AF443339

KNOWLEDGE-BASED INTELLIGENT INFORMATION ENGINEERING SYSTEMS & ALLIED TECHNOLOGIES

Frontiers in Artificial Intelligence and Applications

Series Editors: J. Breuker, R. López de Mántaras, M. Mohammadian, S. Ohsuga and W. Swartout

Volume 69

Volume 1 in the subseries
Knowledge-Based Intelligent Engineering Systems
Editor: L.C. Jain

Previously published in this series:

ISSN: 0922-6389

Knowledge-Based Intelligent Information Engineering Systems & Allied Technologies

KES'2001

Part 1

Edited by

N. Baba

Osaka-Kyoiku University, Kashiwara City, Japan

L.C. Jain

University of South Australia, Adelaide, Australia

and

R.J. Howlett

University of Brighton, Brighton, United Kingdom

IOS
Press

Ohmsha

Amsterdam • Berlin • Oxford • Tokyo • Washington, DC

ISBN 1 58603 192 9 (IOS Press)
ISBN 4 274 90463 6 C3000 (Ohmsha)
Library of Congress Control Number: 2001093968

Publisher
IOS Press
Nieuwe Hemweg 6B
1013 BG Amsterdam
The Netherlands
fax: +31 20 688 33 55
e-mail: order@iospress.nl

Distributor in the UK and Ireland
IOS Press/Lavis Marketing
73 Lime Walk
Headington
Oxford OX3 7AD
England
fax: +44 1865 75 0079

Distributor in the USA and Canada
IOS Press, Inc.
5795-G Burke Centre Parkway
Burke, VA 22015
USA
fax: +1 703 323 3668
e-mail: iosbooks@iospress.com

Distributor in Germany, Austria and Switzerland
IOS Press/LSL.de
Gerichtsweg 28
D-04103 Leipzig
Germany
fax: +49 341 995 4255

Distributor in Japan
Ohmsha, Ltd.
3-1 Kanda Nishiki-cho
Chiyoda-ku, Tokyo 101
Japan
fax: +81 3 3233 2426

LEGAL NOTICE
The publisher is not responsible for the use which might be made of the following information.

PRINTED IN THE NETHERLANDS

Greetings from the General Chairman

On behalf of the Organizing Committee, it is my pleasure to welcome all of you to the 5th International Conference on Knowledge-based Intelligent Information Engineering Systems and Allied Technologies. Thanks to the considerable effort of Professors L.C. Jain and R.J. Howlett, the KES Conference has become quite popular these five years among researchers studying knowledge-based engineering systems, and we welcome you all heartily to share in this exciting conference.

Worldwide conferences such as this one provide superb opportunities for those wanting to observe how top researchers in the field of knowledge-based information engineering systems concentrate themselves in order to deepen the basic theory within each field and/or open up new fields of research.

For this KES-2001 Conference, approximately 300 papers exhibiting high quality and originality have been accepted. In addition, let me direct your attention to three special lectures to be presented by distinguished researchers Professors K. Fukushima, A. P. Refenes, and S. Amari. We hope that as many of you as possible will attend these sessions.

The recent spread of technologies utilizing the Internet has greatly changed the basic structure of our industry and the nature of our daily lives. These recent trends have resulted in the submission to this Conference of a large number of papers dealing with IT related topics. In the special accompanying session scheduled for September 8, active researchers working at leading companies in our country will introduce their recent activities relating to Internet and other new information technologies. I believe that KES-2001 will provide an excellent forum for you to discuss freely among yourselves this variety of topics related to our professions.

In closing, heartfelt thanks are due several individuals who have made important contributions to the success of this conference. First and foremost we salute Professor L.C. Jain (the founder of the KES series of conferences), and Professor R.J. Howlett (Executive General Chairperson of KES 2000), for giving us the opportunity to hold KES 2001 at our university campus and in the Nara Prefecture New Public Hall.

Thanks are also due Professor O. Fujita and the members of the Organizing Committee for their considerable efforts and long hours devoted to the careful organization of this Conference's programs. We would also like to thank all of the invited speakers, the reviewers, the session chairpersons, the authors of the papers, our other colleagues, and Ms. Anne Marie de Rover and Mr. Ron Jansen (IOS Press) for their important contributions to our overall success.

Finally, we would like to express our thanks to FOST, Koei, Ltd., and Matsushita Electric Works, Ltd. whose financial supports have contributed a lot in leading KES-2001 to be a successful event.

It is my personal wish that you all enjoy your participation in KES-2001, using this opportunity to become familiar with each other, to enjoy sharing your ideas, and to work together for progress in our fields in the future.

Norio Baba,
General Conference Chairman,
Professor, Osaka-Kyoiku University.

KES-2001 Organization

General Chairs

Professor N.Baba, Information Science,
Osaka-Kyoiku University, Japan.

Professor L.C.Jain, Knowledge-Based Intelligent Engineering Systems Center,
University of South Australia, Australia.

Professor R.J. Howlett, Intelligent Signal Processing Laboratory,
University of Brighton, United Kingdom.

Local Organization

N.Baba, O.Fujita, I.Hayashi, K.Hirasawa, O.Katai, F.Kojima, M.Kotani, Y.Kuroe,
Y.Maeda, A.Morimoto, N.Nagata, Y.Nara, S.Omatsu, S.Ozawa, Y.Takeuchi, S.Yamawaki

Reviewers

E. Aiyoshi, O. Ahmed, I. Ashino, J. Austin, N. Baba, A. Bargiela, C.T.Chai, Y.W. Chen,
E.D. Damiani, A. Filippidis, C. E. Floyd, O. Fujita, T. Fukuda, R. Hamabe, J. Hasebrook,
Y. Hata, I. Hayashi, K. Hirasawa, A. Hirose, T.P. Hong, M.V. Hull, T. Ichimura, K. Inoue,
F. Ionescu, H. Ishibuchi, N. Ishii, T. Kamano, C. L. Karr, O. Katai, L. Kim, Y. Kinouchi,
I. Kirschning, S. Kitamura, F. Kojima, A. Konagaya, H. Koshimizu, N. Kubota, S. Kunifuji,
C. Kuroda, Y. Kuroe, H.L. Larsen, S.B. Lee, I. Lovrek, Y. Maeda, N. Martin, V.G. Martinez,
L. Medsker, D. P. Mital, A. Morimoto, A. Murgu, M. Nagata, N. Nagata, H. Nagashino,
H. Nakanishi, S. Nishida, T. Nishida, I. Nishizaki, V.A.Niskanen, T. Nomura, K. Obermayer,
Y. Ohsawa, S. Omatsu, S. Ozawa, J. S. Pan, A.S. Pandya, S. Park, R. Price, L. M. Roberts,
M. Sakawa, T. Sawaragi, M. Schmitt, U. Seiffert, I. K. Sethi, C.W. Silva, K. Shimohara,
C.S. Smith, B. Solaiman, C. Sommerer, S. Suzuki, P. Szabo, F. Takeda, H. Takahashi,
Y. Takeuchi, T. Terano, E.Tazaki, C. Tsatsoulis, K. Tsuda, E. Uchino, R. Vaillncourt,
K. Watanabe, S. Watanabe, T. Yamakawa, T. Yamashita, Y. Yamashita, Y. Yoshida.

KES 2001 Technical Session Program Schedule

Thursday, September 6 (Osaka Kyoiku University)

9:45 ~ 9:55	Opening Ceremony (Room 1)			
9:55 ~ 10:55	**Keynote Speech 1:** Recent Advances in the Neocognitron, Prof. K. Fukushima (Room 1)			
Break				

	Room 1 (A-314)	Room 2 (A-215)	Room 3 (A-216)	Room 4 (B3-108-1)
11:05 12:45	**A1:** Fuzzy Systems and Data Mining **Organizer:** Prof. H. Ishibuchi and Prof. T. Nakashima	**A2:** Intelligent Techniques and Software Engineering **Organizer:** Prof. S. Park	**A3:** Biologically Inspired Paradigms in Computational Intelligence **Organizer:** Prof. A. S. Pandya, H. Nagashino and P. Szabo	**A4:** Evolutionary Systems **Organizer:** Dr. K. Shimohara
Lunch				
13:45 Break	**B1:** Neural Networks in Data Mining **Organizer:** Prof. M. Van Hull	**B2:** Applications of Intelligent Systems for Industrial Control, Fault Classification, and Image Processing I **Organizer:** Prof. F. Ionescu	**B3:** Modular Neural Networks: Theory and Applications **Organizer:** Prof. S. Ozawa	**B4:** Conversational Intelligence and Its Applications **Organizer:** Prof. T. Nishida, Prof. M. Koch and Prof. K. Nakata
       18:00	**C1:** Intelligent Data Analysis **Organizer:** Prof. T.-P. Hong	**C2:** Applications of Intelligent Systems for Industrial Control, Fault Classification, and Image Processing II **Organizer:** Prof. F. Ionescu	**C3:** Intelligent Robots Utilizing Soft Computing Techniques **Organizer:** Prof. N. Baba	**C4:** Complex-Valued Neural Networks and their Applications **Organizer:** Prof. A. Hirose
Break				
18:30 ~ 20:00	Reception (Dining Room in the University Hall)			

Friday, September 7 (Osaka Kyoiku University)

9:30 ~ 10:30	**Keynote Speech 2:** Non-linear Models of Financial Market Volatility, Prof. A. P. Refenes (Room 1)			
Break				

	Room 1 (A-314)	Room 2 (A-215)	Room 3 (A-216)	Room 4 (B3-108-1)
10:40 12:20	**D1:** Methodological Aspects of Soft Computing in the Human Sciences **Organizer:** Prof. V. A. Niskanen	**D2:** Application of Soft Computing Techniques for Financial Forecasting **Organizer:** Prof. N. Baba	**D3:** Information Systems for the Access to Digital Services for Wide Community **Organizer:** Prof. I. Kirschning	**D4:** Bioinformatics Towards Post Genome Sequence Era **Organizer:** Prof. A. Konagaya
Lunch				
13:15 Break	**E1:** Digital Image Processing and Analysis: New Developments and Applications **Organizer:** Prof. Le Kim	**E2:** Internet Communication Utilizing Intelligent Agents and Computer Pets **Organizer:** Prof. I. Lovrek and Prof. N. Baba	**E3:** Industrial Application with Artificial Intelligence **Organizer:** Prof. S.-B. Lee	**E4:** Chance Discovery and Its Applications I **Organizer:** Prof. Y. Ohsawa
        16:45	**F1:** Kansei Visual Sensing and Human Interface **Organizer:** Dr. N. Nagata and Prof. H. Koshimizu	**F2:** Granular Computing **Organizer:** Prof. A. Bargiela	**F3:** Artificial Neural Networks and Adaptive Systems for the Information Technologies **Organizer:** Prof. V. Giménez-Martínez	**F4:** Chance Discovery and Its Applications II **Organizer:** Prof. Y. Ohsawa
Break				
17:00 ~ 18:00	Bus Transportation to Nara Hotel			
18:30 ~ 21:00	Banquet (Nara Hotel)			

Thursday, September 6 (Osaka Kyoiku University)

Opening Ceremony (Room 1)				
Keynote Speech 1: Recent Advances in the Neocognitron, Prof. K. Fukushima (Room 1)				
Break				

Room 5 (B3-108-2)	Room 6 (B3-108-3)	Room 7 (C1-105-1)	Room 8 (C1-105-2)	Room 9 (C1-106)
A5: Knowledge-based systems, knowledge discovery I **Organizer:** Prof. K. Tsuda	**A6:** Training Algorithm of Neural Network & Neural Network Model **Organizer:** Prof. N. Baba	**A7:** Data Base and Image Retrieval **Organizer:** Prof. N. Baba	**A8:** Intelligent Knowledge Acquisition Contributed Session	**A9:** SOM and Data Visualization Contributed Session
Lunch				
B5: Knowledge-based systems, knowledge discovery II **Organizer:** Prof. K. Tsuda	**B6:** Intelligent Data Processing in Process Systems and Plants I - Intelligent Monitoring **Organizer:** Prof. C. Kuroda and Prof. Y. Yamashita	**B7:** Knowledge Based Systems I **Organizer:** Prof. N. Ishii	**B8:** Nondestructive Evaluation (NDE) for Diagnostics **Organizer:** Prof. Y. Hata	**B9:** (1) Data Clustering and Classification (2) Intelligent Optimization Using Soft Computing Technologies Contributed Session
C5: Knowledge-based systems, knowledge discovery III **Organizer:** Prof. K. Tsuda	**C6:** Intelligent Data Processing in Process Systems and Plants II - Design, Modeling, Simulation and Control **Organizer:** Prof. C. Kuroda and Prof. Y. Yamashita	**C7:** Knowledge Based Systems II **Organizer:** Prof. N. Ishii	**C8:** Intelligent Systems for Diagnosis and Fault Detection Contributed Session	**C9:** Mathematical Models of Fuzzy Systems Contributed Session
Break				
Reception (Dining Room in the University Hall)				

Friday, September 7 (Osaka Kyoiku University)

Keynote Speech 2: Non-linear Models of Financial Market Volatility, Prof. A. P. Refenes (Room 1)				
Break				

Room 5 (B3-108-2)	Room 6 (B3-108-3)	Room 7 (C1-105-1)	Room 8 (C1-105-2)	Room 9 (C1-106)
D5: Decision Making in Fuzzy and Stochastic Environments **Organizer:** Prof. Y. Yoshida	**D6:** Utilization of Soft Computing Techniques for Intelligent Prediction (Classification) **Organizer:** Prof. N. Baba	**D7:** Intelligent Image Coding and Image Segmentation **Organizer:** Prof. R.Hamabe	**D8:** Perception, Learning and Robotic Systems **Organizer:** Prof. N. Kubota	**D9:** Recurrent Neural Networks and Signal Processing Contributed Session
Lunch				
E5: Emotion Oriented Intelligent Systems I **Organizer:** Prof. T. Ichimura, Prof. T. Yamashita and Prof. E. Tazaki	**E6:** Utilization of Neural Networks and GAs for Intelligent Classification **Organizer:** Prof. N. Baba	**E7:** Knowledge Based Signal and Image Processing I **Organizer:** Prof. P. J.-S. Pan and Prof. Y.-W. Chen	**E8:** Intelligent Control Utilizing Neural Networks I **Organizer:** Prof. N. Baba	**E9:** Utilization of Soft Computing Techniques for Intelligent Knowledge Processing and Data Mining Contributed Session
F5: Emotion Oriented Intelligent Systems II **Organizer:** Prof. T. Ichimura, Prof. T. Yamashita and Prof. E. Tazaki	**F6:** Cognitive Model Contributed Session	**F7:** Knowledge Based Signal and Image Processing II **Organizer:** Prof. P. J.-S. Pan and Prof. Y.-W. Chen	**F8:** Intelligent Control Utilizing Neural Networks II **Organizer:** Prof. N. Baba	**F9:** Recent Trends in Communication Systems Contributed Session
Break				
Bus Transportation to Nara Hotel				
Banquet (Nara Hotel)				

KES 2001 Technical & Special Accompanying Session Program Schedule

Saturday, September 8 (Nara-Ken New Public Hall)

Room 1	Room 2	Room 3	Room 4
9:10 ~ 12:40 **Special Accompanying Session** **(1)** 9:10 ~ 9:55 Matsushita Electric Works, Ltd. **(2)** 9:55 ~ 10:40 KOEI Co., Ltd. Break 10:40 ~ 11:10 **(3)** 11:10 ~ 11:55 Nippon Telenet, Ltd. **(4)** 11:55 ~ 12:40 Hitachi, Ltd.	10:00 ~ 12:40 **G1:** **(1)** Robotic Intelligence by Softcomputing **Organizer:** Prof. K. Watanabe **(2)** Image Processing Contributed Session	10:00 ~ 12:40 **G2:** Evolutionary Computation and Fuzzy Optimisation **Organizer:** Prof. M. Sakawa and Prof. I. Nishizaki	10:30 ~ 13:30 **G3:** **(1)** Internet Application **(2)** Various Soft Computing Techniques Contributed Session (Poster) **(3)** Fuzzy Multiobjective Programming and Evolutionary Games **Organizer:** Prof. A. Murgu (Poster)

Lunch

13:30 ~ 14:30
Keynote Speech 3: Natural Gradient Works Efficiently in Learning, Prof. S. Amari (Room 1)

Break

Room 1	Room2	Room 3	Room 4	Room 5
(5) 14:40 ~ 15:25 Gala, Ltd. **(6)** 15:25 ~ 16:10 Mitsubishi Electric Corporation	14:40 ~ 16:20 **H1:** **(1)** Intelligent Knowledge Processing Utilizing Neural Networks and Expert Systems **Organizer:** Prof. N. Baba **(2)** Hybridization of Various Intelligent Techniques Contributed Session	14:40 ~ 16:20 **H2:** Intelligent Web Tools **Organizer:** Prof. S. Kunifuji	14:40 ~ 16:20 **H3:** Geometry and Statistics in Neural Network Learning Theory **Organizer:** Prof. S. Watanabe	14:40 ~ 16:20 **H4:** A New Approach for Dynamic Document Treatment Contributed Session

Break

16:45 ~ 17:00
Closing Session (Room 1)

Contents

PART 1

KEYNOTE SPEECHES   1

ORGANIZED AND CONTRIBUTED SESSIONS   15

Fuzzy Systems and Data Mining

Intelligent Techniques and Software Engineering

Biologically Inspired Paradigms in Computational Intelligence

PART 2

Internet Application

Various Soft Computing Techniques

Fuzzy Multiobjective Programming and Evolutionary Games

SPECIAL ACCOMPANYING SESSIONS 1607

Part I

Keynote Speeches

KES '01
N. Baba et al. (Eds.)
IOS Press, 2001

Recent Advances in the Neocognitron

Kunihiko Fukushima

Tokyo University of Technology,   Hachioji, Tokyo 192-0982, Japan

Abstract.
The author previously proposed a neural network model *neocognitron* for robust visual pattern recognition. This paper shows recent advances in the neocognitron.

The first topic is handwritten digit recognition by the neocognitron of an improved version. The ability of this network was demonstrated using a large database of handwritten digits (ETL-1). When we used 3000 digits (300 patterns for each digit) for the learning, for example, the recognition rate was 98.5% for a blind test set (3000 digits).

Human beings often can easily recognize a pattern that is partly occluded by other objects, but have difficulty in recognizing it if the occluding objects become invisible. This paper proposes a hypothesis, and consequently a neural network model, explaining why a pattern is easier to recognize when the occluding objects are visible. The model is an extended version of the neocognitron, and can recognize patterns correctly if the occluding objects are visible.

1   Introduction

The author previously proposed a neural network model *neocognitron* for robust visual pattern recognition [1]. It was initially proposed as a neural network model of the visual system. It has a hierarchical multilayered architecture similar to the classical hypothesis of Hubel and Wiesel [2],[3]. It acquires the ability to recognize robustly visual patterns through learning. It consists of layers of S-cells, which resemble simple cells in the primary visual cortex, and layers of C-cells, which resemble complex cells. These layers of S-cells and C-cells are arranged alternately in a hierarchical manner. S-cells are feature-extracting cells, whose input connections are variable and are modified through learning. C-cells, whose input connections are fixed and unmodified, exhibit an approximate invariance to the position of the stimuli presented within their receptive fields. The C-cells in the highest stage work as recognition cells, which indicates the result of the pattern recognition.

This paper shows some recent advances in the neocognitron, such as an improved neocognitron designed for handwritten digit recognition, and a neural network model that can recognize partly occluded patterns correctly.

2   Neocognitron for Handwritten Digit Recognition

This section proposes an improved version of the neocognitron designed for handwritten digit recognition [4]. To improve the recognition rate, several modifications have been applied: such as, the inhibitory surround in the connections from S-cells to C-cells, contrast-extracting layer between input and edge-extracting layers, supervised competitive learning at the highest stage, staggered arrangement of S- and C-cells, and so on. These modifications allow the removal of accessory circuits that were appended to the previous versions, resulting in an improvement of recognition rate as well as simplification of the network architecture. We will demonstrate the ability of this network using a large database of handwritten digits (ETL-1).

2.1   Architecture of the Network

Figure 1 shows the architecture of the proposed network, which has 4 stages of S- and C-cell layers. The stimulus pattern is presented to the input layer U_0. Contrast-extracting layer U_G, which corresponds to retinal ganglion cell layer or lateral geniculate nucleus, follows U_0. Layer U_G consists of two cell-planes: a cell-plane of cells with concentric on-center receptive fields, and a cell-plane of cells with off-center receptive fields. The former cells extract positive contrast in brightness, whereas the latter extract negative contrast from the images presented to the input layer.

The output of U_G is sent to the S-cell layer of the first stage (U_{S1}). Cells of U_{S1} have been trained using supervised learning [1] to extract edge components of various orientations from the input image.

The output of layer U_{Sl} (S-cell layer of the lth stage) is fed to a C-cell layer U_{Cl}, where a blurred version of the response of U_{Sl} is generated. In the conventional neocognitron, the input connections of a C-cell, which converge from a group of S-cells, consisted of only excitatory components of a circular spatial distribution.

An inhibitory surround is newly introduced around the excitatory connections. By the effect of this concentric inhibitory surround, an end point of a line usually elicits a larger response from C-cells than a middle part of the line. This endows the C-cells with the characteristics of end-stopped cells, and C-cells behave like hypercomplex cells in the visual cortex. Bend points and end points of lines are important features for pattern recognition. In the network of previous versions (e.g., [5]), an extra S-cell layer, which was called a bend-extracting layer, was placed after the line-extracting stage. In the proposed network, C-cells, whose input connections have inhibitory surrounds, participate in extraction of bend points and end points of lines while they are making a blurring operation. This allows the removal of accessory layer of bend-extracting cells, resulting in a simplification of the network architecture and an increased recognition rate as well.

The inhibitory surrounds in the connections also have another benefit. The blurring operation by C-cells, which usually is effective for improving robustness against deformation of

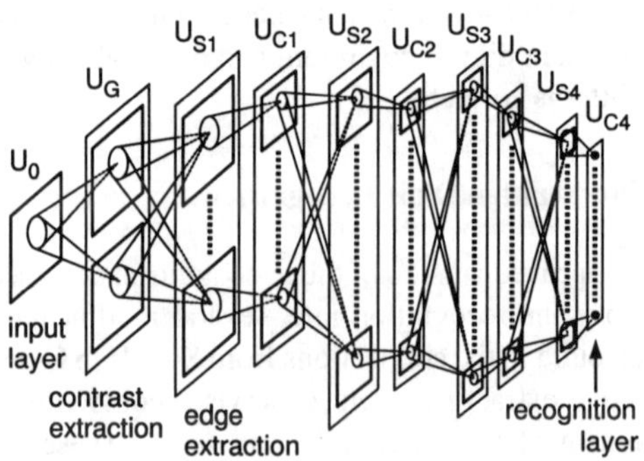

Figure 1: The architecture of the proposed neocognitron.

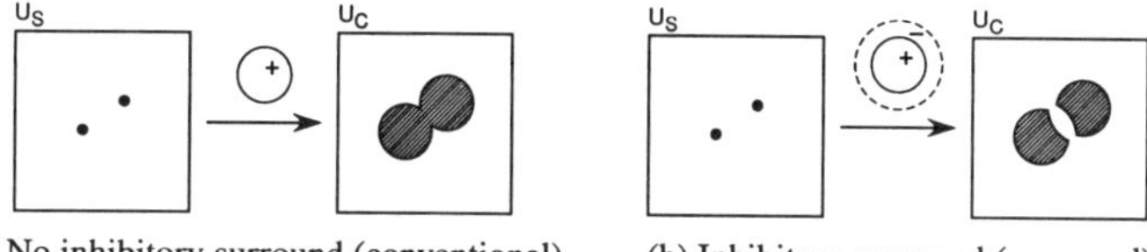

(a) No inhibitory surround (conventional)   (b) Inhibitory surround (proposed)

Figure 2: The blurred responses produced by two independent features can be separated by the inhibitory surround in the input connections to a C-cell.

input patterns, sometimes makes it difficult to detect whether a lump of blurred response is generated by a single feature or by two independent features of the same kind. For example, a single line and a pair of parallel lines of a very narrow separation generate a similar response when they are blurred. The inhibitory surround in the connections to C-cells creates a non-responding zone between the two lumps of blurred responses (Figure 2(B)). This silent zone makes the S-cells of the next stage easily detect the number of original features even after blurring.

The density of the cells in each cell-plane reduces between layers U_{Sl} and U_{Cl}. A staggered arrangement of S- and C-cells is utilized to reduce a harmful side effect of the thinning-out of cells.

The S-cells of intermediate stages (U_{S2} and U_{S3}) are self-organized using unsupervised competitive learning [1]. Every time a training pattern is presented to the input layer, seed cells are determined by a kind of winner-take-all process. Each input connection to a seed cell is increased by an amount proportional to the output of the cell from which the connection leads. Because of the shared connections within each cell-plane, all cells in the cell-plane come to have the same set of input connections as the seed cell. Training of the network is performed from the lower stages to the higher stages: after the training of a lower stage has been completely finished, the training of the succeeding stage begins. The same set of training patterns is used for the training of all stages except U_{S1}. Incidentally, the method of dual threshold is used [6]. Namely, higher threshold values are used for S-cells in the learning phase than in the recognition phase.

Layer U_{S4} at the highest stage is trained through supervised competitive learning. The learning rule resembles the competitive learning used to train U_{S2} and U_{S3}, but the class names of the training patterns are also utilized for the learning. When the network learns varieties of deformed training patterns through competitive learning, more than one cell-plane for one class is usually generated in U_{S4}. Therefore, when each cell-plane first learns a training pattern, the class name of the training pattern is assigned to the cell-plane. Thus, each cell-plane of U_{S4} has a label indicating one of the 10 digits.

Every time a training pattern is presented, competition occurs among all S-cells in the layer. If the winner of the competition has the same label as the training pattern, the winner becomes the seed cell and learns the training pattern in the same way as the seed cells of the lower stages. If the winner has a wrong label (or if all S-cells are silent), however, a new cell-plane is generated and is put a label of the class name of the training pattern.

During the recognition phase, the label of the maximum-output S-cell of U_{S4} determines the final result of recognition. Competition among S-cells occur, and only one maximum output S-cell within the whole layer U_{S4} can transmit its output to U_{C4}.

2.2   Computer Simulation

We tested the behavior of the proposed network using handwritten digits (free writing) randomly sampled from the ETL-1 database. Incidentally, the ETL-1 is a database of segmented handwritten characters and is distributed by Electrotechnical Laboratory, Tsukuba, Japan.

The recognition rate varies depending on the number of learning patterns. When we used 3000 patterns (300 patterns for each digit) for the learning, for example, the recognition rate was 98.5% for a blind test sample (3000 patterns), and 100% for the learning set.

Figure 3 shows a typical response of the network that has finished the learning using 3000 learning patterns. The responses of layers U_0, U_G and layers of C-cells of all stages are displayed in series from left to right, and the responses of S-cell layers are omitted from the figure. The rightmost layer, U_{C4}, is the recognition layer, whose response shows the final result of recognition. It can be seen that the input pattern is recognized correctly as '8'.

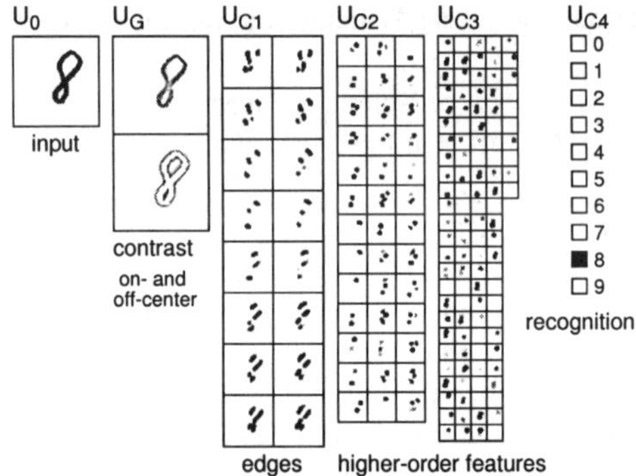

Figure 3: An example of the response of the neocognitron. The input pattern is recognized correctly as '8'.

3   Recognition of Occluded Patterns

Human beings are often able to read or recognize a letter or word contaminated by stains of ink, which partly occlude the letter. If the stains are completely erased and the occluded areas of the letter are changed to white, however, we usually have difficulty in reading the letter, which now have some missing parts. For example, the patterns in Figure 4(a), in which the occluding objects are not visible, are difficult to recognize, but the patterns in (b), in which the occluding objects are visible, are much easier to read.

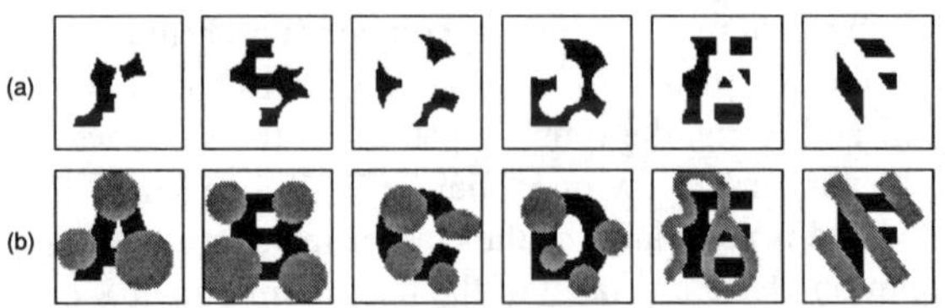

Figure 4: Patterns partly occluded by (a) invisible and (b) visible masking objects. These patterns are used to test the proposed neural network model.

A pattern usually contains a variety of visual features, such as edges, corners, and so on. The visual system extracts, at an early stage, these local features from the input pattern and

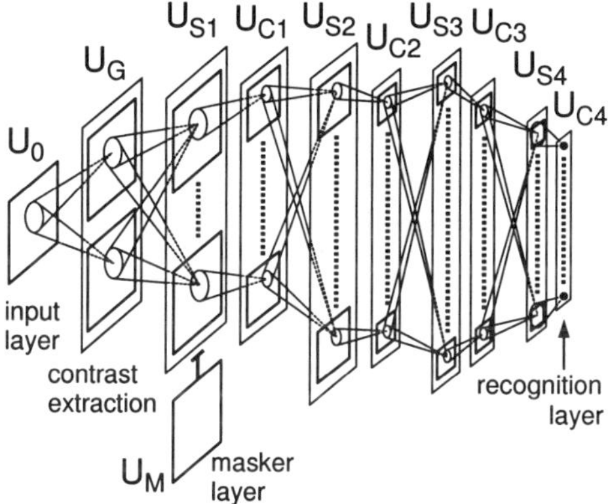

Figure 5: The architecture of the network that can recognize partly occluded patterns correctly.

then attempts to recognize it. When a pattern is partly occluded, a number of new features, which did not exist in the original pattern, are generated near the contour of the occluding objects. When the occluding objects are invisible, the visual system will have difficulty in distinguishing which features are relevant to the original pattern and which are newly generated by the occlusion. These irrelevant features will hinder the visual system in recognizing the occluded pattern correctly.

When the occluding objects are visible, however, the visual system can easily discriminate between relevant and irrelevant features. The features extracted near the contours of the occluding objects are apt to be irrelevant to the occluded pattern. If the responses of feature extractors responsible for the area covered by the occluding objects are suppressed, the signals for the irrelevant disturbing features will be blocked and will not reach higher stages of the visual system. Since the visual system usually has some tolerance to partial absence of local features of the pattern, it should be able to recognize the occluded pattern correctly if the disturbing signals from the irrelevant features are blocked.

3.1   *Neural Network Model*

We have constructed a neural network model based on the above hypothesis [8]. The model is an extended version of the neocognitron. In the hierarchical network of the proposed model, the activity of feature-extracting S-cells whose receptive fields cover the occluding objects is suppressed at a lower stage (U_{S1}).

Figure 5 shows the architecture of the model. Layer U_M, which will be called the *masker layer*, is appended below U_{S1}. Layer U_M, which consists of one cell-plane, detects occluding objects: the occluding objects appears in layer U_M, in the same shape and at the same location as in the input layer U_0. There are topographically ordered and slightly diverging inhibitory connections from layer U_M to all cell-planes of layer U_{S1}. The inhibitory signals from layer U_M suppresses the response of U_{S1} to features irrelevant to the occluded pattern.

Since the irrelevant features generated by occluding objects are thus eliminated, the network, which has some tolerance to partial absence of local features, can recognize occluded patterns correctly.

3.2   Computer Simulation

The network was trained to recognize alphabetical characters. Figure 6 shows the response of the network that has finished learning. The occluding objects, which are visible, is detected by the masker layer U_M. (In the simulation, occluding objects are segmented based on the difference in brightness between occluding objects and occluded patterns.) The responses to local features irrelevant to the target pattern are eliminated from layer U_{S1} by the inhibitory signals from U_M. This can be seen from the response of layer U_{C1}, which is a blurred version of the response of layer U_{S1}. Although the responses of higher layers are slightly sparser than the response to the entire unoccluded pattern, the recognition cell corresponding to 'A' correctly responds in the recognition layer U_{C4}.

If an occluding objects are invisible, however, the masker layer U_M cannot detect the occluding objects and do not send inhibitory signals to layer U_{S1}. As a result, the irrelevant features are transmitted upward without interruption. Disturbed by the irrelevant features, the highest layer of the network fails to recognize the target pattern correctly.

Table 1 summarizes how the model recognizes the images of Figure 4, in both cases when the masker layer U_M is working and not working. Tests were made for two types of input images where the occluding objects are (a) invisible, and (b) visible.

As can be seen from this table, the network with a masker layer correctly recognized all patterns when the occluding objects were visible (Figure 4b). If the occluding objects were invisible (Figure 4a), however, all patterns were erroneously recognized (or rejected). The patterns with invisible occluding objects are hardly recognized correctly even by human

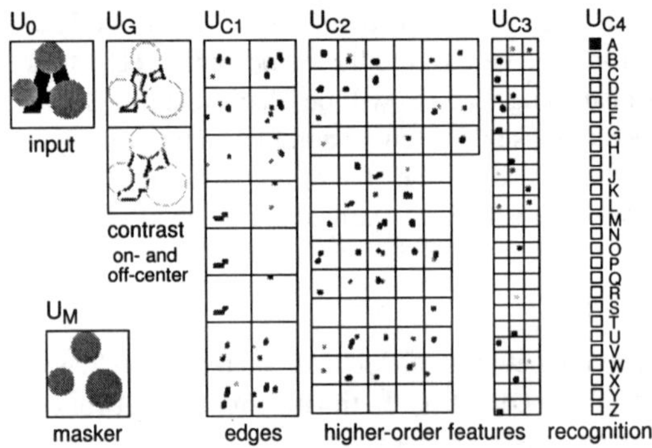

Figure 6: Response of the network when the occluding objects are visible. The occluded input pattern is correctly recognized as 'A'.

Table 1: Result of recognition by the network, when the masker layer is working (proposed model), and not working (neocognitron). This table shows how the images in Figure 4 were recognized.

model	occluding objects	recognized as					
either model	none	A	B	C	D	E	F
	invisible (Fig. 4a)	I	E	—	E	L	G
with masker layer (proposed model)	visible (Fig. 4b)	A	B	C	D	E	F
without masker layer (neocognitron)		Q	B	G	Q	E	X

beings. When the masker layer was removed from the network, the network behaved like the conventional neocognitron, and failed to recognize most of the occluded patterns even when the occluding objects were visible.

References

[1] K. Fukushima, Neocognitron: a hierarchical neural network capable of visual pattern recognition, *Neural Networks* **1**[2] (1988) 119–130.

[2] D. H. Hubel and T. N. Wiesel, Receptive fields, binocular interaction and functional architecture in the cat's visual cortex, *J. Physiology (Lond.)* **106**[1] (1962) 106–154.

[3] D. H. Hubel and T. N. Wiesel, Receptive fields and functional architecture in two nonstriate visual areas (18 and 19) of the cat, *J. Neurophysiol.* **28** (1965) 229–289.

[4] K. Fukushima, Neocognitron of a new version: handwritten digit recognition, *IJCNN'01*, Washington DC, USA (2001), in press.

[5] K. Fukushima and N. Wake, Improved neocognitron with bend-detecting cells, *IJCNN'92*, vol. IV, Baltimore, MD, USA (1992) 190–195.

[6] K. Fukushima and M. Tanigawa, Use of different thresholds in learning and recognition, *Neurocomputing*, 11 (1) (1996) 1–17.

[7] K. Fukushima, Neural network model for selective attention in visual pattern recognition and associative recall, *Applied Optics* **26**[23] (1987) 4985–4992.

[8] K. Fukushima, Recognition of partly occluded patterns: a neural network model, *Biological Cybernetics* **84**[4] (2001) 251–259.

KES '01
N. Baba et al. (Eds.)
IOS Press, 2001

Non-linear Models of Financial Market Volatility

A-P. N. Refenes, W.T.Holt

London Business School,
Sussex Place, Regent's Park,
London NW1 4SA, UK.

ABSTRACT

In recent years neural networks have reportedly achieved considerable successes in a variety of forecasting applications. Although the results are usually accompanied by extensive empirical validation, practitioners and statisticians still remain skeptical: the curse of overfitting is compounded by the lack of rigorous procedures for model identification, selection and adequacy testing. This paper describes a methodology for neural model mis-specification testing.

We introduce a generalization of the Durbin-Watson statistic for neural regression and discuss the general issues of mis-specification testing using residual analysis. We derive a generalized influence matrix for neural estimators which enables us to evaluate the distribution of the statistic. We deploy Monte-Carlo simulation to compare the power of the test for neural land linear regressors. While residual testing is not a sufficient condition for model adequacy, it is nevertheless a necessary condition to demonstrate that the model is a good approximation to the data generating process, particularly as neural network estimation procedures are susceptible to partial convergence. The work is also an important step towards developing rigorous procedures for neural model identification, selection and adequacy testing which have started to appear in the literature.

We demonstrate its applicability in the non-trivial problem of forecasting implied volatility innovations using high frequency stock index options. Each step of the model building process is validated using statistical tests to verify variable significance and model adequacy with the results concerning the presence of non-linear relationships in implied volatility innovations.

Key words: Autocorrelation; Durbin-Watson statistic; Residual diagnostics; Neural networks; Volatility forecasting.

KES '01
N. Baba et al. (Eds.)
IOS Press, 2001

Natural Gradient Works Efficiently in Learning

Shun-ichi Amari
RIKEN Brain Science Institute
amari@brain.riken.go.jp

Abstract. Learning takes place in the parameter space of a system to be trained, and the (stochastic) gradient descent is one of the most popular universal learning methods. However, the parameter space is not Euclidean in general but has a geometrical structure. The natural gradient learning method takes this into account, and its performance has superior to others. The present talk will introduce the natural gradient method by using the multilayer perceptrons, and show how it can be implemented adaptively. The reason why it works so well will be explained by considering the algebraic singularities existing in the parameter spaces of hierarchical systems such as multilayer perceptrons.

1 Multilayer Perceptrons and Its Parameter Space

Let us consider the multilayer perceptron, which receives input signal x and emit output y. When it has h hidden units, the i-th hidden unit calculates the inner product $w_i \cdot x$, and emits its nonlinear function $\varphi(w_i \cdot x)$. Here, φ is a sigmoidal function like hyperbolic tangent. The output unit summarizes all of these outputs with weights v_i linearly, so that it calculates

$$f(x, \theta) = \sum_{i=1}^{m} v_i \varphi(w_i \cdot x), \tag{1}$$

where θ is the vector summarizing all the parameters

$$\theta = (w_1, \cdots, w_h \; ; \; v_1, \cdots, v_h) \tag{2}$$

to specify a perceptron. Learning is carried out by modifying θ adaptively, depending on the current input-output pair from the training examples.

Let us consider a noisy version of the perceptron, where the output is disturbed by the Gaussian noise n whose variance is normalized to $\sigma^2 = 1$. Then the input-output behavior is given by the conditional probability distribution of the output y when x is input,

$$p(y|x, \theta) = \frac{1}{\sqrt{2\pi}\sigma} \exp\left\{-\frac{1}{2\sigma^2}\{y - f(x, \theta)\}^2\right\}. \tag{3}$$

Its logarithm, called the log likelihood, is

$$l(x, y; \theta) = -\log p(x, y; \theta) = -\sum \frac{1}{2}\{y - f(x, \theta)\}^2 \tag{4}$$

which is the negative of the square of the error given by the difference between the teacher signal y and the output of the model having the parameter $\boldsymbol{\theta}$. Hence, the minimizing the square of the error is equivalent to maximize the log likelihood.

It is usual to define the quasi-distance between two probably distributions by the Kullback-Leibler divergence. In the present case, the divergence is written as

$$D\left[\boldsymbol{\theta} : \boldsymbol{\theta}'\right] = \int p(\boldsymbol{x}, y; \boldsymbol{\theta}) \log \frac{p(\boldsymbol{x}, y; \boldsymbol{\theta})}{p\left(\boldsymbol{x}, y; \boldsymbol{\theta}'\right)} d\boldsymbol{x} dy. \tag{5}$$

where E denotes expectation. The square of the distance between two nearby perceptrons specified by $\boldsymbol{\theta}$ and $\boldsymbol{\theta} + d\boldsymbol{\theta}$ is given by the quadratic form

$$D\left[\boldsymbol{\theta} : \boldsymbol{\theta} + d\boldsymbol{\theta}\right] = \frac{1}{2} d\boldsymbol{\theta}^T G(\boldsymbol{\theta}) d\boldsymbol{\theta}. \tag{6}$$

Here the matrix G is the Fisher information matrix (See Amari and Nagaoka, 2000). This introduces the Riemannian metric structure in the parameterspace S of the perceptrons. We call this the neuromanifold.

2   Stochastic Gradient and Natural Gradient

Let $l(\boldsymbol{x}, y; \boldsymbol{\theta})$ be the loss function when the perceptron with parameter $\boldsymbol{\theta}$ is used for processing input $\boldsymbol{x}$, when y should be the true output. The risk function is its expectation,

$$R(\boldsymbol{\theta}) = E\left[l(\boldsymbol{x}, y; \boldsymbol{\theta})\right]. \tag{7}$$

Given a series of examples $(\boldsymbol{x}_1, y_1), \cdots, (\boldsymbol{x}_t, y_t), \cdots$, the stochastic gradient learning rule modifies the current value $\boldsymbol{\theta}_t$ of the parameter as

$$\boldsymbol{\theta}_{t+1} = \boldsymbol{\theta}_t - \eta_t \frac{\partial l\left(\boldsymbol{x}_t, y_t; \boldsymbol{\theta}_t\right)}{\partial \boldsymbol{\theta}}. \tag{8}$$

Here, η_t is a learning rate which may depend on t and

$$\nabla l = \frac{\partial}{\partial \boldsymbol{\theta}} l \tag{9}$$

is the gradient of the loss function l which is believed to be the steepest direction of change in l. This is true when the parameter space is Euclidean and the orthogonal Cartesian coordinates system is taken. However in a Riemannian space where the local distance ds is measured by

$$ds^2 = d\boldsymbol{\theta}^T G d\boldsymbol{\theta}, \tag{10}$$

the true steepest direction is known to be the natural or Riemannian gradient

$$\tilde{\nabla} l = G^{-1}(\boldsymbol{\theta}) \nabla l. \tag{11}$$

Hence, the natural gradient learning method is given by

$$\boldsymbol{\theta}_{t+1} = \boldsymbol{\theta}_t - \eta_t G^{-1}\left(\boldsymbol{\theta}_t\right) \frac{\partial l\left(\boldsymbol{x}_t, y_t; \boldsymbol{\theta}_t\right)}{\partial \boldsymbol{\theta}}. \tag{12}$$

This was proposed in Amari (1998), and applied not only to Multilayer perceptrons but also to Independent Component Analysis (ICA). The natural gradient has been used as a standard method in ICA having a very good performance. In the case of multilayer perceptron, the dynamical behavior of learning in multilayer perceptrons was analyzed by the statistical mechanical method, and is shown to have nearly ideal performance Rattray et al (1998).

However, its implementation looked very difficult, because we need to estimate G and to invert G (Yang and Amari, 1998). However, the adaptive method of directly estimating G^{-1} was proposed in Amari et al (2000) and widened by Park et al (2000), so that it is now easily implemented. The adaptive method is given by the following recurrent estimates of G^{-1}

$$G_{t+1}^{-1} = (1 + \varepsilon_t)\, G_t^{-1} - \varepsilon_t G_t^{-1} \nabla l \left(G_t^{-1} \nabla l\right)^T \tag{13}$$

where ε_t is another learning constant which may depend on t. Simulations show its excellent behaviors, and in many cases, the speed of convergence is more than 1,000 times faster. This is because the natural gradient method can easily escape from plateaus (Amari and Ozeki, 2001).

3 Singular Structure in Hierarchical Systems and Riemannian Metric

An interesting question arises: Why does the natural gradient method work so efficiently? In the above case of the multilayer perceptron, the loss function is given by the log likelihood, and the Riemannian metric is induced also from the same probability model. In such a case, G is the same as the Hessian of the loss function at the optimal point. Therefore, this is equivalent to Newton's method locally, that is, in a neighborhood of the optimal point. Hence the convergence is superlinear. This explains the Fisher efficiency of the natural gradient learning method. However, the natural gradient is different from Newton's method, or its variants, outside a neighborhood of the optimal point. The merit of the natural gradient is not only its local convergence speed but mostly its quick convergence escaping from plateaus. Newton's method does not have such a global property. If the Riemannian structure of the neuromanifold (parameter space) is similar to Euclidean space, the merit would not be so remarkable. This suggests that the parameter space is strongly curved.

Recent studies show that hierarchical systems such as multilayer perceptrons, radial basis function model, Gaussian mixture model, ARMA model in time series, and so on, in general include algebraic singularities in the parameter spaces. On the singularities, the Fisher information matrix degenerates, and its inverse diverges. Therefore, the parameter space is very strongly curved near singularities, where singularities behave like black holes. The dynamics of learning is strongly affected by such singularities. There have been a number of researches concerning the singularities and their effects to parameter estimation and learning (Hagiwara et al, 1993; Hayasaka et al, 1997; Watanabe, 2001; Fukumizu, 2000). One special session is devoted to this topic in KES-2001.

References

[1] S. Amari, Natural gradient works efficiently in learning, Neural Computation **10** (1998) 251–276.

[2] S. Amari and H. Nagaoka, Methods of Information Geometry, New York: AMS and Oxford University press, (2000).

[3] S. Amari and T. Ozeki, Differential and algebraic geometry of multilayer perceptrons, IEICE Tran. **E84-A** (2001) 31–38.

[4] S. Amari, H. Park and K. Fukumizu, Adaptive method of realizing natural gradient learning for multilayer perceptions, Neural Computation **12** (2000) 1399–1409.

[5] K. Fukumizu, Generalization error of linear neural networks in unidentifiable cases, Lecture Notes in Computer Sciences **1720**, eds. O.Watanabe and T. Yokomori, Springer (1999) 51–62.

[6] K. Hagiwara, N. Toda and S. Usui, Nonuniqueness of connecting weights and AIC in multi-layered neural networks, IEICE Trans. **J76-D-II** (1993) 2058–2065.

[7] T. Hayasaka, K. Hagiwara, N. Toda and S. Usui, On the estimation and prediction errors of function representation with orthonormal discrete variable basis in regression model, The Brain and Neural Networks **4** (1997) 18–26.

[8] H. Park, S. Amari and K. Fukumizu, Adaptive natural gradient learning algorithms for various stochastic models, Neural Networks, (accepted 2000).

[9] M. Rattray, D. Saad and S. Amari, Natural Gradient Descent for On-Line Learning, Physical Review Letters **81** 24 (1998) 5461–5464.

[10] H.H Yang and S. Amari, Complexity issues in natural gradient descent method for training multilayer perceptrons, Neural Computation **10** (1998) 2137–2157.

[11] S. Watanabe, Algebraic analysis for non-identifiable learning machines, Neural Computation **13** (2001) 899–933.

Organized and Contributed Sessions

Knowledge Discovery from Local Principal Components Independent of Arbitrary Factors

Chi-Hyon Oh, Katsuhiro Honda, and Hidetomo Ichihashi
Graduate School of Engineering, Osaka Prefecture University,
Gakuen-cho 1-1, Sakai, Osaka, 599-8531 Japan

Abstract. In this paper, we propose a technique of extracting local principal components independent of arbitrary factors chosen. The proposed method takes advantage of Fuzzy c-Regression Models (FCRM) to estimate the parameters of regression models for fuzzy clusters. We decompose the fuzzy scatter matrix of each cluster into two matrices by using the partial regression coefficient matrix obtained by the FCRM. One is closely related to the arbitrary factors and the other is independent of them. Solving the eigen-value problem of the decomposed matrix enables us to extract the local principal components in which influences of arbitrary factors are neutralized. We apply our method to a POS transaction data set in order to discover useful knowledge from it.

1 Introduction

Principal component analysis (PCA) is a technique to compress multidimensional variables into a few indices. Through the observation of those indices, correlations among variables are extracted and we are able to discover some knowledge from the data set. Though we can gain general knowledge from the data set by the PCA, sometimes it could be trivial or valueless because the PCA extract the indices, *i.e.* principal components, so as to have them include the dominant feature of the data set. It might be necessary to analyze data sets from various points of view if one wants to accumulate several characteristics of them. Yanai [1] proposed a technique which has capability of neutralizing influences of arbitrary factors chosen. In [1], Yanai extracted principal components independent of the factors by exploiting the regression analysis technique.

Nonetheless, it is not sufficient to only get rid of influences of arbitrary factors considering lots of real data sets have some substructures. Respective examination of each substructure could lead to appropriate analyses of data sets. Fuzzy clustering is an effective vehicle to partition data sets into some substructures and has a lot of varieties. Fuzzy c-Regression Models (FCRM) [2] proposed by Hathaway *et al.* is one of those algorithms in which parameters of regression models for clusters, *i.e.* partial regression coefficients, are estimated.

In this paper, we propose a technique of extracting local principal components independent of arbitrary factors. Firstly, we implement the FCRM to derive the parameters of c-regression models in our method. The fuzzy scatter matrix is decomposed into two matrices then in the same manner as Yanai's approach. We can obtain local principal components independent of arbitrary factors by solving eigen-value problem of the decomposed fuzzy scatter matrix. We apply our method to a POS transaction data set in which each data point is composed of 20 attributes such as meteorological element and sales of two different supermarkets to gain diverse knowledge from it.

2 Extraction of Local Principal Components Independent of Arbitrary Factors

Yanai's approach [1] to extract principal components independent of arbitrary factors takes advantage of the regression analysis technique. It decomposes the variance co-variance matrix derived in principal component analysis into two parts by using the obtained partial regression coefficient. One is closely related to external criteria and the other is independent of them. In Yanai's approach, the factors one intends to neutralize are regarded as the external criteria. Taking substructures of data sets into consideration, the partial regression coefficient for each of them is estimated in our method. We employ Hathaway's FCRM [2] for the purpose.

In Yanai's approach, it is assumed that the mean value of the data set is zero. We, therefore, introduce the concept of cluster center to the original FCRM. Let Y_c and X_c be sets of response variables, *i.e.* external criteria, and explanatory variables for each cluster c. They can be written as follows:

$$Y_c = (\boldsymbol{y}_{c1}^T, \boldsymbol{y}_{c2}^T, \cdots, \boldsymbol{y}_{cn}^T)^T = \{y_{cki}\}, \qquad y_{cki} = y_{ki} - v_{ci}^y, \quad i = 1, \cdots, t, \tag{1}$$

$$X_c = (\boldsymbol{x}_{c1}^T, \boldsymbol{x}_{c2}^T, \cdots, \boldsymbol{x}_{cn}^T)^T = \{x_{ckj}\}, \qquad x_{ckj} = x_{kj} - v_{cj}^x, \quad j = 1, \cdots, s, \tag{2}$$

where T denotes transposition. t and s are the dimensionalities of response and explanatory variables and n is the number of data points. Note that $\boldsymbol{y}_{ck}^T$ and $\boldsymbol{x}_{ck}^T$ are represented row vectors for convenience sake of notation of the following equations. v_{ci}^y and v_{cj}^x are cluster centers of response and explanatory variables.

Suppose we identify the following C linear models to extract local linear structures.

$$\boldsymbol{y}_c = \boldsymbol{x}_c B_c + \boldsymbol{e}_c, \quad c = 1, \cdots, C, \tag{3}$$

where B_c is the partial regression coefficient matrix for each cluster c and can be written as follows:

$$B_c = \begin{pmatrix} b_{c11} & b_{c12} & \cdots & b_{c1t} \\ b_{c21} & b_{c22} & \cdots & b_{c2t} \\ \vdots & \vdots & \ddots & \vdots \\ b_{cs1} & b_{cs2} & \cdots & b_{cst} \end{pmatrix}$$

B_c is calculated by least squares method in the general linear regression analysis.

The FCRM can be driven by optimization of an objective function. We use Lagrange's method of indeterminate multiplier to derive the objective function for the FCRM. We have the following objective function to be minimized.

$$L = \sum_{c=1}^{C} \sum_{k=1}^{n} u_{ck} \|\boldsymbol{y}_{ck} - \boldsymbol{x}_{ck} B_c\|^2 + T_0 \sum_{c=1}^{C} \sum_{k=1}^{n} u_{ck} \log u_{ck} + \sum_{k=1}^{n} T_k \left(\sum_{c=1}^{C} u_{ck} - 1 \right). \tag{4}$$

u_{ck} is membership of the data point k to the cluster c which represents degree of belonging to clusters. u_{ck} satisfies the following condition.

$$\boldsymbol{u}_c \in \{(u_{ck}) | \sum_{c=1}^{C} u_{ck} = 1, u_{ck} \in [0, 1], c = 1, \cdots, C \}. \tag{5}$$

In (4), the first term represents residual sum of squares. The second term represents entropy maximization as regularization which was introduced in Fuzzy c-Means by

Miyamoto *et al.* [3] for the first time. It enables us to obtain fuzzy clusters. T_0 is the weighting parameter which specifies degree of fuzziness. The remaining terms describe the constraint of the membership u_{ck}, *i.e.* (5). T_k is the Lagrangian multiplier.

From the necessary condition $\partial L/\partial b_{cji} = 0$ for the optimality of the objective function L, B_c can be derived as follows:

$$B_c = (X_c^T D_c X_c)^{-1} X_c^T D_c Y_c. \tag{6}$$

In (6), D_c is the diagonal matrix, diagonal entries of which are memberships of data points.

$$D_c = \begin{pmatrix} u_{c1} & 0 & \cdots & 0 \\ 0 & u_{c2} & \cdots & 0 \\ \vdots & \vdots & \ddots & \vdots \\ 0 & 0 & \cdots & u_{cn} \end{pmatrix}$$

From the necessary conditions $\partial L/\partial u_{ck} = 0$, $\partial L/\partial v_{cj}^x = 0$ and $\partial L/\partial v_{ci}^y = 0$ for L, we also have the following equations for the membership u_{ck}, the cluster center of the explanatory variable v_{cj}^x and the cluster center of the response variable v_{cj}^y.

$$u_{ck} = \exp(A_{ck})/\sum_{a=1}^{c} \exp(A_{ak}), \quad A_{ak} = -\frac{1}{T_0} \sum_{i=1}^{t} \left(y_{aki} - \sum_{j=1}^{s} x_{akj} b_{aij}\right)^2, \tag{7}$$

$$v_{cj}^x = \sum_{k=1}^{n} u_{ck} x_{kj} / \sum_{k=1}^{n} u_{ck}, \tag{8}$$

$$v_{cj}^y = \sum_{k=1}^{n} u_{ck} y_{kj} / \sum_{k=1}^{n} u_{ck}. \tag{9}$$

The solution algorithms for the FCRM are based on an iterative procedure. We remark them below.

FCRM Algorithms

Step 1: Set the number of clusters C, the coefficient of the entropy term T_0 and the terminal condition ϵ. Initialize the membership $u_c k$ randomly so as to have it follow the condition in (12).

Step 2: Update the cluster center v_{cj}^x and v_{cj}^y using (10) and (11).

Step 3: Calculate the partial regression coefficient matrix B_c using (7).

Step 4: Update the membership u_{ck} using (9).

Step 5: If $\max |u_{ck}^{NEW} - u_{ck}^{OLD}| < \epsilon$, then stop. Otherwise, return to Step 2.

We can represent the set of response variables Y_c as in (10) by using the partial regression coefficient matrix B_c.

$$\begin{aligned} Y_c &= X_c B_c + \{Y_c - X_c B_c\} \\ &= Y_{X_c} + Y_{X_c}^-. \end{aligned} \tag{10}$$

In (10), Y_{X_c} is the predicted value by X_c. On the one hand, $Y_{X_c}^-$ is the unaccountable part according to X_c. It is, therefore, able to be said that $Y_{X_c}^-$ is independent of X_c.

We can divide fuzzy scatter matrix S_{fc} into two parts in the same manner as (10).

$$
\begin{aligned}
S_{fc} &= Y_c^T D_c Y_c \\
&= Y_c^T D_c (Y_{X_c} + Y_{\bar{X}_c}) \\
&= Y_c^T D_c X_c B_c + \{ Y_c^T D_c Y_c - Y_c^T D_c X_c B_c \} \\
&= S_{fc}^X + S_{fc}^{X-},
\end{aligned}
\tag{11}
$$

wehre S_{fc}^X is the factor matrix accountable by X_c and S_{fc}^{X-} is independent of X_c. Using (6), we can denote S_{fc}^X and S_{fc}^{X-} as follows:

$$
S_{fc}^X = Y_c^T D_c X_c (X_c^T D_c X_c)^{-1} X_c^T D_c Y_c,
\tag{12}
$$

$$
S_{fc}^{X-} = Y_c^T D_c Y_c - Y_c^T D_c X_c (X_c^T D_c X_c)^{-1} X_c^T D_c Y_c.
\tag{13}
$$

These factor matrices are both symmetric. We can extract principal components independent of explanatory variables by solving the eigen-value problem of the factor matrix S_{fc}^{X-}. We can neutralize the influence of arbitrary factors by setting them as explanatory variables.

3   Knowledge Discovery from a POS Transaction Data Set

We applied the proposed method to a POS (Point of Sales) transaction data set of two different supermarkets to discover useful knowledge from it. The data set consists of 333 data, which have 20 items, the number of customers having come to the store, meteorological element and so on. We enumerate each item and item number below.

Items of the POS transaction data set

> 1: Holiday, 2: Friday, 3: Saturday, 4: Sunday, 5: Average temperature of the day, 6, 7, 8 and 9: Temperature at 6, 12, 15 and 18 o'clock, 10: Humidity, 11 and 12: Weather category during day and night, 13: Precipitation, 14, 15 and 16: Precipitation during 9-12, 12-15 and 15-18 o'clock, 17 and 18: the number of customers of supermarket A and B, 19 and 20: Sales of perishables of supermarket A and B

The items of days of week, Holyday, Friday, Saturday, Sunday, are dummy variables and weather categories are represented by integer values.

We applied Fuzzy c-Varieties (FCV) [4], which is one of the fuzzy clustering algorithm proposed by Bezdek *et al.*, to the POS data set before applying our method to it. In the FCV, prototypes of clusters are multi dimensional linear varieties. Since the linear varieties are spanned by some local principal component vectors, the FCV can be regarded as a simultaneous algorithm of fuzzy clustering and the PCA. Through the analysis of the obtained local principal components, we can discover some knowledge of each cluster. We set the number of clusters to two. The data set was divided into two seasonal clusters, the hot and cold seasons. After analyzing the local principal components obtained by the FCV, we gain the knowledge which says precipitation and the sales of perishables have a close correlation. The less it rains, the more the sales increase.

We applied our method to the POS transaction data set on the basis of the results of the FCV. Since we have obtained the relation with precipitation and the sales of

Table 1: Fuzzy Factor Loading.

Item #	1	2	3	4	5	6	7	8	9	10
Cold season	0.01	-0.48	0.35	0.39	-0.12	-0.12	-0.09	-0.06	-0.10	-0.28
Hot season	-0.01	-0.47	0.30	0.53	-0.46	-0.45	-0.43	-0.46	-0.46	-0.11

Item #	11	12	14	15	16	17	18	19	20
Cold season	-0.33	-0.32	-0.30	-0.22	-0.19	0.78	0.87	0.92	0.88
Hot season	-0.02	0.11	-0.14	-0.15	-0.02	0.75	0.86	0.90	0.92

perishables, we chose precipitation as the factor to be neutralized its influence. We show the value of fuzzy factor loading [5] of each item in Table 1. The fuzzy factor loading quantifies correlations between local principal components and each item. When a certain item has the same sign of the fuzzy factor loading as the other items, it has positive correlation with them. Therefore, if the value of an item increases, that of the other item which has the same sign of the fuzzy factor loading also increases and vice versa. We also specified the number of clusters as two in our method. The data set was divided into the hot and cold seasons. In Table 1, we underlined noteworthy values. Observing the values of fuzzy factor loading, we can mention that the numbers of customers of both supermarkets increase on Saturday and Sunday not Friday and the sales grow in both seasons since Friday has the value of fuzzy factor loading which has the opposite sign of the sales of perishables. Furthermore, when the temperature comes down, the sales go up. In this manner, we can gain diverse knowledge from the data set by using our proposed method.

4　Conclusion

In this paper, we proposed a technique of extracting local principal components independent of arbitrary factors. Firstly, we estimate the partial regression coefficient of each cluster by using the FCRM and decompose it into two matrices then. We can extract the local principal components in which influences of arbitrary factors are neutralized from the decomposed matrix. In the numerical example, we applied the proposed method to a POS transaction data set. Our method has the capability of discovering diverse knowledge from the data set.

References

[1] H. Yanai, Factor Analysis with External Criteria, Japanese Psychological Research 12, 4 (1970) 143-153.

[2] R. J. Hathaway and J. C. Bezdek, Switching Regression Models and Fuzzy Clustering, IEEE Trans. on Fuzzy Systems 1, 3 (1993) 195-204.

[3] S. Miyamoto and M. Mukaidono, Fuzzy c-Means as a Regularization and Maximum Entropy Approach, Proc. IFSA'97 2 (1997) 86-92.

[4] J.C.Bezdek, C.Coray, R.Gunderson, and J.Watson, Detection and Characterization of Cluster Substructure 2. Fuzzy c-Varieties and Convex Combinations Thereof, SIAM J. Appl. Math. 40, 2 (1981) 358-372.

[5] Y.Yabuuchi and J.Watada, Fuzzy Principal Compoment Analysis and Its Application, Biomedical Fuzzy and Human Sciences 3, 1 (1997) 83-92.

KES '01
N. Baba et al. (Eds.)
IOS Press, 2001

Learning Action-Value Functions Using Neural Networks with Incremental Learning Ability

Naoto SHIRAGA   Seiichi OZAWA   Shigeo ABE

Graduate School of Science and Technology, Kobe University, Kobe 657-8501, JAPAN
shiraga@chevrolet.eedept.kobe-u.ac.jp   {ozawa, abe}@eedept.kobe-u.ac.jp

Abstract. When the distribution of given training data is biased and
temporally varied, it is well known that the learning of neural networks
becomes difficult in general. In Reinforcement Learning (RL) problems,
such situations often arise. In this paper, an incremental learning system, which has been devised for supervised learning, is implemented as
an RL agent that can acquire an action-value function properly even
in the above difficult situations. The proposed RL agent is applied to
an extended mountain-car task in which learning domains are temporally expanded. Through computer simulations, we demonstrate that
the proposed agent can acquire a right policy in this task.

1   Introduction

In Reinforcement Learning (RL) problems, an RL agent adapts itself to the environment so
as to maximize the total amount of reward it receives over the long run [1]. Through the
interaction with the environment, the agent learns an appropriate policy that maps a state to
a right action. In general, instead of learning the policy directly, action-values are introduced
into the agent and they are modified based on the error between the estimated reward and
the immediate reward called *Temporal Difference (TD) error*. The action-values are defined
for each state-action pair and they give the estimation of the total reward that the agent
will earn in the future. In tasks with small and finite state sets, it is possible to form these
action-values with a lookup table. In many cases of practical interest, however, there are far
more states than could possibly be entries in this table. In this case, a continuous function is
used for approximating action-values.

Neural networks have been often adopted to approximate action-value functions. However,
it is well known that the learning of neural networks becomes difficult when the distribution
of given training data is biased and temporally varied. In such a situation, the input-output
relations acquired in the past are easily to be collapsed by the learning of newly given data.
This disruption in neural networks is called *interference* that is caused by excessive adaptation
of connection weights for new data. This interference can also arise in RL problems because
similar trials might not be conducted in a short period.

We have proposed an incremental learning system [2], in which Long-Term Memory (LTM)
was introduced into Resource Allocating Network (RAN) [3]. For the notational convenience,
this system is denoted as *RAN-LTM*. In RAN-LTM, storage data in LTM (denoted as *LTM
data*) are produced from inputs and outputs of RAN whose relationships are accurately approximated. When the learning is carried out, some LTM data are retrieved and they are
simultaneously trained to suppress the interference. Although RAN-LTM has been devised
for supervised learning, it can be also applied to RL problems by introducing TD errors.

In Section 2, RAN-LTM is implemented as an RL agent (called *RAN-LTM agent*) that
can acquire proper action-value functions even when the learning domains are temporally
expanded. In Section 3, the RAN-LTM agent is applied to an extended mountain-car task,
in which a car driver (RL agent) has to learn a right policy on the throttle operation to reach
a goal as soon as possible. The agent's ability in approximation of action-value functions is
evaluated through computer simulations. Section 4 presents conclusions of this work.

2   RL Agent with Incremental Learning Ability

The proposed RL agent consists of two parts: *Resource Allocating Network (RAN)* [3] and *Long-Term Memory (LTM)*. RAN is used for approximating an action-value function. The inputs of RAN at time t, x_{ti}, $i = 1, \cdots, I$, are composed of the current agent's states $s_{ti'}$, $i' = 1, \cdots, S$, and the previously selected actions $a_{t-1,i''}$, $i'' = 1, \cdots, A$: i.e. $(x_{t1}, \cdots, x_{tI}) = (s_{t1}, \cdots, s_{tS}, a_{t-1,1}, \cdots, a_{t-1,A})$. Here, S and A are the numbers of agent's states and actions, respectively. The outputs z_{tk}, $k = 1, \cdots, K$, correspond to the action-values, which are utilized for determining agent's actions at time t. Note that the number of input units of RAN, I, is equal to $S + A$ and the number of output units, K, is equal to A. In the followings, the inputs and outputs of RAN are denoted as vectors: i.e. $\boldsymbol{x}_t = \{x_{t1}, \cdots, x_{tI}\}'$ and $\boldsymbol{z}_t = \{z_{t1}, \cdots, z_{tK}\}'$.

While the agent takes actions to achieve a goal, useful pairs of inputs and the associated action-values (called *LTM data*) are automatically produced and accumulated inside the agent. The agent can retrieve these LTM data and learn them in order to sustain its proper action-value function. In learning action-values, the production and retrieval of LTM data are simultaneously carried out inside the agent. The details of these procedures are shown below.

2.1   Resource Allocating Network

RAN is an extended model of Radial Basis Function networks [4]. At the beginning, the number of hidden units is set to one; hence, RAN possesses simple approximation ability at first. As the agent experiences many trials, the approximation ability of RAN is developed by allocating additional hidden units.

The outputs of RAN are calculated as follows:

$$y_{tj} = \exp(-\frac{\parallel \boldsymbol{x}_t - \boldsymbol{c}_j \parallel^2}{\sigma_j{}^2}) \quad (j = 1, \cdots, J) \tag{1}$$

$$z_{tk} = \sum_{j=1}^{J} w_{kj} y_{tj} + \theta_k \quad (k = 1, \cdots, K), \tag{2}$$

where y_{tj} is the output of the jth hidden unit at time t, J is the number of hidden units, $\boldsymbol{c}_j = \{c_{j1}, \cdots, c_{jI}\}'$ is a center vector of the jth hidden unit, σ_j^2 corresponds to variance of the jth radial basis function, w_{kj} is a connection weight from the jth hidden unit to the kth output unit and θ_k is a bias of the kth output unit.

Assume that the kth output for $\boldsymbol{x}_t$ corresponds to an action-value of the kth action, $Q(\boldsymbol{x}_t, a_{tk})$: i.e. $z_{tk} = Q(\boldsymbol{x}_t, a_{tk})$. Then, the agent's action is selected based on the following probability function in Stochastic Action Selector (SAS):

$$\text{Prob}(a_{tk}) = \frac{\exp(Q(\boldsymbol{x}_t, a_{tk})/T)}{\sum_{l=1}^{K} \exp(Q(\boldsymbol{x}_t, a_{tl})/T)} = \frac{\exp(z_{tk}/T)}{\sum_{l=1}^{K} \exp(z_{tl}/T)}, \tag{3}$$

where T is a parameter that controls the randomness of action selection. At the beginning, T is set to a large value, and then it is gradually reduced. Now we assume that the k'th action is selected at time t in SAS, and then the agent's states are changed to $s_{t+1,1}, \cdots, s_{t+1,S}$ and an immediate reward r_{t+1} is given. The following TD error E_{tk} for the kth output is defined:

$$E_{tk} = \begin{cases} r_{t+1} + \gamma \max_l \tilde{z}_{t+1,l} - z_{tk} & (k = k') \\ 0 & (\text{otherwise}), \end{cases} \tag{4}$$

where $\tilde{z}_{t+1,l}$ is the lth output calculated by using the next state $\boldsymbol{x}_{t+1}$ and the current network connections. Based on the value of $E_{tk'}$, either of the following procedures is conducted:

1) If $E_{tk'} > \varepsilon$ and $\|\boldsymbol{x}_t - \boldsymbol{c}^*\| > \delta$, a hidden unit is added to RAN (i.e. $J \leftarrow J + 1$). Here, $\boldsymbol{c}^*$ is the nearest center vector to $\boldsymbol{x}_t$. ε and δ are positive constants. The network parameters for the Jth hidden unit are set to as follows: $\boldsymbol{c}_J = \boldsymbol{x}_t$ and $w_{kJ} = E_{tk'}$.

2) Otherwise, the network connections are modified as follows:

$$w_{kj}^{NEW} = w_{kj}^{OLD} + \alpha E_{tk} y_{tj} \tag{5}$$

$$c_{ji}^{NEW} = c_{ji}^{OLD} + 2\frac{\alpha}{\sigma_j}(x_{ti} - c_{ji}) y_{tj} \sum_k E_{tk} \sigma_{kj}, \tag{6}$$

where α is a positive learning ratio.

2.2 Long Term Memory

Each LTM datum is composed of an agent's state and the associated action-value. They are respectively used as an input and its target in the learning of RAN. Therefore, LTM data should be produced from inputs and outputs of networks whose relationships are accurately approximated. In reinforcement learning, however, the approximation accuracy of an action-value function cannot be measured by only TD errors. Hence, the average number of actions taken to achieve a goal is also considered as a criterion for producing LTM data. The average number of actions is estimated occasionally during the learning of the agent; and if it is small enough, the following procedure is carried out to produce LTM data:

1) If all outputs of hidden units y_{tj}, $j = 1, \cdots, J$, for x_t are less than a threshold value θ_c, then go to Step 5. Otherwise, go to Step 2.

2) Obtain all indices j of hidden units whose outputs y_{tj} are larger than θ_c, and we define a set $\mathcal{I}_1$ of these indices. The variable q_j is updated for all $j \in \mathcal{I}_1$ as follows: $q_j \leftarrow q_j + 2\exp(\rho E_{tk'}) - 1$, where k' means the index of the selected action. $E_{tk'}$ is the TD error for the k'th output $z_{tk'}$ and ρ is a positive constant.

3) If $q_j > \beta$ for $j \in \mathcal{I}_1$, then q_j is initialized and go to Step 4. Otherwise, go to Step 5. Here, β is a positive constant.

4) If no LTM datum for the jth hidden unit had been produced yet, then the jth center vector c_j is given to RAN as its input $\tilde{x}_M$ and the output $\tilde{z}_M$ is calculated. The input-output pair $(\tilde{x}_M, \tilde{z}_M)$ is stored into LTM. The number of LTM data, M, increases by one (i.e. $M \leftarrow M + 1$).

5) $t \leftarrow t + 1$ and go to Step 1.

LTM data are retrieved and utilized for learning the RL agent. In order to learn as efficiently as possible, the suitable number of LTM data should be recalled. Here, based on the activation of hidden units, LTM data to be retrieved are determined as follows:

1) Obtain all indices j of hidden units whose outputs y_{tj} are larger than θ_r, and define a set of these indices as $\mathcal{I}_2$.

2) If $\mathcal{I}_2 \neq \phi$, then go to Step 3. Otherwise, no LTM datum is retrieved. Go to Step 5.

3) Obtain LTM data nearest to the center vectors c_j for all $j \in \mathcal{I}_2$.

4) Recall these LTM data and the output errors are calculated for each LTM datum. Based on the errors, connection weights of RAN are modified.

5) $t \leftarrow t + 1$ and go to Step 1.

3 Simulations

3.1 Extended Mountain-Car Task

To examine the performance of the proposed RL agent, we apply it to an extended version of *Mountain-Car Task*, in which a car driver (RL agent) learns an efficient policy to reach a goal located on the hill between two basins shown in Fig. 1. In the original Mountain-Car Task [1], only the left basin (B_1) is adopted in learning. Here, the right basin (B_2) is also adopted as an additional learning domain. Suppose that a stationary car is positioned in either of two basins at the beginning of an episode. The goal of the RL agent is to successfully drive up the steep incline and to reach a goal state at the top of the hill as soon as possible. The difficulty is that gravity is stronger than the car's engine, and even at full throttle the car cannot accelerate up the steep slope. Therefore, the RL agent has to learn a policy that it first moves away from the goal and up the opposite slope. Then, by applying full throttle the car can build up enough inertia to carry it up the steep slope.

The state of the RL agent is the car's position and velocity. Three actions a_t are available to the agent in each state: full throttle forward $(+1)$, full throttle reverse (-1), and zero throttle (0). The reward is -1 for all state transitions except the transition to the goal state, in which case a zero reward is returned. The car moves according to a simplified physics. Its position x_t and velocity $\dot{x}_t$ are updated by the following dynamics:

$$x_{t+1} = \text{bound}[x_t + \dot{x}_{t+1}], \quad \dot{x}_{t+1} = \text{bound}[\dot{x}_t + 0.001a_t - 0.0025\cos(3x_t)],$$

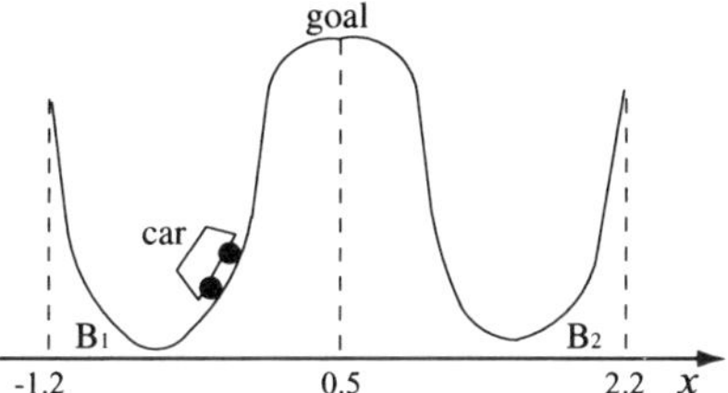

Figure 1: An extended mountain-car task.

Table 1: The average number of steps obtained through 100 episodes and the total time to learn B_1 and B_2. In the column of B_1, 'before' and 'after' respectively correspond to the results of average steps evaluated for B_1 before and after the learning for B_2 converges.

	B_1		B_2	Learning Time
	before	after		(sec.)
RAN agent	811	1595	679	110
RAN-LTM agent	790	541	922	250

where the bound operation enforces $-1.2 \leq x_{t+1} \leq 2.2$ and $-0.07 \leq \dot{x}_{t+1} \leq 0.07$. When x_{t+1} reaches the left and right bound (i.e. $x_{t+1} = -1.2$ and 2.2), $\dot{x}_{t+1}$ is reset to zero. In each episode, the agent starts from a random position with velocity uniformly chosen from the above range. When the agent reaches the goal, the episode is terminated.

3.2   Results

Here, we shall apply the proposed RL agent to the extended mountain-car task in which the learning of the action-value function is conducted only for B_1 at first, and then the learning for B_2 is carried out. That is to say, at first, the RL agent repeatedly experiences episodes in which initial positions are randomly chosen only from B_1, and the learning is continued until the TD error becomes small enough. After the learning for B_1 converges, the RL agent experiences new episodes in which initial positions are randomly chosen only from B_2. Therefore, the episodes experienced in the former learning never occur in the latter learning. After the learning for B_2 is completed, we evaluate the interference by examining the average number of steps needed until the RL agent reaches the goal. Here, the number of steps is equivalent to the number of agent's actions taken in an episode.

Table 1 shows the average number of steps obtained through 100 episodes and the total time to learn B_1 and B_2. In each episode, the RL agent starts from a randomly chosen position in B_1 or B_2. To show the usefulness of the proposed agent (denoted as *RAN-LTM agent*), the performance of the agent using the original RAN (denoted as *RAN agent*) is also examined. The parameters of RAN and Long-Term Memory (LTM) are given as follows:

[RAN] $\varepsilon = 1.0$, $\kappa = 1.0$, $\delta_{max} = \delta_{min} = 0.5$, $\tau = 50$, $\alpha = 0.00001$, $\sigma_j = 0.21$, $\gamma = 0.99$
[LTM] $\theta_c = 0.9$, $\theta_r = 10^{-2}$, $\rho = 1$, $\beta = 0.01$, $\eta = 0.01$.

The average number of steps are evaluated for B_1 before and after the learning for B_2 is carried out. As we can see in Table 1, the number of needed steps in B_1 excessively increases in the RAN agent after the learning for B_2; on the other hand, it decreases in the RAN-LTM agent. Considering that there is not so much difference between the RAN agent and the RAN-LTM agent in terms of needed steps before the learning for B_2, this result suggests that the RAN-LTM agent can sustain proper action-values acquired in the past even if lately presented episodes are quite different from those presented in the past. This result is also certified from Fig. 2, in which the variations of action-values are shown for different positions x in B_1. These variations are given by differences between action-values acquired before and after the learning of B_2. As seen in Fig. 2, due to the interference, the action-values of the RAN agent

          Fuzzy Systems and Data Mining

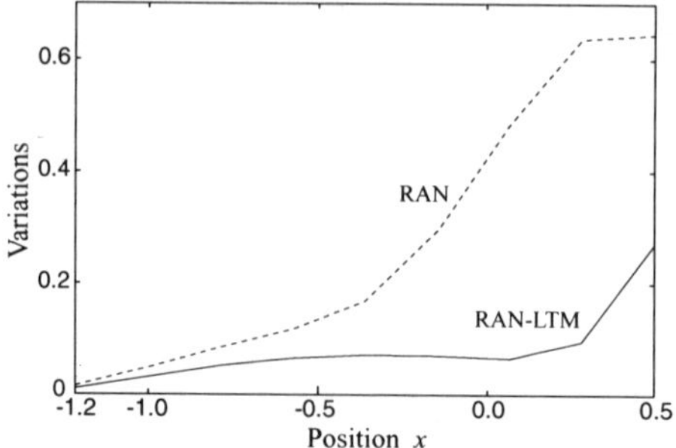

Figure 2: The variations of action-values in B_1.

are varied seriously especially around the boundary between B_1 and B_2 (i.e. $x = 0.5$). On the other hand, the variations are fairly suppressed in the RAN-LTM agent; this suggests that a proper policy for B_1 is maintained.

However, it seems that the larger number of steps in B_2 is needed for the RAN-LTM agent. Since the learning for B_2 was carried out with the fixed number of iterations, it is supposed that the TD error in the RAN-LTM agent did not become small enough within the repetitions, and then the larger number of steps was needed to reach the goal. Furthermore, the learning time of the RAN-LTM agent is two times longer than that of the RAN agent. This is because not only immediate reward but also retrieved LTM data should be learned in the RAN-LTM agent. Therefore, we should devise a more efficient retrieval mechanism for LTM data in order to speed up learning.

4   Conclusions

In this paper, Resource Allocating Network with Long-Term Memory (RAN-LTM) was extended to reinforcement learning. In RAN-LTM, significant pairs of inputs and the associated action-values are automatically stored into Long-Term Memory (LTM). These LTM data are retrieved and utilized for the learning such that the agent holds good approximation accuracy of action-value functions. We evaluated the performance of the RAN-LTM agent in an extended mountain-car task, in which a car driver (RL agent) should learn an efficient policy to reach a goal located on the hill. The learning was conducted only for one basin at first, and then the learning for the other basin was carried out. From the simulation results, we certified that the proposed RAN-LTM agent could suppress the interference and learn a proper policy as compared with the RAN agent.

Acknowledgement

This research has been supported by the Kayamori Foundation of Informational Science Advancement.

References

[1] R. S. Sutton and A. G. Barto: *Reinforcement learning – An introduction*, The MIT Press (1998).

[2] M. Kobayashi, A. Zamani, S. Ozawa, and S. Abe: "Reducing computations in incremental learning for feedforward neural network with long-term memory", *Proc. 2001 Int. Joint Conf. on Neural Networks* (in press).

[3] J. Platt: "A resource allocating network for function interpolation", *Neural Computation*, **3**, 213/225 (1991).

[4] T. Poggio and F. Girosi: "Networks for approximation and learning", *Proc. IEEE Trans. on Neural Networks*, **78**, 9, 1481/1497 (1990).

KES '01
N. Baba et al. (Eds.)
IOS Press, 2001

Feature Extraction from POS Transaction Data by Using Local Independent Components

Katsuhiro Honda, and Hidetomo Ichihashi
Graduate School of Engineering, Osaka Prefecture University,
Gakuen-cho 1-1, Sakai, Osaka, 599-8531, Japan

Abstract. Independent component analysis (ICA), which has been developed mainly in signal processing, is a useful technique for Projection Pursuit as well. For some nonlinearly distributed data, the data set is partitioned into several groups using Fuzzy c-Varieties clustering method before applying the ICA algorithm, which constitute a fuzzy version of Fast ICA by Hyvärinen *et al.*. This paper discusses feature extraction from local independent components by applying the techniques to POS (point-of-sales) transaction data.

1   Introduction

Independent Component Analysis (ICA) is an unsupervised technique, which in many cases characterizes data in a natural way, and is a useful technique for Projection Pursuit as well [1]. In the general formulation of ICA, the purpose is to transform an observed vector linearly into the vector whose components are statistically as independent from each other as possible. The mutual dependence of the components is classically measured by their non-Gaussianity. Maximizing the non-Gaussianity gives us one of the independent components. Projection Pursuit is also a technique developed in statistics for finding "interesting" features of multivariate data. In Projection Pursuit, the goal is to find the one-dimensional projections of multivariate data which have "interesting" distributions for visualization purposes. Typically, the interestingness is measured by the non-Gaussianity. Therefore, the basis vectors of ICA should be especially useful in Projection Pursuit and in extracting characteristic features from natural data.

In spite of its usefulness, the linear ICA models are often too simple to describe real-world data and provide only a crude approximation for general nonlinear data distributions. Karhunen *et al.* proposed local ICA models [2] that were used in conjunction with some suitable clustering algorithms. In the local ICA models, the data are grouped into several clusters based on the similarities between the observed data, ahead of the preprocessing of linear ICA, by using some hard clustering algorithms such as k-Means algorithm. However, the observed data are assumed to be the linear combinations of source signals in linear ICA models. So the clustering methods that partition the data into some spherical clusters like by k-Means are not suitable for the extraction of local independent components, and the data set should be divided into linear clusters.

A technique that uses Fuzzy c-Varieties (FCV) clustering method [3] for extracting local independent components is proposed [4]. The FCV algorithm partitions an observed data set into linear fuzzy clusters based on the similarities of the mixing matrices. Because FCV can be regarded as a simultaneous approach to clustering and Principal Component Analysis (PCA), the FCV algorithm also performs the preprocessing of Fast ICA proposed by Hyvärinen *et al.* [5].

In this paper, we apply the ICA algorithms to a POS (point-of-sales) transaction data and compare the results derived by them.

2   ICA Formulation and Fast ICA Algorithm

Denote that v is the M dimensional observed data vector and s is the N dimensional source signal vector corresponding to the observed data with $N \leq M$. Under the constraints that the elements of source signals $(s_1, s_2, \cdots, s_N)$ are mutually statistically independent and have zero-means, the observed data are assumed to be the linear mixtures of s_i as follows:

$$v = As, \tag{1}$$

where the unknown $M \times N$ matrix A is called the mixing matrix. In ICA, we try to estimate the source signals s_i and the mixing matrix A using only the observed data v.

Fast ICA algorithm proposed by Hyvärinen *et al.* [5] is a useful algorithm that is very simple, does not depend on any user-defined parameters, and is fast to converge to the most accurate solution allowed by the data. Generally, a preprocessing of whitening and sphering by using PCA is applied. In the preprocessing, the observed data v are transformed into linear combinations x,

$$x = Mv = MAs = Bs, \tag{2}$$

such that its elements $(x_1, x_2, \cdots, x_N)$ are mutually uncorrelated and all have unit variance, and $B = MA$ is an orthogonal matrix. Thus we can reduce the problem of finding an arbitrary full-rank matrix A to the simpler problem of finding an orthogonal matrix B, which gives $s = B^{\mathrm{T}} x$.

To derive the matrix B, Hyvärinen proposed the following objective function to be minimized or maximized.

$$J(w_i) = E\{(w_i^{\mathrm{T}} x)^4\} - 3\|w_i\|^4 + F(\|w_i\|^2), \tag{3}$$

where w_i corresponds to one of the columns of the mixing matrix B. The first two terms represent the fourth-order cumulant or kurtosis of the reconstructed signals which is a classical measure of non-Gaussianity. Maximizing the non-Gaussianity of the reconstructed signals gives us one of the independent components. Using the deviation of the objective function, we can obtain the fixed-point algorithm for ICA.

3   Extraction of Local Independent Components by Using Clustering

Even though linear ICA yields meaningful results in many cases, it can provide a crude approximation only for general nonlinear data distributions. Therefore, several techniques where local ICA models were applied to the data grouped by using some suitable clustering methods have proposed.

Karhunen *et al.* proposed local ICA models [2] that were used in conjunction with some hard clustering algorithms such as k-Means. k-Means is a clustering method that groups a data set into C spherical clusters and tries to minimize the criterion,

$$L = \sum_{c=1}^{C} \sum_{v_k \in S_c} \|v_k - m_c\|^2, \tag{4}$$

where m_c represents the mean vector of the cth cluster S_c. The minimum is achieved when the data vectors are divided into C clusters so that the overall mean-square error estimated from the data vectors is the smallest possible. Once we partition the data into C clusters, we apply linear ICA algorithms such as Fast ICA algorithm to each cluster.

Another technique for the extraction of local independent components uses the FCV algorithm [3] which are useful for the grouping of observed data taking the similarities of the mixing matrices into account [4]. The goal of FCV is to determine the memberships of the data in C fuzzy clusters $\boldsymbol{u}_c = (u_{c1}, u_{c2}, \cdots, u_{cJ})^{\mathrm{T}}, c = 1, \cdots, C$ and the M dimensional principal component vectors of each cluster $\boldsymbol{p}_{ci} = (p_{ci1}, p_{ci2}, \cdots, p_{ciM})^{\mathrm{T}}, c = 1, \cdots, C, i = 1, \cdots, N$. Where J represents the number of the observed data. The memberships $\boldsymbol{u}_c$ are constrained as follows:

$$\boldsymbol{u}_c = \{(u_{ck})| \sum_{c=1}^{C} u_{ck} = 1, u_{ck} \in [0,1], k = 1, \cdots, J\}. \tag{5}$$

By using Lagrange's method of indeterminate multiplier, the objective function with entropy regularization [6] is defined as follows:

$$
\begin{aligned}
\min \ L \ &= \ \sum_{k=1}^{J}\sum_{c=1}^{C} u_{ck}\Big\{(\boldsymbol{v}_k - \boldsymbol{m}_c)^{\mathrm{T}}(\boldsymbol{v}_k - \boldsymbol{m}_c) - \sum_{i=1}^{N} \boldsymbol{p}_{ci}^{\mathrm{T}} R_{ck}\boldsymbol{p}_{ci}\Big\} \\
&+ \sum_{c=1}^{C}\sum_{i=1}^{N} \lambda_{ci}(\boldsymbol{p}_{ci}^{\mathrm{T}}\boldsymbol{p}_{ci} - 1) + \alpha \sum_{k=1}^{J}\sum_{c=1}^{C} u_{ck}\log u_{ck} + \sum_{k=1}^{J}\gamma_k\Big(\sum_{c=1}^{C} u_{ck} - 1\Big), \\
R_{ck} \ &= \ (\boldsymbol{v}_k - \boldsymbol{m}_c)(\boldsymbol{v}_k - \boldsymbol{m}_c)^{\mathrm{T}},
\end{aligned}
$$

where λ_{ci} and γ_k are Lagrangian multipliers and α is a positive constant. The entropy term is for fuzzification. The larger the value of α is, the fuzzier the membership assignments become.

Because the FCV algorithm is a simultaneous application of PCA and fuzzy clustering, we can also perform the preprocessing of the Fast ICA algorithm at the same time. After the preprocessing, a modified Fast ICA algorithm named Fuzzy Fast ICA is applied. Fuzzy Fast ICA extracts local independent components from each fuzzy cluster by using following objective function.

$$J_c(\boldsymbol{w}_c) = \frac{\displaystyle\sum_{j=1}^{J} u_{cj}(\boldsymbol{w}_c^{\mathrm{T}}\boldsymbol{x}_{cj})^4}{\displaystyle\sum_{j=1}^{J} u_{cj}} - 3\|\boldsymbol{w}_c\|^4 + F(\|\boldsymbol{w}_c\|^2),$$

where the first two terms represent fuzzy kurtosis defined by considering the memberships $\boldsymbol{u}_c$ derived by the FCV algorithm.

4 Feature Extraction from POS Transaction Data Set

We applied the ICA algorithms introduced in the previous sections to a POS data set to extract meaningful characteristic features and compare the results. The POS data set collected in 1997 at two supermarkets in Osaka is composed of 352 transactions with 9 values: 7 days of the week and the numbers of customers in each supermarket.

First, we applied Hyvärinen's Fast ICA algorithm and extracted two independent components. Figure 1 shows the projection onto the two-dimensional space spanned by the two independent components (IC1 and IC2). In the figure, the horizontal and the

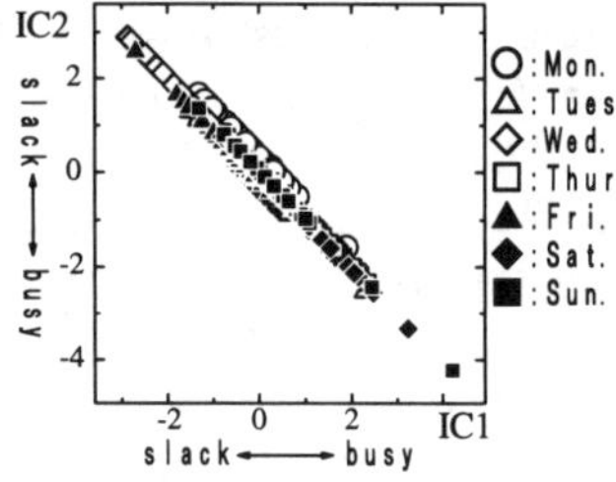

Figure 1: Projection of Independent Components derived by Fast ICA

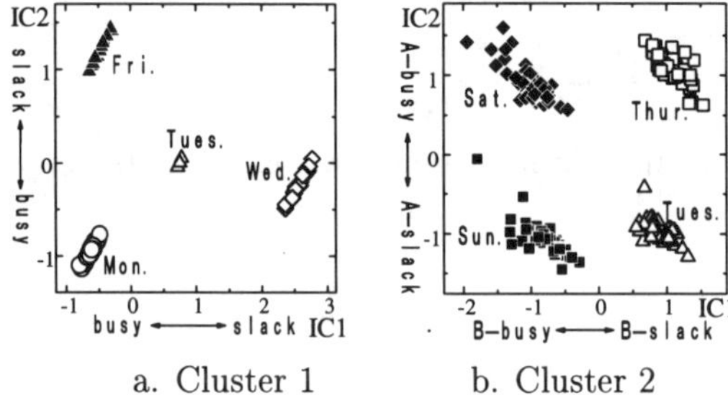

a. Cluster 1   b. Cluster 2

Figure 2: Projections of Independent Components derived by Local ICA

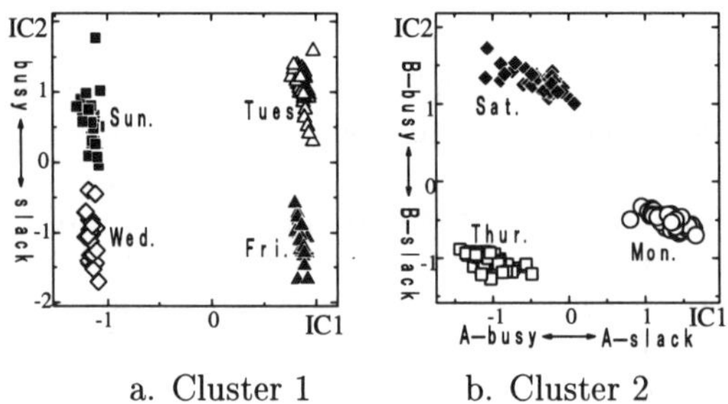

a. Cluster 1   b. Cluster 2

Figure 3: Projections of Independent Components derived by Fuzzy Fast ICA

vertical axes were named based on the correlations between the independent components and the number of customers. For example, the first independent component (IC1) had positive correlation with the number of customers in both the two supermarkets. The data points are distributed on a line and we cannot extract useful knowledge. It is because the distribution of the data is too complicated to analyze with a single model.

Second, we analyzed the data both by local ICA using k-Means and Fuzzy Fast ICA using FCV. Figure 2 shows the projections of independent components obtained by local ICA and Figure 3 shows those by Fuzzy Fast ICA respectively. In the figures, "A-busy" ("B-busy") means supermarket A (B) had many customers while supermarket B (A) didn't have large correlation with the independent component, and vice versa. Figure 2-a and 3-a show the characteristic features that are common to both the two supermarkets. Figure 2-a (3-a) indicates that the number of customers is large on Monday (Sunday and Tuesday) but small on Wednesday and Friday (Wednesday and

Table 1: Average Numbers of Customers for Each Day of Week

day of week	Mon.	Tues.	Wed.	Thur.	Fri.	Sat.	Sun.
supermarket A	613.4	693.7	439.4	827.6	592.0	748.6	720.8
supermarket B	595.9	671.0	478.3	652.0	562.1	750.3	759.9

Friday). On the other hand, Figure 2-b and 3-b show the respective characteristics of each supermarket. In Figure 2-b, we can detect the following features. The number of customers is large at supermarket A but small at supermarket B on Thursday while it is large at supermarket B but small at supermarket A on Sunday. Figure 3-b shows that the number of customers is large at supermarket A but small at supermarket B on Thursday.

Finally, we compared the characteristic features extracted from local independent components with the average numbers of customers. Table 1 shows the average numbers of customers for each day of the week. The number of customers is large on Saturday and Sunday but small on Wednesday and Friday in both the supermarkets while it is large at supermarket A but small at supermarket B on Thursday. These features have apparently appeared from local independent components. Figure 3 shows the features more clearly. The preprocessing using the FCV algorithm is more effective than that using the k-Means algorithm for the feature extraction from the POS transaction data.

5 Conclusion

We have applied the ICA algorithms that extract local independent components by using clustering methods to a POS data set to find some meaningful characteristic features. It is a difficult task to search for interesting projections of multivariate real world data which have nonlinear data distribution, but we could successfully obtain useful knowledge from local models in each cluster. The results showed that the FCV algorithm is more suitable than the k-Means algorithm. The analysis of associations between the number of customers and other elements such as the meteorological elements is left for future work.

References

[1] J. Karhunen, E. Oja, L. Wang, R. Vigario and J. Joutsensalo, A Class of Neural Networks for Independent Component Analysis, *IEEE Trans. on Neural Networks* **8** (1997) 486-504.

[2] J. Karhunen and S. Malaroiu, Locally Linear Independent Component Analysis, *Proc. IJCNN'99*, 1999.

[3] J. C. Bezdek, *Pattern Recognition with Fuzzy Objective Function Algorithms*, Plenum Press, NewYork, 1981.

[4] K. Honda, H. Ichihashi, M. Ohue and K. Kitaguchi, Extraction of Local Independent Components Using Fuzzy Clustering, *Proc. IIZUKA2000*, 2000, 837-842.

[5] A. Hyvärinen and E. Oja, A Fast Fixed-point Algorithm for Independent Component Analysis, *Advances in Neural Information Processing Systems* **9** (1997) 1483-1492.

[6] S. Miyamoto and M. Mukaidono, Fuzzy c-Means as a Regularization and Maximum Entropy Approach, *Proc. IFSA'97* **2**, 1997, 86-92.

KES '01
N. Baba et al. (Eds.)
IOS Press, 2001

Supervised and Unsupervised Fuzzy Discretization of Continuous Attributes for Pattern Classification Problems

Tomoharu NAKASHIMA and Hisao ISHIBUCHI
Department of Industrial Engineering, Osaka Prefecture University
Gakuen-cho 1-1, Sakai, Osaka 599-8531, JAPAN
{nakashi, hisaoi}@ie.osakafu-u.ac.jp

Abstract. In this paper, we examine the performance of two discretization methods of continuous attributes for pattern classification problems. One is a supervised descritization where the class information entropy is used to find threshold values for each attribute. The other is an unsupervised method where each attribute is discretized based on distribution of given training patterns by using a clustering method. Fuzzy and non-fuzzy classification rules are generated from the intervals obtained by the discretization methods. The performance of each discretization method is evaluated by using real-world pattern classification problems, which involve continuous attributes in computer simulations.

1. Introduction

Recently various approaches have been proposed for generating fuzzy if-then rules for pattern classification problems [1-4]. While many sophisticated approaches have been proposed, very simple fuzzy if-then rules were also used as fuzzy classifiers [5,6]. A simple heuristics for generating non-fuzzy classification rules was proposed in [7]. In the generation of fuzzy or non-fuzzy classification rules, one approach is to discretize the domain of each continuous attribute into several intervals. The aim of this paper is to examine the performance of two discretization methods of continuous attributes.

Discretization methods can generally be classified in terms of three aspects [8]. One aspect is supervised or unsupervised discretization. In supervised discretization, class information is used to find threshold values that can effectively divide given training patterns according to their class labels. On the other hand, unsupervised discretization uses no class information, but given training patterns are grouped into several subsets without utilizing clustering.

In this paper, we use an entropy-based partitioning method as supervised discretization and a clustering method as unsupervised discretization. In the minimal entropy partitioning method, threshold values are determined so that given training patterns are maximally divided according to class labels. On the other hand, only the distribution of given training patterns is considered in the clustering method. We specify triangular and trapezoidal membership functions from the intervals obtained by the discretization methods. For comparison, we also generate non-fuzzy classification rules. We compare the performance of each discretization method through computer simulations on read-world pattern classification problems with continuous attributes.

2. Fuzzy Rule-Based Classification Systems

2.1 Fuzzy If-Then Rules

Let us assume that we have m training patterns $\mathbf{x}_1$, $\mathbf{x}_2$, ..., $\mathbf{x}_m$, each of which is described by n continuous attributes a_1, a_2, ..., a_n as $\mathbf{x}_p = (x_{p1}, x_{p2}, ..., x_{pn})$ where x_{pi} is an attribute value (i.e., a real number) of the i-th attribute a_i in the p-th pattern $\mathbf{x}_p$. We assume that a class label is assigned to each of the m patterns. For simplicity, each attribute value is normalized into a real number in the unit interval $[0,1]$. From the given training patterns, we generate fuzzy if-then rules of the following type:

$$\text{Rule } R_q : \text{If } x_1 \text{ is } A_{q1} \text{ and } ... \text{ and } x_n \text{ is } A_{qn} \text{ then Class } C_q \text{ with } CF_q, \tag{1}$$

where R_q is a rule label, A_{qi} is an antecedent fuzzy set, C_q is a non-fuzzy class label, and CF_q is a certainty grade of the fuzzy rule R_q. The generation of fuzzy if-then rules consists of two steps. First, the membership function corresponding to each antecedent fuzzy set is determined. We use triangular or trapezoidal membership functions as in Fig.1. We explain how to generate membership functions in Section 3. Then, we determine the class label C_q and the certainty grade CF_q of the fuzzy if-then rule R_q. A heuristic method for determining the class label and the certainty grade from given training patterns is proposed by Ishibuchi et al.[3].

2.2 Fuzzy Inference

Let us assume that we have K fuzzy if-then rules generated from given training patterns. An input vector $\mathbf{x} = (x_1, x_2, ..., x_n)$ is classified by the single winner rule $R_{q'}$ that has the maximum product of compatibility and the certainty grade:

$$\mu_{q'}(\mathbf{x}) \cdot CF_{q'} = \max\{\mu_r(\mathbf{x}) \cdot CF_r, \ r = 1, 2, ..., K\}, \tag{2}$$

where $\mu_q(\mathbf{x})$ is the compatibility of the input vector $\mathbf{x} = (x_1, x_2, ..., x_n)$ with the fuzzy if-then rule R_q, which is defined by the product operator as follows:

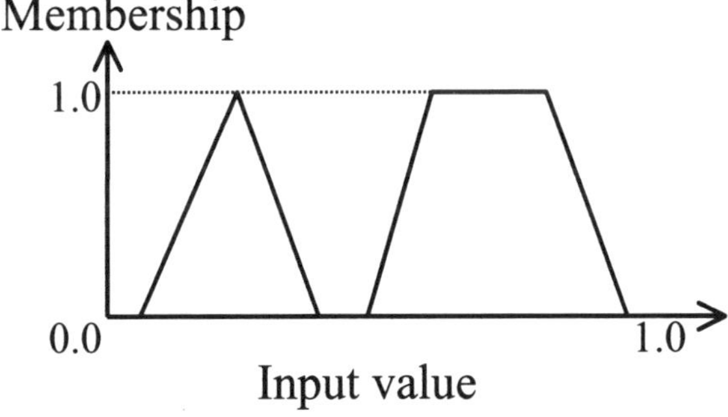

Fig. 1 Examples of triangular and trapezoidal membership functions.

$$\mu_q(\mathbf{x}) = \mu_{q1}(x_1) \cdot \mu_{q2}(x_2) \cdot \ldots \cdot \mu_{qn}(x_n), \tag{3}$$

where $\mu_{qi}(x_i)$ is the membership function of the antecedent fuzzy set A_{qi} defined for the i-th attribute a_i. The input vector $\mathbf{x}$ is classified as the class label $C_{q'}$ of the single winner rule $R_{q'}$ calculated by (2).

When we use fuzzy if-then rules in (1), the membership value takes a real value from 0 to 1. On the other hand, when non-fuzzy if-then rules are used, the membership value takes either 0 or 1. Thus we can define the membership function $\mu'(\cdot)$ by using membership functions as follows:

$$\mu'_{qi}(x_i) = \begin{cases} 1, & \text{if } \mu_{qi}(x_i) \geq 0.5, \\ 0, & \text{otherwise.} \end{cases} \tag{4}$$

3. Discretization methods

3.1 Entropy-based partitioning

This method uses the class information entropy of each partition as an evaluation criterion. Let S denote a set of training patterns. When the set of training patterns S is divided into S_1 and S_2 by a candidate threshold T, the class information entropy $E(S,T)$ induced by the candidate threshold T is calculated as:

$$E(S,T) = \frac{|S_1|}{|S|} \sum_{k=1}^{c} \frac{|S_{1k}|}{|S_1|} \log_2 \frac{|S_1|}{|S_{1k}|} + \frac{|S_2|}{|S|} \sum_{k=1}^{c} \frac{|S_{2k}|}{|S_2|} \log_2 \frac{|S_2|}{|S_{2k}|}, \tag{5}$$

where $|S|$ denotes the number of training patterns in S, S_{jk} is a set of training patterns that belong to class k in a set S_j. We select the threshold that causes the optimal partition with the minimal entropy. Multiple thresholds are selected incrementally. That is, only a single threshold is selected at a time. After determining the first threshold, the next threshold is determined by recursively finding the best threshold with the minimal class information entropy among all the candidate binary partitions of each subset. The number of thresholds to be found is prespecified.

Membership functions are generated based on the thresholds that are determined by the entropy-based partitioning. Let K be the number of intervals for each attribute. In the entropy-based partitioning, $(K-1)$ thresholds should be determined in order to obtain K

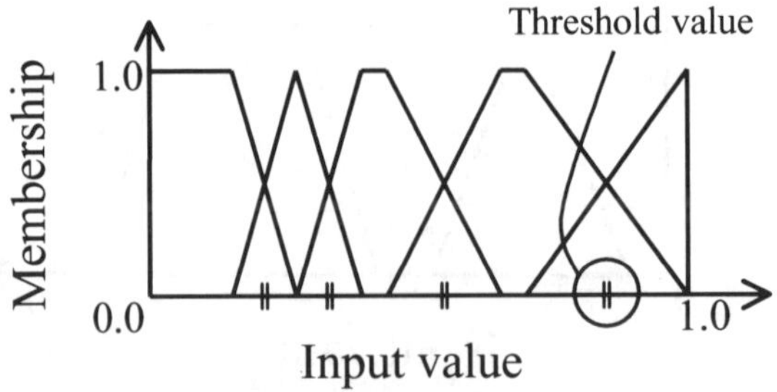

Fig. 2 Generated membership functions from thresholds determined
by the entropy-based partitioning

intervals. Membership functions are generated so that 1) every pair of adjacent membership functions cross with each other at a threshold value and 2) the sum of the membership values is always one (Fig. 2).

3.2 Clustering method

As an unsupervised discretization, we use clustering analysis in this paper. In our clustering analysis, first each training pattern is considered as a cluster. Then the closest two clusters are combined into one cluster. This process is iterated until the number of clusters is reduced to a prespecified number. That is, when the number of intervals is specified as K, the given training patterns are grouped into $(K - 2)$ subsets by the clustering method. As a distance measure between clusters, we use an average linkage method using Euclidean distance. Membership functions are generated from the obtained clusters so that 1) every pair of adjacent membership functions cross exactly at the middle between the means of the corresponding adjacent clusters and 2) the sum of the membership values is always one (Fig. 3).

4. Computer Simulations

We examined the performance of both supervised and unsupervised discretization methods through computer simulations on three real-world pattern classification problems: iris data, appendicitis data, and cancer data. These data sets are available from UCI Machine Learning Databases (ftp::/ftp.ics.uci.edu/pub/machine-learning-databases). The number of classes, dimensionalities, and training patterns is shown in Table 1.

In computer simulations, the number of intervals for each attribute is specified as $K = 2, 3, 4, 5$. We compared the classification ability of fuzzy rule-based classification systems by using the supervised discretization of continuous attributes with the unsupervised discretization. We also examined the performance of non-fuzzy rule-based classification systems. In our computer simulations in this paper, first the domain of each attribute is discretized by the entropy-based partitioning and the clustering method. Then membership functions are generated as described in Section 3. In Table 2 and Table 3, we show the classification results. Table 2 and Table 3 show the performance of each method on the given training patterns. From those tables, we can see that the performance of the supervised discretization was better than the unsupervised discretization for the iris data and the appendicitis data. On the other hand, the unsupervised discretization performed better than the supervised discretization in the case of cancer data. We can also see that the performance of fuzzy rule-based classification systems were better than that of non-fuzzy rule-based classification systems.

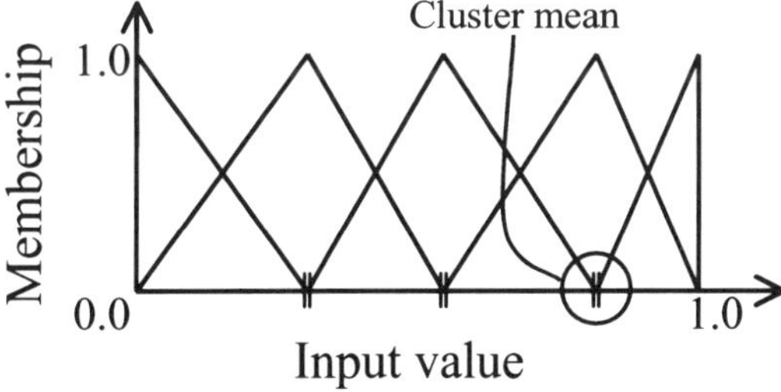

Fig. 3 Generated membership functions from the cluster means obtained by cluster analysis.

Table 1   Data sets used in the computer simulations.

Data set	# of classes	# of dimensionalities	# of patterns
Iris	3	4	150
Appendicitis	2	7	106
Cancer	2	9	286

Table 2   Performance of fuzzy rule-based systems
with supervised and unsupervised discretization.

	Supervised (%)				Unsupervised (%)			
# of partitions	2	3	4	5	2	3	4	5
Iris	80.0	96.0	98.7	98.0	68.7	93.3	93.3	94.7
Appendicitis	89.6	93.4	98.1	98.1	86.8	91.5	91.5	90.6
Cancer	80.4	88.5	88.5	90.2	80.8	94.1	93.4	95.1

Table 3   Performance of non-fuzzy rule-based systems
with supervised and unsupervised discretization.

	Supervised (%)				Unsupervised (%)			
# of partitions	2	3	4	5	2	3	4	5
Iris	73.3	96.7	97.3	98.0	60.0	94.7	85.3	48.0
Appendicitis	90.6	91.5	97.2	99.1	85.9	94.3	92.5	78.3
Cancer	74.1	82.2	88.5	88.1	80.8	86.4	89.2	92.3

5. Conclusions

In this paper, we examined the performance of two discretization methods. One is a supervised method where the class information entropy is used to determine the threshold values for each attribute. The other is an unsupervised method where a given training data set is grouped into several subsets by using a clustering method. Through the computer simulations, we showed that the supervised discretization works better than the unsupervised one.

References

[1] S. Abe and M.-S. Lan, "A method for fuzzy rules extraction directly from numerical data and its application to pattern classification," *IEEE Trans. on Fuzzy Systems*, Vol. 3, No. 1, pp. 18-28, February, 1995.

[2] F.C.-H. Rhee and R. Krishnapuram, "Fuzzy rule generation methods for high-level computer vision," *Fuzzy Sets and Systems*, Vol. 60, No. 3, pp. 245-258, December, 1993.

[3] H. Ishibuchi, K. Nozaki, and H. Tanaka, "Distributed representation of fuzzy rules and its application to pattern classification," *Fuzzy Sets and Systems*, Vol. 52, No. 1, pp. 21-32, November, 1992.

[4] R. Krishnapuram, and F.C.-H. Rhee, "Compact fuzzy rule base generation methods for computer vision," *Proc. of 2nd IEEE International Conference on Fuzzy Systems*, San Francisco, March 28 – April 1, pp. 809-814, 1993.

[5] M. Skubic, S.P. Castrianni, and R.A. Volz, "Identifying contact formations from force signals: A comparison of fuzzy and neural network classifiers," *Proc. of International Conference on Neural Networks*, Houston, June 9-12, pp. 1623-1628, 1998.

[6] H. Ishibuchi, T. Nakashima, and T. Morisawa, "Simple fuzzy rule-based classification systems performed well on commonly used real-world data sets," *Proc. of North American Fuzzy Information Processing Society Meeting*, Syracuse, September 21-24, pp. 251-256, 1997.

[7] R.C. Holte, "Very simple classification rules perform well on most commonly used atasets," *Machine Learning*, Vol. 11, pp. 63-90, 1993.

[8] J. Dougherty, R. Kohavi, and M. Sahami, "Supervised and unsupervised discretization of continuous features," *Proc. of the 12th International Conference of Machine Learning*, Tahoe City, CA, pp. 194-202, 1995.

KES '01
N. Baba et al. (Eds.)
IOS Press, 2001

Interactive Weighing SOM for Data Mining

Masahiro Tanaka†and Yuki Kajitani‡
†Department of Information Science and Systems Engineering, Konan University
Okamoto 8-9-1, Higashinada, Kobe 658-8501, Japan.
‡Graduate School of Information Science and Systems Engineering, Konan University
Okamoto 8-9-1, Higashinada, Kobe 658-8501, Japan.

Abstract. Self Organizing Map (SOM) is a tool to map high dimensional data onto a low (usually 2)-dimensional data. Thus we can visualize the data. We can see the similarity and the difference among the data, but it is severely affected by the normalization. Since there is no automatic mechanism in SOM to tune the weight, the data is usually normalized before computation. However, it is not always reasonable to scale like this. It is not reasonable to give the user-defined weights, either. In this paper, we propose an interactive weighting method by indicating the mapped instance (or equivalently the unit) on the map which the decision maker feels problematic. The weights are increased/decreased based on the difference between the codebook of the unit and the mapped data. A real problem has been treated to emphasize the usefulness of our method.

1 Introduction

Self-Organizing Map (SOM) was desined in early 80's by T. Kohonen [1, 2] and is a paradigm of the neural network to map a high-dimensional space onto a low-dimensional (typically one or two-dimensional) space so that humans can easily understand or visually display the high-dimensional data which is hard to understand as it is.

SOM can be used for data mining in such a way that the resulting 2-D map can be used as a map of preference. Each codebook vector could be used as a representative of a cluster where the mapped data are almost the same and the preference is also at the same category. However this cannot be achieved by simply using the traditional SOM. The problem is that the sensitivity of the preference with respect to the elements are not the same. For some elements of the data, the preference may be very sensitive, while it may not for other elements. When the scale of the items are of different magnitude, this could happen. In this case the scaling can be a useful way. However, rescaling is not a universal tool.

2 SOM for Data Mining

SOM can be designed in arbitrary dimension, typically one or two dimensional, so that humans can directly see the unit information. In this paper, we will treat the 2-Dimensional SOM. The coordinate of the unit is expressed as (i, j), where $i \in \{1, 2, \cdots, p\}$ and $j \in \{1, 2, \cdots, q\}$. Assume the data is n-dimensional. Each unit has a codebook vector $m_{(i,j)} \in \mathbf{R}^n$. The codebook vector is some representative of the data, and each instance of data is mapped onto the unit whose codebook is nearest in Euclidean distance among the whole units.

2.1 Data Mining

The objective of data mining is to find useful information that is not explicitly seen in large databases. There are various ways to express the "knowledge". "IF-THEN" rules are often very useful for the human beings. Rule induction systems ID3, C4.5 and C5.0 developed by Quinlan [4][1] are the methods of extracting decision tree that consists of "IF-THEN" rules. Fuzzy modeling methods are used to extract "IF-THEN" rules which is robust for noise but is not suitable for highly structured rules.

Although SOM is known as a visualization tool, it is important to consider how we want to use SOM. Unlike decision tree systems, SOM does not directly yield rules. It is rather better to use SOM to see the data visually and then use other methodologies such as rule induction methods. Hence most of these tasks have not been formalized, and they are still relied on the users' skill of analysis.

As can be seen in most softwares of SOM (e.g. SOM Toolbox in MATLAB[2].), it is equipped with "scaling" function. This is essential for SOM, because the mapping of SOM is determined by the Euclidean distance and hence the mapping is determined only by the large-scaled variable. Therefore, even when a small difference of some variable is very important, it is hidden in the noise if some other variables are large-scaled.

3 Interactive Weighting

The cases we assume are the following. Our objective is to "mine" the hidden information in the data.

Most of the tools of SOM are desinged to use tne weight for each of the elements. The standard way of weighting is to "normalize" the range to $[0, 1]$. However, it is not always efficient to normalize in this way. The algorithm we propose here is to semi-automatically give the weights by indicating the units of the map that seems inappropriate.

First we show the learning algorithm.

1. Normalize the data in the standard way. Define $\boldsymbol{w} = [1 \ 1 \ \cdots \ 1]$.

2. Use the weight as $\boldsymbol{u}(k) = [w_1 x_1(k) \ w_2 x_2(k) \ \cdots \ w_n x_n(k)]$, $k = 1, \cdots, N$ and train SOM with weighted data $\boldsymbol{u}$.

3. The system shows the data numbers (or some other id labels) mapped to each unit to the DM.

4. The DM chooses a unit which he/she thinks most problematic.

5. If there is no problematic unit, the procedure terminates. Otherwise goto the next step.

$$w_i^{\text{new}} = \frac{w_i^{\text{old}} + \dfrac{c \sum_l |m_i - u_i(l)|}{\sigma_i}}{\displaystyle\sum_j \left(w_j^{\text{old}} + \dfrac{c \sum_l |m_j - u_j(l)|}{\sigma_j} \right)}, \quad i = 1, \cdots, n$$

---

[1]See also http://www.rulequest.com/see5-info.html for a commercial software C5.0.
[2]MATLAB is a software by Mathworks.

where σ_i is the standard deviation of the i-th element of the codebook vector m_i over all the units, and the summation for $|u_i - m_i|$ is over instances that are mapped to this unit i. c is a training parameter.

6. Goto Step 2.

When the algorithm terminates, we have the set of weights which clearly reflects the preference of the DM.

4 An Application

We applied our algorithm to the visualization of "used car sales" problem. The customers have their potential favorite attributes, but it is not always easy to state as words. With this system, it is possible to visually explore their attitudes. The data was from real ones, and the attributes are as follows. In all the experiments, the data is "normalized", that is, each attributes are rescaled so that the range is almost an interval $[0, 1]$ (with some exceptions). The attributes of the data are price, age, mileage, engine size, number of doors, AT/MT and car model.

To apply SOM, we augmented the AT/MT attributes into 2 attributes, and the car model into 14 attributes because there were 14 models of cars. Figure 1 shows the SOM in the first stage; i.e. the weights are all ones. c was set to 0.2.

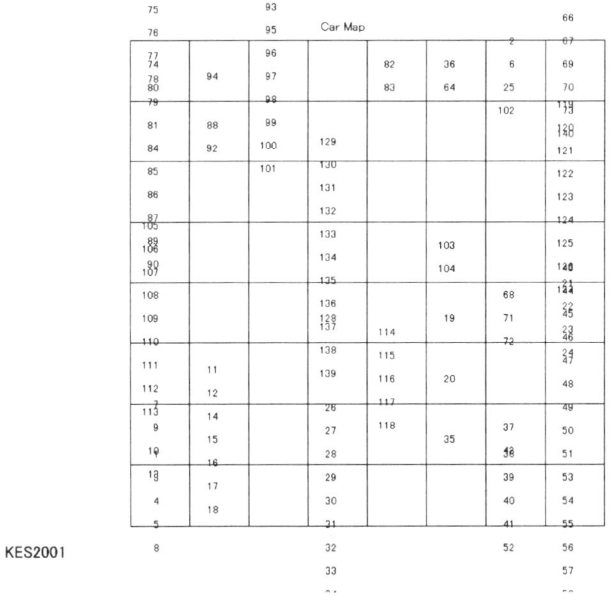

Figure 1: Labels on trained SOM with uniform weights

We can see that data are distributed unevenly, and the clusters are mainly due to the car models (or kinds). Since it is not an interesting phenomenon, we indicated an instance number which is located where the DM has an interest and also within a big mass. It is also effective to indicate a number where some instances in the same unit has different values to the DM, e.g. the price, the age, etc. Figure 2 shows the visualized SOM which was trained after 3 times' interaction. Table 1 was obtained at the final stage.

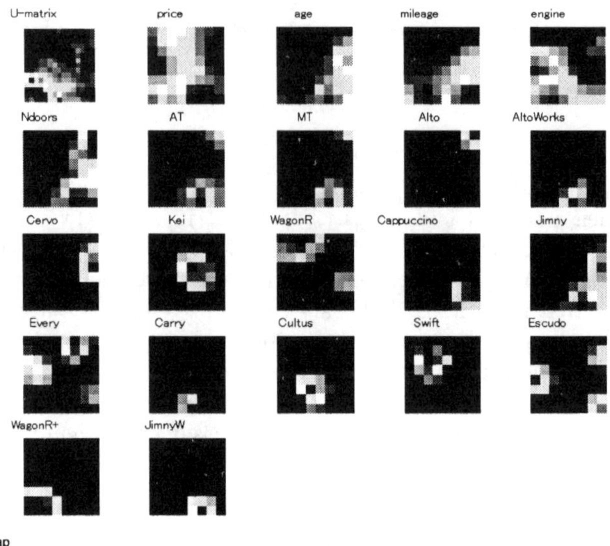

Figure 2: SOM after interaction

The DM decides his attitude also by looking at the Figure 3. Although the image is not very clear, we have much more information as it is dispayed with color. We can see in Table 1 that the learning process is pretty stable. After 3 times interaction, we got the coefficients as shown at the 4th row. Wen can see that the elements 1-3 (corresponding to price, age and mileage) are weighted more heavily than the first trial.

Table 1: Weights transition for 21 elements.

	1	2	3	4	5	6	7	8	9	10	
1	1.00	1.00	1.00	1.00	1.00	1.00	1.00	1.00	1.00	1.00	
2	2.44	1.75	1.70	1.60	0.79	0.79	0.79	0.79	0.79	0.79	
3	2.55	1.94	2.02	1.28	0.97	0.63	0.63	1.36	0.63	0.64	
4	2.48	2.08	1.96	1.29	1.17	0.61	0.61	1.31	0.63	0.62	

	11	12	13	14	15	16	17	18	19	20	21
1	1.00	1.00	1.00	1.00	1.00	1.00	1.00	1.00	1.00	1.00	1.00
2	0.79	0.79	0.79	0.79	0.80	0.79	0.81	0.79	0.80	0.79	0.79
3	0.90	1.67	0.63	0.63	0.65	0.65	0.65	0.63	0.63	0.69	0.63
4	0.86	1.61	0.61	0.67	0.65	0.62	0.62	0.61	0.66	0.67	0.64

5   Conclusions

We proposed a data mining method using SOM. The method is interactive, where the DM can indicate units which seems to be interesting to the person but it still needs

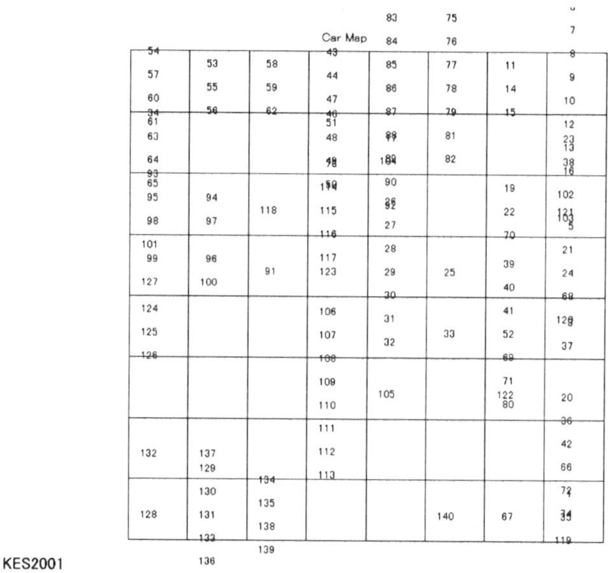

Car Map

KES2001

Figure 3: Mapped labels after interaction

to be separated. By repeating this esperiment, we noticed that DM tends to indicate units which has attracted many instances. Thus, it may be useful to use automatic mechanism to scatter the units with may instances together with the proposed method. It will be our future work.

Acknowledment

We made a program based on (MATLAB's) SOM Toolbox[3], Version 2.0beta created by Esa Alhoniemi, Johan Himberg, Juha Parhankangas and Juha Vesanto.

References

[1] T. Kohonen, Analysis of a simple self-organizing process, Biological Cybernetics, **44** (1982) 135–140

[2] T. Kohonen, Self-organizing formation of topologically correct feature maps, Biological Cybernetics, **43** (1982) 59–69

[3] T. Kohonen, Self Organizing Maps, Springer-Verlag (1995)

[4] J. Ross Quinlan, C4.5: Programs for Machine Learning, Morgan Kaufmann (1992)

[5] M. Tanaka, Y. Furukawa and T. Tanino, Weight Tuning and Pattern Classification by Self Organizing Map Using Genetic Algorithm, Proc. 1996 IEEE International Conference on Evolutionary Computation (1996) 602–605.

---

[3] http://www.cis.hut.fi/projects/somtoolbox/

Uncertainty in Software Engineering Life Cycle: The Very Idea

N. Srinivas
Department of Information and Computer Science, King Fahd University of Petroleum and Minerals,
Dhahran, Saudi Arabia. drsrini@hotmail.com

ABSTRACT

Certain prominent situations in software engineering life cycle that gives room for uncertainty are highlighted. A mathematical framework has been proposed to simulate human decision-making process and is used to handle uncertainty.

Key Words: Uncertainty, Decision Making, Software Engineering,

1. INTRODUCTION

To arrive at a decision in the presence of absolute certainty with respect to all the relevant facts and considerations is rarely affordable to human beings. Uncertainty in any process occurs in many forms, at several stages. For human, the intelligence is an in-built capability. Thus each time, assumptions must be made about data values which are not available, about events which may or may not have occurred, and about consequences likely to flow from a given decision (Pearl, 1988). Many of these assumptions may be made unconsciously or subconsciously. Some may be made explicitly, with whatever degree of justification may be abducted (Zadeh, 1983). Otherwise, only the rules of thumb and acquired experience will serve as a guide.

In the field of software engineering (SE) area, every time, tentative agreement is most important than right. The final judgment with reference to each situation very much depends on the nature of the project at hand. There comes the vital need for a software engineering process (SEP), to monitor and refine the judgments relative to each software project. These processes are complex entities that need to be described and assessed in a precise and unambiguous way (Bandinelli etal, 1995). Inconsistency is inevitable at several stages of this process due to several reasons. Thus there is an acute need to develop a mechanism for an intelligent decision-making. In this peace of work we focus our attention more towards simulating the approach of human decision-making.

During the SE life cycle, one often observes that there are several situations where both the customer and the software engineer were facing uncertainty, due to imprecise or incompleteness of information. Some times it may be also due to a communication gap between the customer and a software engineer (Humphrey, 1989). It is also strongly felt that SE is a study to obtain information about differences between plans and reality in software development (Musa, 1975). Since it involves more human-to-human interaction, than human to machine interaction, leaving a wider scope for human decision-making in reducing inconsistency. At several situations in this paper we adopt the word uncertainty in place of inconsistency, for the purpose of clarity.

We organize this paper as following: In the initial four sections we focus more towards briefing some of the uncertain situations why, how and when they occur in the SE life cycle. In Sections 5 and 6, we present our formalism along with the decision process. Finally in our last section we mention some of the conclusions of our approach.

2. THE CONTEXT

Existing software development process methodologies (SDPM) are complex entities that we have to deal with. The performance of these entities will be in terms of cost, expected quality of delivered software products, and the development time (Genucheten, 1991). This is so, mainly due to the associated uncertainty at several stages of the software development. This necessitates us to identify and if possible remove this uncertainty while we proceed from one stage to the other in a SEP. Inconsistency in a way is an essential ingredient of the evolutionary processes that are found in self-organizing dynamically systems. It allows for variability of cause-effect relations and of interpretation, which is an essential characteristic ingredient of the inductive processes of abstraction, generalization and recognition order embedded in sensory data. We strongly feel that the quality and reliability of any software product is always proportional to that of the SEP, through which it has generated.

Thus reliable processes always produce the reliable products. It is also evident that any software system's "maintenance" activities consume forty-eighty percent of the total system's life cycle cost. This is mainly due to our poor decision making at several stages of the software development. Maintainability is mainly influenced by factors understandability, modifiability and testability [4].

In the context of SE, inconsistency in the customer requirements, at first leads to the lack of proper understanding of the SEP. The understandability thus takes a prominent step in the entire SE life cycle. Initially during the conversation between the "customer" and the staff at "software development organization (SDO).

For example, the customer might initially provide only x% of the requirements and will provide the remaining y% (with x + y = 100%) later at different time intervals. Then this y% of the requirements can roughly be classified as:
(a) Just some additional to the earlier given requirements
(b) Modifications to the existing requirements
(c) Could be the contradictory versions of the earlier provided requirements.

Thus according to the changes in the customer requirement specifications (CRS) the software engineer is forced to modify his design and code, many a times in the entire SDPM. This in turn causes inconsistency to the design, code and even to the testing conditions. Among these the third possibility (c) involves more amount of rework. The following are few reasons in this direction.

1. Some of the CRS could be inconsistent, based on several reasons such as:
   a) Customer's knowledge level might be poor, as the customer may not be aware of all the CRS in the beginning of the project.
   b) Project team's knowledge level while understanding the CRS.
   c) Timely not receiving the clarification to the CRS.
2. Communication gap between customer and the software developers. This is more depends on the degree of clarity of the CRS being communicated. The CRS are often given verbally over teleconferences or across the table over the discussions. Some times the right questions are wrongly posed to the customer, leading to misinterpretation of requirements.
3. Conflicts and confusions among the upper, middle and lower management, on various aspects (such as quality, reliability, risk factors) of the project.

4. Senior personnel leaving the organization for what ever the reason also create inconsistency to the SEP
5. During the cost and size estimation of the project.
6. During staffing: This is a compromise between experience and qualifications of the candidates
7. Inconsistency can also occur when the CRS are transformed into software requirement specifications (SRS).

In all these situations the human decision making process play a vital role in correcting and fixing the issue apart from improving the SEP. Thus, at several situations in SE life cycle a tentative agreement is most important than right. This inconsistency is mainly due to the availability of information "for" and "against" the context, at the same time.

3. WHY TO MEASURE UNCERTAINTY

The necessity for measuring inconsistency in the SE life cycle is aid for an intelligent decision making at several stages. For example, when the CRS are incomplete, one must have a flexible design to transfer these CRS to SRS, prevents project slippage. At the same time the inherent consistency in the design and code (as a result of inconsistencies in the CRS), should not adversely affect the SEP in such a way that the well-defined process is abandoned altogether. In such cases, there needs to be an optimum check/balance between cost, schedule and quality of the final product [13].

Thus measuring uncertainty is critical to have an optimized check over the cost and schedule of the project. In most of the software projects, maintenance accounts for a large part of the system's cost. Correspondingly it also provides large potential for reducing the total system's expenditures, mainly due to the system's uncertainty. There by the total maintenance cost can be reduced by reducing the total system's uncertainty. At the same time, one can also witness the improvement in the quality, reliability and lead time of the project.

Many researchers [1] [2] argue that measuring uncertainty helps the senior management while planning and organizing projects. This can be witnessed especially during their extraction of data and information from the customer as well as from the project leaders.

Measuring uncertainty also gives us a clue for improving the reliability of the organizational process [11].

4.   WHEN TO MEASURE UNCERTAINTY

To have a uniform continuity in the SEP and to stress some of the important key process areas, proposed by SEI CMM [8] for SEP, we need to know when to measure uncertainty. Some prominent situations for measuring uncertainty during the SE life cycle are:

a)   While analyzing the CRS. This will help us in ensuring the clarity to the SRS, for fixing the cost, schedule and team size.
b)   While analyzing the SRS. This is a crucial stage, since it leads to the most crucial phase in the entire SE life cycle, design.
c)   While communicating with the customer. The software engineer has to remove and resolve the communication gaps (if any) while obtaining the requirements and clarifications to the requirements with the customer, at several time intervals. One way of achieving this is to have a clearly defined mechanism in place for communication and ensure that requirements and clarifications are obtained only in the written format [14].

Next one should try to identify the parameters that influence the reliability of the project. These parameters may include, a) clarity in the requirements, b) expertise of the project team c) completeness of the requirements d) availability of adequate man power e) availability of all equipment's and so on.

Any software project does not overrun because of one or two major problems, but rather because of a large number of minor problems. Once if we look into the details of these minor problems, one can realize that uncertainty emerges because of two different factors that influence the customer specifications, at the same time. These are, the success factor (to establish the success of a project) and the failure factor (to oppose the success). There comes the need for an intelligent decision making while handling these individual measures.

We normally assume that all the requirements are feasible, having their success factor to the maximum extent. In reality, one should realize that each of the customer requirements could have their own limitations towards their success and failure, as far as their implementation is concerned.

Initially they give an impression as if they are easily adaptable, but in practice they are not.

Further, measuring uncertainty of an existing (or completed) project is definitely possible, but requires continuous collection of data that the management has determined, together with the data already existing for that particular project.

5.   THE BASIC FORMALISM

We measure uncertainty by associating an uncertain measure (UM), to each of the CRS. UM of a requirement (in simple terms a proposition) is an ordered tuple

$$(a,b) \in [0,1] \times [0,1]$$

As shown in the Fig. 01, where "a" represents the positive measure (or success factor) and "b" represents the negative measure (or failure factor) of a particular CRS [7]. These two measures are independent by nature due to the very fact that any requirement's success/ failure factor depends solely on the several independence sources of evidence that establish them. Thereby the sum of these factors may not always be equal to 1, as is in the case of probability theory.

Thereby UM for a perfectly feasible CRS will be always be (1,0) and is (0,1) that of for non-feasible CRS. We thus introduced an algebra to derive the UM for the compound CRS arising out of logical symbols such as $\neg$, $\wedge$, and $\vee$, and are defined as following:

Let P and Q are two CRS, with UM (P) = (a, b) and UM (Q) = (c, d) then,

(1)   UM $(P \vee Q)$ = {Max (a, c), Max (b, d)}
              if   both a+b > 1, c+d > 1,
         ={Max(a,c),Min(b,d)} Otherwise.
(2)   UM$(P \wedge Q)$ = {Min(a,c), Min(b,d)}
              if both a+b > 1, c+d > 1,
         ={Min(a,c),Max(b,d)} Otherwise.
(3)   UM$(\neg P)$ = {1-b, 1-a} if a+b > 1,
         = {1-a, 1-b} Otherwise.

If the total measure is greater (or less) than 1, we say that we have excess (or lack of) evidence, in either of the cases we are under uncertainty. Here we made a clear distinction between the uncertain points having their total value (i.e., a + b) - which we call it as reliability factor, greater than one (i.e., a + b > 1) from the rest, for the following obvious reasons:

1.   Due to the independent nature of the sources of information.
2.   Compatibility of this algebra.

The following are a few worth noting results in this connection, which can be easily verified.

(a) UM $(T \vee F)$ = UM (T) and UM $(T \wedge F)$ = UM (F) holds for a true proposition T (i.e., UM (T) = (1,0)) and false proposition F (i.e., UM (F) = (0,1)).

(b) UM $(\neg T)$ =UM (F) and UM $(\neg F)$ =UM (T).

(c) UM $(P \vee Q)$=UM (P)+UM (Q) – UM $(P \wedge Q)$, holds for any P and Q.

(d) With reference to Dempster Shafer logic, the interval of Belief (Bl) and Plausibility (Pl), viz., (Bl, Pl) is equals to (a, 1-b). There by, Bl is the Positive evidence, and Pl is that of (1 – Negative evidence).

6. THE ASPECTS OF DECISION MAKING

Once we collect the evidence, next we look for a mechanism to infer our decision, especially in the case of sum of evidences exceeds one. For UM (P), if $a + b > 1$, we say it has an excess evidence otherwise we call it lack of evidence. In either of these cases, to make we normalize P as

UM (P_0) = $\{a/(a + b), b/(a + b)\}$ = (a_0, b_0) = (x, y) with $x + y = 1$.

This is a sure point on the line of demarcation viz., $a + b = 1$ as shown in Figure 02. Since each point on this line obey the basic probability axioms, we call it as the 'reliability' of P. Next for maximizing the success factor, (and minimizing the failure factor), we adopt the following procedure:

Set UM (P) = (a, b) with $a + b > 1$. Using simple geometric principles one realize that

UM (P_1) = $\{(1 + a - b)/2, (1 - a + b)/2\}$ if $a > b$
$\qquad$ = $\{(1 - a + b)/2, (1 + a - b)/2\}$ if $a < b$

By considering the convex combination of the points P_0 and P_1 (as shown in the Fig. 02 we get UM of any general point P_i as

UM (P_i) =(a_i, b_i)=$q\{a/(a + b),b/(a + b)\}$ + $(1 - q)\{(1 + a - b)/2,(1 - a + b)/2\}$ where $0 < q < 1$.

At this junction we can also observe that
a_i = the success factor of [q $\{a/(a + b), b/(a + b)\}$ + $(1-q) \{(1+a - b)/2, (1 - a + b)/2\}$]> $a/(a + b)$

b_i = the failure factor of [q $\{a/(a + b), b/(a + b)\}$ + $(1-q) \{(1 + a - b)/2, (1 - a + b)/2\}$] < $b/(a + b)$

Thus one can also witness that
Max $\{a_i - a_1\}$ = $(a - b)/2(a + b) \{a + b -1\}$ and Min $\{b_i - b_1\}$ = $(b - a)/2(a + b) \{a + b -1\}$

Thereby once we calculate the UM of a project, it is possible to look at maximizing its success factor, and minimizing the failure factor. This improved success factor will give us the improved reliability of the project. Which in turn help in identifying the respective parameters that influence the reliability of the project.

This method has been tested for its consistency both logically and mathematically. It is observed that, the UM obtained from this process (for a particular project) gives us a clue to establish the project's technical quality and reliability.

7. CONCLUSIONS

We represent uncertainty surfacing in a SE life cycle as a bi-directional entity. We conclude that this way of representation resembles more close to human decision making process. Uncertainty handling by this approach for this domain is superior to that of Bayesian approach for the following reasons:

1. Non-availability of the complete information in the beginning of the project.

2. Frequent change requests of the customer may have impact on the entire SE process, which cannot be dealt with the conditional probability.

REFERENCES

1. Anthony C. W, etal. Inconsistency Handling in Multiperspective Specifications, IEEE Tr. SE, Vol. 20, No.8, August 1994.

2. Musa J D., A Theory of Software Reliability and Its Application, IEEE Tr. SE, Vol. SE-1, No. 3, September 1975.

3. Genuchten M. V, Why Software Late? An Empirical Study of Reasons for Delay in Software Development, IEEE Tr. SE, Vol. 17, No., June 1991.

4. Pickard M M and Carter B D., Maintainability: What Is It And How Do We Measure It? Software Engineering Notes, Vol. 18, No., July 1993.

5. Bandinelli S, etal. Modeling and Improving an Industrial Software Process, IEEE Tr. SE, Vol. 21, No. 5, May 1995.

6. Srinivas N, Sarma V.V.S, and George Smith. Representation of Evidence: An Approach to Uncertainty Handling in Knowledge Based Systems, Proc. International Symposium on Artificial Intelligence, Industrial Fuzzy Control and Intelligent Systems, ISAI/IFIS, Cancun, Mexico, Nov. 12 - 15, 1996.

7. Srinivas N, Uncertainty Handling in Knowledge - Based Systems via Evidence Representation, A Ph.D. thesis submitted to Department of Mathematics, Indian Institute of Science, Bangalore, India, May 1996.

8. Basili V. and Green S., Software Process Evolution at the SEL, IEEE Software, July 1994.

9. Pearl J., Probabilistic Reasoning in Intelligent Systems: Networks of Plausible Inferences, Morgan Kaufmann, San Mateo, California, 1988.

10. Zadeh L A., The Roll of Fuzzy Logic in the Management of Uncertainty in Expert Systems, Fuzzy Sets and Systems, Vol. 11, 1983.

11. Blackburn J.D., Scudder G.D., and Wassenhove L.N.V., Improving Speed and Productivity of Software Development: A Global Survey of Software Developers, IEEE Trans. Soft. Engg. , Vol. 22, No. 12, December 1996.

12. Humphrey W. S., Managing the Software Process. Addition - Wesley, Reading, Mass., 1989.

13. Donlad J Reifer, A tale of Three Developers, IEEE Computers, November 1999.

14. Srinivas N. and George Smith, Discontinuities in a Software Engineering Process, ICSE 97, Boston, USA, May 12 – 14, 1997.

Fig.01

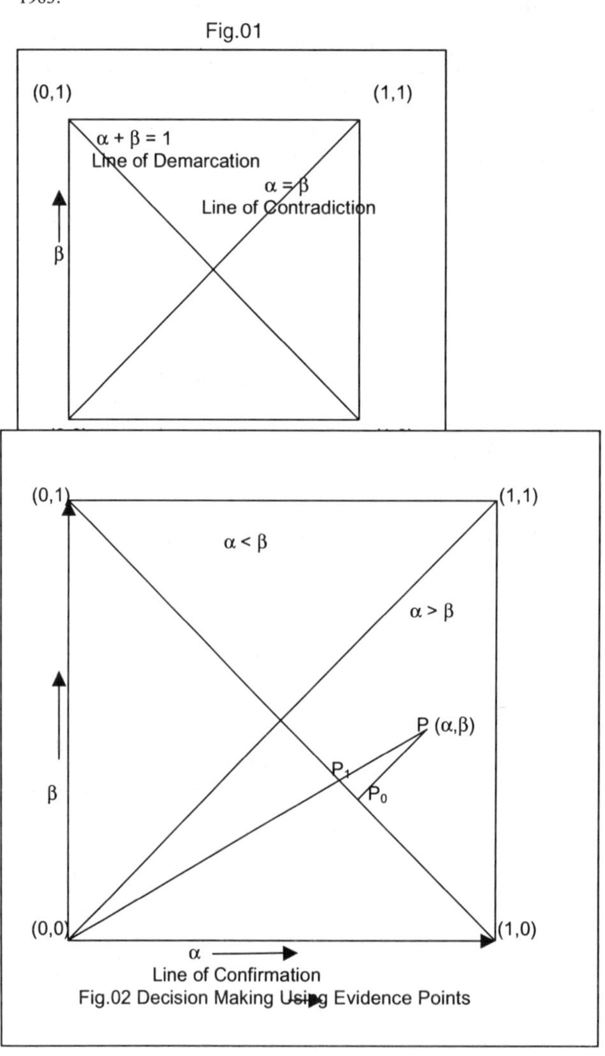

Fig.02 Decision Making Using Evidence Points

Fuzzy Techniques for Software Reuse in-the-Small

C. Bellettini *, E. Damiani ** , M. G. Fugini ***[1]
* Università di Milano, Dipartimento di Scienze dell'Informazione
** Università di Milano, Dipartimento di Tecnologie dell'Informazione
*** Politecnico di Milano, Dipartimento di Elettronica e Informazione
e mail: belletc@dsi.unimi.it,edamiani@crema.unimi.it,fugini@elet.polimi.it

Abstract

In this paper we present a *closed loop*, model-independent approach to software reuse aimed at automatic tuning of software reuse systems. To this end, we rely on polling developers' for their opinions about the reusability value of components, but instead of requiring them to express their views explicitly, our approach relies on observing the user reactions to query answers. The proposed techniques were conceived for *reuse in-the-small*, by fixing some locally measured reuse coefficients for small teams.

1 Introduction

It is widely recognized that taking software reuse systems at the quality level where real benefits can be gained is a difficult task, and requires a great deal of commitment and effort. In some application fields, such as telecommunication, medical, aerospace domains, reuse problems have often been satisfactorily solved exploiting a variety of market place, domain modeling, and product-line approaches. Large organizations are increasingly aware [19] show awareness of the need of *documenting reusable patterns and architectures*, that is, of preserving knowledge and experience of development groups. The same cannot be said for small companies, for which approaches to reuse requiring special purpose environments and specialized user roles [2] are simply unfeasible, to the high rate of change of the environment they operate in. This reason has historically prevented many of them to adopt any kind of reuse policy. Even for well-focused companies, volunteer efforts at knowledge sharing are not enough, due to high personnel turnover. In this paper we try to bridge this gap by exploring software reuse techniques tailored to small companies which are low on the process maturity scale.

A base of reusable artifacts is a crucial tool to preserve the history of projects, lessening the software maintenance burden, and improving the relationships between software producers and customers. Recent research [4] suggests that such tools allow small companies to move forward in their process maturity by encouraging competent work on the design of the code.

The approach presented in this paper relies on the fact that in order for reuse to work in the small, only simple computer-support can be assumed. Even when complex systems considering all kinds of software artifacts [7], including organizational issues, cannot be used due to the small company size, reuse can benefit form simpler tools such as code libraries and component repositories. In our opinion, this kind of support for reuse does not prevent but rather fosters informal communication within the development team. To this aim, we describe a technique suitable for complete automation and automatic adaptation taking into account user feedback, that can be easily coupled with rewarding schemata such as the one described in [29]. We address component based reuse [13], since it is easy to understand and simple to implement, but we do not rely on any specific reuse model: the operations of our approach can be tuned according to the user needs and the organizational constraints.

---

[1] Address for correspondence: Prof. Ernesto Damiani Dipartimento Tecnologie dell'Informazione- Universita' di Milano polo di Crema, Via Bramante 65 26013 Crema Italy edamiani@crema.unimi.it

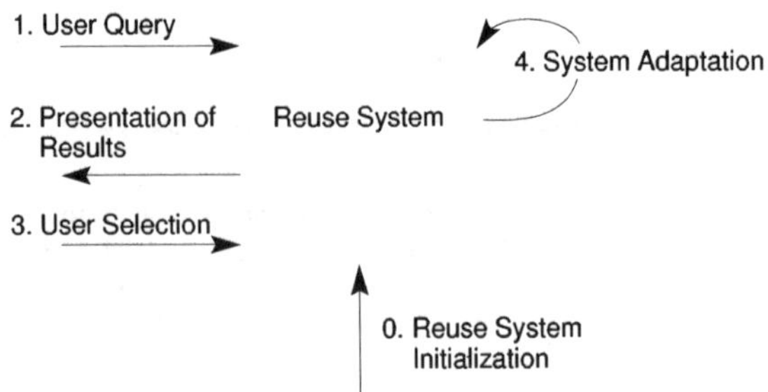

Figure 1: The closed-loop reuse system operation

2   *A closed loop framework for reuse*

Traditionally, large-scale reuse systems rely on repositories where components are classified and organized according to complex company-wide reuse models. Generally speaking, such repositories support two broad approaches to components retrieval. *Information retrieval* techniques are used to search for relevant information based on text, while *software component retrieval* techniques locate software components based on software specifications [28]. In the software component search techniques, several notions of matches were defined, including exact and fuzzy ones . Both categories of reuse systems can be regarded as *open loop* ones [7], since their complexity makes them difficult to tune: the effectiveness of their classification and retrieval of software artifacts depends uniquely on the set of techniques used to initialize the system. In this section we outline how a simple *closed loop* reuse system can be built incrementally, tuning it along the system life cycle according to the views of the user community. This adaptive technique allows one to build a company-wide base of software artifacts without deploying a fully-fledged reuse model. To organize such a knowledge base, we rely on polling developers' for their opinions about the reusability value of components, but instead of requiring them to express their views explicitly, our approach relies on observing the user reactions to query answers. In order to take into account the different levels of skills and awareness of the users about the reuse policy of the organization, we link the importance of feedback to the user typology through a synthetic *user profile*. In Figure 1, we show the framework of a multi-level, model-independent closed-loop reuse system [2] including user feedback collection. Our approach is based on polling user choices as *votes* on the reuse quality of artifacts. Such votes are also used to update user profiles. The operation of our system includes four basic steps.

Step 0.   This step involves the creation of a multi-level reuse system including code and code documentation, design documentation and requirements. We make no assumption on the classification model, if any, used by the system. In small companies it can be as simple as physical storage of artifacts using a conventional file system [23], while in larger organizations it could rely on a more structured approach (e.g., [8]).

Step 1.   As in the previous step we make no assumption on the interaction style between the reuse system and the user. They can vary from simple text retrieval interfaces for very small systems [23] to complex query languages for structured environments [7][8][26].

Step 2.   The reuse system is supposed to have a query interface (possibly a simple text-retrieval one) whose answer is not a single artifact but a set (possibly ranked) of candidate components. Whatever their interaction model, most available reuse systems can provide a ranked list of candidates [1], when answering a user query; even in the absence of such a repository, simple file-system level utilities may provide a similar result.

Step 3.   The user selection is exploited to poll his/her preferences and to log information that will be used for feedback collection, user characterization, and rewarding.

2.1 Assumptions and Prerequisites

We are now ready to list some basic features of our closed-loop reuse techniques. As we shall see, prerequisites are kept to a bare minimum in order to make the approach feasible even for companies with limited resources.

Prerequisites

♦ A simple mechanism of code or class hierarchies browsing is assumed to be available for component search, as in most current programming environments; alternatively, we assume that a retrieval engine exists.

We assume that a cooperative development environment is available that can log some simple development actions (resource access, component selection/modification, project-related intercommunications, etc.).

Features

♦ *Individual profiles.* Users are profiled individually, rather than by user typologies, processes, and/or roles. This seems more reasonable for small groups, where activities scheduling may occur on a day-to-day basis. Instead, in large development teams individual differences matter less, users are better characterized by their role in the team, and the usual process-based measurement is more reasonable [2].

♦ *Local measures.* Required parameters are measured *locally*, that is, for each development environment. Reuse measures are taken automatically through a comparison between reusable components before and after adaptation. In our opinion, reuse in small companies can be effective only if made as most *automatic* as possible.

♦ *Easy initialization.* Even if based on simple tools, such as Unix `grep`, `diff`, or more complex tools such as versioning environments, the reuse system should require only a basic initialization of parameters, to be performed manually and easily by project managers.

♦ *Independence of organizational structure.* Focusing on small companies, we assume that the organizations' structure is flat, highly dynamic, or not precisely defined. Therefore, organizational parameters, such as user roles and project responsibilities, are not applied.

3 Profiling the Reuser

While in big organizations it is an accepted practice to profile software development processes, rather than individuals, small company environments are not always suitable for this kind of analysis. Indeed, in small companies a process involving reuse is often associated with an individual developer, who can change or even work part time [18]. Hence, individual profiles are a critical issue to ensure the success of reuse in the small. It has been widely recognized for nearly a decade [11][24] that empowering the development team has a central role in the management of reuse. However, experience has also shown that empowering developers should not mean to leave them to their own devices: lack of incentive, together with lack of education and awareness on the part of developers are among the primary causes of reuse failure[19]. In a small company setting, a quantitative approach can be taken, and developers' competence and attitude toward reuse can be, if not accurately measured, at least estimated. In this section, we recall nine *profile coefficients* proposed in [29] and aimed at profiling individual users' behavior and competence with respect to the reuse process. All these coefficients take values in the unit interval. Though closely related to classical reusability metrics, the proposed coefficients are *developer-centered*, in that they allow composing a detailed reuse profile for each developer involved in the reuse program. Of course various developer centric metrics can be defined. The metrics proposed have the advantage of being computed automatically, simply by observing the user behavior when interacting with any reuse support system. Therefore, evaluating their validity with respect to a specific organization is straightforward and does not require additional cost or resources. The proposed coefficients can be classified into three classical categories, namely *strategy*, *domain* and *task* [16].

Strategy coefficients

Strategy coefficients estimate the developer's level of general awareness of the company reuse policy. They include *Reuse Percentage*, *After-Delivery Maintenance*, and *Access Level*.

♦ The Reuse Percentage **RP** coefficient is defined as the average reuse percentage attained by the developer, as measured by the development environment tools. Ideally, reuse percentage should coincide with the average *Reuse Merit* value of the artifacts [15], which can be as high as 90% when reuse takes place in the same product line [21]. An upper-bound to this ideal value can be automatically measured as the average difference between the reusable artifacts chosen by the developer before and after adaptation, for instance by means of simple tools like *diff* for requirements and design-level and *sccs* for code-level components [15].

♦ The After-Delivery Maintenance **ADM** coefficient inversely relates to the amount of after-delivery maintenance. It is aimed at estimating the developer's share in the total *Reuse Maintenance Cost* [15]. Again, this coefficient can be transparently measured by the development environment by taking the number of after delivery modification made by the developer, normalizing it with respect to a company-wide maximum value, and then subtracting the result from 1.

♦ Access Level **AL** is the percentage of development resources (in the simplest case, project folders on a network server) to which the developer has access. This metrics is specifically aimed at small companies, where organizational roles are often not formalized and the development environment is local.

The first and the second coefficients express in conjunction the reuse skill of the developer. The first one expresses the reuse rate while the second assesses the quality of his/her reuse choices. Finally, the third can be seen as a rough estimate of the developer's level of responsibility in the company projects.

Domain Coefficients

Domain coefficients allow for making user profiles context-dependent. They include *Project Involvement, Integration Skill* and *Time-to-Market Reduction*.

♦ Project Involvement **PI** is defined as the number of projects in a given domain the developer is involved in, normalized to the company-wide maximum, and is the domain-dependent counterpart of AL.

♦ Integration Skill **IS** is the cardinality of the intersection of the component clusters reused by the developer in recent projects, once again normalized by the company-wide maximum; its upper bound is the systems' *Degree of Commonality* [15] for that domain. IS is the counterpart of the *percentage of components unique to identified projects*, proposed as an organizational-level reuse metrics in [21].

♦ Time-to-Market Reduction **TMR** is a developer-centered, domain-dependent version of the classical decreased level of time-to-market metrics [9].

The first coefficient assesses the experience of the developer in the given application domain while the second expresses the developer's ability in identifying the core component set for a given domain. Finally, the third coefficient takes into account the results of the developer's reuse actions in the domain.

Task Coefficients

Task coefficients aim to measuring the developer's capability at understanding, adapting and reusing components (especially code) at the level of single programming tasks. They include *Programming Tasks Load, Code Reuse Percentage* and *Time-to-Completion Reduction*.

♦ Programming Tasks Load **PTL** is simply the number of programming tasks the developer has performed, normalized to the company-wide maximum value.

♦ Code Reuse Percentage **CRP** is the code-level counterpart of RP, being defined as the average code reuse percentage attained by the developer, ideally coinciding with the *Component Reuse Merit* value [15].

♦ Time-to-Completion Reduction **TCR** estimates time savings obtained by the developer in completing his/her own tasks, thanks to code reuse.

These coefficients are more technical in nature and related to programming and testing capabilities, which, in our opinion, should be evaluated separately from the overall awareness of reuse policies and details.

The first one is a classical measure of programming experience, while the second is a quantitative measure of reuse in the classical terms of "lines of code", rather then in terms of components. In fact, lines of code reused are related to the code comprehension ability. Finally, TCR is a fine grained indicator of the ability to insert lines of reusable code into an application under development.

Though the proposed coefficients allow for a quantitative definition of user profile, their use should by no means considered a "plug-and-play" technique. However, the same difficulty holds when selecting any standardized set of reuse metrics, as evidenced in [19] among the others.

Indeed, all the coefficients discussed above need periodical rescaling to avoid historical data exerting too much influence on their values. Custom initialization techniques, especially to avoid underrating of expert newcomers, are also required. For example, when hiring an expert software designer or programmer, a small company might be bringing in new competencies and skills; in this case, initialization of profile coefficients at the company-wide maximum value could be advisable. Other times, initialization at the company average or minimum may prove a better choice. Furthermore, initialization of profiles for new employees without significant experience will obviously involve lower values. Table1 summarizes the above discussion.

Coefficients	Category	Related Reuse metrics	Expert Newcomer Initialization
RP	strategy	Reuse Merit	Yes, company-wide max
ADM	strategy	Reuse Maintenance Cost	Yes, company-wide max
AL	strategy	-	No
PI	domain	-	Yes, past history based
IS	domain	Degree of Commonality	Yes, company-wide max
TMR	domain	Time-to-Market	Yes, company-wide average
PTL	task	-	Yes, past history based
CRP	task	Component Reuse Merit	Yes, company-wide max
TCR	task	Time-to-Market	Yes, company-wide average

Table 1: Summary of profile coefficients

4   *Computing the User feedback*

User feedback is a learning process leading to permanent modification to the reuse system in order to adapt it to the needs of the company. In our approach, users are considered as *experts* whose opinions are polled simply by registering their selections, as outlined in Section 2. By accumulating the poll results a *Reusability Value* is computed for each component, assessing the views of the user community about the reuse potential of the component. This computation is performed as follows. Suppose that a ranked list of components is presented to the developer by the reuse system. If the developer does not select the first component in the list, but rather the component presented in the k-th position, this is considered as a negative vote against the ranking order. The affected reusability values are those associated to components occupying positions from the first to the $(k-1)$-th in the current query. They are modified in the following way:

$$RV_{new} = (1-\beta)RV_{old} + \beta\,RV_{corr}$$

where the correction factor is $RV_{corr} = -\gamma$. In turn, the chosen component will be affected by the opposite correction ($RV_{corr} = \gamma$). The value of parameter γ expresses system reactivity as well as the total variation scope. In fact, the higher γ, the faster the system can modify the Reusability Value and the broader the total possible variation that can occur as a consequence of user feedback [6]. Moreover, if the Reusability Value is used to modify the system static ranking, tuning γ allows for adapting the feedback to the static system range of values. The confidence granted to user opinion is expressed by parameter β; since variations of the Reusability Value must only occur as a cumulative effect, normally $\beta << 1$. In a ideal user community, where all developers have equal skill and experience, β could be straightforwardly initialized to $1/N$ where N is the number of developers in the team. In the real world, high turnover often results in a workforce having different qualifications, leading to a scenario where users' votes must be weighted with respect to their profiles. Indeed, it is reasonable to assume that senior personnel can estimate the opportunity of reusing a component on the basis of their full awareness of the company's strategic reuse policy, which depends on several (external) factors, such as market issues. On the other hand, if profiles are not used, the number of junior programmers in the team could make their views prevail, potentially endangering the reuse policy.

Each time a user logs on to our system, his/her profile coefficients are aggregated in order to compute the single parameter β used in our feedback mechanism.

4.1 OWA-based aggregation of coefficients

A general way and powerful way of computing coefficients' aggregation is by using a generic *ordered weighted average* operator. This aggregation operations that covers the entire interval between the *min* and *max* operators. Let

$$w = (w_1, w_2, ..., w_n)$$

be a "weighting vector" such that $w_i \in [0,1]$ for all i and

$$\sum_{i=1}^{n} w_i = 1$$

Then, an OWA operation associated with w is the function

$$hw(a_1, a_2, ..., a_n) = w_1 b_1 + w_2 b_2 + ... + w_n b_1,$$

We apply the OWA operator to our coefficients as follows:

$$\beta = \frac{1}{N} \sum_{i=1}^{9} w_i \alpha_i$$

The weights w_i allow one to tune the feedback mechanism w.r.t. the composition of the development team which uses the system.

The OWA operator is a well-known aggregating method suitable for decision problems in which the possible values of rewards are known in linguistic terms, instead of being weighted according to the importance of their information source as it happens with the standard weighted mean (WM) operator.

OWA operators provide us with an adequate framework for representing the optimism degree of the decision when no information is available about the real state.

Thus, since the only difference between the WM and the OWA is the ordering, its adaptation to our qualitative setting is straightforward.

In fact, it is well-known that different OWA operators are distinguished by their weights vector w. In our case, a very simple, intuitive *default setting* for weights w_i, is to slightly privilege skills which appear to be less frequent in the development team. In other words, we adopt a *biased compensative view*, allowing a compensation between the various capabilities and skills expressed by our coefficients. Namely, we set:

$$w_i = \frac{\sum_{j \neq i} \sum_{k=1}^{N} \alpha_{j,k}}{(m-1) \sum_{j=1}^{m} \sum_{k=1}^{N} \alpha_{j,k}} \qquad (1)$$

where N is the total number of users, m is the number of coefficients, and $\alpha_{j,k}$ is the j-th coefficient of the k-th user.

5 Conclusion

We presented a *closed loop*, model-independent approach to software reuse aimed at automatic tuning of software reuse systems.

Our system works by polling developers' for their opinions about the reusability value of components, but instead of requiring them to express their views explicitly, it relies on observing the user reactions to query answers.

References

[1] O. Alonso and W. Frakes. Visualization of reusable software assets. In *Proceedings of the 4th Intl. Conf. On Computer Science and Informatics*, Railegh, 1998.

[2] C. Bellettini, E. Damiani, and M.G. Fugini. User opinions and rewards in a reuse-based development system. In *Proceedings of ACM Symposium on Software Reusability*, Los Angeles, California, 1999.

[3] R.Bellinzona, M. G. Fugini, and B. Pernici. Reusing specifications in OO applications. *IEEE Software*, 12(2):65-75, March 1995.

[4] P. Biggs. A case study in small company software reuse. Technical report, http://students.cs.byu.edu/~pbiggs/smallcomp.html, 1999.

[5] A. Chavez, C. Tornabene, and G. Wiederhold. Software component licensing: A primer. *IEEE Software*, 15(5), September 1998.

[6] E. Damiani and M.G. Fugini. Automatic thesaurus construction supporting fuzzy retrieval of reusable components. In *Proceedings of ACM SIG-APP Conference on Applied Computing (SAC'95)*, Nashville, February 1995.

[7] E. Damiani, M.G. Fugini, and C. Bellettini. A hierarchy aware approach to faceted classification of object oriented component library. *ACM Transactions on Software Engineering and Methodologies*, 8(3):215-262, 1999.

[8] E. Damiani, M. G. Fugini, and E. Fusaschi. A descriptor-based approach to OO code reuse. *IEEE Computer*, 30(10):73-80, October 1997.

[9] J. Favaro. A comparison of approaches to reuse investment analysis. In *Proceedings of the Fourth International Conference On Software Reuse*, pages 136-145, Orlando, FL, April 1996. IEEE.

[10] M.E. Fayad, M. Laitinen, and R.P Ward. Software engineering in the small. *Communications of the ACM*, 43(3):115-118, March 2000.

[11] W. B. Frakes and T. P. Pole. An empirical study of representation methods for reusable software components. *IEEE Transactions on Software Engineering*, 20(8):617--630, August 1994.

[12] R. Johnson. Frameworks = (Components + Patterns). *Communications of the ACM*, 40(10):39-42, October 1997.

[13] C. W. Krueger. Software reuse. *ACM Computing Surveys*, 24(2):131-183, June 1992.

[14] W. Lim. Reuse economics: A comparison of seventeen models and directions for future research. In *Proceedings of the Fourth International Conference On Software Reuse*, Orlando, FL, April 1996. IEEE.

[15] C.McClure. *Software Reuse Techniques*. Prentice Hall, 1997.

[16] P. Mi and W. Scacchi. A knowledge-based environment for modeling and simulating software engineering processes. *IEEE Transactions on Knowledge and Data Engineering*, 2(3):283-294, September 1990.

[17] M. Missikoff and R. Pizzicannella. Interactive and visual environment supporting conceptual modeling of complex OODB applications. In *Proceedings of ACM International Workshop on Advanced Visual Interfaces*, Gubbio, Italy, May 1996.

[18] M. Morisio, M. Ezran , and C. Tully. Introducing reuse in companies: a survey of European experiences. In *Proceedings of ACM Symposium on Software Reusability*, Los Angeles, California, 1999.

[19] N. Nada, D. Rine, and S. Tuwaim. Practices in organizational structure for software Reuse, *The Proceedings of International Conference on Information Systems Analysis and Synthesis*, July 1998

[20] J.S. Poulin. Metrics-guided reuse. *International Journal on Artificial Intelligence Tools (Architectures, Languages, Algorithms)*, 5(1-2), 1997.

[21] D. Rine and R. Sonnemann. Investments in reusable software: a study of software reuse investment success factors. *The Journal of Systems and Software*, 41:17-32, 1998.

[22] D. Schmidt. Why software reuse has failed and how to make it work for you. *C++ Report Magazine*, January 1999.

[23] G. Succi, S. Doublait, C. Uhrik, and F. Baruchelli. Reuse and reusability metrics in a OO paradigm. *International Journal on Applied Software Technology (IJAST)*, 1(3-4):291-303, 1996.

[24] M. Wasmund. Implementing Critical Success Factors in software reuse. *IBM Systems Journal*, 32(4):595-611, 1993.

[25] E. Wiles and F. Bott. Eight steps to your own economic model of software reuse. In *Proceedings of the European Reuse Workshop 98*, pages 123-127, Madrid, Spain, November 1998.

[26] A. M. Zaremski and J. M. Wing. Specification Matching of Software Components. *ACM Transactions on Software Engineering and Methodology*, 6(4):333-369, October 1997.

[27] S. Zuboff, In the Age of the Smart Machine: the Future of Work and Power,: *Basic Books, Inc. Publishers*, New York, 1988.

[28] J-J. Jeng, B.H.C. Cheng. *Specification matching for software reuse: a foundation*. Proceedings of the ACM SIGSOFT Symposium on Software Reusability, ACM Software Engineering Note, Aug. 1995.

[29] C. Bellettini, E. Damiani, M.G. Fugini, *Software Reuse in-the-Small: Automating Group Rewarding*, Information & Software Technology, to appear

[30] V. Torra, *The Weighted OWA Operator*, Int. J. of Intelligent Systems 12 (1997) 153-166.

KES '01
N. Baba et al. (Eds.)
IOS Press, 2001

A Framework for Designing Multi-Agent Systems

Wonyoung Park[1], Sooyong Park[1], and Vijayan Sugumaran[2]

[1] Department of Computer Science, Sogang University

Sinsu-Dong, Mapo-Gu, Seoul, Korea

[2] Department of Decision and Information Sciences, School of Business Administration,

Oakland University

Rochester, MI 48309

Abstract. The era of distributed software environments is emerging and research on Multi-Agent Systems (MAS), which tries to solve complex problems using entities called agents, is on the rise. This paper proposes a framework for a systematic development of MAS. In particular, this approach is geared towards supporting system's properties specially focused on agents' coordination and autonomy. In order to support agent coordination and autonomy, a goal-based approach is utilized for the problem domain analysis, and individual agents are the mapped to the system's refined goals.

1. Introduction

Over the last few years, most of the complex real world problems have been solved using distributed environments, particularly with the advent of the Internet. One approach commonly used to accelerate distributed systems development is to reuse previously developed components with similar functionalities. A large distributed system could be developed through identifying reusable software components, customizing them to meet the new requirements, and integrating them with newly developed software. To manage complex system software development in distributed environments, the research on Multi-Agent Systems (MAS) is on the rise. MAS tries to solve complex problems by focusing on the autonomy of individuals, called 'agents', and on the collaborative interaction that link them together without the human intervention. The effort in designing MAS, however, suffers from lack of systematic approach that is grounded in software engineering paradigm.

In order to develop MAS in a systematic way, we need to analyze the system in terms of its ultimate goals and design the system both in the abstract as well as concrete by mapping the goals and the sub-goals to software agents. The implementation of MAS is only as good as its design; hence, it is critical that correct design decisions are made. A well thought out framework for the system design provides the whole picture of the system and leads to the right implementation with little error. Every system has its own framework and the right establishment of this framework can lead to the right system, and even right analysis and design for extending the system. Furthermore, this is an efficient way to improve the system's reliability and performance. Consequently, in order to generate the right system, it is imperative that a systematic analysis and design process be undertaken.

This paper proposes a framework for the systematic development of MAS. A goal-based approach is used for the problem analysis, and then agents are eventually mapped to the refined goals of the system. Also, identified agents are internally modeled in terms of agent characteristics and designed in terms of agent autonomy. In order to support the coordination needs of agents, which are considered main properties of MAS, architectural styles are utilized in representing the architecture through the agent integration.

The remainder of the paper is organized as follows. Section 2 briefly discusses agent-oriented analysis and design methods as related works. Section 3 describes MAS development framework that consists of problem analysis, agent modeling, agent design, and agent integration. Finally, Section 4 provides summary and future research.

2. Related Research

In recent times there has been a surge of interest in agent-oriented modeling techniques and methodologies.

Gaia [1] is the first agent-oriented software engineering methodology that explicitly takes into account organizations as first-class entities, by providing a coherent conceptual framework for the analysis and design of MAS. In applying Gaia, the analyst moves from abstract to increasingly concrete concepts: abstract – roles, permissions, responsibilities, protocols, activities, liveness properties, and safety properties, and concrete – agent types, services, and acquaintances. Abstract entities are used to conceptualize the system during analysis, but which do not necessarily have any direct realization within the system. In contrast, concrete entities are used within the design process, and will typically have direct counterparts in the run-time system.

Kendall [2] defines a role as an abstraction of agent behavior modeled in terms of responsibilities, possible collaborators, required expertise, and cooperation mechanisms used. The most important advantage of the concept of a role is that it can be freely assigned and reassigned to agents, as long as the agent assigned to the role fulfils the role's requirements. It provides examples of agent role models and explains how role-modeling techniques, found in object-oriented software engineering, can be used to facilitate agent system analysis and design. Also it discusses role model implementations based on two approaches: the Role Object pattern and aspect oriented programming (AOP).

While the above mentioned methodologies provide different approaches to agent system development, they did not include recent software engineering issues including software architecture and component-based development. To complement their work, our proposed approach is architecture-centric and component-based. We discuss the proposed approach in the next section.

3. MAS Development Framework

This section presents our framework for developing MAS. We take a phased approach in generating the MAS architecture, which is shown in Figure 1. Our framework consists of the following phases: problem analysis, agent modeling, agent design, and agent integration. These phases are described further in the following sections.

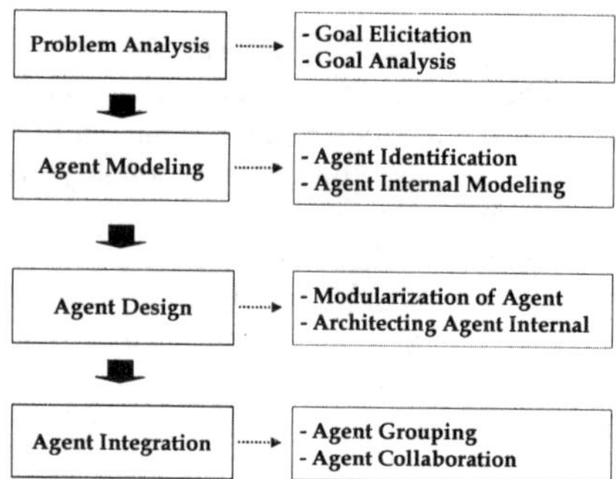

Figure 1. A framework for designing MAS

3.1. Problem Analysis

A prerequisite for designing and implementing systems is to understand the application domain. For understanding the problem domain, there are several techniques available such as use case. Problem analysis is essential for setting up the system's boundary and analyzing user requirements. It is generally agreed that the end users of a system often have difficulty expressing abstract requirements for the system and a goal-directed approach to requirements analysis is an effective mechanism to improve the elicitation process [3]. Thus the concept of goal is central to understanding and designing systems, particularly, agent-oriented systems. There is an increasing trend towards designing agent-oriented software utilizing a goal-directed analysis process. The problem analysis phase in our approach consists of the following activities: a) goal elicitation and b) goal analysis.

A goal could be defined as "a non-operational objective to be achieved by the composite system [4]." In the goal elicitation activity, user requirements are described as goals. The analyst tries to identify itemized goals from the user through interviews or discussions and consults agent-repositories that might have reusable artifacts such as requirements, designs, etc.

The goals identified as a result of the goal elicitation activity are processed using the goal-directed analysis activity to gain better understanding of the requirements for the MAS. This activity may involve clustering, generalization, specialization and decomposition of goals. Clustering of goals entails grouping similar or related goals. Thus, goal analysis mainly involves identifying high level goals, and for each of these high level goals generating sub-goals and also defining the relationships between them. The goals could be categorized into three types: system external goals which are viewed from outside; user goals which are the lowest to be perceived by users; and system internal goals viewed from inside.

3.2. Agent Modeling

We want to represent the target domain using agents based on our understanding of goals in that domain. That is to say, based on the goals, agents need to be identified and modeled to be part of a MAS in this domain. Agent modeling involves identifying and modeling the agents. In particular, this phase consists of the following activities: a) agent identification and b) agent internal modeling.

After gaining an understanding of the domain and the overall goals of the system from the problem analysis, agents are identified to satisfy the goals and their relationships are analyzed. For agent identification, relationship between goal and agents need to be examined. Based on our experience, there could be some heuristic guidelines to map agent to analyzed goals such as "For each identified user's goal, an agent could be identified" and "If the sub-goals become agents, the upper level goal can be a coordinator agent for supporting cooperation among agents."

In the agent identification process, potential objects that can be "agentified," are derived from the goal analysis using agent selection rules. Also, additional agents that need to be added are identified. Then these agents are represented in an Agent diagram that depicts the relationship between various agents and objects in the target problem domain. The notion we use in expressing elements within the Agent diagram is shown below:

Element

- (A) : Agent without mobility
- (M) : Mobile agent
- Class: Established classes

Relation

- Cooperation (–O–): shows the relationship between agents, meaning that cooperation is needed among them.
- Employ (E): shows the relationship between agents and classes, meaning that agents can use established classes.

Figure 2 shows an Agent diagram as an example. There are four agent classes and two object classes. User Info object class is added to make agent adaptive to the specific user.

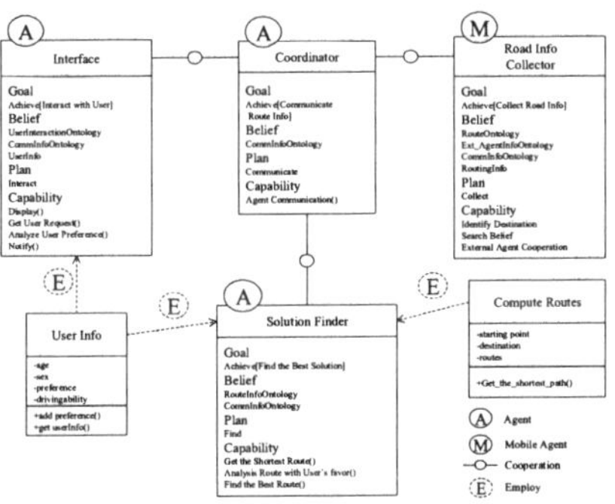

Figure 2. Agent diagram

Once we identify the agents, then the next step is to focus on the internals of the agent. We propose that the internals of an agent consist of the following four parts, namely, goal, belief, plan, and capability [5]. For each agent that is mapped to a refined goal, its internals need to be modeled to reflect that goal. During analysis, agent's belief that represents its knowledge, plan that enable agent's autonomy, and capability are modeled.

3.3. Agent Design

The agent design provides implementation template of each agent. Each modeled agent should have a close relationship with four parts of the agent internal model to achieve its autonomy. The internal agent architecture of agents is represented in the context of agent autonomy. "Autonomy" refers to an agent not depending on the properties or the states of other components for its functionality [6]. That is, an agent has sole control over the activation of its services and may refuse to provide a particular service, or ask for compensation for its services. The autonomy of agents plays a big role in determining the internal architecture of agents and how they react to changes in its surrounding environment. The agent design phase deals with each agent being able to respond to changes that occur in a dynamic environment based on its current state and its set of beliefs. This phase consists of the following activities: a) modularization of agent and b) architecting agent internal.

An agent is composed of the four internal parts: Goal, Belief, Plan, and Capability. These modules are derived from the agent internal models that were generated in the previous activity. They are implemented as plan module, belief module, and capability module to provide inter-agent communication and application services. Finally, This activity results in module definitions.

Once the modules are identified and defined, then we focus on the agent internal architecture. Figure 3 shows how to map the internals of an agent to the elements of agent internal architecture.

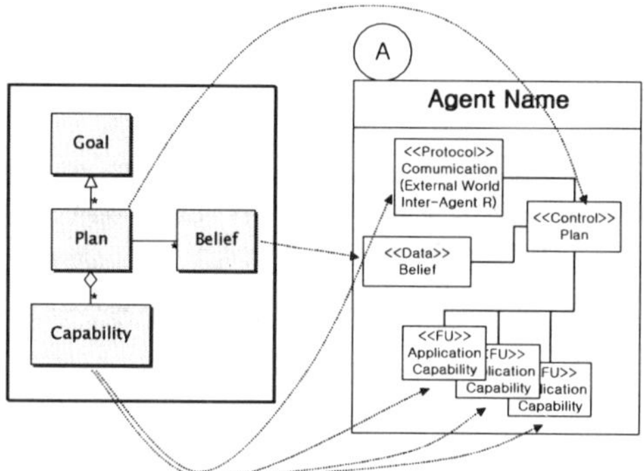

Figure 3. The relationship between the internals of an agent and the elements for agent internal architecture

In MAS environments, an agent tries to achieve its goal and the agent's plan controls other modules to complete that goal. Agent's plan module is regarded as the nucleus of the agent internal architecture. It recognizes the current situation by referencing the belief module, and lets the proper capability module perform its functions. Before executing a task, the agent understands current situation from its goal and the data from the belief. Agent's plan chooses proper strategy and controls its capability based on the strategy. With the architectural styles, various patterns can be applied to represent plan's choice of proper capabilities. State pattern and Strategy pattern [7] can be used for choosing the capability module dynamically and plan's state change can be expressed by State diagram in UML [8].

3.4. Agent Integration

In a MAS, agents work together cooperatively to solve a larger problem than they are capable of solving individually. In this phase we focus on generating the architecture for MAS by integrating the individual agents into a federation. We focus on organizing modeled agents that result from the previous phase in view of agents' coordination. Also we should consider system's reliability and flexibility in terms of agents' interaction. This process involves clearly articulating the definition of essential elements, adjustment of architectural style, architecture representation, and architecture validation. In view of agents' coordination, this phase shows the structure of system organization and the relationship among system elements (agents). Figure 4 shows the process for setting up the agent integration. The agent integration phase consists of the following activities: a) agent grouping and b) agent collaboration.

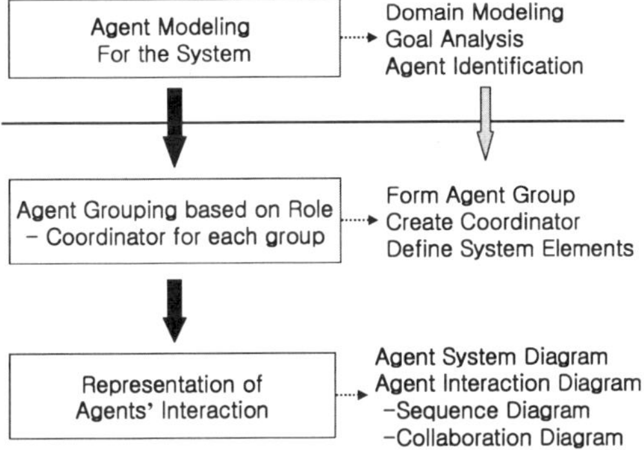

Figure 4. Agent Integration

4. Conclusion and Future Work

Emerging into the era of distributed software environments, the research on Multi-Agent Systems (MAS) is on the rise. This paper has presented a framework for MAS developments. Goal-based approach is used for the problem analysis, and agent is mapped to system's goals and internally modeled. Once the individual agents are identified and modeled, these agents are designed in terms of agent autonomy in order to facilitate their implementation. When we have modeled the agent and created the internal architecture of the agent, we are in a position to create an agent federation or a MAS by integrating these individual agents.

Agent-oriented software needs to be developed based on agent's autonomy and coordination, making it possible to provide complicated services in distributed environments. For this, agent-oriented software development methodology should be systemized and this paper provides a robust framework by suggesting a MAS development method that takes into account system properties such as autonomy and coordination. While we have proposed the initial framework for MAS development, much work remains to be done. We are currently investigating the means for defining agent autonomy, for example, by applying patterns and supporting agent communication and mobility.

References

[1] M. Wooldridge, N.R. Jennings, and D. Kinny. The Gaia Methodology for Agent-Oriented Analysis and Design. *Autonomous Agents and Multi-Agent Systems*, Vol. 3, No. 3, 2000, pp. 285 – 312.

[2] Elizabeth A. Kendall. Role Modeling for Agent System Analysis and Design. *IEEE Concurrency*, April-June 2000, pp. 34-41.

[3] A. v. Lamsweerde and L. Willemet. Inferring Declarative Requirements Specifications from Operational Scenarios. *IEEE Transactions on Software Engineering*, Vol. 24, No. 12, December 1998, pp. 1089-1114.

[4] C. Rolland, C. Souveyet, and C. Ben Achour. Guiding Goal Modeling Using Scenarios. *IEEE Transactions on Software Engineering*, Vol. 24, No. 12, December 1998, pp. 1055-1071.

[5] Sooyong Park, Jintae Kim, and Seungyun Lee. Agent-Oriented Software Modeling with UML Approach. *IEICE Transaction on Information and Communication*, August 2000, pp. 1631-1641.

[6] Onn Shehory. Architectural Properties of Multi-Agent Systems. CMU-RI-TR-98-28, 1998.

[7] Erich G., Richard H., Ralph J., and John V.. Design Patterns: Elements of Reusable Object-Oriented Software, Addison-Wesley, 1995.

[8] Martin Fowler and Kendal Scott. UML Distilled 2nd Edition, Addison-Wesley, 2000.

KES '01
N. Baba et al. (Eds.)
IOS Press, 2001

Proposed Measurement for Complexity of Analysis Classes in UML

Yukyung Kim, Jaenyun Park
Dept. of Computer Science, Sookmyung Women's University
2 Chungpa-Dong, Youngsan-Gu, Seoul, Korea

Abstract Several design metrics have been proposed for evaluating the complexity of object-oriented software. However, most of them discussed only the concepts or properties of the metrics, and have not shown any detailed procedure for computing the specific values. Some metrics requiring more detail than those currently provided by class definitions may well not be computable until implementation is complete and hence, are not practical design metrics. In this paper, we propose a new measurement based on class definitions that would be available during analysis. The metrics is aimed at evaluating the efforts required for understanding classes. It has been analytically evaluated using widely known properties for software metrics. As a result, the proposed metrics is found to satisfy the essential properties, that metrics must have. Finally, experimental evaluation of the metrics is also presented in the paper.

1. Introduction

The measurement of software complexity has been a key interest shared by software engineers in recent years to develop high-quality software. Complexity metrics is used to obtain several benefits including improved estimates of implementation and maintenance effort, as well as early detection of design trouble spots. However, previous research on software complexity has focused on the measurement of software developed using conventional methods. Metrics for any engineered product is governed by the unique characteristics of the product, and object-oriented (OO) software is fundamentally different from conventional software. For this reason, the metrics for OO systems must take into consideration the characteristics that distinguish OO from conventional software [1].

Design-based measurements of psychological complexity have recently sparked considerable interest because they provide earlier feedback to the development project, and earlier feedback means more effective planning and less costly corrective action. Before an OO system is built, the classes that represent the problem to be solved, the manner in which the classes relate to and interact with one another, the inner workings of objects, and the communication mechanisms that allow them to work together must first be defined. All of these things are accomplished during the process of OO analysis [2]. Therefore, the metrics that can measure complexity of the classes during analysis when the first technical activities are performed as part of OO software engineering is necessary. Several design metrics have been proposed to evaluate the complexity of OO software. However, most of them have discussed only the concepts or properties of the metrics, and some metrics requiring more detail than those provided by class definitions may well not be computable until implementation is complete. Hence they did not prove to be practical design metrics [3].

In this paper, we propose a new metrics that can measure the complexity of the analysis classes in UML. A distinguished characteristic of the proposed metrics is that it clearly focuses on using class definitions that would be available during OO analysis as its source of information. It is distinguished from implementation detail by being declarative rather than imperative in nature. Thus, our metrics can be calculated objectively and can provide relatively earlier feedback in the development life cycle.

2. Related works

Complexity metrics for OO software as a research topic just recently emerged. Some metrics that provide an indication of quality at the OO class level have already been presented. The first paper to provide a more substantial treatment of the topic and to propose a suite of metrics was by Chidamber and Kemerer in 1994[4]. Often referred to as the CK metrics suite, the authors proposed six class-based design metrics that focus on class and class hierarchy. The metrics suite also developed metrics for assessing the collaborations between classes and the cohesion of methods that reside within a class. This metrics was analytically evaluated against Weyuker's measurement principles[5]. Lorenz and Kidd divided class-based metrics into four broad categories: size, inheritance, internals and externals [6]. Size-oriented metrics for the OO class focus on counts of attributes and operations for an individual class and average values for the OO system as a whole. Inheritance-based metrics focus on the manner by which operations are reused through the class hierarchy. Metrics for class internals look at cohesion and code-oriented issues, and external metrics examine coupling and reuse. Harrison, Counsell, and Nithi proposed a set of metrics for object-oriented design that provide quantitative indicators for OO design characteristics, called the MOOD metrics suite [3]. Li and Henry showed that a suite of object-oriented complexity metrics can be used as statistically significant predictors of maintenance effort [7]. Lastly, Sharble and Cohen formally defined nine metrics, and applied them in comparison to two different development methods for an object-oriented software [8].

Most of these efforts discussed only the concepts or properties of the metrics, and failed to show the detailed procedure for computing the specific values. Understanding such metrics also proved to be difficult because they were defined informally[9]. Moreover, metrics requiring more details than those provided by class definitions, for example CK metrics suite's "Response for Class" metrics, which requires knowledge of each message send initiated by the class, may well not be computable until implementation is complete. Hence it can not be applied during analysis.

3. Proposed complexity metrics

To develop a precise, understandable, and correct model of the real world, a notation that implements a set of generic components of an OO analysis model must be selected. In UML, a class diagram shows a set of classes, interfaces, and collaborations as well as their relationships. Requirements are assessed and classes are extracted. These classes persist throughout the life of the application. A class defines responsibilities as its behaviour and attributes. Towards fulfilment of a responsibility, a class can collaborate with other classes. Collaboration is defined as the associations and aggregations that are involved in a class diagram. Generalisation, on the other hand, is a relationship between a general and specific class. The specific class inherits everything from the general class, called the parent class[10]. If a class inherits a responsibility of its parent class, such responsibility is called the inherited-responsibility. Total responsibility will be the term used to denote the sum of the inherited-responsibility and the class's unique responsibilities.

3.1 Definitions of new metrics

We define the new metrics, CC(Collaboration Complexity) and IC(Interface Complexity), based on the relationships among classes, because the complexity is related to the interaction between classes as well as the individual complexity of the class.

CC refers to the maximum number of times the collaborations can be achieved with each of the collaborator as indicated by the representation. The maximum number indicates how complex a class can potentially be. Because we do not know exactly, we can only assume that each of the collaborators is used in the mechanisms behind each of the responsibilities that are to be implemented. Thus, CC is computed as the product of the number of collaborators c and the number of the non-inherited responsibilities r.

$$CC = r\,(1 + c)\qquad\qquad\text{(Eqn.1)}$$

Observe that the number of collaborators is $(1+c)$ in (Eqn.1) rather than c because a class can always collaborate with itself. It is quite obvious due to the fact that any method in a class can recursively call itself or any of the other methods. Besides, the complexity of a class includes the external complexity, and the developers of the class have to understand the interfaces of the class itself.

Using IC, we represent the total difficulty associated with understanding the interfaces of each of the collaborators. IC is calculated as the sum of the interface sizes of individual collaborators and the interface size of the class itself. If class E with total responsibilities R collaborates with c other classes E_1, E_2, ... , E_c, IC is given by :

$$IC = IS(E) + \{\, IS(E_1) + IS(E_2) + ..+ IS(E_n)\, \}\qquad\qquad\text{(Eqn.2)}$$

In (Eqn.2), $IS(E_i)$ represents the interface size of class E_i that could be computed as R^2. We assume a non-linear relationship between R and the interface complexity. Accordingly, we choose to measure the interface complexity as R raised to a constant power m. We arbitrarily choose $m=2$ and thus, the interface complexity of a class is given by R^2. At this point we call the interface complexity calculated this way as the interface size. In validating the formulation of the interface size, we will present a result of another investigation. A measure of interface size was chosen as R^2 but the rationale was not entirely clear. To resolve this problem, we used different potential formulations for IS and tested them out on the ATM classes [11]. We have chosen Kendall's rank correlation coefficient to determine the correlation between the values.

Table 1. Formulations of interface size

IS	τ	LOS
R^3	0.65	0.05
R^4	0.66	0.05
R^5	0.66	0.05
R^6	0.66	0.05

The results of the correlation with a Level Of Significance (LOS) of 0.05 are listed in Table 1. The changes shown made little difference in the correlation, proving the sensibility of choosing a simplistic value of R^2.

3.2 Definition of COG

To clarify the computation process using proposed metrics, we use a graph, called COllaboration Graph (COG). An example of COG is shown in Figure 1. From the class diagram [10] which shows analysis classes and their relationships, the graph in Figure 1 is constructed.

COG can represent the information for the relationships among classes and inherited responsibilities of the analysis classes. We construct COG using class diagrams and collaboration diagrams. COG is a directed graph, denoted by COG=(V, E). The elements of a nonempty finite set V that are called vertices can be represented as classes in a class diagram.

We assign a label to each vertex to represent the number of non-inherited responsibilities. The elements of a finite set E are called edges, and represent relationships among classes, denoted by $E= Ec \cup Ei$ where Ec is the set of collaboration relationships and Ei is the set of the inheritance relationships.

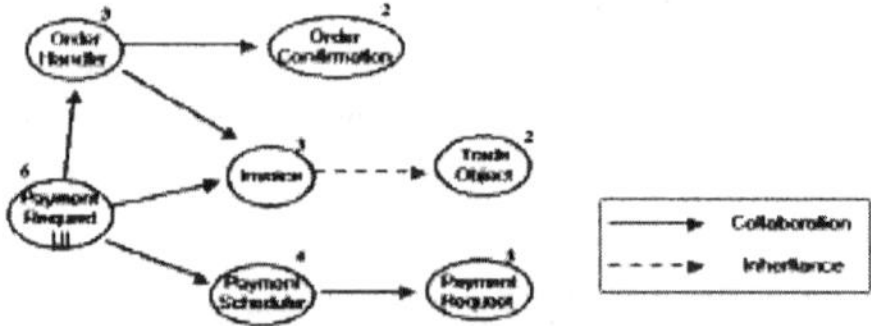

Figure 1. An example of COG

COG provides the information regarding the responsibilities and collaborators of each class, and helps compute complexity. Using Figure 1, we explain the heuristics in calculating the values of the metrics CC and IC. Since the class *"Invoice"* inherits the responsibilities from its parent class *"TradeObject"*, the number of total responsibilities R is 5. Other classes do not inherit in Figure 1, so $r = R$. The number of collaborators c is equal to the outdegree of each vertex. For example, the number of collaborators that interact with class *"Invoice"* and class *"PaymentRequestUI"* is 0 and 3 respectively. Thus, CC(*Invoice*) = r(1+c) = 3(1+0) = 3, IC(*Invoice*) = $R^2 + \sum R^2 = 5^2$, and CC(*PaymentRequestUI*) = 6(1+3) = 24, IC(*PaymentRequestUI*) = $6^2 + 3^2 + 5^2 + 4^2 = 86$.

Comparing the class *"OrderHandler"* with the class *"Invoice"*, CC and IC of class*"OrderHandler"* are higher than the class *"Invoice"*, because class *"OrderHandler"* interacts with three collaborators although they have equal number of responsibilities.

4. Analytical evaluation using Weyuker's properties

In this section, we adopt Weyuker's nine properties [5] to evaluate the new metrics. Weyuker's properties suffer from the following criticisms [12] : First, Zuse pointed out that Weyuker's properties are not consistent with the principles of scaling. Next, Cherniavisky and Smith suggested that we should be careful in using Weyuker's properties, since the properties may give only the necessary conditions for a good measure of complexity. However, since these properties are widely known and accepted, we also adopt them to evaluate the proposed metrics.

Table 2. Evaluation using Weyuker's properties

	Properties								
	1	2	3	4	5	6	7	8	9
CC	O	O	O	O	O	O	X	O	X
IC	O	O	O	O	O	O	X	O	X

Table 2 shows the evaluation results of our proposed metrics with regard to Weyuker's nine properties. In the following analysis, we will exclude discussions on Property2 and Property 8 in this paper. Since the universe of discourse deals with at most a finite set of applications, each of which has a finite number of classes and methods, Property 2 will be met by any metric measured at the class level

[4]. Also, since none of the metrics proposed in this paper depend on the names of the class or methods, they satisfy Property 8.

Properties 1,2, and 3 are exactly the real properties to be satisfied by metrics [5]. All the proposed metrics satisfy these properties. Thus, we can conclude that the proposed metrics have the essential properties of complexity metrics.

Property 7 means changing the order of statements may change the complexity of the class. Chidamber argues that in object-oriented designs, a class is an abstraction of the problem space, and hence the order of statements within the class definition has no impact on eventual execution or use[4]. This will be true only for metrics based on class levels. Therefore, changing the statement order within the class definition does not change the metric involving the class. The proposed metrics are based on classes, and hence, this property is not satisfied. Property 9 states that a combined class will be more complex than its constituent parts. Chidamber suggested that failure in satisfying Property 9 implies that a complexity metric could increase when classes are divided into many more subclasses. Therefore, satisfying Property 9 may not be an essential feature for an object-oriented software design complexity metric[4]. Thus, we do not consider this property for evaluation.

As results shown, the proposed metrics, CC and IC, satisfy Properties 1,2,3,4,5,6, and 8. Cherniavisky contended that Property 7 is not appropriate for object-oriented metrics [12], and in fact the proposed metrics do not satisfy Property 7. Therefore, the result of our evaluation implies that our metrics have the essential properties of a complexity metrics.

5. Case study

We use two commercial systems in order to compare the proposed metrics with the CK design metric suites. One is a well-designed insurance system, the other is WinMarker system (ver. 1.0) that is designed based on the experiences of the developers. For each class that compose the system, we calculate the values of two metrics Coupling Between Objects (CBO) and Response For a Class (RFC) proposed by Chidamber, and our metrics *CC* and *IC*.

The insurance system has been developed using Rational Unified Process (RUP). We use 25 analysis classes, and the corresponding 25 design classes. The values of *CC* and *IC* for the 25 analysis classes, and the values of CBO and RFC for the corresponding 25 design classes, are calculated. WinMarker1.0 has been developed using OO analysis and OO design methodology, together with the Waterfall model, and has been in implemeted by the LG research institute for approximately seven years.

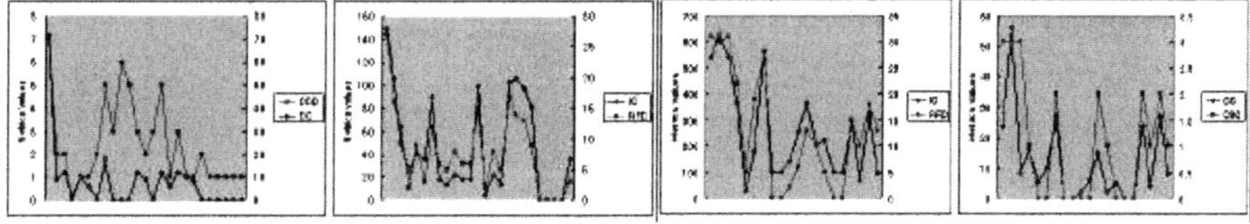

Figure 2. Results of insurance system   Figure 3. Results of WinMarker 1.0

Figure 2 and Figure 3 present the experimental results. As developers depend on their experiences, big classes with various responsibilities emerge in WinMarker1.0. As a result, some classes have very high interface complexity. An interesting aspect is the similarity in the nature of the distribution of the metric values by *IC* and RFC. The higher *CC* and *IC* during the analysis phase, the higher CBO and RFC are during the design phase. Some classes have different results for *CC* and CBO. This is believed to be because CBO counts the number of objects coupled with an object only, whereas CC includes not only the number of collaborators coupled with a class but also the number of responsibilities for their collaborators. Therefore, the value of *CC* is equal to zero for any classes having only attributes, even though its collaborators exist. However, the larger the *CC* is, the larger the CBO will be.

Therefore, measuring the complexity of classes using our metrics *CC* and *IC* during the analysis phase, can allow us to predict the complexity of classes in the course of the remaining processes.

6. Conclusion

In this paper, we proposed a new complexity metrics, called CC and IC, based on the responsibilities and collaborations of analysis classes. Previous research on software complexity metrics has focused on measures of complexity that are code-based. Hence, most of them require more detail than those provided by class definitions. We used the class definitions available during the analysis stage as the source of information to improve this problem. We also proposed some heuristics for computing the actual values of the metric system from class diagrams using COG. We then analytically evaluated our metrics with respect to a set of properties outlined by Weyuker. The result implies that our metrics have the essential properties required of complexity metrics. Finally, the proposed metrics were applied to two practical object-oriented systems, the results of which have likewise been presented in this paper.

We can expect that our complexity metrics may provide us earlier feedback and possibly enable us to predict the efforts and costs needed to develop the system. This would, in turn, allow us to plan as required by the remaining processes. By decreasing the complexity of a system during the analysis stage, the time and cost required to develop the system is reduced. Furthermore, measurements of the complexity during the analysis phase provide earlier feedback to the development project, which translates to more effective and less costly corrective action.

As a result, we expect to develop the cost-effective OO software by reviewing the complexity of analysis classes in the first stage of Software Development Life Cycle (SDLC).

The following are recommended for future research efforts. First, we have to clarify the relationship between the measure of the complexity and the quality of the product. Second, we have to develop the tool, that can collect data needed to measure and compute the values of metrics automatically.

References

[1] B. Henderson-Sellers, Object-Oriented Metrics : Measures of Complexity, New Jersey, Prentice-Hall, 1996.

[2] Roger S. Pressman, Software Engineering-A practitioner' s approach, 5th ed., New York, McGraw-Hill, 2000.

[3] Harrison R., S. Counsell, and R. Nithi, "An Evaluation of the MOOD set of Object-Oriented Software Metrics", *IEEE Trans. on Software Eng.*, vol. 24, no. 6, pp491-496, June 1998.

[4] S. R. Chidamber, and C. F. Kemerer, "A metrics suite for object-oriented design", *IEEE Trans. on Software Eng.*, vol. 20, no. 6, pp476-493, June 1994.

[5] E. J. Weyuker, "Evaluating software complexity measures", *IEEE Trans. on Software Eng.*, vol.14, no.9, pp.1357-1365, 1988.

[6] M. Lorenz, and J. Kidd, Object-Oriented Software Metrics, New Jersey, Prentice-Hall, 1994.

[7] W. Li, and S. Henry, "Object-oriented Metrics That Predict Maintainability", *Journal of System Software*, vol.23, pp.111-122, 1993.

[8] R. Sharble, and S. Cohen, "The Object-Oriented Brewery : A Comparison of Two Object-Oriented Development Methods", *ACM SIGSOFT Software Eng. Notes*, vol.18, no. 2, pp. 60-73, 1993.

[9] E. M. Kim, S. Kusumoto, and K. Kikuno, "A new metric for C++ program complexity and its evaluation in academic environment", IEICE Trans. J79-D-1, vol. 10, pp. 729-737, 1996.

[10] G. Booch, J. Rumbaugh, and I. Jacobson, "The Unified Software Development Process", Massachusetts: Addison-Wesley, 1999.

[11] S. A. Whitmire, "Object-oriented design measurement", New York : John Wiley & Sons, Inc., 1997.

[12] J. C. Cherniavsky and C. H. Smith, "On Weyuker's axioms for software complexity measures", *IEEE Trans. on Software Eng.*, vol. 17, no. 6, pp.636-638, June 1991.

Response Source to Speech and Noise as Revealed by EEG-Based Tomograms

Ali A. Danesh, Ph.D.
*Department of Communication Sciences and Disorders, Florida Atlantic University,
Boca Raton, Florida 33431*

Herbert Jay Gould, Ph.D.
*School of Speech-Language Pathology and Audiology, The University of Memphis,
Memphis, TN 38105*

Abhi Pandya, Ph.D.
*Department of Computer Science and Engineering ,Florida Atlantic University
Boca Raton, FL 33431*

ABSTRACT

Late auditory evoked potentials (LAEPs) were recorded following the binaural presentation of nonsense speech and acoustically similar noise. Data collected from five males and five females from 64 electrodes over their scalps. The major components of LAEP (N1, P2, N2) were detected. Using an EEG mapping technique (LORETA) sources of evoked responses to nonsense speech stimuli were located in the left temporal region. However, for noise stimuli, sources of activity were detected primarily in the fronto-central region. Various scattered areas of activity in the brain were also noticed for both of stimuli.

INTRODUCTION

The new techniques in analysis of physiologic data permit a more precise answer to questions regarding the possible source generators of activity following sensory stimuli. The Late Auditory Evoked Potentials (LAEPs), in response to speech and noise, and their source generators have received a tremendous attention in the current literature. Although previous electrophysiologic and nuclear medicine studies (e.g., Hirano et al., 1997; Krause et al., 1995; Lassen & Friberg, 1988; Lovrich et al., 1988; and Zatorre et al., 1992) have attempted to identify cortical areas in response to speech and non-speech stimuli, they have not used stimuli that were closely matched acoustically. Typically, the previous studies have contrasted speech and either tones, narrow band or wide band noise.

The purpose of this investigation was to determine the differences between the source generators of speech and acoustically similar non-speech (noise) auditory stimuli. The hypothesis was that response to stimuli with linguistic characteristics were generated in different anatomical brain locations than acoustically similar stimuli that lack linguistic characteristics.

METHOD

Stimuli: Two nonsense syllables, DESH and DAUSH (referred to as DEO and DAO, respectively) were used as speech stimuli. A recording of a male native speaker of General American English producing the syllables was obtained using the Computerized Speech Lab (CSL) program (1994). One token of each nonsense syllable was selected and digitized at 16,000 samples per second. Non-speech (noise) stimuli were constructed based on the speech stimuli. In summary the generated noises had: a) stationary spectral shaping and waveform periodicity similar to the original stimulus; b) intensity modulation similar to the original stimulus; and c) a periodic component that was amplitude modulated. The generated non-speech (noise) stimuli were referred to as DAN and DEN for DAUSH and DESH (DAO and DEO), respectively.

Subjects: Ten right-handed adult individuals (five females) with normal hearing served as subjects in the age range of 18-34 years ($M = 23.5$, $SD = 4.522$). All subjects were native speakers of English.

Electrode Placement and Evoked Potential Recording: For each subject meeting the study criteria, an array of 64 surface electrodes was mounted on the subject's head. Fifty-seven of those electrodes were placed via an EEG cap (Electrocap ESR). Five external tin electrodes were attached to the subjects' head at IN, M1, M2, E1, and E2 locations.

Digitization: The location (i.e., X, Y, Z values) of each of the electrodes on the head was measured. A Polhemus 3Dspace Fastrack digitizer was employed and all of the electrodes were digitized.

Subjects' Status and Instruction: Subjects were seated and watched a videotape in a relatively sound treated and electrically shielded room.

Stimulation: Stimuli were presented binaurally through insert earphones (Etymotic ER-3A, 10Ω) at 75dB SPL. Each stimulus was presented 500 times in two blocks of 250. Inter Stimulus Interval was varied to avoid CNV (ISI=1500-2000 msec).

Instrumentation: A NeuroScan (SCAN 3.0) recording system with 64 channels, 486 PC, two Syn-Amps amplifiers (Model 5083), and NeuroScan acquire software, was employed for data collection in this study. Brain electrical activity was collected in a continuous manner. Subsequently, the source generators were detected for grand mean averages of LAEPs employing Low Resolution Electromagnetic Tomography Algorithm (LORETA) technique using EMSE suite.

RESULTS

The Averaged Late Auditory Evoked Potential Waveforms: The ongoing EEG was collected during presentation of the stimuli (DEO, DEN, DAN, and DAO). The EEG was then sectioned into epochs of 100msec pre and 500msec post-stimulus. The epochs were examined for artifact and all records with artifact were rejected manually. The remaining waveforms were averaged based on time of signal arrival and the averaged waveforms for each subject at each electrode were superimposed and displayed. Three major peaks were identified which were labeled N1, P2, and N2 (Figure 1). The waves appear to be inverted from those typically seen. This is because of the position of the reference electrode, 1cm behind Cz, as compared to the typical reference at mastoid, linked ears or nose.

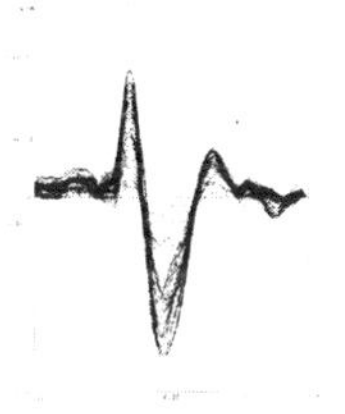

Figure 1.: The superimposed grand average waveforms.

Source Localization of the Linguistic and Non-Linguistic Stimuli: In order to search for the possible locations of the generators for speech and non-speech (noise) stimuli a tomographic source analysis was performed. This analysis is referred to as LORETA. The LORETA method according to Pascual-Marqui et al (1994) "finds a source distribution which accounts for the measured instantaneous field distribution, taking into account both normalized minimum source power and maximum smoothness." Using EMSE suite tomographic display slices, with 1cm thickness, were examined. Analysis started from zero cm at the level of the ear and proceeded up to 10 cm to the level of vertex. Each one of the tomographic displays for every subject and for all of the stimuli was inspected in a window of +/-10 msec around the latencies of major peaks of the individual waveforms. A tomographic analysis between 90-110msec at 3-5 cm for a representative subject (#5, stimulus :DAN) is shown (Figure 2).

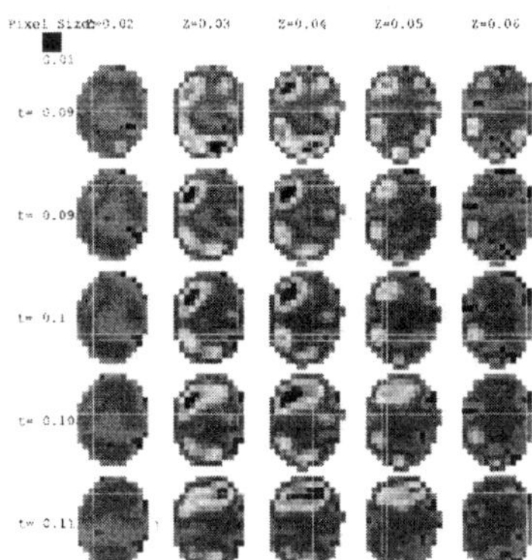

Figure 2.: Tomographic (LORETA) representation of response to DAN between 90msec.

The grand mean average waveforms were used to search for the response generators. Tomographic display slices, with 1cm thickness, in a window of +/- 10msec around the latencies of major peaks were examined. LORETA assumes that adjacent neurons are concurrently and synchronously activated and it is possible to dissociate multiple and scattered sources (Pascual-Marquai et al., 1994). For noise stimuli the bulk of activity was seen in the fronto-central areas. For speech stimuli the temporal areas, remarkably left temporal area, was activated (Figure 3a-d). Scattered areas of activity were also noticed.

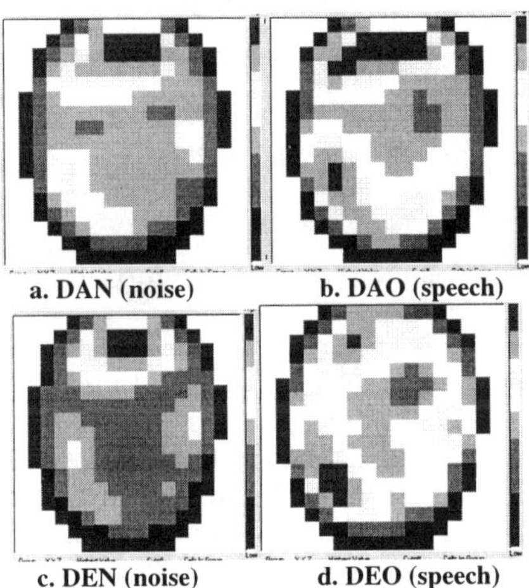

a. DAN (noise)	**b. DAO (speech)**
c. DEN (noise)	**d. DEO (speech)**

Figure3a-d.: LORETA grand averages across subjects to the response to the stimuli shown at 2 cm above the lowest level in the head. Note the foci of activity in the fronto-central region for noise stimuli, and left temporal region for nonsense speech stimuli.

DISCUSSION

The goal of the present study was to investigate the sources of activity following the stimulation with nonsense speech and acoustically similar noise. Previous studies have shown the presence of different foci of activity following speech and noise. However, there was no evidence regarding the comparison of cortical activities following a nonsense speech with acoustically similar noise.

LAEP Waveform Morphology: LAEPs were recorded with the application of a high-density electrode montage using 64 surface electrodes. The major peaks of N1, P2 and N2 were clearly seen for both the nonsense speech and non-speech noise stimuli. The latency of the waveforms was in gross accordance with previous studies (Tonquist-Uhlen et al., 1996; Verkindt et al., 1995). This finding also was compatible with the statements of Hillyard and Woods (1979), as well as Picton and Stuss (1984), that evoked potentials following the presentation of speech stimuli are in essence similar in morphology and scalp distribution to those evoked by non-speech stimuli. Although the waveform morphology and peak latency for the LAEP of all of the stimuli are grossly the same, there is one major difference. That difference is in the latency of the N1 peak for the DAUSH nonsense speech stimulus. This peak was present approximately 25 to 30msec earlier than that of other stimuli. This difference was attributed to the role of the formant transition of this particular stimulus. The transition of the first and second formant (F1 and F2) of this stimulus has a greater frequency extent and consequently sharper slope when compared to F1 and F2 of DEO. This issue was addressed in another paper (Danesh, et al., 1999) and will not be discussed in this presentation.

Source Localization of the LAEPs to Nonsense Speech and Non-Speech Noise Stimuli: The tomographic data from this study that were represented in Figures 5 and 6 reveal differences between linguistic and non-linguistic stimuli. At the latency range of 70-110msec that grossly corresponds to the N1 peak across the LAEP waveforms of this study, it was shown that at 3-5cm above the fiduciary lines, several foci of activity are present in the tomographic slices. It was detected that for non-speech noise stimuli major foci of activity are located in the frontal areas. In addition, foci of activity also were noticed in the temporal regions. On the other hand, the majority of activity and generator sources were in the left temporal area for nonsense speech stimuli. This confirms previous findings regarding the lateralization of speech stimuli toward the left hemisphere (e.g., Lassen & Friberg, 1988; Hirano et al., 1997).

The results of the Zatorre et al. (1992) study showed activation of Heschl's gyri bilaterally following the noise stimulus (Figure 4). This bilateral activation was also seen in the present study at 100msec. However, the activity is more pronounced in the frontal areas in this study. The activity for DEN and DAN are shown directly below the Zatorre et al. (1992) results in Figure 4.

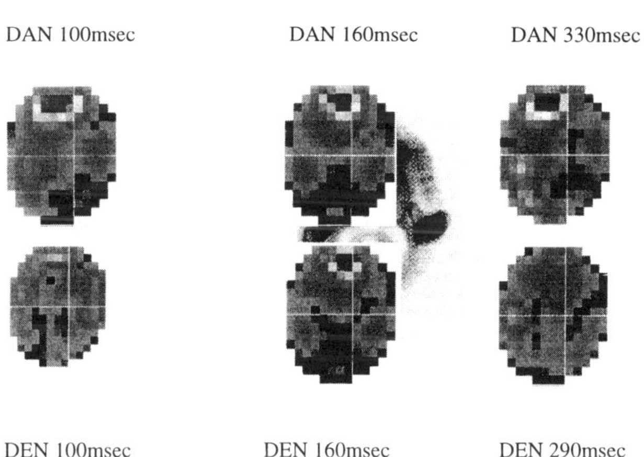

Figure 4.: Comparison between the PET scan using noise stimulus (top) from the Zatorre et al. (1992) study and the tomographic data at 4cm above the fiduciary line in response to the non-linguistic noise stimuli in the present study from a representative subject (#10). Note the bilateral temporal lobe activity in both of the studies. Frontal lobe activity is prominent in the present study.

Zatorre et al. (1992) performed a subtraction procedure with the PET scan so that the response to the noise stimuli was subtracted from the response to the speech stimuli across the average data from all of the subjects. The results of this subtraction, shown in Figure 9, indicated bilateral activity for the speech stimuli that was slightly anterior to the response to the noise stimuli. More specifically, contrasted data showed bilateral activation in the superior temporal gyri anterior to the Heschl's gyri. They also noticed the difference in the left inferior frontal cortex and in the left posterior temporal area.

In the present study, as shown in Figure 4 and 5, points of major activity are plotted for each of the subjects at various periods that can be compared with the Zatorre and colleagues' data. These figures support Zatorre et al. (1992) study by showing bilateral activation along the superior surface of the temporal lobe for both the nonsense speech and non-speech (noise) stimuli.

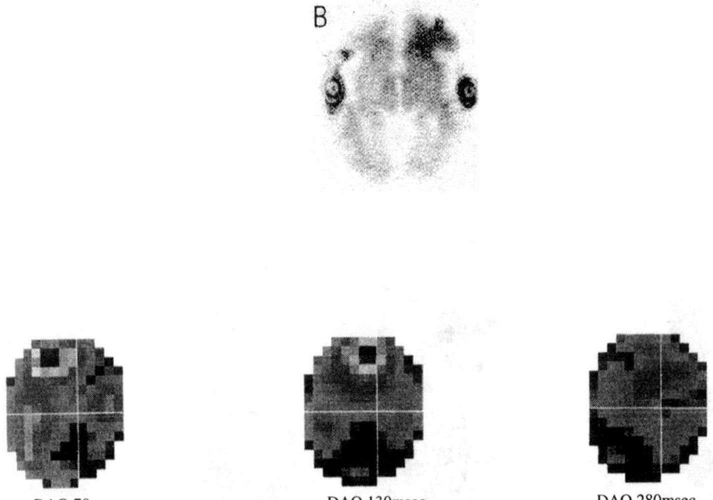

Figure 5.: Comparison between the PET scan showing a subtraction response to speech and noise stimulus (top) from the Zatorre et al. (1992) study, and a linguistic stimulus (DAO) in the present study from a representative subject (#10) at 4cm above the fiduciary line. Note the activity in the bilateral temporal areas in both of the studies.

Examination of the tomographic data of speech stimuli of a representative subject in Figure 9 with the data from Zatorre et al. (1992) reveals some similarities.

First both of the studies revealed activities in the bilateral temporal areas. Second, there is a tendency of extension from temporal areas toward the left frontal area that is more pronounced in the present study. It should be advised that the response from PET studies although encompass fine spatial resolution, possess poor temporal resolution. Therefore, the interpretation of the data should be performed with additional caution.

CONCLUSIONS

The results of this study support findings that the left hemisphere is dominant for speech processing (Lassen & Friberg, 1988; Hirano et al., 1997). Although the stimuli were presented binaurally, the response to the speech stimuli was primarily in the left temporal lobe as demonstrated by LORETA. The stronger left temporal lobe response was not seen for the two non-speech stimuli. A frontal source was also seen for all the stimuli. This focus of activity may be related to an attentional mechanism as the right frontal areas have been related to attention in previous studies (Giard et al., 1988; Näätänen & Michie, 1979; Näätänen, 1975). Frontal sources have also been reported in the past for simple auditory stimuli (Giard et al., 1994; Aliciani et al., 1995). In summary, the findings of this study support the hypothesis of the research that there are separate brain locations activated for the speech and non-speech stimuli.

REFERENCES

Alcaini, M., Giard, M.H., Echallier, J.F., & Pernier, J. (1995). Selective auditory attention effects in tonotopically organized cortical areas: A topographic ERP study. <u>Human Brain Mapping, 2</u>, 159-169.

CSL software, Model 4300B, Software version 5.X (1994). Lincoln Park, NJ: Kay Elemetrics Corp.

Danesh, A.A., Gould, H.J., Rose, D., Buder, E., Mendel, M., Ethington, C. (1999). The N1 response to nonsense syllables and acoustically similar noise. Presented in the XVI Biennial Symposium of International Evoked Response Study Group (IERASG), Trompso, Norway.

EMSE User Manual, 1997. Source Signal Imaging, San Diego: CA.

Giard, M.H., Perrin, P., Echallier, J.F., Thevenet, M., Froment, J.C., & Pernier, J. (1994). Dissociation of temporal and frontal components in the human auditory N1 wave: scalp current density and dipole model analysis. <u>Electroencephalogr Clin Neurophysiol</u>, <u>92</u>, 238-252.

Giard, M.H., Perrin, P., Pernier, J., & Peronnet, F. (1988). Several attention-related wave forms in auditory areas: a topographic study. <u>Electroencephalogr Clin Neurophysiol</u>, <u>69</u>, 371-384.

Hirano, S., Naito, Y., Okazawa, H. Kojima, H., Honjo, I., Ishizu, K., Yenokura, Y., Nagahama, Y., Fukuyama, H., & Konishi, J. (1997). Cortical activation by monaural speech sound stimulation demonstrated by positron emission tomography. <u>Exp Brain Res, 113</u> (1), 75-80.

Krause, C.M., Lang, H., Laine, M., Kuusisto, M., & Porn, B. (1995). Cortical processing of vowels and tones as measured by event-related desynchronization. <u>Brain Topogr, 8,</u>(1), 47-56.

Lassen, N.A., & Friberg, L. (1988). Physiological activation of the human cerebral cortex during auditory perception and speech revealed by regional increases in cerebral blood flow. <u>Scand Audiol</u> (Suppl. 30), 137-176.

Lovrich, D., Novick, B., & Vaughan, Jr., H.G. (1988). Topographic analysis of auditory event-related potentials associated with acoustic and semantic processing. <u>Electroencephalogr Clin Neurophysiol, 71</u>, 40-54.

Näätänen , R. (1975). Selective attention and evoked potentials in humans-a critical review. <u>Biol Psychol, 2</u> ,237-307.

Näätänen , R., & Michie, P.T. (1979). Early selective attention effects on the evoked potential: a critical review and reinterpretation. <u>Biol Psychol, 8</u>, 81-136.

Pascual-Marqui, RD, Michel, CM, & Lehmann, D. (1994). Low resolution electromagnetic tomography: a new method for localizing electrical activity in the brain. <u>Int J Psychophysiol, 18</u> (1), 49-65.

Picton, T.W., & Stuss, D.T. (1984). Event-Related Potentials in the Study of Speech and Language: A Critical Review. In Caplan, D., Roch Lecours, A., & Smith, A. (Eds.).   <u>Biological Perspectives on Language</u> (pp.303-360). Cambridge: MIT Press.

Tonnquist-Uhlen, I. (1996). Topography of auditory evoked long-latency potentials in children with severe language impairment: The P2 and N2 components. <u>Ear & Hearing,17</u>, 314-326.

Verkindt, C., Bertrand, O., Perrin, F., Echallier, J.F., & Pernier, J. (1995). Tonotopic organization of the human auditory cortex: N100 topography and multiple dipole model analysis. <u>Electroencephalogr Clin Neurophysiol, 96</u>, 143-156.

Zatorre, R.J., Evans, A.C., Meyer, E, & Gjedde, A. (1992). Lateralization of phonetic and pitch discrimination in speech processing. <u>Science, 256</u>, 846-849.

Magnetic Measurement System for Multi Object Movement Using Neural Networks and Nonlineair Least-Squares Method

Masatake AKUTAGAWA [1], Xu ZHANG [2], Hirofumi NAGASHINO [3],
Yohsuke KINOUCHI [3] and Rensheng CHE [2]

[1] *School of Medical Sciences, The University of Tokushima,*
3-18-15 Kuramoto-cho, Tokushima, 770-8509, Japan
[2] *Computer Science and Electrical Engineering, Harbin Institute of Technology,*
92 XiDa Road, Harbin,Heilongjiang province, China
[3] *Faculty of Engineering, The University of Tokushima*
2-1 Minamijosanjima-cho, Tokushima, 770-8506, Japan

Abstract. A measurement method for multi object movement using the magnetic field is described in this study. The positions and orientations of the objects are estimated by the magnetic field of small magnets, which are attached at measured objects. It is difficult to obtain its accurate solution rapidly because this problem is one of the inverse mapping problem. In proposed method, the back-propagation neural networks (BPNNs) are used to calculate the position and orientation of the magnets. Furthermore, nonlinear least-squares method are used to obtain more accurate result by using the output of BPNNs as initial value of the iterative calculation. According to results of computer simulations, applicability of the proposed system were confirmed.

1 Introduction

Measurement of object movement is one of the general techniques in various fields, such as industries, medical diagnoses, and academic researches. In several representative methods, the magnetic measurement method has characteristic advantages. It does not have perturbation to the object movement because it has no necessity of any contact between the object and the measurement equipment. In addition, it is capable to measure the movement which is hidden by some structures if they are not ferromagnetic materials. Especially, these features make easy to apply this method to measurement of living body movements, such as movement of eyes[1], jaws[2], tongues[3], locomotion, and beaks[4] etc.

There are two types magnetic measurement method for the moving object. One of them makes use of time variant magnetic field and a small coil attached at the object. In this method, position and orientation of the magnet are encoded to phase transition of the induced current on the coil. Although the fast measurement in high precision is capable, the movement of the object is limited by cables which are connecting between the coil and equipment. The second method is using static magnetic field. A small magnet are attached on the object, and position and orientation of the magnet are estimated from the magnetic field around the object. Using this method, the object movement is not limited because there is no contact to the magnet. Conventionally, this calculation was made by using of pattern matching or iterative calculation method. But it is difficult to carry out fast and precise measurement, and to apply to multi object movement because of non-linearity of the problem.

The purpose of this study is to develop a fast and precise measurement system for multi object movement using magnets and magnetic sensors. In the proposed system, the estimation of the position and orientation of the magnets from the magnetic field were carried out by the multi-layered neural networks[5]. In addition, a non-linear least square (NLS) method were used to obtain more strict measurement accuracy. To examine applicability of the proposed system, computer simulations were performed. Furthermore, we focused on the measurement of the jaw movement as representative application for a living body movement.

2 System Overview

Figure 1 shows the overview of the proposed system. A small magnet is attached to each measured object. Magnetic sensors are fixed on a sensor frame, which is surrounding the measurement area. The sensor frame should be fixed while the measurement in order to minimize the affect of geomagnetic field. Because it performs as DC noise to the measured magnetic field, it can be removed as offset noise.

Each magnetic sensor measures three dimensional magnetic field. Position and orientations of the magnets are calculated by an estimation module. The module has fast mode and precise mode. In the fast mode, the estimation is carried out by only neural networks. Processing time is very fast because they have no iterative calculation. Therefore, it is possible to track the object movement in real time. But the measurement accuracy may be not enough in some condition. At that time, precise mode is used to obtain more precise measurement result. In this mode, the estimation result of the neural network part is used by the NLS part as initial value of the iterative calculation. It needs longer processing time than the fast mode because the iterative calculation are done to gain a convergent solution.

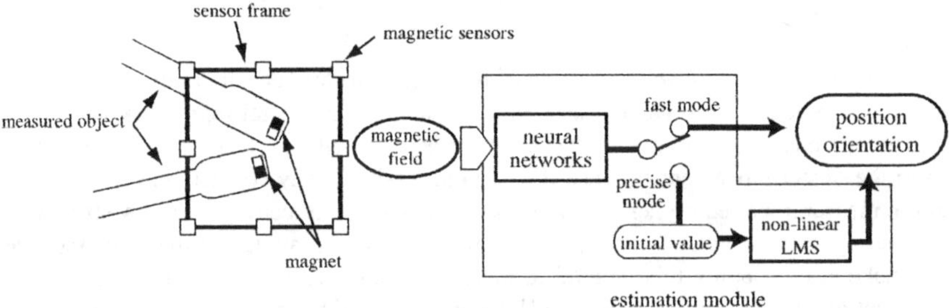

Figure 1: System overview.

The structures of the neural networks were optimized prior to the measurement. They were trained using the back propagation algorithm with a appropriate training pattern set. The training pattern set consists of input vectors and output vectors. The output vectors are positions and orientations of the magnets which are placed in measurement area randomly. The input vectors represent the magnetic field distributions which are calculated for corresponding output vectors.

To calculate exact magnetic field by a magnet, analytical method, such as the finite element method, should be used. In this study, a magnetic dipole are used as a model of a small magnet to simplify calculations of the magnetic field because significant purpose is to examine fundamental characteristics of the proposed method.

The magnetic field B_{ij} at i-th magnetic sensor by j-th magnetic dipole moment m_j at position p_{mj} can calculate by using the Biot-Savart law:

$$B_{ij} = k \left[\frac{3r_{ij}(r_{ij} \cdot m_j) - |r_{ij}|^2 m_j}{|r_{ij}|^5} \right] \tag{1}$$

where $k = \mu_0/(4\pi)$ in which μ_0 is the permeability of free space, $r_{ij} = p_{mj} - p_{si}$ is a position vector of the dipole from i-th sensor at position p_{si}.

The position and orientation of the magnetic dipole can be denoted in six dimensional vector $(p_{mj}, m_j)^T = (p_{mjx}, p_{mjy}, p_{mjz}, m_{jx}, m_{jy}, m_{jz})^T$, and the magnetic field distribution at the sensors is represented in $3S$ dimensional vector $B_j = (B_{1j}, \cdots, B_{Sj})^T = (B_{1jx}, B_{1jy}, B_{1jz}, \cdots, B_{Sjx}, B_{Sjy}, B_{Sjz})^T$ where S is the number of the sensors. The magnetic field distribution observed at the sensors by N magnetic dipoles can be represented as,

$$B = \sum_{j=1}^{N} B_j \tag{2}$$

Therefore the task of the estimation module is to estimate $P = (p_{m1}, \cdots, p_{mN})$ and $M = (m_1, \cdots, m_N)$ from measured magnetic field B_{mea}.

The NLS part refines the estimation result by the neural network part as small iteration steps as possible. Now, the magnetic field distribution B_j for one dipole can be written by using $3S \times 3$ transfer matrix $D(p_{mj})$ as,

$$B_j = D(p_{mj})m_j \tag{3}$$

The element of the $D(p_{mj})$ for one dipole can be expressed as,

$$d_{j,il} = k \left[\frac{3r_{ij}(r_{ij} \cdot e_l) - |r_{ij}|e_l}{|r_{ij}|^5} \right] \tag{4}$$

where $i = 1, \cdots, 3S$, $l = x, y, z$, and e is a unit vector. From Eq. (2), the magnetic field distribution is rewritten as,

$$B = \sum_{j=1}^{N} D(p_{mj}) = A(P)M \tag{5}$$

where $A(P) = (D(p_{m1}), \cdots, D(p_{mN}))$ ($3S \times 3N$ matrix).

To calculate the least-square solution, a measure of fit J_{LS} is defined as,

$$J_{LS} = |B_{mea} - B_{cal}|^2 \tag{6}$$

where B_{cal} is calculated magnetic field by using the magnetic dipole model. The goal of the iterative calculation is to find the set of P and M that minimizes this cost function J_{LS} can expressed as,

$$J_{LS} = \alpha^T W^{-1} \alpha \tag{7}$$

where $\alpha = B_{\text{mea}} - B_{\text{cal}}$ is a noise vector, and W is a covariance matrix. For a given P, the moment M minimizing the J_{LS} is given by,

$$M = A^+ B_{\text{mea}} \tag{8}$$

where $A^+ = (A^T A)^{-1} A^T$ is the Moore-Penrose pseudo inverse solution. In our method, the Gauss-Newton algorithm [6] was applied to solving this problem.

3   Results of Computer Simulations

Figure 2 shows the sensor arrangement and measurement area. They were arranged for measurement of upper and lower jaw movement. Measurement area were two column shape area whose height and diameter were 50mm and 50mm respectively. The orientation of the dipole is limited within 50 degrees. Each region was corresponding to the moving field of upper and lower jaw. 63 magnetic sensor units were arranged on cubic sensor frame. Each sensor unit measured the three dimensional magnetic field. Therefore the magnetic field distribution was represented in 189 (63 × 3) dimensional vector. The input of the estimation module was this magnetic field vector.

The neural network part consisted of four sets of the neural networks show as Figure 3. Each neural network was corresponding to the position and orientation of upper and lower dipoles. Configurations of these neural networks were 189 input units, 2 hidden layers of 80 units, and 3 output units. 50,000 patterns were used to train the networks. After the training, the accuracy of the measurement were tested for 10,000 test patterns. As results of them, average of position and orientation errors were 1.2% (600μm) and 2 degree. Furthermore NLS method was applied to obtain more accurate results. In this case, the averages of the position errors were improved to 0.04498% (22.49μm). In representative case, convergent result were obtained after 3 to 6 iterations.

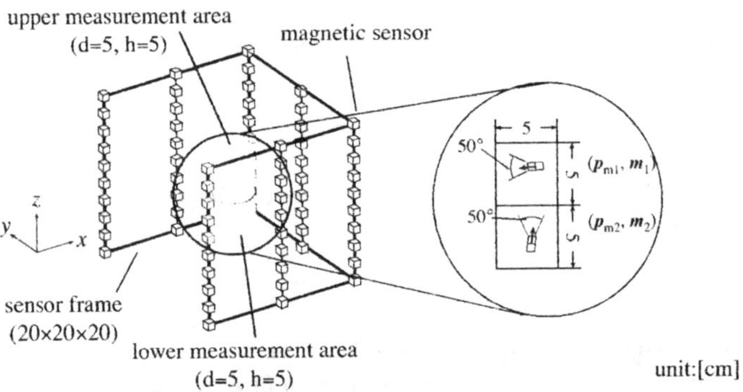

Figure 2: Sensor arrangement and measurement area

Figure 4 (a) and (b)indicates the estimation result for simulated jaw movement. Orientation of the dipoles was fixed to $m_1 = (1, 0, 0)$ and $m_2 = (0, 0, 1)$. The arrows shows position and orientation of estimated dipoles. Almost all part of the locus, the true movement and estimated result were overlapped.

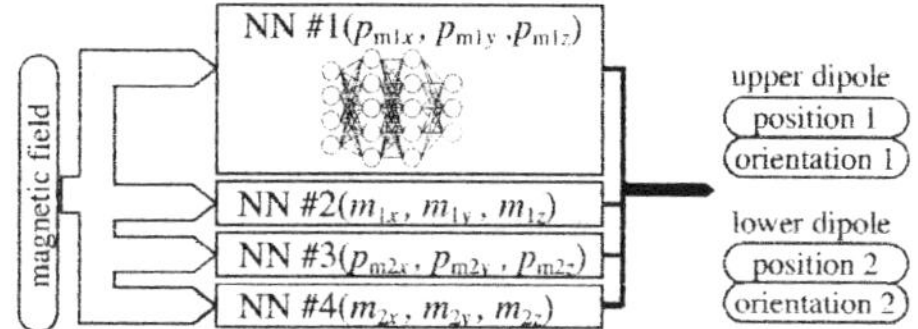

Figure 3: Structure of the neural network part.

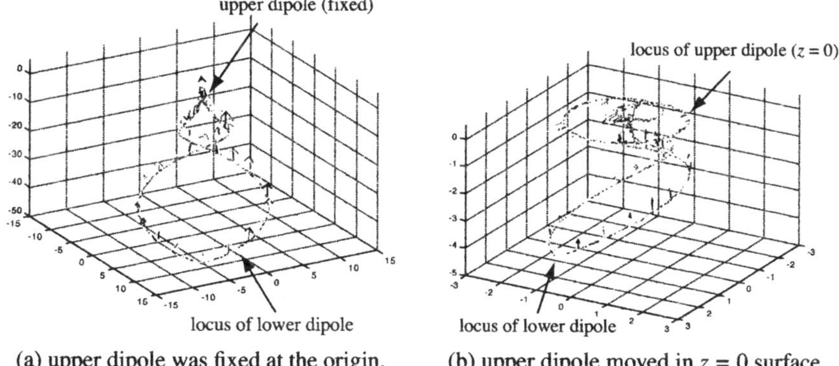

(a) upper dipole was fixed at the origin.　　(b) upper dipole moved in $z = 0$ surface.

Figure 4: Result for simulated jaw movement.

4 Discussion and Conclusions

A magnetic measurement system for multi object movement was proposed. As a representative application of this system, measurement for jaw movement was examined. Results of computer simulations indicated that the proposed system Especially, accurate measurement was capable when both the neural networks and the NLS method were used.

In real measurement system, various factors, such as magnetic noise, will affect the measurement accuracy. We have to assess them, and to improve immunity to the noise.

References

[1] D. A. Robinson, A method of measuring eye movement using a scleral search coil in a magnetic field, IEEE Trans. Biomed. Elect., **10** (1963) 137–145.

[2] B. Jankelson, C. Swain, P. F. Crane, and J. C. Radke, Kinesiometric instrumentation : a new technology, Journal of the American Dental Association, **90** (1975) 834–840.

[3] B. Tuller, S. Shao and J. A. S. Kelso, An evaluation of a alternating magnetic field device for monitoring tongue movements, Journal of the Acoustical Society of America, **88**(2) (1990) 674–679.

[4] J. D. Deich, D. Houben, R. W. Allan, and H. P. Zeigler,, "On-line" monitoring of jaw movements in the pigeon, Physiology & Behavior, **35** (1985) 307–311.

[5] M. Akutagawa, Y. Kinouchi, H. Nagashino, Application of neural networks to a magnetic measurement system for mandibular movement, Proc. of Intl. Conf. of IEEE EMBS, **20**(4) (1998) 1932–1935.

[6] Lennart Ljung, System Identification Theory For the User, Prentice-Hall,Inc, (1990).

KES '01
N. Baba et al. (Eds.)
IOS Press, 2001

Brain Source Localization for Auditory Evoked potentials by Use of Neural Networks

Qinyu ZHANG[1], Ryosuke YAMADA[1], Xiaoxiao BAI[1], Hirohumi NAGASHINO[1],
Yohsuke KINOUCHI[1], Fumio SHICHIJO[2], Shinji NAGAHIRO[2]
2-1 Minamijosanjima-cho, The Faculty of Engineering, The Univ. of Tokushima, Japan
3-18 Koramoto-cho, School of Medicine, The Univ. of Tokushima, Japan

Abstract. The activities in human brain are very complex. Some of them are caused by brain signal sources, e.g. visual, auditory and sensory signals. The properties of the sources can be identified with the technique of brain source localization from EEG. Brain source localization is an important inverse problem for brain diagnosis and functional analysis. The usefulness of BPNN (Back Propagation Neural Networks) to solve this problem in the case of single dipole source localization has been reported. The purpose of this study is to examine the practical effectiveness of BPNN for two dipoles source localization from EEG, then, the developed technique is applied on the clinical EEG data of Auditory Evoked Potentials (AEP).
Keyword. Back propagation neural networks, EEG, Source localization, AEP

1. Introduction

Dipole source localization using Electroencephalograms (EEG's) recorded from the scalp is a classic example of an ill-posed problem in electrophysiology and widely used to make non-invasive estimates of the locations of sources of electrical activity in the human brain[1].

Localization of an electric dipole in the brain has a relatively long history sine Brazier who first suggested a relation between scalp potential distributions and current sources in the human brain. A number of works in the last few decades have attempted to solve the EEG inverse problem by some methods and minimization algorithm such as Simplex Method and Marquardt algorithm [2]. In general, in order to obtain high localization accuracy from the EEG data, long time computer resource and huge computer memory should be requested. A direct method to compute source parameters from a voltage set measured on the scalp such as Moving Dipole [3], Dipole Tracing [4], etc., is therefore proposed. As one of it, the neural networks method is suggested.

Recognizing the potential of neural networks in nonlinear function approximation, the back propagation neural networks (BPNN) has recently been used as an efficient way to solve the inverse problem of electrophysiology [5]. The major advantage of the technique is that once a neural network is trained, it no longer requires iterations or access to sophisticated computers. Since no forward calculation is involved during actual use, any sophisticated head/source model can be used with equal ease irrespective of whether a simple analytical solution or a sophisticated finite element analysis is used. Furthermore, neural networks are capable of capturing the nonlinear dynamics of the source localization problem, while maintaining the noise robustness essential to EEG analysis.

The purpose of this study is to examine the practical effectiveness of BPNN for two dipoles source localization from EEG. Then, this developed technique is applied on the clinical EEG data.

2. Method

The results reported in this paper are based on the following models commonly found in literature: (a) the human brain is represented by the three-concentric-shell model [6], and the biopotential sources are represented by current dipoles [6]. Each dipole is parameterized by its location vector and the dipole moment vector. The set of EEG measuring electrodes are place on the scalp according to the standard international 10/20 system, with a linked-ear reference.

In this study, in order to investigate the usefulness of BPNN for two brain sources localization at the beginning, we therefore set the head model into two regions, the left region and the right region. The head model can be shown in Fig.1. Two dipoles were put randomly in these two regions in order to get training patterns for BPNN.

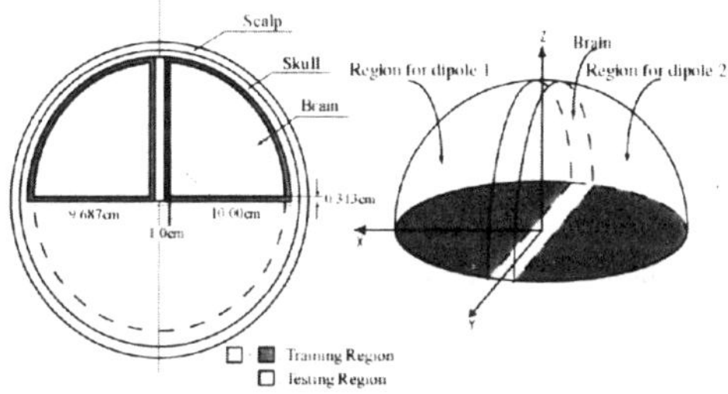

Fig.1. The head model

Once the source model and a head model have been assumed, the next step is to calculate the inverse solution for the location of the source in the model. Because the calculation of an inverse dipole solution is a nonlinear problem, the solution is usually obtained by an iterative process in which the dipolar source is moved about in the head model while its orientation and strength are also changed to obtain the best fit between the recorded EEG and those produced by the sources in the model.

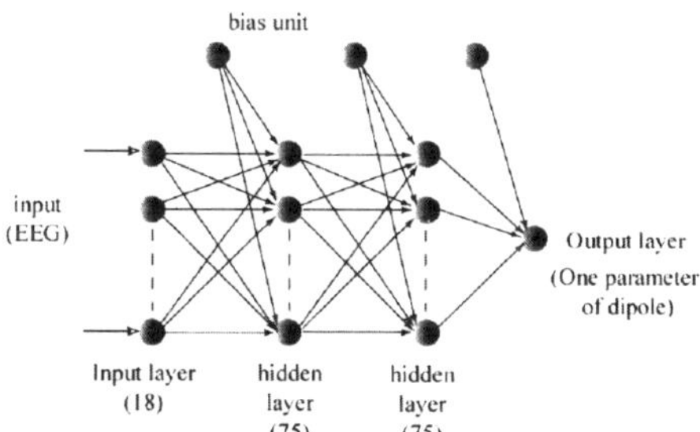

Fig.2. The structure of neural networks

Here, a BPNN (using an error back propagation algorithm) is used extensively for solving inverse problem. The architecture of the neural network used here is shown in Fig.2. If an EEG pattern is fed into the network, the output will be the position of the dipole, and dipole moment. Here, the network consists of four layers (an input layer, two hidden layers, and an output layer). There are 18 neurons to receive the potentials from 18 electrodes in the input layer, 73 neurons in each hidden layer. The network has also bias neurons with a constant output value of 1, to adjust the threshold of each neuron. The output layer just consists of one neuron indicating one of components of dipole position $\mathbf{P}$, i.e., (x, y, z) or moment **M**, i.e., (M_x, M_y, M_z). The twelve same structure networks are therefore used in parallel for position and moment estimations. That is, the network makes use of the independence of twelve source parameters. Neurons in the input layer have a liner input-output function, while other neurons have a sigmoid function.

To get better results for the source localization, we propose to subdivide the head region into 2 parts, one is training region, the other is testing region. The training region is chosen to be lager than the testing region. Here, 30,000 EEG-dipole parameters pairs calculated are used for training BPNN and tested by different 10,000 pairs to examine the localization accuracy.

3. Results

Table 1 showed the localization accuracy based on the models mentioned above. It gave the average position error of 8.7% and direction error of 0.19 degree respectively. Though the localization accuracy was not good while comparing with the case of single dipole source localization, it still examined the usefulness of BPNN for source localization.

Table 1. Localization accuracy

Left		Right	
Position error [%]	Moment error [deg]	Position error [%]	Moment error [deg]
8.3	0.21	7.9	0.17

In order to make the localization accuracy more considerable for clinical application, we therefore defined another head model as shown in Fig.3. Table 2. showed the result of localization.

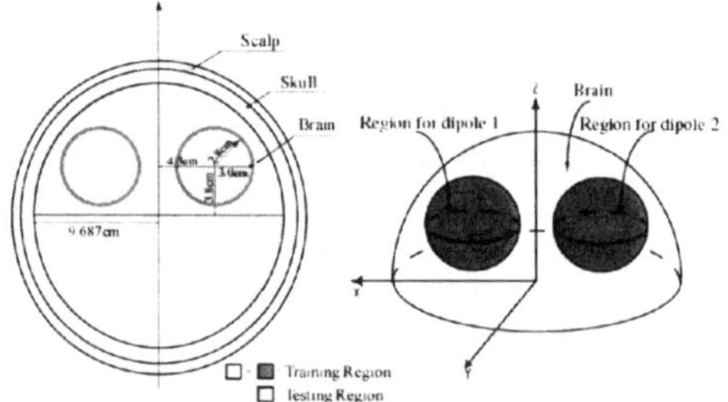

Fig.3 Restricted head model for clinical application

Table 2. Localization accuracy for restricted head model

Left		Right	
Position error [%]	Moment error [deg]	Position error [%]	Moment error [deg]
5.6	0.11	5.4	0.10

4. Clinical Application

The application of the technique developed in this paper to real-world signals is shown below. EEG data have been recorded from a subject who received the external stimulus with a clicking noise, which frequency was 1.5KHz , and the intensity was 80dB. The voltages got from scalp in this way were called AEP (Auditory Evoked Potentials). The estimation result is shown in Fig.4.

In addition, different measures of accuracy have to be employed in clinical situations because of the original position of dipole sources is unknown. The concept of dipolarity, which expresses how close a measured potential field is to the dipole-derived calculated field has been therefore introduced by the expression as follows:

$$Dipolarity(D) = \sqrt{1 - \frac{\sum_{i-1}^{N}\left(v_{m,i} - v_{c,i}\right)^2}{\sum_{i-1}^{N}\left(v_{m,i}\right)^2}} \qquad (1)$$

where $v_{m,i}$ is measured voltage ar scalp electrode 'i'; $v_{c,i}$ is calculated voltage at scalp electrode 'i' with the localized dipole and N is the number of measuring electrodes.

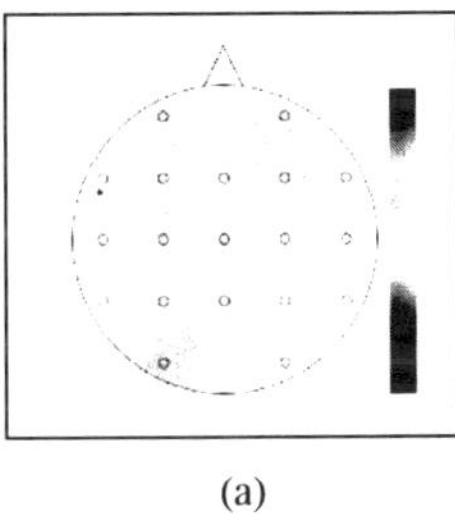

(a)

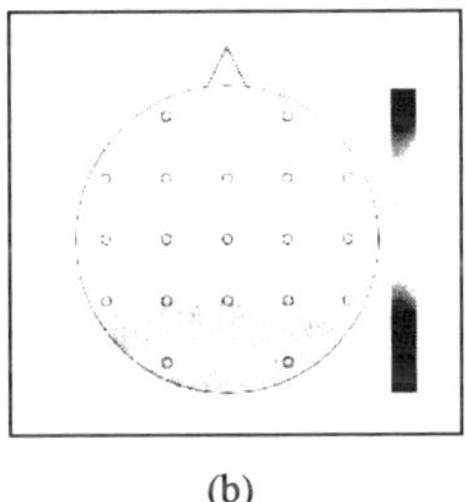

(b)

Fig.4 Estimation results. (a) showed the topography from measured EEG, the two sources could be found near the ears. (b) showed the topography from calculated EEG. The two topographies are almost same

Dipolarity = 100% means that the localization was perfect, the potentials measured on the scalp were truly dipolar, and the localization process was mathematically 100% successful.

Most of the conventional inverse calculations based on iterative algorithms seek to maximize dipolarity in order to find the best dipole parameters. The neural network method does not rely on dipolarity as the estimator in the inverse calculation process, but uses it only to check accuracy.

Here, the dipolarity for the clinical application is 83.6%. Whatever, the result can prove the usefulness of BPNN for two dipoles source localization.

5. Conclusions

We here addressed the problem of estimating two biopotential sources in human brain, based on EEG signals measured on the scalp. This problem is known as the inverse problem of electrophysiology, requires iterative calculation such as Levenberg-Marquardt algorithm. In this paper, we proposed a method of developed back propagation neural networks with the general position error of 7.1% in average that is applicable for clinical use. The application of BPNN on clinical EEG data for AEP revealed evidence in agreement with the above conclusions reached via computer simulations.

References

[1] B.N.Cuffin, EEG Dipole Source Localization, *IEEE Engineering in Medicine and Biology*, pp. 118-122, 1998.

[2] D.W. Marquardt, An Algorithm for Least-squares Estimation of Non-liner Parameters, *J.Soc. Indust. Appl. Math.*, vol. 11,no.2, pp.431-441, 1963

[3] B.N.Cuffin, A Comparison of Moving Dipole Inverse Solutions Using EEG's and MEG's, *IEEE Trans. Biomed. Eng.*, vol. BME-32, no.11, pp.905-910, 1985

[4] Y.Nakajima et.al, Estimation of Neural Architecture in Human Brain by means of The Dipole Tracing Method, *Neuroscience Letters*, 136(1992) pp.169-172

[5] U.R.Abeyratne et.al, Artificial Neural Netwroks for Source Localization in The Human Brain, *Brain Topography*, vol.4, no.1, 1991, pp.3-21

[6] J.Malmivuo and R. Plonsey, *Bioelectromagnetism: Principles and Applications of Bioelectric and Biomagnetic Fields*, Oxford University press, 1995.

KES '01
N. Baba et al. (Eds.)
IOS Press, 2001

Self-Organization of Spatio-Temporal Multi-Rhythms in Neural Networks Stimulated by External Activities

Kenji FUJIMOTO[1], Masatake AKUTAGAWA [2], Naosuke FURUTANI [1],
Hirofumi NAGASHINO [1], and Yohsuke KINOUCHI [1]
[1] *Faculty of Engineering, The University of Tokushima*
2-1 Minamijosanjima-cho, Tokushima, 770-8506, Japan
[2] *School of Medical Sciences, The University of Tokushima,*
3-18-15 Kuramoto-cho, Tokushima, 770-8509, Japan

Abstract. A brain model to memorize periodical signals from external stimuli is described . A living body is always being given various stimuli by the external environment. Some of these stimuli are reconstructed by spatio-temporal pattern generators in the brain. They can be thought as one of the sources of the rhythmic behaviors, such as circadian rhythm. The proposed model is based on the multi-layer type neural network with feedback connections. It can generate arbitrary periodic patterns by training of the network using external input signal. In this study, memorizing of periodic scalar signals and periodic image signals were examined. As results of computer simulations, applicability and capability of this model were confirmed.

1 Introduction

A living body has self-organized information processing system in its nerve system. It has many internal models of various phenomena in external environment. It is possible to create, expand, or modify these models, according to requirement. Rhythms of periodic behaviors, such as circadian rhythm, locomotion and tapping, is supposed to come from some kinds of these internal models. These models are capable to realize arbitrary spatio-temporal patterns from external stimuli autonomously. Once they realize the patterns, they can begin to generate the acquired patterns in some conditions. The trigger of the pattern generation is external stimulus which is similar to the original pattern. Sometimes, the generator is capable to reconstruct the whole spatio-temporal pattern from a part of it. For example, we can recall the whole melody of the favorite music only from listening of its preliminary part.

The purpose of this study is to develop an artificial neural network model which can acquire periodic signal by self organization. The neural network in the real brain has both forward and backward connections[1]. The proposed model is based on the multi-layered neural network with feedback connections. This neural network performs as a periodic pattern generator after the complete training using external signal. In this study, applicability and characteristics of this model is examined for periodic scalar signals and two dimensional raster image.

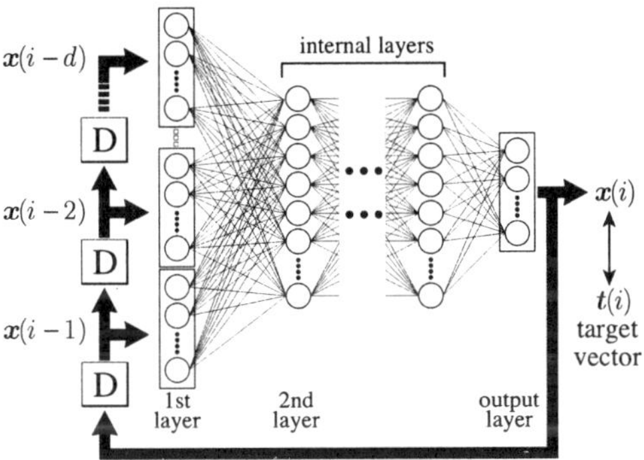

Figure 1: Structure of proposed neural network.

2   Method

Figure 1 represents the structure of the proposed model. This model is based on multi-layered neural networks. The first layer consists of d groups of units. The number of units in each group is M, where is corresponding to the dimension of periodic pattern. Time delay units "D" are used to obtain past output patterns. Now, we have N samples of the periodic pattern $t(1), t(2), \cdots, t(N)$. In case of using d delay units, the number of units in the first layer and that of output layer are Md and M respectively. At i-th iteration, the target value of the neural network is $t(i)$. The input of the first layer is the output of last d iterations, where $x(i - d)$, $x(i - d + 1), \cdots, x(i - 1)$.

There are two phases in following computer simulations, that is, a training phase and a test phase. In the training phase, the connection weights of the neural network are optimized. Our model uses the back propagation method to optimize the connection weights. In this method, it is necessary to use pairs of input and output pattern. The training pattern is given as,

$$
\begin{array}{cccccc}
(& x(1), & \cdots, & x(d) & | & t(d + 1) \ ) \\
(& x(2), & \cdots, & x(d + 1) & | & t(d + 2) \ ) \\
 & \vdots & & \vdots & & \vdots \\
(& x(i - d), & \cdots, & x(i - 1) & | & t(i) \ ) \\
 & \vdots & & \vdots & & \vdots \\
(& x(N - d), & \cdots, & x(N - 1) & | & t(N) \ )
\end{array}
\tag{1}
$$

At i-th iteration, the output $x(i)$ is calculated for $x(i - d), \cdots, x(i - 1)$ using the forward propagation from the first layer to the output layer. Using $x(i)$ and $t(i)$, the error is calculated in accordance with the generalized delta rule [2]. However, in initial several steps, $t(1), \cdots,$ $t(d)$ are used as the input of the first layer instead of $x(1), \cdots, x(d)$ because the previous output of the network cannot be given.

In the test phase, the neural network begins to generate a time series signal independently when initial values are given. At first step, the output $x(d + 1)$ is calculated from the initial values $x(1), \cdots, x(d)$. Next, $x(d + 2)$ is calculated from $x(2), \cdots, x(d + 1)$, and so on.

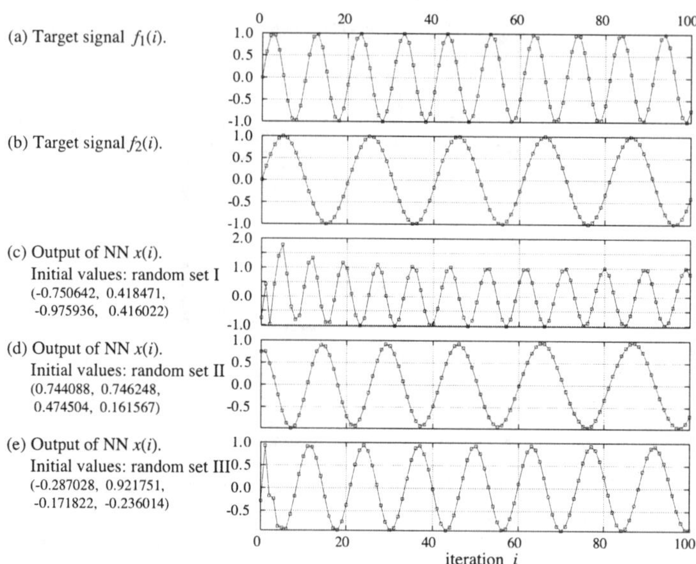

Figure 2: Output of the neural network which was trained to generate two scalar signals.

3   Results of Computer Simulations

We examined applicability of the proposed model for a periodic scalar signal and a periodic image signal.

3.1   *Periodic scalar signals*

Two periodic scalar signals $f_1(i)$ and $f_2(i)$, given as,

$$
\begin{aligned}
f_1(i) &= \sin \omega i \\
f_2(i) &= \sin(\omega/2)i
\end{aligned}
\tag{2}
$$

were used as target signals. The angular frequency was $\omega = 0.62$ [rad/iteration] (Figure 2(a) and (b)). The neural network consisted of 4 layers and 4 delay units. Number of units in each layer was 4, 5, 5 and 1, respectively. At first iteration, $x(1)$, $x(2)$, $x(3)$ and $x(4)$ are given initial values because they can not calculate prior to the first forward calculation. After the network training was completed, several sets of the initial values were examined. The output of the neural network was almost identical to the target signal when the first 4 values of $f_1(i)$ and $f_2(i)$ were presented. In the other hand, Figure 2 (c), (d) and (e) are the output when random value sets (named set I, set II, set III) were presented as the initial values. Figure 3 depicts plots of $x(i)$ versus $x(i+1)$ (Fig. 3(a) shows $f_1(i)$ vs. $f_1(i-1)$, and $f_2(i)$ vs. $f_2(i-1)$). In these figures, thickness of the ellipse represents the signal frequency. In the other word, the thicker ellipse represents the higher frequency signal.

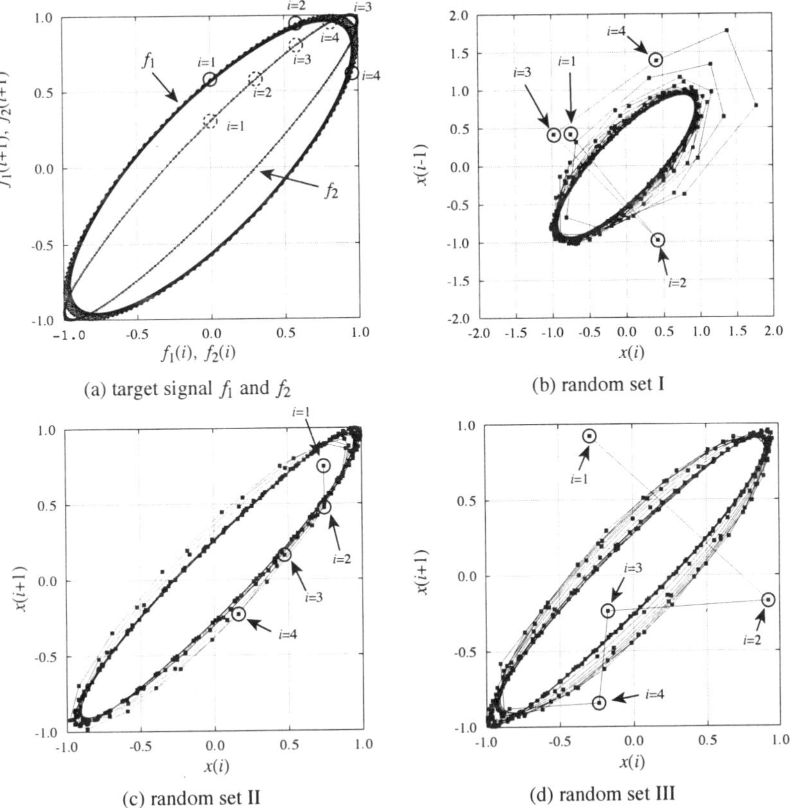

(a) target signal f_1 and f_2

(b) random set I

(c) random set II

(d) random set III

Figure 3: Plots of $x(i)$ vs. $x(i-1)$ for different initial values (except (a)).

Using random set I, the frequency of the output signal converged to that of f_1 after the transition period. In the transition period, range of the signal was larger than original. In case of random set II, transition period was very short because the initial value was close to the original f_2 signal. The last case, thickness of the ellipse in the Figure 3(d) became gradually thin. This means the frequency transition from f_1 to f_2 was occurred.

3.2　Periodic raster images

Raster images of roman alphabet characters were used as the periodic signal. The neural network was trained to generate two series of patterns, where were iterations of A, B, C, D, E, F, and I, J, K, L, M, N. Number of units in the first layer was 224 units (7×8 pixels × 4 delay units). After the training, the neural network generated the corresponding periodic patterns to initial patterns. For example, when a set of A B C D was given, the network generated E F A B C D ····. In the other hand, it generated M N I J K L ··· for the initial patterns I J K L. This

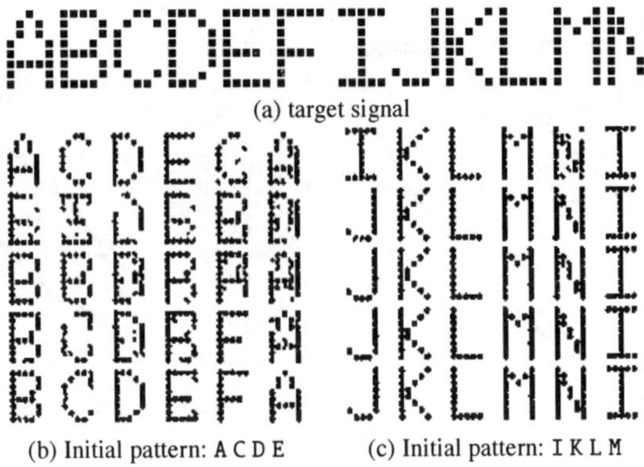

(a) target signal

(b) Initial pattern: A C D E　　　(c) Initial pattern: I K L M

Figure 4: Periodic raster image.

shows the single neural network memorizing two different periodic signals were constructed.

In next stage, we examined the recoverability of unexpected initial patterns. Fig. 4(b)(c) shows examples that recovery of the generated signal was succeeded. After noisy patterns, mixed several patterns or unrecognized patterns were appeared, stable pattern signal was generated.

4　Conclusions

A neural network model to acquire spatio-temporal signals was proposed, and its applicability and capability was confirmed by computer simulations. Using this model, several series of periodic patterns can be acquired by a single neural network. When the appropriate initial values were given to the network, it is capable to generate corresponding patterns. Furthermore, the output neural network converged to one of the acquired signals even if the initial values were unknown, such as random values. This feature is similar to homeostasis of the actual biological system. The future problems are to apply to more complex patterns and to increment the number of periodic patterns.

References

[1]　J. He, T. Hashikawa, H. Ojima and Y. Kinouchi, Temporal integration and duration tuning in the Dosal zone of cat auditory cortex, The Journal of Neuroscience, **17**(7) (1997) 2615–2625.

[2]　D. E. Rumelhart, G. E. Hinton and R. J. Wiliams, Leaning internal representations by error propagation, in D. E. Rumelhart and J. L. MacCleland [Eds.], Parallel Distributed Processing, **1** (1986) 318–362.

KES '01
N. Baba et al. (Eds.)
IOS Press, 2001

Structural Identification of the Multi-layered Neural Networks by using GMDH-type Neural Network Algorithm

Tadashi Kondo[1], Abhijit S.Pandya[2], Thomas Gilbar[3]
*[1]School of Medical Sci., The University of Tokushima,3-18-15
Kuramoto-cho Tokushima 770-8509, Japan
[2]Comp.Sci.& Eng.,Florida Atlantic University, Boca Raton,
FL33431,U.S.A.
[3]Elec. and Comp. Eng., Florida International University,
Miami, FL 33174, U.S.A*

Abstract: In this study, structural identification of the multi-layered neural networks (sigmoid function type neural networks) by using GMDH-type neural network algorithm is developed. The GMDH-type neural network algorithm can generate optimum multi-layered neural network architectures fitting the complexity of nonlinear system so as to minimize the error criterion defined as AIC (Akaike's Information Criterion). The GMDH-type neural networks are organized by using heuristic self-organization method proposed by A.G.Ivakhnenko. The GMDH-type neural networks have abilities of self-selecting the number of layers, the number of neurons in the hidden layers, the useful input variables and the optimum neuron architectures in the hidden layers. Therefore, it is very easy to find out the optimum neural network architectures fitting the complexity of nonlinear system by using the GMDH-type neural network algorithm.

Keywords: GMDH, Neural Networks, Structural Identification

1. Introduction

The GMDH-type neural networks and their applications have been proposed in our early works [1]-[3]. The GMDH-type neural networks can automatically optimize their architecture by using the heuristic self-organization method [4]. They can also determine such structural parameters as the number of layers, the number of neurons in the hidden layers, the useful input variables, and the optimum neuron architectures in the hidden layers. These structural parameters are automatically determined so as to minimize the error criterion defined as AIC [5]. The conventional multi-layered neural networks do not have the structural identification ability of optimum neural network architectures, so applying conventional neural networks to practical complex problems and obtaining good prediction accuracy are difficult.

In this study, multi-layered sigmoid function type neural networks are designed using the GMDH-type neural network algorithm. The algorithm will not only automatically organize the neural network architectures, but also simultaneously estimate the weights that will be used for the individual neurons. The resulting neural network is then applied to the identification problem of a sample nonlinear system, and the results are compared with those obtained by the conventional multi-layered neural networks. These results show that the GMDH-type neural network algorithm is useful for the structural identification of the multi-layered neural networks.

2. Introduction to GMDH

The Group Method of Data Handling (GMDH) algorithm was first presented by the Ukrainian researcher and cyberneticist A.G. Ivakhnenko and his colleagues in 1968. His intent was to develop a rival method to stochastic approximation. Originally designed to estimate higher order regression polynomials, this heuristic self-organization method has been applied to a large variety of fields including medicine, biology, manufacturing, environmental and ecological systems, psychology, economics, etc. [7]

GMDH generates and tests all input-output combinations for a system. Each element of the system implements a nonlinear function of two inputs. Coefficients of the elements are determined using a regression technique. A threshold is specified at each level to determine if the outputs of the elements in a layer are giving acceptable results. If the result from an element is within the threshold, it is passed on to the next layer. Those elements and or variables that are least useful in predicting the proper output are filtered out. Each succeeding layer has more complex combinations. Layers are added until satisfactory results are reached [8]. It's almost a " Darwin" model: Only those elements that are strong and give desired results are allowed to pass on to the next stage. Using this method, the algorithm chooses the optimum set of input variables, the degree of nonlinearity in the final model, and the structure and the degree of interaction terms in the final model [9]. There are four advantages to this method: A small training set is required, the multiple layer architecture of the resulting system results in a feasible way of implementing high degree multinomials, the computation burden is reduced, and inputs/functions of inputs that have little impact on the output are automatically filtered out [8].

3. GMDH-type neural networks with sigmoid functions

The GMDH-type neural networks with sigmoid functions are developed based on the conventional GMDH-type neural networks [2],[3]. The conventional GMDH-type neural networks are organized by the following procedures. First, the original data are separated into training and test sets. The training data are used to estimate the weights of the neural networks, and the test data are used for organizing the neural network architectures. After separating the data, many combinations of the input variables are generated in each layer. For each of these combinations, the optimum neurons fitting the characteristics of the nonlinear system are automatically selected from different neuron architectures. The different neuron architectures used in the GMDH-type neural networks are shown as follows:

1) First type neuron

Σ: (Nonlinear function)

$$z_k = w_1 u_i + w_2 u_j + w_3 u_i u_j + w_4 u_i^2 + w_5 u_j^2 + w_6 u_i^3 + w_7 u_i^2 u_j + w_8 u_i u_j^2 + w_9 u_j^3 - w_0 \theta_1 \tag{1}$$

f : (Nonlinear function)

$$y_k = 1 / (1 + exp(-z_k)) \tag{2}$$

2) Second type neuron

Σ: (Nonlinear function)

$$z_k = w_1 u_i + w_2 u_j + w_3 u_i u_j + w_4 u_i^2 + w_5 u_j^2 + w_6 u_i^3 + w_7 u_i^2 u_j + w_8 u_i u_j^2 + w_9 u_j^3 - w_0 \theta_1 \tag{3}$$

f : (Linear function)

$$y_k = z_k \tag{4}$$

3) Third type neuron

$\sum$: (Linear function)

$$z_k = w_1 u_1 + w_2 u_2 + w_3 u_3 + \cdots + w_r u_r - w_0 \theta_1 \quad (r < p) \tag{5}$$

f : (Nonlinear function)

$$y_k = 1 / (1 + exp(-z_k)) \tag{6}$$

4) Fourth type neuron

$\sum$: (Linear function)

$$z_k = w_1 u_1 + w_2 u_2 + w_3 u_3 + \cdots + w_r u_r - w_0 \theta_1 \quad (r < p) \tag{7}$$

f : (Linear function)

$$y_k = z_k \tag{8}$$

5) Fifth type neuron

$\sum$: (Nonlinear function)

$$z_k = w_1 u_i + w_2 u_j + w_3 u_i u_j + w_4 u_i^2 + w_5 u_j^2 + w_6 u_i^3 + w_7 u_i^2 u_j + w_8 u_i u_j^2 + w_9 u_j^3 - w_0 \theta_1 \tag{9}$$

f : (Nonlinear function)

$$y_k = exp(-z_k^2) \tag{10}$$

6) Sixth type neuron

$\sum$: (Linear function)

$$z_k = w_1 u_1 + w_2 u_2 + w_3 u_3 + \cdots + w_r u_r - w_0 \theta_1 \quad (r < p) \tag{11}$$

f : (Nonlinear function)

$$y_k = exp(-z_k^2) \tag{12}$$

7) Seventh type neuron

$\sum$: (Linear function)

$$z_k = w_1 u_1 + w_2 u_2 + w_3 u_3 + \cdots + w_r u_r - w_0 \theta_1 \quad (r < p) \tag{13}$$

f : (Nonlinear function)

$$y_k = a_0 + a_1 z_k + a_2 z_k^2 + \cdots + a_m z_k^m \tag{14}$$

Here, $\theta_1 = 1$ and w_i $(i = 0, 1, 2, \cdots)$ are weights between the neurons and estimated by applying the stepwise regression analysis [6] to the training data. The output variables y_k of the optimum neurons are called intermediate variables. The optimum neuron architectures fitting the characteristics of the nonlinear system are automatically selected by using the AIC [5]. Therefore, many kinds of nonlinear systems can be automatically modeled by using the GMDH-type neural networks. Then, the L intermediate variables giving the L smallest test errors (calculated using the test data) are selected from the generated intermediate variables y_k. The selected intermediate variables are set to the input variables of the next layer and the same computation is iterated. When the errors for the test data in each layer stop decreasing, the iterative computation is terminated. The complete neural networks that simulate the characteristics of the nonlinear system can be constructed by using the optimum neurons generated in each layer.

The GMDH-type neural networks with sigmoid functions are developed based on the conventional GMDH-type neural networks. This algorithm does not need the separation of the original data into the training and test sets because the AIC can be used as test errors. So all the data are used for training. The procedures for determining the architecture of the GMDH-type neural networks with sigmoid functions conform to the following:

(1) The first layer

$$u_j = x_j \quad (j = 1, 2, \ldots, p) \tag{15}$$

where x_j $(j = 1, 2, \ldots, p)$ are the input variables of the system, and p is the number of input variables. In the first layer, input variables are set to the output variables.

(2) The second layer

Many combination of two input variables (u_i, u_j) are generated. For each combination, the neuron architecture is described by the following equations:

Σ: (Nonlinear function)

$$z_k = w_1 u_i + w_2 u_j + w_3 u_i u_j + w_4 u_i^2 + w_5 u_j^2 + w_6 u_i^3 + w_7 u_i^2 u_j + w_8 u_i u_j^2 + w_9 u_j^3 - w_0 \theta_1 \qquad (16)$$

f : (Nonlinear function)

$$y_k = 1 / (1 + exp(-z_k)) \qquad (17)$$

Here, $\theta_1 = 1$ and $w_i(i=0,1,2,\ldots,9)$ are the weights between the first and second layer. The weights $w_i(i=0,1,2,\ldots,9)$ are estimated by the stepwise regression analysis using AIC[5].

These neuron architectures belong to the first type neuron of the conventional GMDH-type neural networks and the estimation procedure of the weight w_i is as follows:

First, the values of z_k are calculated by using the following equation:

$$z_k = log_e(\phi'/(1 - \phi')) \qquad (18)$$

where ϕ' is the normalized output variables whose values are between zero and one. Then the weights w_i are estimated by using the stepwise regression analysis [6] which selects useful input variables by using the AIC [5]. Only useful variables in (16) are selected by the stepwise regression analysis and neuron architecture can be organized by the selected useful variables.

From these generated neurons, the L neurons which minimize the AIC are selected. The output variables y_k of L selected neurons are set to the input variables of the neurons in the third layer.

(3) The third and succeeding layers

In the third and succeeding layers, the same computation of the second layer is iterated until the AIC values of L neurons stop decreasing. When iterative computation is terminated, the neural network architectures are produced by the selected neurons in each layer.

By using these procedures, the GMDH-type neural networks with sigmoid functions can be constructed.

4. Application to the nonlinear system identification

The nonlinear system is assumed to be described by the following equations:

$$\phi = f_1(x_1, x_2, x_3) / f_2(x_1, x_2, x_3) \qquad (19)$$
$$f_1(x_1, x_2, x_3) = 1.0 + 2.0\, x_1^2 x_2 + 3.0\, x_2^2 x_3 \qquad (20)$$
$$f_2(x_1, x_2, x_3) = 1.0 + 2.0 exp(x_1) + 3.0 exp(x_1 x_2) + 4.0 exp(x_3) \qquad (21)$$

Here, ϕ represents the output variable and $x_1 \sim x_3$ are the input variables. An additional input, x_4, is added as the input variable of the neural networks in order to check that the GMDH-type neural networks can detect and eliminate any useless input variables. The neural networks are organized by using twenty training data. Twenty test data are used to check the prediction and generalization ability.

4.1 The identification results obtained by using the GMDH-type neural networks

(1) The input variables.

Four input variables were used, but the useless input variable x_4 was automatically eliminated.

(2) The number of selected neurons.

Four neurons were selected in the hidden layer.

(3) The architecture of the neural networks.

The calculation of the GMDH-type neural networks was terminated at the fourth layer.

(4) The estimation accuracy.

The estimation accuracy was evaluated by using the following equation:

$$J_1 = \sum_{i=1}^{20} |\phi_i - \phi_i^*| \Big/ \sum_{i=1}^{20} |\phi_i| \qquad (22)$$

where ϕ_i (i =1,2,$\cdots$,20) were the actual values and ϕ_i^* (i =1,2,$\cdots$,20) were the estimated values by the GMDH-type neural networks. The value of J_1 is shown in Table1. In this table, GMDH-NN is the GMDH-type neural networks with sigmoid functions.

Table 1   Identification results of three methods

Method	J_1	J_2
GMDH-NN	0.0122	0.0181
GMDH	0.0405	0.0434
Conventional NN	0.0246	0.0269

(5)   The prediction accuracy.

The prediction accuracy was evaluated by using the following equation:

$$J_2 = \sum_{i=21}^{40} |\phi_i - \phi_i^*| \Big/ \sum_{i=21}^{40} |\phi_i| \qquad (23)$$

where ϕ_i (i =21,22,$\cdots$,40) were the actual values and ϕ_i^* (i =21,22,$\cdots$,40) were the predicted values by the GMDH-type neural networks. The value of J_2 is shown in Table1.

(6)   Variation of AIC and the estimated values.

The variation of AIC is shown in Fig.1. AIC values converged at the fourth layer. The estimated values of ϕ by the GMDH-type neural networks are shown in Fig.2. We can see that the estimated values are very accurate.

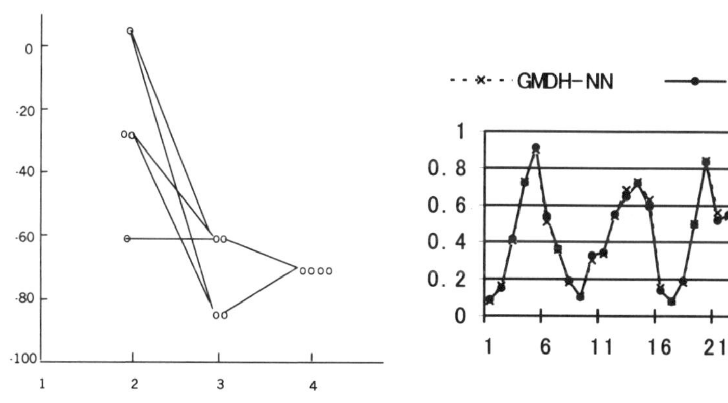

Fig.1: Variation of AIC   Fig.2:   Estimated values by the GMDH-type neural networks

4.2 Comparison of the GMDH-type neural networks and the other models

The identification results of our GMDH-type neural networks were compared with those of the GMDH algorithm and the conventional multilayered neural networks trained using the back propagation algorithm.   In the GMDH algorithm, four input variables were used, but again the useless input variable x_4 was automatically eliminated. Four intermediate variables were selected. The calculation was terminated at the fourth layer. The values of $J_1$and J_2 are shown in Table1.   In the conventional multilayered neural networks, the neural network was developed as a three layered architecture. Four input variables were used in the input layer and four neurons were used in the hidden layer. The weights of the neural networks were estimated by using the back propagation method. The values of J_1 and J_2 are shown in Table1.

From these identification results, both estimation and prediction errors (J_1 and J_2) of the GMDH-type neural networks were the smallest of the three identified models. From this example we can see that the GMDH-type neural network algorithm is a very accurate identification method for the nonlinear system.

5. Conclusion

In this study, the GMDH-type neural networks with sigmoid functions were proposed. This algorithm can automatically organize a multilayered neural network architecture fitting the complexity of the nonlinear system by using the heuristic self-organization method. It is very easy to apply this algorithm to the identification problem of the practical complex system because the structural parameters (such as the number of layers, the number of neurons in the hidden layers, the useful input variables and the optimum neuron architectures in the hidden layers) are automatically determined so as to minimize the AIC. This algorithm was applied to a nonlinear system identification problem and it was shown that this algorithm was a very useful method.

References

[1] T. Kondo, A.S. Pandya, J.M. Zurada,   Logistic GMDH-type neural networks and their application to the identification of the X-ray film characteristic curve,   *Proc. of IEEE International Conference on System, Man, and Cybernetics* (1999) 437-442.

[2] T. Kondo, GMDH neural network algorithm using the heuristic self-organization method and its application to the pattern identification problem, *Proc. of the 37th SICE Annual Conference*, International Session Paper (1998) 1143-1148.

[3] T. Kondo, A.S. Pandya, J.M. Zurada, GMDH-type neural networks and their application to the medical image recognition of the lungs,   *Proc. of the 38th SICE Annual Conference*, International Session Paper (1999) 1181-1186.

[4] A.G. Ivakhnenko,   Heuristic self-organization in problems of engineering cybernetics,   *Automatica*, **6**, No.2 (1970) 207-219.

[5] H. Akaike,   A new look at the statistical model identification,   *IEEE Trans. Automatic Control*, **AC-19**, No.6 (1974) 716-723.

[6] N.R. Draper and H. Smith,   Applied Regression Analysis, John Wiley and Sons, New York, 1981.

[7] K. Chon and   S. Lu ,   A new algorithm for the autoregression moving average model parameter estimation using group method of data handling,   *Annals of Biomedical Engineering*, **29** (2001) 92-98.

[8] F. Chang and Y. Hwang,   A self-organization algorithm for real-time flood forecast,   *Hydrological Processes*, **13** (1999) 123-138.

[9] C. Jiaa and D. Dornfeld,   A self-organizing approach to the prediction and detection of tool wear,   *ISA Transactions*, **37** (1998) 239-255.

HARDWARE IMPLEMENTATION OF RENDEZVOUS INTER-PROCESS COMMUNICATION WITH ROUND ROBIN SCHEDULING :PROTOTYPE TWO PROCESSES

Suryaprasad Jayadevappa, Ravi Shankar and Sam Hsu,
Dept of CSE, Florida Atlantic University, Boca Raton, Florida-33431.

Abstract:
This paper aims to provide a practical approach of porting key Operating System tasks onto the hardware, by extending the Instruction Set Architecture (ISA). We have implemented the rendezvous Inter-Process Communication(IPC) with round robin scheduling. We have used two simple 8-bit RISC processors, with extended ISA. The results obtained from the behavioral simulation has been included.
Inter-Process communication (IPC) is one of the key tasks in any Operating System. In this paper we look into the practical approach of implementing the IPC mechanism between 2 processes. We have implemented the rendezvous approach of IPC , using simple mailbox with zero buffering capacity. To make all this implementable, we have used the round- robin scheduling technique.

Keywords: Inter-Process Communication (IPC), Extended Instruction Set Architecture (EISA), System On Chip (SOC), Operating System (OS), Electronic Design Automation (EDA).

1. Introduction

In the recent years there has been tremendous changes in VLSI design. The reasons for this radical change could be attributed to the availability of better technology, availability of better Electronic Design Automation (EDA) design tools, market demand, early time to market etc.

Migration of applications in software into hardware is becoming more popular in the recent years. The reason for this migration could be many , due to applications type, technology , need of true concurrency, need in higher speed etc. It is very difficult to evaluate one approach against the other. There are always going to be engineering trade-offs.

With the advent of System On Chip (SOC) methodology and improvement in EDA tools, hardware oriented applications nowadays need lesser time to get to the market. One of the reason for this is due to the design reuse capability offered by the SOC methodology. In the System On Chip (SOC) methodology the hardware/software components of a system are identified at the abstract level. An interesting challenge would be in drawing the line between what needs to be in hardware and what needs to get into software. Tanenbaum in his book [3] mentions that the operating system can be in hardware. Also Ian Page discusses similar issues related to converting software into equivalent hardware [4].

To meet the complex requirements of emerging applications, current systems use a combination of general purpose processor with Digital Signal Processing (DSP) processors and application specific integrated circuits (ASIC's) used for specific purposes. Though there has been different approaches followed for different applications, one of the most prominent techniques is the extended instruction set architecture. In this technique the instruction set of the general purpose processor is suitably altered to suit the specific needs of the design.

This paper is our initial step in understanding the issues in porting key OS activities onto a chip. Also to determine if and how we need to suitably change the architecture of a simple general purpose processor to support these key operating system activities. Here we focus on two such important operating system activities, namely IPC and scheduling. We have successfully ported the rendezvous IPC approach with round-robin scheduling algorithm.

We start with a simple strip-down version of an 8-bit RISC architecture. To this we accumulate a few more instructions which would help us in realizing our goal. Further keeping in mind the high level of computational requirements of emerging applications we have two separate processors. According to the applications requirements, these two processors could be used for specific purpose. But in our architecture, one of these processors needs to oversee the IPC and scheduling activity. The simulation (test bench) consisted of two different processes, one for a sort routine and another to perform simple multiplication.

The rest of the paper is organized as follows: section 2 describes the problem, section 3 discusses the methodology and architecture, section 4 provides the simulations and results, and finally section 5 concludes the paper.

2.1 Problem Defined

In this paper we have ventured into porting a simple IPC & scheduling mechanism into hardware. Though the problems seems straight forward, but the difficulty was in looking from the hardware implementation point of view. Some of the important issues in implementing Inter-process Communications are: a) Message Passing , b) Mutual Exclusion during critical activities and c) Proper sequencing when dependencies are present.

Table 1. provides a view of some of the operating system activities and its hardware support. In the recent past more and more of these operating system activities have been supported in hardware on various systems.

OS functions	Hardware Solution	Chip Level	Year	Reference
Dynamic Relocation	Segmentation Registers	Intel 8086	1979	[5]
Memory Management	Virtual Memory	Intel 432	1983	[2]
Process Switching	Register Banks	Berkely's RISC	1983	[7]
System Protection	Supervisory Stack	Motorola 68000	1982	[8]
Memory Management	Memory management Unit, On-chip Cache	Motorola 68000	1984	[8]
Real Time OS	Stack Oriented CPU	SUN's PICO	1999	[9]
Multi-programming	VLIW, Microprogram	Transmeta's Crusoe	2000	[11]
Security	Segmentation Registers	Intel 8086	1979	[5]

Table 1. Various OS activities and the hardware support.

One interesting venture was by Intel in 1983 with iAPX-432, wherein they came out with a chip which supported majority of the operating tasks, was object oriented and was built to support Modula language. The chip was a major failure at that time. The performance was much less than the x286 chips which was available then. The reasons for the failure was multi-dimensional such as, non-availability of technology, trying to put too many complex functionalities onto the chip and an architecture totally oriented to support Ada. But this chip was a very good learning experience for Intel, though a costly one. A lot of features from this failed experience was put to good use in the subsequent chips .

Also as more of these operating system activities migrate into the hardware, the gap between the hardware and the user applications is minimizing. Figure 1., provides a very abstract view of the layers in a typical computer system.

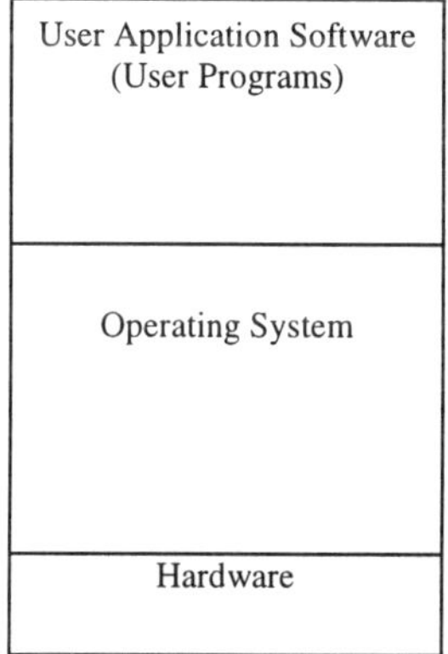

Fig 1.Abstract level diagram of a typical computer system.

The sandwiched layer between the hardware and the user programs is ever so slowly migrating into the lower layer. A major hurdle in this migration has been the challenging user requirements which has further exacerbated this migration. But with better EDA tools and newer hardware design methodologies , the migration process has been hastened to some extent. In the recent operating system design methodology, like that of a microkernel, key functions of the operating system forms a shell and the rest of the operating system support activities are moved up into the higher layer. At present we do not intend to move the complete operating system into the layer below. We intend to identify these important activities like in a microkernel and move this into the hardware. Hence we consider two such important operating system activities, namely inter-process communications and process scheduling. We have implemented the rendezvous IPC technique and the round-robin scheduling algorithm.

Our main interest in this paper is to introduce an appropriate architecture to implement these operating system activities. We intend to evaluate the performance of the various techniques at a later stage.

2.2 Architecture

We have developed and used a simple architecture. Reusability being a key issue in present days VLSI design, we have chosen the available MU0 processor. MU0 is a simple 8-bit RISC based architecture. We have used two of these MU0's in our design. The MU0 reference is designed by students in under-graduate and graduate course of CAD based VLSI design offered in our CSE dept by one of the authors. As MU0 is a very simple 8-bit architecture, enhancements on this can be done easily. The changes that we have to do in-order to accomplish our goals is mainly in the a)Instruction set and b)additional hardware registers.

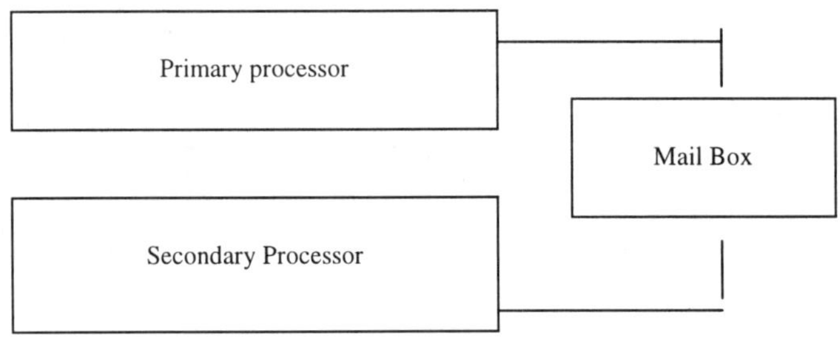

Fig 2. Simple architecture developed.

2.2.1 Extensions in Instruction Set Architecture

The extended Instruction set architecture that we created affected only one of the two processors in our design. The processor with this extended instruction set was named the "primary " and the other "secondary". Typically the primary processor is independent, it has the extra responsibility for the successful implementation of the proposed operating system activities. The process in turn were executed by the secondary processor under the directions of the primary processor.

For example, we added an instruction named LDMBR – which is included in the governor processor. This is similar to any other LOAD instruction, but different in the sense that it is specifically used to load the mail-box address register with the starting address of the process that needs to be executed.

2.2.2 Additional Hardware

We had to make changes in the hardware also. We had to add a few extra registers, with specific purposes:

Mail-box Address Register(MBAR):

This register was a 12 bit address register. It was used by primary processor to communicate with the secondary processor the starting address of the new process to be executed by the secondary processor.

Mail-box Control Register(MBCR):

This is an 8-bit register used by the primary computessor to indicate to the secondary processor that the address present in the MBAR is valid and can be used . At present we are using only one bit called START, to indicate to the secondary processor.

Mail-box Status Register(MBSR):
This register is used to communicate important information about the mail-box status. We have at present used 5 bits in this for various purposes like terminated by primary processor, terminated by secondary processor etc.
 Not all these bits have been used in all the registers except for MBAR in our design. Many have been designed so keeping in mind our future requirements to achieve our long term goal of developing a micro-kernel for the operating system in hardware. There is also another register named TIME which is used by the governor processor to keep track of the time quantum spent by the slave processor on a particular process.

3. Implementation
We have implemented the rendezvous inter-process communication, so that the primary processor can communicate with secondary processor. We also have implemented the round-robin scheduling algorithm to switch between processes.

First, our primary concern was to develop a simple architecture to achieve our goal. So we chose to use two processors, one was assigned to be the primary and the other secondary. Unlike the concept of a real MASTER, our primary processor is an independent processor, with the extra responsibility of ensuring that the communication & scheduling with the secondary processor takes place. The primary processor in our case was just intended for scheduling purpose. For this reason the primary processor had a few extra instruction when compared to that of a secondary processor. Both these processors are simple 8-bit RISC based processors.   They all have fixed length instructions. As can been seen in our implementation, there is option to include more instructions.

We have implemented rendezvous mechanism of message passing. Hence there is no mechanism by which we can buffer other requests. we were able to implement the round-robin scheduling algorithm

The main task of the primary processor is to keep track of the time spent by the secondary processor on a particular task. Whenever there is a new process to be executed, the primary processor gets the starting address. Then the primary processor checks whether the secondary processor is idle by checking the status register. If the secondary processor is idle then the primary processor loads the process address onto the mail-box address register(MBAR) and then sets the START flag in the mail-box control register(MBCR). In the meantime the primary processor also initializes the timer value in the time register and starts counting up. If the secondary processor is busy performing a previous task, then the primary processor checks whether the time quantum has expired. If the time has not expired then the primary processor keeps polling the secondary processor until the time quantum expires. Once the time quantum expires, the primary processor sets the status register indicating that the previous process was suspended by it. In the meantime the secondary processor saves the process information, like address etc.
At the secondary processor end, it checks for the START flag on the MBCR. If the start flag is set, then the secondary processor checks whether it needs to start a new process or a suspended process. If it is a new process then the secondary processor takes the address

present in MBAR and loads it into its program counter. If the time quantum on a particular process has expired then the secondary processor needs to suspend the present process and save the process status information. Hence implementing the round-robin algorithm and the rendezvous inter-process communication technique.

The entire implementation was done at the behavioral level. The language used is VERILOG and the simulator used to verify the results was SILOS-III.

References

[1]   Tannenbaum, A.S., and Woodhull, A.S., Operating Systems, 2nd Edition, Prentice Hall, Upper Saddle River, NJ, 1997

[2].   Intel Staff, Intel 432 System Summary: Manager Perspective, Intel Corp., Santa Clara, CA, 1981.

[3]   Tannenbaum, A.S., Structured Computer Organization, 4th edition, Prentice Hall, Upper Saddle River, NJ, 1999.

[4]   Page, I., Several articles on hardware-software co-design, cosynthesis, and Reconfigurable architectures, Oxford University, http://oldwww.comlab.ox.ac.uk/oucl/people/ian.page.html, 2000.

[5]   TOP500 Supercomputer Sites, Maintained at University of Mannheim, Germany and University of Tennessee, USA. See www.netlib.org/benchmark/top00.html

[6]   Ullman, J.D., Computational Aspects of VLSI, Computer Science Press, Rockville, MD, 1984.

[7]   Wirth, N., "Hardware Compilation: Translating Programs into Circuits," Computer Magazine, Vol. 30, No. 6, IEEE, June 1998, pp. 25-31

[8]   Jean Daniel Nicoud and Andrew Martin Trrell, Ecole Polytechnique, Federale de Lausanne " The Transputer T414 Instruction Set " special feature IEEE 1989..

[9]   Henry Chang, Larry Cooke, Merrill Hunt, Grant Martin, Andrew Mc Nelly, Lee Todd., " Surviving the SOC Resolution –A Guide to Platform- Based Design " . Kluwer press 2000.

[10]   Hennessy, J.L., and Patterson, D.A., Computer Architecture: A Quantitative Approach, 2nd Edition, Morgan Kaufman, San Fransisco, CA, 1996.

[11]   Arnold, M.G,. Verilog Digital Computer Design: Algorithms into Hardware,Prentice Hall PTR, Upper Saddle River, NJ, 1999.

[12]   Inmos Staff, Transputer Architecture, INMOS Corp., Colorado Springs, CO,1985.

[13]   Garber, L., and Sims, D., "In Pursuit of Hardware-Software Co-design," Industry Trends, Computer Magazine, IEEE, June 1998, Pp.12-14.

[14]   Kern, G., and Greenstreet, M.R., Formal Verification in Hardware Design: A Survey, ACM Transactions on Design Automation, Vol. 4, No. 2, April 1999, pp. 123-193.

Proposal of Eco-Evolution

Yoshiji FUJIMOTO
Department of Applied Mathematics and Informatics, Ryukoku University
Seta, Ohtsu, Shiga 520-2194 Japan   fujimoto@math.ryukoku.ac.jp

Katsunori SHIMOHARA
ATR Information Sciences Division
Hikaridai Seika-cho Soraku-gun Kyoto 619-0288 Japan   katsu@isd.atr.co.jp

Abstract. In this study, we propose a new paradigm of evolutionary computation named "Eco-Evolution" for large- scale and complex real problems.The aim of this paradigm is to maintain the diversity of a population and to achieve convergence towards the global optimum and local optima in the evolutionary process. The idea of the paradigm is inspired from Tierra as an evolutionary ecological system that has widespread diversity.
A concept of diversity in terms of applying evolutionary computation to real problems is defined. Several ideas are introduced for the implementation of this feature. They include individual lifetime and individual resource consumption based on fitness, and local crossover for clustering individuals in a population. In this study, Eco-Evolution is confirmed to have the feature of defined diversity.

1. Introduction

In order to apply evolutionary computation to large-scale and complex practical applications, we propose a new paradigm of evolutionary computation named "Eco-Evolution". The practical application of evolutionary computation requires diversity and robustness. Diversity is considered in practical applications to search for the global optimal solution and to obtain as many local optima as possible. Although searching the global optimum is important, however, the global optimum is not always a relevant solution for a practical problem, since many factors can not be reflected to fitness functions such as the ease of manufacture, maintenance, and design, and so on. We therefore have to search for as many alternative solutions as possible. From our standpoint, this means that the local optimal or suboptimal solutions become important as alternative solutions.

Robustness is to search for solutions capable of enduring noise disturbances, errors in manufacturing processes, variations in working circumstances (like temperature changes) and changes with the passage of time. This means finding the solution on the broad peak capable of maintaining the performance against variations in the design parameters, even if it is local optimum. We attempted peak shape identification in the previous paper [1][2] to search for robust solutions.

In this study, we propose a new paradigm of evolutionary computation named "Eco-Evolution" for diversity in evolutionary computation. The aim of this paradigm is to maintain the widespread diversity of a population and to achieve convergence towards the global optimum and local optima in the evolutionary process. This idea is inspired from the evolutionary ecological system Tierra with the widespreading diversity by T. S. Ray [3][4].

In this paper, at first, a concept of diversity that is important for applying evolutionary computation to real problems is described. The paper also introduces an idea of ecological evolutionary systems that include individual lifetime and individual resource consumption for survival based on fitness, resource supply from the environment and local crossover for clustering individuals in a population to achieve diversity.

2. DIVERSITY

The word of "diversity" is often used, but, it is a word with multiple meanings and can be ambiguous.

Considering this, we try to define the concept of diversity in terms of evolutionary computation by the following two conditions.

(1) An individual distribution density in a population is related to a proportional function or monotonic function of the value of the fitness function within the parameter space in the early stage of evolution like the existing probability of a moving particle being proportional to an energy potential function in the Hamiltonian Algorithm [5][6]. Clusters of individuals are formed around the global optimum and local optima as shown in Fig. 1.

(2) The clusters of individuals formed around the global optimum and local optima converge towards optimal solutions in the final stage of evolution as shown in Fig. 2.

When these two conditions are satisfied, we say that "the evolutionary computation has diversity".

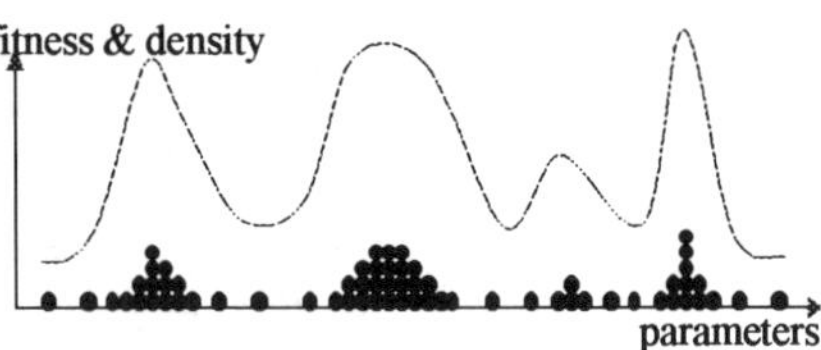

Fig.1 Distribution of individuals based on a fitness function

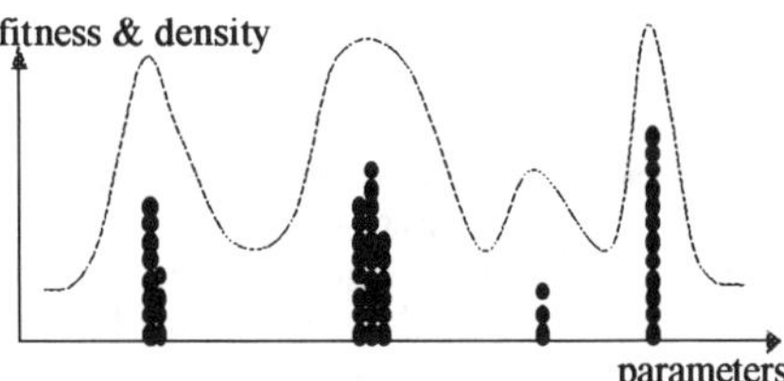

Fig.2 Convergence of individuals towards the global and local optimal points

3. ECO-EVOLUTION

3.1 Basic idea of Eco-Evolution

The basic idea of Eco-Evolution is described in the following.

(1) Each individual is mortal and the lifetime of each individual is determined by its fitness.
An individual with a better fitness has a longer lifetime. In terms of minimization, the lifetime of each individual is inversely proportional to its fitness. Therefore, the population size is variable.

((2) The fitness of each individual means the amount of resources that it needs to live.
Individuals with a smaller consumption of resources (economical) can increase in an environment with limited resources. This means Eco-Evolution aims at solving minimization problems.

(3) New individuals are reproduced by mutation and local crossover (crossover in species).
Mutation is the random generation of individuals and local crossover is the crossover between individuals in a local cluster like a species.

(4) The environment supplies the resources needed for individuals to survive.
The lifetimes of individuals are affected by the ratio of the total consumption by the individuals and the supply from their environment. When the amount of consumption approaches the amount of supply, the resulting exhaustion of resources shortens the lifetime of each individual.

(5) The resources supplied from the environment are gradually reduced when the population size exceeds the proper size able to be fed by the environment.
This increases the selection pressure and prompts the convergence of individuals.

3.2 System Description of Eco-Evolution

This section presents the system descriptions of Eco-Evolution.

(1) The lifetime of an individual
The lifetime of an individual is determined by its fitness as follows.
First, the pseudo-lifetime L_i^P of an individual with $fitness_i$ is defined by the following equations.

$$fitness_i^N = \frac{fitness_i - fitness_{Min}}{fitness_{Max} - fitness_{Min}} \quad \cdots\cdots (1)$$

where $fitness_i^N$ is a normalized fitness and $fitness_{Max}$, $fitness_{Min}$ are the maximum and minimum fitness in a population, respectively.

$$L_i^P = \frac{1}{fitness_i^N + \varepsilon} \quad \cdots\cdots\cdots\cdots (2)$$

$$1 \geq fitness_i^N \geq 0, \quad \varepsilon = \frac{1}{L_{Max}}$$

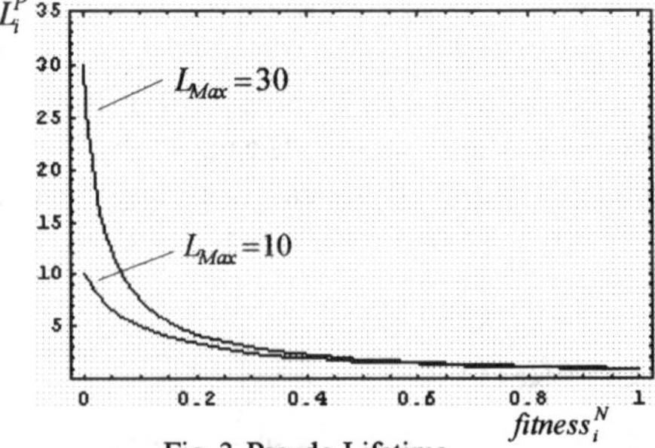

Fig. 3 Pseudo-Lifetime

where L_{Max} is the maximum lifetime.

The longer the maximum lifetime is, the higher the selection pressure is as shown in Fig. 3 because an individual with a better fitness is assigned a longer lifetime.

Next, the lifetime L_i of each individual is normalized so that their average lifetime is $L_{Ave}(t)$ at an evolutionary time t (e-time).

$$L_i = \frac{L_{Ave}(t) \times N_{pop}}{\sum_j L_j^P} \times L_i^P \qquad (3)$$

$L_{Ave}(t)$ is set to a life time a little longer than the equilibrium average lifetime, which is the inverse of reproduction rate P_{Rep} and is multiplied by resource sufficiency coefficient $K(t)$.

$$L_{Ave}(t) = \frac{1.05}{P_{Rep}} \times K(t) \qquad (4)$$

The term equilibrium denotes a state with a constant population size.

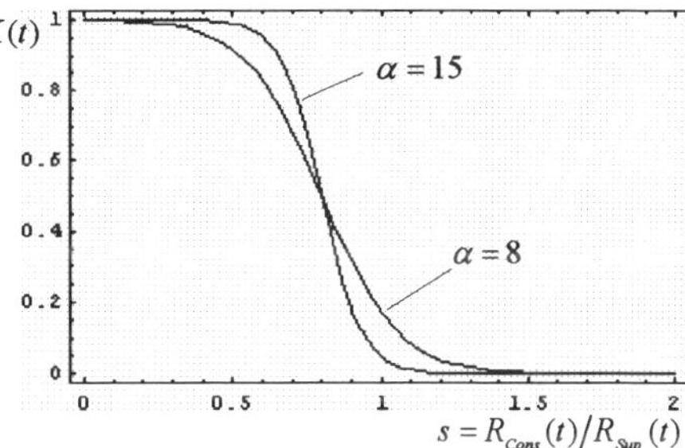

Fig. 4 Resource Sufficiency Coefficient

(2) Resource supply and consumption

The resource supply is the amount of resources supplied from the environment. It is initialized at the value of the worst fitness at the first E-time (Evolutionary time) multiplied by given proper population size N_{Pop}^P of the environment.

$$R_{Sup}(0) = fitness_{Worst} \times N_{Pop}^P \qquad (5)$$

When the population size becomes more than the proper population size, resource supply $R_{Sup}(t)$ decreases gradually with a value that is proportional to the difference between the resource supply and consumption.

$$R_{Sup}(t+1) = R_{Sup}(t) \times$$

$$\left(1.0 - \max\left(\frac{R_{Sup}(t) - R_{Cons}(t)}{R_{Sup}(t)}, 0.05\right) \times \beta\right) \qquad (6)$$

where β is the coefficient that controls the rate decrease.

Resource consumption $R_{Cons}(t)$ is the total fitness of individuals in a population.

$$R_{Cons}(t) = \sum_{i=1}^{N_{Pop}} fitness_i \qquad (7)$$

Resource sufficiency coefficient $K(t)$ is calculated as a function of the ratio $s = R_{Cons}(t)/R_{Sup}(t)$.

$$K(t) = \frac{1 + e^{-0.8\alpha}}{1 + e^{(s-0.8)\alpha}} \qquad (8)$$

α is a constant that determines the steepness of the function as shown in Fig. 4.

(3) Reproduction

New individuals are reproduced by local crossover and mutation with the probabilities of P_{Cross} and P_{Mut}, respectively. A local crossover operation is a crossover between a pair of individuals separated by a relevant distance, i.e., not too close and not too far. It is like a species mating. The mating between kins and the mating between individuals of different species are not allowed.

A local crossover operation is defined by the local crossover probability function of the normalized distance in Eq. 9, which is shown in Fig. 5.

$$P_X(d) = \frac{d}{d_p} e^{\left(1 - \frac{d}{d_P}\right)} \qquad (9)$$

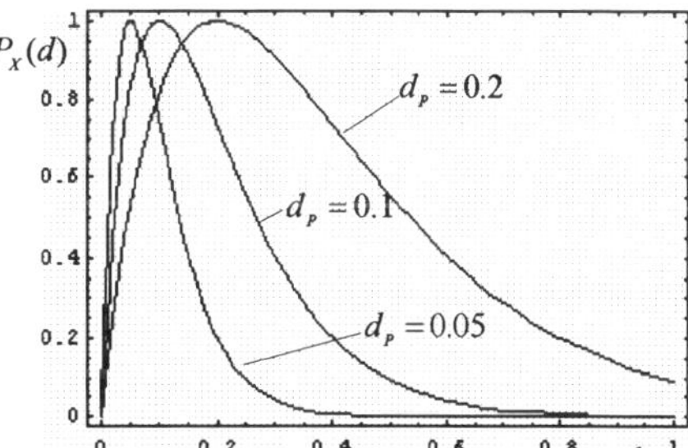

Fig. 5 Local Ccrossover Pprobability Function

where d_p is the most relevant distance for mating.

This means that a pair of parents are allowed to mate with probability $P_X(d)$, when they are separted by a distance d. Local crossover operations are executed on all pairs of individuals along the shuffled order in the population. Crossover probability P_{Cross} is the upper limit of the rate of executed crossover $2N_{Cross}/N_{pop}$.

Therefore, the reproduction rate is calculated by Eq. 10. Then, it is used to calculate the average lifetime.

$$P_{Rep} = P_{Mut} + \min\left(P_{Cross}, \frac{2N_{cross}}{N_{pop}}\right) \quad \cdots \quad (10)$$

where N_{cross} is the number of parent pairs on which crossover operations are executed.

(4) Selection

There is no explicit selection process. Rather, the selection is performed by a lifetime that is inversely proportional to the fitness of each individual, and limited resources supplied from environment by which more individuals with smaller consumption of resources can be reproduced. Only individuals with ages exceeding their lifetimes die.

4. PROCEDURE OF ECO-EVOLUTION

In the Eco-Evolution, the evolutionary computations are executed along the following procedures.

(1) Initialization of population

The e-time is set at zero. The initial population with the given initial population size is formed by generating the individuals randomly.

(2) Evaluation and calculations of lifetime

The fitness values of new individuals are calculated by a evaluation function by Eqs. 1, 2 and 3. The initial value of the resource supply and the resource sufficiency coefficient are determined by Eqs. 5 and 8 and the lifetime of each new individual is calculated.

(3) Modification of the resource supply

If the population size is over the proper population size of the environment, the resource supply is decreased gradually by Eq. 6.

(4) Reproduction

New individuals are generated by local crossovers and mutations with the given rates, independently.

(5) Evaluation and calculation of lifetime

The fitness values of new individuals are calculated by a evaluation function. The resource sufficiency coefficient are determined and the lifetime of each new individual is calculated.

(6) Selection

The individuals of which ages are over lifetime are died.

(7) Check termination conditions

The e-time is counted up and the termination conditions are checked. While the conditions are not satisfied, the procedures from (3) are repeated. When one of the termination conditions is satisfied, the evolutionary process is completed.

5. EMPIRICAL RESULTS

We implemented the Eco-Evolution by Java 2 and tried to apply it to the simple optimization problem of the Himmelblau's function [6] in order to observe the system behavior and confirm the realization of diversity. The Himmelblau's function is given by the following equation and has a global optimum and three local optimums.

$$F_{Him}(x,y) = (x^2 + y - 11)^2 + (x + y^2 - 7)^2 \quad \cdots\cdots \quad (11)$$

$$-5.2 \leq x, y \leq 5.2$$

The experiment is done with the following values of parameters.

- x, y parameters are coded by 16 bits Gray code.
- Initial population size = 500.
- Proper population size = 2000.

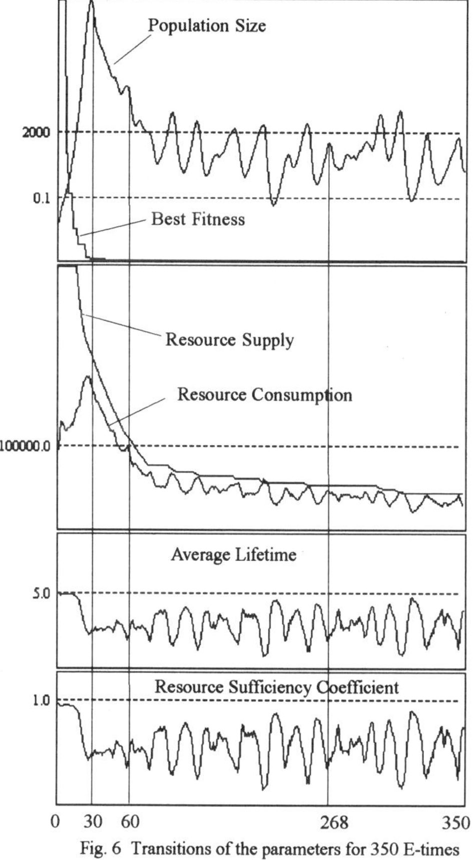

Fig. 6 Transitions of the parameters for 350 E-times

- $\beta = 0.1$ and $\alpha = 7.0$.
- Local crossover rate = 0.15.
- $d_p = 0.065$.
- Mutation rate = 0.05 (selection rate of individuals for mutation as candidates) .
- Bit flip rate of an individual= 0.06.

The results of the experiment are shown in Fig. 6 - 9. The transitions of parameters; the population size, the best fitness, the average lifetime, the resource supply, the resource consumption and the resource sufficiency coefficient for 350 e-times are shown in Fig. 6. The distributions of individuals at the 30th e-time, 62th e-time and 268th e-time are shown in Figs. 7, 8 and 9.

In the first 30 e-times, the population size is increasing rapidly. In this process, the resource is sufficiently supplied for reproducing many offspring. And the reproductions are dominantly done by mutations because the individuals are scattered on the search space as shown in Fig. 7 and have little chance to mate by local crossover. Therefore, multifarious in-

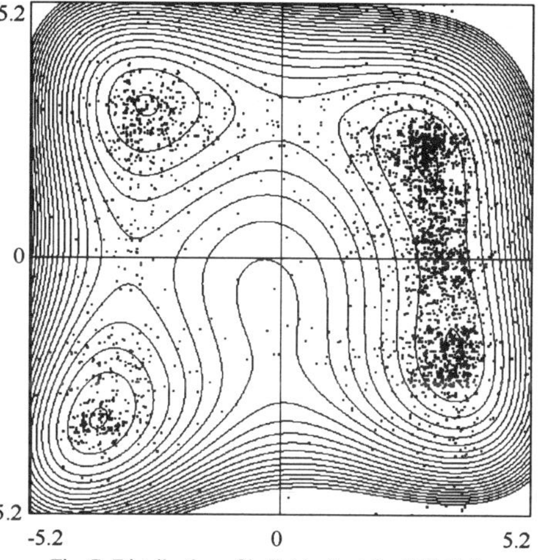

Fig. 7 Distribution of individuals at the 30th E-time

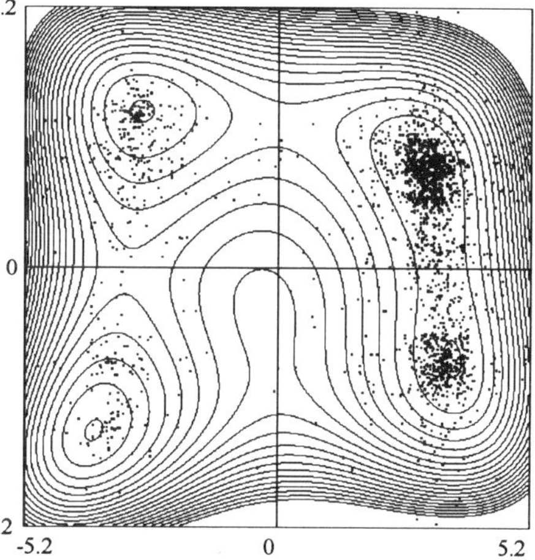

Fig. 8 Distribution of individuals at the 62th E-time

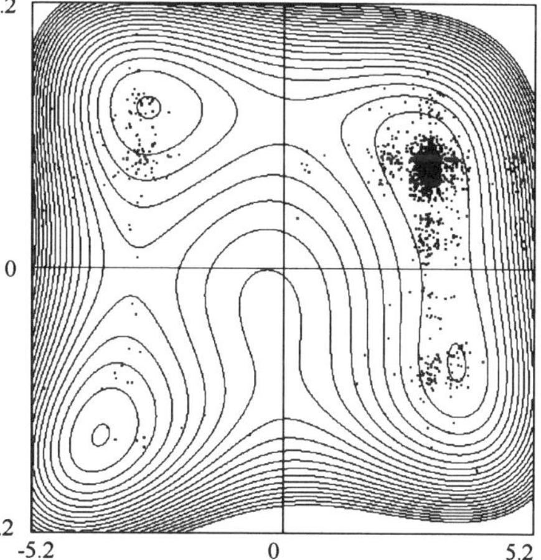

Fig. 9 Distribution of individuals at the 268th E-time

dividuals are reproduced like the Cambrian explosion.

Each individual has the lifetime that is inversely proportional to its fitness determined by the evaluation function. Therefore, the density function of the individual distribution becomes close to the inverse evaluation function and individuals form the clusters around the global optimum and local optimums as shown in Fig. 7. The probability of the local crossover become higher by forming clusters of individuals because the distances between individuals in a cluster are decreasing into the range of local crossover. The local crossover consolidates the clusters of individuals and accelerate them to converge into the global or local optimums.

After the population size is over the proper population size of the environment, the production power of resource in the environment is reduced and the amount of resource supply is decreased. Consequently, the resource sufficiency coefficient is decreasing and the average life time becomes shorter. Shorter average lifetime causes the decrement of population size. This means that the selection pressure become higher and the convergence of individual clusters into their optimal points is prompted. It can be observed in Fig. 8 that individuals converge to each optimum. It is conspicuous for three optima.

In addition, it is also observed that around the more broad peak, the larger clusters is formed. This is expected to be used for searching the robust solutions.

Finally, individuals converge to the global optimum with the elapse of e-time as shown in Fig. 9.

6. CONCLUSION

We propose Eco-Evolution as a new paradigm of evolutionary computation for application to real world problems. We define the concept of diversity in terms of applying evolutionary computation to practical problems and confirm that Eco-Evolution has the feature of diversity defined in this paper. We also find that it has the possibility for searching the robust solutions.

Future work is focusing on the following items.

(1) Multi-resource scheme for improving the capability of the habitant segregations of individuals.

(2) Introduction of the ecological models that cause the periodical oscillations of the environment.

(3) Enhancement to the multi objective optimization and tree-structure representation of an individual like GP.

(4) Applying the Eco-Evolution to real problems and evaluating its performance comparing with the conventional paradigms of evolutionary computations such as GA.

REFERENCES

[1] Y. Fujimoto and S. Tsutsui, "A Peak Shape Identification Genetic Algorithm with a Radial Basis Function", The 1997 IEEE International Conference on Evolutionary Computation, pp. 349-354, April 1997.

[2] S.Tsutsui, A. Ghosh and Y. Fujimoto, "A Robust Solution Searching Scheme in Genetic Search", The Forth International Conference on Parallel Problem Solving from Nature, pp. 543-552, Sept. 1996.

[3] T. S. Ray, "An Approach to the Synthesis of Life", Artificial Life II, pp. 371-408, 1993.

[4] T. S. Ray, "An Evolutionary Approach to Synthetic Biology : Zen and the Art of Creating Life", Artificial Life , Vol. 1, No. 1/2, pp.179-209, 1993.

[5] K. Shinjo and T. Sasada, "Hamiltonian System with Many Degrees of Freedom : Asymmetric motion and intensity of motion in phase space", Physical Review E, Vol. 54, No. 5, pp. 4685-4700, 1996.

[6] K. Shinjo, "Hamiltonian Algorithm : A Method to Solve Optimization Problems", ATR Journal, Vol. 1, pp. 48-49, 1998.

[7] A. Torn and A. Zilmskas, "Global Optimization", Springer-Verlag, pp.183-184 (1989)

Generation of World Model for Mobile Robot based on Adaptive Resonance Theory

Satoshi Shimonaka, Satoru Ishigaki, Masaaki Ida, Osamu Katai
Graduate School of Informatics, Kyoto University
Yoshida Honmachi, Sakyo, Kyoto 606-8501, Japan
shimo@sys.i.kyoto-u.ac.jp

Abstract   We present a world model generation system for behavior based mobile robot. Our robot control system is based on *Subsumption Architecture* which is a class of behavior-based control system. Categorizing the time series of outputs of the robot's behaviors, the system based on *Adaptive Resonance Theory* generates internal symbols corresponding to the features of its external world through the robot's *Sensory Motor Coordination*. Investigating the results of generated world model, we discuss the relations between the robot and human recognition. We present some experimental results to demonstrate the efficiency of our system.

1   World modeling based on behavior

1.1   Sensory-Motor Coordination

In design of control system of a mobile robot, One of the important subject is to generate its internal world model for the real environment which is suitable for intelligent robot control. However, there might be discordance between the generated model and the real environment. The problem is called *Symbol Grounding Problem* [1].

The fundamental concept of environmental modeling in this research is based on *Sensory-Motor Coordination* [2]. The idea of Sensory-Motor Coordination is that the model of external world is generated not only from sensory information but also from the relation of sensory information and action. In this context we proposed a method to generate symbols which is suitable to be coupled with the behavior-based systems. It is based on the idea that categories of landscape objects in the environment are to be extracted from the concepts about the categories of the environmental structure.

Our robot control system is based on *Subsumption Architecture*, that is a class of behavior-based control system [3]. A Subsumption Architecture based control system is constructed by the set of behavior modules, i.e., each module executed asynchronously connects its input information to output information directly. Therefore, the outputs of the behavior modules are regarded as the relation between sensory information and action. Using these behavior modules, our system can be possible to generate a world model that reflects the internal state of the robot concerning the relation of sensory information and action.

1.2   Artificial Resonance Theory

In our previous research, we adopted *Self-Organizing Map* (SOM) [4] to organize world model from the outputs of behavior modules [5],[6]. However, the learning system based on SOM requires much learning time. Moreover it is assumed in the system that the number of organized categories are fixed, so that it is inadequate in case of the addition of new category. Therefore, we adopt the learning system based on *Adaptive Resonance Theory* (ART) [7] in place of SOM in order to generate world model flexibly from the time sequences of behavior activities.

ART is a kind of a competitive-type neural network, which clusters the input information in higher dimensional space to the state space of lower dimension. ART algorithm is presented as follows [7]:

Notation : $\mathbf{I} = (I_1, \ldots, I_M)$ denotes input to the system. Each I_i is in the interval $[0, 1]$. Each category (j) corresponds to a vector $\mathbf{w}_j = \{w_{j1}, \ldots, w_{jM}\}$ of adaptive weight between the input vector and j-th category nodes. The number of potential categories $N(j = 1, \ldots, N)$ is arbitrary. α is positive constant called choice parameter. $\beta \in [0, 1]$ is the learning rate parameter. $\rho \in [0, 1]$ is called vigilance parameter determines the number of clusters to be formed. The operator $\wedge$ is defined as $(x \wedge y)_i \equiv min\{x_i, y_i\}$.

Category choice : For an input vector $\mathbf{I}$ and category j, the choice function T_j is defined by $T_j(\mathbf{I}) = \frac{|\mathbf{I} \wedge \mathbf{w}_j|}{\alpha + |\mathbf{w}_j|}$. The category J is chosen by $T_J = max\{T_j : j = 1 \ldots N\}$

Resonance or reset : Resonance occurs if the match function of the chosen category meets the vigilance criterion, $M_J = \frac{|\mathbf{I} \wedge \mathbf{w}_J|}{|\mathbf{I}|} \geq \rho$, then *Learning* occurs. If $M_J < \rho$, then *Mismatch reset* occurs and the category J is eliminated. A new index J is chosen by *Category choice*.

Learning : The weight vector is updated according to the equation, $\mathbf{w}_J^{(new)} = \beta(\mathbf{I} \wedge \mathbf{w}_J^{(old)}) + (1 - \beta)\mathbf{w}_J^{(old)}$.

The feature of ART is that the rate of learning speed is relatively high and learning stability and plasticity are equipped. Hence ART is expected to be adequate for world modeling in real world situation for resolving *Symbol Grounding Problem*.

2 System architecture

The system proposed in this research consists of the following three stages as shown in Figure 1:

(1) **SA Control System (Aggregation of behavior states)**
The activated state value of each behavior module is calculated while SA control system functioning.

(2) **ART$_1$ (Categorization of behavior states)**
Using the output values of all behavior modules (v_i: velocity, r_i: angular velocity), ART_1 improves its learning parameters to categorize characteristic relations among the realized actions in each time step. The sequential outputs from ART_1 are recorded as time series data after learning process.

(3) **ART$_2$ (Organization of internal symbols)**
Learning of ART_2 occurs from the parts of time series of output data from ART_1. The outputs of ART_2 learning can be regarded as the sets of recognized internal symbols of mobile robot representing the local features of the real environment.

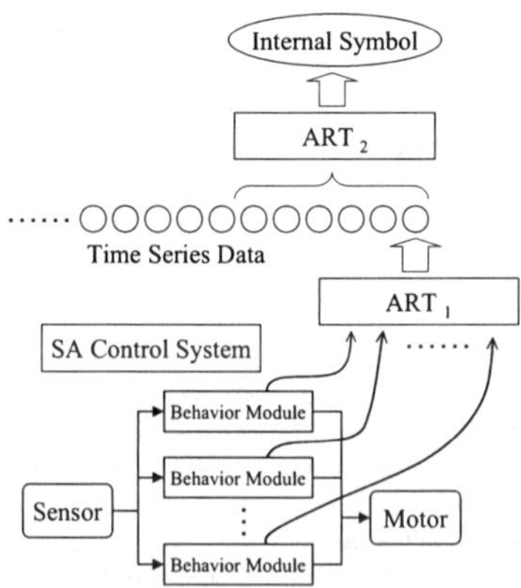

Figure 1: System architecture

The features of our proposed system are summarized as follows: **(1) Filtering and recognition:** There is no guarantee that the representation of world model is not necessarily corresponding to human being's, however it is natural to make the world model which is adequate and rational for the robot that has restricted sensory and action abilities. In our system, only the information which is meaningful for the robot's actions can be extracted from input information. **(2) Expandability:** In case that some behavior modules are newly added or changed in the control system, the system executes learning process in an adequate learning period for readjustment. By improving the robot's behavior, internal symbols are also improved according to the new behavior. **(3) Adaptability:** The system confers the output information from existing behavior modules, therefore any particular type of environment is not presumed for our recognition system. Even when the robot is in different environment, the internal symbols which is suitable for the environment can be obtained by executing learning process again.

3   Experimental results

In this section, we present the exerimental results for demonstrating the efficiency of our system. The dotted lines in each figure represent the robot's trajectories, and numbers represent internal symbols which are generated from learning process. We note that the numbers in each figure are indifferent to the numbers in other figures.

3.1   Behavior modules and internal symbols

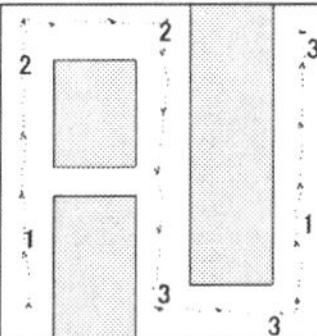
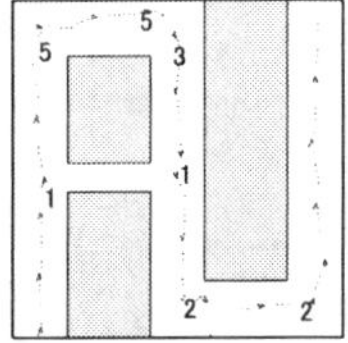
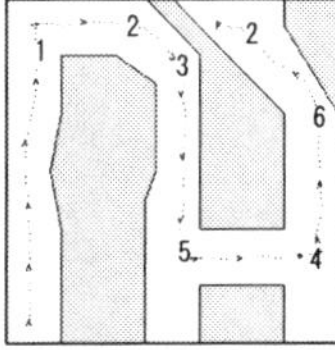

Figure 2: Simple architecture   Figure 3: Augmented architecture   Figure 4: New environment

Figure 2 shows the result of the experiment by simple control architecture which consists of only four behavior modules: "Avoid", "Left wall", "Right wall" and "Forward"". The sequence of activated behaviors is that the robot goes through the straight path while functioning the modules "Forward", "Left wall", and "Right wall", and turns right by "Avoid" and goes straight again and so on. After ten times trials the system generates three kinds of internal symbols as shown in the figure. It can be seen that internal symbol 2 corresponds to right turn and internal symbol 3 corresponds to left turn. Only the meaningful landscape features for the system which have effects on robot's action are filtered and organized as internal symbols.

Next, the system is expanded by adding two behavior modules: "Left turn", "Right turn". The functions of new behavior modules are to alter its direction towards the sides where the wall disappears. Figure 3 shows the result by the augmented control system. The system architecture is not changed from previous experiment without the additions of behavior modules. We obtain four kinds of internal symbols in this case. The figure shows that in addition to turning position of both sides, the place where the robot finds that it goes by a narrow passage has also been recognized as an internal symbol. These internal symbols are indicated as internal symbol 1 in the figure. The result means that the newly generated internal symbols are corresponding to the expanded actions.

Figure 4 shows the result of experiment in more complicated new environment. The behavior control system is the same as the previous experiments. Two different internal symbols such as 4 and 5 in the figure correspond to the place where the robot turned left in a similar manner. The reason of such phenomenon is that because of using the output information of behavior modules, the difference in the robot's internal state in the process of determining action is reflected to generation of internal symbol although the resultant action can be seen as same for human recognition. Internal symbol 2, 3 and 6 in the figure correspond to the gradual corner which does not exist in the former environment. This experiment shows that the internal symbols which adapt to a new environment can be introduced to the same system architecture in another environment since the internal symbol is made of not only the sensory information but also of the relation of sensory information and action.

3.2 Path planning by internal symbols

We can apply the generated internal symbols from our proposed system to path planning problems by using the advantages of our controls system.

Figure 5 shows the result of experiments for path planning in a complicated environment. The system architecture is the same as the above experiments. The path planning process is that the robot obtains the relative positions and activated behaviors corresponding to each internal symbol while wandering in the given environment. Compared to the memorized past sequences of the robot's behavior, the system searches the coincidence between the internal symbol representing the present position and the part of past path sequence that reaches the goal. It is shown in the figure that the tick lines indicate the coincidence points with the past sequence.

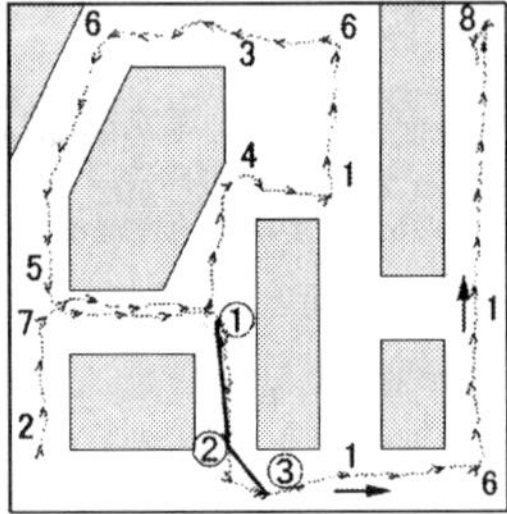

Figure 5: Application to path planning

3.3 Difference in recognition and human interface

Investigating the results of generated world model, we discuss the relations between the robot and human recognition.

As shown in the result of path planning in section 3.2, human being interprets that the internal symbols 1 and 3 correspond to "straight path", internal symbols 5 and 6 correspond to "left corner", and internal symbol 7 corresponds to "right corner". Thus there exist correspondences between the generated internal symbols and the landmark recognition by human being. On the other hand, the system generates internal symbols which can not always be interpreted by human beings, such as internal symbols 2 and 4.

For investigating the relation of internal symbols and the landmark recognition, we should consider the following characteristics of our system:

SA : In our system since the output of behavior modules of SA is utilized as the input of the ART learning system, internal symbols generation is influenced by the number and variety of behavior modules, e.g., if the number of behavior modules is increased, the number of output patterns of behavior modules increase and then this results in the increase of internal symbols.

ART₁ and ART₂ : ART has some system parameters such as ρ, α, β. If the vigilance parameter ρ is increased, more internal symbols which have specific meanings for the place are generated. However, excessive and complicated internal symbols for human being to understand may be generated. On the other hand, if ρ is decreased, then the number of generated internal symbols decreases which is related to more generalized recognition. Both learning rate β and selection parameter α will make effect on internal symbol generation.

Modifying the parameters of ART₁, ART₂ and the behavior modules of SA will make various influence on the internal symbol generation. Figure 6 shows the relations between landmark recognition of human being and internal symbol generation. It is expected that the consideration of the influence by system modification will lead to the suggestion of the development of more effective human machine interface.

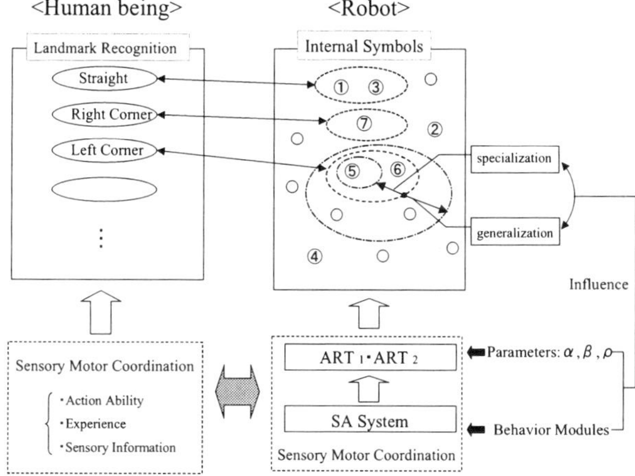

Figure 6: Relation of landmark recognition and internal symbols

4 Conclusion

In this paper we presented the world model generation system based on Adaptive Resonance Theory to generate internal symbols for behavior based mobile robot. In this research, once we abandon the interpretation by human being and adopt internal symbols specified to the robots' style recognition, and after that, we investigate the relationship between world model of the robot and that of human being. Therefore the generated model is regarded as a meaningful world model for the robot itself although it is not necessarily sufficient for the direct interaction between the robot and human being. However, the advantages of our system are that the world model is generated according to the restricted robot's sensing and actuating abilities. Therefore, the generated world model is suitable for intelligent systems with restricted resources under uncertainty such as robot path planning problem. Moreover it is expected that the investigation of the characteristics of such generated world model will lead to the development of consideration on human machine interface.

References

[1] S. Harnad, "The Symbol Grounding Problem," *Physica, D 42*, pp. 335–346, 1990.

[2] R. Pfeifer and C. Scheier, "Sensory-Motor Coordination: the metaphor beyond," *Robotics and Autonomous Systems, Special Issue on "Practice and Future of Autonomous Agents" 20*, 1997.

[3] R. Brooks, "A Robust Layered Control System for A Mobile Robot," *IEEE Journal of Robotics and Automation*, Vol. RA-2, No. 1, 1986.

[4] T. Kohonen, "The Self-Organizing Map," *Proceedings of the IEEE*, 78(9) pp. 1464–1480, 1990.

[5] A. Fukayama, M. Ida, and O. Katai, "Behavior-based Fuzzy Control System for a Mobile Robot with Environment Recognition by Sensory-Motor Coordination," *Proceedings of Fuzz-IEEE99*, pp. 105–110, 1999.

[6] S. Ishigaki, M. Ida, and O. Katai, "Self-Organization of Behavior-based World Model for Autonomous Mobile Robot," *Proceedings of AROB 6th*, pp. 258–261, 2001.

[7] G. A. Carpenter, S. Grossberg and D. B. Rosen, "Fuzzy ART: Fast Stable Learning and Categorization of Analog Patterns by an Adaptive Resonance System," *Neural Networks*, Vol. 4, pp. 759–771, 1991.

KES '01
N. Baba et al. (Eds.)
IOS Press, 2001

Probabilistic Model-building Genetic Algorithms Using Marginal Histograms in Continuous Domain

Shigeyoshi Tsutsui[*1], Martin Pelikan[*2], and David E. Goldberg[*2]
*1 *Department of Management and Information, Hannan University, 5-4-33 Amamihigashi, Matsubara, Osaka 580-8502, Japan, tsutsui@hannan-u.ac.jp*
*2 *Illinois Genetic Algorithms Laboratory, Department of General Engineering, University of Illinois at Urbana-Champaign, 117 Transportation Building, 104 S. Mathews Avenue Urbana, Illinois 61801, USA{pelikan, deg}@illigal.ge.uiuc.edu*

Abstract. Recently, there has been a growing interest in developing evolutionary algorithms based on probabilistic modeling. In this scheme, the offspring population is generated according to the estimated probability density model of the parents instead of using recombination and mutation operators. In this paper, we propose an evolutionary algorithm using a marginal histogram to model the parent population in a continuous domain. We propose two types of marginal histogram models: the fixed-width histogram (FWH) and the fixed-height histogram (FHH). The results showed that both models worked fairly well on test functions with no or weak interactions among variables. Especially, FHH could find the global optimum with very high accuracy effectively and showed good scale-up with the problem size.

1. Introduction

Genetic Algorithms (GAs) [9] are widely used as robust searching schemes in various real world applications, including function optimization, optimal scheduling, and many combinatorial optimization problems. Traditional GAs start with a randomly generated *population* of candidate solutions (individuals). From the current population, better individuals are selected by the *selection* operators. The selected solutions produce new candidate of solutions by being applied *recombination* and *mutation* operators.

Recombination mixes pieces of multiple promising sub-solutions (*building blocks*) and composes solutions by combining them. GAs should therefore work very well for problem that can be somehow decomposed into subproblems. However, fixed, problem-independent recombination operator often break the building blocks or do not mix them effectively [15].

Recently, there has been a growing interest in developing evolutionary algorithms based on probabilistic models. In this scheme, the offspring population is generated according to the estimated probabilistic model of the parent population instead of using traditional recombination and mutation operators. The model is expected to reflect the problem structure, and as a result it is expected that this approach provides more effective mixing capability than recombination operators in the traditional GAs. These algorithms are called the probabilistic model-building genetic algorithms (PMBGAs). In PMBGAs, better individuals are selected from initially randomly generated population like in the standard GAs. Then, the probability distribution of the selected set of individuals is estimated and new individuals are generated according to this estimate, forming candidate of solutions for the next generation. The process is repeated until the termination conditions are satisfied.

According to [16], PMBGAs can be classified into three classes depending on the complexity of models they use; (1) no interactions, (2) pairwise interactions, and (3) multivariate interactions. In models with no interactions, interactions among variables are treated independently. Algorithms in this class work well on problems which have no interactions among variables. These algorithms include the population based incremental learning (PBIL) algorithm [2], the compact genetic algorithm (cGA) [10], and the univariate marginal distribution algorithm (UMDA) [13]. In pairwise interactions, some pairwise interactions among variables are considered. These algorithms include the mutual-information-maximization input clustering (MIMIC) algorithm [7], the algorithm using dependency trees [3]. In models with multivariate interactions, algorithms use models that can cover multivariate interactions. Although the algorithms require increased computational time, they work well on problems which have complex interactions among variables. These algorithms include the extended compact genetic algorithm (ECGA) [11] and the Bayesian optimization algorithm (BOA) [14, 15].

Several attempts to apply PMBGAs in continuous domain have been made. These include continuous PBIL with Gaussian distribution [17] and a real-coded variant of PBIL with iterative interval updating [18]. These algorithms do not cover any interactions among the variables. In the estimation of Gaussian networks algorithm (EGNA) [12], a Gaussian network is learned to estimate a multivariate Gaussian distribution of the parent population. In [4], two density estimation models, i.e., the normal distribution (normal kernel distribution and normal mixture distribution), and the histogram distribution are discussed. These models are intended to cover multivariate interaction among variables. In [5], it is reported that the normal distribution models have showed good performance. In [6], a normal mixture model combined with a clustering technique is introduced to deal with non-linear interactions.

In this paper, we propose an evolutionary algorithm using marginal histograms to model promising solutions in a continuous domain [20]. We propose two types of marginal histogram models: the fixed-width histogram (FWH) and the fixed-height histogram (FHH). The results showed both models worked fairly well on test functions which have no or weak interactions among variables. FHH could find the global optimum with high accuracy effectively and showed good scale-up behavior. Section 2 describes the two types of marginal histogram models. In Section 3 empirical analysis is given. Future work is discussed in Section 4. Section 5 concludes the paper.

2. Evolutionary Algorithms using Marginal Histogram Models

This section describes how marginal histograms can be used to (1) model promising solutions and (2) generate new solutions by simulating the learned model. We consider two typed of histograms: the fixed-width histogram (FWH) and the fixed-height histogram (FHH).

2.1 General description of the algorithm

The algorithm starts by generating an individual population of candidate solutions at random. Promising solutions are then selected using any popular selection scheme. A marginal histogram model for the selected solutions is constructed and new solutions are generated according to the built model. New solutions replace some of the old ones and the process is repeated until the termination criteria are met. Therefore, the algorithm differs from traditional GAs only by a method to process promising solutions in order to create the new ones. The pseudo-code of the algorithm follows:

   0. Set the generation counter $t \leftarrow 0$
   1. Generate the initial population $P(0)$ randomly.
   2. Evaluate functional value of the vectors (individuals) in $P(t)$.
   3. Select a set of promising vectors $S(t)$ from $P(t)$ according the functional values.
   4. Construct a marginal histogram model $M(t)$ according to the vector values in $S(t)$.
   5. Generate a set of new vectors $O(t)$ according to the marginal histogram model $M(t)$.
   6. Evaluate the new vectors in O(t).
   7. Create a new population $P(t+1)$ by replacing some vectors from $P(t)$ with $O(t)$.
   8. Update the generation counter, $t \leftarrow t + 1$.
   9. If the termination conditions have not been satisfied, go to step 3.
   10. Obtain solutions in $P(t)$.

2.2 Marginal fixed-width histogram (FWH)

The histogram is the most straightfoward way to estimate probability density. The possibility of using histograms to model promising solutions in continuous domains was discussed in [4].

2.2.1 Model description

In the FWH, we divide the search space $[\min_j, \max_j]$ of each variable x_j $(j=1,...n)$ into H $(h_j= 0, 1,..., H\text{-}1)$ bins. Then the estimated density $p^j_{\text{FWH}}[h]$ of bin h_j is:

$$P^j_{\text{FWH}}[h] = \left\| \left\{ V[i][j] \ \middle| \ i \in \{0,1,...,N-1\} \wedge h \leq \frac{V[i][j] - \min_j}{\max_j - \min_j} H < h+1 \right\} \right\| \Big/ N \tag{1}$$

where $V[i][j]$ is the value of variable x_j of individual i in the population and N is the population size. Since we consider a marginal histogram, the total size of all the bins should be $H \times n$. Figure 1 shows an example of marginal FWH on variable x_j.

The bin width ε can be determined by choosing appropriate number of bins depending on the required precision for a given problem. For multimodal problems, we need consider some maximum bound on ε value. Let us consider the worst case scenario where the peak with the global optimum of h_1 has smaller width w_1 than ε and the second best local optimum has a much wider peak of width w_2 (Figure 2). Moreover, let us assume that all solutions in each peak are very close to the peak (as it should be the case later in the run). In this case, after the random sampling at generation 0, the mean fitness of individuals in the bin where the second best solution is located is h_2 and mean fitness of individuals in the bin where the best solution is located is $h_1 \times w_1/(2\varepsilon)$. With most selection schemes, if h2 $> h_1 \times w_1/(2\varepsilon)$ holds, then the probability density of the bin where the second best solution is located increases more than the probability density of bin where the best solution is located. As a result, the algorithm converges to the second best optimum instead of the global one. To prevent this, the following relationship must hold:

$$\varepsilon < \frac{w_1 \times h_1}{2 \times h_2} . \tag{2}$$

2.2.2 Sampling methods

In the FWH, new individuals are generated as follows: first a bin is selected according to the probability given by Eq. (1). Then an individual is generated by generating a number from the bin with uniform distribution. This is repeated until all individuals are obtained. The simplest method to sample each bin is

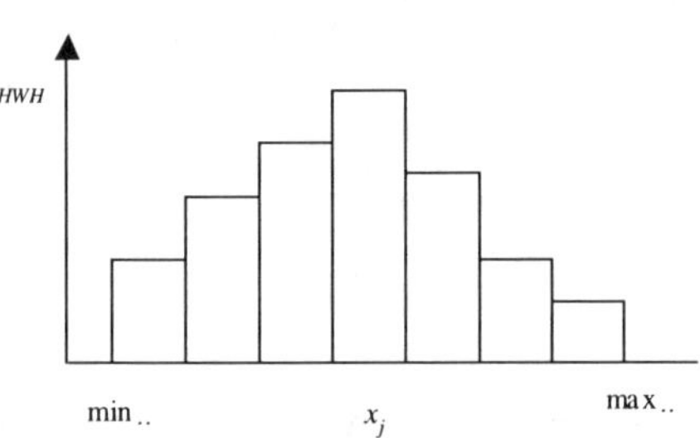

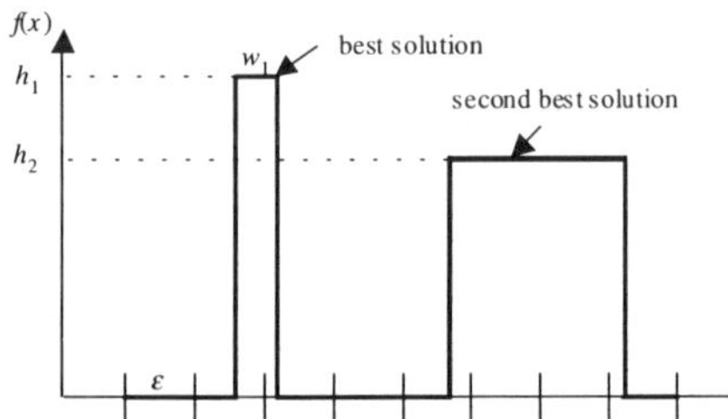

Figure 1: Marginal fixed-width histogram (FWH)

Figure 2 Fitness function with multimodality

to use a roulette wheel (RW). However, RW has a certain amount of stochastic sampling error. Let S be the total number of samples. Then the distribution of R, the number of samples from bin h_j, is

$$P_S^{hj}(R) = \binom{S}{R} \times \left(P_{\text{FWH}}^j[h]\right)^R \left(1 - P_{\text{FWH}}^j[h]\right)^{S-R}, R = 0,1,...,S \tag{3}$$

and the expected value of R is

$$\overline{R} = S \times P_{\text{FWH}}^j[h] . \tag{4}$$

Let us define a "normalized number of samples" $r = R/S$, then distribution of r is obtained as

$$P_S^{hj}(r) = S \times P_S^{hj}(R/S), r = [0, 1] . \tag{5}$$

Figure 3 shows an example of distribution r for various number of S. $P_{\text{FWH}}^j[h] = 0.1$ is assumed. Then the expected value of r should be

$$\overline{r} = \overline{R}/S$$
$$= S \times P_{\text{FWH}}^j[h]/S \tag{6}$$
$$= 0.1.$$

In Figure 3, we can make sure the existence of the stochastic sampling error, although as S becomes larger, the probability density $P_S^{hj}(r)$ around $r = 0.1$ increases.

One well known sampling method to reduce this sampling error is Baker's stochastic universal sampling (SUS), which was proposed for proportional selection operator [1]. In this method, the integer part of the expected number $\overline{R}$ is deteministically sampled and the fraction part of $\overline{R}$ is stochastically sampled without duplicate. We

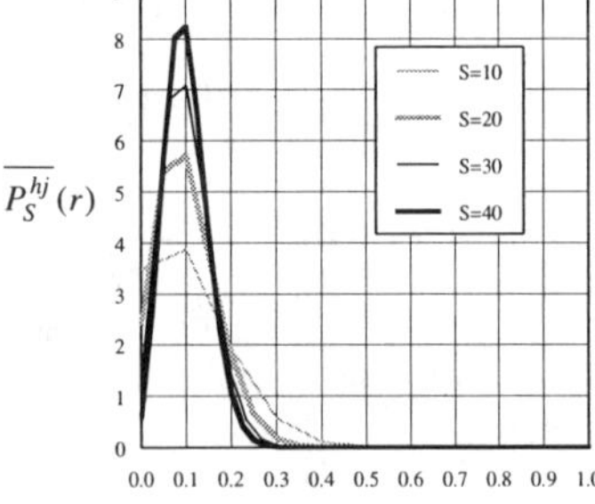

Figure 3: An example of distribution r for various values of S

extend this method to sample from marginal histogram models. We call this method *E-SUS* (extended SUS).

A pseudo C code of E-SUS for FWH is described in Figure 4. To apply this method, The variable "expected" in line 13 is the expected sampling number for bin h of variable x_j. For instance, if the probability of a particular bin is 0.155 and we want to sample 100 individuals, we would expect 15.5 copies of representatives of the bin. The presented algorithm will generate 15 copies and then another copy with probability 0.5. To obtain a random combination among variables, we must shuffle the sampling sequence for each dimension (line 8). In the experiments in Section 3, we use both the RW and the E-SUS.

2.3 Marginal fixed-height histogram (FHH)

2.3.1 Model description

In the FWH each bin has the same width. On the other hand, in the fixed-height histogram (FHH) each bin has the same height. That means that each bin contains the same number of points. See Figure 5 for an example. The important feature of the FHH is that the bins in dense regions are narrower and thus the accuracy of modeling in the important regions increases. In context of evolutionary algorithms, the width of bins around high peaks decreased as more sample points are located in these areas. Since the probability of generating a point from each bin in the FHH is the same, we expect both the density as well as the accuracy improve in promising region of the search space. This is the distinguished characteristics of the FHH. Figure 6 shows this more clearly. Figure 6 (a) is an example of the FHH for a population which has a normal distribution $N(0, 1)$ for variable x_j in the range [-5, 5] as shown in Figure 6 (b). Twenty (20) bins are used. In this example, more points are sampled around $x_j = 0.0$ because the bin width around $x_j = 0.0$ is narrower and therefore there are more bins for the same area.

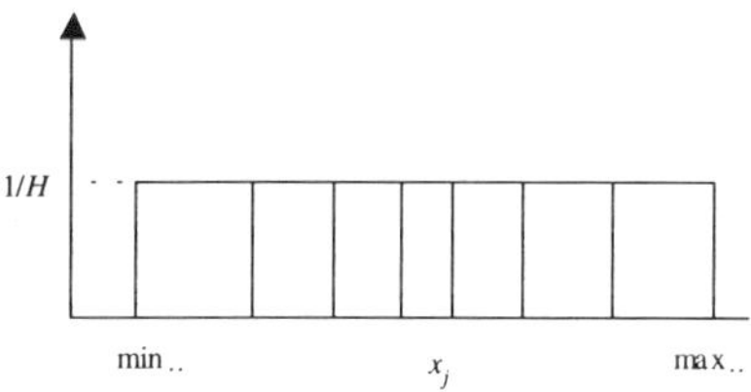

```
1    // S: sampling individual number, n: number of variables, H: number of bins
2    // p[j][h]: probability density of bin h of variable x_j
3    // {j][h]: left edge position of of bin h of variable x_j
4    // Rand(): generate random number from [0, 1.0]
5    V[S][n]; // array for sampled vectors
6    xh[S]={0,1,2,......S-1}; // array for random permutation of bin position
7    for(int j=0; j<n; j++){
8         Shuffle(xh); // get a permutation xh[] by shuffling for each parameter
9         double ptr = Rand();
10        double sum = 0.0;
11        int k = 0;
12        for(int h=0; h<H; h++){
13             double expected = p[j][h]*S;
14             for(sum += expected; sum>ptr; ptr++)
15                 V[xh[k++]][j] =l[j][h]+(l[j][h+1]-l[j][h])*Rand();
16        }
17 }
```

Figure 4: Pseudo C code of the E-SUS for FWH

2.3.2 Sampling methods

Sampling method for the FHH is basically the same as that of FWH except the probability density for FHH has the same value $1/H$ for all bins. The pseudo C code of the E-SUS for FHH is the same with that for FWH except line 13 in Figure 4. The line 13 for the FHH should be:

"double expected = (double)S/H;".

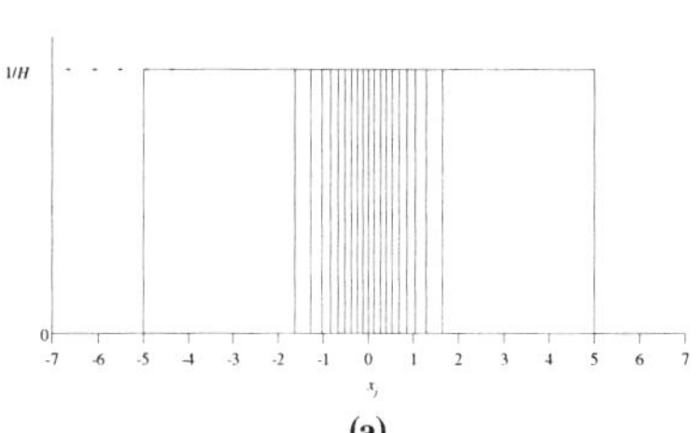

Figure 5: Marginal fixed height histogram (FHH)

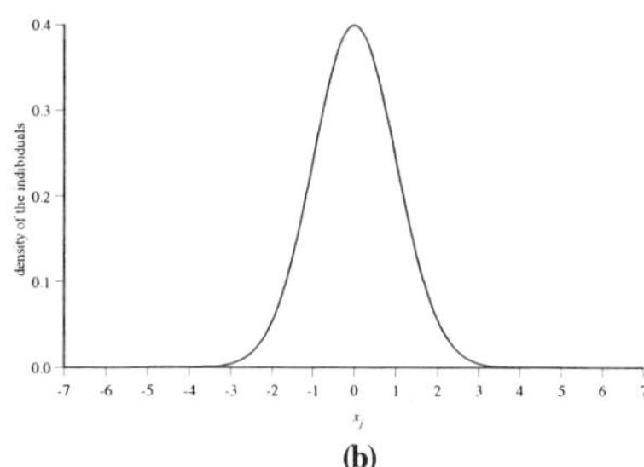

Figure 6: An example of the FHH of normal density

3. Empirical Study

To evaluate the marginal histogram models proposed in Section 2, we ran two histogram models using four test functions. The experimental methodology, test functions used, and experimental analysis are described in this section.

3.1 Experimental methodology

3.1.1 Evolutionary model

The basic evolutionary model we used in these experiments is similar to that of $(\mu+\lambda)$-ES [19]. Let the population size be N, and let it, at time t, be represented by $P(t)$. The population $P(t+1)$ is produced as follows (Figure 7):

1. Marginal histogram model is developed from $P(t)$
2. $K \times N$ new individuals are sampled according to the built model
3. The new individuals are evaluated
4. Individuals in $P(t)$ and $K \times N$ new individuals are ranked and the best N individuals are selected forming $P(t+1)$.

We used $K = 1$ in our experiments in Sub-sections 3.2.1 and 3.2.2.

3.1.2 Test Functions

The test functions we used are summarized in Table 1, which includes the 20-variable two peaks function $F_{TwoPeaks}$, the 20-variable Rastrigin function ($F_{Rastrigin}$), the 10-variable Griewank function $F_{Griewank}$, and the 5-variable Schfewel function $F_{Schfewel}$.

Function $F_{TwoPeaks}$ is the sum of a sub-function similar to the commonly used deceptive function in binary domain and has a broader and lower peak, and a more narrow and higher peak for each variable x_j as shown in Figure 8. This function is designed to see the mixing capability and the scale-up behavior of the algorithm. This function has no interactions among variables. Other three function are commonly used in the evolutionary computation literature. Rastrigin function has also no interactions among variables. Griewank function has weak interactions among variables. Schfewel function has medium-scale interactions among variables.

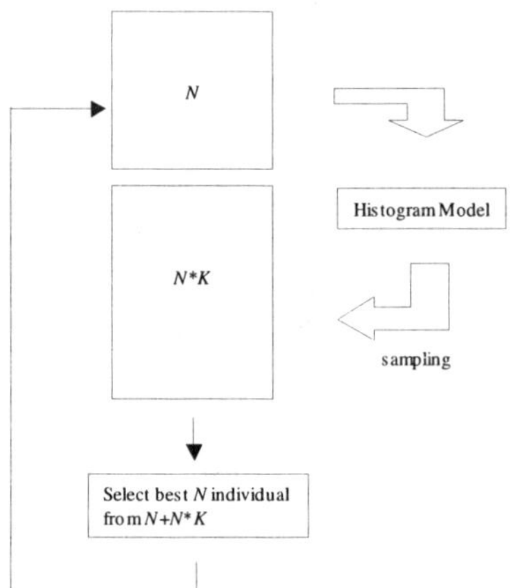

Figure 7: Evolutionary model

Table 1: Test functions

Function	Domain	Interactions*	H^{**}
$F_{\text{Two-peaks}} = n \times 5 - \sum_{i=1}^{n} f_i$	$[0, 12]^n, n = 20$	non	120
$F_{\text{Rastrigin}} = n \times 10 + \sum_{i=1}^{n} (x_i^2 - 10\cos(2\pi x_i))$	$[-5, 5]^n, n = 20$	non	100
$F_{\text{Griewank}} = 1 + \sum_{i=1}^{n} \frac{x_i^2}{4000} - \prod_{i=1}^{n} \left(\cos\left(x_i / \sqrt{i}\right)\right)$	$[-5, 5]^n, n = 10$	weak	100
$F_{\text{Schwefel}} = \sum_{i=2}^{n} \left[\left(x_1 - x_i^2\right)^2 - (x_i - 1)^2 \right]$	$[-2, 2]^n, n = 5$	medium	40

* Interactions among variables ** Number of bins in each variable

As to the number of bins (H) for each variable, we set H so that the initial bin width ε is 0.1. For example, for function $F_{TwoPeaks}$, the number of bins for each variable is $12.0/0.1 = 120$.

3.2 Analysis of results

The experiments in this section focus on (1) convergence properties of the FWH and the FHH with resolution of bin width 0.1, (2) convergence properties of the FHH with high precision, and the (3) scale-up behavior of the FHH on function $F_{TwoPeaks}$ for problem size ranging from 10 to 100 variables.

3.2.1 Convergence properties of the FWH and the FHH with resolution of 0.1

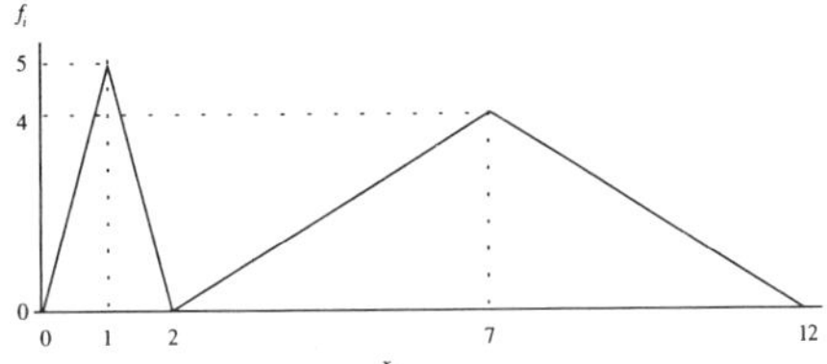

Figure 8: A sub-function of function $F_{TwoPeaks}$

The accuracy of the FWH is at most ε (i.e., 0.1). In our experiments, we evaluated the algorithms by measuring the number #OPT of runs in which the algorithm succeeded in finding the global optimum and the mean number of function evaluations (MNE) to find the global optimum in those runs where it found the optimum. We defined the successful detection of the solution as being within $\pm \varepsilon$ of the actual optimum point. Let $(o_1,...,o_n)$ be the optimal solution of a function. Then, if all values of variables $(x_1,...,x_n)$ of the best individual are within the range $[(o_j-\varepsilon), (o_j+\varepsilon)]$ for all j, we assumed the algorithm to have found the optimal solution. Although the width of each bin for FHH changes during the evolution, we used the same criteria for the FHH in this experiment. Twenty (20) runs were performed. In each run, the initial population $P(0)$ was randomly initialized in the original search space. Each run continued until the global optimum was found or a maximum of 200,000 function evaluations was reached.

Table 2 shows the results of this experiment. We used both the RW and the E-SUS as the sampling method. We also show the results with BLX-α [8], a traditional two-parent crossover operator in continuous domain for comparison.

First, let us see the difference in the performance of the two sampling methods, RW and E-SUS. On almost all experiments with some exceptions, the performance with sampling method E-SUS, which has a smaller stochastic sampling error, is much better than the performance with sampling method RW, which has a larger stochastic sampling error. For example, on function $F_{TwoPeaks}$, the FWH with RW (shown as FWH/RW) found optimal solution 20 times with the population size of 600. The MNE is 14,620.9. On the other hand, the FWH with E-SUS (shown as FWH/E-SUS) found optimal solution 20 times with the population size of 300 and the MNE is 7,178.3, almost the half of the FWH/RW. Next, let us see the difference of performance between the FWH and the FHH. All of the

Table 2: Convergence property of the FWH and the FHH with resolution of 0.1

Model	Function	n	Population size											
			100		200		300		400		600		800	
			#OPT	MNE	#OPT	MNE	#OPT	MNE	#OPT	MNE	#OPT	MNE	#OPT	MNE
FWH/RW	$F_{Two\text{-}peaks}$	20	0	-	1	4,938.0	7	7,852.4	16	9,854.1	20	**14,620.9**	20	18,808.0
	$F_{Rastrigin}$	20	0	-	1	7,851.0	6	9,204.0	12	13,154.2	20	**19,396.6**	20	24,762.7
	$F_{Griewank}$	10	0	-	1	6,527.0	4	9,515.8	14	11,908.6	16	17,884.3	20	**23,198.2**
	$F_{Schwefel}$	5	13	1,188.9	16	1,927.4	19	3,175.5	20	**3,900.2**	20	5,282.0	20	6,828.4
FWH/E-SUS	$F_{Two\text{-}peaks}$	20	1	2,562.0	14	4,926.5	20	**7,178.3**	19	9,609.8	20	14,016.6	20	18,537.9
	$F_{Rastrigin}$	20	1	3,478.0	13	6,770.6	20	**9,720.2**	20	12,616.4	20	18,518.5	20	24,113.5
	$F_{Griewank}$	10	2	2,779.0	7	5,783.3	15	9,220.9	19	12,439.8	19	17,517.2	19	22,502.3
	$F_{Schwefel}$	5	16	1,342.6	19	1,962.3	19	3,302.9	20	**4,012.7**	20	5,924.8	20	7,597.9
FHH/RW	$F_{Two\text{-}peaks}$	20	20	17,952.4	20	9,443.2	20	**8,321.8**	20	10,305.9	20	14,950.5	20	19,442.9
	$F_{Rastrigin}$	20	20	50,099.8	20	46,139.8	20	21,933.1	20	**17,566.2**	20	22,847.3	20	28,990.9
	$F_{Griewank}$	10	17	5,445.1	19	6,561.2	18	9,521.2	16	14,076.5	20	**17,233.4**	19	22,909.5
	$F_{Schwefel}$	5	20	**2,758.9**	20	4,644.6	20	3,821.2	20	5,969.8	20	6,257.1	20	7,101.6
FHH/E-SUS	$F_{Two\text{-}peaks}$	20	20	9,800.4	20	**5,405.8**	20	7,679.9	20	9,982.4	20	14,380.3	20	19,180.1
	$F_{Rastrigin}$	20	20	16,515.7	20	**8,004.2**	20	10,177.6	20	12,690.1	20	18,177.4	20	24,879.6
	$F_{Griewank}$	10	18	3,187.4	18	5,680.6	20	**8,199.6**	20	10,962.5	20	15,782.6	20	20,614.3
	$F_{Schwefel}$	5	20	3,526.7	20	**2,994.3**	20	3,483.3	20	5,020.2	20	6,234.8	20	8,829.2
BLX-α	$F_{Two\text{-}peaks}$	20	0	-	0	-	0	-	0	-	0	-	0	-
	$F_{Rastrigin}$	20	0	-	1	28,521.0	4	47,959.8	11	62,776.0	17	94,994.9	20	**124,825.9**
	$F_{Griewank}$	10	20	**3,316.3**	20	6,347.7	20	9,037.2	20	11,947.3	20	17,483.0	20	23,209.9
	$F_{Schwefel}$	5	20	**3,305.3**	20	3,861.3	20	5,397.8	20	7,141.0	20	10,158.8	20	12,688.0

results of the FHH both with RW and E-SUS showed better performance than FWH. For example, on function $F_{TwoPeaks}$, FHH/RW found optimal solution 20 times with population size of 300 and the MNE is 8,321.8. This is much better than the performance of FWH/RW.

Finally, BLX-α operator showed relatively good performance on functions $F_{Griewank}$ and $F_{Schfewel}$. But it showed poorer performance on functions $F_{Rastrigin}$ and $F_{TwoPeaks}$. Especially on function $F_{TwoPeaks}$, BLX-α could not found optimal solution at all showing very poor mixing capability.

3.2.2 Convergence properties of the FHH

As described in Section 2.2, the bin width of FHH around the global optimum becomes narrower as the evolution proceeds. As a result, the FHH can find much more accurate solution. Figure 9 shows a typical change of bin positions of variable x_1 in a run with FHH/RW, which success to converge to the global optimum, on function $F_{Ranstrigin}$. We can observe that the FHH gathers bin positions to the converged point as evolution proceeds. The population size in the above experiment was set to 800.

To see the convergence properties of the FHH, we ran FHH model on the test functions in Table 1. Population size of 800 was used for all functions in this experiments. Figure 10 shows the changes of mean functional value of the populations until function evaluations reached 100,000. Ten

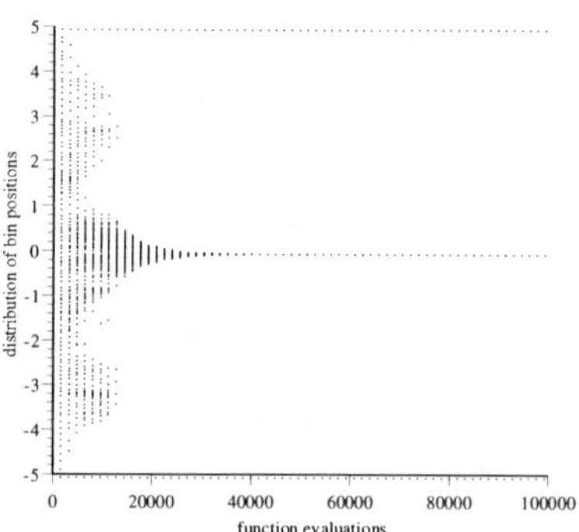

Figure 9: Typical changes of bin positions in parameter x_1 with FHH/RW on function $F_{Griewank}$

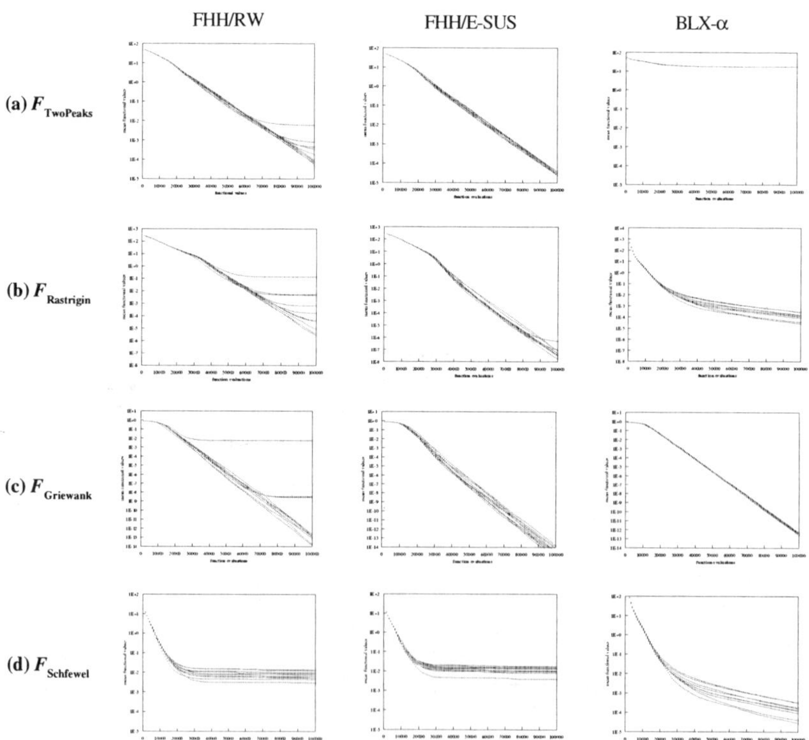

Figure 10: Convergence property of the FHH

(10) runs has been performed.

With RW, the FHH converged to local optima several times on functions F_{TwoPeaks}, $F_{\text{Rastrigin}}$, and F_{Griewank}. However, with E-SUS the FHH converged to global optima 9 out of 10 times on function F_{Rasrigin}. On function F_{Schfewel}, the FHH failed to converge to the global optimum.

Functions F_{TwoPeaks} and $F_{\text{Rastrigin}}$ have no interactions among variables and function F_{Griewank} has weak interactions among variables. Function F_{Schfewel} has medium-scale interactions among variables. With a marginal histogram, it is clearly difficult to solve problems which have medium-scale or stronger interactions among variables.

Results of the BLX-α operator has also shown in the figure. BLX-α performed quite poorly on function F_{TwoPeaks}. On functions $F_{\text{Rastrigin}}$ and F_{Griewank}, it worked similar to the FHH with E-SUS. On function F_{Schfewel} it worked better than the FHH and slowly but gradually converged to the global optimum.

3.2.3 Scale-up behavior of the FHH

This section focus on the scale-up behavior of the FHH with E-SUS using function F_{TwoPeaks} with the number of variables from 10 to 100 with step 10. We used a different evolutionary model, so called MGG. The optimum of function F_{TwoPeaks} is located at $(1.0, 1.0, ..., 1.0)$. We assumed the optimum solution was found if the best individual in the population has a vector value of

$$\forall_{i=1,...,n} \ 1.0 - 0.005 \leq x_i < 1.0 + 0.005 . \tag{7}$$

We did experiment for each number of parameters increasing population size from 200 with step 200 until optimal solution was obtained 10 times in 10 runs.

Figure 11 shows the population size needed to find optimal solution 10 times. The population size increased with $O(n)$. Figure 12 shows the mean number of function evaluations (MNE). The MNE increased with almost $O(n^{1.56})$.

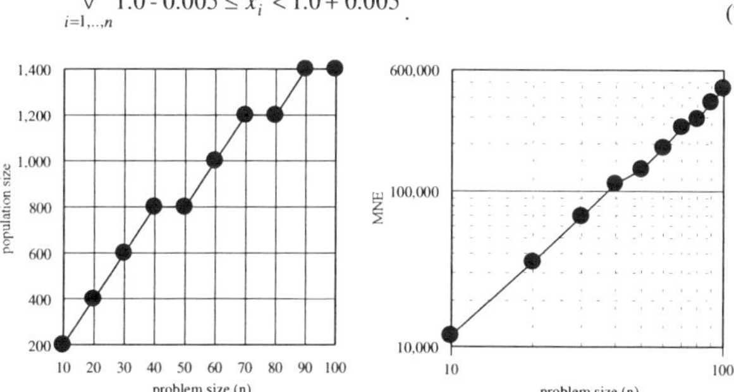

Figure 11: Population size needed to find optimal 10 times in 10 runs

Figure 12: Mean number of function evaluations (MNE) needed to find optimal 10 times in 10 runs

4. Future work

As was observed in Section 3, the marginal histogram models showed good performance on functions which have no or weak interactions among variables. Especially the FHH/E-SUS has a good feature. It can search the solution with high precision and have a good scale-up capability. However, the marginal histogram did not work on functions which have a medium level of linkage among variables. Of course, it will not work well on functions which have strong interactions among variables.

The natural extension of the marginal histogram models is the multivariate histogram models. However, an exponentially growing amount of bins in the form of H^n is required [4]. However, if we can divide a problem into some sub-problems of size m ($m<n$) using multivariate interaction model such as ECGA [10] or BOA [14, 15], we may construct the FHH for each such sub-problem. Many types of FHH for such sub-problems can be considered. For example, let (x_1, x_2) be the parameters of a sub-problem which has a strong interaction between variables x_1 and x_2 and let us assume we can identify the interaction on-line or off-line. Figure 13 is an example of forming a kind of FHH for the sub-problem. In the figure, we first divide the population into two rectangles on x_1 axis so that each rectangle has the same number of individuals. Then, we divide each rectangle into two smaller rectangles on x_2 axis so that again each smaller rectangle has the same number of individuals. We repeat this division recursively while each small rectangle has a number of individuals greater than or equal to defined constant value of K_{min}. Each obtained rectangle corresponds to bin in the marginal histogram. Each rectangle is sampled with equal probability. Individuals are generated by uniform sampling in each rectangle. Figure 14 provides results of both the marginal FHH/RW and the recursively divided FHH/RW on the sub-function of original two dimensional Rosenbrock function

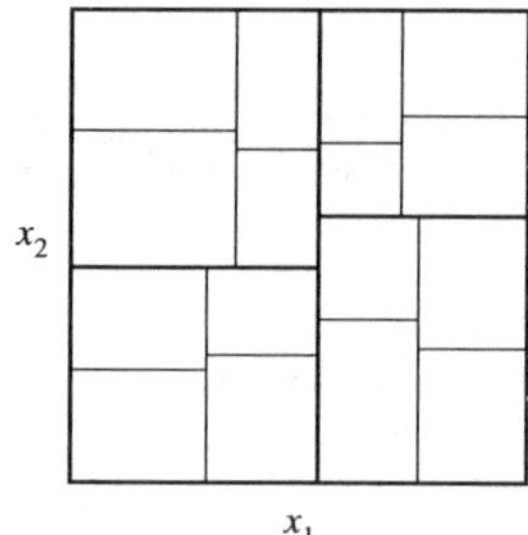

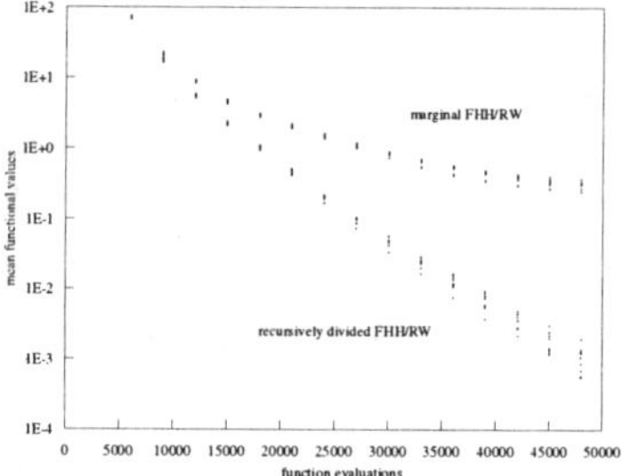

Figure 13: Recursively divided FHH in space x_1, x_2   **Figure 14: Convergence processes on $F_{Rosenbrock}$**

$$F_{Rosenbrock} = 100(x_2 - x_1^2)^2 + (x_1 - 1)^2 ,$$
$$-2.048 \leq x_1, x_2 < 2.047. \tag{8}$$

which has a strong interaction between variables x_1 and x_2. Population size of 3000 and K_{min} value of 2 are used. From this figure, we may see this FHH is working well on this function. Exploring more general FHH models remains a topic for future study.

5. Conclusions

In this paper, we have proposed an evolutionary algorithm using a probabilistic model with marginal histograms in continuous domain. Two types of marginal histogram models were used: the FWH (fixed-width histogram) and the FHH (fixed-height histogram). As the sampling methods, we used both the RW (roulette wheel) and the E-SUS (extended stochastic universal sampling) which is extended from a popular proportional selection operator in GAs.

Both the FWH and the FHH showed good performance on functions which have no or weak interactions among parameters. Comparing two sampling methods, the performance with the E-SUS, which has a smaller stochastic sampling error, showed much better than the performance with the RW, which has a larger stochastic sampling error. The FHH/E-SUS (FHH with E-SUS) showed very good performance. It could find the solutions with a very high precision which is being evolved along with individual solutions. The FHH/E-SUS had also a good scale-up behavior and it mixes solutions very efficiently.

However, the marginal histograms did not work well on function which have a medium (or larger) level of interactions among variables. Introducing models considering interactions among variables as discussed in Section 4 remains a topic for future work. In this paper, although we focused on evolutionary algorithm using probabilistic model with histogram model, it is reported that the probabilistic model using the normal mixture model shows good performance [6]. To combine the histogram model with such a model might enhance the performance of those models because the histogram model has a high mixing capability of sub-solutions.

Acknowledgments

This research is partially supported by the Ministry of Education, Culture, Sports, Science and Technology of Japan under Grant-in-Aid for Scientific Research number 13680469.

David Goldberg's and Martin Pelikan's contribution was sponsored by the Air Force Office of Scientific Research, Air Force Materiel Command, USAF, under grant F49620-00-0163. Research funding for this work was also provided by a grant from the National Science Foundation under grant DMI-9908252. Support was also provided by a grant from the U. S. Army Research Laboratory under the Federated Laboratory Program, Cooperative Agreement DAAL01-96-2-0003. The US Government is authorized to reproduce and distribute reprints for Government purposes notwithstanding any copyright notation thereon.

The views and conclusions contained herein are those of the authors and should not be interpreted as necessarily representing the official policies or endorsements, either expressed or implied, of the Air Force Office of Scientific Research, the National Science Foundation, the U. S. Army, or the U.S. Government.

References

[1] Baker J. E: Reducing bias and inefficiency in the selection algorithm, *Proc. of the 2nd International Conference on Genetic Algorithms*, pp. 14-21 (1987).

[2] Baluja, S.: Population-based incremental learning: A method for interacting genetic search based function optimization and coemptive learning, *Tech. Rep. No. CMU-CS-94-163*, Carnegie Mellon University (1994).

[3] Baluja, S. and Davies: Using optimal dependency-trees for combinatorial optimization: learning the structure of the search space, *Tech. Rep. No. CMU-CS-97-107*, Carnegie Mellon University (1997)

[4] Bosman, P. and Thierens, D.: An algorithmic framework for density estimation based evolutionary algorithms, *Tech. Rep. No. UU-CS-1999-46*, Utrecht University (1999).

[5] Bosman, P. and Thierens, D.: Continuous iterated density estimation evolutionary algorithms within the IDEA framework, *Proc. of the Optimization by Building and Using Probabilistic Models OBUPM Workshop at the Genetic and Evolutionary Computation Conference GECCO-2000*, pp.197-200 (2000).

[6] Bosman, P. and Thierens, D.: Mixed IDEAs, *Tech. Rep. No. UU-CS-2000-45*, Utrecht University (2000).

[7] De Bonet, J. S., Isbell, C. L. and Viola, P.: MIMIC: Finding optima by estimating probability densities, In Mozer, M. C., Jordan, M. I., and Petsche, T. (Eds): *Advances in neural information processing systems*, Vol. 9, pp. 424-431.(1997).

[8] Eshelman, L. J. and Shaffer, J. D.: Real-coded genetic algorithms and interval-schemata, *Foundation of Genetic Algorithms 2*, Morgan Kaufmann, pp. 187-202 (1993).

[9] Goldberg, D. E.: *Genetic algorithms in search, optimization and machine learning*, Addison-Wesley publishing company (1989).

[10] Harik, G., Lobo, F. G., and Goldberg, D. E.: The compact genetic algorithm, *Proc. of the International Conference on Evolutionary Computation 1998 (ICEC 98)*, pp. 523-528 (1998).

[11] Harik, G: Linkage learning via probabilistic modeling in the ECGA, *IlliGAL Technical Report 99010*, University of Illinois at Urbana-Champaign, Urbana, Illinois (1999).

[12] Larranaga, P., Etxeberria, R., Lozano, J.A., and Pena, J.M.: Optimization by learning and simulation of bayesian and gaussian networks, *University of the Basque Country Technical Report EHU-KZAAIK -4/99* (1999).

[13] Mühlenbein, H and Paaß, G.: From recombination of genes to the estimation of distribution I. Binary parameters, *Proc. of the Parallel Problem Solving from Nature - PPSN IV*, pp. 178-187 (1996).

[14] Pelikan, M., Goldberg, D. E., and Cantu-Paz, E.: BOA: The Bayesian optimization algorithm, *Proc. of the Genetic and Evolutionary Computation Conference 1999 (GECCO-99)*, Morgan Kaufmann, San Francisco, CA (1999).

[15] Pelikan, M., Goldberg, D. E., and Cantu-Paz, E.: Linkage problems, distribution estimate, and Bayesian network, *Evolutionary Computation*, Vol. 8, No. 3, pp. 311-340 (2000).

[16] Pelikan, M., Goldberg, D. E., and Lobo, F. G. : A survey of optimization by building and using probabilistic models, *Computational Optimization and Applications*, Kluwer Academic Publishers (in press).

[17] Sebag, M. and Ducoulombier, A.: Extending population-based incremental learning to continuous search spaces, *Proc. of the Parallel Problem Solving from Nature - PPSN V*, pp. 418-427 (1998).

[18] Servet, I. L., Trave-Massuyes, L., and Stern, D.: Telephone network traffic overloading diagnosis and evolutionary computation techniques, *Proc. of the Third European Conference on Artificial Evolution (AE 97)*, pp. 137-144 (1997).

[19] Schefewel, H.-P.: *Evolution and optimum seeking*, Sixth-generation Computer Technology Series, Wiley (1995).

[20] Tsutsui, S., Pelikan, M., and Goldberg, D. E..: Evolutionary algorithm using marginal histogram models in continuous domain, *Proc. of the Optimization by Building and Using Probabilistic Models (OBUPM) 2001*, GECCO-2001 Birds-of-a-feather Workshops (2001).

KES '01
N. Baba et al. (Eds.)
IOS Press, 2001

Asymmetrical Mutations Driven
by Sexual Selection

K. Omori[†], Y. Fujiwara[‡], S. Maekawa[‡],
H. Sawai[‡], and S. Kitamura[*]

[†] *Graduate School of Science and Technology, Kobe University; 1-1 Rokkodaicho,
Nada, Kobe, Japan,* [‡] *Communications Research Laboratory; 2-2-2 Hikaridai,
Seikacho, Soraku, Kyoto, Japan,* [*] *Faculty of Engineering, Kobe University;
1-1 Rokkodaicho, Nada, Kobe, Japan*

Abstract. We propose a model of computation which includes adaptive
mutation rates and sexual selection, and investigate diversified strategies
of adaptive mutation rates between sexes. Proposed model is applied to
a maximum search problem of a 2-dimensional function. The simulation
results show that population can escape from local optima by a runaway
effect of male traits and female preferences. In this phase, asymmetrical
mutation is driven. And the females search conservatively with low
mutation rates, while the males search innovatively with higher mutation
rates. We show how such diversified strategies are brought about in the
process of sexual selection.

1 Introduction

In evolutionary computation, some models which adaptively change the transition rules
are proposed. For example, in the context of an evolutionary model of host-parasite,
Kaneko and Ikegami [1] considered a model with genetic codes including mutation
rates, and investigated the evolution of the self-referential system of code and mutation. Wada et al.[2], for some optimization problems, reported the effectiveness of
asymmetrical mutations motivated by different replication process of double stranded
DNAs in the process of searching. In nature, it is quite interesting to observe that
sexually-reproducing organisms have different mutation rates relative to sexes [3].

In fact, sexual selection [4] can bring about diversified mutation rates between sexes.
In mating, each female observes males' phenotypes and select one. It could be said that
under the sexual selection females have a considerable effect to how males evolve such
as mutation by selecting some direction of traits in the phenotype space, but not in the
genotype space. We are interested in such functional aspects of sexual selection.

In this paper, we propose a model of computation which includes adaptive mutation
rates and sexual selection, and apply it to a maximum search problem of a 2-dimensional
function. We will show how diversified strategies of adaptive mutation rates between
sexes are generated in the process of sexual selection.

2 Sexual Selection

In biology, sexual selection is one of the theories accounting for the species which have
asymmetrical traits between sexes [4]. There are some well-known hypotheses, which
include *runaway hypothesis* and *good gene hypothesis*. The former proposes that female

preferences and male traits mutually affect to each other independently from natural fitness. The latter proposes that female preferences develop to recognize a good gene in male traits, and male traits develop to demonstrate goodness of genes of themselves.

Recently, some studies with sexual selection have been reported in engineering field. Miller and Todd, for example, proposed a sexual selection model in which all individuals have preferences and they mate when they like mutually [5, 6, 7]. Their simulations showed that sexual selection caused acceleration of evolution, escape from local optima, and spontaneous speciations. In these studies, sexuality is defined randomly at mating. In this paper, we focus on asymmetric roles of sexes by proposing a model in which sexuality is defined genetically.

3 Simulation Model

We adopt a real-coded genetic algorithm (GA) as the evolutionary computation model for simplicity of considering functional aspects of sexual selection.

Each individual has sex determined hereditalily, and two phenotypes which are trait vector t and preference vector p. A trait is expressed in both sexes, and determines the individual's fitness for natural selection. On the other hand, a preference is expressed only in females, and works in sexual selection. The preference vector is normalized because only its direction is used for mating in proposed model.

Each individual has sex chromosome, preference chromosome, and trait choromosome and, the mutation rate of an individual is encoded in sex chromosome. However, preference of a male is not expressed but hidden. Although most of living organisms with sexual reproduction are diploid, an individual in proposed model is haploid because we have interest in the interactions between males and females in sexual selection rather than the reproductive structure itself.

The procedure of proposed model is as follows:

1. *Initialization*: Choose initial population randomly whose sex ratio is 1:1.

2. *Natural Selection*: Select according to their fitness separately for each sex.

3. *Sexual Selection*: Make couples according to each female's preference. Each female selects one male by following procedure:

   (a) Sample a certain number of males randomly as a mate candidate, and calculate the average trait t_0 of them.

   (b) Calculate the angle θ between each sampled male's relative trait vector $t' = t - t_0$ and the female's own preference vector p. Then define intensity of preference P by $P = \cos(\theta)$.

   (c) Select one male by the highest intensity, which achieves polygamy.

4. *Genetic Operations*: Apply chromosome-exchange and mutation, then generate one male and one female from each couple.

5. *End Determination*: Repeat the above procedures from Step 2 to Step 4 until the generation reaches a certain number.

Chromosome-exchange is the operation which exchanges parents' chromosomes at recombining them. This operation means the genetic exchange of males and females.

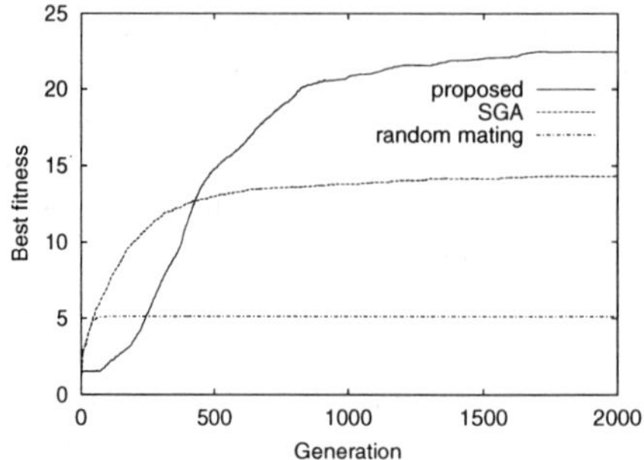
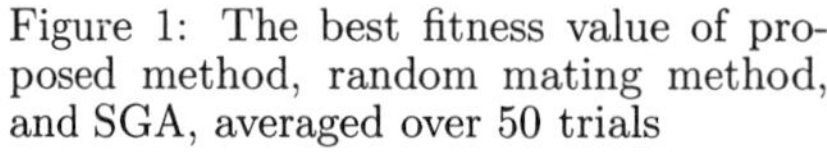

Figure 1: The best fitness value of proposed method, random mating method, and SGA, averaged over 50 trials

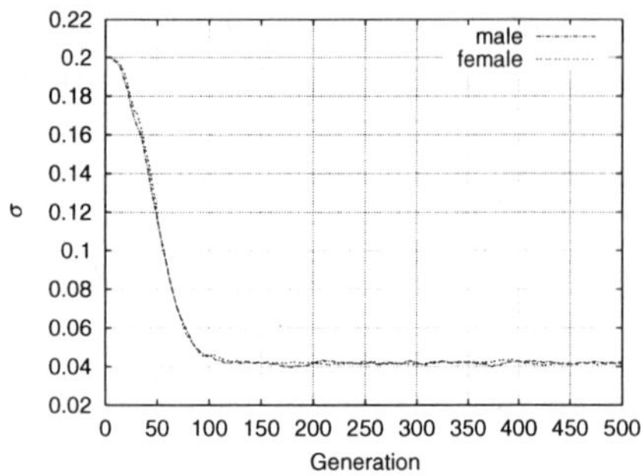

Figure 2: The average value σ of each population of random mating method, averaged over 50 trials

Although females select good males, the traits of the males are not inheritable to females when chromosome-exchange rate (p_c) is 0; in other words, there is no genetic exchange. Therefore, the preferences of females change by only the random genetic drift.

Mutation and its variability are defined for each individual x_i and the magnitude of mutation σ as Gaussian perturbations in the following way:

$$x_i = x_i + \Delta x_i \qquad \Delta x_i \sim \mathrm{N}(0, \sigma) \tag{1}$$
$$\sigma = \sigma + \delta \qquad \delta \sim \mathrm{N}(0, \sigma_0) \tag{2}$$

where σ_0 is constant. Preference vectors are normalized after mutation. The standard deviation σ has the same meaning as mutation rate of standard GAs.

4   Experiments

4.1   Problem

We consider as a toy problem the following maximization of a 2-dimensional function:

$$\max \ f(\boldsymbol{t}) = x_t - 2\sin(\pi x_t) - 0.001 y_t^2 \exp(x_t) \tag{3}$$

where the initial population is arranged around the origin, and trait and preference vectors are denoted by $\boldsymbol{t} = (x_t, y_t)$ and $\boldsymbol{p} = (x_p, y_p)$ respectively. This function has a ridge which becomes narrow exponentially, so the larger local optimum the population arrives, the more difficult it is to to find the desired direction. It can be easily seen that the local optima are neighboring to each other along x_t axis.

4.2   Results

First, we compare proposed model to *random mating method* and *simple-GA* (SGA). Random mating method is equivalent to the proposed one except for that each female selects one male randomly in mating. In SGA, neither of the sexual selection and the adaptive mutation rates are included. In the experiment, the proposed model has the

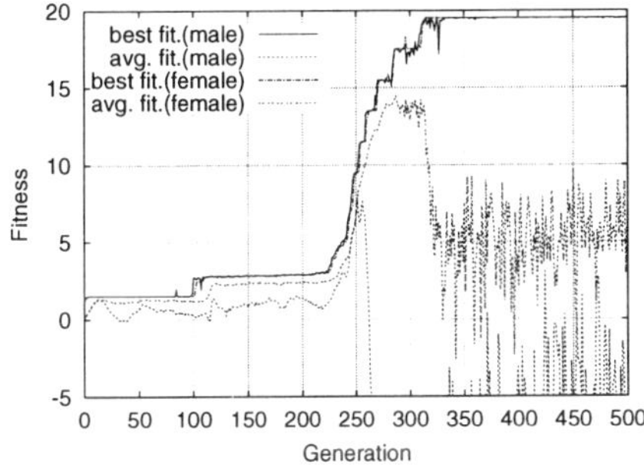

Figure 3: The best and average fitness value of one trial.

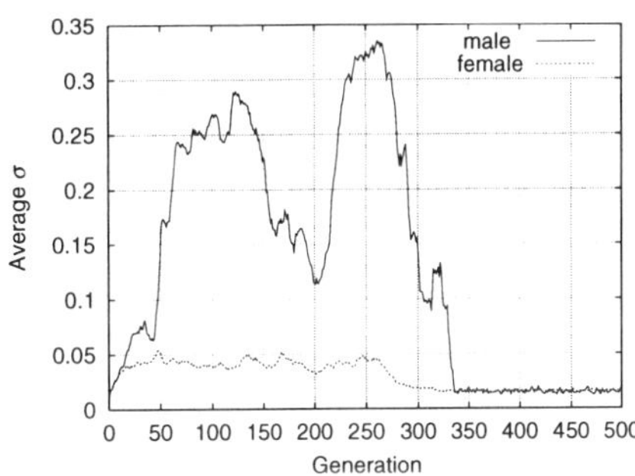

Figure 4: The average value σ of each population of one trial.

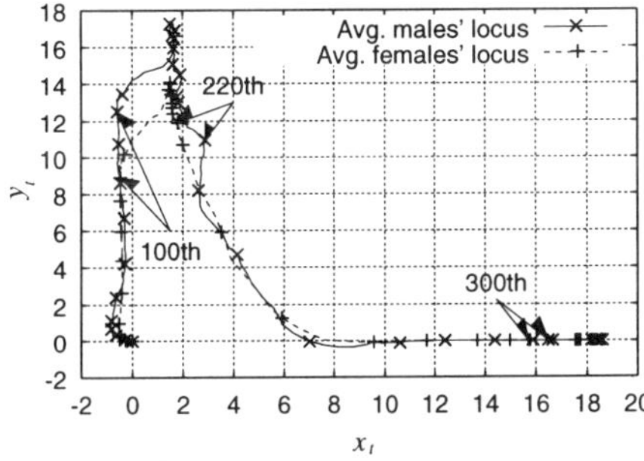

Figure 5: The vector of the average trait of each population in x_t-y_t plane of one trial, plotted every 10 generations.

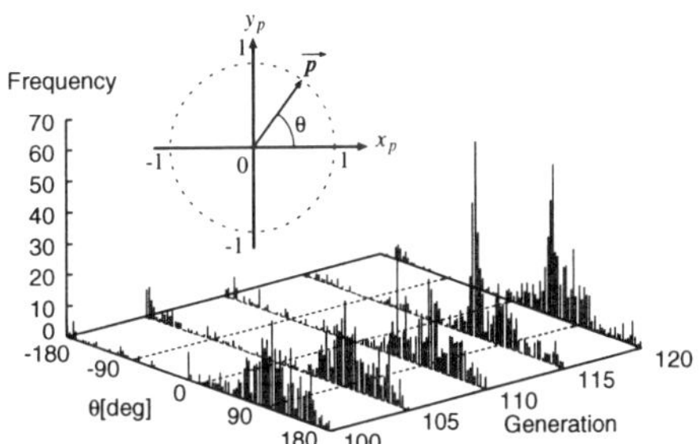

Figure 6: Histogram of preference vectors angles.

following parameters: the population size is 1000, the sampling number of each female is 50, $p_c = 0.1$, the initial value of σ is 0.01, $\sigma_0 = 0.001$. For the random mating: the population size is 1000, $p_c = 0.1$, the initial value of σ is 0.2, $\sigma_0 = 0.001$. And, for the SGA: the population size is 1000, $p_c = 0.1$, the constant value of σ is 0.2.

Figure 1 shows that the proposed model finds larger maxima than the other methods. In particular, the random mating method cannot escape from small local maxima because the values of σ of individuals stay relatively lower (Figure 2). By contrast, in the proposed model, female preferences cause the escape from local optima guiding the male traits in the population as we shall see shortly.

Second, to investigate the details of the process, we focus on the results of one trial. The parameters are the same as above. Figure 3 shows that the searching process have punctuated equilibrium. Note that the averaged fitness value decreases after the 300th generation because even a slight transition for y_t cause a large decrease for the function given by Equation 3. It can be observed from Figure 4 that the average value σ of males is higher than that of females in the evolutionary phases. Figure 5 shows the time-evolution of the average trait of each population during the searching process.

The searching process is regarded as following:

1. When staying in the local optima near the origin where the change of y_t elements remain neutral relatively, The population takes advantage of the slight bias of female preferences by genetic drift to start runaway effect. This preferences' bias means the trend of the generation. Figure 6 shows the the bias of the angles of preferences' vectors. In runaway effect, the more emphatic female preferences become, the larger male traits change.

2. Average value σ of males become higher than that of females Because it is advantageous in mating for males to change one's trait to the direction of the trend. Therefore, males search innovatively with the high mutation rate, and enable to find larger local optima. On the other hand, females search conservatively and keep current local optima. This aspect is considered as the separation of the different searching roles depending on sexes.

3. Females skip to new local optima by the gene exchange for males. Also, females who select some males with higher fitness are so advantageous at next generation that the preferences' trend changes to the positive x_t-axis direction, which leads to the rapid evolution.

4. At the final phase of the searching process, most of transitions caused crucial errors. Therefore, the population keeps the value σ low, and achieves a stable equilibrium. We consider it is the cause of the searching limit that the mutation of traits is restricted to the isotropic transition in a phenotype space.

5   Summary

We have proposed a model of computation with variable mutation rates and sexual selection. In a toy model, we have shown how asymmetrical mutation rates is brought about because of the different roles of females and males under sexual selection. We think that such diversified strategies among sexes provide one of the mechanisms which give a direction of how transition rules themselves should be modified in the context of meta-evolution.

References

[1] K. Kaneko, and T. Ikegami: Homeochaos: dynamic stability of a symbiotic network with population dynamics and evolving mutation rates; *Physica D*, Vol. 56, pp. 406–429 (1992)

[2] K. Wada, H. Doi, S. Tanaka, Y. Wada, and M. Furusawa: A neo-Darwinian algorithm: Asymmetrical mutations due to semiconservative DNA-type replication promote evolution; *Proc. Natl. Acad. Sci. USA* Vol. 90, pp. 11934–11938 (1993)

[3] H. B. Bohossian, H. Skaletsky, and D. C. Page: Unexpectedly similar rates of nucleotide substitution found in male and female hominids; *NATURE*, Vol. 406, 10 August, pp. 622–625 (2000)

[4] Helena Cronin: The ant and the peacock: altruism and sexual selection from Darwin to today; Cambridge University Press (1991)

[5] P. M. Todd, and G. F. Miller: On the Sympatric Origin of Species: Mercurial Mating in the Quicksilver Model; *Proc. of the Fourth Int. Conf. on GA*, Morgan Kaufmann, pp. 547–554 (1991)

[6] G. F. Miller and P. M. Todd: Evolutionary Wanderlust: Sexual Selection with Directional Mate Preference; *From Animals to Animats 2*, MIT Press, pp. 21–30 (1993)

[7] G. F. Miller and P. M. Todd: Biodiversity through sexual selection; *Artificial life V*, MIT Press/Bradford Books, pp. 289–299 (1997)

Para-evolutionary Paradigm of Reasoning

Andrzej BULLER[1], Łukasz KAISER[2] & Katsunori SHIMOHARA[1]

[1] *ATR Information Sciences Division, 2-2-2 Hikaridai, Seika-cho, Soraku-gun, Kyoto 619-0288 Japan*; {buller, katsu}@isd.atr.co.jp

[2] *Faculty of Mathematics, Wrocław University, pl. Grunwaldzki 2/4, Wrocław 50-348 Poland*; kaiser@tenet.pl

Abstract. The paper introduces an unconventional way of reasoning consisting in a selection of entities called memes that inhabit Agent's Working Memory (WM). Agent's concluding belief appears as a population of identical memes that in a certain period of time dominates Working Memory. We discuss mechanisms of meme interactions, early experiments with a simulation model of meme-based Working Memory and its psychological justification, as well as potential possibility of meme-based object recognition and meme-based action planning.

1. Introduction

Let *Agent* be understood as anything that, according to Russel and Norvig [1] "can be viewed as **perceiving** its environment through **sensors** and **acting** upon that environment through **efectors**" and, according to our additional assumption, is equipped with *Working Memory (WM)* inhabited by *memes*.

The term *meme*, understood as an informational counterpart of biological gene, is borrowed from memetics—an emerging discipline searching in a specific way for an answer to the question: how ideas, behaviors, styles or usages spread in societies within a culture (*see* [2]). Although memeticists seem to be strongly interested in collection of facts supporting the thesis that all cultural phenomena appear and disappear as a result of natural selection, they concentrate mainly on creating concepts explaining collective and individual behaviors of humans (*see* [3]). However, the question 'what may meme do inside a single mind' seems to stop being indifferent for memeticists (*see* [4]). Hence the proposal to extend the meaning of the word *meme* in order to discuss a model of a meme-based mental processes in any developed mind, including minds of animals or advanced robots [5][6][7].

Meme, as we assume in this paper, is a **mobile entity** that:

1. represents an ordered pair of elements, where the first element is a proposition, while the second element is either $\varnothing$ or a proposition;

2. inhabits a space called Working Memory (WM) where it may interact with other memes;

3. when (a) its second element is $\varnothing$, (b) it occurs in WM as a member of a population of identical copies, and (c) the population dominates in Working Memory for a certain period of time, then its content may become Agent's temporary belief.

For all p<>∅ a meme ⟨p | ∅⟩ will be called factual meme of f-meme. For all p<>∅ and q<>∅a meme ⟨p | q⟩ will be called r-meme.

An interaction of two memes may take place when they enter at the same time the same compartment of WM. We consider four kinds of meme interactions: *annihilation, positive cross-over, negative cross-over,* and *collision.*

Annihilation takes place in case of an encounter of f-memes ⟨p | ∅⟩ and ⟨q | ∅⟩ when p and q are contradictory propositions.

Positive cross-over takes place when an f-meme ⟨p | ∅⟩ meets an r-meme ⟨q | p⟩. In such case ⟨p | ∅⟩ becomes ⟨p | p⟩ and is annihilated immediately, while ⟨q | p⟩ becomes ⟨q | ∅⟩.

Negative cross-over takes place when an f-meme ⟨p | ∅⟩ meets an r-meme ⟨q | r⟩ and p contradicts r. In such case ⟨p | ∅⟩ becomes ⟨p | r⟩ and annihilates immediately, while ⟨q | r⟩ becomes ⟨s | ∅⟩ such that s contradicts q.

Collision takes place in case of every encounter of two or more memes and consists in a change of directions of memes' movement in a way analogous to a collision of balls on a snooker table.

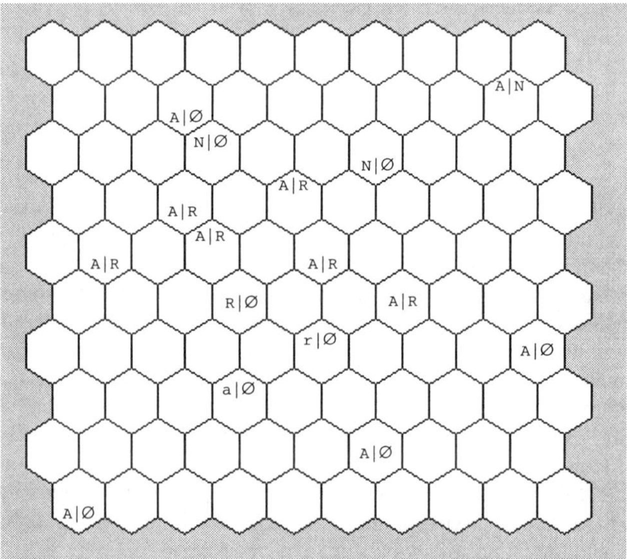

Figure 1. Memes in Cellular Working memory

2. Cellular Working Memory

Working Memory (WM) is a space in which memes can live and interact. It has been implemented as a kind of two-dimensional cellular automaton consisting of hexagonal cells (Figure 1).

It was assumed that a state of a given cell is a function of its own state and states of its three neighbors. State is assumed to be presence of particular memes in particular cell compartments. The function is designed in such a way that if at a given moment only one meme enters to a given cell, the meme in the next moment will disappear from the cell and appear in a neighbor cell (linear movement), while if two memes enter simultaneously the same cell, in the next moment they will disappear from the cell, while their offspring may appear in appropriate neighbor cells (interaction) [5][6][8].

Current implementation of WM facilitating meme interaction is traditional simulation model run in Windows 2000 environment. Next implementation is being prepared on a special-purpose hardware—ATR's CAM-Brain Machine [9].

3. Meme Dynamics in Simulated Working Memory

Dynamics of meme populations has been investigated using a simulation model of Cellular WM. In the framework of one of the experiments the streams of memes being injected to WM represented vagueness of social perception. The uncertainty that perceived person is 'nice' from the point of view of the simulated subject was represented by the existence of streams of contradictory memes: $\langle$'nice'$|\varnothing\rangle$ and $\langle$'not nice'$|\varnothing\rangle$. The uncertainty that perceived person is 'rich' from the point of view of the simulated subject was represented by the existence of streams of contradictory memes: $\langle$'rich'$|\varnothing\rangle$ and $\langle$'not rich'$|\varnothing\rangle$. The subject's criteria of attractiveness were represented by the streams of memes: $\langle$'[I can] agree [to a date with somebody who is]'$|$'nice'$\rangle$ and $\langle$'[I can] agree [to a date with somebody who is]'$|$'rich'$\rangle$. For non-ambiguous data, i.e. when the stream of memes $\langle$'nice'$|\varnothing\rangle$ was much denser than the stream of the memes $\langle$'not nice'$|\varnothing\rangle$ and the stream of memes $\langle$'rich'$|\varnothing\rangle$ was much denser than the stream of the memes $\langle$'not rich'$|\varnothing\rangle$, WM quickly became dominated by the population of memes $\langle$'agree'$|\varnothing\rangle$. For ambiguous data, i.e. when densities of the streams of memes $\langle$'nice'$|\varnothing\rangle$ vs. $\langle$'not nice'$|\varnothing\rangle$ and $\langle$'rich'$|\varnothing\rangle$ vs. $\langle$'not rich'$|\varnothing\rangle$ were comparable, the result was counterintuitive—the populations of $\langle$'agree'$|\varnothing\rangle$ and $\langle$'not agree'$|\varnothing\rangle$ took turned in dominating in WM (Figure 2). In other words, the state of WM went to a 2-state limit-cycle attractor called "strange attractor".

It was shown that the simulated subject hesitated whether to accept the proposal to have a date with the perceived person or to not to accept [6][7]. The discussed model of WM gets strong psychological justification, since experiments with human subjects showed that people's judgment may oscillate between highly positive and highly negative values even in the absence of new data about a person of interest [10].

4. Towards Meme-based object recognition and action planning

Let us assume that Agent's PRS (Perceptual Representation System) detects features f_1, f_2, ..., f_n of observed object and converts their representations onto streams of memes $\langle f_1|\varnothing\rangle$, $\langle f_2|\varnothing\rangle$,..., $\langle f_n|\varnothing\rangle$ respectively. The memes are provided to WM. The density of a particular stream may depend on distinctness of a related feature. Let us also assume that Agent's long-term memories can, for request, provide to WM other memes related to the detected features, as for example, $\langle f_1|o_3\rangle$, $\langle f_1|o_4\rangle$, $\langle f_1|o_7\rangle$, ..., $\langle f_2|o_1\rangle$, $\langle f_2|o_4\rangle$, ..., $\langle f_3|o_1\rangle$, $\langle f_3|o_7\rangle$, ... and others. For all i, j, $\langle f_i | o_j \rangle$ can be interpreted as the fact that an object o_j has a feature f_i, while the density of a stream of the memes $\langle f_i | o_j \rangle$ would depend on strength of evidence that o_j has a feature f_i. As a result of meme interactions, a population of identical memes, say, $\langle o_o | \varnothing \rangle$, where o_o = 'an orange', may achieve domination in WM. It may also happen, that after a while the population lose its domination to the population of memes $\langle o_t | \varnothing \rangle$, where o_t = 'a tangerine', especially when the observed object looks as a small orange and as a big tangerine.

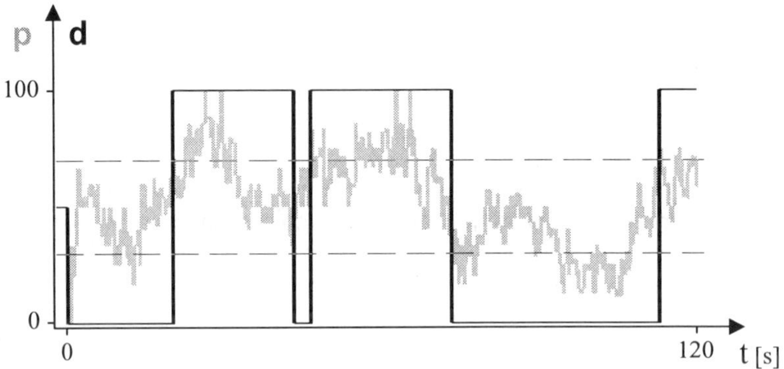

Figure 2. Meme dynamics in simulated Working Memory. Grey plot shows a percentage of given memes in the population of memes of interest. Black plot is a decision yes or not based on the history of population of memes of intrest. Although there was no randomness in data supply, p goes to a 2-state limit-cycle attractor.

Now let us assume that there is a kind of meme denoted $\langle \mathbf{S} | \varnothing \rangle$, that, when occurs in WM in a significant quantity, causes Agent's subjective feeling of satisfaction. Let us also assume, that Agent's PRS can detect states $x_1, x_2, \ldots, x_n$ of observed environment (including he/she/it itelf) and convert their representations onto streams of memes $\langle x_1 | \varnothing \rangle$, $\langle x_2 | \varnothing \rangle$,..., $\langle x_n | \varnothing \rangle$ respectively. Let us also assume that Agent's long-term memories can, for request, provide to WM other sort of memes related to the detected states, as for example, $\langle x_1 | \langle x_5 | u_1 \rangle \rangle$, $\langle x_1 | \langle x_7 | u_2 \rangle \rangle$, $\langle x_1 | \langle x_{11} | u_3 \rangle \rangle$, ..., $\langle x_2 | \langle x_4 | u_1 \rangle \rangle$, $\langle x_2 | \langle x_6 | u_3 \rangle \rangle$, ..., $\langle x_3 | \langle x_4 | u_1 \rangle \rangle$, $\langle x_3 | \langle x_1 | u_2 \rangle \rangle$, ... and others. For all i, j, k, $\langle x_i | \langle x_j | u_k \rangle \rangle$, can be interpreted as the fact that an action u_k can change a state of Agent's environment from x_j onto x_i. The set of state-versus-action-related memes is Agent's practical knowledge about dynamic properties of his/her/its environment. The memes of the type $\langle \mathbf{S} | x_i \rangle$, represent a knowledge about states causing satisfaction.

Let as consider such Agent that has hardwired goal to maintain a reasonable level of satisfaction and, from time to time, to achieve a possibly big level of satisfaction. The hardwired need of satisfaction may result in a spontaneous search for a sequence of actions that could lead to one of the states that can cause satisfaction. An interaction of state-related state-versus-action-related memes in WM is a distributed, only partially controlled generating and testing of more or less complete plans what to do to achieve a desired state of matters. A complete prospective plan is ready when for the first time a sequence of interactions leading to a generation of a non-negligible population of the satisfaction memes (the type $\langle \mathbf{S} | \varnothing \rangle$) took place in a number of places in WM. Provided the initial observed state is x_1, an example of such a sequence may be:

$$\langle x_1 | \varnothing \rangle \text{ and } \langle x_1 | \langle x_5 | u_1 \rangle \rangle \text{ give } \langle x_5 | u_1 \rangle, \quad \langle u_1 | \varnothing \rangle \text{ and } \langle x_5 | u_1 \rangle \text{ give } \langle x_5 | \varnothing \rangle,$$
$$\langle x_5 | \varnothing \rangle \text{ and } \langle x_5 | \langle x_8 | u_7 \rangle \rangle \text{ give } \langle x_8 | u_7 \rangle, \quad \langle u_7 | \varnothing \rangle \text{ and } \langle x_8 | u_7 \rangle \text{ give } \langle x_8 | \varnothing \rangle,$$
$$\langle x_8 | \varnothing \rangle \text{ and } \langle x_8 | \langle x_4 | u_3 \rangle \rangle \text{ give } \langle x_4 | u_3 \rangle, \quad \langle u_3 | \varnothing \rangle \text{ and } \langle x_4 | u_3 \rangle \text{ give } \langle x_4 | \varnothing \rangle,$$
$$\langle x_4 | \varnothing \rangle \text{ and } \langle \mathbf{S} | x_4 \rangle \text{ give } \langle \mathbf{S} | \varnothing \rangle;$$ Conclusion: Agent's satisfaction may be caused
by the sequence of actions: $u_1 \; u_7 \; u_3$.

It is up to Agent whether he/she/it decides to execute the sequence using his/her/its actuators. Some agents may be sufficiently satisfied by the number of satisfaction memes achieved as a result of imagination only. Other agents may want to experience the level of satisfaction caused by a huge number of satisfaction memes that can come only from true perception of the desirable state x_4. Those will try make the idea reality.

5. Conclusions

We introduced an unconventional way of reasoning in which populations of entities called memes try to dominate a living space called Working Memory (WM). The memes belonging to a given resulting population appear as a kind of cross-over of their ancestors who come to WM from Agent's perceptual apparatus. Because of arbitrarily given rules of survival, the process in WM cannot be recognized as evolutionary, hence the qualification *para-evolutionary*. We showed how meme interactions can facilitate a reasoning from uncertain data, as well as object recognition and action planning. The model has strong psychological justification—the intrinsic dynamics of meme population with a 2-state limit-cycle attractor observed during simulation of reasoning in a modeled WM fits well results obtained during experiments with human subjects. The idea of complex memes containing other complex memes as their parts suggests a direction of future investigations of the structure of meme-based semantic memory.

References

[1] Russel S & Norvig P (1995) *Artificial Intelligence. A Modern Approach*, Englewood Cliffs NJ: Prentice Hall.

[2] Blackmore S (2000) The Power of Memes, *Scientific American*, October, 52-61.

[3] Aunger R (Ed.) (2001) *Darwinizing Culture: The Status of Memetics as a Science*, Oxford: Oxford University Press.

[4] Castlefranci C (2001) Towards a Cognitive Memetics: Socio-Cognitive Mechanisms for Memes Selection and Spreading, *Journal of Memetics - Evolutionary Models of Information Transmission*, 5, www.cpm.mmu.ac.uk/jom-emit/2001/vol5/castelfranchi_c.html

[5] Buller A & Shimohara K (1999) Decision Making as a Debate in the Society of Memes in a Neural Working Memory, *3D Forum. The Journal of Three Dimensional Images*, 13 (3), 77-82.

[6] Buller A & Shimohara K (2000) Does the 'Butterfly Effect' Take Place in Human Working Memory? *The Fifth International Symposium on Artificial Life and Robotics (AROB 5th '00), January 26-28, 2000, Oita, Japan*, 204-207.

[7] Buller A, Chodakowski T, Kaiser L, Nowak A & Shimohara K (2001) MemeStorms: Cellular Working Memory and Dynamics of Judgment, *Proceedings of the Sixth International Symposium on Artificial Life and Robotics (AROB 6th ' 01), Tokyo, Japan, January 15-17*, 146-149.

[8] Buller A (2001) Dynamic Fuzzy Sets for Cognitive Modelling, *Proceedings of the Sixth International Symposium on Artificial Life and Robotics (AROB 6th ' 01), Tokyo, Japan, January 15-17*, 150-151.

[9] Buller A, Shimohara K, Chodakowski T & Hemmi H (2000) CoDi Technique: Cellular Automata as Neural Networks, *The Fifth International Symposium on Artificial Life and Robotics (AROB 5th '00), January 26-28, 2000, Oita, Japan*, 192-195.

[10] Nowak A & Vallacher R (1998) *Dynamical Social Psychology*, New York: The Guilford Press.

KES '01
N. Baba et al. (Eds.)
IOS Press, 2001

A Method of Generating Japanese Compound Keywords

Kazuaki ANDO[1], Shigeaki OGOSE[1], Toshinori YAMASAKI[1], and Jun-ich AOE[2]
[1]Faculty of Engineering, Kagawa University, 2217-20 Hayash, Takamatsu 761-0396, Japan
[2]Faculty of Engineering, Tokushima University, 2-1 Minami-josanjima, Tokushima 770-0861,
Japan

Abstract. This paper examines the effectiveness of a derivation method of Japanese compound keywords based on rules for dependency relationship and restrictions on construction, verbal noun, redundant words. Then, we propose a keyword ranking method according to importance with relevance between words (like co-occurrence of common words) in the original text. The proposed method can derive significant keywords that do not appear in the original text.

1. Introduction

Keyword extraction[1],[2],[3],[4] is the important task of various applications such as information retrieval[5], automatic text summarization[6],[7], paraphrasing, text mining and so on. Various methods for keyword extraction[1],[2],[3],[4] have been proposed. In general, almost all methods are based on the hypothesis[4] that words, which express the contents of a document precisely, certainly are appeared in the text. Therefore, in cases in which a word should be a keyword does not occur in the text, there are components of keywords separately in the text, and a keyword is an abstract (subject) word that should be derived from the contents, the methods cannot cope with them[3]. It is important to extract, generate, and derive such keywords for the applications.

This paper examines a method for automatically deriving Japanese compound keywords based on rules of dependency relationship and its restrictions. The proposed method can derive significant keywords that do not occur in the original text. Firstly, to prepare keyword derivation according to rules, we analyzed the keywords assigned to academic papers by the authors (hereafter called author's keyword). Then we define 50 derivation rules based on dependency relationship, restrictions based on syntactic structures, SAHEN NOUN which is a verbal noun in Japanese and redundant words. They could be used for restraining superfluous keyword derivation. Then, we propose a keyword ranking method according to importance with relevance between words (like co-occurrence of common words) in the original text, in order to extract more significant keywords from a set of derived keywords. To verify the validity of the derived keywords, we extract 65 titles and abstracts relevant to computational linguistics from the NACSIS Test Collection for evaluation of Information Retrieval systems (NTCIR-1)[8] as a material for the experiment.

In the following sections, our ideas are described in detail. Section 2 describes the distinctive feature of author's keywords assigned to academic papers. Section 3 describes keyword derivation method based on rules, various restrictions on them, and effective importance measure. Section 4 shows the experimental results with 65 ariticles. Finally, in section 5 the results are summarized and the future research is discussed.

2. Analysis of Author's Keywords

Author's keywords [1] are very useful for indexing for information retrieval and determining whether or not to read the text. However, some of author's keywords do not appear in the original text[3],[4], frequently. People not only extract and select appropriate keywords from the text, but also is capable of generating and deriving more significant keywords freely by understanding the contents of text.

As mentioned above, our purpose in this paper is to generate and derive such compound keywords by rules for applications of keywords extraction and derivation. Thus, we focused on author's keywords, especially: (1) Appear in the document as a different form(synonym, existing part of constituents of the author's keyword, etc.) (2) Do not appear in the text(no components of keywords in the text).

We analyzed author's keywords and abstracts assigned to 50 academic papers related to computational linguistics. The targets were extracted from NACSIS Test Collection for evaluation of Information Retrieval systems [2] (NTCIR-1)[8]. As the result of analysis, nearly 45 percent of whole keywords belong to above 1 or 2.

Typical patterns are shown as follows:
1-1) Dependency Relationship:
　　「音声を認識する → 音声認識」 (recognize a speech→speech recognition)
1-2) Paratactic Construction:
　　「音声の認識および合成 → 音声認識, 音声合成」
　　(recognition and synthesis of speech → speech recognition, speech synthesis)
1-3) Anaphoric and Dependency Relationship:
　　「談話の解析...それを生成する... → 談話生成」
　　(Analysis of discourse ... it is generated ...→ discourse generation)
1-4) Paraphrase:
　　「文書読み上げ → テキスト読み上げ」 (document reading → text reading)
2-1) Association:
　　「推論...知識...コンピュータ... → 人工知能」
　　(inference ...knowledge ...computer... → computer artificial intelligence)」
Especially, since types 1-1, 1-2, and expansion of them can be derived by rules, they will be discussed in the following sections. Confirming limitations of keyword derivation using rules is also another purpose of us in the experiment.

3. Keyword Derivation

3.1 Rules of Keyword Derivation

Miyazaki[9] proposed an automatic segmentation method for compound words using dependency relationships based on rules. Thus, this section defines rules for deriving Japanese compound words by extending Miyazaki's rules.

The following indicates some of rules for deriving Japanese compound words:
[rule 1] : x(common noun)$^+$ を(case-marking particle) y(verbal noun／stem) → xy
Example: 音声／を／認識／する→音声認識
　　　　(recognize a speech→speech recognition)
[rule 2] : x(common noun)$^+$ の(case-marking particle) y(common noun)$^+$ → xy
Example: 情報／の／検索／は→情報検索
　　　　(retrieval of information → information retrieval)

---

[1] In this paper, we deal with only compound words.
[2] The test collection consists of title, abstract, authors, etc, excluding the body.

[rule 3] : x(common noun)$^+$ の(case-marking particle) y(common noun)$^+$ および(case-marking particle) z(common noun)$^+$→ xy, xz

Example:音声／の／認識／および／合成 → 音声認識，音声合成

(recognition and synthesis of speech → speech recognition, speech synthesis)

Above symbol x, y, and z denotes one part-of-speech, respectively. x(*part-of-speech*)$^+$ denotes "one or more concatenations of " x.

3.2 Restrictions on Keyword Derivation

Some needles words may be derived by the rules for keyword derivation because the rules are very simple. This section defines restrictions on keyword derivation, in order to improve the precision of derivation.

Suitable compound words may not be derived by the rules in the case of sentence with syntactic ambiguities. Let us consider that 「人間と計算機を比較する(Man is compared with a computer.)」. Rule 1 may derive an unsuitable keyword 「計算機比較(computer comparison)」 from the sentence. The meaning of the derived keyword is different from the meaning of sentence, because *compare* subcategorizes "と格(one of surface case)" and "を格(one of surface case)".

Thus, we define restrictions on rules for keyword derivation. The following shows an example of restrictions:

[Restriction for rule 1] : If a case-marking particle 「と」 exists in front of the matched sequence, the keyword is not derived by the rule.

[Restriction for rule 2] : If a case-marking particle 「と」 exists in front of the matched sequence, or a case-marking particle 「と」 or 「の」 exists just behind the matched sequence, the keyword is not derived by the rule.

[Restriction for rule 3] : If a case-marking particle 「の」 exists just behind the matched sequence, the keyword is not derived by rule.

When we understand the meaning of a Japanese compound word, we often complement omitted particles between components of the compound word[9] and make phrases based on サ変名詞(verbal noun). Thus, we assume that verbal noun is an important component for semantic understanding of compound words. We assumed that サ変名詞(verbal noun) is an important component for semantic understanding for compound words. Thus, we decided to derive only keywords including verbal noun in this paper[3].

Some words which are very frequent among the texts and do not carry meaning, called redundant(stop) words, could be removed. Two rules for elimination of redundant words are define as follows:

[e-rule1] : Words could not be component of compound words such as 有無(existence), 場合(case), and so on.

[e-rule2] : Compound word derived by frequent appeared expressions in academic papers such as 方法検討(examining methodology), 結果報告(reporting results) and so on.

3.3 Importance Measure for Derived Keywords

In order to extract more significant keywords from derived keywords set, we focused on the number of common morphemes between words in a text. Suppose that "情報 (information extraction)" could be derived as a keyword from a text. If 情報検索 (information retrieval) exists in the text, it should be considered that 情報(information) is relevant to 情報検索 through "情報(information)". Thus, we considered that the weight of

---

[3] We will consider compound keywords without サ変名詞(verbal noun) in the future work.

"information retrieval" should be reflected in the weight of "情報抽出 (information extraction)".

Let $f(w_i)$ be the number of times(frequency) of the word i occurring in a text and k be the derived keyword. The $S(k)$, importance of the derived keyword k, is defined as follows:

$$S(k) = \sum_i \{N(w_i) \times R(w_i, k)\}$$

$$N(w_i) = f(w_i) \Big/ \sum_i f(w_i) \qquad R(w_i, k) = C(w_i, k) / L(w_i)$$

where $R(w_i, k)$: Relevancy between word i and keyword k; $C(w_i, k)$: The number of common morphemes between k and w_i; $L(w_i)$: The number of Morphemes;

Suppose that N(情報(information)) = 0.4, N(抽出(extraction)) = 0.3, N(情報／検索 (information retrieval)) = 0.2. If 情報／抽出(information extraction) could be derived as a keyword, $C(w_i,$ information extraction) are obtained as follows: C(information, information extraction) = 1; C(extraction, information extraction) = 1; C(information retrieval, information extraction) = 1. As for $R(w_i,$ information extraction), R(information, information extraction) = 1 / 1 = 1; R(extraction, information extraction) = 1 / 1 = 1; R(information retrieval, information extraction) = 1 / 2 = 0.5. Therefore, the importance S(information extraction) can be computed as follows: S(information extraction) = $\{N$(information)$\times R$(information, information extraction)$\}$ + $\{N$(extraction)$\times R$(extraction, information extraction)$\}$ + $\{N$(information retrieval) $\times$ R(information retrieval, information extraction)$\}$ = $0.4 \times 1 + 0.3 \times 1 + 0.2 \times 0.5 = 0.8$.

Here, let us consider the importance based on the total number of normalized frequency of components of the keyword. The importance of 情報抽出 (information extraction) could be N(情報(information)) + N(抽出(extraction)) = 0.3+0.4 = 0.7 less than 0.8.

The proposed importance measure can be considered not only importance of the keyword, but also relevancy between words in the text.

Moreover, the importance is compensated by the following considerations: (1) Important words often appear in front and behind of cue expressions such as 「による (by)」 , 「を用いた (using)」 [6];.(2) Rule x(common noun)$^+$ を(case-marking particle) y(verbal noun／stem) $\rightarrow xy$ derives many significant keywords[9].

Thus, final importance, $I(k)$, of derived keywords are defined as follow:

$$I(k) = S(k) \times r_j$$

where r_j denotes the compensation value of jth rule.

4. Evaluation

In order to verify the validity of the derived keywords, we extract 65 titles and abstracts[4], whose the total size is about 37.5KB, relevant to computational linguistics from the NTCIR-1[8] as a material for the experiment. We made 50 rules to derive compound keywords, defined 11 restrictions for deriving needless keywords, and registered 76 redundant words, which were chosen by people under two rules for elimination of them from 50 academic papers used for analysis in section 2. The compensation value of each rule was computed form the 50 papers. The value is shown in Table 1.

Author's keywords are very useful for indexing for information retrieval and determining whether or not to read the retrieved documents. The derived keywords are evaluated by Recall and Precision for author's keywords. Recall is measured as the ratio of the number of correct keywords derived to the total number of correct keywords that exist in the collection including 65 papers, and Precision is measured as the ratio of the number of correct keywords derived to the total number of keywords derived. In this experiment,

---

[4] Experimental data we used here is different from 50 data sets, which are used in section 2.

Table 1 Compensation values for rules

を用いた x(common noun)$^+$ の y(common noun)$^+$ → xy	2.0
による x(common noun)$^+$の y(common noun)$^+$→ xy	2.0
x(common noun)$^+$ を y(verbal noun／stem)$^+$→ xy	1.5
Others	1.0

we defined keywords, which do not appear in the abstract and its title, as correct keywords. The collection consisting of 65 papers includes 76 correct keywords(about 29%) among 263 author's keywords.

The result of the experiment was that Recall and Precision were 21% and 4% respectively. The reason why Precision showed low percentage that one file in the collections contains only one correct keyword, that is, the number of correct keywords is very few against the number of the derived keywords. The correct keywords, which could not be derived, are classified into the following: (A) No information in the title and abstract to derive keywords: 17.5%; (B) 2-1 type in section 2: 47.5%; (C) 1-3 and 1-4 types in section 2: 35.0%.

It seems that analysis, which is more than the method based on rules, are required in order to derive above keywords, that is, to rise Recall.

Then, the derived keywords were evaluated whether they were significant keywords for the text or not. As the results of evaluation by students, 28% of keywords derived were significant. It means that we could obtain about 30% accuracy of keyword derivation using only rules.

5. Conclusion

We focused on that people are capable of freely generating and deriving more appropriate keywords, which do not appear in the original text, by understanding the contents of text. This paper examines the effectiveness of a method of automatically generating and deriving Japanese compound keywords based on rules of dependency relationship and their restrictions. From the experimental results, it turns out that about 30% of total keywords derived were significant. As the future research, we will apply the combination method of keyword derivation using rules with keyword extraction method for surface words to text summarization for retrieved documents by Internet search engine.

References

[1] M. Hara, et al., Keyword Extraction Using a Text Format and Word Importance in a Specific Field, *Journal of Information Processing Society of Japan*, Vol.38, No.2, pp.299-309, 1997.

[2] H. Kimoto, Automatic Indexing and Evaluation of Keyword for Japanese Newspapers, *The Trans. of the IEICE of Japan*, Vol.J74-D-I, No.8, pp.556-566, 1991.

[3] M. Nagata, et al., A Newspaper Keyword Generation Method Based on Key-Concept Extraction, *Proc. of 37th National Convention of Information Processing Society of Japan*, pp.1030-1031, 1988.

[4] M. Morohashi, Automatic indexing survey, *Magazine of Information Processing Society of Japan*, Vol.25, No.9, pp.918-925, 1984.

[5] G. Salton, Introduction to Modern Information Retrieval. McGraw-Hill, 1983.

[6] M. Okumura, et al., Automated Text Summarization: a survey, *Journal of Natural Language Processing of Japan*, Vol.6, No.6, pp.1-26, 1999.

[7] I. Mani et al., Advances in Automatic Text Summarization. The MIT Press, Massachusetts, 1999.

[8] NACSIS Test Collection for IR Systems (NTCIR-1), National Institute of Informatics, 1999.

[9] M. Miyazaki, Automatic segmentation method for compound word using semantic dependent relationships between words, *Journal of Information Processing Society of Japan*, Vol.25, No.6, pp.970-979, 1984.

A Fast Method of Storing Compound Keywords with Related Information for Document Management Systems

Yoshitaka Hayashi[‡] and Hisatoshi Mochizuki[‡‡]

SUMITOMO METAL System Solutions Co., Ltd.[‡]
1-2-6 Doujimahama Kita-ku Osaka, 530-0004, JAPAN[‡]
Dept. of Electrical Engineering and Computer Science
Osaka Prefectural College of Technology[‡‡]
26-12 Saiwai-Cho Neyagawa-Shi Osaka, 572-5872, JAPAN[‡‡]

Abstract. In document management systems, many compound words that are invented freely can be keyword candidates. Since the method should take many costs for processing keyword candidates, it is important to prepare an efficient method to extract keywords with information about the relation between them. This paper presents a technique for storing compound keywords with information about a relation between short based and long based compound keywords by extending Aho and Corasick (AC) string pattern matching machine for a finite number of keyword. By simulation results for 38 Japanese text files, it is shown that the proposed algorithm has performed 8.46ms/KB.

1. Introduction

In document management systems, many compound words that are invented freely can be keyword candidates. In these systems it is important task to process compound words efficiently [1][2][3]. It is impossible method that all compound words appear in a dictionary as keywords. In general we can see methods that managed an individual word appears in a dictionary and information about components of compound words separately. For example, conventional methods extract a word sequence, which is evaluated as a keyword based on its part of speech, or is matched with keyword patterns, as compound keywords [1][3][4][5]. Thus, in extracting keywords, there are two types of criterions by keyword construction. One is to select an individual word, and another is to select a sequence of individual words. In this paper, a keyword consisting of an individual word is called a short component keyword (SC keyword), and a keyword consisting of word sequence is called a long component keyword (LC keyword) [1][2].

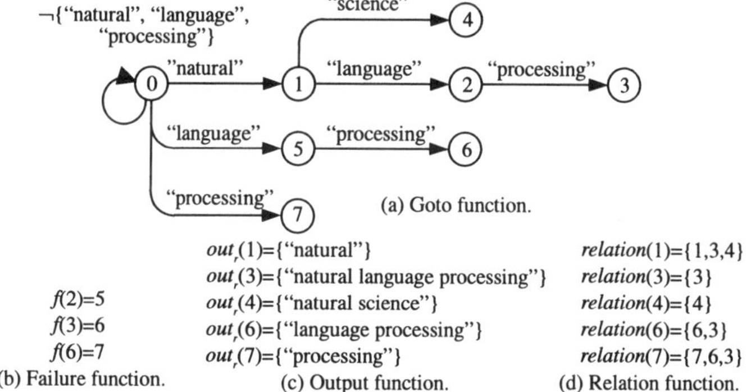

Fig.1 An example of AC-REL machine

Table 1. An example of function *weight*

Function *relation*	SC keyword processing	LC keyword processing
relation(1)={1,3,4}	*weight*(1)=2	*weight*(1)=0
relation(3)={3}	*weight*(3)=0	*weight*(3)=3
relation(4)={4}	*weight*(4)=0	*weight*(4)=1
relation(6)={6,3}	*weight*(6)=1	*weight*(6)=1
relation(7)={7,6,3}	*weight*(7)=2	*weight*(7)=0

Table 2. An example of application for function *weight*

Input keyword	*SCR*	SC keyword processing	LC keyword processing
natural	20	120	20
natural language processing	20	20	170
natural science	80	80	130
language processing	30	80	80
processing	40	140	40

In this paper, it is called *SC keyword processing* that compound keywords with shorter length are given priority. It is easy to detect them by exact matching, but they lose the precise meaning having by LC keywords. On the other hand, it is called *LC keyword processing*, LC keywords can contain SC keywords as their components, but these SC components cannot be detected by exact matching of LC keyword processing. Selection of SC or LC keywords depends on the application systems, so it is necessary to prepare an efficient method to extract both SC and LC keywords. The method should process many operations for appending, separating or comparing substrings for keyword candidates. So, with larger numbers of keywords, the time for processing strings is longer. Therefore, it is important to process them efficiently.

This paper presents a technique for storing compound keywords with information about both SC and LC keywords using the extended Aho and Corasick string pattern matching machine (AC machine) for a finite number of keywords [6]. The extended machine is called *AC-REL (relation) machine*.

2. The AC-REL machine

2.1 Overview

The AC machine consists of constructing a finite state pattern matching machine from the keywords. The AC machine is a simple and efficient algorithm to locate keywords in a text string in a single pass. However, on the application of extracting keywords for compound words, it is necessary to obtain information about the components of each keyword, not locations. The behavior of the AC machine is defined by the three functions: goto, failure and output. In the AC-REL machine, the goto and output function are extended, and the new function *relation* and *weight* are introduced. The notation for the goto function is extended from symbol to string, since the AC-REL machine is storing word sequences. The output function in the AC machine is formalized by associating a set of keywords with every state. However, in the AC-REL machine the output function indicates one keyword, since the new function *relation* substitutes for the associating. Thus, the output function in the AC-REL machine is described as the function out_r.

The new function *relation* has a finite set of nodes. This function is characterized as follows: If $relation(s) \ni t$, then the keyword x where $out_r(s)$ has been indicated is contained in the keyword y where $out_r(t)$ has been indicated. Figure 1 shows an example of the AC-REL machine for the set of keywords K_REL = {"natural", "natural language processing", "natural science", "language processing", "processing"}. For example in Figure 1, since $relation(7) \ni 3$ we see that the keyword "processing"= $out_r(7)$ is contained in the keyword "natural language processing"= $out_r(3)$.

The new function *weight* indicates priority for each states s such that $out_r(s) \neq$empty. The function *weight* is characterized as computing its value dynamically. Table 1 shows an example of function *weight* for the AC-REL machine in Figure 1. In SC keyword processing, suppose c represents the number of the elements of $relation(s)$, then $weight(s)$ is set to c-1. For example in Table 1, $weight(1)$=3-1=2 since $relation(1)$={1,3,4}. Then, we see that $out_r(1)$="natural" is contained in two other keywords: "natural language processing" =$out_r(3)$ and "natural science"=$out_r(4)$. In LC keyword processing, $weight(s)$ is set to the number of state t such that $s \in relation(t)$ and $t \neq s$. For example, $weight(3)$ is set to 3 since state 3 is contained in $relation(1)$, $relation(6)$ and $relation(7)$. Then, we see that $out_r(3)$="natural language processing" contains three other keywords: "natural"=$out_r(1)$, "language processing"=$out_r(6)$ and "processing"=$out_r(7)$. Thus, the function *weight* depends on the function *relation*.

The following section will present algorithms to construct the AC-REL machine. We shall explain algorithms with lines marked by "number>".

2.2 Construction of Goto, Output and Relation Functions

The Algorithm 1 shown in Figure 2 is construction of goto, output and relation functions. The procedure *enter*(y) is given keyword kw_i of K_REL, in the first while-loop at line 2 we have been able to transit the existing states as far as possible, in the for-loop at line 4 keyword kw_i is inserted to the goto graph. At lines 3

```
Algorithm 1: Construction of the goto, output
             and relation functions.
Input: K_REL={kw₁,kw₂,...,kwₙ}.
Output: Functions g, outᵣ and relation.
Method.
  begin
    newstate←0;
    for i←1 until n do enter(kwᵢ)
    for all a such that g(0,a)=fail do
        g(0,a)←0
    relation(0)←φ;
  end

    procedure enter(w₁w₂···wₘ):
    begin
      state←0; j←1;
1>    queue←φ;
2>    while g(state,wⱼ)≠fail do
        begin
          state←g(state,wⱼ); j←j+1;
3>        queue←queue∪{state};
        end
4>    for p←j until m do
        begin
          newstate←newstate+1;
          g(state,wₚ)←newstate;
          state←newstate;
          relation(state)←φ;
5>        queue←queue∪{state};
        end
      outᵣ(state)←{(w₁w₂···wₘ)};
6>    while queue≠φ do
        begin
          let r be the next state in queue
          queue←queue-{r};
          relation(r)←relation(r)∪{state};
        end
    end
end
```

```
Algorithm 2: Construction of the failure and
             relation functions.
Input: Function g and relation
       from Algorithm 1.
Output: Function f and relation.
Method.
  begin
    queue←φ;
1>  for each a such that g(0,a)=s≠0 do
      begin
        queue←queue∪{s}; f(s)←0;
      end
2>  while queue≠φ do
      begin
        let r be the next state in queue
        queue←queue-{r};
3>      for each a such that g(r,a)=s≠0 do
          begin
            queue←queue∪{s}; state←f(r);
            while g(state,a)=fail do
              state←f(state);
            f(s)←g(state,a);
4>          if s∈relation(f(s)) then
              begin
5>              t←s;
6>              while f(t)≠0 do
                  begin
                    relation(f(t))
                    ←relation(f(t))∪relation(s);
7>                  t←f(t);
                  end
              end
          end
      end
  end
```

Fig. 2 Algorithms for the AC-REL machine construction

and 5, a queue of line 1 stores transit states. After inserting keyword kw_i to the goto graph, the function *relation* is constructed in the while-loop at line 6. The constructing function *relation* in Algorithm 1 is associated with common prefixes. In order to process simply the function *relation(s)* is constructed for all states s, not only states s such that $out_r(s)\neq$empty.

For example, we shall explain the AC-REL machine in Figure 1 for the set of keywords *K_REL*. Adding the first keyword "natural" to the graph, $g(0,\text{"natural"})=1$ is constructed. At this point $out_r(1)$ is set to {"natural"}, and *relation*(1) is set to {1} by *queue*={1}. Adding the second keyword "natural language processing", $g(1,\text{"language"})=2$ and $g(2,\text{"processing"})=3$ are constructed. Notice that, when the keyword "natural language processing" is added, there is already an edge labeled "natural" from state 0 to state 1, so this edge does not need to be added. The function $out_r(3)$ is set to {"natural language processing"}. By *queue*={1,2,3}, *relation*(1) is updated to {1,3}, *relation*(2) and *relation*(3) are set to {3}. By adding the third keyword "natural science", the function $out_r(4)$ is set to {"natural science"}. With *queue*={1,4}, *relation*(1) is updated to {1,3,4} and *relation*(4) is set to {4}. When adding the fourth keyword "language processing", the function $out_r(6)$ is set to {"language processing"}, *relation*(5) and *relation*(6) are set to {6} by *queue*={6}. Adding the last keyword "processing", the function $out_r(7)$ is set to {"processing"} and *relation*(7) is set to {7}. The last addition to the graph is a loop from state 0 to state 0 on all input strings other than "natural", "language" or "processing".

2.3 Construction of Goto, Output and Relation Functions

In Algorithm 2 shown in Figure 2, the function *relation* about the substrings except the common prefixes is constructed during the computation of the failure function. This construction depends on the following property of the AC machine [6]: The string x representing state s contains the string y representing state $f(s)$, if $f(s)\neq0$ is determined. So, all the elements of *relation(s)* have to be added to the function *relation(f(s))*. For example in Figure 1 since $f(3)=6$ is determined, we see that the string "natural language processing" representing state 3 contains the string "language processing" representing state 6. Then, all the elements of *relation*(3) are added to *relation(f(3))*. The updating the function *relation* is computed from line 3 to 7 in Algorithm 2. The while-loop at line 6 is updating the function *relation* until the string representing state s does not contain another substrings. The if-statement at line 4 avoids adding the same set.

The failure function is constructed from the goto graph by breadth first search in Algorithm 2. For example in Figure 1, the for-loop at line 1 stores the states 1, 5 and 7 of depth 1 to a queue, and $f(1)=f(5)=f(7)=0$ are determined. The while-loop at line 2 selects the states of depth 1 from a queue. The for-loop at line 3 stores the states 2, 4 and 6 of depth 2 to a queue, and the failure function of depth 2: $f(2)=g(0,\text{"language"})=5$, $f(4)=g(0,\text{"science"})=0$ and $f(6)=g(0,\text{"processing"})=7$ are determined. At this point $relation(f(2))=\{6\}$ is updated to $\{6,3\}$ since f(2)=5 and $relation(2)=\{3\}$, and $relation(f(6))=\{7\}$ is updated to $\{7,6\}$. At the state 3 of depth 3, $f(3)=g(5,\text{"processing"})=6$ is determined. At this point $relation(f(3))=\{6\}$ is updated to $\{6,3\}$. At the state 6, since $f(6)=7$ $relation(f(6))=\{7,6\}$ is updated to $\{7,6,3\}$.

2.4 The Application of AC-REL Machine to Document Management Systems

We shall show an example of application for function *weight*. In general, on the extracting keyword system, a value called score (for example, based on frequency) is introduced to order keyword candidates. The purpose of SC and LC keyword processing is determining this order. For example, Table 2 shows a change of order by SC and LC keyword processing. Let *SCR* be a value called score determined by the extracting keyword system, and let *VAL* be an additional value to *SCR*. *VAL* is defined as $VAL=C{\times}weight(s)$, C is a coefficient. In SC keyword processing, the order of keyword "natural" ranks first from third, since $SCR+VAL =SCR+C{\times}weight(1)=20+2{\times}50=120$, where C is set to 50. Thus, we can see that the SC keywords rank higher. In LC keyword processing, the order of keyword "processing" ranks forth from second. But, we can see that the LC keywords "natural language processing" or "natural science" rank higher.

3. Evaluations

3.1 Theoretical Observations

The worst-case time complexity to construct the AC machine becomes $O(n)$, where n is the sum of the lengths of the keywords [6]. In Algorithm 1 the worst-case time complexity becomes $O(n)$, since all transit states are stored in queue. In Algorithm 2, in the while-loop at line 6, Algorithm 2 makes n-k transitions in the worst case, where k is the number of keywords. However, since $n{>>}k$, the worst-case time complexity of this transition becomes $O(n)$. Therefore, the worst-case time complexity to construct the AC-REL machine becomes $O(n)$.

It is clear that the worst-case time complexity of SC keyword processing becomes $O(k)$. In SC keyword processing, function $weight(s)$ is determined by the number of the elements in $relation(s)$ for state s such that $out_r(s){\neq}$empty. In LC keyword processing, the worst-case time complexity to compute function $weight(s)$ depends on the cost of counting $relation(t)$ including s. Here, $relation(t)$ can be represented by a bit-vector, where state numbers correspond to positions of bits. Therefore, the costs of getting a bit value at an arbitrary position of the bit-vector is constant, then the worst-case time complexity to compute function *weight* becomes $O(k)$, because it is proportional to the number of keywords. Moreover, using a bit vector, it is easy to count the elements in $relation(s)$ in SC keyword processing.

In order to understand the efficiency of the AC-REL, consider computing the function *weight* on the original AC machine. The worst-case time complexity $O(n)$ is required to transit the goto and failure functions in the opposite direction. Therefore, the worst-case time complexity becomes $O(2n)$, except $O(n)$ that the worst-case time complexity to construct the AC machine. Thus, we can see that the time complexity in SC and LC keyword processing on the AC machine becomes $O(3n)$. Using the function *relation*, the AC-REL machine absorbs the time complexity to compute these opposite functions into the time complexity to construct the AC-REL machine, and controls the time complexity to compute function *weight* to $O(k)$. Thus, we can see that the total cost of the AC-REL machine becomes $O(n+k)$ in SC and LC keyword processing.

3.2 Evaluation of Simulation Results

The proposed method was developed on a Microsoft Windows NT 4.0 platform using C++ Language. We selected 38 Japanese texts which are determined into keywords in advance; 8 texts (letter, magazine and etc.), 15 abstracts of papers about information science and 15 abstracts of patents. The total size of these texts was about 37.5K bytes, and the total number of words was 11,250. In our system for extracting keywords, the total number of keyword candidates was of 2,188, a total of 623 SC keywords; and 796 LC keywords.

Table 3 shows the construction time of the AC-REL machine, and SC and LC keyword processing time for each text (TXT in left column represents text, PAP represents paper and PAT represents patent). The average time to construct the AC-REL machine was about 293.6 milliseconds. Specially, after constructing the AC-REL machine, the average time in SC keyword processing was about 25.04 milliseconds, and the average

Table 3. Simulation results for time (milliseconds)

Text (Size:KB)	Construction of AC-REL	SC keyword processing	LC keyword processing	Text (Size:KB)	Construction of AC-REL	SC keyword processing	LC keyword processing
TXT01(1.48)	231.451	35.741	27.891	PAT01(0.92)	280.134	23.286	34.449
TXT02(1.26)	375.630	3.017	2.522	PAT02(0.84)	234.060	28.312	44.579
TXT03(2.26)	376.829	23.965	32.100	PAT03(0.54)	266.335	27.251	24.227
TXT04(1.61)	501.885	3.312	4.754	PAT04(0.74)	184.297	28.738	64.912
TXT05(2.29)	364.712	10.490	9.471	PAT05(0.74)	289.861	6.286	0.632
TXT06(2.28)	390.541	34.527	25.614	PAT06(0.66)	239.448	1.501	9.231
TXT07(1.94)	308.630	20.792	60.018	PAT07(0.60)	239.022	3.312	6.636
TXT08(1.76)	376.840	26.041	22.684	PAT08(0.63)	242.808	22.006	44.598
PAP01(1.25)	248.657	29.319	44.624	PAT09(0.64)	210.780	36.372	21.211
PAP02(0.53)	222.461	1.868	6.197	PAT10(0.60)	1014.512	35.482	84.341
PAP03(0.99)	336.174	36.035	48.243	PAT11(0.64)	191458	32.114	32.930
PAP04(0.99)	329.494	5.883	2.692	PAT12(0.72)	226.719	6.145	0.893
PAP05(0.59)	233.992	5.877	7.430	PAT13(0.63)	246.809	7.204	1.000
PAP06(1.01)	1195.791	22.417	43.544	PAT14(0.63)	341.698	34.826	49.345
PAP07(0.55)	244.716	36.197	43.139	PAT15(0.87)	324.943	2.618	7.372
PAP08(0.86)	1049.373	32.796	33.887				
PAP09(0.99)	247.199	46.698	39.771	Average	293.595	25.037	23.895
PAP10(0.80)	258.861	7.143	29.104				
PAP11(0.71)	268.571	7.599	10.877				
PAP12(0.91)	275.582	8.594	1.111				
PAP13(0.78)	1298.402	10.521	4.930				
PAP14(0.88)	285.376	3.013	4.288				
PAP15(0.89)	299.542	33.218	6.351				

time in LC keyword processing was about 23.9 milliseconds. So, the average time to construct the AC-REL machine and LC keyword processing is 8.46 millisecond per 1 KB in Japanese text. Therefore, the AC-REL machine can extract various keywords effectively.

4. Conclusions

This paper has presented a new method for storing compound keywords with related information as follows.
(1) The SC keyword and the LC keyword have been defined.
(2) An efficient method for storing SC and LC compound keywords with related information has been presented by extending the Aho Corasick string pattern matching machine.
(3) SC and LC keyword processing, a criterion of keyword construction has been introduced.

The presented method has been evaluated in its effectiveness through the simulations. The following future researches remain.
- The better evaluated function for compound keywords, except function *weight* in this paper, is designed.
- The construction of function *relation* is corresponded to the dynamic construction of the AC machine[7].

References

[1] Y.Ogawa, M.Mochinushi and A.Bessho, A Compound Keyword Assignment Method for Japanese Texts (In Japanese), IPS Japan SIG Notes 93-NL-97-15, 1993, pp.103-110

[2] Y.Ogawa, A Bessho, M.Iwasaki, M.Nishimura and M Hirose, A Text Database System based on Simple Words (In Japanese), IPS Japan SIG Notes 92-DBS-90-6, 1992, pp.45-54

[3] S.Ito, H.Niwa, K.Kayashima, S.Maruno and Y.Shimeki, Parametric Keyword Extraction Algorithm and Adaptation Method (In Japanese), IEICE Tech. Report NLC93-53, 1993, pp.41-46

[4] H.Kimoto, Automatic Indexing and Evaluation of Keywords for Japanese Newspapers (In Japanese), IEICE Trans. J74-D-I No.8, 1991, pp.556-566

[5] M.Miyazaki, S.Ikehara and A.Yokoo, Construction of a Dictionary for MT using Combined Word Retrieval (In Japanese), IPS Japan SIG Notes 91-NL-84-12, 1991, pp.87-94

[6] Aho,A.V. and Corasick,M.J., Efficient string matching: An Aid to Bibliographic Search, C.ACM Vol.18 No.6, 1975, pp.333-340

[7] K.Tsuda, H.Iriguchi and J.Aoe, A Method of Constructing Dynamic String Pattern Matching Machines (In Japanese), *IEICE Trans.*, J77-D-I, No.4, 1994, pp.282-289

KES '01
N. Baba et al. (Eds.)
IOS Press, 2001

How to Implement Efficient System for Managing Market Trends and Customer Orientation Transitions

Hisatoshi MOCHIZUKI
Osaka Prefectural College of Technology,
26-12 Saiwai, Neyagawa, Osaka 572-8572, JAPAN

Masayuki KESSOKU
Matsushita Electric CO

Masafumi KOYAMA
Nara National College of Technology

Kazuhiko TSUDA
University of TSUKUBA

Abstract. To carry out public relations activities effectively, it is essential to keep track of orientation transitions of individual customer as well as market trends. However, it is difficult to manage customer orientations using two-dimensional vectors of customer axis and oriented area axis because two dimensions are not enough to represent market trends and customer orientations that change as time passes. To solve this problem, this paper proposes a management method of customer orientations with three-dimensional vectors, which have time axis in additions to customer and oriented area axes. The information volume is rapidly increased in three-dimensional vectors. The proposed method solves this problem by storing only the positions and the values of points where information changes as time passes. In this method, only the past information is stored as positions and values, while the current situation is stored as two-dimensional vectors. This is to immediately understand the most important information, which are the current market situation and customer orientations. This proposed method enables us to conduct sales promotion considering trends in market and orientation transitions of individual customer. It also allows us to grasp the accurate effects of advertisements, which will be an index for more effective advertisement.

1. Introduction

It is essential to keep track of orientation transitions of individual customer as well as the entire market to make public relations activities effective. This is because that there are trends and essential commonsense information in markets and these factors have great impacts on orientation of individual customer. Information systems are receiving recognition as an effective tool to collect these information (orientations of entire market and individual customer) [1]. There is a case that the productivity of sales and marketing has been dramatically improved by utilizing Sales Force Automation (SFA) [2], which is a method to analyze sales activities through marketing activities, in addition to the information gathering.

The marketing in future, especially the marketing in the Internet businesses, requires the way to quickly grasp market trends and orientation transitions of individual customer, utilizing interactivity and immediacy of the Internet. It is also required to grasp the effects of advertisements accurately and to propose the effective timing and contents of advertisements[3].

For the purpose of obtaining information about both market trends and customer orientation at once, one of the effective method to manage customer orientations is the one that use two-dimensional vectors of customer axis and oriented area axis. This method is used in many of Customer Relationship Management (CRM)[2] tools. However, this two-dimensional method is not capable of keeping track of market trends and customer orientations which change as time passes.

In this paper, we propose a method with three-dimensional vectors with time axis in additions to customer and oriented area axes. In three-dimensional vector space, enormous volume of information prevents of maintaining detailed information and slows down the information retrievals. The proposed method solves these problems by storing information of points of changes on customer orientations along with the passage of time and by storing the current situation in two-dimensional vector space.

2. Analysis of Customer Information and One-to-One Marketing

In the progress of IT in recent years, data mining [4] is one of the remarkable methods to analyze customer information . For example, CRM tools are used for storing in-house data, analyzing customer information, and working out customer-oriented business strategies.

Data mining includes various methods such as market basket analysis, memory based reasoning, cluster analysis, decision tree, and neural network. By using the clustering method to find similar records, you obtain the   initial data to start analyses of customer information to grasp market trends and customer orientations. In other words, the valuable information extracted from lump of data by in-house business analysts can be the starting point for planning business strategies.

Two-dimensional vector method with customer and oriented area axes is one of the effective method to grasp market trends and customer orientations from analyses of customer information, and is actually used in many CRM tools. This is because that the vector space in this method is pure image of markets and each customer axis is an individual customer, so that the orientations of individual customer can be kept track by looking at a certain customer axis along with its oriented area axis.

However, this method only allows to manage accumulated strength of orientations in oriented area axis, and it is difficult to keep track of changes in market trends and customer orientations as time passes. Therefore, if an oriented area aroused customers' interest much in the past but does not arouse much in lately, the oriented area still indicate high score. This means that even if an oriented area rapidly arouse customers' interest, it scores low at the initial stage and it is difficult to detect it. However, the highest buying inclination is indicated in the very early stage of oriented areas, which means that the changes of customer orientation must be detected accurately at its early stage to make sales promotions activities effective.

There is a model that considers passage of time among management methods in two-dimension vector space. This method decrements the value of each area element by multiplying a coefficient with the value of less than 1 in every certain period. This results a high scored oriented area in the past to be less scored if the area is losing interests from customers. In this method, oriented areas in the past are considered, however it is impossible to keep track the transitions of oriented areas.

Furthermore, the progress of IT in recent years enables the shift to one-to-one marketing which allows everybody to send and receive information. Since consumers, who may be customers, can obtain advertisements at their own will, conventional style of advertisement that the ads are kept in corporate side may not work well in future. This is why the advertisements to deal with one-to-one marketing is required, which grasp the interest of each one of customers.

One of the new style advertisements is the advertisement system that use WWW [3]. This system requires ad agency WWW server, in addition to advertisers' WWW servers and consumers' WWW clients. This server provide the combination of standard information and advertisements which corresponds to the information obtained by the WWW client just before the advertisement. In this method, the most recent oriented area is considered, however it is difficult to grasp market trends.

3. Information Management System of Market Trends and Customer Orientations

In this chapter, we describe the proposed information management system with three-dimensional vector space of customer, oriented area, and time axes, which is to obtain information of market trends and customer orientations required for data mining to make effective advertisement activities in one-to-one marketing.

3.1 Outline of System

Figure 1 illustrates the configuration of the customer orientation information management system and the outline of storing method for customer orientations. This system is to manage customer orientation information *V* to keep track transitions of customer oriented areas as time passes and to grasp market trends at the same time.

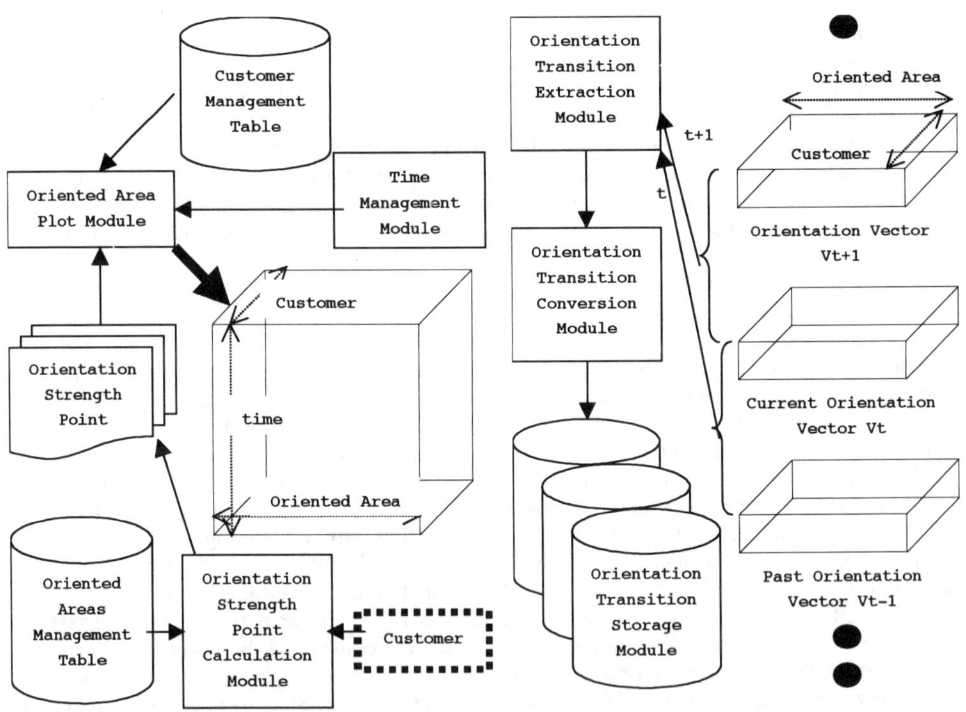

Fig. 1   Configuration of Customer Orientation Information Management System

Fig. 2   Storing Method for Orientation Field by Unit Time

Firstly, we calculate $S_c = \{\ x_1,\ x_2,\ ...\ \}$, a set of orientation strength point x, at the orientation strength point calculation module to keep track the observed behavior of customer c as the customer's orientation. The customer c is managed and maintained in a customer management table that has a list C of customers or customer groups as a database. Each element x of the set S_c is the orientation strength point that corresponds to the oriented areas management table, which maintains a list of oriented areas of customers.

Next, at the oriented area plot module, we plot customer orientation information $V_c = (\ c,\ S_c,\ t\ )$, where the customer c behaves at the time t with the orientation strength points S_c, on to the orientation field which composed of customer, oriented area, and time axes. The customer axis and the oriented area axis correspond to the customer management table and the oriented area management table respectively, and the time t is plotted on the time axis which is a series of unit time managed at the time management module.

3.2 Storing Orientation Field of Each Time-Series

The volume of information will be enormous in customer orientation information system when it is managed in three dimensions of customer, oriented area, and time axes. This proposed system separately manages time axis from the two-dimensional orientation field of customer and oriented area axes (see Figure 2). On the two-dimensional orientation field, the customer orientation information is expressed with orientation vectors. For example, the current customer orientation information is represented as a current orientation vector $V_t = (\ C,\ S_t,\ t\ )$, and the customer orientation information at one unit time past is represented as a past orientation vector $V_{t-1} = (\ C,\ S_{t-1},\ t-1\ )$.

Firstly, the orientation transition extraction module compares a current orientation vector V_t and a past orientation vector V_{t-1}, and extracts the orientation fields with different orientation strength. Then the orientation transition conversion module converts the extracted information into a coordinate point $(\ c,\ s\ )$ on the customer and the oriented area axes and a set orientation strength point x_{t-1} of the past orientation vector, and stores them into the orientation transition storage module. In this way, the past orientation vectors are stored in smaller volume.

When a unit time managed in time management module passes, the current orientation vector V_t in the one unit time past becomes the past orientation vector. Similarly, a new customer orientation strength point calculated based on the customer's behavior at the past unit time becomes a current orientation vector $V_{t+1} = (\ C,\ S_{t+1},\ t+1\ )$.

By repeating this procedure, the past orientation vectors at each unit time are stored in orientation transition module. Consequently, orientation information of unit time is stored as sets of current orientation vectors and orientation transitions in the past. This method enables us to store information for the customer orientation management system in small volume and without omission.

4. Implementation Example

This chapter describes the proposed system with an implementation example. In this example, the customers are the Internet users and the user of this system is an Internet Service Provider.

When a customer use the Internet, the orientation strength point calculation module extracts keywords from the browsed Web pages, looks up the oriented area management table with these keywords, and outputs the orientation strength point x of each element from the table. Then, the module calculates the current orientation vector V_{ct} based on the customer c, the current time t (the unit time when the customer browses the Web page), and a set of orientation strength points S_c. If it is difficult to execute the orientation strength points calculation module in real-time basis, it can be executed afterwards with the historical access records of customers.

Implementing an orientation field in a three-dimensional array requires enormous size of storage. For example, when you have one million customers and a thousand oriented areas, and if a unit time is one week and one year consist of fifty unit times, four bytes of an orientation strength point leads the total volume of an orientation filed to approximately 200 GB.

To prevent this, our method divides an orientation field by unit time and stores the most important current information as current orientation vectors V_t managed in two-dimension of customer and oriented area axes. Then only the differences between past orientation vectors V_{t-1} and current orientation vectors are stored in the orientation transition storage module to maintain the three-dimensional information in small size of storage. In this method, the values of differences and the two-dimensional coordinate positions are only the information to be stored using the orientation transition extraction and conversion modules. MPEG-4[5] can be implemented as a storing method.

The behavior information of customers enables variety of analyses. For example, transitions of customer orientations and market trends can be observed in various time passages by adjusting the length and the cycle of the unit time. It is also possible to link this system to existing statistics tools to make various analyses such as regression analysis and factor analysis.

5. Conclusion

In this paper, we have proposed a information management system to keep track of market trends and customer orientations that changes as time passes. This system accomplishes the environment to determine what is an effective advertisements or sales promotions and when they should be provided to markets and customers. The system enables us to keep track of customer orientations as dynamic and global changes, which is useful and efficient on planning effective business strategies based on transitions of customer orientations and its related factors.

As the importance of information systems in corporate activities grows, it is required to evaluate information systems appropriately. However, it is not easy to evaluate the every effect of information systems on corporate activities. As future work, we need to consider a method to evaluate information systems through customer satisfaction.

It is also important to consider the implementation of ISO/IEC International Standard IS15938(MPEG-7) [5], which is a contents representation standard to search, filter, manage, and process multi-media information. This should increase search efficiency on extracting information from various types of customer orientation data.

We also carry out investigations on information systems to analyze the customer orientation information stored in our proposed system.

References

[1] Moriarity,R.T. and Swartz,G.S., Automation to Boost Sales and Marketing, Harvard Business Review, January-February, pp.100-108, 1989.

[2] Morio Nagata, SFA: Software for Customer Satisfaction, IPSJ Magazine, Vol.40, No.5, pp.528-531, 1999.

[3] Youji Kohda and Susumu Endo, A New Advertising Business Framework in the 1:1 Marketing, IPSJ Journal, Vol.37, No.6, pp1235-1236, 1996.

[4] Michael J.A. Berry and Gordon Linoff, Data Mining Techniques: For Marketing, Sales, and Customer Suppport, John Wiley & Sons, Inc., 1997.

[5] http://www.cselt.stet.it/mpeg.

An Efficient method of Document Management for Sharing the Network Contents

Masafumi KOYAMA(Nara National College of Technology),
(22 Yata-cho, Yamato-Koriyama, Nara, Japan),
Masayuki KESSOKU (Matsushita Electric Industrial Co.,Ltd),
Hisatoshi MOCHIZUKI (Osaka Prefectural College of Technology),
Kazuhiko TSUDA (Tsukuba University)

Abstract. The importance of sharing information on corporate LAN system has been pointed out for years, and many information sharing systems are currently up and running. However, users are not really satisfied with the current systems because 1) it requires a lot of labor to register the data to be shared, and 2) it is difficult to find the registered information efficiently because of mistakes on classifications.

This paper proposes a information sharing system which has resolved these problems. This system allows users to register shared information only by dragging and dropping it into shared folders and automatically rearranges the registered information. To implement these features, we propose the methods to automatically extract and learn the keywords which used to be registered by users manually, to automatically cluster the learned keywords into categories, and to improve the search efficiency. We have evaluated our system using data of a search engine, Hi-Ho e-search, and have accomplished reproduction of 104 categories out of 125 (83.2%).

1. Introduction

The importance of sharing information on corporate LAN system has been pointed out for years, and many information sharing systems are currently up and running. However, not many cases have been succeeded at the viewpoint of sharing information, and even the succeeded cases often spend enormous amount of labor-hours. The possible causes of this situation are that: 1) a lot of labor is required to register the shared data with information sharing systems, and 2) it is difficult for users to find desired information because of mistakes on classifications of shared data.

This paper proposes a information sharing system that has resolved these problems by enabling users to drag and drop shared information into shared folders to register it, and by automatically rearranging the registered information. This system implements the methods 1) to automatically extract and learn the keywords for the data to be registered, 2) to automatically cluster the learned keywords and rearrange the shared data, and 3) to determine effective keywords automatically to improve the search efficiency. By automating these operations, we have solved the problems caused by the differences on criteria among the users when they classify the data, and we have accomplished classifications under the unified criterion. We also expect extensive improvements on maintaining shared data efficiently.

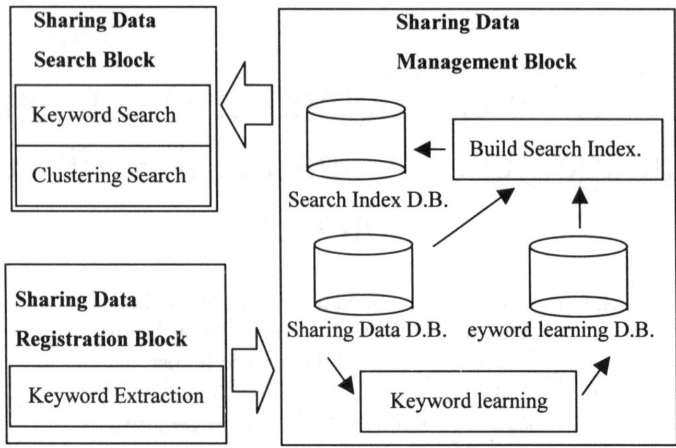

Figure 1. Configuration of Information Sharing System.

2. Configuration of Information Sharing System

This chapter describes the configuration of the proposed information sharing system. This system is composed of three blocks: Registration, Management, and Search blocks, as shown in Figure 1.

2.1 Shared Data Registration Block

Up to now, keywords are manually assigned to text. This way takes long time and huge costs, and moreover the quality is not constant because different person register keywords with different criteria. Here we propose an automatic keywords extraction system. This system extracts keyword candidates from the target text in the Registration block and evaluates these candidates in the Management block. In here, the Registration blocks are distributed to each user and operates separately with the Management block. This is because the data is originally distributed and processing it centrally is not efficient in terms of network traffic and capacity.

Firstly, a user registers a file to be shared by dragging and dropping it to the icon which indicates the Registration block.

Next, keywords are extracted from the shared file in the Registration block. In our method, nouns are selected as keywords, using morphological analysis to separate the text into words and to determine their part of speech. Selecting every noun as keywords lowers the accuracy because irrelevant words are also selected. We eliminate these irrelevant words using morphological dictionary which contains a keyword list and a stop-word list [1]. This keyword list does not contain compound words which will be estimated after morphological analysis.

Lastly, only the keywords and file indexes information of registered data are sent to Management block. The keywords are nouns, such as the words contained in the keyword list and the compound words which are not contained in the stop-word list.

2.2 Shared Data Management Block

Management block is composed of Shared Data Management D.B., Keyword Learning D.B., and Shared Data Search Information. Shared Data Management D.B manages the keywords and file index sent from each user.

Keyword Learning D.B. evaluates keyword candidates. The words used in text have mutual semantic connectivity and they form a word group to describe a topic. We classify the relationships between the words (come before/come after, synonym/quasi-synonym, and categories) within the word group related to a topic, and then we set keyword values for each word. Keyword Learning D.B. manages these keyword values.

Using the Shared Data Management D.B. and the Keyword Learning D.B., we construct Shared Data Search Information. To construct search information which achieves fast keyword searches, we cluster the keywords managed by the Shared Data Management D.B based on the keyword values managed by the Keyword Learning D.B., and then we store the file indexes with the same keyword value in the neighborhood within the file. Then we append filed information of the file indexes to each file index.

To improve the accuracy of the clustering, the keywords within the Keyword Learning D.B. are updated and sophisticated with keywords received from the Shared Data Management D.B.

2.3 Shared Data Search Block

One of the problems of information search is lack of user-entered keywords [3]. One of the other problems is lexical inconsistency, which is that the entered keyword is used in shared data with different semantics or the desired information is indicated with other words than the entered keywords.

This proposed system makes clustering searches using the Shared Data Search Information which is clustered within the Keyword Learning D.B. This is to support acquisition of keywords that specifically represent users' interests from existing shared data. The files of shared data are registered in directory structure based on the categories they belong to. The search result contains not only the corresponding shared data, but also the keywords within the same category, the neighborhood keywords, and main contents of the sets of acquired shared data. Based on these information, the more appropriate keywords are provided for users. Users repeat searches by adding the keywords related to the desired information or by selecting the indicated categories. The Shared Data Search Information is constructed from Shared Data Management D.B. so that the conventional keyword search is also available.

3. Configuration of Information Sharing System

Keywords in Keyword Learning D.B. are updated with combination and division. For example, as the volume of contents associated with a category become larger, the variance of the set of keywords increases (the relation between keywords and its category become ambiguous). When this happens, enormous number of matches are found in clustering searches and it gets more difficult to find the contents that are really desired by users. Therefore, when the variance of the set of keywords exceeds a certain threshold, the category is divided into sub-categories using subordinate keywords of the current keyword (see Figure 2). For example, when "baseball" category becomes too large, it may be divided into sub-categories of "high school baseball" and "professional baseball". Subordinate keywords can be compound words, or "affiliation - dependence" ("Mariners - Ichiro") relation or "agent - object" ("Mariners - Winning the championship") relation in case relation. The word which divides the category into sub-categories with the best balanced variances will be selected [6].

The combination is required when the sets of keywords of two sub-categories divided from different categories are getting similar. Today, the compound area between various field are studied aggressively, which are represented by the words such as "hybrid" or

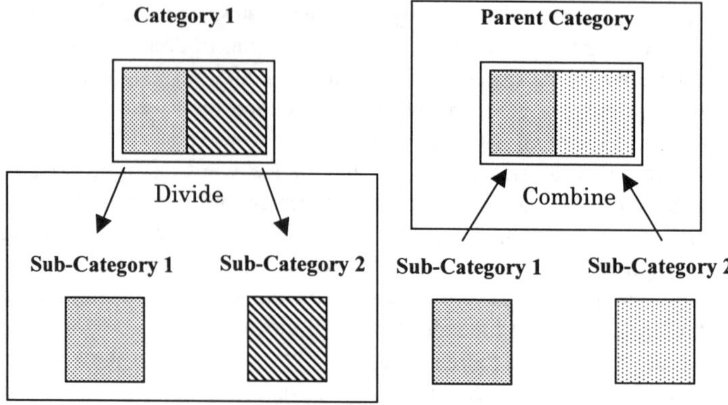

Figure 2. Updating Keyword learning D.B.

"barrier free". This lowers the wall between fields, and consequently, it is essential to have a method to generate intermediate category or to combine categories. In this paper, we use a similar method to the above stated division method to integrate multiple categories, which generate a new category as a superordinate category by deciding superordinate words based on commonality or case relation between keywords.

The above stated methods enable users to narrow down the search candidates efficiently. A method use search history of a user to suggest narrowed down candidates[7], however it has problems such as that:

   - it is passive method and it takes time until the learning effect become visible, and

   - the evaluation criteria for successful histories are ambiguous.

On the other hand, our proposed method determines candidate words automatically based on variances of keywords and excels the conventional methods on its quick effect and clearer selection criteria. Furthermore, this method is highly flexible because it uses case relation and is easy to be applied to keywords other than nouns.

4. Evaluation of Information Sharing System

This chapter evaluates the clustering of Keyword Learning D.B. We use a search engine "Hi-Ho e-search"[4]. Hi-Ho e-search is directory type and consists of category names and Web pages that belong to each category.

Firstly, we have extracted keywords from descriptive text on every Web page for 125 categories. One or two keywords are extracted from each Web page and we constructed a Keyword Learning D.B. from these keywords. We have compared the result of clustering on the Keyword Learning D.B. with the 125 categories of Hi-Ho e-search, and have found that 104 categories are matched, which is 83.2 % of reproduction.

We have conducted preparative experiment on evaluating keyword learning using category division rules of 5 general rules and 20 grammatical rules which compose superordinate/subordinate relation or compound words relation. We have used a method to determine thresholds of divisions which achieve lowest variances with the assortments of frequently appeared keywords. The experiment have been conducted on approximately 15,000 documents and have resulted 23% decrease of variances within each category.

5. Conclusion

This paper has proposed a system to share data on network easily and to automate the clustering of shared information, and has described the result of experiments.

As a future work, we study how to reflect Keyword Learning D.B. to keywords extraction, and we further improve the learning accuracy and the search efficiency of Keyword Learning D.B. using ambiguous words or narrow down information.

Acknowledgements

This research is conducted under the support of Science Research Fund No.12750370 (Intelligent Document Search Method Using Narrow Down Information) from Ministry of Education, Culture, Sports, Science and Technology.

References

[1]   Tetsuya Ishikawa: Japanese Morpholigical and Syntactic Analyzer 'JEMONI' and its Evaluation, Journal of IPSJ,   Vol.32,No.2,pp.220-228 ,1991.(in Japanese)

[2]   Haruo Kimoto: Automatic Indexing and Evaluation of Keywords, Transactions of Natural Language, Transactions of IEICE, Vol.J-74-D-1,pp.556-566,   1991.(in Japanese)

[3]   Wataru Sunayama, *et al.*: Discovery of Chances for Entertainment on Potential Interests,IEEE Proc. of International Conference on Industrial Electronics, Control and Instrumentation(IECON 2000), Nagoya, pp.1648 -- 1651, 2000.

[4]   seach engine Hi-Ho e-search,(URL) http://panasearch.hi-ho.ne.jp/.

[5]   Hisao Mase *et al.*: Experimental Simulation for Automatic Patent Categorization, Transactions of IPSJ, Vol.39, No.7, pp.2207-2216 ,   1998. (in Japanese)

[6]   Ch. Fillmore: A case for case, In E. Bach and R. T. Harms, editors, Universals in Linguistic Theory, pages 1-88. Holt, Rinehart and Winston, New York, 1968.

[7]   Oda, Minami:An Recomendetion Method of Web Documents by The History of References, Transactions of 56th the national conference of IPSJ, pp.203-204, 1998.(in Japanese)

KES '01
N. Baba et al. (Eds.)
IOS Press, 2001

An Improved Approximation of the Error Function of Recurrent Neural Networks for the Real-Time Learning

Keiji Tatsumi†, Yuji Kitano†, Tetsuzo Tanino†, Masao Fukushima‡
† *Graduate School of Engineering, Osaka University,*
Yamada-Oka 2-1, Suita, Osaka 565-0871, Japan
‡ *Graduate School of Informatics, Kyoto University,*
Kyoto 606-8501, Japan

Abstract. Since in real-time leaning methods for recurrent neural networks, the weights is updated at each time, an exact error function and outputs of networks cannot be obtained. Thus, its approximation is often used. When the gap between the exact error function and its approximation widens, the convergence of this method often becomes very slow. Thus, in this paper, we propose a new neural model which can use a closer approximation of the error function by the recalculation of outputs. In addition, we verify its efficiency through some numerical experiments.

1 Introduction

Recurrent neural networks(RNN) have the great capability of handling the temporal signals. They have been successfully applied to many problems involving sequence recognition, sequence reproduction or temporal association[3]. The method of training RNN to obtain desirable properties have been developed in many fields[1, 4, 5, 6, 7]. These methods can be roughly classified in two categories. One is called batch (off-line) training method, which updates the weights after RNN receives all temporal signals. The other is called real-time (on-line) training method, which updates the weights for every current signal. In such case as the adaptive control or the speech recognition, the length of an input or output sequence is often very long or unknown in advance[6]. Then, the real-time training are more suitable.

However, in the real-time training the error function or outputs cannot be exactly obtained since the weight is updated at each time. These methods use an approximate error function or outputs. Thus, because of the gap between an exact value and its approximation, the convergence of this method often becomes very slow. In this paper, we propose the new model of RNN and an improved approximation of error function and outputs of RNN. Further, we verify the efficiency of the proposed model through some numerical experiments.

2 Recurrent Network

In this paper, we consider the general recurrent network (RNN)[3]. In this model, all pairs of units have connections with (synaptic) weights. In addition, there exists a connection even from a unit to itself. Unit j is governed by

$$y_j(t-1) \;\; := \;\; f(h_j(t-1)), \tag{1}$$

$$h_j(t-1) \;\; := \;\; \sum_{i}^{N} w_{ij} y_i(t-1) + x_j(t-1), \tag{2}$$

$$t = 1, 2, \ldots$$

where w_{ij} denotes the weight from unit i to unit j, $y_j(t)$ is the output from unit j at time t and $x_j(t)$ is the external input into the unit j at time t. An initial output $y_j(0)$ of unit j is set to be an appropriate value. The output function f is the logistic function; namely $f(u) := (1 - e^{-\beta u})/(1 + e^{-\beta u})$. Moreover, if there is no input signal, let $x_j(t) := 0$.

The learning of this network is to realize a pair of time sequences, that is an input sequence $x_j(t), t = 0, \ldots, T, j = 1, \ldots, N$, and a desirable output sequence $d_j(t), t = 1, \ldots, T, j \in O(t)$. Here, $O(t)$ is the set of units to which the desirable output $d_j(t)$ is given at time t. Note that for the sake of convenience, we consider the sequcence whose length T is known.

To measure a gap between the desirable outputs and the outputs of RNN obtained by (1) and (2), we define the output error function $E(w)$ by

$$E(w) \;\; := \;\; \sum_{t=1}^{T} \sum_{k \in O(t)} E_k^t(w) \;\; := \;\; \sum_{t=1}^{T} \sum_{k \in O(t)} \frac{1}{2} e_k^t(w)^2, \tag{3}$$

$$:= \;\; \sum_{t=1}^{T} \sum_{k \in O(t)} \frac{1}{2} (y_k(t) - d_k(t))^2. \tag{4}$$

The learning of RNN can be formulated as the problem to minimize $E(w)$. As a real-time methods to solve this minimization problem, Real-Time Recurrent Learning(RTRL) and Extended Kalman Filter(EKF) method are often used[4, 5, 6].

3   Real-Time Learning

In this section, we show the approximation used in real-time learning for a conventional RNN. Since a real-time learning updates weights at each time t, the output of each unit is obtained by the following equalities instead of (1) and (2);

$$y_j(t) \;\; := \;\; f(h_j(t-1)), \tag{5}$$

$$h_j(t-1) \;\; := \;\; \sum_{i=1}^{N} \phi_{ij}^n(t-1)\, y_i(t-1) + x_j(t-1), \tag{6}$$

where the weights at iteration n and at time t is represented by $\phi_{ij}^n(t)$. As an approximation of $\partial e_k(t)/\partial w_{ij}$, we use

$$\sum_{k \in O(t)} e_k^t(\phi^n(t-1))\, g(y_k(t))\, \rho_{ij}^k(t-1), \tag{7}$$

where $g(v) := \beta(1 - v^2)/2$, since f is the logistic function and $f'(u) = \beta(1 - f(u)^2)/2$. $\rho_{ij}^k(t)$ is an approximation of $\partial h_k(t)/\partial w_{ij}$ and defined as follows;

$$\rho_{ij}^k(t) := \begin{cases} 0, & \text{if } t = 0, \\ \delta_{kj} y_i(t) + \sum_{l} \phi_{lk}^n(t) g(y_k(t)) \rho_{ij}^l(t-1), & \text{if } t \geq 1. \end{cases} \tag{8}$$

The real-time training method can be summarized in the following algorithm.

Real-Time Training method, RTT

Step 0 Select initial weights w^0 and $n := 0$.

Step 1 Put $\phi^n(0) := w^n$ and $t := 0$.

Step 2 Compute $y(t+1)$ by (5) and (6).

Step 3 Update weights $\phi^n(t)$ by some procedure utilizing an approximate gradient
(7) and we have $\phi^n(t+1)$.

Step 4 If $t = T$, then put $\phi^n(0) := \phi^n(T+1)$ and go to **Step 5**.

     Else, let $t := t+1$ and go to **Step 2**.

Step 5 If $\|\nabla E(w^{n+1})\| \le \varepsilon$, stop. Otherwise, let $n := n+1$ and go to **Step 1**.

Now, we focus on the approximation of outputs in this algorithm. Although $\phi^n(t-1)$ is updated in Step 3 so as to minimize $\sum_{j \in O(t)}(y_j(t) - e_j(t))^2$ at time t, $y(t+1)$ is calculated by using the previous $y(t)$ at time $t+1$, which is not updated with new $\phi^n(t)$. Thus, the approximation of outputs based on (5) and (6) may not be efficient.

In next section, we propose a new neural model in which $y(t)$ is calculated with updated $\phi^n(t)$ at time $t+1$, again. This model is expected to improve the approximation of the error function or outputs in the real-time training method.

4 New Model

In proposed model, as an approximate output vector of $y(t)$, two vector $y^o(t)$ and $y^c(t)$ are computed as follows;

$$y_j^o(t) := \begin{cases} y_j(0), & \text{if } t = 1, \\ f\left(\sum_{i=1}^{N} \phi_{ij}^n(t-1)\, y_i^o(t-1) + x_j(t-2)\right), & \text{if } t \ge 2, \end{cases} \tag{9}$$

$$y_j^c(t) := f\left(\sum_{i=1}^{N} \phi_{ij}^n(t-1)\, y_i^o(t) + x_j(t-1)\right), \quad t \ge 1, \tag{10}$$

where $y_j(0)$ is an initial value of unit j.

First, at time t, the output $y^c(t)$ is calculated in a way similar to the conventional model. By $y^c(t)$, we have an approximate gradient;

$$\sum_{k \in O(t)} e_k^t(\phi^n(t-1))\, g(y_k^c(t))\, \sigma_{ij}^k(t-1), \tag{11}$$

where $e_k^t(\phi) := y_k^c(t) - d_k(t)$ and $\sigma_{ij}^k(t)$ is defined as follows;

$$\sigma_{ij}^k(t) := \begin{cases} 0, & \text{if } t = 0, \\ \delta_{kj} y_i^o(t) + \sum_{l} \phi_{lk}^n(t)\, g(y_i^o(t))\, \sigma_{ij}^l(t-1), & \text{if } t \ge 1. \end{cases} \tag{12}$$

$\sigma_{ij}^k(t)$ is an approximation of $\partial h_k(t)/\partial w_{ij}$, which corresponds with $\rho_{ij}^k(t)$ in the previous section. Then, weights $\phi_{ij}^n(t-1)$ is updated similarly to RTT with an approximate gradient (11). Next, at time $t+1$ as the updated approximate output of $y(t)$, we have $y^o(t+1)$, which is calculated with the updated weights $\phi_{ij}^n(t)$, again.

The recalculation of outputs can be expected to improve the approximation of the error function or outputs. Moreover, it requires only the computation of (9) in addition to conventional computational complexities. Because the dominated computation (12) in proposed model is exactly equivalent to (8) in the conventional model. Therefore, this model is expected to accelerate the real-time training method.

Further, the real-time training method for proposed model theoretically has the almost same properties as that for conventional model. For example, by a similar proof in [2], it can be easily verified that RTRL for proposed model has global convergence properties under the mild conditions.

5 Simulation Results

In this section, we report numerical results with RTRL methods for the conventional and proposed model. RTRL can be regarded as RTT with the following update;

$$\phi_{ij}^n(t+1) := \begin{cases} \phi_{ij}^n(t) - \eta^n \sum\limits_{k \in O(t)} e_k^t(\phi^n(t))\, g(y_k(t+1))\, \rho_{ij}^k(t), & \text{conventional model} \\ \phi_{ij}^n(t) - \eta^n \sum\limits_{k \in O(t)} e_k^t(\phi^n(t))\, g(y_k^c(t+1))\, \sigma_{ij}^k(t), & \text{proposed model} \end{cases} \quad (13)$$

where η^n is learning rate. We used the Elman-type partial recurrent network[3] having one input unit, one output unit, 16 hidden units and 16 context units. We employed the termination criterion; $E(w) \leq 0.1$. If this termination criterion is not satisfied within 10000 iterations, we judged that the learning was unsuccessful. To measure the efficiency of the algorithms, we consider two indices; the rate of success and the average CPU time. The former is the ratio of the successful trials to the whole trials using different initial weights and the latter is the average CPU time of the successful trials.

We conducted preparatory experiments to find the suitable learning rate η and the suitable parameter β in the logistic function. Initial weights were randomly chosen from 5 intervals, $(-0.1, 0.1)$, $(-0.3, 0.3), \ldots, (-0.9, 0.9)$. As a test sequence, we generated a sequence of 20 points with the sine function; $z(t) := \sin(0.6\, t)$, $t = 0, \ldots, 20$. Test Problem is to predict one-time-ahead of the sequence $\{z(t)\}$.

The results are summarized in Table 1, where "CPU sec" and "# Ite." respectively mean the average CPU time in seconds and the average number of iterations for 300 trials using different initial weights. From this table, we see that RTRL for proposed model outperformed RTRL for conventional model about the rate of success. Further, the former requires less CPU time and iterations than the latter, when they have the almost same rate of success.

6 Conclusion

In this paper, we proposed a new model of RNN with recalculation of outputs, which improves the approximation of error function and outputs. Further, through some numerical simulations we verified the efficiency of the proposed model.

Table 1: RTRL

β	Initial weights	Proposed model			Conventional model		
		Rate of Success	# Ite.	CPU sec	Rate of Success	# Ite.	CPU sec
2	0.1	0.067	8261.5	120.0	0.000	-	-
	0.3	0.067	8485.0	189.0	0.000	-	-
	0.5	0.000	-	-	0.000	-	-
	0.7	0.000	-	-	0.000	-	-
	0.9	0.067	7178.5	109.5	0.000	-	-
2.5	0.1	0.267	8011.1	110.5	0.000	-	-
	0.3	0.400	7243.7	131.8	0.133	6671.8	104.2
	0.5	0.167	6723.2	192.2	0.100	8484.7	129.0
	0.7	0.200	5325.8	129.0	0.100	6505.3	180.3
	0.9	0.233	5821.0	129.6	0.133	6192.8	73.0
3	0.1	0.433	6678.5	108.1	0.400	6776.9	132.8
	0.3	0.367	5450.4	98.4	0.333	7426.9	111.6
	0.5	0.400	7184.4	169.8	0.233	6390.6	94.3
	0.7	0.367	6490.9	164.3	0.300	7857.7	149.0
	0.9	0.500	5722.6	99.5	0.400	5944.8	114.1

Although we focus on the real-time training method, the proposed model can be applied to the hybrid method of the real-time and batch training such as truncated Back Propagation Through Time[7].

References

[1] A.F. Atiya and A.G. Parlos, New results on recurrent network training: unifying the algorithms and accelerating convergence, IEEE Transactions on Neural Networks, **11** (2000) 697–709.

[2] A. A. Gaivoronski, Convergence properties of backpropagation for neural nets via theory of stochastic gradient methods. part 1, Optimization Methods and Software, **4** (1994) 117–134.

[3] J. Hertz, A. Krogh and R.G. Palmer, Introduction to the Theory of Neural Computation, Addison-Wesley, California (1991).

[4] M.W. Mak, K.W. Ku and Y. L. Lu, On the improvement of the real time recurrent learning algorithm for recurrent neural networks, Neurocomputing, **24** (1999) 13–36.

[5] P. Paolo, A. Uncini, F. Piazza and B.D. Rao, On-line learning algorithms for locally recurrent neural networks, IEEE Transactions on Neural Networks, **10** (1999) 253–271.

[6] G.V. Puskorius and L.A. Feldkamp, Neuroconttrol of nonlinear dynamical systems with Kalman filter trained recurrent networks, IEEE Transactions on Neural Networks, **5** (1994) 279–297.

[7] R.J. Williams and J. Peng, An efficient gradient-based algorithm for on-line training of recurrent network trajectories, Neural Computation, **2** (1990) 490–501.

On structure of the residual Jacobian matrix arising in neural-network learning[*]

Eiji Mizutani [1] and   James W. Demmel [2]

eiji@wayne.cs.nthu.edu.tw, demmel@cs.berkeley.edu

1) *Dept. of Computer Science, National Tsing Hua University, Hsinchu 300 Taiwan*
2) *Mathematics and Computer Science, University of California, Berkeley, CA 94720 USA*

Abstract. This paper describes structural features of the Jacobian matrix of the residual vector arising in what we call *neural networks nonlinear least squares problems* for optimizing *neural networks* or *neuro-fuzzy* models. In particular, we shall characterize a sparsity structure of the residual Jacobian matrix that can be found in multiple-output adaptive network models, and describe "layer-independent optimization" algorithms that exploit the special structure. The analysis would be useful to develop efficient algorithms.

1 Introduction

We consider a problem to fit an adaptive *neural network* (NN) (including a *neuro-fuzzy* model) to observations (i.e., a collective data set of many pairs of an input vector and its desired vector) subject to errors, leading to *neural networks nonlinear least squares problems*. We assume our NN model has totally n parameters (denoted by a vector $\boldsymbol{\theta}$ to be optimized) in an N-layered architecture with P_N final output nodes (at terminal layer N); so, the objective function, the "sum of squared errors" measure, can be written as: $E(\boldsymbol{\theta}) = \frac{1}{2}\sum_{q=1}^{d}\sum_{k=1}^{P_N} r_{k,q}^2 = \frac{1}{2}\sum_{q=1}^{d}(\mathbf{a}_q^N - \mathbf{t}_q^N)^T(\mathbf{a}_q^N - \mathbf{t}_q^N) = \frac{1}{2}\mathbf{r}^T\mathbf{r}$, where d denotes the total number of data; $r_{k,q}$ is the kth residual for datum q; $\mathbf{a}_q^N$ is the model's output vector at terminal layer N; $\mathbf{t}_q^N$ is the target output vector; and $\mathbf{r}$ denotes the residual vector composed of r_i, $i=1,\cdots,m$ $(\stackrel{\text{def}}{=} d \times P_N)$. The gradient vector and Hessian matrix of $E(.)$ are given by $\mathbf{g} = \mathbf{J}^T\mathbf{r}$, and $\mathbf{H} = \mathbf{J}^T\mathbf{J} + \mathbf{S}$, respectively, where $\mathbf{J}$ is the $m \times n$ Jacobian matrix of the residual vector $\mathbf{r}$ and $\mathbf{S}$ is a matrix of second-derivative terms. In the following sections, we shall illustrate a sparsity structure found in the residual Jacobian matrix $\mathbf{J}$ using a general multiple-output "multilayer perceptron" (MLP) architecture that has totally N layers (including the first input layer) with $P_s(s = 1, ..., N)$ nodes per layer; hence, the total number of parameters: $n \stackrel{\text{def}}{=} \sum_{s=1}^{N-1} P_{s+1}(P_s + 1)$. We then explain two *layer-independent optimization* methods for adaptive-network learning in a *direct dogleg trust-region algorithm* [5]. Although we shall describe them using an MLP as a concrete example, our descriptions are applicable to many adaptive NN models.

2 Structural Sparsity in Multiple-Output Models

When we consider the residual Jacobian matrix $\mathbf{J}$ associated with a *multiple-output* MLP, we observe a certain *sparsity* structure. The size of sparsity is dependent upon the structure of the MLP; hence, what we call *structural sparsity*. On the other hand,

---

[*] The work was supported in part by the Ministry of Education, Taiwan R.O.C. (grant 89-E-FA04-1-4).

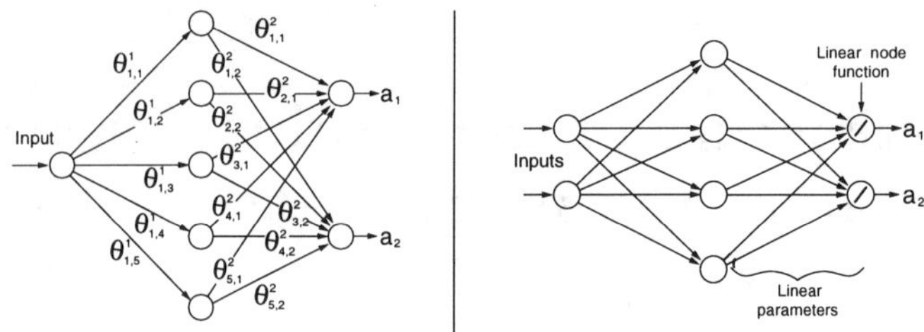

Figure 1: A "1-5-2" MLP architecture (left), wherein our notation θ_{ij}^s denotes a parameter between node i at layer s and node j at layer $(s+1)$, and a "2-4-2" NN model (right), wherein the final output nodes have *linear identity* functions; hence, the parameters between the final output layer and the hidden layer are "linear," whereas the other parameters are "nonlinear." If the hidden nodes have Gaussian-type radial basis functions that may consist of "nonlinear" parameters, then the model is a standard radial basis function network (RBFN).

the Jacobian matrix obtainable from a *single-output* MLP is basically *dense* with no such structural sparsity. As a concrete example, we consider a 1-5-2 MLP that has totally 22 parameters ($n=22$) [see Figure 1(left), wherein no threshold parameters are displayed for simplicity]. On each datum q, two residuals $r_{k,q}$ ($k=1$, 2) are computed, and the corresponding two rows of the Jacobian matrix $\mathbf{J}$ are obtained as shown next:

	$\theta_{1,1}^2$	$\theta_{2,1}^2$	$\theta_{3,1}^2$	$\theta_{4,1}^2$	$\theta_{5,1}^2$	$\theta_{6,1}^2$	$\theta_{1,2}^2$	$\theta_{2,2}^2$	$\theta_{3,2}^2$	$\theta_{4,2}^2$	$\theta_{5,2}^2$	$\theta_{6,2}^2$
row 1 (associated with $r_{1,1}$)	×	×	×	×	×	×	0	0	0	0	0	0
row 2 (associated with $r_{2,1}$)	0	0	0	0	0	0	×	×	×	×	×	×

Notice that "×" denotes some non-zero values, and we omitted the *dense* part associated with $\boldsymbol{\theta}^1$ [i.e., parameters $\theta_{i,j}^1(i = 1, 2;\ j = 1, ..., 5)$ between the first input layer and the next hidden layer]. (In the table, $\theta_{6,1}^2$ and $\theta_{6,2}^2$ denote threshold parameters.) Here, we observe *sparsity* in this part of $\mathbf{J}$ because, in general, the parameters θ_{ij}^{N-1} that have no direct connections to node k ($j \neq k$) have nothing to do with the residual $r_{k,q}$ generated at node k at layer N for any datum q. The sparsity pattern will be clear in the next 6×22 $\mathbf{J}$ (associated with three data) whose rows are permutated appropriately:

$$
\underbrace{\mathbf{J}}_{6\times 22} =
\begin{bmatrix}
\times & \times & \times & \times & \times & \times & \times & \times & \times & \times & \times & \times & \times & \times & \times & \times & 0 & 0 & 0 & 0 & 0 & 0 \\
\times & \times & \times & \times & \times & \times & \times & \times & \times & \times & \times & \times & \times & \times & \times & \times & 0 & 0 & 0 & 0 & 0 & 0 \\
\times & \times & \times & \times & \times & \times & \times & \times & \times & \times & \times & \times & \times & \times & \times & \times & 0 & 0 & 0 & 0 & 0 & 0 \\
\times & \times & \times & \times & \times & \times & \times & \times & \times & \times & 0 & 0 & 0 & 0 & 0 & 0 & \times & \times & \times & \times & \times & \times \\
\times & \times & \times & \times & \times & \times & \times & \times & \times & \times & 0 & 0 & 0 & 0 & 0 & 0 & \times & \times & \times & \times & \times & \times \\
\times & \times & \times & \times & \times & \times & \times & \times & \times & \times & 0 & 0 & 0 & 0 & 0 & 0 & \times & \times & \times & \times & \times & \times
\end{bmatrix}.
$$

In this matrix, the first half of rows all correspond to the residual $r_{1,q}$ ($q=1, \cdots, 3$) produced at the first output node, while the latter half are associated with the residual $r_{2,q}$ at the second output node. Regardless of the number of data, this sparse matrix can be written generally for a multiple P_N-output NN with N layers in the next block form:

$$
\underbrace{\mathbf{J}}_{m\times n} =
\begin{bmatrix}
\mathbf{B}_1 & \mathbf{A}_1 & & & \mathbf{0} \\
\mathbf{B}_2 & & \mathbf{A}_2 & & \\
\vdots & & & \ddots & \\
\mathbf{B}_{P_N} & \mathbf{0} & & & \mathbf{A}_{P_N}
\end{bmatrix},
\tag{1}
$$

where $\mathbf{A}_k(k = 1, \cdots, P_N)$ is a $d \times (P_{N-1} + 1)$ matrix and $\mathbf{B}_k(k = 1, \cdots, P_N)$ is a $d \times n_{dense}$ matrix associated with the residual vector $\mathbf{r}_k$ (at the kth final output node) wherein n_{dense}, the number of parameters located between the first input layer and layer $N-1$ (the last hidden layer), is defined as: $n_{dense} \overset{\text{def}}{=} n - P_N(P_{N-1}+1)$. The first "$n_{dense}$ columns" of $\mathbf{J}$ are the dense part. In a special case where the final output nodes have *linear identity functions* [see Figure 1(right)], all $\mathbf{A}_k$ matrices become the same: $\mathbf{A}_1 = \mathbf{A}_2 = \cdots = \mathbf{A}_{P_N}$. We shall discuss how to exploit this feature in the next section.

Using the above Jacobian matrix, the well-known *Gauss-Newton method* (for $m > n$) can be written as the *Jacobian system*: $\mathbf{J}\Delta\boldsymbol{\theta}_{GN} = -\mathbf{r}$, which is important in nonlinear least squares sense because a pure or *regularized* Gauss-Newton step $\Delta\boldsymbol{\theta}_{GN}$ plays an important role in a trust-region algorithm that attempts to solve a constrained problem: "$\min_{\Delta\boldsymbol{\theta}} \ \|\mathbf{r} + \mathbf{J}\Delta\boldsymbol{\theta}\|_2$ subject to $\|\Delta\boldsymbol{\theta}\|_2 \leq R$ (trust-region radius)" when $\mathbf{S}$ is simply omitted from $\mathbf{H}$. The (modified) Gauss-Newton step, the solution to the *Jacobian system* $\mathbf{J}\Delta\boldsymbol{\theta}_{GN} = -\mathbf{r}$, may be used to form a *dogleg trajectory* to obtain an approximate solution to the posed constrained problem (see refs. [6, 5] for more details). Following this line, we focus on determination of $\Delta\boldsymbol{\theta}_{GN}$ in the remainder. The associated *normal system* can be given by: $\mathbf{J}^T\mathbf{J}\Delta\boldsymbol{\theta}_{GN} = -\mathbf{J}^T\mathbf{r}$, which can be more specifically written as:

$$\begin{bmatrix} \sum_{k=1}^{P_N} \mathbf{B}_k^T\mathbf{B}_k & \mathbf{B}_1^T\mathbf{A}_1 & \cdots & \mathbf{B}_{P_N}^T\mathbf{A}_{P_N} \\ \hline \mathbf{A}_1^T\mathbf{B}_1 & \mathbf{A}_1^T\mathbf{A}_1 & & \mathbf{0} \\ \vdots & & \ddots & \\ \mathbf{A}_{P_N}^T\mathbf{B}_{P_N} & \mathbf{0} & & \mathbf{A}_{P_N}^T\mathbf{A}_{P_N} \end{bmatrix} \Delta\boldsymbol{\theta}_{GN} = - \begin{bmatrix} \sum_{k=1}^{P_N} \mathbf{B}_k^T\mathbf{r}_k \\ \hline \mathbf{A}_1^T\mathbf{r}_1 \\ \vdots \\ \mathbf{A}_{P_N}^T\mathbf{r}_{P_N} \end{bmatrix}. \tag{2}$$

We observe "sparsity" structure (i.e., zero-matrix blocks) even in the Gauss-Newton model Hessian $\mathbf{J}^T\mathbf{J}$. Clearly, if we pay attention to its sparsity structure, then we can form $\mathbf{J}^T\mathbf{J}$ at a smaller cost than mn^2, the dominant cost in the usual normal-equation approach when $m > n$. More specifically, P_N rows of $\mathbf{J}$ are computed on datum q via *backpropagation* process [7], and each row i ($i = q \times k$, where $k = 1, \cdots, P_N$) can be stored into a vector $\mathbf{u}_i$ that has length $n_u [\overset{\text{def}}{=} n_{dense} + (P_{N-1}+1)]$, smaller than n (total number of parameters), by excluding $(P_N-1) \times (P_{N-1}+1)$ zero components (due to the structural sparsity) from the row vector; hence, the total cost for forming $\mathbf{J}^T\mathbf{J}$ reduces to mn_u^2. (For the cost analysis of backpropagation, refer to ref. [7].)

3 Layer independent optimization

In the NN literature, many approximate Jacobian (or Hessian) methods have been discussed: One of them is to optimize the parameters separately "layer by layer." In a three-layered MLP, for instance, this corresponds to *splitting*

$\diamond$ the column space of $\mathbf{J}$ into $[\ \mathbf{B}\ |\ \mathbf{A}\ ]$ (where $\mathbf{B}$ is the *dense* part);

$\diamond$ the parameter space $\boldsymbol{\theta}$ into $\begin{bmatrix} \boldsymbol{\theta}^B \\ \hline \boldsymbol{\theta}^A \end{bmatrix}$ (hence, $n = n_B + n_A$ and $n_B = n_{dense}$); and

$\diamond$ the Jacobian system for the Gauss-Newton step into $\mathbf{B}\Delta\boldsymbol{\theta}_{GN}^B = -\mathbf{r}$ and $\mathbf{A}\Delta\boldsymbol{\theta}_{GN}^A = -\mathbf{r}$. This layer-independent optimization might be of great use when *the final output nodes are linear identity functions*, because $\boldsymbol{\theta}^A$ becomes a *linear parameter** vector; here, we describe two different algorithms in a dogleg trust-region framework:

$\bullet$ *One-step (simultaneous) update method:* solve $\begin{cases} \mathbf{A}\Delta\boldsymbol{\theta}_{GN}^A = -\mathbf{r} \\ \mathbf{B}\Delta\boldsymbol{\theta}_{GN}^B = -\mathbf{r} \end{cases}$ simultaneously to form a

---

*The idea behind Eq.(3) is well-known as *iterative refinement* (see pages 60 and 134 in Demmel [2]; Section 2.9 in Björck [1]) in the linear least squares context. If the nonlinear parameters $\boldsymbol{\theta}^B$ are nearly optimal and thus can be fixed, then the process in Eq.(3) can be readily repeated as an *inner iteration* ($\boldsymbol{\theta}_{\text{now}}^A \leftarrow \boldsymbol{\theta}_{\text{next}}^A$) once matrix $\mathbf{A}$ is factored.

dogleg trajectory to obtain an approximate solution to the trust-region problem;

• *Two-step (sequential) update method*: take the *full* linear least squares (LLS) step (with no trust-region constraint) to optimize linear parameters $\boldsymbol{\theta}^A$ first independently, and then find $\Delta\boldsymbol{\theta}^B_{GN}$ to form a dogleg in parameter space $\boldsymbol{\theta}^B$; i.e., $\mathbf{A}\Delta\boldsymbol{\theta}^A_{LLS} = -\mathbf{r} \implies \mathbf{B}\Delta\boldsymbol{\theta}^B_{GN} = -\mathbf{r}_{\text{next}}$. $\square$ The one-step update method is not restricted to the case where linear and nonlinear parameters are separated into $\boldsymbol{\theta}^A$ and $\boldsymbol{\theta}^B$, whereas the two-step update method is special to it, solving two *linear* and *nonlinear* least squares problems *sequentially* with *different residual vectors* in the following two steps:

(1) *Forward pass step* (for linear parameters) to take the *full* LLS step leading to $\boldsymbol{\theta}^A_{\text{next}}$:

Do forward pass with $\boldsymbol{\theta}^B$ fixed (to compute nodes' outputs) and solve:

$$\mathbf{A}\Delta\boldsymbol{\theta}^A_{LLS} = -\mathbf{r}. \tag{3}$$

Alternatively, since $\mathbf{r} = \mathbf{a} - \mathbf{t}$, this can be expressed with $\boldsymbol{\theta}^A_{\text{next}}(\stackrel{\text{def}}{=} \boldsymbol{\theta}^A_{\text{now}} + \Delta\boldsymbol{\theta}^A_{LLS})$ as:

$$\mathbf{A}\boldsymbol{\theta}^A_{\text{next}} = \mathbf{t}. \tag{4}$$

Then, evaluate the new residual vector $\mathbf{r}_{\text{next}}$ by doing forward pass with $\boldsymbol{\theta}^A_{\text{next}}$.

(2) *Backward pass step* (for nonlinear parameters):

Do backward pass with $\boldsymbol{\theta}^A$ fixed, and obtain $\Delta\boldsymbol{\theta}^B_{GN}$ by solving: $\mathbf{B}\Delta\boldsymbol{\theta}^B_{GN} = -\mathbf{r}_{\text{next}}$ to find a dogleg trust-region (restricted) step $\Delta\boldsymbol{\theta}^B$.

In Step (1), Eq.(3) is less efficient than Eq.(4) since Eq.(3) needs to evaluate a residual vector $\mathbf{r}$. Furthermore, all the components of $\mathbf{A}$ are just *hidden activations*, resulting in: $\mathbf{A}_1 = \mathbf{A}_2 = \cdots = \mathbf{A}_{P_N}(\stackrel{\text{def}}{=} \mathbf{A}_{Lin})$ in Eq.(1). Hence, Eq.(4) can be rewritten as a standard "multiple-output" linear system: $\mathbf{A}_{Lin}\Theta^A_{\text{next}} = \mathbf{T}$, where $\mathbf{A}_{Lin}$ is a $d\times(P_{N-1}+1)$ matrix; $\mathbf{T}$ is a $d \times P_N$ matrix of desired outputs; Θ^A is a $(P_{N-1} + 1) \times P_N$ matrix of linear parameters, wherein the kth column of Θ^A corresponds to the parameters connected to the kth final output node at layer N. For the normal equation approach, $\mathbf{A}^T_{Lin}\mathbf{A}_{Lin}$ is formed and factored only once, and then P_N pairs of triangular systems are solved.

Hrycej called a simple two-step update method "layer-by-layer backpropagation" for optimizing an MLP that has linear identity node functions at terminal layer N, emphasizing its strength using a few examples (see page 174 in Chapter 9 of ref. [3]). In his implementation, the normal equations $\mathbf{A}^T_{Lin}\mathbf{A}_{Lin}\Theta^A_{\text{next}} = \mathbf{A}^T_{Lin}\mathbf{T}$ are solved just once (with no iterative refinement) to optimize the linear parameters $\boldsymbol{\theta}^A$ and a steepest descent-type method (with $\Delta\boldsymbol{\theta}^B = -\eta\mathbf{B}^T\mathbf{r}_{\text{next}}$ or its variant) is employed for optimizing the nonlinear parameters $\boldsymbol{\theta}^B$. However, this two-step update method with a *layer-independent combination of a full LLS step and a steepest descent-type step* is still questionable because those two have different convergence property, and perhaps more importantly because the *nonlinear* parameters $\boldsymbol{\theta}^B$ might be far from optimal values; this is true especially in MLPs since those *nonlinear* parameters between the first input layer and the next hidden layer are usually randomly initialized and therefore far from optimal values; yet, the *linear least squares* method tends to specialize the *linear parameters* $\boldsymbol{\theta}^A$ (between the terminal layer and the next hidden layer) to the current (poor) hidden-node activations. Perhaps, a better framework (than MLPs) for the two-step update method is such a local-tuning NN model as a radial basis function network (RBFN) and a TSK-type (neural) fuzzy system (or ANFIS), in which nonlinear parameters reside only in basis functions (BFs) and fuzzy membership functions (MFs) and those parameters may be well set up and can be assumed "near optimal" due to some problem-specific knowledge. In other words, the goal is to specialize the *linear parameters* to the outputs resulting from current BFs (or MFs). An extreme case is to optimize *linear parameters* alone with BFs (or MFs) fixed. Jang [4] applied the same method as Hrycej's to a single-output neuro-fuzzy ANFIS model, called ANFIS hybrid learning (see pages 668-669 in ref. [4]), wherein the normal equations were solved (with no iterative refinement) by a well-known *Sherman-Morrison matrix inverse formula for* $(\mathbf{A}_{Lin}^T \mathbf{A}_{Lin})^{-1}$ *in the batch mode,* whose dominant cost is $5mn_A^2$; hence, the ANFIS batch-mode hybrid learning works slow and sensitive to round-off errors (although the incremental mode was suggested in ref. [4] but never demonstrated). Overall, this hybrid implementation is not recommendable.

It is worth noting in the normal-equation sense that the layer-independent optimization uses only the diagonal blocks of the Hessian in Eq.(2) for optimization purpose without cross-term matrices $\mathbf{B}_k^T \mathbf{A}_k$ $(k=1,\cdots,P_N)$ that are related to *rotational effects*; so, some Hessian information is certainly lost in this technique. It is thus observable that the "direction for the linear parameters $\Delta\boldsymbol{\theta}^A$" determined by the layer-independent method [with Eq.(3) or (4)] and the "direction" determined by the Gauss-Newton method [e.g., Eq.(2)] can be very different from lack of those cross-term effects. Therefore, in reality, the two-step update method that takes the full LLS step may not work very well, while the one-step update method usually outranks it, effectively restricting iterative steps for optimizing both linear and nonlinear parameters simultaneously. For experimental evidence, refer to ref. [8].

4 Concluding Remarks

Exploiting special structure of the Jacobian matrix may help develop efficient learning algorithms (e.g., with a reduced memory scheme) for optimizing adaptive networks (see ref. [8]). Even for a large-scale problem, full Jacobian information is obtainable through matrix-vector multiplication [6]. We have discussed two algorithms for "layer-independent optimization" that attempt to exploit coexistence of linear and nonlinear parameters: Although the two-step update method is popular in the NN literature, it remains questionable because the model's outputs *depend* upon all parameters across layers. Understanding its weaknesses using Jacobian/Hessian matrices described in this paper would be useful for further research on effective adaptive-network optimization.

References

[1] Åke Björck *Numerical Methods for Least Squares Problems* SIAM, 1996

[2] James W. Demmel *Applied Numerical Linear Algebra* SIAM, 1997

[3] Tomas Hrycej *Modular Learning in Neural Networks: A Modularized Approach to Neural Network Classification* John Wiley & Sons, Inc., New York, 1992

[4] J.-S. Roger Jang "ANFIS: Adaptive-Network-Based Fuzzy Inference Systems" In *IEEE Trans. on Systems, Man, and Cybernetics*, Vol. 23, No. 3, pages 665–685, May 1993

[5] Eiji Mizutani "Powell's dogleg trust-region steps with the quasi-Newton augmented Hessian for neural nonlinear least-squares learning" In *Proceedings of the IEEE International Conference on Neural Networks*, Vol. 2, pages 1239–1244, Washington D.C., July, 1999

[6] Eiji Mizutani and James W. Demmel "On iterative Krylov-dogleg trust-region steps for solving neural networks nonlinear least squares problems" To appear in *Advances in Neural Information Processing Systems*, Vol. 13 (NIPS 13), MIT Press, 2001

[7] Eiji Mizutani and Stuart E. Dreyfus "On complexity analysis of supervised MLP-learning for algorithmic comparisons" To appear in *Proc. of the INNS-IEEE Int'l Joint Conf. on Neural Networks*, Wasington D.C., 2001

[8] Eiji Mizutani and James W. Demmel "On nonlinear least squares algorithms for neuro-fuzzy modular network learning" Submitted to *Neural Information Processing Systems: Natural and Synthetic*, Vancouver (NIPS 2001)

KES '01
N. Baba et al. (Eds.)
IOS Press, 2001

A Neural Network incorporating Coupled Gradient Descent Method

Teijiro Isokawa†, Nobuyuki Matsui† and Ferdinand Peper‡
† *Dept. of Computer Engineering, Himeji Institute of Technology,*
2167 Shosha, Himeji, Hyogo 671-2201, Japan
‡ *Communication Research Laboratory, Ministry of Public Management,*
Home Affairs, Posts and Telecommunications,
588 Iwaoka, Iwaoka-cho, Nishi-ku, Kobe-city, 651-2492, Japan

Abstract. Back Propagation (BP) algorithm for learning method of layered neural network is widely used in many areas of applications. However, BP has several important drawbacks. One of them is "standstill in learning", and it means that the state of network is subject to getting trapped in one of local minima. In this work, we present the Coupled Gradient Descent Method (CGDM) as an efficient way of avoiding local minima in the state of network.

1 Introduction

Since Back Propagation (BP) algorithm[1] was first proposed, it has been applied to many areas of engineering. BP is now representative of the leaning method for layered neural network, but as it involves several drawbacks caused by the gradient descent method it applies, the demand for improving it has been increased. One of the drawbacks is "standstill in learning", which means that the state of network is subject to getting trapped in one of local minima. Local minima in the state of network will increase in accordance with complexity of given tasks. Avoiding local minima is thus necessary for many applications, and researches have been much directed toward overcoming its drawbacks[2, 3, 4, 5].

We present the Coupled Gradient Descent Method (CGDM) as an efficient way of avoiding local minima in the state of network. CGDM is an extension of gradient descent method used in BP, and requires a pair (or group) of gradient parameters combining them explicitly. Combining these parameters makes their operations complementary to each other, and enables them to avoid standstill states.

2 Coupled Gradient Descent Method(CGDM)

2.1 Basic concept of CGDM

CGDM is an extension of conventional Gradient Descent Method(GDM). In order to show the basic concept of CGDM, we first recapitulate a formalism of GDM. For one-dimensional potential space $U(x)$, the dynamics of GDM is expressed as the following

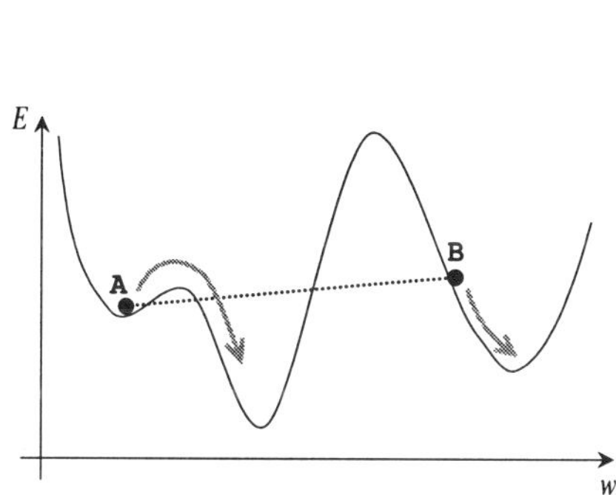

Figure 1: Image of bilateral effects

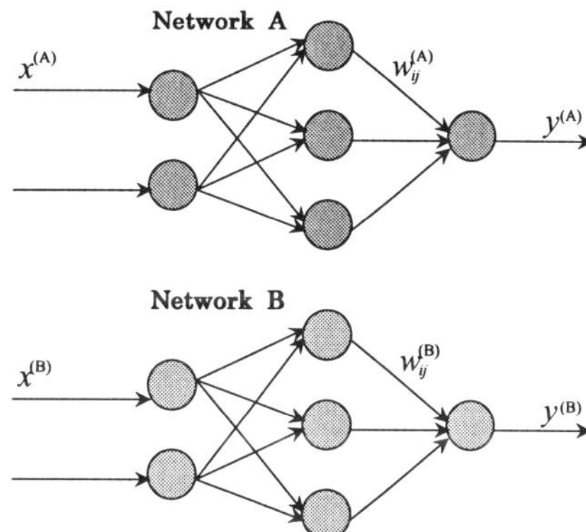

Figure 2: Network structure used in CGDM

equation:

$$\frac{dx}{dt} = -\eta \cdot \frac{dU(x)}{dx} \tag{1}$$

This dynamics leads x to one of the minima of $U(x)$, however, this minimum may not be a global minimum but a local minimum of $U(x)$. Conventional BP also suffers from this drawback because it applies GDM as the learning algorithm.

CGDM uses plural number of xs in single $U(x)$ and updates them by using GDM. When updating one of xs, for example x_i , CGDM uses the gradient of $U(x_i)$ as well as one of $U(x_j)$, where $i \neq j$. To simplify, we consider two xs, x_1 and x_2, which has the different value each other. The dynamics of CGDM in one-dimensional potential space is described below:

$$\frac{dx_1}{dt} = -\eta \cdot \frac{dU(x_1)}{dx_1} - \beta \cdot \frac{dU(x_2)}{dx_2}, \qquad \frac{dx_2}{dt} = -\eta \cdot \frac{dU(x_2)}{dx_2} - \beta \cdot \frac{dU(x_1)}{dx_1} \tag{2}$$

where η and β are constants, β controls mutual influences between the evolution of x_1 and x_2. When $\beta = 0$, x_1' and x_2' evolve independently, therefore in this case CGDM is equivalent with GDM. When $\beta \neq 0$, these settings are expected to cause bilateral effects on x_1 and x_2. Figure 1 shows an example of bilateral effect of x_1 and x_2 in $U(x)$. In Fig.1 the state of x_1 is hauled from the local minimum by the gradient of x_2 and gains the global minimum in $U(x)$.

2.2 BP with CGDM

We introduce this bilateral effect into BP. Two layered neural networks are prepared, where we named these two networks *network A* and *network B* for easy discrimination of these networks. In both of networks, the structures of networks, the characteristics of neurons used in networks are same(See Fig.2). Both of these networks learn the same patterns. The only different point is that the initial values of weights between layers are different in these networks.

One fixed structure of network, characteristics of neurons, and learning patterns creates the output error space $E(\boldsymbol{w})$. Initial different $\boldsymbol{w}$s correspond with different entry points of $E(\boldsymbol{w})$.

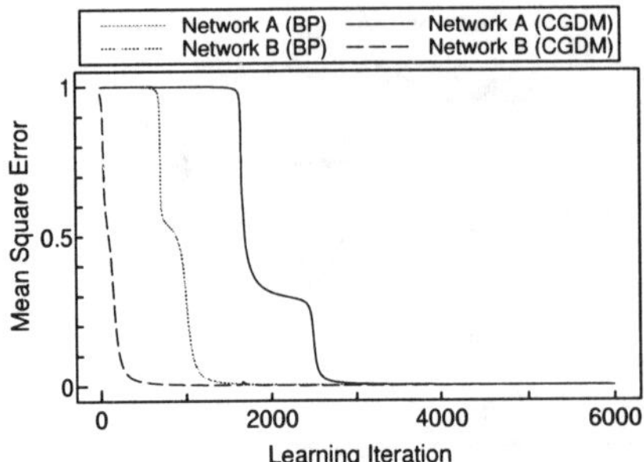

Figure 1: Changes of mean square errors in case using BP and CGDM for combination of two networks with high convergence

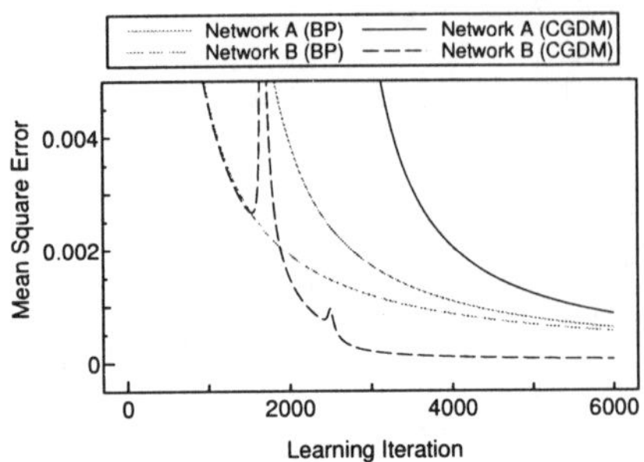

Figure 2: Changes of mean square errors in case using BP and CGDM for combination of two networks with high convergence(magnified)

We adopt updating rules of $\boldsymbol{w}$ as follows.

$$\frac{dw_{ij}^{(A)}}{dt} = -\eta\frac{\partial E_p^{(A)}}{\partial w_{ij}^{(A)}} - \beta\frac{\partial E_p^{(B)}}{\partial w_{ij}^{(B)}}, \qquad \frac{dw_{ij}^{(B)}}{dt} = -\eta\frac{\partial E_p^{(B)}}{\partial w_{ij}^{(B)}} - \beta\frac{\partial E_p^{(A)}}{\partial w_{ij}^{(A)}} \tag{3}$$

where $w_{ij}^{(A)}$ and $E_p^{(A)}$ are the weight connection between neuron i and neuron j and output error of network A, respectively.

3 Experimental Results

In this section we focus on the performance of BP with CGDM, as compared with conventional BP. To make the quantitative comparison between them, we apply these networks to XOR problem. The structure of network we choose is 2-3-1 three layers network. CGDM requires two networks that have same structures each other.

We first show the effect of combination of two networks by CGDM, where convergence of each of networks by BP is known. Figure 1,2 and 3 show the evolutions of mean square errors(MSEs) in the case of applying BP and CGDM. The initial condition(initial weight connection vector) in the case of BP and CGDM are same. From Fig.1, MSEs both in the case of BP and CGDM fall to zero, though trajectories of MSE evolutions are different each other. Figure 2 shows the same result of Fig.1 and magnification of the vicinity of the MSE value 0.0. We see that the MSE of network B in the case of CGDM shows the sharp peak and becomes much more smaller than one in the case of BP.

The initial condition of Fig.3 is different from Fig.1 and 2, the networks in the case of BP get trapped into the standstill states. From Fig.3, we observe that the combination of these networks by CGDM enables both of them to escape the standstill states and get the optimal solutions.

Figure 4 shows β, the mutual control coefficient, dependence of CGDM from the viewpoint of convergence. Convergence of learning is judged from the mean square error $E < 0.001$ after 6,000 iterations of learning is accomplished. Convergence rate(%) of optimal solutions are obtained by executing 500 test trials for each of β values. The dotted line and in the case of $\beta = 0$ show the convergence rate of conventional BP. From

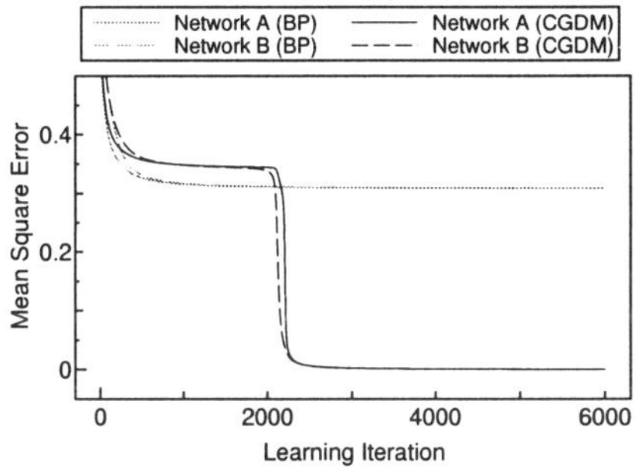

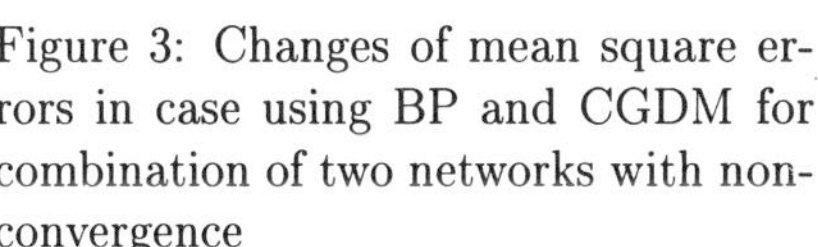

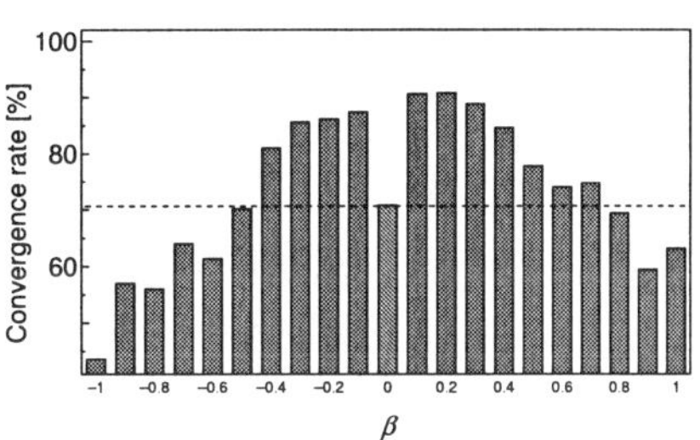

Figure 3: Changes of mean square errors in case using BP and CGDM for combination of two networks with non-convergence

Figure 4: β dependence of convergence rate of CGDM (dotted line represents the convergence rate of BP)

Fig.4, we see that in the region of smaller $|\beta|$ value ($|\beta| < 0.6$) high convergence rate is achieved, it seems to be due to the bilateral effect of learning dynamics. However, in the region of larger $|\beta|$ value, convergence rate becomes lower than conventional BP.

4   Conclusion

In this article, we have proposed the CGDM and explored the performance of this method through solving the XOR problem as a bench mark. From the results of computer simulations, we found that combination of networks by CGDM enables them to achieve efficient leaning, even if both of these two networks could not converge on learning patterns.

It is important to apply CGDM to larger problems and to observe the resultant performances from the viewpoint of engineering applications.

References

[1] D.E.Rumelhart *et al.*: "Parallel Distributed Processing," MIT Press, **1**, pp.295-307 (1988)

[2] T.Rognvaldsson: "On Langevin updating in multilayer perceptrons," Neural Computation, **6**, pp.916–926 (1994)

[3] S.Ma and C.Ji: "Fast training of recurrent networks based on the EM algorithm," IEEE Trans. Neural Networks, **9**, pp.11–26 (1998)

[4] D.R.Hush and J.M.Salas: "Improving the learning rate of backpropagation with the gradient reuse algorithm," Proc. of ICNN, pp.441–447 (1998)

[5] T.Nobori and N.Matsui: "Stochastic Resonance Neural Network and Its Performance," Proc. of IJCNN'2000, TB-1-1 (2000)

KES '01
N. Baba et al. (Eds.)
IOS Press, 2001

Structure Selection of RBF Neural Network Using Information Criteria

Toshiharu Hatanaka Katsuji Uosaki and Toshinori Kuroda
Department of Information and Knowledge Engineering,   Tottori University

Abstract. In this paper, we consider structure selection of neural network of three-layer radial basis function network (RBFN). Three information criteria, Akaike Information Criterion (AIC), Bayesian Information Criterion (BIC) and Takeuchi Information Criterion (TIC), are applied to determine of the number of units in hidden layer of the network. The applicability of these information criteria to structure determination is examined by numerical simulations.

1. Introduction

Recently, there has been a growing interest in modeling of nonlinear relationships since nonlinear characteristics such as saturation, dead-zone, etc., are inherent in many real systems. On the other hand, it has been recognized that a multilayer neural network with the back-propagation algorithm may be viewed as a practical tool for performing a nonlinear mapping. However, an important but difficult problem to determine the optimal number of parameter has not been solved. In other words, we have to determine how many hidden units are needed to nonlinear modeling. The number of hidden units must be sufficient to provide the modeling capability required for the given application; however, if too many units are employed, the network is capable of modeling the training data completely but incapable of generalizing between data that are minor variations of the training data. Moreover, an excessibly large number of hidden units may significantly increase the required training time. Thus we should consider this problem from the viewpoint of the relation between the complexity of the model and the performance for the training and inexperienced data. There have been some researches relating with this [2],[3],[4] for multilayer perceptrons, but very few for multilayer radial basis function network (RBFN) . We address, in this paper, which of the three information criteria, Akaike Information Criterion (AIC), Bayesian Information Criterion (BIC) and Takeuchi Information Criterion (TIC), is suitable to determine the number of hidden units in three-layer RBFN by simulation studies.

2. Radial Basis Function Network (RBFN)

We consider the selection of hidden units number in three-layer radial basis function network (RBFN), one of familiar local basis function networks.
Consider a three-layer feed-forward type RBFN with single output unit (Fig.1).

$$y(t) = \alpha_0 + \sum_{j=1}^{m} \alpha_j \phi_j(\boldsymbol{x}(t)) \tag{1}$$

where basis function $\phi_j(\boldsymbol{x}(t))$ are chosen as

$$\phi_j(\boldsymbol{x}(t)) = \exp\left(-\frac{1}{2}(\boldsymbol{x}(t) - \boldsymbol{\gamma}_j)^T \Sigma_j^{-1}(\boldsymbol{x}(t) - \boldsymbol{\gamma})\right) \tag{2}$$

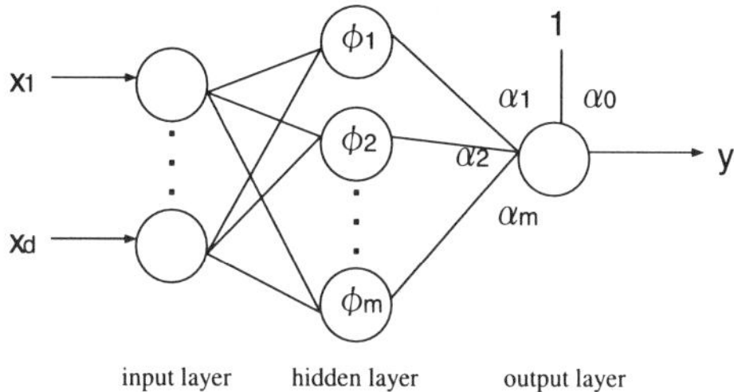

Figure 1: Radial basis function network

$$x(t) = \begin{pmatrix} x_1(t) \\ \vdots \\ x_m(t) \end{pmatrix}, \quad \gamma_j = \begin{pmatrix} \gamma_{1j} \\ \vdots \\ \gamma_{mj} \end{pmatrix}, \quad \Sigma_j = \begin{pmatrix} \sigma_{11}^{(j)}, & \cdots & \sigma_{1m}^{(j)} \\ \vdots & \ddots & \vdots \\ \sigma_{m1}^{(j)}, & \cdots & \sigma_{mm}^{(j)} \end{pmatrix} \tag{3}$$

with $x(t) = (x_1(t), \cdots, x_m(t))^T$, $\gamma_j = (\gamma_{1j}, \cdots, \gamma_{mj})^T$, $\Sigma_j = \sigma^2 I$, and input units d and hidden units m. respectively.

Hereinafter, we assume, for simplicity, $\Sigma_j = \sigma^2 I$ and the value of σ is given. The parameter vector θ defined with the linear weights associated with the output layer and the positions of the centers of the radial basis functions by the following:

$$\theta = (\alpha^T, \gamma_1^T, \cdots, \gamma_d^T)^T \tag{4}$$

with $\alpha = (\alpha_0, \alpha_1, \cdots, \alpha_m)^T$, is unknown and to be estimated. The RBFN is used for nonlinear modeling and the parameter vector θ should be adjusted to make the actual outputs of the network match the desired outputs closely as possible. This can be done by the supervised learning with the cost function defined in terms of deviations of the network outputs $\hat{y}(t)$}from desired outputs $\{y(t)\}$,

$$E(t) = \frac{1}{2}\Big(y(t) - \hat{y}(t)\Big)^2 \tag{5}$$

The correction $\Delta\theta(t)$ applied to $\theta(t)$ is defined by the delta rule with a momentum term,

$$\Delta\theta(t) = -\eta\frac{\partial E(t)}{\partial\theta(t)} + \mu\Delta\theta(t-1) \tag{6}$$

where η $(\eta > 0)$ is called the learning rate, and μ $(0 < \mu < 1)$ is the momentum constant.

3. Determination of Unit Numbers

Generally speaking, hidden units in the multilayer neural network play a crucial role in the operations of the RBFN with back-propagation learning, since they discover significant features of inputs. Hence the choice of hidden units affects the performance of the RBFN. It should be considered from the viewpoints between model complexity and the performance for the training and inexperienced data. This problem can be discussed by using some suitable information criteria. We employ here three information criteria, Akaike Information Criterion (AIC), its Bayesian extension (BIC) and Takeuchi Information Criterion (TIC), and discuss which information criteria is suitable for determination of the hidden units. The information criteria, AIC, BIC and TIC are all relating with Kullback discrimination information,

$$
\begin{aligned}
I[g(x); f(x)] &= \int g(x) \ln\left(\frac{g(x)}{f(x)}\right) dx \\
&= S(g(x); g(x)) - S(g(x); f(x)) \qquad (7) \\
S(g(x); f(x)) &= \int g(x) \ln(f(x)) dx \qquad (8)
\end{aligned}
$$

where $f(x)$ is an approximation to some true density $g(x)$. For a parametric family of density functions $f(x|\boldsymbol{\theta})$, we put $g(x) = f(x|\boldsymbol{\theta}_0)$, $g(x) = f(x|\boldsymbol{\theta}_0)$ with the true parameter set $\boldsymbol{\theta}_0$ and $I[g(x); f(x)]$ can be written by $I[\boldsymbol{\theta}_0; \boldsymbol{\theta}]$. These three information criteria are some approximations of $E[2NI[\boldsymbol{\theta}_0; \boldsymbol{\theta}]]$ and are given by

$$
\begin{aligned}
\text{AIC} &= N \log MSE + 2p \qquad (9) \\
\text{BIC} &= N \log MSE + p \log N \qquad (10) \\
\text{TIC} &= N \log MSE + 2\mathrm{tr}\{\hat{J}^{-1}(F))\hat{I}(F)\} \qquad (11) \\
MSE &= \frac{1}{N} \sum_{t=1}^{N} \{y(t) - \hat{y}(t)\}^2, \qquad (12)
\end{aligned}
$$

where F is a distribution of true model, $\hat{J}(F)$ and $\hat{I}(F)$ are estimates of

$$
\begin{aligned}
J(F) &= -E_G\left[\frac{\partial^2 \log f(\boldsymbol{\theta})}{\partial\boldsymbol{\theta}\partial\boldsymbol{\theta}^T}\right] \qquad (13) \\
I(F) &= E_G\left[\frac{\partial \log f(\boldsymbol{\theta})}{\partial\boldsymbol{\theta}} \frac{\partial \log f(\boldsymbol{\theta})}{\partial\boldsymbol{\theta}^T}\right] \qquad (14) \\
f(\boldsymbol{\theta}) &= \frac{1}{\sqrt{2\pi\hat{\sigma}^2}} \exp\left\{-\frac{(y(t) - \hat{y}(t))^2}{2\hat{\sigma}^2}\right\}, \qquad (15)
\end{aligned}
$$

and N and p are number of training data and parameters, respectively. The model minimizing some information criteria is chosen as the best model which corresponds to the best number of hidden units.

4. Numerical Experiments

To find which information criteria is suitable, we train the RBFN with training data based on certain information criteria and then examine the validity by observing whether the model gives the minimum MSE for validation data.

Simulations are carried out for 10sets of 500 observations. We split the observation data into two parts; first half is used for training and the latter part is used for validation. For each data set, training are started with 10 different initial parameter vector and after 10 times corrections, training is terminated.

[Example 1]

We assume the true RBFN has the following form:

$$
y(t) = \alpha_0 + \sum_{j=1}^{4} \alpha_j \phi_j(\boldsymbol{x}(t)) + w(t)
$$

$$
\phi_j(\boldsymbol{x}(t)) = \exp\left(-\frac{1}{2\sigma^2} \sum_{i=1}^{2} (x_i(t) - \gamma_{ij})^2\right)
$$

where true parameter vector is given by $\boldsymbol{\theta} = (0.5, 1.5, -2.0, -3.0, 2.5, 3.0, 3.0, -3.0, -3.0, 3.0, -3.0, 3.0, -3.0)^T$. and white Gaussian noise $w(t)$ with mean 0 and variance 0.1.

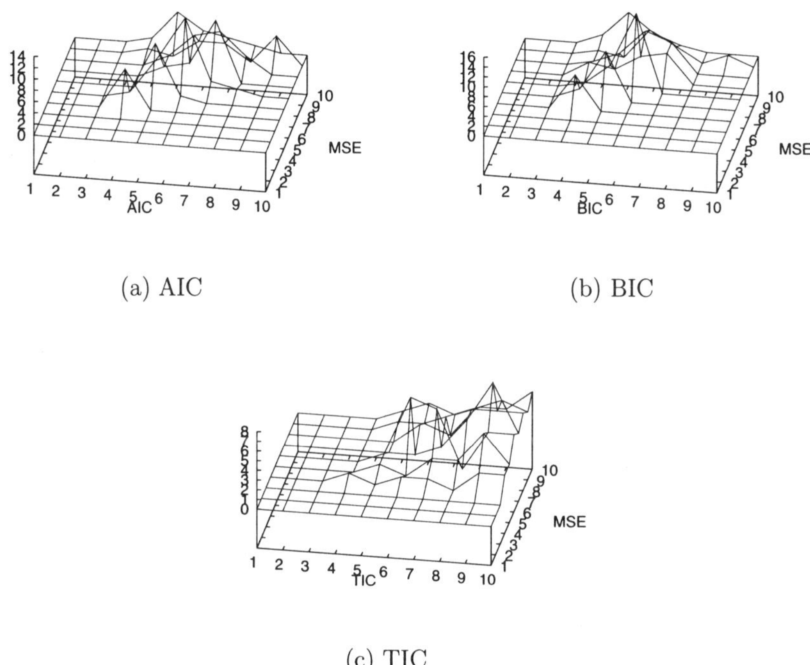

(a) AIC　　　　　　　　　　　　(b) BIC

(c) TIC

Figure 2: Selected number of hidden units (Example 1)

For each information criteria, the relations between the number of hidden units minimizing the information criteria and the number of hidden units minimizing the MSE for validation data are shown in Fig.2. In this case 49% of the hidden units selected by AIC for training data gives the minimum of MSE for validation data, while 40% by BIC and 20% by TIC, respectively, and it is found that TIC is likely to choose larger number of hidden units than AIC and BIC.

[Example 2]

Consider a problem of approximating a function

$$y(t) = 0.5x_1(t) + 3\sin(0.5x_2(t)) + w(t)$$

where input values $\{x_1(t)\}$ and $\{x_2(t)\}$ are random numbers uniformly distributed in [-5, 5]. Figure 3 shows the numbers of hidden units minimizing the information criteria for the training data and MSE for the validation data. In this case 48% of the hidden units selected by AIC for training data gives the minimum of MSE for validation data, while only 15% by BIC and 14% by TIC, respectively. This also shows the tendency to choose larger numbers by TIC than AIC and BIC.

5. Concluding Remarks

We address in this paper the structure selection of neural network of three-layer radial

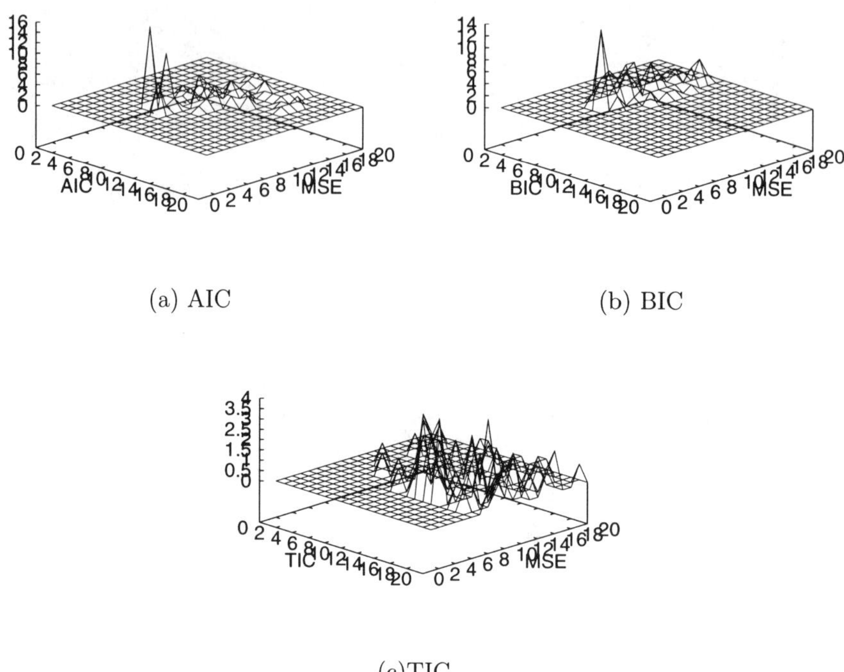

(a) AIC   (b) BIC

(c)TIC

Figure 3: Selected number of hidden units (Example 2)

basis function network (RBFN). We examine, by simulation studies, which of three information criteria, Akaike Information Criterion (AIC), its Bayesian extension (BIC) and Takeuchi Information Criterion (TIC), is suitable for selecting the best number of hidden units for RBFN as a nonlinear modeling system.

References

[1] Vila, J.-P., V. Wagner and P. Neveu, Bayesian nonlinear model selection and neural networks: A conhigate prior approach, *IEEE Trans. on Neural Networks*, Vol. 11, No.2, pp.265–278 (2000).

[2] Fogel, D. B., An information criterion for optimal neural network selection, *IEEE Trans. on Neural Networks*, Vol. 2, No. 5, pp.490–497 (1991).

[3] Murata, N., S. Yoshizawa and S. Amari, Network information criterion—Determining the number of hidden units for an artificial neural network model, *IEEE Trans. on Neural Networks*, Vol. 5, No. 6, pp.865–872 (1994).

[4] Anders, U., and O. Korn, Model selection in neural networks, *Neural Networks*, Vol. 12, No.2, pp. 309–323 (1999).

[5] Konishi, S. and G. Kitagawa, Theory and application of information criteria—Recent development—, *J. Society of Instrument and Control Engineers*, Vol. 38, pp.413–419 (1999). (*in Japanese*)

[6] Schwarz, G., Estimating the dimension of a model, *Annals of Statistics*, Vol. 6, pp.461–464 (1978).

[7] Takeuchi, K., Distribution of information statistics and criteria for modeling fitness, *Mathematical Sciences*, Vol. 15, No. 3, pp.12–18 (1976). (*in Japanese*)

KES '01
N. Baba et al. (Eds.)
IOS Press, 2001

Simultaneous Perturbation Learning Rule for Hopfield Neural Network

Shiro ITONAGA* Yutaka MAEDA Kansai University
Department of Electrical Engineering, Kansai University
3-3-35 Yamate-cho, Suita 564-8680 JAPAN
maedayut@kansai-u.ac.jp

Abstract

Hopfield neural network (HNN) is a recurrent neural network proposed by J.J.Hopfield in1984. The network has a characteristic of the association patterns ahead of time[1]. Originally, the weights of HNN are calculated by patterns to be memorized, then learning rule is not essential[2]. In this paper, we propose a recursive learning method for weight values in HNNs, which gives an optimal weights corresponding to a patterns to be memorized. The learning rule is based on the SP method [3][4][5][6].
Keyword: Hopfield neural network, Simultaneous perturbation learning rule

1. *Introduction*

Usually, the back-propagation (BP) method is widely used as a learning rule in feed-forward neural networks [6]. However, it is relatively difficult to adapt the learning rule for recurrent neural networks like the HNN directly. For the reason, we use the simultaneous perturbation (SP) method for a recursive learning rule of the network.

Usually, it is known that HNN has periodic solutions. Then, it will be more difficult to use the usual BP method. On the contrary, the SP learning rule is relatively easy to implement. In this paper, we propose a learning rule using SP method for HNNs [7]. Some simulation results are also shown.

Therefore, in this paper, we purpose to calculating the optimize weight value using simultaneous perturbation learning rule in order to outputs of HNN to be the periodic solutions.

2. *Hopfield Neural Network*

A model of HNN is depicted in Fig.1. Now, output of HNN and weight matrix is define as follows, respectively,

$$\boldsymbol{o} = \begin{pmatrix} o_1 & o_2 & \cdots & o_n \end{pmatrix}^T \tag{1}$$

$$\boldsymbol{w} = \begin{pmatrix} 0 & w_{12} & \cdots & w_{1n} \\ w_{21} & 0 & & w_{2n} \\ \vdots & & \ddots & \vdots \\ w_{n1} & w_{n2} & \cdots & 0 \end{pmatrix} \tag{2}$$

Where, w_{ij} is a weight value between the i-th neuron and the j-th neuron. w_{ij} is equal to w_{ji}.

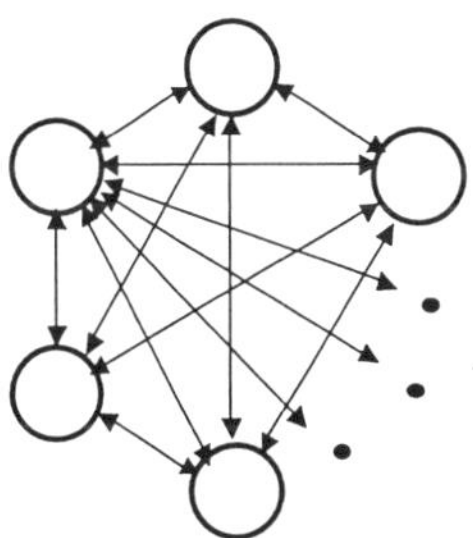

Fig.1 Model of HNN

Behavior of HNN is shown in Fig.2.

```
● Set up initial weights
● Set up initial outputs
-------------------------------------
do
    for i := 1 to n do
        ● Calculate the sum of weighted inputs
    for i := 1 to n do
        ● Decide outputs (threshold process)
until {No change for all outputs?}
```

Fig.2 Behavior of HNN

An output of a neuron in a HNN becomes inputs of all other neurons in the network, except its own neuron. That is to say, a total weighted input of a neuron is

$$net_j(t) = \sum_{i,i\neq j} w_{ij} o_j \tag{3}$$

$$x_j(t+1) = f\left(net_j(t)\right) \tag{4}$$

Where, $f(\cdot)$ is a threshold function. The HNN repeats the same operation until its state becomes stable.

3. Simultaneous Perturbation Learning Rule for HNNs

In this chapter, we describe details of the simultaneous perturbation (SP) learning rule as a learning rule of HNN. In this learning rule, we approximate the first-differential coefficient of an error function using values of the error. Two error values with the perturbation and without the perturbation gives an estimated first-different coefficient using a kind of a difference approximate. The perturbations are added to all weights simultaneously. The procedure has nothing to do with network structure. Therefore, this learning rule can apply to not only ordinary multi layer NN but also recurrent NNs like HNN.

First, we define a weight of HNN again as a vector,

$$w_i = \begin{pmatrix} w_{i1} & w_{i2} & \cdots & w_{in} \end{pmatrix}^T \qquad (5)$$

A sign vector is defined as follows

$$s_i = \begin{pmatrix} s_{i1} & s_{i2} & \cdots & s_{in} \end{pmatrix}^T \qquad (6)$$

s_{ii} denote the i-th element of the sign vector that is 1or -1. The sign of s_{ii} is randomly determined. Moreover, the sign of s_{ii} is independent of the sign of the j-th element s_{ij} of the sign vector. That is,

$$E[s_{ii}] = 0$$

$$E[s_{ii} \cdot s_{ij}] = \begin{cases} 0 & (i \neq j) \\ 1 & (i = j) \end{cases} \qquad (7)$$

E denotes the expectation. Now, we define the i-th element of the vector Δw as follows;

$$\Delta w_{ii} = \frac{J(w_i + cs_i) - J(w_i)}{c} s_{ii} \qquad (8)$$

Where, $J(\cdot)$ denotes the value of the error function. c denotes a magnitude of the perturbation.

We update the weight w_i at time t using the modifying weight Δw_{ii} as follows;

$$w_{(t+1)i} = \begin{cases} w_{max} & \text{if } (w_{ii} - \alpha\Delta w_{ii}) > w_{max} \\ -w_{max} & \text{if } (w_{ii} - \alpha\Delta w_{ii}) < -w_{max} \\ w_{ii} - \alpha\Delta w_{ii} & \text{otherwise} \end{cases} \qquad (9)$$

Where, α is a positive coefficient to adjust a magnitude of a modifying quantity, w_{max} is a maximum value of a weight. We limit the weight value to avoid its extremely larger or smaller value.

We expand the right-hand side of equation (8) at the point w_i, there exist w_s such that

$$\Delta w_i = s_i^T \frac{\partial J(w_i)}{\partial w_i} s_i^i + \frac{1}{2!} cs_i^T \frac{\partial^2 J(w_s)}{\partial w_s^2} s_i s_i^i \quad (10)$$

We take expectation of the above equation (10). Then, we have

$$E[\Delta w_i] = E\left[s_i^T \frac{\partial J(w_i)}{\partial w_i} s_i^i \right]$$
$$+ E\left[\frac{1}{2!} cs_i^T \frac{\partial^2 J(w_s)}{\partial w_s^2} s_i s_i^i \right] \qquad (11)$$

From the conditions of sign vector, the first term the right-hand side of equation (11) is

$$E\left[s_i^T \frac{\partial J(w_i)}{\partial w_i} s_i^i \right] = \frac{\partial J(w_i)}{\partial w_i} \qquad (12)$$

Next, the second term the right-hand side is

$$E\left[\frac{1}{2!} cs_i^T \frac{\partial^2 J(w_s)}{\partial w_s^2} s_i s_i^i \right] = 0 \qquad (13)$$

Thus, we have

$$E[\Delta w_i] = \frac{\partial J(w_i)}{\partial w_i} \qquad (14)$$

Therefore we can find the learning rule (8) a type of a stochastic gradient method. However, the error does not decrease monotonously by this learning rule.

4. Implementation

First, we set up the weights of HNN randomly. Then we repeat learning operation until its state becomes stable. We calculate error function $J(w)$ defined by a difference between its outputs and its patterns to be memorized. Second, we add the perturbations to the all weights of HNN. Then, HNN operates similarly, we calculate the error function $J(w + cs)$ again.

Finally, we calculate the modifying weight Δw using $J(w)$ and $J(w + cs)$ based on equation (8). We update the weights of HNN.

We repeat the series of operations and we obtain optimal weights of HNN which correspond to pattern to be memorized. The flowchart of the learning is shown in Fig.3.

<table>
<tr><td>

●   Set up the random weights
●   Set up the first outputs

do
begin

   ●   Operation of HNN
   ●   Calculate $J(w)$
●   Add the weights to the perturbations and operation of HNN
   ●   Operation of HNN
   ●   Calculate $J(w + cs)$
●   Calculate Δw
●   Modify the weights
end;
until {Evaluate value nearly equal to 0}

</td></tr>
</table>

Fig.3 Flowchart of Learning

5. *Results*

5.1 *Example1*

We show a simulation results for a periodic patterns. The number of neurons in HNN are 25 (5 by 5). Periodic patterns to be memorized are "H" and "S" shown in Fig.4. In the figure, a black square show a +1 output of HNN, and a white square show an −1 output of HNN.

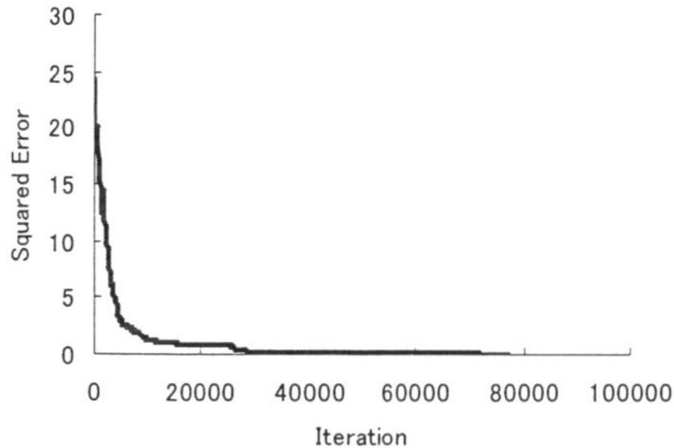

Fig.4 Patterns 1

In the simulation, the maximum perturbation $c_{max} = 0.3$, the minimum perturbation $c_{min} = 0.07$, the maximum value of a weight $w_{max} = 10.0$ and the leaning coefficient $\alpha = 0.05$.

We show a change of the squared error for iterations in Fig.5. As is shown in Fig.5, the squared error decrease as the learning proceeds. After about 70000 times learning, squared error is nearly equal to zero.

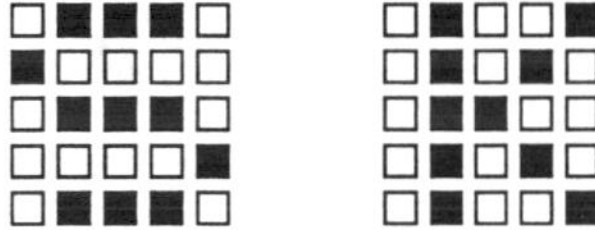

Fig.5 Squared error for patterns 1

Next, we show a change of output patterns at each iterations in Fig.6. As is shown in Fig.6, the squared error decrease as the learning proceeds and the output patterns are closed to the patterns. After 30000 times learning, HNN remembers the proper patterns.

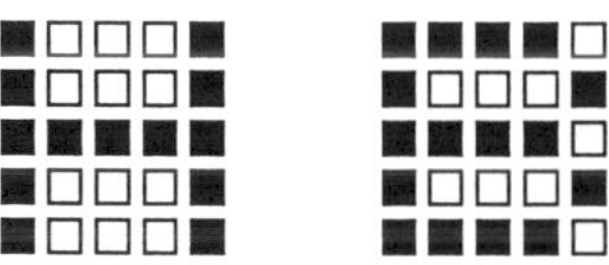

Itteration	Error	Pattern 1	Pattern 2
0	24.1		
10	23.2		
100	19.7		
1000	15.2		
10000	1.28		
30000	0.26		

Fig.6   Change of output pattern for patterns-1

5.2 *Example 2*

Second, using other patterns, we confirm feasibility, we use memorized patterns "S" and "H". The patterns are shown in Fig.7.

Fig. 7 Patterns 2

In the simulation, the maximum perturbation $c_{max} = 0.3$, the minimum perturbation $c_{min} = 0.1$, the maximum value of a weight $w_{max} = 10.0$ and the leaning coefficient $\alpha = 0.05$.

We show the squared error for iteration in Fig.8. As is shown in Fig.8, the squared error decrease as the learning proceeds. After about 1200000 times learning, squared error is nearly equal to zero.

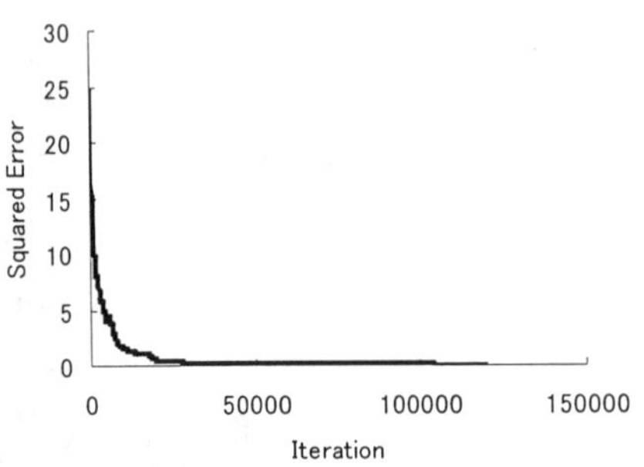

Fig.8 Squared error for patterns 2

Next, we show a change of output patterns at each iterations in Fig.9. Similar to Fig.6, the squared error decrease as the learning proceeds.

Itteration	Error	Pattern 1	Pattern 2
0	24.1		
10	22.9		
100	19.7		
1000	9.97		
10000	1.53		
20000	0.52		

Fig.9 Change of output pattern for patterns-2

As is mentioned above, it is possible to find the optimal weights of HNN for periodic solutions using the learning rule proposed here.

Conclusions

We proposed a learning method to obtain optimal weights for periodic patterns using SP learning rule. Through the simulation, we can confirm a feasibility of learning rule.

Acknowledgments

This research is financially supported by Kansai University Frontier Sciences Center, High Technology Research Center and the Biomechanics and Cybernetics Research Group of the Institute of Industrial Technology of Kansai University.

Reference

[1] J.J.Hopfield, "Neurons with graded response have collective computational properties like those of two-state neurons" Proceedings of the National Academy of Sciences, vol.81, pp.3088-3092

[2] R.Hecht-Nielsen, "Neurocomputing", Toppaninsatsu, 1992 (in Japanese)

[3] Y.Maeda, "Learning rule of neural networks for inverse systems," Transactions of The Institute of Electronics, Information and Communication Engineers, vol.J75-A, pp.1364-1369, 1992(in Japanese), English version of this paper is shown in Electronics and Communications in Japan, vol.76, pp.17-23, 1993.

[4] Y.Maeda and Y.Kanata, "Learning rules for recurrent neural networks using perturbation and their application to neuro-control" Transactions of the Institute of Electrical Engineers of Japan, vol.113-C, pp.402-408, 1993(in Japanese).

[5] Y. Maeda and R. J. P. deFigueiredo, Learning Rules for Neuro-Controller Via Simultaneous Perturbation, IEEE Trans. on Neural Networks, vol.8, no.5, pp.1119-1130, 1997.

[6] D.E.Rumelhart and G.E.Hinton and R.J.Williams, "Learning representations by back-propagation errors" Nature, vol.323, pp.533-536, 1986.

[7] Y.Maeda and H.Tachikawa "Learning Rule for Bidirectional Associative Memory by Simultaneous Perturbation Method" The Institute of Industrial Technology, Kansai University, vol.101, pp.3-7, 1999.

A Data Management Structure for Spatio-Temporal Walkthrough

Kazuo IKEMOTO, Yoshinori HIJIKATA, Mie NAKATANI and Shogo NISHIDA
Department of Systems and Human Science,
Graduate School of Engineering Science, Osaka University
1-3, Machikaneyama-cho, Toyonaka, Osaka, 560-8531, JAPAN

Abstract. This paper deals with data management structures for spatio-temporal walkthrough in virtual 3D environment. In the spatio-temporal walkthrough, it is very important to realize real-time response, that is, efficient search for spatio-temporal data. For this purpose, we propose Adaptive Tree Structure (We call the AT Structure), in which either spatial data structure or temporal data structure is selected adaptively depending on the demand for search. This paper describes an outline of the proposed data structure and evaluation results by computer simulation.

1 Introduction

The management of spatio-temporal data is becoming increasingly important in geographic information systems, computer-aided design, etc. For spatial data management, much research has been done, and structures based on spatial partitioning have been proposed [1]. With these methods, spatial data are handled efficiently and spatial search, such as range search and nearest neighbors search, can be performed rapidly. However, these structures are designed only for managing spatial data with temporal information. On the other hand, in temporal data structures [2, 3], only temporal data with non spatial attribute are considered, and thus these structures do not provide efficient handling methods for spatio-temporal data. To manage spatio-temporal data efficiently, a new data structure is required.

In this paper, the AT structure is proposed for managing spatio-temporal data efficiently. We consider spatio-temporal data which consists of its location and occurrence/extinction time. To manage such spatio-temporal data, we first consider three possible methods based on conventional spatial data structures, and discuss their advantages and disadvantages. Then, based on the analysis, we introduce the AT structure to improve these methods. The AT structure is developed by extending conventional spatial data structures using techniques of searching either structure of the two adaptively. The performance of the AT structure is compared with these methods by computer simulations. We show the AT structure has appropriate properties with respect to spatio-temporal search. Moreover, we propose a method of realizing spatio-temporal walkthrough system by using the AT structure.

2 Management of Spatio-Temporal Data Using Tree Structures

In this section, we consider three possible methods to manage spatio-temporal data by using conventional spatial data structures, which are constructed based on spatial partitioning [4, 5, 6]. We discuss the advantages and disadvantages of each method with respect to both search efficiency and storage cost. The methods considered here are as follows:

(1) MT structure For each version, a distinct tree is constructed. Roots of the trees are managed by array or another tree. We call this the MT (Multi Tree) structure.

(2) ST structure Whole data is managed by a single tree. The tree is constructed based only on location of data. Temporal information is treated as an attribute of each data. We call this the ST (Single Tree) structure.

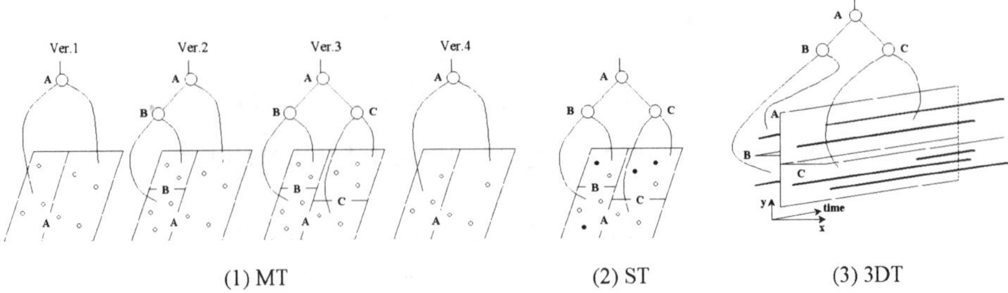

Figure 1: Examples of the MT, ST and 3DT structures

(3) (N+1)-Dimensional Tree structure N-dimensional data with temporal information
is considered as (N+1)-dimensional data. The partition of data space is based on both
location and temporal information of data. We call this the (N+1)-Dimensional Tree
structure.

In the following sections, (N+1)-Dimensional Tree structure is treated as the 3DT for
simplicity. Fig.1 shows examples of the MT, ST and 3DT structures.

The MT structure is suitable for a search for a single temporal version, because every
temporal version of tree is stored separately. In many practical applications, the search for
the current version is most often required. By using the MT, apparently, the search for the
current temporal version is performed rapidly. However, when the access is not confined
to the single temporal version, the searching time increases linearly by extending the time
interval to be searched. On the other hand, the ST structure is not suitable for a search for a
single temporal version. In the ST, temporal version information is not managed by the tree.
Therefore, after the spatial search, temporal version information of each data should also be
examined one by one. This implies the ST cannot perform a rapid spatial search for a single
temporal version.

From a viewpoint of the storage space, the MT is extremely costly. If data are valid
before and after the version update, then they are stored redundantly in the multiple trees.
Therefore, the long-lived data have many copies. It increases the storage cost of the MT.
On the other hand, the ST and 3DT are not costly because each data is stored without
redundancy. In general, there is a tradeoff between the storage cost and the efficiency of the
access to the current data. For example, the MT gives rise to the efficient search method for
a single temporal version with high storage cost.

3   Adaptive Tree Structure

Adaptive Tree structure (AT structure) is proposed to overcome the drawbacks of the ST,
MT and 3DT structures. The ST is constructed based only on spacial information of data,
and temporal information is treated as an attribute of each data. Therefore, by using the ST,
the search for narrow range of space is performed rapidly. However, the search for broader
range of space is not performed efficiently. On the other hand, the MT is constructed based
only on temporal information of data, and it has the same feature as the ST; that is, the
performance becomes worse as the range of temporal search becomes broader. In general,
tree structures have the property that the efficiency of search becomes worse as the range of
search becomes broader. The property cannot be avoided for tree structures.

Based on the above investigation, we propose the AT structure. In this section, we describe
an outline of the AT structure. At first, we prepare two data structures, that is, spatial data
structure and temporal data structure in advance. Then, according to the demand for search,
the method selects one of the data structures that is expected to be searched efficiently. In
short, either spatial data structure or temporal data structure is selected adaptively depending
on the demand for search as shown in Fig.2(1).

In the following subsections, each part of the AT structure is explained in detail.

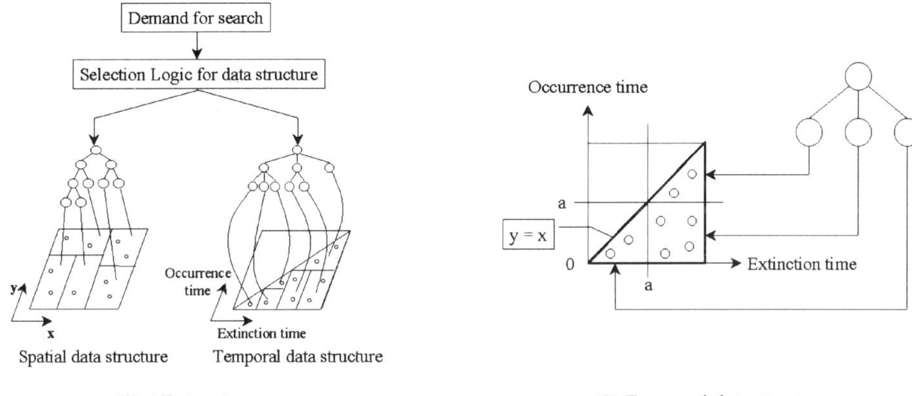

Figure 2: An outline of the AT structure

3.1 Spatial Data Structure

Spatial partitioning for spatial data is done by using k-d tree [4]. Spatial data structure consists of two types of nodes, that is, internal nodes and leaves. An internal node has two children, and it has a discriminator-value to partition the space. Data are stored only in leaves. The maximum number of each leaf is called Capacity. Capacity is determined in advance before data structure is constructed. In the searching process, when arriving at an internal node, the pointer is determined by examining the discriminator-value to decide whether to branch 'left' or 'right' child until arriving at a leaf. In this way, data are obtained from the tree structure.

3.2 Temporal Data Structure

Temporal data is usually treated as a segment in the time coordinate direction. For temporal data, significance exists in occurrence and extinction time. For example, considering management of equipments, it is important when each equipment is started, when repaired and when removed. In our method, temporal data are transformed into points in the occurrence-extinction plane, and these points are managed by tree structures. In the occurrence-extinction plane, all points exist in the area below the line $Y = X$ as shown in Fig.2(2), since occurrence always exists before extinction.

Partitioning of temporal data is done by using ternary trees recursively. Using some threshold value a, triangular data area is divided into three areas, that is, two triangular data areas and one square data area. Each triangular data area is divided into three areas recursively. Each square data area is managed by using k-d tree.

3.3 Selection Logic for Data Structure

In general, the search for narrow range of spatial data is performed rapidly by using spatial data structure. In the same way, the search for narrow range of temporal data is performed rapidly by using temporal data structure. Based on the above property, this logic adaptively selects either spatial data structure or temporal data structure depending on the demand for search. It is necessary for the logic to be simple enough since the total performance is directly affected by selecting time, which is the time needed for the selecting the data structure to be searched.

In this paper, the logic selects the data structure whose ratio between the searching area and the whole area is smaller. This logic is simple, and it is correct when both spatial data and temporal data are uniformly distributed in the whole area.

4   Simulation Results

Computer simulations were carried out to compare the performances of the AT structure, ST and 3DT by using Unix workstation. Data were two-dimensional points (x, y) with temporal information, where $0 \leq x, y \leq 1000$ generated randomly. Temporal data consists of occurrence and extinction data. The occurrence and term data is an integer generated randomly. The extinction data is made up by adding occurrence and term data. In the simulations, the total number of data was 100000, and the leaf capacity was 10 for all structures. The performances of creation of the data structures and spatio-temporal range search are evaluated. In the ST and 3DT, spatial partitioning was done by using k-d tree. We do not consider the MT because it is impossible to create the MT in continuous time.

4.1   Results for Creating Data Structures

Table.1 shows experimental results; CPU-time taken to create a tree, the number of created internal nodes and the number of created internal leaves. The average for 10 sets of data are shown.

The number of created internal nodes of the AT structure was almost twice the case of the ST and 3DT, because the AT structure consists of two data structures, that is, spatial and temporal data structures. The same feature is observed for the number of created internal nodes and CPU-time. In summary, the AT structure is not better than the ST and 3DT with respect to the time cost and the storage cost in creating a data structure.

4.2   Results for Spatio-Temporal Range Search

In the experiments, searching range was supposed to be square in the spatial and temporal plane, whose values are varied from 1 % to 60 % of the whole area. The performances of range search for spatio-temporal data were tested 10 times. The position of each range was generated randomly.

Fig.3(1) shows experimental results of range search under the condition that ratio of spatio searching range is fixed to be 10 % of that of the whole area. As Fig.3(1) indicates, by using the ST, the performance of temporal search is the constant. By using the 3DT, the performance of temporal search becomes worse according to increase of temporal searching range. On the other hand, the AT structure changes the data structure to selected according to the demand for search. Therefore, the AT structure uses temporal data structure under 10 %, and uses spatial data structure over 10 %. For the above mentioned reason, the performance of the AT structure is the same as that of the ST over 10 %.

Fig.3(2) shows experimental results for range search under the condition that ratio of temporal searching range is fixed to be 10 % of that of the whole area. As Fig.3(2) indicates, by using the ST and 3DT, the performance of spatial search becomes worse according to increase of spatial searching range. On the other hand, the performance of the AT structure is the same as that of the ST and 3DT under 10 %. However, the performance of the AT structure is the constant over 10 %.

Fig.4 shows the performances of spatio-temporal search in the three dimensional plane, where X-axis and Y-axis indicate the ratio of temporal and spatial searching range toward the whole area and Z-axis indicates CPU-time. It is confirmed that the performance of the AT structure is better than that of the ST and 3DT in the whole area from the figure.

Table 1: Results for creating the data structures

	ST	3DT	A T	
			temporal	spatial
CPU Time (relative ratio)	1.00	1.01	1.40	1.00
Number of created nodes	16383	16383	16053	16383
Number of created leaves	16384	16384	16185	16384

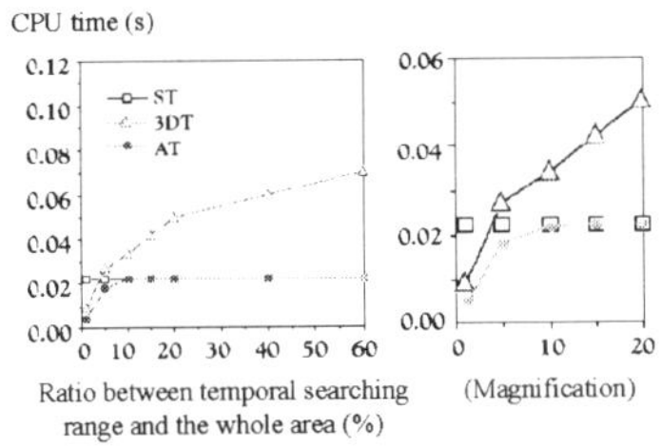

(1) under the condition that ratio of spatial searching range
is fixed to be 10% of that of the whole area

(2) under the condition that ratio of temporal searching range
is fixed to be 10% of that of the whole area

Figure 3: Results for spatio-temporal range search (1)

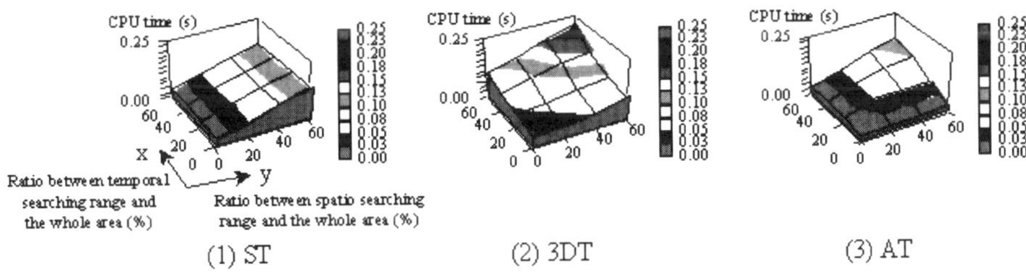

(1) ST   (2) 3DT   (3) AT

Figure 4: Results for spatio-temporal range search (2)

5 Spatio-Temporal Walkthrough System Using AT Structure

In this section, we consider a method of realizing spatio-temporal walkthrough system by using the AT structure. In general, it is important to realize real-time response to continual movement in the walkthrough. Therefor, it is efficient to draw objects which exist not in the whole area but only in the range of vision by using range search.

5.1 System Architecture

Fig.5 shows system architecture of spatio-temporal walkthrough. At first, the AT structure is constructed based on spatial (location) and temporal (occurrence/extinction) information of objects which are managed in virtual 3D environment. Secondly, spatial searching range and temporal searching range are established according to user's input (movement of a point of view, rotation of a direction of one's eyes, change of temporal version). Spatial searching range is established based on a point of view and a direction of one's eyes. Finally, objects obtained by spatio-temporal search are drawn. In the same way, a chain of operation is carried out recursively.

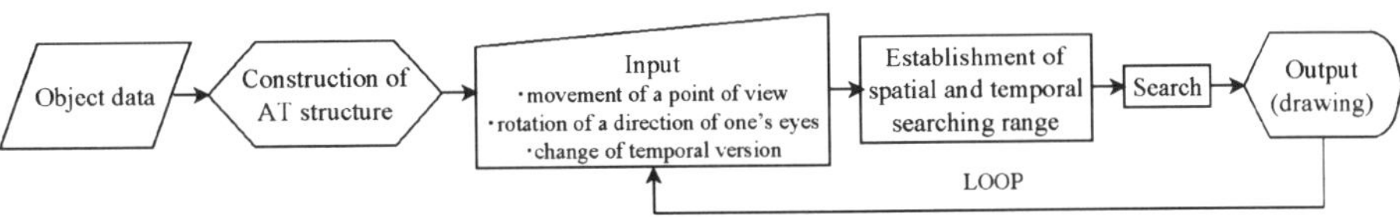

Figure 5: System architecture of spatio-temporal walkthrough

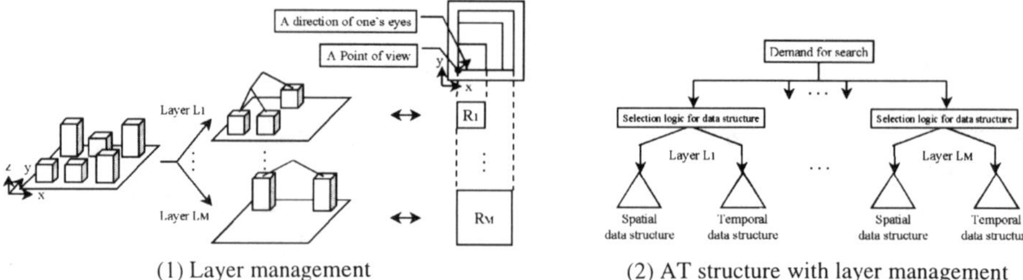

(1) Layer management   (2) AT structure with layer management

Figure 6: Applying management to the AT structure

5.2 *Applying Layer Managemet to AT Structure*

In general, we cannot see low objects which exist far away from a point of view. Therefor, it is efficient to omit drawing low objects in far area. Fig.6(1) shows an outline of layer management. Objects are divided into M layers based on some criterial height, and each layer is managed by using the AT structure as shown in Fig.6(2). In the searching process, each layer structure is searched according to a distinct searching range.

We applied 2-layers management to the AT stucture, and constructed a prototype system. By computer simulation with prototype system, real-time response was confirmed.

6 Conclusions

In this paper, we proposed the AT structure to manage spatio-temporal data efficiently. By computer simulations, the performance of the AT structure was compared with the other possible methods. Experimental results showed that the AT structure has the following advantage; that is, a spatio-temporal range search can be performed more rapidly than the cases of the other possible methods. However, according to the storage cost, the AT structure are not better than the cases of the other possible methods. Moreover, we applied the AT structure to spatio-temporal walkthrough system, and real-time response was confirmed.

The important future works are to consider the interface to control spatio-temporal walkthrough, and also to evaluate our system synthetically.

References

[1] H.Samet, "The Design and Analysis of Spatial Data Structures", Addison-Wesley, 1989.

[2] D.Lomet and B.Salzberg, "Access Methods for Multi version Data", in Proc. ACM SIGMOD'89, pp.315-324, 1989.

[3] R.Elmasri, G.T.J, Wun and Y-J.Kim, "The Time Index: An Access Structure for Temporal Data", in Proc. 16th VLDB Conf., pp.1-12, 1990.

[4] J.L.Bentley, "Multidimensional Binary Search Trees Used for Associative Searching", Commun. ACM, Vol.18, pp.509-517, 1975.

[5] Y.Ohsawa and M.Sakauchi, "The BD-tree - A new N dimensional Data Structure with Highly Efficient Dynamic Characteristics", in Proc. IFIP Conf., pp.539-544, 1983.

[6] Y.Nakamura, S.Abe, Y.Ohsawa and M.Sakauchi, "A Balanced Hierarchical Data Structure for Multidimensional Data with Highly Efficient Dynamic Characteristics", IEEE Trans. KDE, Vol.5, pp.682-693, 1993.

[7] T.Teraoka, M.Maruyama, Y.Nakamura and S.Nishida, "The MP-tree: A Data Structure for Spatio-Temporal Data", Proceedings of the 14th IEEE Phoenix Conference on Computers and Communications, pp.326-333, March, 1995.

[8] T.Tamada, T.Teraoka, M.Maruyama and S.Nishida, "Spatial Data Structure for the 3D Graphical Facility Management System", Proceedings of the 6th International Conference on Human-Computer Interaction, Volume II pp.141-146, July, 1995.

Semiautomatic Color Space Quantization Method Supporting Image Retrieval Using Color Names

Zoran STEJIĆ*   Yasufumi TAKAMA   Kaoru HIROTA
Dept. of Computational Intelligence and Systems Science, Tokyo Institute of Technology
4259 Nagatsuta, Midori-ward, Yokohama 226-8502, JAPAN
**E-mail:* stejic@hrt.dis.titech.ac.jp

Abstract. A semiautomatic color space quantization method is proposed, which models the human perception of color names, and is suited for image retrieval application. Comparison of the proposed Automatic Classification step with the clustering and the manual quantization methods shows 1.5–8.8% increase in the perceptual quality, and 1.7–1,327.0 times decrease in the computational cost. The proposed Manual Reclassification step additionally brings over 30% increase in the perceptual quality. Applied to retrieval from the database of 1,100 color images and compared with the manual quantization method, the proposed method leads to 16% increase in the retrieval precision, enabling a more effective retrieval from large scale image databases.

1. Introduction

The color space quantization problem is considered from the viewpoint of image retrieval application. The focus is on the large scale image databases (e.g. the Internet), where the precision of the retrieval is of primary importance [7].

For querying image colors, content-based image retrieval systems use either the *query-by-example* paradigm (e.g., "Find images with colors similar to *this* image.") or the *textual queries* (e.g., "Find images with a lot of *green* and *blue*."). In both cases, for efficient comparison between database images and the query, color content of an image must be represented in compact form. However, a full-color digital image can contain over 16 millions of colors. The objective of color space quantization is to significantly reduce the dimension of the color space (to a few tens or hundreds of representative colors), with the minimal loss in the color content of an image.

The existing color space quantization methods can be broadly divided in two categories: *clustering* [1, 3, 4, 6], primarily supporting query-by-example, and *manual quantization* [5], suitable for textual queries. The proposed semiautomatic color space quantization method primarily supports textual queries, but is applicable to query-by-example as well.

The proposed method models color names in accordance with the human perception. At first, 64, 216, or 512 representative colors, obtained by *uniform quantization* of RGB color space, are *automatically classified* into 10 color classes (corresponding to basic color names). Then, the colors are *manually reclassified* using the custom color space visualization tool. The result is a mapping from image colors to the named color classes.

The proposed method, compared with the clustering and the manual quantization, brings a decease in the computational cost and, at the same time, an increase in the perceptual quality, resulting in a color space quantization closely approximating the human perception of color names. Applied to image retrieval and compared with the manual quantization method [5], the proposed method leads to 16% increase in the retrieval precision.

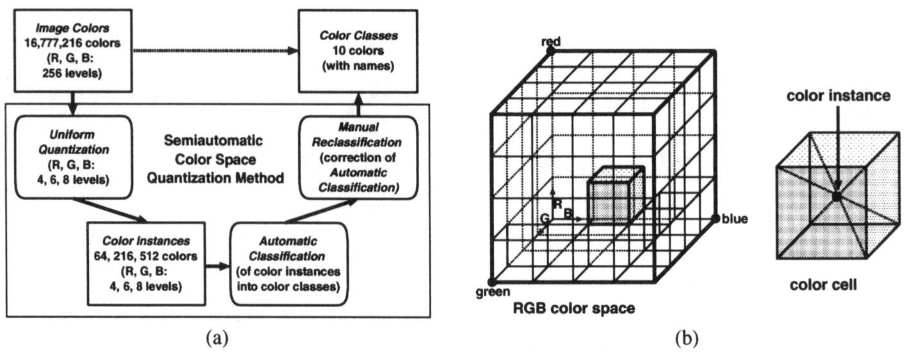

Figure 1: (a) Data flow of the proposed Semiautomatic Color Space Quantization Method; (b) Uniform Quantization (of the RGB color space, each color axis is quantized into 4 levels, resulting in the 64 color cells/instances)

Section 2. describes the conventional color space quantization methods, while Section 3. presents the proposed Semiautomatic Color Space Quantization Method. Finally, Section 4. comparatively evaluates the proposed method, considering its application to image retrieval.

2. Conventional Methods for Color Space Quantization

Among the existing color space quantization methods supporting the query-by-example color queries, clustering techniques are prevailing [1]. The modified *agglomerative hierarchical clustering* algorithm [2], is used for color space quantization in [1] and [6]. The improved version of the *k-means clustering* algorithm is used in [3], while [4] presents the competitive learning, as another improvement of the k-means clustering algorithm. As the references for the comparison with the proposed method, the hierarchical clustering algorithm [1] and the competitive learning algorithm [4] are used, being the representative clustering techniques applied to the color space quantization problem [7].

A representative color space quantization method supporting the textual color queries is the manual quantization of the MTM color space [5] (called here *MTM quantization*). The method defines non-uniform quantization of the MTM color space into 11 color zones (corresponding to the basic color names). The quantization is obtained and verified through intensive perceptual evaluations and observations, involving displaying the MTM color space in sections and incorporating perceptually similar colors by human operators [5]. MTM quantization method is chosen as the reference for comparison with the proposed method because it is expected to model color names with high perceptual quality, since MTM is the closest to the human perception among the known color spaces [7].

3. Semiautomatic Color Space Quantization Method

The purpose of the proposed Semiautomatic Color Space Quantization Method is to define a mapping from image colors to the named color classes, modeling the human perception of color names. The method consists of two steps (Figure 1-a):

(Step 1) Uniform Quantization: The standard RGB color space, with 256 levels (8 bits) per color axis (R, G, and B), is uniformly partitioned into color cells (equal cubes), by uniformly quantizing each axis into 4, 6, and 8 levels, resulting in partitions containing 64, 216, and 512 color cells, respectively. Each color cell is represented by the point in its center (Figure 1-b), called *color instance*. Uniform Quantization step defines the mapping from the original color space containing 16 millions of colors to the three sets of color instances (containing 64, 216, and 512 colors, respectively).

(Step 2) Semiautomatic Classification: The obtained color instances are classified into perceptually uniform classes, corresponding to the basic color names, through the following two steps:

(Step 2-a) Automatic Classification: The standard definitions of the 10 basic color names [8], representing the *color classes*, are taken from the X Window System. In order to classify the color instances into the color classes, the direct classification using the training set, based on the *nearest neighbor rule* (1-NNR [2]) is employed. According to 1-NNR, given the color classes (*training set*), each color instance is assigned to closest the color class. The distance used is the Eucledian distance of the color points in the RGB, HSV, $L^*a^*b^*$, $L^*u^*v^*$, and MTM color spaces.

(Step 2-b) Manual Reclassification: The Automatic Classification is not expected to produce a mapping that perfectly matches the human perfection. Therefore, the classification errors are corrected manually by a human, using the custom tool (*RGB Color Space Visualization Tool*), that visualizes the quantized RGB color space, showing the mapping of the color instances to the color classes.

The number of color instances is a tradeoff between the precision of the quantization and the computational cost of the Semiautomatic Classification step. Even for the images with the rich color content (large number of colors), 512 instances should be sufficient [7]. The number of color classes is arbitrary and depends on the application. In general, the basic color names are sufficient, while in a particular application more color variations (e.g. *light yellow*, *very dark red*) might be necessary. The Manual Reclassification step, enabling the color space quantization to be fine tuned through a questionnaire, guarantees the highest possible correspondence of the proposed color mapping with the human perception.

Uniform Quantization (Step 1) defines a simple mathematical transformation from the RGB coordinates of an arbitrary image color to the coordinates of the color instance representing that color. Semiautomatic Classification (Step 2) defines a lookup table mapping each color instance into the corresponding color class (i.e. color name), in accordance with the human perception. Combination of the two steps represents the overall mapping from the image colors to the named color classes. Once this mapping is defined, it can be used for the image segmentation (by backprojecting the color classes onto the image), or the image retrieval (where the color content of an image is represented as a histogram showing the distribution of the image colors over the color classes).

4. Experiments and Performance Evaluation

4.1. Comparative Performance Evaluation

The performance of the proposed color space quantization method is evaluated through the comparison with the existing methods, with respect to the perceptual quality and the computational cost. Since the Uniform Quantization is a common preprocessing step in the color space quantization methods, and the Manual Reclassification is unique to the proposed method, *only the Automatic Classification step is evaluated* by the comparison with the hierarchical clustering, competitive learning, and MTM quantization methods.

The *perceptual quality* of the quantization refers to how close the mapping of the color instances to the named color classes approximates the human perception. Accordingly, the proposed similarity between the mappings resulting from the different methods and the human-defined mapping (obtained through a questionnaire involving 5 persons) is defined as:

$$Similarity = \left(\frac{\text{number of correctly classified color instances}}{\text{total number of color instances}} \right) \times 100\% \qquad (1)$$

Table 1: (a-b) Comparison of the proposed Automatic Classification step with the existing methods, with respect to the perceptual quality; (c) Comparison of the methods using the color space giving the highest perceptual quality

(a)

Color Space	Similarity (%)											
HSV	70	65	68	68	61	66	68	65	69	69	65	67
MTM	58	50	48	52	61	55	40	52	58	46	45	50
RGB	44	43	44	44	48	32	47	42	36	42	48	42
L*u*v*	38	33	31	34	33	31	26	30	19	31	29	26
L*a*b*	28	27	27	27	23	18	28	23	23	24	23	23
#Instances	64	216	512	Avg.	64	216	512	Avg.	64	216	512	Avg.
Method	Proposed				Hier. Clust.				Compet. Learn.			

(b)

Color Space	Similarity (%)			
MTM	63	61	62	62
#Instances	64	216	512	Avg.
Method	MTM Quant.			

(c)

Method	Similarity (%)			
Proposed (HSV)	70	65	68	68
Compet. Learn. (HSV)	69	69	65	67
Hier. Clust. (HSV)	61	66	68	65
MTM Quant.	63	61	62	62
#Instances	64	216	512	Avg.

Table 2: Comparison of the proposed Automatic Classification step with the existing methods, with respect to the computational cost

Method	Computation Time (ms)			Comp. Time Ratio			
Proposed	0.25	0.61	1.30	1.0	1.0	1.0	1.0
MTM Quant.	0.43	1.02	2.19	1.7	1.7	1.7	1.7
Compet. Learn.	100.00	200.00	310.00	400.0	328.0	238.0	322.0
Hier. Clust.	8.00	340.00	4,400.00	32.0	557.0	3,385.0	1,325.0
#Instances	64	216	512	64	216	512	Avg.

The *computational cost* is defined to be time required to generate the mapping of the color instances to the named color classes (measured under Linux OS, on a 500MHz Pentium 3 machine).

In the experiments evaluating the perceptual quality, different methods are tested using the 5 different color spaces (Section 3.), to examine the effect of the color space selection to the quality of the color quantization. MTM Quantization method is, however, limited to the MTM color space. HSV color space gives the best results for all the methods (Table 1-a, 1-b). The proposed method in average results in the 1.5–8.8% increase in the perceptual quality (Table 1-c).

The computational cost is not affected by the color space selection. As the experiment results for each method show (Table 2), the proposed methods results in 1.7–1,327.0 times decrease in the computational cost.

The objective of the Automatic Classification step is to produce a mapping as-similar-as-possible to the human-defined mapping, so that the human intervention involved in the Manual Reclassification step is minimized. From that point of view, and considering the computational cost as well, the proposed method outperforms the existing ones. The significance of the proposed Manual Reclassification is obvious from the fact that none of the methods results in similarity higher than 70%, meaning that the Manual Reclassification brings over the 30% increase in the perceptual quality.

Table 3: Comparison of the proposed Semiautomatic Color Space Quantization Method with the MTM Quantization, applied to the image retrieval from a database of 1,100 color images

Method	*Retrieval Precision (%)*										
Proposed	98	98	76	98	85	98	96	86	85	100	92
MTM Quant.	97	70	72	92	78	95	59	93	35	85	77
Queried Color	red	orange	brown	yellow	green	blue	purple	white	gray	black	*Avg.*

4.2. Application to Image Retrieval

The proposed Semiautomatic Color Space Quantization Method supports image retrieval using the color names, similarly to the MTM Quantization method [5]. The two methods are compared by incorporating them into an image retrieval system [9]. Color histograms of the database images are computed using both methods, and the retrieval precision [7] is evaluated. As a test database, 1,100 color photographs of fruits, vegetables, roses, skies, and textiles from the *Corel Corporation*'s "Corel Gallery 3.0" image database are used. Relevant images are determined through a questionnaire involving 5 persons. Ten queries were issued, each asking for the images whose dominant color is one of the 10 color classes. Retrieval precision for each of the 10 queries is shown in Table 3. The proposed method in average results in 16% increase in the retrieval precision.

5. Conclusion

The color space quantization problem is considered from the viewpoint of image retrieval from large scale databases (e.g. the Internet). A semiautomatic color space quantization method is proposed, supporting image retrieval using color names. Comparison of the proposed Automatic Classification step with the clustering and the manual quantization methods shows 1.5–8.8% increase in the perceptual quality, and 1.7–1,327.0 times decrease in the computational cost. The proposed Manual Reclassification step additionally brings over 30% increase in the perceptual quality, resulting in a color space quantization closely approximating the human perception of color names. Applied to image retrieval and compared with the MTM quantization [5], the proposed method leads to 16% increase in the retrieval precision, enabling a more effective retrieval from the large scale image databases.

References

[1] Wang, J., Yang, W. J. and Acharya, R. *Color space quantization for color-content-based query systems.* Multimedia Tools and Applications,Vol. 13, pp. 73–91, 2001.
[2] Schalkoff, R. J. *Pattern recognition: statistical, structural and neural approaches.* John Wiley & Sons, 1992.
[3] Coridonni, J. M., Del Bimbo, A. and Vicario, E. *Image retrieval by color semantics with incomplete knowledge.* Journal of the American Society for Information Science, Vol. 49, No. 3, pp. 267–282, 1998.
[4] Uchiyama, T., and Arbib, M. A. *Color image segmentation using competitive learning.* IEEE Transactions on Pattern Analysis and Machine Intelligence, Vol. 16, No. 12, pp. 1197–1206, December 1994.
[5] Gong, Y., Chua, H. C. and Xiayoi, G. *Image indexing and retrieval based on color histograms.* Multimedia Tools and Applications,Vol. 2, pp. 133–156, 1996.
[6] Niblack, W., Barber, R., Equitz, W., Flickner, M., Glasman, E., Petković, D., Yanker, P., Faloutsos, C. and Taubin, G. *The QBIC project: querying images by content using color, texture, and shape.* In SPIE 1908, Storage and Retrieval for Image and Video Databases, pp. 173–181, February 1993.
[7] Del Bimbo, A. *Visual information retrieval.* Morgan Kaufmann Publishers, Inc., San Francisco, California, 1999.
[8] Berlin, B. and Kay, P. *Basic color terms: their universality and evolution.* Berkeley, CA: University of California Press, 1991.
[9] Stejić, Z. *Integrated retrieval of images and text using natural language "kansei" expressions.* M.Sc. thesis, Dept. of Comp. Intelligence and Sys. Science, Tokyo Institute of Technology, Yokohama, Japan, March 2001.

KES '01
N. Baba et al. (Eds.)
IOS Press, 2001

Analysis of Route Choice Behaviour Based on Questionnaire and 3D Traffic Flow Simulator

Toshihiro Hiraoka*, Kouhei Okabe*, Hiromitsu Kumamoto*, Osamu Nishihara*
and Kenji Tenmoku**
*Dept. of Systems Science, Graduate School of Informatics, Kyoto University,
Yoshida Honmachi, Sakyo-ku, Kyoto, Japan*
**Sumitomo Electric Industries, Shimaya 1-1-3, Konohana-ku, Osaka, Japan*

Abstract. To model driver's route choice behaviour, we executed a questionnaire and an experiment with a simplified 3D traffic flow simulator. Questionnaire result shows many drivers prefer the criteria such as cognitive resource demands minimization to travel time minimization when a destination is unfamiliar. Simulation result also shows the existence of the criteria, and that VICS information reduces the travel time.

1. Introduction

Annual traffic congestion losses in Japan amount to about 24 hours per 1 person; in money conversion, about 12 thousand billion yen nationwide. Some traffic management systems in order to ease traffic congestion have been introduced, for example, VICS (Vehicle Information and Communication Systems), signal control systems and so on. VICS provide car navigation systems with the information about congestion, route guidance, predicted travel time and parking area almost real time.

Generally, traffic flow simulator is used in verification of the traffic management systems, but a route choice model embedded in the simulator is very simple one such as choosing the route whose predicted travel time is minimum. That model lacks the reproduciblity of the actual driver's behaviour, so it is difficult to verify what kind of influences or what kind of phenomena happen by the introduction of navigation systems and VICS.

In this paper, we analyse the actual route choice behaviour. At first, the questionnaire result about route choice behaviour is analysed, such as route choice criteria and information collection methods. Experiments are performed using the simplified 3D traffic flow simulator having driver's front view and a navigation screen. We verify the consistency between questionnaire and experiment.

2. Questionnaire about route choice behaviour

2.1 Questionnaire contents

A questionnaire is composed by a question about 1) attributes of examinees (sexuality, age, driving experience, driving skill level, etc.), 2) attitude to navigation system and VICS, 3)

Table 1 Attributes of Examinees

Sex \ Age	Under 19	20 to 24	25 to 29	30 to 39	40 to 49	50 to 59	Over 60	No answer	Total
Male	4	75	41	32	22	21	8	1	204
Female	2	13	14	7	6	6	1	0	49
Total	6	88	55	39	28	27	9	1	253

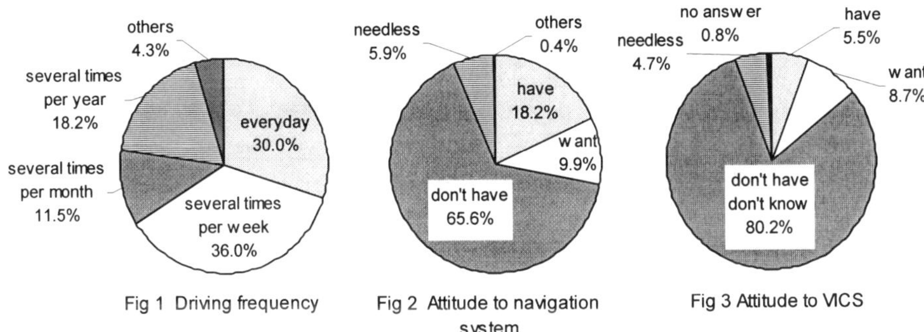

Fig 1 Driving frequency   Fig 2 Attitude to navigation system   Fig 3 Attitude to VICS

route choice criteria in the case where a destination is unfamiliar or familiar, 4) information collection methods for route choice.

2.2 Examinee attributes and attitude to navigation system and VICS

A questionnaire was executed on paper and World Wide Web. Table 1 shows attributes of examinees. This proportion is similar to actual data in Japan, except for the smaller middle age and elder drivers in questionnaire. Driving frequency, attitudes to navigation system and VICS are shown in Fig 1, Fig 2 and 3 respectively.

According to recent researches, the number of passenger cars in Japan amounts to a little more than 50 million, and 7.2 million navigation systems and 2.8 million VICS units are delivered. The ownership rate of navigation system and VICS unit in this questionnaire are similar to actual rate. Fig 3 indicates that the popularity and interest to VICS are very low.

2.3 Cluster analysis

Cluster analysis is applied to the result of following Question 1 and 2. It is an algorithm that can classify the population into finite groups with similar members. There are many methods that calculate similarity. Ward method based on Euclid distance is used in this paper.

2.4 Route choice criteria

Questions about route choice criteria are asked in the both case where the destination is unfamiliar and familiar. Only unfamiliar case is shown in this paper.

Most general route choice criterion is *"travel time"*, in other words, driver chooses the route whose travel time is expected to be minimum. Some researches suggest criteria such as *mileage, runnability* and *comfortableness*. To survey what criteria affect driver's route choice behaviour, Question 1 in Fig 4 was given to examinees.

Four clusters, CLA1 to CLA4, are classified based on a cluster analysis for Question 1 (Table 2). It is shown that runnability and comfortableness are taken seriously in the largest cluster CLA1. Runnability is a driver's desire to arrive the destination certainly, and comfortableness is a desire to drive easily. Therefore, we define CLA1 as *cognitive resource demands minimization type*. And Cluster CLA2, CLA3 and CLA4 are also defined as *travel*

Q1. You are going to drive a car toward unfamiliar destination. What are your route choice criteria in order to decide the route? Please answer the priority percentage of the following route choice criteria. (Sum of the percentages must be 100%)

    a) Travel time   priority:____ %   b) Mileage   priority:____ %

    c) Runnability / comfortableness   priority:____ %   d) Others   priority:____ %

Fig 4   Question 1 (About Route Choice Criteria)

Table 2 Cluster Analysis Result for Route Choice Criteria (Question 1)

Criteria / Cluster	Travel time	Mileage	Runnablity & Comfortableness	Others	Numbers(%)		Type
CLA1	25.4	16.1	**57.1**	1.3	150	(59.3%)	Cognitive resource demands minimization
CLA2	**81.9**	8.4	8.4	1.0	54	(21.3%)	Travel time minimization
CLA3	**42.5**	**42.4**	14.2	1.3	29	(11.5%)	Synthetic evaluation
CLA4	16.9	12.7	15.3	**55.3**	20	(7.9%)	Other

time minimization type, synthetic evaluation type and *other type* respectively.

The cluster size of CLA2, which is regarded as conventional route choice criteria, is 21.3 % and this is about one-third of that of CLA1. This result shows that CLA1 dominates CLA2 because the former is most major route choice criteria when the destination is unfamiliar.

By investigating the correlation between the result of route choice criteria and the attributes of examinees, it is cleared that a driver who is young, beginner and poor at driving tends to belong to CLA1, and that a driver who is elder, veteran and good at driving tends to belong to CLA2 or CLA3.

We have to pay much attention to the followings when route choice criteria are modelled. It's better to use multiple criteria types such as a combination of CLA1 and CLA2, because this questionnaire result reveals that actual route choice criteria are not so simple and uniform. The occurrence ratio of criteria types must be considered.

There is a difficult problem on modelling the route choice criteria CLA1. It is how to formulate the criteria of runnablility and comfortableness, e.g., right / left turn, type of road. As one solution, there is a method to convert the cost of right turn into travel time, but another problem how to set up the cost takes place because it is mental value of each drivers.

2.5 Information collection methods to evaluate route choice criteria

Question 2 in Fig 5 was asked to investigate what information drivers refer to in order to decide the route by evaluating their route choice criteria. It is assumed that all drivers have navigation systems.

Table 3 shows the result of cluster analysis. Four clusters, CLB1 to CLB4, are classified based on the cluster analysis. This result shows that the largest cluster is CLB1, which is 44.7 % of all, where drivers attach weight to map information. We define the information collection method CLB1 as *map oriented type*. CLB2, CLB3 and CLB4 are also defined as *navigation oriented type, navigation dependent type* and *navigation independent type* respectively.

If a driver has a navigation system and a VICS unit, he/she knows congested areas and estimates the travel time more precisely. In this way the navigation system brings many merit

Fig 5   Question 2 (About Information Collection Methods)

Table 3 Cluster Analysis about Information Collection Methods (Question 2)

Method / Cluster	Navigation	Map	Guide sign	Numbers(%)		Type
CLB1	16.8	**62.3**	20.8	113	(44.7%)	Map oriented
CLB2	**53.5**	26.9	19.6	66	(26.1%)	Navigation oriented
CLB3	**89.7**	2.3	8.0	42	(16.6%)	Navigation dependent
CLB4	7.0	36.6	**56.4**	32	(12.6%)	Navigation independent

to drivers, but the cluster CLB1 is a slightly larger than the sum of CLB2 and CLB3. This will be derived from the following reason. 1) Drivers don't notice the merit because of a low ownership rate of navigation system and a low popularity of VICS. 2) The present problem of the navigation system may lower user's trust even though he has it.

The correlation of the result of Question 1 and Question 2 proves that the navigation information is taken more seriously (a rate to belong to CLB2 or CLB3 is higher) from the drivers who belong to CLA2 than those who belong to CLA1, CLA3 and CLA4. This result is thought to reflect that the navigation information is useful for the estimation of travel time.

The questionnaire shows that the navigation information is, at present, almost equivalent to the map information. It is expected that drivers recognize the utility of the navigation information and the rate of drivers who belong to CLB2 or CLB3 increases, when navigation system becomes more popular from now on.

3. Simplified 3D traffic flow simulation

3. 1 Outline of simulation

Fig 6 shows a simulator screen. Experimental subjects choose a route and drive from origin to destination in the simulation area shown in Fig 7 (about 5×5km).

Three types of experiments were executed. In Experiment 1, information about the area map and current position is provided on the navigation screen. In Experiment 2, shortest route information calculated by Dijkstra method is provided in addition to the condition of Experiment 1. In Experiment 3, VICS information (congestion information) is provided in addition to the condition of Experiment 2.

Main congestion sections are shown by dark lines in Fig 7. The congestion situation is renewed every 5 minutes, and experimental subjects can check the situation with 5 minutes maximum delay on the navigation screen.

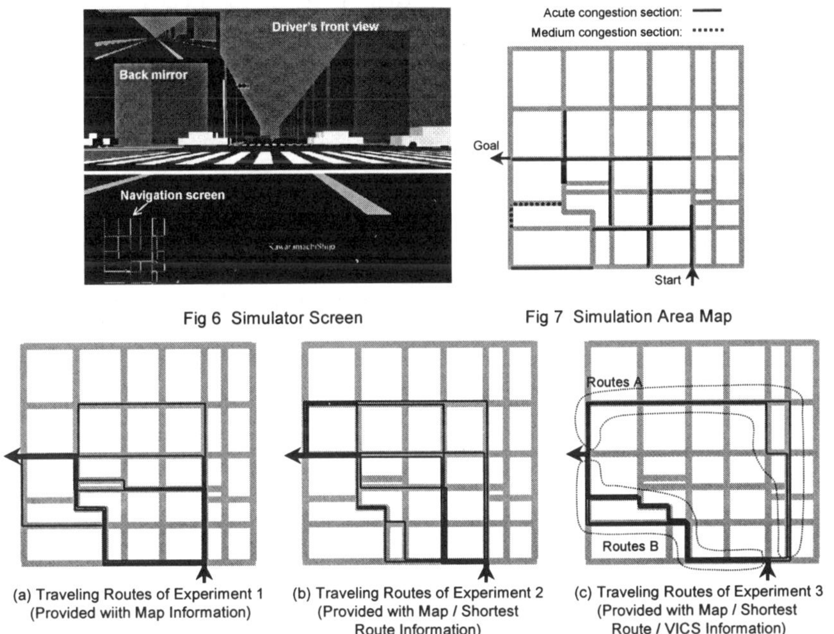

Fig 6 Simulator Screen

Fig 7 Simulation Area Map

(a) Traveling Routes of Experiment 1 (Provided wiith Map Information)

(b) Traveling Routes of Experiment 2 (Provided with Map / Shortest Route Information)

(c) Traveling Routes of Experiment 3 (Provided with Map / Shortest Route / VICS Information)

Fig 8 Simulation Area and Simulation Results

Table 4 Simulation Results

Experimental Subjects	Exp.1	Exp.2	Exp.3
#1	1266	1219	726
#2	486	1766	737
#3	2132	2111	989
#4	1279	1233	739
#5	1626	1233	746
#6	1547	1267	725
#7	1627	1627	750
Ave.	1423	1494	773

Experiments were executed to seven men and women in twenties and thirties who have driver's licence. Route data as well as intentions of behaviours such as congestion avoidance or turns were recorded by direct interviews.

3. 2 Simulation results

Table 4 shows that VICS information reduces travel time almost 50%. In Experiment 3, each experimental subject succeeded to avoid congestion and travelling routes concentrated into Routes A and Routes B shown in Fig 8(c). Routes A is short mileage one with avoiding congestion, and Routes B is simple one with smaller number of turns as well as congestion avoidance. Routes A results from the route choice criteria CLA1 and Routes B from CLA2 or CLA3.

The average number of turns at Experiment 1 is 3.3, and that of Experiment 3 is 6.1. The minimization of the number of turns is one of the criteria satisfying cognitive resource demands minimization. But providing the congestion information changes the order relations between the cognitive resource demands minimization criteria and the travel time minimization criteria, and causes drivers to turn actively. In other words, it represents that drivers change their route choice criteria corresponding to the difference of the information collection methods or the information itself which drivers can get at that time.

Fig 8(b) shows that only one subject chooses the shortest mileage route provided on the navigation screen, because this route intersected many congestion sections.

As mentioned above, the congestion information provision may cause route concentration. This suggests that congestion sections may vary as VICS becomes more popular, and we have to cope with the new congestions.

4. Conclusion

The result of a questionnaire about route choice behaviour shows the criteria such as cognitive resource demands minimization is major one when a destination is unfamiliar. The simulation result shows 1) the existence of the criteria, 2) the effectiveness of VICS information for reduction of travel time and 3) the possibility of route concentration caused by VICS information.

We have to analyse other case, such as where a destination is familiar, and study the method how to formulate the criteria in order to embed in a traffic flow simulator.

References

[1] H. Shimoura *et al.*, Evaluation of the Effect of DRGS using Traffic Flow Simulation System, the Second World Congress on Intelligent Transport Systems, 1995.
[2] T. Saito *et al.*, Analysis about route choice characteristics of the driver by questionnaire investigation, Proc. of IEEJ Meeting on Road Transportation (RTA-98-27), 1998, pp.25-29 [*in Japanese*].
[3] T. Hiraoka *et al.*, A Study of Dynamic Route Selection Based on Simplified 3D Traffic Flow Simulator, Proc. of 40th SICE Annual Conference, CD-ROM, 2001 [*in Japanese*].

Knowledge Re-formulation in Design

Martin DZBOR[*]
Knowledge Media Institute, Open University, Milton Keynes MK7 6AA, UK

Abstract. Many current design theories leave open the question of *'solution talkback'*. Discovery of new design goals and re-formulation of the existing ones is as important as the development of design solutions. This paper looks at knowledge-intensive support to re-formulation in design using tools working with explicit and tacit knowledge. A tool for knowledge acquisition and re-formulation in design is presented drawing upon the methodology developed in house as a part of the ongoing research project.

1   Introduction

Several researchers have investigated engineering design from various perspectives, and they agree on some typical characteristics of design problems, such as initial vagueness and incompleteness of design specification, or iterative solution [1-3]. The amount of combinations of the primitive design elements is usually large and contributes to the exploratory nature of the design problem solving. Below design is viewed as an iterative sequence of elementary actions; some of them are briefly discussed. Our framework uses the 'behavioural' theory of design that sees design as interplay between functional and structural objects [4, 5]. Details of the formal framework can be found e.g. in [6].

2   Summary of Methodology for Reflective Design

The theory of sequential design works with two entities – first order logic as a formal representation of explicit knowledge and unstructured tacit knowledge. Designers have explicit knowledge of various algorithms, methods, mathematical, and physical models. Their tacit knowledge prevails when they talk about 'liking a solution'; they are not able to express what is causing their attitude, only tacitly feel a hidden flaw. Hypothetical set Γ of *all* expectations that may be demanded from, or must not be violated by the designed artefact is *tacit*; i.e. without a definite structure and generally 'unstructurable'. Tacit knowledge complements the explicit one; it is something that is inherently present and used when tackling a problem, though without being expressed or explained explicitly [7]. Paper [6] formalises the position of tacit knowledge in design. One of its corollaries says that a design solution is *acceptable* if it is consistent with both the explicit requirements and designer's tacit expectations!

As argued in [7], designers try to formulate their tacit feelings in explicit terms. Such a formulation is not exactly trivial but unquestionably useful for the creative design; we refer to it as *'tacit inference'* and find it at the bottom of many kinds of design re-formulation. Tacit expectations from the designed artefact help formulating the explicit requirements and constraints thus pushing design process forward. Eventually, designer's *'tacit satisfaction'* with the solution s_i may serve as a stop condition of the iterating design process. Explicit requirements in the step i then act as a *sufficient design specification* and s_i as an *acceptable solution*. Though 'sufficiency' and 'acceptability' are far from the 'optimality', we think the design is more about the 'acceptability' than 'optimality'.

2.1   Basic reasoning techniques in design

Design is ill structured not only in respect to the problem space but also with regard to the reasoning strategies used to navigate in such a space. This section briefly summarises the positions of selected logical operations in reasoning on a knowledge level, and highlights major drawbacks. Details are in paper focusing on the theoretical aspect [6].

I. Logical abduction $(R, A \Rightarrow R) \vdash A$

Suppose there is given a set of desired requirements $F(x) \subseteq R$. In abduction, designers look for an artefact $A(x)$ that implies the desired functionality $F(x)$ consistently with the current design specification.

---

[*] Contact the author on mdzbor@acm.org.

If $F(x)$ can be proven from the current design theory; i.e. $A(x) \vdash F(x)$, in line with the completeness theorem [8] it is possible to conclude that artefact $A(x)$ is sufficient for the satisfaction of the desired requirements $F(x)$, and may be considered a (partial) design solution. Abduction investigates the sufficient means for the achievement of explicit design requirements. The issue with abduction is that a solution sufficiency does not have to hold in a longer run! Abduction is thus a tentative reasoning well suited to the exploratory nature of the design and on-the-fly construction of a design problem space.

II. Logical deduction $(A, A \Rightarrow D) \vdash D$

Abductive reasoning may reveal an artefact $A(x)$ satisfying certain explicit requirements $F(x) \subseteq R$; designers may also want to know the consequences such a tentative commitment. Such consequences may be deduced using the rules of the type $A(x) \Rightarrow D(x)$, or in general terms $A(x) \vdash D(x)$. Deduction is well described in the theory of logic [8]; we note only one corollary. Thanks to the completeness theorem, artefact $A(x)$ is consistent with some property or function $D(x)$ that was deduced and is not among the initial ones (i.e. $D(x) \not\subset R$). Unfortunately, nothing can be said about the consistency between the deduction $D(x)$ and the rest of design requirements R; i.e. holds?$(D(x) \vDash R)$. Even more important assessment of the deduction $D(x)$ is against the tacit expectations Γ; i.e. holds?$(\Gamma \oplus D(x))$.

III. Evaluation of tacit consistency[1] $(R, A, D) \oplus (R \cup D) \vee (R \cup \neg D)$

The *logical* consistency between the deduced consequence $D(x)$ and current explicit requirements R – i.e. $D(x) \vDash R$, can be assessed by some truth maintenance system [9]. However, logical consistency is not sufficient for ending the design process; a design solution has to be consistent also with the designer's tacit expectations. Two interesting results of tacit assessment are:

a. Deduced theorem $D(x)$ refers to a desirable implicit requirement that was originally not mentioned; it is tacitly consistent and extends the current explicit requirements – i.e. $R = R \cup D(x)$.

b. Deduced theorem $D(x)$ is an undesirable feature of a tentative solution and should be avoided in the further design. Its negation is an intuitive necessary condition of solution acceptability.

The former situation extends the current set of design requirements but does not bring any unexpected knowledge into design. It has thus only limited potential for the innovation. On the contrary, the latter one uncovers a new 'goal' $\neg D(x)$ and also introduces inconsistency to the logical theory ($\neg D(x)$ is intuitively desired and $D(x)$ is provable). We see this situation as *tacitly inconsistent* since it was discovered by tacit appreciation of the current solution and design specification. Tacit inconsistency is a powerful vehicle for the construction of a problem space in design because eventually the designer's tacit satisfaction with the solution and its consequences ($\Gamma \oplus A(x)$) may stop the design.

2.2 Design re-formulation $(\Gamma, R, A, D) \oplus R_{NEW}$

Typically, tacit inconsistency may be removed by re-formulating the solution; i.e. backtracking and choosing a functionally alternative artefact $B(x)$ that satisfies the same requirements $F(x) \subseteq R$ but does not exhibit the feature $D(x)$. This is the easiest way but not the only one. Designers may also modify their current interpretation of the available design specification; i.e. the may re-formulate or otherwise restrict the scope of validity of the current design requirements (in other words, $\Gamma \cup R \oplus R_{NEW}$). Another method of re-formulation observed in our experiments was an *assumption-based restriction* of the design. It was observed when the designer wanted to build on top of the current solution $A(x)$ but in the same time tried to restrict it by some conjectured condition $P(x)$ as shown in Eq. 1. The 'translation' of such a conjecture to the plain language says: '*Assuming we could find such $P(x)$ that would guarantee that $D(x)$ is not deduced, the solution $A(x)$ would be again acceptable and consistent.*' An assumption is a theorem with a tentatively accepted validity that can be used in a similar manner as other theorems 'proper'. The tentativeness shows later when a consequence is discovered that is clearly inconsistent with the assumption; assumption in question can be easily identified and cancelled if needed; the consistency would be quickly restored.

$$P(x) \Rightarrow \neg \{ (A(x) \wedge P(x)) \vdash D(x) \} \qquad \textbf{Eq. 1}$$

Paper [6] shows that the conjectured assumption may extend the current design specification as a new requirement or a constraint. Conjectured assumption is usually a condition constraining the scope of validity of an artefact $A(x)$. A requirement can be easily made up from the constraint if in addition

---

[1] Logical consistency is out of scope of this paper; symbol '$\oplus$' denotes tacit consistency, 'Θ' tacit 'inference' [6].

to restricting the solution by a condition $P(x)$, designer also demands a particular value for $P(x)$. In any way, the described conjecture of a new design assumption is clearly a result of non-monotonic reasoning and is closely related to other powerful non-monotonic techniques [10].

Nevertheless, the assumption-based restriction is not powerful enough in all cases of tacit inconsistency. Sometimes there are no feasible functional alternatives or constraining conditions known, and it is necessary to make a more radical reasoning step that would bring into designer's attention an entirely new object or concept he was unaware of before. Cross [11] refers to it as a 'creative leap'; we shall use the term from the theory of logic – i.e. *re-conceptualisation.* In other words, we seek such a set of conceptual objects ψ for which the current theorems $A(\psi)$, $D(\psi)$ or $F(\psi)$ lose their original interpretation leading to the tacit contradiction. In the next section we show briefly how such new objects were identified in our experiments with professional designers, and section 4 looks at the knowledge-intensive support for this non-trivial reasoning operation.

3 Design Re-formulation in Practice

In line with the theoretical framework assuming two complementary types of knowledge involved in design two different tools were used to capture design knowledge. Figure 1 shows a tool for the acquisition of explicit design requirements and management of different design contexts. A tool that was used for acquiring tacit expectations and tentative assumptions is depicted in Figure 2. Both tools are illustrated with snapshots from the experimental session, in which a control strategy of a paper mill extension for smoothing and winding the raw paper was designed [6].

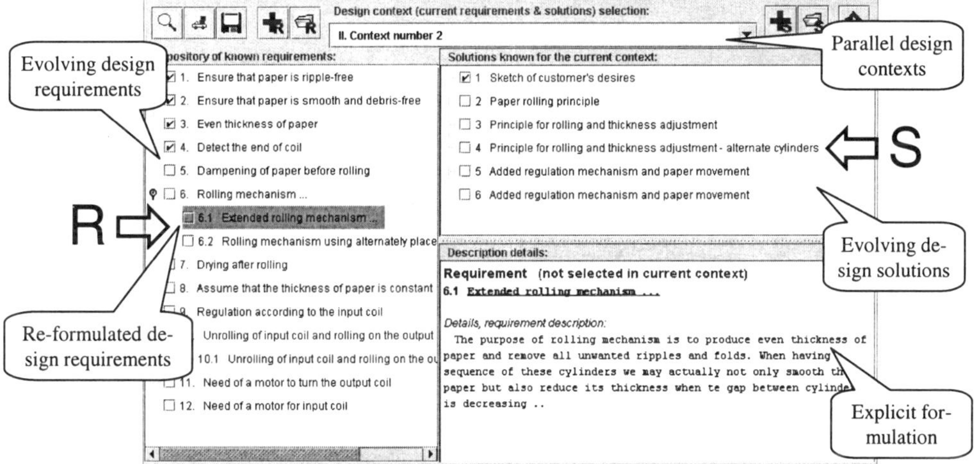

Figure 1. A tool for acquisition of explicit design knowledge

As Figure 1 shows, design requirements are captured in the left-hand panel, the partial solutions are in the top right-hand panel. Those requirements and solutions that belong to the current context are marked with a check. Designer has identified several contexts and he can easily switch between them. Attention shift from one context to the another restores the design state as it existed in the moment the particular context was saved. Such a context storage and restoration supports the pursuit of multiple parallel ideas or alternatives without loosing the local, context-specific information.

It was already mentioned that re-formulation of design requirements is a powerful tool for the construction of a problem solving space of a particular, ill-defined design task. For instance, in Figure 1 requirement number 6 was re-formulated two times – first, a generic rolling mechanism was refined to include pre- and post-processing units; second, the default linear layout of the assembly was modified to an alternate layout in order to address some issues uncovered tacitly. The original requirement 6 appeared in context 3, the first re-formulation (6.1) occurred in context 4 and finally, version 6.2 appeared in context 5. Designer thus recorded all milestones where he made a major tentative decision so that he could backtrack in case the decision showed as unsuitable. The tool keeps the different re-formulations of the same original requirement in different contexts to structure the ill-defined task and emphasise the possibility to cancel any tentative decision if needed in future design.

Explicit design requirements can be acquired straightforwardly; the acquisition of tacit expectations, justifications and reasoning chains leading to such requirements is more difficult. To accommodate the fluidness of tacit knowledge the tool for its acquisition has a form of a threaded 'debate', in which each thread represents a development of a selected idea often drawing upon designer's experience and considering a number of counter-proposals. For example, in Figure 2 we see how the idea of modifying the layout of the rolling drums (marked with '❶') appeared as a result of tacit feeling about low quality of the processed paper ('❷') and an intuitive idea of adding more drums and squeezing them together ('❸'). The particular thread illustrates the situation, in which the designer perceived a potential danger, and what was his initial reaction to avoid the undesired feature ('❷'). The initial remedy did not resolve the issue but it gave the designer enough material to immerse deeper into the design task and investigate closely the reasons why the undesired feature came to his attention and why the initial remedy ('❸') was not sufficient to resolve the deadlock. The thread in Figure 2 also shows the origin of one partial design solution – SOL4 from the previous one – SOL3 ('❹').

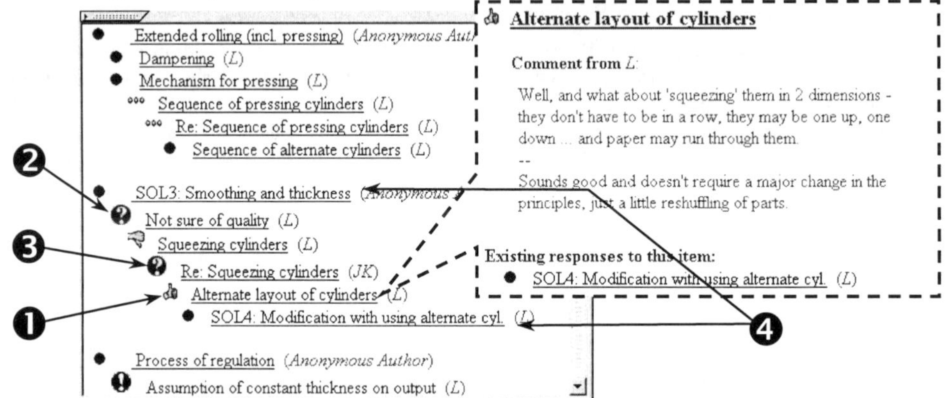

Figure 2. Threaded 'debate' as a tool for acquisition of tacit expectations

Typically, the conclusions of a thread produced a new explicit requirement or a partial solution. For instance, proposition marked as '❶' in Figure 2 leads to the re-formulated requirement number 6.2, whereas the initial idea ('❸') is reflected in the re-formulation 6.1 (see mark 'R⇨' in Figure 1). Thread conclusion is also recorded explicitly in the solution panel (marked '⇦S' in Figure 1). More details of the experiment including other snapshots and sketches of the assembly can be found in [6].

4 Knowledge-intensive Support for Re-formulation in Design

In this section some extensions of the existing acquisition tool with respect to the knowledge-intensive support for design re-formulation are discussed. Due to limited space the techniques mentioned below are not exhaustive, and are described rather superficially. The underlying principle is however, rather simple – since we cannot include a statement about the first-order theory within the theory, we need a complementary reasoning mechanism that can overcome this shortcoming. One of common techniques the expert designers are using whilst tackling ill-structured design tasks is reasoning by analogy. In design, this form of reasoning may take a previous design case, find similarities with the current task and transfer some explicit conclusions from the base to the target [5]. Analogy may help in the deduction filtering, evaluation of the consequences or re-formulation [6] acting on different conceptual levels – below we discuss the abstract analogy for design re-conceptualisation.

Reasoning by analogy knows several ways how to use the existing knowledge in the current task. We identified two large groups in our experiments – a model-based and case-based. The difference is in the usage of base analogs. In the model-based approach an existing design scenario is in the role of a model, and the current problem is seen as instantiation of a selected model. With the case-based approach, instead of instantiation the current task is solved 'similarly' as a previous one. Models for design are often abstracted design scenarios from the past, whereas the cases are concrete tasks with specific decisions and justifications.

Assume there is a base case B with some desired functionality $\Phi(x) \subseteq \Theta$ and known solution $\alpha(x)$. In the current (target) case functionality $F(x) \subseteq R$ is demanded and a solution is sought. The most elementary situations occurs if the cases contain the same requirements; i.e. $F(x) \equiv \Phi(x)$. A solution to the current problem can be either directly $\alpha(x)$ or if $\alpha(x)$ is too abstract it can be *instance-of(A(x), α(x))*. This is however, very rare situation; often the direct knowledge transfer is not possible. Therefore we define an abstracted analogy that uses the structure of the domain knowledge (see Eq. 2).

$$\text{abstracted}(\alpha, \beta) \Leftrightarrow (\exists \beta: \text{type-of}(\alpha, \beta)) \vee (\exists \beta, X: \text{type-of}(\alpha, X) \wedge \text{abstracted}(X, \beta)) \qquad \textbf{Eq. 2}$$

Now we have better chance for finding an analogous match between base and target cases. If the above-mentioned base solution $\alpha(x)$ was an abstraction of a generic prototype $\beta(x)$ then the target solution $A(x)$ may be analogous to $\beta(x)$ or more typical its instantiation; i.e. *instance-of(A(x), β(x))*. And even more abstract knowledge transfer may occur if potential solution $M(x)$ as an instance of $\beta(x)$ is too abstract for the direct usage. Then we may use the finding not as a case but as a model – thus performing a 'double transfer'; the resulting solution $A(x)$ could then be described by schema in Eq. 3.

$$\text{maybe-solution}(A(x)) \Leftrightarrow \exists \beta: \text{abstracted}(A(x), \beta) \vee \text{instance-of}(A(x), \beta) \vee \qquad \textbf{Eq. 3}$$
$$\vee (\text{abstracted}(A(x), M) \wedge \text{instance-of}(M, \beta))$$

An example of such an abstract analogy in our experiment is the identification of a generic rolling mechanism whose functions include 'modification of material surface', which in turn is an abstraction of the originally desired property 'smooth paper surface'. The sequence of rolling drums appears as an instance of a generic principle of rolling as an application of pressure on the surface [6].

5   Discussion and Concluding Remarks

In the paper we discussed the issues with re-formulation of design tasks focusing on two distinctive types of knowledge involved in this non-trivial operation. First, a toolkit for knowledge acquisition and structuring was introduced as a basic vehicle assisting the designer with the construction of a problem solving space for ill-structured design tasks. Second, some extensions of the tool were proposed that would provide the functionality for a computational support for the design. The proposed framework suits particularly well the reflective nature of design problems, and we think that the re-formulation of the design specification is at the bottom of iterative evolution of a design solution.

Similar methodological proposals can be found e.g. in Altshuller's work that proposes an algorithm (or better a heuristic) for resolving apparently insoluble technical problems, in which the required features were mutually contradictory [12]. His theory uses a similar interplay between the model- and case-based reasoning to overcome the deadlock. Nevertheless, we construct a similar approach to account not only for the explicitly inconsistent design specification but also for the situations with only tacit perception of some hidden imperfection of the proposed solution. Such an approach is valid because many potential contradictions do not come to the design explicitly; more typically they are identified through tacit reflection on the current solutions and specifications.

References

[1]   M. Dzbor, Explication of Design Requirements Through Reflection on Solutions. In *Proc. of 4th IEEE Conf. on Knowledge-based Intelligent Engineering Systems*. Brighton, UK, pages 141-144, 2000.

[2]   S. Nidamarthi, A. Chakrabarti, and T.P. Bligh, The significance of co-evolving requirements and solutions in the design process. In *Proc. of 11th ICED*. Finland, 1997.

[3]   H.A. Simon, The structure of ill-structured problems. *Artificial Intelligence*, 4:181-201, 1973.

[4]   T. Bylander and B. Chandrasekaran, Understanding Behavior Using Consolidation. In *Proc. of 9th Intl. Joint Conference on AI (IJCAI'85)*. Los Angeles, California, pages 450-454, 1985.

[5]   L. Qian and J.S. Gero, Function-Behaviour-Structure Paths and Their Role in Analogy-Based Design. *Artificial Intelligence for Engineering, Design, Analysis and Manufacturing*, 10:289-312, 1996.

[6]   M. Dzbor and Z. Zdrahal, Towards Logical Framework for Sequential Design. In *Proc. of 13th ASME DETC Conf. on Design Theory and Methodology*. Pittsburgh, USA, DETC01/DTM-21710, 2001.

[7]   S.D.N. Cook and J.S. Brown, Bridging Epistemologies: The Generative Dance Between Organizational Knowledge and Organizational Knowing. *Organization Science*, 10(4):381-400, 1999.

[8]   M.R. Genesereth and N.J. Nilsson, *Logical Foundations of Artificial Intelligence*. Morgan Kaufmann Publ., USA, 1987.

[9]   J. de Kleer, Problem Solving with the ATMS. *Artificial Intelligence*, 28(2):197-224, 1986.

[10] D. McDermott and J. Doyle, Non-Monotonic Logic. *Artificial Intelligence*, 13(1):41-72, 1980.

[11] N. Cross, Descriptive models of creative design. *Design Studies*, 18:427-440, 1997.

[12] G.S. Altshuller, *Creativity as an Exact Science*. Gordon & Breach Science Publ., USA, 1984.

KES '01
N. Baba et al. (Eds.)
IOS Press, 2001

Adding Event-driven Control Knowledge in a CommonKADS Knowledge Model

Lucile TORRES, Claudia FRYDMAN and Aguinaldo GARRIDO DE CEITA
DIAM-IUSPIM, Av. Escadrille Normandie Niemen, 13397 Marseille Cedex 20, France
lucile.torres@iuspim.u-3mrs.fr, claudia.frydman@iuspim.u-3mrs.fr

Abstract. The CommonKADS methodology became the standard for knowledge-intensive system development. In this methodology, the knowledge model describes the system requirements. It provides a specification of data, functions and control of the system. However, the control specification is only concerned with the function ordering, and it may be defined in the terms of a programming language. This is dedicated to transformational systems, but this does not suit reactive systems which behavior is driven by external events. In this paper, we propose to extend the CommonKADS knowledge model by providing experts with a medium for specifying the reactive part of the system behavior. This is achieved by completing the knowledge model language. Then, operationalizing the control part of the knowledge model into statecharts make possible the dynamics of a reactive system to be validated by simulation.

1. Introduction

The literature on software and knowledge engineering is almost unanimous in recognizing the existence of three dimensions for system modeling [1, 2, 4, 6, 8]:

- a *data dimension* describing the data and the relations between them,
- a *functional dimension* describing the functions in terms of transformations of data (input/output), and their decomposition in sub-functions,
- a *behavioral dimension* defining the temporal evolution and the execution conditions of the functions.

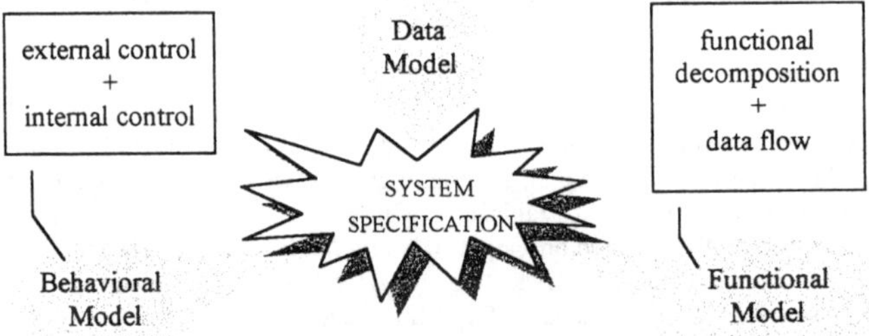

Figure 1: Three complementary views

The three views, represented in Figure 1, are indispensable and complementary. However, one of them is vital, depending on the type of system. Data modeling is preponderant for information systems. The functional model is the principal view for calculation or data processing systems. Finally, the control dimension is essential for supervision or control systems [2] or for graphical user interfaces. This third dimension includes two aspects: the function sequencing (internal control) and the system's reactivity to external events (external control). *"A reactive system is, in contrast with a transformational system, characterized by being to a large extent, event-driven, continuously having to react to external and internal stimuli"* [10]. Thus, the control of a reactive system must be described in terms of *"allowed sequences of input and output events, conditions, actions, perhaps with some additional information such as timing constraints"* [10], instead of specifying temporal order of functions.

In this paper, we are interested with modeling reactive knowledge-intensive systems. Among the methodologies for knowledge-intensive system development, CommonKADS became the leader for transformational systems, being interested especially to represent internal control of the system [8]. We propose to extend CommonKADS to reactive system specification. Since statecharts are the standard for modeling control of such systems, we adapt concepts coming from statecharts to make them useful by the knowledge providers (that are the experts in the application domain), and we will show how these concepts may be introduced in the CommonKADS methodology.

2. The CommonKADS Knowledge Model

The CommonKADS methodology provides a general framework for the development of knowledge-intensive systems[1]. CommonKADS is based on a model suite, where the knowledge model describes the system requirements (except those concerning the interfaces of the system with the environment that are described in another model). Thus, the knowledge model is concerned with problem solving, by describing in an implementation-independent way the role that different knowledge components play in problem solving. It is structured in three knowledge categories: domain, inference and task.

The domain category describes the data dimension of the model.

The inference knowledge describes the lowest level of functional decomposition (in the functional dimension).

The task knowledge completes the functional dimension of the model, and includes also the behavioral dimension. A task is associated with goals to be achieved, and it describes how these goals can be realized in a structural and a control way. The structure of a task is defined as its decomposition in subtasks and ultimately inferences. Its control part describes the dynamic behavior of the task: it represents the internal control of the reasoning process associated with the task by specifying the sequencing of its components (reasoning steps). The emphasis is placed on internal control, by carrying back all aspects of interaction with the outside world into a separate model, the communication model [8].

Internal control may be defined in pseudo-code including statements like conditions, loops, function calls. States and events are considered as a natural medium for describing the control traducing the interaction with environment (that is to say the reactivity of the system to external events): external control may be expressed by means of state-transition diagrams. The CommonKADS methodology recommends employing the UML (Unified

---

[1] "Knowledge-intensive" systems is used in the sense larger than "knowledge-based" systems, for systems integrating knowledge-rich components [8].

Modeling Language) [12] state diagram notation[2] to specify the communication control, or to complete task control in the knowledge model when the behavior of the task is strongly influenced by external events. UML state diagrams are an object-based variant of Harel's statecharts [10], that are themselves an extension of conventional finite-state machines and their graphical representation, state-transition diagrams. In particular, statecharts support a hierarchical decomposition of states (AND/OR decomposition), transitions between states owning to different hierarchical level and communication between concurrent components.

3. Event-driven Control Knowledge

The knowledge providers are traditionally experts in the application domain. They use to describe the behavior of reactive systems by expressing the possible sequence of events to be observed and the relations that bind events to functions to be realized. These experts' knowledge represents, in fact, discrete event abstractions of the dynamic behavior of the considered system. It must be defined in terms of states (of the system) and events (which occurrence provokes state changes of the system). However, states of the system are not explicitly mentioned in the dialogue of experts. Consequently, we propose to represent the experts' knowledge about dynamic system behavior, in terms of events that trigger functions or are produced by functions, instead of by means of state-transition diagrams. This allows experts to validate the knowledge they use to analyze the system since this knowledge is formulated in their natural language.

Table 1: Extension of the knowledge model language

CONTROL-STRUCTURE	::= 'INTERNAL:' PSEUDO-CODE ';' 'EXTERNAL:' 'VARIABLES:' VARIABLE-LIST ';' 'CONTROL:' EVENT-DRIVEN-CONTROL '.'
VARIABLE-LIST	::= variable-name ',' VARIABLE-LIST |
EVENT-DRIVEN-CONTROL	::= task-element EVENT task-element ',' EVENT-DRIVEN-CONTROL | INIT-EVENT task-element ',' EVENT-DRIVEN-CONTROL | task-element EVENT task-element | INIT-EVENT task-element
EVENT	::= event-type '(' CONDITION ')' | event-type | '(' CONDITION ')'
CONDITION	::= COMPARISON BOOLEAN-BINARY-OPERATOR CONDITION | COMPARISON
BOOLEAN-BINARY-OPERATOR	::= 'and' | 'or'
COMPARISON	::= ARITHMETIC-EXPRESSION COMPARISON-OPERATOR ARITHMETIC-EXPRESSION | '(' CONDITION ')' | 'not' CONDITION | 'in' task-element
COMPARISON-OPERATOR	::= '<' | '>' | '$\leq$' | '$\geq$' | '$\neq$' | '='
ARITHMETIC-EXPRESSION	::= TERM '+' ARITHMETIC-EXPRESSION | TERM '-' ARITHMETIC-EXPRESSION | TERM
TERM	::= FACTOR '*' TERM | FACTOR '/' TERM | FACTOR
FACTOR	::= '(' ARITHMETIC-EXPRESSION ')' | variable-name | CONSTANT
CONSTANT	::= integer | string
INIT-EVENT	::= 'init'

The knowledge model language provided by CommonKADS for the task control specification, is concerned with internal control and is defined by pseudo-code. We propose to enrich this language with grammatical structures for specifying event-driven control of tasks, by using familiar terms for experts. The structures we propose are so limited to input/output events of functions that are represented by tasks. These structures

---

[2] CommonKADS provides links with notations of the UML object-oriented development standard (class diagrams, activity diagrams and state diagrams) in any knowledge category of the knowledge model. This makes the knowledge modeling framework more familiar to object-oriented modelers.

allow experts to express when the task will be activated (by the occurrence of what event), whether the task terminates on its own (after a delay, or after an event occurrence), whether two tasks can be carried out in parallel (activated by the same event occurrence). The syntax of the knowledge model language, given in [8], may be completed as given in Table 1, using a BNF notation.

The *PSEUDO-CODE* non-terminal symbol in this grammar remains as defined in [8]. Let us consider the knowledge model with the three main tasks: *avionics-system* for managing an avionics system with a general mode and other concurrent subsystems, such as the radar, *general-mode* for the general mode subsystem, and *radar* for managing the radar subsystem. According to the grammar, the event-driven task control that describes the dynamic behavior of the components of the tasks may be defined in the knowledge model as given in Table 2. This control expresses the experts' knowledge that takes the general form "when the radar is off, the plane is in navigate mode, and the corresponding switch is thrown, the radar is activated". In this example, every external control is concerned with one abstraction level. However, the language allows knowledge engineers to specify task-triggering conditions between tasks of different abstraction levels.

Table 2: Examples of external control of tasks

```
TASK-METHOD
    REALIZES: avionics-system;
    DECOMPOSITION: TASKS: general-mode, radar;
    ROLES: ...;
    CONTROL-STRUCTURE: INTERNAL: ...; EXTERNAL: VARIABLES: ; CONTROL: init general-mode, init radar;
END TASK-METHOD avionics-system;
TASK-METHOD
    REALIZES: general-mode;
    DECOMPOSITION: TASKS: on-ground, navigate;
    ROLES: ...;
    CONTROL-STRUCTURE: INTERNAL: ...; EXTERNAL: VARIABLES: ; CONTROL: init on-ground,
    on-ground take-off navigate, navigate touch-down on-ground ;
END TASK-METHOD general-mode;
TASK-METHOD
    REALIZES: radar;
    DECOMPOSITION: TASKS: on, off,stand-by;
    ROLES: ...;
    CONTROL-STRUCTURE: INTERNAL: ...; EXTERNAL: VARIABLES: ; CONTROL: init off, on switch-off off,
    off switch-on (in navigate) stand-by ; stand-by (delay=5) on;
END TASK-METHOD radar;
```

Finally, the dynamic behavior of problem solving-process can be analyzed by simulating the activation of tasks. The activation conditions of the tasks can be implemented as predicates applied to logical types of event. Such events characterize not only the conditions and the results of the activation of tasks but also the inputs and outputs of the functions of the future knowledge system. Thus the experts are able to validate the dynamic properties of their problem-solving process by simulation. Thus, we have first to make the CommonKADS knowledge model executable. We propose to translate automatically this model into activity charts and statecharts for which efficient simulation tools exist. Indeed, Harel statecharts are the most popular variety of formalisms used for reactive systems behavior specification. They own powerful and flexible features such as nesting of states, concurrency, conditions on transitions, propagated transitions, instantaneous actions (on transitions, on state entry, on state exit), activities occurring as long as a state is active.

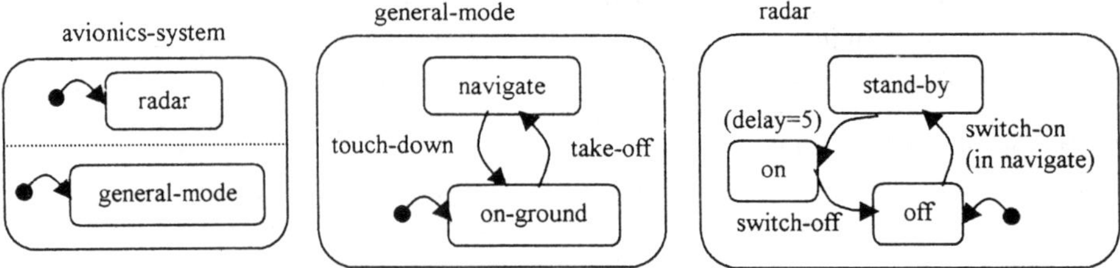

Figure 2: Statecharts for the *avionics-system* task

Activities correspond to functions. So making analogy between a task (in CommonKADS sense) and an activity comes naturally. Then activity charts represent the hierarchy of the tasks. Since a statechart may be associated with one activity, specifying the control of this activity, we propose to translate the external control of each task into a statechart by associating states with the subtasks of the task, and transitions with the events triggering the tasks. Figure 2 describes the statecharts obtained for the example of Table 2.

Finally, statecharts execution allows experts to analyze, simulate step by step and debug the dynamic behavior of the system, at any abstraction level and at any development step (i.e., prior to building the system of course, and even prior to fully specify all its tasks).

4. Conclusion

Operationalizing the task knowledge category in the CommonKADS knowledge model concretizes the dynamic aspects of the reasoning process. It makes possible the validation of these aspects by simulation. Simulation shows when, why, and how the tasks and inference structures are activated and inactivated to solve a problem. This aspect is particularly crucial for the design of a reactive knowledge system.

For this, we proposed to enrich the knowledge model language concerning task category, with specification of event-driven behavior, in respect with the way experts think about it. The next step consisted in operationalizing the behavioral description in the knowledge model into an executable model. So we provided experts with executable specifications of the system behavior that they can exploit more easily than these specifications can be graphically represented and than they correspond to the medium of their knowledge.

References

[1]   D. Harel, H. Lachover, A. Naamad, A. Pnueli, M. Politi, R. Sherman and A. Shtul-Trauring, STATEMATE: A Working Environment for the Development of Complex Reactive Systems, 10th International Conference on Software Engineering, Singapore, April 1988.

[2]   Jean-Paul Calvez, Spécification et conception des systèmes. Une méthodologie,Masson, Paris, 1990.

[4]   Ian Sommerville, Software Engineering, Addison-Wesley, Reading, MA, 1992.

[6]   Lucile Torres and Claudia Frydman, Veryfying and Validating Specifications of Knowledge Based Systems, ECAI 98, Brighton U.K, August 1998.

[8]   Guus Schreiber, Hans Akkermans, Anjo Anjewierden, Robert de Hoog, Nigel Shadbolt, Walter Van de Velde and Bob Wielinga, Knowledge Enhineering and Management, MIT Press, London, 2000

[10] David Harel, Statecharts: A Visual Formalism For Complex Systems, Science of Computer Programming, Vol. 8, pp. 231-274, Elsevier Science Publishers B. V., North Holland, 1987.

[12] Object Management Group, OMG Unified Modeling Language Specification, Version 1.3, http://www.omg.org/, June 1999.

An Information Technology Application to Training Processes

Edson Pacheco PALADINI
Universidade Federal de S. Catarina - CP 476 - 88040 970 - Florianópolis - SC - Brazil

Abstract. This paper discusses a new information technology application to training processes. This application makes feasible a new model of training. As conceptual support to the proposed application, a group of tools of Artificial Intelligence was used, more specifically, interactive modules of a Decision Support Expert System. We have called it "dynamic training" process. In this model, the person being trained does not just possess a passive performance, but, conversely, he acts directly in decision processes that are presented to him. This model is called "intelligent" because (1) it has adaptive characteristics (which means decisions taken according to each situation and to the moment, in an evolutionary way) and (2) because Artificial Intelligence tools are used in the training process. The training is aimed at people that act in decision and managerial processes of industrial organizations. Essentially interactive, the methodology seeks to create a mechanism that makes the person being trained responsible for the presented decisions, changing him into the protagonist of the process. Besides describing the training model, this paper also describes the necessary computational support for its development.

1. Introduction

This paper shows how Artificial Intelligence can be used to design a new information technology application to companies that want to use an interactive process of sharing technical information. So it discusses a new type of training process, i.e., this training process results from a new way to allow people to access information by using a specific technology application. This interactive process makes use of basic techniques of Artificial Intelligence (AI) and other devices of interactive computer science. The training model is structured for "operators". The worker of the factory is called an operator, i.e., the basis of the production process.

This model is known as "dynamic training" because, in this case, the operators being trained do not have just a passive performance, but, on the contrary, he/she acts directly in decision-making processes that are presented to him/her. The training is for operators that will act in decision-making processes of industrial organizations. It is considered also that here there is an intelligent training process, since it has adaptive characteristics, what means a training process for people whose decisions are taken according to each situation and to the moment, in an evolutionary way. The word "intelligent" is used also to refer Artificial Intelligence (AI). In fact, AI tools are used in the training process. There are in technical literature some similar experiments, like the one described by Britz [1] to the statistics area. But there are many differences when comparing these experiments with those conducted in our project. Essentially interactive, the methodology aims at creating a mechanism that makes the operator responsible for decisions when facing some practical problems presented to him/her, giving him/her the lead role in the process (the methodology reproduces the operator's typical activities). The conceptual support to the project includes a group of tools from Artificial Intelligence. More specifically, interactive modules of a Decision Support Expert System have been used. This paper describes the training model and discusses the necessary computational support for its development and the main results obtained. The experiment was conducted along five years at Brazilian industrial companies (1995-2000), including 307 operators of 29 different industries. The main characteristic of the method is the active, dynamic and intelligent participation of the operators in a training process.

2. Intelligent training processes

The main characteristic of the training process described here is the use of Artificial Intelligence tools. These tools can create interactive modules. In particular, an Expert System is used here. Expert Systems (ES) are computational programs that try to solve problems in specific fields of the knowledge. Because of the concepts ES have and because of the diversity and relevance of the applications know, Expert Systems have been used in a great number of practical situations. It is already considerable the number of texts in this area published in Brazil, as, for example, [2]. Classic texts such as [4] are also available.

There are many reasons to justify the use of the ES approach to dynamic and intelligent training. Several good results have shown that this approach is well suited to different industrial process. Useful aspects of Expert Systems should be considered. Some of these aspects refer to the analysis of the efficiency in the treatment of some problems (Expert Systems provide high efficiency in decision processes). Because of the objectivity required for decisions in industry and also because of the safety necessary to the whole process, it is clear that the use of Artificial Intelligence techniques can be extremely useful in obtaining a more reliable information concerning the decision involved, in addition to being useful in making possible to obtain it in a faster way. This aspect can pay off the costs that the structure of the system may bring.

On the other hand, the decision of an Expert System is monitored continually by the user. At this point, the operator begins to play a dynamic role (active part of the process), taking the responsibility for decisions that the System is making, evaluating them and correcting them if necessary. Another analysis can justify better the use of Expert Systems in this project. Since this approach causes intense interaction between the operator and the decision process, it tends to generate a critical behavior for each decision taking and its possible consequences. It is considered also that some authors, when studying the concepts of AI and its more usual tools, emphasized that AI is better suited to situations where the solution to the problem strongly depends on the adaptation of techniques and methodologies to the environment being considered in a given occasion. It can be identify, for instance, the analysis of Waterman [4], for the case of Expert Systems. This analysis is adequate for the objectives of the present project.

3. Practical application

A practical situation where the methodology has been applied is described now. An important problem in industrial process management involves a typical practical decision. The operator has to define how to evaluate the process and to determine the best option from two types of inspection: the inspection developed by attributes or the inspection done by variables. This kind of decision - attributes or variables - was considered as adapted for the application of the methodology. In order to develop it, a module was structured of the Decision Support Expert System that determines the best choice in the case of the decision between quality evaluation by attributes and by variables. It should be clear that this is only one of the several modules of the whole System.

The conceptual basis of the module involves definitions for quality evaluation, such as quality characteristics [3], and the contribution that the evaluation process has to quality. It is important to emphasize that, usually, the evaluation of all the quality characteristics of a product is unfeasible, mainly those of greater complexity. Thus, the control of the quality characteristics tends to be limited to the most important ones. It is obvious that the evaluation concentrates also in quality characteristics that request effective control.

During the preliminary discussion of the issue it is shown that there are two basic forms of applying the quality control to a product, considering the evaluation of its quality characteristic: the control by attributes and the control by variables. The characteristics of each control type are then discussed in general terms. Information about the use of each control is given using practical examples (this introduction with the use of real situations is critical for the whole training process). Thus, for example, it is mentioned that the control by attributes is always done in a discreet scale, and, in general, binomial, where two classifications just define all the variation of quality characteristics. Other facts presented are discussed. After discussing these points, the control analysis by variables must be done, with the characterization of several situations where this kind of control is used. Practical examples of evaluation by variables are presented and also discussed.

The next point in the methodology project is to approach an important subject: which evaluation type to use - attributes or variables. In fact, for the characteristics of each control methodology, for some quality characteristics the control by variables is the most suitable; for others, the control by attributes fits better. Additionally, it could be observed that there are quality characteristics that require a certain control type because of their own nature or for simple convenience reasons. Thus, the selection of the control type to be adopted depends both on the quality characteristic itself and on the particularities of the method.

The several analyses considered for the choice between the two kinds of control are then described. The analyses follow four steps: (1) the importance of correctly selecting the inspection method is the first point to study; (2) as a basis of evaluation of the product quality, a misunderstanding in the selection of the control type to use means the establishment of an incorrect quality level of the product; (3) the methods and techniques of the Statistical Control of Quality, to processes or to products, are specific for each case (attributes or variables); and (4) the inspection by attributes presents great theoretical and practical differences when compared with the inspection by variables.

Most of the differences should be considered, still, in terms of costs when a kind of control is used mistakenly. Differences in costs are a consequence of the fact that it may be executing an expensive control to obtain information that another cheaper type of control would provide in the same way. There are also serious consequences because critical decisions are made based on imprecise information. Finally, from the point of view of the methodology itself of each control type, in general, several practical observations are shown. So far, the operator has listened, attentively or not, to what has been shown. At this moment, the operator gets to know the following: having in mind the specific particularities observed for each kind of control, the next thing to do is to detect the need of structuring a module of the Decision Support Expert System that makes possible to determine which is the best option to adopt in a certain situation, when it becomes necessary to define the most suitable form to evaluate the quality of a product from its quality characteristics.

This is the objective of the present module: to confront the evaluation of the quality made by attributes for a certain situation being studied with that made by variables. Then the module is presented. It is a based on rule Expert System, with the following specifications: The system has 144 rules and 54 qualifiers. The System can list all the qualifiers, as well as the rules where they are being used. There are two choices: attributes or variables.

The system can show all the rules the choices were used in. In this case, the choices appear in all the rules used for making the decision. The decision of the system is: Evaluation by attributes or by variables. The system uses a scale of values between 0 and 10. The adaptation of the option made is made evident by the establishment of values close to 10 to the choice made; the inadequacy is characterized by values close to zero associated to the choice. All possible rules are used in the derivation of data for the selection of the most appropriate choice. The system does not show the rules when they are being used in the execution of the program. The user can alter this option, if so he/she wishes. The System is presented as an interactive process, on a microcomputer screen .The basis of the system is well-known ES software. The user (in the case, the operator) will work with the system selecting options that each qualifier presents to him/her. As an example of qualifier presented to the user, consider the following: *Give an answer to the question below using one of the following procedures: click on the text, with the mouse or write the number of the option in the space. The measurable classification of defects: (1) is necessary. (2) is desired; (3) cannot be used in this case (**write here your answer**).* As an example of a rule used by the Expert System, consider the following: **Rule 17:** *IF the information on the defect should be exact, precise THEN Evaluation by attributes - probability: 2/10; Evaluation by variables - probability: 8/10*

The rules have bibliographical references. They provide conceptual support to the rules. Some rules also have explanatory notes concerning their formulation or concepts they contain . The module is made up of six basic areas. These areas involve relative analyses about the nature of the defects, the results of the inspection, the quality characteristics, the inspection methodology, the inspectors that will work for this kind of evaluation and the productive process as a whole.

In general lines, each area involves the following aspects, among others: (1) As regards the *nature of the defects*: Classification of the defects; occurrence intensity; characterization of the occurrence; information level on the defect (precision, generality, reliability and wideness); frequency of defect occurrence; occasional action of a defect on others. (2) As regards the *results of the inspection*: Forms of expression of the evaluation results; scales for the result representation; forms of obtaining the results. (How the results were obtained). (3) As regards the *quality characteristics* to control: Quality characteristics to control; feasibility of the characteristic for the evaluation; nature of the characteristic and its importance. (4) As regards the *inspection methodology*: Inspection costs and resources; place of inspection; scope of the results of evaluation decisions; analysis of defect causes; emphasis and objectives of the inspection; sensitivity level of evaluation; forms of carrying out the inspection. (5) As regards the *inspectors*: Inspectors' qualification; inspectors' formation; characteristics of the inspectors' action on the evaluation process.(6) As regards the *productive process*: Consequences of the results of the inspection on the productive process and production levels.

4. Results and conclusions

The Expert System described was tested in 102 specific situations, involving 307 operators. In all the contexts, the Expert System showed appropriateness to the cases studied, having in mind the objectives of this project. Being a module of a decision support system, the Expert System was tested initially in practical situations for analysis of its consistency. As an example of its application, the system was used in specific cases where, according to some characteristics, a typical result of the process was expected. The previously defined decision was the same made by the system in all the tests. So the application of the system was done with real data taken from productive processes studied.

Another approach that can be used refers to the evaluation of the results of successive applications of the modules of the system. Experimental use shows that their decisions changed according to well-defined factors. If the results changed, then, it means that some elements also changed.

So these elements should be followed up, because they indicate situations that require control preventive control in most cases. The management of the changes of the results gives the teacher a method to evaluate the operators - if the results change, the decision has been also changed. By evaluating the results, the decisions can be accessed. The Expert System is like a mirror – it shows exactly what the operators have done. Tests in the operation of the modules showed that the sensitivity of the system is high, and its results can be altered with small changes in the decisions of the operators when some qualifiers are presented to them.

The operators have been asked about their confidence in the modules. 86.5% of the operators answered that they found it reliable. In a scale from 0 to 10, their grades ranged from 7.3 to 9.7, with an average of 8.31. The main satisfaction reason was due to the chance of participating in its decisions. In general, they considered it a high motivation factor, and this was evident in the process. When asked to foresee the answers of the system, the operators developed a critical sense about the problem and trained to get the right solution to proposed problem, according to some characteristics of each situation considered, before the system gives its decision. The answers were correct in 81.4% of the situations.

In terms of Expert System operation, it is important to consider the following information: (a) In 88.5% of the cases, the qualifier was understood when first read; in 4.3% of the cases, it was only understood after the second reading; in 4.1% of the cases, it was not understood at all; (b) in 3.1% of the cases, the objective of the module was not understood.

All of these cases had a common element: the participants could not tell whether a given situation was an advantage or a drawback in order to make a decision. For example: if the product has a large number of quality characteristics to be controlled, had they better use inspection by attributes or by variables? In this case, 19 participants (6.1%) showed, at some point, a strange behavior when selecting an alternative from the qualifier. Such behavior was noticed because of such evident signs as: (1) they contradicted previous responses (37% of the cases); (2) they could not justify their responses (33%); (3) they did not proofread their responses, closing the test instead (30%).

For all of these cases, the response the qualifier was absurd. It is concluded that, in these cases, the technician did not know the correct answer and in order to disguise their ignorance, they selected an absurd response.

Finally, it is suggested the application of the system to other situations and the development of more interactive ways in relation to the user as a next step. Since this module is still being tested, it is necessary continually evaluate its use.

Another recommendation to be done regards the development of an interactive evaluation model of the operator's answers. This model, based on Fuzzy Sets, evaluates the operators' answers and gives a grade to the decisions they make. This fuzzy model makes possible to detect weaknesses (that contributed for the taking of mistaken decisions) and potentialities (that determined the selection of correct options) of the decision agents (the operators). Still in an experimental phase, the model should evaluate the progress of the operator according to the decision process under study.

References

[1] Britz, G.; Emerling, D.; Hare, L.; Hoerl, R. and Shade, J. "How to teach others to apply statistical thinking." In *Quality Progress*, ASQ, USA, 1997, **6-1997**, pp. 67-80.
[2] Genaro, S. *Sistemas Especialistas: O Conhecimento Artificial*. Livros Técnicos e Científicos. Rio de Janeiro, 1998.
[3] Paladini, E. P. *Qualidade Total na Prática*. Editora Atlas, São Paulo. 1999
[4] Waterman, D. *Guide to Expert Systems*. Addison-Wesley, Reading, 1985

KES '01
N. Baba et al. (Eds.)
IOS Press, 2001

Neighbor-Interval Based Temporal Logic and its Application in Temporal Knowledge Manipulation

Ali Reza Mansoori and Mohammad Reza Beikzadeh
Iran Telecommunication Reasearch Center,
End of North Kargar, Tehran, Iran
{amansuri, beik}@itrc.ac.ir

Abstract. This paper introduces neighborhood property for temporal relations between temporal intervals and base upon it introduces Neighbor-Interval Based Temporal Logic (NIBTL), a novel temporal logic for temporal knowledge representation and temporal reasoning. Similar to Allen's interval based temporal reasoning, we use a constraint propagation algorithm running on temporal network which represents temporal knowledge but we use some features of NIBTL and some heuristics in our algorithm in order to have a cost effective algorithm and thereby optimized temporal reasoning. Although both Allen's algorithm and our algorithm operate in O(n3) on network with n nodes, our algorithm is faster and in addition to describing the factors of this optimization some results on running and comparing both algorithms for a variety of temporal networks and imposed constraints is represented.

1. Introduction

Time handling is a very important subject in a lot of artificial intelligence applications deal with time. For example in the automated high level synthesis system [1], [2], [3] temporal knowledge manipulation is used effectively and any improvement in its temporal reasoning engine causes whole system improvement [4].

There are some methods for temporal knowledge representation including time interval based and time point based [5][6][7]. In the time interval based methods have been focused in this paper, events are concerned. Each interval has duration, beginning point and ending point.

Allen has introduced interval based temporal logic and temporal network. His temporal network is a simple, transparent and easy to implement and thereby a good candidate for representing temporal knowledge in many applications. His temporal network maintains uncertainty about relations of temporal intervals. Allen has also introduced an algorithm for temporal reasoning. His algorithm is a constraint propagation algorithm in order to satisfying network consistency after imposing any change to the network.

In remaining sections of this paper Allen's temporal logic and temporal reasoning using his temporal network and constraint satisfaction is described, then we introduce neighborhood property of temporal relations and temporal intervals, Neighbor-Interval Based Temporal Logic (NIBTL) and our novel approach for representing temporal knowledge and temporal reasoning based on NIBTL. A new and cost effective constraint propagation algorithm for temporal reasoning using features of NIBTL is concerned and factors of improvement in temporal reasoning using this method will be discussed. Results of some experiences on running our constraint propagation algorithm and Allen's algorithm and comparing them in order to evaluating our algorithm will be presented.

2. Background and Related Works

Allen's interval based temporal logic and maintaining temporal knowledge and temporal reasoning based on his temporal logic has been described in [5]. In brief Allen's work is described in this section. The most essential aspect of every temporal logics is the temporal object on which temporal assertion are interpreted. In the interval based temporal logic assertions are interpreted on time intervals.

2.1. Interval Based Temporal Logic

The relation between any two intervals in real world is one and only one of thirteen possible mutual exclusive relations shown in figure 1. It is very common in reasoning systems that the knowledge about relations between two intervals is incomplete and only possible relations between them are known. The possible relations between two intervals are shown as a vector interpreted as disjunction of its constituent relations, for example (aft, bef, mi).

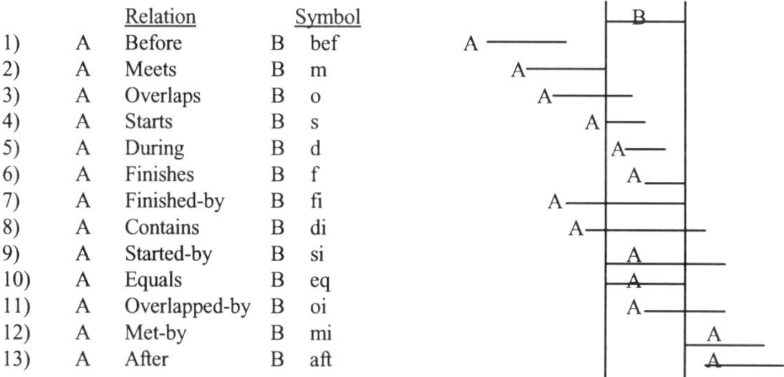

Fig. 1. Thirteen possible relations between temporal intervals

Addition and multiplication are two operations defined on temporal relation vectors and are necessary in temporal reasoning. Addition is intersection of two vectors. For example: (bef, m, o) + (o, s, d) = (o). Inferring the possible relations between two intervals when two distinct information about possible relations between them exist is done by addition.
Multiplication defines transitive relations. Suppose three intervals A, B and C. If the relation between A and B and the relation between B and C are determined for example m and d respectively, multiplication of these two relations specifies the possible relation(s) between them as m*d = (o, d, s). Multiplying two relations is done using a transitivity table offered by Allen[5]. Multiplication is defined not only for temporal relations but also for temporal relation vectors. If R1 is the relation vector between intervals A and B, and similarly R2 for relations between B and C, then R1*R2 is the least restricted possible relations between A and C that is permitted by R1 and R2.

2.2. Temporal Network and constraint propagation

Temporal Network is a tool for representing temporal knowledge. It is based on the temporal logic mentioned in previous section. Temporal network is a directed complete graph. Each node in this graph represents an interval and each arc in this graph that connects two nodes and is labelled by a temporal relation vector containing all possible relations between two intervals connected by that arc. Figure 2 shows a simple temporal network with three nodes representing knowledge about three corresponding intervals.

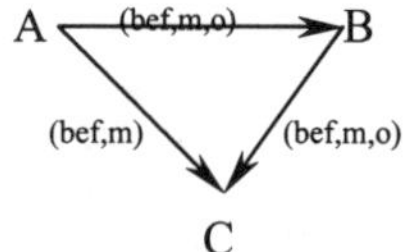

Fig. 2. A simple three node temporal network

Temporal network like any other relation networks should be consistent [8], otherwise it contains contradicted information. In a consistent temporal network, if any vector is restricted to any one of its constituent relations, there is at least one restriction for every other vector to one of their constituent relations such as occurring intervals with remaining relations between them is possible in real world. In figure 2 if we change all relations to (bef,m), the resultant network would be an inconsistent temporal network.

The necessary (and not enough) condition for a temporal network to be consistent is consistency of all its three node subnetworks. Consistency determination in temporal network is NP-hard problem and all n-node subnetworks (n<=network degree) including 3-node subnetworks, 4-node subnetworks, 5-node subnetworks , and so on must be checked for consistency [8].

Imposing a constraint to a consistent temporal network means removing one or more relations from one of edges. This may cause the network to be changed to an inconsistent network. In order to satisfying consistency in the network regarding removed relation(s) some other edges may required to be restricted by removing one or more relations from them. Each consequence restriction is a new constraint and should be propagated in the whole network and this process continues until all restrictions are propagated in the network.

3. Neighbor-Interval Based Temporal Logic (NIBTL)

In our novel temporal logic (NIBTL), the essential object for interpreting time assertions is time interval. In this temporal logic neighborhood property of temporal relation and temporal intervals is used for optimizing temporal knowledge handling.

Suppose two intervals A and B and suppose the aft relation between them. Also suppose we hold one of intervals A or B and slide the other interval on the time axis until the relation between them becomes bef. During this sliding some other relations occur between A and B. Depending on three possible lengthy relations between A and B including LT(less than), EQ(equal) and GT(greater than), the generated relations are different. In figure 3 three groups of generated relations regarding the sequence of generation in each group are shown.

> **Group 1. LA LT LB :**   aft mi oi   f d s   o m bef
> **Group 2. LA EQ LB :**   aft mi oi   eq   o m bef
> **Group 3. LA GT LB :**   aft mi oi   si di fi   o m bef
> **Note:** LA stands for length of A and LB stands for length of B

Fig. 3. Neighbor temporal relations

Two relations are neighbor if they are adjacent relations in one of the three groups in figure 3 (e.g. aft and mi). We define neighbor vector as a vector that all of its constituent relations can be joined together by a neighborhood chain (e.g. (mi, oi, f)) We also define neighbor intervals as interval pairs that temporal relations between them is a neighbor vector.

All intervals in the temporal networks we concern for reasoning are neighbors and thus all relation vectors are neighbor vectors. Properties of neighbor intervals are similar to activities in windows in Devisor, a system for planning in time [9].

In order to have some improvements on two major operations, addition and multiplication we show neighbor intervals as triple (RL, rf, rl) in which RL is one of three possible lengthy relations between intervals including LT, EQ and GT. In this triple rf is the first (leftmost) relation and rl is the last (rightmost) relation in relevant lengthy relation group. For example the vector (mi, oi, f) is shown as (LT, mi, f) (as f occurs in LT group only we conclude that the lengthy relation is LT).

Regarding to Allen's transitivity table in [5], product of any two relations is a neighbor vector. For example aft * o = (aft, mi, oi, f, d) which is a neighbor vector. If the lengthy relation between intervals is specified, the resultant vector could be shown as mentioned triples. For example:

aft * o = (aft, mi, oi, f, d) = (LT, aft, d)   If lengthy relation = LT

or aft * o = (aft, mi, oi) = (EQ, aft, oi)   If lengthy relation = EQ

or aft * o = (aft, mi, oi) = (GT, aft, oi)   If lengthy relation = GT

3.1. Operations in NIBTL

Two operations, addition and multiplication are defined in our NIBTL. Suppose V1=(RL, r1f, r1l) and V2=(RL, r2f, r2l) then V3= V1*V2 = (RL, r3f, f3l) in which r3f is the rightmost relation of r1f and r2f, and r3l is the leftmost relation of r1l and r2l according to order of relations in figure 3. As an example for two vectors addition suppose V1=(LT, mi, d) and V2=(LT, f, bef) then V3=V1+V2=(LT,f,d).

In order to vectors multiplication we need a relation transitivity table similar to Allen's one. For computing V1*V2, in addition to temporal relations of V1 and V2 we need three more parameters including lengthy relation of V1, V2 and V1*V2. Thus our transitivity table has three more entries. For showing the table, rows and columns are regarded for 13 possible relations and each cell at the intersection of relations is like figure 4 containing three major rows and columns for three possible lengthy relations of r1 and r2 respectively, and three vertical atomic cells are dedicated for three possible lengthy relations of resultant vector (LT on top, EQ in middle and GT in bottom). Each atomic cells contains two relations rf and rl of resultant vector. Of course all of 27 atomic cells are not legal cells because some of lengthy relations do not occur. For example suppose three intervals A, B and C, if A LT B and B LT C then the only possible lengthy relation between A and B is LT. The illegal combination cells are dashed in figure 4. Furthermore some other cells are illegal in the table because some relations occur with only one lengthy relation. Relations f, d and s occur with lengthy relation LT, relations si, di and fi occur with lengthy relation GT and relation eq occurs with lengthy relation EQ only.

LT	LT	LT
-----	-----	EQ
-----	-----	GT
LT	-----	-----
-----	EQ	-----
-----	-----	GT
LT	-----	-----
EQ	-----	-----
GT	GT	GT

Fig. 4. Transitivity table cell

Now we can define multiplication of two neighbor vectors based on transitivity table in NIBTL. Table 1 shows a portion of this transitivity table. If we have V1=(R1, r1f, r1l) and V2=(R2, r2f, r2l) then we have V1*V2= (R3, r3f, r3l) in which R3 is resultant vector lengthy relation and r3f is rf in the atomic cell of transitivity table retrieved by entries r1f, r2f, R1, R2 and R3, and similarly r3l is rl in the atomic cell of transitivity table retrieved by entries r1l, r2l, R1, R2 and R3.

3.2. Table lookups improvement in NIBTL

Multiplication cost in NIBTL is two table lookups for any multiplication in order to finding rf and rl of resultant vector. In Allen's temporal logic the cost of multiplying V1 by

V2 with m relations for V1 and n relations for V2 is m*n table lookups in addition to the cost of computing union of m*n vectors. Thus except special cases cost improvement is obvious. In order to have an average case analysis on multiplication we have:

 Number of relations regarding lengthy relations: for LT = 9, for EQ = 7 and for GT = 9
 Number of average relations (regardless lengthy relation) = (9+7+9) / 3 = 8.33
 Average number of relations in temporal vectors = 8.33/2= 4.16
 Average number of table lookups in Allen's method = 4.16*4.16= 17.36

Comparing 17.36 with 2 table lookups in NIBTL shows improvement in multiplication.

```
To Add R(i,j) (* R(i,j) is restricted edge between i and j *)
    If R(i,j)is empty Then signal contradiction
    If R(i,j)+N(i,j) ⊂ N(i,j) Then Begin
      For each Node k Do
        If k<>i and k<>j Then Begin
          pr ← priority(N()i,j),R(i,j));
          If pr<>0 Then Add(<i,j,k>,pr) to queue ToDo;
          pr ← priority(N(i,j),R(i,j),N(k,i));
          If pr<>0 Then Add(<k,i,j>,pr) to queue ToDo;
        End
      N(i,j) ← R(i,j);
    End
    While queue ToDo is not Empty Do Begin
      Get next <i,j,k> with highest priority from queue ToDo;
      R(i,k) ← N(i,k)+(N(i,j)*N(j,k))
      If R(i,k) ⊂ N(i,k) Then Begin
        For each node j Do
          If j<>i and j<>k Then Begin
            pr ← Priority(N(i,k),R(i,k),N(k,j));
            If pr<>0 Then Add(<i,j,k>,pr) to queue ToDo;
            pr ← Priority(N(i,k),R(i,k),N(j,i));
            If pr<>0 Then Add(<j,i,k>,pr) to queue ToDo;
          End
        N(i,k) ← R(i,k);
      End
    End
```

Fig. 5. Constraint propagation algorithm in NIBTL

3.3. Constraint Propagation in NIBTL

Two factors have been focused in order to speed up constraint propagation algorithm shown in figure 5. First is improvement in multiplication cost discussed before and second factor is selecting next edges to be concerned in propagating constraint. Allen's algorithm follows a breadth first strategy and selects a constraint (a restricted edge) and computes all possible multiplications and next layer constraints are added to the end of queue. We use a heuristic for selecting the next edges to be checked for possible new constraints. Our heuristic function computes "maximum probable removed relations" which is computed by two table lookups. When the heuristic function returns 0, the algorithm discards pair of edges for testing new constraints.

In this algorithm, priority is a function for computing the heuristic criterion. ToDo is a priority queue and its elements are pair adjacent edges shown as triples <i, j, k>.

4. Conclusion

Analytically both Allen's algorithm and introduced algorithm run in O(n3) on network with n nodes but introduced algorithm in figure 5 is faster than Allen's. We discussed the factors of this improvement. Here some results of running both algorithms are presented.

We ran both algorithms on a variety of consistent random generated temporal networks and measured some important parameters. The time consumed by NIBTL algorithm was optimized in almost all experiences. As a case study for a network with 100 nodes NIBTL algorithm was three times better (15 sec. vs. 45 sec.). Part of this improvement is the result of optimizing the number of multiplications caused by our heuristic function (1300 vs. 7170) and part of it by rapid multiplication method and less table lookups (17470 vs. 173400 table lookups).

Table 1. A portion of transivity table in NIBTL

		f		si	d	eq	di	S		fi
		LT	EQ	GT	LT	EQ	GT	LT	EQ	GT
aft	LT	>..>		>..>	>..d	>..>	>..>	>..d		>..>
	EQ			>..>			>..>			>..>
	GT			>..>			>..>			>..>
	LT	>..>			>..d			>..d		
	EQ					>..>				
	GT			>..>			>..>			>..>
	LT	>..>			>..d			>..d		
	EQ	>..>			>..oi			>..oi		
	GT	>..>		>..>	>..oi	>..>	>..>	>..oi		>..>
mi	LT	mi..mi		>..>	oi..d	mi..mi	>..>	oi..d		mi..mi
	EQ			>..>			>..>			mi..mi
	GT			>..>			>..>			mi..mi
	LT	mi..mi			oi..d			oi..d		
	EQ					mi..mi				
	GT			>..>			>..>			mi..mi
	LT	mi..mi			oi..d			oi..d		
	EQ	mi..mi			oi..oi			oi..oi		
	GT	mi..mi		>..>	oi..oi	mi..mi	>..>	oi..oi		mi..mi
oi	LT	>..>		>..oi	oi..d	oi..oi	>..oi	oi..d		oi..oi
	EQ			>..oi			>..oi			oi..oi
	GT			>..oi			>..oi			oi..di
	LT	>..>			oi..d			oi..d		
	EQ					oi..oi				
	GT			>..oi			>..di			oi..di
	LT	>..>			oi..d			oi..d		
	EQ	>..>			oi..oi			oi..oi		
	GT	>..>		>..oi	oi..oi	oi..oi	>..di	oi..oi		oi..di

Note: > stands for aft

References
[1] Beikzadeh, M.R., Automated High-Level Synthesis Based On Artificial Intelligence Technique, Univ. of Essex, Colchester U.K., Ph.D. thesis, 1992.
[2] Beikzadeh, M.R. and Mack, R. J., A Primary Planner for High-level Artificial Intelligence Synthesis, Proc. of the 33th MidWest Sym. on Circuits and Systems, Calgray 1990.
[3] Beikzadeh, M.R. and Mack, R. J., The Application of Temporal Logic for Flexible Scheduling within a High-Level Synthesis System, Proc. of the 34th MidWest Sym. on Circuits and Systems, May 1991.
[4] Mansoori, A. R., Intelligent Algorithm for Constraint Satisfaction in Constraint Temporal Reasoning, Sharif Univ. of Tech. Tehran-Iran M.S. Thesis 1996.
[5] Allen, J. F. maintaining Knowledge About Temporal Intervals. Communications of the ACM 26(11): 832-843, November, 1983.
[6] Vilain, M. and Kautz, H., Constraint Propagation Algorithms for temporal Reasoning, Proc. AAAI-86, Phidadelphia, PA (1986): 377-382.
[7] Freksa, C. Temporal Reasoning Based on Semi-Intervals. Artificial Intelligence (54): 199-227, 1992 .
[8] Mackworth, A. K. ,Consistency in networks of relations, Artificial Intelligence (8): 99-118, 1977.
[9] Vere, S. A. , Planning in Time : Windows and Durations for Activities and Goals, Jet Propulation Laboratory, Pasanda, CA, 1983.

KES '01
N. Baba et al. (Eds.)
IOS Press, 2001

Data Mining with Self-Organising Maps (SOM) and Minimal Spanning Tree (MST)

K. Obu-Cann, K. Fujimura, H. Tokutaka, M. Ohkita, M. Inui and S. Yamada
Tottori University, Dept. of Electrical and Electronics Engineering,
Koyama-Minami 4-101, Tottori, 680-8552, Japan

Abstract. Data mining or exploration is part of a larger area of recent research in Artificial Intelligence and Information Processing and Management otherwise known as Knowledge Discovery in Database (KDD). The main aim here is to identify new information or knowledge from database in which the dimensionality or amount of data is so large that it is beyond human comprehension. Self-Organising Maps (SOM) and Minimal Spanning Tree (MST) are used to analyse power transformer database from one of the electric energy providers in Japan. This paper looks at clustering with Minimal Spanning Tree (MST).

1. Introduction

The SOM [1], [2] is a powerful tool for data mining, knowledge discovery and visualisation of high dimensional data. SOM and MST are applied to power transformer database from one of the electric energy providers in Japan. The aim of this work is to apply a data-mining tool based on SOM to learn more about database. The data was analysed to identify the energy consumption pattern of the consumers, season type pattern classification and day type pattern classification. In day type pattern classification the Minimal Spanning Tree (MST) algorithm [3], [4] was applied to augment the efforts of SOM in clustering. Prediction of the oil temperature changes in a power distribution transformer has been practiced using conventional methods based on explicit numerical calculation. Usually associated with such technique of prediction, is a limitation in the prediction accuracy due to assumptions made in the characteristics of the transformer. This paper looks at the application of SOM to the prediction of the oil temperature changes of a power distribution transformer.

2. Clustering with SOM

SOM is usually presented as a two-dimensional grid or map with the units tuned to different input data patterns. This algorithm is based on unsupervised competitive learning, training in this case is entirely data driven and the neurons or nodes on the map compete with each other. The data set contains four measurements (percentage load on the transformer, transformer oil temperature, atmospheric temperature and transformer surrounding temperature) recorded daily every 10 minutes, averaged hourly and normalized by component scaling in order to maximize the differences among the data.

2.1. Season Type Pattern Classification

SOM is applied in the classification of the input patterns into the various seasons of the year namely winter, spring, summer and autumn. The data set for the SOM comprised of data for the months of January, March, July and September 1996. Each month is represented by two different day type patterns. The input patterns are hourly measurements hence each input vector is made up of 96 components, which have been normalised [4]. Figure 1a illustrates the seasonal feature map obtained for the year 1996 after 5000 iterations of training.

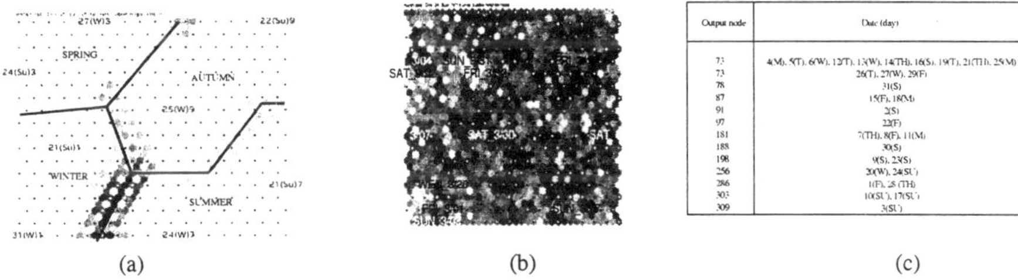

(a) (b) (c)

Output node	Date (day)
73	4(M), 5(T), 6(W), 12(T), 13(W), 14(TH), 16(S), 19(T), 21(TH), 25(M)
73	26(T), 27(W), 29(F)
78	31(S)
87	15(F), 18(M)
91	2(S)
97	22(F)
181	7(TH), 8(F), 11(M)
188	30(S)
198	9(S), 23(S)
256	20(W), 24(SU)
286	1(F), 28 (TH)
303	10(SU), 17(SU)
309	3(SU)

Figure 1: (a) Seasonal SOM for 1996, (b) SOM for transformer oil temperature in March 1996, (c) Summary of SOM for March 1996.

The eight input patterns, which represent the various seasons of the year, were classified into four separate regions on the map. Between winter and summer, the darkest grey separation was identified on the map. This is because during winter, though it is very cold, heating is mostly done with paraffin other than electric energy. In summer a lot of electric energy goes into cooling thereby resulting in a drastic increase in energy consumption. The next in the degree of grey is the difference between winter and autumn followed by winter and spring. The pattern for autumn is closer to spring than summer. This is because in autumn, the heating fuel is also paraffin. Looking at the SOM of figure 1a, a lot can be said about the electric energy consumption patterns of the consumers. Furthermore, this map can be used as a prototype by which recorded data could be compared. If one is confronted with an unlabeled data set, by use of this map, the season within which this data was recorded can be identified by its position on the map.

2.2. Day Type Pattern Classification

SOM is utilised for the classification of daily transformer oil temperature patterns. In SOM, the output units of the network are not specified. It is proposed that the SOM will generate the output based on the different patterns available in the input data space [5]. An input pattern comprises of the 24 hourly transformer oil temperature readings for a day. The 31 input patterns for the month of March 1996 were mapped onto a two-dimensional output network. Figure 1b shows the feature map for March 1996 after 50 iterations of training. 20th March 1996 (Wednesday), the only holiday in March was located at node 256 together with 24th March (Sunday), which was a weekend. Furthermore, the weekend patterns were also classified separately as Saturday and Sunday patterns. Comparing these with results from other months, it was observed that new output nodes, each corresponding to a group of input patterns with a particular feature emerged after the training process. It was also observed that not all the days appeared on the map, this is because days with similar transformer oil temperature patterns are mapped to the same output node as the same day type. The following day types were identified from the feature maps: Weekdays, Holidays and Weekends. Figure 1c is a summary of results from figure 1b.

3. Clustering with MST

Looking at the SOM of figure 1b, the days that belong to a particular cluster cannot be identified easily. In this section, the MST is applied to the SOM to provide a more visual map. The MST method is a kind of tree like structure that links all the units by the shortest path [3], [4]. This structure defines the mutual distance and also describes the similarity relationship of any two points on the map [4], [6]. This algorithm assigns arcs between the nodes in such a manner that all nodes are connected together by single linkages. The lengths of the arcs are defined as the non-weighted norms of the vectorial differences between corresponding reference vectors as follows [4]:

$$\|x - m_c\| = \min_i \{\|x - m_i\|\} \tag{1}$$

Where m_i is the reference vector of node i and m_c is the reference vector of the winner node or unit (*Best Match Unit -BMU*).

3.1. MST 1 (First Winner)

From figure 1b, it is very difficult to tell the number of clusters off-hand. In order to enhance the visual characteristics of the SOM map, the MST 1 was applied. This algorithm is based on figure 2a, the nodes selected to form the neighbourhood in this section comprised of only the first winner nodes.

(a)   (b)

Figure 2: (a) Examples of nodes selected in the MST Topology using only the first winners. The selection process continues until a node, which has already been selected is selected again. (b) Examples of nodes selected using first and second winners.

3.2. MST 2 (Second Winner)

Depending on the degree of similarity among the data, the minimal spanning tree is extended to embrace other winners or nodes in close proximity to the node selected as the winner by the first algorithm. The neighbourhood in MST 1 is extended to include the second winner and its neighbours (see figure 2b). The neighbourhood this time contains far more nodes than that of the former algorithm. The procedure of extending the neighbourhood is continued until there is no appreciable change in the clusters formed by connecting the MST arcs together. Figure 3 illustrates the application of MST to the SOM map for the transformer oil temperature in January 1996. The MST algorithm provided a physical link among the data within a particular cluster thereby providing a more visual map than the previous method. In each of the months analysed, three day-type patterns emerged after the training process. These are Weekdays, Holidays and Weekends. The MST algorithm maintained the same day types that were obtained in section 2.2

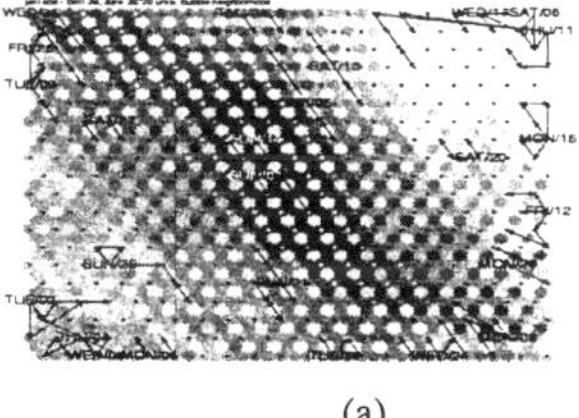

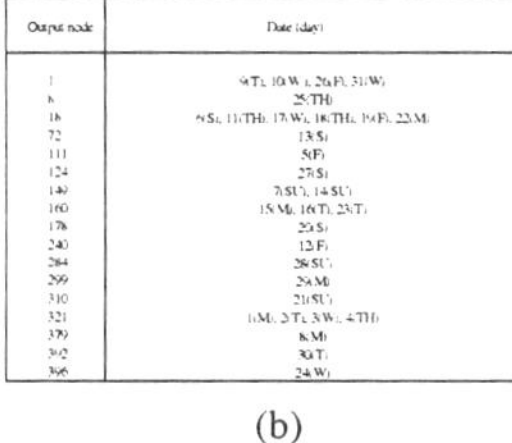

Output node	Date (day)
1	9(T), 10(W), 26(F), 31(W)
8	25(TH)
18	6(S), 11(TH), 17(W), 18(TH), 19(F), 22(M)
72	13(S)
111	5(F)
124	27(S)
149	7(SU), 14(SU)
160	15(M), 16(T), 23(T)
176	20(S)
240	12(F)
284	28(SU)
299	29(M)
310	21(SU)
321	1(M), 2(T), 3(W), 4(TH)
379	8(M)
362	30(T)
396	24(W)

(a)   (b)

Figure 3: (a) MST map for transformer oil temperature in January 1996. The algorithm for this map was extended to include the 2nd winner and its neighbours. (b) Summary of the MST map for January 1996.

4. Transformer oil temperature forecast with SOM

During summer, the power system is put under heavy load. The temperature of the insulation oil for cooling the transformer windings rises. The continuous rise in the oil temperature causes the dielectric strength of the oil to deteriorate. The oil then loses its insulating properties thus creating a short circuit within the windings of the transformer. Under such conditions, if the temperature change of insulation oil of the transformer can be predicted, necessary countermeasures can be implemented to forestall any damage to the transformer. Hence, efficient operation of the transformer can be achieved and also the reliability may be improved.

4.1. Predicting oil temperature from an incomplete data vector

4.1.1 Forecast with 3-month data set

The input data comprised of data for 3 months. The month of the day to be predicted is selected as the centre month and the month before and after are selected to form an input data set. As an example, the input data set for predicting 7th August will be July, August and September. Data for the day to be predicted is made up of the highest value of the atmospheric temperature for the day, the lowest value of the atmospheric temperature for the day and the percentage loading on the transformer. The original oil temperature data for the day to be predicted was presumed unknown and used as evaluation data. Figure 4(a) illustrates the comparison of the original transformer oil temperature readings for 7th August, the SOM forecast values and the Conventional forecast values. The results obtained were compared to the recorded values on 7th of August using equation (2). This approach resulted in a Mean Squared Errors (MSE) of 0.585 (see figure 4(b)).

4.1.2. Forecast with 10-months data

In order to improve the prediction accuracy for the SOM, the input data set was increased from a 3-month data set to a 10-month data set. The input data comprised of data for 10 months, from January to October 1996. The MSE reduced from 0.585 to 0.072 (see figure 4(b)). Increasing the input data set from a 3-month data set to a 10-month data set resulted in a better approximation of the original data.

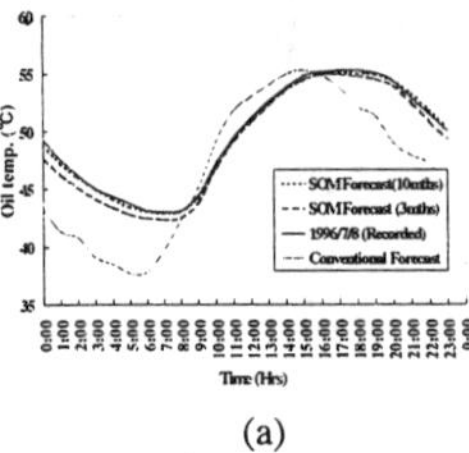

Date (Day)	Mean Squared Error 10-month	Mean Squared Error 3-month	Mean Squared Error Conventional
12/2/96 (Monday)	0.665	1.052	71.724
5/3/96 (Tuesday)	1.659	3.816	63.357
10/4/96 (Wednesday)	0.532	0.956	29.970
8/5/96 (Wednesday)	1.885	1.954	17.855
6/6/96 (Thursday)	18.451	12.475	12.150
17/7/96 (Wednesday)	2.221	1.068	12.958
7/8/96 (Wednesday)	0.072	0.585	16.377
12/9/96 (Thursday)	0.677	2.639	16.844

(a)   (b)

Figure. 4: (a) Recorded transformer oil temperature readings for 7th August, prediction by SOM (SOM prediction was for 10-month and 3-month data sets) and prediction by conventional method. (b) Mean squared error of the predicted results of oil temperature using 10-month data set, 3-month data set and conventional method.

4.1.3. Forecast with Conventional Method

Forecasting oil temperature of a transformer according to the operational guideline of oil filled transformers [7] entails explicit numerical calculation using various constants such as, time constant necessary for the calculation of the corresponding optimal cooling conditions of the installation environment of the transformer. This method resulted in a MSE of 16.38 for the forecast of the 7th August temperature values (see figure 4b). The predicted results obtained from SOM and numerically calculated values, which were based on the conventional method, were compared to the recorded data. The comparison was done based on the mean squared error shown in eqn. (2).

$$Err \; = \; \frac{1}{N} \sum_{j=1}^{N} (r_j - p_j)^2 \qquad\qquad (2)$$

Where, r_j is the value of the recorded oil temperature and p_j is the value of the predicted oil temperature and N is the number of components per data vector. The SOM predictions resulted in a better approximation of the recorded data.

4.2. Predicting a complete data vector omitted from the input data space

In section 4.1, the oil temperature components of the input vector were presumed unknown. In this section the whole input vector representing a particular day was deleted from the input data space. SOM was applied after which the Best Matching Unit (BMU) was identified. Figure 5a shows the feature map for the month of April. In this data set, the data vector for 9th April 1996 (TUE_4/09) was omitted. Using eqn. (3), all the 600 nodes on the grid were compared to the original data for TUE_4/09 and the node with the minimum value selected as the BMU.

$$Err = \sum_{j=1}^{n} (x_j - m_{ij})^2 \qquad\qquad (3)$$

Where x_j and m_{ij} are the j-th component value of the n-th dimensional input data and the i-th unit (node) respectively. Figure 5b compares the transformer oil temperature of the input data vector for TUE_4/09, which was omitted to that which was identified as the BMU after the training process.

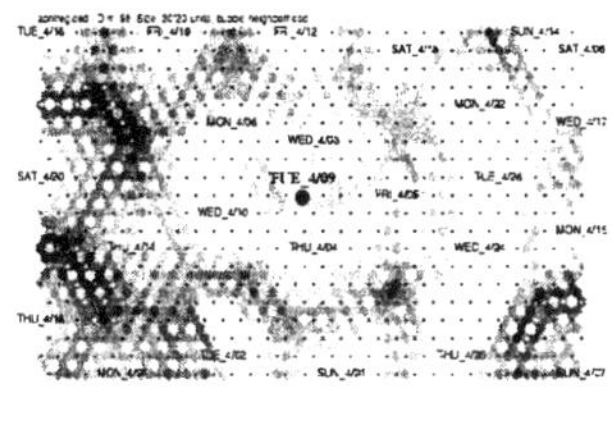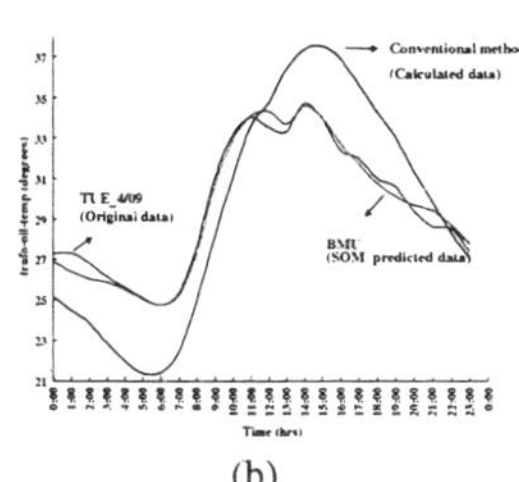

(a)   (b)

Figure 5: (a) SOM for April 1996. The input vector for TUE_4/09 was omitted. The large filled circle was identified as the BMU for TUE_4/09. (b) Predicted values for the transformer oil temperature of TUE_4/09 by SOM and the conventional method using calculation.

Although TUE_4/09 was excluded from the input data, the SOM was able to learn the other input vectors very well so as to predict with a good precision, a data vector not included in the input data space. Furthermore, figure 5b compares SOM to the conventional method currently been used by the service provider for predicting the transformer oil temperature on the TUE_4/09. SOM resulted in a better approximation of the data for TUE_4/09 as compared to the method currently been used by the service provider for data prediction.

5. Conclusion

In its application to the power transformer database, the SOM provides an in-depth knowledge about the consumption pattern of the consumers in a particular day or season. This provides the energy producer with a guide as to the consumption pattern of its consumer in any particular day or season. This could be very vital information for planning engineers in their load forecasting and other prediction and planning activities.

Superimposing the minimal spanning tree onto the SOM also provided a more visual approach in identifying the days that belong to a particular cluster.

SOM was also found to be very effective in regression and this could be a very effective tool for predicting unknown as well as incomplete data vectors in a database.

Forecasting of the oil temperature changes in a power distribution transformer has been done using conventional methods based on explicit numerical calculation. With such techniques of forecasting, the forecasting accuracy is affected due to assumptions made in the characteristics of the transformer. By the use of SOM, this problem is eliminated because the transformer characteristics are not included in the input data set for the SOM.

References

[1] T. Kohonen, Biological Cybernetics, and 43(1): 59-69,1982. Springer-Verlag, 1997.

[2] T. Kohonen, Self-Organizing Maps, IEEE, 78:1464-1480, 1990. Springer-Verlag, 1995.

[3] K. Obu-Cann et al, Clustering by SOM, MCP and MST, NC99-133, pp 121-128, March 2000.

[4] T. Kohonen, Self-Organizing Maps, Springer-Verlag, 1997.

[5] A. Murray, Applications of Neural Network, Kluwer   Academic Publishers, pp 157-189.

[6] K. Obu-Cann et al, Clustering by SOM, MST and MCP, ICONIP'99, vol. 3, pp 986-991, November 1999.

[7] Tech. report of the Institute of Electrical Engineers of Japan, Part 1, No. 143: (1978).

KES '01
N. Baba et al. (Eds.)
IOS Press, 2001

Cerebellar Model Articulation Controller (CMAC) Training by Kohonen Self Organizing Maps (SOM)

Noor Aamir BALOCH, Kunihiko NABESHIMA*, Kazuhiko KUDO
Department of Applied Quantum Physics and Nuclear Engineering, Kyushu University, Hakozaki 6-10-1, Higashi-ku, Fukuoka-shi, 812-8581, Japan

**Japan Atomic Energy Research Institute (JAERI), Tokai-mura, Ibaraki-ken, 319-1195, Japan*

Abstract. Kohonen Self Organizing Maps (SOM) is used to improve the Cerebellar Model Articulation Controller (CMAC) training. It is found that the output of CMAC improved when SOM is employed in training. Further, for a small training sample the results are encouraging.

1. Introduction

CMAC (Cerebellar Model Articulation Controller) network is based on the model proposed by Albus [1] for robot manipulator control and function approximation. It employs simple table lookup technique. CMAC can be regarded as a type of associative memory network. It has a good potential to learn nonlinear function. CMAC is successfully applied to automatic control and function approximation. For a large number of input data the memory requirement of CMAC increases drastically. However, this may be handle by using Hash mapping technique [2]. The important characteristic of CMAC is its build in generalization property. Like other soft computing technique, such as neural networks, CMAC also depend on sufficient training data points for a good learning. In some situations, the availability of large quantity of training data is a difficult task, such as designing the control system of Nuclear Power Plant. CMAC training would not be effective in such cases. In this paper, we have shown that for a small training data, it is possible to train CMAC in such a fashion that the output error decreases.

CMAC is based on a table lookup technique in which the information is stored in the memory location whose addresses are generated by dividing the input space. In order to have better performance, most of the memory location should be train adequately so that output error is minimized. For a small training sample size, the Euclidian distance between the stored weights would be larger, which means that most of the neighborhood of a stored weight would be empty. If we can predict or extrapolate these empty memory locations with the help of known training sample, this will enhance the training process. In this paper, we have used Kohonen Self Organizing Maps (SOM) to fill empty memory regions. Only the neighborhoods of known values are filled. In some methods such as counterpropagation [3], SOM is used as a processing method for input data. In a proposed method, SOM is used only for refinement of weights once it has been trained by ordinary CMAC network.

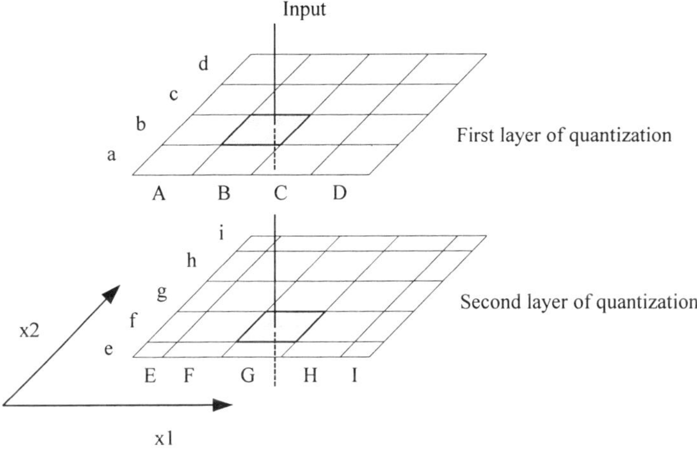

Fig. 1. Simple layout of CMAC network.

2. CMAC Network

Consider Fig. 1, which shows the layout of CMAC network for two input variables x1 and x2. The ranges of both inputs are divided into small discrete intervals called generalization width (GW). These intervals are labeled as A, B, C and D for input variable x1, and a, b, c, and d for input variable x2. The area formed by these intervals, such as Aa, Ab , Ac, Ad, and Ba, are called hypercube. If each variable is shifted by a small interval, different hypercube will be formed, such as Ee, Ef, Eg, Eh, Ei, etc, which lie on the different plane. Similarly, it is possible to add as many layers as desired. We shall call the number of layers in CMAC network as quantization level (QL). In any CMAC network, there exist at least two quantization level. It should be noted that the length of division of input variable is a measure of effectiveness of generalization in CMAC network.

During the training of CMAC network, weights are stored in the hypercube. Each input defines a set of address on the CMAC hypercube plane. Consider an arbitrary input, as shown in Fig. 1, the corresponding hypercube for this input are 'Bb' from first layer, and 'Gf' from second layer. Than the CMAC output would be the sum of weight stored at hypercube 'Bb' and 'Gf'. Mathematically, the CMAC output 'y' is simply given by

$$y = \sum_{j=1}^{ql} w_j \tag{1}$$

Where 'w_j' is the weight stored at the 'jth' quantization level and 'ql' is the total number of quantization level. During the training the weights are updated using least square iterative learning. For a given input, the updating rule can be expressed as

$$w_j^{new} = w_j^{old} + \frac{\alpha}{ql}(\hat{y} - y) \tag{2}$$

Where 'α' is the learning rate, '$\hat{y}$' is the target function value, and '$\hat{y} - y$' is the error for this training sample.

3. Kohonen Self Organizing Maps (SOM) and CMAC Training

The Kohonen's Self Organizing Maps [4] defines a mapping from the input data space R^n onto a two dimensional array of nodes that preserve, as much as possible, the neighborhood relation between the input data. All nodes are connected to an input vector 'v' by weighted connection 'u'. During the training of SOM, the Euclidian distance between the input vector and all weights are calculated as

$$\left\| v - u_w \right\| = \min_i \left\| v - u_i \right\| \tag{3}$$

The weight with the smaller distance is selected as a winner. Then all the weights in the neighborhood Nc around the winner weight 'u_w' are updated as

$$u_i(t+1) = u_i(t) + \gamma(t)\left\| v\text{-}u_i \right\| \ ; \text{if} \in Nc$$
$$= u_i(t) \qquad\qquad ; \text{if} \notin Nc \tag{4}$$

Where '$\gamma(t)$' is the learning rate ($0<\gamma(t)<1$) and it's a monotonically decreasing function of iteration, 't'.

Consider a CMAC network that learns the following target function

$$f(x_1,x_2) = \sin(\frac{\pi\,x_1}{30})\cos(\frac{\pi\,x_2}{30}); \qquad 0 < x_1 < 60, \ \ 0 < x_2 < 60; \tag{5}$$

For applying CMAC network to the above target function, four random samples of size 100, 150, 200, and 250 were selected from the population of size 60×60. Further, the GW was taken as 2, 3, 4, and 5. Quantization level for every GW was taken as 2, 3, 4, 5, 10, 20, 30 and 40. The target function is shown in Fig. 2, whereas Fig. 3 shows the output of CMAC after training with a sample of size of 100, with GW = 3, and QL= 5. Increasing

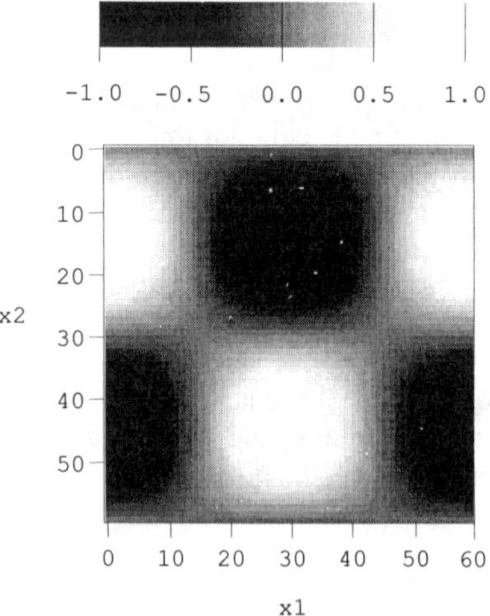

Fig. 2. Target function $f(x_1,x_2)$.

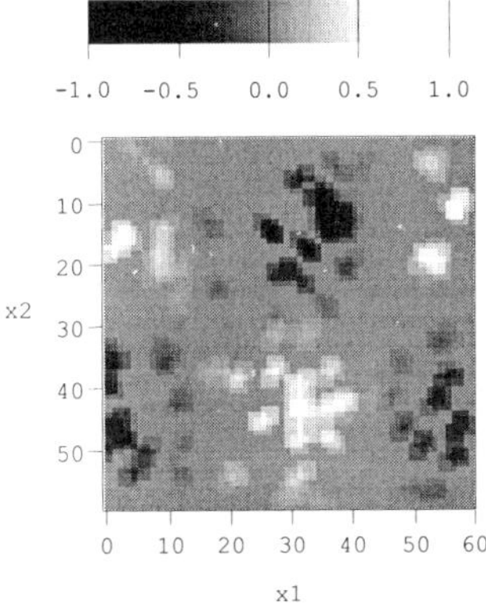

Fig. 3. Image of CMAC output.

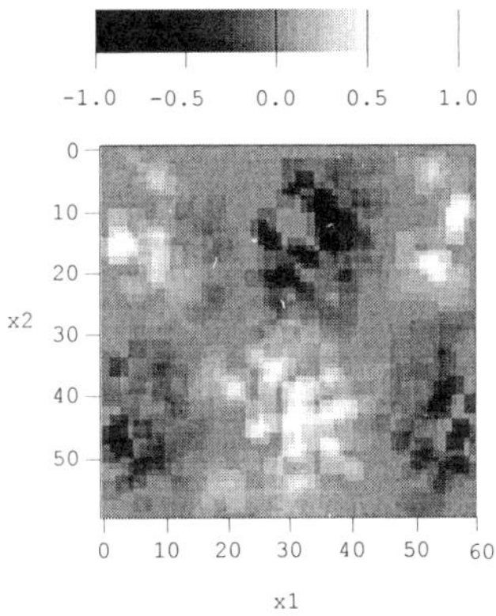

Fig. 4. Image of SOM-CMAC output.

the size of the training sample can reduce this error. Further refinement can be done by selecting the appropriate value of generalization width and quantization level.

It is possible to reduce the output error of CMAC by using Kohonen SOM in the training process. The method is based on the SOM property that during the training of SOM not only the winner weight, but also its close neighbors are trained to some extent to the input vector. After training the CMAC, weights of each quantization layer are modified by using SOM. It can be seen that SOM-CMAC training consists of two phases. In first phase, network is trained as a simple CMAC network, which is supervised. After training, the weights are refined using SOM algorithm, which is unsupervised in nature. Also, note that the counterpropagation network uses the Kohonen SOM in the hidden units, which performs unsupervised learning whereas the output layer uses Grossberg learning which is unsupervised.

The training algorithm for SOM is as follows

(1) Initialization of weights 'w' by random numbers, selected from the weights initially trained by CMAC.
(2) Determination of a winner weight.
(3) The winner weight and its close neighborhood weights (in Euclidian sense) are updated using SOM algorithm as shown in equation 4.
(4) Continue the above process until all weights with untrained neighborhoods are updated.
(5) Repeat the above process for all quantization layers.

Fig. 4 shows the output for SOM-CMAC network for sample of size 100, with GW = 3, and QL = 5. When   compared with Fig. 3, which is the output of ordinary CMAC with same parameters as that of Fig. 4, it can be seen clearly that the neighborhood values are refined by using SOM.

Simulations were carried out for SOM-CMAC training with the random samples of size 100, 150, 200 and 250. The GW was taken as 2, 3, 4, and 5. The quantization level for every GW was taken as 2, 3, 4, 5, 10, 20, 30 and 40. Fig. 5 shows the error between ordinary CMAC and SOM-CMAC network as a functions of quantization level for various training sample size for a fixed value of generalization width (GW =5). As expected the error decreases as the quantization level is increases and reached to its lowest value. Further, the error decreases when CMAC is trained by SOM. In addition, the difference of errors between ordinary CMAC and SOM-CMAC decreases, as the size

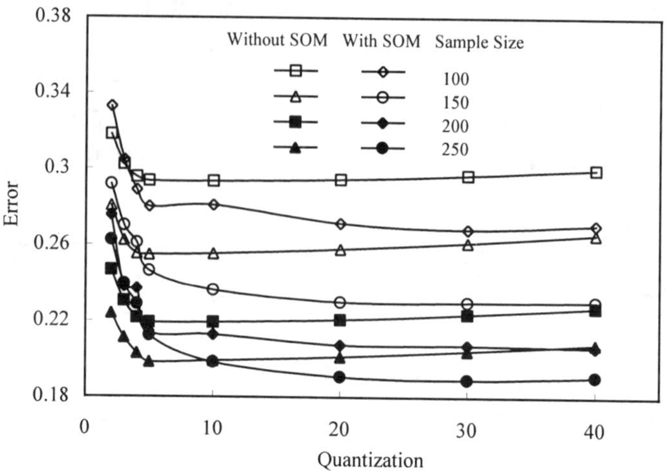

Fig. 5. Comparison of error between ordinary CMAC and SOM-CMAC network.

of the training sample increases. This is because by increasing the size of input sample, more data points would be in the neighborhood of each other; hence, the refinement done by SOM will be reduced.

4. Conclusion

In this paper, training of CMAC network with Kohonen Self Organizing Maps has been investigated, and compared with the ordinary CMAC. It has been found by simulation that the training of CMAC improves by using Kohonen Self Organizing Maps. However, it should be noted that as the size of training sample increases the difference of error between ordinary CMAC and SOM-CMAC network decreases. The method is useful for the cases when the size of training sample is small. For example in case of Nuclear Power Plant control system, where it is difficult to model plant due its complex, time varying and insufficiently known parameters, the size of training would be small or insufficient. It would be interesting to apply SOM-CMAC on these types of complex systems.

References
[1]   J. S. Albus, "A New Approach to Manipulator Control: The cerebellar model articulation controller (CMAC)," *J. Dynamic Syst., Measurement, Contr. Trans. ASME*, pp. 220-227, Sept. 1975.
[2]   Chun-Shin Lin and Ching-Tsan Chiang, "Learning Convergence of CMAC Technique", *IEEE Trans. Neural Networks*, vol. 8, no. 6, pp. 1281-1292, 1997.
[3]   R. H. Nielsen, "Counterpropagation networks," *Applied Optics*, vol. 26, no. 23, pp. 4979-4984, 1987.
[4]   T. Kohonen, "Self Organized formation of topologically correct feature maps, " *Boil. Cybern.* vol. 43, pp. 59-69, 1982.

Comparison of Mapping Methods to Visualize the EC Landscape*

Dilip Kumar Pratihar[†]   Norimasa Hayashida   Hideyuki Takagi

Kyushu Institute of Design, Fukuoka, 815-8540, Japan

takagi@kyushu-id.ac.jp,   TEL & FAX +81–92–553–4555

Abstract. We compare four mapping methods – self-organizing map (SOM), Sammon's non-linear mapping (NLM), topology preserving mapping of sample sets (TOPAS), and VISOR algorithm – to visualize the landscape of evolutionary computation (EC) and accelerate the convergence of EC and interactive EC (IEC). Three experiments are conducted using five benchmark functions and 28 subjects. We compare the computational complexity, the capacity for human visualization, and the effect on convergence for each experiment. These experiments showed that SOM demonstrated the best performance, and the VISOR performed well when CPU time was critical. The other mapping methods, NLM and TOPAS, were far from practical.

1   Introduction

Interactive evolutionary computation (IEC) is a method to optimize tasks based on human preference. In addition to the development of applications in several fields, much research has also been conducted to solve the human fatigue problem [8].

We proposed Visualized IEC to accelerate EC convergence and reduce human fatigue by displaying a 2-D visualized EC landscape mapped from L-D space and allowing the user to intervene in an EC search [1, 7]. Since humans have a superior capability to visualize entire shapes, users estimate a global optimum and use the information as a new elite for the next generation. This method can be combined not only with IEC but also EC as Visualized EC.

An IEC user only evaluates individuals and does not directly join a EC search. Conversely, a Visualized IEC user joins the global optimum search in a 2-D mapped EC landscape in addition to the evaluation of each individual. It is expected that this active intervention with the EC search accelerates the convergence of Visualized IEC/EC. This is especially effective when an IEC where human evaluation is a major part of the working time.

Our objective is to determine suitable mapping methods for the visualization of an EC landscape. There are several projection methods that map the data in L-D space to 2-D space: principal component analysis, least square mapping, projection pursuit mapping, and other linear mapping methods [6], and SOM [2], VISOR algorithm [4], NLM [5], TOPAS [3]and other non-linear methods. Our target mapping methods require practical computational cost, ease of visually estimating the location of the global optimum from the data distribution in 2-D space, and the effect on the acceleration of EC convergence.

In this paper, we review four mapping methods and show the proper mapping methods for the Visualized IEC/EC by evaluating the results from our three experiments.

*This work was supported in part by Matsushita Electric Works Ltd. and Matsushita Electric Works Software Co. Ltd

[†]This work was conducted during his visiting research at Kyushu Institute of Design on leaving from Regional Engineering College, Durgapur, West Bengal, India. He is reached at dilippratihar@lycos.com.

2 Mapping Algorithms

2.1 Self-Organizing Map (SOM)

SOM is a neural network trained by competitive learning [2]. It forms a topological structure of L-D data given from input neurons on its competition layer. When we arrange $n \times n$ neurons in the competition layer spatially in 2-D, the topological structure of L-D data can be expressed in the $n \times n$ neuron arrangement. This feature is used for mapping. The details of SOM has been published in many books.

2.2 Sammon's Nonlinear Mapping (NLM)

The N vectors in an L-D space, say X_i ($i = 1, 2, ..., N$), are to be mapped to a 2-D space, and the mapped N vectors are denoted by Y_i ($i = 1, 2, ..., N$). The N vectors are considered to be random in the 2-D plane. Suppose that d_{ij}^* is the distance between two points X_i and X_j in the L-D space and d_{ij} be the distance between the two mapped points Y_i and Y_j in the 2-D space. The equation, $d_{ij}^* = d_{ij}$, is to be satisfied for the exact mapping.

Let $E(m)$ be the mapping error after m-th iteration and express as $E(m) = \frac{1}{C}\sum_{i=1}^{N}\sum_{j=1(i<j)}^{N}(d_{ij}^* - d_{ij}(m))^2/d_{ij}^*$, where $C = \sum_{i=1}^{N}\sum_{j=1}^{N}d_{ij}^*$ and $d_{ij}(m) = \sqrt{\sum_{k=1}^{D}\{y_{ik}(m) - y_{jk}(m)\}^2}$.

The *steepest descent method* is used to reduce this error to a minimum value. The new 2-D configuration at iteration $(m+1)$ is given by $y_{pq}(m+1) = y_{pq}(m) - (MF)\frac{\partial E(m)}{\partial y_{pq}(m)}/|\frac{\partial^2 E(m)}{\partial y_{pq}(m)^2}|$, where MF is the magic factor that is about $0.3 - 0.4$, $\frac{\partial E}{\partial y_{pq}} = -\frac{2}{C}\sum_{j=1, j\neq p}^{N}(y_{pq} - y_{jq})\{(d_{pj}^* - d_{pj})(d_{pj}d_{pj}^*)\}$, and $\frac{\partial^2 E}{\partial y_{pq}^2} = -\frac{2}{C}\sum_{j=1, j\neq p}^{N}\frac{1}{d_{pj}^* d_{pj}}\left((d_{pj}^* - d_{pj}) - \frac{(y_{pq}-y_{jq})^2}{d_{pj}}(1 + \frac{d_{pj}^*-d_{pj}}{d_{pj}})\right)$.

2.3 Topology Preserving Mapping of Sample Sets (TOPAS)

TOPAS is an improved version of NLM and is based on the rank order evaluation of both the original space (X) and the mapped space (Y) [3]. If the respective rank positions in X- and Y-space are not identical, a correction is used to obtain the proper rank order. To achieve gradual corrections of the points y_i in the visualization plane with regard to y_p in the iterative mapping process, distance information is used in the adaptive rule: $y_{ij}(t + 1) = y_{ij}(t) + \alpha_p(t, \Xi^Y(y_i))(y_{ij} - y_{pj})/dY_{ip}$, where distance, $dY_{ip} = \sqrt{\sum_{j=1}^{D}(y_{ij} - y_{pj})^2}$ and $\alpha_p(t, \Xi^Y(y_i)) = \gamma\beta(t)N_p(\Xi^Y(y_i))$, where $\gamma = sign(\Xi^Y(y_i) - \Xi^X(x_i))$ and $\beta(t) = \eta\exp(-ln(2)\frac{t}{t_h})$, where $sign(x) = 0$ when $x = 0$, $\eta = 0.4$, t is a particular iteration and t_h indicates the total number of iterations, and $N_p(\Xi^Y(y_i), t) = \exp(-\frac{1}{2}(\frac{\Xi^Y(y_i)}{n(t)})^2)$, where $n(t) = n_{start} - \frac{n_{start}-n_{end}}{t_h}t$. The n_{start} and n_{end} will be specified by the user. TOPAS provides a much better structure preservation than the NLM.

2.4 VISOR Algorithm

The pivot-vectors – V_1, V_2, V_3 are determined in the L-D space which provide a convex enclosure of the remaining data points. The determination method is as follows: (1) compute the centroid, M, of all K data points in the L-D space, (2) determine V_1 from the distance, $d(V_1, M) = max_{i=1}^{K}(d(v_i, M))$, (3) determine V_2 from the distance, $d(V_2, V_1) = max_{i=1}^{K}(d(v_i, V_1))$, and (4) determine V_3 from the distance, $d(V_3, V_2) = max_{i=1}^{K}(d(v_i, V_2))$.

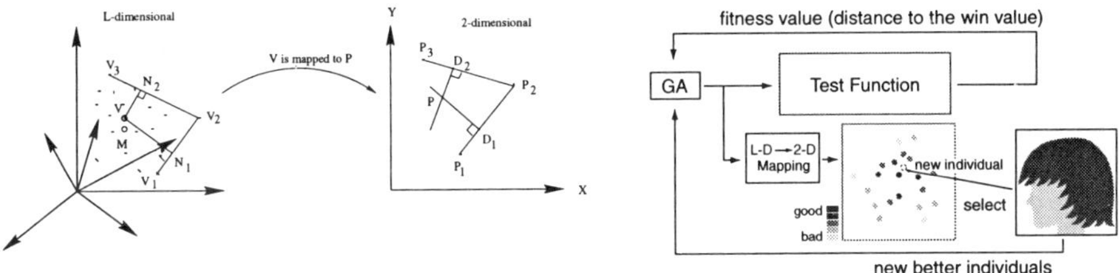

Figure 1: VISOR algorithm.

Figure 2: Experimental systems of the Visualized GA (whole) and conventional GA (upper part).

The pivot-vectors – P_i in 2-D plane correspond to the vectors V_i for $i = 1, 2, 3$.

The mapping procedure of the remaining points, V_i ($i = 4...K$), is: (1) draw perpendicular lines from the V to the lines $\overline{V_1 V_2}$ and $\overline{V_2 V_3}$ and determine the cross points N_1 and N_2, (2) find the points, D_1 and D_2, in 2-D space on the lines $\overline{P_1 P_2}$ and $\overline{P_2 P_3}$ with the same proportion of N_1 and N_2 on $\overline{V_1 V_2}$ and $\overline{V_2 V_3}$, and (3) draw the perpendiculars from D_1 and D_2 to $\overline{P_1 P_2}$ and $\overline{P_2 P_3}$, respectively. The intersection point is P corresponding to V in L-D (Figure 1).

3 Visualized GA as an Experimental System

We compare the mapping methods for visualization using a Visualized GA (genetic algorithm) system (Figure 2). The user of general Visualized IEC is requested to subjectively evaluate individual fitness values and to visually estimate the location of the global optimum. We adopt the Visualized GA instead of the Visualized IGA in our experiment to avoid the former subjective evaluation and objectively evaluate the latter effect.

Reverse mapping from 2-D to L-D space is also required in the Visualized GA. A suitable-sized lookup table is used in our experimental systems except for SOM, since the SOM stores all information during the forward mapping and easily handles the reverse mapping.

Three experiments are conducted in section 4. Mapping methods project 50×50 searching data point in Experiment I and II and 100×100 points in Experiment III from an L-D space to a 2-D space. The five benchmark functions in Figure 3 are commonly used in all experiments.

This experiment is conducted on a 400MHz-Pentium II PC running a Linux OS to directly influence CPU time in section 4.1.

4 Experimental Comparison of Mapping Methods

4.1 Experiment I: Comparison of Time-Complexity

The CPU time to project 2,500 data points using four mapping methods with five benchmark functions is calculated (Table 1). The TOPAS CPU time for the 2,500 data mapping was estimated by extrapolating the CPU time for 11 different sizes of data from 100 to 300.

The VISOR was the fastest mapping method, and the performance of the SOM was acceptable, however, the performance of NLM and TOPAS were far from practical.

4.2 Experiment II: Comparison of Easiness of Visual Inspection

The VISOR and SOM are compared how their 2-D mapped images are visually easier for 28 human subjects to estimate the location of a global optimum using five benchmark functions.

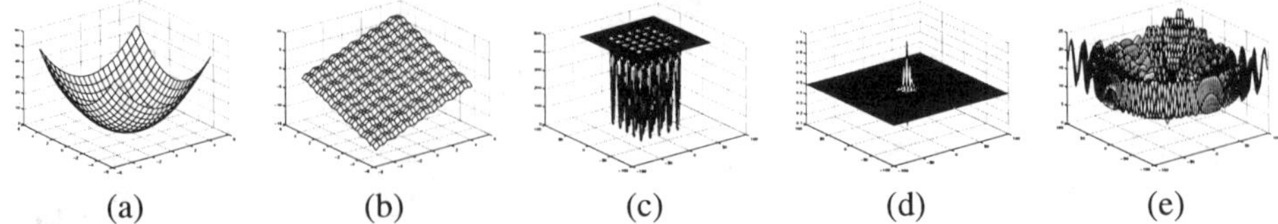

(a)   (b)   (c)   (d)   (e)

Figure 3: Five benchmark functions - (a), (b) and (c) are DeJong's F1, F3, and F5 functions, respectively; (d) and (e) are Schaffer's F1 and F2 functions, respectively.

The left and right boxes in Figure 4 show the mapped data in a 2-D space obtained from an L-D space using SOM and VISOR, respectively. Note that the precision or resolution of each 2-D space is 50×50 points, while the actual number of searched data points is that of dots displayed in the 2-D mapped spaces.

It is interesting to note that the data points are widely distributed in the SOM, whereas their VISOR counterparts accumulate in some region. Subjects reported that this difference of their distribution results made it easier to estimate the location of a global optimum using the SOM.

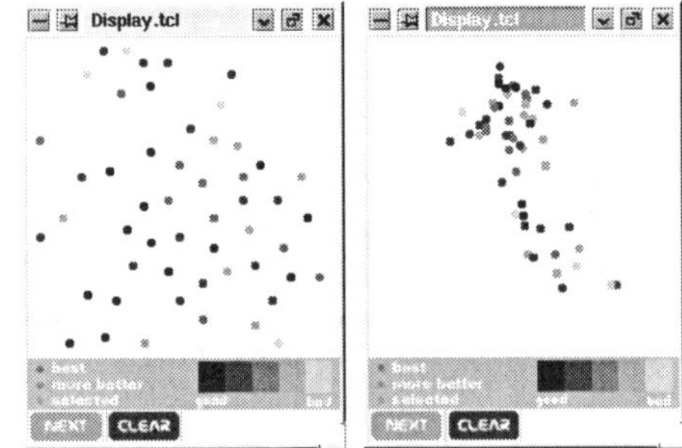

Figure 4: Mapping examples by SOM (left) and VISOR (right). The depth of color shows fitness values.

4.3 Experiment III: Comparison of Visualized GAs' Convergence

The convergences of Visualized GA with SOM and Visualized GA with VISOR for five benchmark functions are compared by 28 subjects. Both mapping methods map 100×100 data, and both GA have 20 population size, 0.9 crossover rate, and 0.02 mutation rate.

Figure 5 shows the average curves of 28 subjects for the two Visualized GAs, and Table 2 shows their sign test results at the fifth generation. These results imply that their superiority depends on the tasks.

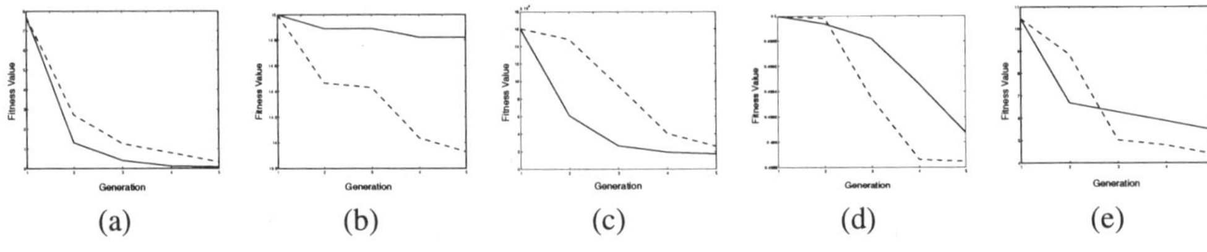

(a)   (b)   (c)   (d)   (e)

Figure 5: Convergence curves of different Visualized GAs for five benchmark functions in Figure 3. Dot and solid lines show Visualized GA with SOM and VISOR, respectively.

The average curves are greatly influenced by the extreme convergence curve among the 28 subjects; the sign test does not consider the difference between two curves. This is why the Figure 5(d) looks different from the results in Table 2.

Table 1: CPU time in seconds of four mapping methods for 2,500 data. DJ and Sc mean DeJong and Schaffer, respectively. The TOPAS CPU time is an estimated value.

	DJ F1	DJ F3	DJ F5	Sc F1	Sc F2
VISOR	1	1	1	1	1
SOM	3	4	3	4	4
NLM	1376	1342	6119	1328	2708
TOPAS	-	-	-	-	3.9×10^7

Table 2: Sign test results of 28 subjects on convergence of two Visualized GAs. SOM and VISOR mean that the number of the case that Visualized GA with SOM or VISOR were faster than another; SAME means that convergence of two Visualized GA were same; ** and * mean a significance with ($p < 0.01$) and ($p < 0.05$)

benchmark function	SOM	VISOR	SAME	sign test
DeJong F1	23	5	0	**
DeJong F3	3	7	18	
DeJong F5	20	8	0	*
Schaffer F1	20	8	0	*
Schaffer F2	8	20	0	*

5   Conclusion

We compared SOM, VISOR, NLM, and TOPAS mapping methods for visualizing an EC landscape to accelerate EC, especially IEC, convergence using five benchmark functions. It was determined that the computational costs of the VISOR and SOM were practical in visualizing an EC landscape, and costs of the NLM and the TOPAS were not, the visualization displayed by the SOM was easier to visualize an EC landscape than that by the VISOR, and the superiority of two mapping methods in convergence depended on the tasks.

We conclude that Visualized IEC/EC with SOM is a suitable combination for both saving time and reducing human fatigue unless the mapping CPU time of SOM is not significant for human fatigue comparable to that of VISOR.

References

[1] N. Hayashida and H. Takagi, "Visualized IEC: Interactive evolutionary computation with multidimensional data visualization," IEEE *Industrial Electronics, Control and Instrumentation (IECON2000)*, Nagoya, Japan, (Oct., 2000) 2738–2743.

[2] T. Kohonen, Self-Organizing Maps, Springer-Verlag, Heidelberg (1995).

[3] A. König, "A survey of multivariate data projection, visualization and interactive analysis," *5th Int'l Conf. on Soft Computing and Information/Intelligent Systems (IIZUKA'98)*, Iizuka, Fukuoka, Japan: World Scientific, Singapore, (Oct., 1998) 55-59.

[4] A. König, O. Bulmahn, and M. Glesner, "Systematic methods for multivariate data visualization and numerical assessment of class separability and overlap in automated visual industrial quality control," *5th British Machine Vision Conf.*, **1**, (Sept., 1994) 195-204.

[5] J. W. Sammon, A nonlinear mapping for data structure analysis, IEEE Trans. on Computers, **C-18**(5), (1969) 401-409.

[6] W. Siedlecki, K. Siedlecka, and J. Sklansky, An overview of mapping techniques for exploratory pattern analysis, Pattern Recognition, **21**(5), (1988) 411-429.

[7] H. Takagi, "Active User Intervention in an EC Search," *Int'l Conf. on Information Sciences (JCIS2000)*, Atlantic City, NJ, USA, (Feb./Mar, 2000) 995–998.

[8] H. Takagi, Interactive Evolutionary Computation: Fusion of the Capacities of EC Optimization and Human Evaluation, Proceedings of the IEEE, (2001) (will appear).

KES '01
N. Baba et al. (Eds.)
IOS Press, 2001

Dynamic Two-Dimensional Feature Map with Modified Ant Clustering for Data Visualization

*Tsuyoshi MIKAMI   **Mituso WADA
*Tomakomai National College of Technology, Nishikioka-443, Tomakomai, 059-1275, Japan
**Graduate School of Eng., Hokkaido University, Kita-13, Nishi-8, Sapporo, 060-8628, Japan

Abstract. The self-organizing networks are useful tools suitable for data analysis in which networks learn the topology of the complicated high-dimensional data structure and can project it on the two-dimensional plane. However, it has been noticed that those methods does not always visualize the high-dimensional clusters adequately on the lower-dimensional plane. In this paper, we proposed the visualization method by the ant clustering algorithm in order to construct the structures on the two-dimensional grid space as feature map.

1. Introduction

Self-Organizing Maps(SOM), proposed by Kohonen, are widely used for many data clustering and its visualization technique for data mining or pattern recognition. The advantage of SOM different from traditional clustering methods (e.g. k-means, isodata and so on) is to visualize the complicated cluster distribution onto the two-dimensional feature map without setting the number of clusters in advance[4]. There have been proposed many applications to massive data analysis by using the SOM. Honkera, et al., proposed the web-document collection and its visualization for browser interface by WEBSOM method, which is advanced version of SOM[3]. Deboeck, et al., applied the SOM to cluster visualization of the finance database so as to analyse it easily[5]. However, the SOM has limitations of the cluster generation because the size of output neurons is fixed continuously. Moreover, the networks cannot learn the topology of the incremental non-stationary data set adaptively, but that of statistical data set. To improve such limitations, many researchers have been proposed the growing self-organizing networks, which is inserting/deleting neurons dynamically. Fritzke proposed the Growing Cell Structures (GCS), which learn the topology of the data structure and is self-tuning the number of cells (output neurons) adaptively and automatically. The GCS is a powerful technique especially for the non-stationary database[7]. As the most recent research, Si, et al., proposed the Dynamic Topology Representing Networks (DTRN), which is a kind of growing self-organizing networks learning the topology of high-dimensional input data with the vigilance test similar to Grossberg's ART2 algorithm. This calculation is faster than the other typical growing self-organizing networks[2].

However, those growing mechanism cannot visualize the high-dimensional topology onto the intuitive space for data analysis. Fritzke proposed the visualization techniques on the two-dimensional map for the GCS but his algorithm is only useful for relative low-dimensional input data (e.g. three-dimension). The real data are usually expressed as high-dimensional attribute vectors, so, the GCS or DTRN algorithm is not always applied to such data. This is a reason that the SOM is still used generally to many applications regardless of its limitations.

This paper describes new visualization methods projecting high-dimensional data onto two-dimensional intuitive space by using the *ant clustering*, proposed by Lumer and Faieta[1]. In this method, each ant random-walks on the two-dimensional grid space and picks/drops each datum having some-dimensional attributes probabilistically according to *dissimilarity* between his held datum and some data existing on his local sight[1]. As a result, it is able to gather the similar data into one or few piles. This method has two advantages as follows:

1. This is able to visualize the high-dimensional data onto two-dimensional plane.
2. The target data is gathered by ant-like foraging mechanism even if the targets are increasing/decreasing dynamically[8].

However, the ant clustering mechanism can deal with only few training data (e.g. less than one thousand data) because it spends a long time until task fulfillment. In this paper, we modified the Lumer's ant clustering algorithm to realize the fast calculation for complicated data analysis.

2. Modification of the Ant Clustering

According to the ecology, real ants follow the simple behavioral rules by their local sight to perform the global task fulfillment. Deneubourg, et al., described the collective mechanism in which ants are foraging and gathering foods by simple behavioral rules and their simulation results are very similar to the work real ants performed. Lumer and Faieta proposed the ant clustering algorithm in which the foods are defined as the numerical data having n-dimensional attributes and the ants are gathering them according to dissimilarity (n-dimensional Euclidean distance) in the local sight. By using this method, it is not necessary to define the number of clusters previously distinct from the traditional clustering, such as k-means, isodata, and so on. Moreover, since each ant does not know the global state but his local sight and his task is done continuously (not converged such as learning), the ants can make the clusters adaptively even if the target data are increasing/decreasing[8]. In the real database, massive data sets are always updated dynamically, so it is difficult to reconstruct the feature map adaptively for such real database by using traditional methods. However, the ant clustering mechanism can deal with only few training data (e.g. less than one thousand data) because it spends a long time until task fulfillment.

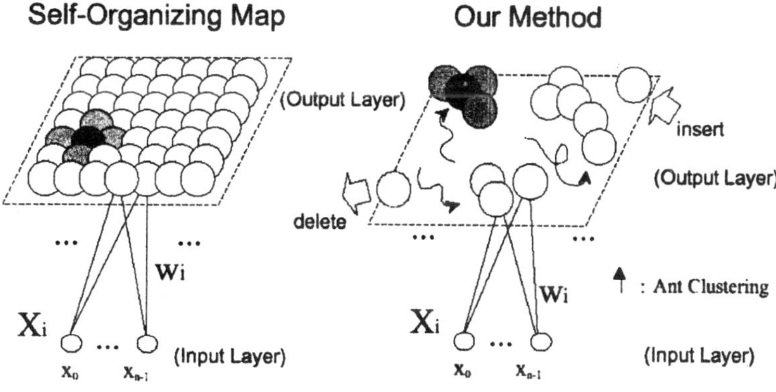

Fig.1 A Comparison between Self-Organizing Map and Our Method

First we considered that it spends a long time for the ants to perform the clustering task fulfillment and reconstructed the ant clustering algorithm to realize faster calculation. In our method, the datum is not carried by the agent (ant) but moving and allocated directly on the feasible location itself. This is not a behavior derived from the ecology, but our method is able to realize the faster algorithm because of the nonuse of random-walk and agent. The datum that should be moved is selected randomly from the index number list of data set. After that, the datum selects some local regions (e.g. 8*8 grids) randomly on the field and finds the best matching region by calculating the follow evaluative function.

$$f(i) = \frac{N^2}{N + \eta \sum_{j \in \pi(i)} d(i,j)}$$

where j is an index of data existing on the local region, $d(i,j)$ is n-dimensional Euclidean distance, η is constant value, N is the number of data on the local sight, and $\pi(i)$ is a set of data existing on the local sight. Each datum is allocated to the best matching region. Our random selection is equivalent to the ant mechanism in which the ant picks the datum and is walking randomly and the local matching mechanism is similar to the mechanism that the ant drops the datum on the adequate region according to its past memory.

3. Experiments

We adopted the four-dimensional artificial data as the training set. Three clusters are distributed on the four-dimensional space where the gravity of clusters is (-2, -2, -2, -2), (0,0,0,0), (2,2,2,2) respectively. The dispersion σ is 1.2 as a common property of three clusters by Gaussian distribution. In our experiments, we applied 400 target data and dispersed them randomly on the two-dimensional field.

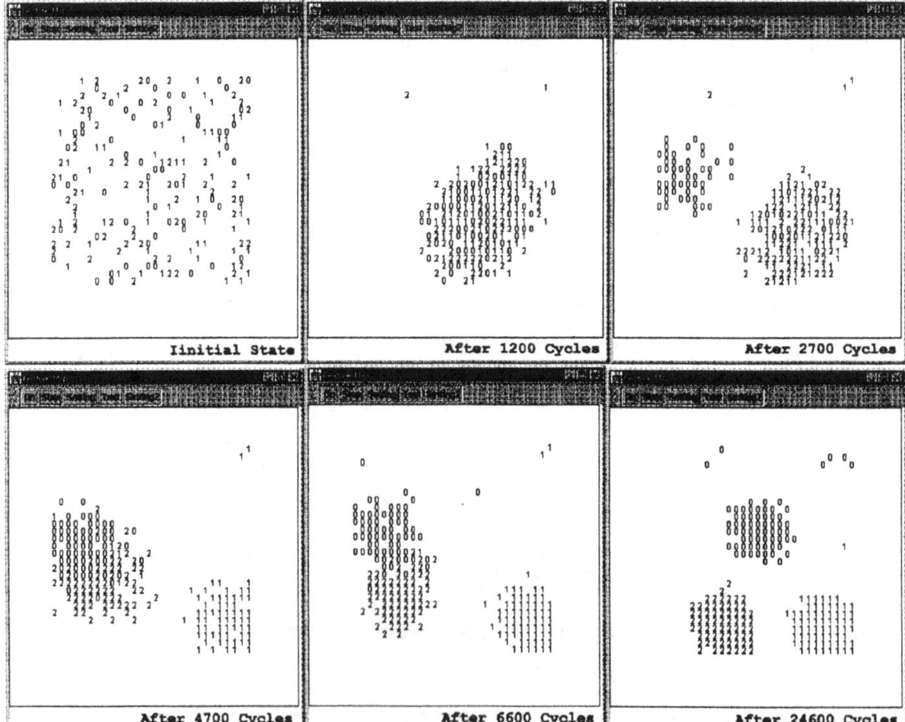

Fig. 2 Time Evolution of the High-Dimensional Data Visualization using Our Method

Figure 1 shows that the four-dimensional data are located on the two-dimensional grid by using our ant clustering method. It is able to understand three clusters intuitively according to two-dimensional feature map. Each cycle is one iteration of our algorithm, namely, 400 cycles mean that 400 data are moving from current location to new location just once. In the experiments, the data are aggregated to only one cluster despite their dissimilarities for few cycles. After 2500 cycles, they are separated from it and reconstructed new clusters involving the similar data. In the traditional ant clustering, proposed by Lumer and Faieta, it is necessary to spend 500,000 cycles until convergence for 400 data set[1]. Kuntz, et al., applied the L-F algorithm to graph partitioning problems for VLSI design, but their results are also needed to over 300,000 cycles[6].

But in our results, it is obvious that the target data are aggregated themselves into three clusters regardless of only 4700 cycles (each datum is moving to new location for only about 12 times). In the experiments, we used the PC/AT compatible computer with Celeron 500MHz CPU and our simulation is executed on the Java Virtual Machine. Nevertheless, it spends only one or two seconds until convergence. Moreover, our result is very similar to SOM in which the grid size is 40*40 (See Figure 2). We carried out some experiments by changing random number, but the results are similar to figure 1 and 2(b) mostly.

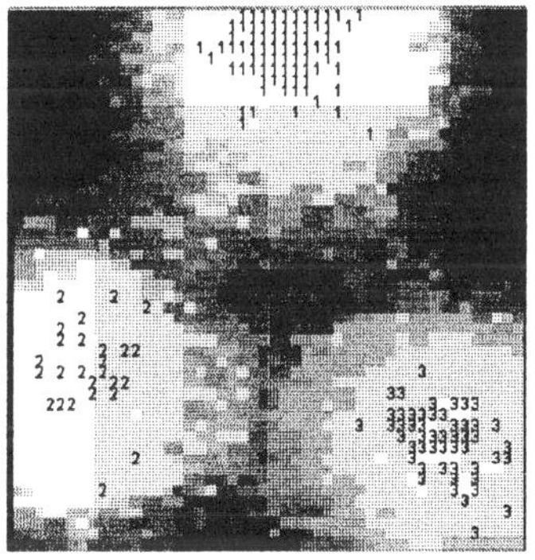 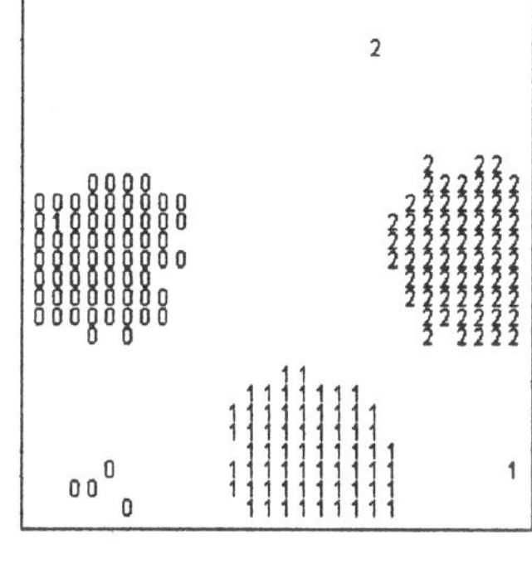

(a). Self-Organizing Map   (b). Our Result

Fig.3 Experimental Results using Self-Organizing Map and Our Method

Moreover, we investigated more complicated data structure, six clusters created from Gaussian distribution. The boundaries among six clusters are not distinct on the four-dimensional space. Fig.4 shows the feature map for those clusters using our method. Six clusters are distributed on the two-dimensional plane and their distribution is influenced by similarity among them. The cluster-2 is combined with cluster-1 and cluster-5, because their boundary data are very similar to each other.

4. Discussion and Conclusion

We proposed the data visualization method based on the ant clustering strategy. In our method, the data set are allocated on the two-dimensional grid and aggregate automatically according to the dissimilarity among them. However, our method should be pre-defined the two-dimensional grid size and is not available if the input data are growing over the size of grid space. It is unknown previously whether the target data are growing over the number of grids or not. In general, it is well-known that the SOM is very lagging calculation algorithm. So, it is non-realistic that large size networks are pre-defined. But in our algorithm, it is not necessary to determine the number of data previously though they are inserting within the limit of the grid size. In the future works, we will investigate the property of feature map for more complicated data, e.g., high-dimensional (e.g. hundred-dimensional) or the massive and non-stationary real data. It is also necessary to compare the SOM with our method in such cases.

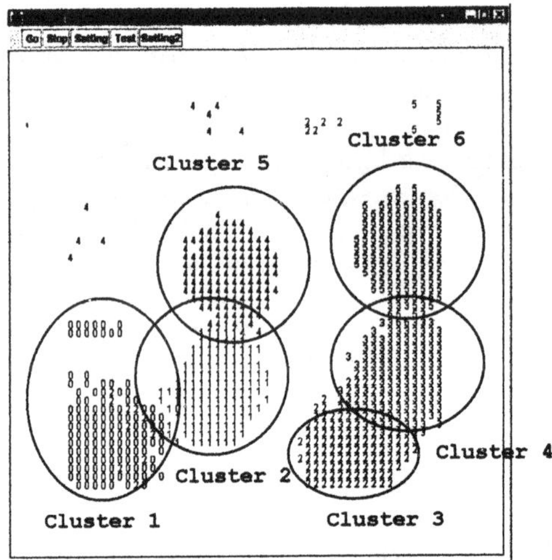

name	gravity
Cluster 1	(2, 2, 2, 2)
Cluster 2	(0, 0, 0, 0)
Cluster 3	(4, 4, 4, 4)
Cluster 4	(6, 6, 6, 6)
Cluster 5	(-2,-2,-2,-2)
Cluster 6	(8, 8, 8, 8)

Gaussian Distribution (σ =0.2)

Fig.4 Visualization of Six Clusters (4-Dimensional Attributes) using Our Method

References

[1] Lumer, E., et al., Diversity and Adaptation in Populations of Clustering Ants, *From Animal to Animats 3: Proceedings of 3rd International Conference on Simulations of Adaptive Behavior*, The MIT Press, 1993

[2] Si, J., et al., Dynamic Topology Representing Networks, *Neural Networks, 13*, 2000

[3] T.Honkela, et al., Newsgroup Exploration with WEBSOM method and browsing interface, *Technical Report A32, Helsinki University of Technology*, 1996

[4] Kohonen, T., Self-Organizing Maps, Springer, 1995

[5] Deboeck, G., et al., Data Explorations in Finance with Self-Organizing Maps, Springer, 1998

[6] Kuntz, P., et al., An Ant Clustering Algorithm Applied to Partitionning in VLSI Technlogy, *Proceedings of European Conference on Artificial Life: ECAL'97*, The MIT Press, 1997

[7] Fritzke, B., Growing Cell Structure - A Self-Organizing Network for Unsupervised and Supervised Learning, *Neural Networks, 7*, 1995

[8] Proctor, G., et al., Information Flocking : Data Visualisation in Virtual Worlds Using Emergent Behaviours, *Virtual Worlds: First International Conference*, Springer, 1998

[9] Fritzke, B., A self-organizing network that can follow non-stationary distributions, *Proceedings of International Conference on Artificial Neural Networks : ICANN'97*, 1997

[10] Fritzke, B., Growing self-organizing networks - why?, *Proceedings of European Symposium on Artificial Neural Networks : ESANN'96*, 1996

[11] Alahakoon, D., Dynamic Self-Organizing Maps with Controlled Growth for Knowledge Discovery, IEEE Transactions on Neural Networks, vol.11, no.3, 2000

[12] Blackmore, J., Visualizing High-Dimensional Structure with Incremental Grid Growing Neural Network, Master Thesis, University of Texas at Austin, 1995

[13] Tanaka, M., et al., Clustering by Using Self-Organizing Map, IEICE Transactions, vol. J79-D-II, no.2, (Japanese), 1996

KES '01
N. Baba et al. (Eds.)
IOS Press, 2001

Budgetary Transfer to Local Governments: Equity, Efficiency and Political Influence

Aiko Shibata   Yu Sakai
Kwansei-Gakuin University2-1 Gakuen, Sanda, Hyogo, 669-1337 Japan

Abstract. Mechanisms of budgetary resource allocations from the Japanese central government to local governments are analyzed in this paper utilizing statistical methods and a Self Organizing Map (SOM). All budgetary transfers to local governments are said to be redistributed on an equitable basis. As a result of the budgetary transfers to local governments that have been carried out for a long time on an equitable basis, we expect inefficient investment in public goods. We also suspect political influences. Using a cross section analysis of fiscal year 1991 and panel data analysis of 18 years between 1977 and 1995, we conclude that the budgetary transfers to the local governments, where political influences were discerned, were carried out equitably and resulted, nevertheless, in inefficient public investment. Though, when cross section data from 46 prefectures are clustered using SOM, and the two largest clusters are analyzed further statistically, the results differ. It was also shown that SOM is useful as a visual data mining method.

1. *Preface*

The Japanese local government's general revenues come from three major sources. They include (1) local taxes,   (2) budgetary transfers from the central government and (3) borrowing. As for the budgetary transfers, there are three kinds of central government transfers to the local governments.   They are allocation taxes (koufuzei), transfer taxes (jyoyozei) and treasury disbursements (kokko-shisyutukin).   These transfers are redistributions from funds collected by the central government, which levies various taxes.

First, we shall use the 1991 data from 46 prefectures, because the 1990 election results are used as variables to show political influence. The 1990 election was the 39th election for the House of Representatives (HR). The so-called " political mechanism of the 1955th year" symbolized by the Liberal Democratic Party's (LDP's) single party rule since 1955 ended with the next election, the 40th election for the HR, which took place in 1993. Second, we shall extend a similar analysis to the period of 1977-1995 for the 46 prefectures. Section 2 analyses a cross-sectional model. In Section 3, panel data of 1977-1995 from the 46 prefectures are analyzed. Section 4 presents our conclusions.

2. Cross sectional analyses of 1991

Let us introduce the following model.

 (1) $Y = a + b X + e$

Y is a dependent variable of the budgetary transfer to local governments, X represents the independent variables, and e is an error term. Data are taken from fiscal 1991 (except for political variables.) X includes population, area, per capita prefectural real income, indexes for prefectural productivity, the accumulated amount of prefectural debts and political variables. We broke down each electoral district by party affiliation and by region (urban versus rural).

Further, clusters are regressed and the results of the regression show that, although some independent variables for political influences are not significant, the other variables present the expected results. However, economic, social and cultural conditions differ among the prefectures. Therefore, we have decided to sort prefectures into clusters based on similarity. We clustered the data with a SOM. Since a SOM creates nodes with vectors similar to those of the input data, it could solve the problem of inadequate data for a statistical analysis after clustering. Figure 1 is the map drawn with a SOM. In the map, two large clusters, Cluster 1 and Cluster 2, are shown. Two main results obtained from cross sectional analyses are as follows.

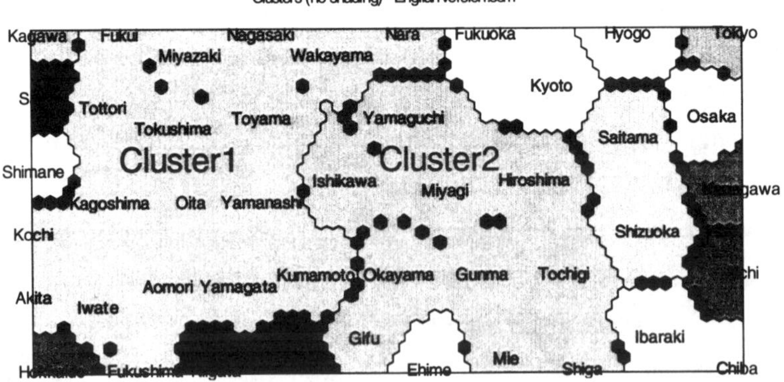

Figure 1. Clustering with SOM

First, if equity and efficiency are considered, the results of the national data, the cluster data and the results between the two clusters differ.   With the national data, it becomes clear that a decrease in per capita income increases the budgetary transfers. This result is expected, as one would assume that governmental policy would try to redistribute the budgetary transfers so as to attain the national minimum standard of living across the country.   However, if productivity decreases, the budgetary transfers to local governments increase.   This result indicates that public investment has been inefficient.

Second, in contrast with the national data, the cluster data show different effects of per capita income and prefectural productivity. Among prefectures in the second cluster, per capita income and productivity are high, while budgetary transfers are low. Among prefectures in this cluster, there is no equitable redistribution of the budgetary transfers. Among prefectures in the first cluster, per capita income and productivity are low. But these prefectures receive much larger budgetary transfers than the national average,   indicating that efficient public investment has been undertaken in these prefectures. We shall extend our analyses to panel data in the following.

3. *Analyses with panel data (1977-95)*

Our panel data are time series data for the period of 1977-95 for 46 prefectures excluding 1992. The independent variables are defined similarly as the 1991 data. And these variables for each prefecture are converted to ratios by dividing them by the national total amount. Let us now introduce the following model that includes two characteristics, i.e., prefectures and time.

$$\bullet \; \bullet \; \bullet \; Y_{it} = a_i + bX_{it} + e_{it} \qquad i = 1, ..., N, \quad t = 1, ..., T \bullet$$

Here, i represents a prefecture and t a time. Y is a dependent variable and a prefectural share of the budgetary transfer, X represents the independent variables mentioned above, and e is an error term We select the fixed effects model after considering three models, the pooled model, the fixed effects model and the random effects model [Green]. The results of the fixed model are shown in column 2 of Table 1. The results are similar to those of the cross sectional analyses. (The adjusted R square is 0.989.)

Now cluster analyses with SOM are performed. The map is drawn with a SOM using the panel data. We obtained 7 clusters. Data were transformed to per capita absolute amounts. And for political variables we have additional variables. Further, clusters are regressed and the results of three clusters, C1, C2 and C3, are shown in Table 1. Significant variables other than political variables have the same signs for parameters. The signs also show the directions as expected. Although the budgetary transfers to local governments are influenced by political variables, the effects differ by clusters.

Table 1. Regression analyses of clusters

Independent	All data: Fixed model	C1	C2	C3
Population	-0.0234	-	-	-
Area	0.2239^{***}	+	+	+
Income	-0.1476^{***}	-	-	-
Productivity	-0.0969^{***}	-	-	-
Local debts	0.0098	+	+	+
Urban- LDP -Ratio		+	-	+[a]
Rural- LDP -Ratio	0.0008^{*}	+	-	-
LDP -Member		-	+	+
Rural-Non-LDP - Ratio	0.0001			

Dependent: transfer. ***,**,* show significance levels of 1%, 5%, 10%, respectively. The signs of parameters of independent variables are listed, if independent variables are significant at a 1%, 5%, and 10% level. "a" indicates a result obtained from nodes created by SOM, when all the signs are identical with a result obtained from actual data.

Let us examine the changes in position over the period for some prefectures. The movement of Gunma Prefecture from 1977 to 1995 is depicted in Figure 2. The white circle is the position of Gunma in 1995. The other end of the line is the position of Gunma in 1977. In 1977 there were 12 prefectures positioned near Gumna shown by black circles. For example, Oita, Aomori, Nagano, Kagoshima, Niigata etc. are in C1. Gunma, Fukushima, Mie, Okayama etc. are in C2. After 18 years in 1995, all prefectures moved to C1 except Kagoshima and Niigata. These two prefectures moved to C2.

Both C1 and C2 have a small number of population, large area and the ratios of senior rural HR members are high. However, C1 has low income per capita and C2 has high per capita income.

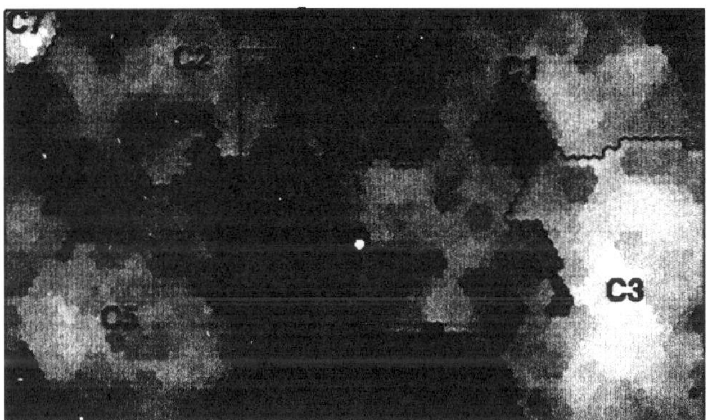

Figure 2. Clustering of panel data with SOM

The index of productivity is low in C1 and is at an intermediate level in C2. C1 receives the largest amount of the transfers per capita. Exploring the reasons for these changes among prefectures over 18 years would be interesting. Thus, a SOM shows advantages in visual data mining.

4. Conclusions

The budgetary transfers to local governments were distributed theoretically for the purpose of attaining a national minimum standard of living. Our findings also support this equitable distribution. However, as a result of the equitable budgetary transfers carried out for a long time, inefficient investment in public goods has occurred. These results are supported by our cross sectional analyses and panel data analyses via the use of statistical methods and a Self Organizing Map (SOM). However, cluster analyses of a cross section of data of 1991 show the different results for equity and efficiency depending on the clusters, although cluster analyses of the panel data did not show such differences. Political influences are detected also.

One of SOM's major advantages is the visual data-mining method that it permits.

5. *Acknowledgements*

We would like to acknowledge for helpful comments from Heizou Tokutaka, Kikurou Fujimura, Katumi Okui. Kagaku Kenkyuhi (No.13630117) of the Ministry of Education and Science of Japan funded the research.

6. *References*

[1] T. Doi, Nihon no Syakai Shihon ni Kansuru Panelbunsek, *Kokumin keizai*, 1998, pp.27-52.

[2] W. Green, *Econometric Analysis, 4th edition*, Prentice Hall, Upper Saddle River, New Jersey, 2000.

[3] Kaizuka, Honma, Takabayashi, Nagamine and Fukuma, Chiho koufuzei no kinou to sono hyoka Part *Financial Review*, vol. 2, August 1986, pp.6-28.

[4] T. Kohonen, *Self Organization and Associative Memory*, Springer-Verlag, 1993.

[5]_____________, *Self-Organizing Maps*, Springer-Verlag, 1995.

[6] S. Meyer and S. Naka, Legislative influences in Japanese budgetary politics", *Public Choice*, 94: 1998, pp.267-288.

[7] S. Reed, *Japan Election Data-The House of Representatives 1947-1990*, Center for Japanese Studies, The University of Michigan, Ann• Arbor, 1992.

[8] H.Tokutaka, S. Kishida and K. Fujimura, *Jikososhikika Map no Ouyou* , Kaibundo, 1999.

[9] M. Van Hulle, *Faithful Representations and Topographic Maps: From Distortion- to Information-based Self-organization*, J. Wiley, New York, 2000.

Parametric and non-parametric regression techniques applied to scoring problems in data mining

Steven Raekelboom
Entropy BVBA
Naamsestraat 84/21
B-3000 Leuven, BELGIUM

Steven Lenaerts
Synes NV
Technologielaan 11
B-3001 Heverlee, BELGIUM
steven.lenaerts@synes.com

Marc M. Van Hulle
Laboratorium voor Neuro- en
Psychofysiologie
Katholieke Universiteit Leuven
Campus Gasthuisberg
Herestraat
B-3000 Leuven, BELGIUM
marc@neuro.kuleuven.ac.be

Abstract. Scoring is one of the most powerful techniques to tackle the bulk of data mining problems encountered in today's real-life business environment. Scoring techniques aim at finding a relation between a set of independent and dependent variables. However, many classic parametric techniques suffer from rather stringent assumptions about the functional format of those relations. This is why non-parametric scoring techniques and, in general, non-parametric statistics attracted the attention of the data mining community. In this paper the well known parametric technique, logistic regression, is compared with a novel non-parametric scoring technique, kMER-based regression, which is based on equiprobabilistic topographic maps.

1. Introduction

Scoring is probably the most interesting application domain within data mining. In practice this technique is used to predict for instance churn behaviour, potential customer value, assess financial risks, etc.[1] From a mathematical point of view, scoring boils down to solving regression problems. The objective is to find a functional relationship between a set of noisy samples and a target variable. Typically a regression problem is formulated as follows:

$$Y = f(X) + \varepsilon$$

The function f needs to be estimated such that the difference between the estimated Y and the actual value of Y, in response to the input X, is minimal. The noise contribution ε is assumed to have zero mean and to be independent from the input X. Such regression problems are often computed on a mean squared error basis.

However, in many practical data mining problems the target variable is binary and the positive targets, usually represented by 1, are underpopulated compared to the negative targets, represented by 0. Following a pure mean square error approach in such cases would lead to regression systems with hit rates that are not considerably better compared to using the a priori distribution of the target variable as decision criterion. One can overcome this problem to a certain extent by bootstrapping the target variable. The disadvantage of this approach is the increase of false positive predictions. Therefore the scoring problem should be formulated differently. The idea is to build a system that is able to filter records with a maximum probability to be a positive target.

$$P(Y = 1 \mid X) = f(X)$$

When building regression models, typically, assumptions are made about the nature of the relationship between the input variables and the output variable. The regression model is then built by computing the parameters of the model. The main shortcoming of such a parametric approach is that the performance of the regression model depends on the validity of the a priori assumptions made about the nature ('shape') of the relationship. The disadvantage of non-parametric models is that they require relatively more samples, more processing power, and that they are more difficult to interpret.

This paper compares the performance of logistic regression[2], as a parametric technique, and that of kMER-based regression, as a non-parametric technique, for data mining problems with dichotomous targets and for which the positive ones are underpopulated. Logistic regression is well known as the extension of multiple regression. KMER-based regression is based on self-organizing topographic map formation.

2. Non-parametric regression with topographic maps

2.1. Topographic Maps

Topographic maps are neural networks with an unsupervised competitive learning scheme. The aim of unsupervised competitive learning is to cluster similar input patterns in the same category. The competitive element in the learning scheme forces the output neurons of the network to compete for being activated by input patterns. This is done in an unsupervised manner, meaning that the categories are found by the network itself from the correlations in the input patterns.

Topographic maps are trained with the same principles but also use the geometrical arrangement of the competitive output neurons to provide information about the input space. Consider two input patterns v_1 and v_2, and two locations l_1 and l_2 of the corresponding activated neurons. As v_1 and v_2 are more similar, l_1 and l_2 should get closer to each other as well. This principal is referred to as topology preservation. Typically the neurons are ordered in a two-dimensional lattice which folds itself into a higher dimensional input space.

One of the first topology preserving map implementations is the SOM algorithm of Kohonen [3]. However the SOM does not guarantee a linear mapping of the data density of the input space to the output space. Recently, Van Hulle [4] embedded the concept of information based learning instead of the SOM's distortion based learning strategy in the kernel-based equiprobabilistic learning rule (kMER). The kMER maps are characterized by many favourable properties that allow superior monitoring of the neighbourhood cooling scheme and a more accurate representation of the input space distribution as kMER guarantees a linear mapping between the density of input space and that of the output space. Initially, the technique was mainly used for clustering and classification purposes. But this technique also provides us with a platform for kernel-based regression, which makes it suited for many generic data mining applications. The next two sections describe the two-phased approach for solving regression problems. During the first phase a topographic map is trained on the input space, generating a data model. In the next phase the target variables are projected onto the topographic map using a one-stage learning algorithm. Note that this approach has attractive advantages from a data mining perspective. First, the data model can be developed separately from the regression application. In this paper four target variables will be projected onto the same data model. Second, since the weight density is proportional to the input density, the regression surface will be more detailed when there is more input data. Finally, training the regression will be fast since only one neuron layer is needed.

2.2. kMER-based regression

The kMER learning rule positions neurons and radii such that each neuron has an equal probability to be active (i.e., a sample falls in its receptive field region). The kernels allocated at the neuron weights can now be used to obtain a continuously differentiable regression surface of which the smoothness depends on the local input density. A neuron output function is defined by supplementing the receptive field region S_i of each neuron with a normalized Gaussian kernel:

$$g_i(x, \rho_s) = \frac{\exp(\frac{-\|x - w_i\|^2}{2(\rho_s \sigma_i)^2})}{\sum_{j=1}^{N} \exp(\frac{-\|x - w_j\|^2}{2(\rho_s \sigma_i)^2})}, \forall i \in A$$

with ρ_s a smoothing factor.

The regression surface is obtained by interpolating between the kernel centres. The output of the regression model can be written as:

$$O(x) = \sum_{j=1}^{N} W_j g_j(x, \rho_s)$$

For binary output values two output neurons are used. This results in a 2 x N weight matrix W. The new weights W_j and the smoothing factor ρ_s are trained such that the mean square error between the actual output values (in this case 0 or 1) and the estimated output values is minimal. The outputs of the regression function for binary targets are activations scores for both positive $Act_1(x)$ and negative $Act_0(x)$ outcome. The probability for a positive target for a record x is then computed as follows:

$$P(1 \mid x) = \frac{Act_1(x)}{Act_1(x) + Act_0(x)}$$

3. Evaluating scoring techniques in data mining

The two methods proposed above will be used to score on four data sets of which the target variable is related to the input variables in several ways. Lift graphs will be used to compare the performance of both techniques [5].

Scoring models that generate good lift graphs are able to make a good ranking of the target probability. Lift graphs (Fig. 1A) however do not evaluate the ability of the model to make accurate estimates about the target likeliness. Calibration graphs (Fig. 1B) enable evaluation of the quality of the target likeliness estimates.

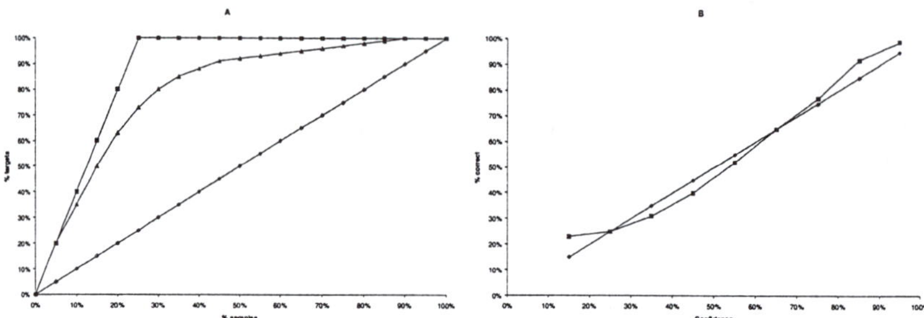

Fig. 1 Evaluation of scoring techniques. (A) Lift graph. The curve with triangular markers reflects a perfect scoring model. In this case the database contains 25% positive targets. The perfect model is able to select all the targets in a subset of 25% of the database. The random model of which the performance is indicated with diamond markers selects only 25% of the positive targets in a subset containing 25% of the total database. A predictive model will score in between. The curve with triangular markers shows that the model is able to filter about 65% of the positive targets in a 25% subset. (B) Calibration graph. The curve with the diamond markers indicates the perfect calibration; the curve with the rectangular markers reflects the calibration of the predictive model. For the 50% to 60% interval on average 55% of those samples should actually have a positive target in case of perfect calibration. The graph shows that the predictive model is slightly over confident for estimates below the 65% level (except the 15% level). Above 65% the predictive model performs better then expected and is therefore slightly under confident.

4. Experiments

Two experiments will demonstrate the performance of logistics regression and kMER based regression. We use a dataset (V) of 5000 records in which each record is characterised by ten variables $v_1, v_2, \ldots, v_{10}$ drawn from a uniform distribution between 0 and 10.

For logistic regression we used the generalised linear models functionality in the statistics toolbox of Matlab 6.0.0.88 release 12. Logistic regression can be performed by invoking B =*glmfit(X, Y, 'binomial', 'logit')* and returns an array (B) of coefficients, applying the scoring model can be done as follows: R=*glmval(B, X, 'logit')*.

For kernel-based scoring we used the Synes Data Prospector™ Suite 1.2. First, a topographic map was trained in batch-mode with following settings: 100 neurons (10 by 10 lattice), a batch size of 1000, 100 batches were presented in the initial run, and a learning rate of $\dfrac{0.2}{batch\,size}$.

In the first experiment, we defined linearly separable target classes as follows:

$$Y = \left\{ y \mid y = 3v_1 - 7v_6 + 6v_7 - 2v_9 + 9, \forall v \in V \right\}$$

$$t = \begin{cases} 1, \forall y \in Y : y < mean(Y) - 0.75stdev(Y) \\ 0, \forall y \in Y : y \geq mean(Y) - 0.75stdev(Y) \end{cases}$$

this results in target distribution of 24,1% positive targets and 75.9% negative targets.

Both techniques score extremely well on the lift graph (Fig. 2A), logistic regression scores nearly perfect while kMER-based regression shows some flaws in determining the last couple of positive targets. Note that in the graph the perfect lift coincides with the lift of logistic regression and kMER-based regression, and therefore markers are difficult to distinguish. When comparing calibration (Fig. 3A) is consistently under confident, while the calibration of kMER-based regression is a rather good fit with the perfect calibration.

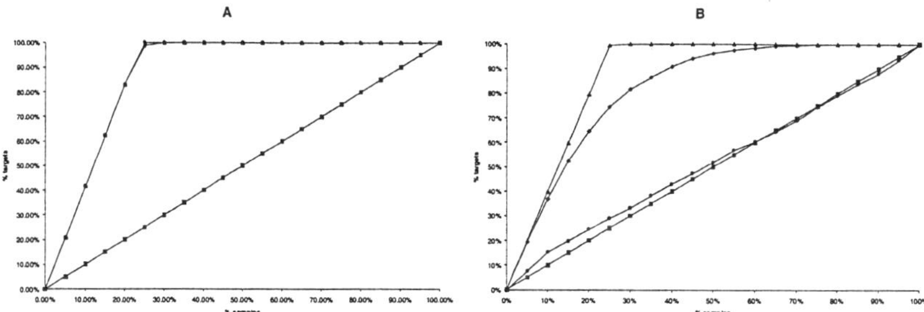

Fig. 2 Lift graphs of the experiments. The curves with diamond markers represent the lift obtained with kMER-based regression, the curves with circular markers represent the lift obtained with logistic regression, the curves with triangular markers indicate the perfect lift, and the curves with rectangular markers stands for the lift achieved with a random selector. (A) first experiment, and (B) second experiment.

In the second experiment, we defined non-linearly separable target classes as follows:

$$t = \begin{cases} 1, \forall v \in V : v_1 > 5, v_9 > 5, v_{10} > 5 \\ \qquad or \ v_1 < 5, v_9 < 5, v_{10} < 5 \\ 0, \forall v \in V : \neg(v_1 > 5, v_9 > 5, v_{10} > 5 \\ \qquad or \ v_1 < 5, v_9 < 5, v_{10} < 5) \end{cases}$$

this results in a target distribution of 25.1% positive targets and 74.9% negative targets.

KMER-based regression achieves a very good lift (Fig. 2B) in this experiment, logistic regression however performs only slightly better than when using a random selector. The calibration (Fig. 3B) of provides a very good fit with the perfect calibration in the confidence ranges from 50% to 70%, but is under confident beyond 70%. For logistic regression, all confidences lay within the 70% to 80% range and the marker coincides with the marker of the perfect calibration. In fact, logistic regression is gambling on negative targets with on average a confidence of 75%, which matches the target distribution.

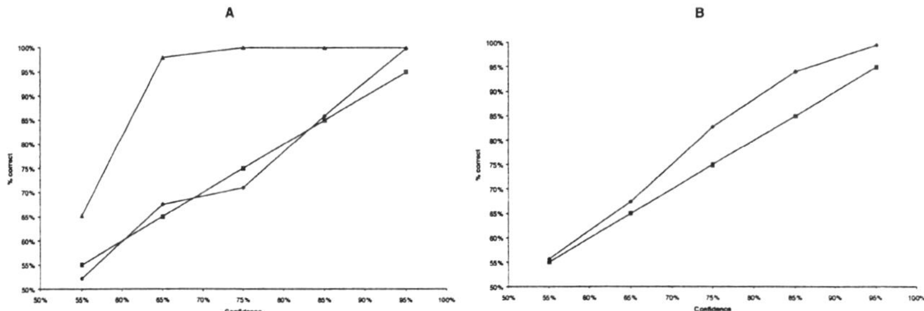

Fig. 3 Calibration graphs of the experiments. The curves with diamond markers represent the calibration achieved with kMER-based regression, the curves with triangular markers reflect the calibration obtained with logistic regression, and the curves with rectangular markers indicate the perfect calibration. (A) first experiment, and (B) second experiment.

In case of the linearly separable experiments, both logistic regression and kMER-based regression scored very well on the lift graph. Calibration-wise kMER-based regression has the edge over logistic regression. In case of the non-linearly separable experiments, kMER-based regression achieves good lifts while logistic regression fails to obtain a reasonable lift. Calibration is good for kMER-based regression, the calibration of logistic regression is quite useless since the models for the non-linearly separable experiments hold minimal predictive power as shown in the lift graph. Hence, when the boundaries between categories become more complex the benefit of not making modelling assumptions becomes apparent, as demonstrated by the better lift graphs of kMER-based scoring.

5. Conclusion

We have introduced a new non-parametric scoring technique and applied it to a dichotomous categorical problem where one category is underpopulated. The technique is based on topographic maps and self-organization, topics that have recently attracted the attention of the data mining community. We have compared our technique with logistic regression, a popular parametric technique for categorical regression. We observe that both techniques perform extremely well when the categories are linear separable. However, many real-life scoring problems cannot be adequately modelled with linear boundaries. This situation occurs quite often in particular when the independent variables are categorical. In such cases, the positive targets can be associated with non-adjacent categories, which leads to non-linear separable scoring problems. With respect to calibration, kMER performs better. Due to the modelling assumptions, logistic regression does not aim at good target probability estimates but is rather optimised to find good probability rankings. This is not the case for our new scoring technique since it directly models the input distribution, which eventually leads to better probability estimates of the underlying scoring function.

6. Acknowledgments

M.M.V.H. is supported by research grants received from the Fund for Scientific Research (G.0185.96N), the National Lottery (Belgium) (9.0185.96), the Flemish Regional Ministry of Education (Belgium) (GOA 95/99-06; 2000/11), the Flemish Ministry for Science and Technology (VIS/98/012), and the European Commission, 5th framework programme (QLG3-CT-2000-30161).

7. References

[1] J. Han and M. Kamber, Data Mining Concepts and Techniques, 2001
[2] C. M. Dayton, Logistic regression analysis, Maryland, 1992
[3] T. Kohonen, Self-organized formation of topologically correct feature maps, Biol. Cybern. **50**,(1982) 35-41
[4] M.M. Van Hulle, Faithful representations and topographic maps, 2000
[5] A. Berson, S. Smith, and K. Thearling, Building data mining applications for CRM, 1999

Automatic FAQ E-mail Classification by Self-Organizing Maps[*]

Sung-Bae Cho

Department of Computer Science, Yonsei University
134 Shinchon-dong, Sudaemoon-ku, Seoul 120-749, Korea

Abstract. Widespread use of computer and Internet leads to an abundant supply of information, so that many services for facilitating fluent utilization of the information have appeared. Accordingly, human operators should respond to users' e-mail questions at the Internet portal services, but the increasing number of users brings serious burden. In this paper, we propose a two-level self-organizing map (SOM) which automatically responds to the users' questions on Internet, and helps them to find their answer for themselves by browsing the map hierarchically. The method consists of keyword clustering SOM which reduces a variable length question to a normalized vector, and document classification SOM which classifies the question into an answer class. The final map is also used for browsing. Experiments with real world data from Hanmail net that is the biggest portal service in Korea show the usefulness of the proposed system.

1. Introduction

Proliferation of computer and increasing interest in Internet lead many people into information-network based services. However, it is difficult for people who are not familiar with such services to get what they want in those services. They suffer from several difficulties, such as installation of program, usage of specific function, and so on. It is time-consuming and difficult to overcome such problems even for expert users as well as beginners. Information-network service corporations, ISP (Internet Service Provider) and IP (Information Provider) companies, operate a call center to help users via telephone, provide answers about questions by web board systems like FAQ (Frequently Asked Questions), and respond to the questions by e-mail. However, the explosive number of users makes service corporations to recognize the needs of automatic response system.

Currently, more than 5 million people use the Hanmail net that is the biggest portal service in Korea and users' questions per day come to about 400 cases. It is redundant and time-consuming to respond to duplicated questions by hand, and even worse users are complaining about the response time. Automatic processing of users' questions might be not only efficient for operators who can avoid redundant work but also satisfactory to users. This paper proposes a two-level self-organizing map (SOM) that automatically classifies the users' questions for Hanmail net. We show that the automatic response system based on the model is plausible for e-mail data of which number and size per class are various.

---

* This research was supported by Brain Science and Engineering Research Program sponsored by Korean Ministry of Science and Technology.

2. Related Works

Generally speaking, automated text categorization (TC) is a supervised learning task, defined as assigning category labels (pre-defined) to new documents based on the likelihood suggested by a training set of labelled documents [1]. The automatic response of the query mails in the Internet is also a sort of the automated categorization of texts into topical categories.

TC has a long history, dating back at least early 60s. Until the late '80s, the most effective approach to the problem seemed to automatic classifiers built manually by means of knowledge engineering techniques. The machine learning paradigm has emerged in the '90s [2]. Within recent paradigm, learning phase builds a classifier that learns the characteristics of one or more categories from the pre-classified training data set, and the classifier is used in categorizing the unseen data. The advantages of this approach are accuracy comparable to human performance and considerable saving in terms of manpower, since no intervention from either knowledge engineers or domain experts is needed [2].

We use the idea of TC to classify the questions for Internet portal services. In particular, SOM is used for machine learning because the current data are not uniformly distributed on the classes, even though we think that it might be useful to apply supervised learning at some stages in the application.

3. Automatic Classification

The system consists of two parts: classification and browsing subsystems. The classification system also consists of two parts. The first part is for preprocessing and keyword clustering which help to encode the input vector for the next classification module by reducing the dimension of the input mail. The dimension of vector is reduced by a SOM for keyword clustering. The second part is to classify and match the queries with the corresponding answers by another SOM called mail classification SOM.

The browsing system is based on the completely learned mail classification SOM. It helps users to search their answer conceptually by navigating the system hierarchically with topology-preserving property of SOM. A user can find answer by changing the resolution of the information of the map. Fig. 1 shows the overall system architecture.

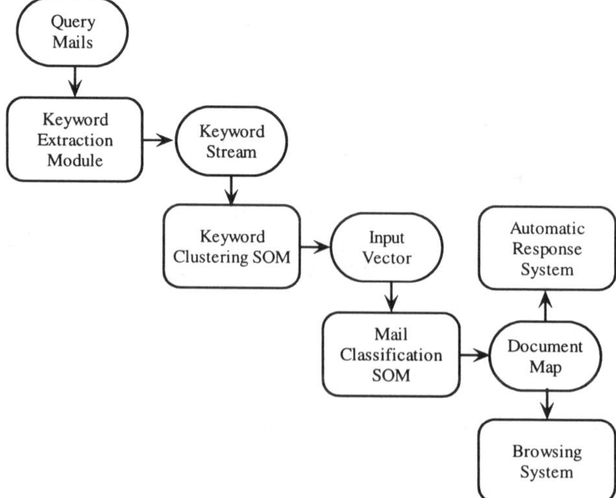

Fig 1. System Architecture

3.1 Keyword Extraction

Query mails are written by natural language, especially in Korean or compound of Korean and English. The query mails must be encoded as normalized to get the input vector for neural network. To vectorize the query mails keyword should be extracted at first, because queries include meaningless components such as auxiliary word, the ending of word, and article. Since the components do not affect the meaning of the mail, they should be removed.

3.2 Keyword Clustering

After the extraction process, the keywords should be converted into vectors of real numbers. There are several methods to vectorize the information, for example vector space model [1]. Due to the computational and spatial reasons the SOM-based approach is adopted as the basic method. This is based on a previous work to encode documents based on their words, such as a neural filter and a neural interest map for information retrieval [2][3].

Keyword clustering SOM is used to reduce and cluster the category of words for query mails. The input of keyword clustering SOM is a keyword identification vector with context information of a sentence that includes the keyword. Because SOM collects and clusters similar data at the neighborhood, similar keywords are mapped into adjacent nodes in SOM [4][5]. Keyword clustering SOM plays the similar role of the thesaurus that discriminates the synonyms in the conventional text processing. Learning algorithm of the SOM is as follows [4][5]:

$$m_i(t+1) = m_i(t) + \alpha(t) \times n_{ci}(t) \times \{x(t) - m_i(t)\}$$

Here, $\alpha(t)$ is learning rate, $n_{ci}(t)$ is neighborhood function, $m_i(t)$ is the weight of the ith node, and $x(t)$ is input of SOM. At the neighborhood function, c means the index of winner node, and winner is obtained by the following equation [4][5]:

$$\|x - m_c\| = \min_i \{\|x - m_i\|\}$$

All the keywords in a query mail are represented by a vector, and each word is represented by an n-dimensional real vector x_i. Input vector for keyword clustering SOM is encoded as follows [6]:

$$X(i) = \begin{bmatrix} E\{x_{i-1} \mid x_i\} \\ \varepsilon x_i \\ E\{x_{i+1} \mid x_i\} \end{bmatrix}$$

Here, E denotes the estimate of the expectation value evaluated over the group, and ε is a small scalar number, e.g., 0.2 [6][7][8][9][10]. The input vector in keyword index i, $X(i)$, consists of one symbol vector x_i and two context vectors of x_{i-1} and x_{i+1}. x_{i-1} means the keyword occurring just before the keyword x_i, and x_{i+1} means the keyword occurring just after the keyword x_i. As mentioned above, all the prior keyword vectors are summed, and then averaged by the number of occurrence of x_i. It is the same for the posterior keywords. The two vectors give the context information of the sentence to the input vector $X(i)$. We train the SOM with these two vectors to cluster the similar keywords.

3.3 Mail Classification

The keyword clustering SOM maps a set of similar keywords into a node of the SOM, because all keywords appeared in a query have a winner node in SOM. According to each winner node, a query mail can be converted into a histogram for keywords that make up a query mail.

The dimension of input vectors is the same with the output number of keyword clustering SOM. Therefore, the size of the map decides the reduction degree and dimension of the input vector of the mail classification SOM. For example, the SOM that has $m \times n$ size map produces *mn*-dimensional input vector. With this method, input vector can be produced, but there is a shortcoming. The vectors produced are very sensitive to the specific query mail, i.e., vectors overfit to query data. Because queries are made by various people with various vocabulary, the vectors to reflect only a specific case are not desirable. To overcome the variation problem, a smoothing process that has frequently used in pattern recognition area is applied. Such process blurs the sensitive integer vectors to non-specific real vectors. Gaussian kernel is used for blurring the histogram [11].

In addition, Shannon's entropy [1][6] that improves the discrimination of keywords is utilized. Entropy converts the discrete frequency to weighted frequency that can be increased by real values. Keywords to reflect the characteristic of specific classes have higher degree of frequency, and others have lower degree of frequency. The entropy weight is calculated as follows:

$$E_w = H_{max} - H(w),$$

$$H_{max} = \log(number\_of\_classes)$$

$$H(w) = -\sum_i \frac{n_i(w)}{\sum_j n_j(w)} \ln \frac{n_i(w)}{\sum_j n_j(w)}$$

Here, $H(w)$ means the entropy value of keyword w, $n_i(w)$ means the occurrence frequency of word w in class i, and H_{max} is the maximum entropy, which is log 67 in our case. V_i is the *i*th component of blurring histogram and calculated as follow:

$$V_i = \sum_w (F_w \times \frac{G_i}{E_w})$$

Here, G_i means the *i*th component of the smoothing kernel, E_w means the entropy weight of keyword w, and F_w means the frequency of keyword w. After vector blurring and entropy weighting processes are finished, a vector of one query mail is produced.

The mail classification map trained with the vectors above maps each query data into a specific node. According to the characteristics of SOM, the query mails that belong to the same class are mapped in the same or neighborhood positions. A new query mail from a user is transformed into an input vector and classified as a certain class, and then automatic response system sends the corresponding answer that matches with the winner node to the user promptly.

4. Experimental Results

Experiments are conducted for 2206 query mails of 67 classes. Since the size of keyword clustering SOM is fixed to 10 by 10, the dimension of the input vector of the mail classification SOM is 100. The first experiment decides the size of mail classification SOM, where entire 2206 data are used for training the map.

The analysis of recognition rate with the keyword clustering SOM of 150×150 is summarized in Table 1. The recognition rates of group A and group D are not good. In case of group A, four classes which do with the e-mail account of Hanmail net are not classified well. Four classes include account change, deletion, duplication, and information confirmation. In these classes, the extracted keywords are so similar that they cannot reflect the attribute of query mail. That is the reason why the recognition rate is so low. Group D includes 16 classes that have trivial information about the remaining 51 classes. It deteriorates the classification rate because the keywords are so similar with the remaining classes.

Next, we have fixed the size of keyword clustering SOM as 150×150, and divided the data set into two groups. One is training data of 1545 query mails, and the other is test data of 661 query mails. Experiments are performed with the change of threshold, which is the quantization error in SOM. Threshold value also affects the rejection rate. The result is shown in Table 2.

Table 1. Recognition rate for each group of data

Group	# of class	Recognition rate	
A	6	954/1002	94.0%
B	7	561/585	95.9%
C	36	124/127	97.6%
D	18	459/492	93.3%
Total	67	2098/2206	95.01%

Table 2. Recognition rate for test data

Threshold	Rejection rate	Recognition rate
0.5	46.4%	58.5%
0.4	60.1%	64.0%
0.3	72.9%	82.7%

5. Concluding Remarks

In this paper, we have proposed an automatic response system of query mails of a portal service corporation in Korea, and shown that the system is promising by conducting several experiments. As shown in experimental result the accuracy rate of training data is high, and the accuracy of test data is plausible but the rejection rate is a little high. We are under way to increase the accuracy while maintaining the low rejection rate. Moreover, we are also attempting to develop a browsing interface appropriately to the novice users.

References

[1] G. Salton, *Automatic Text Processing: The Transformation, Analysis, and Retrieval of Information by Computer*, Addison-Wesley, 1988.
[2] J. C. Scholtes, "Kohonen feature maps in fulltext database: A case study of the 1987 Pravda," *In Proc. Informatiewetenschap*, pp. 203-220. STINFON, Nijmegen, Netherlands, 1991.
[3] J. C. Scholtes, "Unsupervised learning and the information retrieval problem," *In Proc., Int. Joint Conf. on Neural Networks*, pp. 95-100. IEEE Service Center, Piscataway, NJ, 1991.
[4] T. Kohonen, "Self –organized formation of topologically correct feature maps," *Biol. Cyb.*, 43:59-69, 1982.
[5] T. Kohonen, *Self-Organizing Maps*, Springer, Berlin Heidelberg, 1995.
[6] H. Ritter, and T. Kohonen, "Self-organizing semantic maps," *Biol Cyb*, 61:241-254, 1989.
[7] S. Kaski, T. Honkela, K. Lagus, and T. Kohonen, "Creating an order in digital libraries with self-organizing maps," *Proc. World Congress on Neural Networks*, pp. 814-817, 1996.
[8] T. Honkela, S. Kaski, K. Lagus, and T. Kohonen, "Exploration of full-text databases with self-organizing maps," *In Proc., Int. Conf. on Neural Networks*, 1:56-61, IEEE Service Center, Piscataway, NJ, 1996.
[9] K. Lagus, T. Honkela, S. Kaski, and T. Kohonen, "WEBSOM for textual data mining," *Artificial Intelligence Review*, 13:345-364, 1999.
[10] S. S. Kaski, T. Honkela, K. Lagus, and T. Kohonen, "WEBSOM-self-organizing maps of document collections," *Neurocomputing*, 21: 101-117, 1998.
[11] E.Gose, and R. Johnsonbaugh, S. Jost, *Pattern Recognition and Image Analysis*, Prentice Hall PTR, 1996.
[12] K. Lagus and S. Kaski, "Keyword selection method for characterizing text document maps," *Proc. Int. Conf. on Artificial Neural Networks*, **1**, pp. 371-376, IEEE, London 1999.

KES '01
N. Baba et al. (Eds.)
IOS Press, 2001

Generative Mixture Modeling by Autonomous Estimators

Samuel Kaski and Jarkko Salojärvi
Neural Networks Research Centre, Helsinki University of Technology
P.O. Box 5400, FIN-02015 HUT, Finland

Abstract. Generative models of probability density are common in descriptive modeling of data, i.e., summarizing large data sets and exploring the dependencies between their variables. Mixture models are a class of generative models particularly suitable for clustering. They assume that each data sample stems from one of a set of sources that generate data independently. It is then desirable to model the data with a mixture of "experts" corresponding to the sources. The experts cannot be fit to data autonomously, since each of them needs to know the contribution of the others. In this paper we introduce a mixture of autonomous experts model that approximates the plain mixture density model but consists of experts that learn autonomously. Each expert stores simplified representations of the other experts and uses them to judge the contribution of the others. The representations are updated intermittently. During learning the task becomes divided and conquered automatically.

1. Introduction

Mixture density models [6] are a commonly used category of probability density estimators. In mixture models it is assumed that each data sample is generated by one of a set of independently operating mixture components, and that the data set is a mixture of the samples from them.

When fitting a mixture of estimators or "experts" to data, the modeling task is effectively broken into partially overlapping subtasks assigned to the different relatively simple experts. Such division of labor increases analytical tractability and makes the results more easily understandable. This is especially important in descriptive modeling tasks where the goal is to summarize large data sets and gain understanding of their essential characteristics.

Although the experts are independent data generators, experts within a mixture cannot be fitted autonomously to data. The reason is (details are given below) that when fitting each expert the contributions of all the other experts must be known. In this paper we introduce a variant of the mixture model in which the experts can learn autonomously. The experts maintain simplified, periodically updated representations of the other experts, which they use while learning. The experts can then be considered as simple "intelligent agents" which can be implemented as, for example, separate processes in different computers in a distributed fashion.

Our work is related to two topics that have been studied intensively during recent years: the mixture of experts model [3, 4], and committee machines [7]. These models are usually applied to regression and classification tasks, in which the goal is to predict the value of a random variable given the value(s) of other variable(s). The problem of modeling the dependent variable is somewhat more complicated than the density estimation problem considered in this paper. In principle extensions of our methods should be applicable to modeling conditional distributions as well.

2. Mixtures of Autonomous Models

In this chapter the mixture of autonomous experts model will be introduced in the general form that is applicable to any kinds of experts. The goal is to model a data set $\{\mathbf{x}_k\}_k$, $k = 1, \ldots, N$, $\mathbf{x} \in \mathbb{R}^n$.

Denote the prior probability that the jth expert generates a data point by $p(M_j)$, and the probability density function of the samples that the expert generates by $p(\mathbf{x}|M_j)$. The mixture model consisting of K experts then generates the probability distribution

$$p(\mathbf{x}) = \sum_{j=1}^{K} p(\mathbf{x}|M_j)p(M_j) . \tag{1}$$

When the model is fitted to the data the probabilities $p(M_j)$ are considered as parameters and their values, as well as the values of the parameters of the component densities $p(\mathbf{x}|M_j)$, are sought. In this paper we will use the common maximum likelihood criterion.

In order to demonstrate where the simplified representations of the component densities (described by individual experts) are actually needed, let us consider gradient-based optimization. Similar conclusions hold also for the EM-based learning. Denote the parameters of $p(\mathbf{x}|M_j)$ by $\boldsymbol{\theta}_j$. The gradient is then

$$\frac{\partial}{\partial \boldsymbol{\theta}_j} E = \sum_k p(\mathbf{x}_k)^{-1} p(M_j) \frac{\partial}{\partial \boldsymbol{\theta}_j} p(\mathbf{x}_k|M_j) , \tag{2}$$

where $p(\mathbf{x}_k)$ is given by (1). The gradient depends on the parameters of all of the experts through the term $p(\mathbf{x}_k)^{-1}$. The same holds for the gradient of $p(M_j)$.

If the experts are to learn autonomously they need an estimate for the probabilitity $p(\mathbf{x}_k)$. Our solution is to incorporate, in each autonomous expert, simpler representations of the densities generated by the other experts. Denote the representation kept by expert j of the density of expert i by $\hat{p}^{(j)}(\mathbf{x}|M_i)$, and of the prior probability by $\hat{p}^{(j)}(M_i)$. The expert j can then approximate the mixture of all experts by

$$\hat{p}^{(j)}(\mathbf{x}) = p(\mathbf{x}|M_j)p(M_j) + \sum_{l \neq j} \hat{p}^{(j)}(\mathbf{x}|M_l)\hat{p}^{(j)}(M_l) . \tag{3}$$

By assigning the autonomously operating experts to different processors or computers it is possible to distribute the computational load. If the representations the experts store of each other are considerably simpler than the original representations the experts hold of themselves, then the time and memory complexity of each expert is considerably lower than that of the whole mixture. Note that it may be beneficial in distributed computing to fix the prior probabilities of the experts $p(M_j)$ to equal values, so that each expert becomes responsible of approximately the same proportion of the data.

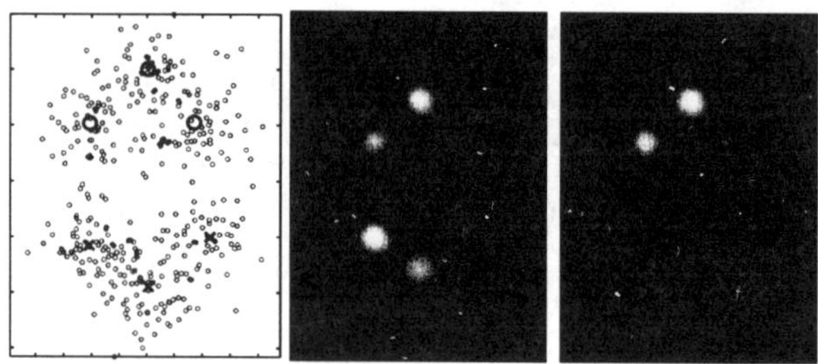

Fig. 1. *A toy example that demonstrates the autonomous experts. The data (the small circles in the leftmost figure) consisted of a total of 400 samples generated from six Gaussian kernels. There were two autonomous experts, each consisting of a mixture of three Gaussians. The means of the Gaussians of the first expert are shown with circles and the means of the second expert with crosses, respectively. The density generated by the mixture of the experts after 54 EM iterations is shown in the center image. The density plot on the right shows how the first experts sees the data. The large Gaussian at the bottom represents the other expert.*

3. Gaussian Mixture Model

In this section the autonomous learning procedure is discussed in the case of relatively simple and widely used Gaussian mixture models [6]. In the plain Gaussian mixture model each expert generates a Gaussian probability density. Each expert may, of course, contain several Gaussians as well. We will assign a set of K_j Gaussians, denoted by $m_i^{(j)}$ to the expert M_j. The mixture of experts then generates the density (cf. 1)

$$p(\mathbf{x}) = \sum_{j=1}^{K} p(M_j) \sum_{i=1}^{K_j} p(\mathbf{x}|M_j, m_i^{(j)}) p(m_i^{(j)}|M_j) . \qquad (4)$$

The mixture can be considered as a kind of a hierarchical Gaussian mixture model. All Gaussian distributions $p(\mathbf{x}|M_j, m_i^{(j)})$ are assumed symmetrical with their means denoted by $\boldsymbol{\mu}_i^{(j)}$ and variances by $(\sigma_i^{(j)})^2$. The variance is the same in each dimension.

The maximum (log) likelihood solution for the parameters can be sought by for example the gradient approach outlined in Section 2 or by the expectation maximization (EM) algorithm [2]. In the case studies of this paper the EM algorithm will be used. The derivation of the update formulas of the EM algorithm follows the standard derivation of the algorithm [6], but instead of $p(\mathbf{x})$ we use the approximation 3 in the expectation step. In this paper we approximate the densities of the other experts by a single Gaussian, denoted by $\hat{p}^{(j)}(\mathbf{x}|M_j)$, which has the nice property that its parameters can be solved in closed form by minimising the Kullback-Leibler distance between the true density and the Gaussian approximation.

The autonomous mixture model has been demonstrated in Figure 1.

4. Experiments

In order to demonstrate and verify the principle of autonomous experts we modeled two real-world data sets with mixture models. The first set was from the machine learning data repository at the University of California at Irvine [1]. It consisted of 690 samples with 14 attributes, derived from Australian credit card applications classified to two groups: accepted or rejected.

The second dataset consisted of 5964 patent abstracts from four patent classes. The document vectors were formed by the standard vector space model of information retrieval, and the dimensionality was reduced to 50 using Latent Semantic Indexing (essentially principal component analysis).

4.1. Results

The autonomous mixtures were compared with plain Gaussian mixtures of the same size. We used 10-fold cross validation with two measures: The likelihood of the test data not used for learning, and the accuracy with which the Gaussian clusters can classify the test data. The classification accuracy was measured by finding the maximum of $p(c|x_k) = \sum_{j,i} p(c|M_j, m_i^{(j)}) p(M_j, m_i^j | x_k)$ over the classes c. If the classification was different than the correct class the classification error was increased.

The Gaussian kernels were initialized with the K-means algorithm, and the variances were initialized to unity. The learning procedure of the autonomous experts was asynchronized by updating a randomly chosen expert at each iteration; the resulting increase in the number of iterations has been compensated for in the figures. The experts updated their estimates of each other after every 10 iterations.

The number of (autonomous) experts was varied while keeping the total number of Gaussians fixed. The prior values $p(M_j)$ were fixed to equal values to simulate optimal work load allocation when the experts are run in different computers. Note that the model reduces to the plain mixture model when the number K of the experts M_j is one.

The results for the two data sets are shown in Figure 2. For both data sets the likelihoods are worse for the autonomous mixture; for the second set the difference is significant (Fig. 2d). The classification accuracy seems, however, equal or even better for the autonomous experts.

5. Conclusions

In this paper we introduced the principle of autonomous experts for mixture density modeling, derived maximum likelihood learning procedures for them, and verified the viability of the principle with simple Gaussian mixture experts. The mixtures approximate the plain mixture models in which the learning is not autonomous, and it was demonstrated in two sets of experiments that the result was close to the original model.

The learning procedure has an interesting interpretation that is perhaps most clearly visible in the stochastic gradient-based learning of the simplest Gaussian mixture model, in which the densities $p(\mathbf{x}|M_j)$ in (1) are single Gaussians. If the mixture learns with stochastic gradient descent, i.e., by moving a short step towards the instantaneous gradient obtained for a single data sample $\mathbf{x}$, then it is straightforward to show that the learning rule for the means of the Gaussians will become

$$\boldsymbol{\mu}_j(t+1) = \boldsymbol{\mu}_j(t) + \frac{\alpha(t)}{\sigma_j^2} p(M_j|\mathbf{x})(\mathbf{x} - \boldsymbol{\mu}_j(t)) , \tag{5}$$

where $\alpha(t)$ is the learning rate that decreases gradually according to the schedules of the stochastic approximation theory [5].

The update rule (5) can be interpreted so that the jth component density of expert i modulates its learning, i.e. movement towards the data, by its estimate of the probability $p(M_j|\mathbf{x})$ of being responsible for the data. It is intuitively appealing that the group of

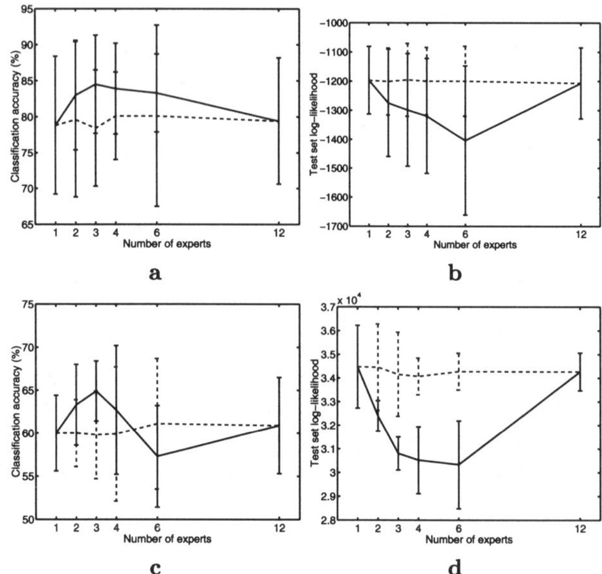

Fig. 2. Comparison of plain Gaussian mixtures and autonomous mixtures for (**a** and **b**) the Australian credit card database and (**c** and **d**) the computer patent database. (**a** and **c**) Classification accuracy, and (**b** and **d**) (log) likelihood for test data. The overall number of Gaussians was 12 in every experiment. The number of EM iterations was 120 for the credit card database and 24 for the document database. Dashed line: plain mixtures of Gaussians; solid line: mixtures of autonomous experts.

autonomous experts learns best if each expert evaluates how good it is in a task, and modulates its learning according to the evaluation. The learning can be considered as a generalization of competitive learning to a group of autonomous experts.

References

[1] C. Blake and C. Merz. UCI repository of machine learning databases, 1998.

[2] A. Dempster, N. Laird, and D. Rubin. Maximum likelihood from incomplete data via the EM algorithm. *Journal of the Royal Statistical Society B*, 39:1–38, 1977.

[3] R. Jacobs, M. Jordan, S. Nowlan, and G. Hinton. Adaptive mixtures of local experts. *Neural Computation*, 3:79–87, 1991.

[4] M. Jordan and R. Jacobs. Hierarchical mixtures of experts and the EM algorithm. *Neural Computation*, 6:181–214, 1994.

[5] H. J. Kushner and G. G. Yin. *Stochastic Approximation Algorithms and Applications.* Springer, New York, 1997.

[6] G. J. McLachlan and K. E. Basford. *Mixture Models. Inference and Applications to Clustering.* Marcel Dekker, New York, NY, 1988.

[7] M. Perrone and L. Cooper. When networks disagree : ensemble methods for hybrid neural networks. In R. Mammone, ed., *Artificial Neural Networks for Speech and Vision*, pp. 126–142. Chapman & Hall, London, 1993.

Data Mining with Self-Organising Maps (SOM) and Minimal Spanning Tree (MST)

K. Obu-Cann, K. Fujimura, H. Tokutaka, M. Ohkita, M. Inui and S. Yamada
*Tottori University, Dept. of Electrical and Electronics Engineering,
Koyama-Minami 4-101, Tottori, 680-8552, Japan*

Abstract. Data mining or exploration is part of a larger area of recent research in Artificial Intelligence and Information Processing and Management otherwise known as Knowledge Discovery in Database (KDD). The main aim here is to identify new information or knowledge from database in which the dimensionality or amount of data is so large that it is beyond human comprehension. Self-Organising Maps (SOM) and Minimal Spanning Tree (MST) are used to analyse power transformer database from one of the electric energy providers in Japan. This paper looks at clustering with Minimal Spanning Tree (MST).

1. Introduction

The SOM [1], [2] is a powerful tool for data mining, knowledge discovery and visualisation of high dimensional data. SOM and MST are applied to power transformer database from one of the electric energy providers in Japan. The aim of this work is to apply a data-mining tool based on SOM to learn more about database. The data was analysed to identify the energy consumption pattern of the consumers, season type pattern classification and day type pattern classification. In day type pattern classification the Minimal Spanning Tree (MST) algorithm [3], [4] was applied to augment the efforts of SOM in clustering. Prediction of the oil temperature changes in a power distribution transformer has been practiced using conventional methods based on explicit numerical calculation. Usually associated with such technique of prediction, is a limitation in the prediction accuracy due to assumptions made in the characteristics of the transformer. This paper looks at the application of SOM to the prediction of the oil temperature changes of a power distribution transformer.

2. Clustering with SOM

SOM is usually presented as a two-dimensional grid or map with the units tuned to different input data patterns. This algorithm is based on unsupervised competitive learning, training in this case is entirely data driven and the neurons or nodes on the map compete with each other. The data set contains four measurements (percentage load on the transformer, transformer oil temperature, atmospheric temperature and transformer surrounding temperature) recorded daily every 10 minutes, averaged hourly and normalized by component scaling in order to maximize the differences among the data.

2.1. Season Type Pattern Classification

SOM is applied in the classification of the input patterns into the various seasons of the year namely winter, spring, summer and autumn. The data set for the SOM comprised of data for the months of January, March, July and September 1996. Each month is represented by two different day type patterns. The input patterns are hourly measurements hence each input vector is made up of 96 components, which have been normalised [4]. Figure 1a illustrates the seasonal feature map obtained for the year 1996 after 5000 iterations of training.

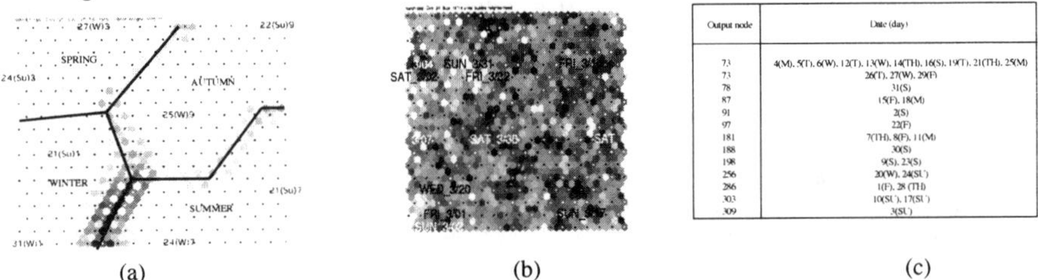

Output node	Date (day)
73	4(M), 5(T), 6(W), 12(T), 13(W), 14(TH), 16(S), 19(T), 21(TH), 25(M)
73	26(T), 27(W), 29(F)
78	31(S)
87	15(F), 18(M)
91	2(S)
97	22(F)
181	7(TH), 8(F), 11(M)
188	30(S)
198	9(S), 23(S)
256	20(W), 24(SU)
286	1(F), 28 (TH)
303	10(SU), 17(SU)
309	3(SU)

(a)   (b)   (c)

Figure 1: (a) Seasonal SOM for 1996, (b) SOM for transformer oil temperature in March 1996, (c) Summary of SOM for March 1996.

The eight input patterns, which represent the various seasons of the year, were classified into four separate regions on the map. Between winter and summer, the darkest grey separation was identified on the map. This is because during winter, though it is very cold, heating is mostly done with paraffin other than electric energy. In summer a lot of electric energy goes into cooling thereby resulting in a drastic increase in energy consumption. The next in the degree of grey is the difference between winter and autumn followed by winter and spring. The pattern for autumn is closer to spring than summer. This is because in autumn, the heating fuel is also paraffin. Looking at the SOM of figure 1a, a lot can be said about the electric energy consumption patterns of the consumers. Furthermore, this map can be used as a prototype by which recorded data could be compared. If one is confronted with an unlabeled data set, by use of this map, the season within which this data was recorded can be identified by its position on the map.

2.2. Day Type Pattern Classification

SOM is utilised for the classification of daily transformer oil temperature patterns. In SOM, the output units of the network are not specified. It is proposed that the SOM will generate the output based on the different patterns available in the input data space [5]. An input pattern comprises of the 24 hourly transformer oil temperature readings for a day. The 31 input patterns for the month of March 1996 were mapped onto a two-dimensional output network. Figure 1b shows the feature map for March 1996 after 50 iterations of training. 20th March 1996 (Wednesday), the only holiday in March was located at node 256 together with 24th March (Sunday), which was a weekend. Furthermore, the weekend patterns were also classified separately as Saturday and Sunday patterns. Comparing these with results from other months, it was observed that new output nodes, each corresponding to a group of input patterns with a particular feature emerged after the training process. It was also observed that not all the days appeared on the map, this is because days with similar transformer oil temperature patterns are mapped to the same output node as the same day type. The following day types were identified from the feature maps: Weekdays, Holidays and Weekends. Figure 1c is a summary of results from figure 1b.

3. Clustering with MST

Looking at the SOM of figure 1b, the days that belong to a particular cluster cannot be identified easily. In this section, the MST is applied to the SOM to provide a more visual map. The MST method is a kind of tree like structure that links all the units by the shortest path [3], [4]. This structure defines the mutual distance and also describes the similarity relationship of any two points on the map [4], [6]. This algorithm assigns arcs between the nodes in such a manner that all nodes are connected together by single linkages. The lengths of the arcs are defined as the non-weighted norms of the vectorial differences between corresponding reference vectors as follows [4]:

$$\|x - m_c\| = \min_i \{\|x - m_i\|\} \tag{1}$$

Where m_i is the reference vector of node i and m_c is the reference vector of the winner node or unit (*Best Match Unit -BMU*).

3.1. MST 1 (First Winner)

From figure 1b, it is very difficult to tell the number of clusters off-hand. In order to enhance the visual characteristics of the SOM map, the MST 1 was applied. This algorithm is based on figure 2a, the nodes selected to form the neighbourhood in this section comprised of only the first winner nodes.

(a)   (b)

Figure 2: (a) Examples of nodes selected in the MST Topology using only the first winners. The selection process continues until a node, which has already been selected is selected again. (b) Examples of nodes selected using first and second winners.

3.2. MST 2 (Second Winner)

Depending on the degree of similarity among the data, the minimal spanning tree is extended to embrace other winners or nodes in close proximity to the node selected as the winner by the first algorithm. The neighbourhood in MST 1 is extended to include the second winner and its neighbours (see figure 2b). The neighbourhood this time contains far more nodes than that of the former algorithm. The procedure of extending the neighbourhood is continued until there is no appreciable change in the clusters formed by connecting the MST arcs together. Figure 3 illustrates the application of MST to the SOM map for the transformer oil temperature in January 1996. The MST algorithm provided a physical link among the data within a particular cluster thereby providing a more visual map than the previous method. In each of the months analysed, three day-type patterns emerged after the training process. These are Weekdays, Holidays and Weekends. The MST algorithm maintained the same day types that were obtained in section 2.2

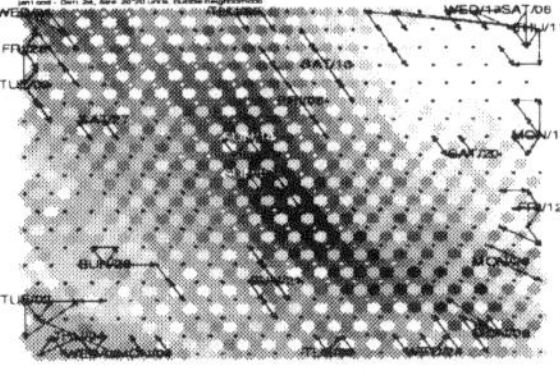

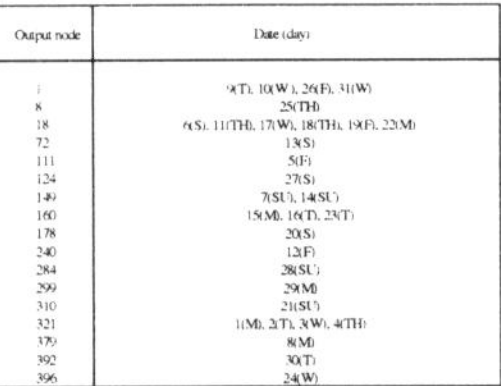

Output node	Date (day)
1	9(T), 10(W), 26(F), 31(W)
8	25(TH)
18	6(S), 11(TH), 17(W), 18(TH), 19(F), 22(M)
72	13(S)
111	5(F)
124	27(S)
149	7(SU), 14(SU)
160	15(M), 16(T), 23(T)
178	20(S)
240	12(F)
284	28(SU)
299	29(M)
310	21(SU)
321	1(M), 2(T), 3(W), 4(TH)
379	8(M)
392	30(T)
396	24(W)

(a)   (b)

Figure 3: (a) MST map for transformer oil temperature in January 1996. The algorithm for this map was extended to include the 2nd winner and its neighbours. (b) Summary of the MST map for January 1996.

4. Transformer oil temperature forecast with SOM

During summer, the power system is put under heavy load. The temperature of the insulation oil for cooling the transformer windings rises. The continuous rise in the oil temperature causes the dielectric strength of the oil to deteriorate. The oil then loses its insulating properties thus creating a short circuit within the windings of the transformer. Under such conditions, if the temperature change of insulation oil of the transformer can be predicted, necessary countermeasures can be implemented to forestall any damage to the transformer. Hence, efficient operation of the transformer can be achieved and also the reliability may be improved.

4.1. Predicting oil temperature from an incomplete data vector

4.1.1 Forecast with 3-month data set

The input data comprised of data for 3 months. The month of the day to be predicted is selected as the centre month and the month before and after are selected to form an input data set. As an example, the input data set for predicting 7th August will be July, August and September. Data for the day to be predicted is made up of the highest value of the atmospheric temperature for the day, the lowest value of the atmospheric temperature for the day and the percentage loading on the transformer. The original oil temperature data for the day to be predicted was presumed unknown and used as evaluation data. Figure 4(a) illustrates the comparison of the original transformer oil temperature readings for 7th August, the SOM forecast values and the Conventional forecast values. The results obtained were compared to the recorded values on 7th of August using equation (2). This approach resulted in a Mean Squared Errors (MSE) of 0.585 (see figure 4(b)).

4.1.2. Forecast with 10-months data

In order to improve the prediction accuracy for the SOM, the input data set was increased from a 3-month data set to a 10-month data set. The input data comprised of data for 10 months, from January to October 1996.   The MSE reduced from 0.585 to 0.072 (see figure 4(b)). Increasing the input data set from a 3-month data set to a 10-month data set resulted in a better approximation of the original data.

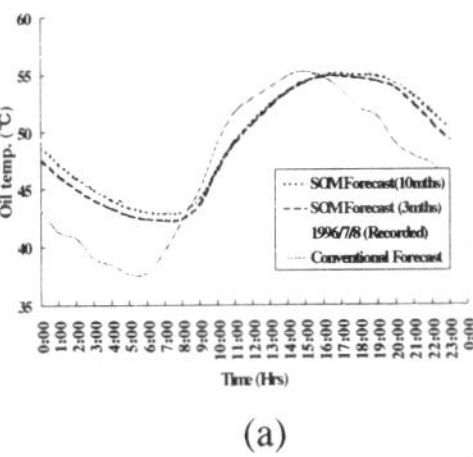

Date (Day)	Mean Squared Error 10-month	Mean Squared Error 3-month	Mean Squared Error Conventional
12/2/96 (Monday)	0.665	1.052	71.724
5/3/96 (Tuesday)	1.659	3.816	63.357
10/4/96 (Wednesday)	0.532	0.956	29.970
8/5/96 (Wednesday)	1.885	1.954	17.855
6/6/96 (Thursday)	18.451	12.475	12.150
17/7/96 (Wednesday)	2.221	1.068	12.958
7/8/96 (Wednesday)	0.072	0.585	16.377
12/9/96 (Thursday)	0.677	2.639	16.844

(a)   (b)

Figure. 4: (a) Recorded transformer oil temperature readings for 7th August, prediction by SOM (SOM prediction was for 10-month and 3-month data sets) and prediction by conventional method. (b) Mean squared error of the predicted results of oil temperature using 10-month data set, 3-month data set and conventional method.

4.1.3. Forecast with Conventional Method

Forecasting oil temperature of a transformer according to the operational guideline of oil filled transformers [7] entails explicit numerical calculation using various constants such as, time constant necessary for the calculation of the corresponding optimal cooling conditions of the installation environment of the transformer. This method resulted in a MSE of 16.38 for the forecast of the 7th August temperature values (see figure 4b). The predicted results obtained from SOM and numerically calculated values, which were based on the conventional method, were compared to the recorded data. The comparison was done based on the mean squared error shown in eqn. (2).

$$Err = \frac{1}{N} \sum_{j=1}^{N} (r_j - p_j)^2 \qquad (2)$$

Where, r_j is the value of the recorded oil temperature and p_j is the value of the predicted oil temperature and N is the number of components per data vector. The SOM predictions resulted in a better approximation of the recorded data.

4.2. Predicting a complete data vector omitted from the input data space

In section 4.1, the oil temperature components of the input vector were presumed unknown. In this section the whole input vector representing a particular day was deleted from the input data space. SOM was applied after which the Best Matching Unit (BMU) was identified. Figure 5a shows the feature map for the month of April. In this data set, the data vector for 9th April 1996 (TUE_4/09) was omitted. Using eqn. (3), all the 600 nodes on the grid were compared to the original data for TUE_4/09 and the node with the minimum value selected as the BMU.

$$Err = \sum_{j=1}^{n} (x_j - m_{ij})^2 \qquad (3)$$

Where x_j and m_{ij} are the j-th component value of the n-th dimensional input data and the i-th unit (node) respectively. Figure 5b compares the transformer oil temperature of the input data vector for TUE_4/09, which was omitted to that which was identified as the BMU after the training process.

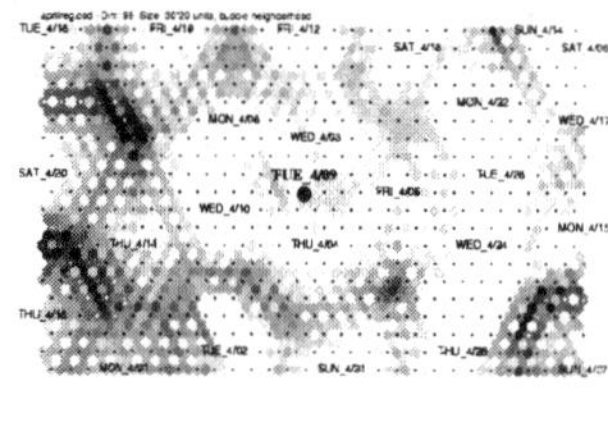

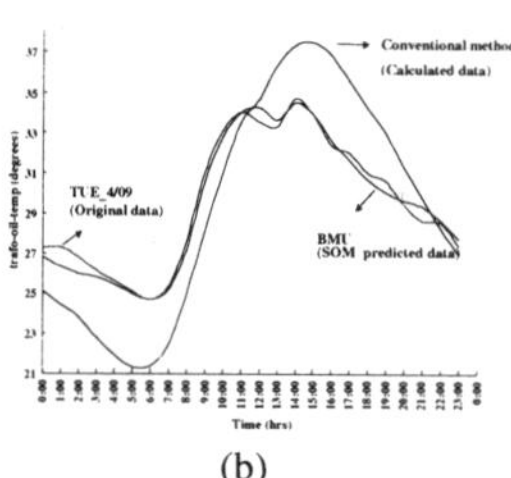

(a)   (b)

Figure 5: (a) SOM for April 1996. The input vector for TUE_4/09 was omitted. The large filled circle was identified as the BMU for TUE_4/09. (b) Predicted values for the transformer oil temperature of TUE_4/09 by SOM and the conventional method using calculation.

Although TUE_4/09 was excluded from the input data, the SOM was able to learn the other input vectors very well so as to predict with a good precision, a data vector not included in the input data space. Furthermore, figure 5b compares SOM to the conventional method currently been used by the service provider for predicting the transformer oil temperature on the TUE_4/09. SOM resulted in a better approximation of the data for TUE_4/09 as compared to the method currently been used by the service provider for data prediction.

5. Conclusion

In its application to the power transformer database, the SOM provides an in-depth knowledge about the consumption pattern of the consumers in a particular day or season. This provides the energy producer with a guide as to the consumption pattern of its consumer in any particular day or season. This could be very vital information for planning engineers in their load forecasting and other prediction and planning activities.

Superimposing the minimal spanning tree onto the SOM also provided a more visual approach in identifying the days that belong to a particular cluster.

SOM was also found to be very effective in regression and this could be a very effective tool for predicting unknown as well as incomplete data vectors in a database.

Forecasting of the oil temperature changes in a power distribution transformer has been done using conventional methods based on explicit numerical calculation. With such techniques of forecasting, the forecasting accuracy is affected due to assumptions made in the characteristics of the transformer. By the use of SOM, this problem is eliminated because the transformer characteristics are not included in the input data set for the SOM.

References

[1] T. Kohonen, Biological Cybernetics, and 43(1): 59-69,1982. Springer-Verlag, 1997.
[2] T. Kohonen, Self-Organizing Maps, IEEE, 78:1464-1480, 1990. Springer-Verlag, 1995.
[3] K. Obu-Cann et al, Clustering by SOM, MCP and MST, NC99-133, pp 121-128, March 2000.
[4] T. Kohonen, Self-Organizing Maps, Springer-Verlag, 1997.
[5] A. Murray, Applications of Neural Network, Kluwer Academic Publishers, pp 157-189.
[6] K. Obu-Cann et al, Clustering by SOM, MST and MCP, ICONIP'99, vol. 3, pp 986-991, November 1999.
[7] Tech. report of the Institute of Electrical Engineers of Japan, Part 1, No. 143: (1978).

Hierarchical Clustering for Data Mining

Anna Szymkowiak, Jan Larsen, Lars Kai Hansen
Informatics and Mathematical Modeling Richard Petersens Plads, Build. 321,
Technical University of Denmark, DK-2800 Kongens Lyngby, Denmark,
Web: http://eivind.imm.dtu.dk, Email: asz,jl,lkhansen@imm.dtu.dk

Abstract. This paper presents hierarchical probabilistic clustering methods for unsupervised and supervised learning in datamining applications. The probabilistic clustering is based on the previously suggested Generalizable Gaussian Mixture model. A soft version of the Generalizable Gaussian Mixture model is also discussed. The proposed hierarchical scheme is agglomerative and based on a $\mathcal{L}_2$ distance metric. Unsupervised and supervised schemes are successfully tested on artificially data and for segmention of e-mails.

1 Introduction

Hierarchical methods for unsupervised and supervised datamining give multilevel description of data. It is relevant for many applications related to information extraction, retrieval navigation and organization, see e.g., [1, 2]. Many different approaches to hierarchical analysis from divisive to agglomerative clustering have been suggested and recent developments include [3, 4, 5, 6, 7]. We focus on agglomerative probabilistic clustering from Gaussian density mixtures. The probabilistic scheme enables automatic detection of the final hierarchy level. In order to provide a meaningful description of the clusters we suggest two interpretation techniques: 1) listing of prototypical data examples from the cluster, and 2) listing of typical features associated with the cluster. The Generalizable Gaussian Mixture model (GGM) and the Soft Generalizable Gaussian mixture model (SGGM) are addressed for supervised and unsupervised learning. Learning from combined sets of labeled and unlabeled data [8, 9] is relevant in many practical applications due to the fact that labeled examples are hard and/or expensive to obtain, e.g., in document categorization. This paper, however, does not discuss such aspects. The GGM and SGGM models estimate parameters of the Gaussian clusters with a modified EM procedure from two disjoint sets of observations that ensures high generalization ability. The optimum number of clusters in the mixture is determined automatically by minimizing the generalization error [10].

This paper focuses on applications to textmining [8, 10, 11, 12, 13, 14, 15, 16] with the objective of categorizing text according to topic, spotting new topics or providing short, easy and understandable interpretation of larger text blocks; in a broader sense to create intelligent search engines and to provide understanding of documents or content of webpages like Yahoo's ontologies.

2 The Generalizable Gaussian Mixture Model

The first step in our approach for probabilistic clustering is a flexible and universal Gaussian mixture density model, the generalizable Gaussian mixture model (GGM) [10, 17, 18], which models the density for d-dimensional feature vectors by:

$$p(\boldsymbol{x}) = \sum_{k=1}^{K} P(k)p(\boldsymbol{x}|k), \quad p(\boldsymbol{x}|k) = \frac{1}{\sqrt{|2\pi\Sigma_k|}} \exp\left(-\frac{1}{2}(\boldsymbol{x} - \boldsymbol{\mu}_k)^\top \Sigma_k^{-1}(\boldsymbol{x} - \boldsymbol{\mu}_k)\right) \tag{1}$$

where $p(\boldsymbol{x}|k)$ are the component Gaussians mixed with the non-negative proportions $P(k)$, $\sum_{k=1}^{K} P(k)$. Each component k is described by the mean vector $\boldsymbol{\mu}_k$ and the covariance matrix Σ_k. Parameters are estimated with an iterative modified EM algorithm [10] where means are estimated on one data set, covariances on an independent set, and $P(k)$ on the combined set. This prevents notorious overfitting problems with the standard approach [19]. The optimum number of clusters/components is chosen by minimizing an approximation of the generalization error; the AIC criterion, which is the negative log-likelihood plus two times the number of parameters.

For unsupervised learning parameters are estimated from a training set of feature vectors $\mathcal{D} = \{\boldsymbol{x}_n; n = 1, 2 \ldots, N\}$, where N is the number of samples. In supervised learning for classification from a data set of features and class labels $\mathcal{D} = \{\boldsymbol{x}_n, y_n\}$, where $y_n \in \{1, 2, \ldots, C\}$ we adapt one Gaussian mixture, $p(\boldsymbol{x}|y)$, for each class separately and classify by Bayes optimal rule (under 1/0 loss) by maximizing $p(y|\boldsymbol{x}) = p(\boldsymbol{x}|y)P(y)/\sum_{y=1}^{C} p(\boldsymbol{x}|y)P(y)$. This approach is also referred to as mixture discriminant analysis [20].

The GGM can be implemented using either hard or soft assignments of data to components in each EM iteration step. In the hard GMM approach each data example is assigned to a cluster by selecting highest $p(k|\boldsymbol{x}_n) = p(\boldsymbol{x}_n|k)P(k)/p(\boldsymbol{x}_n)$. Means and covariances are estimated by classical empirical estimates from data assigned to each component. In the soft version (SGGM) e.g., the means are estimated as weighted means $\boldsymbol{\mu}_k = \sum_n p(k|\boldsymbol{x}_n) \cdot \boldsymbol{x}_n / \sum_n p(k|\boldsymbol{x}_n)$.

Experiments with the hard/soft versions gave the following conclusions. Per iteration the algorithms are almost identical, however, SGGM requires typically more iteration to converge, which is defined by no changes in assignment of examples to clusters. Learning curve[1] experiments indicate that hard GGM has slightly better generalization performance for small N while similar behavior for large N - in particular if clusters are well separated.

3 Hierarchical Clustering

In the suggested agglomerative clustering scheme we start by K clusters at level $j = 1$ as given by the optimized GGM model of $p(\boldsymbol{x})$, which in the case of supervised learning is $p(\boldsymbol{x}) = \sum_{y=1}^{C} \sum_{k=1}^{K_y} p(\boldsymbol{x}|k, y)P(k)P(y)$, where K_y is the optimal number of components for class y. At each higher level in the hierarchy two clusters are merged based on a similarity measure between pairs of clusters. The procedure is repeated until we reach one cluster at the top level. That is, at level $j = 1$ there are K clusters and 1 cluster at the final level, $j = 2K - 1$. Let $p_j(\boldsymbol{x}|k)$ be the density for the k'th cluster at level j and $P_j(k)$ as its mixing proportion, i.e., the density model at level j is $p(\boldsymbol{x}) = \sum_{k=1}^{K-j+1} P_j(k)p_j(\boldsymbol{x}|k)$. If clusters k and m at level j are merged into ℓ at level $j + 1$ then

$$p_{j+1}(\boldsymbol{x}|\ell) = \frac{p_j(\boldsymbol{x}|k) \cdot P_j(k) + p_j(\boldsymbol{x}|m) \cdot P_j(m)}{P_j(k) + P_j(m)}, \quad P_{j+1}(\ell) = P_j(k) + P_j(m) \tag{2}$$

The natural distance measure between the cluster densities is the Kullback-Leibler (KL) divergence [19], since it reflects dissimilarity between the densities in the probabilistic space. The drawback is that KL only obtains an analytical expression for the first level in the

---

[1] Generalization error as as function of number of examples.

hierarchy while distances for the subsequently levels have to be approximated [17, 18]. Another approach is to base distance measure on the $\mathcal{L}_2$ norm for the densities [21], i.e., $D(k,m) = \int (p_j(\boldsymbol{x}|k) - p_j(\boldsymbol{x}|m))^2 \, dx$ where k and m index two different clusters. Due to Minkowksi's inequality $D(k,m)$ is a distance measure. Let $\mathcal{I} = \{1, 2, \cdots, K\}$ be the set of cluster indices and define disjoint subsets $\mathcal{I}_\alpha \cap \mathcal{I}_\beta = \emptyset$, $\mathcal{I}_\alpha \subset \mathcal{I}$ and $\mathcal{I}_\beta \subset \mathcal{I}$, where $\mathcal{I}_\alpha$, $\mathcal{I}_\beta$ contain the indices of clusters which constitute clusters k and m at level j, respectively. The density of cluster k is given by: $p_j(\boldsymbol{x}|k) = \sum_{i \in \mathcal{I}_\alpha} \alpha_i p(\boldsymbol{x}|i)$, $\alpha_i = P(i)/\sum_{i \in \mathcal{I}_\alpha} P(i)$ if $i \in \mathcal{I}_\alpha$, and zero otherwise. $p_j(\boldsymbol{x}|m) = \sum_{i \in \mathcal{I}_\beta} \beta_i p(\boldsymbol{x}|i)$, where β_i obtains a similar definition. According to [21] the Gaussian integral $\int p(\boldsymbol{x}|i)p(\boldsymbol{x}|\ell) \, dx = G(\boldsymbol{\mu}_i - \boldsymbol{\mu}_\ell, \boldsymbol{\Sigma}_i + \boldsymbol{\Sigma}_\ell)$, where $G(\boldsymbol{\mu}, \boldsymbol{\Sigma}) = (2\pi)^{-d/2} \cdot |\boldsymbol{\Sigma}|^{1/2} \cdot \exp(-\boldsymbol{\mu}^\top \boldsymbol{\Sigma}^{-1} \boldsymbol{\mu}/2)$. Define the vectors $\boldsymbol{\alpha} = \{\alpha_i\}$, $\boldsymbol{\beta} = \{\beta_i\}$ of dimension K and the $K \times K$ symmetric matrix $\boldsymbol{G} = \{G_{i\ell}\}$ with $G_{i\ell} = G(\boldsymbol{\mu}_i - \boldsymbol{\mu}_\ell, \boldsymbol{\Sigma}_i + \boldsymbol{\Sigma}_\ell)$, then the distance can be then written as $D(k,m) = (\boldsymbol{\alpha} - \boldsymbol{\beta})^\top \boldsymbol{G}(\boldsymbol{\alpha} - \boldsymbol{\beta})$. Figure 1 illustrates the hierarchical clustering for Gaussian distributed toy data.

A unique feature of probabilistic clustering is the ability to provide optimal cluster and level assignment for new data examples which have not been used for training. $\boldsymbol{x}$ is assigned to cluster k at level j if $p_j(k|\boldsymbol{x}) > \rho$ where the threshold ρ typically is set to 0.9. The procedure ensures that the example is assigned to a wrong cluster with probability 0.1.

Interpretation of clusters is done by generating likely examples from the cluster, see further [17]. For the first level in the hierarchy where distributions are Gaussian this is done by drawing examples from a super-eliptical region around the mean value, i.e., $(\boldsymbol{x} - \boldsymbol{\mu}_k)^\top \boldsymbol{\Sigma}_k^{-1}(\boldsymbol{x} - \boldsymbol{\mu}_k) < const$. For clusters at higher levels in the hierarchy samples are drawn from each Gaussian cluster with proportions specified by $P(k)$.

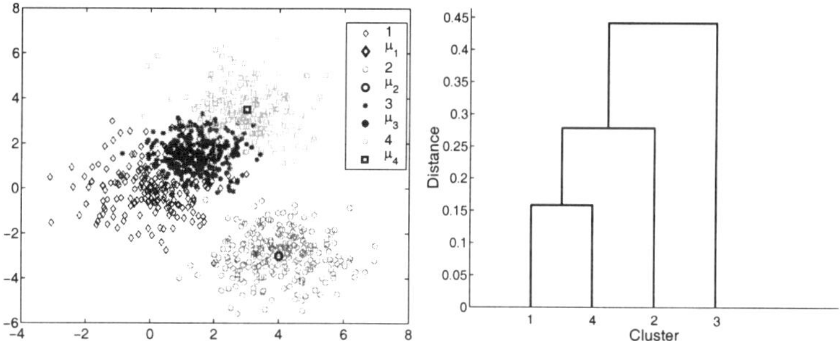

Figure 1: Hierarchical clustering example. Left panel is a scatter plot of the data. Clusters 1,2 and 4 have wide distributions while 3 has a narrow one. Since the distance is based on the shape of the distribution and not only its mean location, clusters 1 and 4 are much closer than any of these to cluster 3. Right panel presents the dendrogram.

4 Experiments

The hierarchical clustering is illustrated for segmentation of e-mails. Define term-vector as a complete set of the unique words occurring in all the emails. An email histogram is the vector containing frequency of occurrence of each word from the term-vector and defines the content of the email. The term-document matrix is then the collection of histograms for all emails in the database. After suitable preprocessing[2] the term-document matrix contains 1405 (702 for training and 703 for testing) e-mail documents, and the term-vector 7798 words. The emails where annotated into the categories: *conference, job* and *spam*. It is possible to model

---

[2] Words which are too likely or too unlikely are removed. Further only word stems are kept.

directly from this matrix [8, 15], however we deploy Latent Semantic Indexing (LSI) [22] which operates from a latent space of feature vectors. These are found by projecting term-vectors into a subspace spanned by the left eigenvectors associated with largest singular value of a singular value decomposition of the term-document matrix. We are currently investigating methods for automatic determination of the subspace dimension based on generalization concepts. We found that a 5 dimensional subspace provides good performance using SGGM.

A typical result of running supervised learning is depicted in Figure 2. Using supervised learning provides a better resemblance with the correct categories at the level in the hierarchy as compared with unsupervised learning. However, since labeled examples often are lacking or few the hierarchy provides a good multilevel description of the data with associated interpretations. Finding typical features as described on page 3 and back-projecting into original term-space provides keywords for each cluster as given in Table 1.

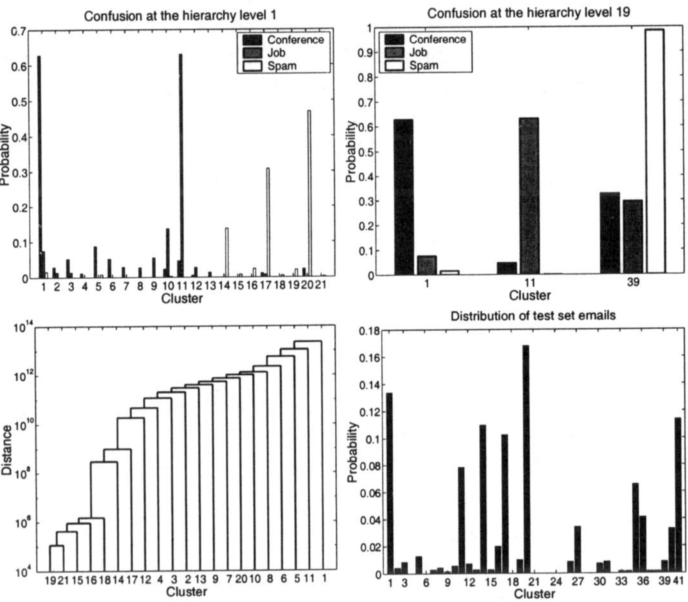

Figure 2: Supervised hierarchical clustering. Upper rows show the confusion of clusters with the annotated email labels on the training set at the first level and the level where 3 clusters remains, corresponding to the three categories *conference*, *job* and *spam*. At level 1 clusters 1,11,17,20 have big resemblance with the categories. In particular *spam* are distributed among 3 clusters. At level 19 there is a high resemblance with the categories and the average probability of erroneous category on the test set is 0.71. The lower left panel shows the dendrogram associated with the clustering. The lower right panel shows the histogram of cluster assignments for test data, cf. page 3. Clearly some samples obtain a reliable description at the first level (1–21) in the hierarchy, whereas others are reliable at a higher level (22–41).

5 Conclusions

This paper presented a probabilistic agglomerative hierarchical clustering algorithm based on the generalizable Gaussian mixture model and a $\mathcal{L}_2$ metric in probabilty density space. This leads to a simple algorithm which can be used both for supervised and unsupervised learning. In addition, the probabilistic scheme allows for automatic cluster and hierarchy level assignment for unseen data and further a natural technique for interpretation of the clusters

Table 1: Keywords for unsupervised learning

1	research,university,conference	8	neural,model	15	click,remove,hottest,action
2	university,neural,research	9	university,interest,computetion	16	free,adult,remove,call
3	research,creativity,model	10	research,position,application	17	website,adult,creativity,click
4	website,information	11	science,position,fax	18	website,click,remove
5	information,program,computation	12	position,fax,website	19	free,call,remove,creativity
6	research,science,computer,call	13	research,position,application	20	mac
7	website,creativity	14	free,adult,call,website	21	adult,government

1	research,university,conference	11	science,position,fax	39	free,website,cal l,creativity

via prototype examples and features. The algorithm was successfully applied to segmentation of emails.

References

[1] J. Carbonell, Y. Yang and W. Cohen, Special Isssue of Machine Learning on Information Retriceal Introduction, *Machine Learning* **39**, (2000) 99–101.

[2] D. Freitag, Machine Learning for Information Extraction in Informal Domains, *Machine Learning* **39**, (2000) 169–202.

[3] C.M. Bishop and M.E. Tipping, A Hierarchical Latent Variable Model for Data Visualisation, *IEEE T-PAMI* **3**, 20 (1998) 281–293.

[4] C. Fraley, Algorithms for Model-Based Hierarchical Clustering, *SIAM J. Sci. Comput.* **20**, 1 (1998) 279–281.

[5] M. Meila and D. Heckerman, An Experimental Comparison of Several Clustering and Initialisation Methods. In: Proc. 14th Conf. on Uncert. in Art. Intel., Morgan Kaufmann, 1998, pp. 386–395.

[6] C. Williams, A MCMC Approach to Hierarchical Mixture Modelling. In: Advances in NIPS 12, 2000, pp. 680–686.

[7] N. Vasconcelos and A. Lippmann, Learning Mixture Hierarchies. In: Advances in NIPS 11, 1999, pp. 606–612.

[8] K. Nigam, A.K. McCallum, S. Thrun, and T. Mitchell Text Classification from Labeled and Unlabeled Documents using EM, *Machine Learning*, 39 2–3 (2000) 103–134.

[9] D.J. Miller and H.S. Uyar, A Mixture of Experts Classifier with Learning Based on Both Labelled and Unlabelled Data. In: Advances in NIPS 9, 1997, pp. 571–577.

[10] L.K. Hansen, S. Sigurdsson, T. Kolenda, F.Å. Nielsen, U. Kjems and J. Larsen, Modeling Text with Generalizable Gaussian Mixtures. In: Proc. of IEEE ICASSP'2000, vol. 6, 2000, pp. 3494–3497.

[11] C.L. Jr. Isbell and P. Viola, Restructuring Sparse High Dimensional Data for Effective Retrieval. In: Advances in NIPS 11, MIT Press, 1999, pp. 480–486.

[12] T. Kolenda, L.K. Hansen and S. Sigurdsson Indepedent Components in Text. In: Adv. in Indep. Comp. Anal., Springer-Verlag, pp. 241–262, 2001.

[13] T. Honkela, S. Kaski, K. Lagus and T. Kohonen, Websom — self-organizing maps of document collections. In: Proc. of Work. on Self-Organizing Maps, Espoo, Finland, 1997.

[14] E.M. Voorhees, Implementing Agglomerative Hierarchic Clustering Ulgorithms for Use in Document Retrieval, *Inf. Proc. & Man.* **22** 6 (1986) 465–476.

[15] A. Vinokourov and M. Girolami, A Probabilistic Framework for the Hierarchic Organization and Classification of Document Collections, submitted for *Journal of Intelligent Information Systems*, 2001.

[16] A.S. Weigend, E.D. Wiener and J.O. Pedersen Exploiting Hierarchy in Text Categorization, *Information Retrieval*, **1** (1999) 193–216.

[17] J. Larsen, L.K. Hansen, A. Szymkowiak, T. Christiansen and T. Kolenda, Webmining: Learning from the World Wide Web, *Computational Statistics and Data Analysis* (2001).

[18] J. Larsen, L.K. Hansen, A. Szymkowiak, T. Christiansen and T. Kolenda, Webmining: Learning from the World Wide Web. In: Proc. of Nonlinear Methods and Data Mining, Italy, 2000, pp. 106–125.

[19] C.M. Bishop, Neural Networks for Pattern Recognition, Oxford University Press, 1995.

[20] T. Hastie and R. Tibshirani Discriminant Analysis by Gaussian Mixtures, *Jour. Royal Stat. Society - Series B*, **58** 1 (1996) 155–176.

[21] D. Xu, J.C. Principe, J. Fihser, H.-C. Wu, A Novel Measure for Independent Component Analysis (ICA). In: Proc. IEEE ICASSP98, vol. 2, 1998, pp. 1161–1164.

[22] S. Deerwester, S.T. Dumais, G.W. Furnas, T.K. Landauer and R. Harshman, Indexing by Latent Semantic Analysis, *Journ. Amer. Soc. for Inf. Science.*, **41** (1990) 391–407.

KES '01
N. Baba et al. (Eds.)
IOS Press, 2001

CONSTRUCTING KNOWLEDGE-BASED ARTIFICIAL NEURAL NETWORK WITH ROUGH SETS

Xiaoshu, Hang   ZhenYu Wang   Fanlun Xiong

(Hefei Institute of Intelligent Machines, The Chinese Academy of Sciences, 230031, Hefei, China)

xshang@mail.iim.ac.cn

Abstract: An approach of constructing knowledge-based artificial neural network based on rough sets is proposed. Crude domain knowledge is extracted from the 0-1 table produced from fuzzy information table of example data by a threshold. The extracted initial rules and their accuracy and coverage are used to configure the fuzzy multilayer perceptron–structure and initial weights. An algorithm of attribute-reduction based on information entropy is also proposed in this paper. Results on diagnosises of rice pests show that the performance of this fuzzy neural system is same with that of conventional multi-layer perceptron.

Keywords: Knowledge-based artificial neural system, rough sets, knowledge reduction, information entropy,

INTRODUCTION

There is a trend in the research of AI that integrates fuzzy sets theory, neural network, genetic algorithms, and rough sets theory for more efficient hybrid system. Fuzzy-neural network, combining the merits of fuzzy set theory and artificial neural networks, can be used to constitute more intelligent systems that premise to handle inexact, uncertainty of knowledge, and the problems of recognition/classification. However, when faced a hybrid system, it is usually much difficult to construct an efficient system because we need a great number of fuzzy rules to describe it.

Rough sets theory, as a new mathematical tool for managing uncertainty that arise from inexact, noisy, or incomplete information, has recently obtained wide attention. It seems to be of fundamental importance to artificial intelligence and cognitive science, especially in the areas of machine learning, knowledge acquisition, knowledge discovery from database, expert system, decision support system, inductive reasoning and pattern recognition[1]. In this paper, rough sets theory is used to extract crude domain knowledge from a 0-1 table produced from fuzzy information table of example data by a threshold. The crude domain knowledge, in the form of fuzzy decision rules, was in turn used to structure knowledge-based fuzzy neural system that is much more efficient than the version of conventional multi-layer perceptron because it less hidden units. According to [2][3], the fuzzy classification rules are mapped into a three layer fuzzy neural system in which the number of hidden units is automatically determined by the number of rules, and the initial weights are also automatically determined by the accuracy and coverage of the fuzzy classification rules.

KNOWLEDGE-BASED ARTIFICIAL NEURAL NETWORK

One major line of research on hybrid intelligent systems is knowledge-based neural network, which concern the use of domain knowledge to determine the initial structure of the neural network. Some researches have demonstrated that a knowledge-based neural network could outperform a standard back-propagation network as well as other related learning algorithms.

The rules-to-network translation is accomplished by establishing a mapping between a rules set and a neural network. The correspondence between rules set and neural network is specified by table1. This correspondence is bidirectional. On one hand, a rules set can be mapped into a neural network, on the other hand, a neural network can be transferred back to a rule-based system.

Table 1 Corresponding between rules set and neural network

Rules set	Neural network
Supporting facts	Input units
Intermediate conclusions	Hidden units
Final conclusions	Output units
Accuracy and Coverage	Weighted connections

There are two constraints on the rules set when it is mapped into a knowledge-based neural network:

- the rules must be in the format of propositional Horn clause:

If $p1$ and $p2$ and ... and pn then r

- the rules can be hierarchical ,but must be non-cyclic;

If rules are not in the format of propositional Horn clause, they can be rewritten into this form by applying the following equivalents:

- *if p or q then r => if p then r and if q then r*
- *if p and (q or s) then r => if p and q then r and if p and s then r*
- *if p then q and r => if p then q and if p then r*

Here is an example from [2]:

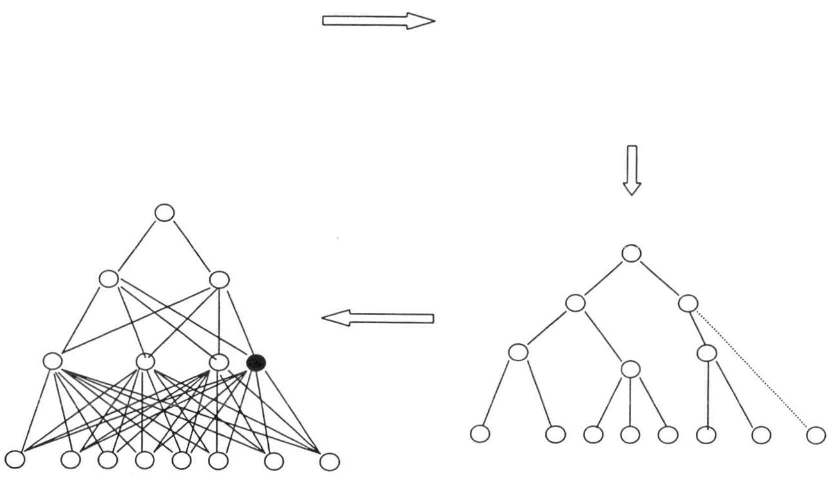

Fig. 1

Each rule has an antecedent consisting of one or more conditions as well as a single consequent. In the neural network configuration, the antecedent is assigned a hidden unit, each condition corresponds to an assigned input unit, and the consequent of rule is assigned an output unit. In general case, we need three-level structure to map a rule: the first level corresponds to the conditions of the antecedent, the next level is the conjunction of these conditions to form the premise, and the third level is the consequent of the rule. If there is only one condition in the premise of a rule, we still use a hidden unit rather than link the condition directly to the consequent

It is an effective way to obtain rough domain knowledge (rules) directly from sample data rather than domain experts to improve the structure and performance of neural network. Rough sets theory is especially suited to knowledge reduction, estimation of data significance, or generation of decision rules from decision table, it can also be used to analysis the relation of dependency between attributes of information table. In this paper, we use rough sets to extract some crude domain knowledge (classification rules) which is used to configure the fuzzy neural network for further obtaining refined fuzzy rules about several kinds of rice pest. In the multi-layer proceptron of conventional version, the link weights are usually initialized by small random values, but the initial link weights of a knowledge-based neural network are determined by the accuracy and coverage of the rules.

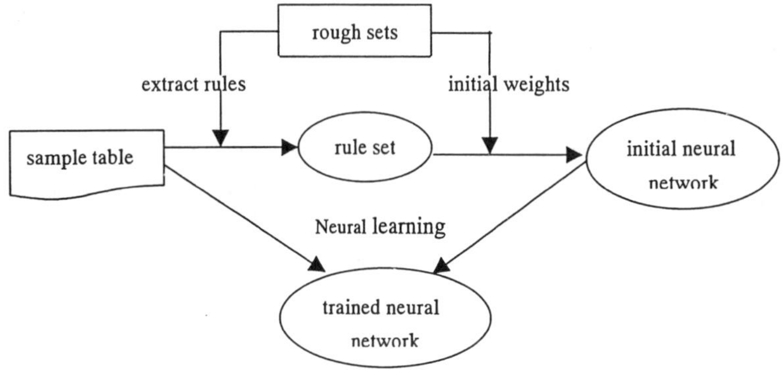

Fig. 2   Flow chart of knowledge-based neural network based on Rough Sets

EXTRACTING CRUDE RULES FROM EXAMPLE DATA WITH ROUGH SETS

(1) Knowledge reduction algorithm based on information entropy theory

We present a knowledge reduction algorithm based on the theory of information entropy, so the meanings of the following definitions are statistics-based.

An information system can be represented as $S=<U, A, V_a>$, in which U is non-empty finite set called universe of objects , A is a non-empty finite set of attributes, and be represented as $A=C\ D$ *when S is a decision table*, V_a *is called the value set of attribute a.*

Definition 1 Let $S=<U, A, V_a>$ be an information table, *attributes subsets* $P\subseteq A$, $Q\subseteq A$, *Let* $P^*=\{p_1,p_2...,p_n\},Q^*=\{q_1,q_2...,q_m\}$ *be the sets of equivalent classes of P,Q on U respectively, the information dependency of Q->P can be measured by conditional entropy*

$$H(Q^*|P^*)=\Sigma_{i=1,n}P(p_i)\Sigma_{j=1,m}P(q_j|p_i)log(q_j|p_i) \quad (1)$$

$H(Q^*|P^*)$ has following properties:

1) $0\ H(Q^*|P^*)\ 1;$

2) if $H(Q^*|P^*)=0,Q's$ classification is completely informational depended on P's classification;

3) if $H(Q^*|P^*)=1,Q's$ classification is completely informational independent from P's classification;

We use $H(D^*|C^*)$ to represent the informational dependency of decision attribute subsets D on conditional attribute subsets C.

Definition 2   Let $S=<U,C\ D,V_a>$be an information table, attribute subset $B\subseteq C$, for any attribute $b\in B$, we call b dispensable in B if $H(B^*)=H((B-b)^*)$,otherwise it is indispensable, let $B'\subseteq B$, we call $B^\wedge$ a reduct of B if $H(B'^*)=H(B^*)$ and for $\forall b'\in B'$, b' is indispensable in B'.

Definition 3   Let $S=<U,C\ D,V_a>$ be an information table , for any conditional attribute $a\in C$, we denote $Sig_{C-a}(a)$ as the significance of a with respect to D:

$$Sig_{C-a}(a)=H(D^*|(C-a)^*)-H(D^*|C^*) \quad (2)$$

if $Sig_{C-a}(a)>0$, attribute a is important with respect to D, otherwise ,it is not important with respect to D.

Definition 4   Let $S=<U,C\ D,V_a>$ be an information table, the set of all attributes in C which are significant in C is called the core of C, we denote $CORE_D(C)$ as the core of C with respect to D.

$CORE_D(C) = \{a \mid \forall a \in C, sig_{C-a}(a) > 0\}$

We propose a relative knowledge reduction algorithm based on information entropy:

Algorithm: RKRABIE(Relative Knowledge Reduction Algorithm Based on Information Entropy)

Input a 0-1 table produced from fuzzy feature table by a threshold, that is $S = <U, C\ D, V_a>, \forall a \in C\ D\ V_a = 0\ or\ 1;$

Output the relative reduct of S;

step1: Computing $H(D^*|C^*)$;

step2: Computing $CORE_D(C)$;

♦ *Let $C_0 = \phi$ $Y = C$;*

♦ *while* $Y \neq \phi$

♦ *let $b \in Y$, computing $sig_{C-b}(b)$;*

♦ *if $sig_{C-b}(b) > 0$ $C_0 = C_0$ {b};*

♦ *$Y = Y - \{b\}$;*

end while

step3 Computing $H(D^*|C_0^*)$ if $H(D^*|C_0^*) = H(D^*|C^*)$ then $B = C_0$, goto step 5;

step4: Let $B = C_0$

while for every c in C–B

♦ *Computing $sig_{B\ c}(c)$; $sig_{B\ c}(c) = H(D^*|B^*) - H(D^*|(B\ \{c\})^*)$;*

♦ *Choose the max $sig_{B\ c}(c)$, $B = B\ \{c\}$;*

♦ *If $H(D^*|B^*) = H(D^*|C^*)$ then goto step5;*

End while;

step5: Output B;

In order to reduce the complexity of computation, Table2 is divided into two sub-tables according to domain knowledge:

$S_1 = <U_1, C_1\ D_1, V_a>, S_2 = <U_2, C_2\ D_2, V_a>$

$S = S_1 + S_2, U = U_1\ U_2, C = C_1\ C_2, D = D_1\ D_2$

S_1: consists of objects: chilo suppressalis, tryporyza incertulas; S2: planthopper, leafhopper, nilaparvata lugens; C_1 : *temperature, humidity, bacterial spike,*

bacterial heart, bacterial sheath. C: temperature, humidity, bacterial leaf, bacterial culm. D_1: the first two columns of part P in table2, D2: the last three columns of table1.

We select 8 features: *temperature, humidity, bacterial spike, bacterial heart, bacterial sheath, bacterial leaf, bacterial culm.* The first two features are the necessary conditions for pest's living, the others are used to diagnose the category of pest. when the feature is numeric, such as *temperature, humidity,* the membership of fuzzy linguistic corresponding respectively to High, Medium, Low is set up by function as follow:

$$\pi(F_j, c, \lambda) = \begin{cases} 2\left(1 - \dfrac{|F_j - c|}{\lambda}\right)^2, & \dfrac{\lambda}{2} \le |F_j - c| \le \lambda \\[2ex] 1 - 2\left(\dfrac{|F_j - c|}{\lambda}\right)^2, & 0 \le |F_j - c| \le \dfrac{\lambda}{2} \\[2ex] 0, & otherwise \end{cases}$$

Table2 λ, c of temperature and humidity

	temperature			humidity		
Param.	H	M	L	H	M	L
c	36	23	10	90	60	30
λ	18	20	20	60	60	30

When the feature is linguistic, such as *bacterial heart,* the memberships corresponding to High, Medium, Low are determined as follow:

$$Low = \left\{\frac{0.95}{L}, \frac{0.55}{M}, \frac{0.05}{H}\right\}$$

$$Medium = \left\{\frac{0.55}{L}, \frac{0.95}{M}, \frac{0.55}{H}\right\}$$

$$High = \left\{\frac{0.05}{L}, \frac{0.55}{M}, \frac{0.95}{H}\right\}$$

Table 1 Fuzzy features information table (size:200)

	Temperature			Humidity			...	Pest				
No.	H	M	L	H	M	L	...	P1	P2	P3	P4	P5.
1	0.94	0.43	0.00	0.56	0.92	0.00	...	0.89	0.34	0.00	0.00	0.00
2	0.47	0.96	0.32	0.46	0.94	0.38	...	0.26	0.94	0.00	0.00	0.00
3	0.00	0.32	0.97	0.96	0.52	0.00	...	0.00	0.00	0.26	0.96	0.00
4	0.57	0.91	0.57	0.52	0.89	0.52	...	0.00	0.00	0.092	0.53	0.00
5	0.00	0.34	0.90	0.00	0.49	0.96	...	0.00	0.00	0.46	0.00	0.91
...		...					...					

(P1: chilo suppressalis, P2: tryporyza incertulas, P3: planthopper, P4: leafhopper, P5: nilaparvata lugens)

R1:if Temp.=m and Humidity=m and bact.spike=h and bact.sheath=l then Pest="tryporyza incertulas"

R2: if Temp.=m and Humidity=h and bact.heart=h and bact.sheath=l then Pest="tryporyza incertulas"

R3:if Temp.=m and Humidity=m and bact.heart=h and bact.sheath=l then Pest=" tryporyza incertulas"

R4: if Temp.=m and Humidity=h and bact.spike=h and bact.sheath=h then Pest =" chilo suppressalis"

R5: if Temp.=m and Humidity=m and bact.heart=h and bact.sheath=h then Pest=" chilo suppressalis"

R6: if Temp.=m and Humidity=m and bact.spike=m and bact.sheath=h then Pest="chilo suppressalis"

R7: if Temp.=h and Humidity=h and bact.heart=m and bact.sheath=h then Pest ="tryporyza incertulas"

R8: if Temp.=m and Humidity=h and, bacterial culm =h   then Pest=" planthopper"

R9: if Temp.=m and Humidity=m and bacterial culm=m   then Pest=" planthopper"

R10: if Temp.=m and Humidity=h and bacterial =h   then Pest =" leafhopper"

R11: if Temp.=m and Humidity=m and bacterial =h   then Pest=" leafhopper"

R12: if Temp.=m and Humidity=h and bacterial leaf =h   then Pest=" nilaparvata lugens"

R13: if Temp.=m and Humidity=m and bacterial leaf =m   then Pest =" nilaparvata lugens"

(2) Accuracy and coverage of rules and configuration of initial link weights

Definition 5 Let $RUL_S=\{R_1,...,R_n\}$ be a set of rules derived from information table S, and $R^{*}(D)=\{D_1,D_2,...,D_n\}$ be a set of equivalent classes of decision attribute D, the equality of a rule R is defined as its accuracy $\alpha_R(D)$ and coverage $\gamma_R(D)$ based on statistical information.

$$\alpha_R(D) = \frac{\left|[x]_R \cap [x]_D\right|}{\left|[x]_R\right|} \qquad (6)$$

$$\gamma_R(D) = \frac{\left|[x]_R \cap [x]_D\right|}{\left|[x]_D\right|} \qquad (7)$$

Where $[X]_R$ represents the set of objects which are indiscernible in discourse U with respect to the antecedent of rule R, and $|*|$ is a operator which represents the cardinality of a set. $\alpha_R(D_j)$ indicates the possibility of an object O_i belonging to decision concept D_j when it's attributes match the antecedent of rule R, a bigger $\alpha_R(D_j)$ means a stronger power of discerning D_j from $\neg D_j$. $\gamma_R(D_j)$ indicates the possibility of an object O_i recognized by rule R when it belongs to decision concept D_j. $\gamma_R(D)=1$ means rule R is the only one that recognizes all the objects O_i belonging to decision concept D_j. In this paper, $\alpha_R(D_j)$ and $\gamma_R(D_j)$ are used to determine the initial link weights of the neural network.

Table 3   Accuracy and coverage of rules

Rule	R1	R2	R3	R4	R5	R6	R7	R8	R9	R10	R11	R12	R13
$\alpha_R(D)$	0.91	0.94	0.97	1.00	0.95	1.00	0.93	0.98	0.95	1.00	1.00	1.00	1.00
$\gamma_R(D)$	0.33	0.34	0.29	0.24	0.26	0.30	0.17	0.61	0.37	0.56	0.40	0.54	0.43

We set up 5 regularities as follow to mapping a rule into the neural network:

1)   One unit is set up corresponding to one rule in the hidden layer, the number of units in the hidden layer equals to the number of the rules.

2)   The logic relation among the conditions of antecedent of rule R is logic "AND", while the logic relation among antecedents of different rules which have the same consequent is logic "OR".

3)   Initial link weights between input layer and hidden layer are set up by $\gamma_R(D)+\varepsilon$, ε is a small random positive value in order to break up the equilibrium.

4)   Initial link weights between hidden layer and output layer are set up by $\alpha_R(D)$;

5)   Other default link weights are randomly set up by small positive value ε.

Rules R1,R2,R3 are mapped into a three-level perceptron as fig.3 according to these regularities.

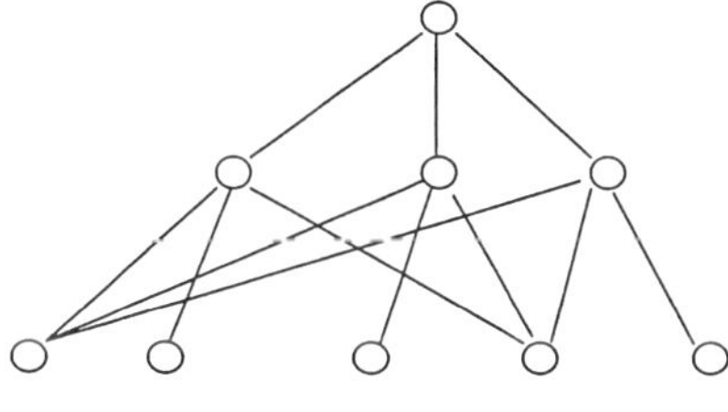

Fig. 3

Neural learning and results

In real life, it is to be noted that human reasoning is somewhat fuzzy in natural. It is impractical to describe exactly the relation between temperature, humidity and rice pests One may incorporate the conception of fuzzy sets into neural network to model the uncertain or ambiguous data.

We constitute a three-layer fuzzy MLP to forecast/diagnose rice pests. There are 3×8 units at input layer since we have features (*temperature, humidity, bacterial spike, bacterial heart, bacterial sheath, bacterial leaf, bacterial culm , *), and each feature is divided into *high, medium, low* fuzzy linguistic variables.

At the output layer, we assign 5 units to the rice pests: chilo suppressalis, tryporyza incertulas, planthopper, leafhopper, nilaparvata lugens.

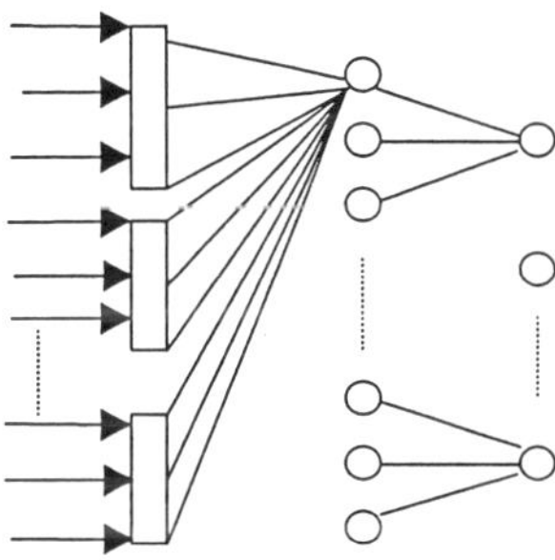

Fig. 4

BP algorithm is used to train the fuzzy perceptron, (8)–(12) are used for updating weights, and O_{pj} represents the actual output value, d_{pj} represents the desired value. η is a learning coefficient.

$$O_{pj} = f\left(net_{pj}\right) = 1 \Big/ \left(1 + e^{-\sum w_{ji}O_{pl}+\theta}\right) \tag{8}$$

$$w_{ji} = w_{ji} + \Delta w_{ji} \tag{9}$$

$$\Delta w_{ji} = \eta \delta_{pj} O_{pl} \tag{10}$$

For units in output layer:

$$\delta_{pj} = (d_{pj} - O_{pj})O_{pj}(1 - O_{pj}) \tag{11}$$

For units in hidden layer:

$$\delta_{pj} = O_{pj}(1 - O_{pj})\sum_k \delta_{pk} w_{kj} \tag{12}$$

Table 4   Results("-"represents no initial value for the feature and processed as "low")

Initial features								Actual values					Desired values				
F1	F2	F3	F4	F5	F6	F7	F8	P1	P2	P3	P4	P5	P1	P2	P3	P4	P5
26	89	H	M	–	L	L	–	.95	.32	.00	.00	.00	.93	.25	.00	.00	.00
23	86	M	H	H	L	L	L	.36	.89	.00	.00	.00	.54	.85	.00	.00	.00
28	72	–	–	M	H	M	L	.00	.00	.94	.24	.00	.00	.00	.89	.00	.00
30	68	–	–	L	M	H	L	.00	.00	.20	.96	.00	.00	.00	.00	.98	.00
33	82	–	–	L	L	L	H	.00	.00	.00	.00	.94	.00	.00	.00	.00	.98

Conclusion

A knowledge-based neural network is designed by integrating rough sets and fuzzy sets into artificial neural network. The results of rice pests classification demonstrate the effectiveness of rough set theory for extracting domain rules and determining initial link weights. The fuzzy MLP of which structure determined by rough sets has less hidden units than the version of conventional MLP, and encode the initial link weights based on knowledge, so the neural network performs an efficient classification. The way of mapping domain knowledge (rule set) into neural network is also introduced. The reduction algorithm of information table is designed based on information entropy.

References

[1] Zdzislaw Pawlak, etc. "Rough set", Communication of the ACM, Vol.38,pp 89-95,1995

[2] Geoffrey G.Towell, Jude W.Shavlik, "Knowledge-based artificial neural network", Artificial Intelligence 70(1994) 119-165.

[3] Li Min Fu "Knowledge-based Connectionism for Revising DomainTheories" trans. System,man and cybernetics, vol 23, 173-182, 1993

[4] Mohua Banerjee etc. "rough fuzzy MLP: knowledge encoding and classification" IEEE trans.Neural networks, vol.9,pp 1203-1215,1998.

[5] Sankar K.Pal, "multilayer perceptron, fuzzy sets and classification" IEEE trans.Neural networks, vol.3,pp 683-695,1992

[6] J.W.Guan, D.A.Bell, "Rough computational method for informational system", Artificial intelligent 105(1998)77-103

[7] Chin-Teng Lin ,C.S.George Lee, "Neural-network-based fuzzy logic control and decision system" trans. On computer, vol 40 ,1320-1336, 1991.

[8] Zeng Huanglin "Rough sets and its application (Chinese)" pressed by Chongqing university, 1998.

[9] Miao Duoqian, Wang Jue, "on the relationships between information entropy and rough ness of knowledge in rough set theory", pattern recognition &artificial intelligence(Chinese), vol. 36, 681-684,1999

KES '01
N. Baba et al. (Eds.)
IOS Press, 2001

ANN for Time Processing

Claudia Lidia BADEA
SIEMENS AG ÖSTERREICH, PSE ECT ENA6, claudia-lidia.badea@siemens.at
Boschstr.10, Vienna, Austria

Abstract: New ANN models for Time Processing based on a new concept of defining the dynamics of a synapse are introduced. One shows conditions for the determination of the stability of equilibrium and an extension of the back propagation algorithm for computing network parameters is described. As particular cases, ANN with week and strong delays are considered.

1. Introduction

In order to provide neural networks with a dynamics, the temporal behaviour of each synapse has been modelled by a linear time invariant filter [1] where the synaptic weight functions were considered as being expressed as the product between some constants- the weights- and the week delay function. The choice of this type of modelling is based on similarities with the RC-circuits and means that the maximum response of the activation rate of the network is to the current activation potential, past activities having a decreasing importance. Yet, there is no mathematical motivation for this type of modelling.

However, another form of defining weight functions – not yet used- could be as the product between the weight constants and the strong delay function, which describes that the maximum influence on the activation rate response to time t is due to the activation potential at time t-T.

The aim of the present paper is to discuss a new concept of defining weight functions by means of modelling the dynamics of synapse with a differential operator. As a consequence, neuro dynamical models will be described with integrodifferential operators [2]. Week and strong delay models are particular cases of these ones.

2. Synaptic weight functions

Let us consider a recurrent neural network with N neurons and assume dynamic of each neuron is described by a differential operator. If we denote neuron with j and synapse with i, function $e_{ij}(t)$ will represent the impulse response i.e. the response of the filter due to unit input impulse at time t=0 on synapse i from neuron j.

In other words, this means, that $e_{ij}(t)$ can be defined as the fundamental solution of the linear differential operator that governs the behaviour of neuron j at synapse i. We shall call e_{ij} synaptic weight functions. Denote with $P_{ij}(D,t)$ the differential operator. According to our definition, e_{ij} will be the solution of the equation

$$P_{ij}(D,t)e_{ij}(t) = \delta_{ij},\tag{1}$$

where we denoted with δ_{ij} the *Dirac* distribution in $t = 0$. The synaptic weight functions depend from the form of operator $P_{ij}(D,t)$ only.

By considering different types of differential operators one can get different forms of synaptic functions. In what follows we will resume our discussion to differential operators with constant coefficients only.

From the Theory of Distributions [3] there is known that the elementary solution for (1) is given by

$$e_{ij}(t) = P_{ij}^{-1}(\delta_{ij}) = Y(t)z_{ij}(t) \qquad (2)$$

where z_{ij} is the solution of the *Cauchy* problem

$$P_{ij}(D)z_{ij}(t) = 0,$$
$$z_{ij}(0) = 0,\dots,z_{ij}^{(m-2)}(0) = 0, z_{ij}^{(m-1)} = 1$$

Using this result we can determine synaptic weight functions for some special forms of $P_{ij}(D)$. Let assume the behaviour of synapse i of neuron j is described by the operator

$$P_{ij}(D) = D + \lambda_{ij}; \lambda_{ij} \neq 0$$

then , we get synaptic weight function of type A,

$$e_{ij}(t) = P_{ij}^{-1}(\delta_{ij}) = Y(t)e^{-\lambda_{ij}t}, t > 0$$

The week delay synaptic function is a particular form of type A.
  For the second order operator

$$P_{ij}(D) = D^2 + \lambda_{ij}^2, \lambda_{ij} \neq 0$$

the elementary solution will be called synaptic weight function of type B,

$$e_{ij}(t) = P_{ij}^{-1}(\delta_{ij}) = \frac{1}{\lambda_{ij}} Y(t)\sin \lambda_{ij}t, t > 0$$

and for the differential operator of order m

$$P_{ij}(D) = (D + \lambda_{ij})^{(m)}, \lambda_{ij} \neq 0$$

the elementary solution will generate another function , which we shall call synaptic weight function of type C

$$e_{ij}(t) = P_{ij}^{-1}(\delta_{ij}) = \frac{1}{(m-1)!} Y(t)t^{m-1}e^{-\lambda_{ij}t}, t > 0$$

The strong delay synaptic function is a particular form of type C.
  As synaptic weight function of type D we shall call any combination of synaptic functions of types A, B or C and will be generated by the general linear differential operator with constant coefficients

$$P_{ij}(D) = D^m + a_1^{ij}D^{m-1} + a_2^{ij}D^{m-2} + \dots + Da_{m-1}^{ij} + a_m^{ij}$$

3. Building ANN Models

Let denote by x_i the stimulus applied to synapse i of neuron j and assume that the dynamics of synapse is described by the differential operator $P_{ij}(D)$. The response $u_{ij}(t)$ of synapse due to the input x_i will be then given by the solution of the equation

$$P_{ij}(D)u_{ij}(t) = x_i(t), t > 0,$$

and is

$$u_{ij}(t) = P_{ij}^{-1}(\delta_{ij}) * x_i(t), t > 0$$

Assume neuron j has M synapse, φ_j is the activation function (hard limiter), v_j is the potential developed across the input of the hard limiter and θ_j an external input bias to this neuron. In order to write the ANN Model we observe that on one side the total activation potential coming to the input of hard limiter is

$$\sum_{i=1}^{M} u_{ij}(t) + \theta_j = \sum_{i=1}^{M} e_{ij}(t) * x_i(t) + \theta_j$$

and on the other side that total flow away from the input node φ_j will be

$$\frac{d}{dt} v_j(t) + v_j(t)$$

The balance equation

$$\frac{d}{dt} v_j(t) + v_j(t) = \sum_{i=1}^{M} \int_{-\infty}^{\infty} e_{ij}(\tau) x_i(t - \tau) d\tau + \theta_j \tag{3}$$

$$x_j(t) = \varphi_j(v_j(t))$$

gives the general form of an ANN Model for Time Processing.
The above computed synaptic weight functions generate following types of ANN dynamic Models.

For week delay:

$$\frac{d}{dt} v_j(t) + v_j(t) = \frac{1}{T} \sum_{i=1}^{M} \lambda_{ij} \int_{-\infty}^{\infty} \exp(-\frac{\tau}{T}) x_i(t - \tau) d\tau + \theta_j \tag{4}$$

$$x_j(t) = \varphi_j(v_j(t))$$

for strong delay:

$$\frac{d}{dt} v_j(t) + v_j(t) = \frac{1}{T^2} \sum_{i=1}^{M} \lambda_{ij} \int_{-\infty}^{\infty} \tau \exp(-\frac{\tau}{T}) x_i(t - \tau) d\tau + \theta_j \tag{5}$$

$$x_j(t) = \varphi_j(v_j(t))$$

for synaptic functions of type A:

$$\frac{d}{dt} v_j(t) + v_j(t) = \sum_{i=1}^{M} \int_{-\infty}^{\infty} \exp(-\lambda_{ij}\tau) x_i(t - \tau) d\tau + \theta_j \tag{6}$$

$$x_j(t) = \varphi_j(v_j(t))$$

for synaptic functions of type B:

$$\frac{d}{dt} v_j(t) + v_j(t) = \sum_{i=1}^{M} \int_{-\infty}^{\infty} \sin(-\lambda_{ij}\tau) x_i(t - \tau) d\tau + \theta_j \tag{7}$$

$$x_j(t) = \varphi_j(v_j(t))$$

and for synaptic functions of type C:

$$\frac{d}{dt}v_j(t) + v_j(t) = \frac{1}{(m-1)!}\sum_{i=1}^{M}\int_{-\infty}^{\infty}\tau^{m-1}\exp(-\lambda_{ij}\tau)x_i(t-\tau)d\tau + \theta_j$$

$$x_j(t) = \varphi_j(v_j(t)) \tag{8}$$

4. Neural Dynamics

Let consider now a recurrent neural network with N neurons and associate to it the dynamics described with (3). The neurons can be input, output or hidden. The input neurons receive external signals and the answer of network is evaluated through the output signals produced by the output neurons. The problem to be solved is: *find parameters of the network- this means find λ_{ij} - such that after receiving an input signal $a=(a_j)$, network stabilizes to an equilibrium state producing a desired response $x_a= (x_{aj})$, $j=1,...,N$.*

The stated problem reduces to an optimization problem: given an input a, find

$$\min_{\lambda_{ij}}\frac{1}{2}\sum_{j\in output}(x_{aj}-x_{ej})^2, \qquad \text{x_{ej} is current state,} \tag{9}$$

subject to (3). In fact, this means: train network in such a way that when it receives an input signal , its evolution in time stabilizes to an equilibrium equal to the desired state. The minimisation of (9) is performed with respect to synaptic coefficients.

4.1. Stability of the Equilibrium

If we denote with x_{ej} the equilibrium state of the model and with v_{ej} the potential, then from the system model (3) we have:

$$v_{ej} = \sum_{i=1}^{M}x_{ei}\int_0^{\infty}e_{ij}(\tau)d\tau + \theta_j$$

$$x_{ej} = \varphi_j(v_{ej}), j = 1,...,N$$

A general methodology to study the stability of an equilibrium for an integrodifferential equation is [4]: linearize equation around the equilibrium and make sure that for the linearized equation the zero solution is asymptotic stable. The linearized form of model (3) is given by (10):

$$\frac{d}{dt}w_j(t) + w_j(t) = \sum_{i=1}^{M}\varphi'(v_{ei})\int_0^{t}e_{ji}(t-\tau)w_i(\tau)d\tau \tag{10}$$

$$w_j(t) = v_j(t) - v_{ej}$$

By applying the above mentioned result we have the following *Lema:*
The zero solution for the model equation (10) is asymptotic stable if and only if the condition

$$M(z) \equiv z + 1 - \sum_{i=1}^{N}\varphi'(v_{ei})L\{e_{ij}\} \neq 0, \operatorname{Re} z \geq 0 \tag{11}$$

where L denotes Laplace transform of e_{ij} is fulfilled.

The stability of the equilibrium depends upon the choice of activation function, synaptic weight functions and the number of neurons in the network.

4.2. Learning Algorithm

Once it has been stated that equilibrium is asymptotic stable, one can start the process of updating network parameters. The standard way is to use a correction rule based on steepest descent,

$$\Delta\lambda_{lk} = -\eta\frac{\partial E}{\partial\lambda_{lk}} = -\sum_{j=1}^{N}\frac{\partial x_{ej}}{\partial\lambda_{lk}}(x_{aj} - x_{ej}) \tag{12}$$

The learning algorithm can be summarised in following steps:
Step1: Compute equilibrium, state if it is asymptotic stable using (11)
Step2: Initialise synaptic coefficients and enter threshold
Step3: Update synaptic weights according to (12)
The implementation of this algorithm has been done for the Week Delay Model (4), Strong Delay Model (5) and for model based on coefficients of type A.

5. Example

As an example let consider the Weak Delay Model (4) and let take a hard limiter the logistic function. The equilibrium given by:

$$v_{ej} = \sum_{i=1}^{M}x_{ei}\lambda_{ij} + a_{j}$$

$$x_{ej} = \varphi_{j}(v_{ej}), j = 1,...,N$$

is asymptotic stable if

$$z+1-\frac{K_{j}}{Tz+1} \neq 0, \operatorname{Re} z \geq 0$$

where

$$K_{j} = x_{ej}(1-x_{ej})\sum_{i=1}^{M}\lambda_{ij}$$

By applying *Routh-Hurwitz* criteria one can show, that condition is fulfilled for $K_j < 1$.

References

[1] Haykin,S. Neural Networks. A comprehensive Foundation, MacMillan College Publ. Comp., New York, 1994
[2] Badea, C.L. Integro-Differentiale Gleichungen. Eine Einführung. Theorie-Anwendungen-Programmierung, ÖCG, R. Oldenburg, München,1990
[3] Schwarz,L. Theorie des Distributions, Gauthier-Villars,1951
[4] Miller,R.K. Asymptotic Stability and Perturbations for linear Volterra Integrodifferential Systems, Ac. Press, N.Y., 1972

KES '01
N. Baba et al. (Eds.)
IOS Press, 2001

Fuzzy Clustering and Time-Frequency Approach for Fault Diagnosis

Dan STEFANOIU [1], Florin IONESCU [2], Dumitru POPESCU [1]

[1] *"Politehnica" University of Bucharest, Dept. of Automatic Control & Computer Science*
313 Splaiul Independentei, Sector 6, 77206-Bucharest, ROMANIA.
Tel.: (+ 401) 410 0400, ext. 327.
E-mails: danny@router.indinf.pub.ro, dpopescu@router.indinf.pub.ro

[2] *FH-University of Applied Sciences-Konstanz, Faculty of Mechanical Engineering / MK2500*
Brauneggerstraße 55, D-78462 Konstanz, GERMANY.
Tel.: (+49) 7531 206-289; Fax: (+49) 7531 206-294
E-mail: ionescu@fh-konstanz.de

Abstract. *A method integrating two very different approaches (fuzzy modeling and time-frequency analysis) is proposed in this paper, regarding the general problem of fault diagnosis by using vibrations. The fuzzy model is aiming to construct and update a primary set of possible faults classifications. The purpose of time-frequency analysis is to refine the classification above by detecting the intimate changes within vibration, with the help of a time-frequency dictionary tuned on the fuzzy model. By learning, a knowledge basis can be built and upgraded depending on vibration signature inside the dictionary, in order to perform diagnosis.*

1. Introduction and problem statement

Nowadays, the trade off between damage costs involved by ignoring fault prevention and costs of hyper-safety of systems is improved. In a complete structure of fault detection and diagnosis [12, 9, 5, 6], a module concerned with monitoring of system symptoms and anticipation for possible failures is included. In general, the symptoms are detected by using two kinds of methods: *analytical* and *heuristic*.

The *analytical methods* are involved with systems for which the characteristic parameters are quantifiable. These parameters are determined by analyzing either some signals or the system itself (like e.g.: the amplitude, the variance, the auto-correlation, the power spectral density, etc.). Basically, the system analysis is founded on a parametric model [10] (such as regressive, state type, concerned with some parity equations). The model parameters are deduced from measured input-output data by system identification techniques. In both cases, a quantitative expertise has to be performed. This consists mostly of comparisons between the measured values and a set of *tolerated* values assigned to normal behavior of system. The malfunction symptoms appear when the parameters start to provide systematically values beyond tolerances. A classification of symptoms can be realized, depending on the difference between the measured and tolerated values.

Sometimes, the analytical approach is not sufficient or cannot be performed (especially since the characteristic parameters are not quantifiable). Moreover, the meaningful of symptoms is important for interpretation of associated faults. Often, this relies on the qualitative expertise of a human operator as expert. For this reason, one says that the detection of symptoms is performed by using *heuristic methods* (from *heuriskein* (gr.) – to investigate). The non-quantifiable information observed from the system could be reflected for example by: colors, smells, tones of noises, etc. However, some quantifiable parameters, but with "fuzzy" values, represented by linguistic terms like: "small", "medium", "large", "about null", etc. belong to this category as well. The human operator integrates this information in a quasi-empirical history of system functioning. Qualitative comparisons are performed between the observed information and the information specified by the history. The history mostly includes information about the maintenance process, repairs, faults types, life-time, fatigue, etc. The decision concerning the symptoms and faults is based on the operator's skills and is affected by uncertainty. However, the experience about the system can be improved by accounting all the events that occurs during its functioning, in a learning mechanism.

Like in medicine, faults prevention remains a demanding task, that requires both *self-anticipation* from the system and intelligent approach from the user. Usually, a self-anticipatory system transmits information about its behavior through some *anticipating signals*. For example, human or animal muscles have different electrochemical activity just before they are damaged, due to high intensity and long effort [11]. Another example is issued from mechanical systems, for which vibrations are anticipating signals. Their intimate structure changes some time before a failure occurs [1]. But this change is so fast and sometimes so difficult to distinguish, that, without special detection and decoding techniques, it could be ignored. In general, the strategy adopted within a fault detection method using vibrations consists of the following: signal acquisition, signal analysis (in order to extract features), features grouping, upgrading faults classification, fault identification (if present). One of the most interesting application of fault diagnosis is concerned with bearings, due to their large integration within mechanical systems [3], [4], [13].

The results obtained so far are based on combinations between time or frequency techniques for signal analysis and neural networks with fuzzy inference rules for faults classification. But, to our best knowledge, a sound joint time-frequency analysis aiming fault prevention is not yet performed and the fuzzy attempt is rather weak. Therefore, in this paper, a different method combining fuzzy and time-frequency approaches is succinctly described. Our attempt is based on the idea that fault classifications are affected by uncertainty due mostly to user's subjectivism and external perturbations, even in case of a rich experience about the system. Moreover, the classifications should be permanently updated, in order to minimize both uncertainty and perturbations effects. In fact, this approach is combining the analytic and heuristic methods.

The method presented briefly hereafter is intending to solve the following general faults diagnosis problem within mechanical systems: *starting from acquired segments of vibration, construct a classification of failures that may occur while the system is functioning, in order to focus on the specific failure the system seems to anticipate.* Four main steps are described next: constructing a fuzzy model of fault classifications; generating a time-frequency dictionary; computing the vibration signature within dictionary; building a knowledge basis of faults.

2. Fuzzy classifications of faults

The most common observable phenomenon when a fault is threatening the mechanical system is vibration energy increasing upon a fixed period of time (although some faults do not produce this effect). Consider that vibration is monitored within a variable length window (up to T_0 - a multiple of sampling period T_s). The acquisition of current segment y is stopped whenever the energy of auto-correlation r_y overpasses a maximum acceptable value σ_{MAX}^2 or a fault announcing value σ_0^2, or the window has reached its maximum duration T_0. Consequently, the time-energy plane can be covered by a network with elementary $T \times \delta^2$ cells, where can lie acquired segments, like in Figure 1(a). The cells, $V_{n,m}$, are grouped into 4 zones and their collection is $\mathcal{V}$. A couple time T_n and energy σ_m^2, taken from the middle is associated to each cell. Also, an occurrence rate of segments within $V_{n,m}$ can be estimated, $v_{n,m}$.

A fault can occur any time during the segment acquisition (zone DA – data acquisition). Three types of faults are considered: malfunctions (when the system still works, but under the desired performances), regular failures (when the system stops itself or it will stop very soon because of internal damages) and severe failures (when the system suffers a brutal and unexpected perturbation damaging it). Malfunctions and regular failures can be detected within NF ∨ FA and FA zones (NF – normal functioning; FA – faults anticipation). In general, severe failures produce either a sudden augmentation of vibration energy to a very high level, or a non-significant change (when the system is stuck, but the engines are still running). The second case is not concerned here, since these failures could be detected by monitoring another parameters. Thus, severe (unexpected) failures are detected within UF zone.

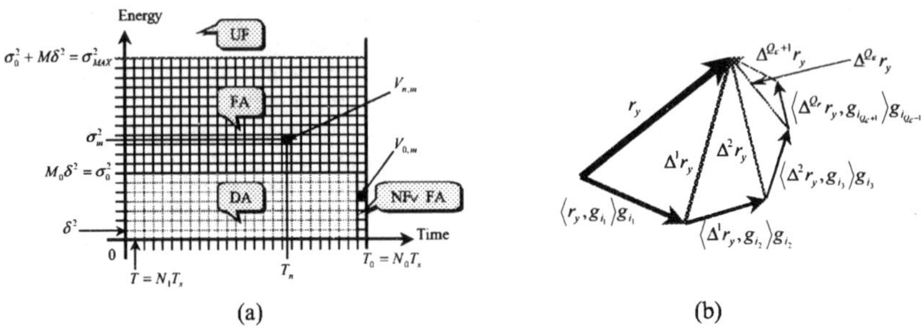

Figure 1. (a) *A time-energy network.* (b) *The matching pursuits principle.*

Define the following crisp relation between cells: V_{n_i,m_i} and V_{n_j,m_j} *are in relation* $R \subseteq \mathcal{V} \times \mathcal{V}$ *if they point to the same fault class*: $V_{n_i,m_i} R V_{n_j,m_j}$. This definition relies on some possible groups of cells, referred to as *clusters*, which seem to describe the same faults class. Several possible *clustering configurations* could cover $\mathcal{V}$, say $K \geq 1$, since no precise classification can be realized. Such a configuration $\mathcal{F}_k$ consists of $P_k \geq 1$ clusters: $\mathcal{F}_k = \left\{ F_{k,p} \right\}_{p \in \overline{1,P_k}}$ ($k \in \overline{1,K}$). It defines a unique relation R_k as follows: $V_{n_1,m_1} R_k V_{n_2,m_2}$ if it exists a cluster $F_{k,p} \in \mathcal{F}_k$ such that $V_{n_1,m_1}, V_{n_2,m_2} \in F_{k,p}$. Better description of R_k is provided by its characteristic (binary) matrix H_k .

The family of crisp relations R_k is used next to construct a set of elementary fuzzy relations $\alpha_k R_k$, where $\alpha_k \in [0,1]$ is a unique membership grade. The membership matrix $\mathcal{M}_k$ of $\alpha_k R_k$ is simply: $\mathcal{M}_k = \alpha_k H_k$. Finally, a fuzzy relation $\mathcal{R}$ is built by aggregating the α - *sharp cuts* $\alpha_k R_k$ through the max fuzzy union, so that its membership matrix is $\mathcal{M} = \max \left\{ \mathcal{M}_k \right\}_{k \in \overline{1,K}}$. A measure that could be used to compute α_k is the *Shannon fuzzy entropy* [7]. In general, the entropy of a set reflects its disorder degree. Thus, α_k should increase/decrease when the entropy of $\mathcal{F}_k$ (denoted by S_k) decreases/increases. The faults classifications will be oriented towards clustering configurations that are more ordered (i.e. with smaller entropy). Thus: $\alpha_k = 1 - S_k$.

In general, $\mathcal{R}$ is reflexive and symmetric, i.e. *proximity* relation and it is not necessarily transitive, i.e. *similarity* relation. But its transitive closure [7], $\overline{\mathcal{R}}$, is a similarity relation. A tree of fault classifications is generated by computing the α -cuts of $\overline{\mathcal{R}}$ and by ordering increasingly the membership grades. For each grade, a collection of more and more refined similarity classes is built. They stand for the fault classifications. The current vibration segment lies in the classification for which the entropy weighted by the membership grade is minimum. The tree is upgraded by updating the occurrence rates and initial configurations for each new segment, according to a sound signal analysis.

3. The time-frequency dictionary

The signal analysis relies on the intimate time-frequency structure of current segment in order to extract useful features in diagnosis. Starting from the similarity class $\overline{F}_{k_0,p_0}$, where lies the current segment, a normalized (unit energy) gaborette-mother g [2] can be defined and used to generate a discrete time-frequency dictionary, by applying 3 operators: scaling, time shifting and frequency modulation, after discretization. An atom within dictionary has the general form: $g_i[l] = g_{[m,n,k]}[l] = s_0^{-m/2} e^{-jkl\omega_0 T_s} g\left((s_0^{-m} l - n) T_s \right)$, where $j^2 = -1$, $i = [m,n,k]$; l is the discrete time; $m \in \overline{0, M_R}$ controls the scaling at scale $s_0 > 0$; $n \in \overline{N_L, N_R}$ is responsible for time shifting with step T_s; $k \in \overline{0, K_R}$ is concerned with the modulation at pulsation $\omega_0 > 0$. The sampling parameters s_0, T_s, ω_0, as well as the left and right bounds of indices are set according to parameters of $\overline{F}_{k_0,p_0}$ and of the current segment. Also, Gabor-Heisenberg Uncertainty Principle [2] should be respected.

4. Vibration signature

In general, the auto-correlation signal r_y is not necessarily included in the subspace spanned by the discrete dictionary. But the dictionary could be constructed such that the distance between r_y and the subspace it spans be as small as possible. Usually, signals provided by physical systems have two components: an *original* (genuine) one and a *parasite* one. Making a clear separation between them is an important and complex problem. The most common assumption is that original and parasite parts are additive. The dictionary is supposed to perform a signal *denoising*. The representation of auto-correlation inside dictionary is the genuine part of vibration segment.

The representation is realized by using matching pursuits method [8]. The main idea is to project the current signal on an atom selected from dictionary such that the 2 signals are the *best matching* ones. The "best matching" means the maximum amplitude of scalar product between r_y and the atoms of dictionary, g_i. If r_y and g_{i_1} are the best matching signals, a signal can be computed by subtracting from r_y the projection of r_y on g_{i_1}: $\Delta^1 r_y = r_y - \langle r_y, g_{i_1}\rangle g_{i_1}$. The residual $\Delta^1 r_y$ becomes the new signal searching for its best matching inside dictionary. The searching for the best matching atom inside dictionary is *pursued* now for $\Delta^1 r_y$, then for $\Delta^2 r_y$ and so forth. If the initial signal r_y is considered a *sentence*, it is decomposed in *words*, by pursuing the best matching between words and subsequent sentences. The best meaningful of sentence is extracted as a *signature* consisting of all coefficients: $\left\{\left\langle \Delta^q r_y, g_{i_{q+1}}\right\rangle\right\}_{q\geq 0}$.

Several properties are verified. The energy of residual signal can be arbitrarily decreased, for example under a fixed level $\varepsilon > 0$, derived from SNR and the signal energy. The energy is dissipated among signature coefficients following a Pythagorean-like relation. This property is due exclusively to employment of projections (which are rectangular) and *it is independent of atoms orthogonality* (i.e. it does not requires that atoms within dictionary be orthogonal each other). This property reflects the decorrelation of signature coefficients and involves that the signature should be more suitable in detection of faults than r_y itself. Finally, a perfect synthesis identity holds, which shows in fact that, once the residual term has reached a fixed minimum level energy, it can be considered as parasite part distorting the original signal.

The principle of matching pursuits is depicted in Figure 1(b). Arrows with decreasing thickness suggest energy dissipation among coefficients. All triangles are rectangular and this also reveals how Pythagoras' relation works, though the atoms within dictionary are in general non orthogonal.

5. Faults knowledge basis

A knowledge basis could be constructed starting from a large number of acquired vibration segments and from their signatures represented by time-frequency images. Each signature offers an image about how the energy of the initial signal r_y is spread in time-frequency plane. The plane is covered by a rectangular network, with respect to the Uncertainty Principle. The network is depicted in Figure 2(a), where N_y is the length of auto-correlation derived from the current vibration segment and 2^0, 2^1, 2^2, ... are frequency resolutions (at constant scales), for $s_0 = 2$.

A time-frequency image consists of 2 pictures: one for the amplitude of signature coefficients and another one for their phase. In Figure 2(b), a time-frequency band of Figure 2(a), at constant scale, is magnified. Each coefficient is localized in a small rectangle of time-frequency plane (for example, those represented in red with a star). The other cells correspond to projections of auto-correlation r_y on the other atoms of dictionary (different from the best matching ones).

Such time-frequency images could be associated to different behaviors of mechanical system within a knowledge basis. A specific fault is reflected in time-frequency plane by an abnormal energy concentration in a zone where, usually, low amplitudes cells are present (e.g. the compact group of red cells with star, towards northeast in Figure 2(b)).

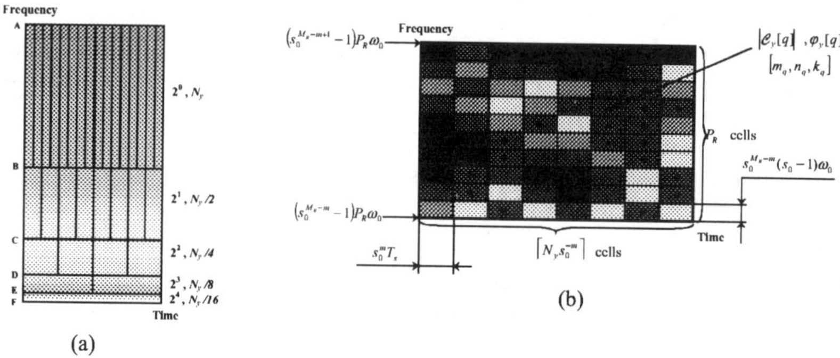

Figure 2. (a) *General shape of a time-frequency image.* (b) *Signature localization.*

The collection of all time-frequency images should be used not only for fault diagnosis purposes, but also to upgrade the fuzzy classifications. Two kinds of clustering configurations can be provided, by using time-frequency images above. On one hand, a set of such configurations is constructible by using some inference mechanism within knowledge basis. On the other hand, direct inspection of time-frequency image related to signature coefficients amplitude could lead to new configurations. The new clusters encompass all partial signals corresponding to cells where an abnormal energy concentration is observed (the red cells with a star, in our example), since these signals actually point to the same fault class.

6. Conclusion

This paper dealt with the problem of faults diagnosis within mechanical systems, starting from their capacity to anticipate failures by means of vibrations. The method proposed here is based on a combination between 2 main types of rationales: fuzzy and time-frequency. The goal of first one is to provide a valid classification of faults, of minimal entropy, starting from subjective observations about system behavior. The second one is meant to refine the previous classification and to point out the possible fault threatening the mechanical system, by using the time-frequency images associated to current segment of vibration.

This method could successfully assist a human operator to detect promptly the failures that may occur in mechanical systems sufficiently complex.

References

[1] **Braun S.** – *Mechanical Signature Analysis,* Academic Press, London, UK, **1986**.

[2] **Cohen L.** – *Time-Frequency Analysis,* Prentice Hall, New Jersey, USA, **1995**.

[3] **Howard I.** – *A Review of Rolling Element Bearing Vibration: Detection, Diagnosis and Prognosis,* Report of Defense Science and Technology Organization, Australia, **1994**.

[4] **Ionescu F., Arotaritei D.** – *Fault Diagnosis of Bearings by using Analysis of Vibrations and Neuro-Fuzzy Classification,* Conference ISMA`23, KU-Leuven, Belgium, September 16-18, **1998**.

[5] **Iserman R.** – *Fault Diagnosis of Machines via Parameter Estimation and Knowledge Processing,* Automatica, Vol. 29, No. 4, 161-170, **1993**.

[6] **Iserman R.** – *Knowledge-Based Structures for Fault Diagnosis and its Applications,* 4-th IFAC Conf. on System, Structure and Control, SSC'97, Bucharest, Romania, October 2-25, 15-32, **1997**.

[7] **Klir G.J., Folger T.A.** - *Fuzzy sets, Uncertainty, and Information,* Prentice Hall, NY, USA, **1988**.

[8] **Mallat S., Zhang S.** – *Matching Pursuits with Time-Frequency Dictionaries,* IEEE Transactions on Signal Processing, Vol. 41, No. 12, 3397-3415, December **1993**.

[9] **Reiter R.** – *A theory of diagnosis from first principles,* Artificial Intell., Vol. 32, 57-95, **1987**.

[10] **Söderström T. Stoica P.** – *System Identification,* Prentice Hall, London, **1989**.

[11] **von Tscharner V.** – *Intensity Analysis in Time-Frequency Space of Modelled Surface Myoelectric Signals by Wavelets of Specified Resolution,* preprint, **2000**.

[12] **Willsky A.S.** – *A Survey of Design Methods for Failure Detection Systems,* Automatica, Vol. 12, 601-611, **1976**.

[13] **Xi J., Sun Q., Krishnappa G.** – *Bearing Diagnostics Based on Pattern Recognition of Statistical Parameters,* Journal of Vibration and Control, No. 6, 375-392, **2000**.

Automated Visual Inspection of Ice - Cream Bars

Theodor BORANGIU

Politehnica University of Bucharest, 313, Spl.Independentei, RO -77206 Bucharest, Romania

Florin IONESCU

FH-University of Applied Sciences, 55. Brauneggerstr., D-78462 Konstanz, Germany

Abstract: The gap between Automated Visual Inspection and Robot Vision applications is nowadays narrowed by the fact that an increasing number of inspection tasks require the manipulation of parts. The paper refers to an integrated approach of image processing and robot control, in the food industry; two technologies are integrated to this purpose: Image Processing and Artificial Intelligence, by exploiting a powerful set of Vision Tools which are available in the V+ programming environment. A complete description of the merged AVI/RV task is included, as well as final implementation recommendations.

1   Introduction

In many **Robot Vision** (RV) and **Automated Visual Inspection** (AVI) tasks, the user must define and include in the application program a number of custom - designed *object features*, using to this purpose **vision tools** [3], [4]. Such features are related to geometric measurements which are applied to the internal and external boundaries, to the body and holes of objects.

The most representative situations which require the utilisation of task - dependent measurements are:

- ❑ **IVA applications** in which one must evaluate *geometric details of parts* (such as edge lengths, angles between edges, distances between centres of holes or other significant points, hole location inside certain zones of parts), or one must check the *correct location of parts*, in particular areas of visualised work places, or the correct relative positioning of different components of an assembling.
- ❑ **RV applications** (visually guided robot motion) in which only the intrinsic features of non-prototyped objects (such as area, perimeter, number of holes, eccentricity, compactness, roundness) or the description of interior and exterior object boundaries in terms of : weighted length of contours and topology of edge points,

are no more sufficient in order to discriminate between objects or to classify parts according to geometric details of the machining of part surfaces [1], [2].

The V+ programming environment offers the possibility to configure such powerful custom designed measurements from a package of vision tools which include Windows for Region Of Interest, linear and circular Rulers and point-and-edge Finders.

2   Windows Region of Interest (WROI)

In many situations, only a small section of the visual field is of interest for the application. In order to process images in a single Region Of Interest with respect to tasks such as: the control of the geometry of part surfaces, the inspection of surface machining and gathering of statistical information in windows of particular shapes, the V+ environment offers the possibility to use special operations like VTRAIN, VWINDOW and VWINDOWI. The (macro)operation:

> **VTRAIN** *prototype, shape, cx, cy, width, height, ang*

will be used in order to create a new prototype of a class of objects, or to modify an existing prototype. Here, *shape* defines the window's form (e.g. rectangular) having respectively the height and width [mm] *height, width,* and being centred at *cx, cy*; the window can be rotated with an angle *ang* [deg] with respect to the x_{vision} axis.

The (macro)operations Window for Region Of Interest (WROI)

> **VWINDOWI** (*cam, type, dmode, sample*) data[ ] = *shape, cx, cy, dx, dy, ang*
> **VWINDOWI** (*cam, type, dmode, sample*) data[ ] = *shape, cx, cy, or, ir, ang0, angn*

perform a general processing of an image taken from the virtual camera *cam*, like in VPICTURE. The information are extracted from a picture window having one of the following possible *shape*s: rectangular, circular disk, ring, ring segment.

Here, *type* specifies the type of statistics on the image which will be computed and returned in the array *data[ ]*, with components such as: total number of pixels in the window, number of nonzero pixels in the window, average level of grey, minimum and maximum levels of grey, total number of pixels belonging to the background, standard deviation of the grey levels in the window relative to the average level, number of edge points in the window.

The argument *sample* expresses the sampling density(e.g. each pixel, or every to lines - pixel is considered). The arguments *cx, cy* locate the window's centre; *dx, dy* are the width and height for a rectangular window possibly rotated with the angle *ang* relative to x_{vision}, whereas *or, ir* specify the outer and inner radius of a circular dik or ring shaped window, having an angular range given by *ang0, angn*.

3   AVI of the Ice - Cream Bars

Figure 1 depicts an AVI application that uses the statistics collected in a WROI and created by the instruction VWINDOWI. The objective is to inspect the wood bars used as ice-cream support. The bars move on a conveyor belt. An image of each bar is taken by a camera; for this down-looking camera, a signal generated by a photocell (which identifies the arrival of the bar in the visual field of the camera) triggers a strobe lighting synchronised with the image acquisition (i.e. with the execution of the VPICTURE operations).

The program presented below is used for this application. The program performs two types of *tests* for the bars moving on the conveyor:

> ➤ a test concerning *the integrity of the bar* (the search of holes with an area greater than 1.2976 mm^2 or of "bays" within the surface of the bar);
> ➤ a test concerning the *presence of dark spots* on the surface of the bar.

A single stationary, down-looking camera is used, which is fixed above the conveyor. The image acquisition is synchronised with the signal generated by a photocell, which detects the presence of a bar in the camera's visual field. The basic parameters of the program and the related significance are described below.

Two **virtual cameras**, "*cam*" and "*cam**" are associated to the physical camera, in order to perform specific tasks:

> • "*cam*" is used for the image acquisition triggered by the external signal from the photocell (this signal raises a V+ rapid digital interrupt line). This signal identifies the occurrence of the external event "*bar arrived in the visual field of the physical camera*". Then, "*cam*" immediately processes in binary mode the image acquired with the recognition disabled, with a unique binarisation threshold V.THRESHOLD[*cam*]. The system computes the centre co-

ordinates and the dimensions of the minimal box, the area and the compactness of the bar's body , as well as the number of holes and checks them against the lower and upper measurements limits: *"ar_min"*, *"ar_max"*, *"cmpct_min"*, *"cmpct_max"*. Taking into account the calibration parameters "mm/pix" = 0.402 and "XYratio" = 1.004 for *"cam"*, it results that the potential detected holes have the area[pixels] greater than $1.2976mm^2$.

- *"cam*"* is used for reprocessing, in binary mode, the previously acquired grey level image (i.e. the bar's image acquired by *"cam"*). *"cam*"* has the same set of switches and system parameters, except the 2nd binarisation threshold, V.2ND.THRESHOLD[*cam*].

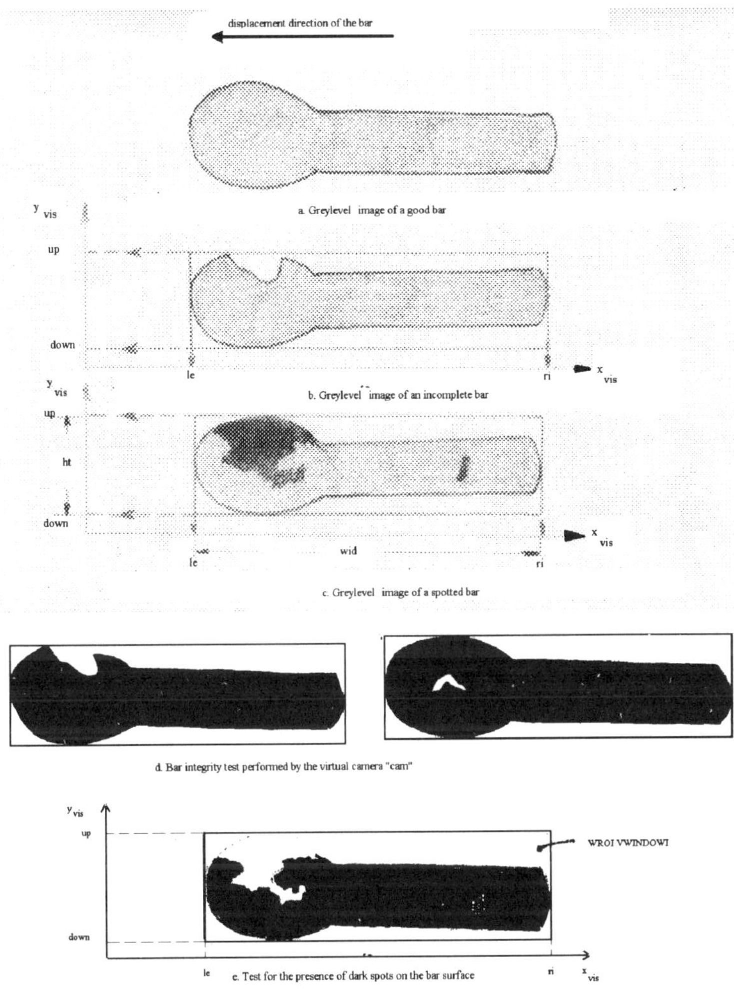

Figure 1 Using a WROI with VWINDOW for the AVI of the ice-cream bars.

"range" represents the grey levels range, which was experimentally evaluated. These grey levels are accepted as "admissible colours" of the bar's visualised surface. Next, a WROI with the dimensions and location respectively identical to the ones of the minimal box will be applied with VWINDOW on the bar image; the system computes the ratio between "the number of non-spotted pixels of the bars surface" and "the number of pixels of the entire bar's area".

The strobe lighting is used to take images of moving objects (here the bars) and to avoid blurred images. In order to get the fastest and the most consistent response at the occurrence of the external event *"activation of the signal from the photocell"*, one uses the rapid digital input interrupt line.

Another problem for the synchronous strobe is the variable time interval between the VPICTURE request (or the occurrence of the external event that triggers VSTROBE) and the effective strobe triggering. The most unfavourable delay (the greatest) measures 16.7 msec and it will be taken into account for the selection of the strobe mode (synchronous or asynchronous), when the objects that have to be visualised are rapidly moving in the visual field of the camera.

Consider now the case of the particular ice-cream bars application, with "mm/pix" = 0.402, in a visual field of 205.824mm = 512 pixels and a speed displacement of the bars of 1m/sec; then the most unfavourable delay of 16.7 msec corresponds to a lagging of:

$$(0.0167sec) * (1000mm/sec) / (0.402mm/pix) \cong 42 \text{ pixels}$$

Hence the problem is to place the photocell so that, taking into account the most unfavourable delay of 16.7msec, the length of the bar and the length of the visual field (the latter being parallel to the displacement direction of the bar), the bar should be still completely visible within the visual field during the image acquisition (i.e. during the execution of VPICTURE).

```
.PROGRAM bar ( )
.........
;Synchronising the image acquisition with an external event that triggers the
;rapid digital input interrupt line
 ENABLE V.STROBE[cam]
 PARAMETER  V.IO.WAIT[cam] = 1
;Area filters are specified for the bars
.........
;Recharging the calibration data for "cam"
 VPUTCAL(cam) cal.arr[]
 pix.to.mm = cal.arr[4]
 scale = SQR(pix.to.mm) * cal.arr[14]
 count.good = 0      ; Counter initialisation for bars
 count.bad = 0
 nonloc = 0
; DO actions sequence
 DO                 ;Until an external signal SIG(1001) is disabled
;Acquisition and processing of an image, synchronised with the rapid interrupt
signal
;from a photocell. No object will be recognised
    VPICTURE(cam) -1, 0
    VWAIT
    VLOCATE (cam,2)"?"
    IF VFEATURE(1) THEN
       cx = VFEATURE(2)
       cy = VFEATURE (3)
       wid = VFEATURE(14) - VFEATURE(13)
       ht = VFEATURE(16) - VFEATURE(15)
       ar = VFEATURE(10)
       per = VFEATURE(41)
       holes = VFEATURE(17)
       ar.mm = ar * scale
       cmpct = SQR(per) / ar_mm
       pass = (holes==0)AND((ar>=ar_min-30)AND(ar<=ar_max+30))AND((cmpct>=0.98* ...
            ...cmpct_min)AND(cmpct<=1.02*cmpct_max))
       IF pass THEN
;If the test of integrity (broken edges - "bays" in the surface), presence of holes
in
;the bar body as suggested in Fig.1(a) and (d)) passes, then reprocessing of the
grey
;level image with the camera "cam*"
       VPICTURE(cam*) 0, 0
       VWAIT
;Binarising  with two thresholds :
       VAUTOTHR (1,PARAMETER(V.THRESHOLD[cam*]),PARAMETER(V.2ND.THRESHOLD[cam*]))
```

```
;A WROI is applied with VWINDOWI\the background is white\the statistic of 3^rd type
is
;returned\the number of pixels in "object"=data[7] represents the surface of the bar
;with accepted grey levels in the non-spotted range:[V.THRESHOLD[cam*],
;V.2ND.THRESHOLD[cam*] + range]
        VWINDOWI(cam*, 3, 0, 1) data[ ] = 1, cx, cy, wid, ht, 0
        ar.not_spot = data[7]
        pass = (ar.not_spot / ar >= 0.98)
        IF pass THEN
;If the test  for the presence of dark spots on the bar surface passes (see Fig.1(c)
;and (e)), then increment the counter of good bars which leave the visual field
            count.good = count.good + 1
        ELSE
;Increment the counter for bad bars and trigger the output signal SIGNAL 1 which
;controls the compressed air to blow the bad bar off.
            GO TO a
        END
    ELSE
a:      count.bad = count.bad + 1
        SIGNAL  1
    END
 ELSE
   TYPE "Bar not located"
   nonloc = nonloc + 1
 END
UNTIL  SIG(1001)
.END
```

4 Conclusions

Experiments have been carried out on a production line consisting of a linear conveyor driven by a DC servo-motor with variable, closed loop controlled speed, a SCARA-type AdeptOne robot with multitasking controller and a video matrix camera with software configurable lighting system [4]. The proposed inspection algorithm has been successfully tested up to a speed of the conveyor belt of 1.34 m/sec.

A statistical analysis is given below, concerning the integrity of ice-cream bars. For such objects, the area deviation expressed in number of black pixels was checked. Bars had a 3 cm head diameter and 2.5 cm body width. The obtained results for the body area deviation checking are represented in the Figure 2.

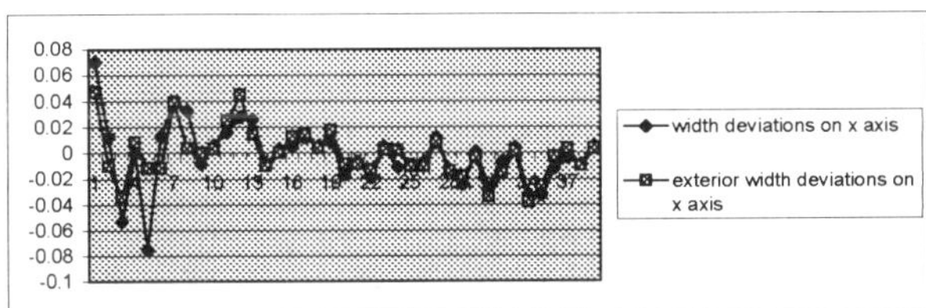

Figure 2 Error values in the area deviation for an ice-cream bar

5 References

[1] Hutchinson, S., Hager, G.D. and P. Corke: "A Tutorial on Visual Servo Control", IEEE Trans. on Robotics and Automation, Vol.12, No.5, Oct.1996.

[2] Arbib, M.A. and A.R. Hanson: "Vision, Brain and Cooperative Computation", MIT Press, Cambridge, MA, 1987

[3] Batchelor, B.: "Intelligent Image Processing in Prolog", Springer Verlag, 1991

[4] ***, "AdeptVision VXL User's Guide", Cat.#7498, Adept Techn. Inc., San Jose, CA, 1998

KES '01
N. Baba et al. (Eds.)
IOS Press, 2001

Dedicated Hardware and Software Architecture for Multipoint Services over ATM Networks

Radu DOBRESCU, Matei DOBRESCU, Daniela ANDONE
POLITEHNICA University of Bucharest, Splaiul Independentei 313, Bucharest, Romania

Abstract. The paper presents the structure and the functional behavior of a server hardware and software architecture able to provide a UNI signaling interface for multipoint communication and to emulate a new rate control mechanism for congestion prevention. The performances of the proposed mechanism were tested through simulations on a point-to-multipoint network model with a binary tree topology.

1. Introduction

One of the most pressing needs for enhanced communications protocols come from multipoint (or group) applications, such as software distribution, replicated database update, command and control systems, audio/video conferencing and distributed interactive simulation. The session parameter negotiation in a multipoint session has to be receiver controlled in order to allow dynamic join and leave. Existing protocols (e.g. IP, CLNP, UDP) are sufficient only for the applications which can be satisfied by a connectionless multicast communication support, but the new generation applications introduce new requirements for the control of time dependencies (transmission delay, ordering, synchronization) and for the definition of session control parameters per participant. In this paper, we offer some improvements in ATM communications: a Multicast ATM Server architecture using UNI 4.0 services based on the Multicast Server Model and a new algorithm for congestion prevention in video dynamic transmissions.

2. A Critical View Over Two ATM Multipoint Communication Models

2.1. The sender controlled model

In ATM architectures, the multipoint communication is connected with the notion of point-to-multipoint virtual circuits (VC's) introduced in the User Network Interface (UNI) [1], [2] for audio conference purposes. Both UNI 3.0. and UNI 3.1. support a source controlled, unidirectional, point-to-multipoint service model defined as the Sender Controlled Model [3]. In this model, when data needs to be sent to multiple destination simultaneously, a source establishes an unidirectional VC that has a single root node (itself) and multiple leaf nodes (the intended recipients). UNI 3.1. does not support the abstraction of a multicast group ATM address to indirectly identify the desired set of leaf nodes, so the source must have an *a priori* knowledge of each leaf node it requires to be on its point-to-multipoint VC. The correct sequence of cells is ensured if we have a single segmentation engine acting as the cell source, considered the root. But if we were to attempt a reverse flow, from leaf nodes to the root, problems arise, because there is no obvious way of differentiating between cells arriving from different leaf nodes. We suggest three possible ways to solve this problem: to use AAL 3/4 end to end and distinguish the AAL_SDU's by a different MID field value in each cell; to modify the switch behavior at merge points when using AAL5, by buffering the cells on incoming links until a complete AAL_SDU has arrived or to ensure the multiple sources do not generate AAL_SDU's that overlap in the time domain.

2.2. The multicast server model

The simplest for a multipoint-to-multipoint service is to independently originate for each source its own point-to-multipoint VC out to the members of the destination group. These leads to a criss-crossing of VC's across the ATM network, denominated usually Multicast Mesh Model (MMM) [4]. Figure 1 shows an example of MMM with three distributed ATM nodes. Due to the results described in [5], we recommend the Multicast Server Model (MSM), a generalization of the AAL_SDU resequencing mechanism mentioned in the previous section. The multicast server attaches to the ATM network and acts as a proxy group member. It terminates VC's from the sources and originates another point-to-multipoint VC out to the actual group members. Its basic function is to reassemble AAL_SDU's arriving from the sources and retransmit them as an interleaved stream of AAL_SDU's out to the recipients. VC's are established to multicast by every endpoint that wishes to send multicast traffic. The paths out to the receivers must be established prior to packet transmission, and multicast server requires a mechanism to identify this receivers. Figure 2 depicts a MSM where each of the three nodes are both active sources and recipients. Quite a MMM network has the opportunity to optimize the path length from each source to its leaf nodes during call setup, potentially minimizing end-to-end latencies for each source's point-to-multipoint VC, the MSM has significant benefits when endpoints resource consumption and signaling load is considered.

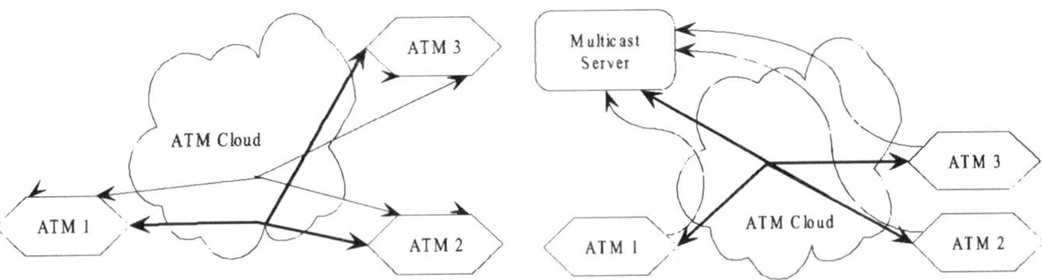

<table>
<tr><td>*Fig. 1. ATM Multicast Mesh Model*</td><td>*Fig. 2. ATM Multicast Server Model*</td></tr>
</table>

In environments where VC's across the UNI or within the ATM cloud are valuable the MSM provides value, as each participant node has at most one VC going out and VC coming in, by contrast with MMM where the number of VC's across the UNI increases linearly with the number of remote sources. MSM provide significantly lower group stability latencies compared to MMM. In addition, the signaling load placed on the ATM network by a group membership change is lower in MSM because in MMM this load is proportional to the number of active sources. As a conclusion, we consider that a Multicast Server is most suitable in UNI support applications. A solution for IP (Internet Protocol) multicasting based on MSM is presented in the next section.

3. Multicast Server Architecture USING UNI 4.0 Service

Our ATM system is a multilevel software architecture having as central software a Java Network Controller called JNC. The JNC knows a number of endpoints of different types (Network Nodes) which can be interconnected. The JNC performs the interconnection of endpoints by configuring the ATM components in an appropriate matter. In addition to the JNC a number of servers are part of the ATM system. The servers provide interfaces to clients that may want to use the ATM network. The UNI-Server, which serves UNI devices, is necessary when the switch devices in the ATM network do not use UNI signaling to set up connections. The UNI-Server provides a UNI signaling interface as documented by the ATM-Forum [2]and consists of six modules: the signaling module, the call handler module, the switch fabric module, the configuration module, the SNMP agent and the backup control module.

The UNI-Server receives requests for connection setup or connection release from the UNI interface. These requests are processed by the UNI-Server and are translated into requests to the JNC. The JNC provides a server interface which is used to setup or release connections in the ATM network. On request the JNC will perform the interconnection of the UNI endpoints. Therefore, transparently to the UNI device, the signaling information is forwarded to the network side signaling stack inside the UNI-Server. The UNI-Server has its own SNMP agent which provides an interface to the Network Management System used to request the status of ports and to manage the UNI connections. Due to the fact that the multicast server constitutes a single point of failure a redundant backup mechanism was installed for better reliability. Switching to the backup server will take place in less than 30 seconds.

4. The Congestion Prevention Algorithm

A closed-loop feedback is utilized to control the overall cell rate (OCR) of the video source. The source periodically transmits a resource management cell (i.e. a forward feedback cell, FFC), that is returned from the destination as a backward feedback cell, BFC. Usually, a single Congestion Indication (CI) bit of this cell indicates the congestion.. The feedback mechanism provide guaranteed service to all video traffic under the Minimum Cell Rate (MCR) and also try to rise up the speed to the Peak Cell Rate (PCR), making use of the unutilized bandwidth. The principle of the control mechanism is the following: for each $N=16$ video cells transmitted, the video source transmits a single FFC, by which the source probes two destinations to obtain the location of the most congested link in the multipoint connection (useful to adjust the PCR) and the congestion state of the entire connection (useful to adjust the OCR). When the source generates the FFC, it indicates the targets for both probes. Since each FFC contains two probe targets, in the return results two BFC, depending of the type of probe the destination is receiving. The feedback polling algorithm requires two algorithms, one to locate the most congested branch (LMCB) in the multicast tree, the other to observe the overall congestion state (OOCS) of this tree.

The LMCB algorithm has two states: SEARCH and PROBE. In the SEARCH state, the source probes each destination, one at a time. Whenever a FBC arrives at a destination, a PT1(Probe Target 1) field is examined and if PT1 matches with the destination's identification, then the destination generates a BFC and sets the LCI (Low-Priority Congestion Indication) bit accordingly. When the source receives a BFC with LCI=1 it enters in the PROBE state. The source probes the destination that generated the LCI and decrements its high PCR by an amount proportional to its current value. Only when the probed destination returns a BFC with LCI=0, the source returns to the SEARCH state. It is obvious that the LMCB algorithm reduces rapidly the high PCR whenever this cell is in danger of being lost, and increases the high PCR only after every destination has indicated a lack of low priority traffic congestion.

The OOCS algorithm observes the Explicit Forward Congestion Indicator (EFCI) bit in the header of the cell being served. When the destination is probed it generates a BFC and marks its CI bit if the last cell arrived to the destination had a marked EFCI. Upon receiving the BFC the source stores the cell's CI bit in a congestion stack that maintains the last N congestion indications received, where N is the number of destinations in the multipoint tree. Then the source calculates a Target Overall Cell Rate (TOCR) as:

$$TOCR = MCR + (N/N_0) \times (PCR - MCR)$$

where N_0 is the number of zeros in the congestion stack. If the current OCR is less (greater) than the target OCR, it is increased (decreased) proportionally. The minimum value (all the paths are congested) is TOCR=MCR. At the other extreme, when all the paths are uncongested, TOCR=PCR. Between these extremes, TOCR is set at an intermediate value corresponding linearly to the degree of congestion.

The adaptive encoding mechanism

The control of the OCR of the video encoder requires an adjustment of the encoder's quantization parameter Q. The video buffer is filled by the encoder and served at an output rate equal to the OCR. When the video buffer's occupancy changes, Q is adjusted to prevent buffer underflow (Q is increased) or underflow (Q is decreased). It is obvious that a decrease of Q leads to the increase of the combined output rate and image quality. The adjustment of the quantization parameter achieves control of the total output rate from the video encoder. In addition, data partitioning is utilized to control the output rate of low and high priority information. The method consists in the division of the encoded video bit stream into two layers with different priorities. The video bit stream appears like a sequence of run-level pairs. To split this stream into two, a priority breakpoint, that specify the number of run-level pairs per block to place into the high priority stream, is introduced. All remaining run-level pairs are placed into the low priority stream. The encoding algorithm ensures for every macroblock that is encoded and packed into ATM cells, the priority breakpoint is adjusted to produce the desired high priority cell rate.

5. Results and Discussions

The simulation model consists of seven ATM switches (S1...S7) interconnected in a binary tree topology. The root switch S1 is considered the video source of the multicast connection, S2 and S3 are placed as primary leaves, while the other four switches (S4...S7) are considered secondary level leaves, each of them having N_D destinations participating to the multipoint communication (4 N_D destinations in all). Links between adjacent switches are assigned a propagation delay D= 5k [μs], where k is the number of kilometers, taken from 2 to 500 in simulations. Each links capacity was considered 100 Mbps. As video data, a stream of 720 frames from a motion pictures was encoded using the proposed rate control and encoding algorithms. At the optimal quantization level (Q=1), for the standard motion picture frame rate of 24 Hz results a mean bit rate of 6.66 Mbps. Interfering traffic is generated according to a Poisson distribution with interarrival rate $\lambda= 6 \times 10^{-5}$ and inserted at each switch. A bottleneck link was simulated using packets with Poisson distribution interarrival rate $\lambda= 3\times10^{-5}$ with an average packet length of eight cells that means a total load of $\rho=0.98$ on a selected bottleneck links. ATM switches are assumed to be nonblocking and output-buffered, each output port having allocated 2000 cells of buffer space. Buffers are served on a first-come-first served discipline. For MCR=1500 cells-per-second (cps) , the worst case queuing delay is about 135 ms, calculated for the given topology with the heuristic formula: *3×(200cells/100Mbps) + (200cells/MCR)*. In the simulations PCR was chosen 20000 cps. The rate adjustment parameters AIR (additive increase rate) and RDF (rate decrease factor) are expressed in cps (AIR), respectively as unitless constant (RDF). In all simulations N was set to 16 and RDF to 1024.

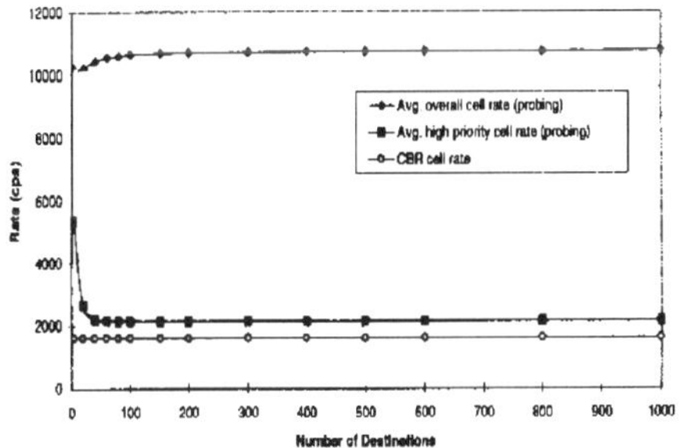

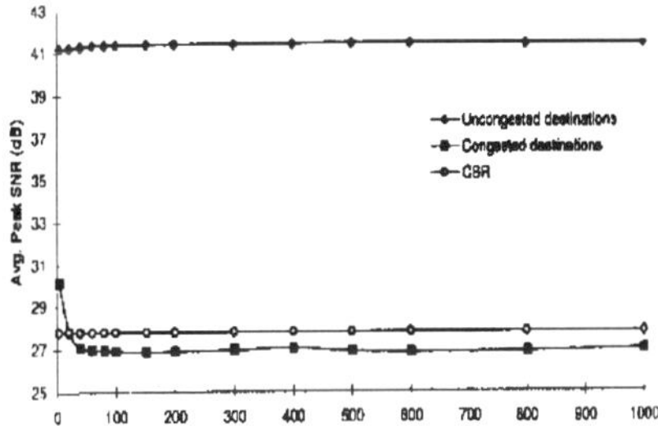

Fig. 3. Average video rate vs. destination number   *Fig. 4. Average peak SNR vs. destination number*

The following performances were tested through simulation:

-- The scalability with the number of destinations of the cell rate, by varying, the number of destinations in the multicast connection while holding all other parameters fixed. In fig.3 one can see that the feedback mechanism's average high PCR decreases rapidly and is dominated by MCR when the number of destinations is greater then 50, quite it ensures an average overall cell rate relatively constant as the number of destinations grows

- The scalability with the number of destinations of the quality of video, compared using as indication the average peak Signal to Noise Ratio (SNR). Fig.4 gives a measure of the quality of video received by two types of destinations participating to the feedback mechanism: congested and uncongested.

- The influence of the propagation delay, obtained with a constant N=100 and interfering high priority traffic introduced on the links from S2 to S4 and S5 (the first 50 destinations). The impact of the propagation delay is assessed by varying the inter-switch propagation delay from 10 μs to 2.5 ms and by measuring the impact on the video quality received by congested (SNR relatively constant 32 dB) and uncongested (SNR decease slowly from 42,3, to 41,8 dB) destinations When the propagation delay increases, the OCR is diminished due to slower feedback response

- The impact of the number of congested path, examined by the simulations of three degrees of congestion: 25% congestion on the link S2/S4, 50 % congestion on the links S2/S4and S2/ S5, 75 % congestion on the links S2/S4, S2/ S5 and S3/S6. Table 1 lists the most representative results.

Table 1: The influence of congestion paths over transmission performance

Congestion degree	Average overall cell rate (cps)	Average high priority cell rate (cps)	Average peak SNR to destinations (dB)
25%	13400	2080	40
50%	11000	2150	35
75%	7600	2060	30

6. Performance Evaluation & Concluding Remarks

By using the proposed feedback mechanism, associated with the encoding algorithm, the UNI-Server is able to handle a number of 500 UNI devices and to set up more than 40 connections per second during tests. The number of simultaneously active connections is more than 1000. Results from the simulations showed that the video feedback mechanism is capable of providing better quality than CBR service at the same amount of available bandwidth, can scale to a large number of destinations, offer an insignificant degree of reduction of the video quality when the propagation delay increases and can adapt to varying degrees of congestion in the network in order to improve video quality when possible.

REFERENCES

[1] ATM Forum User Network Interface (UNI) Specification, Version 3.1, Englewood Cliffs, NJ, Prentice Hall, 1995.

[2] Specification of UNI 4.0 Signaling, ATM Forum afsig-0061.000, 1996.

[3] G. J. Armitage, IP Multicasting over ATM Networks, IEEE Journal on Selected Areas in Communications, vol. 15, no. 3, 1997, pp. 445 - 457

[4] A.S. Sethi – A model for virtual tree bandwith allocation in ATM networks, IEEE INFOCOM, vol.3, no. 4, 1995, pp. 1222 – 1229

[5] R. Dobrescu, Daniela Andone, Anca Molnos, M.Guzu, E. Vitos - Server architecture for UNI signaling over ATM networks, to appear in Proceedings ITC 2001

Module-Based Neural Architectures: Some Effects by the Cooperation of Static and Dynamical Neurons

Kazuyoshi TSUTSUMI
Department of Mechanical and Systems Engineering
Ryukoku University, Otsu, Japan
(*E-mail: tsutsumi@rins.ryukoku.ac.jp*)

Abstract.   This paper introduces two models of module-based neural networks employing static and dynamical neurons, and discusses what effects can be produced due to the integration of mapping and relaxation. In the first model, relaxation is carried out in the internal space warped by mappers or mapping networks. This architecture has the possibility of helping the dynamics avoid inadequate minima. In the second model, two (or multiple) dynamical module-networks are cross-coupled via mapping internetworks. This architecture can produce more stable fixed-point attractors in the network dynamics. Based on some effects in the concrete models, the present paper further discusses the possibility of designing neural network models with more advanced functions via the integration of fundamental principles.

1. Introduction

Among the artificial neural network models proposed so far, let us compare a mapping-type network with static neurons and a relaxation-type network with dynamical neurons. In the learnable mapping-type network, the connection weights between neurons are modified so that the error energy estimating the input-output relationship is minimized [1]. In the relaxation-type network, on the other hand, the thermal energy or the electric power energy is minimized via the overall feedback, and the neuronal outputs themselves are parameters for the energy minimization [2]. The minimization processes in these two kinds of neural networks are apparently similar, but what parameters to be minimized or modified are different.

Each of these fundamental principles seems to have its own plausibility as a model of brain functioning. In this situation, should we understand that these different types of networks function independently in biological brains or neural ensambles? One possibility is the existence of higher mechanisms based on the consistent integration of the fundamental principles. Towards this direction, the author has proposed some integrated network-models with static and dynamical neurons, and has enhanced these models in terms of robotic control, optimization, associative memory, etc., in collaboration with his colleagues.

In the following sections, our two integrated models are introduced, and how mapping and relaxation relate, and what effects are produced through their integration, are discussed.

2. Relaxation in a Warped Internal Space

We assume the following energy function:

$$E \triangleq -\frac{1}{2}\sum_i\sum_j T_{ij}V_iV_j + \sum_i\frac{1}{r_i}\int_0^{V_i} g^{-1}(V)dV - \sum_i I_iV_i. \tag{1}$$

Differentiating it with respect to time, we obtain

$$\frac{dE}{dt} = \sum_i \frac{dV_i}{dt}\left[-\sum_j T_{ij}V_j + \frac{U_i}{r_i} - I_i \right]. \tag{2}$$

U_i and V_i are the internal state and the output of a neuron, respectively. Both are functions of time t, and are related to each other by the following input-output function:

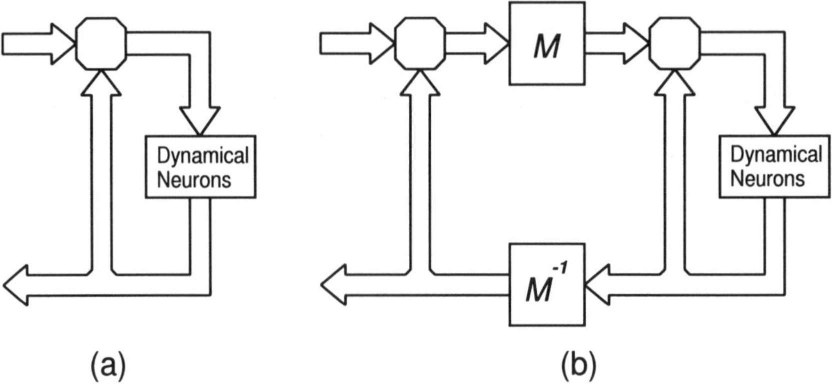

(a)   (b)

Figure 1: (a) Hopfield Network. (b) "Relaxation in a Warped Internal Space" Model.

$$V_i = g\left(U_i \right).$$ (3)

I_i is the electric current from the outside. r_i is the resistance. Here, we construct the following network with the use of capacitance c_i:

$$- c_i \frac{dU_i}{dt} = - \sum_j T_{ij} V_j + \frac{U_i}{r_i} - I_i$$ (4)

Figure 1(a) shows a schematic diagram of the network, which is known as the Hopfield network [3][4][5]. In this network, if the input-output function $g(\cdot)$ is monotonous, the time derivative of the energy function defined by Eq. (1) is kept non-positive, since

$$\frac{dE}{dt} = - \sum_i c_i \left(\frac{dV_i}{dt}\right)\left(\frac{dU_i}{dt}\right) = - \sum_i c_i \left\{ \frac{d}{dt} g(U_i) \right\} \left(\frac{dU_i}{dt}\right)^2 \le 0.$$ (5)

Therefore, the state of the network is ensured to always go to a minimum.

As is known well, however, the network dynamics includes local minima depending on such parameters as T_{ij}, I_i, etc. One method for avoiding these inadequate minima is to have the network relax in a warped internal space. By employing two static mappers M and M^{-1} with an inverse mapping relationship, we can design a network in which dynamical neurons are placed between M and M^{-1} as shown in Fig. 1(b). By simply locating a feedback loop between the two mappers, we can satisfy all of the constraints in the warped internal space. If there are some constraints to be satisfied in the regular external space, we merely place another feedback loop outside of the two mappers. Figure 1(b) shows such a generic architecture. The mappers can be represented by multi-layered neural networks, although it does not matter what structures the mappers have.

Here, we assume that V_i in the external space and $\tilde{V}_i$ in the internal space are mapped to each other at the voltage level as shown below:

$$\tilde{V}_i = M(V_i) \quad and \quad V_i = M^{-1}(\tilde{V}_i).$$ (6)

We also assume that these mappers function at the current level. When only the constraint for $T_{ij}^{(1)}$ is satisfied in the warped internal space and the constraint for $T_{ij}^{(2)}$ is satisfied in the external space, the behaviour of the network depicted in Fig. 1(b) is described by the following equation using only parameters $\tilde{U}_i$ in the internal space:

$$-c_i \frac{d\tilde{U}_i}{dt} = - \sum_j T_{ij}^{(1)} \tilde{V}_j + \frac{\tilde{U}_i}{r_i} + M\left[- \sum_j T_{ij}^{(2)} M^{-1}(\tilde{V}_j) - I_i \right].$$ (7)

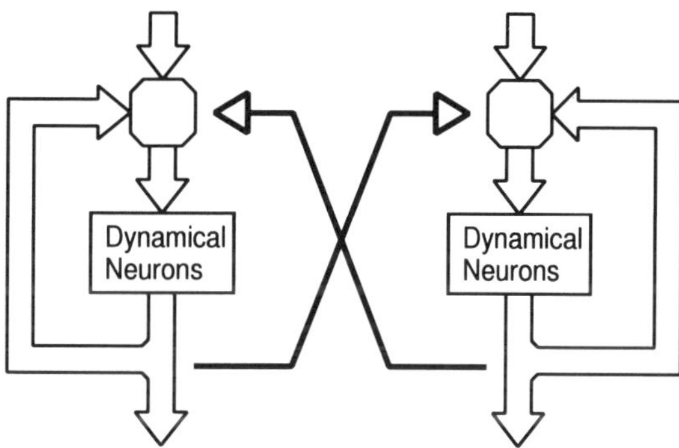

Figure 2: Cross-Coupled Hopfield Nets (CCHN).

Accordingly, the energy processed by the direct feedback loop is minimized based on the measure in the warped internal space, and the energy processed by the indirect feedback loop is minimized based on the measure in the non-warped external space [6][7].

3. Cooperative Relaxation with Module Networks

We assume that two Hopfield networks are cross-coupled according to the conceptual diagram shown in Fig. 2. The output of one of the Hopfield networks is connected to the input of the other, and vice versa. We define the following energy so that it should be minimized and stabilized when the two Hopfield networks relax into their own stable states via the cross-coupling.

$$
\begin{aligned}
E \;=\; & E^{(1)} + E^{(2)} + E^{(cross)} \\
=\; & -\frac{1}{2}\sum_i\sum_j T_{ij}^{(1)} V_i^{(1)} V_j^{(1)} \;+\; \sum_i \frac{1}{r_i^{(1)}} \int_0^{V_i^{(1)}} g^{-1}(V)\,dV \;-\; \sum_i I_i^{(1)} V_i^{(1)} \\
& -\frac{1}{2}\sum_i\sum_j T_{ij}^{(2)} V_i^{(2)} V_j^{(2)} \;+\; \sum_i \frac{1}{r_i^{(2)}} \int_0^{V_i^{(2)}} g^{-1}(V)\,dV \;-\; \sum_i I_i^{(2)} V_i^{(2)} \\
& +\frac{1}{2}\sum_i \left(V_i^{(1)} - O_i^{(2)} \right)^2 \;+\; \frac{1}{2}\sum_i \left(V_i^{(2)} - O_i^{(1)} \right)^2
\end{aligned}
\tag{8}
$$

We assume two-layered linear internetworks without any nonlinear hidden units. Therefore,

$$
O_i^{(\ell)} \;=\; \sum_j W_{ij}^{(\ell)} V_j^{(\ell)} \qquad (\ell = 1, 2) \;.
\tag{9}
$$

From Eqs. (8)(9), the time derivative of the energy function is calculated as follows:

$$
\frac{dE}{dt} =
$$

$$
\sum_i \frac{dV_i^{(1)}}{dt} \left[- \sum_j T_{ij}^{(1)} V_j^{(1)} + \frac{U_i^{(1)}}{r_i} - I_i^{(1)} + \left(V_i^{(1)} - \sum_j W_{ij}^{(2)} V_j^{(2)} \right) - \sum_j W_{ji}^{(1)} \left(V_j^{(2)} - \sum_k W_{jk}^{(1)} V_k^{(1)} \right) \right]
$$

$$
+\sum_i \frac{dV_i^{(2)}}{dt} \left[- \sum_j T_{ij}^{(2)} V_j^{(2)} + \frac{U_i^{(2)}}{r_i} - I_i^{(2)} + \left(V_i^{(2)} - \sum_j W_{ij}^{(1)} V_j^{(1)} \right) - \sum_j W_{ji}^{(2)} \left(V_j^{(1)} - \sum_k W_{jk}^{(2)} V_k^{(2)} \right) \right]
$$

$$
-\sum_i\sum_j \frac{dW_{ij}^{(1)}}{dt} \left[V_j^{(1)} \left(V_i^{(2)} - \sum_k W_{ik}^{(1)} V_k^{(1)} \right) \right]
$$

$$
-\sum_i\sum_j \frac{dW_{ij}^{(2)}}{dt} \left[V_j^{(2)} \left(V_i^{(1)} - \sum_k W_{ik}^{(2)} V_k^{(2)} \right) \right] \;.
\tag{10}
$$

Here, we define $\delta_i^{(1)}$ and $\delta_i^{(2)}$ as follows:

$$\delta_i^{(1)} = V_i^{(2)} - \sum_j W_{ij}^{(1)} V_j^{(1)} \quad \text{and} \quad \delta_i^{(2)} = V_i^{(1)} - \sum_j W_{ij}^{(2)} V_j^{(2)} . \tag{11}$$

If the following network is constructed

$$- c_i^{(1)} \frac{dU_i^{(1)}}{dt} = - \sum_j T_{ij}^{(1)} V_j^{(1)} + \frac{U_i^{(1)}}{r_i^{(1)}} - I_i^{(1)} + \left(V_i^{(1)} - O_i^{(2)} \right) - \sum_j W_{ji}^{(1)} \delta_j^{(1)} \tag{12}$$

$$- c_i^{(2)} \frac{dU_i^{(2)}}{dt} = - \sum_j T_{ij}^{(2)} V_j^{(2)} + \frac{U_i^{(2)}}{r_i^{(2)}} - I_i^{(2)} + \left(V_i^{(2)} - O_i^{(1)} \right) - \sum_j W_{ji}^{(2)} \delta_j^{(2)} \tag{13}$$

and the connection weights of the internetworks are modified according to the following rules,

$$\eta^{(1)} \frac{dW_{ij}^{(1)}}{dt} = V_j^{(1)} \delta_i^{(1)} \tag{14}$$

$$\eta^{(2)} \frac{dW_{ij}^{(2)}}{dt} = V_j^{(2)} \delta_i^{(2)} \tag{15}$$

then Eq. (10) becomes Eq. (16) after substituting Eqs. (11) - (15) for Eq. (10). $\eta^{(1)}$ and $\eta^{(2)}$ represent constants related to the learning speed.

$$\frac{dE}{dt} = - \sum_i c_i^{(1)} \left\{ \frac{d}{dt} g(U_i^{(1)}) \right\} \left(\frac{dU_i^{(1)}}{dt} \right)^2 - \sum_i c_i^{(2)} \left\{ \frac{d}{dt} g(U_i^{(2)}) \right\} \left(\frac{dU_i^{(2)}}{dt} \right)^2$$

$$- \sum_i \sum_j \eta^{(1)} \left(\frac{dW_{ij}^{(1)}}{dt} \right)^2 - \sum_i \sum_j \eta^{(2)} \left(\frac{dW_{ij}^{(2)}}{dt} \right)^2 \leq 0 \tag{16}$$

Therefore, the energy defined by Eq. (8) can be minimized by employing the obtained network architecture and the delta learning rule. In other words, from a Lyapunov function for cross-coupled Hopfield networks, it can be derived that the algorithm necessary for training the two-layered linear internetworks arrives at the delta rule.

If non-linear hidden neurons are assumed in the internetworks, the generalized delta rule can be derived as a learning algorithm for internetworks based on a similar formulation [8][9]. Such a model is named CCHN (Cross-Coupled Hopfield Nets). We can easily expand the model into an architecture that connects multiple Hopfield networks via multiple internetworks [10].

4. Discussion

In the first model, an external space is converted into a warped internal space due to the mapping ability of mappers, and the energy is minimized based on the measure in the internal space. In an example involving a robotic motion planning application, obstacle avoidance was shown to be elegantly actualized [7].

The second model is basically a cross-coupling of Hopfield networks. However, the finally obtained module-based network comprises some non-trivial mechanisms. Each internetwork for the coupling is composed of forward and backward sub-networks with the same connection weights. The errors for calculating the delta values (the inputs to the backward sub-networks) are sent to the inputs of Hopfield networks. The final outputs of the backward sub-networks are also returned to the inputs of these Hopfield networks. Although the backward sub-networks for calculating the delta values are usually assumed as virtual ones, they must really exist for network relaxation. In an example of applying the second model to associative memories, the network had an excellent performance compared with conventional models [10].

The two models introduced here share somthing in common in the sense that mapping (static neurons) and relaxation (dynamical neurons) are integrated. However, it should be noted that the derived architectures and the yielding effects are completely different. In the present paper, the fundamental principles are limited to mapping and relaxation. However, in addition to these two, some frameworks can be considered as other fundamental principles. Accordingly, a greater variety of combinations in terms of integration are expected to yield more advanced mechanisms and to offer more models of brain functioning.

This begs further questioning: Provided each of module-based neural network models introduce here has its own plausibility, should we understand that these different types of "integrated" networks function separately in neural ensambles? One possibility is, in the same way as with more fundamental relationships, that there are more advanced frameworks in which integrated models are consistently unified. Some studies have been done so far on approaches based on integration [11][12][13]. However, there remains a lot to be done. The author believes that "Network Integration" or "Algorithm Integration" approaches will be quite significant.

References

[1] D. E. Rumelhart, G. E. Hinton, and R. J. Williams, Learning Internal Representation by Error Propagation, Parallel Distributed Processing Vol.1, The MIT Press, 1986, pp.318-362

[2] J. J. Hopfield, Neurons with Graded Response Have Collective Computational Properties Like Those of Two-State Neurons, Proc. of the National Academy of Science, U.S.A., **Vol.81**, pp.3088-3092, 1984

[3] J. J. Hopfield & D. W. Tank, 'Neural' Computation of Decisions in Optimization Problems, Biological Cybernetics, **Vol.52**, pp.141-152, 1985

[4] G. A. Carpenter, M. A. Cohen, & S. Grossberg; T. Kohonen & E. Oja; G. Palm; J. J. Hopfield & D. W. Tank, Technical Comments: Computing with Neural Networks, Science, **Vol.235**, pp.1226-1229, 1987

[5] D. W. Tank & J. J. Hopfield, Simple 'Neural' Optimization Networks: An A/D Converter, Signal Decision Circuit, and a Linear Programming Circuit, IEEE Trans. on Circuit and Systems, **Vol.33**, pp.533-541, 1986

[6] K. Tsutsumi & H. Matsumoto, Neural Computation and Learning Strategy for Manipulator Position Control, Proc. of the IEEE First Annual International Conference on Neural Networks, San Diego, **Vol.4**, pp.525-534, 1987.

[7] K. Tsutsumi, A Multi-Layered Neural Network Composed of Back-Prop. and Hopfield Nets and Internal Space Representation, Proc. of International Joint Conference on Neural Networks, Washington, D. C., **Vol.2**, pp.365-371, 1989

[8] K. Tsutsumi, Cross-Coupled Hopfield Nets via Generalized-Delta-Rule-Based Internetworks, Proc. of International Joint Conference on Neural Networks, San Diego, **Vol.2**, pp.259-265, 1990

[9] K. Tsutsumi, Higher Degree Error Backpropagation in Cross-Coupled Hopfield Nets, Proc. of International Joint Conference on Neural Networks, Seattle, **Vol.2**, pp.349-355, 1991

[10] S. Ozawa, K. Tsutsumi, & N. Baba, An Artificial Modular Neural Network and Its Basic Dynamical Characteristics, Biological Cybernetics, **Vol.78**, pp.19-36, 1998

[11] R. Hecht-Nielsen, Counterpropagation Networks, Proc. of IEEE First International Conference on Neural Networks, San Diego, **Vol.2**, pp.19-32, 1987

[12] B. Kosko, Bidirectional Associative Memories, Trans. on Systems, Man, and Cybernetics, **Vol.18**, pp.49-60, 1988

[13] K. Tsutsumi, On Redundancy in Neural Architecture: Dynamics of a Simple Module-Based Neural Network and Initial-State Independence, Neural Networks, **Vol.12**, pp.1075-1085, 1999

KES '01
N. Baba et al. (Eds.)
IOS Press, 2001

A Gaussian Zero-Crossing Discriminant Function for Min-Max Modular Neural Networks

Bao-Liang Lu and Michinori Ichikawa
Lab. for Brain-Operative Device, RIKEN Brain Science Institute
2-1 Hirosawa, Wako-shi, Saitama 351-0198, Japan
{lu;ichikawa}@brainway.riken.go.jp

Abstract. This paper presents a new nonlinear discriminant function called *Gaussian zero-crossing* (GZC) function for min-max modular (M^3) neural networks. The GZC function is designed for solving linearly separable problems, each of which contains only two different data. The GZC function has two different receptive field centers and same receptive field width. Both the receptive field centers and the receptive field width are directly determined by two given training data belonging to two classes. Similar to conventional Gaussian function, the GZC function has locally tuned response characteristics. The objective of introducing the GZC function to M^3 networks is to directly incorporate inherent class relations among training data to emerge M^3 networks.

1 Introduction

In our previous work [2], we have proposed an alternative modular neural network model called *min-max modular* (M^3) neural network for pattern classification. The M^3 network involves three mechanisms: decomposition of a K-class pattern classification problem into a reasonable number of completely independent two-class subproblems as small as the user needs; parallel learning of all of the subproblems by using smaller network modules; and integration of the trained individual network modules into an M^3 network.

By using the proposed task decomposition method [2], a K-class problem can be uniquely divided into a reasonable number of linearly separable problems, each of which contains only two different training data belonging to two classes. Therefore, the task of learning a K-class problem can be easily performed by learning all of the linearly separable problems in parallel. It is well known that a linearly separable problem can be solved by using a linear discriminant function [1] which divides the feature space by a hyperplane decision surface. A fatal weakness of existing linear discriminant functions such as Perceptrons, however, is that they lack locally tuned response characteristics. This deficiency may lead classifiers to produce proper output even when an unknown input is presented.

To overcome the weakness of existing linear discriminant functions, we present a new nonlinear discriminant function called *Guanssian zero-crossing* function. By using the GZC function, an M^3 network for solving a K-class pattern classification problem can be easily designed and constructed according to two simple emergent laws [2]. The M^3 network based on the GZC function has two attractive features. a) very fast learning without error can be guaranteed to any complex pattern classification problems, and b) the entire M^3 network completely maintains the locally tuned response characteristics of the GZC function.

2 Gaussian Zero-Crossing Discriminant Function

The proposed Gaussian zero-crossing discriminant function is defined by

$$f_{ij}(x) = \exp\left[-\left(\frac{\|x - c_i\|}{\sigma}\right)^2\right] - \exp\left[-\left(\frac{\|x - c_j\|}{\sigma}\right)^2\right] \tag{1}$$

where $x \in \mathbf{R}^n$ is the input vector, $c_i \in \mathbf{R}^n$ and $c_j \in \mathbf{R}^n$ are the given training inputs belonging to class $\mathcal{C}_i$ and class $\mathcal{C}_j$ ($i \neq j$), respectively, and are used as two different receptive field centers, $\sigma = \lambda\|c_j - c_i\|$ is the receptive field width, λ is a user-defined constant $(0 < \lambda)$, and the norm $\|z\|$ is the Euclidean norm of vector z. Figure 1(a) shows the GZC function for two-dimensional input.

By using the GZC discriminant function, the output for the GZC classifier is given by

$$g(x) = \begin{cases} 1 & \text{if } f_{ij}(x) > \theta^+ \\ \text{Unknown} & \text{if } \theta^- \leq f_{ij}(x) \leq \theta^+ \\ -1 & \text{if } f_{ij}(x) < \theta^- \end{cases} \tag{2}$$

where θ^+ and θ^- are the upper and lower threshold limits, respectively.

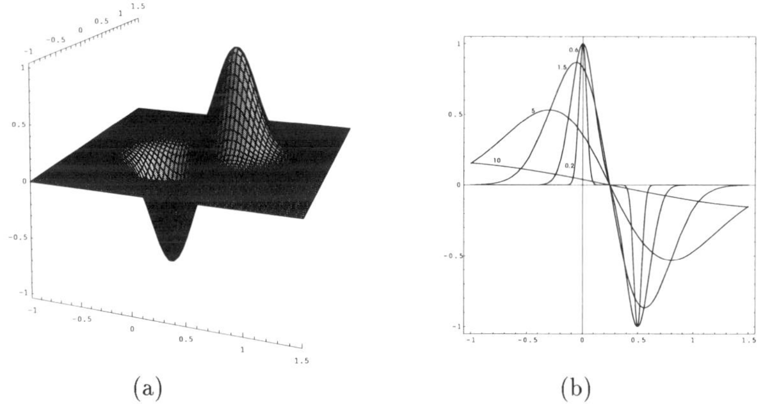

(a)   (b)

Figure 1: (a) The Gaussian zero-crossing discriminant function for two-dimensional input, where $c_i = [0.5, 0.5]^T$, $c_j = [0, 0]^T$, and $\lambda = 0.8$. (b) Variation of the receptive field width with different λ, where $c_i = 0$, $c_j = 0.5$, and $\lambda = 0.2$, 0.6, 1.5, 5, and 10.

The inverse of the GZC function is defined as follows:

$$\bar{f}_{ij}(x) = -f_{ij}(x) = \exp\left[-\left(\frac{\|x - c_j\|}{\sigma}\right)^2\right] - \exp\left[-\left(\frac{\|x - c_i\|}{\sigma}\right)^2\right] \tag{3}$$

From (1) and (3), we can see that the inverse of the GZC function preserves completely the locally tuned response characteristics of the GZC function and satisfies the following relation:

$$\bar{f}_{ij}(x) = -f_{ij}(x) = f_{ji}(x) \tag{4}$$

The above relation is very useful for reducing the number of modules of M^3 networks because $f_{ji}(x)$ can be easily implemented by using the inverse of $f_{ij}(x)$, instead of creating a new discriminant function.

From (1), we see that the receptive field width of the GZC function is determined by two factors: a use-defined constant λ and the distance between two different receptive field centers. Fig. 1(b) illustrates the relationship between the receptive field width and various λ under the same receptive field centers.

3   Problem Decomposition

Let T be the training set for a K-class classification problem,

$$T = \{(X_l, D_l)\}_{l=1}^{L}, \tag{5}$$

where $X_l \in \mathbf{R}^n$ is the input vector, $D_l \in \mathbf{R}^K$ is the desired output, and L is the total number of training data.

We have suggested [2] that a K-class problem defined by (5) can be uniquely divided into

$$\sum_{i=1}^{K} \sum_{\substack{j=1 \\ j \neq i}}^{K} L_i \times L_j \tag{6}$$

linearly separable problems (LSPs), where L_i denotes the number of data belonging to class C_i for $i = 1, \cdots, K$, and $\sum_{i=1}^{K} L_i = L$. The training set for each of the LSPs is given by

$$T_{ij}^{(u,v)} = \{(X^{(iu)}, 1 - \epsilon) \cup (X^{(jv)}, \epsilon)\} \tag{7}$$

where $X^{(iu)}$ and $X^{(jv)}$ are the training inputs belonging to class C_i and class C_j, respectively, ϵ is a small real positive number (e.g., ϵ is set to 0.1), $u = 1, \cdots, L_i$, $v = 1, \cdots, L_j$, $i = 1, \cdots, K$, $j = 1, \cdots, K$, and $i \neq j$.

From pattern classification point of view, $T_{ij}^{(u,v)}$ and $T_{ji}^{(v,u)}$ for $i \neq j$ are the same problem because the difference between $T_{ij}^{(u,v)}$ and $T_{ji}^{(v,u)}$ is only their opposite desired outputs. Suppose that $T_{ij}^{(u,v)}$ has been learned by using a module $M_{ij}^{(u,v)}$ correctly. If we need to learn $T_{ji}^{(v,u)}$, we can get a solution to $T_{ji}^{(v,u)}$ by using the inverse of the output of $M_{ij}^{(u,v)}$, instead of learning $T_{ji}^{(v,u)}$ by using a new module $M_{ji}^{(v,u)}$. The relationship among $M_{ij}^{(u,v)}$, $M_{ji}^{(v,u)}$, and the **INV** unit is given by

$$M_{ji}^{(v,u)} = \mathbf{INV}(M_{ij}^{(u,v)}) \tag{8}$$

where the function of the **INV** unit is to invert its single input. The transfer function of the **INV** unit is defined by

$$q = \alpha + \beta - p \tag{9}$$

where α, β, p, and q are the upper and lower limits of input value, input, and output, respectively. For the GZC function, the values of α and β are 1 and -1, respectively.

According to the above discussion, we see that the number of LSPs that need to be learned to solve a K-class problem can be reduced to

$$\sum_{i=1}^{K} \sum_{j=i+1}^{K} L_i \times L_j \tag{10}$$

4 Emerging M^3 Networks

Suppose that each of the LSPs defined by (7) is solved by a GZC discriminant function. The total number of different GZC discriminant functions for solving a K-class problem is equal to the number of LSPs defined by (10). When λ for each of the GZC discriminant functions is determined, the task of learning a K-class problem is to integrate all of the GZC discriminant functions into an M^3 network according to the following two simple emergent laws [2].

Minimization principle

Suppose a two-class problem $\mathcal{B}$ is divided into P LSPs, $\mathcal{B}_i$ for $i = 1, \cdots, P$, and also suppose that all the LSPs have the same positive training data and different negative training data[1]. If the P subproblems are correctly learned by the corresponding P individual GZC discriminant functions, M_i for $i = 1, \cdots, P$, then the combination of the P GZC discriminant functions with a **MIN** unit produces the correct output for all the training inputs in $\mathcal{B}$, where the function of the **MIN** unit is to find a minimum value from its multiple inputs.

Maximization principle

Suppose a two-class problem $\mathcal{B}$ is divided into P LSPs, $\mathcal{B}_i$ for $i = 1, \cdots, P$, and also suppose that all the subproblems have the same negative training data and different positive training data. If the P subproblems are correctly learned by the corresponding P GZC discriminant functions, M_i for $i = 1, \cdots, P$, then the combination of the P GZC discriminant functions with a **MAX** unit produces the correct output for all the training input in $\mathcal{B}$, where the function of the **MAX** unit is to find a maximum value from its multiple inputs.

5 Computer Simulations

To demonstrate the performance of the proposed GZC discriminant function, a simple two-class problem [3] illustrated in Fig. 2(a) is simulated. From (10), we see that 24 ($L_1 \times L_2 = 6 \times 4$) GZC discriminant functions are required. In the simulation, λ for all of the GZC functions is set to 0.6. The M^3 network for the two-class problem is shown in Fig. 3. The input-output mapping and its contour plot formed by the M^3 network are shown in Fig. 2(b) and 2(c), respectively. Various decision boundaries produced by the M^3 network under different threshold limits are illustrated in Fig. 4.

Examining Figs. 4(a), 4(b), and 4(c), we can see that the most important feature of the GZC function is that the proper generalization performance of the M^3 networks can be easily controlled by adjusting the threshold limits θ^+ and θ^-.

6 Conclusions

We have presented a new Gaussian-like nonlinear discriminant function for pattern classification. We have demonstrated that the min-max modular network based on the proposed discriminant function has locally tuned response characteristics and its proper generalization performance can be easily controlled by adjusting the corresponding threshold limits.

---

[1]If the desired output of a training data is $1 - \epsilon$, then the training data is called *positive* training data. Otherwise, *negative* training data.

References

[1] R. O. Duda, P. E. Hart, and D. G. Stork: *Pattern Classification*, 2nd Ed. John Wiley & Sons, Inc. 2001.

[2] B. L. Lu and M. Ito: Task decomposition and module combination based on class relations: a modular neural network for pattern classification, *IEEE Trans. Neural Networks*, vol. 10, no. 5, pp. 1244-1256, 1999.

[3] B. L. Lu and M. Ichikawa, Emergent on-line learning in min-max modular neural networks, to appear in *Proc. of IJCNN'01*, Washington, D.C., July 14-19, 2001.

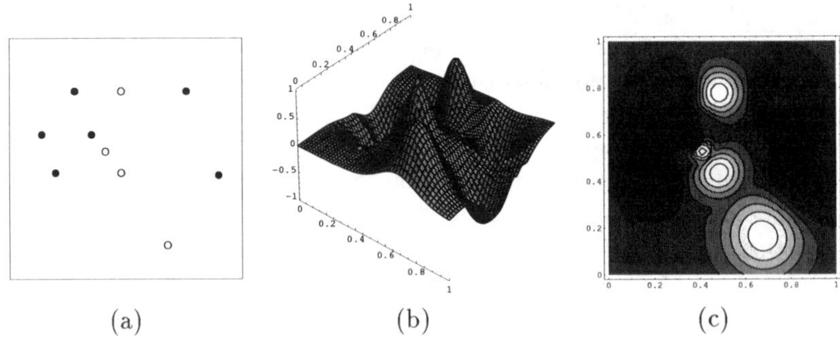

(a)   (b)   (c)

Figure 2: (a) Ten random training inputs for a two-class problem, where point and small open circle represent the inputs whose desired outputs are '0' and '1', respectively. (b) The input-output mapping formed by the M^3 network for solving the two-class problem, (c) The contour plot of the input-output mapping.

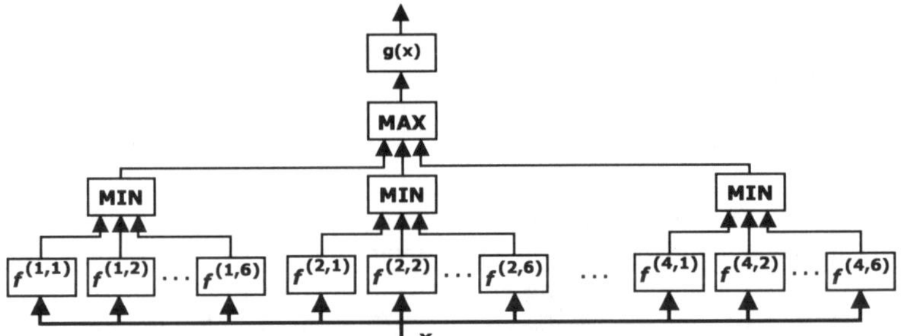

Figure 3: The M^3 network for solving the two-class problem.

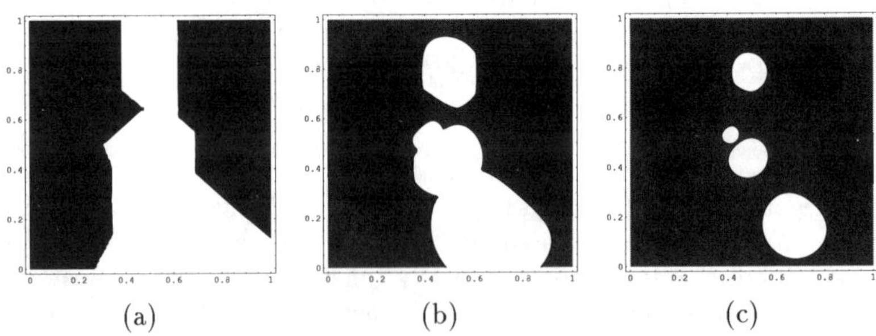

(a)   (b)   (c)

Figure 4: Various decision boundaries. (a) $\theta^+ = 1 \times 10^{-7}$ and $\theta^- = 0$; (b) $\theta^+ = 0.1$ and $\theta^- = -0.1$; (c) $\theta^+ = 0.5$ and $\theta^- = -0.5$. The gray denotes unknown decision regions.

A Study of Brain-like Neural Network Model

Qingyu XIONG, Kotaro HIRASAWA, Jinglu HU, and Junichi MURATA
Intelligent Control Laboratory,
Graduate School of Information Science & Electrical Engineering,
Kyushu University, Japan 812-8581

Abstract. In this paper, a brain-like neural network model is studied, which consists of basic networks and branch control networks. The branch control network can be used to determine which intermediate nodes of the basic network should be connected to the output node with a coefficient of relative strength ranging from zero to one. This determination will adjust the outputs of the intermediate nodes of the basic network. Therefore, by using such a construction, locally functions distributed networks can be realized depending on the values of the inputs of the network. This model is applied to a function approximating problem. The simulation results show that such brain-like model exhibits better performance than the conventional networks with comparable complexity.

1 Introduction

In human's brain, it has been recognized that there are many special intellectual parts related to mathematics knowledge, music knowledge, gymnastic knowledge and so on, and each intellectual part is activated depending on the action of each knowledge. Even in the activated part, the different region is activated with different intensity. This is called "functions are locally distributed in the brain"[1].

But commonly used neural networks, especially layered neural networks, have such a structure that all input nodes and output nodes are connected to the intermediate nodes. Therefore, functions distribution of networks can not be realized in the conventional networks. Here, functions distribution means that only parts of the networks are activated depending on the input values of the network.

In this paper, a brain-like neural network model is proposed using Universal Learning Networks with Branch Control of Relative Strength (ULNs with BR) and it is studied how much ULNs with BR can improve the performance of the networks compared to the conventional networks. This fact has been already verified by using Learning Petri Network [2]-[5] developed in our laboratory.

The remainder of this paper is organized as follows: In Section 2, we present the basic structure and training algorithm of ULNs with BR. In Section 3, we give simulation conditions and simulation results for a function approximating problem Finally, conclusions are presented.

2 Structure of ULNs with BR

2.1 Basic Structure

ULNs with BR is composed of two kinds of networks (see Fig.1). One is a basic network. The other one is a branch control network, which controls the branches from the intermediate nodes to the output node of the basic network in a way that the outputs of the intermediate nodes are multiplied by the coefficients ranging from zero to one, which is calculated by the branch control network. The relationship between the outputs of the branch control network and the branches from the intermediate nodes to the output node in the basic network is one to one correspondence. As a result, the outputs of the branch control network can control the connection of the corresponding intermediate branches of the basic network with a coefficient of relative strength. When the coefficient R_j in Fig.1 takes zero value, the corresponding branch is deleted from the basic network, otherwise, the corresponding output of the intermediate node is multiplied by non-zero coefficient R_j.

In this paper, a commonly used three layered neural network is used as the basic network, while a fuzzy inference network is used as the branch control network, because the localized property of fuzzy inference networks is used to execute the branch control effectively.

- Neural Network

Supposing a static network whose inputs are h_i $(i \in N)$ and output is h_o, then, h_o can be described as follows,

$$h_o = \sum_{j \in JF(o)} \omega_{jo} h_j R_j(Z_j) + \theta_o, \tag{1}$$

$$h_j = f_j(\alpha_j) = z \frac{1 - e^{-\phi_j \alpha_j}}{1 + e^{-\phi_j \alpha_j}}, \tag{2}$$

$$\alpha_j = \sum_{i \in JF(j)} \omega_{ij} h_i + \theta_j, \tag{3}$$

where

$JF(j)$	:	set of suffixes of nodes which are connected to node j,
h_j	:	output value of node j,
f_j	:	nonlinear function of node j,
ω_{ij}	:	weight parameter from node i to node j,
θ_j	:	threshold parameter of node j,
ϕ_j	:	slope parameter of node j.

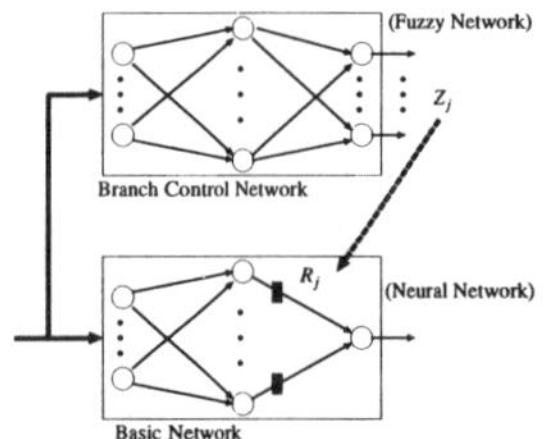

Figure 1: Structure of ULNs with BR

Figure 2: Relative strength R_j calculated by Z_j

$R_j(Z_j) \in [0, 1]$ is the coefficient between the intermediate node j and the output node, which depends on the output Z_j of the fuzzy inference network.

As being stated before, the functions distribution can be realized in ULNs with BR by appropriately calculating $R_j(Z_j)$.

- Fuzzy Inference Network

In this paper, the following fuzzy inference network using Gaussian membership functions is used.

$$Z_j = \frac{\sum_{q \in Q} [f_{hq}(h) \sigma_{jq} \mu_{jq}]}{\sum_{q \in Q} [f_{hq}(h) \sigma_{jq}]}, \tag{4}$$

$$f_{hq}(h) = exp[-\frac{(h - \mu_{hq})^T \sum_{hq}^{-1} (h - \mu_{hq})}{2}], \tag{5}$$

$$f_{jq}(Z_j) = exp[-\frac{(Z_j - \mu_{jq})^2}{2\sigma_{jq}^2}], \tag{6}$$

where

Z_j	:	the output of the fuzzy inference network corresponding to the intermediate node j of the basic network,
$f_{hq}(h)$	:	membership function of the IF part of the qth fuzzy rule,
$f_{jq}(Z_j)$	:	membership function of the $THEN$ part of the qth fuzzy rule,
h	=	$(h_1, h_2, \cdots, h_N)$, input vector,
μ_{hq}	:	vector representing the mean of the IF part of the qth fuzzy rule,
Σ_{hq}^{-1}	:	matrix representing the variance of the IF part of the qth fuzzy rule,
μ_{jq}	:	parameter representing the mean of the $THEN$ part of the qth fuzzy rule, $\mu_{jq} \in [0, 1]$,
σ_{jq}^2	:	parameter representing the variance of the $THEN$ part of the qth fuzzy rule,
Q	:	set of suffixes of fuzzy rules.

The most important coefficient R_j(Relative Strength) is calculated by Eq.(7).

$$R_j(Z_j) = \begin{cases} (\frac{Z_j - Z_j^o}{1 - Z_j^o})^n & \text{if } Z_j \geq Z_j^o \\ 0 & \text{if } Z_j < Z_j^o, \end{cases} \tag{7}$$

where

Z_j^o	:	threshold value to determine whether the intermediate node j is connected to the output node of the basic network;

Therefore, if $Z_j < Z_j^o$, then the connection between the intermediate node j and output node of the basic network is cut, as a result, a functions distributed network is realized. On the other hand, if $Z_j \geq Z_j^o$, then the nodes outputs adjustment is done, that is, the output h_j of node j is multiplied by $R_j(Z_j)$ as shown in Eq.(1).

By adjusting n, we can get different shape of lines about relative strength (see Fig.2) and study what important role the multiplication plays to improve the performance of ULNs with BR.

One of the reasons of using the fuzzy inference network as the branch control network is that neighboring points in the input space should be mapped also to the neighboring points in the output space in the branch control network in order to execute the branch control effectively.

Therefore, by adopting the fuzzy inference network as the branch control network with a coefficient of relative strength, a functions locally distributed network can be realized depending on the input values of the basic network.

The following branch connectivity ρ of the basic network can be adjusted by setting the threshold Z_j^o to an appropriate value:

$$\rho = \frac{l}{L} \tag{8}$$

where l is the number of the connected branches with a coefficient of relative strength between the intermediate nodes and output node of the basic network, and L ($i.e.|JF(o)|$ in Eq.(1)) is the total number of hidden branches of the basic network. The branch connectivity ρ is increased when Z_j^o takes a small value, and vice versa.

2.2 Training of ULNs with BR

As for training of ULNs with BR, parameters of the basic network could be trained in the same way as the commonly used neural networks with supervised learning.

But, as branch connection with a coefficient of relative strength or disconnection of the basic network of ULNs with BR depends on the input values of the network, it is difficult to train the parameters by using the gradient method like back propagation algorithm. So, a kind of random search method named RasID [6] was used to train the parameters of the basic network.

Furthermore, it is noticeable that even though the parameters of ULNs with BR are trained in the following way, ULNs with BR shows distinguished performances: (1) only the parameters of the basic network are trained, (2) the parameters of IF parts of the fuzzy rules such as μ_{hq} and $\sum_{hq}^{-1}$ are set to appropriate values in advance, and the parameters of $THEN$ parts of the fuzzy rules like μ_{jq} are randomly set in $[0, 1]$.

3 Simulations of ULNs with BR

The parameters of the membership functions of the IF parts of the qth fuzzy rules $f_{hq}(h)$ are fixed in such a way that the input space of the network is appropriately covered with 5 fuzzy rules. Here, two dimensional input space is supposed.

$$\mu_{hq} \quad = \quad (0.25, 0.25),\ (0.25, 0.75),\ (0.75, 0.25),\ (0.75, 0.75),\ (0.5, 0.5)$$

$\sum_{hq}^{-1}$ is a diagonal matrix with its element being $\frac{1}{0.03}$, σ_{jq} is equal to 1.0, and μ_{jq} is a random value in $[0, 1]$, which will be fixed during the whole training process . Therefore, it is clear from Eq.(4)-(6) that $0 \le Z_j \le 1$ holds.

As being stated before, it is a distinguished point that ULNs with BR is expected to have higher performance and generalization ability than commonly used neural networks even when $\mu_{jq} \in [0, 1]$ is randomly set.

Simulation is executed by using the following two dimensional function approximation problem which are defined on $0 \le x \le 1, 0 \le y \le 1$.

$$f(x, y) = 0.158((1.5(1-x)+e^{2x-1}sin(3.0\pi(x-0.6)^2)+e^{3(y-0.5)}sin(4\pi(x-0.9)^2))-3)$$

The basic network is a three layered neural network with the number of the intermediate nodes being 20.

Another parameters in the basic network are initialized with random numbers.

Tab.1, Fig.3 and 4 show how the performance and the generalization ability of the ULN with BC change as the connectivity ρ varies by adjusting the threshold Z_j^o using different n (They are the average results over 10 times for each threshold value).Here, ONN in Tab.1 corresponds to an ordinary neural network, and the threshold value Z_j^o ranging from 0.0 to 0.5 corresponds to ULN with BR, i.e., the relative strength $R_j(Z_j)$ is calculated by using Eq.(7).

From our results, we can see clearly that when $n = 1$, ULNs with BR owns the best performance and generalization ability. The reason may be as follows. When n in Eq.(7) gets larger and larger, $R_j(Z_j)$ will get smaller and smaller. After multiplying $R_j(Z_j)$ to the outputs of intermediate nodes in the basic network, the intermediate branches seem to be disconnected from the output node even though they are actually connected. As a result, because of no enough routes for training, the performance will get worse. In another case, if n gets smaller and smaller, ULNs with BR becomes like an ordinary neural network and also, the activated relative intensity of the intermediate nodes in the basic network can not be shown. According to these, it is clear that the relative strength $R_j(Z_j)$ plays a very important role for improving the performance of ULNs with BR.

4 Conclusions

In this paper, a brain-like type of neural network named ULNs with BR is proposed. It consists of two kinds of networks such as basic network and branch control network.

The branch control network can control the branch connection and disconnection of the basic network with a coefficient of relative strength, depending on the input patterns in the input space. From our simulation results, it has been clarified that if the relative strength is appropiately selected, ULNs with BR can own better performance and generalization ability than ordinary neural networks. It means that commonly used neural networks are not always the best in terms of their architectures. This fact is not well known in the neural network community.

Table 1: Mean Squared Error for Different Thresholds using Different n

Z_j^o (n=0.1)	ONN	0.0	0.1	0.2	0.3	0.4	0.5
ρ	1.00	1.00	0.95	0.87	0.78	0.68	0.53
MSE($\times 10^{-4}$)(Training)	8.54	7.69	10.54	22.48	32.74	59.47	73.24
MSE($\times 10^{-4}$)(Testing)	8.93	7.98	11.57	25.31	36.41	74.52	94.28
Z_j^o (n=1.0)	ONN	0.0	0.1	0.2	0.3	0.4	0.5
ρ	1.00	1.00	0.94	0.89	0.77	0.64	0.51
MSE($\times 10^{-4}$)(Training)	8.54	6.64	6.77	6.55	14.27	13.48	17.24
MSE($\times 10^{-4}$)(Testing)	8.93	6.88	7.09	7.07	15.20	15.96	17.31
Z_j^o (n=10)	ONN	0.0	0.1	0.2	0.3	0.4	0.5
ρ	1.00	1.00	0.96	0.90	0.81	0.68	0.55
MSE($\times 10^{-4}$)(Training)	8.54	20.93	25.61	26.19	36.39	74.52	98.52
MSE($\times 10^{-4}$)(Testing)	8.93	22.51	27.71	29.29	40.18	86.89	107.10

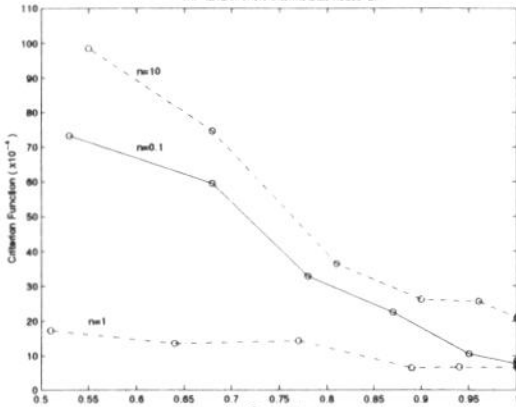

Figure 3: MSE of Training for different thresholds using different n

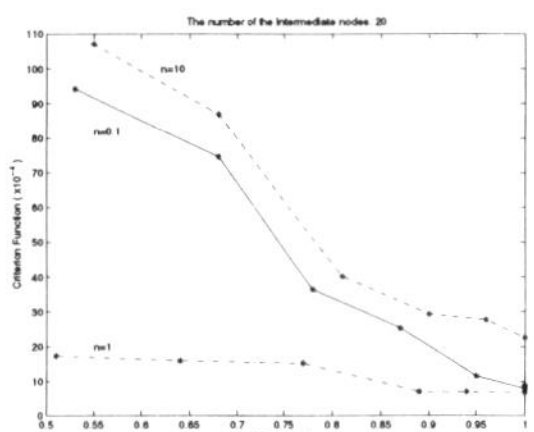

Figure 4: MSE of Testing for different thresholds using different n

References

[1] T. Sawaguchi: *Brain Structure and Evolution*, Kaimeisha, 1989 (in Japanese).

[2] K. Hirasawa, M. Ohbayashi, S. Sakai and J. Hu, "Learning Petri Network and Its Application to Nonlinear System Control", *IEEE Transactions on Systems, Man, and Cybernetics Part B*, Vol. 28, No. 6, pp.781-789, 1998.

[3] K. Hirasawa, S. Oka, M. Ohbayashi, S. Sakai and J. Murata, "Locally Functions Distributed Network -Learning Petri Network- based on Petri Network", *Transactions of the Society of Instrument and Control Engineers of Japan*, Vol. 32, No. 6, pp.241-250, 1996(in Japanese).

[4] M. Ohbayashi, K. Hirasawa, S. Sakai and J. Hu, "An Application of Learning Petri Network to Nonlinear System Control", *Transactions of the Institute of Electrical Engineers of Japan*, Vol. 115-C, No. 6, pp.875-881, 1998(in Japanese).

[5] K. Hirasawa, M. Ohbayashi, J. Murata, C. Jin, S. Oka, S. Sakai, Y.Shitamura, "Modeling of Brain's Functions Distribution by Petri Network", *Transactions of the Institute of Electrical Engineers of Japan*, Vol. 115-C, No. 5, pp.719-727, 1995(in Japanese).

[6] J. Hu, K. Hirasawa and J. Murata, "RasID - Random Search for Neural Network Training", *Journal of Advanced Computational Intelligence*, Vol. 2, No. 4, pp.134-141, 1998.

KES '01
N. Baba et al. (Eds.)
IOS Press, 2001

Application of Modular Neural Networks to Detection of Gas Leakage Under Dynamic Environments

Manabu Kotani, Masanori Katsura and Seiichi Ozawa
Kobe University, Rokkodai, Kobe 657-8501, Japan

Abstract. We examined the application of modular neural networks to a detection of the leakage sound under the dynamic environment. The modular neural network is able to adapt its structure according to the environment. Experiments were performed for an artificial gas leakage device with various experimental conditions for a long term. The discrimination accuracy with the proposed network was about 92.4%, which was nearly the same as that with the conventional network that is, feed-forward network with the error back-propagated algorithm. However, there were some cases in conventional networks where the judging results even for the same data differ between before and after the re-training. In the proposed network, there is nothing such a case. From the results, we confirmed the effectiveness of the modular neural network for the practical use.

1. Introduction

The detection of gas leakage sound from pipes is important in petroleum refining plants and chemical plants, as often the gas used in these plants are flammable or poisonous. Gas sensors can also detect the gas leakage, but it is not always suitable for the early detection of the gas leakage. The ability of the detection is strongly influenced by the direction of the wind, the force of the wind, the density of the gas, and by other factors. On the other hand, if the leakage sound of the gas was used, we could detect the leakage as soon as it happened. Furthermore, the acoustic diagnosis uses the sound detected by microphone, and therefore has the excellent features of non-contact detection and easy handling.

We have already evaluated the detection performance with the neural network for short-term experiment[1]. The applied network is the multi-layered feed-forward model whose learning algorithm is the error back-propagation (BPN). However, the stable diagnosis for the long term is needed the establishment the acoustic diagnosis for the practical use. It is supposed that the background noise varies every season and the deterioration of the equipment affect the background noise. Such change of background noise influences the reliability of diagnosis results. Therefore, it is necessary to examine the diagnostic method that is able to adapt to changes in environment. The simplest method is to re-train the network, but diagnosis result would change before and after the re-training even for the same data. The examination of a novel neural network model is necessary to overcome this difficulty. There are some studies[2]-[4] about the modular neural networks, but the primary difference from these studies is whether diagnosis result may change before and after the adaptation to the environment.

The purpose of this study is to establish the acoustic diagnosis technique for the gas leakage for the practical use. We propose a novel modular neural network to adapt changes in environments and examine the effectiveness of the proposed network for experiments in petroleum refining plants for about 18 months.

2. Outline of Modular Neural Networks

The modular neural network consists of an initial module and some supplementary modules as shown in Figure 1. Outputs of each module are "Normal", "Abnormal", and "Unknown". "Normal" means no leakage, "Abnormal" means the leakage, and "Unknown" means that the input sound differs from the training data and the module can not discriminate to be "Normal" or "Abnormal". The initial module is trained using the data collected in various conditions. The supplementary module is added to the initial module when outputs of the initial module are "Unknown" for many observed data. Furthermore, when the outputs of all modules are "Unknown", a new module is supplemented with the networks.

Hence, the modular neural network adds the new modules to the initial module according to the environment.

Procedure of the construction of the modular network is the following ways:

1) The initial module is trained in the data collected before hand.

2) The data collected in the fixed sampling time is fed to the modular network.

3) If the output of the network is not "Unknown" for the fixed period, go to the 2-nd procedure.

4) The new supplementary module is trained in the data judged "Unknown" and the new module is added to the modular network. Then, go to the 2-nd procedure.

It would be able to diagnose precisely by the training the initial module using acoustic data collected beforehand. Then, in the case that diagnosis result of the "Unknown" continues over the fixed time, the supplementary module trained the data and it is added to the network. The reason that the addition criterion of the module is to continue over fixed time is for preventing the addition of the unnecessary module by the diagnosis of sudden noise, etc.

To attain the output of module to be "Unknown", it is desired that the localization ability for the module is excellent. The localization ability is that the module responses only in the data similar to the training data. Therefore, we apply Radial Basis Function Network (RBF) [5] to the module.

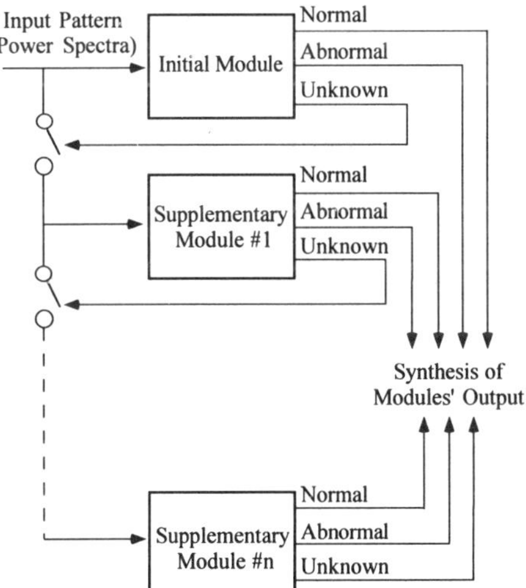

Figure 1: Structure of modular neural network

Here, we evaluated the localization performance of the neural network. Figure 2(a) shows the scatter plot of the training data. Figure 2(b) and (c) show the output of the BPN and RBF, respectively. x and y-axis indicates coordinate of the input data and z-axis indicates the output of the network. Each network has two input units and an output unit. From these results, the BPN could not localize responses around the training data, but the RBF could localize.

3. Experiments

Experiments were performed in the petroleum refining plants using an artificial gas leakage device, with experimental conditions being diameters of artificial defects, gas pressures, and distances between the microphone and the leakage device. Figure 3 shows the experimental setup. A gas cylinder was connected to the artificial gas leakage device through the regulator. Three kinds of defect diameters, which were 0.5, 1.0, and 2.0mm, were artificially processed in the stainless steel pipes of the leakage device. Since the end of the pipe was closed, the gas leaked only from the defect. Gas pressure was 10, 40, and 70kgf/cm^2.

A microphone was placed at distances of 2, 5, and 10m from the leakage device and picked up the sound. The signals from the microphone were recorded on a digital audio tape deck at 48kHz sampling frequency. We have collected the leakage sound and the background noise at the start. Then we have collected only the background noise in 7, 9, 11, and 18 months later.

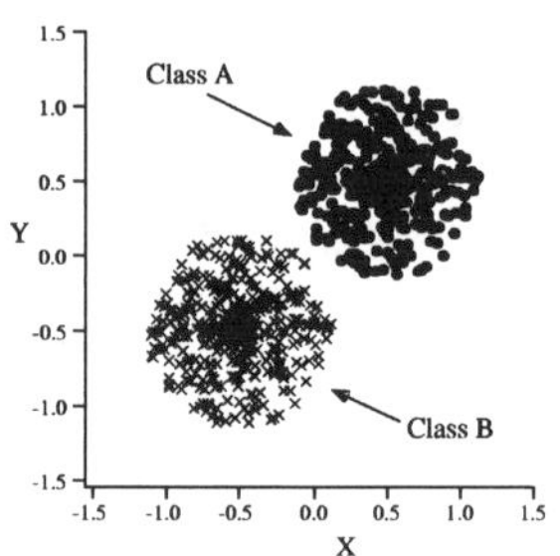

(a) Scatter plot of the training data

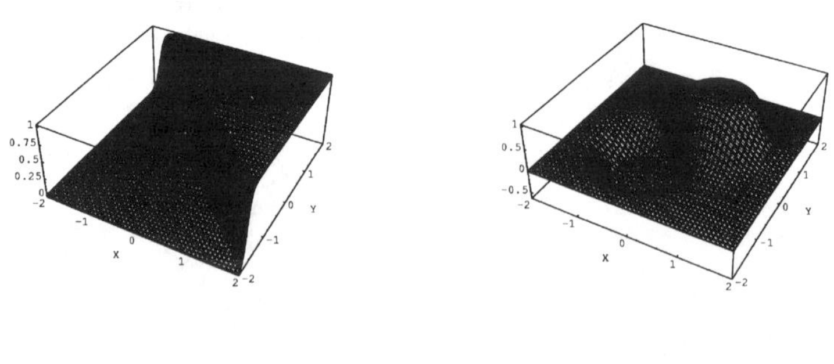

(b) Output of the BPN   (c) Output of the RBF

Figure 2: Localization performance of the neural networks

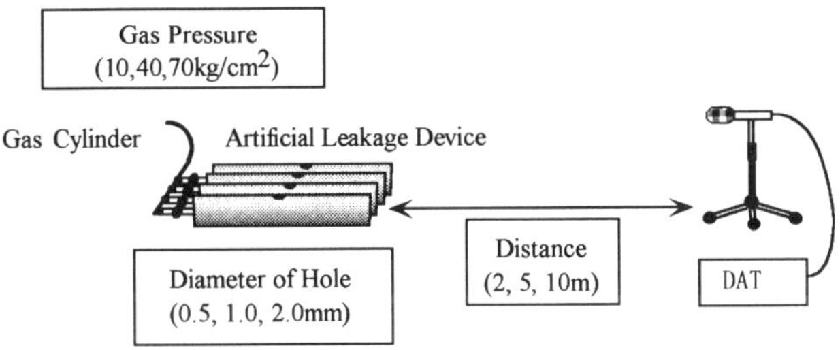

Figure 3: Experimental setup

A pre-processing method is applied Fast Fourier Transform (FFT) for the collected sound and logarithmic power spectra were used as the input information of the neural networks. The FFT computation was carried out every 20ms over Hanning-windowed frames with a 20ms width using FFT analyzer. Consequently, logarithmic power spectra were 200 dimensions.

4. Results

Here, we evaluate the modular neural network using the above mentioned acoustic data. The input information was 200 dimensional logarithmic power spectra and the output of the network was the diagnostic result. The number of input units was 200 and the number of output units was one. The target pattern was +1.0 for the background noise and -1.0 for the leakage sound. The input data was judged "normal" for the output being over +0.3, "abnormal" for the output below -0.3, and "unknown" for the other value.

First, we trained the initial module with the initial data set that consisted of background noise and leakage sound collected at the start. We calculated the discrimination accuracy for the test data of the initial data set. When the data set collected in 7 months later (the second data set) were represented to the module, outputs of the initial module for all the second data set were "unknown". Next, we trained a supplementary module with the above "unknown" data. Table 1 shows results with an initial module and with a supplementary module. Discrimination accuracy is calculated for the test data that is not used for the training of the module. It turns out from Table 1 that the background noise collected in 7 months later are different from the sound of the initial data set. On the other hand, a supplementary module is able to learn the "unknown" data perfectly.

Furthermore, the other data sets were represented to the networks in succession and we constructed the network. Consequently, the modular neural networks consisted of an initial module and three supplementary modules. Table 2 shows results for all data sets using the modular networks.

The discrimination accuracy for the test data in all data set was 92.4%. False rate was 0.3% and unknown rate was 7.3%. We compared the proposed network with the simple network being three-layered BPN with the re-training strategy. The discrimination accuracy with the re-training network was nearly the same as the results with modular networks. However, there were some cases where the judging results even for the same data differ between before and after the re-training.

Table 1: Results of an initial module and a supplementary module

Module	Data	Correct	False	Unknown
Initial	First	93.0%	0.7%	6.3%
Initial	Second	0.0%	0.0%	100.0%
Suppl.	Second	100.0%	0.0%	0.0%

"First" means the initial data set and "Second" means data set collected after 7 months.

Table 2: Results of an initial module and three supplementary modules

Data	Correct	False	Unknown
Initial	93.0%	0.7%	6.3%
7 months later	100.0%	0.0%	0.0%
9 months later	80.0%	0.0%	20.0%
11 months later	87.5%	0.0%	12.5%
18 months later	100.0%	0.0%	0.0%

These results indicate that the proposed modular network is capable of adapting its structure to changes in environments.

5. Conclusion

We have described the acoustic diagnosis for gas leakage from pipes for the long term. We have proposed the modular neural networks suitable for the practical use. Furthermore, we have collected the sound for 18 months. The effectiveness of the modular neural network that can add the new module according to changes in environments was evaluated.

References

[1] M.Kotani, T.Miyata and K.Akazawa: Acoustic diagnosis for gas leakage in pipes under noisy environments, *Proceedings of IMEKO-XV*, Vol. VIII, pp 55-60 (1999).

[2] R.Anand, K.Mehrotra, C.K.Mohan and S.Ranka: Efficient classification for multiclass problem using modular neural networks, *IEEE Trans. Neural Networks*, Vol.6, No.1, pp.117-124 (1995).

[3] C.Rodriguez, S.Rementeria, J.I.Martin, A.Lafuente, J.Muguerza and J.Prez: A modular neural network approach to fault diagnosis, *IEEE Trans. Neural Networks*, Vol.7, No.2, pp.326-340 (1996).

[4] S.Haykin and T.K.Bhattacharya: Modular learning strategy for signal detection in a nonstationary environment, *IEEE Trans. Signal Processing*, Vol.45, No.6, pp.1619-1637 (1997).

[5] T.Poggio and F.Girosi: A theory of neural networks for approximation and learning, *A.I.Memo* 1140, MIT (1989) .

KES '01
N. Baba et al. (Eds.)
IOS Press, 2001

A Modular Constructive Approach to Mobile Robot Navigation

Catarina Silva†‡, Bernardete Ribeiro†
†Departamento de Engenharia Informática, Centro de Informática e Sistemas
Universidade de Coimbra, 3030 Coimbra, Portugal, Email:{catarina,bribeiro}@dei.uc.pt.
‡ESTG - Instituto Politécnico de Leiria - Leiria - Portugal

Abstract

This paper presents a neural modular architecture for navigating mobile robots. The proposed architecture, based on functional task division, has been shown to be efficient in previous work [1].

The traditional difficulties associated with monolithic neural networks are circumvent by introducing a neural modular architecture, developed for the NOMAD mobile robot. The modularity introduced retrieves all the available information, minimizing the incoherence in the training sets, that arises from conflicting training patterns.

After presenting the modular architecture, two additional modules are introduced: Supervisory and Angular Memory. The combination of all defined modules was first tested in simulation and then with the real robot, providing a solution to the problem of navigating NOMAD mobile robot to an objective in an unknown environment.

1 Introduction

The hierarchical level of organization in artificial neural networks may be classified as follows. At the most fundamental level there are synapses, followed by neurons, then layers of neurons in the case of a layered network, and finally the network itself [2]. This description shows the implicit modularity present in monolithic neural networks. However, this modularity can be extended to what are known as *modular neural networks*. Recent years have witnessed a surge of interest in modular approaches that overcome some of the shortcomings presented by monolithic neural networks. There are a number of possible motivations for adopting a modular approach, but the most common is the potential improvement in generalisation involved. In other words, a task could be solved with a monolithic network, but better performance can be achieved when it is broken down into a number of specialist modules [3]. Thus, modularity can be viewed as a manifestation of the *principle of divide and conquer*.

Neural networks have been extensively used in several mobile robot control tasks, exhibiting improvements in performance when compared with more traditional control methods ([4] presents an excellent review). Most of these applications are based on monolithic neural networks, i.e., systems in which a single neural network is responsible for the control task. In this work it will be shown how to use modular neural networks in mobile robot navigation.

The purpose of navigating a robot, i.e., reaching a goal position while avoiding all objects in the path, has often been reached using *a priori* knowledge of obstacle localisation. In these approaches, a map exists or is built along the way. Using this map, a complete trajectory can be determined off-line and, sometimes, on-line. On the other hand, when there is a real application, there's usually an unstructured and dynamic environment where it is very difficult, if not pointless, to plan ahead the path to a goal position. In fact, there are several applications in which there is no meaning in taking the effort to build a map, because the environment is too much changeable.

Traditional approaches to network architecture determination provide *ad-hoc* solutions, often hard to prove optimal or near optimal, because these feed forward neural networks suffer from two major drawbacks. On one hand, it is difficult to decide if a given task can be solved by a given feed forward network. On the other hand, there is no way to determine the most adequate size of a network for a specific application. In this work we explore a class of constructive algorithms that systematically determine a neural network architecture.

Next section will discuss the concepts of modularity and construction and their advantages when applied to neural networks. Section 3 will describe the mobile robot system used as a testbed for this work and section 4 presents the designed architecture. Section 5 shows some results and, finally, in the last section the conclusion is presented.

2 Modular and Constructive Neural Networks

The building block of any neural network is the neuron. The organisation of neurons in layers and interconnection of layers are the next steps in the neural architecture. However, this organisation can go one step higher. Specifically, a neural network can consist of a multiplicity of networks in a modular structure [2].

One of the important improvements achieved by modular neural networks is better generalisation. The convention in neural networks is to use an architecture as small as possible to obtain better generalisation, because the ability of feedforward neural networks to generalise is inversely proportional to the number of weights involved. Some other advantages occur by the use of modular neural networks. First, dividing a task into sub tasks will avoid the problems of temporal and spatial crosstalk [6]. Moreover, by partitioning a complex task, modular neural architectures achieve representations that are more easily understood than those generated by monolithic architectures.

Finally, modularity can enable a learning economy, since some modules can be reused or modified without altering the other modules. This economy can also become crucial when hardware implementations are an objective, since modularity neural can facilitate hardware execution of neural networks [7].

Usually the neural network's architecture is determined using *ad-hoc* techniques that rely on a trial and error approach to determine the topology and the parameters of the net. There are however other algorithms that systematically construct the neural network. There are two general approaches to this optimisation problem: pruning and constructive algorithms. The constructive approach starts with a small network and then grows additional hidden units and weights until a satisfactory solution is found. The pruning approach involves using a larger than needed network and training it until an acceptable solution is found. After this, hidden units or weights are removed if they are no longer actively used.

3 Nomad Mobile Robot

NOMAD is a wheeled cylindrical robot whose diameter is about 46 centimetres, and is shown in figure 1(a). The robot can move forward or backward with varying speed and turn right or left with varying degrees/s. The robot has also a dead-reckoning system for keeping track of the robot's orientation and position.

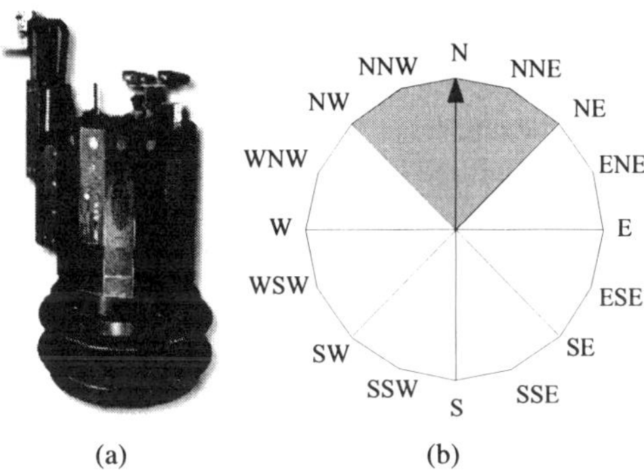

(a)   (b)

Figure 1: NOMAD mobile robot and sensorial directions.

It has 16 ultrasonic sensors and 16 infrared sensors for observing the surrounding environment. Ultrasonic and infrared sensors are evenly distributed around the robot, yielding a 22.5 degrees angle between any two adjacent sensors, distributed according to figure 1(b).

4 System Architecture

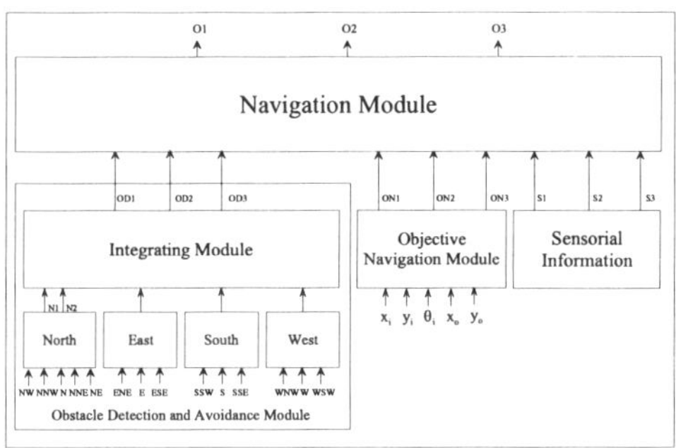

Figure 2: Modular architecture.

The navigation task was initially divided in two main modules: obstacle detection and avoidance and objective navigation. The navigation module was developed to combine the two main modules. Figure 2 shows the modular architecture proposed.

Two additional modules were introduced, taking advantage of the modular structure: supervisory module and angular memory module. They are discussed in section 4.4.

4.1 Obstacle Detection and Avoidance Module

This module is responsible for the task of detecting and avoiding unknown obstacles in the robot's trajectory. This task was further divided into 4 sub-tasks (one for each quadrant), which cooperate in determining the module's outputs. This module will play a central role in the resolution of the problem, since it will receive all the sensorial information available. As referred in section 3, there are two types of sensors aligned in each sensorial direction: an infrared sensor and an ultrasonic sensor. The difference between their ranges can be used to optimise the information in each of the 16 possible sensorial directions.

A pseudo system of cardinal points has been assigned to the sensorial directions available. In the pseudo-system the robot always faces North. Thus, the North direction presents the higher probability of collision. According to this analysis, all sensors in the North quadrant were used as inputs to the North module. The training set of the North module was derived from actual trajectories driven by the supervisor in simulated environments. The other three modules are smaller and identical, taking advantage of the learning economy that underlies the modular structure. To integrate the four modules an integrating module was defined (see figure 2).

4.2 Objective Navigation Module

The Objective Navigation Module computes the heading direction to the goal, considering the goal position and the actual robot's position (x_i, y_i) and orientation, θ_i. The orientation is computed according to (1):

$$\theta_o = atan\frac{y_o - y_i}{x_o - x_i} \tag{1}$$

where (x_i, y_i) is the actual robot's position, θ_i is the actual robots's orientation, (x_o, y_o) is the objectives's position and θ_o represents the angle to the objective. Having the angle that represents the direction, a mapping is made and codified to the cardinal directions.

4.3 Navigation Module

Using the two modules, in which the main task was divided, the navigation module was built to determine the final direction to follow. It also receives three sensorial inputs corresponding to the sensorial values for the heading direction calculated, yielding 9 inputs and 3 outputs, that codify the output direction.

4.4 Supervisory and Angular Memory Modules

Two additional modules were defined to cope with situations were the proposed architecture could be improved. The supervisory module detects obstacles whose dimension is much larger that the robot's dimension and determines the functioning mode: obstacle contour or navigation. The navigation mode is basically the architecture defined in the last section, i.e., the direction is determined by the outputs of the navigation module. The obstacle contour mode uses the outputs of the integrating module to establish the direction to follow.

The Angular Memory module keeps track of the angular directions followed when there is a change in the functioning mode, to detect cycles in the trajectory. Using a circular memory, the transition points are memorised and when a repetition is found an alternative direction is followed, as a way to break the cycle.

4.5 Modules Construction

The topology and parameters of all the modules defined so far were constructed and not determined *ad-hoc*. A *Dynamic Node Creation* [5] method was used with a variant of the back propagation algorithm. Construction algorithms avoid problems usually bound to trial and

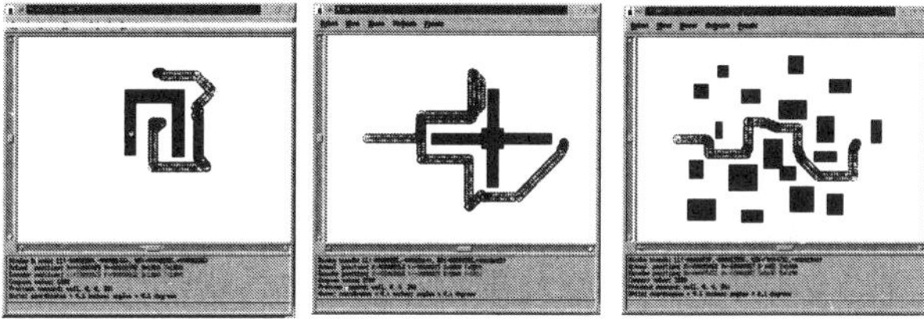

Figure 3: Results in environments with unknown obstacles.

error solutions, as already mentioned in section 2. These algoritms search, not only the correct set of weights for a topology, but also the most apropriate topology for a given problem.

5 Results

The purpose of the simulation was to test the behaviour of the mobile robot using the trained modular neural network system. The simulation was carried in environments where the modular neural network had to control the robot to a pre defined goal position avoiding collisions with obstacles without *a priori* knowledge of the environment. Tests with the real robot were also carried out, strenghtening the confidence in the proposed architecture. Figure 3 presents some of the results obtained.

The modular structure presented proved to be efficient. Neural networks inherent adaptive capabilities enable a good performance as can be seen in the examples presented. The modular architecture avoids robot collisions and produces smooth trajectories, thus showing generalisation after learning.

6 Conclusion

We showed how modular neural networks can be used to navigate a mobile robot among obstacles, without knowledge of the environment. The results presented show the strategy to be robust and adaptive. Future work will include the study of on-line learning as a way to improve the system performance.

This work was partially supported by the portuguese Ministério da Ciência e Tecnologia and the European Union through the R&D Unit 326/94 (CISUC).

References

[1] C. Silva, M. Crisóstomo, B. Ribeiro, "MONODA: A Neural Modular Architecture for Obstacle Avoidance Without Knowledge of the Environment", IJCNN, 2000.

[2] S. Haykin, "Neural Networks - a comprehensive foundation", Prentice Hall, 1999.

[3] A. Sharkey (Ed.), "Combining Artificial Neural Networks", Springer-Verlag, 1999.

[4] B. Krose, J. van Dam, "Neural Vehicles", Neural Systems for Robotics, P.Smagt, O.Omidvar (Eds.), Academic Press, pp. 271-296, 1997.

[5] T. Kwok, D. Yeung, "Constructive Algorithms for Structure Learning in Feed Forward Neural Networks for Regression Problems", in *IEEE Transactions on Neural Networks*, XX(Y).

[6] R. Jacobs, M. Jordan, "Learning Piecewise Control Strategies in a Modular Neural Network Architecture", IEEE Trans. on Systems, Man and Cybernetics, Vol 23, N. 2, pp.337-345, 1993.

[7] G. Auda, M. Kamel, "Modular Neural Networks: A Survey", in *Int. Journal of Neural Systems*, Vol. 9, No. 2, pp. 129-151, 1999.

KES '01
N. Baba et al. (Eds.)
IOS Press, 2001

Adaptive Module Acquisition
In Modular Reinforcement Learning

Hajime Murao[†], Takahiko Kamitsuji[‡], and Shinzo Kitamura[†]
† Faculty of Engineering, Kobe University, JAPAN
‡ Graduate School of Science and Technology, Kobe University
{murao,kitamura}@al.cs.kobe-u.ac.jp

Abstract

This paper proposes an adaptive module acquisition for the modular reinforcement learning, where a learning agent starts with fundamental modules and acquires new modules during the learning if necessary. This relaxes the problem that it is difficult to know suitable module structure to accomplish the task in advance without *a-priori* knowledge of the problem. The criterion to introduce new modules is derived from fundamental aspect of reinforcement learning that the probability of the situation that values of states increase along the greedy policy becomes high after sufficient learning. The proposed method is implemented on Q-learning. It is applied to so-called "pursuit problem" simulated in a computer where two learning agents are navigated to catch a randomly moving object. As a result of computer simulations, the proposed method shows fairly good result in terms of better or the same performance with the less number of states compared to normal Q-learning or modular Q-learning without capability of acquiring new modules.

1 Introduction

Increasing the number of sensors to improve the behavior of a robot causes exponential increase of the number of states in a reinforcement learning, *i.e.* N sensors with resolution R results R^N states. This causes it difficult to acquire suitable behavior to accomplish the task in a feasible learning period. One possible solution for the problem proposed in these days is the modular reinforcement learning[3, 5, 2, 6] where subspaces of the state space are assigned to reinforcement learning modules and their decisions are unified into one by a controle module to determine the behavior of the robot. This decreases the number of states from R^N to RN at most. However, there is still at least one problem: it is difficult to define subspaces over the state space preliminarily. That is, dividing the state space into subspaces more can decrease the total number of states but there possibly be states should be defined accross subspaces so as to be distinguished from other states. If these states are importnt to accomplish the task, this problem becomes critical. This paper proposes adaptive method to unify modules at the states. Where, the number of states increases from RN to R^N through a learning process *pro re nata*. It is applied to so-called "pursuit problem" simulated in a computer where two learning agents are navigated to catch a randomly moving object.

2 Modular Reinforcement Learning

2.1 General Structure of Modular Reinforcement Learning

A modular reinforcement learning is mainly composed of two parts of learning modules and a controlling module.

The learning modules hold policies, each of which is represented as a set of values defined over states and actions as

$$\pi^m = \{q^m(\boldsymbol{s}^m, a) | \boldsymbol{s}^m \in \boldsymbol{S}^m, a \in A\} \tag{1}$$

where π^m is a policy of learning module m, $\boldsymbol{s}^m$ is a state observed by the module, $\boldsymbol{S}^m$ is a subspace of the state space, a is a behavior at single instance of a robot namely "action", and A is a set of possible actions. $q^m(\boldsymbol{s}^m, a)$ represents preference of the action a at the state $\boldsymbol{s}^m$ for which greater value usually means more preferable action.

Let the state space $\boldsymbol{S}$ be defined by a product of fundamental planes as

$$\boldsymbol{S} = S^1 \times S^2 \cdots S^N \tag{2}$$

where S^n is a fundamental plane defined by possible output values of the n-th sensor, N is the number of sensors. A subspace $\boldsymbol{S}^m$ of the state space is defined as

$$\boldsymbol{S}^m = \delta^1 S^1 \times \delta^2 S^2 \cdots \delta^N S^N \tag{3}$$

where $\delta^n \in \{0, 1\}$ indicate what fundamental planes compose the subspace $\boldsymbol{S}^m$ and we here define a string "$\delta^1 \delta^2 \cdots \delta^N$" as "signature" of subspace. A state $\boldsymbol{s}^m$ of a module m is consequently defined as a subset of a state, which is defined by a set of sensor output values as usual, and we introduce a new term "substate" for the state $\boldsymbol{s}^m$ of the module m. In precisely, a state $\boldsymbol{s}$ is defined as follows,

$$\boldsymbol{s} = \{s^1, \cdots, s^n, \cdots, s^N\}, \qquad s^n \in S^n \ \forall n \in \{1, ..., N\} \tag{4}$$

where s^n is an output value of the n-th sensor, and a substate $\boldsymbol{s}^m$ of module m is defined as,

$$\boldsymbol{s}^m = \{\boldsymbol{s}\} \cap \boldsymbol{S}^m. \tag{5}$$

The controlling module manages the learning modules and unifies their output values to a single action of a robot. It also train the learning modules based on the reinfocement signal from outside the robot. There can be many variants of the method to determine an action from the output values of the learning modules. Two typical variants are the cooperative one and the competetive one. For instance, the former trains all of the learning modules at the same time, and the latter trains them individually.

2.2 Q-learning

We here introduce a method based on Q-learning[4] to modify preference values of learning modules. Q-learning is one of the well-known reinforcement learning method, where $Q(\boldsymbol{s}_t, a_t)$, an expected discounted sum of reinforcement signals for the future after taking an action a_t in a state $\boldsymbol{s}_t$ at time t, is updated as follows,

$$Q(\boldsymbol{s}_t, a_t) \Leftarrow (1 - \alpha)Q(\boldsymbol{s}_t, a_t) + \alpha\{r_t + \gamma \max_{b \in A} Q(\boldsymbol{s}_{t+1}, b)\} \tag{6}$$

where r_t is a reinforcement signal given as a result of the action, α and γ correspond the learning rate and the discounting rate respectively.

After a sufficient number of learning, every Q values converge to

$$Q(\boldsymbol{s}, a) = E\{r(\boldsymbol{s}, a)\} + \gamma \sum_{\boldsymbol{s}' \in \boldsymbol{S}} \rho(\boldsymbol{s}, a, \boldsymbol{s}') \cdot \max_{b \in A} Q(\boldsymbol{s}, b) \tag{7}$$

and the greedy policy choosing the action a^* maximizing the corresponding Q value of the state is known to be the best policy, where $E\{r(\boldsymbol{s}, a)\}$ is an expected value of the reinforcement signal r_t in Eq. 6 for the state s and the action a, and $\rho(s, a, s')$ is the probability of transititon from the state s to the state s' by the action a.

During the learning, an action a is usually chosen at a probability $P(s_t, a)$ defined as follows to balance exploratory behavior to find better actions,

$$P(s_t, a) = \frac{\exp(Q(s_t, a)/\tau)}{\sum_{b \in A} \exp(Q(s_t, b)/\tau)} \tag{8}$$

where τ is a parameter controlling the sensitivity of Q values.

2.3   Modular Q-learning

We use the competitive type of controlling module to unify learning modules. That is, a probablity $P(s_t, a)$ of an action a defined by preference values of the learning modules as

$$\begin{aligned}
P(s_t, a) &= \sum_{m \in M} P^m(s_t^m, a) \\
&= \sum_{m \in M} \frac{\exp(q^m(s_t^m, a))}{\sum_{m \in M} \sum_{b \in A} \exp(q^m(s_t^m, b))}
\end{aligned} \tag{9}$$

where M is a set of learning modules and s^m is defined as in Eq. 5.

The preference values are also modified based on Q-learning described in Eq. 6 as follows,

$$q^{m_t}(s_t^{m_t}, a_t) \Leftarrow (1 - \alpha)q^{m_t}(s_t^{m_t}, a_t) + \alpha\{r_t + \gamma \max_{b \in A} q^{m_t}(s_{t+1}^{m_t}, b)\} \tag{10}$$

where m_t indicates the module used to select the action a_t. That is, not only the action at time t but also the module providing the action should be determined simultaniously. It is quite implementation side problem but we would like to notice here because of the purpose the probability $P^m(s_t^m, a)$ is actually used instead of $P(s_t, a)$ to determine both of the action and the module.

2.4   Adaptive Module Acquisition

As a natural characteristic of Q-learning, Q values are settled at a certain equilibrium after the sufficient learning, and the probability of the situation that Q values increase along the greedy policy becomes high. However, if the state space are not suitably organized to distinguish important states to accomplish the task from others, the characteristic of Q values is not satisfied [1]. This is the basic concept of our approach.

In our approach, a condition described as

$$q^m(s_t^m, a_t) \leq \gamma \max_{i \in M, b \in A} q^i(s_{t+1}^i, b) \tag{11}$$

is checked after every greedy actions, and the occurrence of the undesirable condition is counted as

$$d^m(s_t^m) \Leftarrow \frac{k^m \cdot d^m(s_t^m) + d_t}{k^m + 1}, \tag{12}$$

where $d^m(s^m)$ holds the occurrence rate of the undesirable condition in a substate s^m, k^m counts the occurrence of the substate s^m, and d_t is the condition value defined as follows,

$$d_t = \begin{cases} 1 & \text{if Eq. 11 is not satisfied,} \\ 0 & \text{otherwise.} \end{cases} \tag{13}$$

k^m is also updated after updating $d^m(s_t^m)$ as follows,

$$k^m \Leftarrow k^m + 1. \tag{14}$$

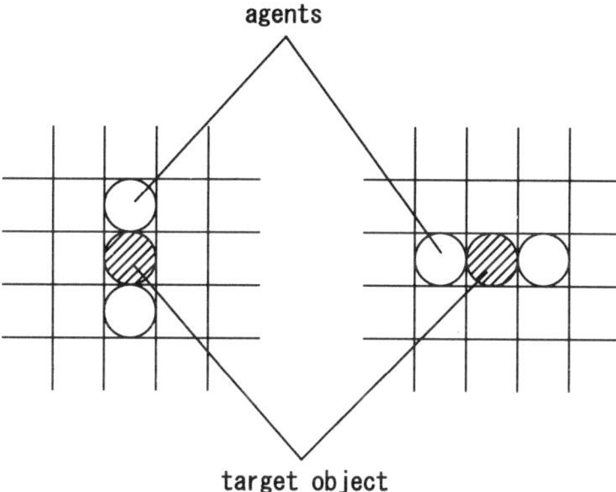

Figure 1: The object is said to be caught by agents only when it is put between them horizontally or vertically.

If $d^m(s^m)$ exceeds a given threshold with enough long observation period k^m, *i.e.* $d^m(s^m)$ and k^m satisfy the following condition

$$d^m(s^m) > \theta_d \quad \text{and} \quad k^m > \theta_k, \tag{15}$$

the substate s^m is removed from the subspace S^m. Where θ_d is the threshold for $d^m(s^m)$ given from $[0, 1]$, and θ_k is the threshold for k^m.

When two substates s^l and s^m of two different modules l and m for the same state s are removed from subspaces S^l and S^m, these two states are combined into a new state s^n of a new module n if it is not existed. This procedure is described as follows,

$$s^n = \{s^l \cap s^m\} \cap S \tag{16}$$

$$S^n \Leftarrow \{s^n\} \cup S^n \quad \text{iff } \{s^n\} \cap S^n = \phi \tag{17}$$

Their preference values $q(s^n, a)$ are initialized to the average of two preference values of original two states s^l and s^m as

$$q(s^n, a) \Leftarrow \frac{q(s^l, a) + q(s^m, a)}{2} \tag{18}$$

3 Experiment

3.1 Simulation Environment

The proposed method is applied to so-called "pursuit problem" navigating two learning agent to catch a randomly moving object. The world in which all characters behave is defined by a 2-dimensional toric lattice. Each agent or the object occupies a lattice point in the world but they cannot be stacked.

They are initially placed at randomly chosen lattice points from the world and each of them can do an action at a time. We say the object is caught by the agents only when the object is put between the agents horizontally or vertically as described in Fig. 1, and positive reinforcement signal namely reward is given to the agents at that time. This process is counted as one trial and is repeated.

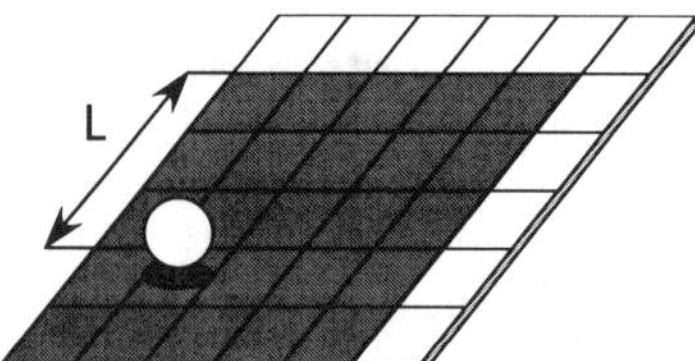

Figure 2: The grayed area indicates the sight of the agent

Table 1: Parameters for Experiments

parameter	value	description
α	0.3	learning rate
γ	0.9	discounting rate
τ	0.1	sensitivity controlling parameter
θ_d	0.95	the threshold for $d^m(s^m)$
θ_k	10	the threshold for k^m

3.2 Learning Agents

The learning agents can move to the direction of up, down, left, and right. With an action to stay the same lattice point, a set A of actions of the learning agents is defined as

$$A = \{\text{up}, \text{down}, \text{left}, \text{right}, \text{stay}\}. \tag{19}$$

Each agent has two sensors measuring relative position of another agents and the target object within a predefined area namely "sight" of agent respectively. The size of the sight is defined by a parameter L as described in Fig. 2. A state of the agent is defined by both of relative position of another agent and the target object. The number of possible sensor values is $(2L+1)^2$ for both sensors and consequently the number of states in conventional monolithic reinforcement learning becomes $((2L+1)^2)^2$.

The proposed method with one controlling module and two learning modules each of which corresponds each sensor is applied to the problem. That is, the state space is composed of two fundamental plane corresponding two sensors and it is decomposed into subspaces for which the signatures can be represented as "10" and "01". This reduces the number of initial states from $((2L+1)^2)^2$ to $2(2L+1)^2$.

3.3 Experimental Results

Computer simulations have been done with paramters represented in Table 1. The number of steps required to accomplish the task has been observed during the learning, and its averages over 20 simulations with various reinforcent learnings are described in Fig. 3.

The number of finally obtained states in the proposed method averaged over 20 simulations is 520. It is quite smaller than the monolithic Q-learning for which the number of states is fixed to 2,500, even though the proposed method finally converges to the same performance with the monolithic Q-learning after 2,000 trials.

The simple modular Q-learning without the ability to acquire new modules cannot learn efficient behavior to accomplish the task. In this case, we completely decompose the state space into subspaces by assigning each module to each sensor. It is slightly different approach from the previous researches such as [2], in which a module is assigned to subspace composed

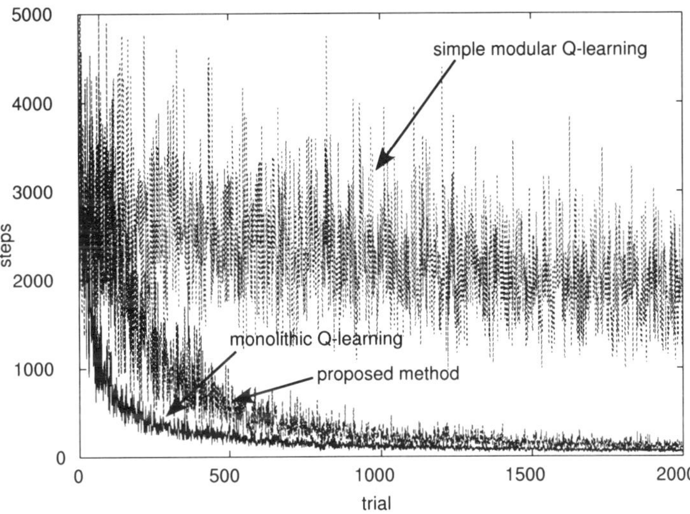

Figure 3: The convergence of required steps to catch the target object with varians of reinforcement learning.

of at least two sensors. However, since it is difficult to know what combinations of sensors are required to accomplish the task *a-prioi*, the situation of our experiments will happen in greater or less degree.

4 Conclusion

This paper has proposed a reinforcement learning with an adaptive modular acquisition. Where a new module corresponding a new subspace of the state space is acquired adaptively during the learning. It has been applied to so-called "pursuit problem" navigating two learning agents to catch the target object. As a result of computer simulations, it was shown that the proposed method could accomplish the task as well as the monolithic Q-learning with the quite smaller number of states.

References

[1] Hajime Murao and Shinzo Kitamura. Q-learning with adaptive state space construction. volume 1545 of *Lecture Notes in Artificial Intelligence*, pages 13–28. Springer-Verlag, 1999.

[2] Norihiko Ono, Osamu Ikeda, and Kenji Fukumoto. Acquisition of coordinated behavior by modular Q-learning agents. In *Proc. of IROS'96*, pages 1525–1529, 1996.

[3] Eiji Uchibe, Minory Asada, and Koh Hosoda. Behavior coordination for a mobile robot using modular reinforcement learning. In *Proc. of IROS'96*, pages 1329–1336, 1996.

[4] Chris Watkins. *Learning from Delayed Rewards*. PhD thesis, Cambridge University, 1989.

[5] Steven Whitehead, Jonas Karlsson, and Josh Tenenberg. Learning multiple goal behavior via task decomposition and dynamic policy merging. *Robot Learning*, pages 45–78, 1993.

[6] Satoshi Yamada. Modular reinforcement learning. *Technical Report of IEICE*, pages 139–149, Mar 1998.

KES '01
N. Baba et al. (Eds.)
IOS Press, 2001

Cooperation of Evolutionary Modular Neural Networks Developed by L-System*

Sung-Bae Cho
Department of Computer Science, Yonsei University
134 Shinchon-dong, Sudaemoon-ku, Seoul 120-749, Korea

Abstract. The optimal design of neural networks by evolutionary algorithms has shown a great possibility to develop sophisticated system for real-world problems. In order to boost up the scalability and utilization, grammatical development has been considered as a promising encoding scheme of the network architecture in the evolutionary process. This paper presents a grammatical development method called L-system to determine the structure of a modular neural network that was previously proposed by the authors, and shows the potential by analyzing the behavior of the network evolved. Experiments of recognizing handwritten numeral digits indicate that the evolved neural network has reproduced a sort of the hierarchical structure composed of subsystems of coarse and fine processing of stimuli in separate pathways.

1 Introduction

There are more than hundred publications that report an evolutionary design method of neural networks [1, 2, 3, 4]. One of the important advantages of evolutionary neural networks is their adaptability to a dynamic environment, and this adaptive process is achieved through the evolution of connection weights, architectures and learning rules [4]. Most of the previous evolutionary neural networks, however, show little structural constraints. In the meantime, there is a large body of neuropsychological evidence showing that the human information processing system consists of modules, which are subdivisions in identifiable parts, each with its own purpose or function.

With this philosophy, we have proposed evolutionary neural networks based on modules as building block [5], and applied a parametric L-system to the development of the network architecture [6]. Each module has the ability to autonomously categorize input activation patterns into discrete categories, and representations are distributed over modules rather than over individual nodes. Among the general principles are modularity, locality, self-induced noise, and self-induced learning. In this paper, we attempts to investigate the emergent properties of the structure and function of the modular neural networks evolved based on grammar, especially L-system.

2 Evolutionary Modular Neural Networks

The basic idea of the evolutionary modular neural networks is to consider a module as a building block resulting in local representations by competition, and develop complex intermodule connections with evolutionary mechanism. In computing terms, an

---

*This research was supported in part by a grant from Brain Science Research under the Ministry of Science and Technology in Korea.

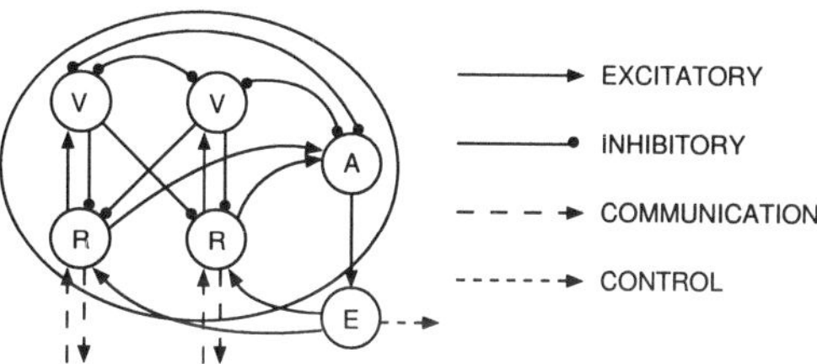

Figure 1: Schematic diagram of the internal structure of a module.

evolutionary algorithm maps a problem onto a set of strings, each string representing a potential solution. In the problem at hand, a string encodes the grammar that can generate network architecture and learning parameters. The evolutionary algorithm then manipulates the most promising strings in its search for improved solutions.

The activation value of each node in the modular neural network is calculated as follows:

$$e_i = \sum_j w_{ij} a_j(t), \tag{1}$$

where w_{ij} denotes the weight of a connection from node j to node i. The effective input to node i, e_i, is the weighted sum of the individual activations of all nodes connected to the input side of the node. The input may be either positive (excitatory) or negative (inhibitory).

The internal structure of each module is fixed and the weights of all intramodular connections are non-modifiable during learning process (see Figure 1). In a module, R-node represents a particular pattern of input activations to a module, V-node inhibits all other nodes in a module, A-node activates a positive function of the amount of competition in a module, and E-node activation is a measure of the level of competition going on in a module. The most important feature of a module is to autonomously categorize input activation patterns into discrete categories, which is facilitated as the association of an input pattern with a unique R-node.

The process goes with the resolution of a winner-take-all competition between all R-nodes activated by input. In the first presentation of a pattern to a module, all R-nodes are activated equally, which results in a state of maximal competition. It is resolved by the inhibitory V-nodes and a state-dependent noise mechanism. The noise is proportional to the amount of competition, as measured through the number of active R-nodes by the A-node and E-node. Evolutionary mechanism gives a possibility of change to the phenotype of a module through the genetic operators.

The interconnection between two modules means that all R-nodes in one module are connected to all R-nodes in the other module. These intermodule connections are modifiable by Hebb rule with the following equation:

$$\Delta w_{ij}(t+1) = \mu_t a_i \left([K - w_{ij}(t)]\, a_j - L w_{ij}(t) \sum_{f \neq j} w_{if}(t) a_f \right), \tag{2}$$

$$\mu_t = d + w_{\mu_E} a_E, \tag{3}$$

where a_i, a_j and a_f are activations of the corresponding nodes, respectively: $w_{ij}(t)$ is the interweight between R-nodes j and i, $w_{if}(t)$ indicates an interweight from a neighboring

R-node f (of j) to R-node i, and $\Delta w_{ij}(t+1)$ is the change in weight from j to i at time $t+1$. Note that L and K are positive constants, and a_E is the activation of the E-node. As a mechanism for generating change and integrating the changes into the whole system, we use evolutionary algorithm to determine the parameters in the above learning rule and structure of intermodule connections indirectly with grammatical encoding.

3   Grammatical Development of MNN

Three kinds of information should be encoded in the genotype representation: the structure of intermodule connection, the number of nodes in each module, and the parameters of learning and activation rules. The intermodule weights are determined by the Hebb rule mentioned at the previous section.

A basic element of the L-system can be defined as a module or a functional group composed of several modules for modular neural networks. A module is denoted as $A(x, y)$ where A identifies the name, x represents the number of nodes and y means the connection pointer of the module, respectively. Positive integer of y means the forward connection and negative one does the backward connection. The functional group is represented by a pair of '[' and ']'. One more addition is a special symbol ',' which is used to represent disconnection between two modules.

In order to see how the grammar generates various network structures, assume that we have the following definition of an L-system for modular neural networks.

- Alphabet $V = \{A, B, C, D, \text{','} \}$,

- Axiom $\omega = A(100, 0)$,

- Productions:

$$A(x, y) \rightarrow A(x, 1)[B(x/10 - 2, y)B(x/10 + 2, -1)]C(x/10, -1) \qquad (4)$$

$$B(x_1, y_1) < B(x, y) > C(x_2, y_2) : x > 10 \rightarrow C(x/2, y)C(x/2, y - 1) \qquad (5)$$

$$B(x_1, y_1) < C(x, y) > C(x_2, y_2) : x > 5 \rightarrow [D(x, y), D(x, 0)] \qquad (6)$$

The sequence of strings generated by the parametric L-system specified as above is like this:

$$
\begin{aligned}
A(100, 0) \quad &\rightarrow \quad A(100, 1)[B(8, 0)\mathbf{B(12, -1)}]C(10, -1) &\qquad (7)\\
&\rightarrow \quad A(100, 1)[B(8, 0)\mathbf{C(6, -1)}C(6, -2)]C(10, -1) &\qquad (8)\\
&\rightarrow \quad A(100, 1)[B(8, 0)[D(6, -1), D(6, 0)]C(6, -2)]C(10, -1) &\qquad (9)
\end{aligned}
$$

4   Simulation Results

In order to confirm the possibility of the proposed model, we have used the handwritten digit database of Concordia University of Canada, which consists of 6000 unconstrained digits originally collected from dead letter envelopes by the U.S. Postal Services at different locations in the U.S. The digits of this database were digitized in bilevel on a 64×224 grid of 0.153mm square elements, giving a resolution of approximately 166 PPI. Among the data, 300 digits were used for training and another 300 digits for testing.

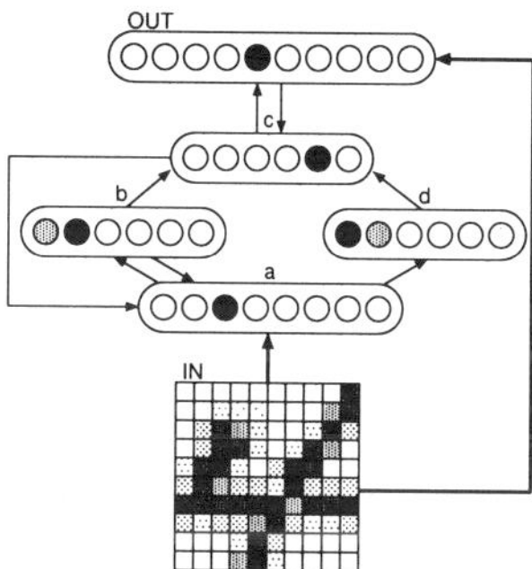

Figure 2: The final modular neural network evolved.

We can see at some representative samples taken from the database that many different writing styles are apparent, as well as digits of different sizes and stroke widths.

The size of a pattern was normalized by fitting a coarse, 10×10 grid over each digit. The proportion of blackness in each square of the grid provided 100 continuous activation values for each pattern. Network architectures generated by the evolutionary mechanism were trained with 300 patterns. A fitness value was assigned to a solution by testing the performance of a trained network with the 300 training digits, and the recognition performance was tested on the other 300 digits. Initial population consisted of 50 neural networks of having random connections. Each network contains one input module of size 100, one output module of size 10, and different number of hidden modules. Every module can be connected to every other module.

Figure 2 shows the final modular neural network evolved. As can be seen, the evolution led to the increase of complexity, and new structures as well as new functionality emerged in the course of evolution: In general, the early networks have simple structures. In the early stages of the evolution some complicated architectures also emerged, but they disappeared as the search of the optimal solution matured. The earlier good specific solutions probably overfitted some of the peculiar training set with lack of generality.

The final network is the architecture producing the best result. This contains four hidden modules of size 8, 6, 6 and 6, implementing different subsystems that cooperatively process input at different resolutions. The direct connection from the input module to the output module forms the most fine-grained processing stream. It is supplemented by a sophisticated modular structure in which a module is globally connected with the input. A sort of hierarchical structure with feedback connections has emerged, and two coarser processing streams as well as local feedback projections support the main processing stream.

It is quite difficult to fully analyze the neural behaviors because they concern with the oscillatory activation dynamics. To make the analysis simpler, we have presented a sample of the class 4 to the final network and obtained a series of snapshots of the internal activations. This system has turned out to produce the correct result with respect to the input as follows:

	(OUT 0123456789	a 01234567	b *	c *	d *)
→	(OUT 0123456789	a 01234567	b 012345	c *	d 012345)
→	(OUT 0123456789	a 01267	b 01345	c 012345	d 0125)
→	(OUT 0123456789	a 01267	b 013	c 2345	d 015)
→	(OUT 0123456789	a 127	b 01	c 45	d 01)
→	(OUT 02456789	a 27	b 01	c 45	d 01)
→	(OUT 4689	a 27	b 01	c 4	d 01)
→	(OUT 49	a 2	b 01	c 4	d 01)
→	(OUT 4	a 2	b 01	c 4	d 01)

In this network, an internal module resolves the competition and induces the correct classification. This shows how the network dynamics finds out the category structure.

In the test of generalization capability, for the patterns that are similar to the trained, the network produced the direct activation through a specific pathway. On the contrary, the network oscillated among several pathways to make a consensus for the strange patterns. The basic processing pathways in this case complemented each other to result in an improved overall categorization. Furthermore, the recurrent connections utilized bottom-up and top-down information that interactively influenced categorization at both directions.

5 Concluding Remarks

We have described a design method of the modular neural networks developed by evolutionary algorithm and a parametric L-system. It has a modular structure with intramodular competition, and intermodular excitatory connections. Compared with the previous work [5], it cannot be asserted that the network evolved based on grammatical encoding gives us a big improvement in recognition performance. However, we hope that this method can give the modular neural network the scalability in complex problems, similarly to the result of [3]. This sort of network will also take an important part in several engineering tasks exhibiting adaptive behaviors. We are attempting to make the evolutionary mechanism sophisticated by incorporating the concept of co-evolution.

References

[1] S.A. Harp, "Towards the genetic synthesis of neural networks," *Proc. Int. Conf. Genetic Algorithms*, pp. 360–369, 1989.

[2] D. Whitley and T. Hanson, "Optimizing neural networks using faster, more accurate genetic search," *Proc. Int. Conf. Genetic Algorithms*, pp. 391–396, 1989.

[3] Kitano, H.: Designing neural networks using genetic algorithms with graph generation system. Complex Systems. **4** (1990) 461–476

[4] X. Yao, "Evolutionary artificial neural networks," *Int. Journal Neural Systems*, **4**, pp. 203–222, 1993.

[5] S.-B. Cho and K. Shimohara, "Evolutionary learning of modular neural networks with genetic programming," *Int. Journal Applied Intelligence*, **9**, pp. 191–200, 1998.

[6] S.-B Cho and K. Shimohara, "Applying L-system to development of evolutionary modular neural networks," *Proc. Asia-Pacific Conf. Simulated Evolution and Learning*, 1998.

Maintaining Knowledge of Conversational Agents

Hidekazu KUBOTA   Toyoaki NISHIDA
Department of Information and Communication Engineering,
School of Engineering, The University of Tokyo

Abstract. In this paper, we discuss how to maintain the knowledge of conversational agents. We have developed a human-style conversational agent called virtualized-ego that represents a real community member. Community members can exchange their knowledge by talking with virtualized-egos. The knowledge of the virtualised-ego is the corpus of past messages of human. The merit and demerit of corpus based knowledge for conversational agents are discussed.

1. Introduction

This paper presents how to maintain the knowledge of conversational agents in creative interaction between humans and agents. A conversational agent that talks humanly can largely help humans because conversation is our daily creative process. It can be good teacher for novices, can guide customers in electronic commerce appropriately [1] and it can also facilitate knowledge creation in a community [2]. In the studies of the conversational agents, how to interact with human is greatly discussed, however how to maintain knowledge of the agents is less discussed. When a conversational agent is practically used in a human community, the knowledge of the agent must be maintained as a status in the community changes. Accordingly, we studies how to handle the maintenance problem about the knowledge of conversational agents.

We have developed human-style talking agents called virtualized-ego that is one's other self which works independently of one's self and can talk on one's behalf. A virtualized-ego stores everyday e-mails and recorded human voices as personal knowledge, and generates a conversation from this knowledge. At first, we describe an evolution of the knowledge of this conversational agent, and next discuss how to maintain the knowledge.

2. Overview of EgoChat

The virtualized-egos work in our system called EgoChat, which is a virtual conversational environment for the community knowledge creation. On EgoChat, community members can talk with virturlized-egos with voices and exchange their knowledge with each other even in some members' absence because virtualized-egos can talk on behalf of absent members.

Figure 1 (a) shows a screen shot of EgoChat. A user is talking with three virtualized-egos, each of which has the face CG of another community member.

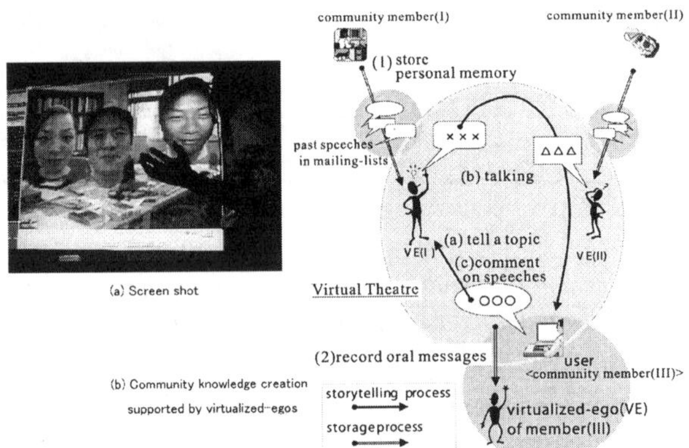

Figure 1. Overview of EgoChat system

The EgoChat system consists of three storytelling processes and two storage processes (Figure 1(b)). In the storytelling processes, community members can share their knowledge by creating a knowledge stream with their virtualized-egos. When a user would like to gain community knowledge, s/he inputs her/his interest into EgoChat by using voice messages(storytelling process(a)). Voice messages of the user are recognized by a commercial speech recognition system, then virtualized-ego(VE)s start talking about related topic(storytelling process(b)). Virtualized-egos will continue talking with each other if the user doesn't interrupt the conversation to comment on speeches of virtualized-egos or change a present topic to another topic(storytelling process(c)).

While talking with virtualized-egos, personal knowledge of a community member is enriched in the storage processes. Before using EgoChat system, the personal knowledge is stored in virtualized-egos by using automated summarizing technology or summarizing humanly from past speeches exchanged on electronic media such as mailing-lists(storage process(1)). Besides text-based speeches, virtualized-egos store newly commented oral messages of a user on EgoChat(storage process(2)), thus virtualized-egos tell community knowledge in past speeches of community members and users add new ideas into their virtualized-egos in a loop of community knowledge creation.

3. Knowledge of virtualized-egos

To manage heterogeneous message such as informal text messages and voices, virtualized-egos have two sets of adaptable representations of the personal knowledge.

3.1 Topics-and-summaries representation

Each virtualized-ego has a set of topics. Past speeches of a community member related to each topic are filed away into personal knowledge of a virtualized-ego. For example, when a virtualized-ego in a community of liquor fans has topics such as "brandy", "beer" and "sake"(Figure 2(a)), some speeches about "brandy" such as "I used to drink V.S.O.P." " I'm fond of diluted brandy."(Figure 2(b)) are stored in knowledge of the virtualized-ego and speeches about other topics are stored in the same way.

All of the past speeches are standardize by labelling with key words that represent contents of the summaries in a word because the speeches are not only summarized text but also recorded voices and they are heterogeneous.

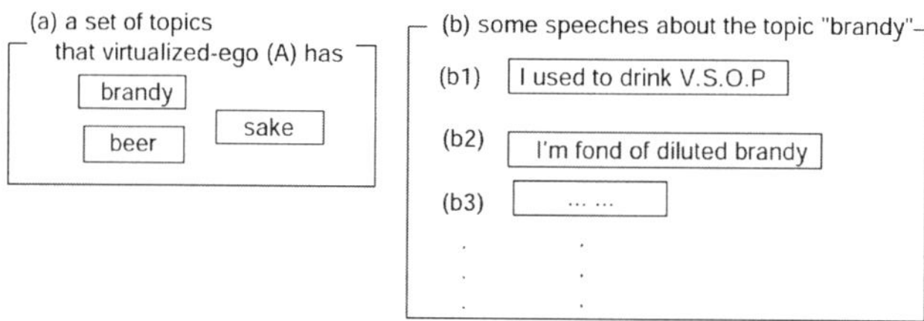

Figure 2. Example of topics-and-summaries

3.2. Flow-of-topics representation

Virtualized-egos change topics at some intervals by referring to an associative representation set of a flow of topics. The associative representation proposed on the work of CoMeMo-Community[3] consists of many-to-many hyperlinks that associate one or more key unit with one or more value unit. The semantics of the associative representation is not defined strictly. Instead, we leave the interpretation of the semantics to human association based on our tacit background knowledge. In the case of virtualized-ego(a)(Figure 3), 'liquor' is a key unit and 'brandy' and 'beer' are value units. This associative representation of virtualized-ego(a) shows a flow of topics from liquor to brandy or beer, and other associative representation such as virtualized-ego(b) has shows other flow of topics.

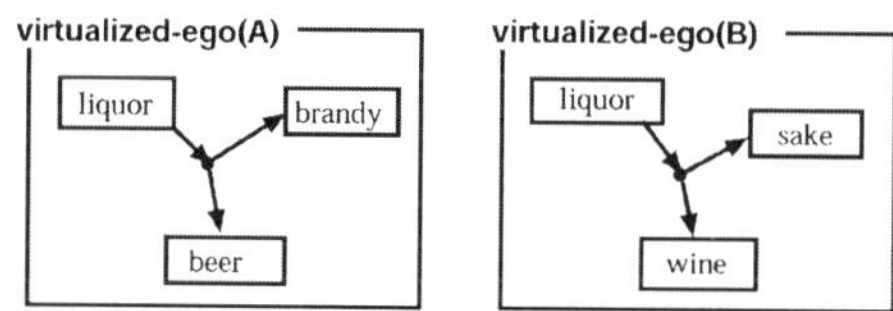

Figure 3. Example of flow-of-topics

Associative representations that shows associations of a community member are stored with ``topics and summaries'' in knowledge of a virtualized-ego. One mediator selected by a user among virtualized-egos selects the next topic that is associated with the current topic, for instance, the message for changing topics is as follows; ``I associate liquor with brandy. Next, let's talk about brandy.'' We suppose that association-based flows of topics make storytelling of virtualized-egos human-like and help users to believe virtualized-egos as community members' other selves that work independently.

3.3. Generating messages from the personal knowledge

In a turn, though all the virtualized-egos select messages associated with a topic at the same time, only one virtualized-ego is selected to speak at a time by comparing priorities of selected messages. Each virtualized-ego generates a priority number when it selects a message in order that the generating conversation may be coherent. In summary, the virtualized-ego that selects the following message after a message mentioned just before is prior in order to exchange messages coherently. A message is regarded as the one that follows the previous message when its key word matches the previous one.

4. Knowledge maintenance of the conversational agent

Difficulty of the knowledge maintenance of the conversational agent is a great difference between the tacit knowledge that appears in a conversation and the explicit knowledge that appears in the most of knowledge representations. The former is too informal and experiential to be generated from the latter that is formal and symbolic. Our approach to solve this difficulty is corpus based. A corpus, in this study, means a set of utterances of a real human member. Virtualized-egos store both literal and oral messages with somes key words and the set of human association. We have made a basic experiment on usability of the EgoChat system to investigate effects of a voice interaction between humans and agents. We summarized a log of the mailing list (15000lines) by hand and stored it into virtualised-egos. Following example is a conversation between virtualized-egos A,B,C and D and a user *(note : the conversation was originally in Japanese)*.

1: A "Let's begin talking about brandy."
2: B "We have more prejudice against brandy than other liquor."
3: A "I often chat over drinking."
4: C "A public bar doesn't deal in brandy because brandy is too expensive."
5: B "Yes, certainly."
6: A "Sure."
7: User "Convenience stores don't deal in brandy too."
8: C "Shouchu *(note : Japanese distilled liquor made from wheat or rice.)*
is more low-priced than brandy."

One merit of corpus based knowledge is easy to add new knowledge into virtualised-egos. We can add our knowledge by simply speaking to virtualised-egos like the line No.7 in the above because the utterance is directly stored in the loose-structured knowledge representation of the virtualised-ego. The other merit is the naturalness of the each speech because the corpus consists of original utterances of human. In the experiment, we interviewed three users who talked with virtualiezed-egos in 3 minutes, then got positive answer about the naturalness of the conversation.

On the other hand, the corpus based knowledge lacks flexibility and is difficult to revise because it largely relies on the original context.

5. Conclusion

In this paper, we discussed how to maintain the corpus based knowledge of conversational agents. We suppose that more conversational maintenance can improve the demerit of this kind of knowledge because our conversational method for adding new knowledge feels casual. Now, we are developing a conversational method for the knowledge maintenance of the virtualised-egos on the EgoChat and planning to apply it to the knowledge creation in a large community.

References

[1]　T. Bickmore and J.Cassell : "How about this weather?" Social Dialogue with Embodied Conversational Agents. In Proceedings of Socially Intelligent Agents : The Human in the Loop, pp.4-8, 2000.

[2]　T. Nishida, N. Fujihara, S. Azechi, K. Sumi, and H.Yano : Public opinion channel for communities in the information age. New Generation Computing, Vol.14, No.4, pp. 417-427, 1999.

[3]　T. Nishida: Facilitating community knowledge evolution by talking vitrualized egos. In H.J.Bullinger and J.Ziegler editors, Human-Computer Interaction Vol 2, pages 437-441.Lawrence Erlbaum Associates, Pub. 1999.

KES '01
N. Baba et al. (Eds.)
IOS Press, 2001

Voice Cafe: Conversation Support System in a Group

Tomohiro FUKUHARA[*1,*3], Toyoaki NISHIDA[*2], and Shunsuke UEMURA[*2]
[*1]Synsophy Project, Communications Research Laboratory
[*2]School of Engineering, The University of Tokyo
[*3]Graduate School of Information Science, Nara Institute of Science and Technology

Synsophy Project, 2-2-2 Hikaridai, Seika-cho, Kyoto 619-0289, Japan
`tomohi-f@synsophy.go.jp`

Abstract. We propose a system for sharing *wet information* in a group. Wet information, which is informal information on group members, is important for collaboration in a group. Wet information facilitates members to know each other. The problem is a lot of time is required for exchanging wet information. Busy members have little time to exchange wet information. We propose a system called Voice Cafe for sharing wet information in a group. In Voice Cafe, conversational agents gossip about group members. Members can easily find other members' conditions, schedules, thoughts and opinions by listening to gossips made by agents. Design concept and an overview of Voice Cafe are described.

1 Introduction

Wet information is important for good collaboration in a group. Wet information here means informal information that is not critical to the tasks of the group[1]. When we have a chat with colleagues in lunch time, we exchange wet information, that are not related to their tasks, such as sport games, plans for holidays, movies and music, and so on. These kinds of wet information facilitates members to understand each other, and enhance relationship among members. The problem is that chat takes a lot of time. Busy groups often have little time to have a chat.

We propose a system called Voice Cafe for sharing wet information in a group. In Voice Cafe, conversational agents (CAs) speak gossips on members. Gossips include members' conditions, schedules, thoughts and opinions, and so on. By listening to the gossips, members can easily find about others.

This paper consists of following sections. In section 2, our design concept for sharing wet information is described. In section 3, an overview of Voice Cafe is described. Section 4 discusses privacy related issues.

2 Design concept

Gossips have advantages to facilitate sharing wet information in a group. CAs are designed to elicit and circulate gossips in a group.

2.1 Gossips for sharing wet inforamtion

Gossips have a role for sharing information among people[2]. We often get information of others from gossips. We also find much about others' reputations, conditions, attitudes, thoughts and opinions from gossips. Although these gossips are not always true, it is easy for us to get information on others.

In this paper, we regard a gossip as a medium that mediates information among people. Fine et al. define a gossip as one's opinion of other's endowments or behaviors[3].

(a) Screen images of Voice Cafe. Right screen shows a display image of Voice Cafe server, and the left screen shows a display image of a client (CA). Each CA's figure is appeared on the server's screen.

(b) CAs are embedded in physical objects. In this case, CAs are embedded in a pot and a cup. Speakers are connected to client PC. Client PC speaks by their TTS system.

Figure 1: Overview of Voice Cafe.

In addition to their definition, we regard a gossip as a medium that mediates information in a group. Although the term "gossip" has negative image, we focus on positive aspects of gossip that facilitate sharing information in a group.

Existing Computer Mediated Communication (CMC) tools such as e-mail, BBS, chat have limitations for sharing wet information in a group. This is because existing CMC tools are not designed for sharing wet information in a group. Exchanging wet information is easier in Face to Face (FTF) situations than communications by existing CMC tools. We need a CMC tools for sharing wet information in a group.

In the field of CSCW, several researches on awareness support systems are proposed. Siio proposed the coffee aroma display that generates coffee aroma when group members are gathered in a meeting room[4]. Users can easily find that people are gathering in the meeting room by coffee aroma. Although this system enables user to have an opportunity to have FTF conversation, users can't know about what members talk in the room. We think to know contents of talking directly is important for sharing wet information.

Brave proposed a haptic communication system that enables to share tactile information over the network[5]. This system enables a user to touch information such as gentle or hard touch each other. Although this system facilitates a user to share nonverbal wet information, it is difficult to understand the meaning of the touch. Much tacit knowledge is required for understanding the meaning of touch. We think verbal communication is also needed to share wet information in a group.

2.2 Implementation design

We design Voice Cafe based on *conversational agent (CA)*. CA is a physical object that gossips on members. CA inquires questions to members. When a member answers, the answer is utilized as a gossip by CAs. Member also add gossips on others by typing in

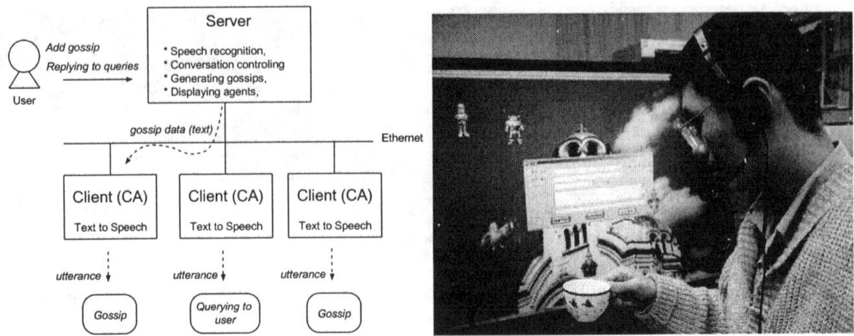

(a) Architecture of Voice Cafe. Voice Cafe consists of a server and several clients (CAs). Server recognizes speeches from users, and generates gossip data that is spoken by clients. Clients speak according to gossip data received from the server by their Text-To-Speech systems.

(b) Conversation with CAs. A cup inquires a question to a user. The user replies to him by using a microphone connecting to the server. Answers from the user is stored in gossip database in the server, and utilized as gossip data.

Figure 2: Architecture and interaction.

the gosipps. CAs gossip according to those gossips.

Gossips are made by CAs. When a member comes to CAs, one CA inquires variety of topics such as his or her condition, whether s/he likes baseball, what kind of music or movies does s/he like and so on. When the member answers, CA recognizes the contents of the answer and store the answer to a database. When another members comes to the CAs, CAs gossip on the topic what they heard from other members. When there is no one in front of CAs, they gossip about members freely. By hearing gossips, members can easily find wet information of others.

3   Voice Cafe

We describe a prototype system called Voice Cafe. Voice Cafe enables users to hear and talk gossips on members. Figure 1(a) and Figure 1(b) show an overview of Voice Cafe. We describe an architecture of Voice Cafe and examples of gossips in this section.

3.1   *Architecture of Voice Cafe*

An architecture of Voice Cafe is shown in Figure 2(a). Voice Cafe consists of a *server* and several *clients (CAs)*. Server recognizes speeches from users, generates and provides *gossip data* to CAs. Gossip data is the contents to be spoken by a CA. CAs speak according to the gossip data received from server by using TTS.

When the server generates a sequence of a gossip, server selects a topic from *topic database* that contains various topics such as baseball players, music, movies, and so on. Then, server retrieves related topics from *members database* that contains members'

Table 1: An example of gossips. Agents gossip about a baseball player and a group member who likes the player.

Talker	Utterance
User A	(*Inputting description of* "*Ichiro*".)
Agent A	I remember "Ichiro".
Agent B	Ichiro plays gardening.
Agent A	You are kidding me.
User A	Do you remember "Ichiro"?
Agent A	Yes, I do. "Ichiro" is a baseball player. Tomohiro is an "Ichiro" fan.
	. . .

(a) Inquiring a question on a baseball player.

Talker	Utterance
User B	Hi, boys and girls.
Agent A	Who are you?
User B	My name is Midori.
Agent A	Hi, Midori.
Agent C	Do you like "Ichiro"?
User B	I don't like him.
Agent C	I remember that you don't like "Ichiro".
Agent B	Tomohiro said that he was an "Ichiro" fan.
	. . .

(b) Gossip on a member.

opinions and preferences on the topics. Finally, server defines which CA to speak gossip data, and send the data to the client.

Figure 2(b) shows a conversation between a user and a CA. In this figure, user listens to gossips made by a cup and a plant. User also replies to the cup by using a microphone connecting to the server.

3.2 Conversation

Examples of gossips are shown in Table 1. In these examples, CAs gossip on a baseball player. In Table 2(a), agent A remembers that a member named Tomohiro likes a baseball player. User can type in what he wants to gossip by agents. In this case, he typed in that he likes the baseball player. Typed data is utilized for gossiping among CAs. Table 2(b) shows gossip on a member. In this example, CAs gossip about Tomohiro when another member named Midori appears.

4   Discussion

In this section, we discuss the following issues.

1. Privacy of members
2. Malicious information

4.1 Privacy of members

Privacy of members must be kept. We suppose that there are two kinds of personal information: (1) innocuous personal information and (2) deep personal information. Former information includes one's condition, tastes for movies and music and so on. Latter information includes his or her age, credit card number, his or her salary and so on. CAs should talk innocuous personal information, however, to distinguish innocuous information from deep personal information is difficult. This is because what is innocuous information is depend on users, i.e., some topics are allowed in some members, on the other hand, those topics are forbidden in the other members.

One approach to keep privacy is to let group members to choice privacy level of CAs manually. There are several levels of personal information from level 1 to level n. In each level, members can set what topics are safe and allowed to talk by CAs. Once topics and levels are set, CAs gossip by obeying those topics and the level. We are planning to implement a function in which users can set topics and privacy level.

4.2 Malicious information

There may occur situations in which some members put malicious information of others to the system such as abuse of others or untrue information. This kind of information may harm members and relationship between them. To identify malicious information from other information is difficult because to define malicious information is different in groups. In some groups, members may allow CAs to talk abuse of others. On the other hand, members are not allowed to talk abuse of others in another groups.

We are planning to implement a function to reveal speaker's name when CAs gossip. By revealing speaker's name, member has to responsibility to their utterances. To prevent malicious information in Voice Cafe is our future work.

5 Conclusion

We described a system for sharing wet information in a group. Wet information is important for good collaboration in a group because wet information facilitates members to know each other. We propose a gossip-based conversation system called Voice Cafe. We are planning to implement a privacy control function in Voice Cafe, and to apply the system to a real group.

References

[1] Azechi, S., "Dry information, communication, and communities", *in Proceedings of Fourth International Conference on Knowledge-Based Intelligent Systems & Allied Technologies (IEEE KES2000)*, 2000, pp.72–75.

[2] Kawakami, Y., "Uwasa ga Hashiru (Spreading Rumors)", Selection social psychology 16, Tokyo, Saiensu-sha, 1997, (ISBN 4-7819-0840-3), (Japanese)

[3] Fine, G.A. et al., "Gossip, gossipers, gossiping", *Personality and Social Psychology Bulletin*, Vol. 4, 1978, pp. 161–168.

[4] Siio, I. et al., "MeetingPot: Informal communication support by an ambient display", *in Proceedings of Interaction 2001*, Information Processing Society of Japan (IPSJ), 2001, pp.163–164. (Japanese).
http://siio.ele.eng.tamagawa.ac.jp/projects/pot/

[5] Brave, S. et al., "Tangible Interfaces for Remote Collaboration and Communication", *The 1998 ACM Conference on Computer Supported Cooperative Work(CSCW'98)*, 1998, pp.169–178.

Communication between Human and Artifact based on Embodiment

Kazunori TERADA[†], Tomohiro FUKUHARA and Toyoaki NISHIDA
Synsophy Project, Communications Research Laboratory
2-2-2, Hikaridai, Seika-cho, Soraku-gun, Kyoto, 619-0298, Japan
E-mail: kazuno-t@crl.go.jp[†]

Abstract. In this paper we argue the communication between human and artifact from two viewpoints. One is the necessity of communicative skill and the other is embodiment aspect of communication. The necessity for the communicative capability of the artifact concerns the activity of the artifact. The function of artifact is defined by embodiment relation between human and artifacts, and varies depending of the change of the embodiment. The active artifact must perceive this functional relation so that it can behave appropriately. We introduce an autonomous mobile briefcase as an example of active artifact to investigate human-artifact communication.
keywords: *active artifact, communication,embodiment*

1 Introduction

Recently, human-machine communication increasingly attract attention in many research areas, and many researchers are studying communicative skill of artifact, such as machine, computer, robot, or agent. Jijo-2 is a mobile robot, which is able to build a probabilistic map of its office environment by acquiring missing location information through conversational dialogues with people[6]. Bischoff et.al. built a humanod robot HERMES which can deliver objects and guide in office-type building with natural language communication[2]. The communicative skill is important for the multi purpose robot such as humanoid robot because it has to execute complex task and interact to human. Therefore, we can say that complexity of communicative skill depends on the complexity of the functional relation which is emerged from interaction between human and agent.

In this paper, we focus on the communication between human and artifacts. The artifact defined in our research does not mean a multi-functional robot such as humanoid but a single functional artifact such as a cup or a chair. Through building such communicative artifact, we will discuss the physically embodied communication. Roy mentioned in his paper that language should be grounded in sensory-motor experience[7]. Machines which process text and spoken language without grounding sensory-motor experience suffer from Symbol Grounding Problem[5]. We consider that a message sent from an artifact to realize its function can be origins of physically embodied message.

In this paper we first present a relation among active artifact, its function and communication, and then describe an architecture of active artifact.

2 Communication between human and artifact

We discuss the communicative skill of the artifact from two viewpoints. One is the necessity of communicative skill and the other is embodiment aspect of communication.

2.1 *Necessity of communicative skill*

The necessity for the communicative capability of the artifact concerns the activity of the artifact. The passive artifact complete its function only by being manipulated by human, and such artifact does not need to communicate to human.

Although the artifact does not need to communicate as long as it is passive, once it becomes active, it needs to communicate to human. Each artifact has its own function. For example, a chair has the function to sit down and a cup has the function to collect water. The active artifact is going to realize its function autonomously and actively. An artifact which realize its function actively should behave appropriately to the situation. The situation varies depending on the functional and mental relation between human and the artifact. In order to know the situation and confirm the humans intention, the artifact should communicate to human.

2.2 *Embodiment aspects of communication*

Some psychological researches have concluded that more than 65 percent of the information exchanged during a face-to-face interaction is expressed through nonverbal information in human-human communication[1]. Recently, in engineering, the importance of embodiment in communication is pointed out and many researches develop communication tool based on embodiment[3][4].

Communicative skill of artifact can be mainly divided into two modes; (1) Verbal mode and (2) Nonverbal mode. Language is a powerful tool for us to communicate each other, and it allows the transmission of meaning from one to another using some physical media such as sound. Similarly, it is considered that embodiment plays a role as a medium in communication. In some case, nonverbal mode is superior to verbal mode because it can transmit non-logical and intuitive information effectively. Nonverbal information means a gesture, physical features of artifact, and haptic motion, etc. It is also considered that even passive artifacts send messages to human through nonverbal information by merely its existence. In this case, nonverbal information is equivalent to physical features, i.e. appearance. Appearance of artifact become an assistant for us to know what the purpose of it is, since each artifact has a peculiar appearance which usually expresses a function. The appearance is perceived by human as nonverbal information. It can be said that this is comparatively negative communication although it may be a kind of communication. What we intend to deal with is positive communication.

The process of translation of embodiment media is considered as that an agent including human perceives embodiment media and it is compared by its embodied experiment. In order to communicate by means of embodiment media, both sender and receiver must assign the same meaning to the common embodiment media. If this assignment differs between sender and receiver, the message is misunderstood. A common understand of embodiment media between human and artifact is important. However, in this case, problem is more difficult

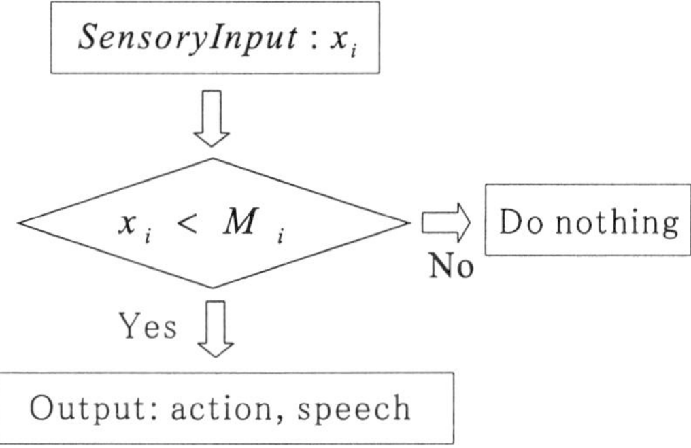

Figure 1: Motivation based system. M_i is threshold of motivation which is predefined by human designer.

because our embodiment is quite different from many artifact's except humanoid robot. If the sender's embodiment is different from receiver's embodiment, receiver hardly understand the embodiment media.

In the process of designing the protocol of embodiment media, the designer of artifact extracts essential embodiment protocol of human-human communication by observation and it is implemented to the artifact. Because almost all artifacts don't have the same embodiment as human, it is difficult to send message through embodiment media. For example, we usually shake our head to send a message of "NO". In order to express such gesture, the message sender should have appropriate embodiment, i.e. head, body, neck, and locomotion ability to realize such action. Almost all artifacts, for example chair or refrigerator don't have such embodiment except humanoid robot. Consequently, such an artifact should use alternative embodiment media to send its message. Fortunately, human can adapt novel or strange phenomenon, human may understand a embodiment media even if it is not completely similar to humans one.

The significance of artifact is to realize its function for human. The function is defined by embodiment relation between human and artifacts, and varies depending of the change of the embodiment. For example, the figure and the size of a computer mouse is meaningful only for humans hand. The active artifact must perceive this functional relation so that it can behave appropriately. The functional relation is perceived through sensors equipped on artifact's body.

3 Active artifact based on motivation

Now we consider an artifact which realize its function actively. The capability of communication of such artifact is used as a method to realize its function appropriately and actively. In this section we describe a concrete system.

3.1 Overview of system

We implement an active artifact as a system based on motivation. That is, the function of artifact is described as motivation, and it realize its function as a result of acting based on

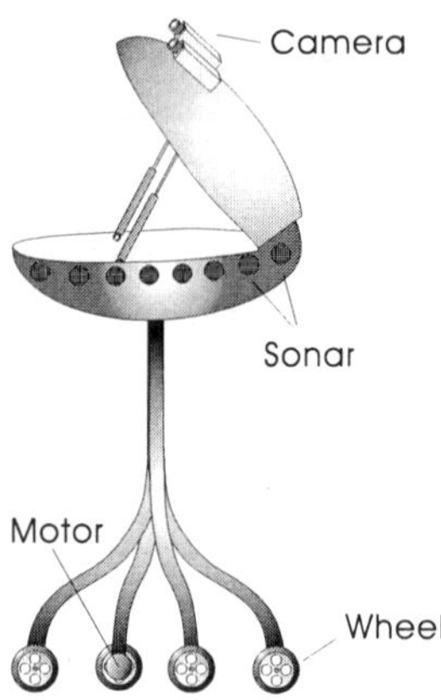

Figure 2: Overview of the autonomous mobile briefcase

the motivation. Figure 1 shows the proposed architecture of active artifact. In other words, motivation is desire. The motivation is described as target value of sensory input. If the value of sensory input correspond to the value of motivation, desire is filled. Therefore, the purpose of the artifact is acting so that desire may be maximized quickly. The artifact can move in physical world and speak to human in order to fill desire. Communication between human and artifact does not need until the artifact would work upon human. We consider that such message is embodied.

The assessment of the situation is important for the active artifact to realize its function efficiently. To work upon human when he does not need is useless.

We are surrounded by many artifact, for example an electronic range, a washing machine, and a chair, etc. These artifact have peculiar functions. A washing machine has a function to wash clothing and a refrigerator has a function to cool foods. The activity is added to the original function. An active chair has wheels to moves around and talk to human in order to make him sit down. The washing machine wants to wash clothing and looks for clothing which should be washed. The washing machine may even talk to human "Isn't is better to wash the dress which you have on soon?" The massage is only advise for human but it is a strategy for washing machine to maximize its desire.

3.2 *Autonomous mobile briefcase*

We propose an autonomous mobile briefcase (AMB) as an example of active artifact to investigate human-artifact communication(See Figure 2). A briefcase is a case in which we put documents and stationary and which is carried by hand. Thus, the briefcase has two functions, i.e. 1)keeping goods and 2)carrying goods. The AMB realize these functions actively and communicate to human. The active artifact need equip actuators and sensors. The most important actuator is wheels to move its body for going to human and bringing goods. The active artifact also need some actuators to open/close door of the case. Sonar sensors are equipped around the body for the purpose of obstacle avoidance. Sensors also have roles of judgment whether desire is filled or not. Motivations of the AMB are desire for keeping objects and desire for carriage objects. The AMB has tactile sensors in the shell so that it can judge whether objects are in the shell or not.

4 Conclusion

In this paper we argued communication between human and artifact from the view point of activity. The message which is sent when the artifact ought to realize its function actively is considered as an embodied message. In the future work, we will built real active artifact and have some experiments.

References

[1] Michael Argyle. *Bodily Communication.* Methuen & Co., 1988.

[2] Rainer Bischoff and Tamhant Jain. Natural communication and interaction with humanoid robots. In *Second International Symposium on Humanoid Robots*, pages 121–128, 1999.

[3] J. Cassell, T. Bickmore, M. Billinghurst, L. Campbell, K. Chang, H. Vilhjálmsson, and H Yan. Embodiment in conversational interfaces: Rea. In *Proceedings of 1999 International Conference on Intelligent User Interfaces*, pages 520–527, 1999.

[4] A. Guye-Vuillème, Tolga K. Capin, Igor Pandzic, Nadia MagnenatThalmann, and Daniel Thalmann. Nonverbal communication interface for collaborative virtual environments. In *Proc. CVE 98, Manchester*, 1998.

[5] Stevan Harnad. The symbol grounding problem. *Physica D*, 42:335–346, 1990.

[6] Toshihiro Matsui, Hideki Asho, and Futoshi Asano. Map learning of an office conversant mobile robot, jijo-2, by dialogue-guided navigation. In *International Conference on Field and Service Robotics*, pages 230–235, 1997.

[7] Deb Roy. Grounded speech communication. In *International Conference on Spoken Language Processing*, 2000.

KES '01
N. Baba et al. (Eds.)
IOS Press, 2001

Motivation for Showing Opinion on Public Opinion Channel: A Case Study

AZECHI Shintaro[*1] and MATSUMURA Ken'ichi[*2]
[*1] *Hokkaido Tokai University, Minami-ku 5-1-1-1, Sapporo, Japan*
[*2] *Synsophy Project, CRL, Hikaridai 2-2, Seika-cho, Kyoto, Japan*

Abstract. How to enhance user's motivation to show her/his own opinion to the communication supporting tools is an important issue. For a user, showing opinion needs much of cognitive resource and has a risk to be abused by some other members, so that they are immediately discouraged to show their opinions. We researched how user's motivation to show her/his own opinion is constructed. We interviewed six Public Opinion Channel users and analysed their propositions. We find four perspectives that construct users' motivation: 1) User's motivation for showing opinions is easily discouraged by non-sense or trash information, 2) Setting clear purpose for using the tool makes users' motivation encourage, 3) Showing the size of user group and keeping anonymity encourages the motivation, 4) Feeling of sharing information among users enhance showing a new opinion. We discuss based on those four findings, how we could improve communication tools.

1. Problem of Motivation for Showing Opinion

How can communication tools enhance users' motivation showing her/his opinions? This is unsolved and important problem for designing communication-supporting tools [1].

For a user, showing her/his opinion is very difficult. There are two problems. 1) Constructing an original, novel, interesting and consistent opinion needs much load of cognitive resources, 2) a user is always exposed to fear of her/his opinion would be refused or ignored by other users. A communication-supporting tool must be designed to solve these users' problems of excessive effort and fear.

For example, it is considered in the development of Public Opinion Channel (POC) [2] how user's motivation for showing her/his opinion should be encouraged. POC is a community-communication supporting broadcast system. POC aims to develop an automatic broadcast system. We suppose that POC gathers daily life opinions in a community. This non-special ignorable information are edited and broadcast to the community by POC. In this enhanced informational cycle, we assume, a community mediated by POC should be activated.

In the POC mediated community, there is also the problem of discouraging users the motivation for showing opinion. When the POC user's motivation is very low, the informational cycle will be not realized. We should design POC to improve to enhance user's motivation for showing her/his opinions.

We start to solve this problem by gathering user's opinion for using POC actually. We research how users' motivation for showing opinion is constructed. We find there are four factors that affect users' motivation. : 1) User's motivation for showing opinions is easily discouraged by non-sense or trash information, 2) Setting clear purpose for using the tool makes users' motivation encourage, 3) Showing the size of user group and keeping anonymity encourages the motivation, 4) Feeling of sharing information among users enhance showing a new opinion.

These findings are useful for not only improving POC but also enhancing communication by CMC, though these are limited findings because our research is a prosaic case study.

2. Method

We interviewed six POC evaluative users who have used POC for more than four weeks. These evaluative users are the development group members including designer, programmer, psychologist and secretary. These interviewees are not naive subjects but experts for POC.

We call them S1 to S6 for convenience to quote her/his answers. Interview carried out for 15-20 minutes for a user.

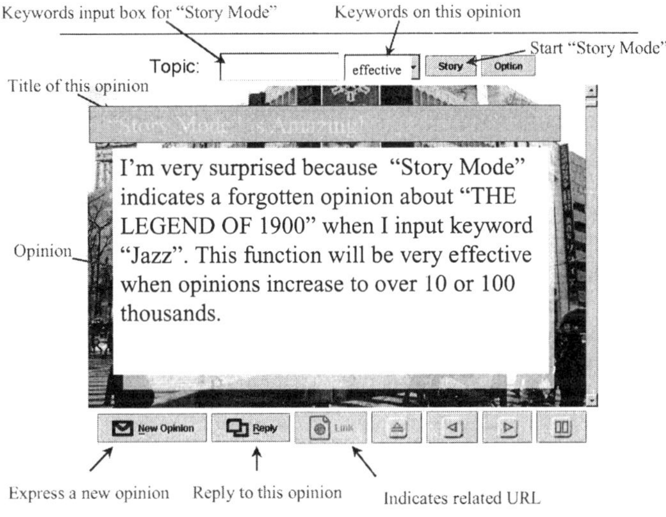

Figure 1: Screen shot of Public Opinion Channel (evaluative beta system): opinions shown by community members (users) are stored and indicated on the main box of the screen in the order of recently first. The "Story Mode" makes a sequence of opinions to be contingent on a inputed keyword.

2-1. Stimulus: POC beta system

POC is a community-communication broadcasting system. Recently, developing team complete to implement POC evaluative beta-version [3].

Although POC intents to be a broadcast system by voice, this version of POC is character-based. And it looks like a bulletin board system on the Web (Figure 1). This system stores opinions shown by community members (users) and indicates those in the stacked order. An opinion is indicated per character in a sequence that imitates and corresponds to the future implementation of voice broadcast unit.

This version of POC has the "Story Mode". When user input a keyword, POC system edits a sequence of opinions corresponding to the key word. So that user can know what other community members are via interested in.

We researched to make clear how users' motivation for showing her/his opinion constructed in via a community by interviewing users of this beta-version POC. Although this beta-version POC is limited implemented as broadcast system, it satisfies fundamental functions of community-communication supporting system. Findings of this research would be useful to improve POC to the broadcast system.

2-2. Questioner

Each interviews carried out like as free discussion not to prevent to draw users' honest impression for POC. There is a questioner containing following five questions that would be convenient for interviewers.

1. How frequent do you use POC on a day and how do you write your opinion in it?
2. How do you use POC? Do you have motivation to use POC?
3. How do you feel about anonymity of POC?
4. Do you think that you can share information among other POC users?
5. Another opinions for using POC.

3. Results

Generally, users say her/his motivation for showing her/his opinion on POC is low. They seem don't have desire to use POC continuously. User's motivation for showing opinion is very low. For examples, S2 (31 / male / cognitive psychologist) said, "I don't have motivation to use POC continuously because there is no special valuable information and opinion", or S4 (26 / male / programmer) said, "I can see some interesting messages on the POC viewer, but all of them isn't useful."

However, forty days has passed since this interview, all six interviewees continuously uses the beta POC. Also at least four users show their opinions actually. Although even the development team members regard POC beta system as a "boring communication system", it seems there is some fascination to facilitate the motivation using the system.

3-1. User's motivation is easily discouraged by trash information

Some users told about the rate of non-sense or trash information of the whole of messages of POC. They think it is very high. They are discouraged to show opinions on POC.

For example, S3 (31 / female / secretary) said, "In this era of the informational flood, reading crowded messages pushed by the system is going against the times".

Filtering trash messages is one of the solving. Removing messages, which are not related to a central topic, would prevent discouraging motivation of the users.

On the other hands, filtering non-sense messages might decrease the amusement of POC. Some user think there are some useful topics in the "trash". For example, S5 (31 / female / social psychologist) said, "I got new knowledge what I hadn't been interested in."

At present, we al least filter the trash information such as repeating messages or abuses., although trash information might also contain valuable information. What is the "trash" is also difficult to be defined. How to filter trash information proposes very difficult problem to develop community-communication tools.

3-2. Setting clear purpose for using the tool makes users' motivation encourage

The users are told that "You can use POC freely, there are no rules for using", because the users are the members of POC developing team and they know the POC system familiarly. Although developers expected the users use POC for discussion, at last, it has been used like as a chat system. S5 said, "Indicated opinion sequence contains many loose topics. It doesn't suit for a discussion along to one subject".

Showing the purpose of using of POC for users, they might be enhanced the motivation for showing opinion on POC. POC has been regarded as a tool for decision-making along a concrete subject. On the other hands, a possibility is appears that POC is appropriate to propose of making friendship and good mood in a community. S6 (30 / male / social psychologist) said, sharing information and opinion by POC "must be some of the base for discussion that is like tacit knowledge [4]".

S4 said "POC is like an instant message system, so I can positively write some of slight messages in it". And S6 said, "I use POC as a memory for thinking aloud". If system's setting allows all users such easygoing sense about message sending, the system should encourage user's motivation. Making good mood in a community preparing an important decision-making discussion should be one of important role of POC.

3-3. Showing the size of user group and keeping anonymity encourages the motivation

Some users have fear to join in the user's group because s\he doesn't infer what a topic would be appropriate for the community. Anonymity of POC seems increase this fear. S1 (28 / male /informational engineer) said, "I'm unpleasant other users say their personal opinions with anonymity".

On the other hands, community closure seem encourage the motivation. S2 said, "I'm interesting that I could infer one write particular opinion in this community, although it is anonymous". In a small and closed community, members are fixed, members could show her/his opinion easily, because there is low possibility to be blame or ignored.

3-4. Feeling of sharing information among users enhance to express new opinion

If users feel they could share information with other users, motivation for presentation should be encouraged. S4 said, "When a response to my opinion is coming, I'm glad and reply to it soon again". Feeling of sharing information would enhance motivation to show new opinion.

4. Discussion

The interviewees have some other ideas encourage motivation for showing their opinion on POC. S2 said that input method by keyboards is complicated, so that POC should be implemented of the input method by voice. S6 proposed some event, for example a user who show 100th or 1000th opinion would be prized, it should enhance users' motivation to show opinions.

Enhancing user's motivation is important to deign a communication-supporting tool. We interviewed POC evaluative users and find four factors that construct user's motivation for showing opinion. Although this is limited research for evaluative user (development team), these results are useful for improving POC and some other community-communication tool design.

References

[1] S. Azechi, Social Psychological Approach to Knowledge-creating Community,. In: Nishida, T. (ed.): Dynamic Knowledge Interaction, CRC Press, 2000, 15-57.

[2] S. Azechi, N. Fujihara, S. Kaoru, T. Hirata, H. Yano, and T. Nishida, Public Opinion Channel: A Challenge for Interactive Community Broadcasting, In T. Ishida (ed.): Digital Cities: Experiences, Technologies and Future Perspectives, Lecture Notes in Computer Science, Springer-Verlag, 2000, 427-441.

[3] S. Azechi, T. Fukuhara, N. Fujihara, K. Matsumura, K. Terada, H. Kubota, and T. Nishida, Public Opinion Channel: A Perspective from Informational Technology and Social Science, Proceedings of the 11th Meeting of Special Interest Group on AI Challenges, 2001, pp. 13-20. (in Japanese)

[4] I. Nonaka and H. Takeuchi, The Knowledge-creating Company: How Japanese Companies Create the Dynamics of Innovation, Oxford University Press, New York,.

KES '01
N. Baba et al. (Eds.)
IOS Press, 2001

Consensus Formation Process in Network Community

Ken'ichi Matsumura

Synsophy Project, Keihanna, Human Info-Communication Research Center, CRL, Japan
Department of Social Psychology, Graduate School of Human Sciences, Osaka University,
Japan

Abstract.   The opinion of minority members will play important role for community to form better consensus in network closed-community. Community members cannot estimate the distribution of the opinion in community precisely. To form better consensus, the opinion of minority had better be reflected. But in labor union, minority members underestimating the attitude of other member and they overestimate the percentage of members supporting minority members. In network community, the same phenomenon will be found. Therefore the opinion of minority will be important for community member to form better consensus.

1. Introduction

In this paper it is discussed the role of minority in network community. And the important factor for community to reflect the opinion of minority will be discussed.

Recently a self-government and inhabit group have a meeting in development of a city and so on to get a consensus between them. Consensus formation between a self-governing body and the inhabitant group can be considered as the consensus formation between two groups. A self-governing body and an inhabitant group consist of many members respectively, and they must form consensus in each group. In this case, they have two consensus formation processes. One is a consensus formation process between the self-governing body and the inhabitant group. The other is the consensus formation process in each group. The strategy to exchange of the information is definitely difference in each process. It is clear that one group sends a message to the other group in process between two groups. In group process, one message from group's member is send to all members in group. We discuss the consensus formation process of the intragroup.

In the network community, it will be able to be said as well. But various communities exist in the network community. In one community, the motive of members is joining a community. In another community, all members have shared goals. In other community, all members resolve some problems with cooperation. People have various motives of joining a community, and the network communities have various goals.

In this paper, the consensus formation process in the network community that does intellectual activities is discussed. Particularly we will discuss the importance of minority' opinion in closed-community.

1) The type of network community
2) For reflecting the opinion of minority
3) False consensus

Based on the result of an investigation in organization, we will discuss these points.

2. Network community

There are mainly two types in network community. They are the open-community and closed-community. Open-community is the community that everyone can participate. It is probably that members who have motives of joining it participate. Participants are unsettled in open-community. Members frequently will get out of community and other members come into it. Therefore the consensus between members will hardly be formed in open-community, it is not need to discuss it.

On the other hand, members are settled, and have a shared goal in closed-community. In closed-community, it is almost that the members know who participates in it one another. Then, information is exchanged here in order to resolve a problem, discuss a problem, and so on. In closed-community, the consensus formation is required in many cases.

2-1. Opened-community

On network, many people often gather with the same purpose. In opened-community, the purpose of the members are gathering and talking about specific topic. The members do not need to performance in it. Therefore, in opened-community, we can't see the consensus formation. Members in it are not form consensus but public opinion. Public opinion is " the opinion or attitude to specific topic that most of people in community express". But the members of opened-community are unsettled in that community. Therefore, public opinion in that community will be always changing. And it is not important in that community. So, opened-community should be called the place that people gather in. The community as the place that people gather in has few settled-member. In this community, many people will not contribute to this community.

2-2. Closed-community

Closed-community is defined as the community that members are settled. In closed-community, it is easy for members to know who is other member. For example, workgroups in organization are closed-community. The members of it know other members in real situation. In closed-community, the community has some goals, and community members share them. And it pursues the best performance. The opinion of minority is important to form better consensus in community. For the above reason, we discuss that the opinion of minority is reflected on the consensus formation in closed-community.

3. Why is the minority important?

In general, the consensus is defined as agreed attitude or opinion expressed. Groups often decide something by majority. But decision by majority is not reflected minority opinion. We doubt that these decide by majority are the best. And we suggest that the opinion of minority must be considered.

There are two reasons that we suggest the opinion of minority must be considered. One is that we find it dubious that decision by majority is better. The other is that minority activate a community. In earlier study, Nemeth argued that the opinion of minority activate divergent thought of majority [1]. Because the new aspects of a subject are provided community members by minority members, they make majority members try to view it from various aspects. In network community, a similar effect of minority will be expected. This effect will make community members send more messages to other member, and the exchange of their opinion will be activated in community. In this point, the opinion of minority is useful for network community. Therefore, minority members must not be ignored. But minority may not be able to express their own opinion.

4. For reflecting the opinion of minority

How does minority opinion influence majority opinion? Matsumura (1999) discussed about the process of minority influence in discussion situation. [2]

4 members of the group discussed atomic power generation with confronting one another. In all session, minority always is agreeing with the use of power generation, and majority is the contrary. In discussion situation, the degree of agreement with minority is important factor for majority to change the opinion. And it is relation with the persuasive minority and the degree of confidence of minority. (Figure1)

The result shows possibility that minority change the opinion of majority. But for majority, the confidence of minority and persuasive of minority are important. If minority members don't show his or her opinion, the opinion of majority will not be changed.

Although it is expected that minority activate divergent thought of majority, minority may not show the opinion. If minority cannot show his or her opinion, they cannot change the opinion of majority and activate community. What make minority be silence?

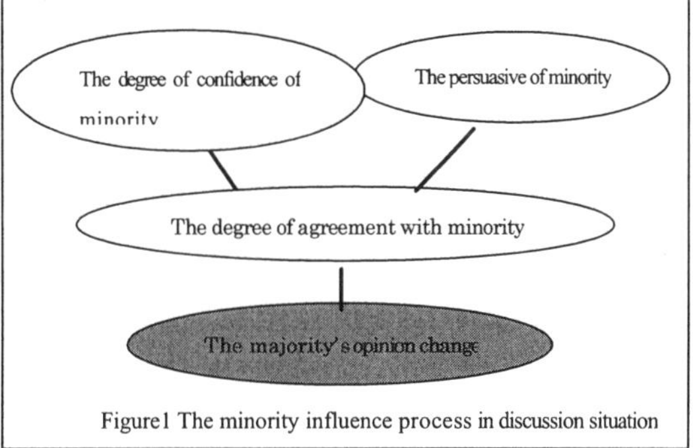

Figure1 The minority influence process in discussion situation

5. False consensus effect

Why cannot minority members show their opinion? It is important factor how does minority guess majority's attitude and behavior. Minority guesses them with two aspects. One of them is the distribution of the community. The other is the guess of other members' attitude.

How do community members recognize the distribution of the opinion? They overestimate or underestimate the percentage of the same their own opinion in groups and society. The members overestimate the proportion of members who has the same attitude. On the contrary, majority members underestimate the proportion of members who has the same attitude. A number of the following members make this difference. Those who endorse an attitude or behavior held (in actuality) only by a minority of others tend to overestimate the percentage of persons who agree with them. In contrast, those who hold a belief actually endorsed by a majority of others tend to underestimate their magnitude of consensus support [3]. Thus members cannot estimate the distribution of the opinion in groups precisely.

5-1. Results.

5-1-1. Minority member's estimation of distribution of the attitude

Minority members were asked to estimate the percentage of others who have positive attitude about the enjoyment of union activity, the attractiveness of union activity, and the behavior of talking about union activity. We calculated the average of the estimation by minority member and actual percentages. Minority member estimate greater consensus or support for his or her own position than the actual distribution of members (See Table1). They overestimate the percentage of other who has the same attitude and behavior. This result corresponds to previous studies [4].

Table1. The estimation by minority and actual distribution

	The enjoyment of union activity	The attractiveness of union activity	Talking about the union activity
Actual distribution	11.36%	12.50%	9.09%
The estimation by minority	14.33%	18.14%	17.41%

5-1-2. Minority members guess of majority's attitude and behavior

Participants were asked to guess the attitudes and behaviors of other members. The average of the attitudes and behaviors that minority guessed and of them that participant estimated were calculated. Table2 shows the average of them. It shows that minority members underestimate the attitudes and behaviors of other members.

Minority overestimate the distribution of the same attitude and behavior that majority member has. On the contrary, they underestimate the attitude and behavior that other members have. Therefore, when we discuss about cognition to the attitude and behavior of others, we must discuss it with two aspects, distribution and estimation.

Table2. The average of the attitude and behavior that minority guessed and of them that participant estimated

	The enjoyment of union activity	The attractiveness of union activity	Talking about the union activity
The average of attitude and behavior	2.45	2.55	2.35
The guess by minority	1.86	2.00	2.27

5-1-3. The effect on behavior of minority

Minority members were distributed to 2groups according to estimation of the attachment and enjoyment. In one group, minority members overestimate the percentage of minority. In the other, they underestimate it.

Minority members underestimating the percentage reduce behavior of talking about labor union and intention of behavior. If minority member underestimate the percentage of minority' attitude, they reduce their intention and behavior. (See Figure 2) In other hand, the attitude of other members guessed by minority don't reduce minority from talking about union activity.

Therefore, minority's intention to act and behavior is affected by the distribution of minority that they estimate particular in attitude.

On contrary, other' attitude guessed by minority don't affect the minority' intention and behavior.

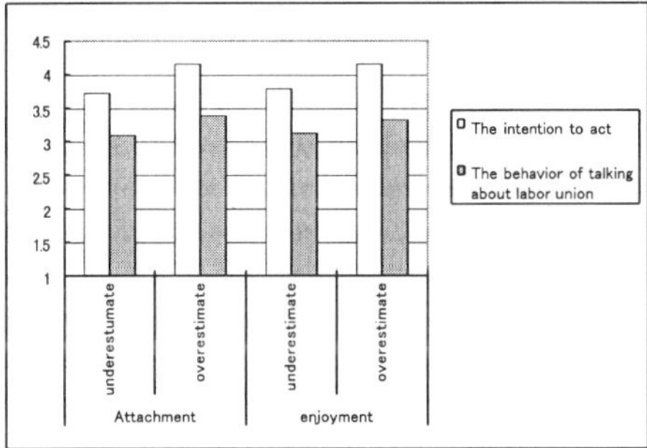

Figure2 The effect of the distribution minority estimate on the intention and behavior

6. The initial distribution of the opinion in network community

The initial distribution of the opinion will greatly influence the consensus formation of the network community. In network community, it is unusual that all members of the community have a shared opinion. Members find majority opinion and minority opinion in community. Will the initial distribution of opinion influence final consensus in community?

In organization, we can find that minority members who underestimate the distribution of minority reduce their behavior Therefore, also in network community, the distribution of the opinion will be important for minority to send massages.

When majority supports a member's opinion, he can justify his opinion. On the contrary, when minority supports a member's opinion, he can't justify his opinion fully. He can choose two strategies to justify his opinion.

Two strategies:
1.   He becomes to insist on his opinion in order to get many supporters.
2.   He becomes silent because insisting on his opinion may make emphasize that he is minority.

Members supported by minority will be able to choose the strategy of insisting on one's opinion, because anonymity is high in network community. As the result, it will activate exchanges of information in community. Therefore it is expected that minority functions as an activator in community. But the final opinion in community can't be predicted.

On the contrary, if minority chooses the strategy of keeping silence, it will not activate exchange of information in community. As the result, majority's opinion will be final opinion in community. But minority member will not assent final opinion in community. Therefore a community can't get sufficiently consensus.

The initial distribution of opinion in community will decide member's behavior, and has influence consensus formation process in network community. In network community, minority opinion, which is often overlooked, will activate exchange of information, and play an important part to make a consensus among community members.

7. Conclusion

It will be difficult for minority members to insist on their opinion in network closed-community. But minority, who propose the opinion that majority don't have, will provide community members new aspects and activate the community [5]. The distribution of the opinion is important for network community to form better consensus, because the distribution of the opinion may make minority be silence. But minority has the useful effect for community. For the community members to form better consensus, it is important that minority must express their attitude or opinion.

Minority has little power in network community. So it is difficult to reflect on the consensus in community. If they find that they are minority, they unlikely to send messages. But that make it more difficult to reflect. If minority members don't send messages, other members cannot consider minority opinion.

In this study, we discussed minority influence on consensus formation process of the network community. But we did not discuss the anonymity. In network community, there is almost the anonymity. This will be the important factor in network community.

Minority member may be able to show his or her opinion by the anonymity. And it will be more difficult that members exactly judge the distribution of the opinion and guess other member's attitude and behavior in network community. Is the opinion we can find majority opinions? The opinion we can find may be the opinion frequently expressed by minority. It is difficult to find which opinion is majority or minority. That may make false consensus effect remarkable. It is reason that community members will try to justify own opinion.

The false consensus effect is not desirable for community to form consensus, because it reduce the action of minority. Therefore, the opinion of minority cannot be reflected on consensus in the community. More members the community has, more difficult it is to judge the actual distribution of the opinion in community. It is important to draw out the opinion we can't find, because the opinion we can find is the opinion that a part of community members have. If we can find own opinions each community members have, we can widely look for better consensus. Therefore it is important to find many opinions including minority opinion in community.

The future issues are that these hypotheses we propose need to be verified by experiment, and two kinds of community is compared. And we must find the factor to form better consensus in network community.

References

[1] Nemeth, C. & Wachtler, J., Creative problem solving as a result of majority versus minority influence. *European Journal of Social Psychology*, **13** (1983), 45-55

[2] Matsumura, K, Minority influence process in discussion situation, *The proceeding of 40th JSSP*, 350-351

[3] Gross, S. R. & Miller, N., The "Golden Section" and bias in perceptions of social consensus, *Personality and Social Psychology Review*, **1** (1997), 341-271.

[4] Mullen, B., & Hu, L+., Social projection as a function of cognitive mechanisms: Two meta-analytic integrations. *British Journal of Social Psychology*, **27**(1988), 333-356

[5] Matsumura, K., Minority influence on Public Opinion Channel, *The proceeding of Jsai'00* (2000), (11)-(12)

How to evaluate application of conversational intelligence

Nobuhiko Fujihara

Research Center of School Education, Naruto University of Education, JAPAN
fujihara@naruto-u.ac.jp

Abstract. It is discussed how to estimate application of conversational intelligence. First, it is insisted that one of the important points for estimation is to set up a control condition. Whether applications of convasatinal intelligence are effective or not, estimations should be executed comparing the case in which tools are used with the case in which tools are not used, that is, a control condition. It is hoped that tools are designed to take setting a control condition into account. Second, possiblitity is discussed whether *network analysis* could be used to estimate application ofconversational intelligence. We report our first trial to use the method for evaluation. Public Opinion Channel (POC) is estimated by network analysis, and discussed the efficacy the method and the effect of POC. As a feature work, we should collect BBS data for baseline to evaluate POC.

1 Introduction

Recently, a various types of network communication tools have been proposed. Some of these tools are based on conversation between users. Probably the most representative tool is a chat system. Video conference system also offers users conversational environment. Public Opinion Channel (POC), an interactive broadcasting system for a community, collects small chats at a part of a community and broadcast them to all over a community [2]. Some of these system aim to support knowledge creation, and some aim to support communication among community members. If tools achieve their goals, the mechanisms of them would apply to later development of new tools. On the other hand, if tools don't achieve their goals, they would be expected to be modified. So, it is very important issue to estimate whether communication system facilitates conversations between users and create novel knowledge.

2 Methodology to evaluate application of conversatoinal intelligence

2.1 Importance of setting a baseline for estimation

To estimate whether tools attain their function, the differences should be compared performances of community or community members between when tools with the function are used and when tools without the function are used. The case where members use tools without the function are called as *control condition*. When control condition is biased, the effect of the function devised on tools cannot be estimated exactly. Thus, it is very important for estimation of tools how control condition should be set up. Tools were sometimes estimated without setting a control condition. But it's not enough to decide whether the tools really support activities of a community and community members. Because it could be denied the possibility that a community and community members achieve performances without the tools as good as with the tools.

How should a control condition be set up? One of the appropriate methods is that tools are designed as composition of a basic part and some additional parts. The case where people use a tool constructed with only a basic part may deal with a control condition, and the cases where people use tools constructed with a basic part and some additional parts may deal with experimental conditions. The effect of a tool would be observed as the difference between a control condition and experimental conditions.

For example, Public Opinion Channel (POC) which is developed and researched by my colleagues and me, is designed in such a way. POC is an interactive community broadcasting system. POC collects information from community members, edits and summarized information, and broadcasts it as a story. Community members listen a story, and respond to it. Repeating the cycle, POC create continuous information circulation in a community. To estimate functions of POC, the case where the system is used which has only basic functions, that is, collecting messages and broadcasting them, is set as a control condition. As experimental conditions, the cases where the systems are used which have basic functions and one or some additive functions. Research issues on POC are "How should information be summarized to facilitate knowledge creation in a community?", "Are anonymous communication systems effective to inhibit troubles in communication like flames?", and so on. One of possible experimental conditions would be POC with summarization function. The effect of the summarization function could be observed when comparing the differences of some measurements, for example, quantities of message circulation, and the increasing rate of users, between the control condition and the experimental condition. In a similar way, some modules which aim to implement a same function can be compared.

Tools are not always designed with modules. So, I'll propose another estimation. As a control condition, typical situations can be used where people use ordinal network communication systems like mailing lists, bulletin board systems, and chats. For the purpose, it is useful to define typical situations and to standardize procedures to collect data and to analyze data. Fujihara & Miura observed search engine users who query information from WWW and analyzed their behavior [1]. In the research, we proposed categories to describe information query behaviors from WWW with search engines. Such research would reveal our common activities in network communities. It will give us a baseline to estimate novel network communication tools to support conversations.

2.2 *Network analysis for evaluation of conversational intelligence*

Recently, we are trying to develop the method "*Network Analysis*" to analyze communications on POC [3]. Originally network analysis is the method to analyze relationships among community members and relationships among companies. It is mainly used in the field of sociology. It describes networks constructed with *nodes* and *links*. Each node means a person or a company, and each link means the relation between people or companies. It is used to investigate the structures of networks, the effect of network structures on community members, and its mechanisms.

We use network analysis to investigate knolwedge structures of communications on POC.

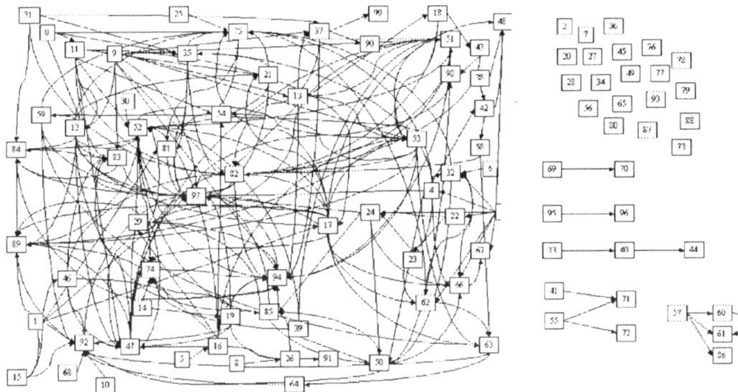

Figure 1: Network structure of POC messages

A network is described with considering each message sent to POC as node. About the ways how to link nodes, there are some possibilities to describe a network. For example, one message and a message replied to it can be linked to describe a network, and messages sharing same topics can be linked. Here, the latter way is adopted, that is, messages sharing two or more content words (which were mainly nouns) are linked.

Figure 1 shows the network with the first 100 messages send to a POC community. Each square represents each message, and the numbers written in squares represent ID numbers of messages. Smaller the ID number is, faster the corresponding message were sent to POC. Twenty messages have no links to other messages, and some construct very simple links. Sixty-five messages construct highly complex network (left part of the figure). This graph is too confused to see, but we can find some nodes which play cores, central roles in the network.

To describe such centrality, some methods for quantification are proposed in network analysis. One of the representative quantification methods is "degree." Degree means the numbers of links each node has. Especially, links which come into each node are called in-degree, and links which go out from each node are called out-degree. Figure 2 shows degrees for each message.

To evaluate whether POC facilitates conversational intelligence, we have to make some assumptions. For example, one assumption is made: the more nodes which have large number of links, the more effective conversations and communications on POC are. On the POC data, the average of total degrees are 8.8 and the numbers of nodes which are larger than the average + 2 standard deviations (in this case, $8.8 + 2 \times 8.9$) are 5. This looks slightly larger than statistically chance level (2.28 nodes).

This analysis is just a first step of our trial. So, we have a lot of issues to discuss as feature works. First, we have to set an appropriate control condition. Now, we analyze BBS data by network analysis. Probably the data can be used as a baseline. Second, there are other possible ways to estimate POC facilitates conversational intelligence. For example, based on the degrees messages could be classified into some clusters. If there were links which connected messages from different clusters, that may indicate the system facilitate our knowledge creation. Also, the numbers of links which connected chronologically separated messages may be one index of knowledge creation. But we believe the network analysis gives

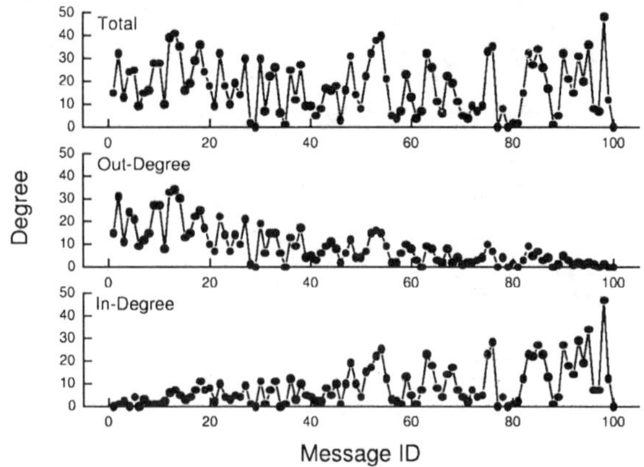

Figure 2: Total degree, in-degree and out-degree for each message

us an interesting viewpoint to evaluate network communication tools.

References

[1] Fujihara, N. & Miura, A. (2000). The effect of the nature of a task on the strategy to search information from Internet. XXVII International Congress of Psychology, Stockholm, Sweden, July 23–28, 2000. (Abstract was published in *International Journal of Psychology, 35 (3/4)*, 84.)

[2] Nishida, T., Fujihara, N., Azechi, S., Sumi, K., & Yano, H. (1999). Public Opinion Channel for communities in the information age. *New Generation Computing, 14*, 417-427.

[3] Yasuda, Y. (1997). *Network Analysis*. Shin'yo-Sha.

KES '01
N. Baba et al. (Eds.)
IOS Press, 2001

A Semantic Representation of Time Expressions Convertible into Numerical Information

Shoji Mizobuchi

Kinki University, Kowakae 3-4-1, Higashiosaka, 577-8502 Japan

Abstract. This paper proposes a semantic representation of time expressions convertible into numerical information. The representation includes both conceptual and quantitative aspects and allows for more precise understanding of time expressions than traditional approaches. Experiments for about 2,000 time expressions show that the proposed method yields correct interpretation in about 93% of the cases.

1 Introduction

In natural language processing systems, it is very important to have the function to understand the meaning of time expressions because they include temporal information which is necessary to grasp the situation of an event [1]. This function is required by a natural language interface suitable to information systems such as temporal database systems [2] and temporal inference systems [3].

The task has been taken up for discussion until now in machine translation systems [4] and information extraction systems [5], but their purpose is to translate or extract a given surface expression into the target expression. For example, "kinou" in Japanese is just translated into "yesterday" in English. Thus, there is no need to exceed the purpose and to interpret the meaning. In contrast, the purpose of this paper presents formalization for a deeper semantic understanding. For example, if today is "March 26th in 1999", the meanings of "tomorrow" include not only conceptual interpretation "the day after today", but also quantitative interpretation "March 27th in 1999."

In this paper, we propose a semantic representation of time expressions convertible into numerical information. The proposed representation is an extension of a series of representations that originate from Leban [6][7][8]. It includes both conceptual and quantitative aspects, and allows for more precise understanding of time expressions than traditional approaches such as those of machine translation systems.

This paper is organized as follows. Section 2 describes time expressions and the properties and construction of them. Section 3 defines a semantic representation of time expressions, called a temporal reference formula. Section 4 evaluates the proposed representation through an experiment on actual time expressions. Finally, section 5 discusses conclusions and future works.

2 Time Expressions

A time expression is a linguistic form which indicates a specific time for the event described by a sentence using the quantitative information involved explicitly or implicitly in the expression. For example, words and phrases such as "September 6, 2001" and "from today until tomorrow" fall under time expressions.

A time expression consists of more than one component. A component is a part of a time expression that makes sense. It can be divided into smaller components if it is not primitive. A primitive component is a minimum element of a time expression.

Seeing time expressions from a semantic standpoint, they are equipped with the properties listed below at least.

- Time expressions indicate a specific time in the span of time.
- Time expressions sometimes indicate more than one time.
- Time expressions indicate time with width.
- Time expressions indicate time which is suited the frame of a calendar.
- Time expressions point to the time which is not settled in the frame of a calendar by combining the time to which it already pointed.

Therefore, a semantic representation of time expressions must be equipped with the form of satisfying these properties.

3 Semantic Representation of Time Epression

The meaning of a time expression is expressed by two kinds of representations: a symbolic representation called a temporal reference formula and a numerical representation called a collection. They express a conceptual meaning and a quantitative meaning of a time expression, respectively. A time reference formula consists of numerical values and symbols, and they are mapped to collections by calendars and operators. The contents of time reference formulas, collections, calendars, and operators are explained below.

3.1 Temporal Reference Formulas

A time reference formula is a symbolic representation which expresses the time to which a time expression points by putting together the times to which its components point.

A temporal reference formula consists of numerical values and symbols. Numerical values are divided into two kinds of ones that represent a position and a length, respectively. In particular, the former corresponds to a collection by means of a calendar. Symbols are divided into three kinds of ones that represent a collection, a calendar and an operator, respectively. In particular, the two latter are connected with more than one adjacent formula and express another time which is different from the time for the formula. The temporal reference formula for time expression "Year 2001" is denoted below as an example.

$$AD \triangleright Y/2001$$

In this formula, "AD" stands for a collection, "Y" stands for a calendar, and "$\triangleright$" and "/" stand for an operator.

3.2 Collections

We assume that time is a dense one-dimensional space consisting of points.

An interval is a convex set of points between two endpoints: a starting point and a finishing point. It is denoted by

$$\langle s, f \rangle,$$

where s and f are a starting point and a finishing point, respectively. There are 13 relationships, introduced by Allen [1], between two intervals.

A collection is a structured and ordered set of intervals. The depth of a collection is a measure of the depth of the structure. A depth 1 collection is an ordered set of intervals and a depth d ($d > 1$) collection is an ordered set of depth $d - 1$ collections. For example, if January 1st, 2001 points to the interval between 0 and 1, depth 1 collections AD and Y are denoted as follows.

$$AD = \{ \langle -780485, +\infty \rangle \}, \quad Y = \{ \dots, \langle -366, 0 \rangle, \langle 0, 365 \rangle, \langle 365, 730 \rangle, \dots \}$$

Moreover, a depth 2 collection $AD \triangleright Y$ is denoted as follows.

$$AD \triangleright Y = \{ \{ \langle -780485, -780120 \rangle, \langle -780120, -779665 \rangle, \dots \} \}$$

3.3 Calendars

A calendar is a mapping that divides time into an infinite number of intervals by assigning a number to one of them without duplication and gap. A calendar Ca is defined as

$$Ca = \{ i_s \mid \forall s \in Z, \text{meets}(i_s, i_{s+1}) \}$$

where Z is the set of all integers, i_s is the interval to which $s \in Z$ is assigned, and the term meets is one of 13 relationships proposed by Allen.

The collection with i_s which is an element of a calendar Ca is denoted as

$$Ca(s) = \{ i_s \}.$$

The collection with an interval which is an element of a calendar Ca and includes the current tick is denoted as

$$\text{now}(Ca).$$

The distance d between two elements in a calendar Ca is denoted as

$$d\,Ca.$$

A calendar can be defined by specifying how it is to be constructed from another calendar. This is denoted by

$$\| Ca; o; w_1, w_2, \dots, w_n \|,$$

where Ca is a calendar, o is the offset between the reference point and the actual starting point of the calendar, and $w_1, w_2, \dots, w_n$ is a list of the number of elements of the calendar Ca which must be grouped together to form an element of the new calendar. In particular, the depth 1 calendar whose intervals are of length 1 is called the base calendar, which is denoted by B.

Suppose that the reference point is January 1st, 2001, days (D), months (M), and years (Y) can be defined as follows.

$$D = \| B;0;1 \| = \{ \dots, \langle -1, 0 \rangle, \langle 0, 1 \rangle, \langle 1, 2 \rangle, \dots \}$$
$$M = \| D;0;31, 28, \dots, 31, 31, 28, \dots, 31, 31, 28, \dots, 31, 31, 29, \dots, 31 \| = \{ \dots, \langle -31, 0 \rangle, \langle 0, 31 \rangle, \langle 31, 59 \rangle, \dots \}$$
$$Y = \| M;0;12 \| = \{ \dots, \langle -366, 0 \rangle, \langle 0, 365 \rangle, \langle 365, 730 \rangle, \dots \}$$

3.4 Operators

Operators combine with more than one adjacent collection, and generate another collection that is different from it. The contents of 7 kinds of operators are explained below.

(1) Dicing Operator ($\triangleright$)

The dicing operator, $C_1 \triangleright C_2$, provides a way of generating new collections from two collections C_1 and C_2. The effect of it is to break up a collection into pieces according to another collection. For example, $Y \triangleright M$ breaks up the collection of years (Y) into months (M). As a result, the depth 2 collection is generated as follows.

$$\left\{ \dots, \begin{cases} \langle 0,31 \rangle & \langle 31,59 \rangle & \langle 59,90 \rangle \\ \langle 90,120 \rangle & \langle 120,151 \rangle & \langle 151,181 \rangle \\ \langle 181,212 \rangle & \langle 212,243 \rangle & \langle 243,273 \rangle \\ \langle 273,304 \rangle & \langle 304,334 \rangle & \langle 334,365 \rangle \end{cases} \begin{cases} \langle 365,396 \rangle & \langle 396,424 \rangle & \langle 424,455 \rangle \\ \langle 455,485 \rangle & \langle 485,516 \rangle & \langle 516,546 \rangle \\ \langle 546,577 \rangle & \langle 577,608 \rangle & \langle 608,638 \rangle \\ \langle 638,669 \rangle & \langle 669,699 \rangle & \langle 699,730 \rangle \end{cases} \dots \right\}$$

(2) Slicing Operator (/)

The slicing operator, C/n, operates on any collection C selecting the n-th interval from each of the contained depth 1 collections. For example, if the application of the $/2000$ operator to $AD \triangleright Y$ generates the following collection of the 2000th year of A. D..

$$AD \triangleright Y/2000 = \{ \langle 0,365 \rangle \}$$

(3) Covering Operator (~)

The covering operator, $C_1 \sim C_2$, generates a collection with minimal intervals covering the corresponding two intervals included in two collections C_1 and C_2. For example, if $AD \triangleright Y/2000 = \{ \langle 0,365 \rangle \}$ and $AD \triangleright Y/2002 = \{ \langle 730,1095 \rangle \}$, then

$$AD \triangleright Y/2000 \sim AD \triangleright Y/2002 = \{ \langle 0,1095 \rangle \}.$$

(4) Uniting Operator (∪)

The uniting operator, $C_1 \cup C_2$, generates a collection with intervals which are in C_1 or in C_2 or in both. For example, if $AD \triangleright Y/2000 = \{ \langle 0,365 \rangle \}$ and $AD \triangleright Y/2002 = \{ \langle 730,1095 \rangle \}$, then

$$AD \triangleright Y/2000 \cup AD \triangleright Y/2002 = \{ \langle 0,365 \rangle, \langle 730,1095 \rangle \}.$$

(5) Sliding Operator (+)

The sliding operator, $C+d$, generates a collection with intervals which are located d unit from each interval in collection C. For example, if $AD \triangleright Y/2000 = \{ \langle 0,365 \rangle \}$, then

$$AD \triangleright Y/2000 + 2Y = \{ \langle 730,1095 \rangle \}.$$

(6) Iterating Operator (∔)

The iterating operator, $C \dotplus d$, generates a collection with intervals which are located at every d unit from each interval in collection C. For example, if $AD \triangleright Y/2000 = \{ \langle 0,365 \rangle \}$, then

$$AD \triangleright Y/2000 \dotplus 2Y = \{ \dots, \langle -731,-366 \rangle, \langle 0,365 \rangle, \langle 730,1095 \rangle, \dots \}.$$

(7) Extending Operator (∔)

The extending operator, $C \dotplus d$, generates a collection with intervals which cover between an interval in collection C and an interval that is located d unit from it. For example, if $AD \triangleright Y/2000 = \{ \langle 0,365 \rangle \}$, then

$$AD \triangleright Y/2000 \tilde{+} 2Y = \{ \langle 0,1095 \rangle \}.$$

4 Evaluation

4.1 Experimental Method

We have conducted an experiment to verify the effectiveness of the proposed representation.

In the experiment, we first collected the primitive components which contain quantitative information implicitly such as "New year's eve" and "spring" because they cannot be converted into formulas directly only by using their notations. Then, the appropriate formulas were made for these notations by hand. After that, the simulation program translates about 2,000 time expressions in Japanese that are extracted from actual documents into the corresponding temporal reference formulas. The formulas that are outputted by the program are then classified into 3 ranks as follows.

A The number of equations is 1 and it is correct

B The number of equations is more than 1 and all of them are correct judging from only time expressions

C Others

4.2 Experimental Result

The experimental result is shown in Table 1. Suppose that the reference point is January 1st, 2000. The list of time expressions and the corresponding formulas for each rank is as follows.

A "1993 nen kara kotoshi made(from 1993 to this year)" $\rightarrow (AD \triangleright Y/1993)-(now(Y))$
B "6 nichi (the 6th day of a month or 6 days)" $\rightarrow M \triangleright D/6, 6D$
C "three years from that day" $\rightarrow$ No formula

In the time expression "6 nichi" of rank B, the simulation program outputs two formulas expressing a position and a length which are correct answers judging from the expression. It is necessary for obtaining more precise interpretations to use global information about the whole sentence and context besides local information, that is, time expressions.

Considering our method as the first step for analyzing time expressions, the formulas that are classified into rank B as much as rank A are included in the correct answers. Since the sum of the rates of rank A and B amounts to 93%, it turns out that the proposed method understands quantitative time expressions well under consideration of local information.

In this experiment, we used time expressions in Japanese. But, the basic idea is available to those in other languages.

Table 1 Experimental result

Rank	A	B	C	A+B	Total
# of time expressions	1,012	848	135	1,860	1,995
Rate(%)	58%	35%	7%	93%	100%

5 Conclusion

This paper has presented a semantic representation of time expressions convertible into numerical information. From the experimental result on actual documents, it proves that our method is useful in the first stage of analysing the time expressions.

To evaluate our method further, we need to apply it to natural language processing systems. We plan to investigate other quantitative expressions such as length, weight and speed.

References

[1] J. F. Allen, Time and Time again: The Many Ways to Represent Time, *Intl J. of Intelligent System* **6** (4), 1991, 341-355.

[2] C. S. Jensen and R. T. Snodgrass, *IEEE Transactions on Knowledge and Data Engineering* **11** (1), 1999, 36-44.

[3] J. F. Allen, Actions and Events in Interval Temporal Logic, *Journal of Logic and Computation* **4** (5), 1994, 531-579.

[4] F. Bond *et al.*, Temporal Expressions in Japanese-to-English Machine Translation, TMI-97 (1997) 55-62.

[5] W. Li *et al.*, Chinese Temporal System and Temporal Components Identification, Proc. 18th Int. Conf. Computer Processing of Oriental Languages, 1999.

[6] B. Leban *et al.*, A Representation for Collections of Temporal Intervals, Proc. 5th Int. Conf. Artificial Intelligence **1**, 1986, 367-371.

[7] R. Chandra *et al.*, Implementing Calendar and Temporal Rules in Next Generation Databases, Proc. 19th Int. Conf. Data Engineering, 1994, 264-273.

[8] P. Terenziani, Integrating Calendar Dates and Qualitative Temporal Constraints in the Treatment of Periodic Events, *IEEE Transactions on Knowledge and Data Engineering* **9** (), 1997, 763-783.

KES '01
N. Baba et al. (Eds.)
IOS Press, 2001

Dimensionality Reduction of Vector Space Model based on Simple PCA

Shingo Kuroiwa, Satoru Tsuge, Hironori Tani,
Tai Xiaoying, Masami Shishibori, Kenji Kita
Tokushima University, Minami-josanjima, Tokushima 770-8506, Japan

Abstract. In this paper, we propose to use the Simple Principal Component Analysis (SPCA)[1] for dimensionality reduction of vector space information retrieval model[2]. The SPCA algorithm is data-oriented fast method, which does not require the computation of the variance-covariance matrix. Experimentally, we show that SPCA-based method offers improvement over the conventional SVD-based method despite of small amount of computation.

1 Introduction

The Vector Space Model (VSM) [2] is a popular information retrieval model, which represents documents and queries by vectors in multidimensional space. The basic idea is to extract indexing terms from a document collection and to represent each document or query as a vector of weighted term frequencies. The similarity comparison among documents, and between documents and queries, is performed via the similarity between two vectors (e.g. cosine similarity). In VSM, the document vectors exhibit the following two properties:

1. Since the size of the indexing terms is typically large, the document vectors are high-dimensional.

2. Since the number of terms in one document is typically far less than the total number of indexing terms, the document vectors are very sparse.

High-dimensional and sparse vectors are susceptible to noise, and are difficult to capture the underlying semantic structure. Additionally, computing resources necessary for the storage and processing of such data is enormous. Dimensionality reduction is a way to overcome these problems. This is a method which maps vectors in high-dimensional space to vectors in a lower dimensional feature space. A variety of dimensionality reduction techniques have been studied extensively in statistical pattern recognition and matrix algebra, such as Singular Value Decomposition (SVD) [3] or Karhunen-Loéve Transform. However, these standard techniques do not work adaptively and so are often much too slow for real-time applications.

In a context of information retrieval, Latent Semantic Indexing (LSI) [4, 5] uses truncated SVD to reduce the dimensionality of the term-document space. LSI has demonstrated improved performance over conventional VSM for several document collections [5, 6, 7]. Despite its effectiveness, however, SVD is computationally expensive for a large document collection. Researchers have tackled this problem using a variety of approximation techniques [8, 9].

In the work described here, we use the Simple Principal Component Analysis (SPCA) [1], which is a data-oriented fast method for producing the principal components approximately, for dimensionality reduction of the vector space model. SPCA has been successfully used for high-dimensional handwritten characters data. Here, we apply the same technique to the vector space information retrieval model.

2 Simple Principal Component Analysis (SPCA)

The SPCA that was proposed by Partridge et al. [1] produces approximate solutions of principal components in an efficient way. In this section, we describe the algorithm according to their paper.

Since SPCA is a data-oriented method like neural network learning [10], we consider an output function y and a threshold function $\Phi_1(y, x_j)$:

$$y = a_i^T x_j, \tag{1}$$

$$\Phi_1(y, x_j) = \begin{cases} x_j & \text{if } y \geq 0 \\ 0 & \text{otherwise} \end{cases}, \tag{2}$$

where x_j is an input vector which was calculated by subtracting the mean of document vectors from each document vector in advance, and a_i is used to an approximate i^{th} eigenvector in the principal component analysis. The first principal component a is estimated iteratively using the following update-function.

$$a^{k+1} = \frac{\sum_j \Phi(y, x_j)}{\left\| \sum_j \Phi(y, x_j) \right\|} \quad ; \qquad y = (a^k)^T x_j, \tag{3}$$

where k is iteration number. Function y is calculated with the previous estimate a^k.

Initially a^0 can be set to any vector. In order to find the next principal component, the effect of the first principal component is removed from the dataset, so as to avoid being found again. The removal of a component from the dataset is called "deflation". This is performed by subtracting the component of each datum which is parallel to the principal component from that datum, leaving only an orthogonal component. Therefore, deflation of datum x_j by the unit principal component direction a can be expressed by the following formula:

$$x_j' = x_j - (a^T x_j)a. \tag{4}$$

The algorithmic complexity of SPCA is $O(dmn)$ for the determination of d principal components of m n-dimensional data. The constants depend on the time required until convergence.

The following three threshold functions were also proposed in addition to Formula (2).

$$\Phi_2(y, x_j) = \begin{cases} +x_j & \text{if } y \geq 0 \\ -x_j & \text{otherwise} \end{cases}, \tag{5}$$

$$\Phi_3(y, x_j) = y x_j, \tag{6}$$

$$\Phi_4(y, x_j) = \frac{1}{\|a^k\|} y x_j. \tag{7}$$

We compare the information retrieval performance using each threshold function in the following experiments. (Although $\Phi_4(y, x_j)$ was only used in an iterative mode in [1], we apply this function to a batch mode.)

3 Experiments and Results

An experimental comparison was made among three information retrieval models:

(1) conventional vector space model (VSM),

(2) latent semantic indexing based on SVD (SVD),

(3) SPCA-based latent semantic indexing (SPCA).

3.1 Conditions

In experiments, we used the MEDLINE collection (see MED collection in [5]), where queries and relevance judgments were available (1033 documents and 30 queries). We first pre-processed documents to eliminate non-content-bearing stop words using a stop list of 439 common English words. Terms occurring in only one document were also removed. The remaining terms were then stemmed using the Porter algorithm [11]. The pre-processing step resulted in 4329 indexing terms.

The ith component of a document vector x_j is calculated by multiplying L_{ij} which is the local weighting of term i and G_i which is the global weighting of the term i. As these term weighting scheme, we used Log-Entropy defined as the following[12]:

$$L_{ij} = \begin{cases} 1 + \log f_{ij} & (f_{ij} > 0) \\ 0 & (f_{ij} = 0) \end{cases}, \tag{8}$$

$$G_i = 1 + \sum_{j=1}^{N} \frac{\frac{f_{ij}}{F_i} \log \frac{f_{ij}}{F_i}}{\log n}, \tag{9}$$

where n is the number of documents in the collection, f_{ij} is the frequency of term i in document j, and F_i is the frequency of term i throughout the entire document collection. We then created 1033 4329-dimensional document vectors. The similarity between two vectors was measured using cosine function.

For the retrieval evaluation, we measured the non-interpolated average precision using the top 50 retrieved documets by using "trec_eval" program[13].

3.2 Results

Figure 1 shows the average precision using SPCA as a function of number of iterations. The number of principal components was fixed to 50, in which the conventional SVD based method indicated the best performance. "Eq.(n)" in the figure correspond to the threshold functions used, which were described in section 2. Regardless of the functions used, SPCA achieved higher precision compared to conventional SVD with only 10 iterations and converged after 30 iterations. The precision of Eq.(2):Φ_1 and Eq.(5):Φ_2 gradually decreased after 10 iterations and converged after 30 iterations. From the point of robustness for the number of iterations, threshold function Φ_2:Eq.(7) seems to have the best performance.

Figure 2 shows the average precision of SPCA with 10 iterations as a function of reduced dimensions. For comparison, we also indicate the SVD's results in the figure. SPCA showed better precision than SVD throughout the all number of dimensions. Eq.(2):Φ_1 and Eq.(5):Φ_2 have achieved high precision against lower dimensions. The best average precision of SPCA

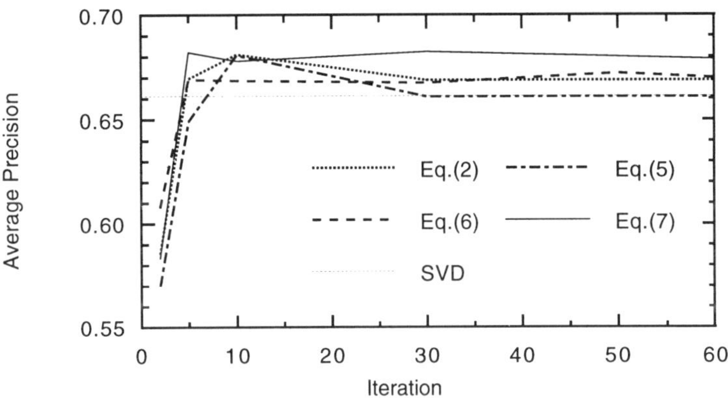

Figure 1: Average precision as a function of number of iteration

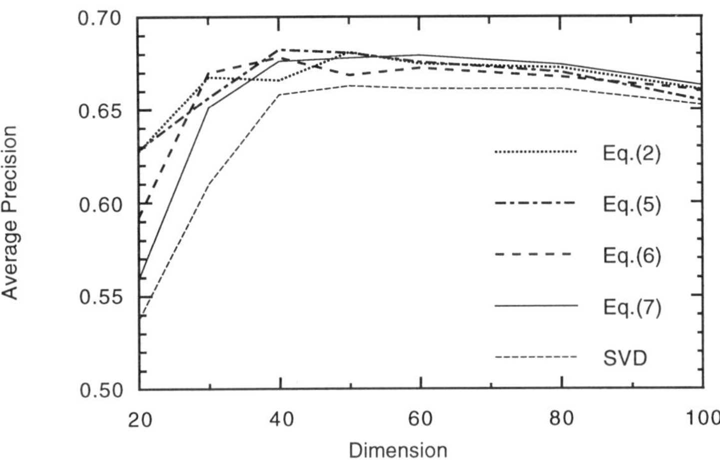

Figure 2: Average precision as a function of dimensions

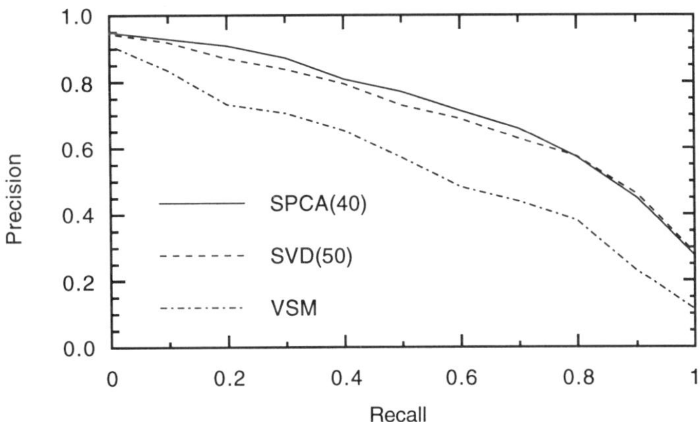

Figure 3: Recall-precision curve

was 0.682 at 40 dimensions with threshold function Φ_2:Eq.(5), while the best average precision of SVD was 0.663 at 50 dimensions and VSM was 0.494, which did not appear in the figure. We have plotted recall-precision curve using all documents retrieved (not top 50) to show the detail of these three results in Figure 3. Although precision of SPCA at high recall did not overcome that of conventional SVD, we consider that SPCA has achieved better performance, since, practically, high precision at low recall is more important to the user of an information retrieval system.

4 Conclusion

We have successfully applied the Simple Principal Component Analysis (SPCA) to dimensionality reduction of the vector space information retrieval model. Despite small amount of computation, SPCA-based method has achieved higher performance than conventional SVD-based method in the experiments. This advantage of SPCA can be attributed to its iterative procedure which is similar to clustering methods such as k-means clustering. We are now planning to apply some clustering techniques to the problem of dimensionality reduction.

References

[1] M. Partridge and R. Calvo. Fast dimensionality reduction and simple PCA. *IDA*, 2(3):292–298, 1997.

[2] G. Salton and J. McGill. *Introduction to Modern Information Retrieval*. McGraw-Hill Book Company, 1983.

[3] G. H. Golub. *Matrix Computations*. The Johns Hopkins University Press, third edition, 1996.

[4] M. W. Berry, S. T. Dumais, and G. W. O'Brien. Using linear algebra for intelligent information retrieval. *SIAM Review*, 37(4):573–595, 1995.

[5] S. Deerwester, S. T. Dumais, G. W. Furnas, T. K. Landauer, and R. Harshman. Indexing by latent semantic analysis. *Journal of the American Society for Information Science*, 41(6):391–407, 1990.

[6] S. T. Dumais. Using LSI for information filtering: TREC-3 experiments. *Overview of the Third Text REtrieval Conference*, pages 219–230, 1995.

[7] T. A. Letsche and M. W. Berry. Large-scale information retrieval with latent semantic indexing. *Information Sciences – Applications*, 100:105–137, 1997.

[8] P. Drineas, A. Frieze, R. Kannan, S. Vempala, and V. Vinay. Clustering in large graphs and matrices. *Proceedings of the 10th ACM-SIAM Symposium on Discrete Algorithms*, 1999.

[9] C. H. Papadimitriou, P. Raghavan, H. Tamaki, and S. Vempala. Latent semantic indexing: A probabilistic analysis. *Proceedings of the 17th ACM-SIGACT-SIGMOD-SIGART Symposium on Principles of Database Systems*, pages 159–168, 1998.

[10] K. I. Diamantaras and S. Y. Kung. *Principal component neural networks : theory and applications*. Wiley, New York, 1996.

[11] W. B. Frakes and R. Baeza-Yates. *Information Retrieval: Data Structures and Algorithms*. Prentice Hall, 1992.

[12] S. T. Dumais. Improving the retrieval of information from external sources. *Behavior Research Methods, Instruments and Computers*, 23(2):229–236, 1991.

[13] TREC homepage. http://trec.nist.gov/.

Dimensionality Reduction of Vector Space Information Retrieval Model based on Non-negative Matrix Factorization

Satoru TSUGE†   Masami SHISHIBORI†   Shingo KUROIWA†
Takashi HIRAI‡   Kenji KITA†
†*Department of Information Science & Intelligent Systems*
Faculty of Engineering, Tokushima University
‡*Hyogo University*

Abstract. The Vector Space Model (VSM) is a conventional information retrieval model, which represents a document collection by a term-by-document matrix. Since term-by-document matrices are usually high-dimensional and sparse, they are susceptible to noise and are also difficult to capture the underlying semantic structure. Additionally, the storage and processing of such matrices places great demands on computing resources. Dimensionality reduction is a way to overcome these problems. Principal Component Analysis (PCA) and Singular Value Decomposition (SVD) are popular techniques for dimensionality reduction based on matrix decomposition, however they contain both positive and negative values in the decomposed matrices. In the work described here, we use Non-negative Matrix Factorization (NMF) for dimensionality reduction of the vector space model. Since matrices decomposed by NMF only contain non-negative values, the original data are represented by only additive, not subtractive, combinations of the basis vectors. This characteristic of parts-based representation is appealing because it reflects the intuitive notion of combining parts to form a whole. Also NMF computation is based on the simple iterative algorithm, it is therefore advantageous for applications involving large matrices. Using MEDLINE collection, we experimentally showed that NMF offers great improvement over the vector space model.

1 Introduction

With the rapid growth of online information, e.g., the World Wide Web (WWW), a large collection of full-text documents is available and opportunity for getting a useful piece of information is increased. Information retrieval is now becoming one of the most important issues for handling a large text data.

The Vector Space Model (VSM) is one of the major conventional information retrieval models that represent documents and queries by vectors in a multidimensional space[1]. Since these vectors are usually high-dimensional and sparse, they are susceptible to noise and are also difficult to capture the underlying semantic structure. Additionally, the storage and processing of such data places great demands on computing resources. Dimensionality reduction is a way to overcome these problems. Principal Component Analysis (PCA) and Singular Value Decomposition (SVD)[2] are popular techniques for dimensionality reduction based on matrix decomposition. However, the cost of their computation will be prohibitive when matrices become large.

This paper propose to apply Non-negative Matrix Factorization (NMF) [3][4] to dimensionality reduction of the document vectors in the term-by-document matrices. The NMF decomposes a non-negative matrix into two non-negative matrices. One of the decomposed matrix can be regarded as the basis vectors. The dimensionality reduction can be performed by projecting the document vectors onto the lower dimensional space which is formed by these basis vectors.

The NMF is distinguished from the other methods, e.g., PCA and SVD, by its non-negativity constraints. These constraints lead to a parts-based representation because they allow only additive, not subtractive, combinations. Also, the NMF computation is based on the simple iterative algorithm, it is therefore advantageous for applications involving large matrices.

2 Non-Negative Matrix Factorization

This section provides a brief overview of Non-negative Matrix Factorization (NMF) [3][4]. Given a non-negative $n \times m$ matrix V, the NMF finds the non-negative $n \times r$ matrix W and the non-negative $r \times m$ matrix H such that

$$V \approx WH. \tag{1}$$

The r is generally chosen to satisfy $(n+m)r < nm$, so that the product WH can be regarded as a compressed form of the data in V.

One of the decomposed matrix W is regarded as the basis vectors. Each basis vectors, i.e., the column of matrix W, can group together semantically related words[4]. Therefore, these vectors can discover the structure that is latent in the data matrix. By projecting document vectors onto new space with lower dimensions using these basis vectors, the NMF may achieve better performance than SVD.

The NMF does not allow negative entries in the matrix W and H. These constraints lead to a parts-based representation because they allow only additive, not subtractive, combination. This characteristic of parts-based representation is appealing because it reflects the intuitive notion of combining parts to form a whole.

2.1 NMF computation

Here, we introduce two algorithms based on iterative estimation of W and H[3][4]. At each iteration of the algorithms, the new value of W or H is found by multiplying the current value by some factor that depends on the quality of the approximation in equation (1). Repeated iteration of the update rules is guaranteed to converge to a locally optimal matrix factorization.

First, we introduce the update rules given in the next equations,

$$\overline{H}_{ij} = H_{ij} \frac{(W^T V)_{ij}}{(W^T W H)_{ij}}, \tag{2}$$

$$\overline{W}_{ij} = W_{ij} \frac{(V H^T)_{ij}}{(W H H^T)_{ij}}. \tag{3}$$

Repeated iteration of these update rules converges to a local minimum of the objective function:

$$F = \sum_i \sum_j \left(V_{ij} - (WH)_{ij}\right)^2 . \tag{4}$$

This objective function is defined as the square of the Euclidean distance between V and WH for a measure. We call these update rules *update rule 1*.

Next, we introduce the update rules which maximize the following objective function:

$$F = \sum_i \sum_j \left(V_{ij} \log((WH)_{ij}) - (WH)_{ij}\right) \tag{5}$$

This objective function is defined as the Kullback-Leibler divergence for a measure. The decomposed matrices W and H are updated as follows:

$$\overline{H}_{ij} = H_{ij} \sum_k W_{kj} \frac{V_{kj}}{(WH)_{kj}}, \tag{6}$$

$$\hat{W}_{ij} = W_{ij} \sum_k \frac{V_{ik}}{(WH)_{ik}} H_{jk},$$

$$\overline{W}_{ij} = \frac{\hat{W}_{ij}}{\sum_k \hat{W}_{kj}}. \tag{7}$$

We call these update rules *update rule 2*.

3 Dimensionality Reduction Using NMF

To apply this NMF to dimensionality reduction of a term-by-document matrix in the information retrieval, we attempt to regard the term-by-document matrix as the data matrix V of the NMF. The following is a resume of the information retrieval by using NMF-based dimensionality reduction:

1. Extract indexing terms from the entire document collection using an appropriate stop list and stemming algorithm. Let we have n indexing terms and m documents.

2. Create m document vectors $d_1, d_2, \ldots, d_m$, where d_j is the i-th component of document vector, a_{ij} is defined $a_{ij} = L_{ij} \times G_i$. Here, L_{ij} is the local weighting for the i-th term in document d_j, and G_i is the global weighting for the i-th term.

3. Apply the non-negative matrix factorization to the term-by-document matrix. The basis vectors W are computed by this process.

4. Project the document vectors onto new r-dimensional space. The columns of W form the axes of this space.

5. Using the same transformation, map a query vector into the r-dimensional space.

6. Calculate the similarity between transformed document vectors and a query vector.

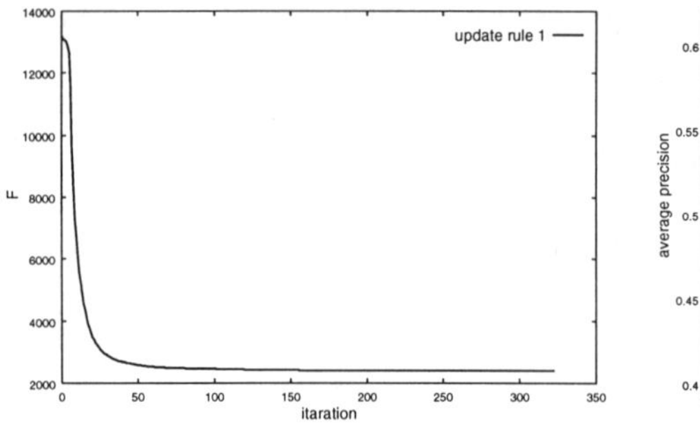

(a) Cost of objective function as a function of number of iterations (update rule 1)

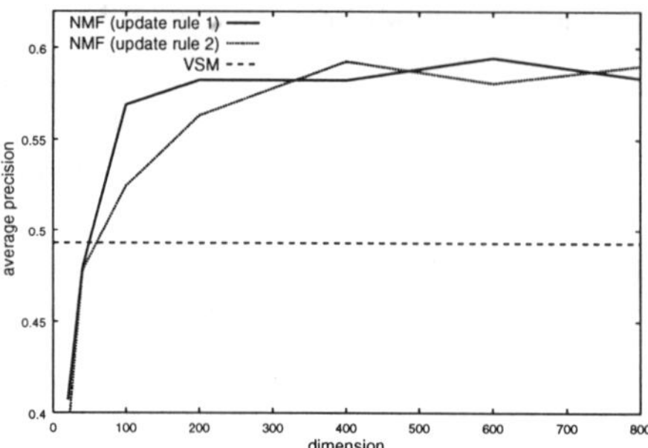

(b) Average precision as a function of dimensions

Figure 1: Experimental results

4   Information Retrieval Experiments

4.1   Conditions

In experiments, we used the MEDLINE collection. This collection consists of thirty queries and 1,033 documents. The average number of relevant documents for each query was 23.2. We first preprocessed documents to eliminate non-content-bearing stopwords using a stop list of 439 common English words. Terms occurring in only one document were also removed. The remaining terms were then stemmed using the Porter algorithm[5]. The preprocessing step resulted in 4328 indexing terms.

By using these documents and terms, the information retrieval system based on vector space model (VSM) was constructed. As a term weighting scheme (see section **3**), we used log-entropy[6]:

Local weight:
$$L_{ij} \;=\; log(1 + f_{ij}), \tag{8}$$

Global weight:
$$G_i \;=\; 1 + \sum_j \frac{p_{ij} \cdot log(p_{ij})}{log(n)}, \tag{9}$$

where n is the number of documents in the collection, and f_{ij} indicates the frequency of the i-th term in the j-th document, and $p_{ij} = \frac{f_{ij}}{\sum_j f_{ij}}$.

For the retrieval evaluation, we measured the non-interpolated average precision, which refers to an average of precision at various points of recall using the top fifty documents retrieved[5][7]. This score was calculated by using "trec_eval" program[8].

4.2   Results

Figure 1 (a) shows that the cost of objective function F given by equation (4) as a function of number of iterations under the condition that the number of the reduced dimensions r was fixed to sixty hundred.

We could see from this figure that the cost of objective function F converged after only 20 iterations. We also confirmed the similarity result under the condition of update rule 2.

Next, Figure 1 (b) shows the average precision as a function of reduced dimensions under the condition that the number of iterations was twenty, in which the cost of the objective function converged. For comparison, the average precision of VSM is also shown in this figure. The dimensions of VSM equaled the number of indexing terms, 4328.

We could see from this figure that the each NMF which had more than 100 dimensions improved the performances compared with VSM. The best performance of the each NMF was achieved at 600 and 400 dimensions using update rule 1 and update rule 2, respectively. These performances were comparable to SVD with same dimensions. As a result, we could consider that the number of the semantic structure in this data was about 500.

We also could see from this figure that the average precision significantly degraded where the dimensions was less than 100. This result implied that only 100 basis vectors were not enough to capture the semantic structure in this data.

5   Conclusions

We have proposed a method for dimensionality reduction of the vector space information retrieval model using the Non-negative Matrix Factorization (NMF). The NMF decomposes a non-negative matrix, i.e., term-by-document matrix, into two non-negative matrices. One of the decomposed matrix can be regarded as the basis vectors. The dimensionality reduction can be performed by projecting the document vectors onto the lower dimensional space which is formed by these basis vectors.

Experimental results on the MEDLINE test collection showed the proposed method gave a better performance than the conventional vector space model. Therefore, we can conclude that the basis vectors, i.e., the columns of the decomposed matrix W, could discover the semantic structure that is latent in the data. We are now planning to analyze the basis vector in order to discover what kinds of semantic structure exist in them.

References

[1] G. Salton and J. McGill. *Introduction to Modern Information Retrieval.* McGraw-Hill Book Company, 1983.

[2] S. Deerwester, S. Dumais, G. Furnas, T. Landauer, and R. Harshman. Indexing by latent semantic analysis. *Journal of the American Society for Information Science*, 41(6):391–407, 1990.

[3] D. Lee and H. Seung. Algorithms for non-negative matrix factorization. *NIPS 2000*, 2000.

[4] D. Lee and H. Seung. Learning the parts of objects by non-negative matrix factorization. *Nature*, 401:788–791, 1999.

[5] I. Witten, A. Moffat, and T. Bell. *Managing Gigabytes: Compressing and Indexing Documents and Images.* Van Nostrand Reinhold, New York, 1994.

[6] E. Chisholm and T. Kolda. New term weighting formulas for the vector space method in information retrieval. *Technical Memorandum ORNL-13756*, 1999.

[7] D. Lewis. Evaluating text categorization. *Proc. of Speech and Natural Language Workshop*, pages 312–318, 1991.

[8] TREC homepage. http://trec.nist.gov/.

KES '01
N. Baba et al. (Eds.)
IOS Press, 2001

A Method to Estimate the Trend of Words by Using a Decision Tree

Makoto OKADA
Department of Mathematics and Information Sciences,
College of Integrated Arts and Sciences, Osaka Prefecture University,
1-1 Gakuen-cho, Sakai, Osaka 599-8531, Japan

El-sayed ATLAM, Masami SHISHIBORI, Jun-ichi AOE
The department of Information Science & Intelligent Systems Faculty of Engineering,
Tokushima University, 2-1, Minami-josanjima, Tokushima, 770-8506, Japan

Abstract. This paper presents an estimation method of the trend of words, which are extracted from text data, considering the time-series variation. To estimate the trend of words, we define five attributes of words in order to construct the decision tree, and the proposed method classifies the words into three classes: Increase, Constant, and Decrease, by using the decision tree. And we evaluate the effectiveness of the proposed method by the experimental results.

1. Introduction

Frequencies of some words in the texts change according to time-series variation. However, traditional document processing systems do not consider the time-series information of words.

Since the results of works by Hisano [1] and Ohkubo et al. [2], it is inferred that some words have a frequency of use that change with time-series variation, and these words often attract the attention of the users in a particular time. Such words are often directly connected with the subject of the text, and are considered as keywords that express the important characteristics of the text.

In this paper, we present an estimation method of the words trend considering the time-series variation by using a decision tree (*DT*), which is constructed by the construction algorithm of decision tree C4.5 [3]. For using decision tree, we define three classes (*Increase, Constant, Decrease*) and five attributes. The words extracted from the text data are classified into these three classes based on the time-series variation of the frequency.

In section 2, we explain a basic concept of the proposed method. In section 3, we describe five attributes used for constructing a decision tree. In section 4, we evaluate the effectiveness of the proposed method by the experimental results. In section 5, conclusion and future works are discussed.

2. Trend of Words

Many words are included the newspaper or texts on Internet, and the frequency of these words sometimes increase, decrease, and change periodically. When we estimate the popularity of these words, we infer the popularity from the change of the words frequency. Fig.1 illustrates these concept.

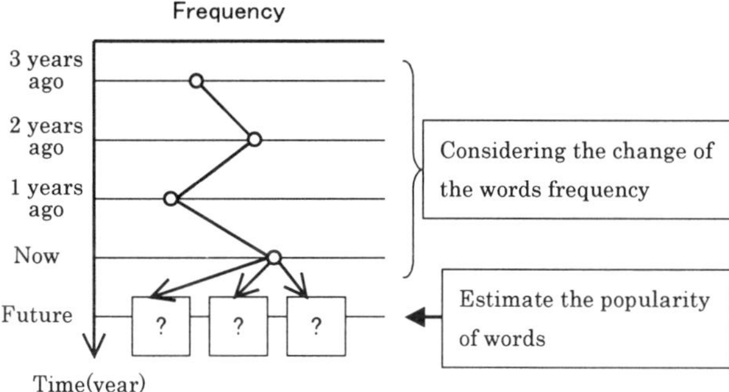

○ : Frequency of this year

Fig.1 An illustration to estimate the trend of words.

As the index used to judge the popularity of words in the time-series variation, and create the stability classes of the words, we defined three classes for judgment words trend other words and the stability of time variation. These classes are defined "*Increasing Class*", "*Constant Class*", and "*Decreasing Class*" in order of the trend of words, and The words belong to each class is called: "*increasing- words*", "*constant-words*", and "*decreasing- words*" respectively. We call these three classes "*stability classes*". The classes of words are decided by the change of appearance frequency of words with time-series variation.

In this paper, five attributes are defined to decide the stability classes, and the words data that are divided into classes beforehand are input into the *DT* automatic algorithm C4.5 [3] as the learning data. Then we use the obtained *DT* to decide automatically the stability class of increasing words. In the next section, the attributes that are used in the *DT* learning to judge the stability class will be described.

3. Attributes

To obtain the characteristics of the frequency change of words quantitatively, we defined five attributes as follows: 1) Slope of regression line (α), 2) Slice of regression line (β), 3) Correlation coefficient (r), 4) Angle between two regression straight lines (ϕ), and 5) Proper Nouns Attributes (*PNA*). The value of each attribute defined here is used as the input data for the *DT* described in section 4.

3.1 Regression Line

Regression analysis is one of the statistical methods, which approximates the change of the sample value with straight line in the two dimension rectangular coordinates, and this approximation straight line is called a regression straight line [4].

In this progress we take the standard years *(x_1= first year, x_2= second year, ...x_i= i year, x_n= n year)* as a horizontal axis, and the corresponding normalization frequency y_i of the words in that years as a vertical axis. The slope segmentation α and the slice β of the equation $y = \alpha x + \beta$ can be calculated by the following formula:

$$\alpha = \frac{\sum_{i=1}^{n} (x_i - \bar{x})(y_i - \bar{y})}{\sum_{i=1}^{n} (x_i - \bar{x})^2} \cdots (1) \qquad \beta = \bar{y} - \alpha \bar{x} \cdots (2)$$

where $\bar{x}$, $\bar{y}$ are the average values of x_i, y_i respectively.

3.2 Correlation Coefficient

Correlation coefficient is used to judge the reliability of the regression straight line. Correlation coefficient is also one of the statistical methods [4], and the calculation equation is shown as follows:

In the above formula $\hat{y}$ is the estimated value by inputting the y_i in the equation of

$$\phi = a\tan(\frac{\alpha_1 - \alpha_2}{1 + \alpha_1 \alpha_2})(\frac{180}{\pi}) \cdots (4)$$

the regression straight line, and α is the slope of the regression straight line. When the absolute value of correlation coefficient r is near to 1, the appearance frequency is concentrated around the regression straight line, and when it is near 0, it means that they are not agreeing on.

3.3 Angle between Two Regression Lines

There are some words for which suitable judgment cannot be achieved by using regression straight line and correlation coefficient only. To solve this problem, we used the angle between the regression straight line that is obtained by both first and second years (old data), versus the regression straight line that is obtained by considering all the data including the newest one (total data of four years).

If the two regression straight lines are given by:

$$y_1 = \alpha_1 x + \beta_1, y_2 = \alpha_2 x + \beta_2$$

then   the formula of the angle ϕ between them is given as follow:

$$r = (sign\ of\ \alpha)\sqrt{\frac{\sum_{i=1}^{n}(\hat{y}_i - \bar{y})^2}{\sum_{i=1}^{n}(y_i - \bar{y})^2}} \cdots (3) \qquad (when \quad 1 + \alpha_1\alpha_2 = 0, \quad \phi = 90)$$

3.4 Proper Nouns Attributes (*PNA*)

In this paper out of the proper nouns attributes, we selected only three kinds of entities: "*Player-personal-name*", "*Player-family-name*", "*Organization-name*", and "*Job / Status-name*" to study the influence of the time-series variation and to obtain the characteristics of the stability classes increasing and decreasing. Also we used "*Ordinary-nouns*", "*Other-proper-noun*" for the constant class. The characteristics of the stability class are easier and correct by using these entities analysis.

4. Experimental Results

In order to confirm the effectiveness of our method, test experiment is made concerning the effectiveness of each attribute and the distribution precision of the improving methods of the *DT* learning data. We will explain more about these experimental data and results in the next sub-sections.

4.1 Experimental data

Table 1 Evaluation Data.

	Period	Total Number	Increasing Words	Constant Words	Decreasing Words
Learning Data (A)	'94 – '97	1069	35	852	182
Test Data (X)	'94 – '98	796	37	675	84

We used *Mainichi-Shinbun* newspapers (*1994-1999*) as an experimental data. We selected the *"professional baseball"* field as a collection data. It was chosen because of the uniqueness of the words in this field are very high, and its tendency to change with the time-series variation. This field has special characteristics than other fields; therefore, we can get information from this field easily.

The data is divided into two groups: learning data and test data. The learning data is the reports of years *1994-1997* and the other includes the reports of years *1994-1998* are used as test data. The data of extracted words is shown in Table 1. The attributes are obtained from the change in frequency of all the words included in the each period. The *DT* is made by using learning data, and the evaluation is made by comparing the results of the test data that is classified by the *DT* and that is decided by human.

In order to get the accuracy of the correct words that are evaluated automatically by *DT* versus the abstract data that are classified by human, we measured: *precision, recall rate* and *F-measure* as follows:

$$Precision\ (P) = \frac{Number\ of\ correct\ words\ extracted\ by\ system}{Total\ number\ of\ words\ extracted\ by\ system}$$

$$Recall\ (R) = \frac{Number\ of\ correct\ words\ extracted\ by\ system}{Total\ number\ of\ correct\ words\ decided\ by\ Human}$$

So, measuring *P, R* are helpful for the purpose of comparison between human abstract decision versus *DT* automatic decision.

4.2 Effectiveness of Attributes

In this section, to evaluate the effectiveness of the attributes that are defined in this paper, the estimation of the precision is calculated with the change of the combination attributes

Table 2 Combinations of Attributes.

Pattern	1			2			3
	a	b	c	a	b	c	
Combination	α, β, r	α, β, ϕ	$\alpha, \beta,$ PNA	α, β, r, ϕ	$\alpha, \beta, r,$ PNA	$\alpha, \beta, \phi,$ PNA	$\alpha, \beta, r, \phi,$ PNA

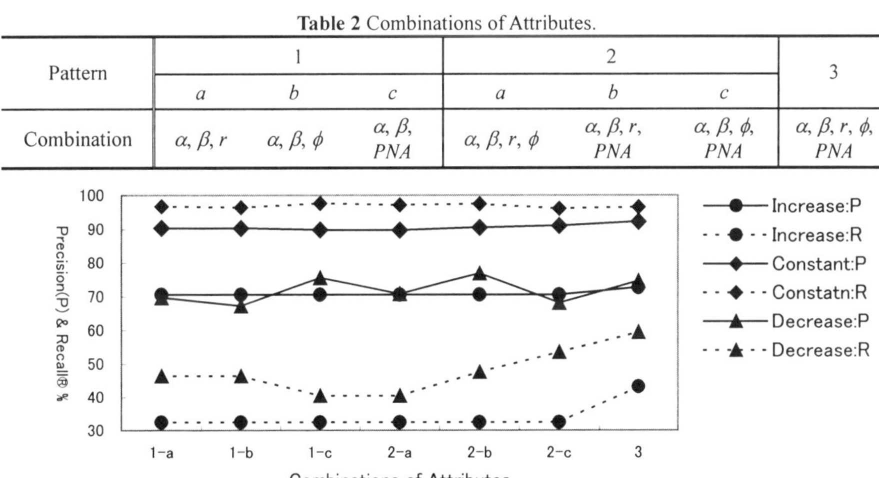

Fig.2 Relationship between the Combinations of the Attributes and the Classification of Precision.

Table 3 Overlapping Words of Test Data X

	Test Data X			Correct Words			Wrong Words		
	Total	Overlap	Non-Overlap	Total	Overlap	Non-Overlap	Total	Overlap	Non-Overlap
Total Number	796	699	97	708	615	93	88	54	4
Increase Words	7	31	6	17	13	4	20	18	2
Constant Words	675	586	89	649	560	89	26	26	0
Decrease Words	84	82	2	42	42	0	42	40	2

used in the *DT* learning data. The combinations of the attributes used in the test are shown in Table 2. The relationship between the combination of attributes and the classification of the precision is shown in Fig. 2.

From Fig. 2, we can recognize that by increasing the number of the used attributes, the classification of the precision is also improved generally. This is because by increasing the attributes, delicate dividing conditions can be made when making the *DT*. Especially, the precision of the case that includes the attribute *"PAN"* is higher generally than before concluding it. This is because the pattern of the attribute value which is mixing together in the learning data of each class can be distinguished clearly by adding the attribute *"pan"*, so that the *DT* learning can be done more correctly and lead to the improvement of the classification precision. Also, we can see that the angle between the two straight lines is efficient to the improvement of the precision of decreasing words as shown in section 3.4. From the testing result, it is confirmed that the defined attributes are effective in the improvement of the classification precision.

4.3 Influence of the Overlap between the Learning Data and Test Data

We confirm the influence of the overlap of words between the learning data A and the Test data X. Table. 3 represent the number of overlapping words and estimation results.

From Table 2, precision of total data, overlap data, and non-overlap data is calculated 88.9(%), 88(%), 95.9(%), respectively. From these results, we confirm that overlap of words between learning data and test data is not influence on to estimate the trend of words by using DT.

5. Conclusion

We proposed the method of estimate the popularity considering the frequency change of words with time-series variation in this paper. The stability class is defined as the index of popularity of words, and five attributes are defined to obtain the frequency change of words quantitatively. The method is proposed to estimate automatically the stability class of the words by having *DT* learning to be done on the extracted attributes from the past text data. We find the effectiveness of the proposed method by the experimental results.

References

[1] Hisano, H., Page-Type and Time-Series Variations of a Newspaper's Character Occurrence Rate, Journal of Natural Language Processing, *7 (2), 2000, pp.45-61.*

[2] Ohkubo, M., Sugizaki, M., Inoue, T., & Tanaka, K., Extracting Information Demand by Analyzing a WWW Search Login, Trans. of Information Processing Society of Japan, 39(7), 1998, pp. 2250-2258.

[3] Quinlan, J.R., C4.5: Programs for Machine Learning, Morgan Kaufmann, 1993.

[4] Gonick, L., Smith, W., The Cartoon Guide to Statistics, HarperCollins Publishers, 1993.

Intelligent Alarm Strategy for Chemical Processes

Chuei-Tin Chang and Chii-Shang Tsai
National Cheng Kung University, Tainan, Taiwan 70101

Abstract. An intelligent alarm system design method, which takes full advantage of the inherent data redundancies in the sensor networks of chemical plants, has been proposed in this study. In particular, systematic procedures have been developed to synthesize corresponding optimal alarm generation logic on the basis of the reconciled data. It is shown in this paper that the proposed approach is superior to any of the existing alarm techniques in the sense that it can be appropriately tailored on-line to minimize the expected loss due to misjudgements in generating alarms.

1. Introduction

Alarm generation is in fact a basic function of the protective system in any chemical process plant. The current practice in the industry is simply to compare measurement data of the variable of interest with a pre-determined threshold value. The decision concerning whether or not to set off an alarm is then made accordingly. Since all measurements are subject to errors, two types of mistakes may be committed in this decision making process. First of all, spurious alarms may be produced due to measurement noises when the variations of the process variables are actually within acceptable limits (type I mistakes). Secondly, the system may fail to detect the existence of hazardous operating conditions and thus no alarms are generated (type II mistakes).

One possible way to improve the current operation is to make use of the *analytical* redundancy embedded in the sensor network of any process plant. Thus, the reconciled data, rather than the raw measurement data, are adopted in this study for alarm generation purpose. In previous studies [1, 2], a systematic method for synthesizing intelligent alarm logics has been proposed according to the design criterion of minimizing expected loss. Although this method was demonstrated to be cost effective, it is applicable only to variables involved in material and energy balances, i.e. the flow rates, the temperatures and the concentrations. Thus, in order to extend this approach to other measurable process variables, it is necessary to perform a more comprehensive study.

2. A Modified Approach to Data Reconciliation

Since the adjusted data are adopted as the basis for alarm generation in this work, the corresponding estimation errors must first be analysed in detail. Traditionally, only the material and energy balances can be used as the constraint equations for data reconciliation [3]. This limitation is relaxed in the present study. In particular, if certain empirically correlated relations are believed to be fairly accurate, it is unreasonable not to include them as the basis for estimating the alarm variables. Let us now express the selected conctraint equations in a general form, i.e.

$$\Phi(\mathbf{v}^t, \mathbf{u}^t, \mathbf{\eta}^t) = 0 \tag{1}$$

where $\mathbf{\Phi}$ denotes the vector of constraint functions, $\mathbf{v}^t$ and $\mathbf{u}^t$ represent respectively the vectors of true values of the measured and unmeasured process variables, and $\mathbf{\eta}^t$ is the vector of unknown parameters. If these constraint equations are linear, the reconciled values of the process variables can be determined analytically. Otherwise, an iterative computation procedure is needed. In order to facilitate prompt remedial actions in emergency situations, an approximate version of the constraint equations is thus needed to produce relatively fair estimates of the process variables. Let us linearize Eq. (1) with respect to the measurement values and then perform row operations to create a linear system of the following form:

$$
\begin{bmatrix} \mathbf{C}_1 & \mathbf{I}_1 & \mathbf{0} & \mathbf{0} \\ \mathbf{C}_2 & \mathbf{0} & \mathbf{I}_2 & \mathbf{0} \\ \mathbf{C}_3 & \mathbf{0} & \mathbf{0} & \mathbf{I}_3 \end{bmatrix}
\begin{bmatrix} \Delta\mathbf{v}_1 \\ \Delta\mathbf{v}_2 \\ \Delta\mathbf{u} \\ \Delta\mathbf{\eta} \end{bmatrix}
= -
\begin{bmatrix} \mathbf{\delta}_1 \\ \mathbf{\delta}_2 \\ \mathbf{\delta}_3 \end{bmatrix}
\tag{2}
$$

where, $\Delta\mathbf{v} = \mathbf{v}^t - \mathbf{v} = [\Delta\mathbf{v}_1 \quad \Delta\mathbf{v}_2]^T$ and $\mathbf{v}$ denotes the vector of measurement values. Similarly, $\Delta\mathbf{u} = \mathbf{u}^t - \mathbf{u}$ and $\Delta\mathbf{\eta} = \mathbf{\eta}^t - \mathbf{\eta}$. It is assumed that $\mathbf{u}$ and $\mathbf{\eta}$ can be determined by solving various subsets of the constraint equations according to the measurement values in $\mathbf{v}$. On the basis of Eq. (2), it can be shown that the reconciled values of the measured variables $\hat{\mathbf{v}}$ can be expressed as

$$\hat{\mathbf{v}} = \mathbf{v} - \mathbf{Q}\mathbf{B}^T(\mathbf{B}\mathbf{Q}\mathbf{B}^T)^{-1}\mathbf{\delta}_1 \tag{3}$$

where $\mathbf{Q}$ is the covariance matrix associated with $\mathbf{v}$ and $\mathbf{B} = [\mathbf{C}_1 \quad \mathbf{I}_1]$. Finally, the estimation errors can be determined on the basis of Eq. (3), i.e.

$$\mathbf{d} = \hat{\mathbf{v}} - \mathbf{v} = [\mathbf{I} - \mathbf{Q}\mathbf{B}^T(\mathbf{B}\mathbf{Q}\mathbf{B}^T)^{-1}\mathbf{B}]\mathbf{e} \tag{4}$$

where $\mathbf{e}$ is the vector of measurement errors.

3. The Performance Function and Indicator Function

In a chemical plant, a subset of the measured variables may be selected as the alarm variables on the basis of process considerations. Each of these variables must satisfy one or more operational constraint. The general form of constraint can be written as

$$G(v'_A) \geq 0 \tag{5}$$

where G is referred to as the performance function and v'_A represents the true value of the A th measured variable.

Notice that an alarm is set off mainly as an indication of constraint violation. Since the true value of each process variable can never be determined, one has to rely on the measurement data to evaluate the performance function. In other words, the values of the following indicator function $G^{(s)}$ must be computed:

$$G^{(s)} = G(v_A^{(s)}) \tag{6}$$

where, $v_A^{(s)}$ denotes the value of alarm variable obtained with the s th independent evaluation method. Apparently, the alarm variable can be monitored directly with sensors. Other than the direct measurements, it is also possible to identify independent evaluation methods to determine the alarm variable indirectly according to the measurement data of other process variables. These indirect methods can be identified mainly by exploiting the inherent analytical redundancy implied in mathematical models. Due to measurement errors, the values of indicator function evaluated with data obtained from different methods are in general not consistent with one another. Nonetheless, one is still required to make a decision concerning whether or not to set off an alarm with these data.

4. Synthesis of Optimal Alarm Generation Logic

As mentioned before, there may be several independent evaluation methods available for the purpose of monitoring the same variable of interest. Let us express these methods with a set of evaluation functions, i.e.

$$v_A^{(s)} = \Psi_A^{(s)}(\mathbf{v}) \tag{7}$$

where $s = 1, 2, \cdots, N_A$ and each function $\Psi_A^{(S)}$ is derived from a subset of the constraint equations. Assuming that such functions are available, one should be able to compute the alarm variable on-line and then substitute the results into the performance function to assess the current operation status. Consequently, a set of binary indicator variables y_s can be determined accordingly, i.e.

$$y_s = \begin{cases} 1 & \text{if} \quad G^{(s)} < 0 \\ 0 & \text{if} \quad G^{(s)} \geq 0 \end{cases} \tag{8}$$

The system alarm can then be generated on the basis of these indicators. The logic for setting off the alarm can be explicitly expressed with a binary alarm function $f(\mathbf{y})$ and $\mathbf{y}$ is a vector whose elements are the indicator variables. The value of this alarm function is set to be one if the system is generating an alarm. Otherwise, its value is zero.

Obviously, the values of the indicator variables y_s s may not be consistent with the true state v_A'. Let us consider the true value of the performance function, i.e. $G' = G(v_A')$. There are two kinds of mistakes that can be identified accordingly, i.e. the indicator y_s is set to be 1 when $G' \geq 0$ (type I mistake) or it is set to be 0 when $G' < 0$ (type II mistake). Similarly, the mistakes committed in generating the system alarm can also be classified into type I and type II. Since both types of mistakes result in financial losses, there are incentives for developing an optimal alarm generation logic which minimizes the expected loss. This loss can be written as

$$L = C_a (1 - P_F) P_a + C_b P_F P_b \tag{9}$$

where P_F is the demand probability which is defined as the probability of violating the constraint, i.e. $G' < 0$, P_a and P_b denote respectively the conditional probability of type I and type II mistakes, and C_a and C_b the corresponding losses. Notice that the conditional probabilities P_a and P_b are affected by the selection of alarm function $f(\mathbf{y})$ and, consequently, there must exist an optimal alarm logic which minimizes the expected loss.

It was shown [1] that the expected loss is minimized if the alarm function can be chosen such that

$$f(\mathbf{y}) = \begin{cases} 1 & \text{if} \quad h(\mathbf{y}) > 0 \\ 0 & \text{if} \quad h(\mathbf{y}) \leq 0 \end{cases} \tag{10}$$

where, $h(\mathbf{y}) = C_b P_F \Pr\{\mathbf{y} \mid G' < 0\} - C_a (1 - P_F) \Pr\{\mathbf{y} \mid G' \geq 0\}$

After obtaining the values of $f(\mathbf{y})$ on the basis of Eq. (10) for all possible $\mathbf{y}$, its functional form can be constructed accordingly. With the functional form given, the logic associated with $f(\mathbf{y})$ can be implemented as a hard-wired circuit or as a computer program.

5. Application Example

The optimal alarm generation strategy has been applied to a realistic process described in the Tennessee Eastman (TE) challenge problem [4]. A simplified process flow diagram is presented in Fig. 1. The on-line measurement data adopted in this example were simulated numerically with the standard FORTRAN program available in public domain.

Let us assume that reactor pressure is the alarm variable in this case. Its high and low alarm limits are chosen to be 2705.6 and 2704.5 KPa guage respectively. Since this variable does not appear in the material and/or energy balance equations, additional relations must be included in the constraint equations for data reconciliation. In particular, other than the mass balances around all process units and the mixing and splitting points, the following empirical formulas for computing the pressure drops across streams 7 and 14 were also treated as constraint:

$$\Delta p_k = \frac{32 f_k l_k m_k^2}{\rho_k \pi^2 d_k^5} \tag{11}$$

where, k=7 or 14; Δp_k is the corresponding pressure drop; f_k is the friction factor; l_k and d_k denote respectively the length and diameter of the pipeline; ρ_k and m_k represent the density and mass flow rate of the fluid. In this study, it is assumed that all flow rates involved in the above constraint equations can be measured on-line and pressure measurements in the reactor, its feeding vessel and the vapour/liquid separator are also available for data reconciliation

Given the measurement data, the first independent method for evaluating the alarm variable is naturally associated with the sensor for directly measuring reactor pressure, i.e. $p_r^{(1)} = p_r$ where p_r denotes the pressure measurement value. The other independent evaluation methods were identified according to Eq. (11). Specifically, these indirect methods can be expressed as:

$$p_r^{(2)} = p_v - 1.726 \times 10^{-7} m_{14}^2 \tag{12a}$$

$$p_r^{(3)} = p_s + 3.097 \times 10^{-8} m_7^2 \tag{12b}$$

where p_v and p_s denotes the measurement values of the vessel pressure and the separator pressure respectively and the constants in these two equations were computed with data obtained from the literature.

The effectiveness of the proposed alarm generation strategy can be demonstrated with simulation studies. The variation in reactor pressure due to an unknown fault was first simulated. Although there are 20 built-in disturbances in the computer program, only one of them, i.e. a step change in the temperature of feed stream D, is considered in the present example due to the limitation of space. The direct measurement data of reactor pressure can be generated with numerical simulation (see Fig. 2). Notice that the dotted lines represent the upper and lower limits of reactor pressure and the line between them is the design pressure.

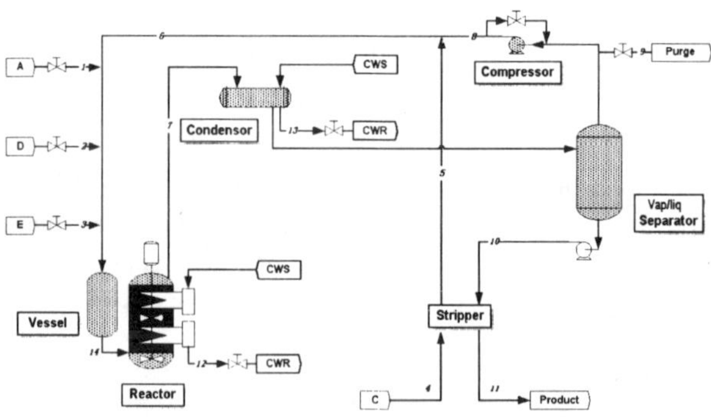

Figure 1. The Flow Diagram of TE Process.

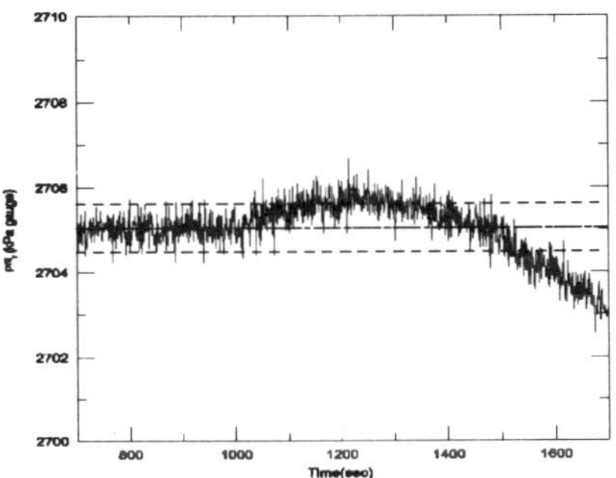
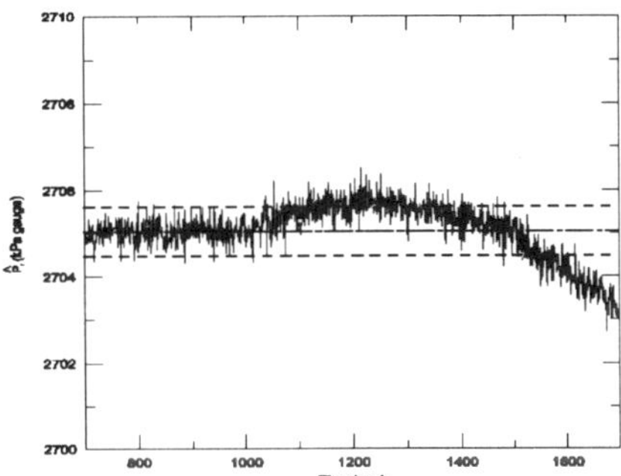

Figure 2. Simulation Results of the Measurement Data of Reactor Pressure

Figure 3. The Reconciled Values of Reactor Pressure

Using the computer-generated measurement data, one can then compute the reconciled reactor pressure. The results are shown in Fig. 3. In this study, the traditional approach, i.e. using the raw measurement data, was taken first to set off alarm. The proportions of type I and type II mistakes in this case were determined to be 0.136 and 0.368 respectively.

Each time a new set of measurement data and the corresponding reconciled pressure are obtained, an optimal alarm logic can be constructed on-line with the proposed synthesis procedure. The computation time needed to build one of these logics is about 0.62 seconds on a Pentium PC. It was found that the proportions of type I and type II mistakes can be lowered to 0.157 and 0.142 respectively by using a C_h-to-C_a cost ratio of 40. Notice that reducing type II errors is usually the first priority in most cases since the purpose for installing an alarm is almost always to protect against certain catastrophic consequences. However, this improvement is often achieved at the expense of committing more type I mistakes. As the ratio C_h/C_a increases, the percentage of type I errors will inevitably increase also. This is nonetheless still acceptable as long as C_h/C_a is consistent with the primary design objectives.

6. Conclusions

From the above discussions, it is clear that the proposed alarm strategy is superior to any of the existing techniques. Such an improvement is brought about mainly by integrating *analytical redundancy* into system design. From experiences obtained in solving the example problem, one can also conclude that the demand for on-line computation is reasonable especially when the sampling interval is in the range of minutes or longer.

References

[1] Tsai, C. S., and C. T. Chang (1997). Optimal alarm logic design for mass-flow networks. *AIChE Journal*, 43, 3021.

[2] Tsai, C. S., C. T. Chang, K. H. Chen, and S. K. Chu (2000). Resilient alarm logic design for process networks. *Ind. Eng. Chem. Res.*, 39, 4974.

[3] Mah, R. S. H. (1990). *Chemical Process Structures and Information Flows*. Butterworth, Boston.

[4] Downs, J. I., and E. F. Vogel (1993), A plant-wide industrial process control problem. *Comput. & Chem. Engng.*, 17, 245.

KES '01
N. Baba et al. (Eds.)
IOS Press, 2001

Process Vectors to Detect Malfunctions in a Process System

Masaru Ishida
Chemical Resources Laboratory
Tokyo Institute of Technology,
4259 Nagatsuta, Midori-ku, Yokohama, Japan 226-8503
Phone:+81-45-924-5254, FAX:+81-45-924-5253, E-mail:ishida@res.titech.ac.jp

Abstract. A chemical process system may have both physical and chemical processes. Hence fault diagnosis should cover both processes simultaneously. Thermodynamics has been applied in the design of a process system. But it may offer a goodtool for detection of malfunctions in a process system. The process vector was proposed as a candidate for such a tool. Its characteristic features and applications were discussed.

1. Introduction

A chemical process system is one of the most complex systems. There are many streams of substances and/or energy. Some streams of substances may have many components. Moreover, various kinds of processes are taking place in a system. There are physical processes such as heat transfer, compression or expansion, and chemical processes such as chemical reaction. Hence, a device to detect a malfunction in a system in its early stage should cover both processes.

The purpose of this paper is to apply thermodynamics to this subject. There are two reasons: (1) Thermodynamics can be applied to any phenomenon covering both physical and chemical processes. (2) Discussions in thermodynamics can be made based on only two quantities, enthalpy and entropy (or exergy).

First, we explain the concept of process vectors, which we proposed for process system design [1]. Then, we show that process vectors can also be applied to the detection of malfunction in a process system. Graphical presentation by process vectors may become a good tool to give us warning against the abnormality in a process system.

2. Process vectors on a thermodynamic compass

2.1. Representation of process features by a vector

A process system consists of many processes. In each process, substances are introduced and some change takes place, yielding output substances. As shown in **Figure 1**, the input substances have enthalpy H_{in} and exergy ε_{in}, whereas the output substances have enthalpy H_{out} and exergy ε_{out}. Then, for this change, we obtain enthalpy change ΔH and exergy change $\Delta\varepsilon$ as follows:

$$\Delta H = H_{out} - H_{in} , \tag{1}$$

$$\Delta\varepsilon = \varepsilon_{out} - \varepsilon_{in} . \tag{2}$$

We can define exergy change $\Delta\varepsilon$ by using entropy change ΔS as follows:

$$\Delta\varepsilon = \Delta H - T_0\Delta S,$$

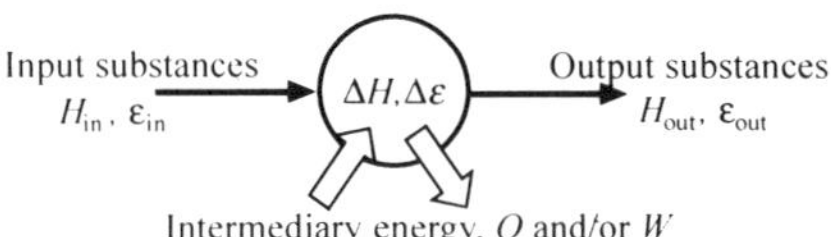

Fig. 1 Energy change ΔH and exergy change $\Delta\varepsilon$ for a process

where $\Delta S = S_{out} - S_{in}$, and $(-\Delta\varepsilon)$ indicates the maximum work obtained from this process by an energy transformation by use of heat sources or sinks at the environmental temperature T_0. Such a change is called a process in this paper. When ΔH is positive, H_{out} is larger than H_{in}, and the process accepts energy of the amount ΔH. On the other hand, when ΔH is negative, it releases energy of the amount $|\Delta H|$ $(= -\Delta H)$. In an entire system, energy released by a process is generally accepted by another process by energy transformation. Hence the energy that is released or accepted by the process is called intermediary energy [2-4].

Then we can present features of a process by a vector on the $(\Delta H, \Delta\varepsilon)$ domain, as shown in **Figure 2**. This domain is called the thermodynamic compass [2-4]. The abscissa and the ordinate, respectively, of the thermodynamic compass indicate ΔH and $\Delta\varepsilon$ for the process. Since a process vector on this diagram gives the values of ΔH and $\Delta\varepsilon$ for the process, this single vector can represent its exergy features as well as energy features simultaneously.

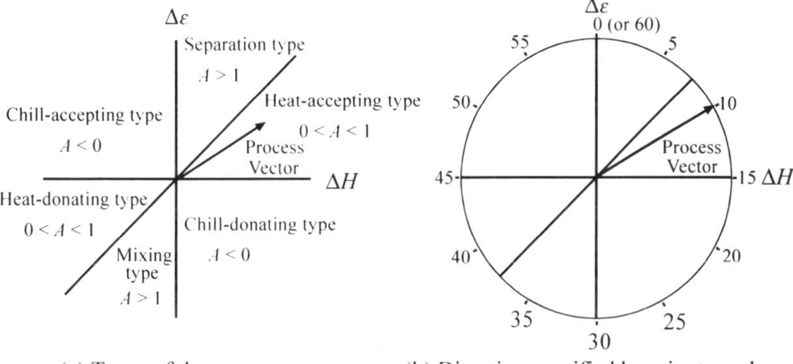

(a) Types of the process (b) Direction specified by minute scale

Fig. 2 Process vector on a thermodynamic compass

2.2 Classification of processes on the compass

ΔH for a process may be positive or negative. Similarly $\Delta\varepsilon$ for a process may be positive or negative. However, simultaneous examination of both ΔH and $\Delta\varepsilon$ gives us much information on the process than the independent examination of each of them. The space on the thermodynamic compass is divided into six regions by three lines, the abscissa ($\Delta\varepsilon = 0$), the ordinate ($\Delta H = 0$), and 45° line ($\Delta H = \Delta\varepsilon$, or $\Delta S = 0$), as shown in Figure 2. Each region represents a specific type of a process, *i.e.* heat-accepting type, separation type, chill-accepting type, heat-donating type, mixing type, and chill-donating type.

For a heating process at $T > 298K$, for example, both ΔH and $\Delta\varepsilon$ are positive, and ΔH is larger than $\Delta\varepsilon$. Hence, its vector is drawn in the region of heat-accepting type. For a cooling process at $T > 298K$, on the other hand, both ΔH and $\Delta\varepsilon$ are negative, and $|\Delta H|$ is larger than $|\Delta\varepsilon|$. Hence, the vector is in the region of heat-donating type. The ratio $\Delta\varepsilon/\Delta H$ indicates the slope of the process vector. This is called energy level A [5,6] and represents quality of the energy released and accepted by the change of substances.

$$A = \Delta\varepsilon \ / \ \Delta H \qquad (3)$$

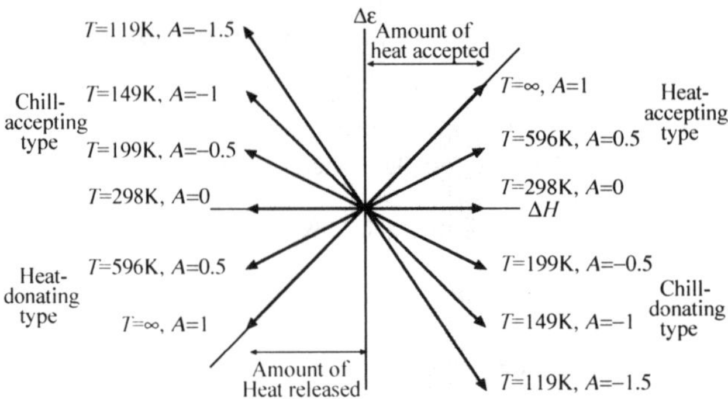

Fig. 3 Process vectors for heating and cooling

For heating or cooling processes, we have $\Delta H = T\Delta S$, and hence the slope of the vector A_T for heat acceptance and release in thermal processes is given by

$$A_T = 1 - T_0 / T_{av} \tag{4}$$

as shown in **Figure 3**, where T_0 is a temperature of the environment and T_{av} is an average temperature for the heating or cooling process. The energy level A_T for such thermal processes is equal to the Carnot efficiency.

As for work-related processes, vectors for gas compression and expansion appear close to the 45° line, as shown in **Figure 4**. For ideal compression or expansion, they are on the 45° line with $\Delta S = 0$, because of $\Delta H = \Delta \varepsilon$,

$$A_W = 1. \tag{5}$$

But due to the loss by friction, the slope of the vector for compression is less than unity, while for expansion greater than unity.

Figure 5 shows that a separation process gives a vector in the separation regime and that a mixing process affords a vector in the mixing regime. This is the case when the output substances and the input substances have the same temperature. When they are different, the vectors may appear in the heat-accepting regime, depending on the condition of the process.

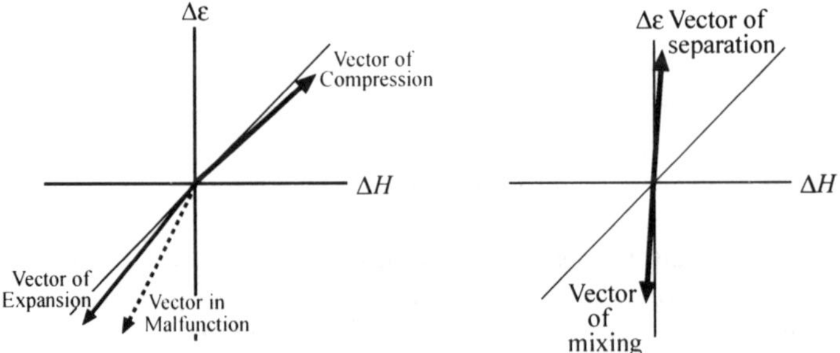

Fig. 4 Vectors for compression and expansion   Fig. 5 Vectors for separation and mixing

Figure 6 shows the process vectors for some chemical reactions. Each chemical reaction gives a specific vector, which may appear in all regimes on the thermodynamic compass. The value of ΔH denotes the heat of reaction. The energy level for a chemical reaction, A_R, indicates the condition under which the reaction can take place. For example, an endothermic process with positive ΔH should accept some kind of energy, say, heat. When A_R for this reaction approaches unity, the slope of the process vector for this reaction is steep; it requires heat at a high temperature, which can be evaluated from the slope of thermal processes in Figure 3.

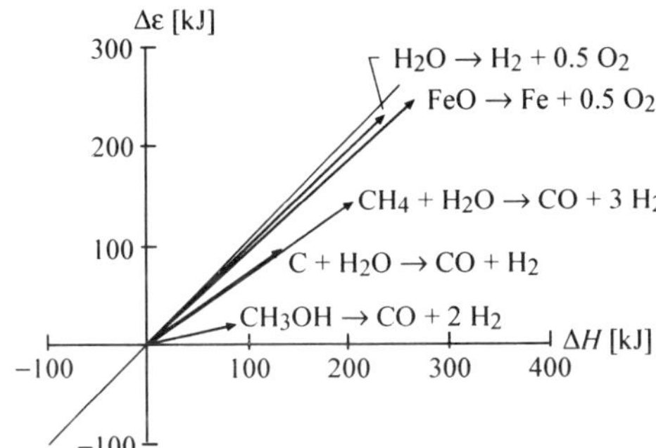

Fig. 6 Examples of process vectors for chemical reaction

Then we can notice the fact that the characteristic feature of a process can be determined not by each value of ΔH or $\Delta\varepsilon$ but by the combination of these two values, *i.e.* the direction of the vector A. This indicates the importance of the simultaneous examination of both ΔH and $\Delta\varepsilon$. The above discussion suggests that the process vector may be a powerful tool to represent the characteristic features of processes. The application of process vectors for creation of functional process structures was discussed elsewhere [1].

3. Application of process vectors to detection of malfunctions

It is very clear that the process vectors can be applied to detect malfunctions in a process system. When we specify the process boundary, the input substances and the output substances can be given. Their temperature and pressure are required to calculate both ΔH ($= H_{out} - H_{in}$) and $\Delta\varepsilon$ ($= \varepsilon_{out} - \varepsilon_{in}$) . The concentrations of each component in the input and output substances are to be measured. Hence, calculation of ΔH and $\Delta\varepsilon$ should be performed, say, by computers. Also we need thermodynamic properties, such as enthalpy and entropy of substances at standard conditions.

When ΔH and $\Delta\varepsilon$ are calculated, the process vector can be presented. Especially, its direction represents characteristic features of the process. When a malfunction occurs in a process, its process vector will give quite unusual direction, for example, the direction of a vector with a dotted line in Figure 4. Hence, we can detect it. The direction of the process vector is sensitive to the wrong values that are used in calculating ΔH and $\Delta\varepsilon$.

The slope, the energy level A, can be used to specify the direction of the vector. Or you may quantify the direction by the minute-scale of the clock, as shown in Figure 2 (b). This has an advantage. When we use the value of A, we must specify the sign of ΔH, too. The processes with positive ΔH and negative ΔH have the same energy level A. However, when we say 10 minutes, this value can specify the process vector in the heat-accepting regime in Figure 2 (b) very clearly.

Another important feature of the process vector is the capability to represent hierarchical characters of a process system. An entire process system, say, zone A in **Figure 7**, has a specific process vector, where ΔH, for example, is given as $H_{exhaust} - (H_{LNG} + H_{Air})$. Some subsystem, say, zone B, in the entire system has also a specific process vector. When the subsystem consists of lower subsystems, each lower subsystem such as zone C has a specific vector. This feature can be applied to detection of malfunctions as follows: For example, we detect abnormal direction of the process vector for some zone, for say zone B. It may suggest the cause of the malfunction. If not, we examine the process vector for each smaller subsystem in it. We can find abnormality in one of the subsystems. In this manner, we can find the location of the malfunction.

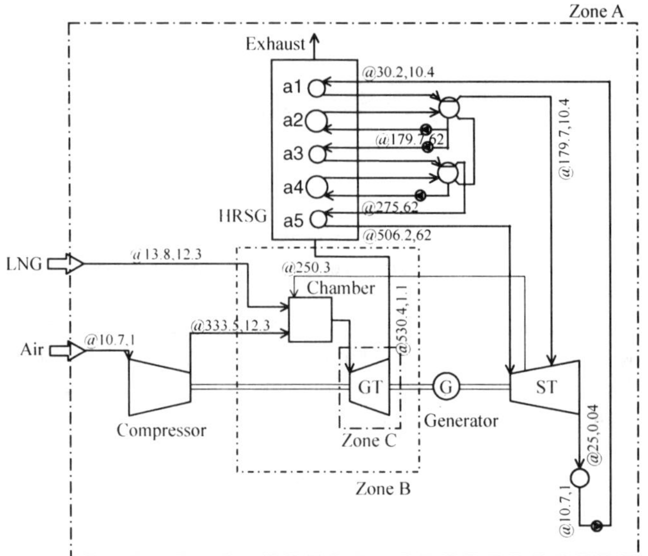

Fig. 7   An example process system, showing three hierarchical zones A, B and C

It is to be noted that the vector is drawn based on thermodynamics and has clear physical meanings. We can grasp various characteristic features of the process by its graphic representation. When we draw also vectors of coupling processes, we can disclose the exergy loss in the process system. Consideration to reduce the exergy loss may lead to new ideas to improve the system [2].

4. Conclusions

Application of thermodynamics to detection of malfunctions was discussed. The process vector (ΔH, $\Delta\varepsilon$) for each process represents both energy and exergy features simultaneously. The direction of the vector represents qualitative feature of the process and is sensitive to the malfunction in the process. The fault diagnosis was presented by applying hierarchical structure of the process system.

References

[1] Yamamoto, M., Ishida M. Process vectors both to create functional structures and to represent characteristics of an entire system In: Proceedings of ECOS99, Tokyo. 1999.p.147-152.

[2] Ishida M. Analysis of hierarchical structure of a process systembased on energy and exergy transformation. In Gaggioli RA, editor. Efficiency and Costing, ACS Symposium Series 235. Washington DC: American chemical society, 1983. p. 179-211.

[3] Ishida M. Thermodynamics-Its perfect comprehension and application (in Japanese) Tokyo: Baifukan, 1995.

[4] Ishida M, Chuang CC. New approach to thermodynamics. Energy Conversion and Management 1997; 38:1543-1555.

[5] Ishida M, Kawamura K. Energy and exergy analysis of a chemical process system with distributed parameters based on enthalpy direction factor diagram.Ind Eng Chem Process Des Dev 1982;21:690-695.

[6] Ishida M, Nakagawa N. Exergy analysis of pervaporation system and its combination with a distillation column based of an energy-utilization diagram. J Membrane Sci 1985;24:271-283.

Application of Wavelet transformation to process diagnosis

Toru Matsuo and Jin Takagaki
Technical Dept., Mitsui Chemicals Inc.; Chigusa-Kaigan 3 Ichihara, Chiba Prefecture, Japan
Hideki Sasaoka and Hirohiko Kazato,
*Research & Development Headquarters, Yamatake Corporation; 1-12-2 Kawana,
Fujisawa, Kanagawa Prefecture, Japan*

Abstract: Analyzing the process behavior of chemical plants, a wavelet transformation tool was developed. The tool could deduce and show certain frequencies in process signals, and it could also suggest correlating signals that exist in different places. The information could provide better tuning for controlling systems and insight regarding the relation between product quality and operational condition.

1. Introduction

By the 1980s, fast Fourier transformation (FFT) analysis became common for many types of frequency analysis. Through progress of computerized numerical processing performance, FFT is now explicitly available not only for frequency analysis of spectroscopy, such as IR, but also for abnormality diagnosis on plant machines, such as bearings equipped in rotating machinery. Although FFT is favorable for detecting stationary waves, it is not very effective for processing non-stationary signals (i.e. temperature, pressure, and flow signals in a plant). Wavelet has capability to analyze these non-stationary phenomena.

A chemical plant includes many controlling systems, and it can be seen as a kind of oscillating system from the viewpoint of control. Applying Wavelet transformation to the signals in control loops, several characteristic frequencies could be seen, and they can be used for monitoring the control system. Wavelet transformation makes diagnosis of a complex plant easier to see through the frequencies. Traditionally, information of signals that we had been using was only absolute values changes over time, but the signal has much more information. The frequency is another source of process information that has not yet been used. The Wavelet transformation can extract the frequency information from signals and show us the meaning of it in relation to the process behavior. Daiguji *et al.* applied Wavelet analysis to fault detection in an oil refinery [1]. Several other applications in chemical engineering are summarized in a book [2].

In this study, a Wavelet transformation tool was developed for process analysis with time-frequency decomposition. The tool is currently being successfully applied to a polymer production plant, to improve process stability and product quality.

2. Method and Case-Study Plant

For analysis of time-frequency decomposition of process signals, a Wavelet analysis tool has been developed by Yamatake Corporation. This Wavelet analysis tool displays the transformed result of the time series data as a visual graph in the time-frequency domain. The characteristics of the data such as the time variant of the frequency are known from the pattern on the graph. In this study, the tool was modified so as to be able to compare the similarities between process variables on the time-frequency graph. By this modification, the more useful information for the understanding and diagnosis of the process is obtained.

For the target of this case study, the Linear Low Density Polyethylene (LLDPE) process, which has one reactor with a latent heat removal system (cooling system), is used in this research (Fig. 1). The data from each measurement (reactor temperature, pressure, gas circulation flow, and polymerization rate) are gathered every minute and stored in memory and analyzed by the Wavelet analysis tool.

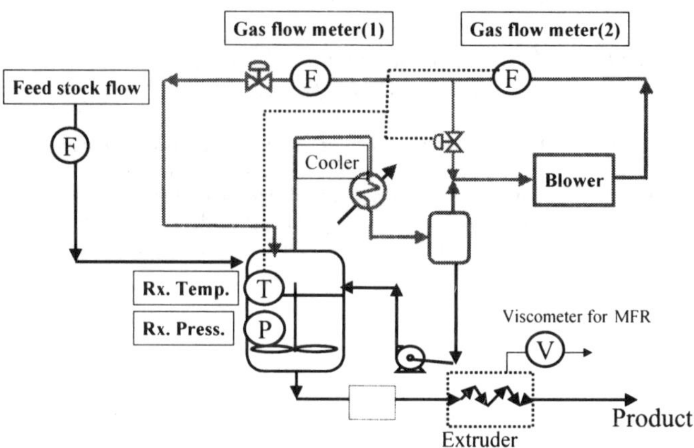

Fig.1   Temp. control system of the LLDPE reactor

3.   Results

3.1 Similarity analysis between process variables

Several process signals in the reactor system are analyzed by the Wavelet analysis tool. Similarities between process variables are easily found from the transformed data, although the original signals are not seen as similar to each other. Those similarities are quantitatively evaluated in frequency domain by the software tool (modified Wavelet transformation tool). Comparing the transformed signals in more detail, several common frequencies are found around 1400, 1100, 220, and 67 µHz, as shown by numbers in Fig. 2, respectively. This information will be useful for understanding and diagnosis of the process.

3.2 Parameters adjustment for a controller

The Wavelet tool can be used for the evaluation of parameter tuning for a controller. After tuning the P, I, D parameters of a temperature controller, the amplitude of signal variations was decreased, and its dominant frequencies were changed (Fig. 3).
These changes can be shown clearly by Wavelet transformation.

3.3 Improvement of product quality

From the Wavelet transformed data of the reactor, the same characteristic frequencies were found in product quality (MFR) and polymer flow rate or Rx temperature. Those frequencies disappeared after tuning of the temperature controlling system. This result implies that the tool provides useful information for the improvement of product quality.

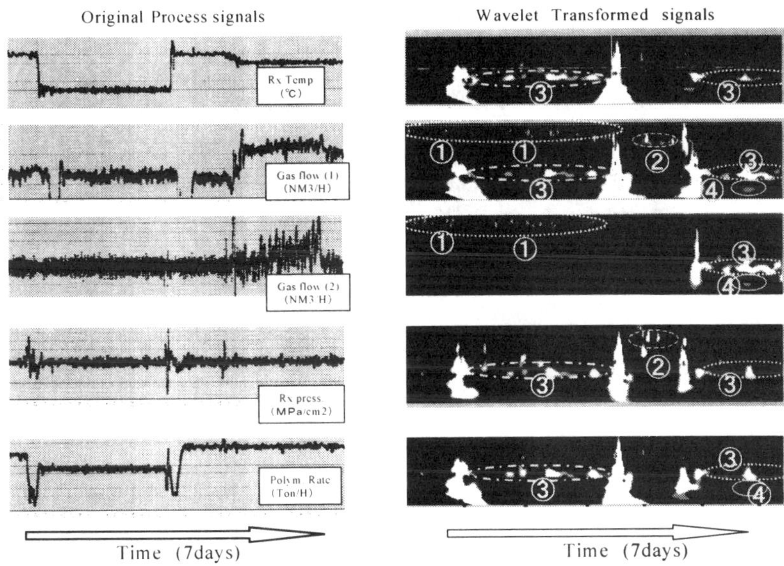

Fig.2   Analysis of the process signals by Wavelet transformation

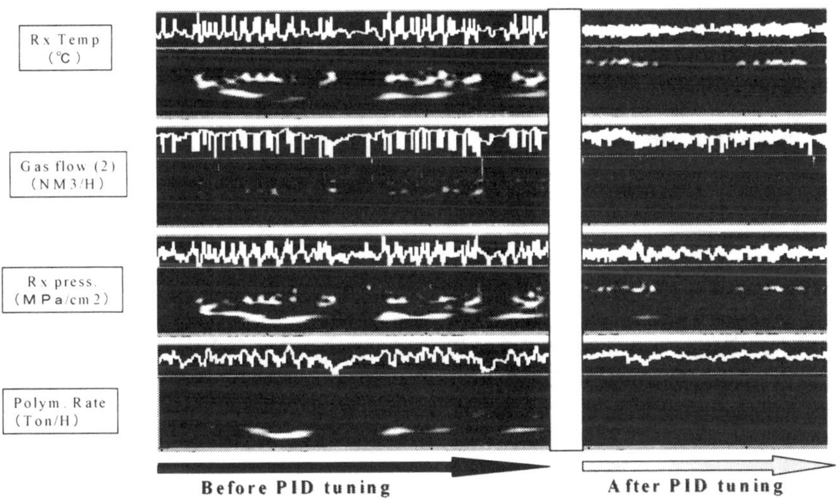

Fig.3   PID tuning and change of the condition in a reactor

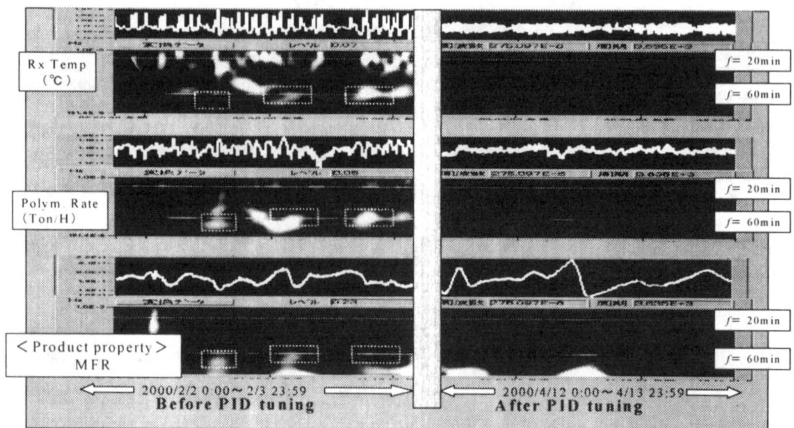

Fig.4   PID tuning and change of the spectrum in product property

4. Conclusion

A Wavelet tool was developed for analysis of process signals.   The tool is currently being successfully applied to a real plant, and new aspects for diagnosing the process condition are obtained.   From the results, we might draw the following conclusions:

- Some aspects of the process operational condition (normal or abnormal) are shown visually.

- Through frequency information, interferential relation in the control system is clarified.

- Effect of P, I, D parameters adjustment is visually evaluated.

- Based on the spectrum, a new adjusting method for the control system is proposed.

5. Acknowledgement

The authors thank Dr. Yoshiyuki Yamashita (associate professor of Tohoku University) for his valuable suggestions, and Intelligent Manufacturing System (IMS) program for financial support.

References

[1] Masaharu Daiguji, Osami Kudo, and Tetsuya Wada, Application of Wavelet Analysis to Fault Detection in an Oil Refinery, *Computers and Chemical Engineering* 21S(1997)1117-1122.
[2] Rodolphe L. Motard and Babu Joseph, eds., Wavelet Applications in Chemical Engineering, Kluwer Academic Publishers, Boston, 1994.

Distributed Adaptive Process Monitoring and Fault Diagnosis based on Data Techniques

Yu Qian [a], Junfeng Wang [a], Xiuxi Li [a], Yaoming Hu [b]
[a] *Chemical Engineering Research Center,* [b] *Department of Automation*
South China University of Technology, Guangzhou, P. R. China, 510640

Abstract. Integrating data analysis, data interpretation and data mining techniques, proposed in this work is a distributed process monitoring and fault diagnosis system. Complex process is partitioned into a number of sub-processes according to fundamental structure of the process. Data analysis and data interpretation techniques are used to monitor sub-processes locally, while data mining to find implicit correlativity between sub-processes. The proposed system is demonstrated in an industrial lubricating de-waxing oil recovery system case.

1. Introduction

Chemical processes are large-scale, multivariate and complicated. Monitoring and fault diagnosis are difficult, slow and inaccurate in many circumstances. Distributed monitoring and fault diagnosis systems are a practical way to deal with it (Mohindra, 1993; Yang, 1997). Process monitoring and fault diagnosis techniques include model-based, knowledge-based and data based methods. In practice, application of the model-based method is limited since accurate fundamental models are not easy to build for many chemical processes. Domain experts and operators knowledge are expensive resources, there are always limitations in applying knowledge-based method such as expert system. Advances in distributed control system techniques allow collection of large quantities of process data. There is definitely a great potential to use such process data to provide valuable information for a better monitoring and fault diagnosis. Data based approaches have recently been intensively researched and applied in chemical processes (Davis, 1996).

However, distributed process monitoring does not facilitate diagnosis of fault sources at the sub-processes level. As a result, an integrated fault diagnosis system may be necessary in industrial application. In general, it may be built by considering both distributed monitoring knowledge base and correlation among the sub-processes found by analyzing fundamental model and expert process knowledge. Since it is difficult to find explicit correlation based on model or knowledge is, data mining techniques are introduced as alternatives. Data mining techniques discover unknown and useful association rules to improve performance of the integrated diagnosis system, which makes the system adaptive.

In this work, common algorithms of data analysis, data interpretation and data mining are reviewed. A distributed process monitoring and fault diagnosis system based on data mining techniques is proposed. At the end, a case study demonstrates the application of the proposed framework in a lubricating de-waxing oil recovery system.

2. Data Analysis and Interpretation

Data analysis and interpretation techniques are basically a pattern recognition methodology. There are two primary tasks: data analysis (or feature extraction), which consists of numerically processing the data to produce numerical features of interest, and data interpretation (or feature mapping), which consists of assigning symbolic interpretations (i.e., labels) to numerical features. The objective of data analysis is to transform numeric inputs in such a way as to reject irrelevant information that may confuse the information of interest and to accentuate information that supports the feature mapping. This is accomplished by numeric-numeric transformation in which the numeric input are transformed into a set of numeric features. Data interpretation is to assign various numerical features to types of labels or descriptions, such as states (e.g., normal or high), trends (e.g., increasing or pulsing), landmarks (e.g., process change), shapes (e.g., skewed, tail), and faults (e.g., flooding or contamination). The resultant symbolic output of a numeric-symbolic interpreter may itself need further interpretation. This requires a symbolic-symbolic mapping such as a knowledge-based system (KBS) that further refines symbolic interpretations.

Current existing algorithms of data analysis and interpretation include: multi-scale wavelet, principal component analysis/ regression (PCA/PCR), partial least squares (PLS), radial basis function network (RBFN), adaptive resonance theory (ART) and so on. The application of these techniques in process monitoring is widely investigated (Whiteley, 1996; Davis, 1996).

3. Data Mining

Data mining techniques have been successfully applied in the finance, business, administration, industrial engineering, and so on. The application in process monitoring and fault diagnosis is also considered and investigated by researchers (Wang, 1999; Yang, 2000; Yamashita, 2000). Data mining as a main part of knowledge discovery from database (KDD), extracts implicit, unknown, and potentially useful information from given data sets. Active investigating areas in data mining include statistical methods, machine learning, neural networks, fuzzy set methods, rough-set methods and so on.

There are a number of data mining algorithms. In considering their function, there are association rules mining, classification/ regression rules mining, clustering rules mining, characteristic rules mining and so on. The popular algorithms among them are: Apriori algorithm, DHP algorithm (Chen, 1996) belonging to association rules mining algorithms; classification and regression trees (CART), simple Bayes, k-nearest neighbor (k-NN) belonging to classification/ regression rules mining algorithms; CLARANS (Chen, 1996) belonging to clustering rules mining algorithms.

4. The Proposed Framework

The data based framework for distributed adaptive process monitoring and fault diagnosis system consists four parts: process decomposition, distributed process monitoring, integrated fault diagnosis and adaptive improvement for knowledge base. It is described in Figure 1.

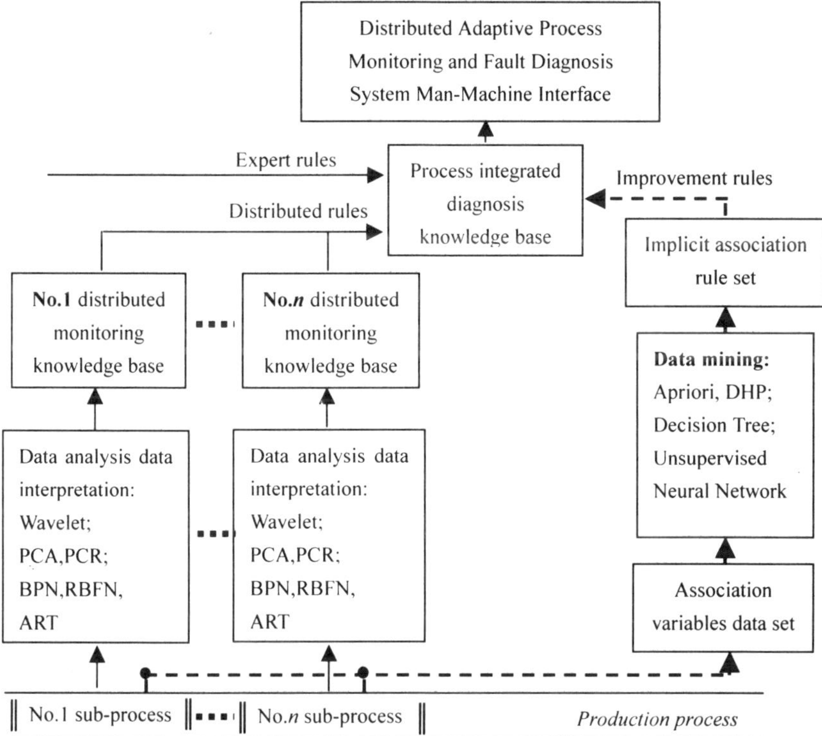

Figure.1. The proposed framework

4.1 Process Decomposition

In a typical chemical process system, an operator console will collect 2000 to 5000 measurements every time increment typically 1 minute or less. Dealing with such large amount of data with any algorithm is extremely difficult. One solution is to partition large-scale, complex process into a number of simple and local sub-processes. When there are n-order of inputs and m-order of outputs in a complex system. It is represented as

$$y_j = \Phi_j(x_1,\ldots,x_n), j = 1,\ldots,m. \tag{1}$$

If the complex system is partitioned into two sub-processes, one with l-order inputs and d-order outputs, the mathematical model is represented as follows:

$$\begin{cases} y_i = \Psi_i(x_1,\ldots,x_l), i = 1,\ldots,d \\ y_k = \Psi_k(x_{l+1},\ldots,x_n), k = d+1,\ldots,m \end{cases} \tag{2}$$

Then the input variables of the two sub-processes are recorded in data sets $\mathbf{X}_1$ and $\mathbf{X}_2$, respectively. Meanwhile the correlative variables between sub-processes are abstracted and recorded in data set Ξ.

4.2 Distributed Process Monitoring

For sub-processes, appropriate data analysis and interpretation algorithms are selected to abstract data features $\mathbf{Z}_i$ from the input data set $\mathbf{X}_i$. Then these data features are assigned to symbolic rules $\mathbf{\Omega}_i$. Finally, by synthesizing symbolic rules and expert experience, distributed process monitoring is implemented. The transforming process is represented as follows:

$$\zeta : \{\zeta' : (x_1, x_2, \ldots, x_d) \rightarrow (z_1, z_2, \ldots z_f)\} \rightarrow \omega_i \tag{3}$$

The implement process is described in Figure 2.

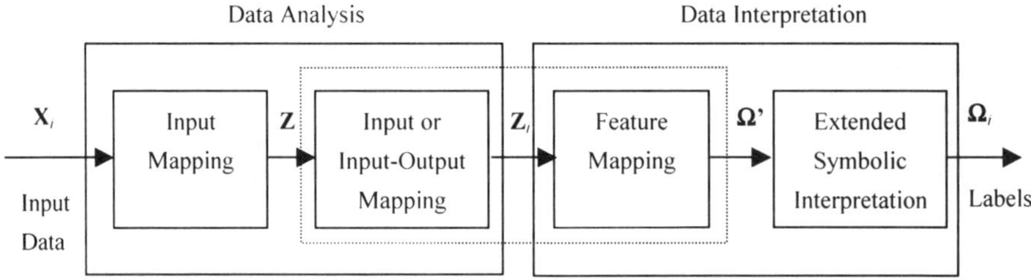

Figure 2. The implement process of distributed process monitoring

Input mapping refers to manipulation of input data, $\mathbf{X}_i$, without considering any output variables or features. This type of mapping is generally used to transform the input data to a more convenient representation, $\mathbf{Z}'$, while retaining all the essential information of the original data. From relating input variables, $\mathbf{X}_i$, output variables, $\mathbf{Y}$, or features of interest, $\mathbf{Z}_i$ at time k, input-output mapping extracts important features, $\mathbf{Z}_i$ at time $k+1$. It aims at the construction of models generally for the purpose of predicting output process behaviors given the input data. Feature mapping assign extracted features, $\mathbf{Z}_i$, to labels, $\mathbf{\Omega}'$, by establishing decision criteria or discriminants based on labeled or annotated data (known and observed) that exhibit sufficiently similar characteristics. Further symbolic interpretations, which is required in many circumstance, is implemented in a symbolic-symbolic mapping from $\mathbf{\Omega}'$ to $\mathbf{\Omega}_i$.

4.3 Integrated Fault Diagnosis

A successful fault diagnosis system cared by operators in the chemical plants should be designed for the whole process. The diagnosis system should have a number of functions in consideration. When find an abnormal situation, the diagnosis system firstly predicts possible fault evolution in future and gives appropriate precautionary measures, secondly discovers possible fault sources and gives possibility degrees of different fault sources, finally gives appropriate suggestion on dealing with these fault sources. So, based on the proposed distributed process monitoring system and the correlative rules among sub-processes, an integrated diagnosis knowledge base is built. The simple and direct correlation among sub-processes is found by analyzing mathematical model and expert experience of the process.

The implicit and unknown correlativity, however, is difficult to find by those methods. So data mining technique is introduced.

4.4 Adaptive Improvement for Knowledge Base

The data source of data mining is the correlative variables data set Ξ among sub-processes and the implement process of data mining is described in Figure 3.

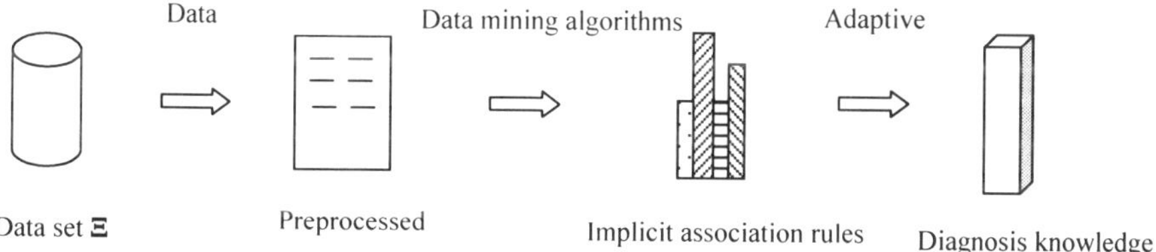

Figure 3.　The implement process of adaptive improvement for diagnosis system

The task of association rules mining is to find rules such as $A_1 \wedge A_2 \wedge, \ldots, A_m \Rightarrow B_1 \wedge B_2 \wedge, \ldots, B_n$, where A_i, B_j represent rules in different sub-process, between data sets with high creditability degree and strong supporting degree, and the implement steps are as follows:

(1) To discover large project sets which supporting degree fulfils the following relation: $S>S_{min}$, where S_{min} is the least supporting degree;

(2) To build appropriate association rules.

Discovered association rules are merged into diagnosis rules to improve the performance of the integrated fault diagnosis system. The process of association rules mining is cycled at intervals, which makes the diagnosis system adaptive.

5. Case Study

The oil recovery process is important in the lubricating oil de-waxing system. The process consists of three stage flash tower (T201, T202, T202/1), furnace (F201), flash tower (T203), stripper (T204) and cooling and heat exchanging process. For an oil recovery system in a southern China refinery, a knowledge based for process monitoring and fault diagnosis system is built by using the fault trees analysis (FTA) method and operators' experience in the abnormal and fault situations of the process (Yu, 2000). The proposed system framework is tested and used to improve the diagnosis system. The oil recovery process, which includes thirty-four variables, is partitioned into six sub-processes. The input space is divided into six data sets X_i and the correlative variables data sets Ξ are abstracted from six data sets. The principal component analysis (PCA) and adaptive resonance theory (ART) are used to implement distributed process monitoring for six sub-processes. Generated monitoring knowledge rules Ω_i are integrated to build one integrate diagnosis system. Finally, the Apriori algorithm is used to discover association rule sets Γ, which adaptively adjusting the performance of the diagnosis system. Table gives a detailed explanation.

Table 1：Variables data sets in the oil recovery process

Sub-process	Input data set	Tag No.	Description	Output knowledge set	Algorithm
T201	X_1	FIC239	T201 input flow	Ω_1	PCA,ART
		TI304	T201 output liquid temperature		
		TI305	T201 output vapory temperature		
		TI306	T201 input temperature		
		LIC233	T201 liquid level		
		PI234	T201 top pressure		
T202	X_2	TI307	T202 bottom output temperature	Ω_2	PCA,ART
		TI308	T202 top output vapory Temp.		
		TI309	T202 input temperature		
		LI276	T202 liquid level		
		PIC206	T202 pressure		
T202/1	X_3	TI310	T202/1 bottom temperature	Ω_3	PCA,ART
		TI311	T202/1 top temperature		
		TIC203	T202/1 input temperature		
		PIC204	T202/1 top pressure		
		LIC223	T202/1 liquid level		
F201	X_4	FIC234	F201 input flow	Ω_4	PCA,ART
		FIC235	F201 input flow		
		PIC201	Fuel oil pressure		
		PIC202	Gas pressure		
		TI348	F201 bottom temperature		
		TI349	F201 bottom temperature		
		TI350	F201 top temperature		
		TI346	F201 output temperature		
		TI347	F201 output temperature		
T203	X_5	LIC224	T203 liquid level	Ω_5	PCA,ART
		PI235	T203 pressure		
		TI314	T203 top temperature		
		TI313	T203 bottom temperature		
T204	X_6	FIC241	T204 input vapory flow	Ω_6	PCA,ART
		LIC225	T204 liquid level		
		PIC236	T204 pressure		
		TI316	T204 top output temperature		
		TI315	T204 bottom output temperature		
Correlativity	Ξ	TI305、TI309、TI310、TI346、TI314、TI315 PIC234、PIC206、PIC204、PI235、PI236 等		Γ	Apriori

6. Conclusion

In this paper, the framework of distributed process monitoring and fault diagnosis is built based data analysis, data interpretation, and data mining. Some of data analysis, data interpretation and data mining methods are shortly reviewed from the perspective of their integration. The introduction of data mining technique into process monitoring and fault diagnosis can adaptively adjust the rules in knowledge bases, which is positive in practical application.

Acknowledgement

Financial support from the National Natural Science Foundation of China (No. 29976015) and China Major Basic Research Development Program (No. G20000263) are gratefully acknowledged.

References

[1] Mohindra, S. and Clark, P. A., A distributed fault diagnosis method based on diagraph models: Steady-state analysis, *Comput. Chem. Eng.*, **17** (1993), 193-198.

[2] Yang Shuzi, The design and application of distributed monitoring and diagnosis system, *J. Huazhong Univ. of Sci. & Tech.*, **25**(1997), 100-105.

[3] Whiteley, J. R., Davis, J. F., Ahmet, M., and Ahalt, S. C., Application of adaptive Resonance theory for qualitative interpretation of sensor data, *Man, Machine and Cybernetics*, **26**(1996), 129-135.

[4] Davis, J. F., Bakshi, B., Kosanovich, K. A., and Piovoso, M. I., Process monitoring, data analysis and data interpretation, "Proceedings, Intelligent Systems in Process Engineering," *AIChE Symposium Series*, **92**(1996a), 312-317.

[5] Chen Mingsyan, Han Jiawei, Philip S Yu., Data mining: an overview from a database perspective, IEEE Transactions on Knowledge and Data Engineering[J], **8**(1996), 866-881.

[6] Wang, X. Z., Data mining and knowledge discovery for process monitoring and control, London: Springer, 1999.

[7] Yoshiyuki Yamashita, Data based approach for intelligent process monitoring, *Proceedings of PSE Asia* 2000, 35-40.

[8] Yoshiyuki Yamashita, Supervised learning for the analysis of process operational data, Comput. Chem. Eng. **24**S(2000),471-474.

[9] Yu Qian, Qiming Huang, Xiuxi Li and Yanbing Jiang, An integrated process operation system platform and fault diagnosis for the lubricating oil de-waxing process, *Proceedings of PSE Asia* 2000, 27-33.

KES '01
N. Baba et al. (Eds.)
IOS Press, 2001

Self-diagnosis of a Neural Net Controller

Takehiro Ohba and Masaru Ishida
Chemical Resources Laboratory
Tokyo Institute of Technology,
4259 Nagatsuta, Midori-ku, Yokohama 226-8503, Japan

Abstract. The neural net controller with self-diagnostic feature is proposed. The proposed network is constructed with small networks that learn the information between the one of the pairs of the controlled and the manipulated variables. From this small network, we can evaluate the learned relations by using the values of weights of the connections. In this procedure, the determinant of the matrix introduced in this evaluation are calculated. The changes in those values suggest whether the network has learnt the characteristics of the process. The proposed self-diagnostic feature is examined by 2-input and 2-output pH control process, and the satisfactory results are obtained.

1. Introduction

A lot of efforts to control a complex chemical process have been done by introducing intelligence to the controller. The use of neural networks is one of those efforts, and the function of the self-learning of the neural network is applied to learn the characteristics of the process.

In the learning procedure, however, the learned information is preserved by a large number of the weights of the connections in the neural network. Hence, it is hard to extract such information from the neural network as an explicit form, and the worse is that we have no way to notice the malfunction of the neural network. These reasons prevent us from applying a neural net controller to control an actual process.

To give a solution to these obstructions is the purpose of this paper. We propose a neural net controller for a MIMO process that has ability to diagnose its condition. The self-diagnostic feature enables the controller to inform us of its trouble, and then we can take steps to meet the situation. This feature can be created by the special network structure of the proposed neural net controller. This network is constructed by gathering small networks, each of which can learn only the relation of a specified pair of the manipulated variable and the controlled variable. This small network uses PENN (Policy- and Experience-driven Neural Network) [1,2] and its training method is based on the indirect learning method proposed [3]. By adopting this structure, we can examine the meaning of each connection and understand the condition of the network. The proposed neural net controller is explained by using a 2-input and 2-output pH control process.

2. Structured PENN controller

2.1. Network structure

The structure of the proposed neural network is the assembly of small networks as shown in Fig.1. It has three layers with M×N hidden-layer units. The input parameters, $\Delta Y_{r,m}$ and

$\Delta Y_{r,p} (1 \le r < N)$, are the changes in the controlled variables, and the output parameters ΔU_q ($1 \le q \le M$) are the changes in the manipulated variables. These parameters will be explained in the following section for the control method. The number of the units in the input layer is twice the number of the controlled variable, M×2, and the number of the output units is equal to the number of the manipulated variable N. The number of the units in the hidden layer is N×M, i.e., the number of the combinations between the controlled and the manipulated variable.

In the general layered neural network, the number of the hidden units is selected as many as the performance of the network can be attained sufficiently. On the contrary, the proposed network uses a fixed number of the hidden units. Hence controlled variable r and manipulated variable q are linked by the partial network shown in Fig. 2. Consequently, only the limited number of the connections, $\upsilon_{r,m}$, $\upsilon_{r,p}$, $\omega_{q,r}$, are used to learn the relation between the controlled variable r and the manipulated variable q. By using this structure, we can evaluate separately the relationship between a specified pair of the controlled variable and the manipulated variable, and this enables the self-diagnostic feature.

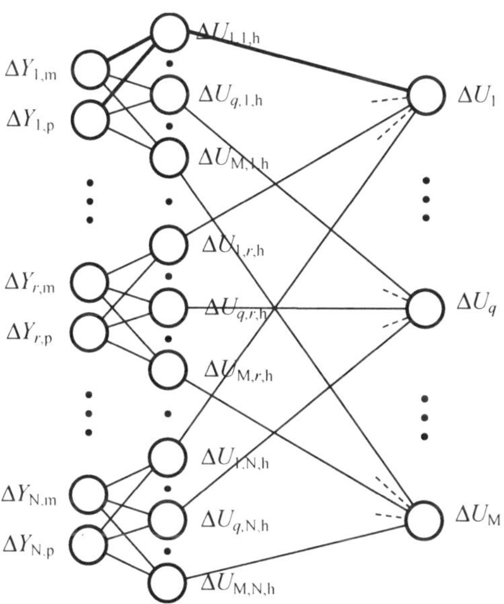

Fig. 1 Network structure of a structured PENN

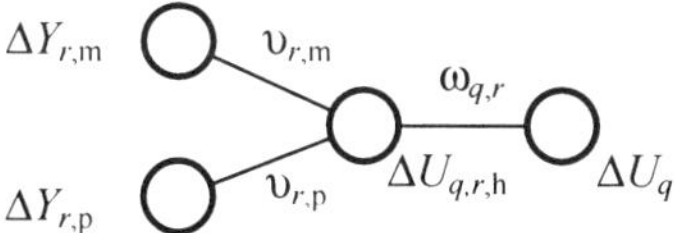

Fig. 2 Partial network structure for manipulated variable q and controlled variable r

2.2. Control method

The inputs for the network, $\Delta Y_{r,m}$ and $\Delta Y_{r,p}$, are defined by Eq. (1).

$$\Delta Y_{r,m} = Y_{r,j} - Y_{r,j-1}$$
$$\Delta Y_{r,p} = R_{r,j+k} - Y_{r,j}$$
(1)

where

$$Y_{r,j} = \frac{y_{r,j} - y_{r,\min}}{y_{r,\max} - y_{r,\min}}$$
(2)

and

$$R_{r,j+k} = \frac{r_{r,j+k} - y_{r,\min}}{y_{r,\max} - y_{r,\min}}$$
(3)

$Y_{r,j}$ is the controlled variable at the instantaneous time j, and $\Delta Y_{r,m}$ is the difference between the present value $Y_{r,j}$ and the value $Y_{r,j-1}$ at the time a sampling period before the present. $\Delta Y_{r,p}$ is the difference between the target value at the time k sampling periods after the present $R_{r,j+k}$ and the value at the present time $Y_{r,j}$, where r_r is the target value of the controlled variable set in advance. The value of k is chosen by considering the delay property of the process and the control interval. The value of k is chosen 1 or 2 for most cases, but for k is set as 3 or 4 when the delay of the process is larger than the control interval.

In this scheme, $\Delta Y_{r,m}$ that shows the change from the past value indicates the factor of the feed back control, whereas $\Delta Y_{r,p}$ reflects the factor of the preview control. The subscript m means 'minus' (i.e., the past from the present time j), and p means 'plus' (i.e., the future from the present).

The output parameter ΔU_q is obtained by forward calculation of the neural network. The input parameters are weighted by υ_m, and $\bullet$ υ_p, respectively. In the hidden and the output units, calculation is performed by the non-linear sigmoidal function is applied.

The new manipulated variable $u_{q,j}$ is obtained by adding the change in the manipulated variable to the previous value $u_{q,j-1}$ at the time a sampling period before.

$$u_{q,j} = u_{q,j-1} + \Delta U_q \cdot \Delta u_{q,\max}$$
(4)

We assume that the constant value of the maximum change in the manipulated variable $\Delta u_{q,\max}$ is given. It has been proved that the performance of the structured PENN controller is much better than that of the traditional PI controller [4].

2.3. Local experience learning

The neural net controller can improve its control performance by learning the responses of the process,. The responses of the process are obtained from the on-line measurement of the previous control results.

One of the input data $\Delta Y_{r,p}$ is the difference between the present temperature $Y_{r,j}$ and the value $Y_{r,j-k}$ at the time k sampling periods before the present. The other input data $\Delta Y_{r,m}$ is the difference between $Y_{r,j-k}$ and $Y_{r,j-k-1}$.

$$\Delta Y_{r,p} = Y_{r,j} - Y_{r,j-k}$$
$$\Delta Y_{r,m} = Y_{r,j-k} - Y_{r,j-k-1}$$
(5)

The target value of the output unit ΔU_q for teaching is obtained as the difference between $u_{q,j-k}$ and $u_{q,j-k-1}$.

$$\Delta U_q = (u_{q,j-k} - u_{q,j-k-1}) / \Delta u_{q,\max}$$
(6)

For the set of these inputs and outputs data as the local experience, the delta rule method is applied to change the weights of the connections.

3. Diagnosis of the neural net controller

3.1. Evaluation of interactions and diagnosis

The self-diagnosis is done during the course of extraction of the process interaction characteristics from the neural network. We need two steps to extract the interactions of the process as the gain constants for the linear process from the structured PENN. First, a representative PENN parameter $p_{q,r}$ is calculated for each pair of the controlled and manipulated variables. Next, the gain constants are obtained based on these PENN parameters.

Step 1: Numerical representation of the structured PENN

In the partial network structure shown in Fig. 1, the output value of the hidden unit $\Delta U_{q,r,h}$ is calculated by the sigmoidal activation function with the input parameters, $\Delta Y_{r,m}$ and• $\Delta Y_{r,p}$, by Eq.(7), since the thresholds are not used in this network.

$$\Delta U_{q,r,h} = f(\Delta Y_{r,p}, \Delta Y_{r,m}) = \frac{1 - e^{-(\Delta Y_{r,p}\upsilon_{r,p} + \Delta Y_{r,m}\upsilon_{r,m})}}{1 + e^{-(\Delta Y_{r,p}\upsilon_{r,p} + \Delta Y_{r,m}\upsilon_{r,m})}} \tag{7}$$

This sigmoidal function gives an output in the range [-1,+1]. Similarly, the output of the partial network ΔU_q is calculated by

$$\Delta U_q = f(\Delta U_{q,r,h}) = \frac{1 - e^{-\Delta U_{q,r,h}\omega_{q,r}}}{1 + e^{-\Delta U_{q,r,h}\omega_{q,r}}} \tag{8}$$

We define the PENN parameter, $p_{q,r}$ as the response of $\Delta Y_{r,p}$ to ΔU_q when some change takes place from the steady state, $\Delta Y_{r,m} = 0$.

$$p_{q,r} \equiv \left.\frac{\partial \Delta U_q}{\partial \Delta Y_{r,p}}\right|_{\Delta Y_{r,m}=0} = \frac{\upsilon_{r,p}\omega_{q,r}}{4} \tag{9}$$

The PENN parameter $p_{q,r}$ indicates linearized relation between the changes in the controlled and manipulated variables:

$$\Delta U_q \approx \sum_r p_{q,r}\Delta Y_{r,p} \tag{10}$$

In the 2-input and 2-output process, the four PENN parameters, $p_{1,1}$, $p_{1,2}$, $p_{2,1}$, $p_{2,2}$, are obtained.

Then we can define the matrix of PENN parameters as follows:

$$P = \begin{pmatrix} p_{1,1} & p_{1,2} \\ p_{2,1} & p_{2,2} \end{pmatrix} \tag{11}$$

The diagnosis of the network condition uses this PENN matrix and its determinant. When the value of the determinant is zero, the PENN matrix is not nonsingular and we cannot solve the simultaneous equation Eq. (10) to obtain forward model of the process. This means that the controller is in ill condition. Also, when the value of the determinant changes dynamically, the value of the determinant tends to become zero, indicating that the controller is unstable.

3.2. Results of diagnosis

The pH control process has been obtained by using the reaction invariants by Gustafsson and Waller[5]. This process has two pumps for supplying acids, one of the pump u_f feeds acid into the alkaline stream and another pump u_t feeds acid into the tank at the lower part of the stream. The pH meters are installed to measure pHs of the inlet stream of the tank y_f and the outlet of the tank y_t. The objective of the control is to maintain the pH of the outlet of the tank at 7 with the fluctuated pH of the alkaline stream.

Figure 3 shows the case when the pump for u_r is broken, and the pump supplies acid at 5 times larger than the command from the controller. In Fig. 3(b), the corresponding p parameter between y_t and u_t to the broken pump for u_r increased. This means that we can know the changes in the process characteristics by watching the changes in the p parameters.

Figure 4 shows the changes in the determinant of the PENN matrix when the pH meter for y_t always reports wrong value. The determinant became 0 and the diagnosis of the controller is accomplished.

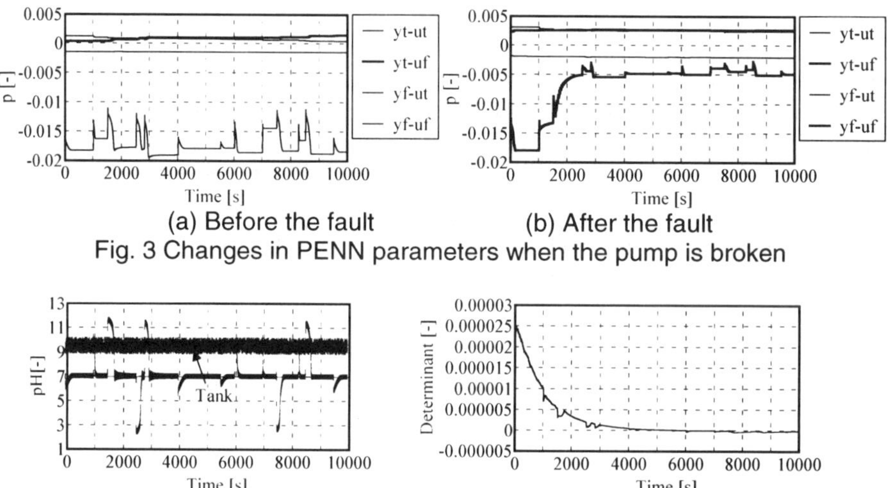

Fig. 3 Changes in PENN parameters when the pump is broken

Fig. 4 Mulfunction of the pH meter and the changes in the determinant of the
PENN matrix

4. Conclusions

The structured PENN (Policy- and Experience- driven Neural Network) controller is the assembly of partial networks which link controlled variable y_r and manipulated variable u_q.

It can perform self-diagnosis by evaluating the PENN parameters from the weights of the connection of the neural network and the determinant of the PENN parameters.

References

[1] Ishida M. Control by a policy- and experience-driven neural network. J. of Chem. Eng. of Japan 1992;25:108-111

[2] Ishida M, Ohba T. Control by a new policy- and experience-driven neural network to follow a desired trajectory. Journal of Chem Eng of Japan 1994;27:137-138

[3] Psaltis D, Sideris A, and Yamamura A. A multilayered neural network controller. IEEE Control Systems Magazine 1989; April:17-21

[4] Ohba T, Ishida M. Application of a neural network to evaluation of interactions in a MIMO process. AIChE Journal 1998;44:2018-2024

[5] Gustafsson, T.K., and K. V. Waller, "Dynamic Modeling and Reaction Invariant Control of pH," Chem. Eng. Sci.,38(3),389(1983)

More flexible user interface for support system of marine engine operation

Kazuhiko NAGAO, Kuniyuki MATSUSHITA
Yuge National College of Maritime Tech.
1000,Yuge Ehime, Japan

Masayoshi NUMANO
National Maritime Research Institute
6-38-1,Shinkawa Mitaka, Tokyo,Japan

Naohiro ISHII
Nagoya Institute of Tech.
Gokiso Nagoya, Aichi, Japan

Abstract. We propose a new concept of a support system for marine engine operation onboard ship. In the concept an agent system is introduced, who has a man-machine interface of speech communication and acts as the skilled engineer by using chief engineer's knowledge. Near future, the marine engine system will be operated by a deck officer at the bridge who has not enough experience. To cope with the emergency and normal start-stop situation, several intelligent control systems are developed, but it can only change the engine load and indicate the failure parts.
We have proposed and been developing the Chief Engineer System (CE-SYS) which instructs the crew about the preferable operation during each step of the engine conditions. CE-SYS is a computer agent system, which has the experienced chief engineer's behaviour. CE-SYS creates a message from the condition of engine in automatically and tells it to the crew as a instruction. In this report, we analysis about man-machine interaction part, and improve the more flexible user interface.

1. Introduction

A ship sailing in the ocean becomes independent and the natural environment of the ocean influences its operation. It is important to maintain a safe level of operation with a limited crew and equipment even if the ship encounter emergency situations such as collisions, running aground, and/or sinking [1]. Moreover, this is the maritime affairs business manager's (equal to Captain and Chief Engineer) responsibility to maintain this level of safety. Generally status of ship engine is divided into two forms of operation "Regular State" and "Non-regular State" like as a "Starting Engine". Currently the M0 (machinery space zero) standard ship and the super-automation ship can only operate in the "Regular State" for several hours or more in nighttime, without someone on duty in the engine room[3]. The technology of the M0 and other ship engine automation systems are not designed to handle emergencies and other out of the ordinary situations, which include engine start and stop. The demand to have a system in place to handle these situations with a limited crew and equipment has grown. This demand has grown because a majority of the engine accidents and breakdowns occurred during the "Non-regular State" [2,4]. There are two reasons, crew has not enough skills, and to err is human.

The promotion of the experienced crews has become difficult in recent years. The reason for this change has come about because economic reason and reducing the size of the crews, the aging and the internationalizing of crews, etc [5,6].

In the man-machine systems, the miss-operation and miss-judgement causes malfunction of whole system, this mistake is called "human error". A human error occurs because of

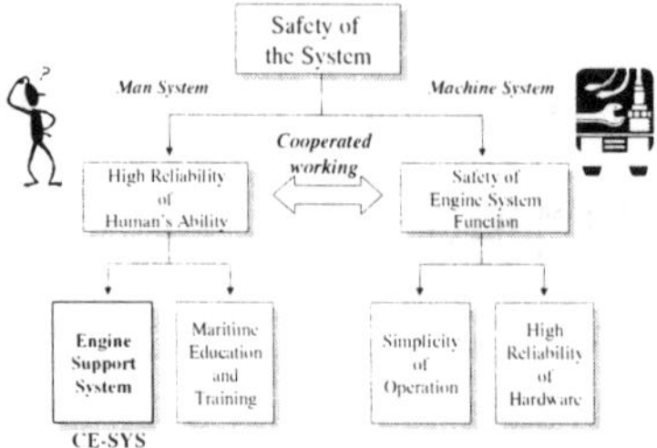

Fig. 1 The safe level of maritime engine system

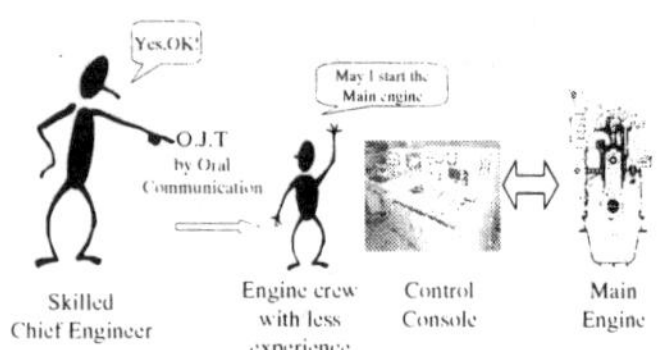

Fig. 2 Flow of engine system

operator's tiredness or less-skill.

The purpose of this research is to develop a system that will prevent the miss-operation and the misjudgment of the crew when working with the engines in a "Non-regular state". This is especially important due to the complexity of the engine system start and stop operation. The new operation support system was developed based on the knowledge of expertized chief engineers [12,13,14]. This report adds approach by an agent function to CE-SYS. The agent system always exists in the support system, and it maintains man-machine interaction. This provides the explanation function and user profiling function.

2. Integrated safety of the marine engine system

The block chart of Fig.1 shows the means to maintain the ships engine system at a safe level of operation. This level of safety is the integration of the engine system (machine system) operation and the reliability of the human element (man system) that is responsible for the operation of the equipment. In other words you cannot have one without the other.
So far, the determination of the operation was being entrusted all by man and trained the expert who had the ability by the education and training. It is difficult that the machine system decide the operation. Such systems may be become a large-scale and is not practically. It is appropriate that man decides the operation and takes all the responsibilities at the accident generation on operating man finally.

However, man has the fault like tiredness and the miss-operation, etc. and no matter how the person trains, this fault is not canceled completely.

This system provides the traditional education and the training method and it supplements man's decision making from the machine side. That is, to avoid the miss-operation and the misjudgment by the inattentive and tiredness, the machine generates an appropriate instruction in each condition.

3. Prototype of CE-SYS

3.1. Agent in CE-SYS

Agents represent software programs that independently perform requests on behalf of a user in a networked and digital world, for example, the search for information, the support for collaboration, user interface, etc. In the marine engine system, crew should watch many complex meters and decide the next operation immediately. The support system, which monitors meters and decides the appropriate operation, is required.

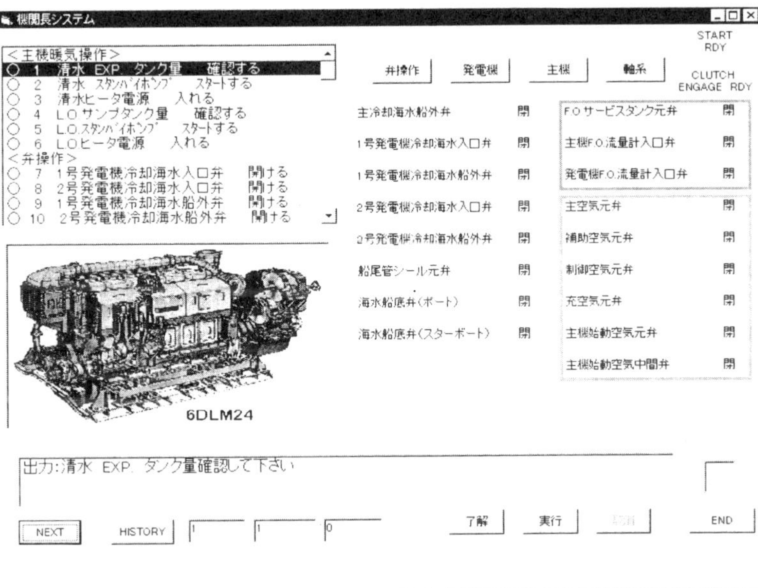

Fig.3 Prototype of CE-SYS

On the training ship, when an inexperienced crew operates the hardware directly, it is necessary to confirm the operation by another experienced crew (i.e. skilled chief engineer) in close proximity giving voice commands. In this system the safe operation of the engine system rest with the chief engineer. The chief engineer can explains the particular condition of the engine and teaches appropriate operations.(Fig.2)

The CE-SYS, which we proposed, is the computer agent system which act as the chief engineer.

3.2. The main function of CE-SYS agent

The authors showed integrated safety of a marine engine and proposed design of chief engineer system (CE-SYS) in the related work [12,13,14]. The proposed CE-SYS is applied college training ship "Yuge-maru". So we build the knowledge database by experience of a skilled chief engineer about marine engine operation and proposed the situation of the support system by talking with the crew. In addition, it was produced the prototype-1(which has the voice recognition function) as the first model ship "Yuge-maru"(shown in Fig.3). The experiment about recognition rate by human sound correspondence by prototype-1 executed at a research voyage of the summer in the last year. As the result it is enough to practical use clear on oral-communication by using some inputting devices such as touch panel and a mouse jointly.

This report adds approach by an agent function to CE-SYS prototype-1. The agent system always exists in the support system, and this is a protocol to have will decision ability of autonomous problem solution. Main improvement point of this time did a divide by class of task that we build already (78 items of preparations works support before the departure from a harbor) with built-in helping function.

4. Improvement for man-machine interface

```
                              Skill Level

         Rule 1                    A

         Rule1.1                   B

   Rule1.1.1    Rule1.1.2          C
```

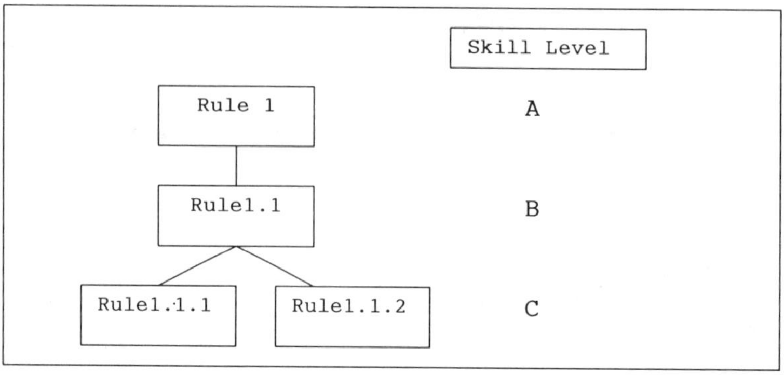

Fig.4 Classified rule for skill level

The prototype-1 of CE-SYS has enough to training purpose for students, but in the practical use, there still has several problems. For example, it can not create most preferable explain for each users or situations. In this section, we focus about these problems.

4.1 Multiple messages for explanation

Prototype-1 of CE-SYS has script of operations, and each operation has only one message. In this report, we resolve this restriction, new type of script permits several messages for each operation. Prototype-2 of CE-SYS can display most preferable explanation for each level. Moreover, if a user fails one operation, CE-SYS particularly mentions that again.

4.2 Support for user profiling

If an agent in CE-SYS wants to explain about the engine condition or next operation, CE-SYS has to know about a skill level of each user (operator, student). So we append the maintain function for the user profile. This function holds a skill level of user and operation log book (which may include the errors).

4.3 Classification of man-machine interface

By using the above extensions, we can control the message for skill of users on each operation. However, CE-SYS follows the course of the script, we can not control the execution of rule. To cope with this problem, we classify the operation to structured rules.

By using the classified rule, we are able to control the executing of rules for explanation level. Fig.4 shows that there is no need to execute the rule-1.1 for skill level A, but the other user (skill level is B or C) have to execute this rule, and so.

4.4 Evaluation of man-machine interface

We developed the rule-based system for man-machine interface functions by using the developing tool for export system BABYLON [15]. We confirm that this system indicates the most preferable messages for each user's skill, and if a user encounters to particular mistake, System displays the message about that problem more particularly.

5. Conclusion

In this research, we propose the agent of the chief engineer which support and the education of engine room crew's decision making in place of the chief engineer. We improved the CE-SYS prototype, which has the data base for the engine start procedure and voice interactive system on the personal computer. The effectiveness of this system is confirmed by the experiment on training ship "Yuge maru". Moreover, the system can be constructed only with a normal computer system, so the cost can be saved.

In the future, we develop the mechanism for the acquirement of knowledge of the chief engineers and crews, the verification of mental model of the crew, and develop the microphone device under the noisy condition.

References

[1] Y.Murayama, "Factor Analysis of Marine Engine Troubles", Journal of The Marine Engineering Society in Japan, No.11,Vol.1,1976.
[2] Y.Murayama, "A Data Bank of Marine Engine Failures and Its Application to Improve Availability", Journal of The Marine Engineering Society in Japan, No.14,Vol.12,1979.
[3] Mitsubishi Heavy Industries(LTD), "Research Report of Navigation Support Systems", 1998.
[4] H.Iwakiri, "An Overview of Marine Engine Failures", Journal of The Marine Engineering Society in Japan, No.34,Vol.5,1999.
[5] Y.Itou,M.Numano, "A Study on a Method of Safety Assessment by Using a Simulator for a High-Speed Vessel", Ship Research Institute Ministry of Transport Report,1998.
[6] L.Baarman, "Engine Control Room Simulator with Open Software Architecture – A Long Term Development Project", Proc. of ICERS4, 1999
[7] Sheridam,T.B, "Telerobotics, Automation, and Human Supervisory Control", MIT Press, 1992.
[8] W.Brenner, "Intelligent Software Agents – Foundations and Applications", Spinger, 1998.
[9] O.Katai, "Human-System Interactions and Intelligent Support", Journal of Japanese Society for Artificial Intelligence,Vol.13.No.3,1998.
[10] A.Gofuku, "Modeling of Function and Structure of Man-Machine Systems", Journal of Japanese Society for Artificial Intelligence, Vol.13.No.3,1998.
[11] "ViaVoice Home Page", IBM, http://www-4.ibm.com/software/speech/
[12] Nagao, Matsushita, "A Concept of agent system for marine operation on the training ship", Proc. of KES2000, Vol.1.2000.
[13] Matsushita, Nagao, "Support System of Marine Engine Operation based on skilled Chief engineer's knowledge",Proc. of ISME2000, 2000.
[14] Matsushita, Nagao, "The Personal Computer-aided Engine Operation Support System based on Oral Communication between Human and Machine", Proc. of IMLA 11, 2000.
[15] T.Christaller, Primio, "THE AI WORKBENCH BABYLON",1992.

KES '01
N. Baba et al. (Eds.)
IOS Press, 2001

A Concept of a Teacher-Support System for Symbolic Calculation

Takayoshi YOSHIOKA, Satoshi MATSUI, Yasuto KAJIWARA, Hitoshi NISHIZAWA
Toyota National College of Technology
2-1 Eisei-cho, Toyota, 471-8525, JAPAN
Tel: [+81] (565) 36-5849, Fax: -5925
{yoshioka, nisizawa}@toyota-ct.ac.jp

Abstract. In this paper, we proposed a concept of a teacher-support system for symbolic calculation. The teacher-support system provides the teacher with the candidates of possible operation-rules his students used for their symbolic calculations by analyzing the students' expressions used in the calculations. The operation-rules include not only proper mathematical operation-rules but also bug-rules caused by students' misconceptions. The system optimizes its method by learning the selection of the teacher among the proposed candidates by the system.

1. Introduction

Intelligent CAI systems using AI technologies, such as machine learning or reasoning so on, have been developing in various field to provide each learner with adequate materials to his/her level of learning. It would attempt to imitate intelligent teachers to help ones learning at the special domain. In general the system consists of 3 elements; man-machine interface, student model that presents the condition of learners, teacher model that presents teaching strategies with knowledge representations of the special subjects [14].

The buggy-model is one of the typical student models, and it stores bug-patterns collected from learners' exercises to the database. Some CAI system using buggy model had reported in the learning programming language system [4][5].

CAS (Computer Algebra System) [1], which we focus on our study, is one of the fields of CAI, and it has the symbolic calculating software such as MATHEMATICA [15] on its backend process [2][3][6][10]. We have been developing a WWW-based exercise system of algebra, and using it as a tool of remedial education [7][8][9][11][12][13]. The system uses WWW technology for communicating with the students and takes advantage of a CAS for symbolic calculations necessary to analyze the students' inputs, and to return the adequate response depending upon it.

In this paper, how we make teacher-support system to be more intelligent using buggy-model is described.

2. Students' Misconceptions in Calculations

When a student makes an exercise of symbolic calculation, for example, an addition of two fractional expressions without common denominators, he types his new expressions with the keyboard into one of the input-fields on the Web page. He is allowed to fill as many fields as he likes for the intermediate expressions from the problem toward the answer, and the expression in the last field is thought to be his answer. All the expressions, sent to the server for evaluation, are compared with the answer and analyzed independently. Every expression receives a comment telling the equality to the answer and appropriateness as the answer. The comment to a non-equivalent expression is "includes a calculating error".

The comments returned from the system help many students to find their mistakes during the exercises, and correct them by themselves. But some of the students, slow learners and novices, claimed it difficult to find the cause of their mistakes because the comments the system returns to an non-equivalent expression tells nothing about the kind or cause of the mistakes they did.

Besides the unkindness of the comments, there exists another reason of difficulty for those students. They have misconception about the operation-rules related to the calculations. Careless mistakes are fairly easy for the students to recognize when the position of the error is notified. But the errors caused by misconception of the students are sometimes quite difficult to accept even if the position of the error is pointed exactly.

In the case of misconception, a short comment is not enough to guide him. A detailed explanation telling why the particular operation leads him to a wrong expression must be given with the description of the alternative operation. A numerical example, which apparently shows the inappropriateness of his operation, helps him to recognize the mistake. Giving such a detailed explanation is currently the task only human teachers or TAs can accomplish. In this paper, we would like to propose a system, which is able to support the teachers by analyzing the students' expressions and picking up some features that might be hard to find for the teachers.

3. Needs for the Support System

Although exercise is essential for mastering symbolic calculation, the amount necessary for the students differs widely. Able students learn the entire lesson within a few exercises, but slow learners not only make many mistakes during the exercises but also continue to make same mistakes in every exercise. Because some of their mistakes are caused by their misconception, it is not easy to let them recognize the mistakes. General tendency that slow learners have difficulty connecting abstract explanations done at lectures to the concrete example worsen the matter.

For providing them with more specific explanation based on the concrete example in front of them is necessary. The problem is that such specific explanations consume many hours of the teachers. In many cases, it exceeds the capacity of usual teachers.

The support system proposed in this paper is an answer to the issue. If the system returns enough information for providing the adequate specific explanation for students' mistakes, it reduces the teacher's load and gives them the opportunity to help more students in trouble.

4. Characteristics of the students in trouble

One of the authors has been teaching algebra for the fresh students (age: 15 - 16) of our institution for five years, and collected the common characteristics of the students that have difficulty in symbolic calculation;
1) The students tend to write down only their final answers on the exercise-sheets. Some of them even calculate in their mind without using any pencils and papers. Others use papers just for the temporal memo of the manual calculations that are disposed just after the calculations. The answer-sheets are full of vacant spaces and lack in description.
2) Explaining how they calculated to others is a difficult task for them. Such a request "Explain the calculating procedure!" sounds quite new to them. Even if they can calculate, they do not explain how they did. Maybe, they do not know how they did by themselves.

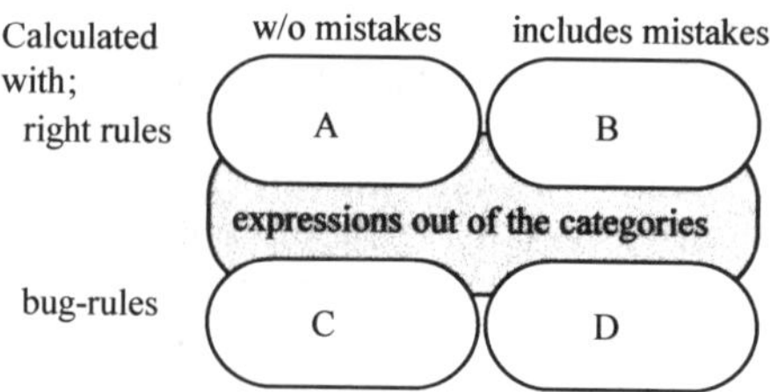

Figure 1 Categories of Expressions

They seem to treat their mind as a black box for the calculation or they are just making a kind of pattern matching.

3) Symbolic calculation, especially fractional calculation, is too complicated to adapt their pattern matching technique. There are too many similar but non-equivalent patterns to

4) Select the adequate pattern. As the result, their performances are deteriorated in calculations of fractional expressions.

These characteristics of the students produce much kind of calculating errors on their exercises, especially in the calculation of fractional expressions.

5. Categories of Students' Expressions

Finding two equivalent expressions is a simple task with the help of a CAS. By confirming the mathematical equivalence of a calculated expression with the original expression, we can separate the correctly calculated expression of category A in Figure 1 with the other expressions. However, categorizing an improperly converted expression becomes a complicated task.

In this paper, we categorize the improperly calculated expressions two dimensionally as shown in Figure 1. One of the measures is the operation-rule, which controls the calculation. Not only right mathematical operation-rules but also bug-rules often created by the students' misconception control the actual calculations. If we convert the original expression with the same bug-rule the student used, the resulting expression must be equivalent with his expression. For this process, we have to collect as many bug-rules as possible in our daily educational activity.

The other measure is the existence of mistakes. When a student's calculation includes a mistake other than the improper conversion lead by misconception, we cannot create any equivalent expressions even if we use any operation-rules or bug-rules. Now, the expression is not equivalent to any converted expressions but is similar to one of them or similar to some of them.

If it is similar to the expression of category A, the calculation supposed to have a "careless

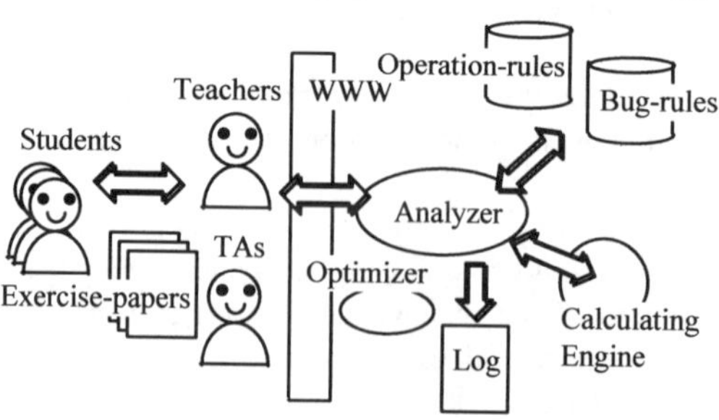

Figure 2
Structure of the
Support System

mistake" and the expression belongs to category B. If it is similar to an expression converted by a bug-rule, which belongs to category C, the calculation must be also controlled by the bug-rule but also includes a "careless mistake". Such an expression belongs to category D.

Because similarity is an ambiguous measure, there remains the room for the expressions that are not categorized into any of the four groups. The usefulness of the support system depends on the method of measuring the similarity.

6. The structure and processes of the support system

The system has the WWW-based interface, an expression analyzer, a database storing operation-rules and bug-rules, a calculation engine (a CAS in our system), and an optimizer as shown in Figure 2. For getting the aid of the system, the user, the teacher or the TA, sends a list of symbolic expressions to the system through the WWW interface. The expression-list in the form of {LHS, RHS} includes two expressions, LHS as the original expression before the student's operation and RHS after the operation. Since the student's operation includes some kind of a mistake, RHS are generally not equal to LHS mathematically.

The expression-list is processed as shown in Figure 3.

1) The expression analyzer confirms the inequality of the two expressions, LHS and RHS. If they are equal, the following processes are bypassed and a simple comment "The expressions are equal" is returned to the user.

2) The analyzer finds the influenced parts in the expressions by the student's operation. An operation may change only one of the terms in LHS and leaves the other terms untouched.

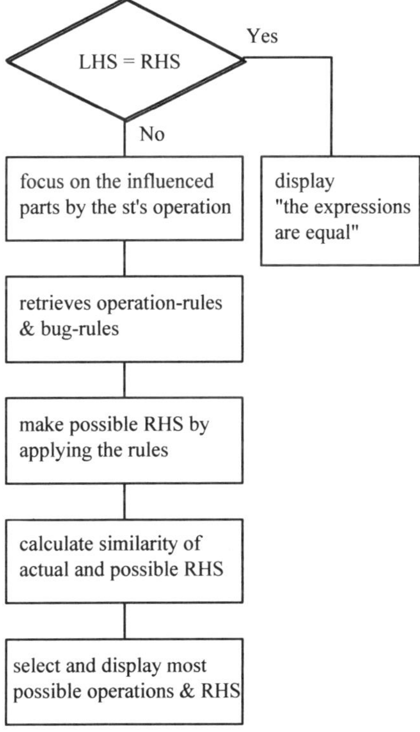

Figure 3 Process on the Server

Another operation may affect only the numerator of a fraction, and other operation may convert the whole LHS expression, etc. The following analysis is focused on the influenced parts of the expressions.

3) After finding the focused parts, the analyzer retrieves the operation-rules and the bug-rules that match the structures of the LHS and RHS from the database. The structure of an expression in this paper is expressed in RPN (reverse Polish notation), where an expression is described with multilayered sets of an operator and the elements. Each element may be an expression also described as a set of child operator and the child elements. Because the operation, factorization, for example, convert the outermost operator of an expression from "Plus" to "Multiply", it is only referred only if the LHS has the outermost operator "Plus" and the RHS has "Multiply".

4) It produces a list of possible RHS expressions by applying the retrieved rules to the LHS of the student's expression.

5) Then, the analyzer calculates the structural similarities of the actual RHS expression and the possible RHS expressions. The structural similarity is expressed as a number from 0.0 to 1.0. The number basically means the ratio of common elements under common operators in two expressions. Since the matching starts from the root to the leaves of an expression and stops at the uncommon branches, similarity number is calculated as zero if the most outer operator of the two expressions is different. When all the operators and the elements of the two expressions match perfectly, the number should become 1.0.

6) A few operation-rules of highest similarity are selected and sent back to the user with the possible RHS expressions. They are displayed on the user's screen as the most possible operations selected by the student and the corresponding RHS expressions.

If the displayed rule is a right mathematical operation-rule, it suggests that a careless mistake may be happening in the student's calculation. If there is a bug-rule, it suggests a misconception of the student. Reviewing the displayed information, the teacher selects one of the rules as what his student tried to use. The user's selection is stored in a log file on the server, and will be used to optimize the parameters in the similarity-calculating procedure.

7. Calculating similarity of expressions

Using the structure of the symbolic expressions as shown in Figure 4, we could calculate the similarity of two expressions. Three expressions in Figure 4 have common outermost operator "Plus" and three elements. The expression a) and b) also shares the same operators and an element as the three outermost elements of the expressions. They are different in only

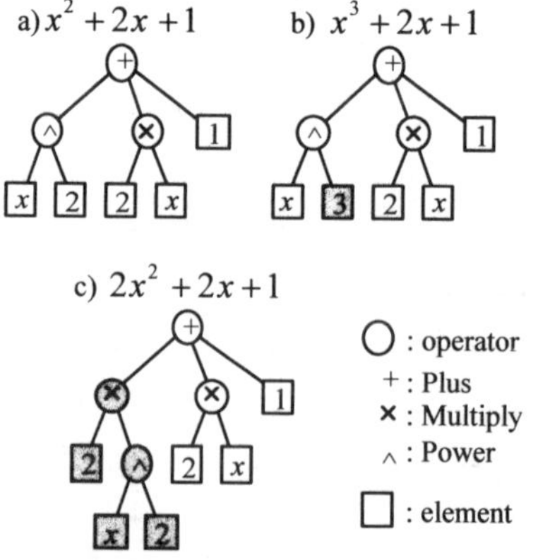

Figure 4
Structure of
Symbolic Expression

one of the child element of the first element. The difference between the expression a) and c) is larger. The operator of the first outermost element of the expression c) "Multiply" is different from the operator "Power" of the expression a). In this case, we conclude that the expression b) is closer to the expression a) than the expression c).

8. Conclusions

We proposed a concept of a teacher-support system for symbolic calculation, which aids the teacher to make suggestions to his students. The system analyzes the operation-rules controlling the students' calculations and calculates the similarity of the resulting expressions to the possible expressions. Thus, the system lists up possible operation-rules that the students might be selecting during their calculation. The operation-rules include not only proper mathematical operation-rules but also bug-rules caused by students' misconceptions. The system optimizes its method of calculating the similarity by learning the decision of the teacher who selects the operation-rules from the provided candidates.

References

[1] Buchberger B., Should Students Learn Integration Rules?, SIGSAM Bulletin 24(1), 1990, pp.10-17.
[2] Dominik A. & Fuchs K.J., MATHEMATICA Palettes - A Structured Environment for Training Mathematical Skills and Strategies, Proc. of ICTMT4, 1999.
[3] Dorfel G., Doring B., Distelmaier H., A Concept for Modelling Dynamic Natural Settings, Proc of Fourth Int. Conf. on Knowledge-Based Intelligent Engineering Systems & Allied Tech. Vol 1, University of Brighton Sussex U.K, 30th & 31st August, 1st September 2000, pp. 257-260.
[4] Kalayar M., Ikematsu H., Hirashima T., Takeuchi A., Intelligent Learning support System for Algorithm Learning, Advanced Research in Computers and Communications in Education IOS Press, (Proc. of ICCE99), Vol.1, pp.840-843 (1999).
[5] Konishi T., Suzuki H., Itoh Y., A Method of Automated Evaluation of Learners, Programs for Assisting Teachers (in Japanese), IEICE Trans. D-I Vol.J83-D-1No.6 pp.682-692, (2000).
[6] Matsubara Y. & Nagamachi M., Construction of factorisation ITS based on the domain knowledge Construction System (in Japanese), Trans. of Japanese Soc. for Information and Systems in Education, 13(3), pp.137-150 (1996).
[7] Nishizawa H., Nakayama K., Kanaya T., A remote tutoring system for algebra enhanced by Mathematica, Proc. of IMS '97, 1997, pp. 361-368.
[8] Nishizawa H., Saito T., Pohjolainen S., Evaluation Methods of Mathematical Expressions in On-line Exercises on Expanding Polynomials and Simplifying Rational Expressions, ICTMT-4, 1999, p52.
[9] Nishizawa H., Saito T., Pohjolainen, An Implementation of a Hypermedia Learning Environment for a Small Group, Trans. of Japanese Soc. for Information and System in Education, vol.15 no.4, 1999, pp. 249-253.
[10] Strickland P. & Al-Jumeily D., A Computer Algebra System for Improving Student's Manipulation Skills in Algebra, IJCAME, 6(1), 1999, pp. 17-24.
[11] Yoshioka T. Fuchs K., Nishizawa H., Dominik A., Step-by-step Instruction of Symbolic Calculation, ATCM2000, 2000, pp. 186-192.
[12] Yoshioka T., Higuchi Y., Nishizawa H., A Web-based Interactive Exercise System for Learning Mathematical Functions, ICCE2000, 2000, pp. 1574-1576.
[13] Yoshioka T., Nishizawa H., Tsukamoto T., Method and Effectiveness of an Individualized Exercise of fundamental Mathematics, Community College Journal of Research and Practice, 25, Taylor & Francis, 2001, pp. 373-378.
[14] Wenger, E., Intelligent CAI (in Japanese), Ohmsya (1990).
[15] Wolfram S., The MATHEMATICA Book, Wolfram Media and Cambridge University Press (1996).

KES '01
N. Baba et al. (Eds.)
IOS Press, 2001

Writer Recognition by means of Fuzzy Membership Function and Local Arcs

Masahiro OZAKI[1], Yoshinori ADACHI[2], Naohiro ISHII[3] and Mitsu YOSHIMURA[4]
*[1]Nagoya Women's University, [2]Chubu University, [3]Nagoya Institute of Technology,
[4]Nagoya City University*
[1]3-40, Shioji-cho, Mizuho-ku, Nagoya, JAPAN   467-8610   ozaki@nagoya-wu.ac.jp

Abstract: Writer recognition was studied by means of local arc, curvature, through the experience obtained from the previous work using the Fuzzy membership functions. The combination of similarity values of each chord length gives good results even for the characters could not be recognized by the previous method.

1. Introduction

We had proposed the off-line writer recognition system[1][2][3][4] for Japanese "hiragana" characters. The proposed system was obtained 98.7% recognition ratio, but there are a few problems still remained.
1. Because of long period for collecting characters, a difference between characters increases and consequently characters scatter.
2. In spite of the same kind of characters, some writers write differently every time.

To overcome these problems, character normalization has been done beforehand. However, as in case 1, normalizing characters make the same writer's characters have different stroke widths. As in case 2, these quite different characters were omitted for recognition as inadequate characters, so in the case of too few characters, the recognition process happen to be abandoned.

In the previous work, we tried to make a writer recognition system as simple as possible by using two-dimensional Fuzzy membership functions which is obtained from simple summation of characters.   However, sticking to the method, the above problems could not be overcome.   Then, we introduced the idea of local arcs, which is proposed by one of our co-authors[5][6] and is not influenced so much by the difference of the size and shape of characters.

2. The previous method by using fuzzy theory

In the previous study, we used fuzzy membership functions to construct dictionaries for each type of characters and each writer.   Eight types of "hiragana" "は(ha)","ま(ma)",""に(ni)","す(su)","の(no)","と(to)","を(wo)","る(ru)", often appeared letters, were used for investigation.   The written characters on the framed sheets were input through a scanner and were normalized as 50 x 50 dots square after centered the gravity.   Then to make a dictionary, we eliminated characters that far from the average character.   The average and variance are obtained from the following equations,

$$\overline{f_{ij}}(x, y) = \frac{1}{n} \sum_{k=1}^{n} f_{ij,k}(x, y) \quad (x, y = 1,..., 50) \quad (1)$$

where $f_{ij,k}(x,y)$ was a function indicating an existence of character strokes as follows,

$$f_{ij,k}(x,y) = \begin{cases} 1 & \text{if stroke exists at } (x,y) \\ 0 & \text{others} \end{cases} \tag{2}$$

and subscripts i, j, k were indicating a writer, a type of character, and character number, respectively.

$$\sigma^2_{ij,k} = \frac{1}{2500} \sum_{x,y=1}^{50} (f_{ij,k}(x,y) - \bar{f}_{ij}(x,y))^2 \tag{3}$$

To emphasize a writer's feature, a principal component analysis was used for each writer and each type of character. And 2D membership functions $\mu_{ij}(x,y)$ was expressed by the following equation,

$$\mu_{ij}(x,y) = \sum_{l=1}^{m} \lambda_{ij,l} X_{ij} \vec{e}_{ij,l} / \sum_{l=1}^{m} \lambda_{ij,l} \tag{4}$$

where $X_{ij} = (\vec{x}_{ij,1}, \vec{x}_{ij,2}, \cdots, \vec{x}_{ij,n})$ is a character matrix and $\vec{x}_{ij,k}(50 \times 50 = 2,500 \text{ elements})$ is a character. $\lambda_{ij,l}(l = 1, \cdots, n)$ is an eigenvalue and $\vec{e}_{ij,l}(l = 1, \cdots, n)$ is an eigenvector, which are obtained from the covariant matrix $Q_{ij} = X_{ij}^T X_{ij}$.

Writer recognition was done by the following similarity evaluation function,

$$\eta_{ij,k} = \sum_{(x,y)\in A_{\alpha,ij}} \mu_{ij}(x,y) f_{ij,k}(x,y) \tag{5}$$

where $\eta_{ij,k}$ indicates the k-th character's similarity value to the i-th writer.

$$A_{\alpha,ij} = \{(x,y) \mid \mu_{ij}(x,y) > \alpha \tag{6}$$

where α was set to 0.5 empirically.

Handwritten characters usually fluctuated and those variances are depending on writers and types of characters. Then to absorb scattering of characters and to characterize one's handwriting features, a combination of four types of characters could work effectively. The following equation is used to obtain a similarity,

$$\eta_i = 1 - (1 - \eta_{ij_1 k_1})(1 - \eta_{ij_2 k_2})(1 - \eta_{ij_3 k_3}) \tag{7}$$

which is a product of three different kinds of characters, where k_1, k_2 and k_3 are indicating character numbers.

In **Table 1**, the average recognition ratios of 4 characters with changing the number of characters used for the dictionaries, which are made of 4, 5, 6, 7, and 8 characters, are listed. Where characters have low similarity value, 0.4, were omitted as inadequate characters for the writer recognition. The results show fairly good agreement. However, in the case of 8 characters dictionary of writer no. 6 and except 4 characters dictionaries of writer no. 19, 4 characters were not selected for evaluation because of most characters became inadequate character. Then evaluations were not carried out. The characters of these writers were varied those size and shape each time they wrote. Examples of those characters are depicted in **Figure 1**. Which are the characters "は(ha)"of writer no. 19, the distances between the first stroke and the other strokes change very much.

Table 1: Effect of number of type of characters on recognition ratio

Writer no.	No. of characters				
	4	5	6	7	8
6	1.00	1.00	1.00	1.00	*
13	1.00	1.00	1.00	1.00	.999
19	1.00	*	*	*	*
Average of others	1.00	1.00	1.00	1.00	1.00

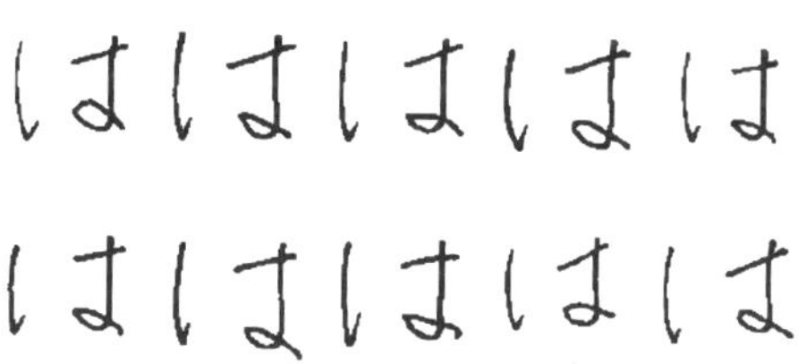

Figure 1: Characters "ha" of writer no.1

3. Proposed method

3.1 Calculation of curvature

To overcome the problems occurred from the normalization and variance of character shapes mentioned in the previous section, curvature of strokes was used to figure out the feature of writes. The curvature of local arc was shown to be useful for writer recognition by Yoshimura [5][6]. Yoshimura used mask patterns of arc to obtain the stroke's curvatures with the normalized characters. In this study, we calculate the curvature directly with the original characters. The following processes obtained the curvature.

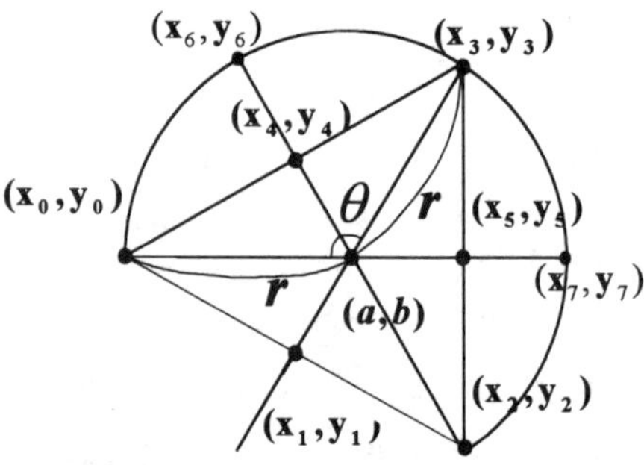

Figure2: The arc of the chord

(1) First two points making chord are fixed.
   The length of the chord is 5, 9, 13, and 17 dots and the direction is changed 15 degrees interval, in total 12 directions are investigated.
(2) The arc of the chord is found out by the following procedure as shown in **Figure 2**.
   a. The tip of the chord with co-ordinates (x_0,y_0) is placed on the stroke.
   b. It is checked that the other end of the chord with co-ordinates (x_1,y_1) is on the stroke or not. If not on, then the chord is changed.
   c. The perpendicular bisector l is decided that is pass the midpoint (x_2,y_2).
   d. The point (x_3,y_3) on l and on a stroke is searched within the distance from the midpoint less than 4 times of the length of the chord.
   e. The circle which passes those three points, (x_0,y_0), (x_1,y_1), and (x_3,y_3), is obtained as follows:

$$(x-a)^2 + (y-b)^2 = r^2 \tag{8}$$

where

$$a = \frac{(y_0 - y_1)(y_1 - y_2)(y_3 - y_0) + (x_1 + x_3)(x_1 - x_3)(y_0 - y_1) - (x_0 + x_1)(x_0 - x_3)(y_1 - y_3)}{2\{(x_1 - x_3)(y_0 - y_1) - (x_0 - x_1)(y_1 - y_3)\}} \tag{9}$$

$$b = \frac{1}{2}\left\{(x_0 + x_1 - 2a)\frac{x_0 - x_1}{y_0 - y_1} + y_0 + y_1\right\} \tag{10}$$

$$r^2 = (x_0 - a)^2 + (y_0 - b)^2 \tag{11}$$

(3) The arc of the chord is divided into quarters. The points $(x_6,y_6),(x_7,y_7)$ are the quadruple points. It is examined that these two points are on the stroke or not. These points are obtained as follows:
   a. The midpoints of points (x_0,y_0) and (x_3,y_3), and (x_1,y_1) and (x_3,y_3) are obtained as (x_4,y_4) and (x_5,y_5) respectively.
   b. The quartered points $(x_6,y_6),(x_7,y_7)$ are obtained from the ratio as follows:

$$x_6 = x_4 + \frac{1-\cos\frac{\theta}{2}}{\cos\frac{\theta}{2}}(x_4 - a), \quad y_6 = y_4 + \frac{1-\cos\frac{\theta}{2}}{\cos\frac{\theta}{2}}(y_4 - b) \tag{12}$$

$$x_7 = x_5 + \frac{1-\cos\dfrac{\theta}{2}}{\cos\dfrac{\theta}{2}}(x_5 - a), \quad y_7 = y_5 + \frac{1-\cos\dfrac{\theta}{2}}{\cos\dfrac{\theta}{2}}(y_5 - b) \qquad (13)$$

where

$$\cos\frac{\theta}{2} = \sin\alpha = \sqrt{1-\cos^2\alpha} \qquad (14)$$

$$\cos\alpha = \frac{\sqrt{(x_0 - x_3)^2 + (y_0 - y_3)^2}}{2r} \qquad (15)$$

(4) If 2 points $(x_6, y_6), (x_7, y_7)$ are both on a stroke, then radius of the curvature is calculated. Otherwise, next condition is examined.

The above procedure (1) to (4) is carried out for every character types.

3.2 Dictionary of curvature

The curvatures of randomly selected 5 characters for each type of character were calculated. The obtained curvatures were classified for 12(12 directions) times 11(curvature value from −5 to 5) matrix. From the principal component analysis of these matrixes, eigenvalues and eigenvectors were obtained to satisfy that the cumulative proportion was larger than 0.9. By weighting sum using these eigenvalues and eigenvectors, dictionaries were prepared for each type of character and each writer.

3.3 Results of writer recognition

Eight types of characters, "は(ha)","ま(ma)",""に (ni)","す(su)","の(no)","と(to)","を(wo)", and "る (ru)", for 21 writers were used for the recognition tests, and the results are listed in **Table 2**. The recognition ratio is strongly depending on the length of the chord and type of characters, and generally the chord length 9 or 13 gives better results. In the average speaking, the recognition ratio of the chord length of 5 is 79%, those of 9, 13, 17 are 86%, 87%, and 87% respectively. However, it depends on the

Table 2: Effect of number of lengths of chord on recognition ratio

Writer no.	lengths of chord			
	5	9	13	17
1	0.48	0.68	0.69	0.78
2	0.89	0.93	0.96	0.98
3	0.88	0.95	0.93	0.80
4	0.80	0.90	0.84	0.85
5	0.85	0.88	0.91	0.90
6	0.73	0.86	0.80	0.79
7	0.94	0.88	0.96	0.96
8	0.69	0.79	0.83	0.83
9	0.81	0.85	0.88	0.88
10	0.55	0.70	0.76	0.83
11	0.78	0.85	0.85	0.90
12	0.85	0.93	0.89	0.91
13	0.61	0.70	0.75	0.68
14	0.95	0.90	0.95	0.95
15	0.71	0.80	0.75	0.74
16	0.86	0.91	0.96	0.88
17	0.86	0.91	0.91	0.91
18	0.76	0.94	0.98	0.95
19	0.98	0.99	0.96	0.94
20	0.85	0.80	0.90	0.91
21	0.78	0.88	0.89	0.88
Av.	0.79	0.86	0.87	0.87

size of character and stroke shape. In the table, recognition ratios of the chord length 5 of the writer no. 1,10, and 13 are very low, 48%, 55%, and 61%, because they have habits to write large characters. On the contrary, those of writer no. 14 and 19 are quite high, 94% and 98%, because they write comparatively small characters.

3.4 Compilation of chord length

The above results shows that the chord length very much affects to the writer recognition in accordance with the size of characters. Then same as the previous study, we introduced the combination of four types of chord lengths. The similarity is evaluated by the following equation:

$$\eta = 1 - (1 - \eta_5)(1 - \eta_9)(1 - \eta_{13})(1 - \eta_{17}) \qquad (16)$$

where $\eta_5, \eta_9, \eta_{13}, \eta_{17}$ indicate the similarities obtained from the chord length 5, 9, 13, and 17 respectively.

In **Table 3**, the effects of number of chord length type are listed. The average writer recognition ratio of the combination of 2 chords, 3 chords, and 4 chords are 91%, 95%, and 97% respectively. In the case of the proposed method, there were no inadequate characters exist which have to be specially mentioned. The results of writer no. 6 and 19, whose characters were mostly inadequate characters in the previous work, were fairly improved like as 88%~94%,99%~100% recognition ratios respectively.

Table 3. Result obtained recognition ratio

combination of chords	recognition ratio
5 and 9 dots	0.962
5 and 13 dots	0.957
5 and 17 dots	0.948
9 and 13 dots	0.958
9 and 17 dots	0.976
13 and 17 dots	0.980
5,9 and 13 dots	0.984
5,9 and 17 dots	0.982
5,13 and 17 dots	0.985
9,13 and 17 dots	0.992
5,9,13 and 17 dots	0.994

4. Conclusion

In this work we proposed the hybrid writer recognition method, which is a combination of the previous method and the proposed curvature method. The previous method is very simple using 2D Fuzzy membership functions but has high recognition ratios. However there also exclude characters as inadequate characters. The local arc gives writers' characteristic features fairly well and analysis of curvature gives good writer recognition ratio. Therefore only small number of characters are required to perform the writer recognition, usually only three types of characters and 5 characters for each type of characters. In the future we are going to extend the idea to Roman characters.

References

[1] M. Ozaki, Y. Adachi, N. Ishii, and T. Koyazu: Fuzzy CAI System to Improve Hand Writing Skills by Using Sensuous (1996) Trans. of IEICE Vol.J79-D-II NO.9 pp.1554-1561

[2] S. Watanabe, T. Furuhashi, K. Obata, and Y. Uchikawa: An Off-Line Signature Recognition Using a Fuzzy Net (1994) Trans. of IEE Japan Vol.114-D No.6 pp.674-679

[3] M. Ozaki, Y. Adachi, and N. Ishii: Writer Recognition by means of Fuzzy Similarity Evaluation Function (2000) Proc. KES 2000, pp.287-291

[4] M. Ozaki, Y. Adachi, and N. Ishii: Study of Accuracy Dependence of Writer Recognition on Number of Character (2000) Proc. KES 2000, pp.292-296

[5] M.Yoshimura and I.Yoshimura : Writer recognition the state-of the art and issues to be addressed (1996) TECHNICAL REPORT of IEICE PRMU96-48 pp.81-90

[6]I.Yoshimura, M.Yoshimura: Writer Identification Using Localized Arc Pattern Method, Trans. of IEICE Vol.J74-D-II NO.2 pp.230-238

Study on the Visual Evoked Potential using Correlation Filter Method

Hiroshi Sasaki*, Hideharu Tsubota* and Naohiro Ishii**
Fukui University of Technology, 3-6-1 Gakuen, Fukui 910-8505, Japan
E-mail: h-sasaki@ccmails.fukui-ut.ac.jp
***Nagoya Institute of Technology, Gokiso-cho, Showa-ku, Nagoya 466-8555, Japan*
E-mail: ishii@ics.nitech.ac.jp

Abstract. Many colors are shown in our surrounding, and we can discriminate the difference of colors easily. The organs which perform information processing with regard to such a visual processing are the retina and the cerebrum. The information (visual stimulus) inputted into the retina reaches the visual cortex of cerebrum finally, and is processed there. The electroencephalogram of a very small potential produced at this time is called visual evoked potential. Since it superposes on the original electroencephalogram (basic rhythm), this evoked potential cannot be seen directly. Therefore, if we can remove the basic rhythm of the electroencephalogram by data processing, we can extract an evoked potential only. In this study, the three primary colors of light (red, blue and green) are used as a visual stimulus. And the viewing pattern is made by choosing two kinds of colors from them and combining. As the method of extracting the visual evoked potential in case of giving such a visual stimulus, we use correlation filter method. And, it is compared with average processing method which has been used until now about the result of a latency, a potential and FFT processing.

1. Introduction

We respond to a visual information first in the retina which exists in the eye ball. In the retina, the form of an object and the kind of a color, etc. are inputted. The color information from the retina reaches the visual cortex of cerebrum, finally it is summarized as a visual information, and the difference among colors information can be seen as the evoked potential. However, since the potential of the basic rhythm is very large compared with the evoked potential, it is difficult for us to extract the changed part of the evoked potential directly. Then, the conventional method used for the analysis of the evoked potential is average processing method. This method is to get the wave by carrying out the division by N, after N-times electroencephalograms are added. Thereby, the signal-to-noise ratio of the evoked potential hidden by the basic rhythm is improved. However, the actual evoked potential does not have an definite time-locked relation to the stimulus time starting by the fluctuation phenomenon. Moreover, the basic rhythm is not necessarily generated regularly at random. Therefore, even if it increases the number of times of addition without any restriction, the signal-to-noise ratio is not improved.

Then, in this study, correlation filter method is devised as the method for improving the fluctuation phenomenon which is the weak point in average processing method. By using this method, it is considered that the fluctuation of the time (latency) in the evoked potential wave form can be canceled, and the effect of average processing method can be pulled out to the maximum extent, and the very small evoked potential can be taken out definitely.

2. Correlation Filter Method

Correlation filter method is the method which combine the average processing and the correlation processing (Fig.1). As for the process, for each RAW data (100 electroencephalogram data), every ten average processing is performed as pre processing, and ten data groups are created. Next, the 1st data group is made into the template, and the correlation processing with this template and the 2nd data group is performed. When the fluctuation is observed, the data group is modified as shown in the following. The average processing is performed to the 1st data group and the 2nd data

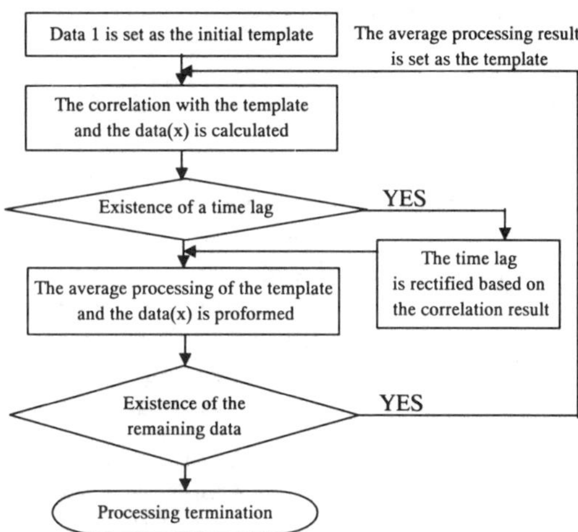

Fig.1   Flow chart of correlation filter method

group, and this result is set as the new template. This series of processing is performed one by one for ten data groups.

Here, before applying correlation filter method to the actual electroencephalogram, it is necessary to verify the effect of this processing method quantitatively. Therefore, we evaluated correlation filter method using the computer simulated potential waves, called here pseudo-evoked potential wave. The pseudo-evoked potential wave used in the simulation is expressed with the following numerical expression.

$$W(t) = 0.125\sin(t + \tau)\exp(0.35t) + 0.5\sin(t) \cdot r$$

(τ : fluctuation of a latency,   r : function of random numbers)

100 pseudo-evoked potential waves which added τ at random by a maximum of 20 msec is created using this numerical expression. An example of the used pseudo-evoked potential wave form is shown in Fig.2. We performed the simulation of average processing and correlation filter processing to 100 pseudo-evoked potential waves (Fig.3). From Fig.3, in the conventional average processing, the error has arisen in the potential and the latency

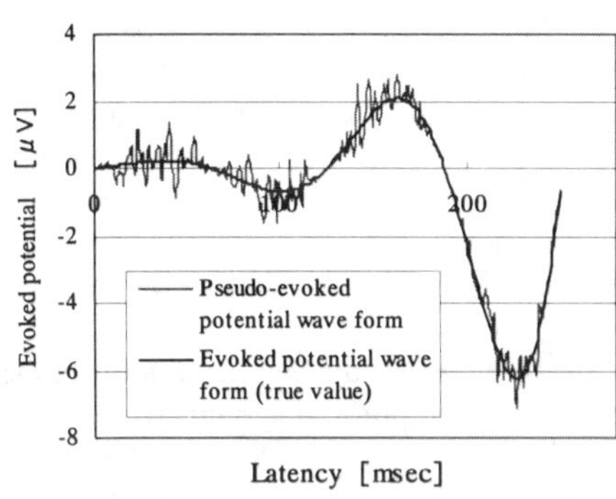

Fig.2   Example of the pseudo-
evoked potential wave form

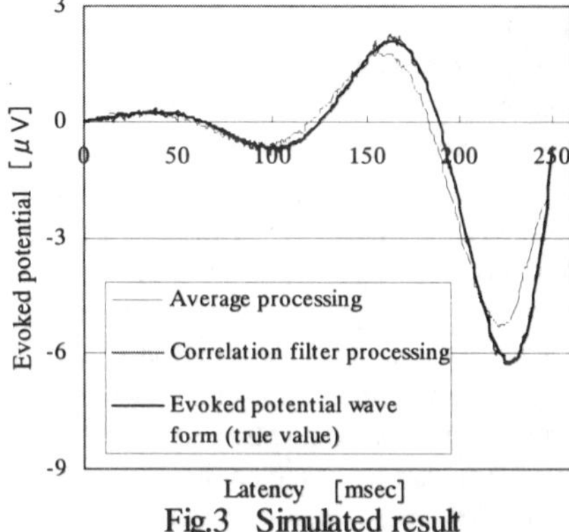

Fig.3   Simulated result

under the influence of the artifact and the fluctuation of a latency, compared with the evoked potential wave form (true value). On the other hand, in correlation filter processing, the evoked potential wave form (true value) could be reproduced faithfully, and we could check that correlation filter processing was effective.

3. Experimental Method

Ag-AgCl electrodes with a diameter of about 1cm are attached in the subject (color vision normality, 22 years old, male). One of the attachment positions is MO position which only 5cm distance separated from the inion. And it is LO and RO position which only 5cm distance separated from MO position to the left and the right, respectively. Both the earlobe juncture electrodes are used as the reference electrode, and the body ground is attached to a wrist. And the potential between each electrode of LO, MO, RO position and the reference electrode are measured. We use the display equipment (background color: black) of a personal computer as a stimulus unit and the round shape with a diameter of about 20cm is displayed to show the stimulus color pattern of Fig.4. Measurement is performed by giving a subject the visual stimulus of these patterns. However, each stimulus is given at 1 second interval.

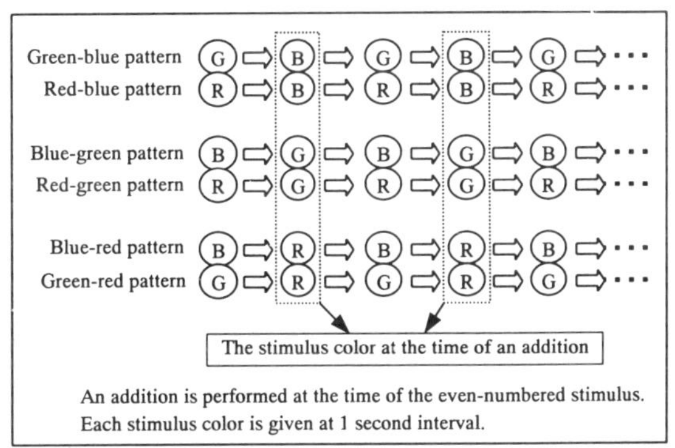

Fig.4   Stimulus color pattern

4. Result

An example of measured result about latency and potential are shown in Table 1 and Table 2 respectively. P1 and N1 are the first upper and lower peak point respectively in the measured wave form. P2 and N2 are the 2nd upper and lower peak point respectively. P1-N1 shows the potential between P1 and N1, P2-N2 shows the potential between P2 and N2. Moreover, Fig.5 is an example of the measured wave form. A vertical axis shows the dimension of an evoked potential and the horizontal axis shows the elapsed time (latency) after giving a stimulus. Fig.6 is the result of FFT processing in the wave form of Fig.5, the vertical axis shows the power spectrum (the maximum is set to 100), and the horizontal axis

Table 1   Example of the result (latency)

Processing method	Electrode position	P1 [msec]	N1 [msec]	P2 [msec]	N2 [msec]
Average processing	LO	30	83	114	208
	MO	32	83	114	207
	RO	32	83	114	204
Correlation filter processing	LO	41	91	141	208
	MO	45	88	142	200
	RO	38	88	141	206

Table 2   Example of the result (potential)

Processing method	Electrode position	P1-N1 [μV]	P2-N2 [μV]
Average processing	LO	2.603	4.110
	MO	2.588	3.998
	RO	2.400	4.882
Correlation filter processing	LO	2.897	3.991
	MO	2.648	4.012
	RO	2.760	4.346

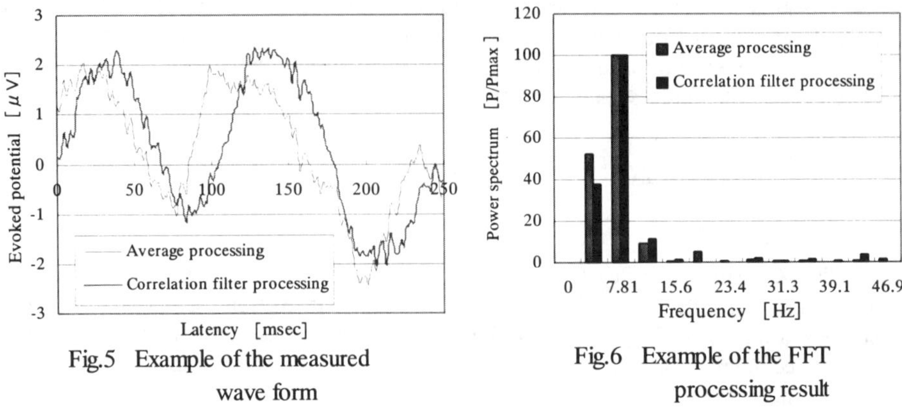

Fig.5   Example of the measured
wave form

Fig.6   Example of the FFT
processing result

shows the frequency. The stimulus color pattern of these examples in the measured result is blue-green pattern (an addition color is green).

5. Consideration

From the result of a latency (Table 1), the latency tends to become late in correlation filter processing compared with average processing. (The genesis of each peak becomes slow.) As for this, in the average Processing, it is thought that the wave form is distorted under the influence of the artifact, and then the latency becomes early. The difference of this latency is confirmed also in the simulated result. Next, from the result which compared the latency for every stimulus color pattern, the latency tends to become early, when an addition color is green (red-green pattern, blue-green pattern).

And then, from the monitoring result (Table 2) of potential, we can confirm that P1-N1 potential increases in the correlation filter processing of almost all the stimulus color pattern. In the case of average processing, in the phase of an electroencephalogram measuring, the fluctuation of a latency has arisen to each 100 RAW data, and it is thought that potential will appear low as a result. Moreover, as a result of comparing both processing to P2-N2 potential, a great difference which was seen at the time of P1-N1 potential did not appear. This result shows that P2, N2 potential with the comparatively large amplitude have little influence of fluctuation also at the case of only average processing which is just the same as the result of a latency. As a result compared by the stimulus color pattern, when an addition color is blue (red-blue pattern, green-blue pattern), the potential is in the low trend similar to P1-N1 potential.

From the FFT processing result, in the average processing, the maximum peak exists at 7.81Hz and the 2nd peak exists at 3.91Hz. When only average processing is performed, in order to amplify even the noise by mixing of the artifact, and the fluctuation of the latency, this FFT processing result shows that the excessive frequency component is intermingled besides the main frequency component of the evoked potential. This is considered to have been the hurdle for obtaining the exact evoked potential wave form. On the other hand, in the correlation filter processing, although the frequency of the maximum peak is the same compared with that of the average processing, but the value at 3.91Hz is decreased. This shows that the correlation filter processing can obtain the more exact evoked potential wave form rather than the average processing.

6. Conclusion

In this study, we devised the correlation filter method which combined the average processing and the correlation processing instead of the conventional average processing method. The merit of this method is that the effect of the average processing is demonstrated to the maximum extent by performing correlation processing. And from the result of simulation, measured wave form and FFT processing, correlation filter method could not be easily influenced of the artifact and the fluctuation of latency compared with average processing method, and we were able to confirm the effect in the evoked potential analysis. Since the value of peak potential was small around P1 and N1, it was difficult to use as measured data especially until now. However, the value of the latency and the potential of P1, N1 could be known clearly this time, and it became an effective result for advancing a study from now on. By the way, since correlation filter method consists of three phase of processing process (average processing of pre processing, correlation processing and average processing), the week point of this processing is that it needs much time and effort. Therefore, now, this method unsuitable for data processing of many stimulus color patterns. In order to solve this problem, the introduction of the method which can be done by batch processing is considered as a future subject.

References

[1]   M.G.Larimore, C.R.Johnson, J.R.Treicher, "Theory and Design of Adaptive Filters", 268 pages, Prentice Hall, March 2001

[2]   Woody,C.D., "Characterization of an adaptive filter for the analysis of variable latency neuroelectric signal", Med. & Biol. Eng., Vol.5, pp.539-553, 1967

[3]   McGillem,C.D. et al., "Improved waveform estimation procedures for event-related potentials", IEEE Trans., BME, Vol.32, No.6, pp.371-379, 1985

[4]   A.Calway, S.Kruger, D.Tweed, "Motion Estimation Using Correlation and Local Directional Smoothing", Proc. IEEE Int. Conference on Image Processing, pp.1-5, 1998

[5]   B.Esf ari, J.J.Little, "Cepstral Analysis of Optical Flow", Technical Rept. of Computer Science, Univ. of British Columbia, TR-92-06, Nov. 1992

KES '01
N. Baba et al. (Eds.)
IOS Press, 2001

Neural Network Interstitial Lung Disease Analysis

Takaharu KOUDA and Hiroshi KONDO
Electrical Engineering Department, Kyushu Institute of Technology
kondou@ele.kyutech.ac.jp

Abstract. Computer-aided diagnosis for pneumoconiosis using Neural Network is presented. The rounded opacities on the pneumoconiosis X-ray photo are picked up quickly through a back propagation (BP) neural network with several typical training patterns. The training patterns from 0.6 mmØ to 4.0 mmØ are made as simple circles. The neck problem for an automatic pneumoconiosis diagnosis has been to reject the unnecessary part like ribs and vessel's shades. In this paper such unnecessary parts are rejected well by the special technique called "moving normalization".. The new technique called moving normalization is developed here in order to made an appropriate bi-level ROI image. The total evaluation is done from the size and figure categorization. Mary simulation examples show that the proposed method gives much reliable result than traditional ones.

1. Introduction

In these several years computer-aided system are extremely popular in a medical field. The aim of computer-aided diagnosis is to alert the radiologist by indicating potential lesions and/or providing quantitative information as second options. Since the middle of 1980, a number of computerized schemes for computer-aided diagnosis have been developed for chest radiography, mammography, angiography, and bone radiography. Especially in chest radiography, many computerized schemes have been applied to the detection and classification of pneumoconiosis, because the quantitative analysis of it has been required from the viewpoint of workmen's accident compensation insurance. Pneumoconiosis is a lung disease caused by, for example, the long-term inhalation of coal dust and the local tissue reaction to the accumulated dust particles. The first radiological symptom in the development of simple pneumoconiosis is the appearance of small opacities, either rounded of somewhat irregular, in the chest X-rays. According to the profusion of small opacities, categories 0-3 have been established to indicate the severity of the disease where category 0 means normal case and category 3 means very numerous small opacities. The early studies of computer pneumoconiosis analysis have been done by several groups [1][2][3]. In their studies the texture analysis for the X-ray photo has been taken. Recently, however, the study trends toward the detection of the small rounded opacity itself because of the extremely development of the computer hard and software [4]. They have utilized a special filtering for dropping off the unnecessary part like rib shade in the X-ray. The performance of the filtering is not satisfied sometimes due to the vagueness of the X-ray. In this paper the neural network is introduced to pick the rounded opacities up from the X-ray photo with no filtering. A neural network is powerful for pattern matching. Here a back propagation neural network with three layers is used.

2. Pneumoconiosis categorization

Pneumoconiosis is one of the serious lung occupational disease. Hence the diagnosis result of a medical doctor gives a big implication for the workmen's accident compensation insurance. Even such doctor's diagnosis results, however, are not often consistent with each other. For this reason it has been required that the quantitative analysis of pneumoconiosis is established. According to the classification scheme of the International Labor Office (ILO), there are two kinds of categories: one is number and area density classification and the other is size-figure one. The former one has three ranks from 0 to 3, where 0 means normal case and rank 3 means very serious case. The size-figure classification has also three types as P, Q, and R, where P means the equivalent diameter d of the opacity is less than or equal to 1.5 mm, Q means 1.5mm< d⌋ 3.0mm, and **R** means 3.0< d⌋ 10.0 mm. Figure 1 shows a normalized X-ray photo with 3000x3000x8bit. The normalization is made as setting the minimum gray level 0 (black) and the maximum value 255 (white). The other gray levels are transformed linearly between the above two value. Figure 1 is the photo in (3,P) category. Usually in the analysis of pneumoconiosis chest X-ray image the original image like Fig 1 is divided into three blocks from the top to the bottom shown in Fig 1. We call them high lung field, middle lung field, and low lung field respectively. From these fields the region of interest (ROI) is quarried for the analysis (See Fig.1). The size of the ROI is 512x512 pixel here. These divisions for the ROIs are done automatically. The opacity figures are almost rounded. And such rounded opacities in the pneumoconiosis chest X-ray image appear so often in the high and middle lung fields because of the position near bronchial tubes. Figure 2 shows an example of ROIs. This is the right high lung field from (2,Q) categorized image. From such ROI image the rounded opacities must be detected. The field evaluation for the classification is done by calculating the number density and the area density of the rounded opacities, and by comparing those values of the ILO standard images.

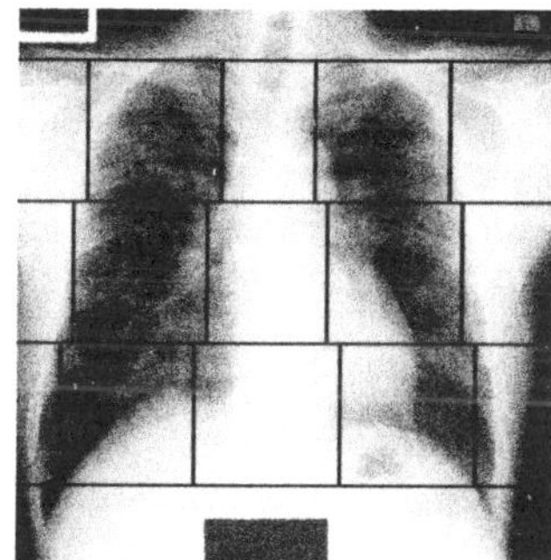

Fig.1 X-ray photo (3,P) 3000x3000x8bit

Fig.2 Example of ROI (3,P)

3. Procedure of pneumoconiosis analysis

The coincidence rate with a medical doctor in the evaluation results depends upon the exactness of rounded opacity detection. It is the most important part of the pneumoconiosis analysis to detect each rounded opacity. In this paper a neural network is utilized for detecting a rounded opacity. A neural network is abbreviated to NN here. An NN has an excellent property for a pattern matching. Here we utilized a back propagation NN with three layers. The training patterns are circular figures with various radiuses like the rounded opacities.

The important step of this procedure is moving normalization of the ROI. Chest X-ray image has often a big variance in its gray levels.

Especially tiny area average value is quite different with that of another area. Hence only a

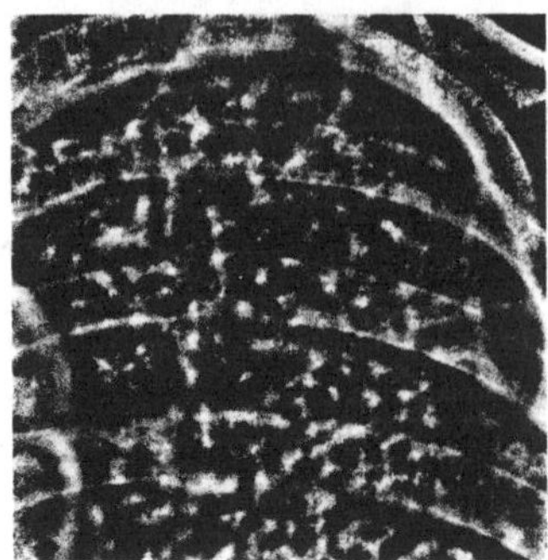

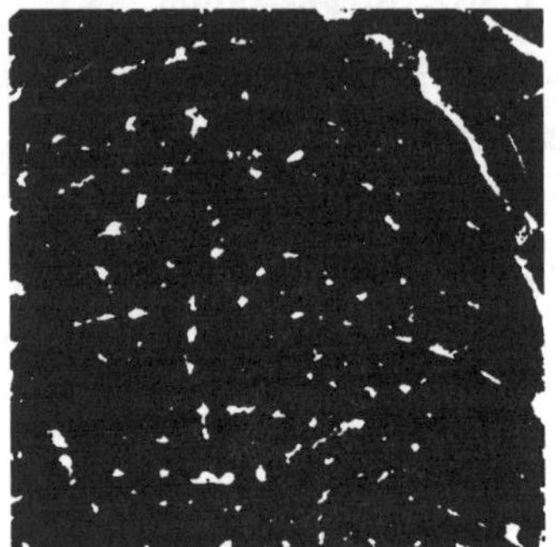

Fig. 3 Moving normalization   **Fig. 4 Bi-level image**

full-picture simple normalization of the ROI is not necessarily sufficient for processing. We introduce here a new technique in order to get a high contrast image and we call it a moving normalization that is shown in the following section. From the result of the moving normalization we can make a bi-level image by using a threshold value. The input of NN is taken from the bi-level image and its size is 30x30 pixel. If the center pixel of the taken input has 0 (black) then it is not needed to input because only the center image of the input is the object image for NN. If the center pixel of the input has 1 (white) then other isolate whites are deleted i.e., changing from 1 to 0 and take it as an input to NN. This processing is called here an exclusive deleting one. It is important to delete the back ground because NN output reacts also for background of the input. The detection of rounded opacities is the next step. This is the main topic of this paper and also shown in the following section. Finally calculating the number density and area density we get a classification result in comparison with these value of the ILO standard X-ray images.

3.1. Moving normalization

In a chest X-ray image usually the contrast is low but the local area average gray level is often much different with the other local area one. Hence the full picture normalization in the gray level gives the unsatisfactory results for our processing. Here we employ special technique for the normalization. First, the tiny area R_e with 32x32pixel is taken from the ROI at left top corner. Let the minimum and the maximum values in the area Re, f_{min} and f_{max} respectively the gray level $f(i,j)$ of R_e is changed as follows.

$$g(i,j) = (f(i,j) - f_{min}) \frac{F_{Max}}{f_{max} - f_{min}} \qquad (1)$$

where F_{Max} is the maximum value: If we employ 8bit for the quantization then $F_{Max}=255$. The result $g(i,j)$ is quantized from 0 to 255. After transforming by using Eq. (1), only the central pixel is left and all other pixels are disposed. And we repeat this procedure by shifting one pixel to the left. The procedure is done from left to right and from top to bottom. The marginal 15 pixels of the ROI are also disposed. If we need such margins then it is sufficient to take ROI wider by 15 pixels. Figure 3 is one example of the moving normalization. Very high contrast image is gotten in comparison with the original. Fig.3 has the same local mean and local variance at any point. Hence taking one threshold value we can make a desired bi-level image (Fig.4). In a moving normalization we tried to take 8x8, 16x16, 32x32, and 64x64. And 32x32 is the best one of all.

4. Detection of Rounded Opacities

Using a neural network we detect the rounded opacities in the X-ray photo. Here we utilize a back propagation neural network (NN). Back propagation NN requires training patterns. Figure 6 shows the training patterns for pneumoconiosis- rounded opacities with several sizes. The last five stick-like figures are for unnecessary parts like vessel and rib shade. These are actually not rounded but thin and long ones. Hence such training patterns are effective for reducing the false positive. The number of NN output is 23 which includes 18 different rounded opacities and 5 stick-like ones shown in Fig.5, the hidden layer neuron number is 171 determined by heuristic way, and the input layer neuron number is 900 (30x30).

Figure 6 shows a status of detecting a rounded opacity by NN. The target equivalent diameters are 1.2 mmØ, 1.2 mmØ, 1.4 mmØx2, 1.4 mmØx2, and 1.4 mmØ. The horizontal axis is the position and the vertical axis is the output value percent.

Form this figure we can see easily that our NN can detect and identify the position and the opacity size exactly.

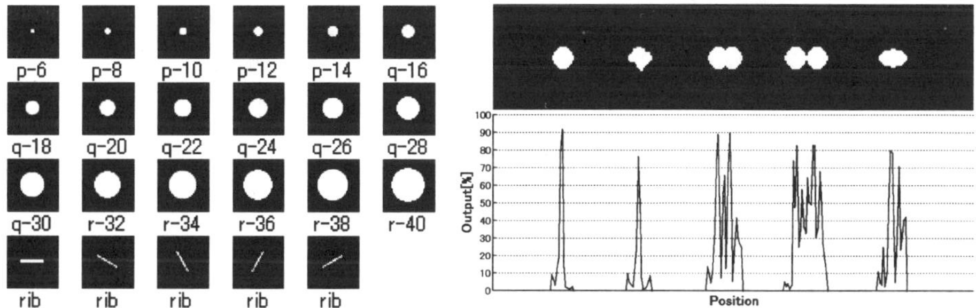

Fig. 5 Several training patterns	Fig.6 Neural Network Detection

5. Simulation

Figure 7 and 8 are simulation examples. Figure 7 shows a category type 1 with not so many P-size rounded opacities. The detecting affected part is marked as the equivalent black circle. Of course, Q-size and R-size of the rounded opacities are also there. But the major size is P-size, so we call this ROI (1,P). Figure 8 shows a detected result with (3,R) type. Many big rounded opacities are detected. Even when the figure of the opacity is not rounded our NN may catch such opacity (See Fig.6). In such a case the equivalent radius is detected. After detecting the opacities we calculated the densities:

$$D_N = N_r / N_E \qquad (2)$$

$$D_A = N_a / N_E \qquad (3)$$

where N_r and N_a are the number and the area of rounded opacities respectively. N_E is the back ground all area. D_N is called the number density and D_A is called the area density. The classification is done in comparison with the densities of ILO standard pneumoconiosis X-ray photo.

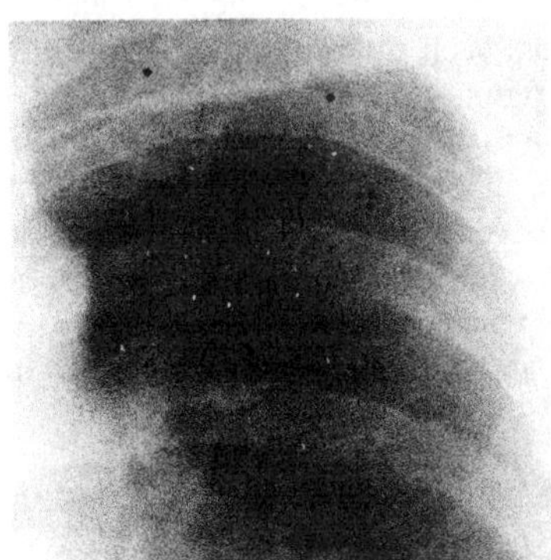

Fig.7 Detected result (1,P) type ROI

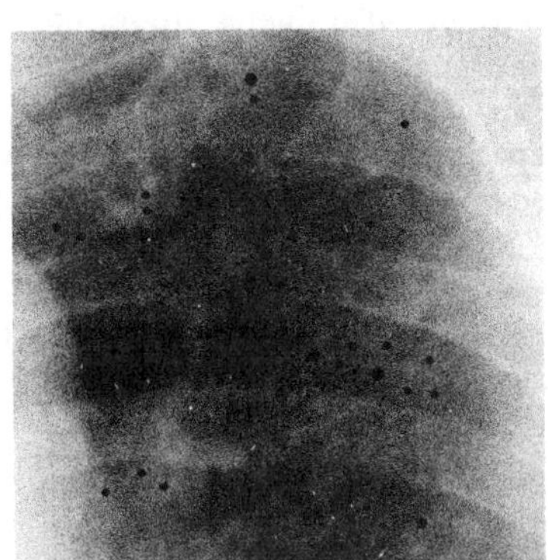

Fig.8 Detected result (3,R) type ROI

6. Conclusions

We have presented a new automatic diagnosis for pneumoconiosis radiographs using neural network. Rounded opacities are caught by our NN from tiny to big one. The detecting rate is higher than the several traditional method [3][4][5]. As a pre-processing for NN-input we have developed a moving normalization method that is very effective for getting a high contrast image of a chest X-ray photo. It is important to modify an input suitably for NN. From upper left side to lower right side of the ROI the input scan is performed successively one pixel by one pixel.

Although NN has a merit in real time processing, our proposed system is not a real time one because it takes a little bit much time for pre-processing and so many time repeating of NN processing by scanning. Now the processing time is about 30 minutes for our personal computer (700MHzCPU). The left problem is to confirm that the high consistency with a doctor's opinion is obtained in a large amount of the chest X-ray photos.

7. References

[1] Katsuragawa S., Doi K., Macmabon H., Nakamori N., Sasaki Y., and Fennelsy J., "Quantitative Computer-aided Analysis of Lung Texture in Chest Radiographs", RSNA annual meeting of the kurt Rossmann Laboratories for Radiologic Image Research, Dept. of Radiology. Unv. of Chicago Radio Graphics vol 10, 1988, pp. 257-269.

[2] Savol.M.A., Li.C.C,. and Hoy.R.J., "Computer-Aided Recognition of Small Rounded Pneumoconiosis Opacities in Chest X-rays", IEEE Trans. on PAMI, VOL.PAMI-2, No.5, 1980, pp. 470-482

[3] Sasaki Y., Katsuragawa S., and Yanagisawa T., "Quantitative Analysis of Pneumoconiosis in Standard Chest Radiographs", Dept. of Radiology, vol.52-10, 1992, pp. 1385-1393

[4] Chen X., Hasegawa J., and Toriwaki J., "Recognition of small Rounded Opacities for Quantitative Diagnosis of Pneumoconiosis Radiographs", Trans. on IEICE, D-II vol. J72-D-II, No.6, 1989, pp. 944-953

[5] X.Chen, J.Hasegawa, and J.Toriwaki, "Recognition of Small Rounded Opacities for Quantitative Diagnosis of Pneumoconiosis Radiographs", Trans. on IEICE, vol. J72-D-II, No. 6, 1989, pp.944-953

KES '01
N. Baba et al. (Eds.)
IOS Press, 2001

Cross-Validation for Neural Discriminant Models

Masaaki TSUJITANI and Takashi KOSHIMIZU
*Faculty of Information Science & Arts, Osaka Electro-CommunicationUniversity, Osaka, JAPAN
and Product Development, Bayer Yakuhin, Ltd., Osaka, JAPAN*

Abstract. The usage of cross-validation is considered in nonlinear discriminant analysis by a feed-forward neural network model. We illustrate that the criterion based on the cross-validation can be an alternative to Akaike's information criterion (*AIC*) and *EIC* computed from the log likelihood when selecting the optimum number of hidden units for neural network model. We also provide the bias correction of error rate associated with a discriminant rule constructed by fitting to the training sample in neural network model. We additionally propose model checking in order to identify influential observations and examine distributional assumptions in neural network model fitting.

1. Introduction

In this article, the role of cross-validation is highlighted for nonlinear discriminant analysis using a feed-forward neural network model when the data consist of binary response and a set of predictor variables, which is closely related to statistical pattern recognition. In attempting to cope with these issues, the statistical tools for the inference presented here include *i*) selection of the optimum number of hidden units that contribute to the overall network performance by use of cross-validation to circumvent the identification problem for the misspecified models [1], *ii*) cross-validation estimate of the bias of excess error in prediction [2], and *iii*)influential analysis in order to identify outliers and examine distributional assumptions in neural network model fitting [3]. These methods are applied through the analyze of diabetes data[4].

2. Neural discriminant models
(1) Determination of Hidden Units

An alternative model selection strategy to *AIC* and *EIC* for the bias correction of the log likelihood is "leaving-one-out" cross-validation for neural network model, which is asymptotically equivalent to *TIC* [5]. Let the initial (training) sample $X = \left\{ X^{<1>}, X^{<2>}, ..., X^{<D>} \right\}$ with $X^{<d>} = \left\{ x^{<d>}, t^{<d>} \right\}$, $x^{<d>} = \left\{ x_1^{<d>}, ..., x_I^{<d>} \right\}$ be independently distributed in an unknown distribution. The cross-validation algorithm is then given as follows:

Step 1. Generate training samples $X_{[i]} = \left\{ X^{<1>}, X^{<2>}, ..., X^{<i-1>}, X^{<i+1>}, \cdots, X^{<D>} \right\}$, $i = 1, 2, ..., D$. The notation $[i]$ written as a subscript to a quantity stands for the deletion of the i-th data point $X^{<i>}$ from the initial sample $X = \left\{ X^{<1>}, X^{<2>}, ..., X^{<D>} \right\}$.

Step 2. Using each training sample, fit a model and estimate unknown parameters denoted by $\theta\left(X_{[i]} \right)$ and predict the output $o_{[i]}^{<i>}$ for the deleted sample point $X^{<i>}$.

Step 3. As a matter of convention the cross-validation criterion is often stated as that of minimizing

$$CV = -2\ln \sum_{i=1}^{D} \ln\left[\{o_{[i]}^{<i>}\}^{t^{<i>}} \left(1 - o_{[i]}^{<i>}\right)^{1 - t^{<i>}} \right].$$ (1)

See also [6] for additional related material with bootstrapping.

(2) Excess error estimation

Let $\eta_{X_{[i]}}$ be the discrimination rule by training sample $X_{[i]}$. Given a patient with medical measurements $x^{<i>}$, we predict presence/absence of the disease by $\eta_{X_{[i]}}\left(x^{<i>}\right)$. The cross-validation algorithm for estimating of excess error rate when fitting neural network models is given as follows:

Step 1. Generate the training sample $X_{[i]}$, and construct the realized discrimination rule

$\eta_{X_{[i]}}$ based on $X_{[i]}$. Then define

$$Q\left(t^{<i>}; \eta_{X_{[i]}}\left(x^{<i>}\right)\right) = \begin{cases} 1 : \text{incorrect discrimination} \\ 0 : \text{otherwise} \end{cases}.$$

Then proportion of incorrect discrimination is given by

$$\frac{1}{D}\sum_{i=1}^{D} Q\left(t^{<i>}; \eta_{X_{[i]}}\left(x^{<i>}\right)\right)$$

Step 2. The estimator of the bias is given by

$$Bias = \frac{1}{D^2}\sum_{i=1}^{D}\left\{\sum_{j=1}^{D} Q\left(t^{<i>}; \eta_{X_{[j]}}\left(x^{<i>}\right)\right)\right\} - \frac{1}{D}\sum_{i=1}^{D} Q\left(t^{<i>}; \eta_{X_{[i]}}\left(x^{<i>}\right)\right).$$ (2)

Step 3. Using bias correction, the estimator of actual error rate can be obtained as

apparent error rate $-$ Bias of Eq.(2), (3)

3. Influential Analysis

To assess the discrepancies between a target value and the corresponding fitted log likelihood at the d-th observation by use of deviance, influence measure provides guides and suggestions that may be carefully applied to neural network model [3]. The effect of the d-th observation on the deviance can be measured by computing

$$\Delta Dev_{[d]} = Dev - Dev_{[d]}$$ (4)

where

$$Dev = 2\left[\ln L\left(\max; X\right) - \ln L\left(\hat{\theta}; X\right)\right], \quad Dev_{[d]} = 2\left[\ln L\left(\max; X_{[d]}\right) - \ln L\left(\hat{\theta}; X_{[d]}\right)\right].$$

The distribution of $\Delta Dev_{[d]}$ will be approximated by χ^2 on d.f.=1 when the fitted model is correct. An index plot is a reasonable rule of thumb for graphically presenting the information contained in the values of $\Delta Dev_{[d]}$.

References
[1] Anders and O.Korn, "Model selection in neural network," *Neural Networks*, Vol.12, pp.309-323,1999.
[2] G. Gong, "Cross-validation, the jackknife, and the bootstrap: Excess error estimation in forward logistic regression," *J. Amer. Statist. Assoc.*, Vol.81, pp.108-113, 1986.
[3] W.Johnson, "Influence measures for logistic regression: Another point of view," *Biometrics*, Vol.72, pp.59-65, 1985.
[4] L.Prechelt, *Proben 1-A set of neural network benchmark problems and benchmark rules*, Universität Karlmatik, Techical Report 21/94, 1994.
[5] R. Shibata, "Statistical aspects of model selection," In *From Data to Model*(edit. By J.C. Willems), pp.215-240,Springer, 1989.
[6] M.Tsujitani and T. Koshimizu, "Neural discriminant analysis," *IEEE Trans. Neural Networks*, Vol.11, pp.1394-1401, 2000.

KES '01
N. Baba et al. (Eds.)
IOS Press, 2001

An Automated Tissue Discrimination Based on Fuzzy Analysis of Ultrasonic Wave

Keisuke SUGANO[1], Kouki NAGAMUNE[1], Syoji KOBASHI[1], Yutaka HATA[1],
Toshiyuki SAWAYAMA[2] and Kazuhiko TANIGUCHI[3]

[1]*Department of Computer Engineering, Himeji Institute of Technology,
2167, Shosha, Himeji, 671-2201, Japan*
[2]*NEW SENSOR Inc., 9-39, Hiradohri, Nose-Cho, Toyono District, 563-0356, Japan*
[3]*KINDEN Co.,Ltd., 3-1-1, Saganakadai, Kizu-Cho, Souraku District, 619-0223, Japan*

Abstract. This paper proposes an automated procedure for discriminating tissues using ultrasonic waves. Generally, the property of the echoes varies at each tissue. It is useful to discriminate the tissues according to difference of the property for diagnosis of diseases such as cancer. But, it is difficult to discriminate the tissue from the property because of its inter- and intraobserver variability. Therefore, the proposal method is aided by fuzzy inference and consists of two stages. The first stage constructs the expert system. In this stage, the fuzzy membership functions are automatically constructed at each characteristic value from pre-experimental data. The second stage predicts the tissues from another waves using the constructed expert system. Finally, the method was applied to six tissues. The experimental result indicated that all the tissues were discriminated (96.7%) with high accuracy than "C4.5" (93.3%).

1. Introduction

The ultrasonic wave has become more important diagnostic technique for the human body due to its low cost and non-invasive property. The information of hardness of tissues supports the diagnosis of diseases such as cancer. The ultrasonic wave acquired from an ultrasonic device includes echoes of target tissues. Many researchers have heretofore been observed the property of the echo using ultrasonic wave [1]-[3]. However, they did not predict the tissues from the obtained property of ultrasonic wave because of its inter- and intraobserver variability. Therefore, this paper proposes the discriminating method of the tissues based on fuzzy analysis of ultrasonic wave.

This paper describes an automated discriminating method of the tissues by fuzzy inference. The property of the echo, which depends on tissues, has the deviation [4]. Since this deviation prevents us from predicting the tissue, we develop the procedure for controlling the deviation by fuzzy inference [5][6], and for constructing the fuzzy membership function automatically from the property of the echo.

The proposal method consists of two stages. In first stage, the fuzzy membership functions are constructed automatically at each characteristic value from pre-experimental data. This method is using two echoes as the characteristic value that we called the first echo and the second echo. The first echo is the echo of back of the target tissue, and the second echo is the echo of shuttle path between surface and back of the target tissue. The second stage predicts the tissues from another waves using the constructed expert system to evaluate the performance of this method. Finally, the method was applied to six tissues. The experimental result indicated that all the tissues were discriminated (96.7%) with high accuracy, compared with the method by "C4.5 [7]" (93.3%).

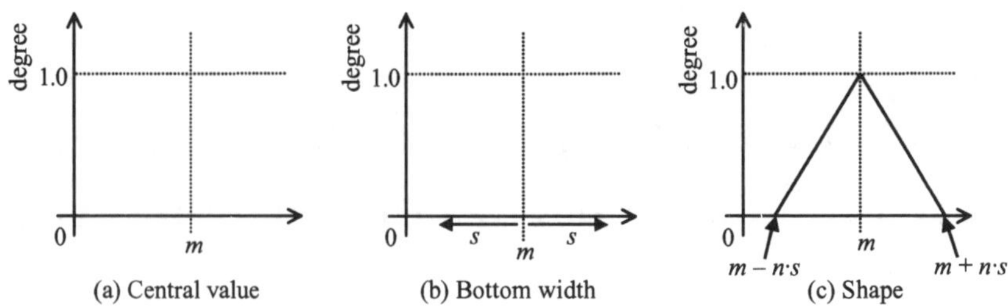

Figure 1 Construction procedure of membership function.

2. Automated Construction Procedure of Fuzzy Membership Function

In our method, the fuzzy membership functions are constructed automatically from the mean and standard deviation of the characteristic values of the acquired ultrasonic wave. As shown in Figure 1, the mean of the characteristic values, m, is used as the center value of the membership function (Fig. 1(a)), standard deviation of the characteristic values, s, is used as the value of bottom width of this (Fig. 1(b)) and the shape of constructing membership function is the triangular function (Fig. 1(c)). According to the value of n, the bottom width of the membership function is changeable. This membership function is constructed at each characteristic value determined in our method.

3. Ultrasonic Discriminating System

The ultrasonic discriminating system consists of an ultrasonic probe (KARL DEUTSCH, TS 6 WB 4-20) whose frequency band is from 4MHz to 20MHz and whose crystal size is 6mm, an ultrasonic pulsar receiver (Krautkramer Japan Co., Ltd., HIS-AD2-C), an A/D converter (Pico Technology Ltd., ADC-200) and a personal computer for data processing. Figure 2 shows the overview of the system. As shown in this figure, the ultrasonic pulsar receiver transmits and receives ultrasonic waves via the preamplifier, and the ultrasonic waves are acquired by the personal computer using the A/D converter. The system obtains the data under the following conditions that direct contact and immersion testing.

The ultrasonic wave includes the plural echo. Figure 3 illustrates the traveling path of the ultrasonic wave. The surface echo is the echo of surface of the probe (A: Fig. 3(a)). The first echo is the echo of back of the target material (B: Fig. 3(b)). The second echo is the echo of shuttle path between surface and back of the target material (C: Fig. 3(c)).

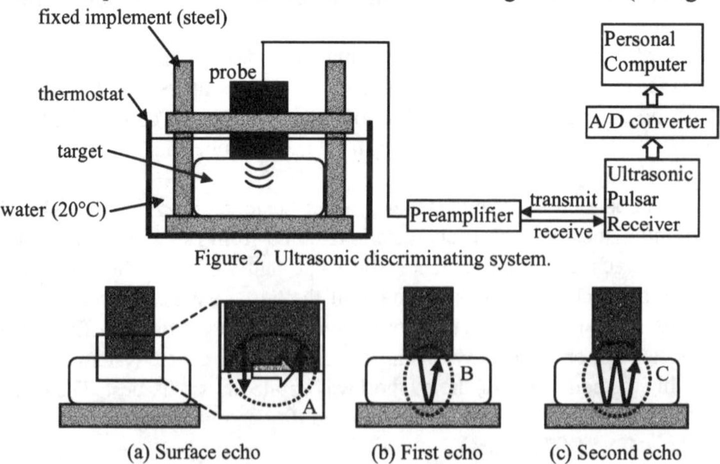

Figure 2 Ultrasonic discriminating system.

Figure 3 Traveling path of the ultrasonic wave.

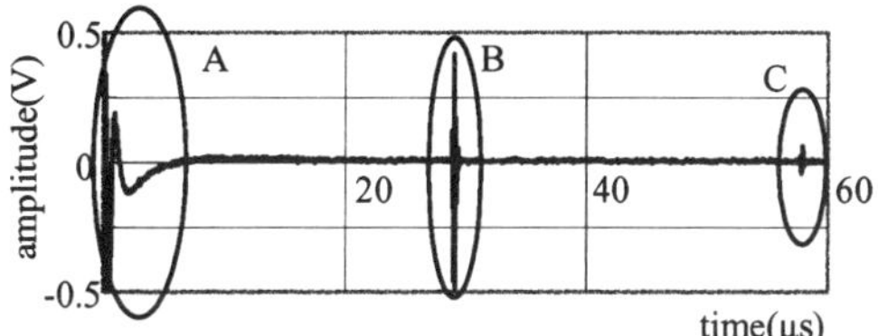

Figure 4 An example of the ultrasonic wave.

Figure 4 illustrates an example of the acquired ultrasonic wave.

4. The Discriminating Method by Fuzzy Analysis

4.1 Construction of The Expert System

Our method consists of two stages (i.e. Construction stage and Discrimination stage). In the pre-experiment, we acquire the ultrasonic wave twenty times for each target material as the reference echo, and extract amplitude value of first and second echo as the characteristic values of the material. We calculate the mean, m, and standard deviation, s, from each characteristic value for constructing fuzzy membership function.

We consider only two characteristic values (i.e., amplitude value of first and second echo) of the acquired ultrasonic wave for fuzzy inference. Figure 6 shows the amplitude value of the first echo. As shown in this figure, amplitude value is the difference between maximum value and minimum value on an echo.

The dependence of the echo on the property of the material causes the characteristic of the echo. We can derive knowledge from the characteristic of the echo. This knowledge can be expressed by the if-then rule as follows,

IF $A1_Y$ is $A1_X$ AND $A2_Y$ is $A2_X$, THEN Y is X,

where, X denotes the material of the reference echo; Y denotes the unknown material; $A1_X$ (or $A1_Y$) denotes amplitude value of the first echo of X (or Y); $A2_X$ (or $A2_Y$) denotes amplitude value of the second echo of X (or Y). These rules can be represented by fuzzy membership functions shown in Figure 7 ($n=3$, in our method).

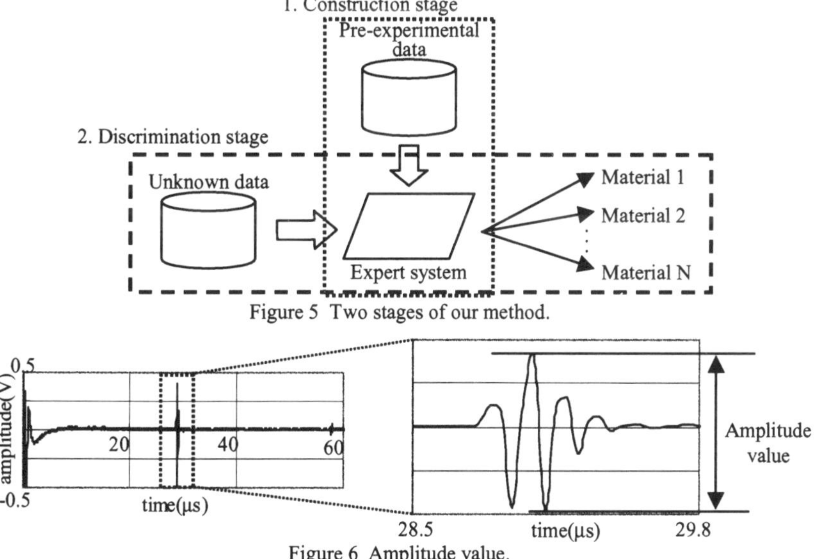

Figure 5 Two stages of our method.

Figure 6 Amplitude value.

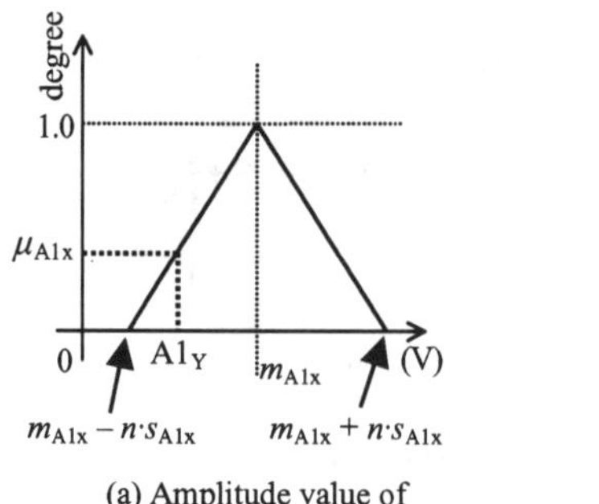
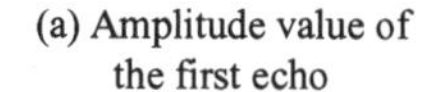
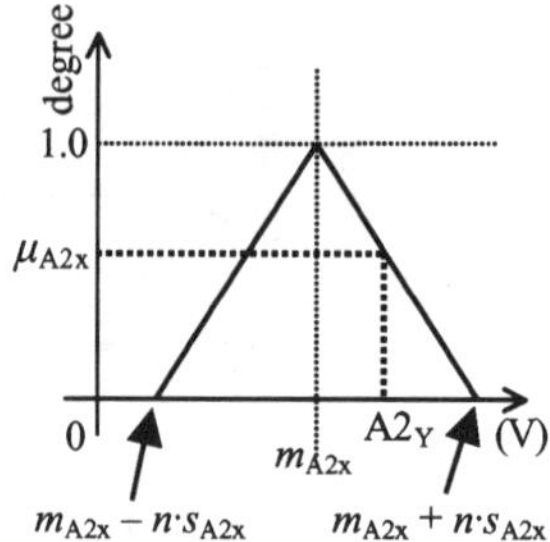

(a) Amplitude value of the first echo   (b) Amplitude value of the second echo

Figure 7   Membership functions.

4.2 Tissue Discrimination

Applying this method to characteristic values of the inferred ultrasonic wave, we can obtain the membership degree of characteristic value of reference echo. We can obtain membership degree of material X by the following equation,

$$\mu_X(Y) = \frac{\mu_{A1_X}(A1_Y) + \mu_{A2_Y}(A2_Y)}{2} ,$$

where, $\mu_X(Y)$ denotes a total degree of material X. When $\mu_X(Y)$ is maximum, the material Y is discriminated as the material X.

5. Experimental Results

To evaluate the performance of the proposed method, we applied to six materials ((A): chest of chicken, (B): gizzard, (C): liver of cattle, (D): *momen-tofu* (Momen means cotton.), (E): *goma-tofu* (Goma means sesame.) and (F): konjac). Table 1 shows mean and standard deviation of characteristic values of each material in pre-experiment. Figure 8 shows the constructed membership functions.

The example of an obtained total degree of the material is shown in Table 2. In this table, column is the materials of the inferred echo (for the purpose of evaluation, material of inferred echo is known), and row is the materials of the reference echo. As shown in this table, we confirmed that degree of same material is highest.

In this experiment, we attempted to discriminate of sixty inferred echoes and could get desirable result with accuracy of 96.7%. On the other hand, the result of another algorithm to classify "C4.5" is 93.3%. Hence, the availability of this method was indicated.

Table 1   The mean and standard deviation.

	First echo		Second echo	
	mean(m)	standard diviation(σ)	mean(m)	standard diviation(σ)
Chest of chicken	0.6199	0.0679	0.0914	0.0111
Gizzard	0.6238	0.0639	0.0202	0.0095
Liver of cattle	0.4690	0.0534	0.1477	0.0105
Momen-tofu	0.3258	0.0175	0.0384	0.0043
Goma-tofu	0.3044	0.0131	0.0411	0.0048
Konjac	0.3052	0.0196	0.0639	0.0196

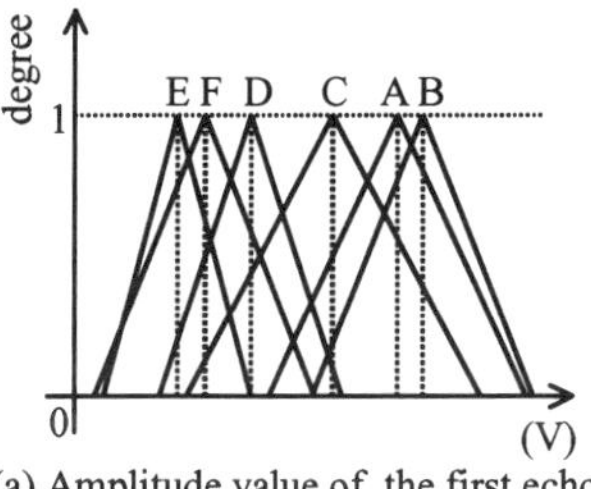

(a) Amplitude value of the first echo

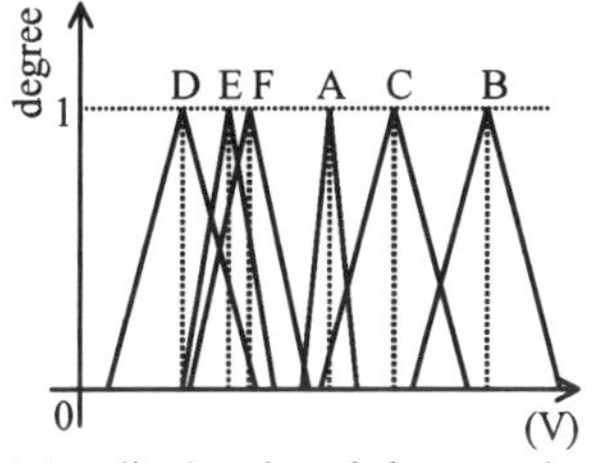

(b) Amplitude value of the second echo

Figure 8 Constructed membership functions.

Table 2 Total degree of each material.

inference \ reference	Chest of chicken	Gizzard	Liver of cattle	*Momen-tofu*	*Goma-tofu*	Konjac
Chest of chicken	0.5838	0.1381	0.4499	0	0	0
Gizzard	0.3477	0.8103	0	0	0	0
Liver of cattle	0.2320	0.2049	0.8077	0	0	0
Momen-tofu	0	0	0.1066	0.8135	0.3938	0.1796
Goma-tofu	0	0	0.0050	0.8056	0.9013	0.4569
Konjac	0.0994	0	0	0.2395	0.4237	0.8990

Additionally, when we applied this method to ultrasonic wave of unused material at pre-experiment, we obtained the membership degree of each known material from the membership function. That is, our method can obtain the degree of similarity between unused material and known material.

6. Conclusions

In this paper, we proposed an automated discriminating method of tissues based on analyzing ultrasonic wave by fuzzy inference. This method used the knowledge that characteristic values are different among target materials. We used the first echo and the second echo, and discriminated of materials by using amplitude value of two echoes. As a result, the expert system was constructed automatically and we successfully discriminated of the target materials of the inferred echo using the constructed membership functions.

The future works are to apply this method to other materials.

References

[1]　R. C. Chivers and R. J. Parry, "Ultrasonic Velocity and Attenuation in mammalian tissues," J. Acoust. Soc. Am, vol. 63(3), pp. 940-953, 1978.

[2]　S. A. Gross, R. L. Johnston and F. Dunn, "Comprehensive Compilation of Empirical Ultrasonic Properties of Mammalian tissues," J. Acoust. Soc. Am, vol. 64(2), pp. 423-457, 1978.

[3]　S. A. Gross, R. L. Johnston and F. Dunn, "Comprehensive Compilation of Empirical Ultrasonic Properties of Mammalian tissues 2," J. Acoust. Soc. Am, vol. 68(1), pp. 93-108, 1980.

[4]　L. W. Schmerr, Jr., *Fundamentals of Ultrasonic Nondestructive Evaluation*, Pleum Press, 1998.

[5]　L. A. Zadeh, *Fuzzy sets and applications*, John Wiley and Sons, 1987.

[6]　S. Kobashi, N. Kamiura, Y. Hata and M. Ishikawa, "Automatic Robust Threshold Finding Aided by Fuzzy Information Granulation," Proc. of IEEE 1997 Int. Conf. on Image Processing, vol. 1, pp. 711-714, Oct. 1997.

[7]　J. R. Quinlan, *C4.5: PROGRAMS FOR MACHINE LEARNING*, Morgan Kaufmann Publishers, 1993.

KES '01
N. Baba et al. (Eds.)
IOS Press, 2001

A Medical Image Segmentation Method Using K-means Clustering and Rough Sets

Takehiko MATSUURA[1], Syoji KOBASHI[1] and Yutaka HATA[1]
[1]*Department of Computer Engineering, Himeji Institute of Technology,*
2167, Shosha, Himeji, 671-2201, Japan

Abstract. Image segmentation is one of the fundamental techniques to develop a computer-aided diagnosis (CAD) system in the medical field. This paper first introduces rough sets into image segmentation method. In this method, attribute values of each pixel of an image of interest are given by using K-means clustering, and the attribute values divide the image into many regions. By applying value reduct, which is one of the typical concepts of rough sets, to the attribute values, dissimilarities between regions are calculated. Final clustering result is obtained by merging similar regions. To evaluate the performance of the proposed image segmentation method, it was applied to an artificial generated image, and a human brain Magnetic Resonance (MR) image. The results were also compared with convention K-means clustering.

1. Introduction

Image segmentation is one of the fundamental techniques to develop a computer-aided diagnosis (CAD) system in the medical field. For example, it is very important to develop methods that can automatically extract regions of interest (ROIs). Extracting methods based on image segmentation techniques are the powerful tools to assist in extracting shapes of ROIs. Image segmentation technique includes a process that divides the whole into several regions. Clustering methods are often used to segment an image that has ambiguous boundary. Clustering method divides a universal set into adequate number of clusters according to a criterion such as similarity between objects. Conventional clustering methods are K-means method [1], fuzzy c-means (FCM) method [2], and tournament clustering method [3], et al. However, these methods have the problem to estimate equally about every attribute.

Rough sets [4], proposed by Pawlak, have been remarkable attention in the field of knowledge discovery [5], since they provide tools to mathematically treat roughness of knowledge. In rough sets, the roughness of knowledge can be expressed by equivalence relation. Since this equivalence relation can classify a universal set into several categories, rough sets would provide a powerful framework for clustering. Many measures for estimating roughness have been proposed [6]-[8].

This paper shows a segmentation method of medical image acquired roughness of knowledge of attribute based on rough sets. First, we apply K-means clustering to the original image with increasing the number of clusters. The results represent the attributes for each pixel of the image. Second, the attributes of all pixels consider the equivalence relation, and image is segmented into small regions by an indiscernibility relation using all equivalence relation. Third, value reduct, which is one of the typical concepts of rough sets, is calculated as the dissimilarity between the regions. Finally, these dissimilarities between regions determine the equivalence relation, and final segmented result is obtained by integration of this equivalence relation. We applied this method to an artificial generated image and a human brain Magnetic Resonance (MR) images. This method could then produce good results.

2. Preliminaries

2.1 Equivalence Relation

The fundamental concepts of rough sets are classification and approximation. Let $U\,(\neq\phi)$ be a universe of discourse and X be a subset of U. An equivalence relation, R, classifies U into a set of subsets $U/R = \{X_1, X_2,..., X_n\}$ in which following conditions hold:

$$(1)\ X_i \subseteq U,\ X_i \neq \phi, \quad \text{for any } i,$$
$$(2)\ X_i \cap X_j = \phi, \quad \text{for any } i, j,\ i \neq j$$
$$(3)\ \bigcup X_i = U, \quad \text{for any } i.$$

2.2 Indiscernibility Relation

Any subset X_i, called a category, represents an equivalence class of R. A category in R containing an object $x \in U$ is denoted by $[x]_R$. For a family of equivalence relations $\mathbf{P} \subseteq \mathbf{R}$, an indiscernibility relation over $\mathbf{P}$ is denoted by $IND(\mathbf{P})$. Moreover,

$$[x]_{IND(\mathbf{P})} = \bigcap_{R \in \mathbf{P}}[x]_R.$$

2.3 Value Reduct

Let $\mathbf{R}$, $\mathbf{P}$ and $\mathbf{Q}$ denote a set of equivalence relations, a subset of $\mathbf{R}$ and a proper subset of $\mathbf{P}$, respectively. Here we define a reduct r_i of object x_i as

$$r_i = \left\{A_i(\mathbf{P})\,\middle|\,[x_i]_\mathbf{P} \subseteq [x_i]_\mathbf{R}, [x_i]_\mathbf{Q} \not\subseteq [x_i]_\mathbf{R}\right\}, \quad \text{for all } \mathbf{P} \subseteq \mathbf{R}, \mathbf{Q} \subset \mathbf{P},$$

where $A_i(\mathbf{P})$ denotes a set of attribute values of x_i associated with a set of relations $\mathbf{P}$.

3. Segmentation Method of Medical Image

We propose a segmentation method of medical image using rough sets. The procedure of our method is summarized as follows:

Step 1) Do K-means clustering based on intensity then let the clustering result attribute values. Next, we find indiscernibility relation using every attribute.

Step 2) Calculate weights of attribute based on concept of value reduct, and determine dissimilarity between the regions using weight of attribute.

Step 3) Classify into similar regions and dissimilar ones about each region based on dissimilarity between the regions. Next, integrate similar regions.

3.1 Extraction of Attribute Based on K-means Clustering

The first step, we determine some attributes for whole pixels. To perform K-means clustering based on intensity gives an attribute, and we change number of clusters to obtain some attributes for whole pixels. This step produce clustering results as follows: $\{U/R_{x_1}, U/R_{x_2},..., U/R_{x_n}\}$. We find an indiscernibility relation of $\mathbf{R}$ that means set of all equivalence relations. We regard pixels that belong to same category as same region. We then call each category as "region".

3.2 Dissimilarity Between the Regions Based on Value Reduct

The second step determines weight of attributes and dissimilarity between the regions. Weight of attributes is calculated by value reducts of given attribute. Every value reduct provides a decision rule to classify regions into a category on a set of equivalence relations. This measure finds decision rules by assigning every attribute based on an equivalence relation. A decision rule is driven by a set of all equivalence relations. An employing appearance frequency of attribute value determines weight of corresponding attribute.

Weight of attribute is defined as follows:

[Definition 1] Weight of attribute

Let $\mathbf{R}$ be a set of all equivalence relations associated with every attribute, $x_i \in U$ be region of the universe, a_k be the k-th attribute of a set of all attributes, v_i be a value reduct of x_i on $\mathbf{R}$ and $v_i(a_k)$ be a attribute value a_k on a value reduct v_i, l be the total number of value reducts for the universe. Weight of attribute a_k, w_k, is defined as follows:

$$w_k = 1 - \frac{\sum_{x_i \in U} \omega(v_i, a_k)}{l},$$

$$\omega(v_i, a_k) = \begin{cases} 1, & \text{if } v_i(a_k) \text{ is indispensable value} \\ 0, & \text{if } v_i(a_k) \text{ is dispensable value} \end{cases}.$$

□

By assigning weight for every attribute, dissimilarity between the regions can be calculated. Dissimilarity between the regions is defined as follows:

[Definition 2] Dissimilarity between the regions

Let $U = \{x_1, x_2, \ldots, x_n\}$ be the universal set, $A = \{a_1, a_2, \ldots, a_m\}$ be a set of attributes and $a_k(x_i)$ be an attribute value of region x_i on attribute a_k. For region x_i and x_j, the dissimilarity between the regions, $d_N(x_i, x_j)$, is defined as follows:

$$d_N(x_i, x_j) = \sum_{k=1}^{m} (w_k \times \lambda_{ij}^k),$$

$$\lambda_{ij}^k = \begin{cases} 0, & \text{if } a_k(x_i) = a_k(x_j) \\ 1, & \text{if } a_k(x_i) \neq a_k(x_j) \end{cases},$$

where w_k denotes weight of attribute a_k derived by Definition 1. $d_N(x_i, x_j) = d_N(x_j, x_i)$ and $d_N(x_i, x_i) = 1$ hold.

□

3.3 Combination of Equivalence Relations

The third step assigns equivalence relations with a dissimilarity threshold *Th*. An equivalence relation classifies the universe into two categories: one contains similar regions, and another contains dissimilar regions. Regions contained in a category are considered indiscernible each other. An equivalence relation $R_i \in \mathbf{R}$ for a representative region x_i is defined as follows:

[Definition 3] Equivalence relation based on dissimilarity

For a representative region $x_i \in U$, equivalence relation $R_i \in \mathbf{R}$ classifies U as

$$U / R_i = \left\{ \left\{ x_j \big| s(x_i, x_j) \leq Th \right\}, \left\{ x_j \big| others \right\} \right\} \text{ for all } j (1 \leq j \leq n),$$

where *Th* is a dissimilarity threshold for determining whether regions x_i and x_j are similar under R_i or not. Note that the same threshold value is used for every relation in $\mathbf{R}$.

□

Finally, an indiscernibility relation over **R** using every equivalence relation is defined as follows:

$$[x]_{\text{IND}(\mathbf{R})} = \bigcap_{R_i \in \mathbf{R}} [x]_{R_i}.$$

We define the same region as the region that belong to the same category in an indiscernibility relation over **R**. Every region is a final segmentation result.

4. Experimental Results

We first applied to an artificial generated image shown in Figure 1(b). This image was obtained by smoothing a Figure 1(a) image 50 times. We applied our method to this image as number of attribute was 3(C=2,3,4), *th*, which was threshold of dissimilarity was 2. This result is shown in Figure 1(c) that has 4 clusters. Moreover, Figure 1(d) shows a result of K-means clustering as number of clusters is 4. The segmented images consist of four gray levels, each denoting a different region. In spite of the results of the same number of clusters, the boundaries were obtained the different portion by this method and the K-means method.

Next, we obtained a human brain MR slice image. Figure 2(a) shows one slice of this MR image. We applied our method to this image as number of attribute was 3(C=3,4,5), *th*, which was threshold of dissimilarity was 1.33. This result has 6 clusters. Each cluster is shown in Figure 2(b)-(g). For example, Figure 2(b) includes background region, Figure 2(c) includes the white matter region, Figure 2(d) includes cerebrospinal fluid (CSF) region, and Figure 2(e) includes the gray matter. To compare with the proposal method, Figure 3 shows a result of K-means clustering as number of clusters is 6. The white matter was segmented as many discontinuous by K-means (Figure 3(f), (g)). On the contrary, the white matter was segmented into one region by the proposal method (Figure 2(c)). Therefore, we were able to regard the proposal method produces a better result than K-means clustering.

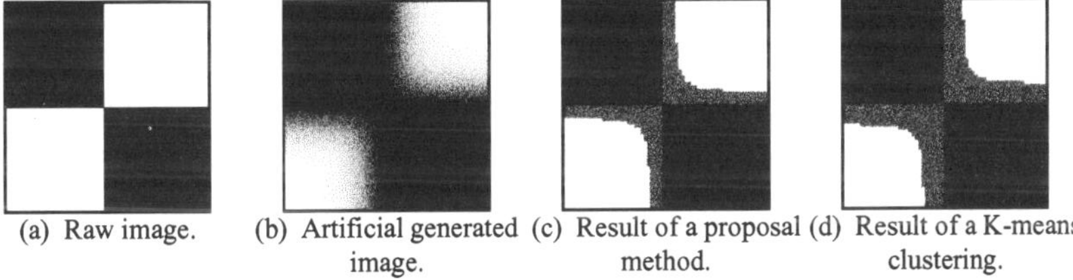

(a) Raw image. (b) Artificial generated image. (c) Result of a proposal method. (d) Result of a K-means clustering.

Figure 1 Experimental result of an artificial generated image.

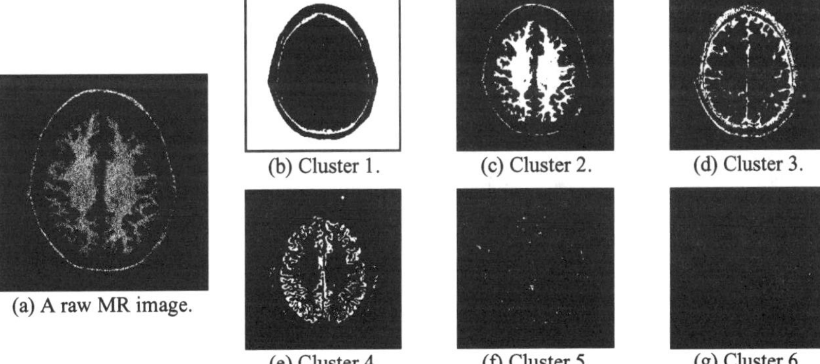

(a) A raw MR image. (b) Cluster 1. (c) Cluster 2. (d) Cluster 3.

(e) Cluster 4. (f) Cluster 5. (g) Cluster 6.

Figure 2 Each cluster for proposal method.

    Nondestructive Evaluation (NDE) for Diagnostics

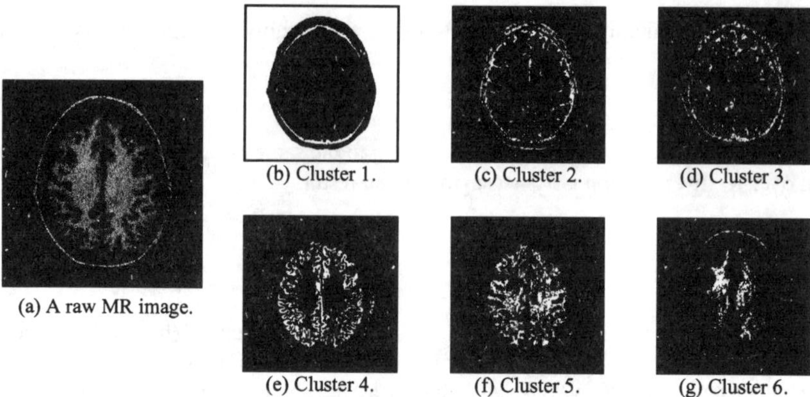

(a) A raw MR image.

(b) Cluster 1.

(c) Cluster 2.

(d) Cluster 3.

(e) Cluster 4.

(f) Cluster 5.

(g) Cluster 6.

Figure 3 Each cluster for K-means clustering method.

5. Conclusions

In this paper, we proposed a segmentation method of medical image based on rough sets. In this method, K-means clustering based on intensity gave attribute to each pixel, and it was assumed equivalence relation of each pixel. Next, we found dissimilarity between the regions using concept of value reduct. Finally, we defined equivalence relation that was separated regions into high dissimilarity and low one, and got a segmentation result by finding indiscernibility relation using every equivalence relation. In our experiment, we applied a proposal method to an artificial generated image and a human brain MR image. Our method was able to segment them, compared with K-means clustering method, and confirmed good results. The future works are to consider how to determine first attribute and to improve the performance.

References

[1] S. Z. Selim and M. A. Ismail, "K-means-type Algorithm," *IEEE Trans. Pattern Anal. Mach. Intell.*, Vol. 6, No. 1, pp. 81-87, 1994.

[2] J. C. Bezdek, *Pattern Recognition with Fuzzy Objective Function Algorithms*, Plenum Press, New York, 1981.

[3] M. R. Anderberg, *Cluster Analysis for Applications*, Academic Press, 1973.

[4] Z. Pawlak, *Rough sets, Theoretical aspect of reasoning about data*, Kluwer Acad. Publ., Dordrecht, pp. 1-8, 116-132, 1991.

[5] A. Nakamura, "Rough sets – its theory and applications," *Journal of Japan Society for Fuzzy Theory and Systems*, Vol. 8, No. 4, pp. 594-603, 1996. (in Japanese).

[6] S. Tsumoto and H. Tanaka, "Induction of disease description based on rough sets," *1st Online Workshop on Soft Computing*, pp. 19-30, Aug. 1996.

[7] S. Tsumoto, "Automated extraction of medical expert system rules from clinical databases based on rough set theory," *Information Science*, Vol. 112, pp. 67-84, 1998.

[8] A. Øhrn, S Vinterbo, P. Szymansski, and J. Komorowski, "Modelling cardiac patient set residuals using rough sets," *the AMIA Annual Fall Symposium Nashville*, TN, USA, pp. 25-29, Oct. 1997.

KES '01
N. Baba et al. (Eds.)
IOS Press, 2001

441

Fuzzy Rule Based Segmentation of CT Knee Images

Makoto SHIBATA*, Syoji KOBASHI*, Yutaka HATA*,
Yasuhiro TOKIMOTO** and Makoto ISHIKAWA**
**Department of Computer Engineering, Himeji Institute of Technology,
2167, Shosha, Himeji, 671-2201, Japan*
***Ishikawa Hospital, 784, Bessho, Bessho-cho, Himeji, 671-0221, Japan*

Abstract. In this paper, we propose an automated method for segmenting the cruciate ligament and the meniscus from CT knee images. The method first finds the candidate region of interests (ROIs) and the bone region by using intensity thresholding. The obtained bone region is decomposed into the femur and the tibia by watershed segmentation. To eliminate the cartilage and the cortical bone from the candidate region we can express these tissues by using the fuzzy if-then rules. To segment the ROIs we employed physician's knowledge; 'the cruciate ligament and the meniscus are wedged between the femur and the tibia', 'the shape of the meniscus is half-moon' and 'the cruciate ligament are located near the center of the knee'. The knowledge is converted to fuzzy if-then rules, and then the rules can compute the fuzzy degree for ROIs. To evaluate our method, it was applied to 5 normal subjects. Quantitative evaluation of the resultant images by a physician shows that our method can give interesting 2D reconstructed and 3D surface rendering images. These results would help us to understand 3D shape and to evaluate the condition of the cruciate ligament and the meniscus.

1. Introduction

Computed tomography (CT) and magnetic resonance imaging (MRI) can obtain detailed information inside a human body. A three dimensional (3D) representation of the cruciate ligament and the meniscus is essential for diagnosing the injuries of the tissues. The method segmenting the meniscus from MR images was proposed [1]. However, segmentation method of the cruciate ligament was not described yet because the discernment is difficult in MRI images. Since the manual delineating of the regions of interest (ROIs) is time-consuming, automated procedures are required from CT knee images.

This paper proposes an automated segmentation method of the ROIs from CT knee images, where the ROIs are the cruciate ligament and the meniscus. First, we extract the candidate region and the bone region by intensity thresholding. Secondly, we segment the bone region into the femur and the tibia using watershed segmentation [2]. Thirdly, we eliminate the cartilage and the cortical bone from the candidate region. We employ the fuzzy if–then rules [3] to express the cartilage and the cortical bone, determine the voxels belonging to the cartilage and the cortical bone. Finally, to segment the ROIs we employ physician's knowledge; 'the cruciate ligament and the meniscus are wedged between the femur and the tibia', 'the shape of the meniscus is half-moon' and 'the cruciate ligament are located near the center of the knee'. The knowledge is converted to fuzzy if-then rules, and then the rules can compute the fuzzy degree for ROIs. We determine the voxels belonging to the ROIs according to the degree. We applied our method to 5 normal subjects. Quantitative evaluation of the resultant images by a physician shows that our method can provide interesting 2D reconstructed and 3D surface rendering images. These images would help us to understand 3D shape and to evaluate the condition of the cruciate ligament and the menisci.

2. Preliminaries

We used transaction CT images acquired from HiSpeed CT scanner (GE Medical Systems). Field of view (FOV) was 170mm. Matrix was 512 by 512. Thickness of the slice was 2.0mm. Each of the volume data was composed of 59 separate slices. Figure 1 shows the coordinate system for segmenting the ROIs. For CT images, we constructed the voxel data, which consisted of 512 by 512 by 59 voxels with coordinate (x, y, z) as shown in this figure. The intensity of all data sets ranged between –1500HU and 4096HU. Figure 2 shows a raw CT image. In this image, the brighter voxels imply higher intensity.

3. Segmentation of the Cruciate Ligament and the Meniscus

Figure 3 shows chart of our method. Figure 4 illustrates a coronal image of the knee. As shown in this figure, the knee consists of the cruciate ligament, the meniscus, the bone (the femur, the tibia, etc), the cartilage, the cortical bone, the muscle, etc. Furthermore, the ROIs (i.e., the cruciate ligament and the meniscus) are wedged between the femur and the tibia, and the cartilage and the cortical bone are located round the bone as shown in this figure. Therefore, to extract the femur and the tibia is essential for segmentation of the ROIs. Figure 5 shows the intensity range of each tissue. We firstly extract the candidate region and the bone region by using intensity thresholding as shown in this figure. Figure 6 shows the extraction result. As shown in this figure, the candidate region involves whole ROIs, and the bone region involves the whole bone. In the constriction between the femur and the tibia, the Euclidean distance from the voxel to closest voxel belonging to the bone region is short. We segment the bone region into the femur and the tibia by watershed segmentation using the distance. The candidate region extracted by using the intensity thresholding involves the cartilage and the cortical bone. To eliminate these tissues from the candidate region, we employ the fuzzy if-then rules. This processing is explained in detail in Section 3.1. Finally, to segment the ROIs from the candidate region that eliminated the cartilage and the cortical bone, we employ the fuzzy if-then rules. This processing is explained in detail Section 3.2.

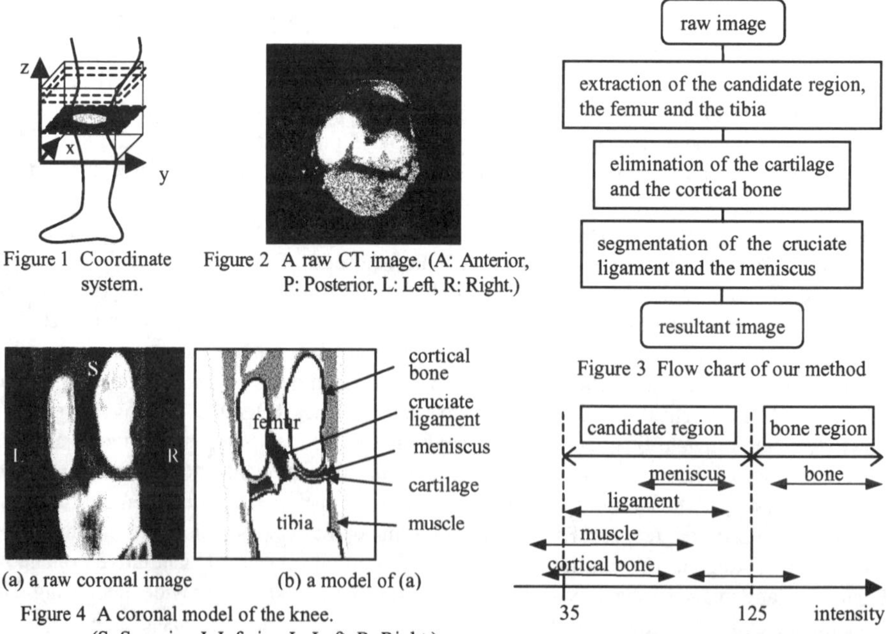

Figure 1 Coordinate system.

Figure 2 A raw CT image. (A: Anterior, P: Posterior, L: Left, R: Right.)

Figure 3 Flow chart of our method

(a) a raw coronal image   (b) a model of (a)

Figure 4 A coronal model of the knee. (S: Superior, I: Inferior, L: Left, R: Right.)

Figure 5 The intensity range of each tissue.

3.1 Elimination of the Cartilage and the Cortical Bone from the Candidate Region

We express the cartilage and the cortical bone by using the fuzzy if-then rules, and eliminate these tissues from the candidate region. To express these tissues we consider the knowledge; 'in the cartilage and the cortical bone, the gradient magnitude of the intensity is large' and 'the cartilage and the cortical bone anatomically are located near the bone'. We can derive the following fuzzy if-then rules from the knowledge.

If g is "Large" and db is "Short" then μ_A is "High",

where g denotes the gradient magnitude of the intensity and db denotes the Euclidean distance from the voxel to the closest voxel belonging to the bone region; μ_A denotes the degree for the cartilage or the cortical bone. μ_A is calculated by the following expression:

$$\mu_A = \mu_g \times \mu_{db},$$

where μ_g (or μ_{db}) denotes the degree of the fuzzy membership functions for g (or db) as shown in Figure 7. Figure 8(a) shows an example of the inference results. After giving the total degree to the voxels of the candidate region, the voxels whose degree is more than a threshold 0.4 are classified into the cartilage or the cortical bone, and we eliminate their voxels from the candidate region. Figure 8(b) shows the classification result of the cartilage and the cortical bone. Figure 8(c) shows the elimination result, where the white voxels imply the candidate region after the elimination process.

3.2 Segmentation of the Cruciate Ligament and the Meniscus

In candidate region that eliminated the cartilage and the cortical bone, to segment the ROIs we employed the following physician's knowledge.

Knowledge 1. The ROIs are located in the neighbourhood of both the femur and the tibia, and the meniscus is near than the cruciate ligament.

Knowledge 2. The shape of the meniscus is half-moon, and the cruciate ligament is located near the inside of the meniscus's position.

Knowledge 3. The cruciate ligament is located near the central sagittal section of the knee.

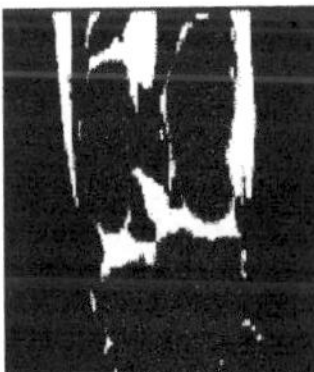

Figure 6 Extraction of the candidate region and the bone region. (The white voxels imply the candidate region and the gray voxels imply the bone region.)

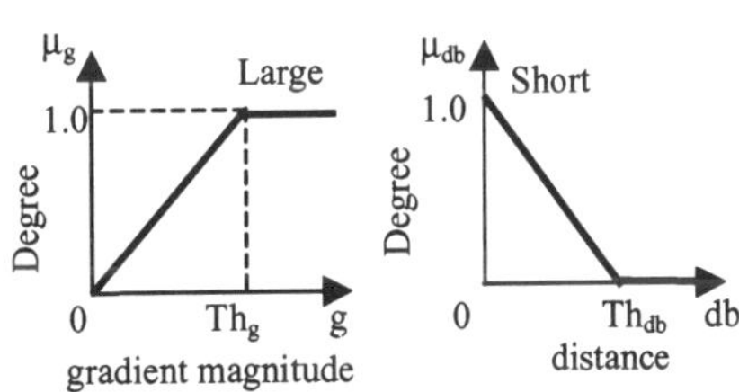

Figure 7 Membership functions for g and db.

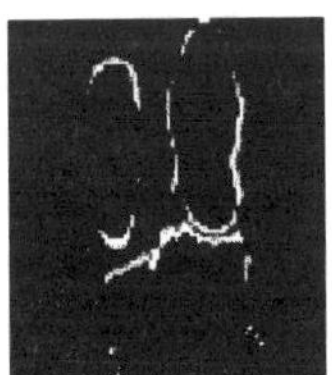

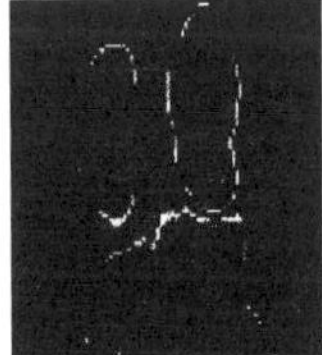

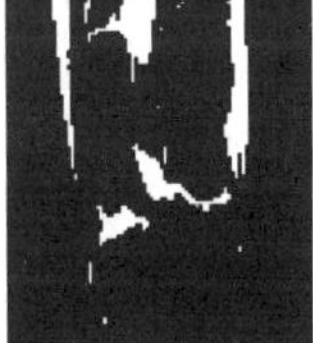

(a) total degree for the cartilage and the cortical bone, μ_A

(b) the cartilage and the cortical bone

(c) elimination result

Figure 8 Elimination of the cartilage and the cortical bone. (The brighter voxels imply higher degree in (a).)

This knowledge is converted to fuzzy if-then rules, and then the rules compute the fuzzy degree for ROIs. Firstly, we can derive the following fuzzy if-then rules from Knowledge 1.

If df is "Short" and dt is "Short", then μ_s is "High",
else if df is "Middle" and dt is "Middle", then μ_m is "High",

where df (or dt) are denoted the Euclidean distance from the voxel to the closest voxel belonging to the femur (or the tibia). μ_s and μ_m are calculated by the following expression:

$$\mu_s = \mu_{dfShort} \times \mu_{dtShort}, \qquad \mu_m = \mu_{dfMiddle} \times \mu_{dtMiddle},$$

where μ_{df} (or μ_{dt}) denotes the degree of the fuzzy membership functions for df (or dt) as shown in Figure 9. Secondly, we give the elliptical membership functions for the ROIs from the Knowledge 2. When the coordinates of the centerline are (x_o, y_o), we define dl(x, y) by the following expression:

$$dl(x, y) = \sqrt{\left(\frac{x - x_o}{2}\right)^2 + (y - y_o)^2} \ .$$

The membership functions are given according to dl as shown in Figure 9. μ_{dl} denotes the degree of the fuzzy membership functions for dl. When we calculate the total degree for the cruciate ligament, the degree acquired from the black function is used in this figure. Thirdly, we can derive the following fuzzy if-then rules from Knowledge 3.

If ds is "Short", then μ_{ds} is "High",

where ds denotes the Euclidean distance from the voxel to central sagittal section of the knee and μ_{ds} denotes the degree of the fuzzy membership functions for ds as shown in Figure 9. Finally, the total degree for the cruciate ligament and the meniscus, μ_C and μ_M, are calculated by following expression:

$$\mu_C = \mu_s \times \mu_{dl} \times \mu_{ds}, \qquad \mu_M = \mu_m \times \mu_{dl}.$$

After giving the total degree to the voxels of the candidate region, the voxels whose μ_C is more than a threshold 0.3 are classified into the cruciate ligament and the voxels whose μ_M is more than a threshold 0.6 are classified into the meniscus. Figure 10 shows an example of the inference results and segmentation result.

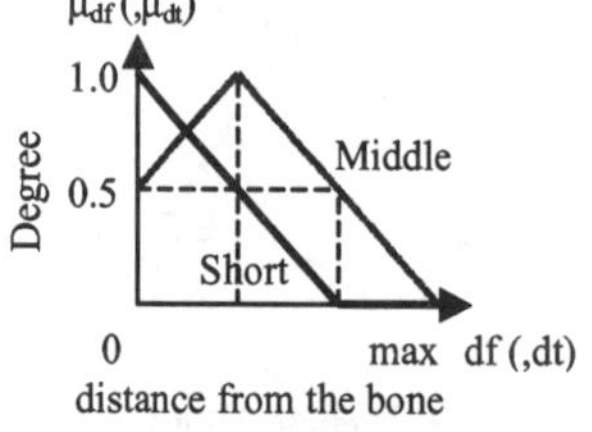

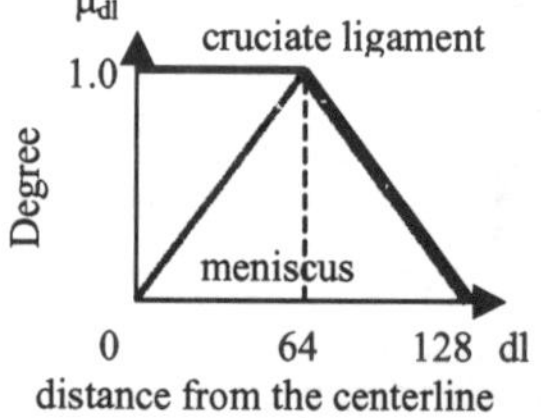

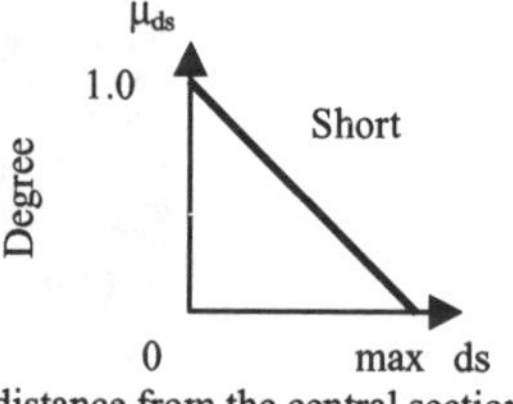

Figure 9 Membership functions for df, dt, dl and ds.

(a) total degree for the cruciate ligament, μ_C (b) total degree for the meniscus, μ_M (c) segmentation result

Figure 10 Segmentation of the cruciate ligament and the meniscus. (The brighter voxels imply higher degree in (a) and (b). The white voxels imply the cruciate ligament and the gray voxels imply the meniscus in (c).)

4. Experimental Results and Conclusions

We applied our method to 5 normal subjects. On our experiment, we set 2 parameters as follows: $Th_g = 1000$, $Th_d = 5mm$. For each image, we successfully segmented the cruciate ligament and the meniscus. The computational time required about 12 minutes on each CT knee images on SGI O2 (R10000, 255 MHz). Qualitative evaluation by a physician showed that this method could successfully segment them. Figure 11 shows the 3D visualization of the ROIs that we segmented in 5 normal subjects of the right knee by our method. In this figure, the rotation angle is -35° and the elevation angle of them is 30°. For one of them, we showed the segmentation result of each of the cruciate ligament and the menisci and both of them. Thus, this method can produce useful visualization for them.

In this paper, we proposed a segmentation method for CT knee images. The method first found the candidate region and the bone region by using intensity thresholding. The obtained bone region was segmented into the femur and the tibia by watershed segmentation. We employed the fuzzy if-then rules to eliminate the cartilage and the cortical bone from the candidate region. The physician's knowledge was converted to fuzzy if-then rules, and then the rules computed the fuzzy degree for ROIs. We determined the voxels belonging to the ROIs according to the fuzzy degree. The experimental results showed that our method segmented the cruciate ligament and the meniscus. It remains as the future work to apply this method to the disease person's subjects.

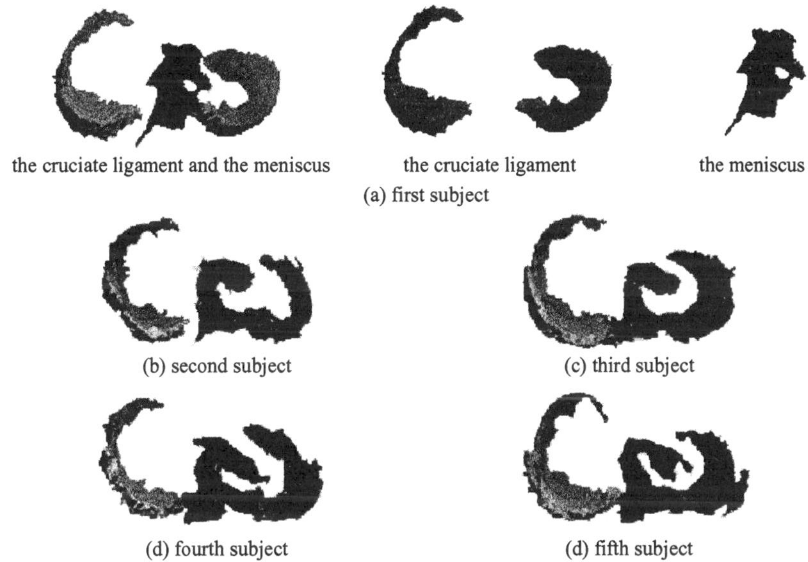

the cruciate ligament and the meniscus   the cruciate ligament   the meniscus

(a) first subject

(b) second subject   (c) third subject

(d) fourth subject   (d) fifth subject

Figure 11 3D visualization of segmentation results.

Acknowledgement

Research supported in part by the BISC Program of UC Berkeley.

References

[1] Y. Hata, T. Sasaki, Y. Tokimoto, Y. Ando, M. Ishikawa and H. Ishikawa, "Automated Segmentation of Magnetic Resonance Image Using Fuzzy Logic," *Proceeding of the Eighth International Fuzzy Systems Association World Congress*, pp. 274-278, Aug. 1999.

[2] J. C. Russ, *The image processing handbook. 2nd edition*, CRC Press, Inc., 1994.

[3] L. A. Zadeh, *Fuzzy Sets and Applications*, CRC Press, Inc., 1992.

KES '01
N. Baba et al. (Eds.)
IOS Press, 2001

3-D Visualization of the Cholecyst and the Bile Duct Using Fuzzy Clustering

Chihiro YASUBA[*], Syoji KOBASHI[*], Yutaka HATA[*],
Yasuhiro TOKIMOTO[**] and Makoto ISHIKAWA[**]
**Department of Computer Engineering, Himeji Institute of Technology,
2167, Shosha, Himeji, 671-2201, Japan
**Ishikawa Hospital, 784, Bessho, Bessho-cho, Himeji, 671-2201, Japan*

Abstract This paper proposes a method for extracting the cholecyst and the bile duct from magnetic resonance cholangiography (MR-C) volumetric images. We propose weighted fuzzy c-means clustering to classify an MR-C image into some clusters in which voxels have similar intensity and similar position. Then, the method finds the clusters corresponding to the cholecyst and the bile duct by evaluating the center vectors. Our experimental result on six subjects showed that this method could extract both the cholesyst and the bile duct.

1. Introduction

Magnetic resonance cholangiography (MR-C) is an imaging technique, which non-invasively depicts the bile duct with high contrast [1]. MR-C has been using for diagnosing the shape and the direction of the bile duct and the pancreatic duct to reconstruct 3-D projection from MR-C volumetric image. Maximum intensity projection (MIP) technique is widely used [2]. The technique can provide interesting images, and can help us understand the location of the cholecyst and the bile duct called regions of interests (ROIs). However, the given images usually contain unnecessary tissues (e.g., stomach) with high intensity. They prevent us to discriminate between the ROIs and the unnecessary tissues. To reduce such unnecessary tissues, it is required to extract ROIs from MR-C volumetric images. This paper proposes a weighted fuzzy c-means and applies it to extract the cholecyst and the bile duct from MR-C volumetric images.

This method is based on the fuzzy c-means (FCM) clustering algorithm introduced by Bezdek [3][4]. For each slice the FCM classifies the pixels into some clusters whose voxels have similar intensity and similar position. First, a processing region is limited to around ROIs by an operator. For the specified region, thresholding technique is applied. The threshold value is previously given so that it is lowest value of the intensity distribution of the ROIs. The remained voxels are passed to next clustering process. The clustering is performed with the FCM with respect to both of the intensity and the position. In the FCM clustering, weights are employed in order to converge the center of clusters within the ROIs.

Finally, some clusters are selected as ROIs by evaluating the intensity of center vectors. To evaluate the performance of the proposed methods it was applied to six subjects with some diseases. In any subjects ROIs were correctly segmented. The given 2-D reconstructed images and 3-D surface rendering images are enough to help radiologists to diagnose the 3-D shape and the connective of the cholecyst and the bile duct.

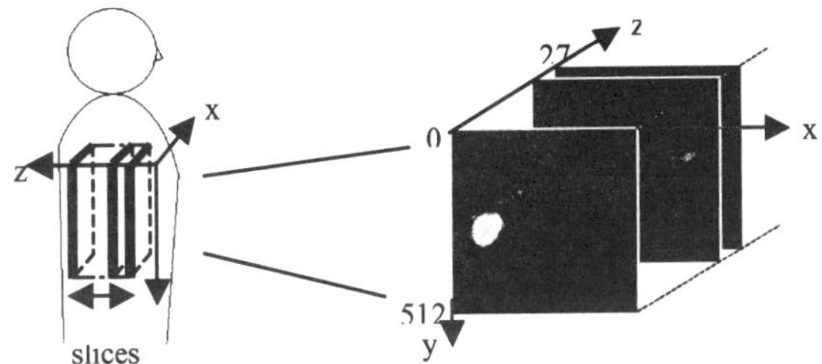

Figure1 Coordinate system.

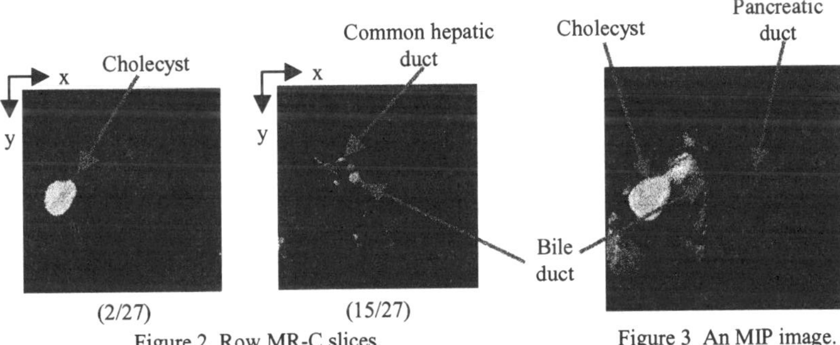

Figure 2 Row MR-C slices.

Figure 3 An MIP image.

2. Preliminaries

We obtained MR-C images from Genesis Signa 1.5 Tesla MRI scanner (General Electric Medical Systems). The image acquisition method was single slice first spin echo with TR = 75790.7 msec and TE = 188.2 msec. Field of view was 240 mm. Matrix was 512 by 512. Thickness of the slice was 3.0 mm. Voxel size was $0.47 \times 0.47 \times 3.0$ mm^3. Each of the volume data was composed of about 30 separated slices. The intensity value for all voxels of all intracranial structure ranged between 0 and 4095. Figure 1 shows our coordinate system. Figure 2 shows raw MR-C images. Figure 3 shows the MIP image. In these images the cholecyst and the bile duct are appeared as rather light gray.

3. Segmentation method based on weighted FCM Clustering

First fuzzy c-means (FCM) clustering algorithm proposed by Bezdek is introduced. The FCM minimizes the following objective function J_{FCM} with respect to fuzzy membership u_{ik}.

$$J_{FCM}(U,V) = \sum_{k=1}^{n} \sum_{i=1}^{c} (u_{ik})^m \|w_k - v_i\|^2, \qquad (1)$$

where

$$w_k = (x_{k1}, x_{k2}, \cdots, x_{kp}), \quad k = 1, 2, \ldots, n \qquad (2)$$

$v_i (i = 1, 2, \ldots, c)$ is a vector of cluster centers, p is the dimension of the vectors w_k, c is the number of clusters, n is the number of vector, and $m < 1$ is the fuzziness index. The FCM algorithm is executed in the following steps.

Step 1) Initialize memberships u_{ik} of w_k belonging to cluster i that

$$\sum_{i=1}^{c} u_{ik} = 1, \quad k = 1, 2, \ldots, n. \qquad (3)$$

Step 2) Compute the vector of fuzzy cluster centers v_i for $i = 1, 2, \ldots, c$ using

$$v_i = \frac{\sum_{k=1}^{n}\{(u_{ik})^m w_k\}}{\sum_{k=1}^{n}(u_{ik})^m}. \tag{4}$$

Step 3) Update the fuzzy membership u_{ik} using

$$u_{ik} = \left[\sum_{j=1}^{c} \left(\frac{\|w_k - v_i\|^2}{\|w_k - v_j\|^2} \right)^{\frac{1}{m-1}} \right]^{-1}. \tag{5}$$

Step 4) Check the stopping criterion. If it is not satisfied, go to Step 2.

In our method proposed here, a processing region is manually set to around the cholecyst and the bile duct. To reduce the processing time, background region (e.g., fat, air) are removed by thresholding technique. For whole volume data only one threshold intensity is given by an operator so that it is the lowest value of intensity distribution of the cholecyst and the bile duct. In our method, setting the processing region and finding the threshold value need operator's interaction. In the following we call the remained voxels target region. Figure 4 shows a raw MR-C image and the target region. For the target region, we apply the FCM method where the attribute values are the intensity and the position, i.e. coordinate values, x and y. All of the attribute values are normalized by using normalizing transform, and the attribute value of intensity is multiplied by 1.5.

Intensity change among neighbor voxels inside each tissue is small because voxels consisting ROIs have homogenous intensity. Otherwise, intensity change on boundary of each tissue is big. Therefore, to move the center vectors into inside the tissues we rewrite the function to calculate calculating the center vectors of clusters shown in Eq.(4) as

$$v_i = \frac{\sum_{k=1}^{n}\{(u_{ik})^m \times w_k \times a_k\}}{\sum_{k=1}^{n}\{(u_{ik})^m \times a_k\}}, \tag{4$'$}$$

where a_k is weight of voxel k. The weight is given by

$$a_k = 1.0 - \int_{-\infty}^{z_k} \frac{1}{\sqrt{2\pi}} \exp(-k^2/2)\,dk, \tag{6}$$

where z_k is the standard deviation (SD) of voxel k [5]. We call this weighted FCM clustering. In our experiment, 11×11 kernel was used to compute the SD. Figure 5 shows the relation between z_k, and the weight, a_k.

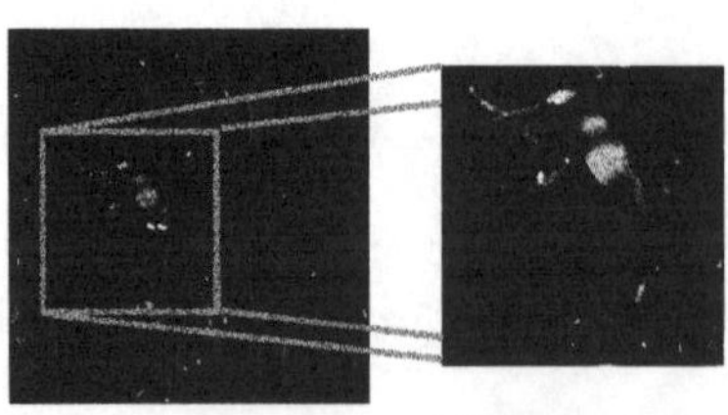

(a) A raw MR-C image   (b) Target region

Figure 4 Target region.

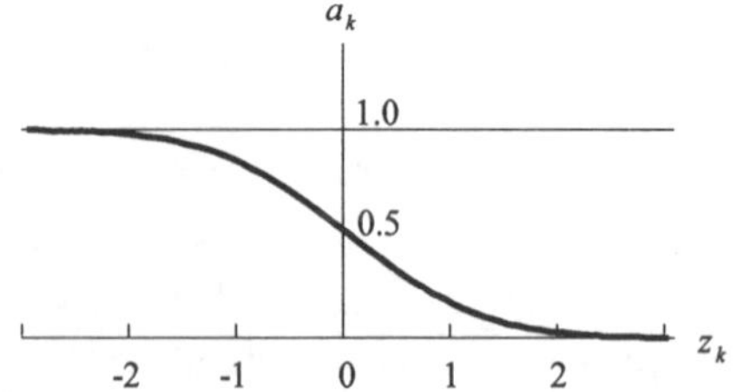

Figure 5 Relation between intensity change and weight.

Next, our method selects the clusters corresponding to the cholecyst and the bile duct by evaluating center vectors of the clusters. The cholecyst and the bile duct have high intensity in MR-C images. Let v_i^{int} be center vectors of the clusters arranged according to high intensity. We divide v_i^{int} into two groups, and we calculate means of each groups as follows,

$$m_{c'}^{up} = \frac{\sum_{i=1}^{c'} v_i^{\text{int}}}{c'}, \qquad m_{c'}^{low} = \frac{\sum_{i=c'+1}^{c} v_i^{\text{int}}}{c - c'}, \quad 0 < c' < c, \tag{7}$$

where c' is number of cluster arranged. Next, we find c' when $m_{c'}^{up}$ has most different value from $m_{c'}^{low}$. Finally, we select clusters which values of center vectors are not lower than value of $v_{c'}^{\text{int}}$.

4. Experimental Results

We applied our method to six volume data for six patients. In this examination, number of clusters is fixed by five. Figure 7 and Figure 8 show the resultant images. In these figures, (a) target regions specified by an operator, (b) results of clustering, (c) extracted region by our method, and (d) the results delineated by a radiologist. As shown these figures our method can correctly segment the ROIs with compared to a radiologist segmentation. Figure 9 and Figure 10 show 3-D surface rendering images of regions of resultant image each of Figure 7 and Figure 8, respectively. In these figures, (a) MIP images, (b) of our method results, and (c) the ones delineated by a radiologist. These figure show that the bile duct can be observed easily by using our method, and our results are almost equivalent to the results by a radiologist. Processing of each subject required about fifteen minutes, running on SGI O2 (R10000, 225MHz).

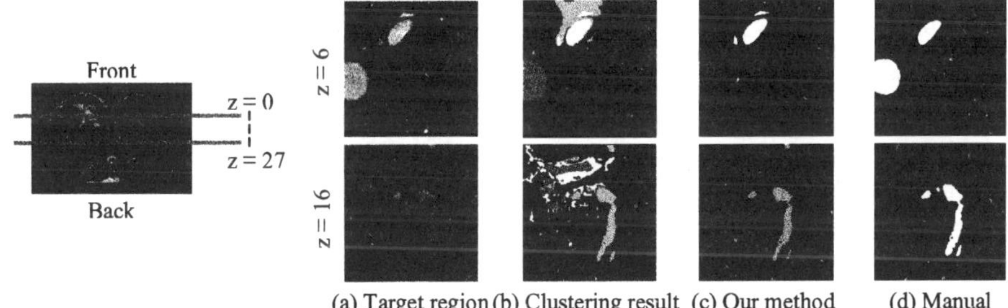

Figure 7 Experimental results 1 (sex : M, age : 75).

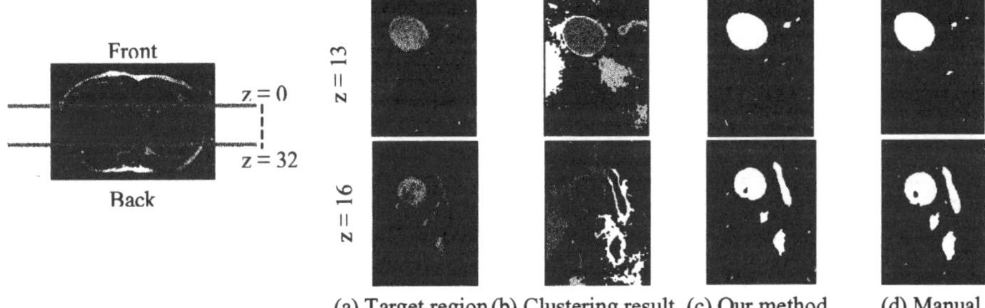

Figure 8 Experimental results 2 (sex : F, age : 59).

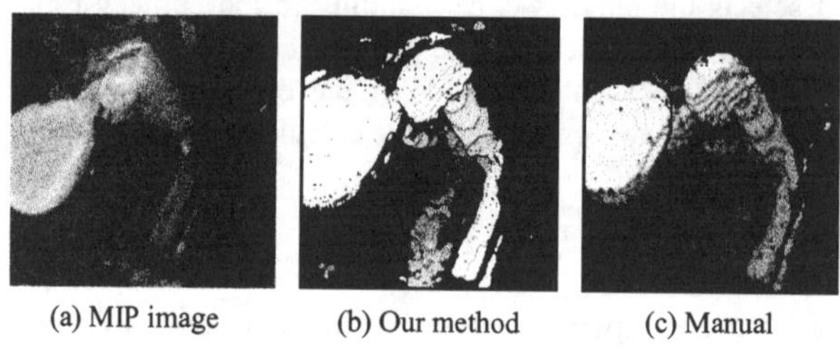

(a) MIP image   (b) Our method   (c) Manual

Figure 9 3-D images of result (Figure 7).

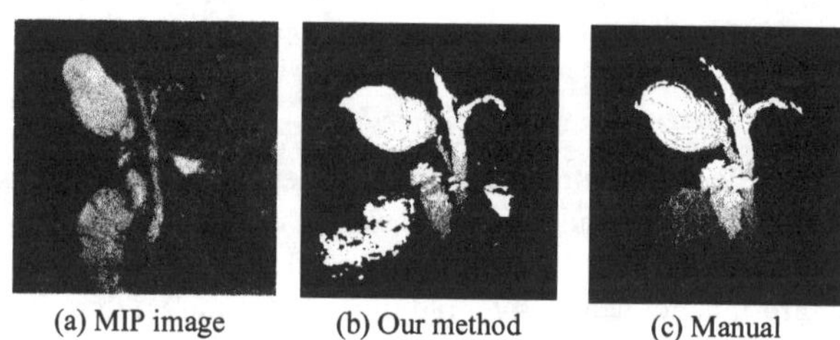

(a) MIP image   (b) Our method   (c) Manual

Figure 10 3-D images of result (Figure 8).

5. Conclusion

In this paper, we described an automated method for extracting the cholecyst and the bile duct from MR-C volumetric images. Generally, at the outline of an ROI the differential value among the neighbor voxels or pixels (e.g., sobel value, laplacian value) has high value. In the contrast, the value at the inner of the ROI is small. Following this fact, the proposed method gives high weights to the voxels with low SD. In FCM clustering we called this weighted FCM clustering. The weights move the center of clusters into the ROIs. The experimental result on six subjects showed that our method was able to extract both the cholecyst and the bile duct. The future work is to apply the proposed method to other portions such as the pancreatic duct and visualization for the bile stone.

Acknowledgement

This work was supported in part by the BISC Program of UC Berkeley.

References

[1] Y. Watanabe, M. Dohke, T. Ishimori, Y. Amoh, K. Oda, A. Okumura, K. Mitsudo, and Y. Dodo, "High-resolution MR Cholangiopancreatography," Critical Reviews in Diagnostic Imaging, vol. 39, no. 2-3, pp. 115-258, 1998.

[2] B. Lichtenbelt, R. Crane, and S. Naqvi, "Introduction to Volume Rendering," Hewlette-Packard Company, pp. 49, 50, 135, 136, 1998.

[3] L. O. Hall, A. M. Besaid, L. P. Clarke, R. P. Velthuizen, M. S. Silbiger and J. C. Bezdek, "A Comparison of Neural Network and Fuzzy Clustering Techniques in Segmenting Magnetic Resonance Images of the Brain," IEEE Transactions on Neural Networks, vol. 3, no. 5, pp. 672-682, 1992.

[4] C. Bezdek, and S. K. Pal, *Fuzzy Models for Pattern Recognition Methods that Search for Stryctures in Data*, IEEE PRESS, pp. 219-220, 1992.

[5] E. Gose, R. Johnsonbauch, and S. Jost, *Pattern Recognition and Image Analysis*, A Simon & Schuster Company, p. 42, 1996.

KES '01
N. Baba et al. (Eds.)
IOS Press, 2001

On the Similarity Measure of Sample Data for Clustering based on the Mixture Model

Osamu FUJITA and Norio BABA
Osaka Kyoiku University, 4-698-1 Asahigaoka, Kashiwara, Osaka, 582-8582, Japan

Abstract. A similarity measure based on multiple clustering is proposed. The similarity between two sample data is defined as the probability that the two samples are classified into the same cluster. In practice, assuming the various mixture models and estimating each set of model parameters by using the EM algorithm, every sample is classified into each appropriate cluster for every reasonable model, and then the similarity is evaluated. This similarity measure can be consistent with the categorization and clustering based on the same mixture model.

1. Introduction

Probabilistic approach is a way of great promise in a variety of fields such as pattern recognition, pattern generation, information retrieval and data mining. Classification based on Bayes decision theory and data clustering based on the mixture model are fundamental tools in connection with supervised and unsupervised learning from examples [1,2]. The theory has been developed on the basis of statistical properties of sample data. In practice, however, there is a need to look into the detailed relationship between individual samples. For example, the document retrieval requires the similarity measure of documents which is consistent with the results of document classification and clustering. The consistency is important also from a methodological point of view.

Traditional data clustering is done on the basis of the similarity or dissimilarity measure between individual samples. The dissimilarity is generally defined as the distance between samples. The clusters are formed so that the distance between samples within a cluster is less than that of inter-clusters. There are wide variations in the definitions of the similarity and distance measure. The clustering results such as the shape and the member of clusters can vary widely according to the definition. In any way, the metric space of samples provides a universal basis for data processing.

There are two types of the clustering methods: hierarchical clustering and non-hierarchical clustering. In the hierarchical clustering, the nearest pair of samples or clusters is merged into one cluster, and this process is repeated recursively until the number of clusters decreases to a desired value. In this procedure, it is necessary to evaluate the distance of every pair of samples or clusters, and so it is time-consuming when the number of samples is very large. In the non-hierarchical clustering, a typical example is the k-means algorithm. The first thing to do is setting the unknown number of clusters even if there is no a priori knowledge about sample data, and then distributing all samples to clusters almost equally. Then the two steps, 1) determine the center of each cluster and 2) reassign each sample to a cluster whose center is nearest from it, are repeated until there is no change. The computation time is usually less than that of hierarchical clustering. In either case, the distance between samples should be measured before clustering.

The probabilistic approach of data clustering based on the mixture model is similar with k-means algorithm [2]. However, the sample space is not necessarily a metric space. The procedure is designed to maximize the likelihood of model parameters with respect to sample data, and so there is no need of the distance measure. When the probability is modeled by a multinomial distribution, in a sense, the Kullback-Leibler divergence can be used for measuring the deviation of the model of a cluster distribution from a sample, but it is improper to the distance measure between individual samples due to asymmetry. In such a situation, the problem is what is the best concept of the distance measure between samples. We introduce one of possible solutions in Chapter 3.

2. Data Clustering with Mixture Models

The mixture model provides a probabilistic approach to data clustering in a natural way. It is based on the assumption that sample data are generated from the mixture of many individual distribution functions. Each individual distribution can be regarded as a cluster. The shape of the clusters and the membership of samples are characterized by model parameters of the distribution functions. The best assignment of samples is considered to maximize the likelihood of model parameters with respect to samples.

Let $X = \{\mathbf{x}_i\}_{i=1}^{S}$ denotes the sample data set. It is usually assumed to be generated independently from a mixture density

$$P(\mathbf{x}_i) = \sum_{j=1}^{Nc} P(\mathbf{x}_i \mid w_j; \theta) P(w_j) \tag{1}$$

where each component of the mixture is denoted w_j and parameterised by θ. We assume w_j and Nc indicate the label of the j-th cluster and the number of clusters respectively, for convenience's sake, though w_j does not necessarily have one-to-one correspondence with a class or cluster in general.

The maximum likelihood estimate of θ can be obtained by the EM algorithm [3]. In practice, however, the log likelihood using Eq. (1) involves the log of a sum so that it is not easily maximized numerically. To avoid this difficulty, using the binary indicator variables $\mathbf{z}$, $\mathbf{z}_i = (z_{i1}, \ldots, z_{iNc})$, where $z_{ij} = 1$ iff $\mathbf{x}_i$ belongs to the j-th cluster, a "complete-data" log likelihood function can be written as

$$l_c(\theta \mid X, Z) = \sum_{i=1}^{Ns} \sum_{j=1}^{Nc} z_{ij} \log[P(\mathbf{x}_i \mid \mathbf{z}_j; \theta) P(\mathbf{z}_j; \theta)]. \tag{2}$$

where X and Z denote the sets of $\mathbf{x}_i$ and $\mathbf{z}_i$, respectively. The EM algorithm estimates the parameter θ for unknown $\mathbf{z}$ by iterating the following two steps:

E-step: $\quad Q(\theta \mid \theta_k) = E[l_c(\theta \mid X, Z) \mid X, \theta_k],$

M-step: $\quad \theta_{k+1} = \arg\max_{\theta} Q(\theta \mid \theta_k).$ $\tag{3}$

The cluster that $\mathbf{x}_i$ belongs to can simply be decided by

$$C(\mathbf{x}_i) = \arg\max_{z_{ij}} P(z_{ij} \mid \mathbf{x}_i; \theta). \tag{4}$$

where the a posteriori probability, $P(z_{ij} \mid \mathbf{x}_i; \theta)$, can be computed by using Bayes rule.

There are some problems in this method in practice. First, the maximum likelihood estimate of θ cannot be determined uniquely in most cases. There can be many solutions. This is primarily because the number of clusters is unknown before the clustering. To obtain the optimal solution, it is necessary to examine various models that are different in the number of clusters. Even if the optimal number is determined, however, the likelihood function generally has many local maxima for mixture models. Second, the number of samples is not always enough for parameter estimation, and so high accuracy cannot be expected. It is very difficult to judge which is the best model when the values of local maxima of the likelihood are close to each other. Lastly, as the number of iteration of EM steps increases and the likelihood approach to the maximum point, the convergence speed becomes very slow. In this situation, the likelihood little increases and the assignment of samples little changes, though each EM step consumes a regular time. It might be worthwhile to search another solutions from different initial conditions rather than to waste time to compute for obtaining less information.

3. Similarity Measure based on Multiple Clustering

The similarity between samples in the mixture model is influenced by the above-mentioned problems. If it is very difficult to choose the optimal model parameters among many solutions, it is better to take reasonable models as many as possible into consideration. To meet this requirement, we introduce a new similarity measure based on the probability that two samples are classified into the same cluster. Let the similarity and the distance between two samples of $\mathbf{x}$ and $\mathbf{y}$ be $S(\mathbf{x}, \mathbf{y})$ and $D(\mathbf{x}, \mathbf{y})$, respectively, and defined as follows:

$$S(\mathbf{x},\mathbf{y}) = P(C(\mathbf{x}) = C(\mathbf{y})),$$
$$D(\mathbf{x},\mathbf{y}) = P(C(\mathbf{x}) \neq C(\mathbf{y})) = 1 - S(\mathbf{x},\mathbf{y}), \tag{5}$$

where $C(\mathbf{x})$ denotes a class that $\mathbf{x}$ belongs to. This definition obviously satisfies the following three properties of the distance.

i) $D(\mathbf{x},\mathbf{y}) \geq 0, \quad D(\mathbf{x},\mathbf{y}) = 0 \Leftrightarrow \mathbf{x} = \mathbf{y}$,
ii) $D(\mathbf{x},\mathbf{y}) = D(\mathbf{y},\mathbf{x})$, $\tag{6}$
iii) $D(\mathbf{x},\mathbf{y}) \leq D(\mathbf{x},\mathbf{v}) + D(\mathbf{v},\mathbf{y})$.

The third property can be proved by considering

$$P(C(\mathbf{x}) \neq C(\mathbf{y}))$$
$$= P(C(\mathbf{v}) \neq C(\mathbf{x}) \neq C(\mathbf{y}) \neq C(\mathbf{v})) + P(C(\mathbf{x}) \neq C(\mathbf{y}) = C(\mathbf{v})) + P(C(\mathbf{v}) = C(\mathbf{x}) \neq C(\mathbf{y}))$$

and

$$P(C(\mathbf{x}) \neq C(\mathbf{v})) + P(C(\mathbf{v}) \neq C(\mathbf{y})) - P(C(\mathbf{x}) \neq C(\mathbf{y}))$$
$$= P(C(\mathbf{v}) \neq C(\mathbf{x}) \neq C(\mathbf{y}) \neq C(\mathbf{v})) + 2P(C(\mathbf{v}) \neq C(\mathbf{x}) = C(\mathbf{y})) \geq 0$$

where equality holds iff $P(C(\mathbf{x}) \neq C(\mathbf{v}) \neq C(\mathbf{y})) = 0$.

In the mixture model, it is assumed that there are many models that approximately fit to sample data. Let M be the set of such considerable models. As $C(\mathbf{x})$ varies with the models, the similarity can be evaluated as

$$S_{MC}(\mathbf{x},\mathbf{y}) \sim \frac{1}{|M|} \sum_{m \in M} \delta(C(\mathbf{x},\theta^{(m)}), C(\mathbf{y},\theta^{(m)})) \tag{7}$$

where $|M|$ means the number of models,

$$\delta(a,b) = \begin{cases} 1 & for\ a = b \\ 0 & for\ a \neq b \end{cases}$$

and

$$C(\mathbf{x},\theta^{(m)}) = \arg\max_{w_j} P(w_j \,|\, \mathbf{x},\theta^{(m)}). \tag{8}$$

The function, $C(\mathbf{x}, \theta^{(m)})$, is a kind of the discriminant function based on Bayes decision theory. The drawback of this definition is to ignore the probability distribution that $\mathbf{x}$ belongs to the other clusters. This information loss can be regarded as a kind of quantization error.

The similarity depends greatly on what kind of model is included in the set M. In principle, even random partition of sample space by using such as hyper-planes is available for the function $C(\mathbf{x}, \theta^{(m)})$. However, if the number of samples is much less than the dimension of the sample space, i. e., sparse samples in a very high dimensional space, the S_{MC} of almost all pairs will gather around 0.5. It is better to select reasonable models by using maximum likelihood estimation.

4. Comparison with the Different Types of Similarity Measures

There can be a variety of different definitions of the similarity measure. However, it seems there is no perfect definition for the mixture model. In the following examples, model parameters are assumed to be determined uniquely in any case, and therefore they cannot solve the above-mentioned problems.

In the field of document retrieval and categorization, for example, the following similarity function has been derived based on the Fisher kernel [4].

$$K(\mathbf{x},\mathbf{y}) = \sum_{i=1}^{Nc} P(w_i \mid \mathbf{x}) P(w_i \mid \mathbf{y}) / P(w_i) \tag{9}$$

where w_i means the latent class variable. According to this definition, if $\mathbf{x}$ and $\mathbf{y}$ belong to the same cluster w_i, the smaller $P(w_i)$ is, the larger the similarity between them is. This qualitative property seems to be very reasonable, but it brings about a somewhat bad property quantitatively. The point is that $K(\mathbf{x}, \mathbf{x})$ varies with $\mathbf{x}$ and $P(w_i)$, and that it does not always have the maximum value. It is possible that $K(\mathbf{x}, \mathbf{x}) < K(\mathbf{x}, \mathbf{y})$ for $\mathbf{x} \neq \mathbf{y}$. This means that there is no appropriate distance measure corresponding with $K(\mathbf{x}, \mathbf{y})$. It does not seem to be a good property theoretically, though it is unclear how this affects the results of document categorization and retrieval in practice.

From the viewpoint of the consistent relationship between the similarity and the distance, the following two types of definitions can easily be derived intuitively.

$$S_V(\mathbf{x},\mathbf{y}) = \sum_{i=1}^{Nc} \left[P(w_i \mid \mathbf{x}) P(w_i \mid \mathbf{y}) \right]^{\frac{1}{2}},$$

$$D_V(\mathbf{x},\mathbf{y}) = \frac{1}{2} \sum_{i=1}^{Nc} \left[\sqrt{P(w_i \mid \mathbf{x})} - \sqrt{P(w_i \mid \mathbf{y})} \right]^2 = 1 - S_v(\mathbf{x},\mathbf{y}), \tag{10}$$

$$S_K(\mathbf{x},\mathbf{y}) = \prod_{i=1}^{Nc} \frac{P(w_i \mid \mathbf{x})^{P(w_i \mid \mathbf{y})} P(w_i \mid \mathbf{y})^{P(w_i \mid \mathbf{x})}}{P(w_i \mid \mathbf{x})^{P(w_i \mid \mathbf{x})} P(w_i \mid \mathbf{y})^{P(w_i \mid \mathbf{y})}},$$

$$D_K(\mathbf{x},\mathbf{y}) = \sum_{i=1}^{Nc} \left[P(w_i \mid \mathbf{x}) \log \frac{P(w_i \mid \mathbf{x})}{P(w_i \mid \mathbf{y})} + P(w_i \mid \mathbf{y}) \log \frac{P(w_i \mid \mathbf{y})}{P(w_i \mid \mathbf{x})} \right] = -\log S_K(\mathbf{x},\mathbf{y}), \tag{11}$$

where $D_V(\mathbf{x}, \mathbf{y})$ means Euclidian distance in the Nc-dimensional vector space such that $\sqrt{P(w_i \mid \mathbf{x})}$ is the i-th component of the vector. On the other hand, $D_K(\mathbf{x}, \mathbf{y})$ can be derived by analogical inference based on Kullback-Leibler divergence so as to be symmetric with respect to $\mathbf{x}$ and $\mathbf{y}$. Unfortunately, in either case, there is no solid reasoning based on the theory of probability and statistics.

The above three definitions have a common problem. In a special case, for example, that $P(w_i|\mathbf{x}) = P(w_i|\mathbf{y})$ for every i, $\mathbf{x}$ and $\mathbf{y}$ are regarded as identical, even if $\mathbf{x}$ is not actually the same with $\mathbf{y}$. This situation may be probable when the dimension of $\mathbf{x}$ is much larger than the number of clusters. This problem generally occurs whenever the dimension of the feature space is considerably reduced from that of the sample space. The same holds for S_{MC} too, but it can be improved by examining as many models as possible.

5. Application of S_{MC} to the Model Selection

The information represented by S_{MC} reveals the relationship between individual samples based on their collective properties. In an application to document retrieval, for example, the similarity scoring of documents can be consistent with document categorization and clustering based on the mixture model. Furthermore, S_{MC} provides the way of the hierarchical clustering and the results can be fed back to the mixture model for the model selection. The procedure is as follows:

1) Obtain many reasonable mixture models by using EM algorithm,
2) Classify all samples into clusters based on each mixture model,
3) Compute S_{MC} for all pairs of samples,
4) Do hierarchical clustering based on S_{MC},
5) Select the best mixture model that is most similar with the result of the above hierarchical clustering.

Although it seems to be much time-consuming, it must be necessary for the optimal model selection to take time in any way. There must be no efficient method, even if such a criterion as Akaike's Information Criterion (AIC) and Minimum Description Length (MDL) is used.

It is interesting that the similarity of sample data is evaluated after data clustering or categorization, which is based on the complementary relationship between the distance of samples and the partition of the sample space. The similarity measure of S_{MC} is widely applicable not only to the mixture model but also to more complex models. For example, $C(\mathbf{x})$ can be such a complex non-linear function as neural networks. If a certain class of neural network models can be defined, the metric of a transformed feature space will be analysed based on S_{MC}.

6. Summary

The similarity measure based on the multiple clustering is proposed. The similarity between two sample data is defined as the probability that the two samples are classified into the same cluster. In the mixture model, this similarity is measured after the maximum likelihood estimation of model parameters by using the EM algorithm, and then it is used for hierarchical clustering and the selection of the best mixture model. This method is applicable to document clustering and retrieval for example.

References

[1] R. O. Duda & P. E. Hart, Pattern Classification and Scene Analysis. John Wiley & Sons, 1973.
[2] T. M. Mitchell, Machine Learning. McGraw Hill, 1997.
[3] A. P. Dempster, N. M. Laird, & D. B. Rubin, Maximum Likelihood from Incomplete Data via the EM Algorithm. Journal of the Royal Statistics Society, Series B, **39** (1991) 1-38.
[4] T. Hofmann, Learning the Similarity of Documents: An Information-Geometric Approach to Document Retrieval and Categorization, In: S. A. Solla, T. K. Leen, & K. –R. Muller (ed.), Advances in Neural Information Processing Systems 12. MIT Press, 2000, pp. 914-920.

KES '01
N. Baba et al. (Eds.)
IOS Press, 2001

Experiments with an Ensemble Self-Generating Neural Network

Hirotaka INOUE and Hiroyuki NARIHISA

Okayama University of Science, 1-1 Ridai-cho, Okayama-shi, Okayama, 700-0005, Japan

Abstract. In an earlier paper, we introduced an ensemble model called ESGNN (ensemble self-generating neural network) which can be used to reduce the error for classification and chaotic time series prediction. Although this model can obtain the high accuracy than a single SGNN, the computational cost increase in proportion to the number of SGNN in an ensemble. In this paper, we propose a new pruning SGNN algorithm to reduce the memory requirement for classification. We compared ESGNN with nearest neighbor classifier using a collection of machine-learning benchmarks. Experimental results show that our method could reduce the memory requirement and improve the accuracy over the nearest neighbor classifier's accuracy.

1 Introduction

Neural networks have been widely used in the field of intelligent information processing such as classification, clustering, prediction, and recognition. Generally, these neural networks have to be decided the network structure and some parameters by human experts. It is quite tricky to choose the right structure of neural networks suitable for a particular application at hand.

Self-generating neural networks (SGNN) [1] have abilities of the simplicity of networks design and the high speed learning. SGNN are some kinds of extension of the self-organizing maps (SOM) of Kohonen [2] and utilize the competitive learning algorithm which is implemented as self-generating neural tree (SGNT).

In order to improve the accuracy of SGNN, we proposed ensemble self-generating neural networks (ESGNN) for classification [3] and time series prediction [4]. ESGNN apply ensemble averaging [5] to SGNN and fully utilize the characteristics of the high speed convergence. Although ESGNN are improved the accuracy by using various SGNN, the computation time and the memory capacity increase in proportion to increase the number of SGNN in ensemble.

In this paper, we propose a new SGNN's pruning method to reduce the computational cost for classification. The pruning is composed of two phases, the merge phase and the dead leaf removing phase. After pruning, pruning SGNT keeps the classification accuracy of unpruned SGNT. We compare pruned ESGNN with nearest neighbor (NN) classifier in view point of the computational cost and the accuracy using ten problems in UCI repository [6].

2 Pruning Self-Generating Neural Networks

Self-generating neural networks (SGNN) are based on SOM and implemented as a self-generating neural tree (SGNT) architecture. Moreover, this neural tree can be constructed directly from the given training data without intervening human tricky effort. The SGNT algorithm is defined as a tree construction problem how to construct a tree structure from

```
1 begin    initialize j = the height of the SGNT
2    do for each subtree's leaves in the height j
3       if all leaves belong to the same class,
4       then merge all leaves to parent node
5       if all subtrees are traversed in the height j,
6       then j ← j − 1
7    until j = 0
8 end.
```

Figure 1: The merge algorithm.

```
1 begin    initialize j = 0, N =# of leaves on the merged tree
2    do j ← j + 1; for each attribute x_j
3       search the nearest leaf of x_j
4       if the nearest leaf different from the original leaf,
5       then prune the original leaf as the dead leaf
6    until j = N
7 end.
```

Figure 2: The dead leaf pruning algorithm.

the given data which consist of multiple attributes under the condition that final leaf neurons correspond to the given data.

After all input data are inserted into SGNT as leaf neurons, weights of each node neuron are averages of the corresponding weights of all its children. Whole network reflect the given feature space by its topology.

SGNT has the capability of high speed processing, but the accuracy of SGNT are inferior to nearest neighbor classifier because SGNT is not guarantee to reach the nearest node for unknown data. Hence, we take the majority of plural SGNT's outputs to improve the accuracy.

In order to reduce the memory requirement, we propose a pruning algorithm. This algorithm has two parts, the merge phase and the dead leaf pruning phase. The merge phase is performed to reduce dense leaves (Figure 1). This phase use the class information to decide the merge or not. If leaves, which have the same parent node, are same class, then these leaves are pruned and the parent node represents these leaves. The pruned tree has the same classification capability to the unpruned tree. The dead leaf pruning phase is performed to prune the dead leaves in the current tree. The dead leaf is the leaf which has never referred. The algorithm is given as Figure 2. We repeated the merge phase and the dead leaf pruning phase until the tree become stable.

We show an example of our pruning algorithm in Figure 3. This is a two-dimensional classification problem with two equal circular Gaussian distribution. The shaded plane is the decision region of class 0 and the another plane is the decision region of class 1 by SGNT. The dotted line is the Bayes decision boundary. The number of training samples is 200 (See Figure 3(a)). The unpruned SGNT is given in Figure 3(b). In this case, 200 leaves and 118 nodes are automatically generated by the SGNT algorithm [1]. The height of the unpruned SGNT is 8. First, we execute the merge phase for the unpruned SGNT. Figure 3(c) shows the pruned SGNT after the merge phase once. 199 leaves and 106 nodes are pruned away. Second, we execute the dead leaf pruning phase. The only one original leaf which remain the merge phase is pruned. Third, we execute the merge phase and the dead leaf pruning phase again. Finally, we obtain the final pruned tree in Figure 3(d). Although this tree has only one root node and two leaves, decision regions are same as the unpruned tree.

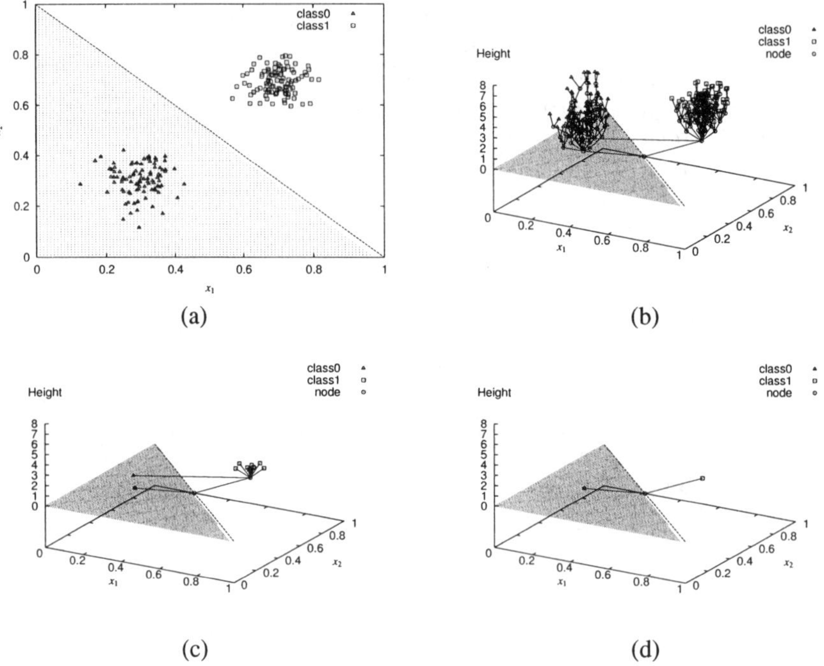

Figure 3: An example of the SGNT pruning, (a) a two dimensional classification problem with two equal circular Gaussian distribution, (b) the structure of the unpruned SGNT, (c) the structure of the pruned tree after the merge phase, and (d) the structure of the final pruned SGNT. The shaded plane is the decision region of class 0 by the SGNT and the doted line shows the Bayes decision boundary.

In the above example, we use all training data to construct SGNT. The structure of SGNT is changed by the order of the training data. Hence we can construct ESGNN from the same training data by changing the input order. We call this approach as "shuffling".

3   Experimental Results

We investigate the computational cost (the memory capacity and the computation time) and the accuracy of single SGNN, ensemble SGNN, nearest neighbor, and 3-nearest neighbor with two sampling method, shuffling and bagging. Shuffling use all training data by changing the order of the training data randomly on each classifier. Bagging [7] is a resampling technique which permit the overlapping the data. We evaluate 10-fold cross-validation for ten benchmark problems of UCI repository. In this experiment, we use a modified Euclidean distance measure for all classifiers as follows:

$$d(\boldsymbol{x}, \boldsymbol{y}) = \sqrt{\sum_{i=1}^{m} w_i \cdot (x_i - y_i)^2} \,, \tag{1}$$

where m is the number of attributes, x_i, y_i are the i-th attribute values of $\boldsymbol{x}$ and $\boldsymbol{y}$ and w_i is the weight for the i-th attribute as follows:

$$w_i = \frac{1}{\max_i - \min_i} \,, \tag{2}$$

Table 1: The average memory requirement and the computation time (sec.) of ten trials for shuffled SGNN, bagged SGNN, and nearest neighbor.

Dataset	shuffled SGNN		bagged SGNN		nearest neighbor	
	memory	time(sec.)	memory	time(sec.)	memory	time(sec.)
balance-scale	0.384	0.08	0.304	0.07	0.664	0.4
breast-cancer-w	0.125	0.11	0.093	0.10	0.703	0.63
glass	0.582	0.03	0.459	0.02	0.654	0.05
ionosphere	0.423	0.10	0.299	0.09	0.697	0.26
iris	0.196	0.02	0.153	0.01	0.657	0.03
letter	0.290	15.91	0.239	14.14	0.677	979.23
liver-disorders	0.690	0.06	0.483	0.05	0.670	0.15
new-thyroid	0.270	0.02	0.235	0.02	0.654	0.05
pima-diabetes	0.570	0.15	0.428	0.14	0.672	0.72
wine	0.238	0.02	0.179	0.01	0.703	0.05
Average	0.376	1.65	0.282	1.47	0.675	98.16

Table 2: The average classification accuracy of ten trials for single SGNN, ESGNN, nearest neighbor (1-NN), and 3-nearest neighbor (3-NN) with shuffling. The standard diviation is given inside the bracket.

Dataset	SGNN	ESGNN	1-NN	3-NN
balance-scale	0.781(0.053)	0.843(0.059)	0.771(0.057)	0.816(0.049)
breast-cancer-w	0.954(0.020)	0.967(0.023)	0.954(0.025)	0.963(0.024)
glass	0.632(0.102)	0.692(0.075)	0.669(0.086)	0.697(0.078)
ionosphere	0.858(0.042)	0.883(0.043)	0.872(0.034)	0.858(0.044)
iris	0.953(0.055)	0.967(0.047)	0.960(0.047)	0.960(0.047)
letter	0.893(0.007)	0.958(0.003)	0.960(0.004)	0.956(0.005)
liver-disorders	0.574(0.086)	0.635(0.055)	0.626(0.066)	0.623(0.064)
new-thyroid	0.944(0.070)	0.953(0.049)	0.958(0.040)	0.940(0.063)
pima-diabetes	0.687(0.078)	0.711(0.070)	0.692(0.084)	0.737(0.058)
wine	0.938(0.042)	0.960(0.039)	0.949(0.032)	0.960(0.027)
Average	0.821	0.856	0.841	0.851

where $\max_i$ and $\min_i$ are the maximum and minimum value of the i-th attribute in the training set respectively. All computations of these classifiers are performed on an IBM PC-AT machine (CPU: Intel Pentium II 450MHz, Memory: 323MB).

Table 1 shows the average computational cost (the memory requirement and the computation time (sec.)) of ten trials for the SGNN and the nearest neighbor. It is found that bagged SGNN have a higher performance in the memory compression and the computation time than shuffled SGNN. Both SGNN are fast compared with the nearest neighbor method. Especially, a remarkable difference appear in letter (#data: 20000, #input dimensions: 16). This supports the fact that pruning SGNN can use for large scale problem effectively.

Table 2 and Table 3 show the average classification accuracy of ten trials for shuffled and bagged SGNN, ESGNN, nearest neighbor, and 3-nearest neighbor respectively. We use 25 SGNN for ESGNN. It is clear that, over 10 datasets, both shuffling and bagging ESGNN lead to obviously more accurate classifiers. These are superior to nearest neighbor and comparable to 3-nearest neighbor on the average.

At the comparison to shuffling and bagging, bagging leads to greater improvement in the classification accuracy and superior to shuffling on 5 of the 10 datasets. In short, bagging is

Table 3: The average classification accuracy of ten trials for single SGNN, ESGNN, nearest neighbor (1-NN), and 3-nearest neighbor (3-NN) with bagging.The standard diviation is given inside the bracket.

Dataset	SGNN	ESGNN	1-NN	3-NN
balance-scale	0.768(0.055)	0.850(0.047)	0.786(0.054)	0.822(0.044)
breast-cancer-w	0.959(0.030)	0.973(0.017)	0.957(0.021)	0.964(0.024)
glass	0.593(0.107)	0.702(0.060)	0.603(0.110)	0.589(0.093)
ionosphere	0.849(0.042)	0.869(0.041)	0.863(0.038)	0.835(0.036)
iris	0.960(0.056)	0.960(0.047)	0.953(0.063)	0.953(0.063)
letter	0.874(0.009)	0.955(0.003)	0.948(0.003)	0.939(0.005)
liver-disorders	0.574(0.077)	0.614(0.051)	0.617(0.077)	0.583(0.053)
new-thyroid	0.945(0.065)	0.967(0.050)	0.949(0.047)	0.963(0.037)
pima-diabetes	0.671(0.056)	0.719(0.050)	0.715(0.067)	0.706(0.062)
wine	0.938(0.063)	0.960(0.039)	0.933(0.035)	0.944(0.037)
Average	0.813	0.856	0.819	0.830

better than shuffling in view point of the computational cost and the classification accuracy.

4 Conclusions

In this paper, we proposed a new pruning method for ESGNN and evaluate the computation cost and the accuracy. Moreover, we investigated the difference of two sampling method. Experimental results show that the memory requirement is reduced remarkably, and the accuracy increase by using ESGNN superior to the nearest neighbor's accuracy. Ensemble model is good for parallel computation because each classifier behave independently. Therefore, this model is useful and practical for data mining. As the future work, we will consider that the comparative study of ESGNN with C4.5 and kernel methods (i.e., support vector machines).

References

[1] W. X. Wen, V. Pang, and A. Jennings. Self-generating vs. self-organizing, what's different? In P. K. Simpson, editor, *Neural Networks Theory, Technology, and Applications*, IEEE Technology Update Series, pages 210–214. IEEE Technical Activities Board, Piscataway, NJ, 1996.

[2] T. Kohonen. *Self-Organizing Maps*. Springer-Verlag, Berlin, 1995.

[3] H. Inoue and H. Narihisa. Improving generalization ability of self-generating neural networks through ensemble averaging. In Takao Terano, Huan Liu, and Arbee L P Chen, editors, *The Fourth Pacific-Asia Conference on Knowledge Discovery and Data Mining*, volume 1805 of *LNCS*, pages 177–180, Kyoto, Japan, Apr. 18–20 2000. Springer-Verlag.

[4] H. Inoue and H. Narihisa. Predicting chaotic time series by ensemble self-generating neural networks. In *The IEEE-INNS-ENNS International Joint Conference on Neural Networks*, volume 2, pages 231–236, Como, Italy, July 24–27 2000.

[5] S. Haykin. *Neural Networks: A comprehensive foundation*, chapter 7. Prentice-Hall, Upper Saddle River, NJ, second edition, 1999.

[6] C.L. Blake and C.J. Merz. UCI repository of machine learning databases, Irvine, CA: University of Califomia, Department of Information and Computer Science, 1998 [http://www.ics.uci.edu/~mlearn/MLRepository.html].

[7] L. Breiman. Bagging predictors. *Machine Learning*, 24:123–140, 1996.

Mercer Kernels and 1-Cohomology

Bernd-Jürgen Falkowski
University of Applied Sciences Stralsund
Zur Schwedenschanze 15
D-18435 Stralsund
Germany
Email: Bernd.Falkowski@fh-stralsund.de

Abstract: For the construction of support vector machines so-called Mercer kernels are of considerable importance. Since the conditions of Mercer's theorem are hard to verify some mathematical results are collected here to provide a complete description of Mercer kernels enjoying certain invariance properties under a group action. Two well-known types of Mercer kernels are seen to arise as special cases of this description and a less well-known one is exhibited. In addition an interesting connection to the Levy-Khinchine formula of probability theory comes to light. These results have, in essence, been known for quite some time but are rather spread out in the literature and hence possibly not so familiar to Neural Network researchers.

Introduction

Recently support vector machines, cf. e.g. [1], have received much attention. In this context Mercer kernels, cf. e.g. [1], p. 35 for Mercer's theorem, which are important building blocks of such machines, have frequently been used. Thus it seems somewhat surprising that, apart from three basic types of kernels, cf. e.g. [5], p. 333, and some construction rules, cf. e.g. [1], pp. 42-44, very little appears to be known about such kernels amongst Neural Network researchers. This is all the more surprising since the conditions of Mercer's theorem are not easily verifiable in general.

Hence it seems worthwhile to collect together some mathematical results which have, in essence, been known for quite some time, cf. [4], [7], [2], to provide a complete description of all Mercer kernels possessing certain invariance properties under a group action: Besides establishing an interesting connection to the Levy-Khinchine formula of probability theory (which roughly speaking describes the logarithms of characteristic functions that are "infinitely divisible") it will be shown that two of the three basic kernel types mentioned above arise as special cases of the general description. In addition an apparently less well-known kernel type will be exhibited.

Note that due to lack of space only sketch proofs will be given and only continous kernels will be considered here. For details and further background information we refer to [2], [3], [4], [7].

1. Conditionally Positive Definite Kernels and Mercer Kernels

Definition 1.1: Given a topological space X and a continuous function $K : X \times X \to C$. Then K is called a <u>positive definite (p.d.) kernel on $X \times X$</u>, if it satisfies

   a) $K(x, y) = \overline{K}(y, x) \; \forall x, y \in X$

   b) $\sum_{i,j=1}^{n} \alpha_i \overline{\alpha}_j K(x_i, x_j) \geq 0 \; \forall \alpha_i \in C, \forall x_i \in X$

If a) above holds and b) holds under the additional condition

c) $\sum_{i=1}^{n} \alpha_i = 0$

then K is called <u>conditionally positive definite (c.p.d.)</u>.

Hence, by remark 3.7 in [1], p. 35, a Mercer kernel is just a real-valued positive definite kernel.

Example 1: If X is a complex Hilbert space with inner product denoted by $<.,.>$, then the kernel K defined by $K(x,y) := <x, y>$ is positive definite.

Example 2: If K is c.p.d. on $X \times X$ then for any $z \in X$ the kernel L_z defined by $L_z(x,y) := K(x,y) - K(x,z) - K(z,y) + K(z,z)$ is p.d.: Given arbitrary points $x_1, x_2, ..., x_n$ and scalars $\alpha_1, \alpha_2, ..., \alpha_n$, set $x_0 = z$ and $\alpha_0 = -\alpha_1 - \alpha_2 - ... - \alpha_n$, then the claim follows from the definition of conditionally positive definite.

Example 2 shows how, given a c.p.d. kernel, one can immediately construct a p.d. kernel. There is, however yet another way to construct p.d. kernels from c.p.d. ones. It is described in essence by the following (well-known) lemma.

Lemma 1.2: Suppose that a p.d. kernel K and a polynomial p with positive coefficients are given. Then $p(K)$ is also a p.d. kernel.

Proof (sketch): Linear combinations of p.d. kernels with positive coefficients are obviously p.d. It remains to show that products of p.d. kernels are again p.d. This may be achieved by considering p.d. matrices $\mathbf{A} := [a_{ij}]$ and $\mathbf{B} := [b_{ij}]$ as covariance matrices of two independent normally distributed random variables $\mathbf{X} := [X_1, X_2, ..., X_n]$ and $\mathbf{Y} := [Y_1, Y_2, ..., Y_n]$ with mean vector zero. Then the matrix $\mathbf{C} := [a_{ij}*b_{ij}]$ is the covariance matrix of $\mathbf{Z} := [X_1*Y_1, X_2*Y_2, ..., X_n*Y_n]$ and hence p.d.   Q.E.D.

Corollary to 1.2: For every c.p.d. L and every $t>0$, the kernel defined as $K(x,y) := exp[tL(x,y)]$ is p.d.

Proof (sketch): Given L consider for an arbitrary z the kernel L_z as defined in example 2. This is p.d. and it follows from lemma 1.2 and continuity that $exp[tL_z]$ is p.d. for every $t>0$. Multiplying this with the p.d. kernel $exp[tK(x,z)]*exp[tK(z,y)]$ and again appealing to lemma 1.2 gives the result.   Q.E.D.

The converse to the corollary is true as well. In fact one has the following lemma.

Lemma 1.3: If the kernel $K_t(x,y) := exp[tL(x,y)]$ is p.d. for every $t>0$, then L is c.p.d.

Proof (sketch): It is easy to see that $(1/t)*(K_t(x,y)-1)$ is c.p.d. for every $t>0$. Hence, taking limits, L is seen to be c.p.d.   Q.E.D.

Thus there clearly is an intimate connection between c.p.d. kernels and p.d. kernels. It is the aim of this article to exhibit a class of c.p.d. kernels enjoying certain invariance properties under a group action on X. In fact they will be described in terms of the 1-cohomology of the group in question. This description will give a systematic characterization of certain well-known Mercer kernels via the above correspondence as well as of other kernels that appear to be rather less well-known.

2. Group Actions and Invariant Positive Definite Kernels

Definition 2.1: Let G be a topological Group with identity e and X be a topological space, as before. G is said to act continously on X if

1. for every fixed $g \in G$, the map $g \to gx$ is one to one and onto.
2. $ex = x$ for all $x \in X$
3. $g_1(g_2x) = (g_1g_2)x$ for all $g_1, g_2 \in G$, $x \in X$
4. $(g,x) \to gx$ is continous
5. for every fixed $g \in G$, the map $x \to gx$ is a homeomorphism of X

A p.d. (c.p.d.) kernel K (L) is said to be invariant under G if $K(gx,gy) = K(x,y)$ $(L(gx,gy) = L(x,y))$.

Theorem 2.2: Let X be a topological space and let G be a group acting continuously on it. Suppose that K is a p.d. kernel on X×X invariant under G.

Then there exists a complex Hilbert space H and a weakly continuous unitary representation $g \rightarrow Ug$ of G in H (i.e. we have that $Ug_1g_2 = Ug_1Ug_2$ and that the map $g \rightarrow$ $<Ug\,v_1,v_2>$ is continuous for every v_1, $v_2 \in H$) and a continous map $v\colon X \rightarrow H$ such that the vectors $v(x)$ span H and

    1. $K(x,y) = <v(x),v(y)>$
    2. $v(gx) = Ug\,v(x)$

Proof (sketch): This is essentially a consequence of the Kolmogorov consistency theorem, for details see e.g. [7], theorem 1.2.   Q.E.D.

Corollary to 2.2: If X= G and G acts on itself by left multiplication, then a G-invariant p.d. kernel K may be viewed as a p.d. function φ in the usual sense. This is obtained by setting $\varphi(g)\colon= K(g,e)$ $(K(g_1,g_2) = K(g_2^{-1}g_1,e))$. On setting v(e) := v_0 in 2.2, a well-known result on p.d. functions due to Gelfand and Raikov (analogous to the GNS-construction for C*-Algebras) follows: Every p.d. continuous function φ may be written as $\varphi(g)= <Ugv_0,v_0>$.

3.   1-Cohomology and Invariant Conditionally Positive Definite Kernels

Definition 3.1: Let G be a topological group acting continuously on the topological space X and let $g \rightarrow Ug$ be a weakly continuous unitary representation of G in a Hilbert space H. A map $\delta\colon X \rightarrow H$ is called a <u>first order cocycle with origin x_0</u> if

    $Ug\,\delta(x) = \delta(gx) - \delta(gx_0)$

If $G = X$, G acts on itself by left multiplication, and $x_0 = e$ then δ is just called a <u>first order cocycle (or just a cocycle) associated with U</u>.

Example 3: Suppose that v is a fixed vector in H. Then a trivial cocycle (<u>coboundary</u>) associated with U may be defined by setting

    $\delta(g)\colon= Ug\,v - v.$

Theorem 3.2: Let G be a topological group acting continuously on the topological space X and let L be a c.p.d. kernel on X×X invariant under G. Then for any fixed $x_0 \in$ X there exists a weakly continuous unitary representation $g \rightarrow Ug$ of G in H and a continuous cocycle $\delta\colon X \rightarrow H$ with origin x_0 such that

    $< \delta(x), \delta(y)> = L(x,y) - L(x,x_0) - L(x_0,y) + L(x_0,x_0)$ for all x,y $\in$ X --------------(*)

Conversely, if $g \rightarrow Ug$ is a weakly continuous unitary representation of G in H and δ is a continuous cocycle with origin x_0 and values in H and if L is a kernel such that (*) holds, then L is a G-invariant continuous c.p.d. kernel.

Proof: See Corollary 1.4, theorem 3.4, and remark 3.5 in [7].

Corollary to 3.2: Taking again X= G and G acting on itself by left multiplication, then a G-invariant c.p.d. kernel L may be viewed as a c.p.d. function $\psi(g)\colon=L(g,e)$ in the usual sense and by 3.2 all c.p.d. functions arise in this manner. If $\psi(e) = 0$ (i.e. ψ is normalized) then it satisfies

    $< \delta(g), \delta(h)> = \psi(h^{-1}g) - \psi(g) - \psi(h^{-1})$ for all g,h $\in$ G

From this it immediately follows that

    $Re\,\psi(g) = -1/2< \delta(g), \delta(g)>.$

Hence it becomes clear that G-invariant continuous c.p.d. kernels are completely described by the 1-cohomology of G. As a consequence, using the corollary to 1.2 and example 2, a large class of Mercer kernels is completely described in this manner.

Of course, the 1-cohomology of groups is by no means easy to compute in general. However, for R^n, a case of particular interest, it is completely known: R^n, being an abelian group with respect to addition, is a type 1 group, cf. [6] for further information. Hence every weakly continuous unitary representation may be written as a direct integral (the continuous analogue of a direct sum) over irreducible representations (1-dimensional characters) in an essentially unique manner. A similar statement is true for the associated cocycles, cf. [7], p.83 for the details. The cocycles associated with irreducible representations of R^n are easy to describe: If the representation is a non-trivial character, then the associated cocycle can only be a coboundary. If, on the other hand, the representation is trivial (the identity) then an associated cocycle must be a homomorphism, cf. [7] p.94. Since an arbitrary weakly continuous unitary representation of R^n can be written as a direct sum of a trivial representation and a representation which does not contain the trivial representation the mentioned results may be summarized in the following theorem.

Theorem 3.3: Let U be an arbitrary weakly continuous unitary representtion of R^n in a Hilbert space H. Then there exists a continuous homomorphism η: $G \to H$, a measure space (Ω, μ) and a measurable map χ: $\Omega \to \hat{G}$ (the character group of R^n, which is of course isomorphic to R^n itself), such that a first order cocycle associated with U may be written as

$$\delta(x) = \eta(x) + \int c(\omega)[<\chi(\omega),x> - 1]\, d\mu(\omega) \quad \text{for } x \in R^n.$$

Here $<\chi(\omega),x>$ denotes the value of the character $\chi(\omega)$ at the point x and the meaning of the other symbols has just been discussed.

Corollary to 3.3: Every real-valued c.p.d. function ψ on R^n may be written as

$$\Psi(x) = -1/2<Ax,x> + \int |c(\omega)|^2 [Re <\chi(\omega),x> - 1]\, d\mu(\omega) \text{ -----------------------}(**)$$

where A is a positive definite matrix.

Proof: This follows from 3.3 and the corollary to 3.2.   Q.E.D.

Remark 1: The corollary to 3.3 contains an abstract version of the well-known Levy-Khinchine formula, which describes the logarithms of characteristic functions of infinitely divisible probability measures. In fact the measure in question can be characterized in rather more detail, see e.g. [7], p. 99.

Remark 2: By varying the parameters in the corollary to 3.3 the well-known (and some not so well-known) Mercer kernels may be derived.

4. Some Mercer Kernels

We use the corollary to 3.3 to derive various Mercer kernels.

4.1 The Inner Product Kernel

Taking the second summand in (**) equal to zero and the Matrix A = I gives as c.p.d. function

$$\Psi(x) = -1/2<x,x>.$$

Applying example 2 gives (with K(x,y) = $\Psi(x-y)$, $z = 0$)

$$L_0(x,y) := <x,y>$$

is positive definite and hence a Mercer kernel, as is well-known.

4.2 The Gaussian Kernel

The same choice of parameters in (**) and consequently the same c.p.d. function gives when combined with the corollary to 1.2 the Gaussian kernel $G(x,y) := \exp$ -t$<x-y,x-y>$ as positive definite and hence a Mercer kernel for any t>0. Again this is well-known.

4.3 Kernels derived from Coboundaries

Note that $\exp$ -t$<x,x>$ - 1 = $1/[2^n*(\pi t)^{n/2}] \int [\exp(i<x,s>) - 1] \exp[-<s,s>/4t]\, ds$ and thus that this defines a c.p.d. function derived from coboundaries. Hence we are led to Mercer kernels using the methods of 4.1 and 4.2, which must be quite well-known as well.

4.4 Subordinate Kernels

Of course positive multiples of the kernels derived in 4.3 may be added for various non-negative values of t without changing the positive definiteness. Now it is well-known, cf. [4], p. 188, that there exist measures μ_α on $[0, \infty)$ such that $\int (1 - \exp\text{-}tr^2)d\mu_\alpha(t) = |r|^\alpha$, for $0 < \alpha < 2$. And hence, taking limits, it clearly follows that $\Psi(x) = -|<x,x>|^\alpha$ is c.p.d. From this then one can derive Mercer kernels as in 4.1 and 4.2.

4.5 Other Kernels

Clearly other kernels may be obtained by using an explicit version of Ψ as described in 3.3. An interesting worked out example may be found in [4], p. 135.

5. Conclusion

The results given above have, in essence, been known for quite some time. Seeing that they do not seem to be well-known amongst Neural Network researchers (even Wahba in [9] does not mention them) it still seems worthwhile to point them out.

6. References

[1]   Cristianini, N.; Shawe-Taylor, J.: An Introduction to Support Vector Machines and other Kernel-Based Learning Methods. Cambridge University Press, (2000)

[2]   Erven, J.; Falkowski, B.-J.: Low Order Cohomology and Applications. Springer Lecture Notes in Mathematics, Vol. 877, (1981)

[3]   Falkowski, B.-J.: Levy-Schoenberg Kernels on Riemannian Symmetric Spaces of Non-Compact Type. In: Probability Measures on Groups, Ed. H. Heyer, Springer Lecture Notes in Mathematics, Vol. 1210, (1986), pp. 58-67

[4]   Gangolli, R.: Positive Definite Kernels on Homogeneous Spaces. In: Ann. Inst. H. Poincare B, Vol. 3, (1967), pp. 121-225

[5]   Haykin, S.: Neural Networks, a Comprehensive Foundation. Prentice-Hall, (1999)

[6]   Mackey, G.W.: Induced Representations of Groups and Quantum Mechanics. Benjamin, (1968)

[7]   Parthsarathy, K.R.; Schmidt, K.: Positive Definite Kernels, Continuous Tensor Products, and Central Limit Theorems of Probability Theory. Springer Lecture Notes in Mathematics, Vol. 272, (1972)

[8]   Schölkopf, B.; Burges, J.C.; Smola, A.J. (Eds.): Advances in Kernel Methods, Support Vector Learning. MIT Press, (1999)

[9]   Wahba, G.: Support Vector Machines, Reproducing Kernel Hilbert Spaces, and Randomized GACV. In [8].

KES '01
N. Baba et al. (Eds.)
IOS Press, 2001

Simulation Based Optimization Using SVM and GA

Hirotaka Nakayama and Koji Washino

Department of Information Science and Systems Engineering, Konan University
8-9-1 Okamoto Higashinada-ku Kobe Japan 658-8501
E-mail:nakayama@konan-u.ac.jp

Abstract. In many practical engineering design problems, the form of objective function is not given explicitly in terms of design variables. Under this circumstance, it usually takes a lot of time to obtain the value of objective function by some analysis such as structural analysis, fluid mechanic analysis, and so on. In order to make the number of analyses as few as possible, we suggest a method by which optimization is performed in parallel with predicting the form of objective function. In this paper, support vector machine (SVM) is employed in predicting the form of objective function, and genetic algorithms (GA) in searching the optimal value of the predicted objective function.

1. Introduction

Our aim in this paper is to optimize objective functions whose forms are explicitly unknown in terms of design variables. In many engineering design problems under this circumstance, it takes a lot of time to obtain the value of objective function by some complicated analysis of large scale such as structural analysis or other analysis. Under this situation, the number of necessary analyses should be as few as possible. So, we suggest a method consisting of two stages predicting by SVM and optimizing by GA while the some stop conditions are not satisfied. The most important thing in this method is how to get a good approximation of the objective function based on as few training data as possible, that is, how to add effective training data as new training data.

2. Support Vector Machine(SVM) for Regression

Support Vector Machine(SVM), based on the statical learning theory, was developed by V.Vapnik originally for classification problems. But it is possible to adopt to solve regression problems. The given training data (x_i, y_i) $(i = 1, ..., l)$ in the input space are mapped into a high dimensional feature space in order to be capable of providing good generalization ability by nonlinear transformation function $\phi(x)$ such as radial basis function.

$$\phi(x) = [\ exp(-\frac{\|x - x_1\|^2}{r^2}), ..., exp(-\frac{\|x - x_l\|^2}{r^2})\ ]$$

Here, r can be given by $r = d_{max}/\sqrt[b]{a * b}$, where d_{max} is the maximum distance between training data, a is the dimension of training data in the input space, and b is the number of training data. The regression function in a high dimensional feature space is expressed by

$$f(x, w) = w * \phi(x).$$

The regression formulation for the SVM uses ε-insensitive loss function called Vapnik's loss function. The linear ε-insensitive loss function $L^\varepsilon(\boldsymbol{x}, y, f)$ is defined by

$$L^\varepsilon(\boldsymbol{x}, y, f) = |y - f(\boldsymbol{x})|_\varepsilon = max\,(0, |y - f(\boldsymbol{x})| - \varepsilon),$$

where f is a real-valued function. Let γ denote the margin which measures the amount by which the training and test accuracy differ, and ϑ denote test accuracy. A training point is considered as a mistake if its accuracy is less than $\vartheta - \gamma$. That is, any training data lying outside a band of size $\pm(\vartheta - \gamma)$ around the approximate function are considerd to be training mistakes. It is immediate that the margin slack variable $\xi((\boldsymbol{x}_i, y_i), f, \vartheta, \gamma)$ satisfies

$$\xi((\boldsymbol{x}_i, y_i), f, \vartheta, \gamma) = L^{\vartheta - \gamma}(\boldsymbol{x}_i, y_i, f)$$

which uses an ε-insensitive loss function with $\varepsilon = \vartheta - \gamma$, if the loss function is compared with the margin slack vector $\xi(\boldsymbol{S}, f, \vartheta, \gamma)$. The margin slack vector of a training set $\boldsymbol{S} = ((\boldsymbol{x}_1, y_1), ..., (\boldsymbol{x}_l, y_l))$ with a function f, and loss margin γ

$$\boldsymbol{\xi} = \xi(\boldsymbol{S}, f, \gamma) = (\xi_1, ..., \xi_l)$$

contains the margin slack variables

$$\xi((\boldsymbol{x}_i, y_i), f, \vartheta, \gamma) = \xi_i = max(0, |y_i - f(\boldsymbol{x}_i)| - (\vartheta - \gamma))$$

In our regression, therefore, the sum of the linear ε-insensitive losses

$$\frac{1}{2}\|\boldsymbol{w}\|^2 + C\sum_{i=1}^{l} L^\varepsilon(\boldsymbol{x}_i, y_i, f)$$

should be minimized. The equivalent primal optimization problem is as follows.

$$\frac{1}{2}(\boldsymbol{w} \cdot \boldsymbol{w}) + C\sum_{i=1}^{l}(\xi_i + \acute{\xi}_i) \;\to\; MIN$$

subject to constraints

$$\begin{cases} y_i - \sum_{i=1}^{l} w_i * \phi(\boldsymbol{x}_i) \leq \varepsilon + \acute{\xi}_i \\ \sum_{i=1}^{l} w_i * \phi(\boldsymbol{x}_i) - y_i \leq \varepsilon + \xi_i \\ \xi_i \,,\; \acute{\xi}_i > 0 \;(i = 1, ..., l) \end{cases}$$

where ε controls the width of insensitive zone. If the deviation between the actual and predicted value is less than ε, the regression function is not considered to be in error. C, given large number, affects trade-off between complexity and the training error.

3. How to Select Additional Data

Until some stop conditions are satisfied, we add some data in order to improve the approximation of objective function. We add two training data at each additional learning in order to add both global information and local information. The first additioal training data for giving local information is taken from neighborhoods of the current optimal point. Let the neighborhood S be given by a square whose center is the current optimal point having the length of a side l. The length of l is given by $l = l_0 * \frac{1}{C_\# + 1}$. $C_\#$ is the number counting how many training data have already been in the square S_0 whose center is also the current optimal point with the fixed length of a side l_0. The second additional training data for giving global information is taken from a part in which the existing learning data are sparse. A part with sparse existing data can be found by the following steps: First, a certain number of data $p_i(i = 1, ..., N_{rand})$ are generated randomly. Denote by d_{ij} the distance between random data p_i and the existing learning data $q_j(j = 1, ..., N)$. Second, select the shortest k distances $d_{ij}(j = 1, ..., k)$ for each p_i, and sum up these k distances. Finally, take p_t which maximizes $D_i = \sum_{j=1}^{k} d_{ij}(i = 1, ..., N_{rand})$ as a second additional training data in a part in which the density of training data is low.

The algorithm is summarized as follows:

Step.1   Predict the form of objective function by SVM.

Step.2   Estimate an optimal point for the predicted objective function by GA.

Step.3   Count $C_\#$ which is the number of optimal points appeared in S_0 so far.

Step.4   If $C_\# \geq C_0$, terminate the iteration. Otherwise add new two training data, and go to Step.1.

4. GA for Optimizing Predicted Objective Functions

We employ GA to optimize the predicted objective function, because it is simple and easy to use. It is also able to provide an approximate solution to the global optimum. We apply BLX-α method which can treat continuous design variables in a simple way. It uses continuous values of design variables as codes of individuals as they are. In crossover, children are generated randomly inside a hyper box including their parents. It has been observed that BLX-α method without mutation is effective in problems with continuous design variables.

5. A Numerical Example

Consider an example given by

$$f(x_1, x_2) = 10 * exp(-0.01(x_1 - 10)^2 - 0.01(x_2 - 15)^2)sinx_1$$
$$(0 \leq x_1 \leq 15, \ 0 \leq x_2 \leq 20)$$

This function has a maximum value $f = 9.5585$ at $x_1 = 7.8960$ and $x_2 = 15.0000$. The five training data, $(x_1, x_2) = (0, 0), (15, 0), (0, 20), (7.5, 10), (15, 20)$ are taken for initial training data as shown in Figure 1. Set $C = 20000$, $e = 0.5$ in SVM, and *population* $= 10$, *generation* $= 200$ in GA, and $l_0 = 4.0$, $C_0 = 20$, $N_{rand} = 20$, $k = 3$. The results of simulation are shown in table 1. Figure 2 and 3 are intermediate process of the predicted function in simulation 1. Figure 4 shows the final stage: training data are shown by:○, the correct optimal point by:□, and the predicted optimal point by: ×.

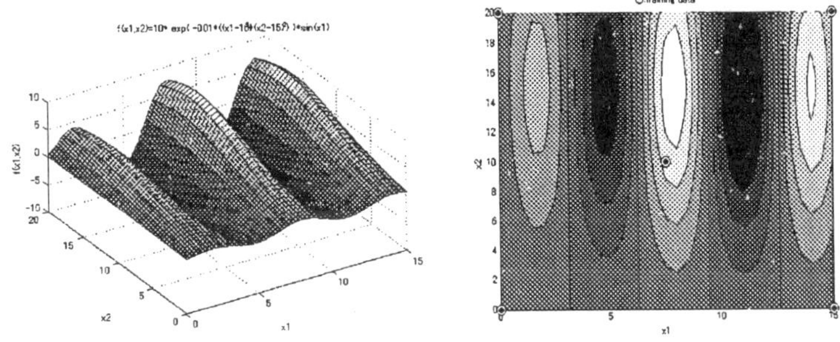

$$x_1 = 7.8960, \ x_2 = 15.0000, \ f(x_1, x_2) = 9.5585$$

Figure 1 : Example

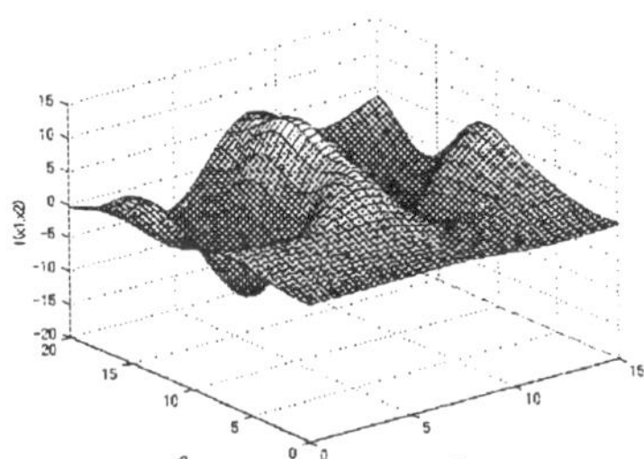

Figure 2   : Result of simulation
(25 training data)

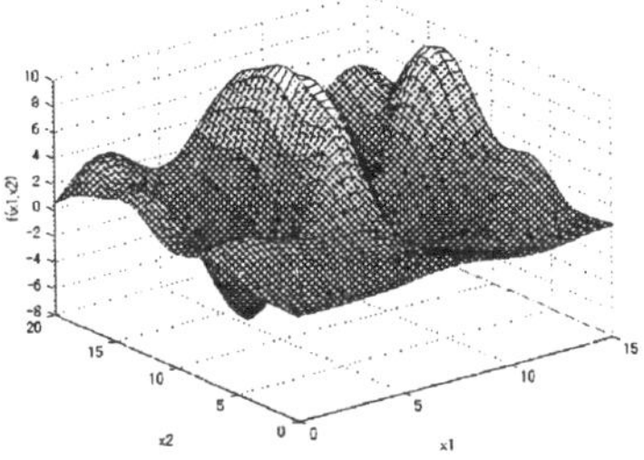

Figure 3   : Result of simulation
(35 training data)

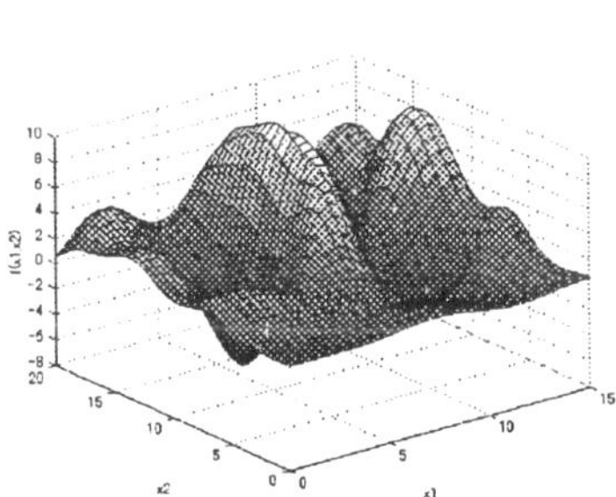

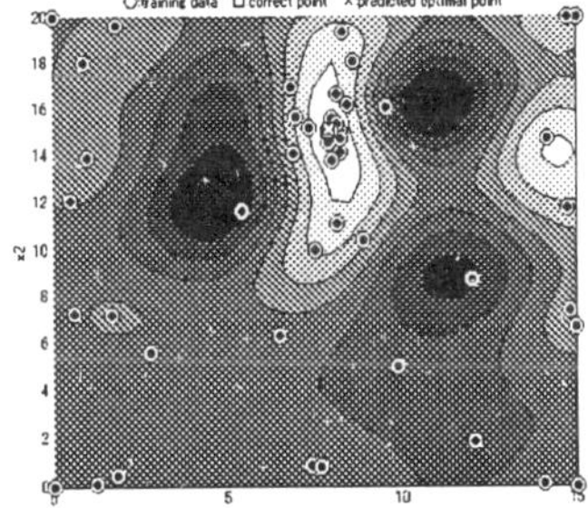

$$x_1 = 7.9265, \ x_2 = 15.1034, \ f(x_1, x_2) = 9.5529$$

Figure 4 : Result of simulation (49 training data)

Table 1 : The result of simulations by suggested method

	#analysis	x_1	x_2	$f(x_1, x_2)$
Theoretical value	–	7.8960	15.0000	9.5585
1	49	7.9265	15.1034	9.5529
2	63	7.6640	15.5812	9.2672
3	77	7.7341	14.4875	9.4066
4	63	7.7846	16.7767	9.2030
5	71	7.8056	15.6532	9.4782
average	64.6	–	–	–

For the sake of comparison, the result obtained by only GA is given in Table 2. The population is 10 which is the same as in the proposed method. When the top 10 individuals with high ranks among parents and children converge within a given fixed range, the calculation is stopped. Table 2 shows the total number of generated individuals (represented by #analysis) until the calculation is stopped. The symbol $*$ shows the case converging to a local maximum point different from the exact solution. In this table, one may see that the average number of evolution of objective function in GA is almost twice as the suggested method to get similar precision.

Table 2 : The result of simulations with only GA
(Necessary analysis corresponding to training data)

	#analysis	x_1	x_2	$f(x_1, x_2)$
1	100	8.008	14.678	9.4869
2	60	7.980	13.635	9.3532
3	100*	14.093	13.851	8.3383
4	90	8.000	12.650	8.9946
5	80	7.902	14.937	9.5580
6	780	6.912	14.634	5.8333
7	70	7.876	14.845	9.5542
8	70*	1.762	14.754	4.9773
9	60*	7.675	6.310	4.9773
10	130	8.066	13.864	9.2966
average	154	-	-	-

6. Concluding Remarks

In optimizing objective functions which are not given exlicitly in terms of design variables, we suggested a method which uses SVM and GA. Although we did not mention in this paper, we compared the regression ability of SVM to that of RBF networks separetly. It has been observed from our numerical experiments that both methods have almost the same ability in regression. Utilization of support vector in selecting additional training data will be subjected to future research.

References

[1] Nello Cristianini, John Shawe-Taylor : An Introduction to Support Vector Machines and other Kernel-based learning method, Cambridge, 2000

[2] Hirotaka Nakayama, Masao Arakawa, Rie Sasaki : Optimization with Implicitly Known Objective Function Using RBF Networks and Genetic Algorithms, Proc. of ICANNGA, pp.387-390, 2001

Genetic Algorithm with Self-Adaptive-Inversion Operator

Nidapan SUREERATTANAN

Computer Science and Information Management,
School of Advanced Technologies,
Asian Institute of Technology,
P.O. Box 4, Klong Luang, Pathumthani 12120,
Thailand
nidapan@cs.ait.ac.th

Abstract. Genetic algorithm (GA) has been known as a classical search algorithm for a large-space data. It works so well in the vast majority of applications. However, a problem concerning to selective pressure still remains. It is a challenge topic for many researchers. High selective pressure can speed up the convergence as well as cause to lost of diversity in a population. In this paper, we introduce a self-adaptive-inversion operator to give a survival chance for the worse chromosomes even high selective pressure is applied. Results of applying the proposed GA are presented and compared to the results of applying the conventional GA, on the optimization functions and real-world data.

1 Introduction

Genetic algorithm (GA) [1][2][3][4][5][6] is a search method simulated from the theory of natural evolution. It was developed by Holland [4] in the early 1960s. Recently, it becomes the well known search algorithm in wide use. The algorithm has received consideration over the years, as a population-based search method, for many applications [1][7][8][9][10][11][12]. GA requires a scalar measure of chromosome's fitness as a guidance for searching solutions. Based on a bias chance of survival, according to a high selective pressure, the better chromosomes always occur in dominating proportion whereas the worse are more likely to die off. Subsequently, a loss of population diversity can be occured when the chromosomes are too alike or identical. The event is a cause of premature convergence. However, a low selective pressure is absolutely apart from wasting time. To avoid a complete loss of population diversity even a high selective pressure is applied, the self-adaptive-inversion operator is proposed in this study. The operator performs a survival chance for the worse chromosomes. This approach has been implemented on a set of optimization functions [13][14], and a set of real-world data in feature selection [15][16][17][18] problem. The experiments illustrate the performance of the proposed GA on both sets comparing to the performance of the conventional GA.

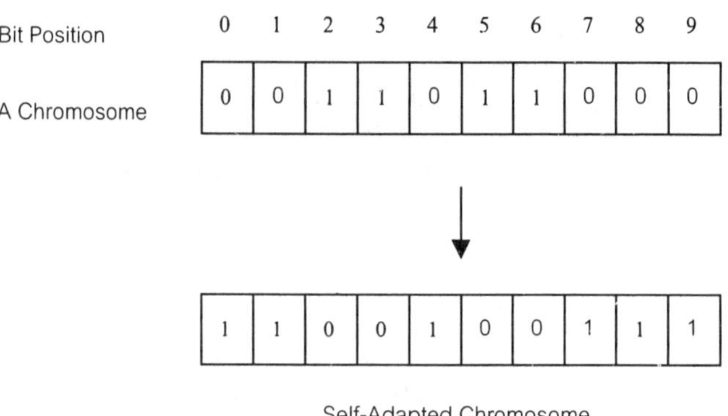

Figure 1: Self-Adaptive-Inversion Operator: An Example

2 Proposed GA

Keeton [19] said that natural selection not only plays a creative role in leading to new gene combinations and in giving direction to evaluation, it also plays an extreme conservative role. The key concept of the proposed GA mainly concerns to a conservation of the worse chromosomes, which are likely to die off. Learning behaviors of living things motivates the interest to develop the self-adaptive-inversion operator. In nature, organisms learn to perform self-adaptation for survival competition. Opposite action is a kind of self-adaptive learning behaviors. By our proposed self-adaptive-inversion operator, there is an attempt to extend a survival chance for the worse chromosomes. The operator simulates opposite action by inverse of all gene values on the worst chromosome, at each generation, from '0' to '1' or vice versa (See the example of self-adaptive-inversion operator in Fig. 1). After evaluating its fitness, the survival chance is re-determined whereby the higher the fitness, the higher the survival probability.

3 Experiments

The experiments attempt to evaluate the performance of the conventional GA and the proposed GA (GA_S), on the optimization functions and real-world data for feature selection problem. In order to evaluate the capabilities of different strategies, the parameters are defined as followings:

- *Population size:* $\ln(2^n)$

- *Crossover rate: 0.6*

- *Mutation rate: 0.001*

where n is a length of string representing a search space (However, the minimum population size is 10). The binary string representation is used in all experiments.

3.1 Evaluated Performance on Optimization Functions

Table 1 (modified from [20]) summarizes all functions used. Table 2 illustrates the performance of the GA and the GA_S on the test functions. For function F_1, F_3, F_4, and F_5, GA_S can find the optimized solution at less generation than GA. Although both methods cannot find the optimized solution for function F_2 within 5,000 generations, GA_S still illustrates the better performance than GA in solution found and generation used as well. The execution time to return the best solution of different methods is shown in Table 3. For all functions, the execution time used by GA_S is more satisfactorily than the execution time used by GA.

Table 1: Standard Set of Test Functions

Function	Search Space	
	Bounds	Resolution
$F_1(x) = \sum_{i=1}^{3}(x_i)^2$	$-5.12 \leq x_i \leq 5.12$	$\Delta x_i = 0.01$
$F_2(x) = 100 * (x_1^2 - x_2)^2 + (1 - x_1)^2$	$-2.048 \leq x_i \leq 2.048$	$\Delta x_i = 0.001$
$F_3(x) = \sum_{i=1}^{5}[x_i]$	$-5.12 \leq x_i \leq 5.12$	$\Delta x_i = 0.01$
$F_4(x) = \sum_{i=1}^{30} ix_i^4 + GAUSS(0,1)$	$-1.28 \leq x_i \leq 1.28$	$\Delta x_i = 0.01$
$F_5(x) = \left[\frac{1}{500} + \sum_{j=1}^{25} \frac{1}{j+\sum_{i=1}^{2}(x_i - a_{ij})^6}\right]^{-1}$	$-65.536 \leq x_i \leq 65.536$	$\Delta x_i = 0.001$

Table 2: Comparative Performance: GA vs GA_S

Function	Minimum Value	GA		GA_S	
		Solution	Generation	Solution	Generation
F_1	0	0	429	0	366
F_2	0	1.002	4134	0.001	2756
F_3	-30	-30	834	-30	218
F_4	0	0	1885	0	1319
F_5	1	1	92	1	48

3.2 Evaluated Performance on Feature Selection Problem

The experiment is conducted to compare the effectiveness of GA_S and GA in feature selection problem. In this study, the proportion of training set is 50% of its entire set. However, the maximum is set to 500 for any huge data set. All data sets are randomly divided into training and testing set. The training set is used for the genetic search whereas the testing set is used for measuring the performance of the selected solution. The predefined maximum number of generation, 20, is the stopping criteria used. The fitness function using multiple correlation coefficient [8] is used for guiding the genetic search. Table 4 and Table 5 illustrate that GA_S is faster than GA on Satellite and Texture data. In term of accuracy, GA_S is also more satisfactorily than GA on Cleveland Heart, Satellite, and Texture data. Although GA can reduce more features than GA_S on Satellite data but the feature set selected by GA_S can be the better representative for classification.

Table 3: Execution Time Comparison on Function Optimization

Function	Time Required (hh:mm:ss)	
	GA	GA_S
F1	00:00:01	00:00:01
F2	00:00:15	00:00:11
F3	00:00:11	00:00:02
F4	00:14:09	00:09:16
F5	00:00:01	00:00:01
Total	00:14:37	00:09:31

Table 4: Performance Evaluation of GA on Feature Selection Problem

Data (Feature: Accuracy)	Selected Feature	Accuracy	Execution Time (hh:mm:ss)
Cleveland Heart (13: 81.52)	8	82.18	00:00:02
E. Coli (7: 84.23)	7	84.23	00:00:04
Iris (4: 93.33)	1	96.00	00:00:02
Satellite (36: 78.10)	17	79.05	00:08:12
Texture (40: 76.55)	22	78.24	00:12:02
Wisconsin Breast Cancer (9: 96.14)	4	96.28	00:00:02

Table 5: Performance Evaluation of GA_S on Feature Selection Problem

Data (Feature: Accuracy)	Selected Feature	Accuracy	Execution Time (hh:mm:ss)
Cleveland Heart (13: 81.52)	8	83.50	00:00:02
E. Coli (7: 84.23)	7	84.23	00:00:04
Iris (4: 93.33)	1	96.00	00:00:02
Satellite (36: 78.10)	19	79.54	00:04:47
Texture (40: 76.55)	22	78.89	00:09:22
Wisconsin Breast Cancer (9: 96.14)	4	96.28	00:00:02

4 Conclusion

In this paper, the self-adaptive-inversion operator is proposed for GA. The operator is originated for a survival opportunity of the worse chromosomes. The experimental studies illustrate that the proposed GA can speed up the convergence on the set of optimization functions. The proposed GA can also achieve in real-world feature selection problem. The overall performance of the proposed GA is more satisfactorily than the overall performance of the conventional GA in all experiments in this study.

References

[1] Y. Davidor, Genetic Algorithms and Robotics. *Robotics and Automated Systems*. Singapore: World Scientific, 1991.

[2] M. Gen and R. Cheng, Genetic Algorithms and Engineering Design. A Wiley-Inter Science Publication, 1997.

[3] D. E. Golberg, Genetic Algorithms in Search, Optimization, and Machine Learning. Addison–Wesley, New York, 1989.

[4] J. H. Holland. Outline for a Logical Theory of Adaptive Systems, *Journal of the Association for Computing Machinery* 3 (1962) 297–314.

[5] C. G. Langton, Artificial Life: An Overview. MIT Press, 1995.

[6] M. Mitchell, An Introduction to Genetic Algorithms. MIT Press, 1996.

[7] K. Kristinsson and G. A. Dumont, System Identification and Control Using Genetic Algorithms, *IEEE Transactions on Systems, Man, and Cybernetics* **22** (1992) 1033–1046.

[8] N. Chaikla and Q. Yulu, Feature Selection using Domain Relationship with Genetic Algorithms, *Knowledge and Information Systems Journal* **1** (1999) 257–268.

[9] N. Morad and A. Zalzala, Genetic Algorithms in Integrated Process Planning and Scheduling, *Journal of Intelligent Manufacturing* **10** (1999) 169–179.

[10] J. K. Parker and D. E. Goldberg, Inverse Kinematics of Redundant Robotics Using Genetic Algorithms, *Proceedings 1989 IEEE International Conference on Robotics and Automation*, IEEE Computer Society, Washington, D.C., 1989, pp. 271–276.

[11] C. S. Pattichis and C. N. Schizas, Genetic–Based Machine Learning for the Assessment of Certain Neuromuscular Disorders, *IEEE Transactions on Neural Networks* **7** (1996) 427–439.

[12] T. P. Runarsson and M. T. Jonsson, Genetic Production Systems for Intelligent Problem Solving, *Journal of Intelligent Manufacturing* **10** (1999) 181–186.

[13] K. A. De Jong, An Analysis of the Behavior of a Class of Genetic Adaptive Systems, *Ph.D. Thesis*, University of Michigan, Michigan, 1975.

[14] J. J. Grefenstette, Optimization of Control Parameters for Genetic Algorithms, *IEEE Transactions on Systems, Man, and Cybernetics* **SMC-16** (1986) 122–128.

[15] M. Dash and H. Liu, Feature Selection for Classification, Elsevier Science, 1998. URL: http://www-ast.elsevier.com/ida/browse/0103/ida00013/article.htm

[16] A. K. Jain and B. Chandrasekaran, Dimensionality and Sample Size Considerations. In: P. R. Krishnaiah and L. N. Kanal (eds.), Pattern Recognition in Practice **2** (1982) 835–855, North–Holland.

[17] W. S. Meisel, Computer–Oriented Approaches to Pattern Recognition, Academic Press, New York, 1972.

[18] J. T. Tou and R. C. Gonzalez, Pattern Recognition Principles, Addison–Wesley, United State of America, 1974.

[19] W. T. Keeton, Elements of Biological Science, W.W. Norton and Company, New York, 1969.

[20] L. Booker, Improving Search in Genetic Algorithm, Genetic Algorithm and Simulated Annealing, Morgan Kaufmann, San Mateo, California, USA, 1987, pp. 61–73.

KES '01
N. Baba et al. (Eds.)
IOS Press, 2001

Using Centripetal Force to Solve SAT by Lagrange Programming Neural Network

Masahiro NAGAMATU and Makio HOSHIURA
Kyushu Institute of Technology, Kitakyushu, Fukuoka 804 Japan
Tel, Fax : +81-93-884-3257, E-mail : nagamatu@brain.kyutech.ac.jp

Abstract. We proposed an artificial neural network called LPPH for the satisfiability problem (SAT) of propositional calculus. The dynamics of the LPPH is based on the Lagrangian method and have the following properties: 1) it is not trapped by any point which is not the solution of the SAT, and 2) when it comes near a solution it converges to the solution. Experimental results show that the LPPH can find the solution more efficiently than already proposed combinatorial algorithms even if it is numerically simulated on conventional computers. The dynamics of the LPPH consists of the differential equations of the neuron outputs and the weights of connections between neurons. In this paper we introduce centripetal force to the dynamics, and investigate the effectiveness of the force for solving the SAT.

1. Introduction

The satisfiability problem (SAT) of propositional calculus is an important basic problem in many fields of computer science. However the SAT is a famous NP-complete problem, and no efficient algorithm is known. There are two types of algorithms for the SAT, complete ones and incomplete ones. The complete ones can decide whether the given SAT problem has a solution or not, while incomplete ones cannot decide when the given problem has no solution. However, when the size of the given problem becomes large, complete algorithms cannot solve in general, while incomplete ones can find the solution efficiently as long as there is a solution.

Algorithms using neural network are generally incomplete ones. The dynamics of Hopfield type neural network[1] is a gradient descent one, and inevitably, trapped by local minima. Nagamatu[2] proposed a Lagrange programming neural network with polarized high-order connections (LPPH). For the basic version of the LPPH, the following two properties are proved theoretically[3]. 1) Every equilibrium point of the dynamics is a solution of the continuous version of the SAT, called CONSAT, and *vice versa*. 2) Almost all equilibrium points have some kind of asymptotical stability. These properties are used to solve the CONSAT. After starting from randomly generated initial point, it moves around in the state space without getting trapped by any point which is not the solution of the CONSAT, and when it comes near a solution it converges to the solution. Above two properties do not exclude the possibility that it moves around forever. However experimental results show that such a case does not occur and the LPPH can solve the SAT effectively [4].

Several revisions of the LPPH are proposed to improve the efficiency[5]. In this paper we introduce "centripetal force" to the LPPH dynamics. This force works so that it makes the value of each variable gathers to the center of 0(false) and 1(true). Experimental result show, with respect to the speed of searching solution, the improvement is obtained for most problems, and there are problems for which more then 10 times improvement is obtained.

2. Satisfiability Problem

Let $X=\{x_1,x_2,...,x_n\}$ be a set of Boolean variables. A Boolean expression in conjunctive normal form (CNF) is a conjunction of clauses:

$$E = C_1 \wedge C_2 \wedge ... \wedge C_m,$$
$$C_r = L_{r1} \vee L_{r2} \vee ... \vee L_{rl_r} \quad r = 1,2,...m,$$

where L_{rk} is the kth literal of the rth clause, and l_r is the number of literals in the clause C_r. Satisfiability problem (SAT) for an expression E is stated as follows:

$$\text{(SAT)} \quad \text{find } \boldsymbol{x}$$
$$\text{such that} \quad \boldsymbol{x} \text{ satisfies } C_r \quad r = 1,2,...,m,$$
$$\boldsymbol{x} \in \{0,1\}^n.$$

Without loss of generality, we can make the following assumption:

Assumption 1: No variable appears more than once in each clause.

The SAT is a discrete valued problem. Here we will convert the SAT into an arithmetic continuous valued problem. Let us consider $\boldsymbol{x}=(x_1,x_2,...,x_n)$ to be a vector of arithmetic continuous valued variables. For each $i=1,2,...,n$, and $r=1,2,...,m$, a function $g_{ir}:[0,1]^n \rightarrow [0,1]$ is defined as follows:

$$g_{ir}(\boldsymbol{x}) = \begin{cases} x_i & \text{if } x_i \text{ appears in } C_r \\ & \quad \text{as a negative literal,} \\ 1-x_i & \text{if } x_i \text{ appears in } C_r \\ & \quad \text{as a positive literal,} \\ 1 & \text{otherwise.} \end{cases}$$

Corresponding to C_r ($r=1,2,...,m$) a function $h_r:[0,1]^n \rightarrow [0,1]$ is defined as follows:

$$h_r(\boldsymbol{x}) = \prod_{i=1}^{n} g_{ir}(\boldsymbol{x}).$$

The continuous valued satisfiability problem (CONSAT) is stated as follows:

$$\text{(CONSAT)} \quad \text{find } \boldsymbol{x}$$
$$\text{such that} \quad h_r(\boldsymbol{x}) = 0 \quad r = 1,2,...,m,$$
$$\boldsymbol{x} \in [0,1]^n.$$

It is obvious that the SAT and the CONSAT are equivalent if we consider x_i such that $0<x_i<1$ in solutions of the CONSAT to be Boolean "don't care" value.

3. Lagrange Programming Neural Network LPPH

A neural network called LPPH (*Lagrange programming neural networks with polarized high-order connections*)[2] was proposed for solving the CONSAT. In the LPPH neuron i corresponds to variable x_i. Hereafter let x_i denotes output of neuron i. Each connection corresponds to the clause. The connection r, which corresponds to clause C_r, connects neurons which correspond to variables included in C_r. Each neuron has a polarity for each connection. If variable x_i is in C_r as a positive (negative) literal, the polarity of neuron i for the connection r is negative (positive, respectively). Each connection r has a weight $w_r>0$.

Fig.1 shows an example of the LPPH for the following CNF:

$$E = (\bar{x}_1 \vee \bar{x}_2 \vee x_3) \wedge (\bar{x}_1 \vee x_2) \wedge (x_1 \vee \bar{x}_3).$$

In the figure circles represent neurons, and squares represent connections. If neuron i is negative for connection r, the output of neuron i to the connection r is converted to $1-x_i$. The input to neuron i from connection r is $- \partial h_r(x)/\partial x_i$.

Let $w=(w_1,w_2,\ldots,w_m)$. Dynamics of the LPPH are defined as follows:

$$\frac{dx_i}{dt} = -x_i(1-x_i)\frac{\partial F(x,w)}{\partial x_i}, \quad i=1,2,\ldots,n. \tag{1}$$

$$\frac{dw_r}{dt} = h_r(x), \quad r=1,2,\ldots,m, \tag{2}$$

$$\text{where } F(x,w) = \sum_{r=1}^{m} w_r h_r(x).$$

The differential equations (1) and (2) can be considered to be a kind of Lagrangian method for the following optimization problem:

(OPT)　　minimize 0　(objective function has no meaning),

subject to　$h_r(x) = 0, r = 1,2,\ldots,m,$

$$x \in [0,1]^n.$$

For the LPPH we have the following properties:

Property 1[3]: For any $x \in [0,1]^n$ and $w \in (0,\infty)^m$, $(x;w)$ is an equilibrium point of the LPPH if, and only if, x is a solution of the CONSAT. Hence the dynamics does not trapped by any point which is not the solution of the CONSAT.　　□

Property 2[3]: Almost all equilibrium points are asymptotically stable. Hence when the dynamics comes near some solution, it converges to the solution.　　□

From experiments it is known that the speedup for finding solutions can be achieved when we use the following differential equations instead of (1) and (2) [4].

$$\frac{dx_i}{dt} = -b(x_i)\frac{\partial F(x,w)}{\partial x_i}, \quad i=1,2,\ldots,n. \tag{3}$$

$$\frac{dw_r}{dt} = -\alpha w_r + h_r(x), \quad r=1,2,\ldots,m, \tag{4}$$

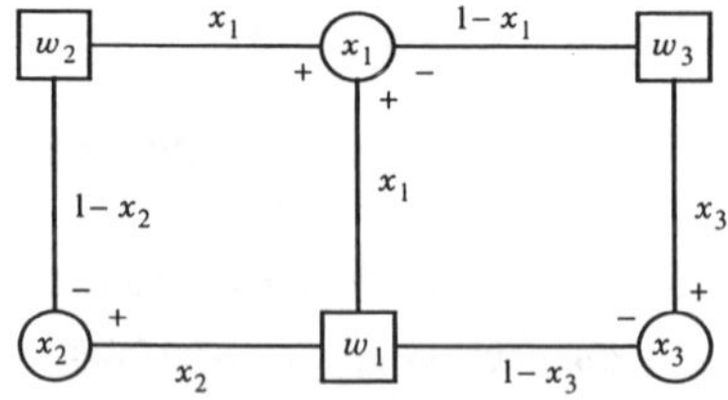

Fig.1. An example of the LPPH.

$$\text{where } b(x_i) = \begin{cases} 1, & 0 < x_i < 1, \\ 0, & \text{otherwise,} \end{cases}$$

$$\alpha \geq 0 : \text{decay factor of } w_r\text{'s.}$$

4. LPPH with Centripetal Force

In this section we introduce an additional force called "centripetal force" to the LPPH dynamics as follows:

$$\frac{dx_i}{dt} = b(x_i)\left(-\frac{\partial F(x,w)}{\partial x_i} + \beta(0.5 - x_i)\right), \quad i = 1,2,\ldots,n. \tag{5}$$

The coefficient β represents the strength of the centripetal force. The role of the term $\beta(0.5 - x_i)$ is to makes the value of x_i close to 0.5.

Fig. 2 shows experimental results. In the experiment, the LPPH dynamics of (5) and (4) is simulated by Euler method. The step size of the Euler method is set to considerably large to accelerate the solution search. In Fig.2 the horizontal axis indicates the value of β, and the vertical axis indicates the ratio (the number of steps of Euler method to find the solution when the value of horizontal axis is used for β)/(the number of steps when $\beta=0$). Namely the vertical axis indicates the ratio of reduction of CPU time when the centripetal force is used. EXP00,...,EXP19 are the 3-SAT problems which are randomly generated and have 50 variables and 215 clauses. The CPU time and the number of steps to find the solution when $\beta=0$ are shown in Table 1.

From Fig. 2 we can see that for all problems the speedup is achieved when the best value of β is used for each problem. For some problems the speedup factor of about 10 is achieved. Although the best value of β depends the problem, if the value of the range from 0.5 to 2 is used, good results are obtained for most problems.

5. Conclusion

The centripetal force is introduced to the LPPH dynamics to accelerate the solution search. The speedup factor of about 10 is obtained for some problems. Now we are investigating why the centripetal force is useful. One reason is conjectured that: if the forces from other variables are very small for some variable x_i (this means that any requirement does not exist for the value of x_i), it is desirable that x_i has value close to 0.5 to prepare for the later requirement for its value. We are planning experiments to prove this conjecture.

References

[1] J. J. Hopfield and D. W. Tank, 'Neural' computation of decisions in optimization problems, *Biological Cybernetics*, vol. 52, pp. 141-152, 1985.

[2] M. Nagamatu, and T. Yanaru, Lagrange programming neural networks with polarized high-order connections for satisfiability problems of propositional calculus, *IIZUKA'94, The 3rd International Conference on Fuzzy Logic, Neural Nets and Soft Computing*, pp. 233-235, August 1994.

[3] M. Nagamatu, and T. Yanaru, On the stability of Lagrange programming neural networks for satisfiability problems of propositional calculus, *Neurocomputing*, vol. 13, no.2-4, pp.119-133, 1996.

[4] M. Nagamatu, and T. Yanaru, Lagrangian method for satisfiability problems of propositional calculus, *ANNES'95*, pp.71-74, November 1995.

[5] M Nagamatu and T Yanaru, Solving SAT by Lagrange Programming Neural Network with Long and Short Term Memories, in "Information Modelling and Knowlege Bases XI," Edited by Eiji Kawaguchi, et al., IOS Press, pp.289-301, 2000

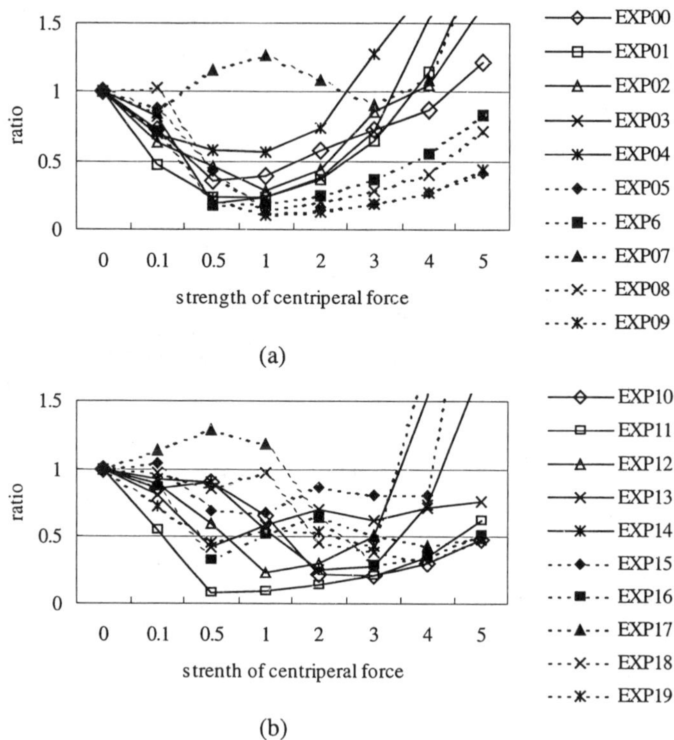

(a)

(b)

Fig.2. Time reduction obtained by the centripetal force.

Table 1. CPU time and the number of steps for each problem.

Problem	CPU(sec)	#step	Problem	CPU(sec)	#step
EXP00	0.008	36.2	EXP10	0.025	124.8
EXP01	0.006	29.5	EXP11	0.020	101.1
EXP02	0.010	51.4	EXP12	0.020	101.6
EXP03	0.009	48.0	EXP13	0.045	226.1
EXP04	0.008	38.9	EXP14	0.032	163.1
EXP05	0.019	93.0	EXP15	0.066	342.1
EXP06	0.012	59.2	EXP16	0.026	131.4
EXP07	0.021	110.8	EXP17	0.044	226.1
EXP08	0.028	142.6	EXP18	0.082	413.0
EXP09	0.034	165.8	EXP19	0.138	707.5

Incremental Update of Large Itemsets by Sampling Techniques

Hsin-Tai Yang and Shie-Jue Lee
Department of Electrical Engineering
National Sun Yat-Sen University
Kaohsiung 804, Taiwan

Abstract. In the paper, we focus on the problem of finding large itemsets in a large database that is updated incrementally. We present an algorithm that scans the full database as seldom as possible by sampling the original database into a much smaller database and then use this sampled database to estimate the possibility of the sizes of itemsets. As a result, subsets of candidate itemsets can be pruned efficiently from the search space. Experimental results have shown that our method performs efficiently with middle-sized sampling rates.

1 Introduction

In recent years, the problems of data mining have attracted a lot of interest in the database community [1, 2, 3]. In particular, among the data mining techniques, discovery of association rules becomes a major research topic. The interest arises from the fact that many organizations have collected massive digitized data that might be very informative and useful for their business. For instance, supermarkets collect electrically the purchase data of their customers from receipts. And the managers might want to know what products are often bought together in a basket. The goal of mining association rules is to analyze these large data sets and to discover their associations that are useful for decision support.

The size of the data can be regarded as the most influential factor to execution time in mining association rules because with m items, there are 2^m combinations of potential associations. In general, association algorithms require multiple passes over the whole database. As a result, how to reduce the frequency of access to the large database by pruning as many candidates as possible is a critical issue.

In this paper, we focus on the problem of finding large itemsets in a large database that is updated incrementally. We present an algorithm that scans the whole database as seldom as possible. The proposed method is based on the ideas of negative/positive borders and database sampling which help finding large itemsets and guiding the pruning strategy to terminate the mining process earlier, respectively.

2 Problem Description

Agrawal et al [4, 5] defined the problem of discovering association rules in databases. Let $I = \{I_1, \ldots I_m\}$ be a set of items (attributes), and DB be a set of transactions. Let X, called an itemset, be a set of items in I. An itemset X is called a k-itemset if it contains k items in I. The support of an itemset X in DB, $sup_{DB}(X)$, is the number of transactions in DB that contain X. An itemset is called a large itemset if $sup_{DB}(X)/|DB|$ exceeds a minimum fractional support threshold, *minsup*, and a small itemset otherwise. An association rule is an implication of the form $X \Rightarrow Y$, where $X \subset I, Y \subset I$, and $X \cap Y = \emptyset$.

The problem of finding association rules can be decomposed into two phases. The first phase is to generate all combinations of items with fractional support above a certain threshold, *minsup*. The second phase is to use the large itemsets to generate associations. In general, the first phase is very time-consuming, and is the main concern of this paper.

In practice, there are situations that large databases are updated incrementally. That is, a small amount of transactions *db* are added into the whole large database at a time. Suppose we are given an original database DB with a set of large items L under the minimum support, *minsup*, and an added database *db* with a set of large item L' under the same minimum support, the incremental update problems can be defined as finding the collection L_u of large items in $DB \cup db$ under the minimum support, *minsup*.

3 Our method

Cheung's approach [7] uses a so-called level-wise technique to discover large itemsets. The approach is based mainly on the observation that if an itemset is small, then its superset could not be large. As a result, the candidates they generate in the higher levels are only the ones whose subsets are sure to be large. However, this kind of approach has the drawback that when a large itemset contains many items, all the subsets of the underlying itemset should be explored before the large itemset itself being discovered. For example, let $L = \{A, B, C, D, E\}$ be a large itemset. Level-wise algorithms need to check all of its subsets, i.e., $\{A\}, \{B\}, \ldots, \{B, C, D, E\}$, before discovering L to be large.

To solve the problem described above, we propose a heuristic algorithm that can estimate the possibility of some itemset to be large in advance. With this look-ahead property, redundancy can be avoided. Our algorithm for finding large itemsets incrementally is based on the negative border algorithm proposed in [6], and can be mainly divided into two phases. In what follows, we first describe the concept of the negative/positive borders and then discuss our approach in detail.

3.1 Concept of Negative/Positive Borders

The concept of the negative border can be defined as follows [6].
Definition: Let R be a set of items and $P(R)$ denote the power set of R. Given a collection $S \subseteq P(R)$ of itemsets, closed with respect to the set inclusion relation, the negative border NB(S) of S consists of the minimal itemsets $X \subseteq R$ not in S.
Consider an example. Let $R = \{A, B, \ldots, F\}$ and assume the collection of itemsets

$$S = \{A\}, \{B\}, \{C\}, \{F\}, \{A, B\}, \{A, C\}, \{A, F\}, \{C, F\}, \{A, C, F\}.$$

The negative border of this collection is $NB(S) = \{\{B, C\}, \{B, F\}, \{D\}, \{E\}\}$.

Note that, conceptually, any set in the negative border will not be a large itemset, but all its subsets are. Accordingly, we can define the positive border such that any set in it will be large, but all its supersets are small.

3.2 Pruning with a Sampling Technique

Assume that the set L of the large itemsets in the original DB is given, and the added database and updated database are denoted as *db* and $DB \cup db$, respectively. In the beginning, we sam-

Algorithm
Input: A database DB with the increment db, a minimum support threshold $minsup$.
Output: The collection of large itemsets in $DB \cup db$.
Method:

> L:=the large itemsets in DB
> L':=the large itemsets in DB - the losers found by using the technique in [7]
> DB_s:=Sample DB with a sampling rate α
> Repeat until $PB(L')$ unchanged./*Shrinking*/
> > For each l in $PB(L')$
> > > if $(sup_{DB_s \cup db}(l) < minsup)$
> > > > $L' := L' - \{l\}$
> > >
> > > Endif
> >
> > Endfor
>
> Endrepeat
> Repeat until $NB(L')$ unchanged. /*Expanding*/
> > For each l in $NB(L')$
> > > if $(sup_{DB_s \cup db}(l) >= minsup)$
> > > > $L' := L' \cup \{l\}$
> > >
> > > Endif
> >
> > Endfor
>
> Endrepeat
> L' is the collection of all large itemsets

Table 1: The algorithm for finding large itemsets.

ple DB with a sampling rate α to construct a new database DB_s. After that, the sizes of itemsets can be estimated from the two smaller databases DB_s and db. At this time, the derived result is not exactly the final one. Next, two operations are performed on the positive border and negative border, respectively: *Shrink* and *Expand*. The operation *Shrink* is to narrow down the positive border by removing large itemsets from it. Those itemsets removed are estimated to be large in $DB_s \cup db$ but are actually small in $DB \cup db$. Similarly, the operation *Expand* is to expand the negative border by adding itemsets to the negative border that are estimated to be minimally small in $DB_s \cup db$ but is actually large in $DB \cup db$. After the two operations are performed, the large itemsets for $DB \cup db$ can be discovered.

3.3 Algorithm

The algorithm for discovering large itemsets in our method is shown in Table 1.

3.4 An Example

Let the number of items be 5, denoted as $\{A, B, C, D, E\}$, $minsup = 0.5$, and sampling rate $\alpha = 0.5$. Suppose we are given databases DB and db as shown in Figure 1, and the collection L of the large itemsets in DB. Note that the transactions in the databases are represented by binary sequences, with '1' in the first place denoting the presence of A and '0' denoting the absence of A, and so on. For example, 01010 denotes the transactions DB.

1. Prune the large itemsets in $L = \{B\}$ that will be no longer in $DB \cup db$. This step can be easily done by only scanning the added database [7]. The resulting itemsets are denoted as $L' = \{B\}$.

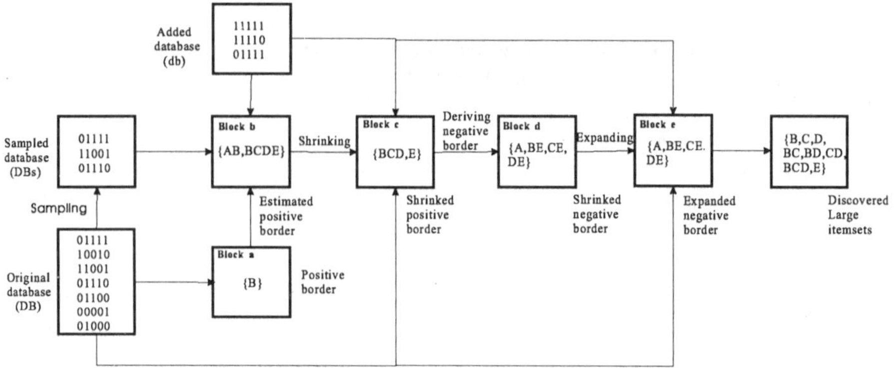

Figure 1: An example.

2. Find the set $PB(L')$ of the positive border with respect to L'. We have $PB(L') = \{\{B\}\}$.

3. Sample DB with the sampling rate $\alpha = 0.5$ to get the sampled database DB_s, as shown in the block a of Figure 1.

4. Use DB_s to guide the possible new negative border of the large itemsets in the updated database $DB \cup db$ to get the estimated positive border. The result is shown in the block b of Figure 1.

5. Perform the *Shrink* operation to skip out the itemsets that no longer belong to the positive border. The result is shown in the block c of Figure 1.

6. Derive the negative border from the resulting positive border. The result is shown in the block d of Figure 1.

7. Perform the *Expand* operation to add the itemsets that are outside the negative border. The result is shown in the block e of Figure 1.

8. Discover the large itemsets, i.e., $\{\{B\}, \{C\}, \{D\}, \{B, C\}, \{C, D\}, \{B, D\}, \{B, C, D\}, \{E\}\}$.

4 Experiments and Discussions

To verify the effectiveness of our approach, we conduct experiments to show how the sampling rates affect the execution time. In our experiment, the numbers of transactions in the original database and the added database are 10,000 and 2,000, respectively. The minimum support, $minsup$, is 0.3 and the sampling rates are changed from 10% to 100%. The experimental result is shown in Figure 2. In the figure, the x-axis represents the sampling rates and the y-axis the execution time with rescaling. The figure reveals that when the sampling rates are small, the execution time is large. This is because the sampled database does not correctly model the original database. As the sampling rates increase constantly, the execution time decreases because the sampled database models the original database well. The sampling rates continue to increase, and the sampled database is much more alike with the original one and the overhead comes out because two large databases, the sampled and the original ones, need to be processed simultaneously.

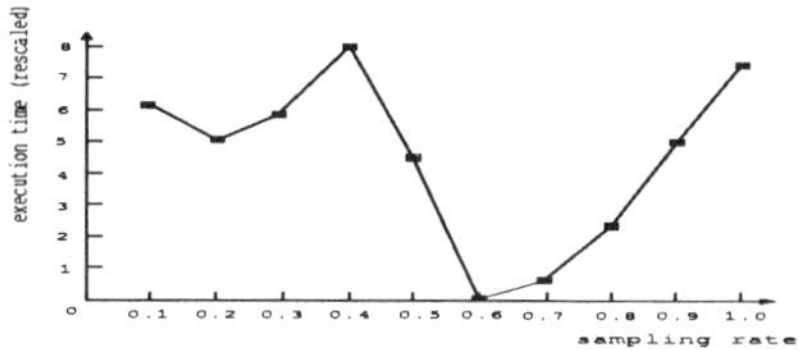

Figure 2: Experiments on sampling rates

Our method performs well, compared with other methods. In particular, our approach works pretty well in the case when the newly added large itemsets contain a lot of items in contrast to the original ones. Under this circumstance, we can estimate correctly the size of the underlying itemsets to be large. Therefore, our method executes much more efficiently than other approaches, such as level-wise algorithms.

5 Conclusions

We have described a method that can find large itemsets in a large database that is updated incrementally. Our method scans the full database as seldom as possible by sampling the original database into a much smaller database and then use this sampled database to estimate the possibility of the sizes of itemsets. As a result, subsets of candidate itemsets can be pruned efficiently during the search of the whole database.

References

[1] G. Piatetsky-Shapiro and W. J. Frawley in *Knowledge Discovery in Databases*, (AAAI Press, Menlo Park, CA), 1991.

[2] U. M. Fayyad, G. Piatetsky-Shapiro, P. Smyth, and R. Uthurusamy in *Advances in Knowledge Discovery and Data Mining*, (AAAI Press, Menlo Park, CA), 1996.

[3] T.-P. Hong, C.-Y. Wang, and Y.-H. Tao, "Incremental data mining based on two support thresholds," in *Fourth International Conference on Knowledge-Based Intelligent Engineering Systems and Allied Technologies.*, pp. 436–439, 2000.

[4] R. Agrawal, T. Imielinski, and A. Swami, "Mining association rules between sets of items in large databases," in *Proceedings of ACM SIGMOD'93*, pp. 207–216, May 1993.

[5] R. Agrawal and R. Srikant, "Fast algorithms for mining association rules," in *Proceedings of VLDB'94*, pp. 487–499, 1994.

[6] H. Mannila and H. Toivonen, "On an algorithm for finding all interesting sentences," in *Cybernetics and Systems, Volume II, The Thirteen European Meeting on Cybernetics and Systems Research*, pp. 973–978, 1996.

[7] D. W. Chueng, V. T. Han, J. Ng, and C. Wong, "Maintenance of distributed associated rules in large databases: An incremental updating technique," in 12th *IEEE International Conference on Data Engineering*, pp. 106–114, 1996.

KES '01
N. Baba et al. (Eds.)
IOS Press, 2001

Mining Similar Association Rules from Transaction Databases

Shyue-Liang WANG, Chun-Yin KUO and Tzung-Pei HONG
Department of Information Management, I-Shou University
Kaohsiung, 84008, Taiwan, R.O.C.

Abstract. This paper introduces the problem of mining similar association rules from transaction databases. Mining of association rules among items in transaction databases has been studied extensively in recent years. Various forms of data types such as quantitative and categorical data in databases as well as domain knowledge such as taxonomies have been considered in the mining algorithms. In this work, we present a data mining algorithm to discover similar association rules where similarity relations are assumed among database items. The proposed algorithm is a generalization of the apriori mining algorithm of association rules. The results developed here can be applied to cross-marketing, customer segmentation and prediction.

1. Introduction

Data mining, also known as knowledge discovery in databases, means a process of nontrivial extraction of implicit, previously unknown and potential useful information from databases [4]. It has been recognized as a new area for database research and received considerable interests in recent years.

The problem of mining association rules was introduced in [1]. Let $I=\{i_1, i_2, ..., i_m\}$ be a set of literals, called itmes. Given a set of transactions D, where each transaction T is a set of items such that $T \subseteq I$, an association rule is an expression $X \Rightarrow Y$ where $X \subseteq I$, $Y \subseteq I$, and $X \cap Y = \phi$. Many algorithms for efficient mining of association rules have been proposed. Most of these algorithms identify the relationships among transactions using binary values. However, transactions with quantitative values, categorical values or even fuzzy values are commonly seen in real–world applications. In addition, domain knowledge such as taxonomies (*is-a* hierarchies) over the transaction items may be available. Various generalizations of association rules and algorithms to handle these data types and constraints are thus developed [1][2][3][5][6]. For example, if taxonomy such as Jackets *is-a* Outerwear *is-a* Clothes exists, then a generalized association rule might be *Outerwear* $\Rightarrow$ *shoes*. An example for quantitative association rule might be *(Age, 30..39) and (Married, Yes)* $\Rightarrow$ *(Car, 2)*. An example for a fuzzy association rule might be *(Age, old)* $\Rightarrow$ *(Wealth, high)*.

In this work, we consider a new type of constraints on transaction items where similarity relations exist among the items. Introduced in [7], similarity relation has been used for describing how similar two elements from the same domain are, as the name implies. For example, CPU products from different companies such as Intel and AMD are similar to each other due to their functionalities and specifications. Another example is that certain printers from HP and EPSON might be similar. Similarity relations may then be defined on the CPUs and printers from different companies or categories. Given a set of transactions and a set of similarity relations on the items, an example of such association rule, which we call similar association rule, might be *(Intel or AMD)* $\Rightarrow$ *(HP or EPSON)* with minimum similarity 0.8. An algorithm based on the apriori mining algorithm [1] to discover similar association rules is proposed.

The rest of our paper is organized as follows. Section 2 reviews the similarity relation. Section 3 gives a formal definition of similar association rule. Section 4 presents the similar associations rule algorithm. A conclusion is given at the end of the paper.

2. Similarity Relation

The concept of a similarity relation is essentially a generalization of the concept of an equivalence relation [7]. It is useful for describing how similar two elements from the same domain are, as the name implies. For a given domain Σ, a similarity relation S is a mapping of every pair of elements in the domain onto the unit interval [0, 1]. More specifically:

Definition: A similarity relation, S, in domain Σ is a binary relation in Σ, such that for any $a, b, c \in \Sigma$, (1) $S(a, a) = 1$, (2) $S(a, b) = S(b, a)$, (3) $S(a, c) \geq \max [\min (S(a, b), S(b, c))]$, for all $b \in \Sigma$.

Example: Assuming that a similarity relation S on the domain CPU is shown in Figure 1. The domain of CPU consists of six elements: I850, I933, A700, A850, C600, C700. The grade of membership $\mu_S(I850, A850)$ is 0.8, meaning that the similarity between the domain elements $I850$ and $A850$ is the degree of 0.8.

	I850	I933	A700	A850	C600	C700
I850	1	0.4	0.4	0.8	0.4	0.4
I933	0.4	1	0.4	0.4	0.4	0.4
A700	0.4	0.4	1	0.4	0.6	0.8
A850	0.8	0.4	0.4	1	0.4	0.4
C600	0.4	0.4	0.6	0.4	1	0.6
C700	0.4	0.4	0.8	0.4	0.6	1

Figure 1. Similarity relation of CPU

For any α in (0, 1), an α-level value (α-cut) on a similarity relation creates an equivalence relation S_α that forms a partition on Σ. Each pair of elements in an equivalence class thus has similar degrees equal to or larger than α. Let π_S^α denote the partition corresponding to the equivalence relation S_α. Clearly, two elements d_1 and d_2 will belong to the same equivalence class of this partition if and only if $S(d_1, d_2) \geq \alpha$. Therefore, the set of all possible domain partitions for a similarity relation S defined on domain Σ is formed as: $DP(S) = \{ \pi_S^\alpha \mid \alpha \in (0, 1) \}$. These partitions are nested in the sense that $\pi_S^{\alpha_1}$ is a refinement of $\pi_S^{\alpha_2}$ iff $\alpha_1 \geq \alpha_2$.

3. Similar Association Rules

Let $I = \{i_1, ..., i_m\}$ be a set of items, D be a set of transactions T_i, $T_i \subseteq I$, $1 \leq i \leq n$. Consider that a set of similarity relations, $\{S_A, S_B, ...\}$, is defined on item sets $A, B, ...$, where $A \subset I$, $B \subset I$, ... For a user-specified minimum similarity level value α, it induces a partition on each similarity relation, $\pi_{S_A}^\alpha$, $\pi_{S_B}^\alpha$, ... etc. For each partition, there exists a set of equivalence classes $A_{i_k}^\alpha$ that contains item $i_k \in A$.

A similar association rule is an implication of the form $XY \Rightarrow Z$, where $X, Y, Z \subset I$ and X,

Y, Z are subsets of different equivalence classes, induced by a similarity level value α. The rule $XY \Rightarrow Z$ holds in the transaction set *D* with confidence *C* if *C%* of transaction in *D* that supports *XY* also supports *Z*. The rule $XY \Rightarrow Z$ has support *S* if *S%* of transaction in *D* supports $X \cup Y \cup Z$.

The problem of mining similar association rules can be stated as follows. Given a set of transactions *D* and a set of similarity relations on the transaction items, the problem of mining similar association rules is to find all similar association rules that have support, confidence and similarity greater than user-specified minimum support (called *min-sup*), minimum confidence (called *min-conf*), and minimum similarity (called *min-α*)

4. The Similar Association Rule Algorithm

This section describes the proposed data mining algorithm based on the apriori algorithm [1] to discover similar association rules from transaction data. The proposed mining algorithm first determines the partitions and equivalence classes for each similarity relation according to the user-specified minimum similarity level value *min-α*. The algorithm then calculates the largest *k*-EC (equivalence class) sets that have supports greater than *min-sup*. It then generates similar association rules that have confidence greater than *min-conf*. Details of the proposed mining algorithm are described below.

The similar association rule

INPUT: A set of transactions, $D=\{T_1, T_2, ..., T_n\}$, a set of similarity relations, S_A, S_B, ... on item sets *A, B, ...etc, min-α, min-sup, min-conf.*

OUTPUT: A set of similar association rules.

STEP 1 :   For each given similarity relation S_A, calculate the partition $\pi_{S_A}^{\alpha_A}$ and its equivalence classes $A_{i_k}^{\alpha_A}$, $i_k \in A$, where α_A is the least α value in S_A greater than *min-α.*

STEP 2 :   (a) Find all the candidate 1-EC (Equivalence Class) sets $A_{x_i}^{\alpha}$ by grouping items, in all transactions, belonging to the same equivalence class and put item in C_1.

(b) Calculate the cardinality of each 1-EC set in C_1 for the transaction *D*, denoted by $\| A_{x_i}^{\alpha} \|$

STEP 3 :   Determine the large 1-EC sets by selecting those 1-EC sets with $\| A_{x_i}^{\alpha} \| \geq$ *min-sup* and put them in L_1.

STEP 4 :   Set *r=1*, where *r* is the number of equivalence class sets kept in the current equivalence class sets.

STEP 5 :   Generate the candidate set C_{r+1} from L_r. in a way similar to that of the apriori algorithm[1]. It first joins L_r and L_r assuming that *r-1* EC's in the two EC sets are the same other the other one is different. It then keeps in C_{r+1} the EC sets, which have all their sub-EC sets of *r* sets existing in L_r.

STEP 6 :   (a) Calculate the cardinality of each *(r+1)*-EC set in C_{r+1}

(b) Determine the large *(r+1)*-EC set by selecting those with cardinality greater than *min-sup* and put them in L_{r+1}.

STEP 7 :   IF L_{r+1} is null, then do the next step; otherwise, set *r=r+1* and repeat STEPs 5 to 7.

STEP 8 :   Construct the similar association rules for all large *q*-EC's with $\{s_1, s_2, ..., s_q\}$, *q≥2*, using the following sub-steps:

(a) Form all possible similar association rules as follows:
$$s_1 \wedge ... \wedge s_{w-1} \wedge s_{w+1} \wedge ... \wedge s_q \rightarrow s_w, \text{w=1 to } q.$$
(b) Calculate the confidence values of all similar association rules using the

formula:

$$\frac{\parallel S_1 S_2 ... S_q \parallel}{\parallel S_1 S_2 ... S_{w-1} S_{w+1} ... S_q \parallel}$$

STEP 9 : Check the output rules with confidence values larger than or equal to the predefined *min-conf*.

After STEP 9, the rules output can act as the meta-knowledge for the given transactions.

5. Conclusion

In this work, we have introduced a new type of association rule, called similar association rule, and presented a data mining algorithm to discover this type of rules from a transaction database. The proposed rule is a kind of fuzzy association rule in the sense that similarity relations are permitted on transaction items. The proposed algorithm is based on the apriori algorithm [1] for discovering association rules. More works need to be done on improving the efficiency and scalability of the proposed approach.

References

[1] R. Agrawal, T. Imielinksi and A. Swami, "Mining association rules between sets of items in large database", Proc. of the ACM SIGMOD Conference on Management of Data, Washington DC, May 1993, 207-216.

[2] R. Agrawal, R. Srikant, "Fast Algorithms for Mining Association Rules", Proc. of the 20th Int'l Conference on Very Large Databases, Santiago, Chile, September 1994, 487-499.

[3] T. P. Hong, C.S Kuo, and S.C Chi, "Mining association rules from quantitative data", Intelligent Data Analysis, vol.3, 1999, 363-376.

[4] G. Piategsky-Shapiro, "Discovery, analysis and presentation of strong rules", Knowledge Discovery in Databases, AAAI/MIT press, 1991, 229-248.

[5] R. Srikant, R. Agrawal, "Mining Generalized Association Rules", Proc. of the VLDB conference, Zurich, Switzerland, September 1995.

[6] R. Srikant, R. Agrawal, "Mining Quantitative Association Rules in Large Relational Tables", Proc of the ACM SIGMOD Conference on Management of Data, 1996, 1-12.

[7] L.A. Zadeh, "Similarity relations and fuzzy orderings", Inform Sci., vol. 3, no.1, 1971, 177-200.

KES '01
N. Baba et al. (Eds.)
IOS Press, 2001

Two-phase Data Types Transformation Framework in Data Mining

Mon-Fong JIANG, Shian-Shyong TSENG, Shang-Yi LIAO, Wei-Chou CHEN
Department of Computer and Information Science, National Chiao Tung University
Hsinchu 300, Taiwan, ROC

Abstract. As we know, the data processed in data mining may be obtained from many sources in which different data types may be used. However, no algorithm can be applied to all applications due to the difficulty for fitting data types of the algorithm, so the selection of an appropriate mining algorithm is based on not only the goal of application, but also the data fittability. Therefore, transforming the non-fitting data type into target one is also an important work in data mining, but the work is often tedious or complex since a lot of data types exist in real world. Merging the similar data types of a given selected mining algorithm into a generalized data type seems to be a good approach to reduce the transformation complexity. In this work, a two-phase data types transformation framework including merging and transforming phases is proposed. With the data type transformation framework, the user can select appropriate mining algorithm iterative and interactive for the goal of application without considering the data types.

1. Introduction

In recent years, the amount of various data grows rapidly. Widely available, low-cost computer technology now makes it possible to both collect historical data and also to institute on-line analysis for newly arriving data. Automated data generation and gathering leads to tremendous amounts of data stored in databases. Although we are filled with data, we lack for knowledge. Data mining [1, 2, 3, 5, 7, 8, 9, 10, 15] is the automated discovery of non-trivial, previously unknown, and potentially useful knowledge embedded in databases. Hence, the importance of data mining is increasing.

As we know, each kind of previous data mining methods and algorithms [5, 10] has its own advantages and suitable application domains. However, it is difficult for users to choose a suitable one by themselves without prior knowledge about data mining [15]. Since determining the suitable data mining algorithm may be considered as an iterative and interactive process, series of try-and-error are indispensable and the preprocessing for each mining algorithm will become tedious. Actually, what kind of data mining methods should be applied depends both on the characteristics of the data to be mined and the kind of knowledge to be found through the data mining process [13]. Hence, the types of data stored in databases are the key factors to determine the data mining methods. This causes "the data types fittability problem"[13] for data mining arises. To solve this problem, we have an idea to investigate the relationships between the characteristics of the data to be mined and the characteristics of data mining techniques. With the relationships, we can clearly analyze the data types fittability problem and further determine whether the data types transformation can be performed or not.

In this work, a data type transformation framework including merging and transforming phases is proposed. In the merging phase, a knowledge acquisition process for soliciting data mining algorithms characteristics is executed to construct a mining algorithm characteristic table (MACT) to indicate the relations between the mining algorithms and the data types. According to the MACT, the similar data types will be merged into the generalized ones stored in the data types mapping table (DTMT) by the constructing algorithm. In the transforming phase, the DTMT is then used to convert the original data types of data sources to be mined into the generalized one for the selected mining algorithm. Notice that, the DTMT is constructed once in merging phase and is reusable in the transforming phase, even the different data mining algorithm is selected in the different iteration. As we know, the preprocessing work of data types transformation is often tedious or complex since a lot of data types exist in real world. With the two-phase data type transformation framework, the preprocessing work is finished in the first phase, users only need to determine which kinds of mining algorithms will be used in the following phase. With the data type transformation process, the user can select appropriate mining algorithm just for the goal of application without considering the data types.

2. Related Work

As we know, none of the data mining techniques can be used to solve all the problems in real world. Therefore, users have to know which kind of data mining problems their cases belong to before choosing an appropriate mining algorithm. Because data mining algorithms can usually be applied to the suitable data types, users may transform data types of source data before the selected algorithm has been executed. Therefore, data types transformation can be seen as a preprocessing of data mining. Other preprocessings such as data selection, data cleaning, dimension (attribute) reduction, missing data handling may also need to be performed before running the selected data mining algorithm [4]. Kerbert et al. [14] proposed an Active Template research to create a single, unified environment that a data analyst can use to carry out a knowledge discovery project, and to deliver the resulting solution in the form of an Active Template. In summary, the whole process of data mining is the so-called knowledge discovery in databases (KDD) [5, 6, 8].

Among a variety of data mining algorithms provided, a suitable one should be chosen for solving the data mining problem user specified. However, if the data types transformation is not included in the tool, some mining methods (algorithms) could not be applied to the specified data. For example, the ID tree method is not suitable for numeric data. If it is used to find classification rules from a relation which contains some numeric attributes (i.e. real type), the data types transformation must be performed before running the selected data mining algorithm.

3. Two-phase Data Types Transformation Framework

As we know, the preprocessing of data mining is often tedious or complex since a lot of data types exist in real world. The cost of rerunning the whole data mining process is high. The data types transformation framework we proposed is a two-phase process to solve above problem in conventional data mining process. The tedious preprocessing is done only once in the first phase, and the data types are actually transformed in the second phase.

In the merging phase, the similar data types will be merged into the generalized ones by knowledge acquisition process for data mining algorithms characteristics and by constructing DTMT algorithm. In the transforming phase, according to the DTMT, the original data types of data sources are converted into the generalized one for the selected mining algorithm. Notice that, the DTMT is constructed once in merging phase and is reusable in the transforming phase, even the different data mining algorithm is selected in the different iteration. With our data type transformation framework, the preprocessing work is finished in the first phase, and users only need to determine which kind of mining algorithms will be used in the second phase. The architecture of the two-phase data types transformation framework is illustrated in Figure 3.

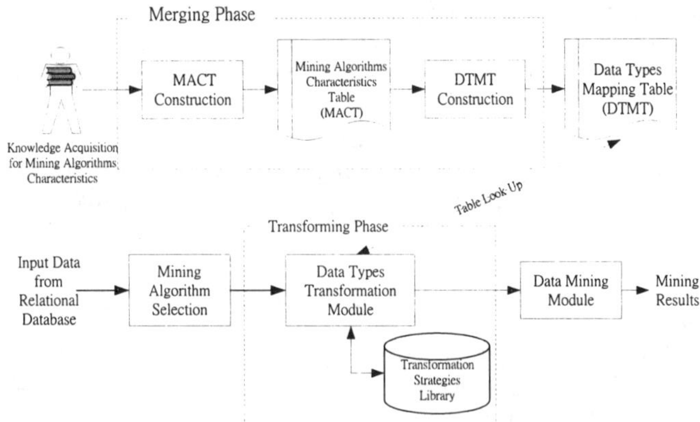

Figure 1: Two-phase data types transformation framework.

According to the architecture showed in Figure 1, the algorithm of two-phase data type transformation is proposed as follows.

Algorithm 3.1: *Two-phase Data Type Transformation Framework*
> **Phase 1:** Merging phase.
>> **Step 1:** Construct MACT by Algorithm 3.2.
>> **Step 2:** Construct DTMT by Algorithm 3.3.
> **Phase 2:** Transforming phase (Algorithm 3.4).
>> **Step 1:** Generalize the original data type according to the DTMT.
>> **Step 2:** Transform the generalized data type into the suitable data type for the selected mining algorithm according to the transformation strategies library.

3.1 Merging Phase

The merging phase of data types transformation framework is to merge the data types with similar characteristics into a generalized one. The first step is to construct MACT by the following algorithm.

Algorithm 3.2: *Constructing MACT*
> **Output:** The MACT which is a two dimensional array storing the relationships of each available data mining algorithm and each data type, where the value of each MACT [data-type, mining-algorithm] entry may be labeled 'Y,' 'N,' or 'X.'
> **Step 1:** Determine the set of data mining algorithms which may be used. Place the name of the algorithms across the top of the array.
> **Step 2:** Determine the available data types.
> **Step 3:** Fill the entry. 'Y' means the data type is suitable to the mining algorithm, 'N' means the data type has no relationship to the mining algorithm, and 'X' means don't care in the case.

In the first two steps of the algorithm, we may understand two main characteristics of the available data mining algorithms: what kind of knowledge to be found and what suitable data type to be used. However, the Step 3 is usually performed by domain experts. In the previous work [13], some data mining algorithms have been collected and tested to know the relationship between the data mining algorithms and corresponding data types. Based upon the relationship, we may decide whether the data type is suitable to the mining algorithm.

In the above algorithm, the MACT is represented by a two dimensional array, the advantage is to simplify the process of knowledge acquisition. The real storage format may use the triples (data-type, mining-algorithm, Y/N), and the absence of a triple would indicate a 'don't care.' According to the MACT, the similar data types can be merged into the generalized ones by the following DTMT constructing algorithm.

Definition 3.1: More Specific
> We say that data type D_2 is more specific than data type D_1, if MACT$[D_2, A_i]$ = MACT$[D_1, A_i]$ or 'X' for every data mining algorithm A_i.

Algorithm 3.3: *Constructing DTMT*
> **Input:** MACT
> **Output:** DTMT, which is a two-fields table indicating the relation between the original data type and the generalized ones.
> **Step 1:** For each two data types D_1 and D_2, if D_1 is more specific than D_2, remove the row of D_1.
> **Step 2:** List and rename all generalized data types with G_i in the first column and fill the corresponding data types which is more specific than G_i in the second column of DTMT.

According to constructed DTMT, the transforming phase then converts the original data types of data sources to be mined into the generalized one for the selected mining algorithm. Notice that, the DTMT only has to be constructed once in merging phase and is reusable in the transforming phase.

3.2 Transforming Phase

After the merging phase, we then transform the generalized data types of the original data into the suitable one. In Algorithm 3.4, we first map the original data types of the input data into the generalized one and then transform the generalized data type into the suitable one according the selected mining algorithm.

Algorithm 3.4: *Transforming Data Types*
> **Input:** Original data types of the input data and DTMT.
> **Output:** Suitable data type for the selected data mining algorithm.
> **Step 1:** Generalize the original data type according to the DTMT.
> **Step 2:** Transform the generalized data type according to the transformation strategies library.

In Step 2, the transformation strategies library should at least include the strategies for transforming each one generalized data type in DTMT into another, but it is difficult to build the library automatically.

Algorithm 3.5: *Constructing Transformation Strategies Library*
> **Input:** DTMT and the selected data mining algorithm.
> **Output:** Transformation strategies library.
> **Step 1:** Determine which kind of generalized data types of DTMT that the original data type belongs to.
> **Step 2:** Select one of the following four suggestions:

(1) From discrete to continue: define distance or similarity using available domain knowledge.

(2) From continue to discrete: clustering on the attribute.

(3) From discrete with numerous values to discrete: reduce the distinct values by generalization on the attributes.

(4) From discrete with numerous values to continue: define similarity using data format.

4. Concluding Remarks

In this paper, we first mentioned the fact that none of data mining algorithms can be applied to all kinds of data types. We then proposed the viewpoint of the iterative and interactive data mining process. After that, we proposed a two-phase data types transformation framework to provide the transformation strategies to reduce the cost of the rerunning in data mining. This work was supported in part by MOE program of Excellence Research under Grant No. 89-E-FA04-1-4.

References

[1] R. Agrawal and J. C. Shafer, "Parallel mining of association rules," IEEE Tran. Knowledge and Data Engineering, Vol. 8, No. 6, pp. 962-969, Dec., 1996.

[2] R. Agrawal, T. Imielinski, and A. Swami, "Database mining: a performance perspective," IEEE Tran. Knowledge and Data Engineering, Vol. 5, No. 6, pp. 914-925, Dec., 1993.

[3] C. Apte and S. Weiss, "Data mining with decision trees and decision rules," Future Generation Computer Systems, Vol. 13, pp. 197-210, 1997.

[4] M. S. Chen, J. Han and P. S. Yu, "Data Mining: An Overview from Database Perspective," In IEEE Trans. on Knowledge and Data Engineering, Vol. 8, No. 6, pp. 866-883, Dec., 1996.

[5] D. W. Cheung, A. W. C. Fu, and J. Han, "Knowledge discovery in databases: A rule-based attribute-oriented approach," Proc. 1994 Int'l Symp. On Methodologies for Intelligent Systems, pp. 164-173, Charlotte, North Carolina, Oct., 1994.

[6] D. W. Cheung, V. T. Ng, A. W. Fu, and Y. Fu, "Efficient mining of association rules in distributed databases," IEEE Tran. Knowledge and Data Engineering, Vol. 8, No. 6, pp. 911-922, Dec., 1996.

[7] U. Fayyad, and P. Stolorz, "Data mining and KDD: Promise and challenges," Future Generation Computer Systems Vol. 13, pp. 99-115, 1997.

[8] J. Han, "Mining knowledge at multiple concept levels," Proc. of 4th Int. Conf. on Information and Knowledge Management, pp. 19-24, Baltimore. Maryland, Nov., 1995.

[9] J. Han, "Data mining techniques," ACM-SIGMOD'96 Conference Tutorial, June, 1996.

[10] J. Han, Y. Cai, and N. Cercone, "Data-driven discovery of quantitative rules in relational database," IEEE Tran. Knowledge and Data Engineering, Vol. 5, No. 1, pp. 29-40, Feb., 1993.

[11] J. Han and Y. Fu, "Dynamic generation and refinement of concept hierarchies for knowledge discovery in databases," Proc. AAAI'94 Workshop on knowledge Discovery in Databases (KDD'94), pp. 157-168, Seattle, Wa, July 1994.

[12]. A. K. Jain and R. C. Dubes, Algorithms for clustering data, Prentice-Hall Inc., pp. 58-89, 1988.

[13] M. F. Jiang, S. S. Tseng, and S. Y. Liao, "Data Types Generalization for Data Mining Algorithms," *IEEE SMC'99 Conference*, Tokyo, Japan, 1999.

[14] R. Kerber, H. Beck, T. Anand, and B. Smart, "Active Templates: Comprehensive Support for the Knowledge Discovery Process," Proceedings of the fourth international conference on knowledge discovery and data mining, pp244-248, 1998.

[15] C. Westphal and T. Blazton, Data Mining Solutions: Methods and Tools for Solving Real-World Problem, Wiley Computer Publishing, 1998.

Web Mining for Browsing Patterns

Tzung-Pei HONG[†], Kuei-Ying LIN[‡] and Shyue-Liang WANG[‡]
[†]*Department of Electrical Engineering, National University of Kaohsiung*
[‡]*Department of Information Management, I-Shou University*
Kaohsiung, Taiwan, R.O.C.

Abstract. In this paper, we use the data mining techniques to discover relevant browsing behavior from log data in web servers. The browsing time of a customer on each web page is used to analyze the retrieval behavior. Since the data collected are numeric, fuzzy concepts are used to process them and to form linguistic terms. A sophisticated web-mining algorithm is thus proposed to find relevant browsing behavior from the linguistic data. Each page uses only the linguistic term with the maximum cardinality in later mining processes, thus making the number of fuzzy regions to be processed the same as the number of the pages. Computational time can thus be greatly reduced.

1. Introduction

Techniques of web mining have recently been requested and developed to increase efficient and effective web retrieval. Cooley et. al. divided web mining into two classes: web-content mining and web-usage mining [6]. Web-content mining focuses on information discovery from sources across the World Wide Web. On the other hand, web-usage mining emphasizes on the automatic discovery of user access patterns from web servers. In the past, several web-mining approaches for finding sequential patterns and user interesting information from the World Wide Web were proposed [2-5]. Chen and Sycara proposed the WebMate system to keep track of user interests from the contents of the web pages browsed. It can thus help users easily search data from WWW [3]. Chen et. al. mined path-traversal patterns by first finding the maximal forward references form log data and then obtaining the large reference sequences according to the occurring numbers of the maximal forward references [2]. Cohen et. al. sampled only portions of the server logs to extract user access patterns, which were then grouped as volumes [4]. Files in a volume could then be fetched together to increase the efficiency of a web server.

Recently, fuzzy set theory has been used more and more frequently in intelligent systems because of its simplicity and similarity to human reasoning [8][9]. The theory has been applied in fields such as manufacturing, engineering, control, diagnosis, economics, among others [10]. Hong et al. proposed a fuzzy mining approach for finding association rules from transactions [7]. In this paper, we propose a novel web-mining algorithm to find linguistic browsing behaviors from data logs on web servers. The browsing time of a customer on each web page is used to analyze the retrieval behavior of a web site. Since the data collected are numeric, fuzzy concepts are used to process them.

2. Review of Agrawal and Srilant 's Mining Approach

Agrawal and Srikant proposed a mining algorithm to discover sequential patterns from a set of transactions [1]. Five phases are included in their approach. In the first phase, the transactions are sorted first by customer ID as the major key and then by transaction time as the minor key. This phase thus converts the original transactions into customer sequences. In the second phase, the set of all large itemsets are found from the customer sequences by comparing their counts with a predefined support parameter α. This phase is similar to the process of mining association rules. Note that when an itemset occurs more than one time in a customer sequence, it is counted once for this customer sequence. In the third phase, each large itemset is mapped to a contiguous integer and the original customer sequences are transformed into the mapped integer sequences. In the fourth phase, the set of transformed integer sequences are used to find large sequences among them. In the fifth phase, the maximally large sequences are then derived and output to users.

3. Notation

The following notation is used in our proposed algorithm:

n: the total number of log data;

m: the total number of files in the log data;

c: the total number of clients in the log data;

n_i: the number of log data from the i-th client, $1 \le i \le c$;

D_i: the browsing sequence of the i-th client, $1 \le i \le c$;

D_{id}: the d-th log transaction in D_i, $1 \le d \le n_i$;

I^g: the g-th file, $1 \le g \le m$;

R^{g_k}: the k-th fuzzy region of I^g, $1 \le k \le |I^g|$, where $|I^g|$ is the number of fuzzy regions for I^g;

v_{id}^g: the browsing duration of file I^g in D_{id};

f_{id}^g: the fuzzy set converted from v_{id}^g;

f_{id}^{gk}: the membership value of v_{id}^g in region R^{g_k};

f_i^{gk}: the membership value of region R^{g_k} in the i-th client sequence D_i;

$count^{gk}$: the scalar cardinality of region R^{g_k};

$max\text{-}count^g$: the maximum count value among $count^{gk}$ values;

$max\text{-}R^g$: the fuzzy region of file I^g with $max\text{-}count^g$;

α: the predefined minimum support value;

λ: the predefined minimum confidence value;

C_r: the set of candidate sequences with r files;

L_r: the set of large sequences with r files.

4. Fuzzy Web Mining for Browsing Patterns

Log data in a web site are used to anallyze the browsing patterns on that site. Many fields exist in a log schema. Among them, the fields *date, time, client-ip* and *file name* are used in the mining process. Only the log data with .asp, .htm, .html, .jva and .cgi are considered home pages and used to analyze the mining behavior. The other files such as .jpg and .gif are thought of as inclusion in the home pages and are omitted. The number of files to be analyzed is thus reduced.

The log data to be analyzed are sorted first in the order of client-ip and then in the order of date and time. The duration of each web page browsed by a client can then be calculated from the time interval between the page and its next page. Since the time durations are numeric, fuzzy concepts are used here to process them and to form linguistic terms. In the log data, each transaction contains only one web page. The mining process can thus be simplified when compared to that for multiple-item transactions in Agrawal and Srilant 's mining approach [1].

The proposed web-mining algorithm then calculates the scalar cardinality of each linguistic term on all the log data, and adopts an iterative search approach to find large itemsets. Each item (web page) uses only the linguistic term with the maximum cardinality in later mining processes, thus making the number of fuzzy regions to be processed the same as the number of original web pages. The algorithm therefore focuses on the most important linguistic terms, which reduces its time complexity. The mining process based on fuzzy counts is then performed to find fuzzy browsing patterns from these large itemsets. The detail of the proposed web-mining algorithm is described as follows.

The web mining algorithm:

INPUT: A server log, a set of membership functions, a predefine minimum support value α.

OUTPUT: A set of linguistic browsing patterns.

STEP 1: Select the transactions with file names including .asp, .htm, .html, .jva .cgi and closing connection from the log data; keep only the fields *date, time, client-ip* and *file-name*. Denote the resulting log data as D.

STEP 2: Transform the client-ips into contiguous integers (called encoded client ID) for convenience, according to their first browsing time. Note that the same client-ip with two closing connections is given two integers.

STEP 3: Sort the resulting log data first by encoded client ID and then by date and time.

STEP 4: Calculate the time durations of the web pages browsed by each encoded client ID from the time interval between a web page and its next page.

STEP 5: Form a browsing sequence D_j for each client c_j by sequentially listing his/her n_j tuples (web page, duration), where n_j is the number of web pages browsed by client c_j. Denote the *d-th* tuple in D_j as D_{jd}.

STEP 6: Transform the time duration v_{id}^g of the file name I^g appearing in D_{id} into a fuzzy set f_{id}^g represented as

$$\left(\frac{f_{id}^{g_1}}{R^{g_1}} + \frac{f_{id}^{g_2}}{R^{g_2}} + \;....\; + \frac{f_{id}^{g_l}}{R^{g_l}} \right)$$

using the given membership functions, where I^g is the g-th file name, R^{g_k} is the k-th fuzzy region of item I^g, f_{id}^{gk} is v_{id}^g's fuzzy membership value in region R^{g_k}, and l is the number of fuzzy regions for I^g.

STEP 7: Find the membership value f_i^{gk} of each region R^{g_k} in each browsing sequence D_i as

$$f_i^{gk} = \underset{d=1}{\overset{|D_i|}{MAX}} f_{id}^{gk},$$

where $|D_i|$ is the number of tuples in D_i.

STEP 8: Calculate the scalar cardinality of each region R^{gk} as:

$$count^{gk} = \sum_{i=1}^{c} f_i^{gk},$$

where c is the number of browsing sequences.

STEP 9: Find $max\text{-}count^g = \underset{k=1}{\overset{l}{MAX}}\left(count^{gk}\right)$,

where $1 \leq g \leq m$, m is the number of files in the log data, and l is the number of regions for file I^g. Let $max\text{-}R^g$ be the region with $max\text{-}count^g$ for file I^g. $max\text{-}R^g$ will be used to represent the fuzzy characteristic of file I^g in later mining processes.

STEP 10: Check whether the value $max\text{-}count^g$ of a region $max\text{-}R^g$, $g = 1$ to m, is larger than or equal to the predefined minimum support value α. If a region $max\text{-}R^g$ is equal to or greater than α, put it in the set of large 1-sequences (L_1). That is,

$$L_1 = \left\{max - R^g \middle| max - count^g \geq \alpha, 1 \leq g \leq m\right\}.$$

STEP 11: If L_1 is null, then exit the algorithm; otherwise, do the next step.

STEP 12: Set $r=1$, where r is used to represent the length of sequential patterns currently kept.

STEP 13: Generate the candidate set C_{r+1} from L_r in a way similar to that in the aprioriall algorithm [1]. Restated, the algorithm first joins L_r and L_r, under the condition that r-1 items in the two itemsets are the same and with the same orders. Different permutations represent different candidates. The algorithm then keeps in C_{r+1} the sequences which have all their sub-sequences of length r existing in L_r.

STEP 14: Do the following substeps for each newly formed (r+1)-sequence s with contents $\left(s_1, s_2, ..., s_{r+1}\right)$ in C_{r+1}:

(a) Calculate the fuzzy value f_i^s of s in each browsing sequence D_i as:

$$f_i^s = \underset{k=1}{\overset{r+1}{Min}} f_i^{s_k} ,$$

where region s_k must appear after region s_{k-1} in D_i. If two or more same subsequences exist in D_i, then f_i^s is the maximum fuzzy value among those of these subsequences.

(b) Calculate the scalar cardinality of s as:

$$count^s = \sum_{i=1}^{c} f_i^s ,$$

where c is number of browsing sequences.

(c) If $count^s$ is larger than or equal to the predefined minimum support value α, put s in L_{r+1}.

STEP 15: IF L_{r+1} is null, then do the next step; otherwise, set $r=r+1$ and repeat STEPs 13 to 15.

STEP 16: Output the maximally large q-sequences, $q \geq 2$, to web-site mangers as browsing patterns.

5. Conclusions

In this paper, we have proposed a novel web-mining algorithm, which can process web-server logs to discover fuzzy sequential browsing patterns among them. The duration of each web page browsed by a client is calculated from the time interval between the page and its next page. Since the time durations are numeric, fuzzy concepts are used here to process them and to form linguistic terms. Each web page uses only the linguistic term with the maximum cardinality in later mining processes, thus making the number of fuzzy regions to be processed the same as the number of original web pages. The algorithm therefore focuses on the most important linguistic terms, which reduces its time complexity. A fuzzy mining process has then been performed to find fuzzy browsing patterns. The mined rules are expressed in linguistic terms, which are more natural and understandable for human beings.

References

[1] R. Agrawal, R. Srikant, Mining Sequential Patterns, The Eleventh International Conference on Data Engineering, 1995, pp. 3-14.

[2] Ming-Syan Chen, Jong Soo Park and Pjilip S. Yu, Efficient Data Mining for Path Taversal Patterns, IEEE Transactions on Knowledge and Data Engineering, Vol. 10, 1998, pp. 209-221.

[3] Liren Chen and Katia Sycara, WebMate: A Personal Agent for Browsing and searching, The Second International Conference on Autonomous Agents, ACM, 1998.

[4] Edith Cohen, Balachander Krishnamurthy and Jennifer Rexford, Efficient Algorithms for Predicting Requests to Web Servers, The Eighteenth IEEE Annual Joint Conference on Computer and Communications Societies, Vol. 1, 1999, pp. 284 –293.

[5] R. Cooley, B. Mobasher and J. Srivastava, Grouping Web Page References into Transactions for Mining World Wide Web Browsing Patterns, Knowledge and Data Engineering Exchange Workshop, 1997, pp. 2 –9.

[6] R. Cooley, B. Mobasher and J. Srivastava, Web Mining: Information and Pattern Discovery on the World Wide Web, The Ninth IEEE International Conference on Tools with Artificial Intelligence, 1997, pp. 558 -567

[7] T. P. Hong, C. S. Kuo and S. C. Chi, A data mining algorithm for transaction data with quantitative values, Intelligent Data Analysis, Vol. 3, No. 5, 1999, pp. 363-376.

[8] A. Kandel, Fuzzy Expert Systems, CRC Press, Boca Raton, 1992, pp. 8-19.

[9] L. A. Zadeh, Fuzzy logic, IEEE Computer, 1988, pp.83-93.

[10] H. J. Zimmermann, Fuzzy Set Theory and Its Applications, Kluwer Academic Publisher, Boston, 1991.

Fuzzy Rule Induction for Classification

Te-Min CHANG and Kun-Hsien CHEN
Department of Information Management, National Yat-Sen University,
Kaohsiung, Taiwan, R.O.C.

Abstract. Nowadays, knowledge discovery in databases has become an important issue to help business acquire its intelligence that assists operational and managerial work. Among many types of knowledge, classification knowledge is widely used. Most classification rules learned by induction algorithms are in the crisp form. Fuzzy linguistic representation of rules, however, is much closer to the way human reasons. The objective of this paper is to propose a method to acquire classification knowledge with fuzzy linguistic descriptions. An experiment is conducted based on several databases to show advantages of this work. The proposed method is justified with good system performance. It can be easily implemented in various business applications on classification tasks.

1. Introduction

Modern industries collect data and store them in databases. Massive raw data are, however, of little use because they can only provide limited information for decision-making. Therefore, technology to transfer data into knowledge becomes essential for business applications. Among many types of knowledge, classification knowledge is widely used. Typical algorithms used to acquire classification knowledge include tree and rule induction (TRI), and artificial neural networks (ANN). Since ANN is deemed as a black box, results from TRI are much easier for users to comprehend [8].

Two types of knowledge representation are commonly seen in TRI techniques: decision trees and "if-then" rules. The most famous algorithm for decision tree structure is ID3 [6] and its successors such as C4.5 [7]. On the other hand, typical algorithms to directly generate if-then rules include AQ [4] series, CN2 [1], HCV [9], and RITIO [10]. This latter approach does not generate decision tree during the induction process. It was shown that it usually outperformed the decision tree approach [1][9][10].

The outputs of rule induction are a set of if-then rules that are often in crisp forms. A crisp rule has the disadvantage that a rule condition has to be exactly matched for firing. Furthermore, such an expression is unnatural to human beings because people used to think and reason in a fuzzily linguistic way. This research is then motivated by using the concept of fuzzy set and fuzzy logic in rule induction to generate linguistic rules for classification tasks.

2. Fuzzy Rule System

In this study, we proposed a four-step method to generate fuzzy rule systems. Step 1 is data preparation where data are collected and preprocessed. The collected data are then partitioned into training data set and test data set. Fuzzy regions and membership functions are determined in step 2. According to the defined fuzzy regions, training data are transformed into fuzzy examples in step 3 and a rule inductive algorithm is employed to generate fuzzy rules in step 4. Detailed descriptions are presented in the following.

<u>Step1. Data Preparation</u>

Data acquired from any specific database comprise a set of examples and each example E_i is in the form as

$$E_i = (X_1, \ldots, X_v, Y_k),$$

where v is the number of input attributes, X_j is the value of the j'th attribute, and Y_k is the output class of E_i. That is, an example E_i contains multi-input attributes and a single-output class. Data usually contain noises if they are collected directly from databases of real cases and thus they should be preprocessed to render meaningful results. Data may be noisy in that (1) they contain wrong attribute values or misclassifications, (2) contradiction is caused by examples with the same attribute values but different classes, (3) examples with missing values lead to incompleteness, and (4) the same examples result in redundancy. In case (1), inductive learning is robust to such errors. Unlike traditional inductive learning, contradictory examples and redundant examples resulted from case (2) and (3) can be retained in the database as long as an appropriate inference mechanism is employed to yield reasonable results. In case (4), missing values are replaced with the most frequently occurring values.

After data are collected, they are partitioned into two data sets: a training data set and a test data set. Training data are used to generate the fuzzy rule system. Test data will be utilized to examine the generalization ability of the generated system.

<u>Step 2. Fuzzy Region and Membership Function Determination</u>

Fuzzy regions of each input attribute are determined by discretizing its domain interval into several regions with each region fuzzified accordingly. An efficient way to discretize the input is to find several possible cut points and perform partition for several times. Fayyad and Irani [2] have shown that with examples sorted in ascending order by their values of a continuous attribute, possible cut points exist in the boundary where two adjacent examples belong to different classes. A cut point with the most information gain is then chosen. This partition is employed iteratively until the user-defined number of intervals is reached.

Each crisp region is then fuzzified with a fuzzy membership function associated with it. Typical shapes of membership functions are triangular, trapezoidal, and bell-shaped. Without losing the generality, the membership function used in our study is of trapezoidal shape.

<u>Step 3. Generating Fuzzy Examples</u>

After fuzzy regions and membership functions are determined, training data are transferred into fuzzy examples. For each crisp value, its non-zero membership degrees in all possible fuzzy regions are determined. The crisp value is then converted to the fuzzy term whose region has the maximum membership degree. A potential problem for such conversion is example conflicts. All conflicting examples, however, are retained to form a conflicting rule set. They can be dealt with by our proposed inference mechanism.

<u>Step 4. Inductive Learning</u>

Non-conflicting examples are further generalized to yield fuzzy inductive rules by the rule induction algorithm, RITIO, as proposed by Wu and Urpani [10]. It was shown that RITIO achieved higher levels of predictive accuracy over HCV and C4.5 in their work. RITIO starts by copying the whole training data to a rule matrix (RM). Rules for classification knowledge are generated in the way of deleting one attribute at a time in RM. The least relevant attribute determined by the most entropy value is about to remove. RITIO will check whether inconsistency occurs if the attribute is removed. Should any inconsistency occur, RM is divided into remove-group and retain-group. The least relevant attribute is removed in the remove-group but retained in the retain-group. Employing RITIO algorithm renders the following advantages in our study: (1) there is no overhead to transfer trees to rules. (2) generated rules are clear and usually short. (3) the generated rules can be simply combined with conflicting rules as long as appropriate inference mechanism is applied.

3. Fuzzy Inference Mechanism

After the above procedure, a fuzzy rule system that contains a non-conflicting rule set from RITIO and a conflicting rule set is generated. Fuzzy inference mechanism is then employed to infer the output class for a given input. The inference process will include the fuzzy matching and the output reasoning. Fuzzy matching is a process to match the given input and a specific rule in the rule base. A rule is matched if all its fuzzy terms in the antecedent have non-zero membership degrees for the input values. We define the fitness of a match between the input I_i and a fuzzy rule R_j as:

$$Fit(I_i, R_j) = min(D_1, ..., D_v)$$

where D_k stands for the membership degree of kth fuzzy term in the rule antecedent. The fitness value measures the degree of similarity between the input and the antecedent of the rule.

After fuzzy rules are matched, three outcomes may be resulted: single-match, multiple match and no-match. In the case of single-match, it is obvious that the output class is inferred successfully. In the case of multiple-match, a weighting scheme is employed to judge which class is closer or more reliable to the input. The *estimate of possibility* (*EP*) that measures the possibility of a class being satisfied by the given input is proposed to deal with such situations. The *EP* value of a rule R given input I is defined by the fitness value, i.e., $EP(R,I)$ $= Fit(I,R)$. The *EP* value of a class c_i is then assigned by operating fuzzy *algebraic sum* on all the matched fuzzy rules that have the same class c_i where the fuzzy algebraic sum $\oplus$ is defined as A $\oplus$ B = A + B − AB [3]. For example, in the case of two matched rules R_1 and R_2 with the same output c_i, *EP* is calculated as:

$$EP(c_i, I) = EP (R_1, I) + EP (R_2, I) − EP (R_1, I) EP (R_2, I)$$

The formula is used recursively if there are more than two matched rules for the class c_i. After this procedure is applied to every class, the class with the highest *EP* value will be chosen as the final result.

In the case of no-match, we utilize the membership degrees of fuzzy terms in the rule antecedent to measure the "similarity" between the rule and the input. In the case of no-match, however, there must be some zero membership degrees in each rule's antecedent. Only those rules with the least number of zeros are considered similar to the input and we drop those zeros as if the rules were matched by the input. The result would be a single-match or a multiple-match case, which in the latter case the same procedure described above is applied to yield the final output class.

Performance of classification system is measured by the class prediction accuracy (CPA) of inputs defined as:

$$CPA = \frac{nr(IN, RU)}{n(IN)} \times 100\%$$

where $nr(IN,RU)$ stands for the number of right classification of an input set IN against the rule set RU, and $n(IN)$ represents the number of inputs in the set IN. Class prediction accuracy with respect to training data represents the system's mapping ability for the given input-output data. On the other hand, class prediction accuracy with respect to unseen data or test data stands for the system's generalization ability.

4. Experiment and Results

An experiment is conducted to examine the feasibility of the proposed method. Databases used in the experiment are acquired from the University of California at Irvine machine learning database repository [5]. Five databases are chosen in a way that they exhibit different characteristics. In this experiment, system performance from the proposed method was compared with that resulted from using RITIO algorithm.

Table 1. Comparison of Two Methods

Database	CPA from RITIO for training data	CPA from Fuzzy RITIO for training data	CPA from RITIO for test data	CPA from Fuzzy RITIO for test data
Breast Cancer	92.4	**97.1**	89.5	**94.7**
BLD	71.1	**73.0**	67.0	**68.9**
Heart	**76.7**	75.7	69.1	**72.8**
Iris	96.0	**97.3**	96.0	96.0
LN	75.0	**77.5**	70.6	**76.5**

The results of comparison are shown in Table 1. Better results for each database are highlighted with **boldface** font. Of all the five databases, fuzzy rule systems overall perform better than systems resulted from the RITIO algorithm with respect to both mapping ability and generalization ability. The improvement of system performance results from the fuzzy boundary that average out possible noises and the fuzzy inference based on similarity measure of rules to any given inputs. This result further shows the applicability of using our proposed method.

5. Conclusions

Most of classification knowledge extracted is expressed by crisp rules. Crisp rules have less description power than fuzzy rules in the way human thinks and reasons. The purpose of this research therefore is to propose a method to mine classification knowledge represented by fuzzy rules. The proposed method includes four steps. An experiment is conducted to show the performance of fuzzy rule systems generated using our proposed method. It is shown that overall system performance using our proposed method is better than that using RITIO algorithm that outperforms HCV and C4.5 as reported. Consequently, the feasibility of our proposed method to mine classification knowledge is justified.

References

[1] Clark, P. E., and R. Boswell, "Rule Induction with CN2: Some Recent Improvements," *Proceeding of Fifth European Working Session on Learning*, pp.151-163. Porto, Portugal: Springer-Verlag, 1991.

[2] Fayyad, U. and K. B. Irani, Multi-interval discretization of continuous-valued attributes for classification learning, *Proceeding of the 13th International Joint Conference on Artificial Intelligence*, Morgan Kaufmann, pp. 1022-1027, 1993.

[3] Klir, G. J. and B. Yuan, *Fuzzy sets and fuzzy logic: theory and applications*, Prentice Hall, 1995.

[4] Michalski, R. S., I. Mozetic, J. Hong, and N. Lavrac, "The Multi-Purpose Incremental Learning System AQ15 and Its Testing Application to Three Medical Domains," *Proceeding of Fifth National Conference on Artificial Intellegence*, pp. 1041-1045, 1986.

[5] Murphy, P. M. and D. W. Aha, UCI repository of machine learning databases, machine-readable data repository, Department of Information and Computer Science, University of California, Irvine, 1995.

[6] Quinlan, J. R., Induction of Decision Trees, *Machine Learning*, vol. 1, pp.81-106, 1986.

[7] Quinlan, J. R., *C4.5: Programs for Machine Learning*. Morgan Kaufmann, 1993.

[8] Spangler, W. E., J. H. May, L. G. Vargas, Choosing Data-Mining Methods for Multiple Classification: Representational and Performance Measurement Implications for Decision Support, *Journal of Management Information Systems*, vol. 16, pp. 37-62, 1999.

[9] Wu, X., Dealing With Noise and Real-Valued Attributes, in *Knowledge Acquisition from Databases*, Ablex, 1995.

[10] Wu, X. and D. Urpani, Induction by attribute elimination, *IEEE Transactions on Knowledge and Data Engineering*, Vol. 11, Issue 5, pp. 805 –812, 1999.

KES '01
N. Baba et al. (Eds.)
IOS Press, 2001

Data Mining of Inclusion Dependency from Relational Database

Shyue-Liang Wang, Yu-Ching Chen and Tzung-Pei Hong
Department of Information Management, I-Shou University
Kaohsiung, 84008, Taiwan, ROC

Abstract. This paper presents a set of data mining methods to facilitate the discovery of inclusion dependencies (IND) from database relations. An inclusion dependency states that values in column of one relation must appear as values in column of some other relation. The mining of inclusion dependencies consists of two basic tasks: validation and searching. Many set and non-set oriented data mining techniques of inclusion dependencies for static database relations have been proposed. In this work, we propose three new set-oriented (SQL Language) validation techniques and one heuristic searching method of inclusion dependencies for static and dynamic databases. Computational complexities of the new validation techniques are analyzed. Numerical simulation of the heuristic searching method is performed. The results show good computational efficiencies at the cost of minor space requirements. The methods developed here can be applied to database maintenance and
database reverse engineering.

1. Introduction

Data dependencies, such as functional dependency and inclusion dependency, describe the data semantics between attributes and represent constraints on the possible relations in relational database design. Usually, data dependencies are invented by the designer for the conceptual schema design. However, automated discovery of data dependencies from existing relations can simplify the design process and can be of great help when relationships among the attributes of the relation are not obvious. This may help for optimal design of relation schemas.

An inclusion dependency, $R(X) \subseteq S(Y)$ states that values in column X of one relation R must appear as values in column Y of some other relation S. An unary inclusion dependency restricts this definition to single attribute column. A discovered inclusion dependency may be treated as a potential candidate of foreign key in schema design. Research on discovering inclusion dependencies have been presented [1-9]. Basically the discovery process consists of two tasks: validation and searching. Most of the methods proposed utilize SQL statements for validation and pair-wise companions for searching. Although heuristics [2-3] and inclusion dependency transitivities [1] have been introduced for pruning the search space, these methods are still computationally expensive. One reason is that the SQL validation statements are all based on JOIN operation which is computationally expensive and can be improved. Another reason is that the number of SQL statement calls is minimized only under IND transitivity properties. It can be further reduced by considering the ordering and testing sequence of the attributes. In addition, most of the proposed methods consider only static databases. For dynamic databases such as adding new records or attributes to the databases, the discovery process need to be repeated without utilizing current information about dependencies.

In this work, we propose three new SQL language based validation methods and two heuristic searching methods of inclusion dependencies for both static and dynamic databases. The rest of our paper is organized as follows. Section 2 proposes one new validation method of IND on static database and two new validation methods of IND on dynamic database. Section 3 proposes a heuristic searching method of IND's for static databases. Section 4 presents the analyses of computational complexities of the proposed validation and searching methods. Section 5 presents the numerical simulation of the proposed searching heuristics. Finally a conclusion is given at section 6.

2. Validation Methods of Inclusion Dependencies

This section first proposes a union-based validation method of inclusion dependency on static database. Two simple validation methods, IND-based and Value-number-based methods, under different given informations are then introduced when new attribute values are added to the database.

2.1 Union-based Validation Method

The time complexity of the SQL statement for calculating the intersection of $R_i(A_1)$ and $R_j(A_2)$ in Bell & Blockhausen [1] is $O(n^2)$, where n is the largest number of attribute values. Calculating intersection basically requires a JOIN operation and is time consuming. However, the time complexity of calculating the number of distinct values from the union of $R_i(A_1)$ and $R_j(A_2)$ is only $O(n)$. Therefore, we propose the following union-based validation method for IND. Let $R_m(A_k)$ be the union of $R_i(A_1)$ and $R_j(A_2)$.

<u>SQL Statement</u>

   US1. SELECT COUNT(DISTINCT $R_m.A_k$)
        FROM R_m $=:e'$
   US2. SELECT COUNT(DISTINCT A_1)
        FROM R_i $=:e_1$
   US3. SELECT COUNT(DISTINCT A_2)
        FROM R_j $=:e_2$

<u>Validation Statement</u>

UJ1. $e'=e_2$ $\Rightarrow$ $R_i(A_1) \subseteq R_j(A_2)$
UJ2. $e'=e_1$ $\Rightarrow$ $R_j(A_2) \subseteq R_i(A_1)$
UJ3. $e'=e_1=e_2$ $\Rightarrow$ $R_i(A_1) = R_j(A_2)$
UJ4. $e'\neq e_1, e'\neq e_2$ $\Rightarrow$ No IND

The computation time for the SQL Statement US1 is only $n+n = 2n$. However, additional space of $2n$ is required by $R_m(A_k)$. The total computation time for the three SQL statements is $4n$.

2.2 IND-based Validation Method

This subsection proposes a validation method of IND when the inclusion dependency between the attributes is known and new attribute values are added to the databases. Let the IND between $R_i(A_1)$ and $R_j(A_2)$ be known and new attribute values t_2 is added to $R_j(A_2)$. Table 1 shows the IND-based validation method that determines the inclusion dependency between the two attributes. It can be seen that only $3n+2$ running time is required to validate the IND when a new attribute value is added. The number of SQL query required is 3.

Table 1. IND-based validation method

Input	Validation Statement / Output	Computation Complexity	Query
$R_i(A_1) \subseteq R_j(A_2)$	$R_i(A_1) \subseteq R_j(A_2')$	None	None
$R_i(A_1)=R_j(A_2)$	if $(t_2 \in R(A_2))$ then $R_i(A_1)=R_j(A_2')$ else $R_i(A_1) \subseteq R_j(A_2')$	n	1
$R_i(A_1) \supseteq R_j(A_2)$	if $(t_2 \notin R(A_1))$ then *No IND* else if $(e_1=e_2')$ then $R_i(A_1)=R_j(A_2')$ else $R_i(A_1) \supseteq R_j(A_2')$	n $n,\ n+1$	3
No IND	if $(e_2'=e'')$ then $R_i(A_1) \subseteq R_j(A_2')$ else *No IND*	$n+1,\ 2n+1$	2
Total		$3n+2$	3

2.3 Value-number-based Method

This subsection proposes another validation method of IND when the numbers of distinct attribute values are known and new attribute values are added to the database.

Let e_1, e_2, e' be the numbers of distinct attribute values of $R_i(A_1)$, $R_j(A_2)$ and $R_m(A_k) = R_i(A_1) \cup R_j(A_2)$ respectively. Assuming that a new attribute value t_2 is added to $R_j(A_2)$ and the numbers of distinct attribute values become e_1, e_2', e''. The following algorithm calculates the values e_1, e_2', e'' from e_1, e_2, e'.

Procedure Value-number-based validation:

Input: E=$(e_1$, e_2, $e')$, new attribute value t_2 added to $R_j(A_2)$

Output: E=$(e_1$, e_2', $e'')$

Begin

1. $E := (e_1, e_2, e')$
2.   if $(t_2 \in R_j(A_2))$ then $E := (e_1, e_2, e')$
3.   else if $(t_2 \in R_i(A_1))$ then $E := (e_1, e_2+1, e')$
4.   else $E := (e_1, e_2+1, e'+1)$

End

The validation statements based on E=$(e_1$, e_2', $e'')$ are as follow:

VJ1. $e''=e_2'$ $\Rightarrow$ $R_i(A_1) \subseteq R_j(A_2')$

VJ2. $e''=e_1$ $\Rightarrow$ $R_i(A_1) \supseteq R_j(A_2')$

VJ3. $e''=e_1=e_2'$ $\Rightarrow$ $R_i(A_1)=R_j(A_2')$

VJ4. $e'' \neq e_1, e'' \neq e_2'$ $\Rightarrow$ No IND

3. Searching Method of Inclusion Dependencies

This section propose a heuristic searching method of inclusion dependencies on databases. For a given set of attributes of the same data type, the searching method finds all inclusion dependencies between the attributes. The following bisection searching method of IND's is an extension of Bell & Blockhausen's [1] searching algorithm. Besides pruning by transitivities of inclusion dependency, we consider the ordering and the testing sequence of the attributes. The saving of number of queries by the proposed bisection method will be simulated numerical and presented in Section 5.

Algorithm: Bisection searching method:

Input: A list of all attributes of one type

Output: A list of all INDs between attributes of one type
1.   Compute all candidate attributes where the attribute value interval is a subset or superset of any other attributes.
2.   Sort and number all candidate attributes from A_1 down to A_n according to interval restriction.
3.   Construct the testing sequence B_1, B_2, ..., B_n from the list A_1, A_2, ..., A_n in preorder traversal fashion as follows:
     (a)   Select the midpoint of the list of attributes as B_1, and bisect it into left and right sublists,
     (b)   Repeat (a) recursively on left sublist, i.e., select the midpoint of left sublist as next B_i and bisect it into left and right sublists,
     (c)   Repeat (a) recursively on right sublist, i.e., select the midpoint of right sublist as next B_i, and bisect it into left and right sublists.
4.   Construct the unidirectional list structure is as follows:

$$[[B_1 : [\underline{B_2}],[\underline{B_3}],...,[\underline{B_{i-1}}],[\overline{B_i}],[\overline{B_{i+1}}],...,[\overline{B_n}]]$$

$$[B_2 : [\underline{B_3}],[\underline{B_4}],...,[\overline{B_{i-1}}],[\overline{B_i}],...,[\overline{B_n}]]$$

$$\vdots$$

$$[B_{n-1} : [\overline{B_n}]]]$$

$\underline{B_i}$ and $\overline{B_j}$ respectively are symbols for the INDs $B_k \subseteq B_i$ and $B_j \subseteq B_k$.

5.   For all B_i with $1 \le i < n$ do:
     (a)   Check and prune by inclusion independent transitivity.
     (b)   Check and prune by inclusion dependent transitivity.

4. Analysis

This section describes the computation complexity analysis of the validation methods we proposed in section 2. Table 2 shows the time space complexities and number of queries for Bell & Blockhausen [1], union-based, IND-based, value-number-based validation methods of IND. It is observed that union-based method requires less computation time than Bell & Blockhausen's method. However, additional space is required for storage. It is also observed that when new attribute values are added, IND-based and value-number-based methods should be used if IND or e_1, e_2, e' of all attribute values are known respectively.

Table 2. Complexities of validation methods

	Time Complexity	**Query**	**Space**
Bell & Blockhausen Method	$n^2 + 3n + 1$	3	None
Union-based Method	$4n + 2$	3	$2n + 1$
IND-based Method	$3n + 2$	3	$2n + 2$
Value-number-based Method	$2n$	2	3

5. Experiments

This section describes the numerical simulation of the bisection method and compares it with different testing sequences of the attributes. Table 3 shows the average numbers of SQL queries required by the (1) bisection method, (2) testing sequence from the largest to the smallest attributes (LtoS), and (3) testing sequence from the smallest to the largest attributes (StoL). For n attributes, the total possible number of IND's is $n(n-1)/2$. For the simulation, we assume that $A_n \subseteq A_{n-1} \subseteq .... \subseteq A_1$. We simulate the cases that number of known IND's range from 0 to $n(n-1)/2$ randomly. The results show that the bisection method requires less SQL queries on the average. In addition, when the number of attributes increases and the number of known IND's decreases, the proposed bisection method shows better saving of SQL queries.

Table 3. Average numbers of SQL queries for 3 testing sequences

Attribute	Method	0	1	2	3	4	5	6	7	8	9	10
3	Bisection	2.00	1.33	0.67	0							
	LtoS	3.00	1.67	0.67	0							
	StoL	3.00	1.67	0.67	0							
4	Bisection	4	3.17	2.33	1.70	1.07	0.50	0				
	LtoS	6	4.33	3.00	1.95	1.13	0.50	0				
	StoL	6	4.33	2.93	1.95	1.13	0.50	0				
5	Bisection	6.00	5.20	4.44	3.77	3.10	2.46	1.90	1.30	0.84	0.40	0
	LtoS	10.00	8.00	6.33	4.95	3.80	2.90	2.17	1.36	0.87	0.40	0
	StoL	10.00	8.00	6.33	4.95	3.79	2.88	2.18	1.37	0.87	0.40	0

6. Conclusion

We have presented three new set-oriented (SQL language based) validation methods and a heuristic searching method of inclusion dependencies for static and dynamic databases. Computational complexities and query requirements of the new validation techniques are analyzed. Numerical simulation of the heuristic searching method is performed. The results show good computational efficiencies at the cost of minor space requirements. The methods developed here can be applied to database maintenance and database reverse engineering.

For future work, other types of database dependencies, such as functional dependencies and multi-valued dependencies, need to be considered.

References

[1] S. Bell and P. Brockhausen, "Discovery of data dependencies in relational databases", Tech. Rep. LS-8, Report-14, University of Dortmund, Apr. 1995.

[2] R.H.L. Chiang, T.M. Barron and V.C. Storey, "Reverse engineering of relational databases : Extraction of an EER model from a relational database", Data & Knowledge Engineering 12(2), 107-142, 1994.

[3] R.H.L. Chiang, "A knowledge-based system for performing reverse engineering of relational databases", Decision Support Systems 13, 295-312, North-Holland, 1995.

[4] H. Ito, T. Kajiyama, T. Fukumura, "Discovering Conceptual Structure of Relations", Proceedings of the IEEE SMC'96, 591-596, 1996.

[5] J.H. Jahnke, W. Schäfer, and A. Zündorf. "Generic Fuzzy Reasoning Nets as a Basis for Reverse Engineering Relational Database Application", Proceedings of ESEC/FSE, LNCS1302, 1997.

[6] M. Kantola, H. Mannila, K.J. Räihä, and H. Siirtola, "Discovering Functional and Inclusion Dependencies in Relational Databases", International Journal of Intelligent Systems, Vol. 7, 591-607, 1992.

[7] J-M. Petit, J. Kouloumdjian, J-F. Boulicaut, F. Toumani, "Using Queries to Improve Database Reverse Engineering", Proceedings of the 13th International Conference on the ER Approach, volume 881 of LNCS, 369-386, Oct. 1994.

[8] J-M. Petit, F. Toumani, J. Kouloumdjian, "Relational Database Reverse Engineering: A Method Based on Query Analysis", International Journal of Cooperative Information Systems, 4(2,3), 287-316, 1995.

[9] J-M. Petit, F. Toumani, J-F. Boulicaut, J. Kouloumdjian, "Towards the Reverse Engineering of Denormalized Relational Databases", Proceedings of the 12th International Conference on Data Engineering, New Orleans, Louisiana, USA, IEEE Press, 218-227, Feb. 1996.

KES '01
N. Baba et al. (Eds.)
IOS Press, 2001

Experimental Mechatronic System for Failure Detection and Selection in Laminate and Composite Structures

Radyk HRADYNARSKI, Florin IONESCU*, Mitko MIHOVSKI, Kostadin KOSTADINOV
Institute of Mechanics, Acad. G. Bonchev St., Block 4, Sofia 1113, Bulgaria
**University of Applied Sciences - Konstanz, Brauneggerstr. 55, Germany*

Abstract: This paper deals the development of an experimental mechatronic system for failure control, particularly, in honeycomb structures applied in aviation industry for "unsticking" discontinuities control. The approach is based on theoretical and experimental studies using an electromagnetic actuator for excitation and acoustic emission technique for signal propagation registration. The system is consisted of the exciter, acceleration sensor for mechanical feedback, one chip microcomputer HC11P2, three acoustic emission sensors for linear and plane localization; acoustic emission analyzer, and personal computer for signal processing and failure control. Obtained information from the acoustic sensors for defined failures of structure's models is analyzed to determine the optimal excitation, optimal sensor localisation, failure control and selection rules.

1. Introduction

Composite materials manufactured by glueing, soldering, welding or by using other technologies are popular in a number of industrial areas. Note that several are the profits gained by using such materials - their introduction into staructures makes the latter lighter, their mechanical characteristics are comparable to these of metal alloys and they are corrosion resistant. These qualities make the composites indispensable for use in aviation industry, automobile transport, chemical-mechanical enginnering and other industrial branches. However, manufacture of these materials implies the use of precise technologies. Any violation of the technological prescriptions results in the formation of structure defects inadmissible for further material use. In addition, structure discontinuities may occur during the process of material exploitation [1], too.

Optical, thermal, radiation and acoustic methods turn out to be perspective and applicable for the failure control of composite structures. In particular, the acoustic methods of control are mostly developed as theory and practice. They prove to be effective for measurements in various frequency ranges. The most popular technique used for low frequency studies comprises impedance, velocimeter, amplitude, spectral, resonance and topographical methods of measurement and control[2,3]. Furthermore, impulse, reverberation and resonance methods are used for ultrasonic studies. Recently the acoustic emission method grows popular, being used for the control of structures undergone of mechanical impacts.

The aim of this paper is to develop a mechatronic system for the failure control of composite structures applied in aviation industry with controlled low frequency impulse excitation. Two methods are employed for this purpose - a modified low-frequency method for the excitation of elastic oscillations and a method of acoustic emisson for signal record and analysis.

2. Object of Study. Method for Excitation

The object of study are structure composite materials, type "honeycomb", applied in aviation industry. They are characterized by typical discontinuity e.g. unsticking of an aluminium alloy casing from the structure.

Propagation of bending oscillations in an infinite plate. Let's consider the propagation of mechanical oscillations within an infinite plate without discontinuities[4,7]. For the case of cylindrical coordinate system (r, φ) and the symetry of wave propagation, the propagation problem can be solved considering the coordinate r, only. The equation of wave propagation through an isotropic material has the form:

$$DV^4 w + \rho_0 \frac{\partial^2 w}{\partial t^2} = P(r.t) \tag{1}$$

where $D = Eh^3/[12(1-\upsilon^2)]$, E - Young's modulus, h is material thickness, ρ_0 - densinty per unit area of the plate, υ - Poisson's coefficient, $w(r.t)$ - displacement, $P(r.t)$- load acting over a unit area, t - time, $\nabla^4 = \frac{1}{2}\frac{\partial}{\partial r}\left[r\frac{\partial}{\partial r}\left(\frac{1}{r}\frac{\partial}{\partial r}(r\frac{\partial}{\partial r}) \right) \right]$.

The solution for the displacement for load given as Dirak's function[4],: is

$$w(r.t) = w(\beta) = \frac{I_a}{4\pi D}\left(\frac{\pi}{2} - \int_0^{\beta^2/4} \frac{\sin x}{x}dx \right) \tag{2}$$

where I is the load amplitude, $a^2 = D/\rho_0$,, $\beta^2 = r^2/at$, r -distance to the point where the displacement is assessed, t-time of impulse propagation recorded from the moment of load application and lasting till the moment when the load reaches a point with a coordinate r, a - material parameter, β - similar parameter characterizing the material area under consideration. Eqn. (2) for large values of β at points far from the spot of applying the excitation is:

$$w \sim \frac{I.a\cos(\beta^2/4)}{4\pi D\beta^2/4}. \tag{3}$$

The analysis of relation (3) shows that the maximum amplitude of the propagating wave decreases by mode $1/r^2$, whereas the wave shape significantly changes with the increase of the distance. If eqn. (2) is solved for $w(r,t) = 0$, it is useful to introduce the relation

$$r^2 = 4\alpha\left(\frac{D}{\rho_0} \right)^{0.5}.t \;;\; \text{where } \alpha = \frac{r^2}{4t}\left(\frac{\rho_0}{D} \right)^{1/2} \tag{4}$$

The zero amplitudes $w(r.t) = 0$ are obtained for the following values of α_i: $\alpha_1 = 1.9265$, $\alpha_2 = 4.8938$, $\alpha_3 = 7.9727$ and $\alpha_4 = 11.084$ etc.

The phase velocity of the wave propagation V for a definite point r, is:

$$V = \frac{\partial r}{dt} = \left(\frac{\alpha h}{t} \right)^{0.5} \cdot \left(\frac{E}{12(1-\upsilon^2)} \right)^{\frac{1}{4}} \tag{5}$$

The propagating acoustic wave is characterized by a dispersion[4], and the wave length is determined by the expression, where f is the wave frequency:

$$\lambda = V/f \tag{6}$$

Calculations are performed for an aluminum alloy plate used with the following characteristics: $\rho_0 = 2,700$ kg/m^2, $E = 7.10^{10}$Pa, $h = 1.10^{-3}$ m, $v = 0.338$. The displacements distribution of w (r.t = 5 μ s) is shown in Fig. 1.

Low frequency-bending oscillations of a plate with finite dimensions. Oscillation of a model specimen with a cylindrical discontinuity is considered. To introduce the oscillations, it is assumed that the excitation of elastic transversal waves in the discontinuity area is symmetric. All considerations are based on the classical theory of circular elastic plate oscillation firmly fixed along plate perimeter. Plate displacements are given in the form [7]:

$$W = \frac{P}{16\pi D}(R^2 - r^2 - 2r\ln\frac{R}{r}, \quad W_{max} = 0.224\frac{R^2 P}{h^3},$$

where: r is the current radius, R - radius of the discontinuity within the plate, h-plate thickness within the discontinuity area, □- applied concentrated load.

The eigen frequencies of oscillation of the model discontinuity are: $f_{nm} = A_{nm} \sqrt{\dfrac{Eh^2}{12R^4 \rho_0 (1-\upsilon^2)}}$,

where A_{nm} is the coefficient determining

Table 1

N	n	m	f_{nm},kHz
1	0	0	159
2	0	1	620
3	0	2	1389
4	1	0	331
5	1	1	952
6	1	2	1883

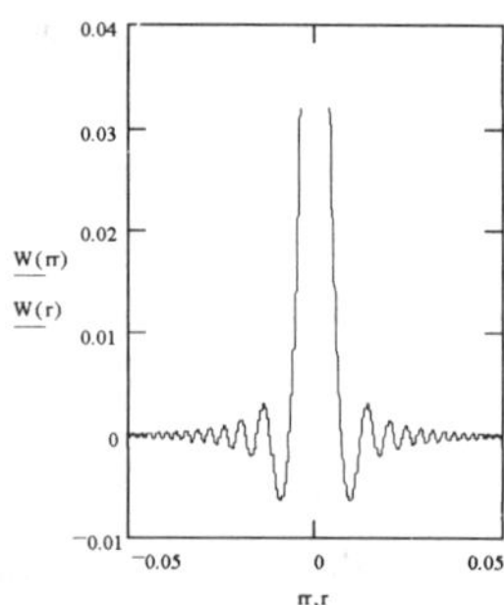

Fig.1. Displacement distribution of the w ($r.t$ = 5 µs).

the plate oscillation modes, m - the number of the main circles, n - number of the main diameters. The A_{nm} values for the first several oscillation forms are as follows: $\square_{00}$=3.19; $\square_{10}$=4.61; $\square_{20}$=4.9; $\square_{01}$=6.30; $\square_{02}$=9.43 etc. [7]. The results of numerical calculations (Table 1) for the determination of the oscillation eigen frequencies f_{nm} are performed for a model cylindrical discontinuity with radius R = 5 mm and thickness h = 1 mm (whereas $\square$=1). The excited oscillations within the discontinuity area can be appropriately registered by using acoustic emission. The signals are registered by means of ultrasonic sensors mounted by following a definite scheme of ordering. A more precise oscillation analysis seems useful, when, for instance, a model plate, freely supported along its perimeter, is considered.

3. Design of the Mechatronic System for Failure Detection in Laminate and Composite Structures.

To assess the casing - "honeycomb" contact, we propose a mechatronic system for failure detection in laminate and composite structures which block scheme is shown in

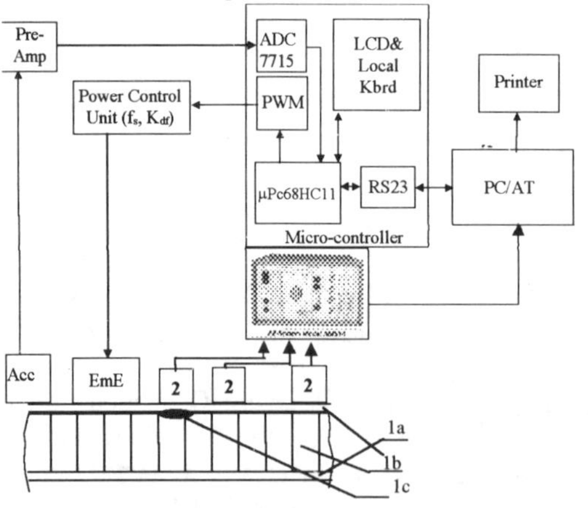

Fig. 2. Structure of the Mechatronic System

Fig. 3. General View of the Exciter.

It consists of the following elements: electromagnetic actuator(EmE) as a low frequency impulse exciter with a controllable amplitude and with impulse frequency; acceleration sensor (Acc) for mechanical feedback, microcontroler 68HC11P2 for excitation control, 1-object of study, comprising casing (1a), a "honeycomb" structure (1b) and a discontinuity area(1c); "Vallen-systeme" acoustic emission analyser, type AMSY-4; 2- three acoustic emission sensors for linear and plane localization; PC for additional processing of information.

The exciter is developed as an electromagnetic actuator generating force impulses with frequency from 1 to 150 Hz for a maximum load of about 60 N/cm^2. To ensure uni-axial translator motion of the end-effector and to enable high force nearly constant to be delivered to suit individual test specifications, severe constraints are placed on the design of the electromagnetic actuator. In such a manner the electromagnetic actuator works in the motion range of 2 mm with nearly constant force. The general view of the electromagnetic actuator is shown in Fig.3.

To guarantee the equal conditions during the measurements of mechanical object with different characteristic the acceleration feedback through the object studied is included. The accelerator is mounted on the object points belonging to the its kinematic and dynamic sensitivity directions [4]. Acceleration signal is used to form feedback signal while controlling the duty cycle of the PWM. In such a manner, the uniform excitation has been realized independence of the mechanical characteristics of the object studied. To obtain various features of the actuator we control the actuator by two parameters - system frequency f_s in the range 0-500Hz and duty cycle of the PWM. The controlled generated oscillations of the end-effector are in the ratio 1-150Hz, limited by the moving end-effector mass m.

4. Experimental Studies

The experimental studies are performed by testing model plates and a helicopter wing.

Model experiments. Two groups of aluminum alloy plates with dimensions 150x50x3 [mm] were studied: - plates without any discontinuity and plates with a cylindrical flat-bottom discontinuity of diameter 10 [mm]. Plate thickness within the discontinuity area is h = 1 mm. Acoustic oscillations are excited under three operation regimes of the electromagnetic exciter, i.e.: I - signal low frequency and high amplitude; II -signal medium frequency and amplitude; III -signal high frequency and low amplitude. The acoustic emission analyzer operates in a regime of linear localization, while the exciter is located as shown in Fig. 2. The results for a plate without a discontinuity (Fig.4.) and for a plate with a discontinuity (Fig.5.) are obtained. The following notations are used: Amp-amplitude of the signals registered, Freq-number of the counted sinusoid crosses in an AE impulse for a definite level,

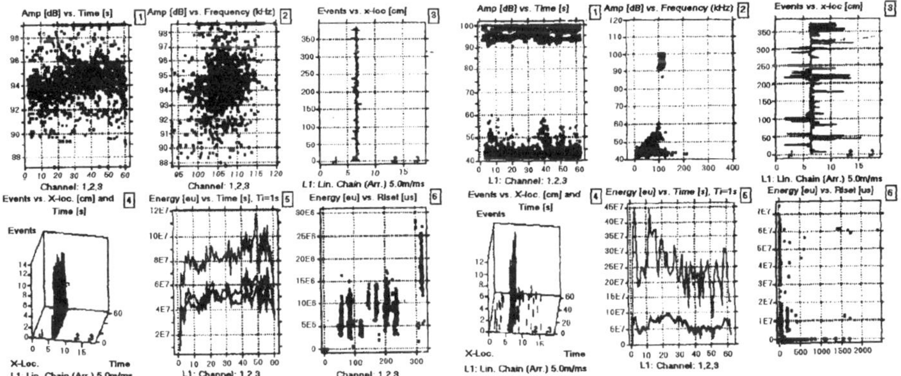

Figure 4 Results for the plate w/o discontinuity.   **Figure 5** Results for the plate with discontinuity.

Energy-quantity proportional to the impulse energy and given in absolute energy units [eu], Events - number of registered events, Time - current time of study, Riset - time, starting from the moment of the first cross of a definite impulse level and lasting till the moment of attaining the maximal amplitude, □-loc-location of the AE source for a given localization scheme. The analysis shows that signals with amplitudes within ranges 40-60dB and 80-100dB and frequency within the 0-200 kHz range are registered. The maximal energy of the acoustic emission signals is within ranges 4.10^7-40.10^7[eu]. The time of signal growth is within ranges 0-2500µs. Regarding definite excitation conditions, the performed analysis proves that the energy of acoustic emission signals can be considered as an advisbale information signal. Signals with higher energy are registered under decrease of the excitation frequency and for increase of the shock amplitude. The excitation of signals with maximal energy is attained in the case when the exciter is located within the discontinuity area. The rest of the parameters are less informative as compared to the energy parameter. µs

Studies under real conditions of a helicopter Mi-8 wing. Employing an ultra sonic impedance method, it is proved that there are no discontinuities of the contact between the aluminium sheet and the "honeycomb" composite. Fig.6 shows results for a composite wing

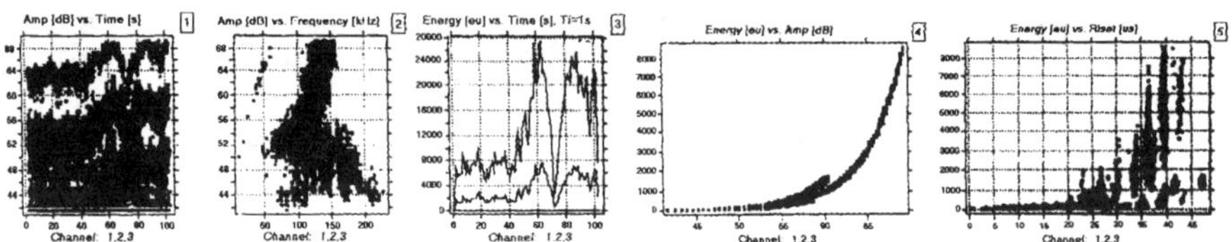

Figure 6. Results for a composite wing with an artificially "insticking" flaw.

with an artificially created "insticking" flaw. The presence of a flaw yields a significant increase of the amplitude and energy of the registered AE impulses, but for insignificant frequency change(Freq). Signals with greater energy are registered for the unsticking case and AE signals with longer growth time prevail.

5. Conclusions

The paper proposes a mechatronic system for failure detection in laminate and composite structures. Particularly, the control of honeycomb structures applied in aviation industry for "unsticking" discontinuities is considered.

The approach used is based on theoretical and experimental studies and is developed by a joint use of an intelligent electromagnetic actuator for low frequency excitation of mechanical oscillation and acoustic emission technique for the registration of signal propagation.

The comparison between the obtained data and these of the model study confirms the plausibility of the adopted theoretical and experimental model. The obtained results presented here show high reliability of failure detection in structures subject to permanent exciting.

Acknowledgements: The authors gratefully acknowledge the partial support by the Bulg. Nat. Science Foundation under the Projects TH708/97 & TH819/98 and DFG.

References
[1] Fallström K. E., A nondestructive method to detect delaminations and defects in plates, NDT and E International, 1991, 24, 2, 67-76.
[2] Dapkus R., A. Daugela, Smart Acoustic Impedance Sensors for the testing of Engineering Materials, Proc. of MOVIC'98, Zurich, 25-28.8.1998, 1, 225-230.
[3] Broch J. Tr., Mechanical vibration and shock measurements, K. Larsen & Son A/S, 1980.
[4] Boiadjiev G., *Kinematic Sensible Directions of Manipulating Systems,* Proc. of 6th Int. Symp. Measurement and Control in Robotics, Brussels, 1996, pp. 77 - 82.

KES '01
N. Baba et al. (Eds.)
IOS Press, 2001

Sensor Data Fusion Using H-Infinity Filters

Jitendra R. RAOL

Flight Mechanics and Control Division, National Aerospace Laboratory, Bangalore, India

Florin IONESCU

FH- University of Applied Sciences, Brauneggerstr. 55, D-78462 Konstanz, Germany

Abstract. Sensor data fusion algorithms based on H-Infinity filters are investigated for two-sensor situation. Their performance is evaluated when there is data loss in either of the sensors. Results of validation are presented for two local filters tracking a moving object.

1. Introduction

Sensor fusion, a short notation for multi sensor data fusion (MSDF) is defined as the process of integrating information from multiple sources to produce the most specific and comprehensive unified data about an entity, activity or event. The data fusion achieves improved accuracy and more specific inferences than could be achieved by the use of a single sensor alone. The applications of MSDF include remote sensing, monitoring of manufacturing processes, mechatronics, robotics and medical diagnosis [1]. As a technology, MSDF is the integration and application of many disciplines & new areas including communications and decision theory, estimation theory, digital signal processing, numerical methods and neuro-fuzzy techniques. At a basic level, the processing operations in sensor fusion are dominated by numeric procedures involving linear estimation techniques and statistical methods. In this paper, two H-infinity filter [2,3] based fusion algorithms are evaluated, when there is specific data loss in either of the two sensors used for tracking a moving object. The sensor data fusion results using H-Infinity filters are relatively new. The performance is evaluated in terms of state errors and H-infinity based norm using simulated data.

2. Sensor Data Fusion

Each sensor employs an estimator to obtain the state vector of an object's trajectory and associated covariance matrices from respective sensor measurements. These state vectors are transmitted over a data link to a fusion center where state vector fusion is performed to obtain a composite state vector. This composite state vector represents fusion state of the object. The sensor stations employ individual H-infinity filters to create two sets of track files. The kinematics model of a moving object is described by:

$$x(k+1)= Fx(k) + Gw(k) \tag{1}$$

Where the object state vector has two components: position and velocity and

$$F = \begin{bmatrix} 1 & T \\ 0 & 1 \end{bmatrix}; \quad G = \begin{bmatrix} T^2/2 \\ T \end{bmatrix}$$

Here T is the sampling interval. Also we have for w as the white Gaussian noise:

$$E\{w(k)\} = 0; \, Var\{w(k)\} = Q$$

The measurements at each sensor are given by (i = 1, 2, ... , m; number of sensors). The measurement noise is assumed to be white Gaussian with the statistics:

$$z_i(k) = Hx(k) + v_i(k) \tag{2}$$

$$E\{v_i(k)\} = 0; \; Var\{v_i(k)\} = R_{vi}$$

3. H-Infinity *A Posteriori* Filter Based Fusion Algorithm (HIPOFA)

The estimates are obtained for each sensor (i=1,2) using H-infinity A posteriori filter [2]:

Covariance Time Propagation:

$$P_i(k+1) = FP_i(k)F' + GQG' - FP_i(k)[H_i^t \; L_i^t] R_i^{-1} \begin{bmatrix} H_i \\ L_i \end{bmatrix} P_i(k)F' \tag{3}$$

$$R_i = \begin{bmatrix} I & o \\ 0 & -\gamma^2 I \end{bmatrix} + \begin{bmatrix} H_i \\ L_i \end{bmatrix} P_i(k)[H_i^t \; L_i^t] \tag{4}$$

Filter Gain:

$$K_i = P_i(k+1)H_i^t \, (I + H_i P_i(k+1)H_i^t)^{-1} \tag{5}$$

Measurement update of states is obtained by :

$$\hat{x}_i(k+1) = F\hat{x}_i(k) + K_i(y_i(k+1) - H_i F\hat{x}_i(k)) \tag{6}$$

The conditions for the existence of the H-Inifinty filters are given in [2]. The fusion of the estimates from the two sensors then can be obtained by:

$$\hat{x}_f(k+1) = \hat{x}_1(k+1) + \hat{P}_1(k+1)(\hat{P}_1(k+1) + \hat{P}_2(k+1))^{-1}(\hat{x}_2(k+1) - \hat{x}_1(k+1)) \tag{7}$$

$$\hat{P}_f(k+1) = \hat{P}_1(k+1) - \hat{P}_1(k+1)(\hat{P}_1(k+1) + \hat{P}_2(k+1))^{-1}\hat{P}_1^t(k+1) \tag{8}$$

The fused state vector and the covariance of the fused state utilize the individual estimate state vectors (of each sensor) and covariance matrices.

4. H-Infinity Global Fusion Algorithm (HIGFA)

This filtering algorithm is based on [3].

The local filters are given for each sensor (i=1,2,...,m):

State/Covariance Time Propagation

$$\tilde{x}_i(k+1) = F\hat{x}_i(k) \tag{9}$$

$$\tilde{P}_i(k+1) = F\hat{P}_i(k)F' + GQG' \tag{10}$$

Covariance update

$$\hat{P}_i^{-1}(k+1) = \widetilde{P}_i^{-1}(k+1) + [H_i^t \; L_i^t]\begin{bmatrix} I & 0 \\ 0 & -\gamma^2 I \end{bmatrix}^{-1}\begin{bmatrix} H_i \\ L_i \end{bmatrix} \tag{11}$$

Local filter gains:

$$A_i = I + 1/\gamma^2 \hat{P}_i(k+1)L_i^t L_i \quad ; \quad K_i = A_i^{-1}\hat{P}_i(k+1)H_i^t \tag{12}$$

The measurement update of local states:

$$\hat{x}_i(k+1) = \widetilde{x}_i(k+1) + K_i(y_i(k+1) - H_i\widetilde{x}_i(k+1)) \tag{13}$$

Time Propagation of Fusion State/Covariance:

$$\widetilde{x}_f(k+1) = F\hat{x}_f(k) \tag{14}$$

$$\widetilde{P}_f(k+1) = F\hat{P}_f(k)F' + GQG' \tag{15}$$

Measurement update of the Fusion States/Covariance:

$$\hat{P}_f^{-1}(k+1) = \widetilde{P}_f^{-1}(k+1) + \sum_{i=1}^{m}(\hat{P}_i^{-1}(k+1) - \widetilde{P}_i^{-1}(k+1)) + \frac{m-1}{\gamma^2}L'L \tag{16}$$

Global Gain:

$$A_f = I + 1/\gamma^2 \hat{P}_f(k+1)L'L \tag{17}$$

Global (measurement update) Fused State:

$$\hat{x}_f(k+1) = [I - A_f^{-1}\hat{P}_f(k+1)H_f^t H_f]\widetilde{x}_f(k+1)$$

$$+ A_f^{-1}\hat{P}_f(k+1)\sum_{i=1}^{m}\{\hat{P}_i^{-1}(k+1)A_i\hat{x}_i(k+1) - (\hat{P}_i^{-1}(k+1)A_i + H_i^t H_i)F\hat{x}_i(k)\} \tag{18}$$

5. Results and Discussions

The simulated data are generated (in PC MATLAB) using eqns. (1) and (2) with associated F and H matrices. The sampling interval is 0.5 sec. The normalized (with standard deviation) random noise is added to the state vector. Also the measurements of each sensor are corrupted with random noise. The sensors could have dissimilar measurement noise variances (e.g. Sensor 2 having higher variance than Sensor 1). The initial condition for the state vector is given as x(0)=[200 0.5]. The performance of the fusion filters is also evaluated in terms of

$$\frac{\sum_{i=0}^{N}(\hat{x}_f(k) - x(k))'(\hat{x}_f(k) - x(k))}{(\hat{x}_{0f} - x_{0f})'P_{0f}(\hat{x}_{0f} - x_{0f}) + \sum_{k=0}^{N}w'(k)w(k) + \sum_{i=1}^{m}\sum_{k=0}^{N}v_i'(k)v_i(k)} \tag{19}$$

Basically this ratio (the H-infinity norm) should be less than square of gama, which can be considered as an upper bound on the maximum energy gain from the input to the output. It can be observed from this norm that the input to the filter consists of energies due to the error in the initial condition, state disturbance (process noise), and measurement noise in two sensors. The output energy of the filter is due to the error in fused state. Tables 1-3 give performance indices for these two filters(with Rv1=Rv2): under normal condition and with data loss in either sensor for about a few seconds. The theoretical covariance norm of the fused state was found to be lower than that of the individual filters. Figs. 1 & 2 show state error time histories with their bounds [4] for HIPOFA and HIGFA when there is data loss in Sensor 1. It can be observed that the two fusion algorithms are fairly robust to the loss of data.

6. Concluding Remarks

The results of performance evaluation of H-Infinity filters-based sensor data fusion algorithms have been presented when there is loss of data in either sensor. The filters are fairly robust.

References

[1] Bar-Shalom Y. and Fortman T.E. Tracking and Data Association. Academic Press, Anaheim, CA. 1988.
[2] Hassibi B., Sayad A.H. and Kailath T. Linear estimation in Krein spaces-Part II: Applications. IEEE Trans. On Autom. Contrl., Vol. 41, No. 1, January 1996.
[3] Kyung-Keun Kim, Jin-Bae Park, Seung-Hee Jin and Tae-Sung Yoon. Design of decentralized H-Infinity filter via Krein space state model. IEE Electronics Letters, U.K., 2000.
[4] Candy J.V. Signal Processing-The Model-Based Approach. McGraw Hill Book Company, N.Y. 1986.

Table 1: Percentage Residual Fit Errors

	Normal	Data loss In Sensor 1	Data loss In Sensor 2
HIPOFA-F1	0.443	0.442	0.443
HIPOFA-F2	0.435	0.435	0.427
HIGFA-F1	0.443	0.442	0.443
HIGFA-F2	0.436	0.436	0.427

Table 2: Percentage State Errors

	Normal		Data loss in Sensor 1		Data loss in Sensor 2	
	Posi	vel	posi	vel	posi	vel
HIPOFA-F1	0.210	5.56	0.202	5.55	0.210	5.55
HIPOFA-F2	0.210	5.99	0.207	5.99	0.188	5.98
HIPOFA	0.151	5.54	0.146	5.54	0.142	5.53
HIGFA-L1	0.211	5.55	0.203	5.55	0.211	5.55
HIGFA-L2	0.210	5.94	0.207	5.94	0.188	5.92
HIGFA	0.065	6.24	0.066	6.24	0.064	6.24

F1& F2: individual sensor filters; L1& L2: Local Filters

Table 3 : H-Infinity norm (fusion filter)

	Normal	Data loss in sensor 1	Data loss in sensor 2
HIPOFA	0.0523	0.0523	0.0523

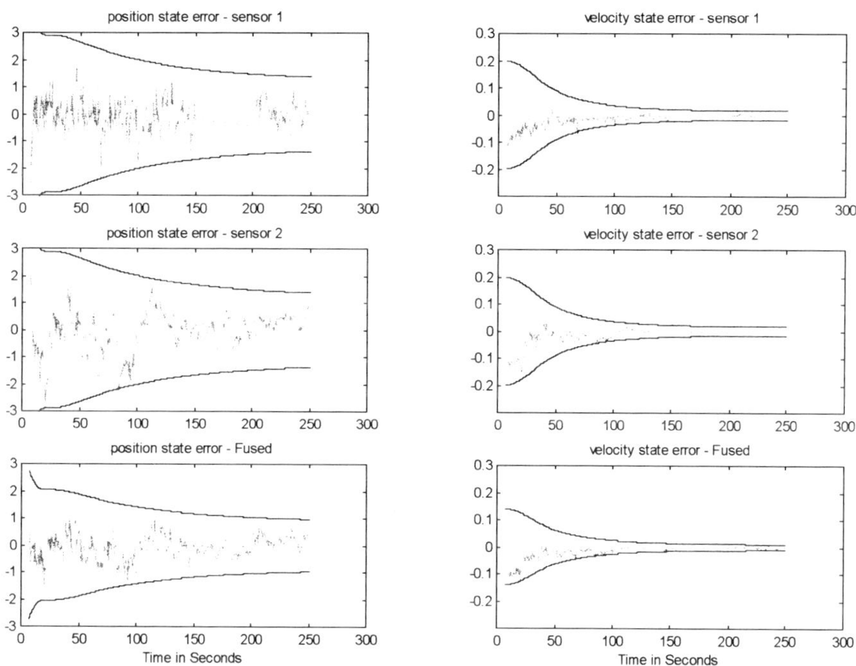

Fig. 1. State Errors with Bounds for HIPOFA (var(v2)=9var(v1) ; Data Loss in Sensor 1)

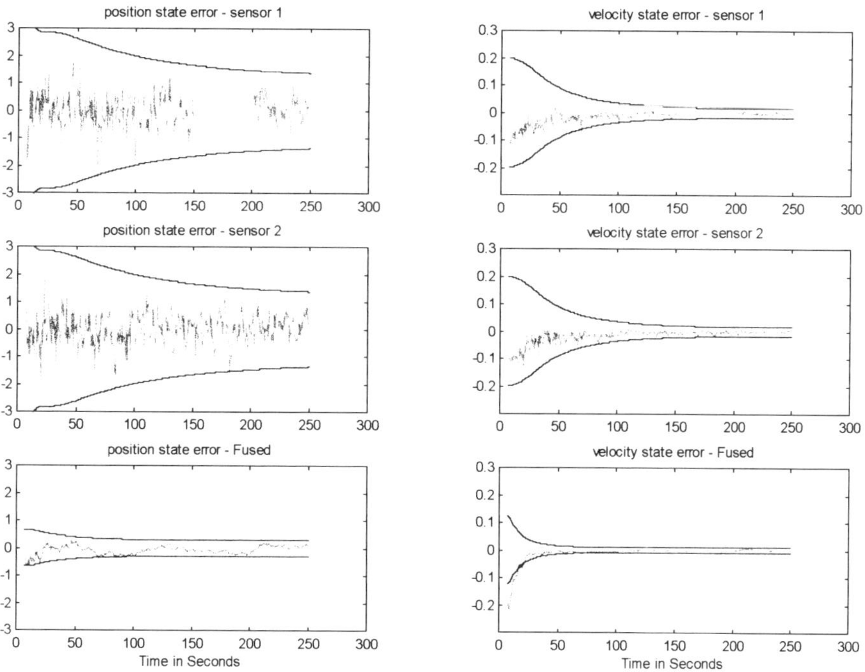

Fig. 2: State Errors with Bounds for HIGFA (var(v2)=var(v1) ; Data Loss in Sensor 1)

KES '01
N. Baba et al. (Eds.)
IOS Press, 2001

A Proposal for Adaptive Routing using Neural Predictor in a Distributed Environment

Florin IONESCU
University of Applied Sciences Konstanz, Postfach 10 05 43, D-78402 Konstanz, Germany

Dragos AROTARITEI
Aalborg University Esbjerg, Niels Bohrs Vej 8, 6700 Esbjerg, Denmark

Mihai Horia ZAHARIA
"Gh. Asachi" Technical University, B-dul Mangeron Nr. 53. A Iasi 6600, Romania

Uffe Koch WIIL
Aalborg University Esbjerg, Niels Bohrs Vej 8, 6700 Esbjerg, Denmark

Abstract: The internet applications continuously increase their development. This increasing complexity makes us to believe that the classical approach in data flow control offers only partial solutions. In this paper adaptive router architecture is proposed, based on the load predictions in order to optimize the routing using this supplementary criteria

1. Introduction

The continuous growth of Internet has made necessary to find and implement innovative solutions in order to improve routing performance. New router architectures with a processing capability of millions of packets/second are being used now. Although recent researches in transmission technology have aided in improvements in transmission bandwidth, other elements of a network also play a vital role in the performance improvement.

The area of distribution for networks hit new ground every day and the traffic over the Internet are escalating. One can easily understand that there must be some factors that limit the amount of traffic that can be sent. Traditionally, the bottleneck has been the routers. The strategy of the router vendors have been to "use a bigger hammer" to scale up the performance. As a result, there is a common misbelieve in the network industry is that routing is inherently slow. Some say "switch where you can, route where you have to".

2. The Routing Problem

The routing problem is summarized shortly as follow. User-level benchmarks, that record the performance of data transfers across the network provide some information, but suffer from poor scaling and the inability to determine network topology.

A network performance prediction system must have the ability to give predictions for any combination of machines selected from a large set of possible machines at many sites across the network.

Topology information is extremely important for performance-based machine selection. For example, without topology information, it is impossible to determine when multiple communication paths within the same application will be sharing a single network link.

Failure to consider this intra-application sharing can result in an overestimation of the network performance an application can achieve. An alternative to benchmarking is obtaining performance information through direct queries to the components making up the network. There are many methods used for packet switching in routing technology and the most used are:

- IP Switching: Traffic-based, per-hop, downstream originated;
- CSR: Traffic-based, originated by downstream/upstream/both;
- Tag switching: Topology based, one VC per route;
- ARIS: Topology based, one VC per egress router;
- MPLS combines various features of IP switching, CSR, Tag switching, ARIS.

An ideal router has certain characteristics as follow:
- high internal bandwidth for intra processing of packets;
- high processing power or high bandwidth (at rate of several million packets/sec);
- specific processing of incoming packets that requires manipulation on a number of relevant fields of a packet header.
A centralized CPU is normally responsible for the per packet processing.

Differentiated Services demands the following operations to be supported:
- packet classification according to requirements;
- buffer management. i.e. how much buffer space to be given for certain kinds of network traffic and what packets to discard;
- packet scheduling, which is made more efficient by the new developments in the areas of fair queuing and per flow scheduling.

3. Adaptive Routing

Usually by adaptive routing is seen as a concept that helps in lowering average message latency, increasing network throughput but has received less importance due to the overhead involved with it. This is primarily due to the use of virtual channels and more complex routing schemes. A virtual channel slows down the clock by 25-30 % but can be improved by a pipelined design. A number of solutions for increasing the cost effectiveness have been proposed. Path selection mechanisms (for example: Least Recently used, Least Frequently used) can improve efficiency. Adaptive routing table look up can be implemented in a number of ways:

Full Table implementation where size is proportional to the number of nodes in the network.

Meta-table that partitions the networks into groups. The use of meta-table reduces memory requirements.

In general the look ahead adaptive scheme serves good for low and high workloads. This approach is insufficient if we take into account the variety that exists in the Internet structure. As is previous mentioned most proposed solutions, either hardware or software approach, are efficient only to solve some subclasses from the involved topics. Became clear that the mean of "adaptive", must be understand a little beat different. The approach of traffic load prediction can provide a good optimization in the packet routing. Usual the chosen path is the one that is functionally or has the minimum number of hopes until the destination is reached. But the instantaneous load on a path is dynamic so it will be most efficient if, by the use of traffic prediction, the selection will be made by taking into account this parameter in front of path depth.

The most used prediction techniques, for network performances are:

- **Application based:** the history of application run is the analyzed parameter.
- **Benchmark based:** the application itself made his local benchmark's in order to made an optimal decision. This techniques have some important limitations:
- **The scaling problem:** the usual numbers of machines that are grouped into a LAN make the problem to become too complex from communication and/or data analyses.
- **The invasiveness:** the local benchmark introduce himself a supplementary overload in the network
- **The topology:** the involved algorithms can have O(P!) complexity.

It appears, now, that is simpler to move the prediction process at the router level. One of the classic methods is to use the neural networks. Here some observation can appear relative to this choice:

Disadvantages:

- great complexity of algorithm;
- lose time with training;
- large amount of memory is needed;

Advantages:

- the router is self-reconfigurable for optimal performance ;
- the impact of scalability problem is reduced;
- the topology problem does not exist;

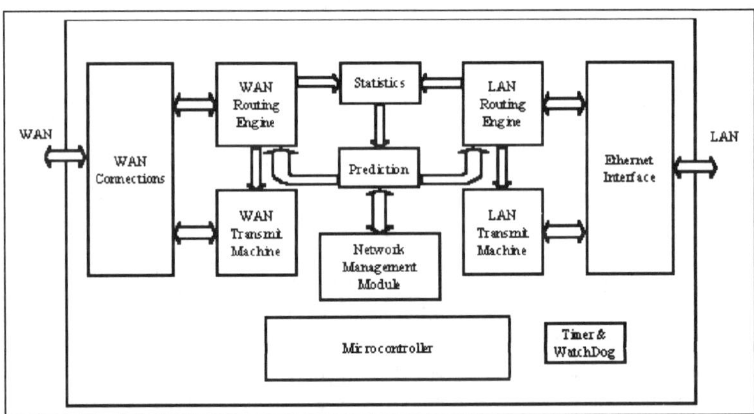

Figure 1. Adaptive router — internal architecture.

The system have two distinct phases in his work: **the training phase** (when the classic methods are used to make packet routing) and **the optimal phase** (when the prediction is used). The training phase is reached every time when the error rate in prediction is over a predefined value, by a long time – the non-optimal phase.

This method can be implemented both at software or hardware level. The last one is recommended because minimize some of the method disadvantages. The prediction module can de implemented into a FPGA area coupled with an high speed RAM which are controlled by micro controller as is presented in Figure 1.

4. Neural Network Predictor

Neural networks (NN) are used widely for time series prediction (also named temporal sequence processing -TSP). Various architectures and learning algorithms are present in a very rich literature. Multilayer perceptron (MLP), recurrent NN, Radial basis Functions NN based on ARMAX models, Temporal NN, TDD (Time Delay NN) and so on are only ones of the different types of NN [4-6] used in a vary large area of applications.

Examples of this approach are market predictions, meteorological and network traffic forecasting, chaotic time series prediction for modeling chaotic systems. Two important criteria are taken into account in these applications: the frequency with which data should be sampled, and the number of data points which must be used in the input representation.

Often these NN are feed-forward architecture [5] and employ a sliding window over the input sequence. In prediction one-step ahead, usually, the present or the future value of a time series is a nonlinear function f of its k past values:

$$y(t) = f(y(t-1), y(t-2), ..., y(t-k))$$

A measure of error, often used in performance evaluation, is Error Mean Square:

$$E = 1/2 \cdot (\hat{y}(t) - y(y))^2 \text{ - where } \hat{y} \text{ denote the predicted value.}$$

5. Experimental results

In order to prove the soundness of the proposed idea a number of simulations was made. The data sets used in this simulations where taken directly from *www.tuiasi.ro* server statistics for at a day level for three different sub-servers which make routing in three different heterogeneous LAN's at the inbound and outbound traffic rate in Kb/s (Figures 2-4).

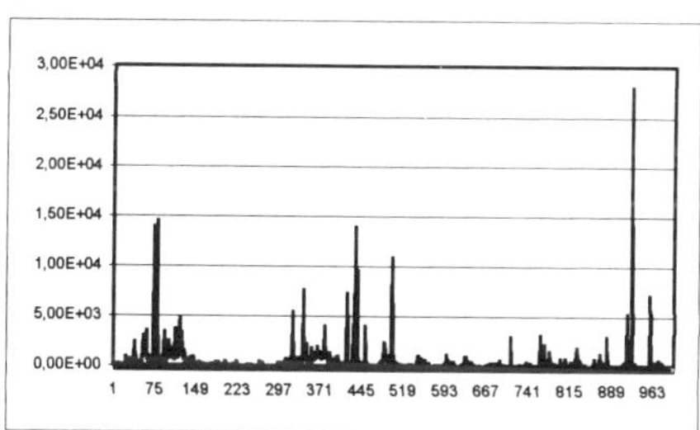

Figure 2. The inbound

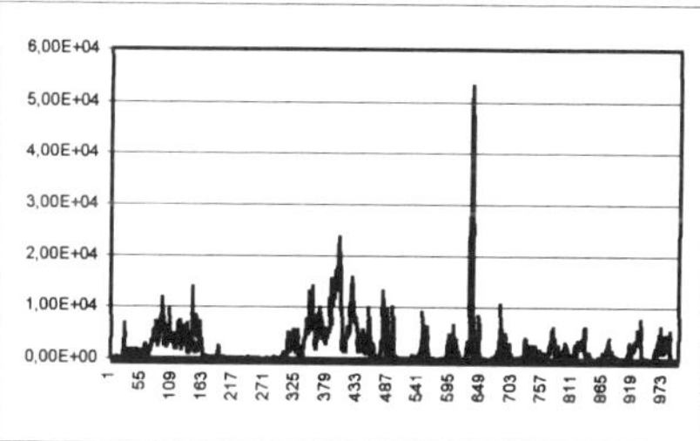

Figure 3. The outbound

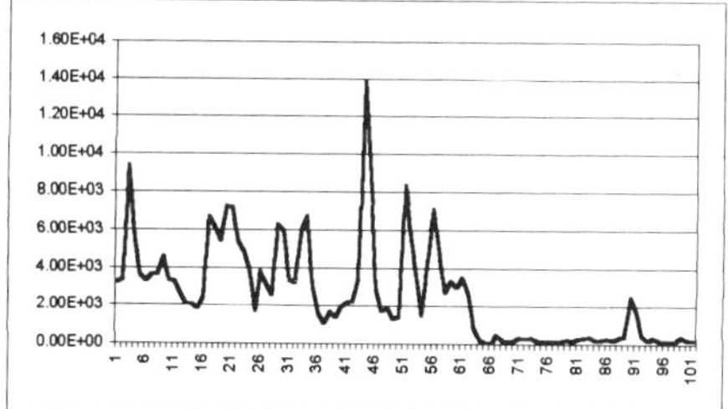

Figure 4. The inbound (100 samples selected)

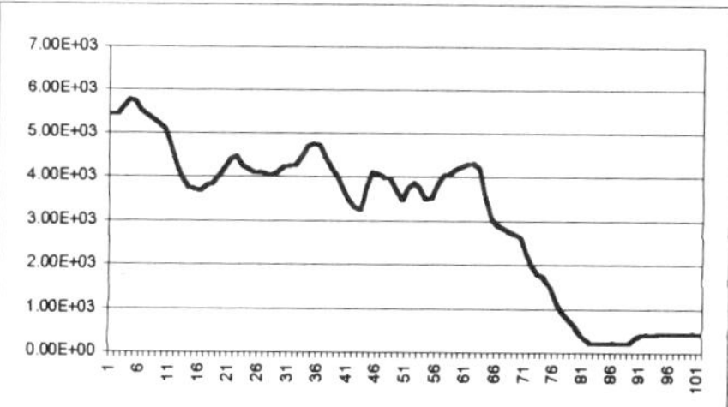

Figure 5. The inbound averaged

We propose a typical MLP structure with 24 neurons on the firs layer, 16 on second layer and one on the last layer. The activation function is sigmoidal for the neurons from the first two layers and linear for the neuron (or neurons) in the last layer.

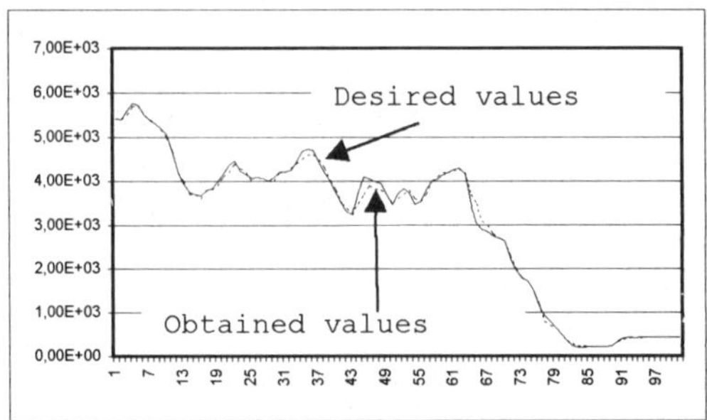

Figure 6. The desired and predicted values.

A classic back propagation algorithm has been used for training NN. A dynamic variant architecture (so namely recurrent MLP) with modified back propagation algorithm has been tested, also.

It is clear that the shape is very abruptly and hard to be predictable. In our experiments we used an average over last k samples (figure 5). The NN predictor was trained with series with k =3, 5 10, and 20 samples. The result in the testing phase (after a learning phase over a 100.000 cycles over all the samples), for k = 10 is given in Figure 6. The dotted line represents the predicted value and the continue line represents the desired values.

6. Discussion and Conclusions

The simulation results, which are in good agreement with the proposed model, prove that this approach can be efficient for a large area of routers. An adaptive standard in packet routing is in progress. Other experiment can use more performing NN for prediction: using standard FIR in synapses, modular NN and additional techniques of preprocessing the signal at the NN predictor input. Technologies for p-steps ahead are also very promising.

Time series data mining area of research could give some solutions for prediction of high peaks or prediction with other .

References

[1]   B. Lowekamp, O'Hallaron, Th. Gross, Direct Queries for Discovering Network Resources Properties in a distributed Environment, *Cluster Computing*, 2000, pp. 1-13.

[2]   R. Morris, E. Kohler, J. Janoti, M.F. Kaashoek, The click modular router, *Operating system review* 34 (5), December 1999, pp. 217-231.

[3]   J. Touch, Board for Realizing Intelligence in Network, *USC /Information Science Institute*, May 1997.

[4]   A. Cichocki, and R. Unbehauen, Neural Networks for Optimization and Signal Processing, John Wiley & Sons, 1993.

[5]   R.P. Lippmann, An Introduction to Computing with Neural Nets, *IEEE ASSP MAGAZINE*, April 1987, pp 4 –22.

[6]   S. Haykin, Neural Networks, A Comprehensive Foundation, Prentice-Hall Inc., 1999.

[7]   Internet resources: http://www.rad.com/products/family/rj-018/rj-018.htm, http://www.tuiasi.ro.

On Syntactic Character Recognition Using Attributes

Mihail Prundaru
Siemens Dematic AG, Buecklestrasse 1-5, D-78467, Konstanz, Germany

Ioana Prundaru
Swiss Stock Exchange, Selnaustrasse 30,CH-8021, Zurich, Switzerland

Marius Bosica
Compaq Computer GmbH, Leinfelderstrasse 60, D-70771, Leinfelden, Germany

Abstract: The paper presents some aspects of the syntactic pattern recognition paradigm using attributes for character recognition. The input characters are identified, thinned to a one pixel width pattern and a feature-based description is provided. The method's features are the set of *terminals* (or terminal symbols) for the application. Each feature has a set of three *attributes*. The combined feature and attribute description for each input pattern preserves in a more accurate way the structure of the original pattern. The grammar inference engine uses the feature-based description of each input pattern from the training set to build as many grammars as classes of patterns. For each input pattern from the training set, the production rules are derived together with all the necessary elements: the *nonterminals* , branch and testing conditions. Since the grammars are regular, the process of deriving the production rules is simple. All the productions are collected together providing the tags to be consecutive, without gaps. The size of the class grammars is reduced at an acceptable level for further processing using a set of Evans heuristic rules. These algorithm identifies the redundant productions, eliminating those productions and the correspondent nonterminal symbols. The stop criteria for the Evans *thinning* algorithm makes sure that no further reductions are possible. The last step of the grammar inference process enables the grammar to identify class members which were not in the training set: a cycling production rule. The above built grammars are used by the syntactic (character) classifier to identify the input patterns as being members of *a-priori* known classes.

1. INTRODUCTION
1.1. Parsing

The task of the system is to sort the input patterns in m given classes (for numerals m is 10, being the digits from 0 to 9). The terms used in this paper are presented in [5]. We'll call these classes $\omega_0 \dots \omega_m$. Each input pattern is "checked" and the basic *features* (primitives) are identified; for each primitive a set of attributes is computed. The features set is a-priori determined by other application. The features set is shown in figure 1 [6]:

Figure 1. The features set

The features are used as the terminals for the pattern grammars , one grammar for each class.

Let $G_1, G_2 \dots G_m$ denote the pattern grammars [5] where:

$$G_j = (V_{N,j}, V_{T,j}, S_j, P_j)\ \ 1 \le j \le m \tag{1}$$

Let us call the features set as :

$$F = \{\ f_1,\ f_2,\ f_3,\ f_4,\ f_5,\ f_6,\ f_7,\ f_8,\ f_9\}\qquad\qquad(2)$$

and the attributes set :

$$A = \{\ a_1,\ a_2,\ a_3\}.\qquad\qquad(3)$$

Each input pattern Π_i is regarded as a string of concatenated features , each having the attributes.

Each attribute has a finite set of allowed values; let's call V_1 as the set of all possible values for the attribute a_1. Analog we define V_2 and V_3 as the sets of values for the attribute a_2 and a_3 respectively.

With the notations made above ((2) and (3)), an input pattern Π_k can be represented as a string of attributed features as follows:

$$\Pi_k = \{\ f_i,\ [v_{i1};\ v_{i2};\ v_{i3}]\}\ \oplus\ \{\ f_j,\ [v_{j1};\ v_{j2};\ v_{j3}]\}\ \oplus\ ....\ \oplus\ \{\ f_m,\ [v_{m1};\ v_{m2};\ v_{m3}]\}\qquad(4)$$

where f_n is one of the given features (for the numeral case is 10), and the $v_{n1};\ v_{n2};\ v_{n3}$ are one of the allowed values of attribute a_1, a_2, and a_3 respectively. The $\oplus$ stands for the concatenation symbol. For simplicity we'll use the string representation for features of an input pattern without the attribute information; we'll use the semantic information (attributes) when necessary. Therefore the input pattern Π_k can be seen as a string of features $\Pi_k = f_i\ f_j .... f_m.$

Let W_j denote the language generated by G_j over $V_{T,j}$ set. We assign the input pattern Π_k to class j if

$$\Pi_k \in W_j\qquad\qquad(5)$$

The method that decides whether or not a given word belongs to the language generated by a class grammar is implemented as the parser. We process in a top-down manner in order to parse the input pattern; the other approach uses the bottom-up manner. We start with the start symbol and , using the production rules from each given grammar, try to obtain the terminal given word.

1.2 Syntactic distance

Ideally the grammars should be built to have the following properties: the languages generated are disjoint, and each input pattern belongs to one of the languages. In reality there are patterns which can be generated by more than one class grammar, and there are patterns which are generated by no grammar at all. In the first case with more grammars recognizing the input pattern, the solution is simple use the first Grammar which recognized the pattern. For the second case a syntactic distance is introduced. The second case is handling the "noisy" input patterns, with other words the imperfections in our methods.

The first distance introduced is the Levenshtein distance [3] between the input pattern and each class grammar. This distance assumes that the input pattern is a noisy word; the possible errors induced by noise can be [1] substitution, deletion and insertion.

Due to these error types the input pattern, seen as a word, distanced itself from the words generated by the class grammar. The distance is given as the minimum of the weighted sums of the above error types for each word from each language. Let n_{Si} be the number of substitution errors for word w_i, n_{Di} the number of deletion errors and n_{Ii} the number of insertion errors respectively; then the Levenshtein distance d_L is computed for each word w_i as it follows:

$$d_L (\Pi_k, w_i) = k_1 \, n_{Si} + k_2 \, n_{Di} + k_3 \, n_{Ii} \tag{6}$$

where the k_1, k_2 and k_3 are the weighted coefficients for each possible type of errors.

The method looks for the minimum of these distances for each class grammar.

1.3. Attributed distance

Since we use attributes for each feature, a second distance is computed: the distance between attributes[7]. This distance is a weighted sum of the *distance* between the features attributes. The attributes have a different contribution to the pattern description.

For each word belonging to a given grammar, a combined attributed and syntactic weighted distance is computed.

2. GRAMMAR INFERENCE

The method uses the concept of a *Programmed Attributed Regular Grammar* to build the classifier [5]. The process of finding the production rules which generate each input pattern from the training set is called the grammar inference. The typical grammar inference process steps are shown below:

a. collect data

- each input sample , input word, is parsed into the basic elements.

- based on its decomposition, the production rules are derived.

- the trainer will make the correct assignment of the input pattern.

- the non-representative patterns will be rejected.

After all the samples were parsed, the second step can begin.

b. the trainer provides some "good" samples for each class [8].

c. create all the class grammars

- for each grammar, all the production rules are collected and made continuous modifying the jump tags.

- the redundant productions are replaced in order to appear only once.

d. optimization of the production rules using the Evans heuristic rules [5]:

- the production rules are inspected in all the possible combinations;

- the redundant productions are identified; for these productions , one non-terminal is replaced by the other one throughout the production rules.

- the residue productions are used to   generate recursive productions [1]; in many practical cases the residue productions exist and the grammar is recursive: it generates not only the samples from the training set, but also similar patterns.

- all non-reachable productions, due to replacements, will be deleted.

e. the stop criterion is based on the maximum of the *Grammar's power* [5].

3. SYNTACTIC RECOGNITION OF NUMERALS USING ATTRIBUTES

The numeral recognition is used in many image understanding applications such as: mail sorting, bank forms, industrial items tracking etc.

The input patterns are scanned (usually from paper) and stored as a binary rectangle dot matrix for further processing. As preprocessing we use the *deskew* of the character's dot matrix and the *thinning*. What remains after thinning of the input character is a skeleton which has a width of one pixel, as shown in figure 3.

Each thinned character is *inspected* to find the important points for its description: end, branch or corner points. This is the first level segmentation for an individual character. Then, the *features* or *primitives* are identified while *inspecting* the line which connects two consecutive important points. That's the second level segmentation for a character. Each feature is identified together with its attributes; the attributes are the *relative position* inside the character's rectangle, the *relative size* and a *flag* which indicates *any rotation* from the ideal feature.

Figure 3. Post code

Each attribute has a set of discrete allowed values which are calculated relatively. The method's results are presented in figure 4.

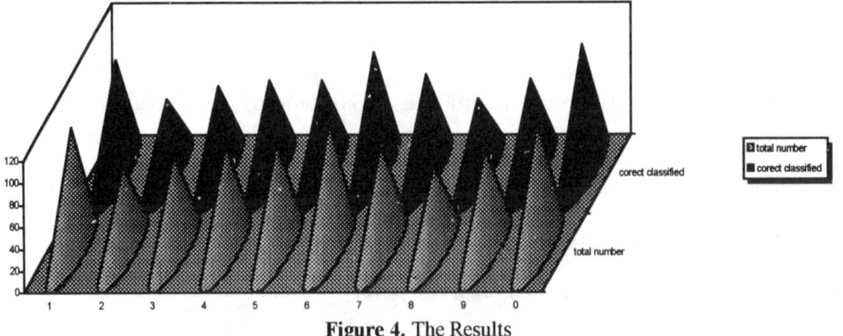

Figure 4. The Results

4. CONCLUSIONS

The paper is an attempt to automate the process of grammar inference for numeral recognition. The algorithms used to build the grammars incorporate several search algorithms. The computational effort is minimal and is mainly directed towards searching through production rules. The authors used real data for training and for recognition. The training set had 1000 samples, and the input set the same.

The main parts of a syntactic pattern recognition application are: determining the grammars based on the existing samples - the grammar inference, and deciding which pattern belongs to which class - the classification. Both problems were approached by this paper and solutions were given. The usage of attributes for each feature eased the pattern description, but added complexity to the grammar design.

References

[1] S. Banks, *Signal Processing, Image Processing and Pattern Recognition,* Prentice Hall International Ltd., 1990

[2] A. Conway, *Page Grammars and Page Parsing - A Syntactic Approach to Document Layout Recognition,* Proceedings of the Second International Conference on Document Analysis and Recognition, Tsukuba, Japan, 1993

[3] K.-S. Fu, *A Step Towards Unification of Syntactic and Statistical Pattern recognition,* IEEE Transactions on PAMI, May 1986.

[4] K.-S. Fu and T. L.Booth, *Grammatical Inference: Introduction and Survey-Part I & II,* IEEE Transactions on PAMI, May 1986.

[5] M. Prundaru and I. Prundaru, *Inference process of programmed attributed grammars for character recognition,* SPIE's 13th Annual Symposium on Electronic Imaging, San Jose, California, USA, 2001.

[6] M. Prundaru and I. Prundaru, *Syntactic Handwritten Numeral Recognition by Multi-dimensional Grammars,* SPIE's 12th Annual International Symposium on Aerospace/Defence Sensing, Simulation and Controls, Orlando, USA, 1998.

[7] M. Prundaru and I. Prundaru, *Syntactic Numeral Recognition Using Programmed Attributed Grammars,* Fifth Joint Conference on Information Sciences, vol.1, pp.867-870, Atlantic City, NJ, USA, 2000.

[8] Y. Xu and G. Nagy, *Prototype Extraction and Adaptive OCR,* IEEE Transactions on PAMI, Vol.21, No.12, pp.1280-1296, December 1999.

[9] V. Wu, R. Manmatha and E. Riseman, *TextFinder: An Automatic System to Detect and Recognize Text in Images,* IEEE Transactions on PAMI, Vol.21, No.11, pp.1224-1229, November 1999.

KES '01
N. Baba et al. (Eds.)
IOS Press, 2001

Semiotic Learning for Social Robot Using Evolutionary Computing

Tetsuo SAWARAGI and Takuji SHIMOKAWA

Dept. of Precision Eng., Graduate School of Eng., Kyoto University
Yoshida Honmachi, Sakyo, Kyoto 606-8501, Japan

Abstract. For realizing a naturalistic collaboration between the human and the robot, we have to establish the intention-sharing from the series of motion data that are observed and exchanged between the human and the robot. In a word, this is a problem to detect "meanings" out of the digitized data stream. In this paper, we propose a novel approach based on the semiosis, and present a number of ways for implementing the ideas using recurrent neural networks and interactive evolutionary computing method.

1. Introduction

Motor control for a social robot poses challenges beyond stability and accuracy. Human observers will perceive motor actions as *semantically* rich, regardless of whether the robot intends the imputed meaning. Such perception on intent and emotional state transparent at an intuitive level is constrained by the robot's physical appearance and movement, and is also formed by those with whom it interacts (i.e., a human). We attempt to make a robot regulate its interactions so that they can suit its perceptual and motor capabilities with what humans naturally perceive in their natural interactions among people. For this purpose, we have to design the robot's behaviors readable to humans and variable in its allowable operations. This will let both of robot and human participate in natural and intuitive *social* interactions. In this paper, at first we describe about the background ideas of our work. Next, we introduce an idea of visual segmentation of a stream of continuous motor actions, and based on this we design a mobile robot that can generate a variety of semantically rich movements. By getting the feedback of how the human observer feels in seeing those, the robot evolves its behaviors so that a human observer can form consistent emotional ontologies in a cooperative fashion with the robot.

2. Semiosis as a Framework for Bridging between Biology and Machine Intelligence

Engineers who want to make semantic machines are commonly faced with the task of defining "meaning", which at present exists only in brains, and then with the task of learning how to make or cause "meaning" in machines. Shannon-Weaver information theory, which is representational, has divorced meaning from information and therefore does not apply to brains [1].

Peirce instead formulated his semiology from a more general definition of semiosis which introduces the interaction of three abstract subjects: the *sign*, its *object* and its *interpretant* [2]. His theory is moreover not limited to intentionally transmitted and artificial signs, but covers all phenomena that eventually can be perceived by our *senses* and interpreted by our *mind*. The core of the semiosis is the problem of the relationships between *meanings* and *representations*: A representation, as a material object or process, has no meaning in itself, but is mostly signified through usage in practice being enclosed within an actor. The importance of the above triadic relationships in any sign system has been repeatedly stressed by many in the context of biology and genetics.

From a computational perspective, semiosis is a combination of the following subprocesses [3]:

- encoding of the sensations by using elementary signs
- associating encoded sensation with codes of actions
- constructing strings of codes for "states-actions-..." stored in the memory assigning to these strings of signs values of goodness interpretable under specific goals, which allow interpretation of "experiences"
- discovering classes of experiences (generation of the concepts)
- forming hypotheses of new behavioral rules using previously stored results of prior processes of semiosis

In the followings of this paper, we attempt to apply the above semiotic process into a human-robot system design, where a robot tries to build emotional concepts through the interactions with a human.

3. Social Robots and Social Learning

In the curret market, there appeared a distinguished shift of preferences of the consumers seeking for the artifactual products. Typical examples of that are "pet robots" and "cohabitant breeding game software", both of which are based upon the interactions between the products and the human user and the users seek for enjoying the conversation, both in verbal and nonverbal communication. The common properties embedded in those products are;

- lack of technology is supplemented by the user's proactive engagement.
- users can actually feel that they are committing with the artifacts.
- entrainment of the human user within the interactions with the artifacts is designed.
- design of human-intervening "events", rather than the "products"
- they are based upon the bi-directional communication via analog media.
- target of the consumers are people who seek for experiencing, rather than the people whose needs and goals expected to the products are explicit and preexist.

In a word, those products are oriented towards the design of "relations" and/or "process" between the artifacts and the human user, rather than the products isolated from the user's participation. In order to keep the users in their active and continuous commitments, we have to desin the emergence of novel views and perspectives out of the increased interactions between the human user and the artifacts, and the prerequisite therein is the user's continuous *participation*; becoming engaged, and being embedded in a situation.

This is a typical style of a *social* interaction, and is typically seen in *turn-taking* between a mother and her baby in humans. Such an interaction emerges without any mechanism which is explicitly controlling turn-taking between them: a mother responds to her baby's pauses in sucking with jiggling in order to encourage the infant to resume sucking. The success of this emergent turn-taking (jiggling, sucking) relies fully on the mother's *interpretation* of the baby's behavior, although the mother received a message that was never sent.

Herein we can relate sociality with semiosis. Namely, a behavior like turn-taking may actually seem to be "social" from an external observer point of view, but there only exists two *self-enclosed* processes in parallel; the process of selectively expressing a behavior which is already a part of specific repertoire and the other process of interpreting that.

Social learning is to improve such appropriate selections based on reinforcement that is brought about by the individual perception of progress relative to the *current* goal (not the absolute one), and/or by evaluateing performance according to its own perspective based on the use of its resources. For a behavior-based system, resources are the behaviors available for controlling the actions of the robot. Assuming that behaviors are programmed with an initial policy, the objective is to try to find some regularities in the interactions between the robot and the human reacted in the robot's use of behaviors over time.

With respect to the meanings that the social robot gets to have, theoretical model is needed to explain how an autonomous agent may originate new meanings. The agent is autonomous in the sense that its ontology is not explicitly put in by a designer, nor is there any explicit instruction. Meaning is defined as a conceptualisation or categorisation of reality which is relevant from the viewpoint of the agent. Meanings need not be expressed through language, but may take many forms depending on the context and nature of the situation concerned as mentioned in the previous section.

In this paper we assume that the origins of meaning are based on construction and selection processes embedded in human-robot interactions. Each individual agent is assumed to be capable to reconstrict its memory generating diversity and variation, which is subjected to selection pressure coming from the human partner.

4. System Architecture

An architecture of our system is illustrated in Fig.1. An agent shown in Fig.2 is a kind of software, but is distinguished from the usual programs in that that learns by observing interactions with a human user and by operating a robot. A robot functions as an interface with an external world for an agent. It collects data from an environment for an agent, and executes a command given by an agent, that brings about some effects on the environment. A human observes a robot's behaviors and returns some responses reflecting what he has felt and/or obtained from a robot's behavior. Having these responses from a human, an agent makes efforts to know about the relations between its exerting actions and how a human has received and deemed them. Sometimes, it has to modify the rules to generate actions that it has obtained so far tentatively. Herein, techniques of machine learning are used to acquire knowledge from human inputs.

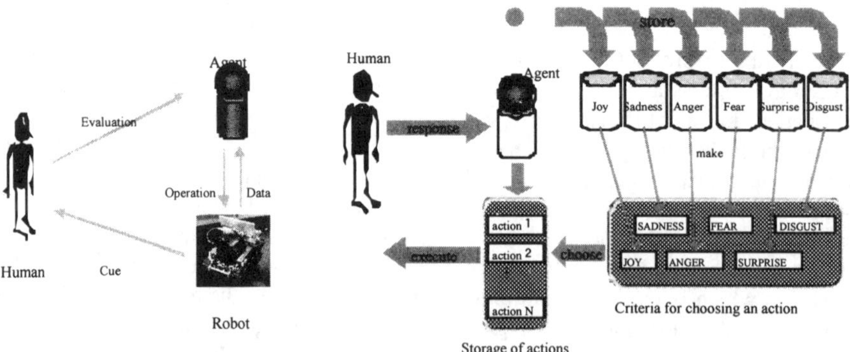

Fig.1 Triadic of Human-Agent-Robot   Fig.2 Details of interactions

For the purpose of enriching and diversifying the behavior of the robot, we introduce the theory of the boundary of visual motion into the description of actions that the agent generates and makes the robot execute [4]. This theory focuses on boundaries to theorize how a motion of an object is seen by the human. We introduce this idea because it is consistent with the human recognizing ability for visual motions in general.

Fig.3 shows the template that our agent uses to generate an action. Actions generated by an agent can be processed as follows. One unit of actions consists of series of motion states and motion boundaries arranged in a line. And each motion state has a time duration parameter that specifies the time the state lasts. A motion state corresponds to a motor state of a robot, while a motion boundary specifies the process of transition between two motion states. This is represented as a linear bit string (i.e., genotype representation).

The agent's goal of learning is to get to represent its emotions properly by the robot's "acting". When the agent executes an action, it selects one from the storage of actions it possesses. In selecting an action, the agent evaluates all actions in the storage under a particular criterion, and then it chooses one with the highest evaluation from action alternatives. An agent's learning process consists of two phases. One is to improve a repertory of actions, and the other one is to improve the criteria used in choosing an action. These criteria are prepared as many as the number of emotions that are felt by a human who watches the behaviors of a robot (described later). We use the *interactive genetic algorithm* to modify the actions in the storage. The criteria for choosing a specific action are dynamically formed in referring to a human's responses that are stored in the storage paired with the actions that bring about those responses. The agent forms those criteria using statistical method.

The learning process progresses as follows: First, the agent selects a specific action pattern to execute, and the robot moves around in conformity with that pattern. Observing that behavior, a subject guesses the class of emotions it indicates and returns the result to the agent as his/her responses. Then, the agent modifies the current contents of the storage of actions, and renews normative action patterns corresponding to the designated class of emotions that the agent has formed so far. The above interaction is iterated in turn and the storage in the agent is kept being reformed.

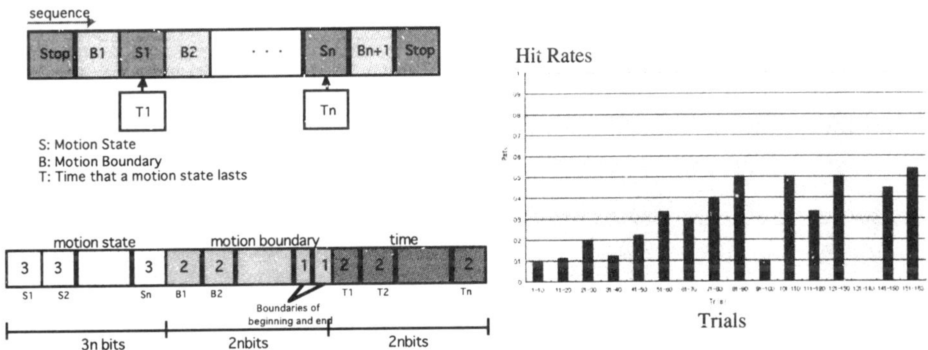

Fig.3 Genotype representation　　　Fig.4 Rate of the agreement between human and robot

5. Experimental Results

We show the results obtained from a particular subject. The setting of this experiment is as follows.

- The agent distinguishes six classes of emotions: joy, sadness, anger, fear, surprise, and disgust.

- At the beginning of the interaction, the agent has randomly-generated fifty patterns of actions in its storage.
- 163 times of interactions were done in the experiments.
- In an action, there are 4 motion states between 'Stop' states of both ends of it. So n = 4 in Fig.3.

Fig.4 shows the transition of the rate of the agreement between what the human judged and what the agent actually intended. At the beginning of the interaction, the rate is increasing. But it stops increasing and gets unstable in the latter half.

More detailed investigation into how the meanings of each emotional concept within the agent are reconstructed will be presented elsewhere [5], but in brief it was found that the representation of "joy" is quite different from those of other emotions, and those of surprise and disgust are not very similar to each other. On the other hand, there are strong correlations among the representations of sadness, anger, fear, and surprise or disgust. It was found that the subject has come to rely on a particular parameter like a changes of the throttle of the robot to judge the internal state of the agent.

6. Conclusions

In this study, a software agent was developed which is capable to evolve its own behavior through interactions with a human using a robot as a mediator. To realize this, at first a robot was developed, and then for letting the agent learn from human responses, interactive genetic algorithm and statistical technique were adopted. The results of the experiment show that a partial success of acquiring behavior patterns representing emotions was attained and we can expect to extend our system to the one that can further activate human-robot interactions in sharing their emotions. We have to note that not only robots but also a human himself/herself may change during this interactions; a human may become extremely sensitive and keen in finding even faint cues to guess and distinguish the agent's internal state from the behavior of the robot. Such human-robot co-evolution is of extreme importance in the design of social robots in general.

References

[1] Freeman, W.J. (1997). A Neurobiological Interpretation of Semiotics: Meaning, Representation, and Causality, *Proc. of the 1997 International Conference on Intelligent Systems and Semiotics: A Learning Perspective*, pp.487-492.

[2] Peirce, C.S. (1955). *Phylosophycal Writings of Peirce*, Buchler, J. (Ed.), NY., Dover Pub..

[3] Meystel, A.M. (Ed.) (1997). *Proc. of the 1997 International Conference on Intelligent Systems and Semiotics: A Learning Perspective*, sponsored by NIST, IEEE, NSF, ARO, Gaithersburg, US.

[4] Rubin, J.M. and Richards, W.A. (1985). Boundaries of Visual Motion, AI-Memo No. 835, Massachusetts Institute of Technology Artificial Intelligence Laboratory.

[5] Shimokawa, T. and Sawaragi, T. (2001). Acquiring Communicative Motor Acts of Social Robot Using Interactive Evolutionary Computation, to be appeared in 2001 IEEE Conf. On System, Man and Cybernetics.

KES '01
N. Baba et al. (Eds.)
IOS Press, 2001

Knowledge Sharing Using Belief Measure in Multi Robot System

Futoshi Kobayashi, Shiro Sakai and Fumio Kojima
Graduate School of Science and Technology, Kobe University
1-1 Rokkodai, Nada, Kobe 657-8501, Japan

Abstract. In this paper, we consider the problem of sharing knowledge in multi-robot exploration. Recently, some research works of multi-robot exploration have studied. However, in traditional research works, it is very difficult to share each robot's knowledge of explored area. We use the belief measure as the expression of sensing information in each robot for exploring an unknown environment. Then, multiple robots share the knowledge of the environment considering the degree of trust for other robots. The effectiveness of our approach is demonstrated by a real experiment for the case of two mobile robots.

1 Introduction

Recently, some research works with multiple robots are studied in various field because various types of robots are developed and produced. In research works, the exploration of an environment with multiple robots have received much attention. The use of multiple robots have several advantages over single robot systems. Firstly, multiple robots can explore an environment faster than a single robot. Secondly, using some cheap robots can be expected to be more fault-tolerant than using one powerful and expensive robot. Finally, multiple robots can localize themselves more efficiently by exchanging sensing information about the environment. Though the exploration problem has been dealt with for single robot[1], there are few approaches for multiple robot. Rekleitis et al. proposed a multi-robot exploration of reducing the odometry error during exploration[2, 3] and Burgard et al. consider a collaborative multi-robot exploration in order to minimize the overall exploration time[4]. Moreover, Singh et al. proposed a technique for heterogenous robots[7]. In this technique, the robots share a common map which is built during the exploration. However, traditional approaches of exploring by multiple robots can not share sensing information efficiently in multi robot system.

In this paper, we consider the problem of sharing knowledge in collaborative exploration of an unknown environment by multiple robots. This approach uses grid maps expressed by the belief measures for sensing information of the individual robots. The robots have not a common map, but a map for sharing sensing information. Accordingly, the knowledge of each robot can share considering the difference of sensing information by the sensing ability and the motion ability of each robot. For showing the effectiveness, our approach is implemented on real robots in real environment.

2   Knowledge Sharing Using Belief Measure

2.1   Expression of Knowledge by Belief Measure

In order to explore an unknown environment, we use grid maps to represent the environments. The concept of grid maps is to use a grid of equally spaced cells and to store in each cell the degree of confidence that this cell is occupied by an obstacle. In this paper, we represent the belief measure[5, 6] as the degree of confidence. Here, we define the 2 types of sets as follows:

O: Cell without Obstacle,
N: Cell with Boundary of Obstacle.

Accordingly, each cell has the belief measures $Bel_O(x, y, r, t)$, $Bel_N(x, y, r, t)$ and $Bel_{O \cup N}(x, y, r, t)$. Here, x and y represents the location of the cell, r represents the number of the mobile robots, respectively. Then, the basic basic assignments are calculated as follow:

$$
\begin{aligned}
m_O(x, y, r, t) &= Bel_O(x, y, r, t) \\
m_N(x, y, r, t) &= Bel_N(x, y, r, t) \\
m_{O \cup N}(x, y, r, t) &= Bel_{O \cup N}(x, y, r, t) - Bel_O(x, y, r, t) - Bel_N(x, y, r, t)
\end{aligned}
\tag{1}
$$

Besides, the belief measures are expressed by the basic assignments as follows:

$$
\begin{aligned}
Bel_O(x, y, r, t) &= m_O(x, y, r, t) \\
Bel_N(x, y, r, t) &= m_N(x, y, r, t) \\
Bel_{O \cup N}(x, y, r, t) &= m_O(x, y, r, t) + m_N(x, y, r, t) + m_{O \cup N}(x, y, r, t)
\end{aligned}
\tag{2}
$$

Here, following equation is composed:

$$
Bel_{O \cup N}(x, y, r, t) = 1.
$$

Initially, we define the basic assignments for all cells as follows:

$$
\begin{aligned}
m_O(x, y, r, 0) &= 0 \\
m_N(x, y, r, 0) &= 0 \\
m_{O \cup N}(x, y, r, 0) &= 1
\end{aligned}
$$

2.2   Temporal Knowledge Sharing

In this subsection, we explain the knowledge sharing method of sensing data in each robot. When the evidences $m_{\{O, N, O \cup N\}}$ are obtained from sensing data, $m_{\{O, N, O \cup N\}}(t)$ are calculated from the Dempster's rule of combination[5] as follows (x, y and r are overleaped):

$$
m_O(t) = \frac{m_O(t-1) \cdot m_O + m_O(t-1) \cdot m_{O \cup N} + m_{O \cup N}(t-1) \cdot m_O}{1 - m_O(t-1) \cdot m_N - m_N(t-1) \cdot m_O}
\tag{3}
$$

$$
m_N(t) = \frac{m_N(t-1) \cdot m_N + m_N(t-1) \cdot m_{O \cup N} + m_{O \cup N}(t-1) \cdot m_N}{1 - m_O(t-1) \cdot m_N - m_N(t-1) \cdot m_O}
\tag{4}
$$

$$
m_{O \cup N}(t) = \frac{m_{O \cup N}(t-1) \cdot m_{O \cup N}}{1 - m_O(t-1) \cdot m_N - m_N(t-1) \cdot m_O} = 1 - m_O(t) - m_N(t)
\tag{5}
$$

Here, when the robot can detect the obstacle by its sensors, the evidences m_O and m_N for corresponding cells are 0.05 and 0.0, respectively. On the contrary, when the robot cannot detect, the evidences m_O and m_N are 0.0 and 0.05, respectively.

2.3 Spatial Knowledge Sharing

In this subsection, we explain the knowledge sharing method of sensing data between multiple robot. Each robot share the sensing information of itself and the sensing information of other robots as shown in Figure 1. Here, the degree of trust for other robots $p_{rr'}$ is defined as follows:

$$p_{rr'} = 1 - \frac{1}{1 + \exp(-d_{rr'} + \alpha/\beta)} \tag{6}$$

where, r' and $d_{rr'}$ represent the number of other mobile robots and the distance between the robot r and the robot r', respectively. By using the degree of trust, the basic assignments $m'_{O,N,O\cup N}(r)$ after sharing information between robots are calculated from the Depster's rule

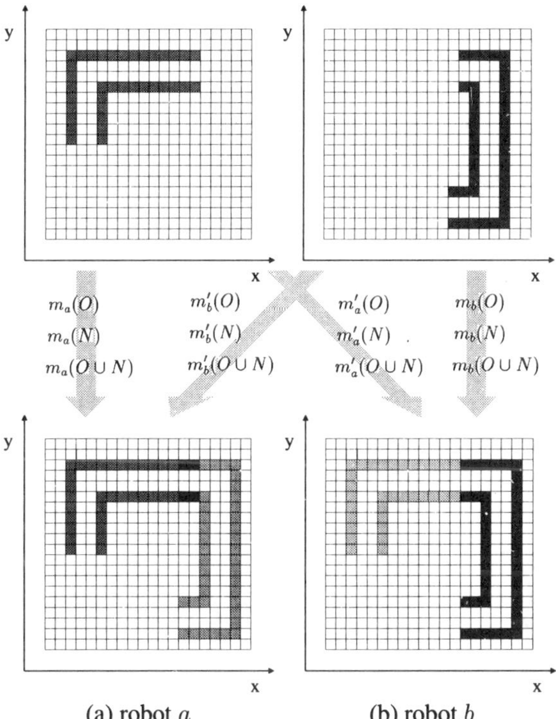

Figure 1: Sharing of Sensing Information by Individual Robot

of combination as follows (x, y and t are overleaped):

$$m'_O(r) = \frac{m_O(r) \cdot p_{rr'} m_O(r') + m_O(r) \cdot p_{rr'} m_{O \cup N}(r') + m_{O \cup N}(r) \cdot p_{rr'} m_O(r')}{1 - m_O(r) \cdot p_{rr'} m_N(r') - m_N(r) \cdot p_{rr'} m_O(r')} \quad (7)$$

$$m'_N(r) = \frac{m_N(r) \cdot p_{rr'} m_N(r') + m_N(r) \cdot p_{rr'} m_{O \cup N}(r') + m_{O \cup N}(r) \cdot p_{rr'} m_N(r')}{1 - m_O(r) \cdot p_{rr'} m_N(r') - m_N(r) \cdot p_{rr'} m_O(r')} \quad (8)$$

$$m'_{O \cup N}(r) = 1 - m'_O(t) - m'_N(t) \quad (9)$$

Accordingly, each robot acquire the grid map for the unknown environment by sharing the sensing information of other robots.

3 Exploration by Multi Khepera Robot

Our approach has been implemented on real robots and in real environment. In this experiment, we use two Khepera robots (Figure 2) equipped with 8 Infra-red proximity sensors in an environment as shown in Figure 3. The diameter of the Khepera robot is $52.5mm$, the size of the environment to be explored in this experiment is $492 \times 820mm^2$, and the size of a grid cell is $2 \times 2mm^2$. Each Khepera robot moves along the right side of obstacles.

Figures 4 and 5 show the individual grid map of each robot for Bel_O and Bel_N, respectively. Figures 6 and 7 show the shared grid map of each robot for Bel_O and Bel_N, respectively. The shading of gray for each cell in these figures indicates belief measures. As shown in each sharing grid map, cells that the robot measures by its sensors is dark, and cells that other robots measures is light. By this experiments, two robots can share their sensing information efficiently and explore the unknown environment.

4 Summary

In this paper, we proposed a method of sharing knowledge in multi robot system for exploring an unknown environment. The key idea of our approach is that the degree of confidence expressed by the belief measure is shared between multi robots. Our approach has been implemented on real robots. The experiment in this paper presents that our approach can share the knowledge of each robot.

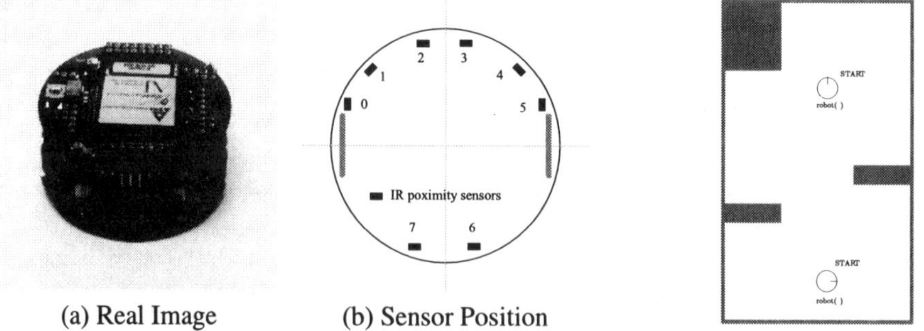

(a) Real Image   (b) Sensor Position

Figure 2: Khepera Robot   Figure 3: Exploration Environment

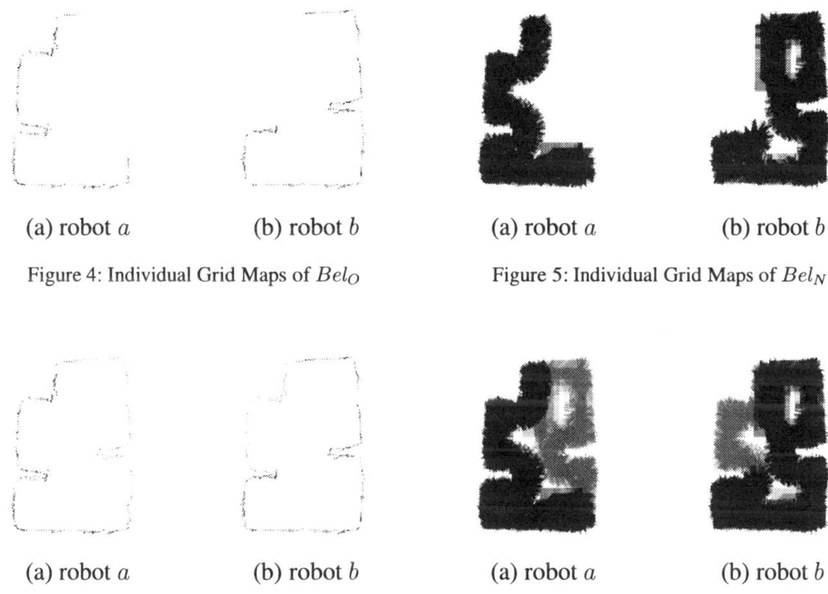

(a) robot a (b) robot b (a) robot a (b) robot b

Figure 4: Individual Grid Maps of Bel_O Figure 5: Individual Grid Maps of Bel_N

(a) robot a (b) robot b (a) robot a (b) robot b

Figure 6: Shared Grid Maps of Bel_O Figure 7: Shared Grid Maps of Bel_N

References

[1] H. Gonzalez-Banos and J.C. Latombe, Planning Robot Motions for Range-Image Acquisition and Automatic 3D Model Construction, Proc. of AAAI Fall Symposium (1998).

[2] I.M. Rekleitis, G. Dudek, and E.E. Milios, Graph-Based Exploration using Multiple Robots, Proc. of 5th International Symposium on Distributed Autonomous Robotic Systems (2000) 241–250.

[3] I.M. Rekleitis, G. Dudek, and E.E. Milios, Multi-Robot Exploration of an Unknown Environment, Efficiently Reducing the Odometry Error, Proc. of International Joint Conference in Artificial Intelligence **2** (1997) 1340–1345.

[4] W. Burgard, M. Moors, D. Fox, R. Simmons, and S. Thrun, Collaborative Multi-Robot Exploration, Proc. of the 2000 IEEE International Conference on Robotics and Automation (2000) 476–481.

[5] A.P. Dempster, Upper and lower probabilities induced by a multi-valued mapping, Ann. Math. Stat **38** (1967) 325–339.

[6] G. Shafer, A Mathematical Theory of Evidence, Princeton Univ. (1976).

[7] K. Singh and K. Fujimura, Map Making by Cooperative Mobile Robots, Proc. of IEEE International Conference on Robotics and Automation (1993) 254–259.

KES '01
N. Baba et al. (Eds.)
IOS Press, 2001

Application of Recurrent Neural Network via Simultaneous Perturbation Rule to Flexible Arm System

Norimichi Kubo, Yutaka Maeda
Department of Electric Engineering, Kansai University
3-3-35 Yamate-cho, Suita 564-8680 JAPAN
maedayut@kansai-u.ac.jp

Abstract

This paper presents a method to control a robot arm made of a flexible material, using a neural network. The simultaneous perturbation learning rule as a learning rule is utilized. If we use the gradient method as a learning rule of the neural network, Jacobian of the plant is essential. However, this learning rule does not require Jacobian of an objective plant so that the neural network uses only outputs of an objective system. Simulation results and actual control results of a real flexible arm system are described to confirm a feasibility of the proposed method.
Keywords : Neural networks, Flexible arm, Simultaneous perturbation

1.Introduction

Neuro-controller(NC) by a direct inverse control scheme(see Fig.1) is one of promising approaches in non-linear control problems.

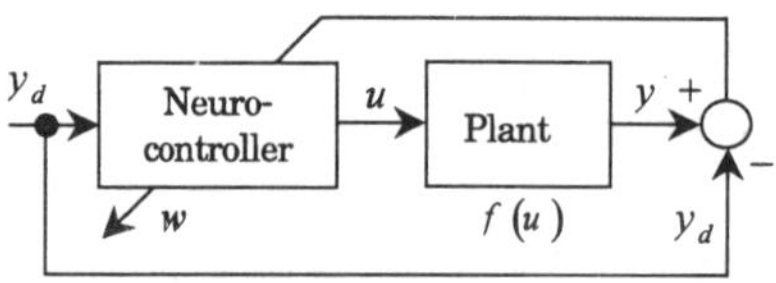

Fig.1 A basic scheme for a direct neuro-controller.

In order to use a NC as a direct controller, the NC must be an inverse system of an objective plant, that is, the NC must learn an inverse system in so-called indirect inverse modeling. Then, the learning rule plays a practical role, because it will be related to an arrangement of overall system.

In the case of indirect inverse modeling, generally, we need a plant model or a sensitivity function of the plant to acquire the derivatives needed for learning such as the backpropagation(BP) method, because the error function is usually defined not by an output of the NN but by that of the plant.

In this paper, we propose a NC using the simultaneous perturbation learning rule for a flexible arm system. This learning rule does not require a derivative of an error function but only values of the error function itself. Therefore, without knowing Jacobian of the objective arm system, we can design a direct neuro-controller. Then the neuro-controller can learn a changing environment about the system as well.

Ordinarily, an error function $J(w)$ is defined by a squared error of the plant. When we use a usual gradient method as a learning rule of the NC, we must know the quantity $\partial J(w)/\partial w$.

Then, we have

$$\frac{\partial J(w)}{\partial w} = \frac{\partial J(w)}{\partial y}\frac{\partial y}{\partial w} = (y - y_d)\frac{\partial f(u)}{\partial u}\frac{\partial u}{\partial w} \quad (1)$$

where y_d denotes the desired output of the plant., so $(y - y_d)$ is known. Moerover, we can calculate $\partial u/\partial w$ like the back-propagation learning rule. However, we don't know the sensitivity function. As a resutlt, we can not obtain a proper modifying quantities for weights in the NC.

On the other hand, we can introduce an idea of the simultaneous perturbation learning rule[1],[2]. In this case, there is no need to know the sensitivity function of the unknown plant. Only using the values of an error function, the learning rule can estimate a gradient of the error function with respect to adjustable parameters, weights of the NC in this case.

The idea of the simultaneous perturbation was proposed by J.C.Spall as an extension of Kiefer-Wolfowitz stochastic approximation[2][3]. J.Alespector et al. and G.Cauwenberghs also proposed the same idea[4][5]. Independently, Y.Maeda introduced the same algorithm as a learning rule of neural networks[1][6][7]. J.C.Spall et al. and Y.Maeda reported some applications of the simultaneous perturbation method in control problems[6][8][9][10].

2.   Simultaneous   perturbation learning rule

Now, we describe the simultaneous perturbation(SP) leaning rule.

Define a weight vector and a sign vector as follows;

$$w(t) = (w_1(t), w_2(t), \cdots, w_N(t))^T$$
$$s(t) = (s_1(t), s_2(t), \cdots, s_N(t))^T \quad (2)$$

Where t denotes an iteration, superscript T is transpose of a vector. $s(t)$ is a sign vector whose components are $+1$ or -1.

The i-th component of the modifying vector of the weights $\Delta w_i(t)$ is defined as follows;

$$\Delta w_i(t) = \frac{J(w(t)+cs(t)) - J(w(t))}{c} s_i(t) \qquad (3)$$

Where c is a magnitude of the perturbation. The weights are updated as follows;

$$w(t+1) = w(t) - \alpha \Delta w(t) \qquad (4)$$

Where α is positive learning coefficient.

Note that only two values of the error function, $J(w(t))$ and $J(w(t)+cs(t))$ are used to update the weights in the network. Any information about the objective plant does not included in the learning rule.

3. Simultaneous perturbation for flexible arm system

Now, we consider a one-freedom flexible beam shown in Fig.2 as an objective plant.

The feedback control scheme is applicable to this problem. Using proper states of the plant, we can design a state-variable feedback control.

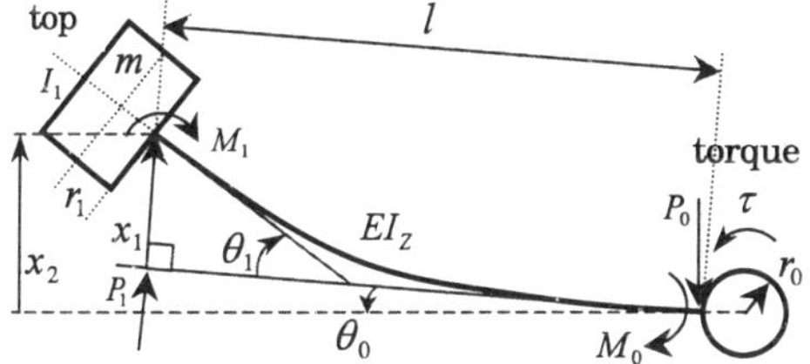

Fig.2 A flexible arm

However, when a system parameter such as a mass of the top of the arm changes, we must recalculated the feedback gain in the state-variable feedback control scheme to maintain suitable control accuracy by using the state equation. Then, we must know the new state equation. This means that we must observe changes of the plant exactly.

That is, in usual control scheme, we must have an exact model of an objective plant. Controllers are basically designed based on the identified model. Therefore, when some characteristics of the plant change, we must detect the change to compensate the controller.

On the other hand, our scheme used here works without this information, because only values of the error function is required. Without a model of the plant or information about the plant, the NC via the simultaneous perturbation can generates proper input for the plant.

Moreover, proper modification of the controller is

carried out under operation. That is, on-line learning is possible in this learning scheme.

3.1. Model of the flexible beam

Mass of the top of the arm is 0.145[kg], and length of the arm is 0.472[m]. The system is basically nonlinear. Therefore, it is relatively difficult to have a proper model of the system. However, neglecting higher modes we can obtain the following state equation and the output equation. Parameters in this equation is determined by actual parameters of the plant such as a mass of the top and so on.

$$\dot{x} = \begin{pmatrix} 0 & 1 & 0 & 0 & 0 & 0 \\ 0 & -9.18 & -1.25 \times 10^4 & 0 & 2.20 \times 10^3 & 0 \\ 0 & - & 0 & 1 & 0 & 0 \\ 0 & -5.02 & -8.54 \times 10^4 & 0 & 2.54 \times 10^4 & 0 \\ 0 & 0 & 0 & 0 & 0 & 1 \\ 0 & -9.18 & 5.13 \times 10^6 & 0 & -1.59 \times 10^6 & 0 \end{pmatrix} x + \begin{pmatrix} 0 \\ 90.5 \\ 0 \\ 49.5 \\ 0 \\ 90.5 \end{pmatrix} u \qquad (5)$$

Then, x is the six dimensional state variable as follows;

$$x = \begin{pmatrix} \theta_0 & \dot{\theta}_0 & x_1 & \dot{x}_1 & \theta_1 & \dot{\theta}_1 \end{pmatrix}^T \qquad (6)$$

u denotes a torque input τ as a command against the plant. Inputs of the NC are four states and their derivatives of Eq.(6) as shown in Fig.3. We know that these six states are necessary to produce a proper command input for the system.

Using these states, the neural network outputs a torque τ.

3.2. Neural network

The objective plant has a dynamics. Therefore, simple multi-layered neural network is not appropriate to control the plant, since the network can not have any memories. Thus we used a multi-layered neural network with feedback.

The network used here is shown in Fig.3. Basic construction is a simple multi-layered network. However, the network has time-delayed feedback inputs from output of itself. This feedback gives dynamics to the network.

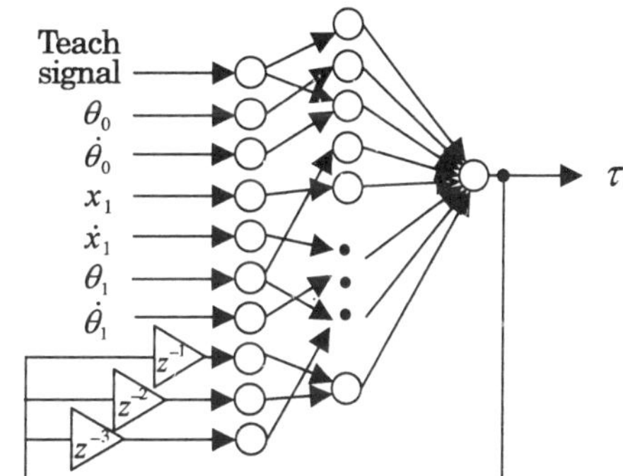

Fig.3 Neural network used here

4. Simulation results

Fig.4 shows a configuration of the system. The error function is defined as follows;

$$J(u(w)) = \sum_i \left(x_{2,i} - d_i\right)^2 \qquad (7)$$

Where, d_i denotes a desired position of the top of the arm at the i-th sampling time. That is, the error is a sum of the squared error of the position of the end for ten seconds. Every trial gives a value of the error function.

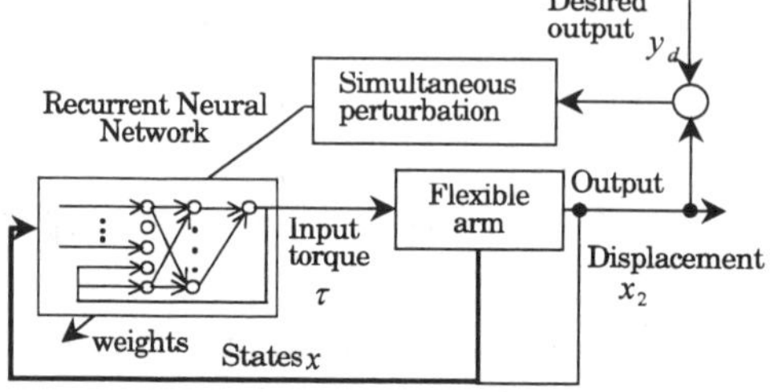

Fig.4 Overall configuration of the system

Without perturbation, we make a trial and obtain a value of the error in Eq.(7). Next, we add perturbations to all weights simultaneously and make a trial. Then, we have a value of the error, i.e., $J(u(w+cs))$. By using Eq.(3),(4) we can update all weights in the NC. We repeat this procedure.

4.1 Tracking control 1

We consider a random motion for the top of the arm. The perturbation $c = 2\times 10^{-5}$ and the gain coefficient $\alpha = 5\times 10^{-5}$ in the learning rule.

We obtain a result shown in Fig.5. This is a result after 35000 times learning. The locus of the top of the arm is very close to the desired locus.

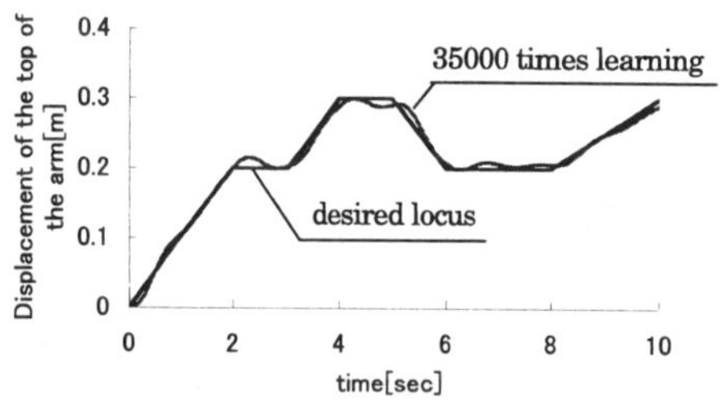

Fig.5. A simulation results of a tracking control

4.2 Tracking control 2

We used the same NN. Initial weights values are those of the NN after 35000 times learning in the previous simulation. Every trial, desired locus are

generated randomly.

We obtain a result shown in Fig.6. This is a result after 2000 times learning. The perturbation $c = 5\times 10^{-5}$ and the gain coefficient $\alpha = 2\times 10^{-5}$ in the learning rule.

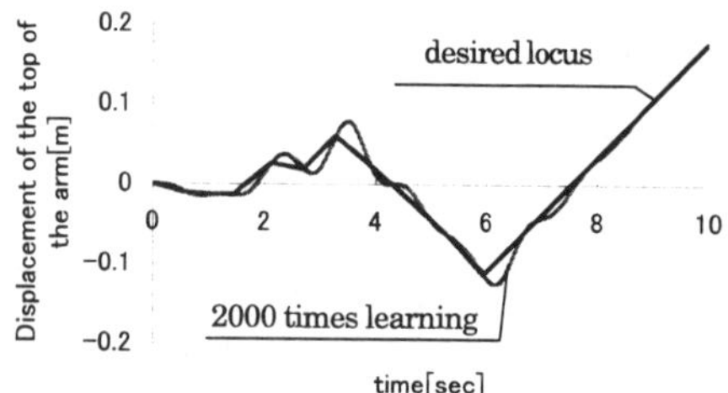

Fig.6. A simulation results of a tracking control

5. Experimental results by on-line learning

Through experiments using an actual flexible arm system, we confirm a feasibility of this learning scheme by the simultaneous perturbation rule.

Total flowchart is shown in Fig.7. First, we need a simulation as a pre-training of the neural network, because it is difficult and ineffective for the neural network to learn suitable weight values from randomly determined weight values in actual operation.

After the pre-training of the neural network, the network is utilized as a controller of the plant.

Based on the measurements, a position of the top of the arm, angles θ_1 and θ_2 and their derivative are calculated. These measured data are fed into the NC.

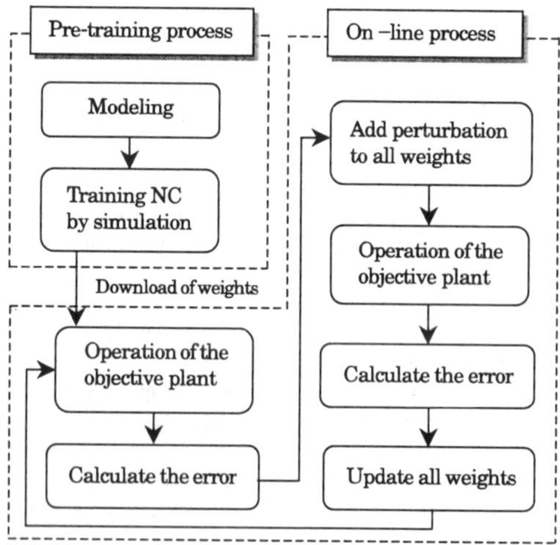

Fig.7 Flowchart

5.1 Vibration control

From a certain initial state, we would like to reduce vibration of the top of the flexible arm. We cannot perfectly realize the initial state as same as this by simulation.

Fig.8 is a result after 30 times on-line learning by our learning rule. The NC controls the actual flexible arm system.

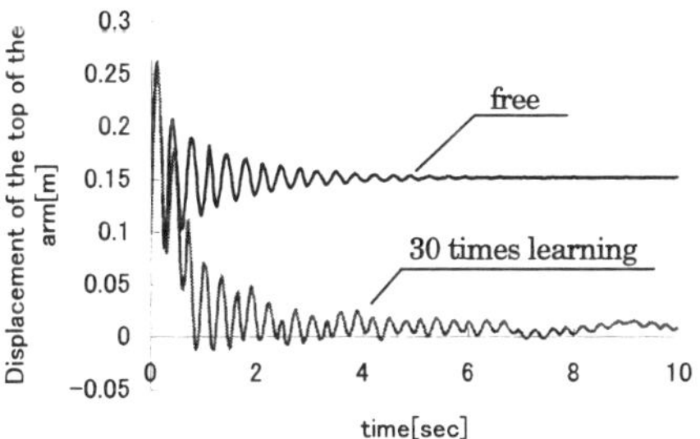

Fig.8 A vibration control of the flexible arm

5.2 Tracking control

Without the pre-training, we use the same NC. Fig.9 is a result after 100 times, 160 times and 280 times on-line learning by our learning rule. Desired locus is sinusoidal wave which amplitude is 0.1[m], period is 20[sec]. Moreover the wave is moved 0.1[m] in the top direction and $\pi/2$[rad] moved in the right direction.

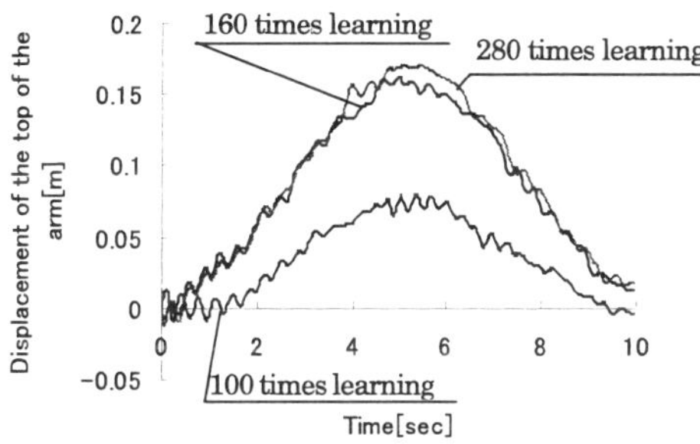

Fig.9

6. Summary and Conclusions

In this paper, a flexible arm system controlled by a NC using the simultaneous perturbation learning rule is described. Moreover, we could apply this scheme to a real time system. The back-propagation learning rule is not applicable to this. However, by using the simultaneous perturbation learning rule, the NC can learn an inverse of the object plant without any information on the objective plant.

Acknowledgement

This research is financially supported by Kansai University Frontier Sciences Center and High Technology Research Center.

References

[1]Y.Maeda, "Learning rule of neural networks for inverse systems," Transactions of The Institute of Electronics, Information and Communication Engineers, vol.J75-A, pp1364-1369, 1992(in Japanese), English version of this paper is shown in Electronics and Communications in Japan, vol.76, pp.17-23, 1993

[2]J.C.Spall, "A Stochastic approximation technique for generating maximum likelihood parameter estimates," Proceedings of the 1987 American Control Conference, pp.1161-1167, 1987.

[3]J.C.Spall, "Multivariable stochastic approximation using a simultaneous perturbation gradient approximation," IEEE Trans. Automatic Control, vol.37, pp.332-341, 1992.

[4]J.Alespector, R.Meir, B.Yuhas, A.Jayakumar and D.Lippe, "A parallel gradient descent method for learning in analog VLSI neural networks," in S.J.Hanson, J.D.Cowan and C.Lee(eds.), Advances in neural information processing systems 5(pp.836-844), San Mateo, CA : Morgan Kaufmann Publisher, 1993.

[5]G.Cauwenberghs, "A fast stochastic error-descent algorithm for supervised learning and optimization," in S.J.Hanson, J.D.Cowan and C.Lee(eds.), Advances in neural information processing systems 5(pp.244-251), San Mateo, CA : Morgan Kaufmann Publisher, 1993.

[6]Y.Maeda and Y.Kanata, "Learning rules for recurrent neural networks using perturbation and their application to neuro-control," Transactions of the Institute of Electrical Engineers of Japan, vol.113-C, pp.402-408, 1993(in Japanese).

[7]Y.Maeda, H.Hirano and Y.Kanata,"A learning rule of neural networks via simultaneous perturbation and its hardware implementation," Neural Networks, vol.8, pp.251-259, 1995.

[8]J.C.Spall and J.A.Cristion, "Nonlinear adaptive control using neural networks : Estimation with a smoothed form of simultaneous perturbation gradient approximation," Statistica Sinica, vil.4, pp.1-27, 1994.

[9]J.C.Spall and D.C.Chin, "A model-free approach to optimal signal light timing for system-wide traffic control," Proceedings of the 1994 IEEE Conference on Decision and Control, pp.1868-1875,1994.

[10]Y.Maeda and R.J.P.deFigueiredo, Learning Rules for Neuro-Controller Via Simultaneous Perturbation, IEEE Trans. on Neural Networks, vol.8, no.5, pp.1119-1130, 1997.

KES '01
N. Baba et al. (Eds.)
IOS Press, 2001

Reinforcement Learning of Robots with Context-specific Formation of Fuzzy Control Rules

Hideki Yamagishi*, Hiroshi Kawakami*, Tadashi Horiuchi** and Osamu Katai*
**Graduate School of Informatics, Kyoto University,*
Yoshida Honmachi, Sakyo-ku, Kyoto 606-8501, Japan
E-mail: {yama, kawakami, katai}@sys.i.kyoto-u.ac.jp
***Dept. of Information Engineering, Matsue National College of Technology*
14-4, Nishi-ikuma, Matsue, 690-8518, Japan
E-mail: horiuchi@it.matsue-ct.ac.jp

Abstract. Recently, reinforcement learning methods have been successfully applied to various problems where latent rules cannot be observed nor acquired manually. Q-learning is one of the effective methods for reinforcement learning. One of the simplest ways to estimate Q-values is to look up a Q-table, but it cannot deal with continuous-valued inputs and outputs. We have already proposed a framework of reinforcement learning with Condition Reduced Fuzzy Rules (CRFRs) where Q-values are interpolated by the use of fuzzy inference. In this paper, we apply C4.5 algorithm to integrate some fuzzy rules learned by reinforcement learning and introduce the notion of "boundaries of motion" for chunking motion sequences into an action.

1 Introduction

Recently, reinforcement learning methods have been successfully applied to various problems where latent rules cannot be observed nor acquired manually. Q-learning is one of the basic methods for reinforcement learning. One of the simplest ways to estimate Q-values is to look up a Q-table, but it cannot deal with continuous-valued inputs and outputs. We have already proposed a framework of reinforcement learning with Condition Reduced Fuzzy Rules (CRFRs)[4] where Q-values are interpolated by fuzzy inference.

Q-learning requires only experienced conditions/actions pairs and rewards combinations. Therefore, Q-learning can be applied to problems where meaningful I/O sets cannot be specified beforehand. When we applied Q-learning with CRFRs to those types of problems, fuzzy rules cannot be specified beforehand by referring to domain-specific heuristics. Namely, the initial rule set has to be made of all the combinations of conditions and actions. In this paper, we will apply Q-learning with CRFRs to learn fuzzy control rules and will show a way to get a set of simple latent rules using C4.5 algorithm from the acquired useful rules.

2 Q-learning with Condition Reduced Fuzzy Rules and its Properties

2.1 Q-learning with Condition Reduced Fuzzy Rules

Fuzzy inference has been successfully used to interpolate the values (components) of a Q-table [3]. This method estimates Q-values for conditions/action pairs by fuzzy inference (QL+Fuzzy). The framework of learning process is almost the same as the standard Q-learning algorithm [2] except for the ways of (1) estimating Q-values by using Takagi-Sugeno Model[7], of (2) selecting an action, and of (3) revising the Q-value of each fuzzy rule.

Table 1: Parameters for generating the initial fuzzy rules

N_c:	the total number of conditions
N_r:	the number of conditions which are included in each CRFR
N_a:	the total number of actions
N_f:	the number of fuzzy sets for each condition and action
N_g:	the number of fuzzy sets having nonzero grades for an arbitrary input

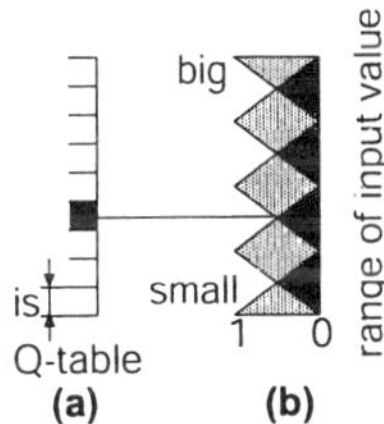

Figure 1: Fuzzy interpolation

We have proposed fuzzy inference with Condition Reduced Fuzzy Rules (CRFRs) and have incorporated it to Q-learning. The antecedent part of each CRFR consists of all the possible actions and the selected conditions, and the consequent part shows the Q-value of the corresponding rule. Table 1 shows the parameters related to generate the initial fuzzy rules.

Let us assume that each condition and action involves fuzzy sets and that their numbers are equally set to be N_f. Figure 1(b) shows the case $N_f = 5, N_g = 2$ and the way how fuzzy membership function interpolates Q-tables shown in Figure 1(a). In case of $N_c + N_a = 3$ and $N_f = 5$, the normal form of fuzzy rules will be represented as the following form:

$$\text{if } (X_1 \text{ is } P_{1i_1}) \wedge (X_2 \text{ is } P_{2i_2}) \wedge (X_3 \text{ is } P_{3i_3}) \quad \text{then} \quad (Q_k = C_{k_j}),$$

where P_{ij}, Q_k and C_{k_j} denote fuzzy set, Q-variable and its value, respectively. The combination of i_1, i_2 and i_3 such that $1 \le i_1, i_2, i_3 \le 5$ yields $125 (= 5^3)$ fuzzy rules in total. In this case, $8 (= 2^3)$ fuzzy rules will be revised at each learning step. On the other hand, in case of $N_r + N_a = 2$, the CRFRs will be given as

$$\text{if } (X_1 \text{ is } P_{1i_1}) \wedge (X_2 \text{ is } P_{2i_2}) \quad \text{then} \quad (Q_l = C_{l_j}),$$
$$\text{if } (X_2 \text{ is } P_{2i_2}) \wedge (X_3 \text{ is } P_{3i_3}) \quad \text{then} \quad (Q_m = C_{m_j}),$$
$$\text{if } (X_1 \text{ is } P_{1i_1}) \wedge (X_3 \text{ is } P_{3i_3}) \quad \text{then} \quad (Q_n = C_{n_j}),$$

where the total number of CRFRs is $75 (= 3 \times 5^2)$ and the number of revised fuzzy rules at each step is $12 (= 3 \times 2^2)$. Generally, the total number of rules is $_{N_c}C_{N_r} N_f^{N_r+N_a}$, and the number of revised rules is $_{N_c}C_{N_r} N_g^{N_r+N_a}$. When $_{N_c}C_{N_r} N_f^{N_r+N_a} < N_f^{N_c+N_a}$ and $_{N_c}C_{N_r} N_g^{N_r+N_a} < N_g^{N_c+N_a}$, the total number of CRFRs is smaller, and the number of CRFRs revised at each learning step is larger than those of Q-learning with normal fuzzy rules. The framework of Q-learning with CRFRs is shown in Table 2.

Table 2: Framework of Q-learning with CRFRs

---

1. Initialize C_i (value of Q_i) for $i = 1, 2, \cdots, N_c\, C_{N_r}\, N_f^{N_r + N_a}$

2. Repeat forever

   (a) Repeat T times

      i. Assume (select) a set of actions at random.

      ii. Calculate the matching grade of each rule (ω_i) under the current conditions and assumed actions.
         Matching grade ω_i of each fuzzy rule R_i is given by the algebraic product of the grades of its antecedents.

      iii. Estimate the Q-value for the current conditions/actions pair, i.e.,
         $Q = (\sum_i \omega_i C_i)/(\sum_i \omega_i)$.

   (b) (after the second cycle)

      i. Calculate ΔQ for the last conditions/actions pair by the standard way of Q-learning, i.e.,
         $\Delta Q = \alpha\{r_t + \gamma \max_b Q(x_{t+1}, b) - Q(x_t, a_t)\}$.

      ii. Distribute ΔQ to CRFRs so as to ΔC_i be proportional to ω_i, i.e.,
         $\Delta C_i = (\sum_j \omega_j \omega_i \Delta Q)/(\sum_j \omega_j^2)$.

   (c) Select one of the actions by roulette wheel selection and execute it.

   (d) Observe the next state and the reinforcement signal.

2.2 Synchronic firing structure

In the standard Q-learning, the controller learns Q-values in an $N_c + N_a$ dimensional space, using all the variables in the antecedent part of fuzzy rules. In Q-learning with CRFRs, the controller learns Q-values in $_{N_c}C_{N_r}$ "r.d.s." (reduced dimensional spaces) (N_r dimensional conditions + N_a dimensional actions) [5]. Q-values in a N_c dimensional space are estimated by fuzzy inference using back-projection of the $_{N_c}C_{N_r}$ r.d.s.

Figure 2 illustrates a conceptual image of learning Q-values in four r.d.s.. In Q-learning with CRFRs, the learning of Q-values proceeds on the four planes rather in the original space. As fuzzy rules synchronically fire on the four planes (r.d.s.'s) at each step, the actions with high Q-values are selected stochastically. The learning process makes some regions on the four planes emerge where fuzzy rules fire synchronically.

3 Experimental Analysis

3.1 Experimental Environment

Figure 3 illustrates an experimental problem. Learning methods in this case will yield rules for controlling boats to go around the racing track. The state variables (condition for selecting an action) of this system are the current location (x, y), the velocity (v_x, v_y), the direction (r) and the angular velocity (w) of the boat, and the action is the combination of "steering wheel operation (hdl)" and "acceleration lever operation (acc)". In this case, we have $N_c = 6$ and

$N_a = 2$. We set time constant large enough yielding the controlling task quite difficult.

Fuzzy membership functions were set as shown in Figure 1 (b). Rewards in this case were set to be inversely proportional to the distance between the boat and the nearby local target as shown in Figure 3. Each time a boat collides with a fence, it will be penalized by receiving a certain negative value as the reinforcement.

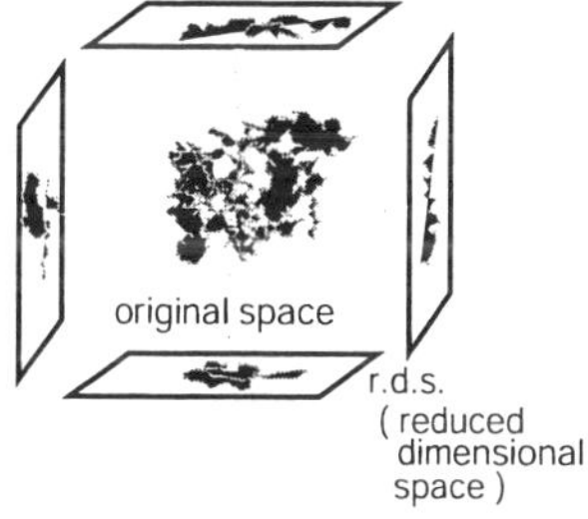

Figure 2: Conceptual image of the
synchronic firing structure

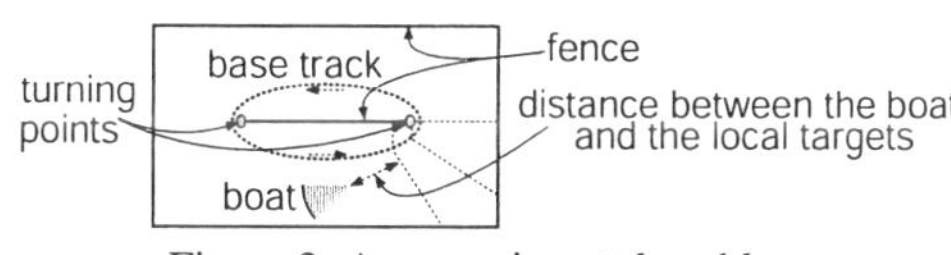

Figure 3: An experimental problem
(boat racing track)

3.2 *Experimental results*

3.2.1 Learning performance

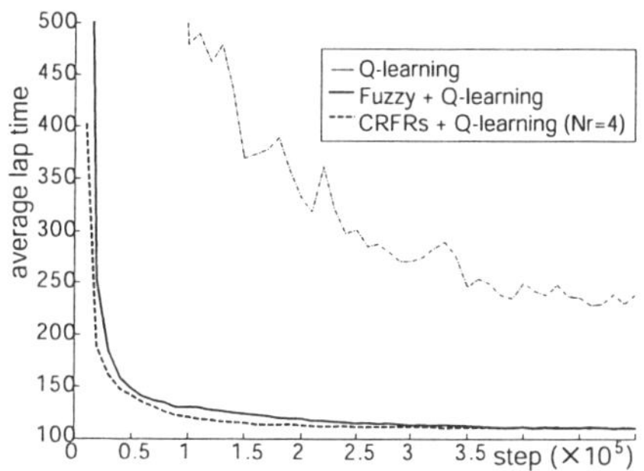

Figure 4: Comparison of the learning under
various Q-learning methods

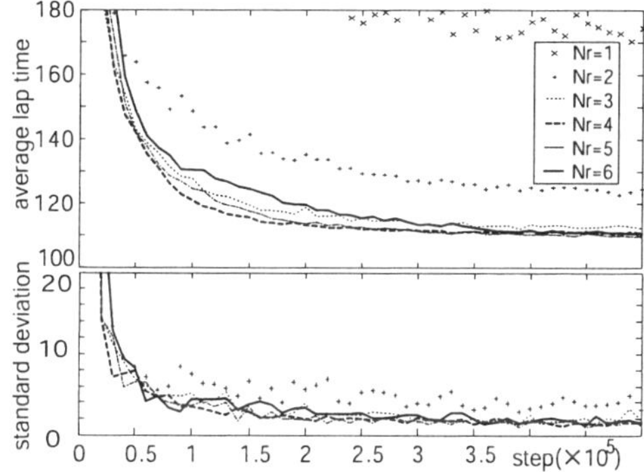

Figure 5: Learning performances of our method in
various setting of the antecedent parts of fuzzy rules

Figure 4 shows the learning performance of Q-learning with CRFRs, Q-learning with Fuzzy reasoning and normal Q-learning. Fundamental difference between QL+Fuzzy and the proposed method is the number of variables in the antecedent part of the rule. If all the variables are used as the antecedent part (i.e., $N_r = N_c$), the proposed method is the same as QL+Fuzzy.

The upper part of Figure 5 shows the average lap times of twenty trials, and the lower part of the figure shows their distributions for each $N_r(1 \sim N_c)$. Generally speaking, it is expected that the rules using all the conditions will yield better performance than the rules with reduced number of conditions. However, Figure 5 reveals that the performance of CRFRs with an appropriate value of N_r is equivalent to or even better than that of the most detailed rules (i.e., $N_r = 6$). We have shown in section 2.1 that we can set the total number of rules smaller than QL+Fuzzy, while the number of revised rules at each learning step is larger than QL+Fuzzy. Therefore, it is expected in general that the learning proceeds faster than in the case of Q-learning+Fuzzy, and this expectation is realized in the case of $N_r = 4$ in this experiment.

3.2.2   Separation of rules

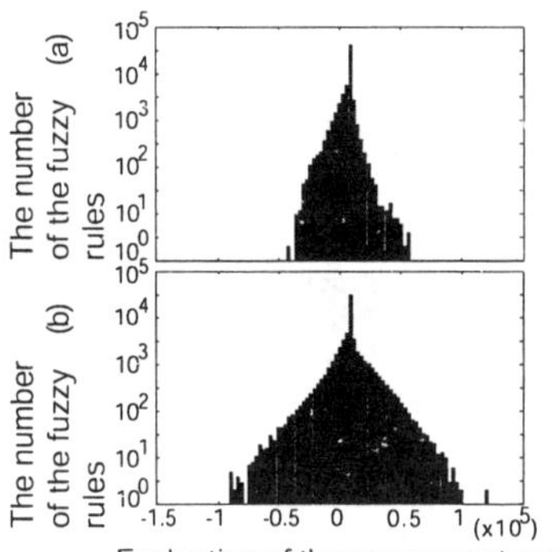

Evaluation of the consequent part

Figure 6: Histograms of Q-values in the proposed method

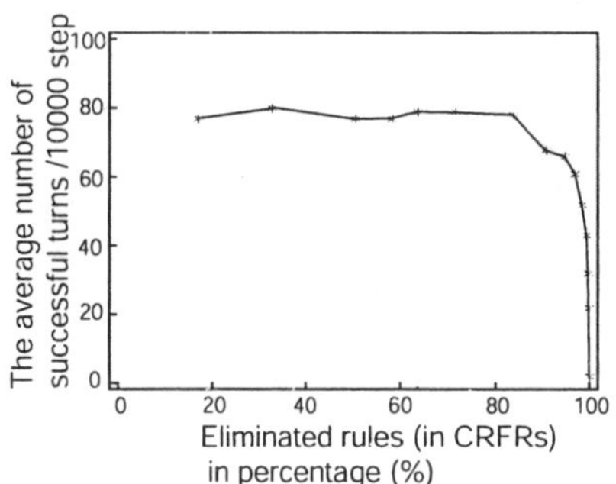

Eliminated rules (in CRFRs) in percentage (%)

Figure 7: Elimination of rules

Figure 6 shows an process of division of "valid" rules having high Q-values and "harmful" rules having low Q-values (not necessarily negative Q-values). The upper part of this figure shows a histogram of the rules at an early stage of learning, and the lower part shows that of the rules at the final stage of learning. From this figure, we can see that most of the rules have not changed their Q-values and thus have worked little. Only a few rules have contributed at the action selections. Thus, we can considerably reduce the number of rules deriving a set of valid rules.

For an agent to execute so complicate task that the agent should make plans, it is generally assumed that motion sequences should be chunked into an action. The idea of "boundaries of motion" is one of the ways to chunk and to label motions[8]. Thus we should put together many rules to create boundaries of motions and further reduce the number of rules which are acquired by the reinforcement learning.

We employ C4.5 algorithm to integrate useful rules acquired after learning. Defining classes is important since C4.5 algorithm classifies rules into classes. The classes may not reflect the Q-values of the rules, for considering merely Q-values may yield rules without action variables, i.e., they may be of no use for selecting actions if there are rules which have the same conditions and different actions. We cannot apply C4.5 by focusing merely on the actions. This is because penalty-avoidance-rules cannot be composed in this case. Then, we should apply C4.5 by focusing on "actions and Q-values" pairs, and the conditions become nodes of the decision trees.

Table 3 shows a part of the results, where "acc ", "hdl", "sn", "st", "sr" and "sl" stand for acceleration, handling, slightly strong, straight, slightly right and slightly left, respectively. The number of rules acquired by this method is around 440 ,i.e., 0.1% of the initial rules. We can interpret these integrated rules as "behavior modules" because inputs of rules are conditions and outputs are actions. Further, a transition from one integrated rules to another

Table 3: Decision trees acquired by C4.5 algorithm

situation	attribute-value pairs		class
accessing a turning point	vx=3:	x=1:	acc=sn,hdl=sr,+
		x=2: y=3: w=2:	acc=sn,hdl=sl,−
departing from a turning point	r=0:	w=2:	acc=sn,hdl=st,+
		w=3:	acc=sn,hdl=st,+
		w=1:	acc=sn,hdl=sr,+
	r=4:	w=2:	acc=normal,hdl=sl,+
		w=3:	acc=normal,hdl=sl,+
		w=1:	acc=sn,hdl=st,+

denotes a transition of boundaries of motions because a transition brings about a change of motions, and thus each integrated rules represent boundaries of motion.

4 Conclusion

In this paper, we have applied Q-learning with CRFRs to a control problem and shown its learning performance. The proposed method separates useful control rules from huge number of initial rules, i.e., only useful rules are assigned relatively high/low Q-values and other rules are ignored. Furthermore, through analyzing the Q-values of each rule by using C4.5 algorithm, the number of rules are considerably reduced (only 0.1% of the number of initial rules) by integrating the rules that are acquired by Q-learning with CRFRs. In addition, we have shown that we can categorize rules via the notion of "boundaries of motion" and chunk motion sequences into an action.

References

[1] R. S. Sutton and A. G. Barto, *Reinforcement Learning: An introduction*, The MIT Press, 1998.

[2] C. Watkins and P. Dayan, "Technical Note: Q-learning", *Machine Learning*, 8-3/4, pp.279-292, 1992.

[3] T. Horiuchi, A. Fujino, O. Katai and T. Sawaragi, "Fuzzy Interpolation-Based Q-Learning with Continuous Inputs and Outputs "(in Japanese), *Trans. of the Society of Instrument and Control Engineers*, 35-2, pp. 217-279, 1999.

[4] H. Kawakami, T. Konishi and O. Katai, "A Reinforcement Learning with Condition Reduced Fuzzy Rules", *Proc. of the 2nd Asia-Pacific Conference on Simulated Evolution and Learning*, LNAI 1585, Springer, pp. 198-205, 1998.

[5] H. Yamagishi, H. Kawakami, T. Horiuchi and O. Katai, "Separation of Behavior Modules by Reinforcement Learning with Condition Reduced Fuzzy Rules," *Proc. of the 4th Japan-Australia Joint Workshop*, pp. 114-121, 2000.

[6] H. Yamagishi, H. Kawakami, T. Horiuchi and O. Katai, "Reinforcement Learning based on Resonance between Agent and Field", *Proc. of the 6th Int. Symp. on Artificial Life and Robotics(AROB 6th' 01)*, pp. 405–408, 2001.

[7] M. Sugeno, *Fuzzy Controls* (in Japanese), Nikkan Kogyo Shinbun, 1988.

[8] J. M. Rubin and W.A. Richards, Boundaries of visual motion, A.I.Memo 835, MIT 1985.

KES '01
N. Baba et al. (Eds.)
IOS Press, 2001

Uniqueness of Feedforward Complex-valued Neural Network with a Given Complex-valued Function

Tohru Nitta

Mathematical Neuroinformatics Group, Neuroscience Research Institute,
National Institute of Advanced Industrial Science and Technology,
AIST Tsukuba Central 2, 1-1-1 Umezono, Tsukuba-shi, Ibaraki, 305-8568 Japan

Abstract. In this paper, we will prove the uniqueness theorem for complex-valued neural networks: the 3-layered complex-valued neural network is uniquely determined by its corresponding complex-valued function, up to a finite group, provided that the complex-valued neural network is irreducible. Moreover, we will investigate the order of the finite group, and make clear that the redundancy of the learning parameters of the complex-valued neural network is the same as the one of the real-valued neural network. We will also derive a sufficient condition for a given 3-layered complex-valued neural network to be minimal.

1   Introduction

It is expected that complex-valued neural networks, whose parameters (weights and threshold values) are all complex numbers, will have applications in fields dealing with complex numbers such as telecommunications. Independently of Georgiou et al [3] and Benvenuto et al [1], Nitta et al [5, 6] proposed a (multi-layered type) complex-valued back-propagation network and demonstrated its characteristics [5, 6, 7]; (a) properties greatly different from those of the real-valued back-propagation network [8] including 2D motion structure of weights and the orthogonality of decision boundary of a complex-valued neuron, (b) the learning property superior to the real-valued back-propagation, and (c) an inherent 2D motion learning ability (an ability to transform geometric figures), which was more outstanding than (a) and (b).

In the meantime, Sussmann proved that the 3-layered (real-valued) neural network was uniquely determined by its input-output map, up to an obvious finite group, provided that the real-valued neural network was irreducible (*Uniqueness Theorem*) [9]. Based on Sussmann's result, Fukumizu et al elucidated the mechanism of local minima of (real-valued) feedforward neural nets [2]. Hagiwara et al showed that AIC (Akaike Information Criterion) could not be derived for (real-valued) three-layered networks due to the nonuniqueness of the connection weight[4]. We can find from Sussmann's result that the asymptotic behavior of the maximum likelihood estimator of the parameters (weights) of the (real-valued) feedforward neural network is not clear. As we have described above, Sussmann's result is the basis to investigate the property (especially, statistical property) of real-valued neural networks. Similarly, it is very important to make certain the corresponding fact on complex-valued neural networks for the purpose of investigating the properties of the complex-valued neural nets.

In this paper, we will prove the uniqueness theorem for complex-valued neural networks: the 3-layered complex-valued neural network is uniquely determined by its corresponding complex-valued function, up to a finite group, provided that the complex-valued net is irreducible, where in addtion to the two generators of the finite group for real-valued neural nets, two new generators related to the rotation of complex numbers are added to the finite group. Moreover, we will investigate the order of the finite group, and make clear that the redundancy of the learning parameters of the complex-valued neural network is the same as the one of the real-valued neural network. We will also derive a sufficient condition for a given 3-layered complex-valued neural network to be minimal. All the proofs are omitted due to space limitations.

2 The Complex-valued Neural Network

This section describes the complex-valued neural network used in the analysis. First, we will consider the following complex-valued neuron. The input signals, weights and output signals are all complex numbers (note that the threshold parameter is not defined.). The net input U_n to a complex-valued neuron n is defined as:

$$U_n = \sum_m W_{nm} X_m, \tag{1}$$

where W_{nm} is the (complex-valued) weight connecting complex-valued neurons n and m, and X_m is the (complex-valued) input signal from complex-valued neuron m. To obtain the (complex-valued) output signal, convert the net input U_n into its real and imaginary parts as follows: $U_n = x + iy = z$, where i denotes $\sqrt{-1}$. The (complex-valued) output signal is defined to be

$$\sigma(z) = \tanh(x) + i \tanh(y) \tag{2}$$

where $\tanh(u) \stackrel{\text{def}}{=} (\exp(u) - \exp(-u))/(\exp(u) + \exp(-u)), u \in \boldsymbol{R}$ ($\boldsymbol{R}$ denotes the set of real numbers) and is called *hyperbolic tangent*. Note that $-1 < Re[\sigma], Im[\sigma] < 1$. Note also that $\sigma(z)$ is not holomorphic, because the Cauchy-Riemann equation does not hold.

A complex-valued neural network consists of such complex-valued neurons described above. The network used in the analysis will have 3 layers: $m - n - 1$ network. We will use $w_{ij} = w_{ij}^r + iw_{ij}^i \in \boldsymbol{C}$ for the weight between the input neuron i and the hidden neuron j ($\boldsymbol{C}$ denotes the set of complex numbers), $c_j = c_j^r + ic_j^i \in \boldsymbol{C}$ for the weight between the hidden neuron j and the output neuron. Let $y_j(\boldsymbol{z}), h(\boldsymbol{z})$ denote the output values of the hidden neuron j, and the output neuron for the input pattern $\boldsymbol{z} = {}^t[z_1, \cdots, z_m] \in \boldsymbol{C}^m$ where $z_k = z_k^r + iz_k^i \in \boldsymbol{C} (1 \leq k \leq m)$, respectively. Let also $\nu_j(\boldsymbol{z})$ and $\mu(\boldsymbol{z})$ denote the net inputs to the hidden neuron j and the output neuron for the input pattern $\boldsymbol{z} \in \boldsymbol{C}^m$, respectively. That is, $\nu_j(\boldsymbol{z}) = \sum_{i=1}^m w_{ij} z_i$, $\mu(\boldsymbol{z}) = \sum_{j=1}^n c_j y_j(\boldsymbol{z})$, $y_j(\boldsymbol{z}) = \sigma(\nu_j(\boldsymbol{z}))$, and $h(\boldsymbol{z}) = \sigma(\mu(\boldsymbol{z}))$. The set of all $m - n - 1$ complex-valued neural networks decribed above will be denoted by $N_{m,n}$, which is the object of the analysis.

3 Uniqueness Theorem for Complex-valued Neural Networks

This section gives the uniqueness theorem for complex-valued neural networks described in Section 2. Several definitions and propositions are prepared.

Definition 1 *For a fixed m, two complex-valued neural networks $N_1 \in N_{m,n_1}$ and $N_2 \in N_{m,n_2}$ are called* **I-O equivalent** *if their corresponding complex-valued functions are same.* $\qquad\square$

Definition 2 *Two complex-valued linear affine functions $\alpha, \beta : \mathbf{C}^m \to \mathbf{C}^1$ are called* **rotation-equivalent** *if one of the following conditions holds:*

$$\forall \mathbf{z} \in \mathbf{C}^m; \ \alpha(\mathbf{z}) = \beta(\mathbf{z}) \ (= \exp[i0] \cdot \beta(\mathbf{z})), \tag{3}$$

$$\forall \mathbf{z} \in \mathbf{C}^m; \ \alpha(\mathbf{z}) = -\beta(\mathbf{z}) \ (= \exp[i\pi] \cdot \beta(\mathbf{z})), \tag{4}$$

$$\forall \mathbf{z} \in \mathbf{C}^m; \ \alpha(\mathbf{z}) = i\beta(\mathbf{z}) \ (= \exp[i(\pi/2)] \cdot \beta(\mathbf{z})), \tag{5}$$

$$\forall \mathbf{z} \in \mathbf{C}^m; \ \alpha(\mathbf{z}) = -i\beta(\mathbf{z}) \ (= \exp[i(3\pi/2)] \cdot \beta(\mathbf{z})). \tag{6}$$

$\qquad\square$

(Remarks)

1. A net input to a complex-valued neuron i, $\nu_i(\mathbf{z}) = \sum_{j=1}^{m} w_{ji} z_j$, is a complex-valued linear affine function on $\mathbf{C}^m$.

2. In addtion to the two conditions of *sign-equivalence* that Sussmann defined to prove the uniqueness theorem for real-valued neural nets in [9], two new conditions (eqns (5) and (6)) related to the rotation of complex numbers are added.

Definition 3 *A complex-valued neural network $N \in N_{m,n}$ is called* **reducible** *if one of the following three conditions holds:*

1. *One of the weights between the hidden layer and the output layer is zero:* $1 \leq \exists j \leq n; \ c_j = 0.$

2. *There exist two hidden neurons such that the net inputs to them are rotation-equivalent: $1 \leq \exists j_1, \exists j_2 \leq n; \ \nu_{j_1}$ and ν_{j_2} are rotation-equivalent.*

3. *There exists a hidden neuron such that the net input to it is a constant:* $1 \leq \exists j \leq n; \ \nu_j$ *is a constant.*

$\qquad\square$

We can easily see the following proposition on *reducibility*.

Proposition 1 *If a 3-layered complex-valued neural network is reducible, then it is I-O equivalent to another 3-layered complex-valued neural network with fewer hidden neurons.* $\qquad\square$

Definition 4 *A 3-layered complex-valued neural network is called* **irreducible** *if it is not reducible.* $\qquad\square$

The four transformations described in the following Proposition 2 are needed to state the redundancy of the learning parameters of complex-valued neural networks.

Proposition 2 *There are four transformations that can be applied to a 3-layered complex-valued neural network without changing its corresponding complex-valued function, which we will now state:*

1. θ_j^{-1} changes the sign of the weights between the input layer and the hidden neuron j ($\{w_{ij}\}_{i=1}^m$) and the sign of the weight between the hidden neuron j and the output neuron (c_j).

2. θ_j^i multiplies the weights between the input layer and the hidden neuron j ($\{w_{ij}\}_{i=1}^m$) by $-i \in \mathbf{C}$, and the weight between the hidden neuron j and the output neuron (c_j) by $i \in \mathbf{C}$.

3. θ_j^{-i} multiplies the weights between the input layer and the hidden neuron j ($\{w_{ij}\}_{i=1}^m$) by $i \in \mathbf{C}$, and the weight between the hidden neuron j and the output neuron (c_j) by $-i \in \mathbf{C}$.

4. $\tau_{j_1 j_2}$ interchanges two hidden neurons j_1 and j_2.

Let us use $\theta_j^{-1}(N), \theta_j^i(N), \theta_j^{-i}(N)$ and $\tau_{j_1 j_2}(N)$ to denote the complex-valued neural networks resulting from the above four transformations for a 3-layered complex-valued neural network $N \in N_{m,n}$, respectively. □

(Remark) In addtion to those of the case of the real-valued neural net in [9], two new transformations (θ_j^i and θ_j^{-i}) related to the rotation of complex numbers are added.

Proposition 3 *The transformations $\theta_j^{-1}, \theta_j^i, \theta_j^{-i}$ and $\tau_{j_1 j_2}$ described in Proposition 2 generate a finite group $W_{m,n}$ of transformations of the set $N_{m,n}$.* □

Definition 5 *Two 3-layered complex-valued neural networks $N_1, N_2 \in N_{m,n}$ are called* **equivalent** *if they are related by a transformation in the group $W_{m,n}$.* □

All the preparations have been done. The following Theorem 1 is one of the main results obtained in this paper.

Theorem 1 (Uniqueness Theorem) *Let N_1, N_2 be irreducible I-O equivalent 3-layered complex-valued neural networks in N_{m,n_1}, N_{m,n_2}, respectively. Then (i) $n_1 = n_2$ and (ii) N_1 and N_2 are equivalent.* □

Theorem 1 states that the redundancy of the learning parameters of an irreducible complex-valued neural network to approximate a given complex-valued function is determined up to a finite group.

Theorem 2 *The order of the finite group $W_{m,n}$ described in Proposition 3 is $2^{2n} \cdot n!$.*

□

Consider a $m - 2n - 1$ real-valued neural network and a $m - n - 1$ complex-valued neural network, the number of learning parameters of which is both $2(mn + 2n)$. Then, the order of the finite group for the real-valued neural network $|W_{m,2n}^{(real)}|$ is equal to $2^{2n} n!$ from Sussmann's result [9], and the one for a complex-valued neural network $|W_{m,n}^{(complex)}|$ is equal to $2^{2n} n!$ from Theorem 2. Thus, they are same. In other words, the redundancy of the learning parameters of the complex-valued neural network is the same as the one of the real-valued neural network.

Definition 6 *A 3-layered complex-valued neural network with n hidden neurons is called* **minimal** *if it is not I-O equivalent to any 3-layered complex-valued neural network with fewer hidden neurons.* □

Corollary 1 *An irreducible 3-layered complex-valued neural network is minimal.*　□

Corollary 1 means that if only we check the three simple conditions in Definition 3, we can find if a 3-layered complex-valued neural network is minimal. Since it is generally difficult to check the minimality of a given 3-layered complex-valued neural network, Corollary 1 is very useful.

4　Conclusions

We have clarified the redundancy of the learning parameters of 3-layered complex-valued neural networks through theoretical analyses. The main results may be summarized as follows. (i) We proved the uniqueness theorem for 3-layered complex-valued neural networks. (ii) Considering the order of a finite group, we showed that the redundancy of the learning parameters of complex-valued neural networks was equal to the one of real-valued neural networks. (iii) We derived a sufficient condition for a given 3-layered complex-valued neural network to be minimal. We believe that these results will be effectively used to elucidate the properties of complex-valued neural networks.

In future studies, based on the results of this paper, we will analyze the asymptotic behavior of the maximum likelihood estimator of the learning parameters and the mechanism of local minima of feedforward complex-valued neural networks.

Acknowledgements
We wish to thank the members of the Mathematical Neuroinformatics Group for discussing this study. We are also grateful to the anonymous reviewers for their many helpful suggestions.

References

[1] N. Benvenuto and F. Piazza, On the Complex Backpropagation Algorithm, *IEEE Trans. Signal Processing*, Vol.40, No.4, 1992, pp.967-969.

[2] K. Fukumizu and S. Amari, Local Minima and Plateaus in Hierarchical Structures of Multilayer Perceptrons, *Neural Networks*, Vol.13, No.3, 2000, pp.317-327.

[3] G. M. Georgiou and C. Koutsougeras, Complex Domain Backpropagation, *IEEE Trans. Circuits and Systems–II: Analog and Digital Signal Processing*, Vol.39, No.5, 1992, pp.330-334.

[4] K. Hagiwara, N. Toda and S. Usui, On the Problem of Applying AIC to Determine the Structure of a Layered Feed-forward Neural Network, *Proc. of International Joint Conference on Neural Networks*, Vol.3, 1993, pp.2263-2266.

[5] T. Nitta and T. Furuya, A Complex Back-Propagation Learning, *Transactions of Information Processing Society of Japan*, Vol.32, No.10, 1991, pp.1319-1329 (in Japanese).

[6] T. Nitta, An Extension of the Back-Propagation Algorithm to Complex Numbers, *Neural Networks*, Vol.10, No.8, 1997, pp.1392-1415.

[7] T. Nitta, An Analysis on Fundamental Structure of Complex-valued Neuron, *Neural Processing Letters*, Vol.12, No.3, 2000, pp.239-246.

[8] D. E. Rumelhart *et al.*, Parallel Distributed Processing, Vol.1, MIT Press, 1986.

[9] H. J. Sussmann, Uniqueness of the Weights of Minimal Feedforward Nets with a Given Input-Output Map, *Neural Networks*, Vol.5, 1992, pp.589-593.

KES '01
N. Baba et al. (Eds.)
IOS Press, 2001

Qualitative Analysis of a Class of Complex-Valued Associative Memories

Yasuaki Kuroe   Naoki Hashimoto
Takehiro Mori
Department of Electronics and Information Science
Kyoto Institute of Technology
Matsugasaki, Sakyo-ku, Kyoto 606-8585, Japan
kuroe@dj.kit.ac.jp

Abstract. This paper presents a model of associative memories using complex-valued continuous neural networks and studies its qualitative behaviors theoretically. The model is an extension of the conventional real-valued associative memories of self-correlation type. We investigate the structures and asymptotic behaviors of solution orbits near each memory pattern based on the center manifold theory. We also discuss a recalling condition of each memory pattern, that is, a condition which assures that each memory pattern is correctly recalled.

1 Introduction

In recent years, there have been increasing research interests of artificial neural networks and many efforts have been made on applications of neural networks to various fields. One of the most useful and most investigated areas of applications of neural networks addresses implementations of associative memories. As applications of the neural networks spread more widely, developing neural network models which can deal with complex numbers is desired in various fields. Several models of complex-valued neural networks have been proposed and their abilities of information processing have been investigated [1, 2, 3, 4, 5, 6].

The purpose of this paper is to present a model of associative memories using complex-valued continuous neural networks and to investigate its qualitative behaviors. The model is an extension of the conventional real-valued associative memories of self-correlation type. It is shown that, in the model, each memory vector is an equilibrium point of the network, but is not isolated one and a point of an equilibrium set which is one-to-one corresponding to the memory vector. We investigate the structures and asymptotic behaviors of solution orbits near each memory pattern based on the center manifold theory. We also discuss a recalling condition, that is, a condition that assures that each memory pattern is correctly recalled.

In the following, the imaginary unit is denoted by i ($i^2 = -1$). The n-dimensional complex (real) space is denoted by $C^n (R^n)$ and the set of $n \times m$ complex (real) matrices is denoted by $C^{n \times m} (R^{n \times m})$. For $A \in C^{n \times m}$ ($a \in C^n$), its real and imaginary parts are denoted by A^R (a^R) and A^I (a^I), respectively.

2   Complex-valued Associative Memory of Self-Correlation Type

2.1   Model

Let m be the number of memory patterns to be stored, and each memory pattern is an N dimensional complex vector, denoted by $s^{(\gamma)} \in C^N$, $\gamma = 1, 2, \cdots, m$. Suppose that the set of memory patterns satisfies the following orthogonal relations.

$$s^{(\gamma)*} s^{(l)} = \begin{cases} N, & \gamma = l \\ 0, & \gamma \neq l \end{cases}, \qquad l = 1, 2, \cdots, m \tag{1}$$

$$|s_j^{(\gamma)}| = 1, \qquad j = 1, 2, \cdots, N \tag{2}$$

where s^* is the conjugate transpose of s and $s_j^{(\gamma)}$ denotes the jth element of vector $s^{(\gamma)}$. Note that there is a difference between the complex domain and the real domain in choosing a set of vectors satisfying the above conditions. In the complex domain there always exist N vectors which satisfy (1) and (2) in any N dimensional complex space. On the other hand, in the real domain such N vectors exist only in 2^n dimensional real space where n is a natural number, and if $N \neq 2^n$, there exist only less than N vectors which satisfy (1) and (2).

We consider a complex-valued and continuous-time neural network described by differential equations of the form:

$$\begin{cases} \tau \dfrac{du_j(t)}{dt} = -u_j(t) + \displaystyle\sum_{k=1}^{N} w_{jk} x_k(t) \\ x_j(t) = f(u_j(t)) \end{cases}, \qquad j = 1, \cdots, N \tag{3}$$

where $u_j(t) \in C$ and $x_j(t) \in C$ are the state and the output of the jth neuron at time t, respectively, τ is the time constant (positive real number), $w_{jk} \in C$ is a connection weight from the kth neuron to the jth neuron and $f(\cdot)$ is an activation function which is a nonlinear complex function ($f : C \to C$).

Let us determine the weight matrix $W = \{w_{jk}\} \in C^{N \times N}$ and the activation function $f(\cdot)$ so that the neural network (3) can store the set of memory vectors $\{s^{(\gamma)}\}$. Taking account of the orthogonal structure of the set of memory vectors $\{s^{(\gamma)}\}$, we determine the weight matrix W as the sum of the autocorrelation matrix of each memory vector $s^{(\gamma)}$:

$$W = \frac{1}{N} \sum_{\gamma=1}^{m} s^{(\gamma)} s^{(\gamma)*}. \tag{4}$$

As a candidate of the activation function $f(\cdot)$ we will choose a complex function which satisfies the conditions: (i) $f(\cdot)$ is a smooth and bounded function, by analogy with the sigmoidal function of real-valued neural networks, and (ii) each memory vector $s^{(\gamma)}$ becomes an equilibrium point of (3). The condition (ii) is accomplished by choosing a function $f(\cdot)$ which satisfies

$$f(s_j^{(\gamma)}) = s_j^{(\gamma)}, \qquad j = 1, 2, \cdots, N, \quad \gamma = 1, 2, \cdots, m. \tag{5}$$

In regard to the condition (i), we recall the Liouville's theorem, which says that 'if $f(z)$ is analytic at all $z \in C$ and bounded, then $f(z)$ is a constant function'. Since a suitable $f(z)$

should be bounded, it follows from the theorem that if we choose an analytic function for $f(z)$, it is constant, which is clearly not suitable, as is discussed in [1]. We choose

$$f(z) = \frac{\eta z}{\eta - 1 + |z|}, \quad \eta - 1 > 0, \ z \in C \tag{6}$$

as the activation function in which η is a tunable parameter satisfying $\eta > 1$. The function (6) is not analytic, but has the continuous partial derivatives $\partial f^R / \partial z^R$, $\partial f^R / \partial z^I$, $\partial f^I / \partial z^R$ and $\partial f^I / \partial z^I$ and is bounded:($|f(z)| < \eta, \forall z \in C \ (|z| < \infty)$). Note also that, the function (6) satisfies (5).

2.2 Memory Patterns as Equilibrium Sets

In the model of complex-valued associative memories (3), (4) and (6), each memory vector $s^{(\gamma)}$ becomes an equilibrium point of the network. However $s^{(\gamma)}$ is not an isolated equilibrium point as shown below. For each memory vector $s^{(\gamma)}$, we consider the vector $e^{i\alpha} s^{(\gamma)}$ where α is a real number ($\alpha \in R$). By using the function (6) we can check that $e^{i\alpha} s^{(\gamma)}$ satisfies

$$e^{i\alpha} s_j^{(\gamma)} = f\left(\sum_{k=1}^N w_{jk} e^{i\alpha} s_k^{(\gamma)}\right), \qquad j = 1, 2, \cdots, N \tag{7}$$

for any real number α. This implies that $e^{i\alpha} s^{(\gamma)}$ is also an equilibrium point of the network. Hence each memory vector $s^{(\gamma)}$ is not an isolated equilibrium point but a point in the equilibrium set $\Phi^{(\gamma)}$ defined by

$$\Phi^{(\gamma)} = \{e^{i\alpha} s^{(\gamma)} : \forall \alpha \in R\} \subset C^N \tag{8}$$

which is a closed curve in C^N. Considering this fact, we identify all the points in the equilibrium set $\Phi^{(\gamma)}$ and regard $\Phi^{(\gamma)}$ as a memory pattern for each memory vector $s^{(\gamma)}$.

3 Analysis of Qualitative Behaviors Near Memory Patterns

The qualitative behaviors of a nonlinear dynamical system near an equilibrium point can be studied via linearization with respect to that point. Note that the activation function $f(z)$ given by (6) is not differentiable with respect to z, but its real and imaginary parts, f^R and f^I, are continuously partially differentiable with respect to z^R and z^I. It is, therefore, possible to derive the linearized model by separating the model (3) into its real and imaginary parts. Choose a real number α arbitrarily and let $q^{(\gamma)} = e^{i\alpha} s^{(\gamma)}$. We separate the model (3) into its real and imaginary parts and linearize them around each equilibrium point $q^{(\gamma)}$. Let $\Delta x_i(t) := x_i(t) - q_i^{(\gamma)}$ and define $y(t) \in R^{2N}$ by

$$y(t) = (\Delta x_1^R(t), \Delta x_2^R(t), \cdots, \Delta x_N^R(t), \Delta x_1^I(t), \Delta x_2^I(t), \cdots, \Delta x_N^I(t))^T \tag{9}$$

where $(\cdot)^T$ denotes the transpose of $(\cdot)$. The linearized model is given by

$$\frac{dy(t)}{dt} = D(q^{(\gamma)})y(t), \quad \gamma = 1, 2, \cdots, m. \tag{10}$$

The expression of the coefficient matrix $D(q^{(\gamma)})$ is omitted for lack of space.

We now analyze the qualitative behavior and structure of the solution orbits near each memory pattern. This can be done by investigating the eigenvalues and eigenvectors of the coefficient matrix $D(q^{(\gamma)})$ of the linearized model (10).

Theorem 1. *For an arbitrary real number β, the matrices $D(\boldsymbol{q}^{(\gamma)})$ and $D(e^{i\beta}\boldsymbol{q}^{(\gamma)})$ are orthogonally similar, that is, there exists an orthogonal matrix T such that*

$$D(e^{i\beta}\boldsymbol{q}^{(\gamma)}) = TD(\boldsymbol{q}^{(\gamma)})T^{-1}. \tag{11}$$

The theorem implies that all the eigenvalues of the matrix $D(\boldsymbol{q}^{(\gamma)})$ are the same for all point $\boldsymbol{q}^{(\gamma)} = e^{i\alpha}\boldsymbol{s}^{(\gamma)} \in \Phi^{(\gamma)}$ and the corresponding eigenvectors relate with each other by the orthogonal matrix. The theorem can be proved by the choice of the matrix T as

$$T = T(\beta) = \begin{bmatrix} I_N \cos(\beta) & -I_N \sin(\beta) \\ I_N \sin(\beta) & I_N \cos(\beta) \end{bmatrix} \tag{12}$$

where I_N is the N dimensional unit matrix.

Theorem 2. *Let $\lambda_i(D(\boldsymbol{q}^{(\gamma)}))$ be the ith eigenvalue of $D(\boldsymbol{q}^{(\gamma)})$. $\lambda_i(D(\boldsymbol{q}^{(\gamma)}))$, $i = 1, 2, \cdots, 2N$ are all real and their real parts are all less than or equal to zero:*

$$Im\{\lambda_i(D(\boldsymbol{q}^{(\gamma)}))\} = 0 \ \ and \ \ Re\{\lambda_i(D(\boldsymbol{q}^{(\gamma)}))\} \le 0, \quad i = 1, 2, \cdots, 2N \tag{13}$$

for all $\gamma = 1, 2, \cdots, m$. Furthermore $D(\boldsymbol{q}^{(\gamma)})$ has at least one eigenvalue 0 and at least one eigenvalue $-1/(\tau\eta)$. It also has $2(N-m)$ eigenvalues $-1/\tau$. The corresponding eigenvector to the eigenvalue 0 is $\boldsymbol{p} = ((\{i\boldsymbol{q}^{(\gamma)}\}^R)^T, (\{i\boldsymbol{q}^{(\gamma)}\}^I)^T)^T$ and that to the eigenvalue $-1/(\tau\eta)$ is $\boldsymbol{r} = ((\boldsymbol{q}^{(\gamma)R})^T, (\boldsymbol{q}^{(\gamma)I})^T)^T$.

The proof of Theorems 2 is omitted. Note that $-1/(\tau\eta) < 0$ and $-1/\tau < 0$ because $\tau > 0$ and $\eta > 1$. It is important to note that Theorems 1 and 2 hold for all $\Phi^{(\gamma)}$, $\gamma = 1, 2, \cdots, m$. The proposed associative memory is homogeneous in this sense.

It is known that qualitative behavior of solutions near an equilibrium point of a nonlinear dynamical system is determined by its linearized model at the point if it is hyperbolic (the coefficient matrix of the linearized model has no eigenvalues with zero real part) [7]. From Theorem 2, however, all the equilibrium points $\boldsymbol{q}^{(\gamma)}$, $\gamma = 1, 2, \cdots, m$ are not hyperbolic. In such a case, the center manifold theory can be utilized to study the structures and asymptotic behaviors of solution orbits near the equilibrium points.

It follows from the center manifold theory that, from Theorem 2, each point $\boldsymbol{q}^{(\gamma)}$ in each memory pattern $\Phi^{(\gamma)}$ has a stable manifold W^s and a center manifold W^c, and has no unstable invariant manifold W^u. Let c be the number of eigenvalues 0 which $D(\boldsymbol{q}^{(\gamma)})$ has. Then there exist a $2N-c$ dimensional stable invariant manifold W^s and a c dimensional center manifold W^c near the equilibrium point $\boldsymbol{q}^{(\gamma)}$.

Consider the case that $c = 1$. It can be shown that, if $c = 1$, that is, $D(\boldsymbol{q}^{(\gamma)})$ has only one eigenvalue 0, the eigenspace associated with the eigenvalue 0, denoted by $E^{c(\gamma)}$, is given by

$$E^{c(\gamma)} = \{a\boldsymbol{p} : \forall a(\in \boldsymbol{R})\} \subset \boldsymbol{R}^{2N} \tag{14}$$

and $E^{c(\gamma)}$ is stable manifold of the linearized model (10). Here we consider the neural network (3), (4) and (6) to be a $2N$ dimensional real dynamical system by separating it into its real and imaginary parts. Denote the expressions of the equilibrium point $\boldsymbol{q}^{(\gamma)}$ and the memory pattern $\Phi^{(\gamma)}$ in the $2N$ dimensional space by $\boldsymbol{q}^{(\gamma)2N}$ and $\Phi^{(\gamma)2N}$, respectively. They are given by $\boldsymbol{q}^{(\gamma)2N} = ((\boldsymbol{q}^{(\gamma)R})^T, (\boldsymbol{q}^{(\gamma)I})^T)^T$ and

$$\Phi^{(\gamma)2N} = \{T(\beta)\begin{pmatrix} \boldsymbol{q}^{(\gamma)R} \\ \boldsymbol{q}^{(\gamma)I} \end{pmatrix} : \forall\beta \in \boldsymbol{R}\} \subset \boldsymbol{R}^{2N}. \tag{15}$$

Theorem 3. *If $D(\boldsymbol{q}^{(\gamma)})$ has only one eigenvalue 0, the equilibrium point $\boldsymbol{q}^{(\gamma)2N}$ has one dimensional center manifold, denoted by $W^{c(\gamma)}$, and $2N - 1$ dimensional stable invariant manifold, denoted by $W^{s(\gamma)}$, and $W^{c(\gamma)} = \Phi^{(\gamma)2N}$.*

Proof. The first half of the theorem is obvious. We will show $W^{c(\gamma)} = \Phi^{(\gamma)2N}$. Note that $\Phi^{(\gamma)2N}$ is invariant for the dynamical system (3), (4) and (6) because it is its equilibrium set. Calculating the gradient of $\Phi^{(\gamma)2N}$ with respect to β at $\beta = 0$, we obtain

$$\left. \frac{d\Phi^{(\gamma)2N}}{d\beta} \right|_{\beta=0} = \begin{pmatrix} -\boldsymbol{q}^{(\gamma)I} \\ \boldsymbol{q}^{(\gamma)R} \end{pmatrix} = \boldsymbol{p}. \tag{16}$$

Then $\Phi^{(\gamma)2N}$ is tangent to $E^{c(\gamma)}$ given by (14) at $\beta = 0$. This completes the proof. $\qquad\square$

Theorem 3 implies that if $D(\boldsymbol{q}^{(\gamma)})$ has only one eigenvalue 0, the memory pattern $\Phi^{(\gamma)2N}$ itself is the center manifold of the equilibrium point $\boldsymbol{q}^{(\gamma)2N} \in \Phi^{(\gamma)2N}$. Note that this theorem holds for all points in the memory pattern $\Phi^{(\gamma)2N}$ if it holds for any one point $\boldsymbol{q}^{(\gamma)2N}$ in $\Phi^{(\gamma)2N}$.

4 Discussion

We have shown that, if $D(\boldsymbol{q}^{(\gamma)})$ has only one eigenvalue 0 for a point $\boldsymbol{q}^{(\gamma)}$ in a memory pattern $\Phi^{(\gamma)}$, the structure of solution orbits near each point in $\Phi^{(\gamma)2N}$ consists of $2N - 1$ dimensional stable invariant manifold and one dimensional center manifold. Furthermore the memory pattern $\Phi^{(\gamma)2N}$ itself is the center manifold which is tangent to the stable manifold $E^{c(\gamma)}$ of the linearized model (10). Considering these facts, we come to the recalling condition of each memory pattern $\Phi^{(\gamma)}$ that, if '$D(\boldsymbol{q}^{(\gamma)})$ has only one eigenvalue 0', each memory pattern $\Phi^{(\gamma)}$ is correctly recalled, that is, every solution starting in the neighborhood of $\Phi^{(\gamma)}$ approaches $\Phi^{(\gamma)}$ as $t \to \infty$. To verify the condition we have carried out numerical experiments. Note that for each $\Phi^{(\gamma)}$ it is enough to evaluate the eigenvalues of $D(\boldsymbol{q}^{(\gamma)})$ at any one point in $\Phi^{(\gamma)}$.

First we let $m = 5$ and choose the vectors $\boldsymbol{s}^{(1)}, \boldsymbol{s}^{(2)}, \cdots, \boldsymbol{s}^{(5)}$ satisfying (1) and (2), and construct the network (3), (4) and (6). We evaluate the eigenvalues of $D(\boldsymbol{q}^{(\gamma)})$ for all $\gamma = 1, 2, \cdots, 5$. It turns out that each $D(\boldsymbol{q}^{(\gamma)})$ has one eigenvalue 0, one eigenvalue $-1/(\tau\eta) = -5.0$, eight eigenvalues $-1/\tau = -10.0$ and eight eigenvalues with negative real parts. Since each $D(\boldsymbol{q}^{(\gamma)})$ has only one eigenvalue 0, we can determine that all the 5 memory patterns are correctly recalled. To verify this, the recalling test was performed. We ran the network (3) starting from the various initial conditions in the neighborhood of each memory pattern. Figure 1 shows examples of the obtained recalling process. In the figure, the directional cosine $a(t)$ between $\Phi^{(\gamma)}$ and the solution $\boldsymbol{x}(t) := [x_1(t), x_2(t), \cdots, x_N(t)]^T$ of the network (3) versus time t is plotted for $\gamma = 1$, where the directional cosine $a(t)$ is defined by $a(t) := ||\Phi^{(\gamma)*}\boldsymbol{x}(t)||/(||\Phi^{(\gamma)}|| \, ||\boldsymbol{x}(t)||)$. $a(t) \to 1$ implies that the memory pattern is correctly recalled. It is seen that, if we choose the initial condition $\boldsymbol{x}(0)$ close enough to the memory pattern $\Phi^{(1)}$, the solution $\boldsymbol{x}(t)$ approaches $\Phi^{(1)}$, that is, the memory pattern $\Phi^{(1)}$ is correctly recalled. Similarly, we have checked that all the other memory patterns are correctly recalled.

Next, we increase the number of the vectors to be stored to $m = 6$ and construct the network (3), (4) and (6). We evaluate eigenvalues of $D(\boldsymbol{q}^{(\gamma)})$ for all $\gamma = 1, 2, \cdots, 6$. It turns out that each $D(\boldsymbol{q}^{(\gamma)})$ has three eigenvalues 0, three eigenvalues $-1/(\tau\eta) = -5.0$, six eigenvalues $-1/\tau = -10.0$ and six eigenvalues with negative real parts. Hence in this case, we cannot determine if each memory pattern is correctly recalled. We performed the recalling test. Examples of the results for the memory pattern $\Phi^{(1)}$ are shown in Fig. 2. It is seen from the figure that, no matter how close we choose the initial condition $x(0)$ to the memory pattern $\Phi^{(1)}$, the solution $x(t)$ does not approach the memory pattern $\Phi^{(1)}$, that is, the memory pattern is not recalled. These experimental results support the recalling condition.

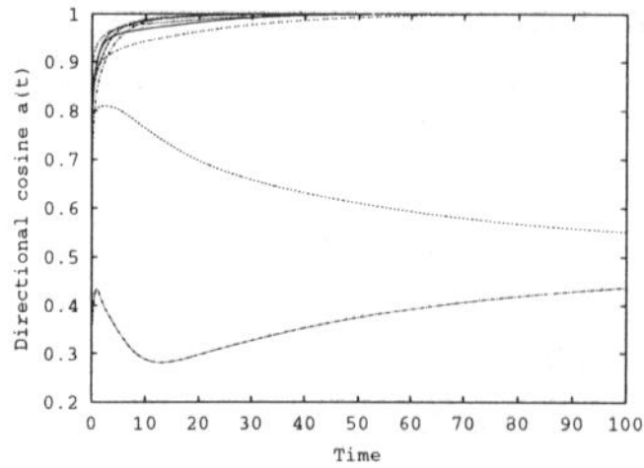

Figure 1: Time evolution of directional cosine $a(t)$ between $\Phi^{(\gamma)}$ and $x(t)(N = 9, m = 5)$.

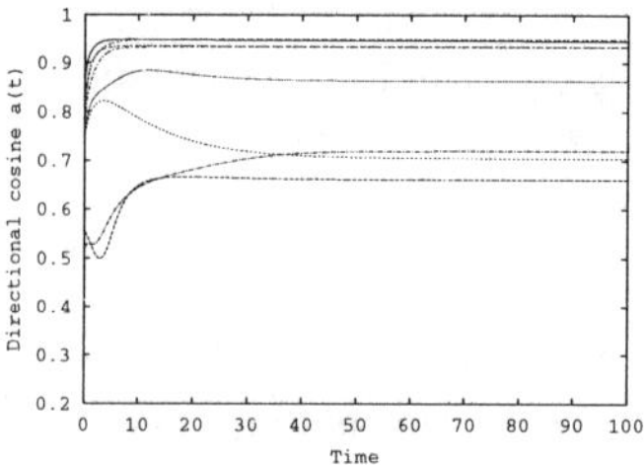

Figure 2: Time evolution of directional cosine $a(t)$ between $\Phi^{(\gamma)}$ and $x(t)(N = 9, m = 6)$.

5   Conclusion

In this paper we have presented a model of associative memories using complex-valued continuous neural networks and studied its qualitative behaviors theoretically. The model is an extension of the conventional real-valued associative memories of self-correlation type. In the model, each memory pattern is not an isolated equilibrium point of the network but an equilibrium set, which is a closed curve in the complex state space. We investigated the eigenstructures and asymptotic behaviors of solution orbits near each memory pattern and discussed its recalling condition. Similar analysis can be done for a model of complex-valued associative memories using discrete-time neural networks [8].

References

[1] G. M. Georgiou and C. Koutsougeras, "Complex domain backpropagation," IEEE Trans. on Circuits and Systems, vol. 39, no. 5, pp. 330–334, May 1992.

[2] N. Benvenuto and F. Piazza, "On the complex backpropagation algorithm" IEEE Trans. on Signal Processing, vol. 40, no. 4, pp. 967–969, Apr. 1992.

[3] I. Nemoto and T. Kono, "Complex-valued neural network," The Trans. of IEICE, vol. J74-D-II, no. 9, pp. 1282-1288, 1991 (in Japanese).

[4] A.J.Noest, "Phaser neural network," Neural Information Processing Systems, D.Z.Anderson,ed., pp. 584–591, AIP, New York, 1988.

[5] A. Hirose, "Dynamics of fully complex-valued neural networks," Electronics Letters, vol.28, no.16, pp.1492-1494, 1992.

[6] S.Jankowski, A.Lozowski, and J.M.Zurada, "Complex-valued multistate neural associative memory," IEEE Trans. Neural Networks, vol. 7, no. 6, pp. 1491-1496, Nov. 1996.

[7] J. Guckenheimer and P. Holmes, Nonlinear Oscillations, Dynamical Systems, and Bifurcations of Vector Fields, pp. 16–22, Springer-Verlag, New York, 1986.

[8] N.Hashimoto, Y. Kuroe and T.Mori, "Theoretical Study of Qualitative Behaviors of Self-Correlation Type Associative Memory on Complex-valued Neural Networks," Trans. of IEICE, vol.J83-A, no.6, pp.750-760, June 2000 (in Japanese).

Complex–Valued Associative Memory
Storing Grayscale Images
with Attribute Information Attached

Hiroyuki AOKI
Department of Electronic Engineering, Tokyo National College of Technology,
Hachiouji 193–0997 Japan

Yukio KOSUGI
Interdisciplinary Graduate School of Science & Engineering, Tokyo Institute of Technology,
Yokohama 226–8502 Japan

Abstract. In view of applying Complex–Valued Associative Memory (CAM) for grayscale image processing in various fields, attaching attribute information to each image to be stored in the memory is more significant for practical use. This paper proposes a CAM capable of storing grayscale images with attribute information attached, and examines a recalling ability of the memory. This paper also presents a useful technique for improving the performance of the memory.

1. Introduction

In many practical signal–processing fields, complex–valued expressions are widely used to analyse data. In the field of neural network applications, the development of complex–valued models is also expected. Complex valued neural networks [1][2] are well suited to deal with multi–valued information such as grayscale images, since introducing complex–valued units enables us to make use of phase information. As an application of complex–valued models, Lee et al.[3] have introduced an example of grayscale pattern recognition. We also have proposed a grayscale image association system, which has been realized by applying Complex–Valued Associative Memory (CAM), combined with a 2-dimensional discrete Fourier transform (2–D DFT)[4][5]. This approach makes possible to apply CAM easily for grayscale image processing, and extends application fields of CAM. A few applications of CAM for image processing have been reported. They are, for example, the system for recognizing the current position of a vehicle in a town by performing a scene image association, and the system for identifying the tip position of fiberscopes under the endoscopic surgery by using an endoscopic image association [6][7]. In view of building more practical systems, it would be more useful, if each image to be stored in a CAM has its some attribute, which provides more information about itself. This paper is intended to build a CAM storing grayscale images with attribute information expressed by characters and to examine the recalling ability of the system.

2. Preliminaries

2.1 Construction of CAM

We consider fully connected complex–valued neural networks as shown in Fig.1 (a). Each neuron can take K states of the unit circle in the complex plane as shown in Fig.1 (b).

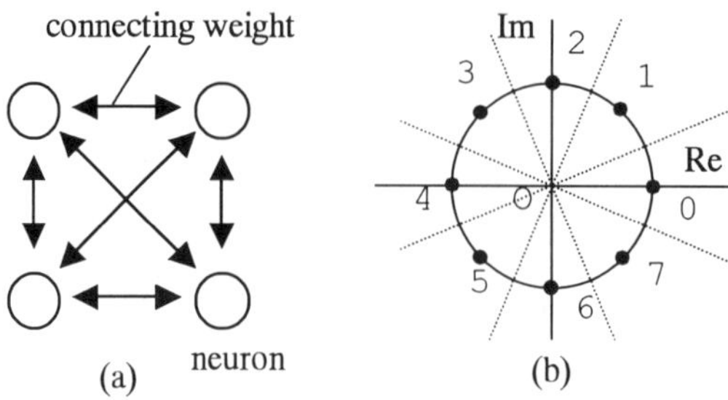

Fig.1 (a) Composition of CAM. (b) States of complex-valued neuron defined in complex plane (an example of K=8).

The state of k th complex-valued neuron, denoted by $x(k)$, is given by

$$x(k) = \exp\left(i\theta_K s(k)\right), \quad \theta_K = 2\pi/K, \quad s(k) = 0, 1, \cdots, K-1. \tag{1}$$

Let $\mathbf{W} = \left(w_{kj}\right)$ be the complex-valued weight matrix among neurons, where w_{kj} denotes the connecting weight from the j th to the k th neuron. The fully connected CAM state transition equation can be given by

$$x_{next}(k) = \mathrm{csgn}\left(y(k)\right) \ , \ y(k) = \sum_{j \neq k} w_{kj} x(j), \tag{2}$$

where

$$\mathrm{csgn}\left(y(k)\right) = \exp\left(i\,\theta_K s_{next}(k)\right) \ , \ s_{next}(k) = \left[\left\{\arg(y(k)) + \tfrac{1}{2}\theta_K\right\}/\theta_K\right] \ (\mathrm{mod}\ K) \tag{3}$$

and the symbol $[c]$ presents the maximum integer not exceeding c .

2.2 Phase Matrix Image Representation

A phase matrix image representation has already been proposed to represent a grayscale image using CAM [4]. For the representation, a 2-dimensional discrete Fourier transform (2-D DFT) and an inverse DFT (2-D IDFT) are employed. Let the 2-D DFT pair of an $M_L \times N_L$ image $\mathbf{u} = \{u(m,n)\}$ be denoted by

$$\mathbf{v} = \{v(m',n')\} = DFT\left(\mathbf{u}\right) \tag{4}$$

$$\mathbf{u} = \{u(m,n)\} = IDFT\left(\mathbf{v}\right) . \tag{5}$$

Then, let

$$\left|v_{max}\right| = \max\left\{\left|v(m',n')\right|; \ \ 0 \leq m' \leq M_L - 1, \ 0 \leq n' \leq N_L - 1\right\} \tag{6}$$

$$\alpha(m',n') = \arg\left(v(m',n')\right), \quad \gamma(m',n') = \cos^{-1}\left(\left|v(m',n')\right|/\left|v_{max}\right|\right). \tag{7}$$

The phase matrix $\mathbf{x}_{phase} = \left\{x_{phase}(m',n')\right\}$ of an image $\mathbf{u}$ is defined as

$$x_{phase}(m',n') = \exp\left(i\left(\alpha(m',n') + \gamma(m',n')\right)\right). \tag{8}$$

On the other hand, the original image $\mathbf{u}$ is restored from the phase matrix $\mathbf{x}_{phase}$ by the following inverse transform

$$\mathbf{u} = \left|v_{max}\right| \mathrm{Re}\left\{IDFT\left(\mathbf{x}_{phase}\right)\right\}. \tag{9}$$

In the above equation, $\left|v_{max}\right|$ is not always required to restore the image, since $\left|v_{max}\right|$ is a constant value. In Eq.(8), for each element $x_{phase}(m',n')$, one complex-valued neuron is allocated, resulting in CAM construction. In the process, the number of neurons in CAM can be reduced by removing high frequency elements from the phase matrix $\mathbf{x}_{phase}$ at the degree in which the image degradation is not conspicuous. Let us here $\mathbf{x}_{img}$ be a vector obtained by removing high frequency elements from the phase matrix $\mathbf{x}_{phase}$. The operation given by Eq.(8) and the removal process of the high frequency elements are carried out in the pre-processing shown in Fig.2. The image restoration process given by Eq.(9) is carried out in the post–processing shown in Fig.2.

3. Construction of CAM Storing GrayScale Images with Attribute Information Attached

3.1 Representation of Characters by CAM

In this paper, let attribute information be expressed by several characters. Suppose that we use a convention known as ASCII code, one character is expressed by the length of 8 binary digits. This allows us to represent one character with one complex-valued neuron, if the number of its states K is greater than or equal to $2^8 = 256$. In order to attach attribute given by N_{cha} characters to each image to be stored in CAM, N_{cha} complex-valued neurons should be added. Let $\mathbf{x}_{cha}$ be a complex-valued neuron state vector generated from N_{cha} characters.

3.2 Calculation of the Weight Matrix

A complex-valued state vector $\mathbf{x}$ of CAM is composed of the vectors $\mathbf{x}_{img}$ and $\mathbf{x}_{cha}$. Let

$$\mathbf{x}^{\alpha} = \begin{pmatrix} \mathbf{x}^{\alpha}_{img} \\ \mathbf{x}^{\alpha}_{cha} \end{pmatrix}, \quad (\alpha = 1, 2, \cdots, P) \tag{10}$$

be P complex-valued state vectors to be stored in CAM. The weight matrix $\mathbf{W}$ can be given by the following equation according to the idea of the optimized associative mapping [8]. Let N_{img} and N_{cha} be the numbers of dimension of the vectors $\mathbf{x}^{\alpha}_{img}$ and $\mathbf{x}^{\alpha}_{cha}$, respectively. Let $N = N_{img} + N_{cha}$ and $N \times P$ matrix $\mathbf{S} = \left(\mathbf{x}^1, \mathbf{x}^2, \cdots, \mathbf{x}^P \right)$, then

$$\mathbf{W} = \mathbf{S}\,\mathbf{S}^{+}, \quad \mathbf{S}^{+} = (\mathbf{S}^{*t}\mathbf{S})^{-1}\mathbf{S}^{*t} \tag{11}$$

where the notation $\mathbf{S}^{+}$ indicates the Moore–Penrose pseudoinverse matrix of $\mathbf{S}$

4. Phase Adjustment

A CAM built by Eq.(11) has the following property. When memorizing a vector $\mathbf{x}^{\alpha}$ into the CAM, then the phase-shifted vectors of the $\mathbf{x}^{\alpha}$ including $\mathbf{x}^{\alpha}$ itself also become fixed points of the CAM. In Eqs.(2) and (3), the following relation holds.

$$\mathbf{x}^{\alpha}\exp\left(i\theta_K s\right) = \mathrm{csgn}\left(\mathbf{x}^{\alpha}\exp\left(i\theta_K s\right)\right) \quad (s = 0,1,2,\cdots,K-1) \tag{12}$$

Therefore, the CAM may retrieve a phase-shifted vector $\mathbf{x}^{\alpha}\exp\left(i\theta_K s\right)$ instead of $\mathbf{x}^{\alpha}$ itself, in response to the presentation of an incomplete or noisy version of the vector $\mathbf{x}^{\alpha}$. This property causes a problem of preventing us from obtaining the clear memorized images and their correct attribute information. To solve the problem, we propose a phase adjustment process, which is carried out as soon as the CAM finishes a recalling operation. As long as the value of s in Eq.(12) can be found, the vector $\mathbf{x}^{\alpha}$ is obtained by shifting the phase of the recalled vector by $-\theta_K s$.

For finding the value of s, we make use of the following property on the phase matrix image representation described in the previous section. The element of the phase matrix at $m' = 0$ and $n' = 0$, or $x_{phase}(0,0)$, always takes the following constant value. That is

$$x_{phase}(0,0) = \exp\left(i\pi/2\right) = \exp\left(i\theta_K K/4\right), \tag{13}$$

where K is assumed to be a multiple of 4. Eq.(13) is derived as follows, 1) the $v(0,0)$ taken from Eq.(4) is called the Direct Current (DC), which means the brightness or bias of the whole image, is real value so that $\alpha(0,0) = 0$ is obtained, and 2) we set $\gamma(0,0) = \pi/2$ in the process of the phase matrix generation, which means to get rid of the effect of DC component from an input image. Then, we pay attention to the neuron corresponding to the element $x_{phase}(0,0)$ after a recalling process is finished. The phase difference between the recalled final state of the neuron and the state $\exp\left(i\pi/2\right)$ is expected to indicate the value of $\theta_K s$.

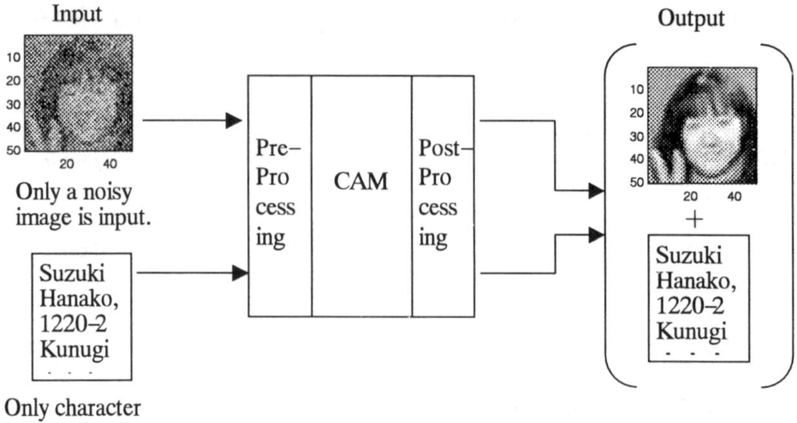

Fig.2 Schematic diagram of the simulation

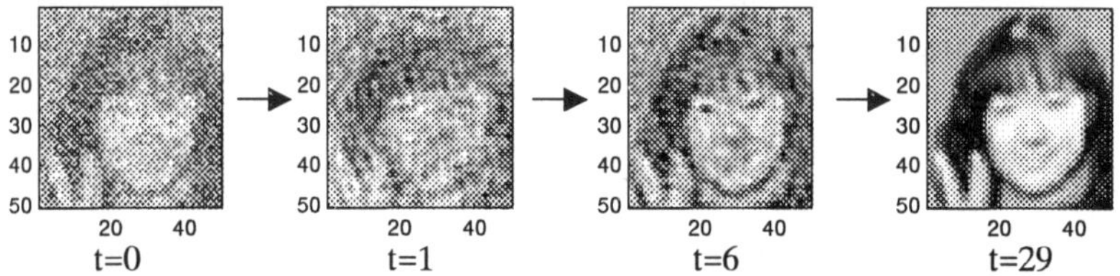

Fig.3 A recalling process of an input image degraded by 2 dB SNR additive, Gaussian white noise.

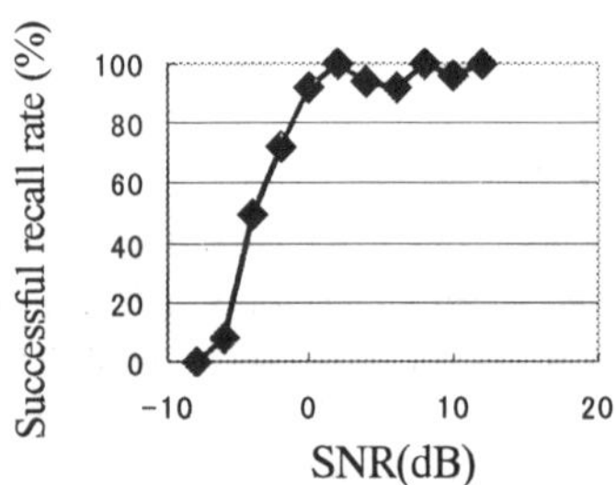

Fig.4 SNR Dependence of successful recallrate

5. Simulation Results

Let us build a CAM, which deals with 50×50 images with $[0, 255]$ integer-valued gray levels. The human face images are mainly collected as test images. The attribute information expressed by 200 characters is attached to each memorized image. The CAM consists of 1213 complex-valued neurons, in which the number for images $N_{img} = 1013$, and for characters $N_{cha} = 200$. The characters are randomly generated. The number of states of each neuron is set at 256. Let 170 images be memorized into the CAM. The storage capacity (P/N) is 0.14. The synchronous operation is used in the following simulations.

Two kinds of simulations were carried out to examine the recalling ability of CAM. 1) The case in which a noisy image degraded by additive Gaussian white noise to the memorized image is input. 2) The case in which the only attribute information is input.

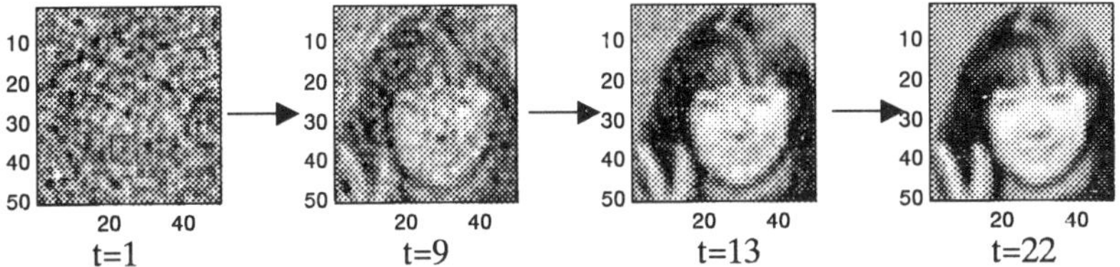

Fig.5 A recalling process, where only the character information is given as the initial state..

The signal-to-noise ratio (SNR) by additive Gaussian white noise is defined in (dB) as

$$SNR = 10\log_{10}\left(\sigma^2/\sigma_n^2\right) \tag{14}$$

where σ^2 and σ_n^2 are the variance of the original image and the noise, respectively.

Fig.3 displays a recalling process, in which case only an image degraded by 2 dB SNR without attribute information is input. The memorized image and attribute information associated with the image are perfectly recalled in 29 iterations. Shown in Fig.4 is the SNR dependence of successful recall rate. We count each recalling process successful if and only if the memorized data are perfectly recalled. Fig.5 displays a recalling process, in which case only attribute information is input. The 95% of all the memorized images including the case of Fig.5 image are successfully recalled from only their attribution information. The successful case means the whole image and attribute data can be recalled from only the 16% part of it. The phase adjustment described in Sec. 3.3 is carried out in all these simulations.

6. Conclusion

A content addressable memory storing grayscale images with attribute information attached has been realized using CAM. In order to realize the memory, the following techniques are required: 1) the phase matrix image representation to represent images using CAM, 2) the removal operation of high frequency components of images to reduce the number of neurons, 3) the phase adjustment proposed in this paper to increase the successful recall rate. The development of the system capable of dealing with large-scale image data, for example, for medical use and geographical use, etc., is a future subject of this study.

References

[1] A.Hirose, Proposal of fully complex-valued neural networks , Proc. IJCNN 92, Vol V, pp.152-157, Baltimore, 1992.

[2] Igor Aizenberg, Naum. Aizenberg, Joos Vandewalle, Multi-Valued and Universal Binary Neurons , Kluwer Academic Publishers, 2000

[3] Dong-Liang Lee and Wen-June Wang, A multivalued bidirectional associative memory operating on a complex domain , Neural Networks 11, pp.1623-1635, 1998.

[4] H. Aoki, MR.Azimi-Sadjadi and Y. Kosugi, Image Association Using a Complex-Valued Associative Memory Model , IEICE Trans., Vol.E83-A, no.9, pp.1824-1832, Sep.2000.

[5] H. Aoki and Y. Kosugi, An Image Storage System Using Complex-Valued Associative Memoies , in Proc. International Conference on Pattern Recognition Barcelona, Vol II, pp.626-629, Sep.2000.

[6] H. Aoki and Y. Kosugi, Image Guided Positioning by Using a Complex-Valued Associative Memory for Intelligent Navigation Systems , in Proc. 2001 IEEE Intelligent Vehicle Symposium Tokyo, to be published.

[7] H. Aoki, E. Watanabe, A. Nagata and Y. Kosugi, Rotation-Invariant Image Association for Endoscopic Positional Identification Using Complex-Valued Associative Memories , in Proc. International Work-conference on Artificial and Natural Neural Networks Granada, to be published.

[8] T.Kohonen, Self-Organization and Associative Memory , 2nd Ed, Springer-Verlag, New York, 1987.

KES '01
N. Baba et al. (Eds.)
IOS Press, 2001

Complex-Valued Nagumo-Sato Model of a Single Neuron

Iku NEMOTO and Kei SAITOH
Department of Mathematical Sciences, Tokyo Denki University
Hatoyama,   Saitama 350-0394,   Japan

Abstract. We propose a comple-valued version of the Nagumo-Sato model of a single neuron. Its weights and membrane potential u take complex values. The output is either 0 or $e^{i\arg u}$ depending on the modulus of u. Simulation results suggest that the model dynamics involve both continuous and discontinuous types of chaos. Some of the basic properties such as fixed points and period-two orbits of the model dynamics have been also studied.

1. Introduction

The Nagumo-Sato model of a single neuron has been studied in detail by Hata[1]. The model is simple but the dynamics is rich having discontinuous chaos. This paper proposes a complex-version of the model and reports some of its behavior.

Some researchers have proposed complex-valued neural networks. We proposed complex-valued associatiative memory and showed its capacity by simulation[2,3]. Complex-valued neurons allow the description of both mudulus and phase, thus approximating better the neural signals of the biological neurons than the conventional two-valued model. The complex version of the Nagumo-Sato model has complex weights and complex membrane potential u.   The output is either 0 or $\exp\{i\arg u\}$ depending on the modulus of u.   The behavior of the proposed model is quite interesting and appears to have both discontinuous and continuous chaos.   Discontinuous chaos comes from the step-function-like behavior of the output function with a sort of threshold and continuous chaos results from the nonlinearity of the output function above the threshold.

This report proposes the model and shows its very basic properties and some of the results of numerical experiments with emphasis on behavior which appears to be chaotic.   We present the experimental ground to support the conjecture that it is chaos.

2. The Model

The model is a complex version of the Nagumo-Sato model.   Let the output of the complex neuron model at time n be denoted by

$$\xi_n = \Theta(\eta_{n-1}) \qquad (1)$$

where the complex membrane potential η_n is given by

$$\eta_n = A - \frac{1}{\alpha}\sum_{r=0}^{n}\beta^r\xi_{n-r} \qquad (2)$$

where A and β are complex and $\alpha > 0$.   As β stands for the effect of the past outputs, we set $|\beta| < 1$. β is expressed both in polar and orthogonal coordinates as $\beta = b\exp(i\theta) = \beta_R + i\beta_I$. Θ is a complex version of the Heaviside step   function:

$$\Theta(u) = \begin{cases} 0, & |u| < 1 \\ e^{i \arg u} & |u| \geq 1 \end{cases} \qquad (3)$$

We can make the above model into a dynamical system by the following variable conversion:

$$z_n = 1 + \alpha\beta A - \sum_{r=0}^{n} \beta^r u_{n-r} \qquad (4)$$

We let $c = 1 - \alpha A(1 - \beta)$ and then we get the dynamical system we deal in this report:

$$z_n = f(z_{n-1}) \qquad (5)$$

$$f(z) = \begin{cases} 1 + \beta(z - c), & (|z - c| < \alpha) \\ 1 + \beta(z - c) - \dfrac{z - c}{|z - c|}, & (|z - c| \geq \alpha) \end{cases} \qquad (6)$$

We treat $\alpha \geq 0, \beta, c$ as independent constants.

3. Some Basic Properties

We first note that the dynamics of the model is bounded; it is easily shown that for any initial value z_o and any $\varepsilon > 0$,

$$|z_n - c| \leq \frac{1 + |1 - c|}{1 - b} + \varepsilon \qquad (7)$$

for sufficiently large n. Therefore, we only have to deal with orbits within this bound.
Another convenient property of the model is its two kinds of topological conjugacy:

$$f(\bar{z} \mid \bar{c}, \bar{\beta}) = \overline{f(z \mid c, \beta)} \qquad (8)$$

$$f(T_\tau z \mid T_\tau c, \beta) = T_\tau f(z \mid c, \beta) \qquad (9)$$

where the bar stands for complex conjugate and T_τ is an operator to rotate around 1 by angle τ. The topological conjugacy allows us to restrict the parameter space to

$$S = \{\alpha \geq 0, \beta \in \mathbf{C}, c \leq 1 : b \leq 1, 0 \leq \arg \beta \leq \pi\} \qquad (10)$$

We let $C_r \equiv \{z : |z - c| < r\}$, $\Gamma_R \equiv \{z : |z - 1| < R\}$ and $\partial C_r, \partial \Gamma_R$ their boundaries. We call ∂C_α the critical circle abbreviated as **CC** hereafter. The mapping function of the model can be rewritten as

$$z = c + r e^{i\phi} \qquad (11)$$

$$f(c + r e^{i\phi}) - 1 = \begin{cases} \beta r e^{i\phi}, & (r < \alpha) \\ e^{i\phi}(\beta r - 1), & (r \geq \alpha) \end{cases} \qquad (12)$$

This shows that a circle ∂C_r is mapped to one of the Γ_R's. This circle-to-circle correspondence is a very basic feature of the dynamics.

4. Some Numerical Experiments

Fig. 1 shows some typical orbits. $\{f^n(z_0): n = 1, 2, \ldots, 20000\}$ starting from the initial value $z_0 = 0.5 + i0.5$ are plotted. In (a), $\beta = 0.2 \exp(i\pi / 3), c = 0.3$ and $\alpha = 0.1$. The orbit consists of two series of points which quickly converge to two points (arrows). These two points are period-two points and both outside the **CC**. In (b) $\beta = 0.5 \exp(i\pi / 3)$ and the other parameter values are the same as (a). The orbit appears to form a dense ring. In (c) $\alpha = 0.5$

and the other parameters are the same as (b). The orbit converges to one of the three sets of period-three points depending on the initial value. To avoid confusion, the first 1000 points of each orbit are eliminated. In (d), $\beta = 0.98\exp(i\pi/18)$, $c = 0.3, \alpha = 0.1$. The orbit did not show any sign of periodic behavior and appears to be chaotic, though not proved to be so.

We study the last example more closely. Fig. 2(a) is a close-up view of the part of the orbit roughly indicated by the arrow in Fig. 1(d). Here 2 million iterations were performed to obtain enough points in this very small area. Further iterations made the orbit denser and so far no sign of periodicity was observed. In (b), $\partial f''(C_{0.3}), f''(\partial C_{0.3})$ are shown for $n = 1, 2, \ldots 20$. It is seen that 20 iterations of the map of the **CC** and its inner lining reveal the basic feature of the orbit of a single point shown in Fig. 1(d). It is easily seen that for this set of parameter values, $f^2(\overline{C_{0.3}}) \subset f(\overline{C_{0.3}})$ holds so that the orbit lies within $f(\overline{C_{0.3}})$ if it starts there. In fig. 2(c), $f^{20}(\overline{C_{0.3}})$ is shown which again looks quite close to Fig. 1(d). This suggests that the ω-limit set of 0.5+i0.5 (and probably of any point) coincides with the invariant set of the map. Fig. 2(d) shows the high sensitivity of the dynamics on the initial value. The initial values are 2000 equally spaced points on the circle with radius 0.0001 centered at 1. The 150-th images of the 2000 points are plotted which show close resemblance to Fig. 1(d). All these features strongly suggest that we are looking at chaos.

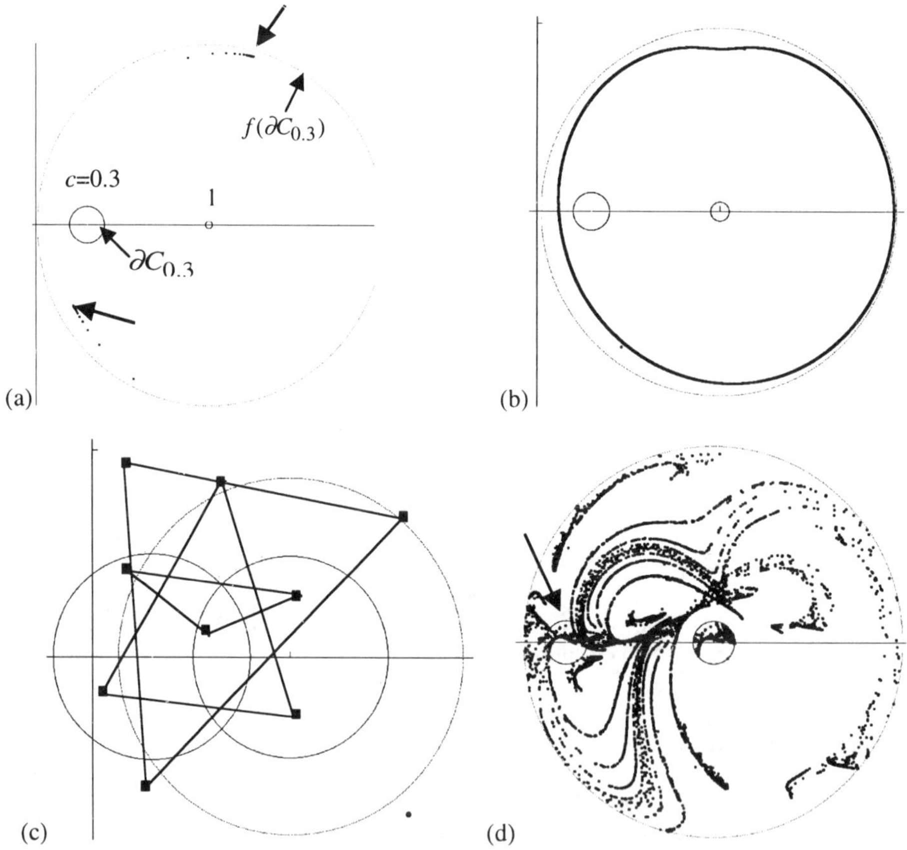

Fig. 1 Typical orbits of the model. $c=1$ and 1 are shown to be on the horizontal(real) axis. The large circle centered at 1 is $f(\partial C_{0.3})$. See text for parameter values.

Fig. 3 shows a bifurcation diagram for $\phi_n = \arg(z_n - c)$ as function of $10° \leq \theta \leq 14°$. The other parameter values are: $b = 0.98, c = 0.3, \alpha = 0$. The dynamics takes place always outside **CC**(its radius is 0). In this quite narrow range of $\arg\beta$, the dynamics changes its behavior in a very complicated manner.

5. Concluding Remarks

The present article reports the result of our numerical experiment with the complex version of the Nagumo-Sato model. The result suggests strongly that this model can cause chaotic behavior in the phase of the output. The phase of neural impulse has recently received much attention as a biological means to associate objects in our memory. The use of complex-valued models of neural networks will be of much use in modeling such kind of phenomenon. Also this is an interesting model for itself and should be studied more theoretically. Some theoretical aspects of the model such as fixed points and period-two points are discussed elsewhere[4].

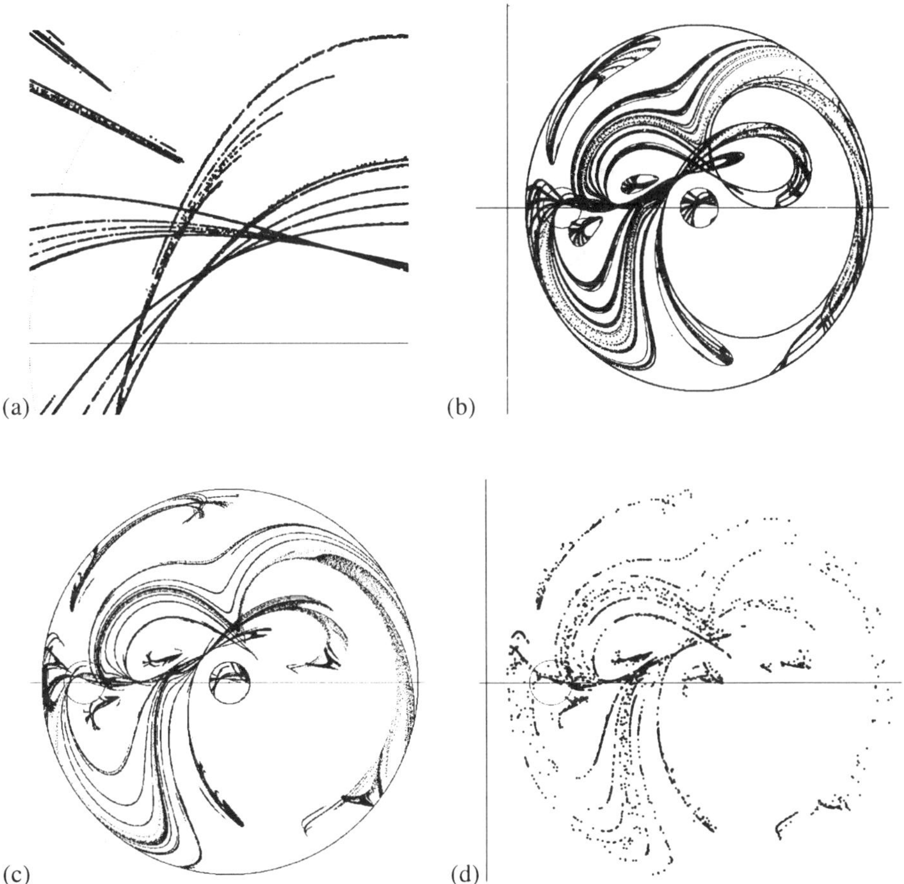

Fig.2.   Experimental ground suggesting that Fig.1(d) shows chaotic behavior.

References
[1]   M. Hata, Dynamics of Cainiello's Equation, *J. Math. Kyoto Univ.*, **22**(1982) 177-173.
[2]   I. Nemoto and T.Kono, Complex Neural Networks, *Systems and Computers in Japan.*, **23**, 75-84 (1992).
[3]   I. Nemoto and M.Kubono, Complex Associative Memory, *Neural Networks*, **9**, 253-261 (1996).
[4]   I. Nemoto and K. Saito, A Complex-Valued Version of Nagumo-Sato Model of a Single Neuron and its Behavior, *submitted for publication.*

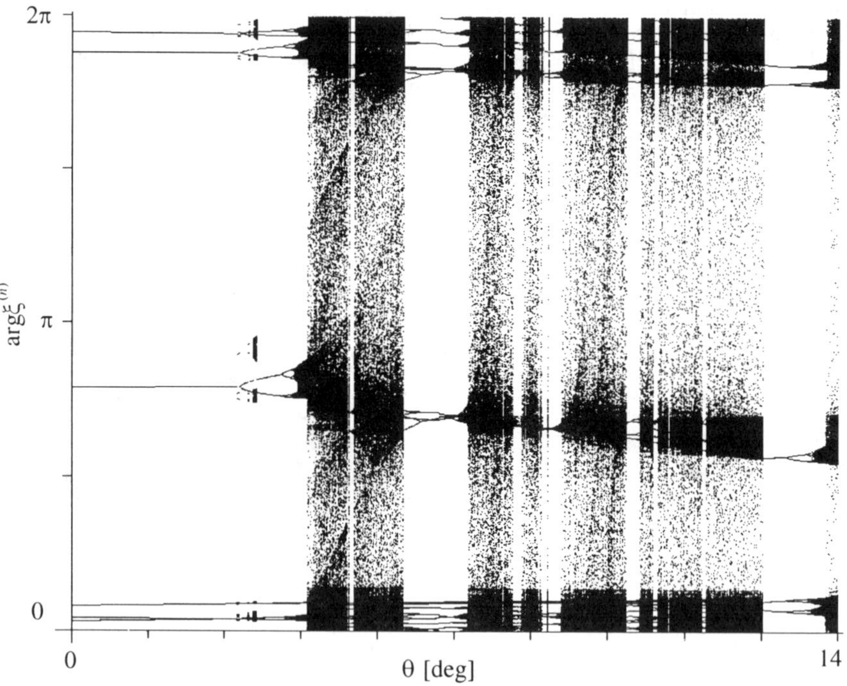

Fig. 3   Bifurcation diagram for $\arg \xi^{(n)}$ as function of $\arg \beta$.

KES '01
N. Baba et al. (Eds.)
IOS Press, 2001

An Application of a Phasor Model with Resting States to Multiuser Detection

Teruyuki MIYAJIMA and Kazuo YAMANAKA

Department of Systems Engineering, Ibaraki University, 316-8511, Japan

Abstract. We had proposed a phasor model of neural networks where the state of each neuron possibly takes the value at the origin as well as on the unit circle. In this paper, we apply the phasor model to data detection in code-division multiple-access (CDMA) communications. In the CDMA system considered, transmitted data take complex values, and users are allowed to be the inactive mode as well as the active mode. The parameters of the phasor model are determined by comparing the energy function of the phasor model and the likelihood function for the optimum detection. Simulation results show that the receiver using the phasor model has the possibility to outperform a conventional receiver.

1 Introduction

It is a natural selection to adopt complex-valued neural networks in order to process complex-valued signals. In most of complex-valued neural networks proposed earlier, the state of each neuron takes the value on the unit circle of the complex domain [2, 3, 4]. Such a neural network is called phasor model since the state can be considered as a phase of pulse sequences from a firing neuron. On the other hand, we proposed a phasor model, which is referred to as phasor model with resting states, where neurons are allowed to rest as well as fire[1]. In this model, the state of each neuron takes either the value on the unit circle or zero. In [1], a stability property of equilibria was studied.

In this paper, we consider an application of the phasor model with resting states to communications. We focus on CDMA systems since they have been applied in many wireless communication systems such as IS-95 and IMT-2000. Real-valued Hopfield networks were used for data detection in CDMA systems where the data take real values and all users are always active[5]. On the other hand, in the CDMA system considered in this paper, the data take complex values and users are allowed to be inactive as well as active. Then, if a phasor model with resting states is applied to the data detection, the complex-valued data of an active user can be represented by the state of an active neuron, and an inactive user can be represented by the resting state of a neuron. We will show that mimicking the optimum multiuser detection can derive a receiver using the phasor model.

2 Phasor Model with Resting States

2.1 Updating Rule

Consider a neural network with N neurons. The local state (output) of the jth neuron is denoted by x_j. A local state is allowed to be either at the origin, $x_j = 0$, or on

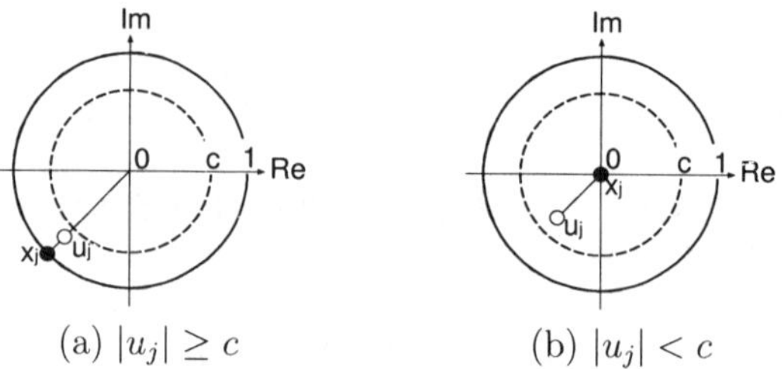

(a) $|u_j| \geq c$ (b) $|u_j| < c$

Figure 1: Updated state.

the unit circle, $x_j = \exp(i\phi_j)$, where ϕ_j denotes the argument of x_j. A local state is assumed to be asynchronously updated based only on its membrane potential given by

$$u_j = \sum_{i=1}^{N} w_{ji} x_i \tag{1}$$

where w_{ji} represents the complex-valued connection weight between the jth and ith neuron. It is assumed that the Hermitian property $w_{ji} = w_{ij}^*$ holds where $(\cdot)^*$ stands for the complex conjugate. When a neuron is updated the new local state is determined by the following rule:

$$x_j = \begin{cases} \exp(i\phi_j), & \phi_j \stackrel{\triangle}{=} \arg(u_j) & (|u_j| \geq c) \\ 0 & & (|u_j| < c) \end{cases} \tag{2}$$

where $c > 0$ is a given threshold. Figure 1 illustrates the relation between the destination of the state and the membrane potential of a neuron.

2.2 Stability of Equilibria

Here we briefly summarize the results concerning with the stability of equilibria presented in [1]. The energy function is defined by

$$U(\mathbf{x}) \stackrel{\triangle}{=} -\frac{1}{2} \sum_{j=1}^{N} \sum_{i=1}^{N} x_j^* w_{ji} x_i \tag{3}$$

where $\mathbf{x} \stackrel{\triangle}{=} (x_1\ x_2\ \cdots\ x_N)$ denotes a global state. Consider an equilibrium $\bar{\mathbf{x}} = (\bar{x}_1\ \bar{x}_2\ \cdots\ \bar{x}_N) = (\exp(i\bar{\phi}_1)\ \exp(i\bar{\phi}_2)\ \cdots\ \exp(i\bar{\phi}_N))$. Let the set of indices of the local states on the unit circle be denoted by I, i.e., $I \stackrel{\triangle}{=} \{j|\ |\bar{x}_j| = 1\}$, and the number of these neurons by $|I|$. We assume $0 < c < \min_{j \in I} |\bar{u}_j|$ where $\bar{u}_j$ is jth neuron's membrane potential at $\bar{\mathbf{x}}$. This assumption is needed for $\bar{\mathbf{x}}$ to be an equilibrium. Define δ-neighborhood of a state $\mathbf{x}_0 = (x_{01} \cdots x_{0N})$ as

$$\Omega \stackrel{\triangle}{=} \{\mathbf{x}|\ x_j = 0,\ j \notin I\ \text{ and }\ |\arg(x_j) - \arg(x_{0j})| < \delta,\ j \in I\}$$

for arbitrary real number $\delta > 0$. Considering the energy change in Ω, we can obtain the following result.

 Lemma 1: The function U decreases so long as the state transition occurs in a neighborhood Ω of an equilibrium. $\square$

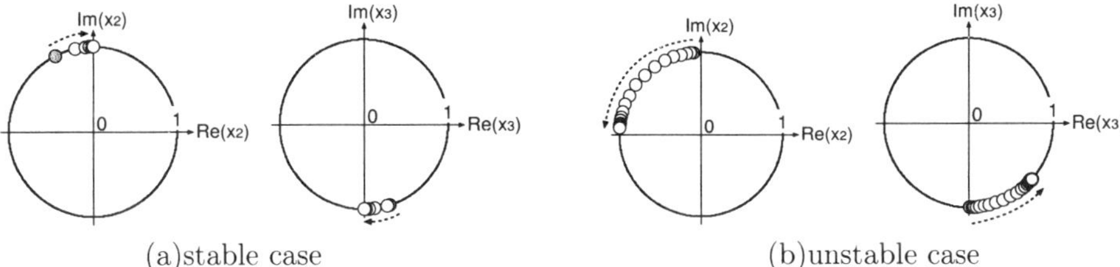

(a)stable case (b)unstable case

Figure 2: State transition.

Furthermore, we get the following result concerning with the stability of the equilibrium:

Proposition 2: Choose an arbitrarily $i \in I$. Suppose that ϕ_i is fixed to $\bar{\phi}_i$ and U is convex in a neighborhood Ω as a function of the other $|I| - 1$ variables ϕ_j. Then, there exists a neighborhood $\Omega_0 \subset \Omega$ such that

$$\mathbf{x}_k \to \bar{\mathbf{x}} \quad (k \to \infty)$$

whenever the sequence of the global states starts from the inside of neighborhood Ω_0.□ Hence, we can conclude that the equilibrium is asymptotically stable if the energy is locally convex. A sufficient condition for the local convexity of the energy is that all the $(|I| - 1) \times (|I| - 1)$ principal submatrices of the Hessian matrix are positive definite. A sufficient condition for the positive definiteness of the principal submatrices is

$$\mathrm{Re}(\bar{x}_j^* w_{ji} \bar{x}_i) > 0 \quad i, j \in I. \tag{4}$$

We show simulation examples. We consider two networks consisting of five neurons. Both networks have an equilibrium at $\bar{\mathbf{x}} = (1\ i\ -i\ 0\ 0)$. The connection weights of the first network are chosen to satisfy the condition in (4). On the other hand, the weights of the second network do not satisfy the condition. Figure 2 shows the behavior of x_2 and x_3 with small initial disturbance when the state of the 1st neuron was fixed at $x_1 = \bar{x}_1$. From this figure, one can observe that the disturbed state approaches the equilibrium asymptotically if the energy is locally convex. Moreover, although the initial state is very close to the equilibrium the disturbed state goes far away from the equilibrium if the energy is not locally convex. We observed that the states x_4 and x_5 remain at 0.

3 Application to Multiuser Detection

3.1 *Communication System*

Consider a synchronous QPSK/DS/CDMA system[6]. The system structure is shown in Fig.3. The complex baseband representation of the received signal is given by

$$r(t) = \sum_{k=1}^{K} A_k b_k s_k(t) + n(t) \tag{5}$$

where $b_k(i) \in \{0, 1, e^{j\pi/2}, e^{j\pi}, e^{j3\pi/2}\}$ is the complex-valued information symbol, $A_k \in R$ is the signal amplitude, $n(t)$ is a zero mean white complex Gaussian noise process with

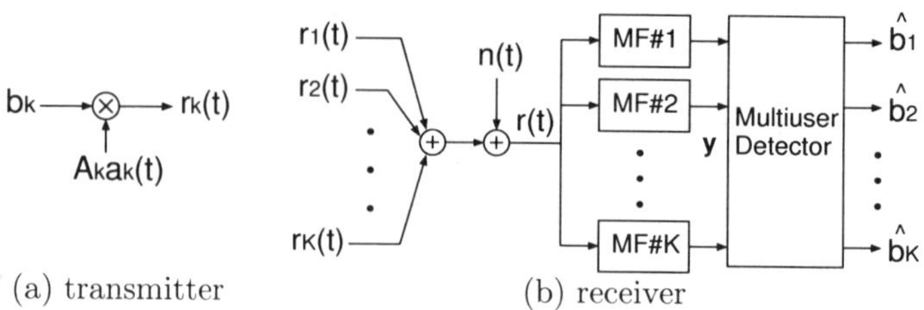

Figure 3: System structure.

power spectral density $N_0/2$, K is the maximum number of users, and $s_k(t)$ is the spreading waveform of the form

$$s_k(t) = \sum_{l=0}^{L_c-1} a_{kl} P_{T_c}(t - lT_c),\ 0 \le t < T_b \tag{6}$$

where $a_{kl} \in \{\pm 1\}$ is the spreading sequences, L_c is the length of $\{a_{kl}\}$, T_b is the symbol duration, $T_c(= T_b/L_c)$ is the chip duration, and $P_{T_c}(t) = 1(0 \le t < T_c), 0(\text{otherwise})$. Users are allowed to be either active or inactive. If the ith user is inactive, both the transmitted symbol and amplitude are zero, $b_i = 0, A_i = 0$.

The optimum multiuser detection rule selects $\hat{\mathbf{b}} = [\hat{b}_1 \ \cdots \ \hat{b}_K]$ which maximizes the following likelihood function[7]

$$L(\hat{\mathbf{b}}) = 2\text{Re}\{\mathbf{y}^H \mathbf{A}\hat{\mathbf{b}}\} - \hat{\mathbf{b}}^H \mathbf{ARA}\hat{\mathbf{b}} \tag{7}$$

where H denotes the Hermitian transpose operation, $\mathbf{A} = \text{diag}[A_1 \ \cdots \ A_K]$, $\mathbf{R}$ is the correlation matrix of the spreading waveforms given as

$$\mathbf{R} = \frac{1}{T_b} \int_0^{T_b} \mathbf{s}(t)\mathbf{s}^T(t)dt \tag{8}$$

where $\mathbf{s}(t) = [s_1(t) \ \cdots \ s_K(t)]^T$, and

$$\mathbf{y} = \frac{1}{T_b} \int_0^{T_b} r(t)\mathbf{s}(t)dt \tag{9}$$

is the output of the conventional matched filter receiver. The conventional matched filter receiver experiences performance degradation caused by the interference that is due to non-zero cross-correlations between the spreading waveforms. On the other hand, the complexity of the optimum detection grows exponentially with increasing the number of users.

In this paper, we propose to use a phasor model with resting states for low-complexity and sub-optimum multiuser detection. If there is the 0th neuron whose output is always 1, the energy function can be rewritten as

$$U(\mathbf{x}) = -\frac{1}{2} \sum_{j=1}^{N-1} \sum_{i=1}^{N-1} w_{ji} x_j^* x_i - \sum_{i=1}^{N-1} \text{Re}(w_{0i} x_i) \tag{10}$$

Comparing the energy in (10) and the likelihood function in (7), the parameters of the phasor model can be determined as follows:

$$N = K + 1,\ w_{ij} = -2h_{ij}A_iA_j\ (i,j \ne 0),\ w_{0i} = 2y_i^* A_i\ (i \ne 0),\ x_0 = 1 \tag{11}$$

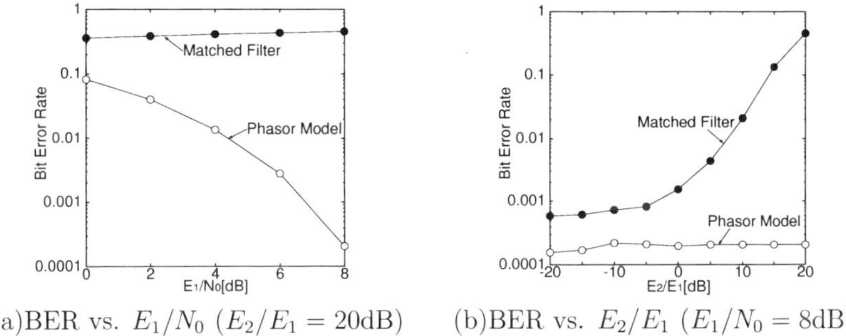

(a)BER vs. E_1/N_0 ($E_2/E_1 = 20$dB)   (b)BER vs. E_2/E_1 ($E_1/N_0 = 8$dB)

Figure 4: Performance comparison.

The output of the kth neuron, x_k, after the dynamics converged is the estimation of the kth user's symbol, $\hat{b}_k$.

3.2 Simulation Results

The maximum number of users is $K = 4$. The 4th user is assumed to be inactive, i.e., $A_4 = b_4 = 0$. The signal energy per bit of the 1st user is equal to that of the 3rd user, i.e, $E_1 = E_3$. The spreading sequence used is the Gold sequence of length 7. The threshold c is set 0.01. In Fig.4, bit error rate performances for the 1st user are shown. It can be observed from these figures that the performance of the conventional receiver degrades as the interference becomes strong, and the proposed receiver is superior to the conventional one. We observed that the state x_4 always converges to 0.

4 Conclusions

In this paper, we have considered an application of the phasor model with resting states to multiuser detection in a CDMA system in which the data take complex values or zero. Simulation results show that the proposed receiver can reduce the effect of interference from other users unlike the conventional matched filter receiver.

References

[1] T.Miyajima, F.Baisho, K.Yamanaka, K.Nakamura, and M.Agu, A Phasor Model with Resting States, *IEICE Trans. on Inform. & Syst.* **E83-D** (2000) 299–301.

[2] A.J.Noest, Associative Memory in Sparse Phasor Neural Networks, *Europhys. Lett.* **6** (1988) 469–474.

[3] A.Hirose, Dynamics of Fully Complex-Valued Neural Networks, *Electronics Lett.* **28** (1992) 1492–1494.

[4] M.Agu, K.Yamanaka and H.Takahashi, A Local Property of the Phasor Model of Neural Networks, *IEICE Trans. on Inform. & Syst.* **E79-D** (1996) 1209–1211.

[5] T.Miyajima, T.Hasegawa and M.Haneishi, On the Multiuser Detection Using a Neural Network in Code-Division Multiple-Access Communications, *IEICE Trans. on Commun.* **E76-B** (1993) 961–968.

[6] S. Verdú, Multiuser Detection, Cambridge Univ. Press, Cambridge, 1998.

[7] M.K.Varanasi, Group Detection for Synchronous Gaussian Code-Division Multiple-Access Channels, *IEEE Trans. on Inform. Theory* **41** (1995) 1083–1096.

Coherent Neural Network Architecture Realizing A Self-Organizing Activeness Mechanism

Akira HIROSE, Chiharu TABATA and Dai ISHIMARU
Dept. Frontier Informatics, Graduate School of Frontier Sciences / RCAST
The University of Tokyo, 4-6-1 komaba, Meguro-ku, Tokyo 153-8904, Japan

Abstract. The coherent neural networks deal with amplitude and phase information of carrier wave signals. Their dynamics is carrier-frequency sensitive so that the self-organizing, learning and signal-processing behavior is controllable by a carrier-frequency modulation. We present a coherent network architecture for realizing a self-organizing activeness mechanism which is one of the most important functions of future brain-type information processing systems. The carrier-frequency is used as an internal state signal used for activeness expression such as context-dependent behavior and active attention. Complex-valued Hebbian rule generates stably equilibrium behavior points with an adaptive metric in the frequency domain. We report experimental results on the self-organization of the network that expresses the activeness mechanism.

1. Introduction

The complex-valued neural networks are often classified into two types in terms of the activation functions (undifferentiable nonlinearity) and the behavior dynamics, i.e., the real-imaginary parts type and the amplitude-phase type. Though they are almost the same in mathematics, there is some difference in their dynamics features.

The authors have been interested in the amplitude-phase type networks to investigate ultimate neural devices / subsystems, electromagnetic wave / lightwave signal processing as well as future brain-type information processing systems [1]. Complex-amplitude signals are processed adaptively and nonlinearly in our systems such as coherent lightwave associative memory systems [2] and interferometric radar image processing systems [3]. In such real-world applications, the amplitude-phase complex-valued neural networks have an advantage since they are free from the coordinate system. That is to say, the real and imaginary parts of complex signals are dependent on the reference coordinate (e.g., direction of 'real'), whereas the amplitude and phase can be physically meaningful themselves (squared amplitude = energy, phase = temporal delay or advance) and hence essentially effective to realize useful systems. The nature does not have any coordinate system before we observe it.

Therefore, we regard the amplitude and the phase of the signals and weights as the primary variables. The frequency (time derivative of the phase) is also an important variable, especially when we process signals having carrier such as radio, television and most in the electromagnetic wave / lightwave communication and broadcasting systems.

The coherent neural networks deal with such carrier signals. They have a carrier-frequency dependent behavior and, therefore, realize a behavior control by a carrier frequency modulation as well as a frequency-domain parallel operation [4]. (A similar dynamics have ever been reported also in biological observation [5] where neural behavior is modulated by oscillation frequency variation of a neural loop.) Their learning and self-organizing dynamics are similar to the Fourier synthesis (integration of orthogonal sinusoidal functions) [6]. Therefore, theoretically speaking, the generalization characteristics in the frequency domain can be arbitrarily formed adaptively to environment at a certain condition [7]. This fact leads to a wider applicability than conventional networks with additional control input terminals because such networks can realize only a gradual behavior variation attributed to the identical transform relation of the control signals with that of the activation function.

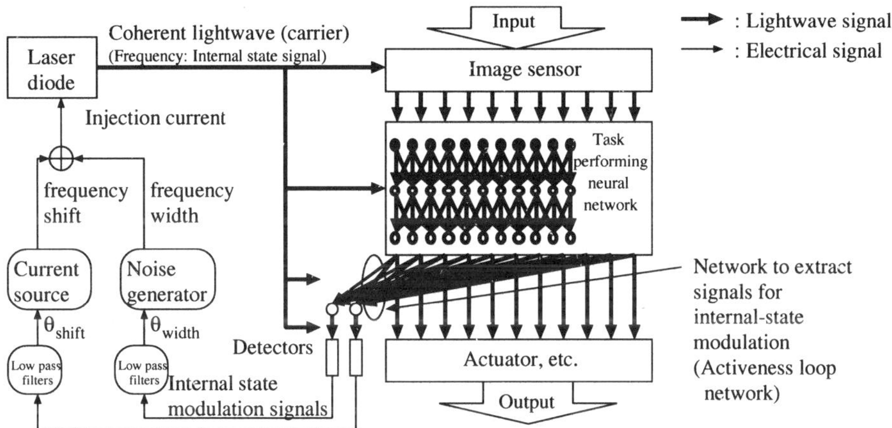

Figure 1. Physical construction using laser lightwave to control the coherent neural network to make the activeness mechanism network self-organize.

One of the most important application aiming at the future brain-type information processing systems is the activeness mechanism that self-organize to perform context dependent activities and active attention for example. Such mechanism has conventionally been modeled by preparing beforehand plural neural modules and switches (or integrator for a population coding) to realize an active modal behavior [8]-[11]. Though the models are capable of presenting the activeness, we have to determine at the initial stage the number of the neural modules and the structure of the switches. In this sense, the activeness mechanism cannot self-organize.

In this paper, we present an architecture of self-organizing activeness mechanism utilizing the carrier frequency controllability of the coherent neural network. The essence of the architecture lies in the phase-sensitive dynamics of the network in self-organization and task processing. However, in this paper, a physical system using coherent laser lightwave is also presented as an example of substantial construction for an easier understanding of operation.

Complex-valued Hebbian rule makes the network self-organize to form the activeness expressing network as well as the task performing one. In the coherent network, each connected neuron pair has plural synaptic connections in order to realize an adequately adaptive frequency-domain generalization characteristics. Experiments demonstrate that the network forms itself to have stably equilibrium points in the frequency domain at which the behavior converges depending on the history. Thereby a frequency-domain metric is constructed in the information space so that the carrier frequency can be used as an internal state signal. A context dependent behavior is reported.

2. Architecture and dynamics

2.1 System architecture

Figure 1 shows a possible physical construction of the system using coherent lightwave carrier. For example, a color image is input where its brightness and hue modulate the amplitude and phase of the carrier lightwave, respectively, to generate optical parallel signals. The signals are processed (e.g., memory and recall) in the coherent lightwave network [2] and yield output signals.

The output is branched into another network (activeness loop network) and generate an internal-state modulation signals. They generate currents to modulate the laser diode frequency. Consequently the coherent network behavior is modulated and the total system forms a feedback loop. In a certain condition [6], the activeness network self-organizes to generate stably equilibrium points of behavior in the frequency domain by a complex-valued Hebbian rule. The input-output-relation enhancing dynamics performs the creation of such stable behavior points and basins around them.

2.2   Complex-valued Hebbian rule in coherent neural dynamics

The Hebbian rule for connection weights $\mathbf{W} \equiv [w_{ji}]$ is obtained for a μ-th input $\boldsymbol{x}_\mu \equiv [x_{i\mu}]$ and output $\boldsymbol{y}_\mu \equiv [y_{j\mu}]$ signal pair (point attractor in this paper) as

$$\tau \frac{dw_{ji}}{dt} = -w_{ji} + K \sum_\mu y_{j\mu} \, (x_{i\mu})^* \tag{1}$$

where τ and K denote learning time constant and gain, respectively, and $(\cdot)^*$ stands for Hermitian conjugate. To write the rule in terms of the amplitude and phase values, we take the polar form by expressing the input signal $x_{i\mu}$ by its amplitude $|x_{i\mu}|$ and phase $\alpha_{i\mu}$, as well as the output $y_{j\mu}$ by its amplitude $|y_{j\mu}|$ and phase $\beta_{j\mu}$, as: $x_{i\mu} \equiv |x_{i\mu}| \exp[i\alpha_{i\mu}]$ and $y_{j\mu} \equiv |y_{j\mu}| \exp[i\beta_{j\mu}]$. Then (1) is written as

$$\tau \frac{dw_{ji}}{dt} = -w_{ji} + K \sum_\mu |y_{j\mu}| \, |x_{i\mu}| \exp[i(\beta_{j\mu} - \alpha_{i\mu})] \tag{2}$$

This rule is rewritten in terms of the transparency $|w_{ji}|$ and the phase $\arg[w_{ji}]$ of the weight as

$$\tau \frac{d|w_{ji}|}{dt} = -|w_{ji}| + K \sum_\mu |y_{j\mu}| \, |x_{i\mu}| \cos[\beta_{j\mu} - \alpha_{i\mu} - \arg[w_{ji}]] \tag{3}$$

$$\tau \frac{d(\arg[w_{ji}])}{dt} = -\arg[w_{ji}] + K \sum_\mu \frac{|y_{j\mu}| \, |x_{i\mu}|}{|w_{ji}|} \sin[\beta_{j\mu} - \alpha_{i\mu} - \arg[w_{ji}]] \tag{4}$$

In the coherent lightwave network, the phase shift $\arg[w_{ji}]$ is obtained by the change of connection delay time (optical length) τ_{ji} realized by an optical phase modulator; i.e., we have the relation $\arg[w_{ji}] = 2\pi f \tau_{ji}$. By converting (4), we obtain the delay-time variation rule as

$$\tau \frac{d\tau_{ji}}{dt} = -\tau_{ji} + \frac{K}{2\pi f} \sum_\mu \frac{|y_{j\mu}| \, |x_{i\mu}|}{|w_{ji}|} \sin[\beta_{j\mu} - \alpha_{i\mu} - 2\pi f \tau_{ji}] \tag{5}$$

Though the product term may seem complicated, it can simply be generated physically by a square-law detection of the output term $|y_{j\mu}| \exp[i\beta_{j\mu}]$ and the input term $|x_{i\mu}| \exp[i(\alpha_{i\mu} + 2\pi f \tau_{ji})]$ with the delay time taken into consideration.

2.3   Use of carrier frequency and spectrum as internal state

Figure 2(a) is a conceptual illustration of the frequency dependence of the network behavior. On the other hand, Fig.2(b) shows a possible use of the carrier frequency to realize various attention modes. The carrier frequency can be broad in its spectrum. In general, the behavior is determined by a synthesis process expressed as follows. The neural weight w_{ji} consists of a transparency $|w_{ji}|$ and a delay time τ_{ji}: $w_{ji}(f) = |w_{ji}| \exp[i2\pi f \tau_{ji}]$ where the weight value is a function of the carrier frequency f. An output signal of a neuron $y_j(f)$ is hence a density function of f determined by input signals x_i, weights w_{ji}, and a spectral density of the carrier wave $S(f)$. The consequent output signal y_j is an integral of the density $y_j(f)$ in the frequency domain expressed as

$$y_j = \int S(f) \, y_j \left(w_{ji}(f), x_i \right) df \tag{6}$$

In the physical coherent system, the spectral integration is performed as a temporal integration of the instantaneous output signal sequence where the spectral density is determined by a quick frequency modulation of the laser diode.

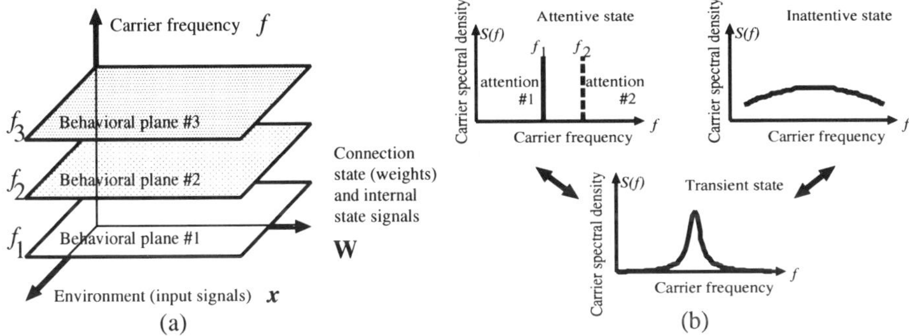

Figure 2. (a)Frequency dependent behavior planes and (b)possible relation between carrier spectra and internal states (attention states, etc.).

3. Simulation experiments

We choose associative recall as the processing task in the experiment. (See Fig.1.) The complex-valued Hebbian rule is applied to the task performing neural network (single layer, recurrent for recall, three parallel connections with different initial delays including various $2n\pi$ terms for each connected neuron pair [7]). The Hebbian rule is adopted also in the internal-state modulation-signal extracting network to generate stably equilibrium points in the carrier frequency domain. Here the carrier frequency is modulated roughly as $f = f_0 + f_{\max} \times \sin\theta_{\text{shift}}$ without the noise term.

To clarify the dynamics of the self-organization of the extracting network, we assume the following condition for simplicity. (i)The associative memory learns complex-amplitude point attractors with a given set of input and output images and carrier frequencies. (ii)After the learning of the associative memory, a global feedback loop is formed and the frequency is set free-running to make the extracting network self-organize. Simultaneously the recalling process is performed with (noiseless) input images presented in turn (i.e., on-line self-organization). Therefore, the interaction of the recalling behavior and the internal-state modulation performs the self-organization of the extracting network. (It is further expected that both the memorizing process and extracting-network self-organization can be performed simultaneously with the closed loop and the free-running frequency. The detail will be reported elsewhere.)

Figure 3 shows the evolution of the carrier frequency f and the overlap between output and target (memorized) images $\mathbf{Re}[\boldsymbol{x}^* \cdot \boldsymbol{s}_\mu]$. At the initial state (time step = 0), the associative memory network already completed to learn almost orthogonal three input images with different frequencies. On the other hand, the extracting network has random initial weights at first. With the memory weights fixed, the network system starts self-organizing where the carrier frequency is free-running and the extracting network weights are changing with the Hebbian rule.

At the first time steps in Fig.3(a) with input 1 presented to the task network, the carrier frequency shows an oscillatory variation to converge at a certain frequency (which is incidentally the same frequency as that used in the associative learning). The corresponding overlap value in Fig.3(b) varies largely at first and converges at almost unity, which means a stable attention mode to recall one of the memorized information. The oscillatory behavior suggests that the basin is formed both in the frequency domain and the complex-amplitude signal information domain.

When the input image is changed (input 2), we observe a similar (but different) oscillatory behavior and a convergence. Again the overlap becomes almost unity, showing a successful recall. However, in the input 3 case, the overlap is far from unity and the recalling process fails, though the frequency converges at a certain value different from that used in the learning process of the associative memory.

Finally input 1 is presented again. The overlap convergence value is almost unity but a little lower than the former one. The convergence frequency is completely different

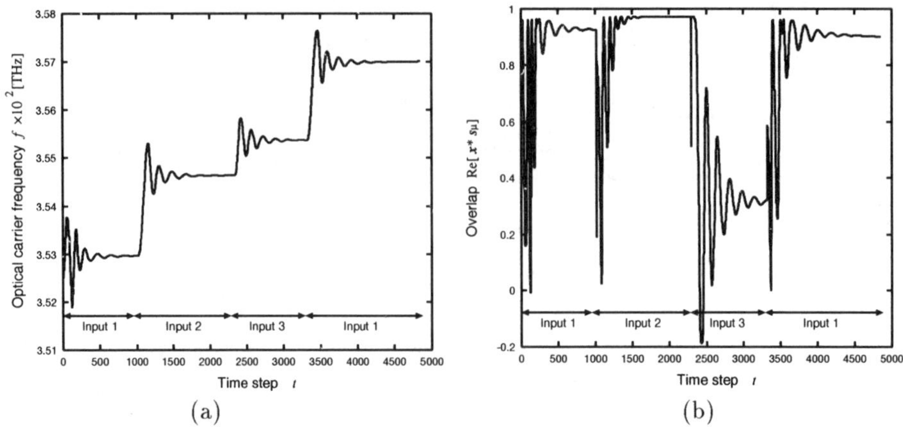

Figure 3. Evolutions of (a)optical carrier frequency f and (b)overlap between recalled output and intended signals $\mathbf{Re}[\boldsymbol{x}^* \cdot \boldsymbol{s}_\mu]$.

from the first input 1 case. The result means that, though the network has a sufficient attention to recall, its internal state is completely different from that in the first input 1 case. Therefore, if we have another task neural network in parallel to the associative memory, we expect that the network behavior should be different. (The convergence curves are similar but different from one another.)

Figure 3 also shows a context dependent behavior. That is to say, the convergence frequencies (internal states) for the two input 1 cases are different and dependent on the preceding states. The phenomena is another expression of an active behavior.

4. Conclusion

We have presented a coherent network architecture for realizing self-organizing active-ness mechanisms which is one of the most important functions of the future brain-type information processing systems.

References

[1] A.Hirose, Coherent neural networks and their applications to control and signal processing, In: S.G.Tzafestas (ed.), Soft Computing in Systems and Control Technology, World Scientific Publishing Co. (1999) 397-422.

[2] A.Hirose, M.Kiuchi, Coherent optical associative memory system that processes complex-amplitude information, *IEEE Photon. Technol. Lett.*, **12** (2000) 564-566.

[3] A.B.Suksmono, A.Hirose, Adaptive complex-amplitude texture classifier that deals with both height and reflectance for interferometric SAR images, *IEICE Trans. Electron.*, **E83-C** (2000) 1912-1916.

[4] A.Hirose, R.Eckmiller, Behavior control of coherent-type neural networks by carrier-frequency modulation, *IEEE Trans. on Neural Netw.*, **7** (1996) 1032-1034.

[5] E.Koerner, U.Koerner, Concurrent parallel-sequential processing in gamma controlled cortical-type networks of spiking neurons, *International Conference on Artificial Neural Networks (ICANN) '97 Lausanne*, Proc. (1997) 91-96.

[6] A.Hirose, Self-organizingly emerging activeness architecture realized by coherent neural networks, *ICANN'99 Edinburgh*, Proc. 2 (Sept. 7-10, 1999) 726-731.

[7] C.Tabata, Neural networks having self-organizing activeness mechanisms, *Graduation thesis*, Dept. Electronics Engineering, The University of Tokyo (2001) (in Japanese).

[8] T.Omori, A.Mochizuki, K.Mizutani, M.Nishizaki, Emgergence of symbolic behavior from brain like memory with dynamic attention, *Neural Networks*, **12** (1999) 1157-1172.

[9] D.M.Wolpert, M.Kawato, Multiple paired forward and inverse models for control, *Neural Networks*, **11** (1998) 1317-1329.

[10] J.E.Lewis, W.B.Kristan, Jr., A neuronal network for computing population vectors in the leech, *Nature*, **391** (1998) 76-9.

[11] P.Hartono, S.Hashimoto, Temperature switching in neural network ensemble, *Journal of Signal Processing*, **4** (2000) 395-402.

Improvement of the Hierarchical Compact Patricia Trie for a Dynamic Large Key Set

Masami Shishibori, Minsoo Jung, Satoru Tsuge, Jun-ichi Aoe
Department of Information Science & Intelligent Systems,
The University of Tokushima,
2-1 Minami-Jhosanjima-Cho, Tokushima-Shi, 770-8506 Japan

Abstract. A Patricia trie is the fastest access method in binary tries, because it has the shallowest tree structure. But, it needs the storage memory as large as the size of registered keyset. In order to solve this problem, we have already proposed a method to compress the hierarchical Patricia trie into the compact bit stream, which is called the Hierarchical Compact Patricia trie (HCPat tree). This HCPat tree requires very small storage memory, but it takes a lot of times to insert a key if the size of keyset becomes large. This paper proposes a construction algorithm for improving the updating times is applied to only one of separated trees in the new HCPat tree.

1 Introduction

A Patricia trie (Pat tree) does not have any single descendant nodes, which have only one arc in binary tree. Manber et al.[1] proposed the method to compress the Patricia tree into one array structure, which is called a PAT array. This PAT array is created by mapping the number of each external node of the Patricia tree into the array in order, and the key is searched by doing the binary search on the PAT array. Certainly, the PAT array is a compact data structure, however many disk accesses are required for the key retrieval. We proposed the algorithm to compress the Pat tree to a Compact Patricia trie (CPat tree), which consists of very compact bit stream. However, if the bit length of this CPat tree becomes long, it requires much time expense that search keys allocated on the end of the bit stream.

This paper proposes modified a Pat tree's structure that can avoid that the time efficiency is worsened even if the key set became very large. It is separates from a CPat tree to small tree and composes hierarchically. The speed of operation in each small tree detached is fast. On this account, it can extract unnecessary scanning for each tree detached in CPAT tree. This new tree structure is called the Hierarchical Compact Patricia trie (HCPat tree). For the dynamic key set, moreover, a composition method of HCPat tree is the important problem, because a Pat tree usually has the eliminated nodes. For this reason, if information of eliminated nodes is changed by the insertion of a new key, all child nodes of the internal node that has the corresponding eliminated nodes should be modified. For this problem, we propose the method to improves insertion time of the HCPat tree.

2 A Pat Tree represented by a Compact Binary Tree

A Pat tree is eliminated single arc node, and in order to avoid the false matches, each node of the Pat tree must have either the counter for the number of the eliminated nodes and a pointer of record storing eliminated symbols. In the case that the number of eliminated nodes is stored into each node, every time each node is traversed to retrieve a key, the binary sequence of the key must be skipped over by the stored number before the next inspection. Table-1 shows each binary sequence of the key set K. Figure-1 (a) indicates the corresponding Patricia trie to the key set K.

$$K = \{air, art, bag, bus, tea, try, zoo\}$$

Since all nodes in the Pat tree are always guaranteed to have two leaves without the dummy leaf, this tree can be transformed into the CB tree[2] of bit string base. As a result, *Leafmap* of the CB tree becomes unnecessary, furthermore also the number of bits of *Treemap* decreases by the double number of dummy leaves in the CB tree. The information about eliminated nodes included into the Patricia trie, however, must be represented as the compact data structure. Shishibori et al.[3] has proposed the corresponding CB tree to a Pat tree which consists of *Treemap*, *B_TBL* and *Nodemap* instead of *Leafmap*. This new CB tree is called a Compact Patricia trie (CPat tree). *Treemap* and *B_TBL* are obtained in the same way as Jonge's method[2], however the bit length of *Treemap* is shorter by the double number of dummy leaves. *Nodemap* indicates whether the corresponding node includes the number of eliminated nodes or not. *Nodemap* is obtained by traversing the tree in pre-order, then if the internal node stores the number of eliminated nodes, emit '1' the same times as the number of eliminated nodes, after that emit '0' one time. Figure-1 (b) shows the CPat tree for the key set K. The corresponding internal node number and the eliminated node sign (symbol 'e') are shown within the round '()' above *Nodemap*.

Table 1. Binary sequences of a key set K

Keys	Internal codes	Binary sequences
air	0 / 8 / 17	00000 01000 10001
art	0 / 17 / 19	00000 10001 10011
bag	1 / 0 / 6	00001 00000 00110
bus	1 / 20 / 18	00001 10100 10010
tea	19 / 4 / 0	10011 00100 00000
try	19 / 17 / 24	10011 10001 11000
zoo	25 / 14 / 14	11001 01110 01110

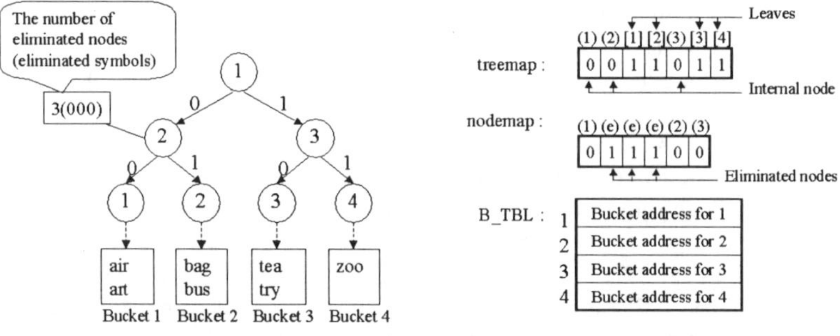

(a) The Pat tree for the Key Set K.   (b) The CPat tree for the Key Set K.

Figure 1. The Pat tree and CPat tree for the key set K.

3 A Hierarchical Pat Tree represented by a Compact Binary Tree

3.1 A hierarchical Compact Pat tree

We have already proposed a new CPat tree that proves the time-cost to reduce this time expense. This CPat tree consists of the separated trees of fixed depth, and each tree is linked each other by pointer. Figure-3 shows a Pat tree for the key set K, where a depth of tree separated is two, and the bucket is 1. A compact Pat tree for this hierarchical Pat tree is called Hierarchical Compact Pat tree (HCPat tree). Figure-2 (a) shows a HCPat tree based on the HPat tree of Figure-2 (b).

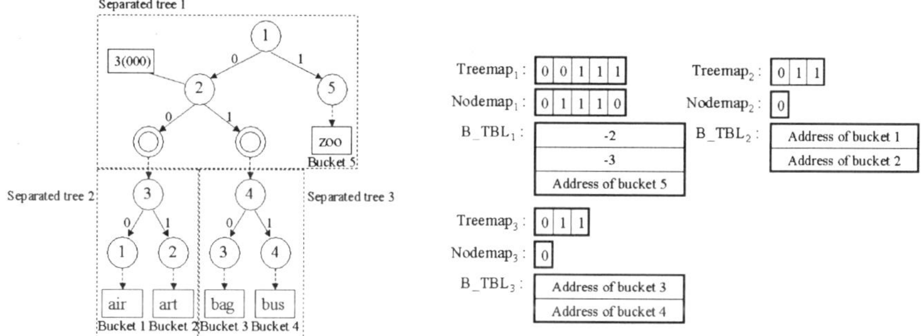

(a) The hierarchical Pat tree based on Fig-1 (a).   (b) The HCPat tree based on Fig-1 (b).

Figure 2. The hierarchical Pat tree and HCPat tree based on the Pat tree of Figure-1.

3.2 A problem of the insertion process for the HCPat tree

An insertion process of HCPat tree is happened when a new node is added at external nodes or internal nodes. This insertion process is classified into the following three cases:
  (1) The case when the bucket is partially full.
  (2) The case when the bucket is full.
  (3) The case when the key cannot belong to the bucket, that is, the key happens the false drop.
In the first case and second case, we can utilize the algorithm of the hierarchic CB tree. But, in the last case when the false drop occured by a new key, we must add new nodes in internal node, which stored the eliminated nodes, and modify the eliminated nodes. But, one problem is created, that is, if this internal node is modified, all lower child nodes than the internal node should be also modified.
Let's consider the insertion process that a new key is inserted in the HCPat tree at Figure-3. First of all, we can find the bucket that corresponds to a new key "eat" (=00100...). Next, we can know that the false drop happens when we compare the bit string of this key "air" and a new key "eat". In this case, we must a add new node in internal node 2 where the false drop happens. Then, this process causes the problem that all lower child nodes of internal node 2 must be updated, off course separated tree 2 and 3 also must be modified. So we propose the new insertion algorithm to avoid these time expense problem.

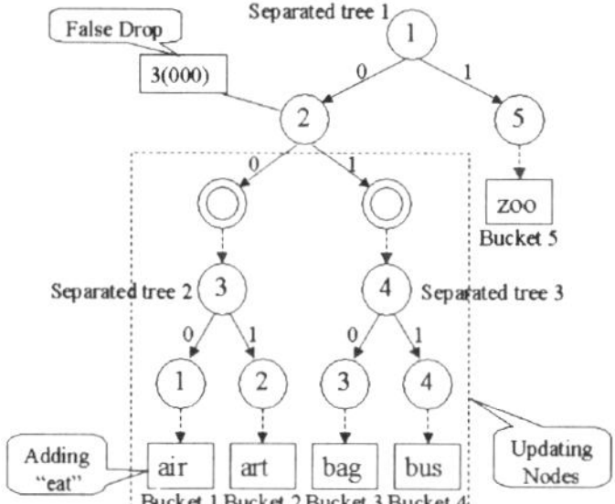

Figure 3. Insertion problem in the HCPat tree

4 Improvement of the Insertion Algorithm for the HCPat tree

In the case when the false drop happens, it is required the all child nodes must be modified. We propose the method to avoid the modification of low rank nodes. Figure-4 (a) indicates that this proposed method does not need to modify low child nodes. This method modifies the only one separation tree which includes the false drop node, and

attaches an original node and a new tree to pointer. This requires only two. Figure-4 (b)
is the HCPat tree after inserting a new key "eat" in hierarchic Pat tree of Figure-3.
Next we indicate the new insertion algorithm when the false drop happens.

[A new insertion algorithm for the HCPat tree]

num   : the number of tree detached that is during current retrieval operation.
keypos   : the current position in bit strings of key.
treepos   : the current position in $Treemap_{num}$
nodepos   : the current position in $Nodemap_{num}$

Step-1: [Store the position of the false drop node]
Store the position of *keypos, treepos, nodepos*, num in the false drop node.

Step-2: [Product the new separated tree]
Product the new $Treemap_{num}$, $Nodemap_{num}$, and B_TBL_{num} that have the lowest
num to empty.

Step-3: [Transpose the $Treemap_{num}$ to the new separated tree]
Obtain the bits from false drop node until below child node in $Treemap_{num}$, and
these bits are transposed to the *Treemap* of the new separated tree, and replace
moved bits of $Treemap_{num}$ by '1'.

Step-4: [Transpose the $Nodemap_{num}$ to the new separated tree]
Obtain the bits from false drop node until below child node in $Nodemap_{num}$, and
these bits are transposed to the *Nodemap* of the new separated tree.

Step-5: [Verification of the bit value of the key]
If the bit of the key pointed to by *keypos* is '0', proceed to step-6, otherwise to
step-7.

Step-6: [Appending the external node to left subtree]
Replace the first bit '0' of *Treemap* in the new separated tree by '010', and
proceed to step-8.

Step-7: [Appending the external node to right subtree]
Replace the first bit '0' of *Treemap* in the new separated trie by '001', and
proceed to step-8.

Step-8: [Modification the *Nodemap* of new separated tree]
Replace bits of false drop position in *Nodemap* of the new separated tree by '0'.

Step-9: [Transpose the B_TBL_{num} to the new separated tree]
Transpose the buckets to the new separated tree, these buckets are included
below child node of the false drop node.

Step-10: [Linkage to the new separated tree]
Append the pointer of the new separated trie to B_TBL_{num} by minus.

Step-11: [Appending the new key to the new separated tree]
Append a new key to the B_TBL_{num} of the new separated tree.

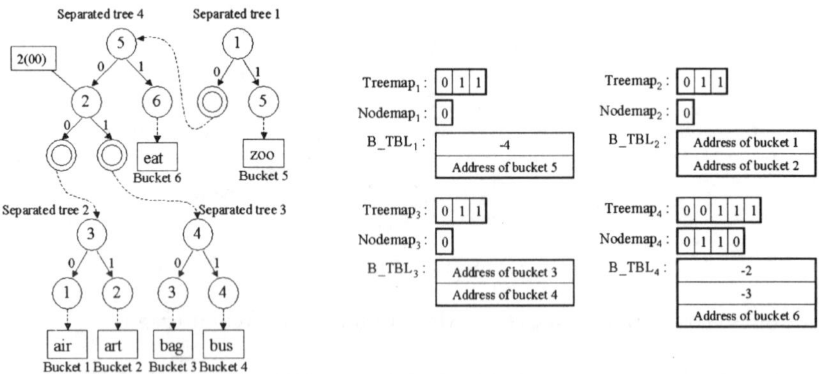

(a) The HPat tree after inserting a new key "eat".   (b) The HCPat tree after inserting a new key "eat".

Figure 4. The HPat tree and HCPat tree after inserting a new key "eat" in Figure-2.

5   Evaluations

This algorithm is implemented by C++ and we tested in console mode of Windows 2000 that has an Intel 300MHz CPU. The experimental data is 70,000 nouns in Japanese with a length of 6~100 Byte and 120,000 word in English with a length of 1~50 Byte. As for the evaluation, we compare the time and space efficiency between the Pat tree, CPat tree, and HCPat tree, where the depth of HCPat tree is 3 and the bucket size is 10. Table-2 and Table-3 show the experimental results. The Pat tree is with very fast process time, as all nodes are linked by pointer, and the CPat tree is less memory expense by large key set, as all nodes are consisted of bit array. On the other hand, the HCPat tree can get fast operation result into small memory capacity. From these result, it is found that, as for the time efficiency, the HCPat tree is about 40 times faster than the CPat tree, and as for the space efficiency, it requires 1/3 memory space of Pat.

Table 2. Experimental result of the Pat tree, CPat tree, and HCPat tree for Japanese words.

	Pat tree	CPat tree	HCPat tree
Information of tree			
Number of nodes	25,857	25,857	30,696
Internal nodes	12,928	12,928	12,928
External nodes	12,929	12,929	17,768
Times			
Registration (s)	10.3	928.0	26.7
Retrieval (ms)	0.084	12.711	0.219
Insertion (ms)	0.147	13.257	0.382
Storage			
Treemap (Kbytes)	103.4	3.2	3.8
Nodemap (Kbytes)	51.7	3.4	3.4
Bucket table (Kbytes)	0.0	51.7	71.0

Table 3. Experimental result of the Pat tree, CPat tree, and HCPat tree for English words.

	Pat tree	CPat tree	HCPat tree
Information of tree			
Number of nodes	44,517	44,517	52,976
Internal nodes	22,258	22,258	22,258
External nodes	22,259	22,259	30,718
Times			
Registration (s)	23.5	3981.4	52.6
Retrieval (ms)	0.109	30.294	0.245
Insertion (ms)	0.195	33.178	0.438
Storage			
Treemap (Kbytes)	178.1	5.6	6.6
Nodemap (Kbytes)	89.0	7.9	7.9
Bucket table (Kbytes)	0.0	89.0	122.8

6   Conclusion

A CPat tree is the compact data structure and can retrieve keys in order, but the time-efficiency becomes bad for a large key set. This paper proposed a new HCPat tree and the improved registration algorithm. For evaluations, we compared the HCPat tree with traditional Pat tree and CPat tree. As for the time efficiency, the HCPat tree is about 40 times faster than the CPat tree. As for the space efficiency, it requires 1/3 memory space of Pat tree that is consisted of pointer structure. As for the future works, we will propose *Treemap*'s compression method that increases more memory efficiency.

References

[1] Gonnet G.H., Baeza-Yates R.A.and Snider T., New indices for text : Pat trees and Pat arrays. In Information Retrieval -Data Structures & Algorithm, New Jersey.: Prentice Hall, 1992, pp.66-82.

[2] Jonge W. D., Tanenbaum A.S. and Reit R.P., Two Access Methods Using Compact Binary Trees, IEEE Transactions on Software Engineering, 13(7), 1987, pp.799-809.

[3] Shishibori M., Koyama M., Okada M. and Aoe J., Two Improved Access Methods on Compact Binary (CB) Trees, International Journal of Information Processing & Management, 36(3), 2000, pp.379-399.

KES '01
N. Baba et al. (Eds.)
IOS Press, 2001

Knowledge Discovery
for Document Classification

Sangkon Lee, Akihiro Tanaka, Masao Fuketa, Kazuhiro Morita, and Jun-ichi Aoe
2-1 Minami Josanjima, Tokushima,
Dept. of Information Science and Intelligent Systems,
University of Tokushima 770-8506, Japan

Abstract. We describe what is field-associated terms and how to discover field-associated terms, which exist in any text, and define five levels to classify them. All of them are stored to field tree to make use of extraction of field-coherent passages for document classification. Percentages of levels for Japanese FTs are 60% of FTs belong to level 1, which are perfect FTs, and 72% are in either level 1 or level 2, which are semi-perfect FTs. We believe this number was large enough to validate this study.

1. Introduction

To extract and classify field-coherent passages in document we have focused on morphological evidence of text. Passage retrieval system, which divides a text into several passages by using field-associated terms to find field of text. Hoenkamp and Groot indicated that, "many researchers share the intuition that noun phrases are a richer and more precise representation of meaning then keywords: 'horse' and 'race' may be related, but 'horse race' and 'race horse' carry a more circumscribed meaning than the words in isolation. Intuitively, words forming phrases present a more precise description of content than a sequence of keywords"[1]. A field-associated term is a single or compound word that occurs in any document, and which enables field identification. For example, we immediately recognize the field of document as <baseball> or <politics>, when we encounter the word 'pitcher' or 'election.' Tsuji et al. proposed an efficient method for organizing Japanese field-associated terms, which allowed field specification in accordance with a fixed document classification scheme[2, 3]. In addition, by using field-associated words Fuketa et al. introduced a document classification method by using field-associated terms[4]. In this study, we present a knowledge discovery method to accomplish English document classification.

2. Identification of Field-Associated Term

As Hoenkamp and Groot concluded that, "Evidence that phrases would be more effective for information retrieval is inconclusive," although conventionally, not much attention has been paid to words or their structures in passage retrieval. However, it is natural for people to identify the field of document when they notice specific words. We shall refer to these specific words as *field-associated terms* (FTs); specifically they are words that allow us to recognize intuitively a field of text or field-coherent passage. Therefore, FTs can be used to identify the field of a passage, and can be also used to classify different fields among passages. For these reasons, we can use FTs as a clue to identify a passage field. As mentioned in previous section, FTs can be either words or phrases.

To classify fields of passages in a document, first of all, we need to construct a so-called *field tree* with levels (that is, a detailed subject based classification system), which are described in following sub-sections. In addition, because FTs may change with time, we also need to define a so-called *stability rank*. For example, the words "pitcher" or "hitter" are stable words when used as FTs in the field of *sports*, however, winning team or famous player will change with time, so that these FTs would be unstable. According to the degree of stability of FTs, each FT is assigned its stability rank manually. Because the length of a single FT (i.e., a word-form FT) is generally quite short, and their numbers are finite, we could collect FTs manually, and create a field dictionary according to a topic statement and a document classification scheme. Compound FTs (i.e., phrase-form FTs) could be unlimitedly generated, and by constructing a *combination-pattern* of single FTs, compound FTs can be built.

2. 1 Field-Associated Term

We define a minimum unit (or a word), as one which cannot be further divided without losing its semantic meaning, as a single FT. Compound FTs are defined to consist of two or more single FTs. Both terms are expressed by enclosing them within quotation marks. A compound FT is regarded as being single if it loses its field information when divided; for example "consumption tax," "nuclear weapon," or "global warming." In addition to this, proper nouns with the exceptions of people's names, such as "South Burlington" – the name of high school baseball team should be regarded as a single FT. When a proper noun contains a person's name, it is also regarded as a single FT. i.e., "Babe Ruth"; but if it also contains a person's title, for example, "Coach R. Dawson," we divide it into two singles FTs: "coach" and "R. Dawson." These singles FTs belong to baseball field having different levels. A computer is taught these humanly selected FTs, which are then saved in the field tree as our knowledge base. This combination pattern is made after carefully examining Imidas '99, a Japanese language terminology book [5].

A FT could be a word (e.g., "game"), a phrase (e.g., "victory and defeat"), or a right-truncated word (e.g., "single:"), which represents that matches the letters to the left of the colon to deal with a FT "singles." FTs are the words that indicate each subject matter category in the classification scheme. Since the basic concept behind FTs involves the choice of a limited set of words that match a given document best, they describe a set of *discriminating words*. Moreover, FTs are not always the same as subject words. In the sense that FTs are obtained from a document but subject words may not shown in a document, FTs could be better discriminator words than subject words. On the other hand, many of the FTs are non-subject bearing words, such as "case" or "use." While the semantic differences among FTs are small, FTs choices are used in a document is mainly a matter of personal style. In addition, FTs seem to be terms with high *idf* in general so they play an important role in passage retrieval.

2.2 Field Tree

A field tree is a *field scheme*, which represents the relationships between fields using a tree structure. A *leaf field* (or leaf node) in a tree is a *terminal field*, and other nodes are called *medium fields*. In this study, the field tree was constructed based on Imidas '99[5]. It contained 443 fields, and among these there were 50 medium fields and 393 terminal fields:

There are 174 terminal fields at 2nd level, 208 at the 3rd level, and 11 the 4th. In addition, each field is referred to as *parent* or *child* in accord with its relative position. <Path> was chosen to designate a field, the root name was omitted, and provided no conflict occurred, only the terminal field was used, not the full path. For example, if a path <Path> is <Sports/Ball Games/Tennis>, it describes field <Tennis> as a sub-field of <Ball Games>, which is also a sub-field <Sports>. Based on the result of morphological analysis for a given document, FTs that match with morphemes are extracted. Concentration rates of these FTs were calculated using Tsuji et al.[3]. These extracted FTs are single words registered in morphological dictionary. Conversely, compound FTs are constructed semi-automatically – in the sense that their stability rank were based on field inheritances of single FTs. Note that, in this study, unregistered words were not included in the morphological dictionary.

2.3 Level (Depth) of FTs

As it mentioned in the previous section, according to the scope of a field, it can be classified as one of the following three groups: a terminal field, a medium field or a multiple fields (i.e., set of terminal and/or medium fields). Each FT belongs to one of those groups according to their strength of indicating a specific field. Since each FT has different scope to associate with a field, assigning proper levels (i.e., depths in a document classification scheme or strength) is necessary. In this study, we define five different levels in terms of the FT's reminding ability of a field. The following criteria determine on which fields an FT belongs to.

Definition 1.   Levels for Field-Associated Term w

   Level 1. **Perfect FT:**

   FT w belongs to a unique terminal field.

   Level 2. **Semi-Perfect FT:**

   FT w is associated with some limited set of terminal fields sharing parent node.

   Level 3. **Medium FT:**

   FT w is not a perfect or semi-perfect FT, and needs to be associated with unique medium field.

   Level 4. **Multiple FT:**

   FT w is not a perfect, semi-perfect, nor medium, and needs to be associated with several multiple fields.

   Level 5. **Non- (or Anti-) FT:**

   Term w is not a FT, and cannot be associated to a field, this group includes stop words, such as articles, prepositions, pronouns, sentential marks etc.

For instance, a term "Yokozuna[1]" is a perfect FT with a level of 1 in the tree, and is limited only one terminal node <Sumo>. In level 2, the semi-perfect FT "singles," which has same parent node <Sports>, also belongs to several terminal fields <Badminton>, <Table Tennis>, or <Tennis>. The other FTs, "game," "play," "contest," or "tournament," are not restricted to a terminal node, but they are limited to a unique medium field <Sports>. Meanwhile, the multiple FT "victory and defeat" is limited to two terminal

---

[1] *Yokozuna* means that a grand champion who is the highest rank of Sumo wrestling. Power, skill and class are required for dignity to the rank. If there is no wrestler competent to be a Yokozuna, the position will be left vacant. There is no demotion from Yokozuna. Once a wrestler becomes a Yokozuna he will never return to the lower ranks.

nodes, <Hobby & Entertainment/Games/Chess> and <Politics & Law/Elections>, or limited to two terminal nodes with the same medium field, namely <Sports/Martial Arts/Sumo> or <Sports/Ball Games/Baseball>. Finally, the words "case" and "use" do not specialize any fields, and level 5 is assigned.

3. Evaluation

We describe a collection of Japanese and English FTs in this section. In Japanese, 3,248 FTs were collected for all nodes (or categories) in the field tree. Most of documents used for collecting FTs were obtained from the CD-Asahi Newspaper 1995~97 Text Collection although in certain cases FTs were obtained from the Internet. For each of the collected FTs, a level labels, which were assigned, as Section 2.3.

To be frank with you, we had a plan to organize the following 11 fields – <Business & Economy>, <Culture & Fine Arts>, <Education>, <Environmental Problems>, <Health & Medical Science>, <Hobby & Entertainment>, <Industry>, <International>, <Nature>, <Politics & Law>, <Science & Technology>, <Social Life>, <Sports>, and <Study> the following four were omitted since only a small number of Japanese FTs found: <Education>, <Health & Medical Science>, <Industry>, and <Social Life>. For the same reason, five fields for English FTs were omitted, namely <Earth Sciences & Geography>, <The Arts>, <History>, <Domestic Life>, and <General Knowledge & Philosophy> among <Earth Sciences & Geography>, <Life Sciences>, <Physical Sciences>, <Technology>, <Mathematics & Measurements>, <Religion>, <History>, <Society & Social Institutions>, <Business & Economics>, <The Arts>, <Domestic Life>, <Sports and Recreation>, and <General Knowledge & Philosophy>. It should be noted that FT collection is an on-going process. We expect that the omitted fields can be added in future research. A histogram of Japanese FTs will be shown in presentation. Number of FTs for <Sports> is relatively large, not only because sports documents usually have many teams or famous player's names, and it is easy to find machine readable texts in Internet, but also because there are more specific FTs available in sports field than in other fields. We showed the percentages of levels for Japanese FTs are 60% of FTs belong to level 1, which are perfect FTs, and 72% are in either level 1 or level 2, which are semi-perfect FTs. We believe this number was large enough to validate this study.

4. Conclusion

In this paper, single and compound field-associated terms have been defined, on the basis of the characteristics of a single word, which is stored in the morphological dictionary. A document classification scheme, which can be helpful to a classifier to decide a field of document, is also presented by a tree structure. In order to extract relevant passages, we first defined levels of field-associated terms, which provided a scope of restricting a document's field, focusing on changes in field-associated terms (FTs) with time, as well as their stability rankings. Because the length of FTs is the shortest meaningful distance and their number is finite, FTs are collectable. It is the combination of FTs that gives a strong clue to the document's field. In this research, to accomplish a new passage retrieval technique using FTs as clues to a specific passage theme's field. Our knowledge discovery will well prevents separation or field duplication problems between passages. Similar to human thought processes, our system can determine the field of a passage. In conclusion, the proposed system can extract the passages that are not only the most relevant in terms of user's query but also well represent the semantic information of a text.

Further research is needed on the following topics to enhance our system. First of all, since language is not static, the field information must be continually updated to remain current language. *Innovative Multi-Information Dictionary* Imidas '99 is a general information source that does not contain relationships of all fields. Therefore, further work needs to be done to identify other sources of word knowledge, such as domain-specific thesauri, dictionaries, and statistical word usage information that might be integrated with this work. We also need to collect more Japanese FTs in the four fields mentioned in previous section, and may need to find a domain specialist for English FTs. Because level 4 FTs, which indicate multiple fields, usually have different occurrence frequencies in different fields, it is possible to enhance recall by setting different weights to each field. In addition, more researches on relationship among FTs are of necessary. Studies about the inheritance of phrase FTs are in process. Secondly, collocation or case information can be assigned to FTs because there are certain collocation relationships among words. This facilitates decision on the correct field of passages, which contains level 2, 3, and 4 FTs but no level 1 FTs. In addition to additional work on FTs, attention should be paid to connecting words, conjunctions, and signal words.

References

[1] Hoenkamp, E., and Groot, R., Finding Relevant Passages Using Noun-Noun Compounds: Coherence vs. Proximity. The Twenty-Third Annual International ACM Special Interest Group on Information Retrieval Conference on Research and Development in Information Retrieval, 2000.

[2] Tsuji, T., Nigazawa, H., Okada, M., and Aoe, J., Early Field Recognition by Using Field Association Words. Paper Presented at the Proceedings of the 18th International Conference on Computer Processing of Oriental Language, 1999.

[3] Tsuji, T., Fuketa, M., Morita, K., and Aoe, J., An Efficient Method of Determining Field Association Terms of Compound Words. Journal of Natural Language Processing, Vol. 7, No. 2 (2000), 003-026. *(in Japanese)*

[4] Fuketa, M., Lee, S., Tsuji, T., Okada, M., and Aoe, J., A Document Classification Method by Using Field Association Words, *An International Journal of Information Sciences,* Elsevier Science, Vol. 126, No. 1-4 (2000), 57-70.

[5] Dozawa, T. (Editor), Innovative Multi-Information Dictionary, Imidas '99, Annual Series, Japan: Zueisha Publication Co., 1999 *(in Japanese)*

KES '01
N. Baba et al. (Eds.)
IOS Press, 2001

An Efficient Construction Method of Compound Field Association Words

El-Sayed Atlam, Akihiro Tanaka, Kazuhiro Morita, Masao Fuketa, and Jun-ichi Aoe
*Dept. of Information Science and Intelligent Systems, University of Tokushima
Tokushima,770-8506, Japan.   E-mail: atlam@is.tokushima-u.ac.jp*

Abstract. This paper presents two main point: the first one is a strategy for   building Morphological machine dictionary of English efficiently to infer meaning of derivations from a simple word by considering morphological affixes and their semantic   classification. .The second one, although there are many kinds of research about text classification depend on term information in the whole text,   humans can recognize the filed of a text by finding a minimum number of specific words in that text. In this paper, such terms are called a field association (*FA*) word . Generally, it is easy to collect single *FA* word because the number of it is finite, but there are some difficulties: how t o select useful and important compound *FA* word from a huge number of combinations of single *F A* word. For *FA* words, five-association levels are defined and two kinds of ranks based on stability and inheritance are presented. Around 88%   redundant candidates of compound *FA* words can be removed remarkably by using the level and the rank. The proposed methods are applied to 20,000 relationships between verbs and nouns extracted   from a large tagged corpus.

1.   Introduction

With recent growth of the Internet systems, the popularity of the Internet has caused an exponential increase in the amount of on-line text and in the number of people who create and use this text. As the amount of documents and the number of users rise, automatic document classification becomes an increasingly important tool for helping people organize this vast amount of data. Also, many document retrieval systems have been developed and a variety of facilities are increasingly needed in every areas such as keyword retrieval, similar file retrieval, automatic document classification, document summarization and so on [3,4,7,8]. Considering early semantic analysis [2,10] in natural language processing, a morphological dictionary should contain as much semantic information as possible in order to be acceptable at the level of morphological analysis. Particularly, major factors in the transmission of meanings among words can be related to processes of morphological word modification and their semantic classification.
The first aim of this paper is to formalize the type of meaning inheritance based on word formation and to utilize it as one strategy of lexical acquisition and computational lexicology and to propose a practical retrieval technique for semantic representation. As mentioned above, document classification and summarization independent of existing classified information is certainly important.   But, it is natural for people to know the field of document when they see specific words in the document. Field-Association (*FA*) word is a single or compound words whose terms occurs in any document, and which makes possible to recognize a field of text by using common knowledge of human. Generally, it is possible to collect single *FA* words because the number is finite, but there are some difficulties to select the useful and important compound *FA* words from a large number of combinations of single *FA* words. Thus, the second aim of this paper is to propose an efficient construction method of compound filed association words for specializing field in accordance with a fixed document classification scheme from well-classified document database. Redundant candidates of compound *FA* words can be removed remarkably by using the level and the rank. the presented methods is verified to 20,000 relationships among verbs and nouns extracted from a large corpus .

2.   Field Association Words and Levels

2.1 Field Association Words

A single word is defined as a minimum unit (or word), which cannot be dividable anymore without losing a semantic meaning and called single *FA*. For compound *FA*, it consists of two or more single *FA* words [9, 11]. We regard a compound *FA* words as single if they are easy to lose its field information when they are divided, for example "nuclear weapon", "image processing" or "global warming". In addition, the proper noun except for person's name, such as "Atlanta Braves" – the name of baseball team as a single *FA* word. When a proper noun contains a person's name, it is regarded as a single *FA* word. If the proper noun is a person's name in a whole such as, "Sammy Sosa"; but if it contains a person's title, for example, " Coach R. Neilson" we divide it with two single: "Coach" and "R. Neilson". These two singles belong to baseball field with different levels.
 For easier explanation, hierarchical field categories will be introduced by defining super-fields and sub-fields. For example, a word "pitcher" can be associated with the sub-field <baseball> of <SPORTS>, so it is denoted by <SPORTS\baseball> [6,11].
Consider relationships between the structure of compound words and field classification.   About compound words consisting of two nouns, usually the right side one is the grammatical head in general [12]. Similarly, when the right side one becomes the syntactic head and is also the lexical head, it is called taxonomic class term [12]. In this case, it can be also available to field classification because the left one just modifies the right one [6]. In order to know the meaning of compound nouns and connect them with field classification, we have to know the three natures of compound nouns. The first is that the right one or the left one is the lexical head and is directly related to field classification. The second is both of them are needed to understand the meaning of the compound word. The third is that one side modifies the other and it narrows the associative field.

2.2   Levels of Single *FA* words

 FA words have its strength and scope. Particularly, the scope is ambiguous. Hence, we define five different levels considering *FA* word strength to associate with a field with the following critration:
Definition 1 (Levels of *FA* words)
(Level) Perfect-*FA* words: That associated with one sub-field only.
(Level 2) Semi-Perfect *FA* words: That associated with a few sub-fields in one super-field.
(Level 3) Super-*FA* words: That associated with one super-field uniquely.
(Level 4) Multiple-*FA* words: That associated with a few sub-field of some different super-fields.
(Level 5) Non-*FA* words: do not specify the fields and also includes stop words, as articles, repositions, etc.

3.   Classification of Compound *FA* word

3.1 Word Formation (Word Compounding)

A COMPOUND word is made up of two or more words that together express a single idea. There are three types of compounds. An *open compound* consists of two or more words written separately, such as *salad dressing*, or *April Fools' Day*. A *hyphenated compound* has words connected by a hyphen, such as *age-old, force-feed*. A *solid compound* consists of two words that are written as one word, such as *keyboard* or *typewriter*. In addition, a compound may be classified as permanent or temporary. A *permanent compound* is fixed by common usage and can usually be found in the dictionary, whereas a *temporary compound* consists of two or more words joined by a hyphen as needed, usually to modify another word or to avoid ambiguity. In general, compounds begin as *temporary compounds* that become used so frequently they become established as permanent compounds. Although the dictionary is the first place to look when you are trying to determine the status of a particular compound, reference works do not always agree on the current evolutionary form a compound, nor do they include *temporary compounds*. Generally, in this study we will concentrate on a *permanent compound* word.

3.2   Relation between Derivation Frames and Semantic Representation

3.2.1   Derivation Frames

Morphology consists of inflectional morphology and derivational morphology. The former deals with word modification such as in person, number, and tense for verbs; in number and case for nouns and so forth. The other deals with word modification through affixation, conversion and compounding. In this study we consider all thing about compounding and call a word a derived word, if it has come through a morphological modification to its stem. To encode derived words and their semantic classification into a dictionary, suffixal derived words are grouped in the same derivation frames, and the prefixal ones grouped in different derivation frames, provider that they satisfy the following conditions:

a)　*Beside stems, a derived word in derivation frame belongs to only one category.*

b)　*A derived word retrieved from the reduced dictionary will be a unique string.*

3.2.2　Semantic Primitive

It is very essential to have systematic study on the verbs and noun, to have a deep knowledge. Due to implicit in the events, a generic knowledge is necessary. By building the relationships between verb and noun in the case frame. For example, the verb *"Process"* in the following sentence demands a noun associated with one of the semantic primitive *"Computer"* as the agent of the verb by using relationships among words as follows:

1-　*Information is processed by Computer at Laboratory*

2-　*Data is Processed by Computer at Laboratory*

3-　*Image is Processed by Computer at Laboratory*

4-　*Language is Processed by Computer at Laboratory*

However, frequency semantic representations for other categorical expressions are given a simpler semantic definition (i.e. semantic primitives [1]). For example, it is possible to define the detailed semantic representation of the verb "process" in terms of a case frame as follows:

    (Case Frame of "process"

    (ACTOR:　Data)

    (OBJECT:　Computer)

    (LOCATION:　Laboratory)

    (TOOL:　　Software))

However, as long as a derived word "processor" is given a simple semantic definition "computer", the semantic details that PROCESS carries will not be carried over to the derived word. Let OR be a semantic categorization code for a suffix "-or", then we understand the semantic representation for "processor" as function SUFFIX (PROCESS, OR), which indicates the OBJECT of PROCESS. Hence, one aim of this paper is to formalize the type of meaning inheritance based on word modification and to utilize it as one strategy of lexical acquisition and computational lexicology.

3.2.3　Implicit Inference from the Semantic Stem

Let us consider a derivation frame of *"process"* as follows:

(**Derivation Frame** of *"process"*

(**Morphological Stem**: process)

(**Semantic Stem**: process)

(**Category**: Verb)

(**Inflectional Modification**

    (**Singular**: processes)

    (**Present Participial**: processing)

    (**Past**: processed)

    (**Past Participial**: processed))

(**Derivational Modification**

  (**Suffixal**

    (processor

      (**Suffix Code**: +or)

      (**Semantic Code** : INDICATE_OBJECT)

      (**Semantic Code** : INDICATE_INSTRUMENT))

  (**Prefixal**

    (reprocess

      (**Prefix Code** : re+)

      (**Semantic Code**: AGAIN)

(**Derivation Frame of** "reprocess"

  (**Morphological Stem**: reprocess)

  (**Semantic Stem**: reprocess)

  (**Category**: Verb)

  (**Inflectional Modification**

    (**Singular**: reprocesses)

    (**Present Participial**: reprocessing)

In this frame, only the semantic stem is accessible to semantic knowledge. Semantic representation of each derivation word in the derivation frame of "process" can be inferred in the following manner:

i)   "processor"

A surface word "processor" can be retrieved from a derivation frame of "process". The semantic representation produced by the code INDICATE_OBJECT and INDICATE_INSTRUMENT, respectively being into prominence the OBJECT and INSTRUMENT slots of the knowledge representation of the semantic stem. For Both OBJECT and INSTRUMENT the primitive of "processor" becomes computer, but the implicitly includes knowledge of the stem means " human used software to manage some data".

   ii)   " reprocess"

a surface word "reprocess" can be retrieved from the a derivation frame of " reprocess", but the semantic representation is produce from knowledge associated with the semantic stem of derivation frame, "process" as follows: a) obtain the code of "*reprocess*" as AGAIN

   b) access to the representation of "*process*"

   c) produce the representation include AGAIN , that is to say : "ACTOR *manage again*......".

4.   Concepts and Determination of Compound FA Words
4.1 Concepts of Compound *FA* Words

Some words become more restricted and give us more information retrieval when it becomes compound *FA* word that reminds some limited set of terminal fields. We regard these compound *FA* words as single if they are easy to lose its field information when they are divided as in Fig. 1. From Fig.1 we find the compound *FA* words "*American League*" and "*Perfect Game*" related to the specific field *baseball* with higher rank level, but if we divide them to single words they will lose their field information and they will belong to different lower field levels.

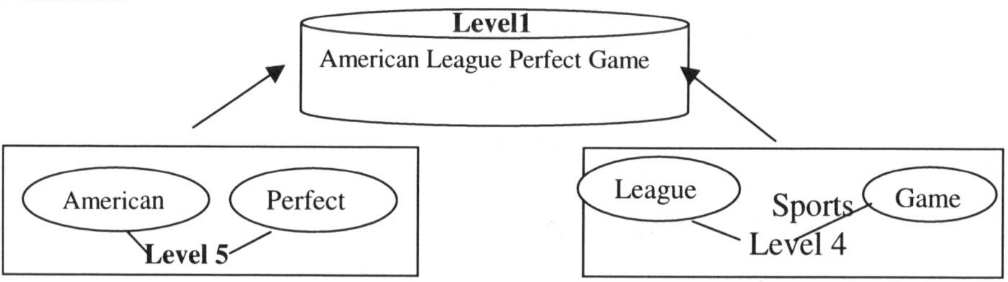

Fig. 1 ranking of compound *FA* word

4.2 Removed Redundant Compound word

Redundant candidates of Compound *FA* words can be removed remarkably by using the level and the rank. There are some compound *FA* word that do not remind a human by any more information than single *FA* words because their levels are some, such words will be remove. For Example the compound words *Cancer Virus* is not consider as a compound *FA* words because its level is equal the level of word *Cancer*

4.3 Determination of Compound *FA* Words

By using the precise single *FA* words. We can determine a compound *FA* word $z = x\,y$ structured by two single words by the Following Algorithm:

***Algorithm1* : (Case1):** z is a *FA* word if LEVEL (x) = 5, LEVEL (y) = 5

(Case2): z is a *FA* word if LEVEL (x) = 5, LEVEL (z) < LEVEL (y) <= 4 (1,2,3,4), or if LEVEL (y) = 5, LEVEL (z) < LEVEL (x) <= 4 (1,2,3,4).

(Case3): z is a *FA* word if z associate with fields except fields by x and y.

(Case4): z is not *FA* word if LEVEL (x) < LEVEL (y), and LEVEL (x) <= LEVEL (z) and if z is associated to the same field.

4.4 Compound *FA* rule:

FA rules are extracted from *FA* words based on compound words. Let CONCEPT (x) be the concept of x. For example, CONCEPT ("Gorge Push") is [name of USA president]. Let PREFIX (y) be a set of elements x for *FA* words $z = x\,y$.

***Algorithm 2:* (Step1):**Extract common concept μ such that CONCEPT (x) = μ for x in PREFIX (y).

(Step2): Determine candidates for *FA* rules such that $[\mu] + y$.

(Step3): Generate $w\,x$ for w such that CONCEPT (w) = μ, and determine the rule if $w\,x$ is associated with the same field as $z = x\,y$.   **(Step4):** For SUFFIX (x), Steps 1-4 are repeated.

5. Evaluation Results

20,000 relationships extracted from a large tagged corpus that have different features are involved in the experiment. Details about this data are shown in Table 1. Algorithm 1 take around 15,600 candidate of compound FA words consisting of two nouns and has reduced them to 1439 FA words. it turns out that Algorithm 1 can remove 88% of redundant compound FA words .

Table 1. Ban Tree Bank Corpus Data

Data		Number
Number of pair:	*Verb- Subject*	*16,970*
	Verb – Object	*2,214*
	Verb- Location	*679*
	Compound Word	*1439*

Fig.2 shows the total frequencies of all verbs and their semantic primitives, we find all main semantic primitive going increase with verbs "sell" and "buy" because we can find many relations of these two verbs in many fields, while the OBJECT is very high with verbs "Process" and "exchange", because you can find their relations in some specific fields only.

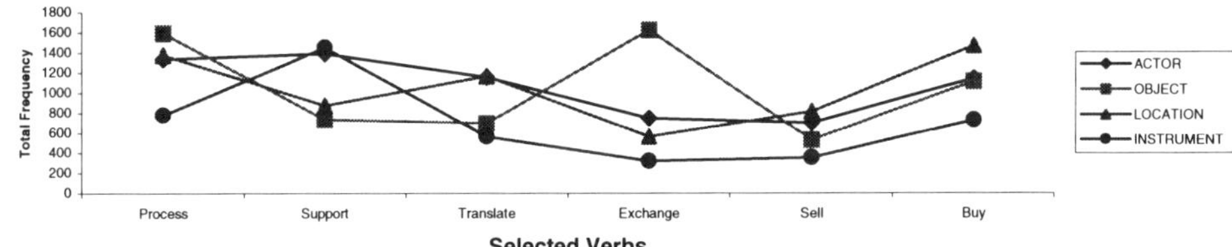

Fig.2 Total Average Frequencies of Some Verbs with Main Semantic Primitive

6. Conclusion

In this paper, we have defined a relation between derivation frames based on morphological and knowledge bases of verb lexicons in order to obtain a deeper representation of a text the technique has been applied to intelligent machine translation that can read documents as input and can produce output in the stylistic form of a skeleton or paraphrase. Also, an efficient method of determining compound *FA* words has been presented and around 88% of redundant candidates of compound *FA* words removed remarkably by using the level and the rank.

REFERENCES

[1] J.Aoe and M. Fujikawa, "A Efficient Representation of Hierarchical Semantic Primitives"- An Aid to Machine Translation Systems-, *In Proceeding of the second International Conference on Supercomputing, Santa Crala, USA*, pp.361-370, 1987.

[2] R. Cohen, "Analyzing the Structure of Argumentative Discourse", Computational Linguistics, 13, 1-2,pp. 11-24, 1987.

[3] N. Fuhr, "Models for retrieval with probabilistic indexing", *Information processing and Retrieval,* Vol.25, No.1, pp.55-72 (1989).

[4] F. Fukumoto and Y. Suzuki, "Automatic clustering of articles using dictionary definitions". *In Proceeding of The 16th International Conference on Computational Linguistic (COLINGO'96)*, pp.406-411, (1996).

[5] M. Fuketa, S. Mizobuchi, Y. Hayashi and J. Aoe, "A fast method of determining weighted compound keywords from text databases". *Information Processing & Management,* Vol.34, No.4, pp.431-442 (1998).

[6] K. Kawabe and Y. Matsumoto, "Acquisition of normal lexical knowledge based on basic level category". *Information Processing Society of Japan, SIG note*, NL125-9, pp.87-92 (1998). (in Japanese).

[7] J. Kupiec, J. Pedersen and F. Chen, "A trainable document summarizer" . *In Proceeding of Annual internationa lConference on Research and Development in Information Retrieval(SIGIR'95)*,pp.68-73 (1995).

[8] H. Kimoto, "Automatic indexing and evaluation of keywords for Japanese newspapers". *Transaction on Information and Systems, IEICE of Japan,* Vol.J74-D-I, No.8, pp.556-566 (1991) (in Japanese).

[9] T. Mamiki, "New English Grammar Word Formation" Vol.2, Taisyukan (1985). (in Japanese)

[10] C.S.Mellish, *Computer Interpretation of Natural Language Description, Ellis Horwood*, 1985.

[11] G. Salton and M. J. McGill:, Introduction to Mordern Infromation Retrieval, McGraw-Hill (1983).

[12] T. Tsuji, H. Nigazawa, M. Okada and J. Aoe, "Early field recognition using field association words". *In Proceeding of 18th International Conference of Oriental languages (ICCPOL'99)*, pp.301-304 (1999).

KES '01
N. Baba et al. (Eds.)
IOS Press, 2001

Similarity Measurement using Two Objects Weight Function

El-Sayed Atlam, Masaki Oono, Kazuhiro Morita, Masao Fuketa, and Jun-ichi Aoe
Dept. of Information Science and Intelligent Systems
University of Tokushima,Tokushima, 770-8506, Japan.
E-mail: atlam@is.tokushima-u.ac.jp

Abstract. A "term weighting" is an useful technique for keyword extraction and document classification. The traditional approach depend on high frequency terms, called positive weight (PW) function. In this paper a new "term weighting" function called, negative weight inverse verb frequency (NWIVF) function that depends on low frequency terms has been defined, and new similarity measurement is presented by combining the NWIVF and PWIVF functions. By using this new method both recall and precision have improved by 35% and 23% respectively over the positive weight method.

1. Introduction

Recently, many types of Natural Language Processing (NLP) systems have been developed and there are many individual knowledge such as case frame knowledge [7], synonyms [2], co-occurrences [6] and so on. "Term weighting" has been widely investigated in information retrieval [8] . It is one of the important issue in keyword extraction, text categorization [5], and classification [1]. The usage of term weighting is used in similarity measurement for this application. In this paper, we introduce word similarity based on term weight to obtain similarity measurement among relational information by the case frame [3]. A well-known approach weighted inverse verb frequency (WIVF) [10], has been widely used, but the disadvantage of this method is that it depends on high frequency terms, called positive weight (PW) function, therefore, we propose a new "term weighting" method called Negative Weight Inverse Verb Frequency (NWIVF) function which depends on the low frequency terms. The presented method is verified to 11,000 relationships among verbs and nouns data extracted from a large tagged corpus

2. Term Weighting

2.1 Positive Weight Inverse Verb Frequency

Takenobu [9] defined a factor called *weight inverse verb frequency (WIVF)* to incorporate the object frequency over the collection of verbs, in this paper we call this factor a *positive weight inverse verb frequency (PWIVF)* function. For verb x_i ($1 \leq i \leq n$) and for object y_h ($1 \leq h \leq m$) that related to this verb, the function $PWIVF(x_i, y_h)$ is given by:

$$PWIVF\ (x_i, y_h) = \frac{TF(x_i, y_h)}{\sum_{h=1}^{m} TF(x_i, y_h)} \quad \cdots\cdots\cdots\cdots\cdots (1)$$

2.2 Two Objects Weight Function

To focus the similarity between two objects we introduce a new term called a *positive two term weight* $PTTW((x_i, y_h), (x_i, y_k))$ function ($h \neq k$) by extending Eq. (1) for using two objects as follows:

$$PTTW((x_i, y_h), (x_i, y_k)) = (PWIVF(x_i, y_h), PWIVF(x_i, y_k)) \quad \cdots\cdots\cdots\cdots (2)$$

To apply this function we used a part of speech dictionary to extract **verb-object.. etc.** with its frequency from large tagged corpus, we took some typical data from this tagged corpus and used it as shown in Table 1.

Table1 positive and negative weight functions of verbs and objects

X / y	Eat	Drink	Draw	Write	Buy
Orange					
TF (x, y)	154	60	1	2	95
TF- Average(x)	154-52 = 103	60 -23 = 37	1 -21 = - 20	2 - 60= -58	95 -123= -28
PWIVF (x, y)	154/416=0.37	60/184=0.32	---	---	----
NWF (x, y)	---	----	1/186 =0.005	2/480=0.004	95/984=0.095
NWIVF(x, y)	---	----	20/186 =0.12	58/480=0.12	28/984 =0.02
Apple					
TF (x, y)	142	50	2	1	90
TF- Average (x)	142-52=90	50-23= 27	2-21= -19	1-60=-59	90 -123 = - 33
PWIVF (x, y)	142/416=0.34	50/184= 0.27	----	---	---
NWF (x, y)	---	---	2/168=0.01	1/480=0.002	90/984=0.09
NWIVF (x, y)	---	---	19/168=0.11	59/480=0.13	33/984 =0.03
Shirt					
TF (x, y)	2	5	11	4	101
TF- Average(x)	2-52=- 50	5-23= -18	11 21- -10	4-60=-56	101-123= -22
PWIVF (x, y)	---	---	---	---	----
NWF (x, y)	2/416= 0.004	5/184=0.03	11/168= 0.059	4/480=0.008	101/984= 0.101
NWIVF (x, y)	50/416=0.13	18/184=0.1	10/168=0.06	56/480=0.11	22/984=0.022
Jacket					
TF (x, y)	1	7	12	3	102
TF- Average(x, y)	1-52= -51	7-23=-16	12 -21= - 9	3-60=-57	102-123= -21
PWIVF (x)	---	---	---	---	-----
NWF (x, y)	1/416= 0.002	7/184=0.04	12/168 = 0.06	3/480=0.006	102/984=0.102
NWIVF (x, y)	51/416= 0.12	16/184=0.09	9/168 =0.05	57/480=0.12	21/984 = 0.021
Book					
TF (x, y)	4	14	52	190	105
TF- Average(x)	4-52= -48	14 -23= -9	52-21= 20	190-60= 130	105-123= -18
PWIVF (x, y)	---	---	52/168=0.31	190/480=0.39	----
NWF (x, y)	4/416= 0.009	14/184=0.08	---	---	105/984=0.105
NWIVF (x, y)	48/416=0.11	9/184=0.085	----	-----	18/984 =0.018
Paper					
TF (x, y)	3	9	46	170	107
TF- Average(x)	3-52 =-49	9 -23= -14	46-21=25	170-60=110	107 -123= -16
PW (x, y)	---	---	46/168= 0.27	170/480=0.35	-------
NWF (x, y)	3/416=0.007	9/184= 0.03	---	----	107/984 = 0.107
NWIVF (x, y)	49/416= 0.12	14/184=0.09	----	-----	16/984 =0.016
Duck					
TF (x, y)	55	20	26	57	191
TF- Average(x)	55-52=3	20-23= -3	26-21= 5	57-60=-3	191- 103= 88
PWIVF (x, y)	55/416=0.134	---	26/168=0.16	---	191/984=0.19
NWF (x, y)	---	20/184=0.1	-----	57/480=0.11	-----
NWIVF (x, y)	---	3/184=0.01	----	3/480=0.005	------
Goose					
TF (x, y)	54	18	25	53	194
TF- Average(x)	54-52=2	18-23= -5	25-21= 3	53-60=-7	194 -123= 71
PWIVF (x, y)	54/416=0.13	--	25/168 =0.15	---	194/984=0.194
NWF (x, y)	---	18/184=0.097	---	53/480=0.10	----
NWIVF (x, y)	---	5/184=0.03	---	7/480=0.006	------
$\sum_{h=1}^{m} TF\,(x_i, y_h)$	416	184	168	480	984
Average (x_i)	416/8=52	184/8 =23	168/8 = 21	480/8 = 60	984/8= 123

We take all pairs of objects with the same verb that are having the highest *PWIVF* function value and extract the highest *PWIVF* function. From Table 1, we extract verb *"Eat"* with two objects *"Orange"* and *"Apple"*, and apply Eq. (2), which measures the *PTTW* function for objects with their related verbs as follows:

$$PTTW\left(("Eat", "Orange"), ("Eat", "Apple")\right)$$

$$= \left(\frac{TF("Eat", "Orange")}{\sum_{h=1}^{h=8} TF("Eat", y_h)}, \frac{TF("Eat","Apple")}{\sum_{h=1}^{h=8} TF("Eat", y_h)} \right) = \frac{154}{416}, \frac{142}{416}) = (0.37, 0.34)$$

By using the same method we can measure the *PTTW* function for all other objects with their related verbs, as in Table 1.

3. Similarity Measurement
3.1 Using Positive Weight Function

Using a term weighting method, the object weighted function *PWIVF* (x_i, y_h) of verb x_i with object y_h represented by term vectors $V(y_h)$ of the form:

$$V(y_h) = (\, PWIVF(x_1, y_h), PWIVF(x_2, y_h), \cdots, PWIVF(x_n, y_h)\,) \quad \cdots\cdots\cdots\cdots (3)$$

Giving the vector representations of verbs as in Eq. (3), similarity between two objects can be obtained by element-wise comparison of the vectors.

There are many ways to measure similarity between two vectors. We adopt the *Jaccard* function ($Sim\,(V_i, V_j)$), which is the normalized inner product of two vectors. In this study we rename the *Jaccard* function as *P-Sim* ((V (y_h), $V(y_k)$) as shown in the following equation:

$$P-Sim(V(y_h), V(y_k)) = \frac{\displaystyle\sum_{i=1}^{n} PWIVF\,(x_i, y_h) \cdot PWIVF\,(x_i, y_k)}{\displaystyle\sum_{i=1}^{n} PWIVF\,(x_i, y_h) + \sum_{i=1}^{n} PWIVF\,(x_i, y_k) - \sum_{i=1}^{n} PWIVF\,(x_i, y_h) \cdot PWIVF\,(x_i, y_k)} \quad \cdots\cdots (4)$$

where $PWIVF\,(x_i, y_h)$ means *i-th* element of the term vector $V\,(y_h)$. The greater the value of $P\text{-}Sim\,(V\,(y_h),\,V\,(y_k))$ is, the more similar to these two objects are.

For example, if we take two objects *"Orange" and "Apple"* with all verbs and apply Eq. (4), we measure the similarity between two objects as follows:

$P-Sim\,(V\,("\,Orange\,"),V\,("\,Apple\,"))$

$$= \frac{((0.37 \times 0.34) + (0.32 \times 0.27)}{((0.37 + 0.32 + 0.005 + 0.004 + 0.095) + (0.34 + 0.27 + 0.01 + 0.002 + 0.09) - ((0.37 \times 0.34) + (0.32 \times 0.27)}$$

$$= \frac{0.2122}{1.506 - 0.2122} = \frac{0.2122}{1.2938} = 0.16401$$

But if we apply the same equation for Objects *"Apple" and "Shirt"*, we obtain different results as follows:

$P-Sim\,(V\,("\,Apple\,"),V\,("\,Shirt\,"))$

$$= \frac{((0.34 \times 0.004) + (0.27 \times 0.03)}{((0.34 + 0.27 + 0.01 + 0.002 + 0.09) + (0.004 + 0.03 + 0.059 + 0.008 + 0.1) - ((0.34 \times 0.004) + (0.27 \times 0.03))}$$

$$= \frac{0.00946}{0.913 - 0.00946} = \frac{0.00946}{0.90354} = 0.01046$$

As we mentioned that the greatest value of the $P\text{-}Sim\,(V\,(y_h),\,V\,(y_k))$ is, the more similar these two objects are. Therefore, we can decide that objects *"Orange" and "Apple"* are more similar than objects *"Apple"* and *"Shirt"*.

Problems for using the *PWIVF* function only:

Although the *PWIVF* function is very useful to measure similarity between words, but it could not concern many low values in Table 1 as follows:

PWIVF ("*Write*", "*Orange*") = 0.004 and   *PWIVF* ("*Write*", "*Apple*") = 0.002

Due to the advantages of using the lowest weight function, we propose a new method by combining the highest and the lowest weight functions. It is more useful to perform similarity on the basis of both functions.

3.2 Using Negative Weight Function

In this part, a low weight function was defined as a *Negative Weight Inverse Verb Frequency (NWIVF)* function, we found that *NWIVF* function has many advantages:

1-   We clarified that by using the *NWIVF* function, similarity measurement has a significant value.
2-   When we combine the *NWIVF* function with the *PWIVF* function, we obtained a significant amount of similarity measurements as compared with the *PWIVF* function only.

Definition 1 (*Negative Weight Inverse verb Frequency*):

For verb x_i ($1 \leq i \leq n$), for object y_h ($1 \leq h \leq m$), and for the number of objects m. The average of these objects is given by:

$$\text{Average}\ (x_i) = \frac{\displaystyle\sum_{h=1}^{m} TF\,(x_i, y_h)}{m}$$

and the *negative weight inverse verb frequency NWIVF* (x_i, y_h) function is given by:

$$NWIVF(\,x_i, y_h) = \frac{\left|\,TF(x_i, y_h) - \text{Average}(x_i)\,\right|}{\displaystyle\sum_{}^{m} TF(x_i, y_h)} \quad \cdots\cdots\cdots\cdots\cdots (5)$$

Definition 2 (*Negative Two Term Weight Function*)

We can also define a *negative two term weight NTTW* $((x_i, y_h), (x_i, y_k))$ function for two objects to extract a similarity as follows:

$$NTTW((x_i, y_h), (x_i, y_k)) = ((NWIVF(x_i, y_h), NWIVF(x_i, y_k)) \cdots\cdots\cdots\cdots\cdots (6)$$

Suppose we have same objects as in Table 1. We take all pairs of objects with the same verb that are having the lowest *NWIVF* function and extract the lowest *NWIVF* function. For example, from Table 1 we extract verb *"Write"* with two objects *"Orange"* and *"Apple"*, and we apply Eq. (6) which measures the *NTTW* function for objects with their related verbs as follows:

$NTTW$ (("Write", "Orange"), ("Write", "Apple")) = $((NWIVF$ ("Write", "Orange"), $NWIVF$ ("Write", "Apple"))

$$(\frac{|TF\ ("Write\ ","Orange\ ") - Avreage\ ("Write\ ")|}{\displaystyle\sum_{h=1}^{8} TF\ ("Write\ ", y_h)}, \frac{|TF\ ("Write\ ","Apple\ ") - Avreage\ ("Write\ ")|}{\displaystyle\sum_{h=1}^{8} TF\ ("Write\ ", y_h)})$$

$$= (\frac{|2 - 60|}{480}, \frac{|1 - 60|}{480}) = (0.12, 0.13).$$

By using the same method we can measure the *NTTW* function for all other objects with their related verbs, as in Table 1. As we mentioned before that we are using the *Jaccard* function, which is the normalized inner product of two vectors, here we developed a new modification to be able to extract the similarity using the *XWIVF*, as shown in the following equation:

$$X - Sim(V(y_h), V(y_k)) = \frac{\displaystyle\sum_{i=1}^{n} XWIVR(x_i, y_h) \cdot XWIVR(x_i, y_k)}{\displaystyle\sum_{i=1}^{n} XWIVR(x_i, y_h) + \sum_{i=1}^{n} XWIVR(x_i, y_k) - \sum_{i=1}^{n} XWIVR(x_i, y_h) \cdot XWIVR(x_i, y_k)} \cdots\cdots\cdots\cdots (7)$$

where $X = N$ *(Negative)* or P *(Positive)*, and *XWIVF* (x_i, y_h) means *i-th* element of the term vector $V(y_h)$. The greater the value of X-Sim $(V(y_h), V(y_k))$ is, the more similar these two objects are.

Consider the following cases as examples of Eq.(7):
Case (1) *Using negative weight function only:*
We define similarity measurement by using the *NWIVF* function only. We check every two objects with all verbs, then we extract the lowest *NWIVF* function, e.g. we checked objects *"Shirt" and "Jacket"* with all verbs, we found that these objects have the *NTTW* function with two verbs *"Eat" and "Write"* as follows:
$NTTW$ (("Eat", "Shirt"), ("Eat", "Jacket")) $\quad = \quad (0.13, 0.12)$,
$NTTW$ (("Write", "Shirt"), ("Write", "Jacket")) $\quad = \quad (0.11, 0.12)$,
Then we calculate similarity measurement by applying equation (9), which measures the similarity using the *NWIVF* function only as follows:

$N - Sim(V("Shirt"), V("Jacket"))$

$$= \frac{0.0288}{0.409 - 0.0288} = \frac{0.0288}{0.3802} = 0.075749$$

Case (2) *Using combination of positive and negative weight functions:*
Check every two objects with all verbs, then we extract the highest *PWIVF* function, e.g. we checked objects *"Orange" and "Apple"* with all verbs, we found that these objects have the *PTTW* function with two verbs *"Eat" and "Drink"* as follows:
$PTTW$ (("Eat", "Orange"), ("Eat", "Apple")) $\quad = \quad (0.37, 0.34)$, $PTTW$ (("Drink", "Orange"), ("Drink", "Apple")) $\quad = \quad (0.32, 0.27)$, at the same time these objects have the *NTTW* function with verb *"Write"* as follows: $NTTW$ (("Write", "Orange"), ("Write", "Apple")) $\quad = \quad (0.12, 0.11)$,
By combination of both *NWIVF* and *PWIVF* functions, we obtain similarity measurement of applying equation (7) as follows:

$X - Sim\ (V\ ("Orange"),\ V\ ("Apple"))$

$$= \frac{0.241}{1.506 - 0.241} = \frac{0.241}{1.265} = 0.19051$$

By using the *NWIVF* function only we can measure similarity between objects as we mentioned in case (1), section (3.2). The value we obtained by this method (*N-Sim* = 0.075749) is small but has a great value when compared with a positive method (*P-Sim* = 0) (not considered). This shows that our procedure has a significant development in similarity measurement by using the *NWIVF* function.
Similarity measurement using combination of PWIVF and NWIVF functions:
By using the combination of *PWIVF* and *NWIVF* functions we can also improve similarity measurement as we mentioned in case (2), section (3.2). If we compare the results that we obtained by combination of the two weight functions, we can find it higher similarity measurement than using the *PWIVF* function only, i.e
X-Sim(V("Orange"), V("Apple"))=0.19051 > *P-Sim* (V("Orange"), V("Apple")) = 0.16401 in *PWIVF* method.

3. Simulation Results

99,714 sentences from tagged corpus (Peen Treebank), which have different features, (612 Verbs & 7,000 Nouns) are involved in this experiment. Precision and Recall shown in Table 2.

For M is all possible pair of objects, N' number of combinations of objects and verbs, and N is the number of positive weight function, we defined these parameters as follows: $M = (M-1) + (M-2) + . + 2 + 1$. $N' = n . m$, for the number of verbs n and the number of objects m. By using these three parameters M, N, and N' with the typical data we define the precision and Recall [8] for similarity pair in traditional and new method as follows: ***Precision   P = N/ N'   and   Recall R= N/ M*** ....(8)

From Table 2, if we consider column 6, which contains 14 objects and 10 verbs, $M = 91$ (13+12+11+10+9+8+7+6+5+4+3+2+1), and $N' = 140$, and the number of positive weight function in *PWIVF* method $N_{trad} = 50$. Then we calculate precision and recall by using Eq. (8), we find $R_{trad} = 50/91 = 0.55$, $P_{trad} = 50/140 = 0.35$, by the same manner we calculate precision and recall, when combining the number of positive and negative weight function are $N_{new} = 82$, then $R_{new} = 82/91 = 0.90$ and $P_{new} = 82/140 = 0.58$. Then the new *NWIVF* function can improve the recall and precision by 35% and 23% respectively over the *PWIVF* function.

Table 2. Precision and Recall by traditional and new method

	6 obj. 4 verb*	*8 obj.* 6 verb*	*10 obj.*8 verb*	*12obj.*10 verb*	*14obj*10 verb*
M (all possible pairs of objects)	15	28	45	66	91
N' (number of combinations of objects and verbs	24	48	80	120	140
N_{trad} (number positive weight function)	3	7	12	30	50
R_{trad}(Recall in Traditional)	3/15 = 0.2	7/28 = 0.25	12/45 = 0.28	30/66 = 0.46	50/91 = 0.55
P_{trad} (Precision in traditional)	3/24 =0.12	7/48 =0.14	12/80 = 0.15	30/120 =0.25	50/1140 =0.35
Number negative weight function.	2	5	8	20	32
N_{new} (number of positive & negative weight function)	5=3 +2	12=7 +5	20= 12 +8	50 =30 +20	82 =50 +32
R_{new} (Recall in New method)	5/15 = 0.33	12/28 = 0.43	20/45 =0.45	50/66 = 0.76	82/91 = 0.90
P_{new} (Precision in new method)	5/24 =0.20	12/48 =0.24	20/80 = 0.25	50/120 = 0.41	82/140= 0.58

trad = traditional method using the PWIVF function only.
new = new method using combination of PWIVF and NWIVF functions.

4. Conclusion

A new "term weighting" function called, *negative weight inverse verb frequency (NWIVF)* function has been definend, new similarity measurement is presented by combining the *NWIVF* and *PWIVF* functions. Recall and precision values have also improved compared with the *PWIVF* method. We verified the proposed method, it was found to be in excellent agreement by the simulation results.

The presented method in this paper is applied only for extracting the similarity between words and in future study, this method will be applied to text categorization.

REFERENCES

[1] Biber, M. J., Hebrail, G., Moteil M. G., and Penot, N. "Automatic Document Classification". *In proceedings of the Annual International ACM SIGIR Conference, 199, pp. 51-58.*

[2] Darling, B. C. A dictionary of selected synonym in the principal language. *Ph.D. thesis, University of Chicago, 1988.*

[3] Fillmore,C.J. " The case for case". In Universal in Linguistic Theory, New York: Hort, Reinhart and Winston, 1968.

[4] Fuhr. N. "Models for Retrieval with probabilistic indexing" *Information Processing & Management, Vol.25, No. 1, 1989,pp. 55-72.*

[5] Lewis, D. D. "An Evaluation of Phrasal and Clustered Representations of a Text Categorization". *In proceedings of ACM SIGIR International conference, 1992, pp. 37-50.*

[6] Li, H., & Abe, N. "Clustering Words With the MDL Principle". *Journal of Natural Language Processing, Vol. 4,o.2, 1997, pp 71-88.*

[7] Oishi, A., & Matsumoto, Y.. "A Method for Deep Case Acquisition Based on surface Case Pattern Analysis", *NLPRS., Vol.34, No.2, 1995, pp. 678-684.*

[8] Salton, G., & McGill, M.J. Introduction to Modern Information Retrieval, *McGraw-Hill, 1983.*

[9] Takenobu, T. & Makoto, I. "Text Categorization Based on Weight Inverse Document Frequency", *SIG-IPSJ, 1994, pp. 33-39.*

[10] Utsumi, A., Hori, K.,& Ohsuga, S. "An Affective- Similarity-Based Method for Comprehending Attribution Metaphors", *Journal of Natural Language Processing, Vol.5, No.3, 1998, pp. 3-30.*

Automatic Generation of Operating Procedures for Batch Process Plants

Kenji Hoshi, Kousuke Nagasawa, Yoshiyuki Yamashita and Mutsumi Suzuki

Department of Chemical Engineering
Tohoku University, Sendai 980-8579, Japan

Abstract. A method for automatic generation of operating procedures for batch process plants is presented. Material-procedures graph and plant-structure graph are used for the representation of the problem. By using these two graphs interactively, a recursive search algorithm is described to obtain detailed operating procedures. The methodology is successfully illustrated by a case study.

1. Introduction

For the production of low-volume, high value products, the importance of operating procedures synthesis is increasing. The operations must be carried out with satisfying constraints such as safety, efficiency, energy consumption and so on. Computer support for operating procedure synthesis is desirable since it is currently carried out almost manually, which often require considerable amount of time and effort. It is also required for improve optimality and safety of plant operations.

The problem of operating procedure synthesis was originally treated by a pioneering work of O'Shima [2]. Foulkes *et al.* [1] utilize pattern matching and search to find all paths for a specific route of materials in a process network. Viswanathan *et al.* [4, 5] proposed an approach based on a discrete event modeling concept.

In this paper, a methodology for the automatic generation of operating procedures for a batch chemical plant is presented. In addition to the material-procedure graph, plant structure graph are used for the mapping between unit operating procedures and process equipment units.

2. Problem Definition

In this paper, off-line synthesis of operating procedures for a batch process plant is treated. The planning can be defined as searching allowable sets of sequential operations and an initial state to realize a given final state without violating constraints.

Required knowledge to solve the problem include material conversion procedures, plant structure, and constraints. For the purpose of explanation, let us consider the problem obtaining material E from given materials A and C by a plant shown in Fig. 1.

2.1 Material-procedures graph

Based on the material and/or operating procedures graph [4], reactions and mixing equations for materials are represented by a digraph. Figure 2(a) shows the material-procedure graph for the example problem.

Symbol	Type	Description
UP1	Reaction	$^{i}A \rightarrow B$
UP2	Reaction	$^{i}C \rightarrow D$
UP3	Mixing	$B + D \rightarrow E$

^{i}: indicate source materials

Fig. 1: Flow sheet of the example

2.2　Plant-structure graph

Any plants would be modeled by means of 'fragments' of pipeline through which flow can occur. It was found to be convenient to model the fragments of line using valves as the endpoints — unless a vessel provided a natural endpoint[1]. In this study, pipeline fragments around a junction among valves are treated as a single 'fragment', although Foulkes *et al.*[1] treated them as individual fragments. Structure of the example plant is represented by a structure graph as shown in Fig. 2(b), where FR0, FR1, $\cdots$, FR6 are seven 'fragments' as shown in Fig. 1. By using these 'fragments', it is naturally assured that only the valves along the graph-path between two equipments should be opened to transfer a material from one equipment to another.

3.　Method

3.1　Architecture

Based on these graph representations of knowledge of material conversion procedures and plant structure, a methodology to generate detailed operating procedures is considered with constraints. The schematic architecture of the algorithm is shown in the Fig. 3.

At first, routes of a material transfer through complex piping system among processing equipments are considered. Then, equipment connectivity and valve connectivity are analyzed with a plant structure graph, where the equipment connectivity is a matrix representation of direct connection of equipments, and the valve connectivity is a list of

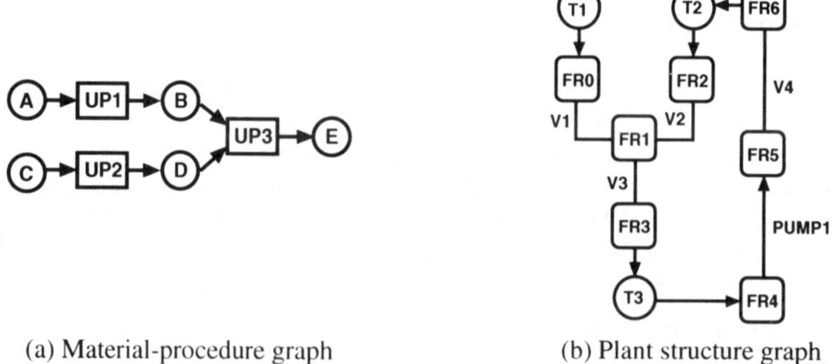

(a) Material-procedure graph　　　　　　　　(b) Plant structure graph

Fig. 2: Graph representations of the example

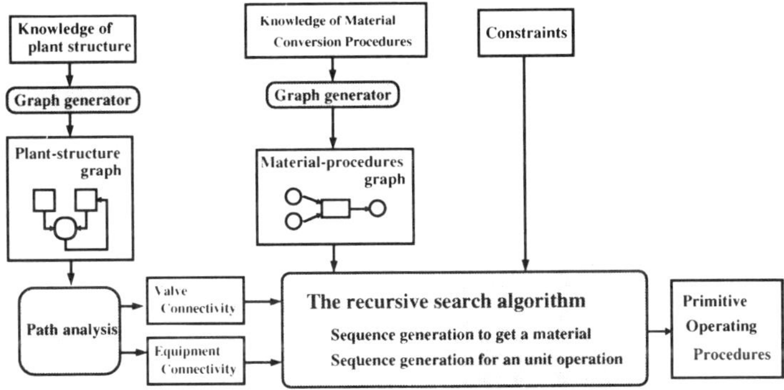

Fig. 3: Schematics of the algorithm

Symbol	Type	Description
M1	Mixing	$^iR1 + {}^iR2 + {}^iR3 \rightarrow I1$
F1	Filtration	$I1 \rightarrow I3$
R1	Reaction	$^iR4 + {}^iR5 + I3 \rightarrow I4$
De1	Decantation	$I4 \rightarrow I5$
M2	Mixing	$^iR1 + {}^iR4 + I5 \rightarrow I5'$
F2	Filtration	$I5' \rightarrow I7$
D1	Distillation	$I7 \rightarrow Material - A$

^{i}: indicate source materials

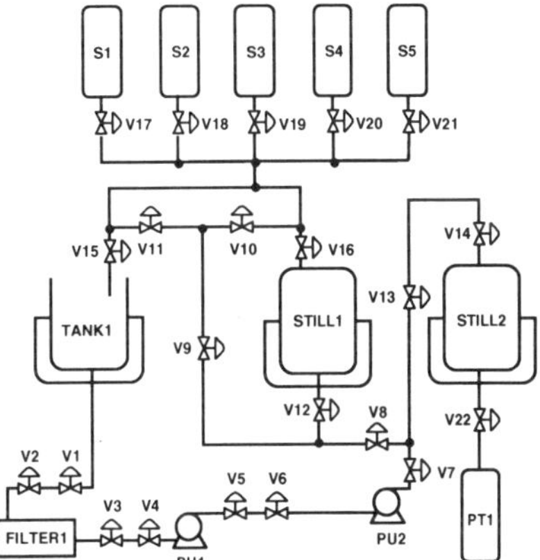

Fig. 4: Case Study Plant

connected valves between the equipments. Combining the above knowledge with constraints, equipments and primitive operating procedures are assigned for each other to generate allowable series of detailed operating procedures.

3.2 Implementation

In this study, the assignment algorithm is implemented recursively by using two program functions. The first function is to generate a sequence of primitive operating procedures to obtain a specific material in a specific equipment. The second function is to generate primitive operating procedures to realize the specified operation in a specified equipment. After activating the first function with the target material and its container equipment, both of the functions are recursively called interactively to generate possible series of primitive operating procedures.

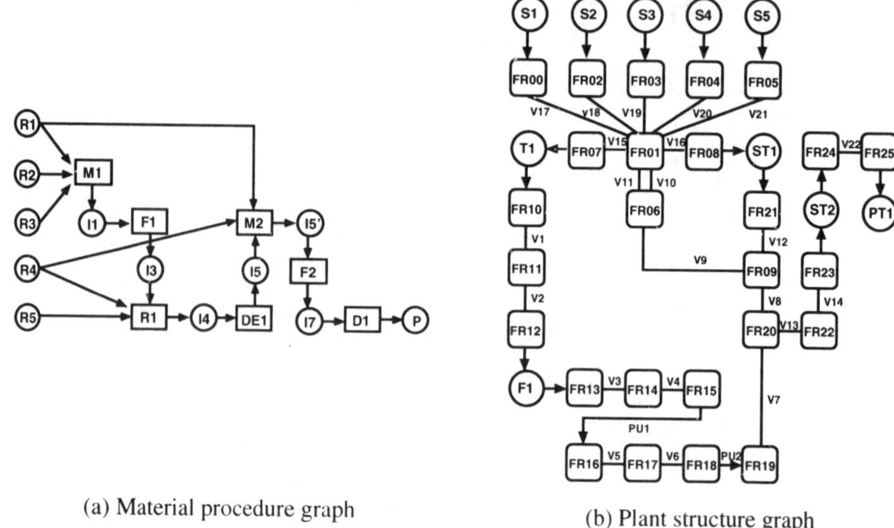

(a) Material procedure graph

(b) Plant structure graph

Fig. 5: Graph Representations of the Plant

4. Case Study

To illustrate the methodology, the above mentioned method is applied to an example problem, which is similar to an example in Viswanathan *et al.*[5]. Flow sheet and material conversion procedures are shown in Fig. 4. Applicable equipments for each specific conversion operations are specified as constraints. For example, equipments Tank1, Still1 and Still2 can be used for M1 (mixing) operation but Filter1 is the only equipment for F1 (filtration) operation. Constraints such as order of operations of a specific pump and valves have not been used, but they could be easily added, if necessary.

Material-procedures graph and plant structure graph of this example are shown in Fig. 5. By applying the proposed algorithm, 112 sets of detailed operating procedures are generated. Table 1 shows an example set of the procedures to obtain material A in tank $PT1$. In this table, close of valves are not shown explicitly. The generated series of operation seems to be reasonable. If necessary, order of operations of a specific pump and valves could be easily added by heuristic constraints.

5. Discussions and Conclusions

Let us suppose that N steps of unit operations are required to obtain the target material, where each of unit operation has M possible candidates of primitive operating procedures. Then the order of calculation of the proposed method is $O(M^N)$ to get all sets of operating procedures. From this simple consideration, the proposed method is relatively tough for the increase of plant complexity, but it is not free from the problem of combinatorial explosion. Heuristic approach will give some kind of solutions for this problem.

Concurrent operations have not been considered yet in this study. By following the idea of Uthgenannt [3], such operations could be allowed by locking used equipments after getting operating procedures for single operation.

Table 1: An example of generated operating procedure

Operation	Position	
PUT R1	S1	
PUT R2	S2	
PUT R3	S3	
PUT R4	S4	
PUT R5	S5	
MOVE	S1 → TANK1	(Open V17,V15)
MOVE	S2 → TANK1	(Open V18,V15)
MOVE	S3 → TANK1	(Open V19,V15)
EXEC M1	TANK1	
EXEC F1	TANK1 → FILTER1 → STILL1	(Open V1,V2,V3,V4,PU1,V5, V6,PU2,V7,V8,V9,V10,V16)
MOVE	S4 → STILL1	(Open V20,V16)
MOVE	S5 → STILL1	(Open V21,V16)
EXEC R1	STILL1	
EXEC DE1	STILL1	
MOVE	STILL1 → TANK1	(Open 12,9,11,15)
MOVE	S1 → TANK1	(Open V17,V15)
MOVE	S4 → TANK1	(Open V20,V15)
EXEC M2	TANK1	
EXEC F2	TANK1 → FILTER1 → STILL2	(Open V1,V2,V3,V4,PU1,V5, V6,PU2,V7,V13,V14)
EXEC D1	STILL2	
MOVE	STILL2 → PT1	(Open V24,V25)

In conclusion, based on the knowledge of given plant structure and material operating procedures, a method to generate allowable series of primitive unit operating procedures are proposed. The method is successfully applied to an example problem and confirmed the ability to generate reasonable operating procedures without conflicting constraints.

References

[1] N. R. Foulkes, M. J. Walton, P. K. Ansow and M. Galluzo, Computer-Aided Synthesis of Complex Pump and Valve Operations, *Computers and Chemical Engineering*, **12** (1988) 1035–1044.

[2] E. O'Shima, Safety supervision of valve operations, *Journal of Chemical Engineering of Japan*, **11** (1978) 390–395.

[3] J. A. Uthgenannt, Path and Equipment Allocation for Multiple, Concurrent Processes on Networked Process Plant Units, *Computers and Chemical Engineering*, **20** (1996) 1081–1087.

[4] S. Viswanathan, C. Johnsson, R. Srinivasan, V. Venkatasubramanian and K. E. Ärzen, Automatic operating procedure synthesis for batch process: Part I. Knowledge representation and planning framework, *Computers and Chemical Engineering*, **22** (1998) 1673–1685.

[5] S. Viswanathan, C. Johnsson, R. Srinivasan, V. Venkatasubramanian and K. E. Ärzen, Automatic operating procedure synthesis for batch process: Part II. Implementation and application, *Computers and Chemical Engineering*, **22** (1998) 1687–1698.

KES '01
N. Baba et al. (Eds.)
IOS Press, 2001

A PCA Based Output Integrated Recurrent Neural Network for Dynamic Process Modeling

Yu Qian, Huanong Cheng, Xiuxi Li, Qinghua Yin and Yanbin Jiang
Chemical Engineering Research Center, South China University of Technology,
Guangzhou, P. R. China, 510640

Abstract. Combining data analysis and neural network techniques, this paper proposes a novel modeling method, the principal component analysis (PCA) based output integrated recurrent neural network (OIRNN). The modeling and training algorithm are presented and demonstrated for a typical chemical process. It is characteristic of rational model structure and good prediction of the non-linear dynamic behaviors.

1. Introduction

Most of real industrial chemical processes are non-linear complex systems. Dynamic behavior depiction and time responses are relatively rigorous. Development of precise first principle models is difficult and time-consuming, while fuzzy logic models depend heavily on expert knowledge. Conventional statistical regressive models are incapable to model the underlying structure with significant non-linear characteristics.

To remedy the defects of single models, integrate models are actively researched recently. One kind of hybrid models [1] combines part of first principle equations with ANN, in which ANN is used to determine parameters of the first principle model. Fuzzy logic approach [2] is used for representing imprecision and approximation of the relationship among process variables. It is successfully incorporated into conventional process simulators. Several efforts [3] have been made to integrate statistical analysis within non-linear regression, which are polynomial, spline function and ANN. Among them, ANN is a non-parametric method with better prospects in integrated modeling.

A lot of operation data is available in the industries. Among many approaches for data analysis and data interpretation to extract process features and to predict process tendency, PCA is a popular statistical approach for data analysis [4]. With PCA, input data are projected onto a lower-dimensional space. Useful information is mined up, while redundant information is filtered out. By using principal components, modeling complexity is greatly reduced. For many nonlinear processes, conventional linear principal component regression (PCR) may bring distinct errors. Feedforward network (FFN) was later proposed to improve linear PCR. The main idea is to derive components by PCA and replace the linear regression with FFN. It is basically a static system and not good at high-order dynamic chemical processes. A large number of relevant past inputs are needed in FFN, which would entail an impractical large networks size.

Recently, Recurrent neural networks (RNN) are proposed to simulate dynamic processes.

Just mention a few, there are Hopfield neural network, Elman neural network and Jordan neural network. They are good at single order system, while not suited for high-order systems. Zhu[5] and Wu[6] modified Elman neural network, proposed a new RNN with multiple sub-feedback-layers and a state-integrated RNN, respectively. The modified RNNs reserve past time sequence information as much as possible to facilitate simulating high-order dynamic non-linear behavior.

Complex structure of MIMO systems would limit the application of RNN in modeling of actual processes. In ordinary conditions, nodes in the hidden layer of RNN are much more than those in the input layer. When dealing with too many input variables, complex structures occur in multiple sub-feedback-layers RNN and state-integrated RNN.

2. PCA based OIRNN and the algorithm

Two approaches in this paper are proposed to solve this problem. First, the original input variables are transformed into principal components by PCA. These principal components reflect features of original data space. The principal components are used as input nodes. Second, Jordan RNN is modified with integrating the output values of a certain steps of past times.

The modified RNN is improved as follows:

$$\begin{cases} \dot{X} = F(X,U) \\ Y = G(X) \end{cases} \tag{1}$$

where X, U, Y are the state vector, the input vector and the output vector, respectively. $\dot{X} = dX / dt$. Equation (1) is developed in difference equation:

$$\begin{cases} X_{t+1} = X_t + F(X_t, U_t) \\ Y_{t+1} = G(X_{t+1}) \end{cases} \tag{2}$$

Equation (2) is modified to an output integrated recurrent function,

$$Y_{t-1} = G^*(W_t Y_t, W_{t-1} Y_{t-1}, \ldots, W_{t-k} Y_{t-k}, U_t) \tag{3}$$

where, $Y_t, Y_{t-1}, \cdots, Y_{t-h}$ are values of output nodes of RNN of h past times. The h is the number of integrated state of the virtual nodes. The weights, $W_t, W_{t-1}, \cdots, W_{t-h}$, are decided by networks training. The improved Jordan neural network is named the output integrated recurrent neural network (OIRNN). The modified RNN simulates high-order non-linear dynamic system.

Based the work above, this paper proposed a novel integrated model, PCA based OIRNN. The model consists of two parts: The first part is PCA from the input layer to the feature layer, while the second part is OIRNN from the feature layer to the output layer including virtual nodes. Model structure is shown in Figure 1. Original input data from the input layer is disdimensioned to the feature layer by PCA. The number of nodes in the feature layer is the number of the principal components. The output values from the output layer are feed back to the virtual nodes. The inner structure of virtual nodes is shown in Figure 2. As shown in Figure 2, a virtual node integrates output states of h time steps. The PCA for the original data space is similar to a network structure. Thus the PCA based OIRNN is in nature an integrated network. The node number of the input layer, the feature layer, and the hidden layer and the output layer are p, m, m_h and q, respectively.

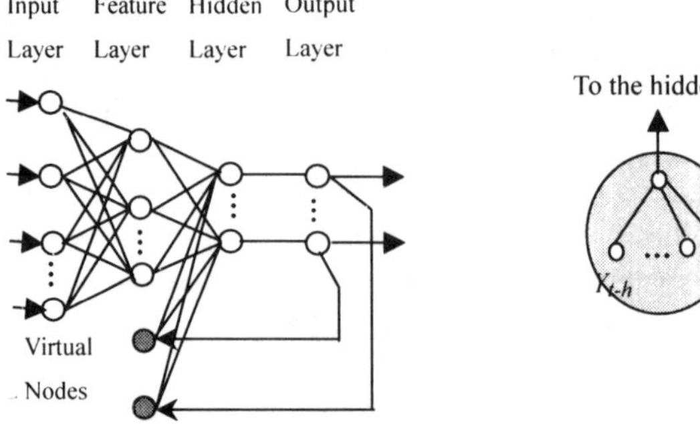

Figure 1.   PCA based OIRNN   Figure 2.   The inner structure of the virtual node

Training algorithm for the PCA based OIRNN is divided into two steps: the first step is to train from the input layer to the feature layer, whose weights are obtained with PCA for the original data space. The second step is to train from the feature layer to the output layer with genetic algorithm.

2. 1. PCA algorithm from the input layer to the feature layer

First, the input data matrix X is standardized. For convenience, the standardized X is still marked as X. The covariance matrix V of X is calculated by, $V = X'X / n$. Then the eigenvalues of V λ_1, λ_2, $\cdots, \lambda_m$ ($\lambda_1 \geq \lambda_2 \geq \cdots \geq \lambda_m$) and responsive eigenvectors a_1, a_2, $\cdots$, a_m are calculated. These vectors are orthogonal.

Principal components F_i $(i=1,...,m)$ are calculated by:

$$F_i = Xa_i = \sum_{j=1}^{p} a_{ij} x_j$$

The number of principal components m is determined by: $Q_m \geq T$. Where T is a preset precision. Q_m is ratio of the m-dimension eigenvector space explaining the original data space:

$$Q_m = \sum_{h=1}^{m} Var(F_h) / Var(x_j), Var(\bullet) \text{ is variance arithmetic operator.}$$

For standard data, $\quad Q_m = \dfrac{1}{p} \sum_{h=1}^{m} \lambda_h$

2.2. The OIRNN algorithm from the feature layer to the output layer

Training of the networks is to minimize the deviation of prediction from the network and the real output. When the real output is $D=[d_1,d_2...d_q]^T$, and the output of model is $Z=[z_1,z_2,...,z_q]^T$, objective function is the prediction error sum of squares (*PRESS*):

$$PRESS = \sum_{i=1}^{n} [\sum_{j=1}^{q} (z_{ij} - d_{ij})^2].$$

Mathematical representation of neural node in each layer is addressed as Table 1:

Table 1　Mathematical representation of neural node in each layer

	Feature Layer $(1 \le i \le m)$	Virtual Nodes $(1 \le i \le q)$	Hidden Layer $(1 \le i \le m_h)$	Output Layer $(1 \le i \le q)$
U	$U_{if} = F_i$	$U_{il} = \theta_{il} + \sum_{j=0}^{h-1} w_{ij} z_{il}$	$U_{ih} = \theta_{ih} + \sum_{j=1}^{m} w_{ij} O_{jf} + \sum_{j=1}^{q} w_{ij} O_{jl}$	$U_{io} = \theta_{io} + \sum_{j=1}^{m_h} w_{ij} z_{ij}$
O	$O_{if} = U_{if}$	$O_{il} = U_{il}$	$O_{ih} = 1/(1 + \exp(-U_{ih}))$	$O_{io} = 1/(1 + \exp(-U_{io}))$

In the Table 1, θ_{ik} is the domain value of the ith node in the kth layer. w_{ij} is the weight from the jth node in the previous layer to the ith node in the following layer. U_{ik} is the input of the ith node in the kth layer. O_{ik} is the output of the ith node in the kth layer.

In the PCA base OIRNN algorithm, principal component analysis from the input layer to the feature layer (PCA) is a linear conversion of the multi-dimensional matrix, while from the feature layer to the output layer (OIRNN) is a non-linear transformation. Computation expenses of PCA are much smaller than that of OIRNN. Thus the PCA based OIRNN not only enables model to simulate the non-linear system, but also reduces the complexity of the process modeling.

3. Case Study

Used as the case study is Tennessee-Eastman process proposed by Downs and Vogel [7], which is currently used widely as a test bed for new approaches and methodologies for process control and optimization. TE process consists of a series of units: a continuous stirring tank reactor (CSTR), a vapor-liquid separator, a stripper, a compressor, a condenser, a reboiler, etc. There are 12 operation variables in TE process. The flow sheet is showed in Figure 3. The data come from a TE simulator by part of the authors' previous work [8].

The product of TE process is sent to downstream refine process. Changes of the product stream Q and composition G outside $\pm 5\%$ are unacceptable. Control objectives are to minimize variability of the product stream Q and the composition G. The proposed PCA based OIRNN is used to simulate the flow rate Q and the change of composition G with the operation conditions under a preset disturbance.

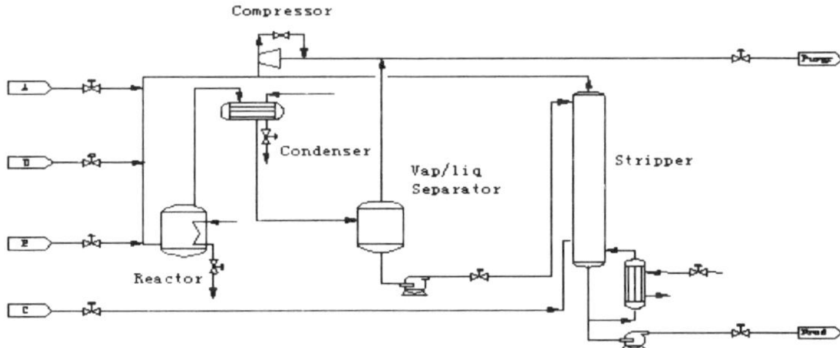

Figure 3.　Flow sheet of TE process

The 12 manipulated variables of TE process are components flow of D, E, A, A+C in the feed stock, compressor recycle valve, purge valve, separator recycle valve, purge valve, separator pot liquid flow, stripper liquid product flow, stripper stream valve, reactor cooling water flow, condenser cooling water flow and agitator speed. There would be rather complex structure if these variables are all used and put into the input nodes of OIRNN. With PCA, 3 new variables (principal components) are found. These principal components explain 95%

message of the original data, as shown in Table 2.

Table 2 The cumulative contribution of principal components (PC) for original data

Principal Component	PC1	PC2	PC3
Cumulative Contribution	0.42220	0.74860	0.94638

PCR, PCA based FFN, and PCA based OIRNN methods are separately used for this dynamic process modeling. The preset dead error is 5. The input node and the output node number of FFN are 3 and 2, respectively. In OIRNN model, there are 3 input nodes, 2 virtual nodes and 2 output nodes. The virtual nodes integrate the 5 output states of past time. Number of the hidden node of the FFN or OIRNN is adjusted to satisfy the dead error. 100 sets of data are used for training and another 100 data are used for test. Table 3 shows the result of model training, where *Er* is the relative mean error.

Table 3.   Models comparison results

	For trained data			For prediction		
	Er(%)		*PRESS*	*Er(%)*		*PRESS*
	G	Q		G	Q	
PCR	0.4132	0.6735	4.753	8.317	6.451	1949
FFN	0.3930	0.3032	4.331	1.321	0.912	48.02
OIRNN	0. 3312	0.3210	3.694	0.7341	0.6168	15.49

Under the preset dead error, the hidden node number of FFN and OIRNN is 30 and 7, respectively. The number of OIRNN parameters is 68, smaller than that of FFN, 182. Comparison of the model testing shows that *Er* and *PRESS* of OIRNN are smaller than those of PCR or PCA based FFN. With its linear characteristics in PCR model, there is big error for this nonlinear system. FFN model doesn't perform as well as OIRNN because of its networks structure. From the dynamic responses of the flow-rate Q shown in Figure 4, PCA based OIRNN shows a good ability to predict the general dynamic trend.

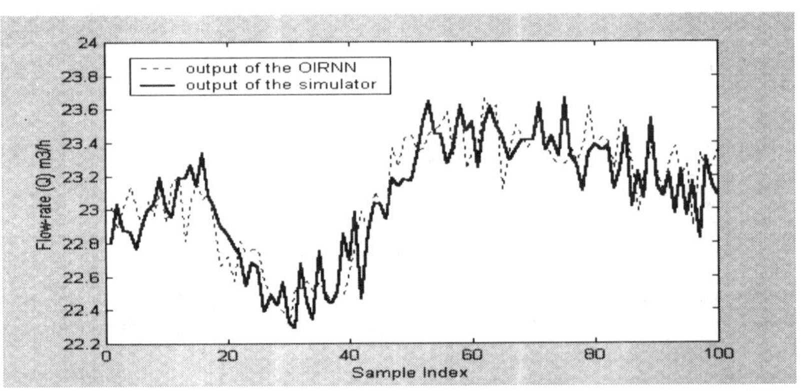

Figure 4.   Prediction of the trend of the flow-rate from the PCA based OIRNN

When developing the model for TE dynamic process, effects of the node number of the hidden layer and the integrated state number on PRESS is investigated. In general condition, the node number of the hidden layer is 2~3 times of the input node number, while the integrated state number of 5~7 is better than enough.

4. Conclusions

Comparing with other modeling methods, the PCA based OIRNN is characteristic of rational and simplified structure. With PCA, redundant information in input data is filtered, and new principal variables are obtained. The number of the principal components is smaller than that of original variables, which reduce computation expenses of OIRNN. By integrating the information of past time, OIRNN is able to simulate the high order dynamic processes. The proposed PCA based OIRNN performs satisfactorily in modeling of the complex dynamic chemical systems.

Acknowledgements

Financial support from the National Natural Science Foundation of China (No. 29976015) and China Major Basic Research Development Program (No. G20000263) are gratefully acknowledged.

References

[1] Qi, Haiyu et al., A hybrid neural network --first principals model for fixed-bed reactor, *Chemical Engineering Science*, 54, 1999, 2521-2526

[2] Qian, Yu et al., Fuzzy rule-based modeling and simulation of imprecise units and processes, *Canadian J. Chemical Engineering*, 77(1), 1999, 186-190

[3] Baffi G. et al., Non-linear projection to latent structures revisited: the quadratic PLS algorithm, *Computers and Chemical Engineering*, 23, 1999, 395-412

[4] Qian Yu, et al., Nonlinear dynamic principal component analysis for on-line process monitoring and diagnosis, *Computers and Chemical Engineering*, 20(Supplement), 2000, 423-429

[5] Zhu Qunxiong, Application of RNN. *Journal of Beijing University of Chemical Technology*. 25(1), 1998, 86-89.

[6] Wu Jianfeng et al., A New Kind of Recurrent Neural Networkssss used in Modeling Chemical Process, *Proceedings of PSE Asia 2000*, 2000, 417-422

[7] Downs J.J. et al., A Plant-wide industrial process control problem, *Computers and Chemical Engineering*, 17(3), 1993, 245-255

[8] Qian Yu et al., An object/agent based environment for the computer Integrated Process Operation System, *Computers and Chemical Engineering*. 24(supplement), 2000, 457-462

KES '01
N. Baba et al. (Eds.)
IOS Press, 2001

Agent-Oriented Dynamic Process Simulation based on Ordinary and Partial Differential Equation Models

Kazunori Matake, Hideyuki Matsumoto and Chiaki Kuroda
Department of Chemical Engineering
Graduate School of Science and Engineering
Tokyo Institute of Technology, Tokyo 152-8552, Japan

Abstract: A main objective is to develop an efficient and precise dynamic process simulation using novel hybrid computing techniques. In development of a hierarchical hybrid simulation system for chemical process design and operation, a new simulation framework, that is an agent-oriented dynamic simulation system combined with the O.D.E model and the P.D.E. model, is proposed. Two kinds of interfaces between the above two communicable model agents are also intelligent communicable agents, which can easily adapt to changes in the simulation circumstances. A specific system was applied to dynamic simulation in a bulk polymerization reactor of polystyrene, and effective and precise simulation was realized in a start-up operation.

1. Introduction

Process information based on computer simulation using various process models is required for many problems in process industries. Appropriate process models make it easy to estimate undetectable information in process equipment. Also in a scale-up stage of process design, easy and efficient tasks can be achieved by using simulation tools, and it is very important to select an appropriate simulating method in the complicated hierarchical modeling framework. A target of this study is to create an effective process-simulation manager whose task would be to coordinate modeling activities composed of hierarchical networks for the purpose of process evaluation. It is very important to clarify that the characteristics information should flow through interfaces among various models.

Considering environmental and safety factors in process industries, the modeling of fluid process with chemical reactions is one of the most important activities. As an example, dynamic phenomena of temperature and conversion (concentration) in a continuous polymerization reactor have been generally simulated by numerical integration of ordinary differential equations concerned with the heat and mass balance. However, the assumption of a perfect mixing tank reactor produces inaccuracy, because spatial distributions of state variables and properties are not neglected in the polymerization reactor [1]. Therefore, the inner volume is required to be divided into some perfect mixing regions and some inhomogeneous regions, and all dynamic data concerned with reaction, heat transfer, mass transfer and mixing are concurrently needed by combining simulation in all regions [2, 3, 4].

Dynamic and spatial phenomena in a process can be simulated by many commercial flow simulators (e.g. FLUENT) based on the P.D.E. (partial differential equation) model.

In this model, vectors of momentum transfer are characteristic data.   This simulation can provide detailed process information, but the simulated area must be restricted within a small region to achieve rapid convergence.   Also, the dynamic simulation based on the O.D.E. (ordinary differential equation) model, i.e. the perfect mixing model, can be generally used using many commercial process simulators (e.g. Hysys).   However, the combined simulation of O.D.E. and P.D.E. models has not been attempted yet by any researcher.   We expect efficient and precise results from the combined simulation in complicated chemical processes.   A main objective of this study is to develop an efficient and precise dynamic process simulation using novel hybrid computing techniques.

2. Agent-Oriented Simulation

Sometimes the hypothesis of perfect mixing doesn't exist in a part of simulating region, where a partial differential equation model is required to know the spatial distributions of variables in detail.   Therefore, the combination of ordinary differential equation models and partial differential equation models is important.   The whole system is simulated using an O.D.E. model, but some parts need to be analyzed in detail using a P.D.E. model.   These two models have to exchange the simulated data through common interfaces that transform boundary data between the two models.   Efficient and precise dynamic simulation is possible using such hybrid computing techniques.

It is important to design an interface between an O.D.E. model without spatial distributions and a P.D.E. model with spatial distributions.   At the interface from the P.D.E. model to the O.D.E. model, it is necessary to integrate various flows, and to average various properties and variables based on the area, the volume and the time.   At the interface from the O.D.E. model to the P.D.E. model, it is necessary to get boundary conditions with spatial distributions through the differentiation based on some assumptions, e.g. the linearity of distributions.   These interfaces need to be intelligent communicable modules, which can easily adapt to changes in the simulation circumstances. **Figure 1** shows such a transferring method of data through the interfaces based on an agent-oriented model.   The time management among four kinds of agents is synchronized by a common clock.

3. Implementation and Simulation

Agent-oriented dynamic simulation was implemented for the dynamic behavior in a bulk polymerization reactor of polystyrene shown in **Figure 2**.   In the simulation, the inner space of the reactor was separated into 15 regions shown in **Figure 3**.   In all regions, the simulation based on each O.D.E. model was executed all together, and at the same time the simulation based on a P.D.E. model was executed in 1 – 7 regions.   In the boundary between the regions 1 and 15, the boundary conditions of the region 1 were replaced as constant values (an assumption) each calculating step based on the calculating results of the region 15.   In the 1 – 7 regions, the calculating results by each O.D.E. model were replaced each calculating results with the average values over the volume obtained from the P.D.E. model.   **Figure 4** shows one cycle of calculation for the iteration time Δt.   Two interface agents installed between the multiple O.D.E. model agents and the P.D.E. model agent (FLUENT) are independent modules, and can easily adapt to changes in the simulation circumstances of region dividing.   In this simulation, the PI controller was also combined. Such calculation and data exchange were continuously repeated to simulate the dynamic behavior of the whole reactor.

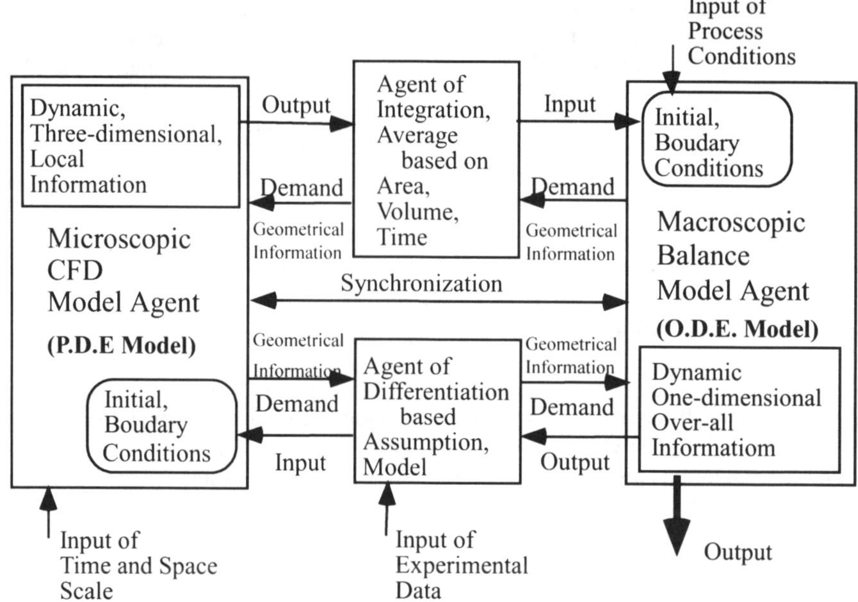

Figure 1   Data transferring through interfaces based on an agent-oriented model.

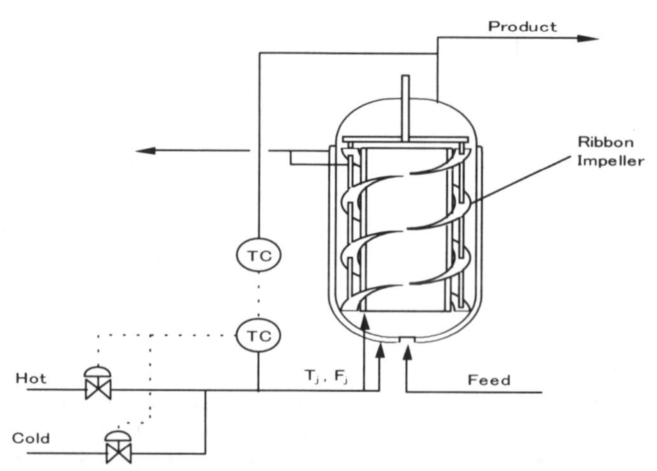

Figure 2 Schematic diagram of a bulk polymerization reactor of polystyrene.

Figure 5 shows the comparison between the above-mentioned agent-oriented dynamic simulation and the single O.D.E. model simulation.   We can find much difference between them, and we think that the result of the agent-oriented dynamic simulation is reasonable from the viewpoint of real operations.   The difference of calculation is caused by the calculative accuracy in each region, 1, 2, 4, 6, namely it means that the effect of the input flow to the region 1 can't be neglected.   It is very important to find such non-uniform regions in the whole system.

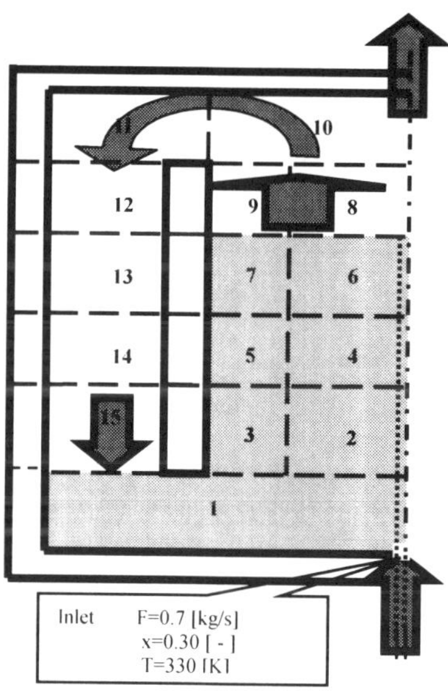

Figure 3   Division of inner space into 15 regions.

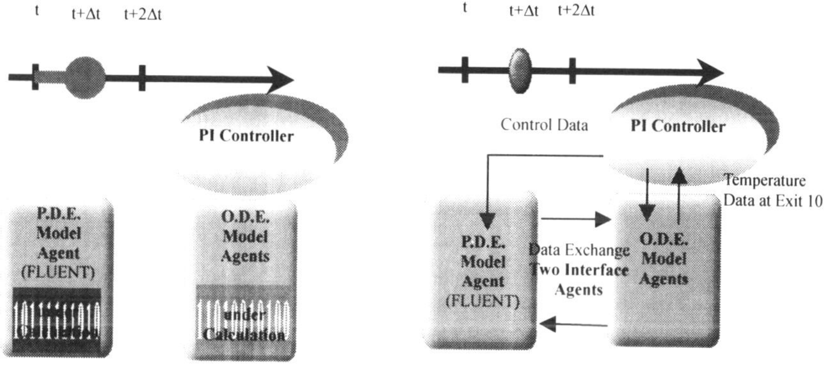

Figure 4 One cycle of calculation and data exchange for the iteration time Δt.

In this study, it doesn't much matter to compare quantitatively the calculation efficiency between the agent-oriented simulation and the single P.D.E model simulation, because it is hard to simulate efficiently the well-mixed complicated behavior in the regions 12, 13, 14, 15 using the P.D.E. model.   And the single P.D.E. model simulation certainly requires far longer time because of far more precise calculation.   The comparison of simulation with the real results is now impossible because it is very difficult to get the real dynamic data experimentally inside a bulk polymerization reactor.   As already mentioned about Fig. 5, the normality of control is a criterion for the normal dynamic simulation.

Moreover, this agent-oriented technique can be easily extended to the more complicated multistage polymerization reactors by introducing the concept of multi-agents.

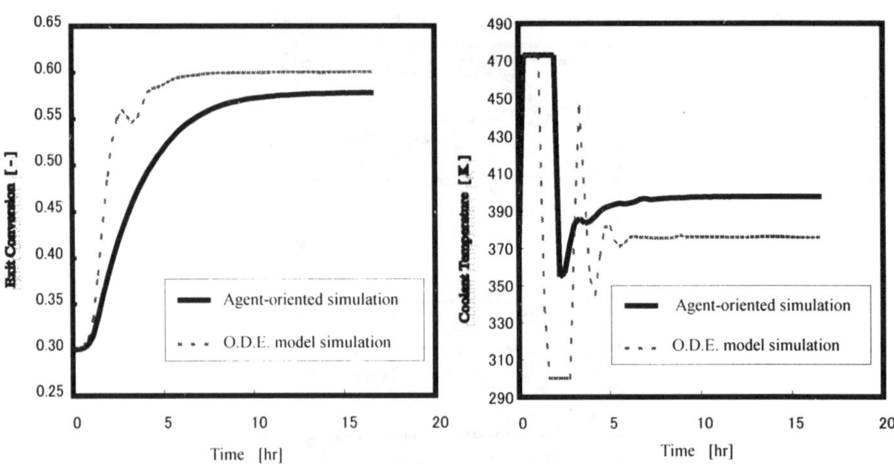

Figure 5 Comparison between the agent-oriented dynamic simulation and the single O.D.E. model simulation.

4. Conclusions

In development of the hierarchical hybrid simulation system for chemical process design and operation, a new simulation framework, that is an agent-oriented dynamic simulation system combined with the O.D.E model and the P.D.E. model, was designed and implemented. In this study, the system was applied to the dynamic simulation in a bulk polymerization reactor, and the efficient and precise simulation was realized. This simulation method is thought to develop more by combining hierarchically with other simulation systems.

Acknowledgements

This work was supported by the project fund from Japan Society for the Promotion of Science. The authors are grateful to Prof. Yuji Naka, who is the leader of our future research project, Mr. Masahiko Abe and Mr. Naoki Kimura, who are graduate students, for fruitful discussions and advices.

References

[1] Tosun, G. and A. Bakker; "A Study on Macrosegregation in Low-density Polyethylene Autoclave Reactors by Computational Fluid Dynamic Modelling", *Ind. Eng. Chem. Res.*, **36**, [2], 296-305 (1997)

[2] Marini, L. and C. Georgakis, "Low-density Polyethylene Vessel Reactors", *AIChE J.*, **30**, [3], 401-407 (1984)

[3] Ochs, S., P. Rosendorf, I. Hyanek, X. Zhang and W. H. Ray, "Dynamic Flowsheet Modeling of Polymerization Processes using Polyred", *Computers Chem. Engng*, 20, [6/7], 657-663 (1996)

[4] Villa, C. M., J. O. Dihora and W. H. Ray, "Effects of Imperfect Mixing on Low-density Polyethylene Reactor Dynamics", *AIChE J.*, **44**, [7], 1646-1656 (1998)

KES '01
N. Baba et al. (Eds.)
IOS Press, 2001

Neural Network Simulation of Formation of Constant Pattern in Column Adsorption

Seppo PALOSAARI and Hajime TAMON
Department of Chemical Engineering
Kyoto University, Kyoto 606-8475, Japan

Abstract. Artificial neural network is applied to simulation of formation of constant pattern in column adsorption, based on the data obtained by solving the column equations. The results show that the neural network is able to analyze the behaviour of the system and to predict the length of column required for establishing the constant pattern.

1. Introduction

One of the practical objectives is to estimate the breakthrough curve which informs about the length of the operation time before the feed starts to appear in the output flow. With a long column this requires a long computing time. Generally, column adsorption is a difficult operation to simulate by solving the fundamental equations because of the number of the phenomena participating. From the engineering viewpoint it is useful to develop an expression which does not require solutions of partial differential equations. In this work the approach is based on the idea that the operation of a column can be divided into two parts. Firstly, there is the period where the constant concentration pattern in coordinates of concentration versus distance from feed point is being established. Secondly, the concentration wave proceeds in the column as a constant pattern the velocity of which is easy to calculate. Therefore, the difficult part is to simulate the formation of constant pattern. It is studied in the following.

2. Basic Equations

Mass balance in an adsorption column for plug flow with dispersion, in constant temperature and pressure, is as follows:

$$\frac{\partial c}{\partial t} + u\frac{\partial c}{\partial z} + \frac{\rho_b}{\varepsilon}\frac{\partial q}{\partial t} = D_L\frac{\partial^2 c}{\partial z^2} \tag{1}$$

where c is the concentration in liquid, D_L is the axial dispersion coefficient, q is the solid loading in the particle, i.e. the concentration of the adsorbent expressed as weight fraction in the particle, u is the flow velocity, ε is the void fraction in the column, and ρ_b is the bed density. In solving the above equation the following general simplifying assumptions are made so that, the particles are of equal shape and size, usually spheres, temperature is constant throughout the bed, void fraction and density of the bed are the same throughout the bed and there is only one component adsorbing.

The term $\partial q/\partial t$ is the mass flux from the surface of the particle into the particle, and it must be equal to the mass flux from the bulk liquid to the surface of the particle:

$$\frac{\partial q}{\partial t} = a_k k_s\left(q_i - \bar{q}\right) = \frac{a_k}{\rho_b} k_f\left(c - c_i\right) \tag{2}$$

where

$$q_i = K_c c_i^{1/e} \qquad \text{(Freundlich adsorption isotherm)} \tag{3}$$

and where q_i is solid loading or "concentration" on the surface of the particle. It is related to the concentration in the liquid film adjacent to the surface. This relationship can be described by the adsorption isotherm as is done above. Further, in Eq.(2) a_k is the total surface of the particles for a volume unit of bed, k_s is the mass transfer coefficient in solid between the surface and the mean concentration in the particle, and k_f is the mass transfer coefficient in liquid between the bulk and the surface of the particle. In Eq.(2) the linear driving force assumption is applied. Then, the diffusion in the particle is expressed in terms of mass transfer coefficient. The relationship between the mass transfer coefficient and the effective diffusivity in the particle is expressed as follows:

$$k_s = \frac{10\,D_s}{d_p} \tag{4}$$

The bed length required for the attainment of constant pattern can be calculated from the solutions of the above equations. This was done by observing the concentration versus location curves in the column, and detecting the curve where the similar shape of two adjacent curves was found. The location where the constant wave was at two-percent relative concentration was then chosen as the location where the constant pattern is attained.

However, solving the partial differential equations and the doing necessary calculations for detecting the attainment of constant pattern is not an easy task. Therefore, in this work it was decided to calculate a number of solutions, and then calculate from the solutions a neural network to describe the length of the bed required for attainment of constant pattern. The work is explained in the following.

Solution of the Column Equation

In order to produce data for finding an expression for the required distance in the column to attain constant pattern it is necessary to solve the column equations presented above. The equations describing the operation of the column, as presented above, were solved by the NUMOL method presented by Schiesser [1]. In differentiation of the transfer equations between the particles and the fluid the mass flux through the surface is calculated by Eq. (2). The surface concentration of the particle is iterated so that the flux from the fluid to the surface, and from the surface into the particle becomes equal. We have verified that the method presented above to gives correct results by comparing the calculated values with values obtained by analytical methods under such conditions where analytical solutions work, i.e. with linear isotherm. The LDF approximation limits the use of the method to cases where the adsorption isotherm is not very non-linear in case of favourable isotherm. When Eqs. (1)-(4) were solved it was found that following eight variables are sufficient to describe liquid adsorption in a column from a water solution, namely $\rho_b \times F_c$, Fe, ε, D_s, D_f, d_p , v_S, and c_{feed}. The eight variables were used as inputs for neural network training, and the respective outputs were the lengths required to attain the constant pattern.

3. Detection of the Formation of the Constant Pattern

The concentration waves in liquid concentration versus location co-ordinates were calculated by means of the width of the mass transfer zone. Near to the feed point the mass transfer zone is narrow, and then widens towards the end of the column. When the mass transfer zone was 95 percent of the end value in such a long column where the steady state could be assumed to be attained the constant pattern was taken to be established. The reason for choosing the 95 percent value was that above it a small increase in the percentage requires a long distance in the column. Then a small inaccuracy in numerical solution causes a large error in column length required for constant pattern.

4. Training of the Neural Network

A three layer artificial feedforward neural network with one hidden layer was used in training. The training method was to minimize the square difference between the values calculated from Eqs(1)-(3) and the values given by the neural network. The values of the weight coefficients were adjusted to attain the minimum which was found by the use of 20 hidden nodes. We applied the method presented by Lanouette et al [3]. In that method different neural networks are trained to fit the data by finding the starting values randomly. The final output in the use of the neural network is calculated as the mean value of the outputs produced by all the networks. Our experience shows that the method is useful and improves the fit in most cases. In some cases the attainable improvement requires only three to five networks whereas in some case the improvent requires up to 15 trainings. In the present case the effect of the Lanouette training was not very good, but it did improve the result as shown in Table 1. The final results accepted to be used in the final simulations were those of the five trainings.

Table 1. The results of the Lanouette training. The validation or test data are such that the network is not trained to those data but predicts the output for them.

NUMBER OF TRAINING	REL. MEAN DEVIATION FOR TRAINING DATA, PERCENT	REL. MEAN DEVIATION FOR VALIDATION DATA, PERCENT
1	3.50	3.87
2	3.17	3.61
3	2.97	3.46
4	2.90	3.43
5	3.00	3.46

With eight inputs the data space has eight dimensions, and if all the inputs induce non-linear behaviour to the output, then the minimum number of training data sets is $3^8 = 6561$. However, in many practical applications change in some of the inputs induces linear change in the output. Then the number of data sets needed becomes much lower. Assume that all of the inputs induce a linear change in output. Then the minimum number of data sets needed may be approximated to be $2^8 = 256$. There were eight input variables, as shown in Table 2. Training was carried out by 2900 training data sets with 928 test data sets. Though this is not a sufficient number of data to describe the behaviour of the output in case of all inputs showing a non-linear effect on the output, it was found that in this that six of the variables had a linear or nearly linear effect on the output. Therefore the 2900 data sets were considered sufficient.

Table 2. The variables and their ranges used in training the neural network.

$\rho_b \times F_c$	Fe	ε	D_s	v_S	d_p	D_f	c_{feed}
300–660	0.25–0.7	0.4–0.55	1×10^{-12} – 1×10^{-10}	0.4×10^{-3} – 0.7×10^{-3}	0.5×10^{-3} – 1.25×10^{-3}	0.8×10^{-9} – 1.2×10^{-9}	0.05–5.0

Table 3. The results of neural network training of the 2900 training data sets, and 928 test data sets.

NUMBER OF HIDDEN NODES	TRAINING ERROR, PERCENT	TEST DATA ERROR, PERCENT
23	3.24	3.67
20	3.00	3.46
17	3.41	3.89

The result of neural network training are shown in Table 3. The result is not very strongly dependent on the number of hidden nodes but in tests carried out twenty hidden nodes produced the lowest test data error. Consequently, the 20 hidden node neural network was chosen to be used in the present work. Figure 1 shows the training results. Though the result is not very accurate it gives sufficient basis to study the relationships between the variables.

Table 4. The values of the parameters used in Figs. 2-3.

$\rho_b \times F_c$	Fe	ε	v_s	d_p	D_f	c_{feed}
500	0.45	0.45	0.6×10^{-3}	0.001	1×10^{-9}	1.0

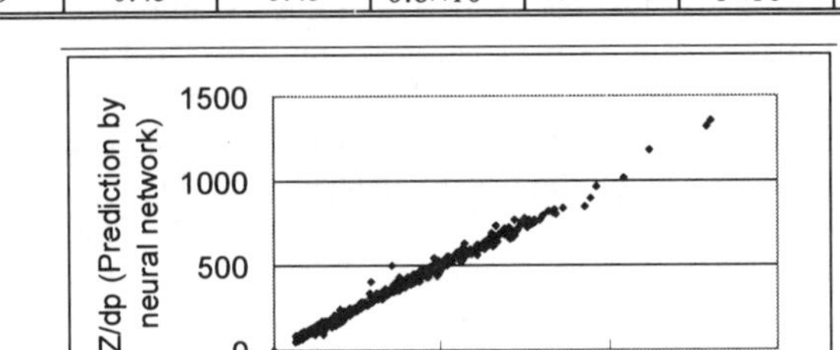

Figure 1. Neural network for the length required for the establishment of constant pattern expressed as z/d_p where d_p is the particle diameter. The data are those presented in Table 3. The number of hidden nodes was 20.

5. Simulation by Neural Network

The neural network obtained was then used in simulation where the effect of each variable is presented in turn over the whole interval of calculation. Because the particle diffusivity has a non-linear effect, it was chosen to be the parameter in the calculations. The results for the two non-linear input-output relations are shown in Figs. 2-3. The values of the parameters are shown in Table 4. The information given Fig. 3 is now explained in more detail. The curve in the figure is produced by the neural network. The horizontal lines are calculated by the low accuracy used in producing the input data for the neural network. The triangles are calculated by nearly accurate solution of Eqs. (1)-(4). Solution of these equations tended to take computing time because great accuracy was needed for the reason that the attainment of constant pattern was detected as a difference between consecutive concentration waves along the length of the column at each time moment. The inaccurate solution, to produce the 3828 solutions used here, took 14 days of CPU time with 600 MHz Pentium III processor. The nearly accurate solution would have taken over 50 days CPU time. The number of node points along the column was 1000 for the 1.6 meter column in the inaccurate solution. The nearly accurate solution required 4000 nodal points. The figure shows that the inaccurate solution is sufficient in accuracy because the error in neural network prediction is greater than the difference between the accurate and inaccurate solution. It appears that the determining factor to high neural network accuracy is a great number data sets. As the conclusion it may be said that if solution of fundamental equations are used as inputs to neural network then it is advisable to produce a great number of relatively inaccurate solutions.

The results show that the effect of all other parameters except the Freundlich exponent and the particle diffusivity has a linear or nearly linear effect on the Z/d_p. The result is unexpected. Furthermore, the results show the strong effect of the particle diffusivity when it approaches the value of 1×10^{-12}. Then, the principal mass transfer resistance is in the particle.

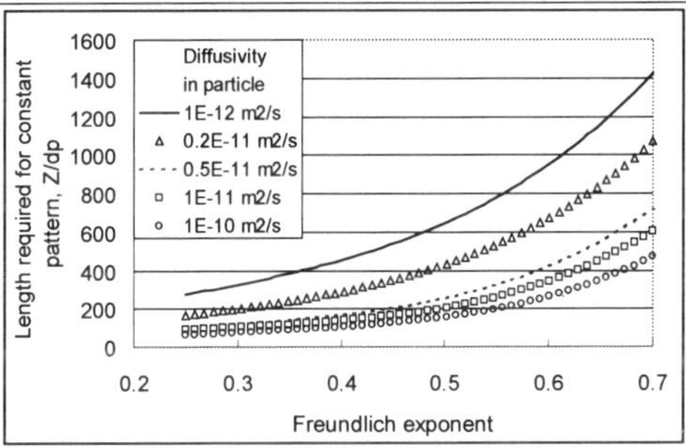

Figure 2. The effect of the Freundlich exponent on the column length required to attain constant pattern.   The values of the parameters used are shown in Table 3.

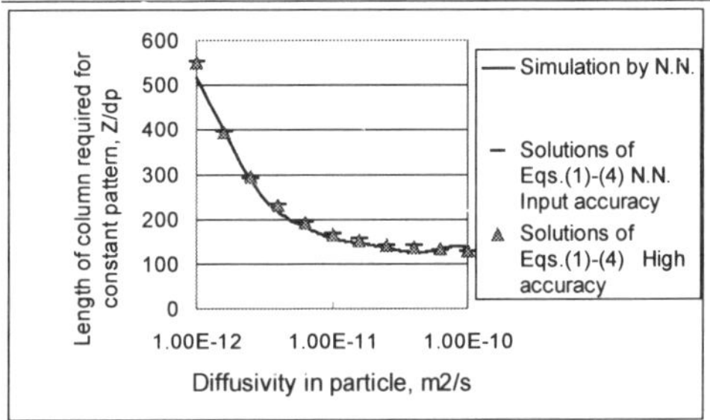

Figure 3. The effect of the particle diffusivity on the column length required to attain constant pattern. The values of the parameters used are shown in Table 3.

6. Conclusions

The above results show that neural network can be applied successfully in finding out the behaviour of the adsorption system, namely the column length required for constant pattern to become established. With a simple simulator using the neural network it was easy to find relationships in the system.

The above results show that out of the eight parameters chosen to describe the behaviour of the column, the effect of six variables is nearly linear or linear to the length required to attain the constant pattern whereas the Freundlich exponent and the particle diffusivity display a very non-linear effect on the length required to attain the constant pattern. For this reason with adsorbent of high value of the Freundlich exponent, i.e. above 0.5, and with low value of the diffusion coefficient in the particle, near to 1×10^{-12} m^2/s, the adsorption column should be long. Adsorption may not be a practical operation in those cases.

References
[1] W.E. Schiesser, *The Numerical Method of Lines*, Academic Press, San Diego, 1991
[2] W.H. Press, B.P. Flannery, S.A. Teukolsky and W.T. Vetterling, *Numerical Recipes,* Cambridge University Press, Cambridge, USA, 1987.
[3] Robert Lanouette, Jules Thibault and Jacques L. Valade, *Process modeling with neural networks using small experimental datasets*, Computer and Chemical Engineering 23(1999)1167-1176.

KES '01
N. Baba et al. (Eds.)
IOS Press, 2001

Design of a pH Controller using a Neural Network Optimized by Genetic Algorithm

Masahiko Abe, Hideyuki Matsumoto and Chiaki Kuroda

Department of Chemical Engineering

Graduate School of Science and Engineering

Tokyo Institute of Technology

2-12-1 Ookayama, Meguro-ku, Tokyo 152-8552 Japan

Abstract: In new design of chemical processes, it is required to design both a reactor and a control system concurrently based on a few process informations. Applications of neural networks (NN) are very attractive in control of the non-linear chemical reactors. Here it is considered that genetic algorithm (GA) is a powerful tool in optimizing NN controllers without teacher data.

The purpose of this study is to investigate on a GANN (NN optimized by GA) pH control system from the viewpoint of both control performance and flexibility. In results, there is found an appropriate value as to the number of input time series data and hidden units. GA can automatically optimize the number of hidden units, and therefore GANN is effective in design of powerful NN controllers.

1. Introduction

In new design of chemical processes, it is required to design both a reactor and a control system concurrently in view of safety and efficiency of design. There are, however, many unknown data of flow conditions, disturbances, etc. in a reactor at the initial stage of design. Therefore, it is desirable to design a control system using a process simulator that models dynamic behaviors of the reactor.

Applications of neural networks (NN) are very attractive in control of non-linear chemical reactors. The typical optimization algorithm of NN is a back propagation (BP) method using teacher data, however it is difficult to get teacher data in the initial stage of design. Here it is considered that genetic algorithm (GA) is a powerful tool in optimizing NN without teacher data. On the other hand, it is important to determine the structure of NN such as the characteristic and the number of neural units in order to get good performance of control and its flexibility for various requirements [1]. GA gives a solution for also this problem.

The purpose of study is to investigate on a GANN (NN optimized by GA) pH control system, because pH control is one of the most popular non-linear problems [2]. Here we think that the NN structure has a significant effect on the performance and the flexibility of non-linear control.

2.　The pH Control System

As shown in Figure 1, a neutralization process is a continuous stirred tank reactor (CSTR) with reactions among acetic acid (AcH), propionic acid (PrH) and sodium hydroxide solution (NaOH *aq.*). C_{0ACH}, C_{0PrH} and C_{0NaOH} are the inlet concentrations of AcH, PrH and NaOH *aq.*. Its process simulator was developed referred to the previous study[3], where the reaction kinetics showed the strong non-linearity.

Figure 1 shows the whole schematic diagram of this pH control system. The NN used is a three-layered network, which has 6-10 input units, 1-10 hidden units and an output unit. Some time series data of operational, controlled and set variables are supplied into the input layer. Output of NN is the value of the next operational variable. The weights of connections and the number of units are off-line optimized by GA based on the evaluation of controlling results.

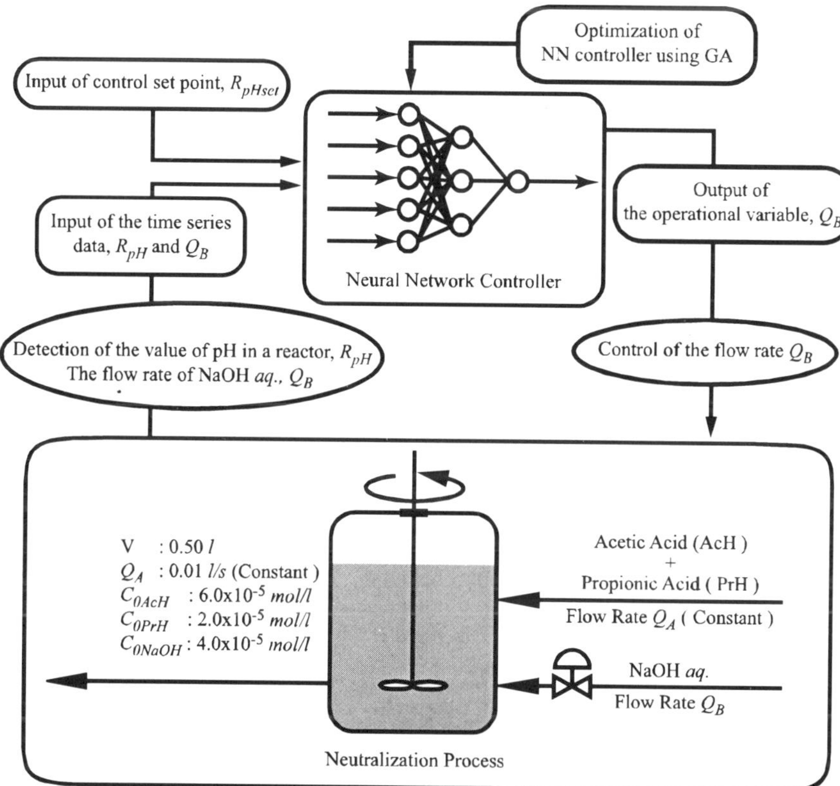

Figure 1. The structure of pH control system

3.　Optimization of NN Controllers

In order to optimize the connection weights using GA, the weight is encoded to four binary bits, which make a basic group of gene. An individual (a NN) is consisted of some basic groups of gene. The optimization of NN is carried out by mutation, crossover, etc. among multiple individuals [4]. GA optimizes the group of individuals, and the best NN is used as a pH controller.

In this study, each NN is evaluated by the control performance. The fitness function is defined as follows;

$$f = \frac{1}{rms} \qquad (1) \qquad\qquad rms = \sqrt{\frac{\sum_{1}^{n}\left(R_{PHset}(n) - R_{pH}(n)\right)^2}{n}} \qquad (2)$$

where f is the fitness of a NN, $R_{pHset}(t)$ is the control set point at time t, $R_{pH}(t)$ is the value of pH in the reactor and n is the number of sampling data. In this study, the test time is 1000 seconds and the sampling time is 2 seconds.

4. Results and Discussion

In this study, the following two problems are mainly investigated, that is the effects of the number of input and hidden units on control performance and flexibility.

In the simulations for evaluating flexibility, the following disturbances and time-delay were added to the detected signals.

- The error for Q_A and Q_B ... ±1%.
- The time-delay of Q_B ... 2 seconds.
- The error for R_{pH} ... ±2%.

In order to evaluate the performance and flexibility of a NN, the total utility function is defined as follows;

$$U = \left(\frac{u_f}{u_{f,\max}}\right)^{W_1} \left(\frac{u_g}{u_{g,\max}}\right)^{W_2} \qquad (3)$$

where u_f is the average value of f for applicable control samples, $u_{f,max}$ is the maximum of u_f, u_g is the number of applicable control samples, $u_{g,max}$ is the total number of control samples (225), W_1 and W_2 are the weights of the control performance and flexibility. In this study, W_1 and W_2 are fixed 0.4 and 0.6 respectively.

4.1   Effects of The Number of Input Units

A NN has three kinds of process variables as inputs, that is R_{pH}, Q_B and R_{pHset}. In this study, the numbers of R_{pH} and Q_B time series data are fixed three. The effects of the number of R_{pHset} time series data are investigated by changing from zero to four as shown in Table 1.

Figure 2 shows a result of averaged U for the number of R_{pHset} time series data by twenty NNs with the same input structure. These twenty NNs were optimized based on test controls of R_{pHset} = from 7 to 8. It was found that the averaged U showed the maximum value in case of three time series inputs, and that it was important to determine proper treatment of time series data.

Table 1. The number of time series input of the NN

The number of R_{pHset} time series data	Time series data
0	(*None*)
1	$R_{pHset}(t+1)$
2	$R_{pHset}(t+1)$, $R_{pHset}(t)$
3	$R_{pHset}(t+1)$, $R_{pHset}(t)$, $R_{pHset}(t-1)$
4	$R_{pHset}(t+1)$, $R_{pHset}(t)$, $R_{pHset}(t-1)$, $R_{pHset}(t-2)$

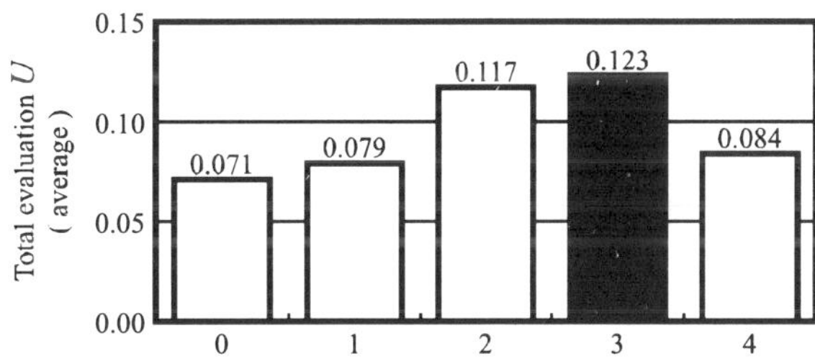

Figure 2. A result of the evaluation of control performance and flexibility

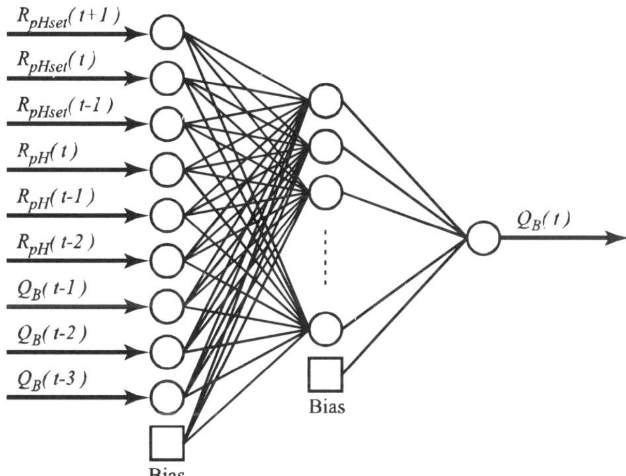

Figure 3. The structure of a NN controller investigated in section 4.2

4.2   Effects of The Number of Hidden Units

In this section, a NN as shown in Figure 3 was used to investigate on the effects of hidden units. The number of hidden units was changed in one, seven and ten. Each control result for each hidden unit (1, 7, 10) was shown in Figure 4 (a, b, c) respectively. There are an appropriate number of hidden units for stable control, and it is about seven in this study. On the other hand, we tried validation tests using NN with the different number of hidden units and in result, the less the number of hidden units become, the more adaptability appears.

GA can automatically optimize the number of hidden units, and therefore GANN is effective in design of powerful NN controllers.

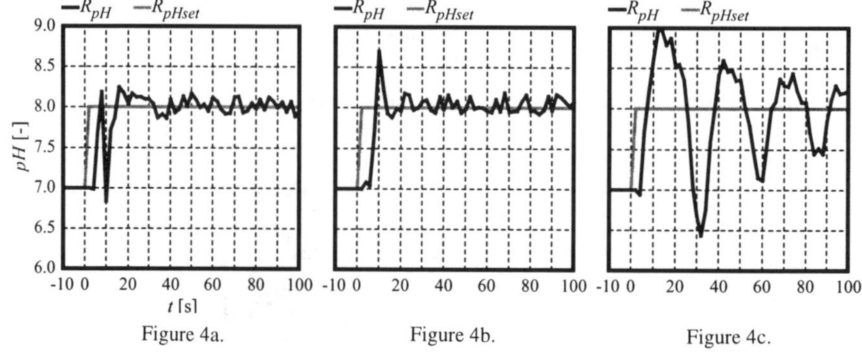

Figure 4a.   Figure 4b.   Figure 4c.

Figure 4. Some results of a pH control

5. Conclusions

A GANN pH controller system was studied from the viewpoint of both control performance and flexibility. In results, there was found an appropriate value as to the number of input time series data and hidden units. GA can automatically optimize the number of hidden units, and therefore GANN is effective in design of powerful NN controllers. However, the problem about selecting the number of time series data is left in future.

Acknowledgements

This work was gratefully supported by a Grant-in-Aid for Scientific Research (No. 10555263) from the Ministry of Education, Science and Culture of Japan.

Reference

[1]   Dirion,J.L., M.Cabassud, M.V.Le Lann and G.Casamatta; "Design of a Neural Controller by Inverse Modelling", *Computers & Chemical Engineering,* **19**, Suppl, S797-S802 (1995)

[2]   Behera, L. and K.K.Anand; "Guaranteed Tracking and Regulatory Performance of Nonlinear Dynamic Systems Using Fussy Neural Networks", *IEE Proceedings: Control Theory & Applications,* **146**, 5, 484-491 (1999)

[3]   Palancar,M.C., J.M.Aragon, J.A.Miguens and J.S.Torrecilla; "Application of a Model Refernce Adaptive Control System to pH Control. Effects of Lag and Delay Time", *Industrial & Engineering Chemistry Reseach,* **35**, 11, 4100-4110 (1996)

[4]   Abe,M., H.Matsumoto and C.Kuroda; "An Artificial Neural Network Optimized by a Genetic Algorithm for Real-time Flow-shop Scheduling", *International Conference on Knowledge-Based Intelligent Engineering Systems,* **Proceedings,** 1, 329-332 (2000)

Expert Systems for Controlling a KCP (Kureha Crystal Purifier)

K. Otawara

Kureha Techno Eng, 135 Ochiai, Nishiki, Iwaki, Fukushima 974-8232, Japan

Abstract. A KCP (Kureha Crystal Purifier) continuously produces pure crystal from crude crystal. KCPs have been practiced around thirty years in chemical industries. Even though the KCP requires little operation, it had not been automated previously because its dynamic model was not obtainable. Various complicated phenomena are involved in the KCP, and it is extremely difficult to obtain its dynamic model. Therefore, an expert system is adopted for automating the KCP. The system requires no operation and the crystal yield has been improved by a few percent.

1. Introduction

Kureha Crystal Purifier (KCP) is one of systems for continuous fractional crystallization. This KCP has been practiced more than thirty years for purifying crystalline chemicals. It features high purity, high yield, energy saving, little operation, little maintenance, small installation area, long stable operation, low capital cost, short start-up time, no scaling or clogging in the column, continuous process, flexible selection of a crystallizer which can be independent of the KCP, etc. It has been applied to industries as listed in table 1.

2. Principle of a KCP

Figures 1 and 2 illustrate the schematic diagram and structure of a KCP, respectively. The KCP mainly consists of columns, screw conveyors, a slurry feeder, and a melter. The crude crystal is fed at the bottom of the column and pushed upward by double screw conveyors. In the course of being conveyed upward, the crude crystal contacts countercurrently with the

Table 1　KCP's operated in chemical plants

Year [y]	Diameter [inch]	Unit	Capacity per unit [T/y]	Product	Melting point [°C]	Purity [%]
1969	18	1	4,600	*p*-dichlorobenzene(DCB)	53	99.998
1974	18	1	4,600	*p*-DCB	53	99.998
1978	18	2	1,700	Bromine compound	83	99.99
1980	20	1	5,800	*p*-DCB	53	99.998
1987	20	1	5,800	*p*-DCB	53	99.998
1991	20	2	5,000	Unsat. carboxylic acid	16	99.99
1994	20	1	5,800	*p*-DCB	53	99.998
1997	22	1	6,800	*p*-DCB	53	99.998
1998	30	2	12,800	*p*-DCB	53	99.998
1998	30	1	10,000	Unsat. carboxylic acid	16	99.99
2001	3	1	N/A	Multi purpose	N/A	N/A

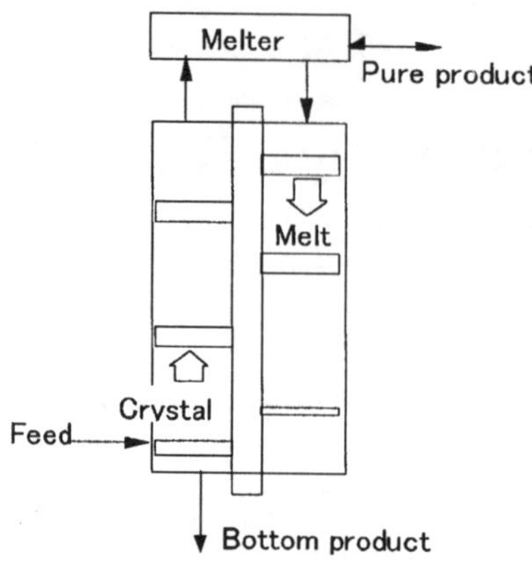

Fig. 1. Schematic diagram of a KCP.

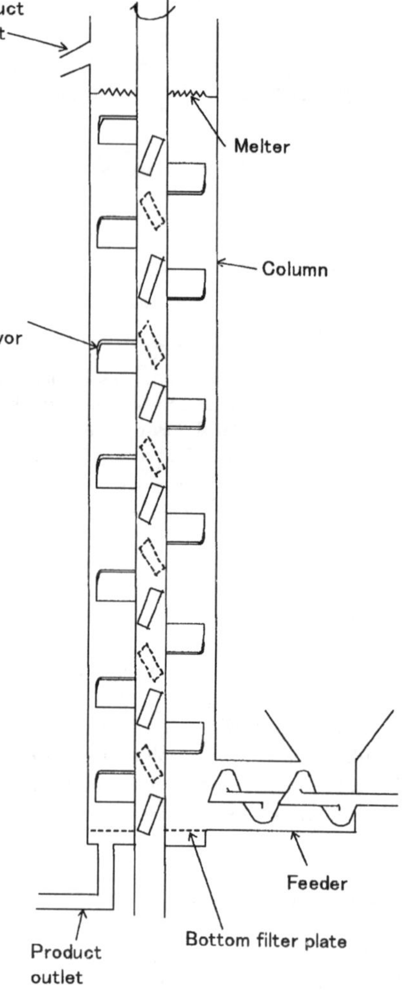

Fig. 2. Structure of a KCP.

reflux melt: This melt is produced by the melter at the top of the column and trickles downward among the crude crystal. Then, the crude crystal from the bottom is washed by this melt and sweats. The upper part of the column, the purer of the crystal. Eventually, the crystal is purified at the melter. Hence, the pure product exits as crystal from the top of the column.

3. Mathematical Model of a KCP

As elucidated in the previous section, phenomena in the KCP is extraordinary complicated. For scaling up the KCP, the following two phase counter current model has been adopted as the steady state model.

$$E_x \frac{d^2 C_x}{dz^2} + V_x \frac{dC_x}{dz} + kaC_y = 0 \tag{1}$$

$$E_y \frac{d^2 C_y}{dz^2} - V_y \frac{dC_y}{dz} - kaC_y = 0 \tag{2}$$

and the boundary condition for its open system is expressed as

$$-E_x\left(\frac{dC_x}{dz}\right)_{z\to L} = V_x(C_{x_{z\to L}} - C_{xT})\quad z=L \tag{3}$$

$$\frac{dC_x}{dx} = 0 \qquad z=0 \tag{4}$$

$$-E_y\left(\frac{dC_y}{dz}\right)_{z\to 0} = V_y(C_{yB} - C_{y_{z\to 0}})\quad z=0 \tag{5}$$

$$\frac{dC_y}{dy} = 0 \qquad z=L \tag{6}$$

where C, V, E, k_a, x, and y are concentration of impurities, velocity, axial dispersion coefficient, sweating rate, liquid, and solid, respectively [1-2]. This model already simplified many aspects of phenomena in the KCP: For example, temperature profile is not taken into account. Unsteady state modeling had also been tried to derive before, but it appears excessively difficult. Even if it is possible, to use such a model for controlling the KCP may not be practical on the online realtime operation. Thus, unsteady state model for a KCP is not available and it appears vital to resort to an expert system for controlling the KCP.

4. Expert System for Controlling a KCP

As mentioned earlier, a KCP requires relatively few operators. Nevertheless, to maximize the productivity from KCP's installed to purify para-dichlorobenzene (p-DCB) in Kureha Nishiki Factory, an expert system was developed in 1990. G2, a real-time expert system developed by Gensym in U.S.A., has been adopted as a tool for the expert control. The values of sensors are transmitted from a Distributed Control System (DCS) to a Sun workstation (SPARC Station 1) via modems every minute. A KCP is controlled by the heat fgiven to the KCP, specifically, by the flow rate and temperature of the hot water supplied to the melter placed at the top of the KCP. The set points of the melter flow rate and temperature are determined by the expert control system and they are transferred back to the DCS. An example of the user interface of the system is illustrated in Fig. 3.

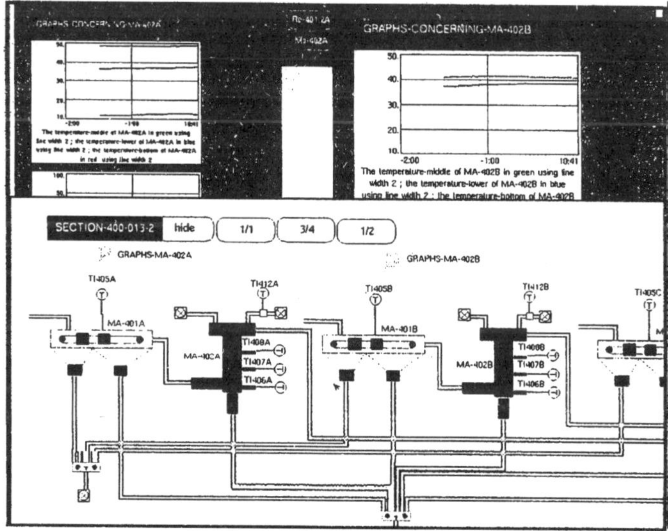

Fig.3.Example of the expert control system.

The principle of controlling a KCP is based on two factors: One, if heat given to the melter of the KCP is large, then the crystal yield is small. Two, if heat is small, then product crystal is impure. The state of the KCP can be empirically estimated by the temperature sensors placed in the axial direction. Therefore, the heat can be optimized by the temperature profile along the axis of the KCP. The rules implemented in the expert control system are mainly divided into five groups:

Group 1) Determine the melter flow rates, if necessary.
Group 2) Check if the state of the KCP improved after the flow rate had been changed according to group 1: If improved, do not invoke group 1, otherwise, invoke Group 1.
Group 3) Determine the melter temperature based on the state of the KCPs and the melter flow rates: Since all KCPs share a hot water supply system, changing the melter temperature affects all KCPs.
Group 4) Determine the melter flow rates when the operating conditions of the equipments prior to KCPs change.
Group 5) Offer operational guidance.

Group 1 or 2 is applied to a KCP, alternatively. The interval of invoking rules is one minute. These rules have been obtained through several interviews with an expert operator. After implementing the rules on G2 and the expert system was installed in the controlling room of the factory, the rules have been tuned up by gradually adopting the system.

The expert control system has totally automated KCPs. Since then, no manual operation has essentially been required except critical situations. Furthermore, the yield of p-DCB from a KCP has been improved by a few percent and KCPs are more stable than before: The system watches and controls KCPs every minute, which is naturally impossible for human operators, thus rendering possible to minimize the heat given to the melters of the KCPs. The system is also reliable. So far, only hardware problems, but not G2, have caused the system to halt.

G2 has proven to be very effective to develop an expert system. Before introducing G2, another tools based on LISP language were adopted at earlier stage of the development. These tools did not have online realtime functions, therefore, building the expert control system was very tedious and time-consuming. G2, however, has online realtime functions. It also features natural language programming, user friendly graphical tools, object oriented system, intelligent editor, etc. These programming environments have saved much effort involved in the development of the expert system.

It was attempted to control a KCP with conventional feedback controllers, previously. It was found that these controllers were not adequate for controlling a KCP since the time response of the KCP is excessively long.

Before developing the expert control system, the methods for controlling a KCP had depended on individual operators. Now, the rules for controlling a KCP are written explicitly and the operators share the rules. Thus, the expert system can also be an excellent tool to explicitly derive empirical knowledge.

5. Concluding remarks

A KCP is one of systems for continuous column fractional crystallization. This KCP has been practiced more than thirty years for purifying crystalline chemicals. Even though the KCP requires little operation, it was not automated before because its dynamic model was not readily obtainable. Thus, an expert system is adopted for automating the KCP. G2 and SPARC Station 1 have been adopted as software and hardware, respectively. This system requires no operation any more and the crystal yield has been improved by a few percent. G2 has proven efficient for developing online realtime applications.

References

[1] T. Miyauchi and T.Vermeulen, *Ind. Eng. Chem. Fundam.* **2** (1963) 113.

[2] K.Otawara, T. Matsuoka, and S. Saito, Proc. Int. Sym. Industrial Crystallization: an overview of the present status and expectations for the 21st century, Tokyo, September 17-18, 1998, pp.598-605.

KES '01
N. Baba et al. (Eds.)
IOS Press, 2001

Recognition of Finger Character
or Sign Language

Mowey CHIN[1], Yoshinori ADACHI[1], Masahiro OZAKI[2] and Naohiro ISHII[3]
[1]Chubu University, [2]Nagoya Women's University, and [3]Nagoya Institute of Technology
[1]1200 Matsumoto-Cho, Kasugai, Aichi, Japan 487-8501   adachiy@isc.chubu.ac.jp

Abstract: A proposed method using LUV color space and ellipse extraction was an extension of the method proposed by Adachi et al. Furthermore, by introducing compensation process, the extraction improved quite much. Extracted hand images were used to make an automatic finger character recognition system and got an adequate accuracy about 80% recognition ratio.

1. Introduction

In the field of man-machine interface, besides the voice, the gesture is also important communication tool. In this study, as the part of the sign language recognition, the finger character was noticed, and the automatic recognition of the finger character was examined. The sign language is not only universally common language, but also difficult for the healthy person to obtain the skills to communicate through the sign language efficiently. Furthermore, it is very difficult to transmit ones will and thought to the others who completely does not know the sign language. It seems to be one of the important themes to make a interpretation system of the easy finger character for the complete understanding. On the sign language recognition, the many researches have been made[1]-[4]. However, in many cases, the data glove was used or the special situation that the shape of the hand was easy to extract had been set. There seldom exist researches using the automatically extracted hand images for the sign language recognition.

In this study, in order to carry out the recognition without being dependent on the background, as a pretreatment of the recognition, the new method to extract the flesh color region was proposed by improving the technique in the previous study. And the matching between the extracted hand and dictionary of the finger character was examined.

2. Proposed method

The proposed method mainly consists of two parts. The extraction division and the recognition division are those. The flesh color image (image of the hand) is extracted in the extraction division and the matching is carried out with the dictionary in the recognition division.

2.1 Extraction of flesh color regions

In the previous study, we proposed the extraction method by using the ellipse region on the UV plane in the LUV space, and showed that the better results were obtained comparing with the existing methods[5]. In this study, further improvement was added on the basis of the previous technique. Concretely, the flesh color region is extracted by next procedure.

(1) To begin with, the RGB color system is converted into the XYZ color system, and in addition, it will convert into the LUV color system. The images of the hand of 133 sheets photographed under various conditions were projected and were superimposed in the UV space in order to limit appearance region of the flesh color. The part of the images is shown in **Figure 1**.

The result of superimposing flesh color appearance region is shown in **Figure 2**. As shown in this figure, it is proven that the flesh color is distributed in the specific region in the UV plane, and it is proven that the flesh color region may be extracted by limiting to this region.

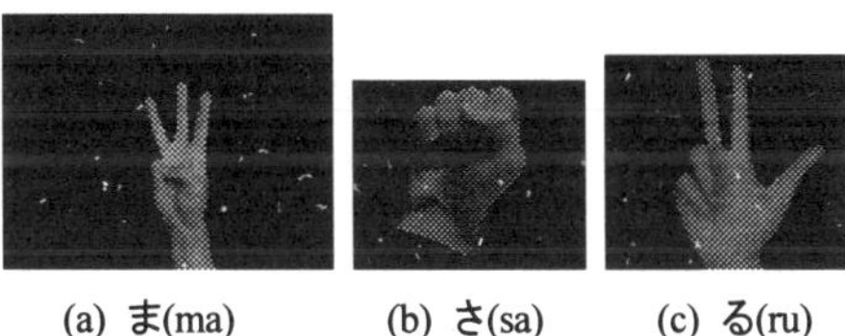

(a) ま(ma)　　　(b) さ(sa)　　　(c) る(ru)

Figure 1. The part of sample image of the 133 sheets used in order to limit the flesh color appearance region.

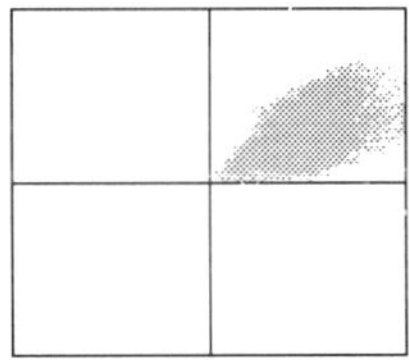

Figure 2. The region in UV plane which superimposed the 133 flesh color images.

(2) The target image is projected in the UV plane, and the peak in the flesh color appearance region decided in (1) was searched, and it was made to be a candidacy for the flesh color region that should be extracted by the ellipse centered around the peak. The ellipse is decided here as the following.

a. In search of pixel distribution on the straight line that connects the peak on the UV plane of flesh color with the origin, the boundary with the region less than 0.05 times the peak value continues is found, and it is made to be the length of the major axis.

b. In search of pixel distribution on the straight line that is orthogonalized to straight line that connects the peak with the origin and passes through the center of the major axes, the length of the minor axis is obtained in the same way.

c. The center of the ellipse is decided as the center of the minor axis and the lengths of the major and minor axes are already obtained in the above.

Decided ellipse is shown by the following equation.

$$\left[\frac{(U-P)\cos\theta+(V-Q)\sin\theta}{A/2}\right]^2+\left[\frac{-(U-P)\sin\theta+(V-Q)\cos\theta}{B/2}\right]^2=1 \qquad (1)$$

where (P, Q) is the center of the ellipse and θ is a slope of the line connected the peak and the origin. A and B are the lengths of the major and minor axes respectively.

Using the image of finger character "い(i)", decision of the ellipse and example of the extracted flesh color image are shown in **Figure 3**.

(3) In extraction result of the flesh color region by the ellipse, deficiency and excess exist in the extraction result. This will be corrected as following.

a. The pixel in the ellipse region is projected in the L space. And by leaving region with large peak, small peak is deleted. This can mostly remove the close color noise from the flesh color. This example is shown in **Figure 4**.

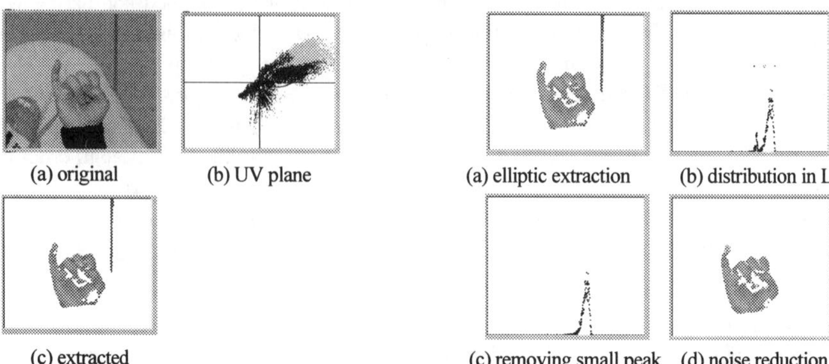

(a) original   (b) UV plane   (a) elliptic extraction   (b) distribution in L

(c) extracted   (c) removing small peak   (d) noise reduction

Figure 3. Example of elliptic extraction of flesh color region.   **Figure 4**. Example of noise reduction by removing small peak.

b. The boundary of the hand is obtained by the Sobel method, and the flesh color part is expanded to the boundary in order to compensate the part that is lacking in the flesh color region. The search direction of the boundary is made to be 8 directions of the 45-degree interval. The results of the boundary image got by the Sobel method and compensated flesh color image are shown in **Figure 5**.

(4) The margin part of extracted and corrected image is deleted, and the region is narrowed. The processes of (1) to (5) obtain the image of the hand. The example of the image got in this way is shown in **Figure 6**.

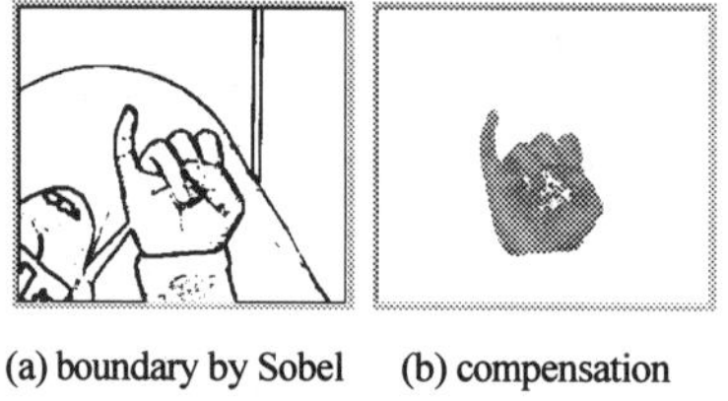

(a) boundary by Sobel   (b) compensation

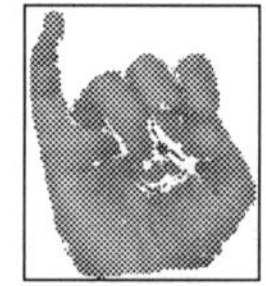

Figure 5. Example of boundary image by Sobel method and compensated flesh color image.   **Figure 6**. Example of extracted image using for matching.

2.2 Recognition of finger character

Comparing the agreement with prepared dictionary, the recognition of the finger character carried out. According as the shape of the finger character, the horizontal-vertical ratio of the region of the dictionary is different, and the unknown image was adjusted to the dictionary for the matching, the scaling was done. The procedure of recognition process is described in the following.

(1) As a dictionary, the each finger character image is photographed, and the image up to the wrist that manually removed the background is registered. In order to decrease the margin utmost, the region outside the extracted and corrected image has deleted. Therefore, the horizontal-vertical ratio is different for the every finger character. The example of the finger character of the dictionary is shown in **Figure 7**.

(2) The similarity between a dictionary of 41 finger characters and extracted and compensated unknown flesh color image was calculated for matching.   As usual, inner product was used as the similarity evaluation function, and the dictionary that took the largest value indicated the finger character.   The similarity is obtained by the following equation

$$\eta = \frac{\left(\vec{a}\Box\vec{b}\right)}{|\vec{a}||\vec{b}|} \qquad (2)$$

where η is the similarity value. $\vec{a}$ and $\vec{b}$ are dictionary and extracted images.

(3) Before the recognition, in order to obtain the summary value of the ratio of the scaling for the matching, the images of hand vertical position and horizontal position were input, and the magnification ratio is calculated.

(4) Extracted rectangle is adjusted to which length in length and width of the dictionary. Adjusting edge is the edge that takes small value of the magnification ratio, and it is correspondent to the part that designated expanding extraction image in the every finger character of the dictionary, and the similarity is calculated.   This aspect is shown in **Figure 8**.

(5) The finger character in which the agreement takes largest value is made to be the recognized result.

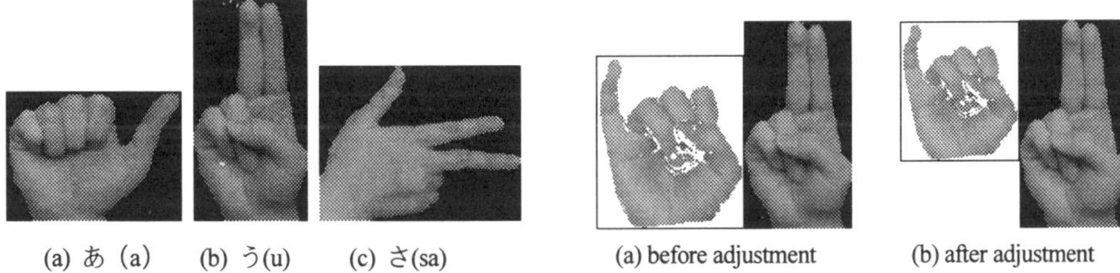

<table>
<tr><td>(a) あ (a)   (b) う(u)   (c) さ(sa)</td><td>(a) before adjustment   (b) after adjustment</td></tr>
</table>

Figure 7. Example of finger characters using as a dictionary.

Figure 8. Example of matching.

3. Result and discussion

In this study, without carrying out the examination on many images yet, the reliability of the results do not become sufficiently high enough.   Though in many cases, the recognition was generally satisfactory.   Some of the results are listed in **Table 1**.   The similarity values of recognized images were about 0.65 to 0.90, and it became that the agreement was considerably high.   For the contrariety, the followings were considered on the image of which the recognition was failure.

(1) The recognition process was done on the image with not sufficient extraction.

(2) The misunderstanding was occurred between the images in which only the shape of the finger differs.   Especially in this case, the individual difference is big for bended degree and extended angle, and the finger is not possible to identify the finger character of which the shape only of the finger differs.   This example is shown in **Figure 9**.

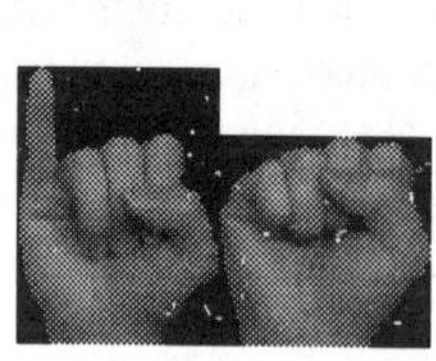
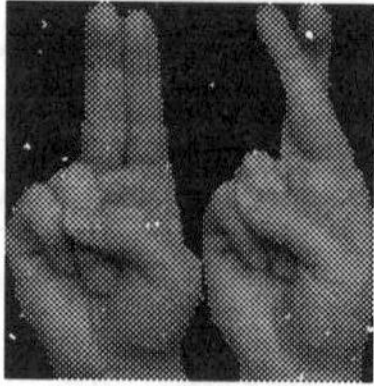

(a) い(i) and さ(sa)   (b) う(u) and ら(ra)

Figure 9. Example of similar finger characters.

(3) This time, only one person's finger characters were used as a dictionary, but multiple average finger characters seem to be necessity.

(4) To obtain the similarity value, the edge designated in the every finger character was used for the positioning. It seems to be necessary to decide the position by more sophisticated method such as the dynamic programming matching method.

4. Conclusion

This time, it would be possible to obtain the considerably accurate extraction image by extraction and compensation of the flesh color region. And the recognition ratio is about 80%. However, it seems to obtain the better recognition rate by adopting dynamic programming matching method, on the matching. In the future, such point will be examined.

Table 1. Some results of matching.

Finger character	Similarity //	* indicates failure
あ (a)	0.838	
い (i)	0.781	
う (u)	0.827	
え (e)	0.847	
お (o)	0.805	
か (ka)	0.712	* (tu) 0.654
き (ki)	0.646	
く (ku)	0.877	
け (ke)	0.860	
こ (ko)	0.733	
さ (sa)	0.838	
し (si)	0.847	* (ku) 0.838
す (su)	0.866	
せ (se)	0.888	
そ (so)	0.795	
た (ta)	0.848	
ち (ti)	0.829	
つ (tu)	0.794	
て (te)	0.853	
と (to)	0.856	
な (na)	0.762	
に (ni)	0.819	
ぬ (nu)	0.806	
ね (ne)	0.736	
は (ha)	0.798	
ひ (hi)	0.757	* (ta) 0.751
ふ (hu)	0.867	
へ (he)	0.717	
ほ (ho)	0.860	
ま (ma)	0.797	* (na) 0.755
み (mi)	0.813	
む (mu)	0.813	
め (me)	0.807	
や (ya)	0.805	
ゆ (yu)	0.781	* (ho) 0.668
よ (yo)	0.799	* (i) 0.737
ら (ra)	0.867	* (u) 0.739
る (ru)	0.835	
れ (re)	0.839	
ろ (ro)	0.837	
わ (wa)	0.868	

References

[1] Sagawa H., Sakou H., Oohira E., Sakiyama A., and Abe M.: "Sign-Language Recognition Method Using Compressed Continuous DP Matching", Trans. of IEICE D-II, J77-D-II, 4, pp.753-763 (1994).

[2] Uchida M., Ishikawa K., and Ide S.: "Finger Character Recognition System and Application to Signed Language", Trans. of IEE Japan, 114-C, 10, pp.995-1000 (1994).

[3] Kiyasu S. and Fujimura S.: "A Method for Connected Word Recognition of Sign Language Using an Image Sequence", Trans. of SICE, 35, 8, pp.1099-1104 (1998).

[4] Imagawa K., Taniguchi R., Arita D., Matsuo H., Lu S., and Igi S.: "Recognition of Local Features for Camera-based Sign-Language Recognition System", J. of ITE, 54, 6, pp.848-857 (2000).

[5] Adachi Y., Imai A., Ozaki M., and Ishii N.: "Study on Extraction of Flesh-Colored Region and Detection of Face Direction", Intl. J. of KES, 5, 2, pp.112-117 (2001).

Recognition of Face Direction

Toshiyuki KAMIYA[1], Yoshinori ADACHI[1], Masahiro OZAKI[2], and Naohiro ISHII[3]
[1]Chubu University, [2]Nagoya Women's University, and [3]Nagoya Institute of Technology
[1]1200 Matsumoto-Cho, Kasugai, Aichi, Japan 487-8501   adachiy@isc.chubu.ac.jp

Abstract: From the observations of flesh-colored pixel distributions on UV plane in the LUV color space, an ellipse successfully extracted a face region.   A proposed method using ellipse extraction and compensation with Soble boundary was an extension of the method proposed by Adachi et al.   Extracted rotate face images by 15 degrees interval were used to make an automatic face direction detecting system. The face direction was estimated through two processes. The first process was a projection into an eigenspace, and the second process was an estimation of direction through analytical interpolation. At first, a seven-dimensional eigenspace was obtained from the principal component analysis of thirteen face images. Then, from the distances in a eigenspace, the face direction was detected analytically.   The accuracy of detection of face directions was adequate.

1. Introduction

There are many researches on the field of human face recognition and face direction recognition [1][2]. Those are expected from the security or man-machine interface point of view. Most of the works related to the human face recognition are using characteristics of face parts such as the relative locations or the shapes of eyes, mouse, nose and so on. Therefore it is very important to face parts correctly.

The most popular methods to detect the locations of face parts are using edge information. For example, an edge-congested region is judged as an eye region [3]. However, because of the environmental light changes or the human face rotations, the edge image is not always made sharply.

There is a method using mosaic template matching [4]. But this has also difficulty in lighting and size changes. Therefore many restrictions are imposed on the face image taken. Usually a full face is taken for this purpose.

The three-dimensional (3D) object recognition or its direction detection from two-dimensional (2D) images is important in industry and has been studied by many researchers [5][6]. Most of them are based on the 3D structure of the object, because appearances of the 3D objects change so much by the object's direction or the lighting direction.

In this work, we discussed about the flesh- colored region extraction and the face direction detection from the 2D images.   At the beginning, the extraction and the compensation of the flesh-colored region was examined. Secondly, an eigenspace was constructed from the rotate 2D images by principal component analysis and all images were projected in the space. Thirdly, face direction of unknown image was analytically obtained from the distances in the eigenspace.

2. Proposed method

The proposed method mainly consists of two parts. The extraction division and the recognition division are those.   The flesh-colored region (face image) is extracted and compensated in the extraction division and the matching is carried out with the dictionary in the recognition division.

2.1 Extraction of flesh-colored region

In the previous study[7], we proposed the extraction method by using the ellipse region on the UV plane in the LUV space, and showed that the better results were obtained comparing with the existing methods.   In this study, further improvement was added on the basis of the previous technique.   Concretely, the flesh-colored region is extracted by next procedure.

(1) The RGB color system is converted into the XYZ color system, and in addition, it will convert into the LUV color system.   The images of the hand of 130 sheets photographed under various conditions were projected and were superimposed in the UV plane in order to limit appearance region of the flesh color.   The part of the images is shown in **Figure 1**.

The result of superimposing flesh color appearance region is shown in **Figure 2**.   As shown in this figure, it is proven that the flesh color is distributed in the specific region in the UV plane, and it is proven that the flesh-colored region may be extracted by limiting to this region.

(2) The target image is projected in the UV plane, and the peak in the flesh color appearance region decided in (1) was searched, and it was made to be a candidacy for the flesh color region that should be extracted by the ellipse centered around the peak.   The ellipse is decided here as the following.

Figure 1. Sample face image to limit appearance region.

Figure 2. The region in UV plane which superimposed the 130 flesh color images.

a. In search of pixel distribution on the straight line that connects the peak on the UV plane of flesh color with the origin, the boundary with the region less than 0.05 times the peak value continues is found, and it is made to be the length of the major axis.

b. In search of pixel distribution on the straight line that is orthogonalized to straight line that connects the peak with the origin and passes through the center of the major axes, the length of the minor axis is obtained in the same way.

c. The center of the ellipse is decided as the center of the minor axis and the lengths of the major and minor axes are already obtained in the above.

Decided ellipse is shown by the following equation.

$$\left[\frac{(U-P)\cos\theta+(V-Q)\sin\theta}{A/2}\right]^2+\left[\frac{-(U-P)\sin\theta+(V-Q)\cos\theta}{B/2}\right]^2=1 \qquad (1)$$

where (P, Q) is the center of the ellipse and θ is a slope of the line connected the peak and the origin. A and B are the lengths of the major and minor axes respectively.

Using the face image, decision of the ellipse and example of the extracted flesh-colored image are shown in **Figure 3**.

(3) In extraction result of the flesh-colored region by the ellipse, deficiency and excess exist in the extraction result.   This will be corrected as following.

a. The pixel in the ellipse region is projected in the L space. And by leaving region with large peak, small peak is deleted.   This can mostly remove the close color noise from the flesh color.   This example is shown in **Figure 4**.

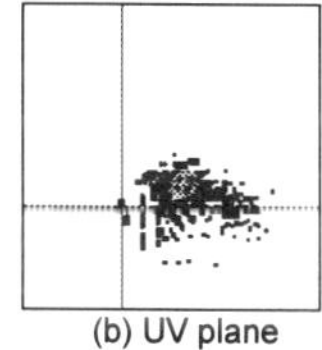

(a) original (b) UV plane (c) extracted

(a) before noise reduction (b) after noise reduction

Figure 3. Example of elliptic extraction of flesh-colored region.

Figure 4. Example of noise reduction by removing small peak.

b. The boundary of the face is obtained by the Sobel method, and the flesh color part is expanded to the boundary in order to compensate the part that is lacking in the flesh-colored region.   The search direction of the boundary is made to be 8 directions of the 45-degree interval.   The results of the boundary image got by the Sobel method and compensated flesh color image are shown in **Figure 5**.

By the process of (1) to (3), the face image is obtained.   The example of the image got in this way is shown in **Figure 6**.

(a) boundary by Sobel (b) compensation

(a) 30 degrees (b) 150 degrees

Figure 5. Example of boundary image by Sobel method and compensated flesh color image.

Figure 6. Example of extracted image using for matching.

2.2 Recognition of face direction

The recognition of the face direction was carried out by the method for specifying the angle by forming the eigenspace from rotational face image, and projecting the image on this space for the recognition, and comparing the distance with the original images. Because of the skin color and hairstyle, extracted face region is different one by one. Then in order to allow to some extent difference, many images were examined.   The procedure of recognition processing is described in the following.

(1)   13 face images of the 15 degree interval is photographed from the right side to the left side for several people, and each eigenspace is formed from the image which removed the background in the manual.   The eigenspace space was formed from the peculiar face, which corresponded to the eigenvector, by the principal component analysis. The number of the eigenvector was automatically decided from the contribution ratios, in this case 0.9, and in most cases it became 6 to 7 dimensional space.   The eigenvectors are depicted in **Figure 7.**

(2)   The matching was carried out by projecting the extracted unknown image in the eigenspace, and choosing the angle in which there are most small distances with face image of the original 13 images, and interpolating in the before and after.

(3) Originally, the size of the face must be adjusted, before it is projected in the eigenspace. In this study, this process has been omitted for the simplicity, since the photographing distance was made to be being constant.

(4) However, in order to detect face direction accurately, extension and reduction processing which put the size of the face together are required.   And it is possible to absorb a little difference in the interpolation process.

(5) There are various techniques in interpolation; a neural network is one of it. But this time, most simple method, the proportional distribution, was used.

3. Results and discussion

This time, without carrying out the examination on many images yet, the reliability of the results is not sufficiently high enough, though in the many cases, this method gave good results.   The extraction rate of extracted face image showed the value of about 0.75 to 0.95, and most of the flesh-colored region would be able to be extracted and compensated.   And, it was confirmed that in the direction recognition, there was the error in the range of few degrees in most cases, and the eigenspace technique worked fairly well.   The part of the results is depicted in **Figure 8**.

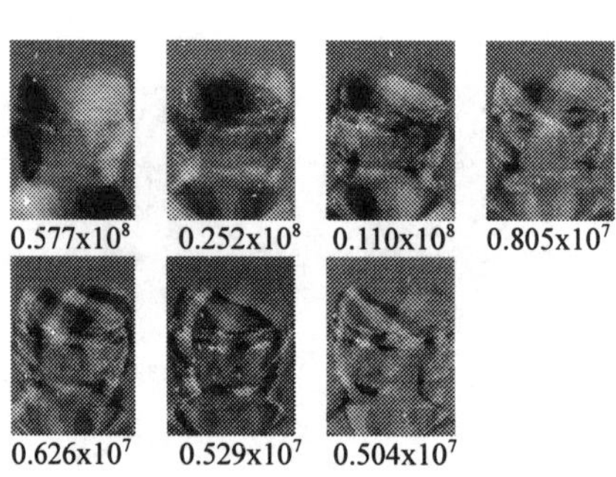

Figure 7. Eigenvectors corresponding to the seven largest eigenvalues.

Figure 8. Comparison of experimental and estimated face directions.

However, there exists large error cases and the cause of the large error is considered as followings:

(1) First case is that the image that is not sufficient extracted is used.

(2) Second case is that there is a difference at the height and the direction of the face differs perpendicularly large.   In this case, it is necessary to form the eigenspace of the perpendicular rotation image also.

(3) Third case is that the size of the face largely differs.   The size of the flesh color region greatly affects the generation of the eigenspace.   Then, area adjusting is necessary.

(4) Fourth case is that the exposure condition of the skin is different by the hair.

4. Conclusion

This time, it would be possible to obtain the considerably accurate extraction image by extraction and compensation of the flesh color region.   It is considered that however, condition setting of the recognition object must be changed on the direction recognition, as long as the eigenspace is used.   And, it is regarded as also examining the method for guessing from relative configuration condition of the face parts, in order to be able to cope even in what kind of condition.   In the future, such point will be examined.

References

[1] SONG X., LEE C.-W., XU G., and TSUJI S.: "Extraction of Facial Organ Features Using Partial Feature Template and Global Constraints", Trans. of IEICE D-II, J77-D-II, 8, pp.1601-1609 (1994).

[2] DOI M., CHEN Q., MATANI A., OSHIRO O., SATO K., and CHIHARA K.: "Lock Control System Based on Face Identification", Trans. of IEICE D-II, J80-D-II, 8, pp.2203-2208 (1997).

[3] AOYAMA K., YAMAMURA T., OHNISHI N., and SUGIE N.: "Estimating Face-and-Eye Direction from Images Taken with a Camera", Technical Report of IEICE, PRU95-233, pp.131-136 (1996).

[4] KOSUGI M.: "Human-Face Recognition Using Mosaic Pattern and Neural Networks", Trans. of IEICE D-II, J76-D-II, 6, pp.1132-1139 (1993).

[5] CHIN R. T. and DYER C. R.: "Model-Based Recognition in Robot Vision", ACM Computing Surveys, 18, 1, pp.67-108 (March 1986).

[6] BESL P. J. and JAIN R. C.: "Three-Dimensional Object Recognition", ACM Computing Surveys, 17, 1, pp.75-145 (1985).

[7] ADACHI Y., IMAI H., OZAKI M., and ISHII N.: "Study on Extraction of Flesh-Colored Region and Detection of Face Direction", Intl. J. of KES, 5, 2, pp.112-117 (2001).

KES '01
N. Baba et al. (Eds.)
IOS Press, 2001

Implementation of the Protocol to Avoid Deadlocks in π-Calculus

Kazunori Iwata and Naohiro Ishii
Dept. of Intelligence and Computer Science, Nagoya Institute of Technology,
Gokiso-cho, Showa-ku, Nagoya, 466-8555, Japan
E-mail kazunori@egg.ics.nitech.ac.jp, ishii@egg.ics.nitech.ac.jp

Abstract. We have implemented the protocol for the multi-threaded processes with choice written in π-calculus[1]. We have shown the outline of the protocol. Moreover, we have proposed the language which can be used on a computer more easily than the mathematical notations of π-calculus and explained how to implement the interpreter by employing JAVA. Finally, we have implemented the system to execute the language and shown the example for the executions.

1 Introduction

We implement the protocol for multi-threaded processes with choice written in π-calculus[1]. π-calculus is a process calculus to describe a channel-based communication among distributed processes, which consist of sequential processes and choice processes. We focus on the property of π-calculus, which can describe the distributed processes, to implement an agent describe language in which the communication among agents and distributed agents are easily described.

In π-calculus, each process concurrently communicates with other processes by executing its own processes. If a process is a choice process, the process must choose one process from concurrent processes. It executes the chosen process after blocking other concurrent processes. However, when it cannot execute the chosen process and release the blocked processes, it runs into the deadlock. Hence, we adjust the situations in the communication and implement the protocol to avoid deadlock[2, 3, 4, 5, 6]. We employ JAVA to implement the system and assign a thread to each process in π-calculus.

2 π-calculus

π-calculus is a process calculus which is able to describe dynamically changing networks of concurrent processes. π-calculus contains just two kinds of entities: process and name. Processes, sometimes called agents, are the active components of a system. The syntax of defining a process is in Figure 1. The first name in output or input process(eg. x in $\overline{x}[y]$) is called channel on which the process communicates with other processes.

Processes interact by synchronous rendezvous on channels, (also called names or ports). When two processes synchronize, they exchange a single value, which is itself a channel.

The output process $\overline{x}[y].P_1$ sends a value y along a channel named x and then, after the output has completed, continues to be as a new process P_1.

$$
P \quad ::= \quad \begin{array}{ll}
\overline{x}[y].P & \text{/* Output */} \\
x(z).P & \text{/* Input */} \\
P \mid Q & \text{/* Parallel composition */} \\
(\nu x)P & \text{/* Restriction */} \\
P + Q & \text{/* Summation */} \\
0 & \text{/* Nil */} \\
!P & \text{/* Replication */} \\
[x = y]P & \text{/* Matching */}
\end{array}
$$

Figure 1: The syntax of process: P and Q are processes. x and y are names.

Conversely, the input process $x(z).P_2$ waits until a value is received along a channel named x, substitutes it for the bound variable z, and continues to be as a new process $P_2\{y/z\}$ where, y/z means to substitute the variable z in P_2 with the received value y. The parallel composition of the above two processes, denoted as $\overline{x}[y].P_1 \mid x(z).P_2$, may thus synchronize on x, and reduce to $P_1 \mid P_2\{y/z\}$.

Fresh channels are introduced by restriction operator ν. The expression $(\nu x)P$ creates a fresh channel x with scope P. For example, expression $(\nu x)(\overline{x}[y].P_1 \mid x(z).P_2)$ localizes the channel x, it means that no other process can interfere with the communication between $\overline{x}[y].P_1$ and $x(z).P_2$ through the channel x.

The expression $P_1 + P_2$ denotes an external choice between P_1 and P_2: either P_1 is allowed to proceed and P_2 is discarded, or converse case. Here, external choice means that which process is chosen is determined by some external input. For example, the process $\overline{x}[y].P_1 \mid (x(z).P_2 + x(w).P_3)$ can reduce to either $P_1 \mid P_2\{y/z\}$ or $P_1 \mid P_3\{y/w\}$. The null process is denoted by 0. If output process (or input process) is $\overline{x}[y].0$ (or $x(z).0$), we abbreviate it to $\overline{x}[y]$ (or $x(z)$).

Infinite behavior is allowed in π-calculus. It is denoted by the replication operator $!P$, which informally means an arbitrary number of copies of P running in parallel. This operator is similar to the equivalent mechanism, but more complex, of mutually-recursive process definitions.

π-calculus includes also a matching operator $[x = y]P$, which allows P to proceed if x and y are the same channel.

The output, input and restriction primitives of π-calculus are monadic: exactly one channel is exchanged during each communication. Polyadic π-calculus[1] is a useful extension of π-calculus which allows many channels exchanged during each communication.

3 Outline of the Protocol

In this section, we explain the outline of the protocol for multi-threaded processes with choice. The processes concurrently communicate each other. Firstly, we introduce basic concepts concerning the communication.

3.1 Basic Concepts

The relationship among these elements(Processes, CMs, CHMs) is in Figure 2.

Process: Processes are units of concurrent execution of our concurrent and distributed system. Processes are implemented as threads. If processes meet a choice process, they make new threads for each process in the choice process.

Communication Manager:

Communication Managers(CMs) manage communication requests on channels from processes. They make possible for processes to communicate with one another. They have two queues consisting of the communication requests from processes.

Choice Manager: Choice Managers(CHMs) manage choice processes on processes. They observe the threads made from the choice, and decide which process should be chosen.

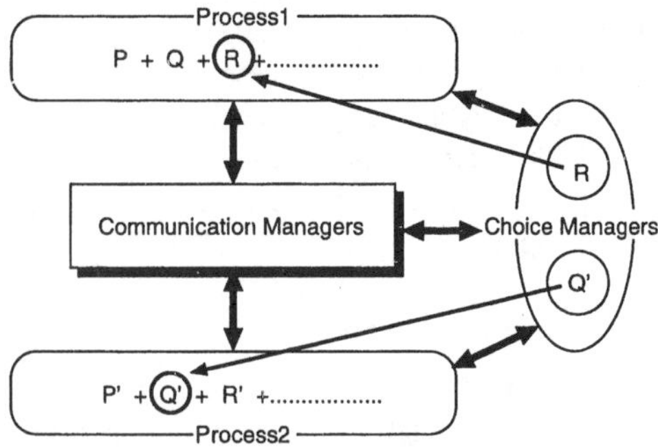

Figure 2: The relationship among the elements: Each element communicates with each other. If there is no choice process, processes and CMs do not communicate with CHMs.

3.2 *Behavior of Elements*

Next, we explain the behavior of the elements in Figure 2.

Behavior of Processes: Processes perform a output process (resp. an input process) through CMs. To perform a output process (resp. an input process), processes send the information, which includes the channel and the values, to CMs and wait for the answer from CMs.

When processes encounter a choice operator, they send the information about it to CHMs and assign a new thread to each process in a choice. Each thread interacts CMs independently each other. One of these thread is selected by CHM, and the others are stopped.

Behavior of Communication Managers: Each CM maintains a queue that stores the communication requests from processes. CMs divide the communication requests according to the channel name. When the output request and the input request are gathered on the same channel, CMs send processes the results of communication.

Behavior of Choice Managers: Each CHM controls the choice operator. It gets the information from the processes and interact CMs to decide which thread should be chosen. It send the result of selection to processes and CMs.

4 Implementation of the Protocol

We employ JAVA to implement the system. In this system, we assign the thread, which is concurrently executed, to each process in π-calculus.

4.1 *Design Language*

Here, we design a primitive language instead of mathematical notations of π-calculus for this system to implement an easily on computer in Table 1.

Table 1: The comparison between π-calculus and Our Language

Functions	Polyadic π-calculus	Our language
Output	$\overline{x}[y_1 \ldots y_n].P$	$(\text{out } x \; y_1 \ldots y_n).P'$
Input	$x(z_1 \ldots z_n).P$	$(\text{in } x \; z_1 \ldots z_n).P'$
Parallel composition	$P_1 \mid \ldots \mid P_n$	$P'_1 \mid \ldots \mid P'_n$
Restriction	$(\nu \, x_1 \ldots x_n)P$	$(\text{new } x_1 \ldots x_n).P'$
Summation	$P_1 + \cdots + P_n$	$P'_1 + \cdots + P'_n$
Nil	0	@
Replication	$!P$	$!P'$
Matching	$[x = y]P$	$[x = y].P'$

4.2 Define Classes

We define the classes which are related to the model in Figure 2 to implement the system.

Analysis Class reads the processes according to Table 1. It checks the number of parentheses and decomposes the parallel composition to the single process and treats the restriction.

Assistance Class parses the processes in order that they may easily execute the processes and checks the existence of summations. It decomposes the process according to the signs which are "(", ")", "+", ".", "[", "]" and "!".

Process Class executes the processes which have been already analyzed by Analysis Class and Assistance Class. It gets the processes from Analysis Class and sends them to Assistance Class. When it gets the parsed processes from Assistance Class, it executes them by communicating with CM Class and CHM Class.

CM Class plays a role in CM in subsection 3.1. It makes new thread related to the channel name.

CHM Class plays a role in CHM in subsection 3.1. It makes new thread related to the process which has summations. The thread has an id called CHid.

5 Experiments to Verify System

We experiment some cases in π-calculus to verify the system.

(1) $a(b) \mid \overline{a}[c] \; \left(\equiv \; P \mid Q \right)$

This case is the basic case to communicate on the channel. The processes P and Q communicate each other on the channel named a.

(2) $(\nu \, a)(a(b).a(b) \mid \overline{a}[c]) \mid \overline{a}[e] \; \left(\equiv (\nu \, a)(P.P \mid Q) \mid \overline{a}[e] \right)$

This case checks the restriction of the name a. The restricted name $(\nu \, a)$ has an effect on the processes $P.P \mid Q$. The name a in the process $\overline{a}[c]$ is different from the name a in the processes $P.P \mid Q$. Hence, the communication is not established among the processes $\overline{a}[c]$ and $P.P \mid Q$.

(3) $(a(b) + a(d)) \mid \overline{a}[c].\overline{a}[e] \; \left(\equiv (P + a(d)) \mid Q.\overline{a}[e] \right)$

This case means the choice process can select the process $a(b)$ or $a(d)$. The communication is established between the processes $a(b)$ and $\overline{a}[c]$ or the processes $a(d)$ and $\overline{a}[c]$. The processes not to be chosen is discarded.

We show the result of the experiments (1) and (2) in Figure 3, (3) in Figure 4.

The result of the experiment (3) has two answers. Because it has the choice process. The results mean the choice processes are successfully executed. In Figure 4, the results show the processes P and $a(b)$ are selected, respectively.

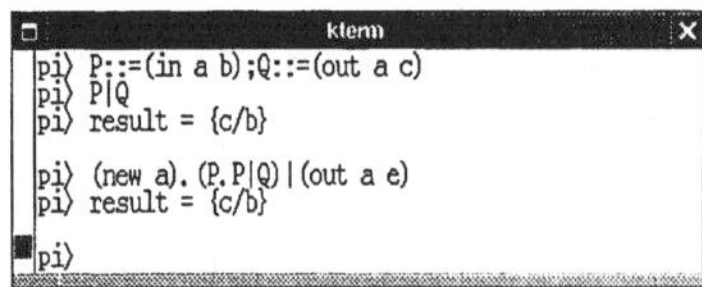

Figure 3: The results on experiments (1) and (2): The results mean the communication between the process P and Q or only in the process $(P.P \mid Q)$ is established. Because the operation *new* makes new local name in $(P.P \mid Q)$.

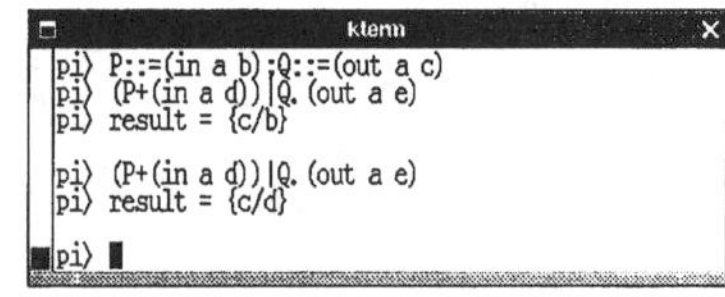

Figure 4: The results on experiment (3): The result means the process P or $(in\ a\ d)$ is chosen in the process $(P + (in\ a\ d))$.

6 Conclusion and Future Works

In this paper, we have implemented the protocol for multi-threaded processes with choice written in π-calculus. We have explained the three elements: processes, Communication Managers(CMs) and Choice Managers(CHMs) to implement the protocol. Processes are units of concurrent execution of our concurrent and distributed system. CMs manage communication requests on channels from processes. CHMs manage choice processes on processes.

We have defined the language instead of using the mathematical notations of π-calculus and explained how to implement the interpreter by employing JAVA.

Finally, we have implemented the interpreter and experimented to verify the system.

One of the future works is to implement the compiler for our language and build it into any system like as process systems, distributed object systems and so on.

Acknowledgments

A part of this research result is by the science research cost of the Ministry of Education.

References

[1] R. Milner. Polyadic π-calculus:a Tutorial. LFCS Report Series ECS-LFCS-91-180, Laboratory for Foundation of Computer Science, 1991.

[2] G.N.Buckley and A. Silberschatz. An effective implementation for the generalized input-output construct of CSP. In *ACM Transactions on Programming Languages and Systems*, volume 5, No.2, pages 223–235, 1983.

[3] C.A.R. Hoare. Communicating sequential processes. In *Communications of the ACM*, volume 21, No.8, pages 666–677, 1985.

[4] E. Horita and K. Mano. Nepi: a network programming language based on the π-calculus. In *Proceedings of the 1st International Conference on Coordination Models, Languages adn Applicationos 1996,*,volume 1061 of *LNAI*, pages 424–427. Springer, 1996.

[5] E. Horita and K. Mano. Nepi^2: a two-level calculus for network programming based on the π-calculus. In *IPSJ SIG Notes, 96-PRO-8*, pages 43–48, 1996.

[6] E. Horita and K. Mano. Nepi^2: a Two-Level Calculus for Network Programming Based on the π-calculus. ECL Technical Report, NTT Software Laboratories, 1997.

KES '01
N. Baba et al. (Eds.)
IOS Press, 2001

A Genetic Algorithm for the Classification Problem and Its Applications

Shan DING, Naohiro ISHII
Department of Intelligence and Computer Science, Nagoya Institute of Technology,
Gokiso-cho, Showa-ku, Nagoya 466, Japan

Abstract. The classification problem is one of the typical problems in data mining and machine learning. A genetic algorithm (GA) for the classification problem under an undetermined environment is proposed, which is based on a fuzzy distance function. The GA is to calculate the attribute weights and attach the attribute weights to modify the distance of similarity degrees. Furthermore, the attribute weights calculated by the proposed GA are applied to the discretized value difference metric (DVDM), and a new distance metric called weighted discretized value difference metric (WDVDM) is proposed. Computational experiments are conducted on benchmark problems, downloaded from UCI machine learning databases. Experimental results, compared with the C4.5 algorithms, show the efficiency of the proposed algorithm, and the WDVDM improves the discretized value difference metric (DVDM).

1. Introduction

The classification problem is one of the typical problems in data mining and machine learning. There are many methods that have been researched to solve this problem. Piatetsky-Shapiro [1] presented some rules that partition the given data into disjoint groups. Yang and Honavar [2] selected a subset of attributes of features from a much larger set to represent the patterns to be classified. Cover and Hart [3] proposed the nearest neighbor (NN) methods learn by storing examples as points in a feature space, which requires some means of measuring distances between examples.

The NN method is the base of many applications of case-based reasoning systems [5,6]. The process is as follows: Given a set of classified examples, which are described as points in an input space, a new unclassified example is assigned to the known class of the ``nearest" example. The ``nearest" relation is computed by using a (similarity) metric defined on the input space. Ricci and Avesani [7] presented a class of local similarity metrics. They start from an initial set of stored cases, improve the retrieval accuracy by modifying the local definition of the metric. Their learning procedure is reinforcement learning algorithm and can be run as a block box since no particular setting is required. With the aid of classical test sets it was shown that their proposed metric can improve the accuracy of nearest neighbor method in many cases. The NN method is extremely simple to implement and leaves itself open to a wide variety of variations. On the other hand, it suffers from the existence of noisy attributes. Simpler metrics may fail to capture the complexity of the problem domains, and as a result, may not perform well. To decrease the influence of noisy attributes, one way is to calculate the attribute weights and attach weights to the attributes to further modify the distance of similarity degrees. In this paper, we present genetic algorithms (GA) to calculate attribute weights from similarity information which is generated from training sets. The simulation experiments show the effectiveness of the proposed GA. Furthermore, our proposed learning methods are applicable not only to nonlinear similarity functions, but also any other similarity functions.

The remainder of this paper is organized as follows: in Section 2 the problem is defined and the preparation knowledge about the similarity function and data normalization is described. Section 3 discusses in details our GA approach. In Section 4, a new distance metric based on discretized value difference metric (DVDM) is proposed and experimental tests are carried out. Some concluding remarks follow in Section 5.

2. Problem Description
2.1 Qualitative Similarity Information and Condition

If the case x and y are in the same output class, we say that the case x is similar to y. We call the given similarity information as *Qualitative Similarity Information (QSI)* [9] which means the case *x* is similar or not to case *y*. Let $w = (w_1, w_2, \ldots, w_n)$ denotes a weight vector, where w_i (i=1, 2, ... , n) is the weight assigned to feature f_i. Let $s_w(x, y)$ be a similarity function which calculates the degree of similarity between cases x and y using the weight vector *w*. The problem is to find a weight vector *w* and a threshold α ($0 < \alpha < 1$), which satisfy the following *Qualitative Similarity Condition(QSC)*. For each chosen case pair (x, y), if x is similar to y then $s_w(x, y) \geq \alpha$; otherwise $s_w(x, y) < \alpha$

It should be noticed that there does not exist a single *w* and α which satisfies the condition. We add the following condition to *w* to reduce the search space. $\forall x$, y and any *w*, $\exists w'= \{ w_1',$ $w_2', \ldots, w_n'\}$, $s_w(x, y)=s_{w'}(x, y)$ where $w'=w/\sum_{i=1}^{n} w_i$, note that almost all of the similarity functions have this type of property.

2.2 Fuzzy Similarity Measure

In order to learn feature weights for similarity, we use a nonlinear similarity function. It is based on fuzzy integrals [4]. The importance of attributes is reflected on the attribute weights which will be used in deciding the fuzzy measure for calculating the fuzzy integrals. This nonlinear similarity function for training set is used for experiments. In fact, the attribute weights learned by GA also can be used in Euclidean distance functions [5].

2.3 Normalization

If one input attribute has been changed in a relatively large range, it would overpower the other attribute's influence on the distance function. For example, if an application has just two attributes A and B, and A can have values from 1 to 1000, whereas B has values only from 1 to 10, then B's influence on the distance function will usually be overpowered by A's influence. Therefore, the attribute should be normalized by its range (i.e., maximum-minimum), so that the normalized value for each attribute is in the approximate range [0, 1].

$$V_a= (V_{ta} - V_{mina}) / (V_{maxa} - V_{mina}),$$

where
   V_{ta} is the value for attribute a;
   V_{maxa} is maximum value for attribute a;
   V_{mina} is minimum value for attribute a;

Using the nonlinear function, we can obtain the similarity distance of two recorders by calculating the distances of each attribute of two recorders respectively. If attribute values of two recorders are all unknown, return a distance of 1 (i.e., a maximal distance). If either of the attribute values is unknown(for example we know one attribute values is v_a), return the larger value either v_a or $1-v_a$.

3. The Genetic Algorithm for QSI
3.1 The Genetic Algorithm (GA)

In this subsection, we describe the genetic algorithm to calculate *w* and α for QSI.

Coding weight vector

A weight vector can be encoded by a string of real numbers where the *i*th one is the weight assigned to the *i*th feature. The threshold of the individual is added to the end of the string. Fig.1 shows an example of an individual.

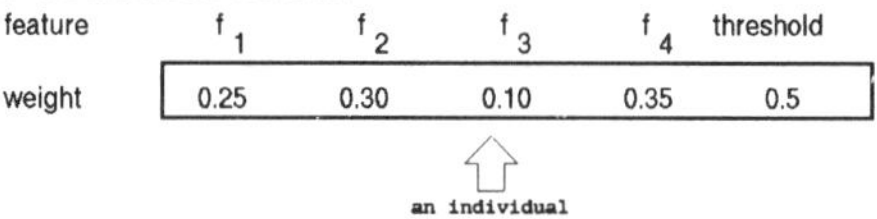

Fig1. Example of an individual

Evaluation

The fitness value of an individual is taken as the satisfaction degree of QSI, i.e., N_{QSI}/N, where N_{QSI} is the number of the sets of two recorders which satisfy QSC using weight vector w and α representing the individual, and N is the number of chosen sets of two recorders. The fitness value of a population is the maximum fitness value of individuals in it.

Selection

All individuals in the population are sorted by their fitness values, and the first individual is the best. In our method, all individuals are selected for mating as described elsewhere [10] in details.

Crossover

Two selected individuals are called parent-A and parent-B, and assume the fitness value of parent-A is higher or equal to that of parent-B as described elsewhere [10] in details.

Mutation

For two genes chosen to be mutated, if both are weights then randomly exchange a gene for another different one, or replace a gene pair (g_1, g_2) with a new one (g_1', g_2'), where $g_1 + g_2 = g_1' + g_2'$ for keeping the sum of weights unchanging.

Differentiation of the same individuals

After crossover and mutation, the differentiation of the same individuals will be carried out for individuals in the population as described elsewhere [10] in details.

GA parameters

The population size is $2N_w$, where N_w is the number of attributes. The probability of mutation is 0.8. The stopping criterion is taken to be 200 N_w individual evaluations.

4. Application of the weight vector

4.1 Value Difference Metric (VDM) and Discretized VDM

The attribute weight that calculated by the proposed learning methods is applicable to not only nonlinear similarity functions but also any similarity functions. In this section, we apply the calculated weight vector to the discretized value difference metric (DVDM) described in paper [8].

The Value Different Metric(VDM) [5] is defined as follow:

$$VDM_a(x, y) = \sum_{C=1}^{C} | Pa, x, c - Pa, y, c |^2$$

where
1. C is the number of output class in the problem domain;
2. $P_{a, x, c}$ is the conditional probability that the output class is c given that attribute a has the value x, i.e., $P(c|x_a)$.

$$P_{a, x, c} = N_{a, x, c}/N_{a, x}$$

1.$N_{a, x}$ is the number of instances in the training set T that have value x for attribute a;
2. $N_{a, x, c}$ is the number of instances in T that have value x for attribute a and output class c;

From the description above, VDM can only be applied in discretization attributes. In order to allow VDM to be used directly to continuous attributes, Wilson and Martines developed the DVDM. In the DVDM, continuous values are discretized into s equal-width intervals, where s is an integer supplied by the user. Similar to that in Wilson and Martinez[8], we use a heuristic to determine s automatically. Let s be the larger one between 5 and C, where C is the number of output classes in the problem domain. The width w_a of a discretized interval for attribute a is given by

$$w_a = |\max_a - \min_a|/s,$$

where $\max_a$ and $\min_a$ are the maximum and minimum value, respectively, occurring in the training set for attribute a. The discretized value v of a continuous value x for attribute a is an integer from 1 to s. and is given by

$$v = d_a(x) = \begin{cases} x, & \text{if } a \text{ is discrete esle} \\ s, x = \max_a, \text{ else} \\ \lfloor (x - \max_a) \rfloor / w_a + 1 \end{cases}$$

After deciding upon s and finding w_a, the discretized values of continuous attributes can be used just like discrete values of nominal attributes in finding $P_{a, x, c}$. For details to calculate $P_{a, x, c}$, see Wilson and Martinez [8].

As its generalization, the nearest neighbor classifier algorithm can use the distance function DVDM, which is defined as follows:

$$DVDM(x, y) = \sum_{a=1}^{m} |\text{vdm}_a(d_a)(x_a), d_a(y_a))|^2$$

4.2 Weighted Discretized Value Difference Metric

According to the different role (weight) which each attribute has in the classification problem (in training data set), the weight of each attribute in distance function is decided. Cost and Salzberg [12] developed a weight nearest neighbor algorithm for learning symbolic features. Based on their algorithm and DVDM, we propose a new distance metric call Weighted Discretized Value Difference Metric (WDVDM).

$$WDVDM(x, y) = \sum_{a=1}^{m} |w_a \text{vdm}_a(d_a)(x_a), d_a(y_a))|^2$$

where w_a is calculated from similarity information by GA. vdm_a is described as before.

4.3 Simulation and Test

The proposed GA and the WDVDM are implemented using C language on the SunOS 5.6 with Pentium III 600MHz and tested on 10 datasets also downloaded from UCI machine learning databases. The datasets are listed in Table.1. Table.1 also presents the number of training datasets in each database ("INST"), the number recorders of test database ("TEST") and C4.5 algorithms including before pruning algorithm ("BEFORE"), after pruning algorithm ("AFTER") and rules algorithm ("RULES").

Experimental results of Table.1 show the efficiency of the proposed GA and WDVDM is more accurate accuracy than the other algorithms on average. That means that the proposed WDVDM improves the accuracy of the DVDM.

DATASET	INST	TEST	WDVDM	DVDM	C4.5 (BEFORE)	C4.5 (AFTER)	C4.5 (RULES)
Annealing	400	150	94%	93.3%	97%	95%	94%
Dermatology	256	110	94.4%	94.5%	93.6%	93.6%	93.6%
Cleveland	149	303	74.5%	77.6%	70%	69%	62.7%
Glass	150	64	82.1%	81.3%	85.9%	85.9%	85.9%
Golf	7	7	78.6%	85.7%	42.9%	42.9%	42.9%
Monk1	124	432	100%	87.9%	76.6%	71.4%	100%
Monk2	169	432	74.9%	74.5%	65.3%	60.3%	66.2%
Monk3	122	432	94%	93.8%	92.6%	82.9%	96.3%
Iris	105	46	92.2%	91.3%	93.3%	93.3%	93.3%
Vote	300	135	96.7%	96.3%	94.8%	93.1%	94.8%
AVER			88.14%	87.62%	81.2%	78.74%	82.97%

Table. 1 Results of test

5. Concluding Remarks

In this paper, a genetic algorithm (GA) that learns attribute weight from the similarity information is proposed. A nonlinear similarity function based on fuzzy integrals is used as the distance function. Furthermore, the attribute weights calculated by the proposed GA are applied to the discretized value difference metric (DVDM), and a new distance metric called weighted discretized value difference metric (WDVDM) is proposed. 10 sets of benchmark problems downloaded from UCI machine learning databases are tested. Computational experiments show the efficiency of the proposed GA and the WDVDM is superior to the DVDM and the C4.5 algorithms. The other applications are also under considerations.

References

[1]G. Piatetsky-Shapiro (Editor), Knowledge Discovery in Database, AAAI/MIT Press, 1991.

[2] J. Yang and V. Honavar, ``Feature Subset Selection Using a Genetic Algorithm'',   Proceedings of the Genetic Programming Conference, GP' 97, pp 380--385, Stanford University, CA, 1997.

[3] T. Cover, and P. Hart, ``Nearest neighbor pattern classification'', IEEE Transactions on Information Theory, 12:1, 21-27, 1967.

[4] M. Sugeno, ``Fuzzy measure and fuzzy integral''(in japanese), Trans. of the Society of Instrument and Control Engineers, 8(2): 218-226, 1972.

[5] C. Stanfill and D. Waltz, ``Toward memory-based reasoning'', Communication of ACM, 29:1213-1229, 1986.

[6] S. Wess, K. D. Althoff, and G. Derwand, ``Using k-d trees to improve the retrieval step in case-based reasoning'', In Topics in Case-Based Reasoning, First European Workshop, EWCBR-93, pp 167-181 Berlin, 1993.

[7] F. Ricci and P. Avesani, ``Learning a Local Similarity Metric for Case-Based Reasoning'', First International Conference, ICCBR-95, pp 301-312. 1995.

[8] D. R. Wilson and T. R. Martinez, ``Improved heterogeneous distance functions'', Journal of Artificial Intelligence Research, 11:1-34,1997.

[9] Y. Wang and N. Ishii, ``Learning Feature Weights from Similarity Information'', International Journal on Artificial Intelligence Tools, Vol. 7, No. 1, pp 31-41,1998.

[10] S. Ding, N. Ishii, ``An Online Genetic Algorithm for Dynamic Steiner Tree Problem'', 2000 IEEE International Conference on Industrial Electronics, Control and Instrumentation. pp. 812--817, 2000.

[11] J. R. Quinlan, C4.5: Programs for Machine Learning, The Morgan Kaufmann Series in Machine Learning, Pat Langley, Series Editor October 1992

[12] S. Cost and S. Salzberg, ``A Weighted Nearest Neighbor Algorithm for Learning with Symbolic Features'', Machine Learning, 10, 57--78, 1993.

Robust Diagnosis Method of Plant Machinery by Hidden Markov Model

Peng CHEN*, Toshio TOYOTA** and Tomoya NIHO*

*Faculty of Computer Science and System Engineering, Kyushu Institute of Technology

680-4 Kawazu, Iizuka-shi, Fukuoka-ken, Japan.

E-mail: chen@mse.kyutech.ac.jp

**Japan Condition Diagnosis Technology Laboratory Incorporated

Takasuhigashi 3-11-1, Wakamatu-ku, Kitakyushu, 808-0144 Japan

E-mail: toyo99@lily.ocn.ne.jp

Abstracts: *This paper proposes a robust condition diagnosis method for plant machines using hidden Markov Model (HMM). The merit of the method proposed in this paper is that the condition diagnosis of machinery can be carried out even if the specifications and operating condition of the machine are unknown. The method has been proved by applying it to the real condition diagnosis of plant machinery. In this paper, a diagnosis example of rolling bearing has been shown.*

1. INTRODUCTION

In the field of plant machinery diagnosis, in order to identify fault types of plant machinery precisely, not only the signals measured in normal state and abnormal state of the diagnosed machine are required, but also the specification and the operating conditions of the machine must be known beforehand[1][2]. For example, when diagnosing rolling bearing fault precisely, the outer race diameters D, inner race diameter d, the number of rolling elements z and the rotating speed must be known.

However, because there are innumerable machines in the large scale of plant, it is not easy to measure and record normal signals at all inspecting points, and it is very difficult to know all detailed specifications and operating conditions of innumerable parts, especially the parts sealed up in these machines.

In order to overcome the difficulty, we proposed a robust diagnosis method by using Hidden Markov Model (HMM) [3][4]. By the method, the fault type can be sensitively identified without the specifications and the operating conditions. In this paper, we show practical diagnosis examples of rolling bearing using sound signals and vibration signals to verify the efficiency of the method.

2. DECISION OF HMM SYMBOLS

2.1 HMM

HMM is an undecided statistical automaton having finite states. Fig. 1 shows an example of HMM. The pattern of the state transition in the HMM is not unique even if the inputted and outputted symbol series are same. The HMM used to diagnose is left-to-right having initial and final states. Because the state transition is not unique and only the symbol series can be observed, it is called "Hidden Markov Model"

In the case of Fig.1, a_{ij} is the transition probability from state i to state j. {a,b} is the symbol series, and the numerical value if [·] is output probability of the symbol when state transits from state i to state j. These values are acquired by learning the data measured from plant machine to be diagnosed.

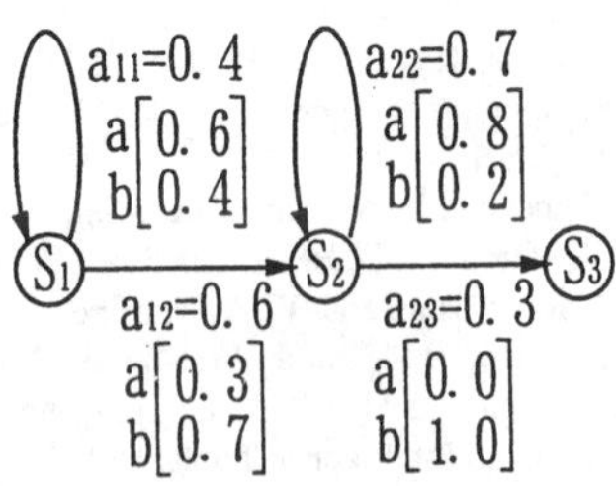

Fig. 1   an example of HMM

The numbers of symbol series and states are decided by trail and error as the case may be. After the values are decided by learning as shown in Fig.1, the probability outputting a symbol series can immediately calculated. For example, when state transits from state i to state j, the transition patterns for outputting symbol series {abb} are $S_1 \rightarrow S_1 \rightarrow S_2 \rightarrow S_3$, $S_1 \rightarrow S_2 \rightarrow S_2 \rightarrow S_3$. The probability can be calculated as follows.

$$P(\{abb\} \mid S_1 \rightarrow S_1 \rightarrow S_2 \rightarrow S_3) = 0.4 \times 0.6 \times$$
$$0.6 \times 0.7 \times 0.3 \times 1.0 = 0.03024 \tag{1}$$

$$P(\{abb\} \mid S_1 \rightarrow S_2 \rightarrow S_2 \rightarrow S_3) = 0.6 \times 0.3 \times$$
$$0.7 \times 0.2 \times 0.3 \times 1.0 = 0.00756 \tag{2}$$

We have,

$$P(\{abb\}) = 0.03024 + 0.00756 = 0.0378 \tag{3}$$

2.1 Decision of Symbol Series for Rolling Bearing Diagnosis

The symptom parameters, which can reflect the feature of time signal, are relative crossing information (Iq), relative Kullback information (Ip) and Kurtosis factor (γ). They are defined as follows.

The frequency q_i crossing over some level i of vertical coordinate of time signal y(t) with plus slope in unit time can be calculated from power spectrum P(f) of y(t) as follows[5];

$$q_i = \frac{\sigma_v}{2\pi\sigma_x} e^{-(i/2\sigma_x)^2} \tag{3}$$

Here,
$$\sigma_x^2 = \int_0^\infty P(f)df$$
$$\sigma_v^2 = \int (2\pi f)^2 P(f)df$$

The value p_i of probability density function in level i of vertical coordinate of time signal y(t) can be calculated by

$$p_i = \frac{\sqrt{2\pi}}{\sigma_v} q_i \tag{4}$$

Here, we divide the time signal y(t) into N parts as shown in Fig. 2. The vertical coordinate of the first part $y_{sj}(t)$ is divided into M sections from maximum $y_{sj}(t)$ (=max{ $y_{sj}(t)$ }) to minimum $y_{sj}(t)$ (=min{ $y_{sj}(t)$ }). The vertical coordinate value y_i of section i (i=1~M) is

$$y_i = \min\{ y_{sj}(t) \} + i \times [\max\{ y_{sj}(t) \} - \min\{ y_{sj}(t) \}]/M \tag{5}$$

If we measure the signal by K times, then the average of the q_{si} and p_{si} calculated with $y_{sj}(t)$ (j=1~K) are

$$q_{si} = \sum_{j=1}^{K} q_{sij} / K \tag{6}$$

$$p_{si} = \sum_{j=1}^{K} p_{sij} / K \tag{7}$$

The time interval of the parts $y_{zk}(t)$ behind $y_{sj}(t)$ is t_{zk} (k=1~N). $q_{i,zk}$ and $p_{i,zk}$ of $y_{zk}(t)$ can similarly calculated by formula (1) and (2), and the relative crossing Information (Iq) is defined as

$$I_{qk} = \sum_{i=1}^{M} \left|\log(q_{i,zk} / q_{si})\right| / M \tag{8}$$

The relative Kullback information (Ip) [6] is defined as

$$I_{pk} = \sum_{i=1}^{M} \left|\log(p_{i,zk} / p_{si})\right| / M \tag{9}$$

The Kurtosis factor (γ) of $y_{zk}(t)$ is defined as

$$\gamma_k = \frac{\sum_{i=1}^{N_k} (y_{zk}(t)_i - \overline{y_{zk}(t)})^4}{N_k \sigma^4} \tag{10}$$

Here, $y_{zk}(t)_i$ is the descrete sampling value of $y_{zk}(t)$. N_k is the total number of $y_{zk}(t)_i$. $\overline{y_{zk}(t)}$ is the mean value of $y_{zk}(t)_i$. σ is the standard division of $y_{zk}(t)_i$.

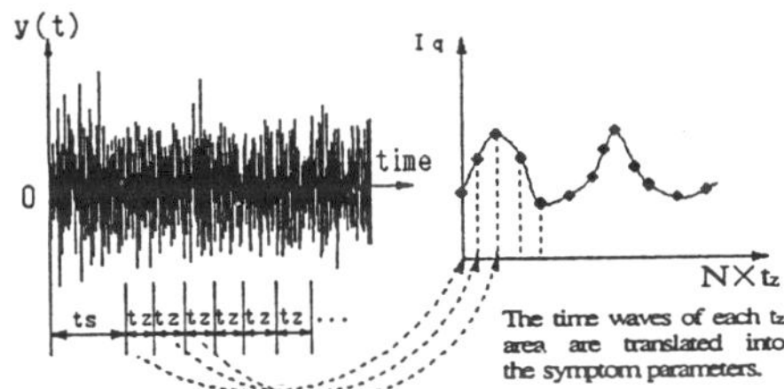

Fig. 2 Translating time signal
into symptom parameter waveform

We use the three parameters (I_{qk}, I_{pk} and γ_k) to diagnose faults. To rising the sensitivity of the diagnosis, the I_{qk} and I_{pk} are calculated only using the values of $y_{zk}(t)_i > |1.5\sigma|$ in the time signals $y_{zk}(t)_i$.

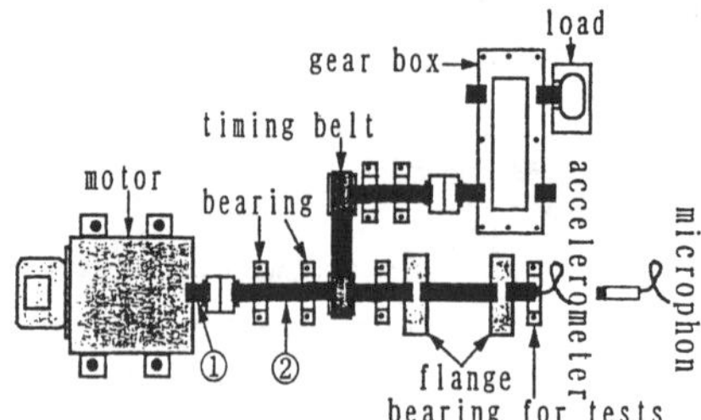

Fig. 3 The rotating machine

Fig. 3 shows the rotating machine and the sensors for tests. The rolling bearing states in the machine for the tests are normal state, outer race flaw, inner race flaw and rolling element flaw. The vibration signals are measured by microphone and accelerometer.

The accelerometer is on the bearing box, and the distances from the microphone and the bearing are 500cm, 1500cm, 2500cm and 3500cm. The rotating speeds are 400rpm, 800rpm, 1200rpm and 1600rpm as shown in Table 1.

Signals used for learning or tests are shown in Table 1. Sampling frequency is 20kHz, and the sampling data number in one file is 40960.

Table 1 data specifications for leaning and tests

b \ a	400	800	1200	1600
50	L	T	L	T
150	T	L	T	L
250	L	T	L	T
350	T	L	T	L

a: rpm, b: measuring distance,
L: for learning, T: for test

Fig. 4 shows the vibration signals measured by the accelerometer. Fig. 5 shows the sound signals measured by the microphone. The waveform examples of parameters (I_q, I_p, γ) calculated by formulae (8), (9) and (10) are shown in Fig. 6-8.

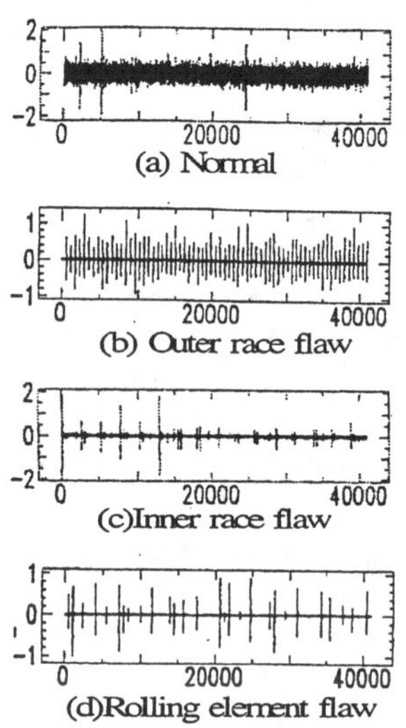

(a) Normal

(b) Outer race flaw

(c) Inner race flaw

(d) Rolling element flaw

Fig. 4 Vibration signal (1200 rpm)

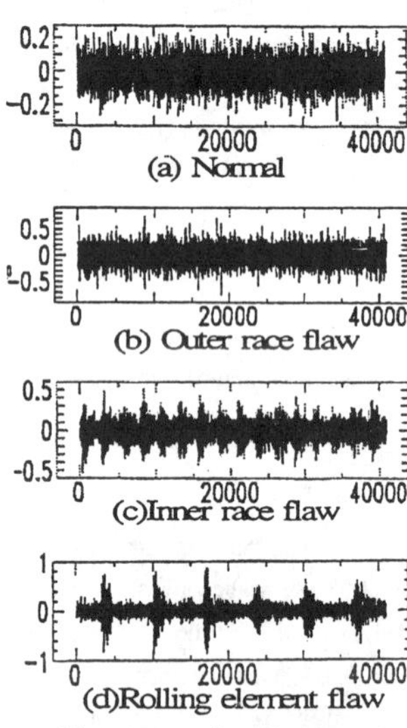

(a) Normal

(b) Outer race flaw

(c) Inner race flaw

(d) Rolling element flaw

Fig. 5 Sound signal (1200 rpm)

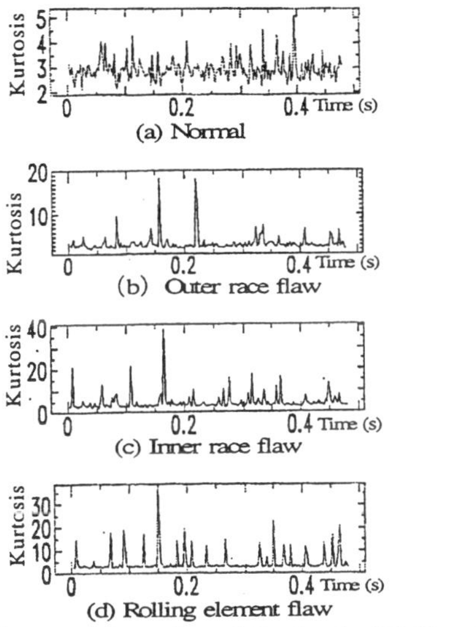

Fig. 6 Kurtosis waveform of vibration signal (1200 rpm)

These parameter waveforms are translated into symbol series {a, b, c, ···} by clattering. Because there is not a theory by which the number of symbol and HMM states can be decided yet, we try to select several types of HMM, and chose one of them that has the highest detection rate of faults to practical diagnosis. Fig. 9 shows an example of HMM which has 10 states and 9 symbols.

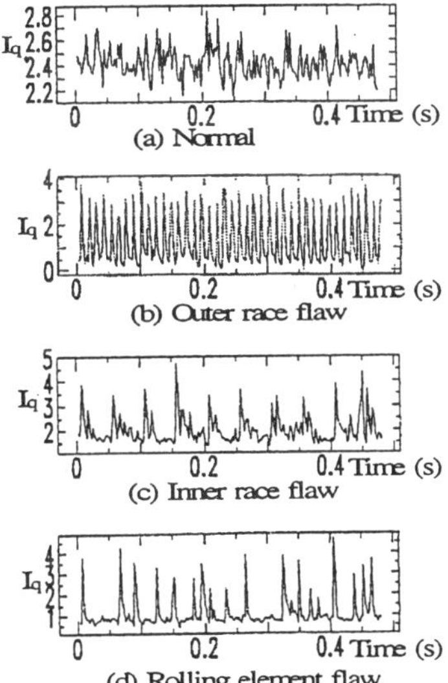

Fig. 7 I_q waveform of vibration signal (1200 rpm)

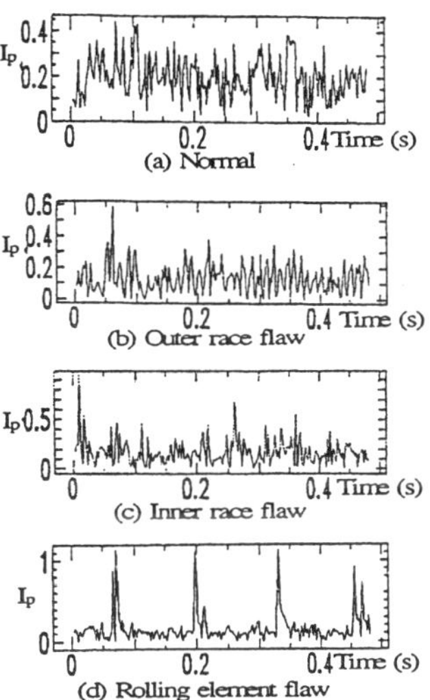

Fig. 8 I_p waveform of vibration signal (1200 rpm)

3. VERIFICATION

The result of distinguishing normal state from abnormal states (outer race flaw, inner race flaw and rolling element flaw) is shown in Fig. 10. Fig. 10 (a) and (b) are the test results using vibration signals and sound signals respectively.

In the case of vibration signals, the distinguishing rates (D.R) are 100%, because the noise including in the signals is smaller than in the sound signals. In the case of sound signals, the distinguishing rates (D.R) are also not so low (>90%).

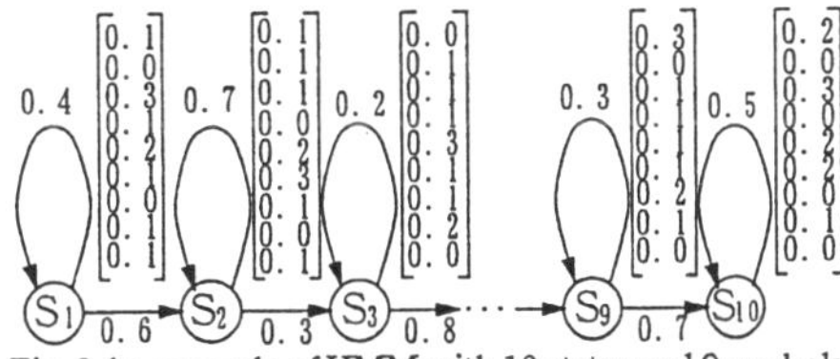

Fig. 9 An example of HMM with 10 states and 9 symbols

D.R : distinguishing rate (correct No./taltal test No.)

No. of states	No. of symbols	D.R		No. of state	No. of symbols	D.R
10	9	80/80		10	9	58/64
15	14	80/80		15	14	57/64
21	20	80/80		21	20	59/64

(a) the case of vibration signals   (b) The case of sound signals

Fig. 10 Results of discriminating normal and abnormal

D.R : distinguishing rate (correct No./taltal test No.)
O.R.F: Outer Race Flaw
I.R.F: Inner Race Flaw
E.R.F: Rolling Element Flaw

No. of states	No. of symbols	D.R		
		O.R.F	I.R.F	R.E.F
10	9	26/26	26/26	25/26
15	14	26/26	26/26	25/26
21	20	25/26	24/26	23/26

(a) the case of vibration signals

No. of states	No. of symbols	D.R		
		O.R.F	I.R.F	R.E.F
10	9	26/26	26/26	24/26
15	14	26/26	25/26	25/26
21	20	25/26	24/26	25/26

(b) The case of sound signals

Fig. 11 Results of discriminating abnormal types

Fig. 11 shows the results of distinguishing fault types. In the case of vibration signals, the distinguishing rates (D.R) are larger than the rates in the case of sound signals. From Fig. 10 and 11, the HMM which has 10 states and 9 symbols is better for the bearing diagnosis.

4.CONCLUSIONS

In this paper, we have proposed a robust condition diagnosis method for plant machines. The merit of the method proposed here is that the condition diagnosis of machinery can be carried out without the specifications and operating condition of the machine. The method has been proved by applying it to the real condition diagnosis of machinery. In this paper, the diagnosis example of rolling bearing has been shown.

When noise including in the signal is strong, the accuracy of the diagnosis may become lower. Therefore, it is necessary to eliminate the noise before diagnosis by HMM as much as possible. There are many methods of the noise elimination which can be used to the diagnosis[7][8]. Other symptom parameters may be used for the HMM besides that shown in formulae (8), (9) and (10). We are trying to select the better symptom parameters and studying on the automatic decision method of the optimum construction for HMM.

References

[1] H. MATUYAMA: Diagnosis Algorithm, Journal of JSPE,Vol.75 No.3, pp.35-37, 1991.

[2] Toshio TOYOTA, Tomoya NIHO and Peng CHEN : Condition Monitoring and Diagnosis of Rotating Machinery by Gram-Charlier Expansion of Vibration Signal, Proc. of Fourth International Conference on Knowledge-Based Intelligent Engineering System & Allied Technologies (KES2000), pp.541-544, 2000.

[3] Levinson S.E., Rabiner L.R., and Sondhi M.M. : "An Introduction to the Application of the Theory of Probabilistic Functions of a Markov Process to Automatic Speech Recognition " ,Bell Syst. Tech.J.,Vol.62,No.4,pp.1035-1074(1983)

[4] Rabiner L.R. : "A Tutorial on Hidden Markov Models and Selected Application in Speech Recognition" , Proc.IEEE,77-2,257-285(1989)

[5] J.S.Bendat: Probability Function for Random processes, NASA Rep. CR-33,(1969).

[6] S. Kullback : Information Theory and Statistics ·, John Willy & Sons, Inc, (1959).

[7] Toshio TOYOTA, Peng CHEN : STATISTICAL FILTER FOR EXTRACTION OF PURE FAILURE SIGNAL, Proc. of SICE'92, pp1029-1032, 1992.

[8] Peng CHEN, Toshio TOYOTA : Extraction Method of Failure Signal by Genetic Algorithm and the Application to Inspection and Diagnosis Robot, IEICE TRANSACTIONS on Fundamentals of Electronics, Communications and Computer Science VOL.E78-A, No.12, pp.1622-1626, 1995.

Fault Monitoring And Compensation In Hydraulic Systems

C. ANGELI

Department of Mathematics and Computer Science
Technological Education Institute of Piraeus
Konstantinoupoleos 38, N. Smirni
GR-171 21 Athens, Greece
E-mail: angeli@compulink.gr

and

A. CHATZINIKOLAOU

Automation Systems S.A.
S. Patsi 62
GR-118 55 Athens, Greece
E-mail: chatzi@compulink.gr

Abstract. Fault monitoring and compensation functions add effectiveness and reliability to a fault detection process when they are included in a diagnostic method. In this work, the sensor data from an actual electro-hydraulic system are recorded, analysed and presented in a suitable format to obtain the information needed for the decision making process. In these systems some kind of faults can be compensated for a period of time by developing suitable diagnostic procedures. The presentation of the fault compensation process in a hydraulic system using the data acquisition and control software DASYLab is the overall goal of this paper.

1. Introduction

Fault monitoring and detection techniques have been developed for automated processes during the last years. These methods include numerical methods, artificial intelligence methods or combination of the two methodologies. Fault detection using numerical techniques is a well established subject and a lot of survey papers have been published, such as [1], [2], [3], [4], [5].

In addition, the usage of coupled techniques for fault detection is considered as an important research area [6], [7] and the implementation of coupled systems to on-line processes [8], [9], [10] offers diagnostic functions that increase the productivity of technical systems.

Faults in hydraulic systems are often caused in many cases by leakage. There is a period of time that the leakage is kept in acceptable limits but it is necessary for safety reasons to detect, monitor and compensate the consequences of this fault until the defective component is repaired or replaced. The establishment of a monitoring system for fault detection and compensation is very useful in terms of safety, reliability and higher operational efficiency of these systems. Considering that the complexity of these systems and the high operational pressures involve a significant risk in detecting a defective component, the implementation of a reliable monitoring and data acquisition process for detecting and compensating faults has a particular importance for these systems. Previous diagnostic and monitoring methods for hydraulic systems was contributed, among other researchers, by [11].

2. The hydraulic system

In the actual hydraulic system used for the experimentation process of this work a hydraulic motor is controlled by a proportional 4-way valve with a cyclical routine which requires a high speed for a short period and then returns to a low speed. This is achieved by applying a command voltage $U_2 = 6$ V to the proportional 4-way valve for a period of 0,8 s and after that period by reducing the command voltage to 1 V. The total period of the cyclical routine is 15 s. The main elements of the hydraulic system for the control of the hydraulic motor are presented in Figure 1.

A data acquisition and control system is used for the measurement of the output signals of the actual system, their interpretation into an accepted format by the computer as well as the analysis and presentation of the signal information. The main measurable variables of the hydraulic system are the angular velocity of the hydraulic motor and the pressures between the components of the system.

A model must be derived and validated for the hydraulic system. Figure 1 shows the main part of the hydraulic system. Regarding that the system pressure p_o at the P-port of the proportional 4-way valve is constant (it is achieved through a pump with a higher capacity than needed by the hydraulic motor and the accumulator (1.5)) we have to model only the

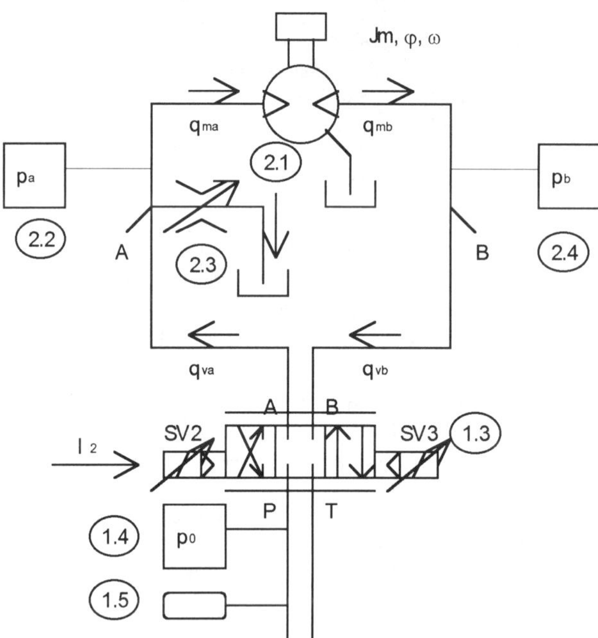

Figure 1. The main part of the hydraulic system for the control of the hydraulic motor.

hydraulic motor with the coupled mass (2.1), the proportional 4-way valve (1.3) and the connection pipes between them.

A common fault in a hydraulic system is a leakage due to a worn motor. This results to a decrease of the rotation speed of the motor. The flow regulator (2.3) is used for the simulation of this fault.

3. The data acquisition

For the data acquisition the DASYLab software was used. The DASYLab (Data Acquisition System Laboratory) is a graphical programming software that takes the advantage of the features and the graphics provided by Microsoft Windows. Among the module functions provided are A/D and D/A converters, digital I/O, mathematical and statistical functions, digital filters of several types, logical connectors, counters, I/O files, digital displays, bar graphs, analogue meters and more. It obtains real-time logging at a rate of up to 800 KHz and real-time on-screen signal display at a rate of up to 70kHz. The acquired data and process results can also be saved into files so that they can be retrieved for further processing at a later time. Using DDE (Dynamic Data Exchange), data can be transferred directly to other Windows application supporting the DDE protocol.

The focus of the data acquisition system for this application is the measurement of the pressures p_a , p_b and the angular velocity ω as well as the monitoring of the digital signals that indicate the correct operation of the electric motor and other electrical devices of the system as filter, oil level switch e.t.c.

The data acquisition system runs in parallel to the actual system and acquires data for the angular velocity and the pressures while the speed is changing. Figure 2 shows the data acquisition process and the definition of the motor speed (outct3).

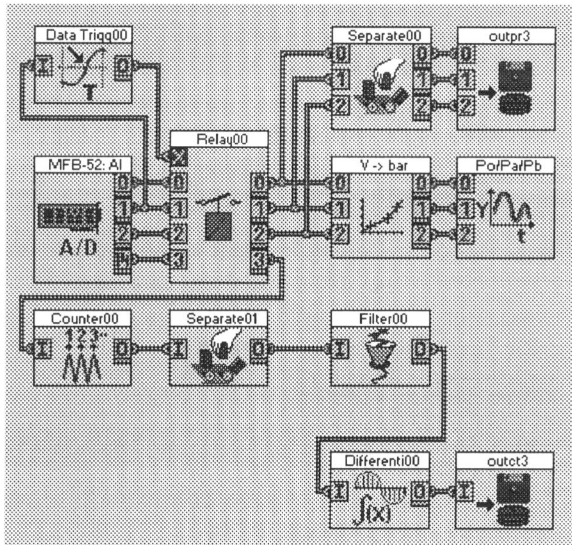

Figure 2. The implementation of the data acquisition process

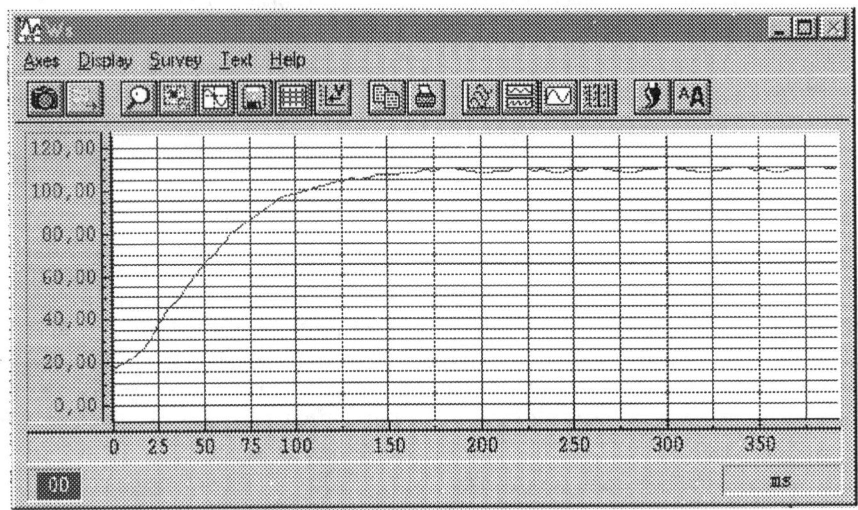

Figure 3. Angular velocity ω

The sampling rate must be chosen as high as possible especially for the accuracy of the measurement of the angular velocity. The global experiment sampling rate was chosen 2000 Hz (800 samples at 0,5 ms interval with a total time of 0,4 s) because this is the highest possible frequency under which the DASYLab experiment is working. A higher sampling rate would overload the data acquisition system and affect the performance of the experiment.

A simulation program is running simultanously and calculates the values of the angular velocity and the pressures for the same period of time (0,4 sec). The comparison of the measured and the calculated values is then performed through another DASYLab "experiment" and the diference is written to an output file.

Since the acquired data refer to the dynamically changing state and the actual system is operating with a periodical voltage U_2 that is applied every 15 sec there is time enough for the running of the modules of the system. The data acquisition system runs for 3 sec, the simulation for 0,4 sec and the expert system for 4 sec.

As already mentioned, when the command input value U_2 to the system changes from 1 to 6 V, the speed of the hydraulic motor increases as shown in Figure 3. In this Figure the angular velocity ω in rad/s for the time-period of 400 ms is presented.

The curve was initialy stepped. To eliminate the steps and make the comparison with the simulation result easier a DASYLab "Filter" module was used. The filter was a second order "Bessel" lowpass filter with a cut-off frequency of 8 Hz.

4. The control and compensation process

The control of the hydraulic motor and the compensation process are realised by means of the DASYLab "experiment" that is shown in Figure 4.

The combination of the modules "U0", "TTL(=5V)" and "U2" produce a periodical square-wave form voltage signal that takes the values $U_0 = 1$ V and $U_2 = 6$ V, corresponding to the low and high speed of the hydraulic motor.

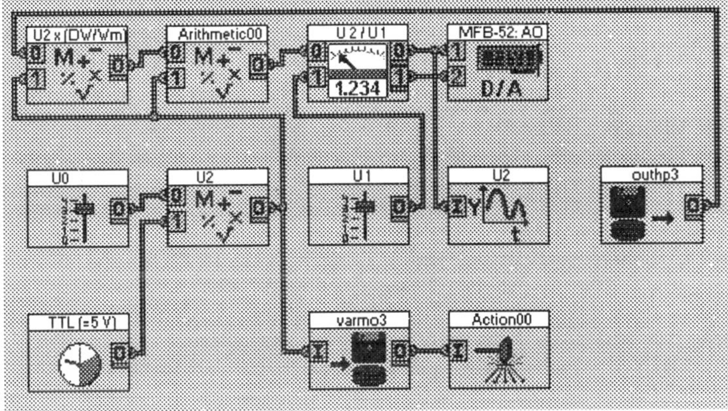

Figure 4. Control and fault compensation of the hydraulic motor

In another part of the DASYLab "experiment" the measured speed in the steady state of the hydraulic motor is compared with the theoretical value that comes from the simulation of the fault free system. By this the difference between these values is acquired and written to the file "outhp3". The data of this file are used for the compensation of the motor speed reduction caused by the motor leakage.

To compensate the reduction of the motor speed, the correction voltage U_2 x $\Delta\omega/\omega_m$ is calculated by the module "U2x(DW/Wm)" and added to the command voltage U_2. The result then influences the motor speed in a fault compensating way.

The "U1" module gives the possibility for experimentally varying the system pressure p_o with the voltage U_1. The "U2" chart module displays the voltage U_2 to the proportional 4-way valve. The "U2/U1" analogue meter module shows the voltage U_2 for the regulation of the speed of the hydraulic motor and U_1 for the regulation of the system pressure. The "MFB-52:AO" module corresponds to two analogue outputs of an I/O card that is built in the computer and transfers the command voltage signals to the amplifier of the proportional 4-way valve for the regulation of the oil flow to the hydraulic motor and the amplifier of a proportional pressure valve for the regulation of the system pressure. The "varmo3" module is used to store the value of the voltage U_2 that is needed by the simulation program, so that the simulation runs evry time with the updated command value. The "Action00" module is used to switch to the expert system after finishing the fault compensation process.

5. Conclusion

The main importance of embedding fault compensation functions in a diagnostic system lies in the economical benefits gained for hydraulic systems by preventing unexpected shutdowns in a production procedure as well as in the offered higher degree of safety and reliability for industrial systems. In this paper, the monitoring and fault compensation procedure in a hydraulic system using the DASYLab software is presented. The experimentation results show that the control of the electro-hydraulic motor and the fault compensation system using the DASYLab software is reliable and the method is applicable to real world situations.

6. References

[1] R. Isermann, Process Fault Detection Based on Modelling and Estimation Methods - A Survey. *Automatica* (1984),Vol 20, N. 4. pp 387-404.

[2] R. Patton, P. Frank and R.N. Clark, *Fault diagnosis in dynamic systems, Theory and application,* Prentice Hall, 1989.

[3] P. Frank, Fault Diagnosis in Dynamic Systems Using Analytical and Knowledge-based Redundancy - A Survey and Some New Results. *Automatica* (1990)Vol. 26, No. 3. pp. 459-474.

[4] R. Isermann and P. Balle, Trends in the Application of Model Based Fault Detection and Diagnosis of Technical Processes. *IFAC World Congress 1996*, San Francisco.

[5] R. Patton, Fault-Tolerant control: The 1997 situation, In IFAC Symposium, *Fault Detection, Supervision and Safety for Technical Processes*, Kingston Upon Hull, UK, 26-28 August 1997, pp. 1029-1051.

[6] J. Zhang and A. Morris, A comparison of on-line fault diagnosis systems for a CSTR. IFAC Workshop on on-line Fault Detection and Supervision in the Chemical Process Industries, Newcastle, 1996, June 12-13.

[7] P. Frank, Robust model-based fault detection in dynamic systems. In IFAC symposium, *On-line fault detection and supervision in the chemical process industries* Newark, Delaware, Eds. P. Dhurjati and G. Stephanopoulos. No.1, Pergamon Press, 1993.

[8] M. Pfau-Wagenbauer and W. Nejdl, Integrating model-based and heuristic features in a real-time expert system. *IEEE Expert* (1993), August, pp.12-18.

[9] R. Rengaswamy and V. Venkatasubramanian, An integrated framework for process monitoring, diagnosis, and control using knowledge-based systems and neural networks. In IFAC symposium, *On-line fault detection and supervision in the chemical process industries,* Eds. Dhurjati P. and G. Stephanopoulos. No.1, Pergamon Press, 1993.

[10] Harris T., C. Seppala, P. Jofriet and B. Surgenor, Plant-wide feedback control performance assessment using an expert system framework. *On-line fault detection and supervision in the chemical process industries* . Eds. Morris A. and E. Martin, Pergamon, 1996.

[11] C. Angeli and A. Chatzinikolaou, Fault Prediction and Compensation Functions in a diagnostic knowledge-based system for hydraulic systems. *Journal of Intelligent and Robotic systems* (1999), Vol. 25, No.2, pp. 153-165.

A Codeword-Based Approach to Diagnose the Thermal Imaging of Printed Circuit Boards

Shih-Yuan Huang*, Chi-Wu Mao*, Kuo-Sheng Cheng**

*Department of Electrical Engineering,
**Institute of Biomedical Engineering,
National Cheng Kung University, Tainan 701, Taiwan, R.O.C.
(No. 1 Ta-Hsueh Road, Tainan City, Taiwan, 701, R.O.C.)
Tel: (886)6-2757575 ext 62346 Fax: (886)6-2761785
E-mail: syhuang@chinyi.ncit.edu.tw

Abstract. In this paper a new thermal image fault detection technique called Codeword-Based Comparison Technique (CBCT) is presented. First, a normal reference thermal image denoted as 'gold' thermal image is compressed into codebook, and then each block in thermal image of the Board Under Test (BUT) is compared with its corresponding codeword. Since each block is represented as the codeword in the codebook during the encoding phase, every codeword can be compared accordingly with each specific block in BUT. The algorithm is thus highly structured into modules, hardware independent, and easy to work. The experimental results demonstrate that the performances of the proposed method are satisfactory.

1. Introduction

There are several non-destructive diagnostic techniques which have been developed to augment the traditional Automatic Test Equipment (ATE). However, the thermal image of Printed Circuit Board (PCB) can be affected by faults such as internal device damage, short loop, open circuit nodes, and power supply failure. Therefore, many thermal image processing techniques have been proposed to locate a rough range of faults in PCB [1-6]. In this paper the major feature is to encode the gold thermal image into codebook, then it is compared with the BUT to identify the fault blocks.

The reference thermal image for the normal operation of the printed circuit board is denoted as the 'gold' thermal image. For the gold thermal image, the peak temperature is used as a thermal feature in our experiment. Then, each divided block of BUT is compared with its corresponding codeword in the codebook. Since vector quantization has the potential to represent the main feature of an image into a codebook with a minimum distortion, we apply an unsupervised learning approach called compensated fuzzy Hopfield neural network (CFHNN) [7] to design a codebook for the gold thermal image. This paper is organized as follows. The proposed Codeword-Based Comparison Technique (CBCT) is described in Section 2. The experimental results are presented in Section 3, and finally the conclusion is made in Section 4.

2. Codeword-Based Comparison Technique

First, the gold thermal image is divided into n blocks. Each block z_x contains b elements. Let blocks $\{z_x\}$ be the input, and apply the CFHNN [7] to generate a codebook having c codeword $\{w_i\}$. The optimal index can be found by minimizing the squared Euclidean distance between codeword w_i and input vector z_x as Eq. (1)

$$ED_{x,i} = \min_i \{ED_{x,j}\} \quad for \ j = 1,2,\ldots,c, \tag{1}$$

where

$$ED_{x,j} = \sqrt{\sum_{s=1}^{b}(z_{x,s} - w_{j,s})^2} \ . \tag{2}$$

Second, denote m_x as the mean value of z_x and m_i as the mean value of w_i. According to the Cauchy-Schwarz inequality [8], the difference between m_x and m_i for the Euclidean distance $ED_{x,i}$ between z_x and w_i can be derived as

$$\left| m_x - m_i \right| \leq \frac{ED_{x,i}}{\sqrt{b}} = \Delta_{x,j} \ . \tag{3}$$

Because the gray level of a pixel in the gold thermal image represents the maximum operating temperature in the same position for all normal function boards, the mean temperature of block z_x in gold thermal image would be greater than or equal to the mean temperature of the corresponding block z_y in the thermal image of BUT.

If a mean temperature in a block of a BUT is defined as m_y in accordance with Eq. (3), m_y in normal BUT will be limited by no more than the upper mean temperature m_U as

$$m_U \leq m_i + \frac{ED_{x,i}}{\sqrt{b}}, \tag{4}$$

Finally, the thermal image of BUT is also divided into n blocks and each block z_y contains b elements (i.e. all the sizes of w_i, z_x, z_y are equal). Thus, the thermal information of the gold thermal image is included into a codebook. Accordingly, we can directly measure the mean difference between z_y and its corresponding codeword w_i.

To increase the diagnostic sensitivity and alleviate the problem arising from the large $ED_{x,j}$, we propose an adaptive threshold as follows. Let $DM_i = m_{i+1} - m_i$. If any one of these criteria described in the following is matched, z_y is then identified as a defect block.

Criterion 1: $m_y > m_U$ *and* $\Delta_{x,i} \leq DM_i$.

Criterion 2: $m_y > m_{i+1}$ *and* $\Delta_{x,i} > DM_i$.

The detailed description of the proposed method to identify the faulty blocks is illustrated in Figure 1. The algorithm may be summarized as follows.

Step 1: Divide the gold thermal image of PCB into n blocks (z_x represents the x's block) with the size of each block b .

Step 2: Use the CFHNN to generate the codebook.

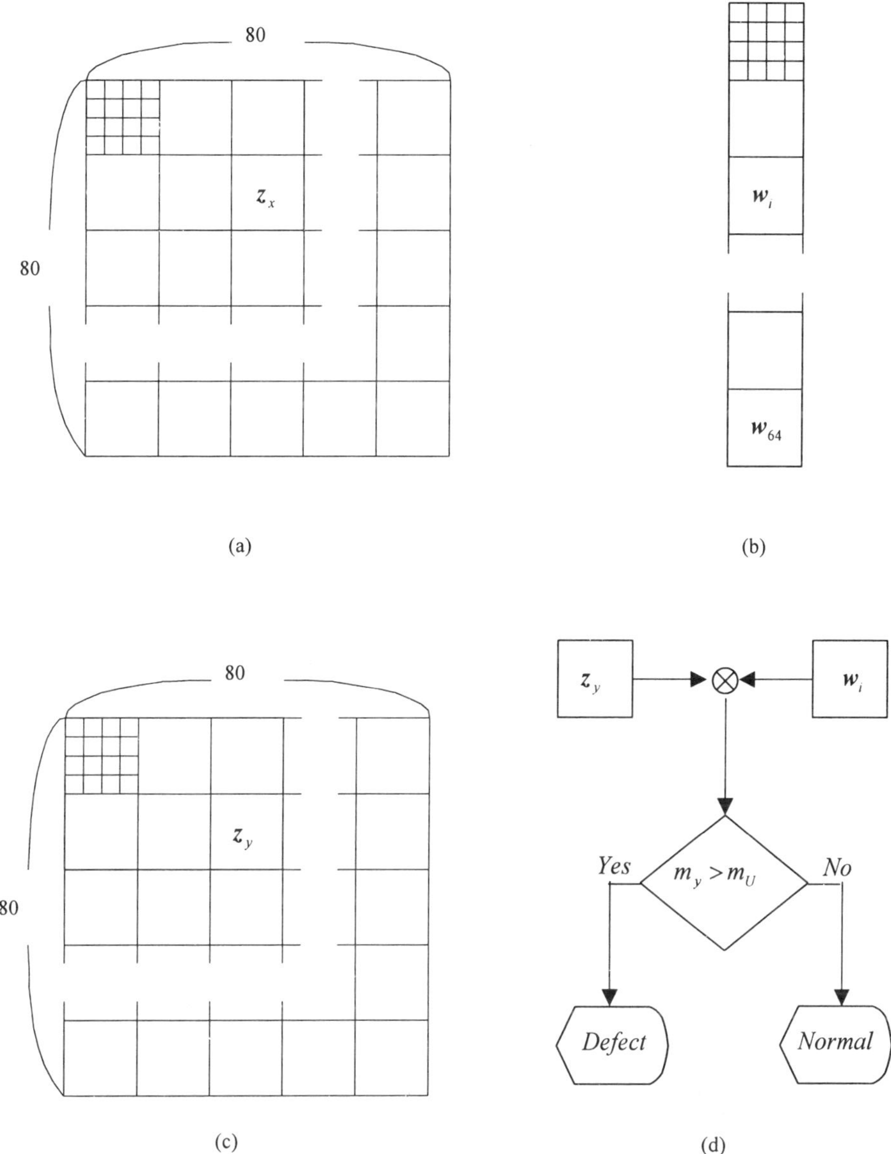

Figure 1. Illustration of the experimental procedure. (a).Step 1: Divide gold thermal image into 400 blocks, (b). Step 2,3: Generate codebook and encoding, (c). Step 4: Divide BUT image into 400 blocks, and (d). Step 5: Identify the faulty block

Step 3: Apply Eq. (1) to search each block z_x and map to i's codeword in codebook. Then, record $\Delta_{x,i}$ as defined in Eq. (3), and the mean value m_i for each codeword w_i. Compute the distance between the mean value and the difference between two consecutive codeword DM_i, and the upper limit of mean value m_U as Eq. (4).

Step 4: Divide the thermal image of BUT into n blocks. The size of the block is the same as the block z_x and codeword w_i.

Step 5: Identify the faulty blocks using adaptive threshold criterion technique.

3. Experimental Results

We select a monitor control board as a BUT, in which an Altera CPLD and two buffer chips are used. To verify the novel algorithm proposed in this paper, five thermal images for five normal operation boards are acquired to extract the gold thermal image and the thermal images for some other boards driven by the wrong clocks are obtained for diagnosis. In the experiments, the 40-MHz clock signal was the correct clock for normal operation of this board. While the wrong clock frequencies such as 50 MHz and 80 MHz are used in the fault boards denoted as BUT1 and BUT2, respectively. The thermal image for the gold image is shown in Figure 2, and BUT1 and BUT2 are shown in Figure 3. In the analysis, the gold thermal image as shown in Figure 2(a) was divided into 400 non-overlapping blocks with a 4x4 size to generate a codebook containing 64 codewords. The image reconstructed from the codebook is shown in Figure 2(b).

For performance comparison, an different approach called Block Based Comparison Technique (BBCT) is also developed. In the BBCT method, the blocks of the BUT are directly compared to the mean values of those corresponding blocks in the gold thermal image. Figures 4 show the experimental results in fault conditions for 50 MHz and 80 MHz. It is demonstrated that the proposed codeword-based diagnostic system performs better.

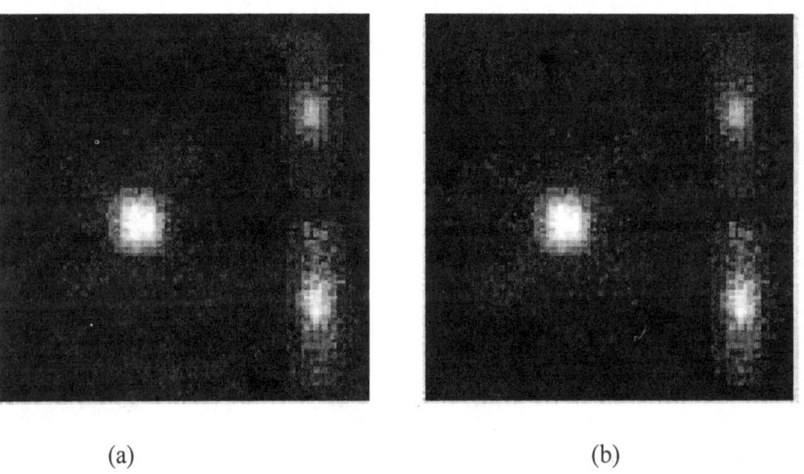

(a)   (b)

Figure 2. (a) The gold thermal image extracted from normal operation printed circuit board. (b) The image reconstructed from the codebook.

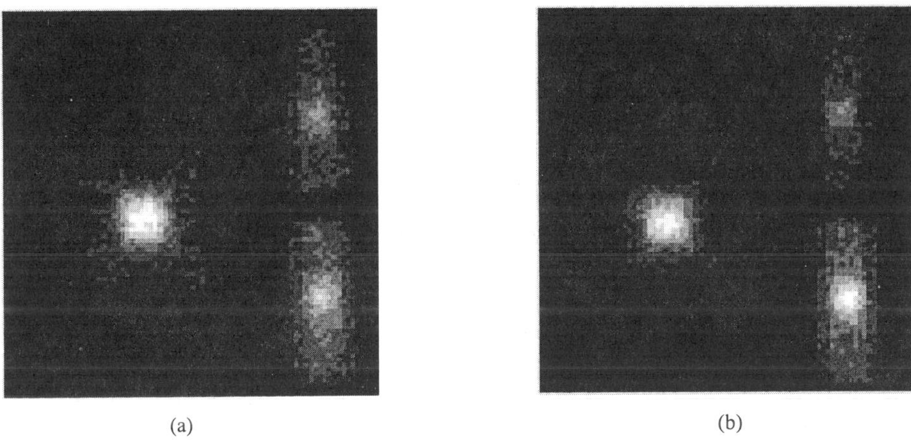

(a)   (b)

Figure 3. Two examples of the thermal images for PCB faults. (a) BUT1 worked in 50 MHz, and (b) BUT2 in 80 MHz.

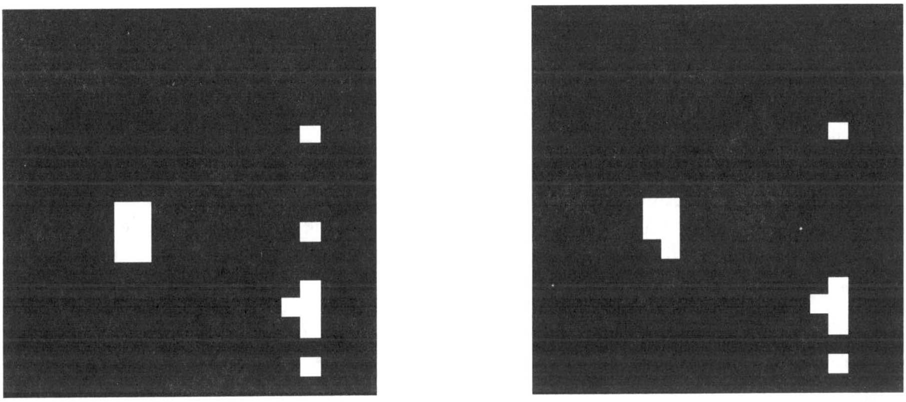

(a)   (b)

(c)   (d)

Figure 4. The identified fault regions for two cases. (a) and (b) for the case 1 using BBCT and CBCT, respectively, and (c) and (d) for the case 2 Using BBCT and CBCT, respectively.

4. Conclusions

In this paper, a novel thermal image analysis system for printed circuit board diagnosis is presented. The system basically works on the codewords generated from a gold thermal image. The proposed algorithm, which is highly modularized, can be further implemented to run on a multiprocessor system or a pipelined architecture for speeding up the computation. Since only the mean and comparing operations are involved in the diagnostic phase of the proposed method, it may be easily realized in hardware design approach.

Acknowledgements

This work was partly supported by the National Science Council, ROC under the Grant No. NSC 89-2213-E-006-095. The authors would also like to thank Prof. Herchang Ay at the Department of Mechanical Engineering, Southern Taiwan University of Technology, Tainan, Taiwan, ROC.

References

[1] Hugh F. Spence and Daniel P. Burris, "An artificial neural network printed circuit board diagnostic system based on infrared energy emissions," *Proc. Autotestcon*, pp. 41-45, 1991.

[2] Lloyd G. Allred and Gary E Kelly, "A system for fault diagnosis in electronic circuits using thermal imaging," *Proc. Autotestcon*, pp. 455-458, 1992.

[3] Stepen A. Merryman and R. M. Nelms, "Diagnostic technique for power systems utilizing infrared thermal imaging," *IEEE Trans. Industrial electronics*, pp. 615-628, vol. 42, no. 6, Dec. 1995.

[4] Delores Washburn and Michael Nguyen, "Commercial off the shelf (COTS)/Non developmental item (NDI) testing for department of defense using infra-red technology," *Proc. Autotestcon*, pp. 455-458, 1995.

[5] Lloyd G. Allred and Thomas R. Howard, "Thermal imaging is the sole basis for repairing circuit cards in the F-16 flight control panel," *Proc. Autotestcon*, pp. 418-424, 1996.

[6] Lloyd G. Allred and J. Paul Harames, "Guidelines for successful implementation of infrared thermography for repairing electronic circuit cards," *Proc. Autotestcon*, pp. 410-417, 1996.

[7] S.–H. Liu and J.–S. Lin, "A compensated fuzzy Hopfield neural network for codebook design in vector quantization," *International Journal of Pattern Recognition and Artificial Intelligence.* Vol. 14, No. 8, pp. 1067-1079, Dec. 2000.

[8] Murray R. Spiegel, *Mathematical Handbook of Formulas and Tables*, New York: McGraw-Hill, 1968.

Diagnosis and Treatment of Machinery Based on Case of Typical Fault Modes, Features and Countermeasure Analysis (TFMFCA)

Zhang Jinyu, Zhang Youyun, Xie Youbei   (Xi'an Jiaotong University, Xi'an, China)

Abstract: Intelligent diagnosis system based on case-based reasoning (CBR) is a kind of new applied system in diagnostic area. In this paper, firstly, a rationalizing case knowledge acquisition method based on TFMFCA is proposed to meet the problems in CBR knowledge capture. Then, a common relative database (RDB) to a multi-attribute fuzzy RDB is extended, and the concepts of feature knowledge cell (FKC) and FKC dictionary and the case knowledge representation method based on RDB is proposed. The third, case retrieval models and strategy depending on FKC is developed. Finally, a retrieval method and applied example based on fuzzy similarity measurement is presented. The example shows that the troubleshooting based on TFMFCA is feasible.

Index Words: Fault Diagnosis, FMECA, CBR, Retrieval, Knowledge System, AI, Database, and Fuzzy Theory.

1 INTRODUCTION

Intelligent diagnosis based on CBR has been developing [1,2] along with progress of CBR research in artificial intelligence (AI) area over the last years. This kind of reasoning method can well imitate cognition activity of human as fault diagnostician, for example, by this method the previous typical case in memory could be used to analyze, diagnose and treat current failure. At the same time, the new failure case of diagnosis and treatment could be saved into the memory for later diagnosis and treatment. Therefore it is regarded by human being.

CBR is different from traditional rule-based reasoning (RBR). The latter needs a mass of rule knowledge and a rule-based reasoned. The former needs a great deal of fault case knowledge and a similarity-based case retrieval mechanism.

The research of diagnosis and treatment based on CBR mainly focuses on case retrieval mechanism [3-5], integrating CBR and other theory [5-7], and system implementation *et al.* [8] now. The study object of case retrieval is similar with traditional fault diagnosis [9-12] to investigate the corresponding relations between fault features and fault causes mainly. But diagnostic CBR system has better flexibility and more available information. It is shown that efficient case retrieval is successful key of the diagnosis and treatment, and that efficient case knowledge acquisition and management is diagnostic foundation in CBR.

It is unfavorable for the development of diagnosis and treatment that case knowledge acquisition and management mainly depends on people's handwork, as we know. Some important investigations in diagnostic RBR by means of FMECA were presented in Ref. [13-17]. These results establish a good foundation for FMECA in diagnostic CBR system.

According to the requirement of intelligent diagnostic CBR system and RDB, some important improvement and extension of FMECA are made, a set of case knowledge models and methods of acquisition, representation, management and retrieval based on TFMFCA are proposed in this paper.

2 TFMFCA

The fault mode, effects and criticality analysis (FMECA) is an important method in reliability and security engineering. This method is essentially a "from bottom to top" systemic analysis discipline. The effects of every fault mode of item to system performance are analyzed and determined one by one from basic parts (called as item) to equipment or system. The criticality of every fault mode to person and equipment is evaluated by consequence analysis.

A new method is developed by integrating diagnosis feature set, logic resolution with FMECA in order to represent and manage case knowledge of rotate machinery diagnosis effectively. This method is fully different from traditional FMECA. Firstly, this method is diagnosis and treatment oriented. It focuses on the typical fault modes (TFM), reduces the analysis task. Secondly, this method emphasizes to analyze the fault features and treatment countermeasure of TFM. Therefore, we define this method as "Typical Fault Mode, Feature and Countermeasure Analysis (TFMFCA)". Following tasks are carried out in TFMFCA:

1. The peculiarity and consequence of every fault mode from bottom basic parts of equipment to system (include error operation and circumstance) are analyzed. The TFM would be fond out which have brought major hazard in history or has grave consequence.

2. The mechanism, causes and peculiarities of TFM are investigated. The taking place, growing, inner rule and outer behavior of fault should be determined.

3. The corresponding relation between fault causes and features should be discovered by use of fault diagnosis methods in existence.

4. The effects and consequences of TFM on the equipment should be analyzed.

5. The maintenance countermeasure of TFM will be decided by using a progressive logic diagram.

3 CASE KNOWLEDGE ACQUISITION BASED ON TFMFCA

It is known from section 2 that we are able to find out various contents such as fault mode, fault cause, fault feature, fault consequence and countermeasure of fault treatment from the definition of TFMFCA. All these contents are the knowledge needed in failure diagnosis and treatment. Thereby, they may be regarded as the case knowledge of diagnosis and treatment.

Four kinds of tables for acquiring and analyzing case knowledge are suggested in this paper In order to generalize the analysis process and content of fault case knowledge efficaciously based on TFMFCA. Taking turbine-generator for example, we set up a case knowledge analysis and acquisition computer-aided system based on TFMFCA, and applied it to a 200MW steam turbine. There are several typical cases of the rotator in Table 1-4.

1. Partial important item and TFM

Equipment always consists of many elements, parts and subsystems, which possess some special functions and have some link relations between each other such as machine, physics or chemistry. All the elements, parts and subsystems are generally called as item. The function and the significance of every item are commonly inequitable. Consequently, the important items should be firstly determined, and the functions, subjection relationship and typical fault information of item should be considered too. So the contents of Table one are shown in table 1.

Table 1 Partial Important Item and Its TFM

No.	Item Name	Father Item	Functions	TFM	remark
1	Rotor	Rotor System	Energy Transformation	1. Rotor Unbalance; 2. Liquid Accumulation Inside Rotor	
2	Shafts	Rotor System	Energy Transfer	1.Out of Alignment	
3	Bearings	Rotor System	Supporting	1.weared; 2.High Temperature	
...	...	...	...	...	

2. Causes and peculiarities of TFM

In order to analyze the mechanism, direct cause, indirect cause, taking place, growing, inner rule and outer behavior of every TFM, the Table two is designed as table 2:

Table 2 Causes and Peculiarities of Partial TFM

No.	Name of TFM	Fault Mechanisms	Fault Causes	Peculiarities	Types
1-1	Rotor Unbalance	Mass Eccentricity	Original unbalance; Parts fall off; Bend of shaft; Encrusting matter on the blades.	Increasing with the square of rotating speed; Gradually growing.	
1-2	Liquid Accumulation Inside Rotor	Excitation of centrifugal force and friction of accumulated liquid.	Bug of Design and manufacture; Coagulated Liquid	Unstable	
2-1	Out of Alignment	Out of center alignment or angle alignment.	The Error in installing; Coupling error; Distortion of box or base; Abnormal pipeline force.	Appearing after the rotor is installed, steadily gradually growing.	
...	...	...	...	...	

3. Features of TFM

It is the foundation of fault diagnosis to analyze the typical behavior features of equipment in the fault condition. These features usually come from several different aspects including vibration time-domain features, vibration frequency-domain features and other features such as noise, heat, oil, electromagnetism etc. According to the specialty of rotating machinery, the diagnosis features are divided into three categories in this paper: 1) frequency features based on amplitude spectrum; 2) independent condition features; 3) knowledge features. The Table three is designed as table 3:

Table 3 Feature of Partial TFM

No.	Name of TFM	Spectrum Feature	Independent Features	Knowledge Feature Based on Rule
1-1	Rotor unbalance	5:0.9, 6:0.05, 7:0.05	3:1, 11:1	1. Work frequency is the main component. 2 Axis track is positive ellipse. 3.Axis vibration is nearly to 0 at low velocity.
1-2	Liquid accumulation Inside Rotor	3:0.2, 4:0.5, 5:0.3	1:3, 2:3, 3:1, 4:1	1. The component of less than work frequency is the main component and the work frequency is secondary. 2 Axis track is positive ellipse. 3. Vibration is disorder with strong randomicity.
2-1	Out of Alignment	5:0.4, 6:0.5, 7:0.1	3:1, 4:1, 6:1, 8:1, 9:1,	1.The duple frequency is the main component and work frequency is secondary. 2. Axis track is positive "banana" or "8". 3. One of the oil-film pressures of two adjacent bearings becomes bigger, and another becomes smaller. 4. Vibration is sensitive to charge. 5.The relative phase of two rotors beside coupling is near $180°$.
...	...	...	...	...

4. Consequence and countermeasure of TFM

There may be various faults in equipment's life. Effect and consequence of every fault also differ in hundreds ways. Some faults are catastrophic or critical; they affect safety of person and equipment. Some are major; they result in placing out of service. Some are minor. So we should carefully analyze the effects and consequences of the faults in order to diagnose and treatment the faults correctly.

Then, how to decide the consequences and countermeasures? The reliability centered maintenance (RCM) thought and maintenance steering group (MSG) decision logic is adopted in this paper. That is to say, in the light of the logical gradation of RCM, the assured question is answered and the maintenance strategy determined one by one. The Table four is designed as table 4:

Table 4 Consequences and Countermeasures of Partial TFM

No.	Name of TFM	Fault Effects			Consequences	Counter-measure	Treatment
		Local	upper	Final			
1-1	Rotor unbalance	Rotor is acted by alternate charge, tend to occur fatigue demagnificati on.	Bearing load is increased; Shafting vibration is excited.	Reliability and life of equipment are strongly reduced.	Economy	Predicted Maintenance	Repair and correct balance
1-2	Liquid accumulat ion inside rotor	The vibration modality of rotor is changed.	Unsteady vibration is excited.	Equipment security is affected.	Security	Predicted Maintenance	Reduce or prevent from the liquid.
2-1	Out of Alignment	Rotor, bearing and coupling are acted by alternate stress; the fault probability is increased.	Abnormal vibration of shafts is resulted in.	The life and security of shafts are decreased.	Economy	Predicted Maintenance	Adjust the alignment, and consider effect of work state.
...	...	...	...	...	...	...	...

4 REPRESENTATION OF TFMFCA CASE BASED ON RDB

TFMFCA case knowledge is recorded in 2-D table from section 2. Relational database (RDB) can help us to manage and operate the case knowledge in high efficiency.

4.1 RDB Model

RDB system mostly is applied to manage object and its attribute data, e.g. someone and his/her personal information, equipment and their fault information.

DEF 1: Suppose $X = \{x_1, x_2, \cdots, x_N\}$ and $H = \{a_1, a_2, \cdots, a_M\}$ are finite sets, note

$$X \otimes H = \{(x,a); \quad x \in X, a \in H\},$$

$I \subset X \otimes H$ is a relation from X to H, $(x,a) \in I$ denotes x has attribute a, and I has property as follows:

(I) $\forall x \in X$, there are $a \in H$ and $a' \in H$, make $(x,a) \in I, (x,a') \notin I$;

(II) $\forall a \in H$, there are $x \in X$ and $x' \in X$, make $(x,a) \in I, (x,a') \notin I$.

If P is a probability measure on X we call (X,P,H,I) as a relational database system.

We might find out that RDB commonly represent data by OO-table from DEF1. This is similar to TFMFCA case knowledge representation.

In order to build knowledge system based on RDB, first of all problems to be solved is representation of TFMFCA case knowledge in database table. Practical researches demonstrate there are three kinds TFMFCA case knowledge mainly: (i) descriptive knowledge (DK); (ii) feature knowledge (FK); and (iii) algorithm knowledge (AK). The DK may be directly saved in database. The AK may be implemented in a kind of computer language. But the FK can not be directly represented in database.

4.2 Extending RDB

FK of equipment faults usually is some experiential mapping relations between features and faults, and is correlative to the equipment type. For example, fault feature sets of different equipment, such as rotator system, gearbox and reciprocating machinery, are different each other. So the representation methods of fault FK are also not the same. Generally there are three kinds of the fault features in rotating machinery: (i) feature vector; (ii) independent feature; (iii) rule based knowledge (refer to table 3).

It is clearly that to represent the knowledge in an ordinary (X,P,H,I) RDB directly is difficult. In order to represent and manage these feature sets and attribute relations effectively in ordinary database, the RDB must be extended.

DEF2: Suppose $X = \{x_1, x_2, \cdots, x_N\}$ and $H = \{a_1, a_2, \cdots, a_M\}$ are finite sets, the value set of $a_i (1 \le i \le M)$ is also finite set $Y_i (1 \le i \le M)$. Note:

$$Y = Y_1 \otimes Y_2 \otimes \cdots \otimes Y_M$$

$V : X \rightarrow Y$ · a mapping, where

$$V(x) = (a_1(x), a_2(x), \cdots, a_M(x)) \in Y$$

Means that x has attribute $a_i (1 \le i \le M)$, viz. $a_i(x) \in Y_i$ is value of attribute a_i to x.

If P is a probability measure on X, we call (X,P,H,Y,V) as a multi-attribute relational database system (MARDB).

Deduction 1: In a MARDB system, if $Y_i = \{0,1\}(i \le M)$, then a MARDB system become ordinary RDB.

DEF3: Suppose $F = \{f_1, f_2, \cdots, f_N\}$ and $A = \{a_1, a_2, \cdots, a_M\}$ are the finite sets, they respectively represent TFM sets and fault feature sets. The value sets of $a_i (1 \le i \le M)$ are feature vector, and finite sets $Y_i \in Y (1 \le i \le M)$, denote that f has feature vector of attribute sets $a_i (1 \le i \le M)$. $V_i (1 \le i \le M)$ is attribute set of Y_i and B_i is fuzzy measure of the feature vector. Note:

$K_i(a_i) = (Y_i, V_i, B_i)$ to be a feature knowledge cell (FKC). Then fault feature set may be represented in FKC set: $A = (K_1, K_2, \cdots, K_M)$.

If P is a probability measure on F, Y_i and V_i has fuzzy relation on A, we call (F,P,A,Y,V) as a multi-attribute fuzzy relational database system (MAFRDB).

Deduction 2: If FKC $K_i(a_i)$ is a common set of a_i, then a MAFRDB system becomes a

MARDB system.

4.3 FK Representation

There are several ways to extend a RDB system to a MAFRDB. One of the effective methods is to represent FKC by extending a string field of RDB. For example, the FKC of independent feature may be represented as "feature name: amplitude: membership value." The FKC of spectrum vector of rotating system in nine bands may be represented as follows:

name	amplitude	membership value
the first band spectrum f1	A1	$\delta 1$
the second band spectrum f2	A2	$\delta 2$
……	……	……
the ninth band spectrum f9	A9	$\delta 9$

Rule based knowledge may be represented as

name	threshold value	membership value
rule1	Tv1	$\Delta 1$
rule2	Tv2	$\Delta 2$
……	……	……
rule N	Tvn	Δn

Through such transformation, the case FK based on TFMFCA could be represented, managed, edited and saved expediently in ordinary RDB.

Unfortunately, DBMS or program can not make a right judgement to two very close conceptions or sentences that human can easily identify. Therefore, it is necessary to standardize and code the FKC that need enter database. In other words, a special dictionary of FKC should be designed for comparing, managing and editing FKC conveniently. Meanwhile, the space requirement of database table may also be reduced in this way. We code FKC in three decimal numbers $(b_3 b_2 b_1)$. The b_3 denote the type of

FK: 1 denotes feature vector; 2 denotes independent feature; 3 denotes rule based knowledge. The $b_2 b_1$

denotes serial number of feature name.

5 RETRIEVAL MODEL BASED ON FUZZY SIMILARITY MEASURE

It is well known that a fault of rotating machinery is usually corresponding to multi-group features, a group of features is sometime corresponding to several faults. Fault and feature sets are not simply corresponding in one by one relation, but in more complicated fuzzy relation. Such relation makes retrieval more difficult. So rational retrieval model and strategy should be built.

Suppose $F = \{f_1, f_2, \cdots, f_N\}$ is a set of case, $A_i = (K_{i1}, K_{i2}, \cdots, K_{iM}) \subseteq A$ is a set of feature knowledge of case i. Here $K_{ij} = (Y_{ij}, V_{ij}, B_{ij})$ is FKC, $A = A_1 \cup A_2 \cup \cdots \cup A_L$ is the dictionary of FKC. For Single FKC $O_k = \{K_k\} \subset A$, the retrieval model of case may be denoted as a mapping between finite set F, A, O_k, similarity measure M and threshold value λ:

$$F^k = f(F, A, O_k, M, \lambda) \subseteq F \qquad (1)$$

The retrieval mechanism as follows:

(1) Find out the target case set $F^o = \{f_1^o, f_2^o, \cdots, f_P^o\} \subseteq F$ that possesses searched FKC O_k, make $O_i \cap A_i \neq \varnothing$.

(2) Compute the fuzzy similarity measure $S_i (0 \leq S_i \leq 1, 1 \leq i \leq P)$ between FKC O_k and the FKC of every target case f_i^o according to the similarity measure M.

(3) Order the fuzzy similarity measure S_i of FKC in descending.

(4) For a given threshold level $\lambda(0 < \lambda \leq 1)$, take $F^I = \{f_i^o, S_i \geq \lambda, 1 \leq i \leq P\}$ as the cases which meet the retrieval condition.

6 RETRIEVAL METHOD OF ROTATING MACHIBERY

Suppose the fault case set of a rotating machine is $A = \{A_1, A_2, \cdots, A_N\}$, the i th case vector of spectrum feature is $A_i = \{A_{i1}, A_{i2}, \cdots, A_{i9}\}$, its weight vector is $W = \{w_{i1}, w_{i2}, \cdots, w_{i9}\}$. Criterion spectrum vector is $B = \{B_1, B_2, \cdots, B_9\}$. The feature vector of pending fault mode is $O = \{O_1, O_2, \cdots, O_9\}$. Then standardized case feature vector $a_i = \{a_{i1}, a_{i2}, \cdots, a_{i9}\}$ may be defined as a membership value vector of spectrum feature vector vs. criterion spectrum vector, viz.:

$$a_{ij} = 1 - e^{-k(\frac{B_j}{A_{ij}})^n} \qquad (i = 1, 2, \cdots, N ; j = 1, 2, \cdots, 9) \qquad (2)$$

Where k is control coefficient, n is rank.

Basing on the same reason, standardized feature vector $o = \{o_1, o_2, \cdots, o_9\}$ of pending fault mode may be also computed by using equation (6).

Then the fuzzy similarity measure based on Manhaton distance is:

$$\delta_E(A_i, O) = \sum_{k=1}^{9} w_{ik} |a_{ik} - o_k| \qquad (i = 1, 2, \cdots, N) \qquad (3)$$

Well then, the retrieval based on fuzzy similarity measure is to find the one or several cases corresponding to the biggest $\delta_E(A_i, O)$.

7 APPLICATION IN STEAM TURBINE DIAGNOSIS

7.1 Case Study

To check the method described in section 6 and 7, a real fault case of 200MW steam turbine was investigated. This steam turbine started service in 1987. More than 30 times of unwanted vibration appeared during Jan and May of 1989. Early unwanted vibration can decrease automatically. Afterward the vibration can't decrease until decreasing charge or increasing oil temperature. When the charge of steam turbine reaches 100MW, the rotor vibration with half of rotating frequency is quite obvious. When the charge reaches 160MW the bearing vibration of steam turbine rapidly increase to 120 μm or so, the charge has to be released, the steam turbine has to be switched off. When the unwanted vibration appears, the vibration amplitude of no.2 and no.3 bearing generally reach 70~100 μm, the maximum one exceeds 145 μm. Fig. 1 is the vibration spectrum of no 2 bearing. The upper figure shows the vibration in horizontal direction. The lower figure shows the vibration in vertical direction.

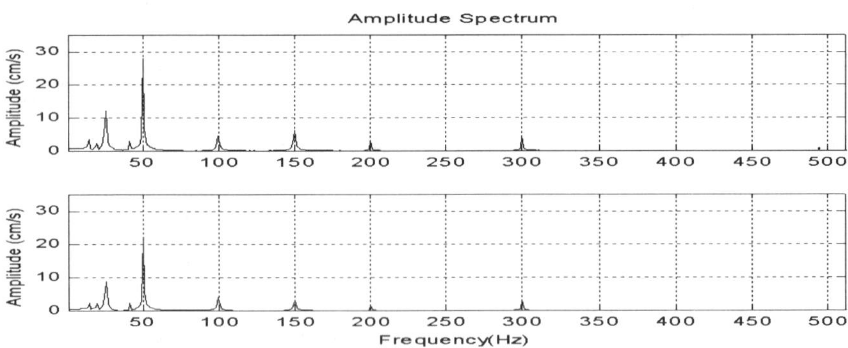

Fig. 1 Vibration spectrum of NO. 2 bearing

From the vibration spectrums, it is shown that:

1) The vibration can be quickly excited and the amplitude fluctuates very much.

2) The 25Hz vibration sometimes appears, sometimes disappears, but its max amplitude does not exceed 10% of total band in general. While the strong vibration abruptly emerges, the 25Hz vibration increases greatly, the 50Hz vibration decreases a little. Sometimes, except 25Hz vibration there are plenty of frequency components in the spectrum.

The feature vectors of the spectrum as follow:

$$O_1 = \{3.37, 2.43, 2.01, 2.96, 28.06, 4.49, 6.09, 2.84, 4.64\},$$

$$O_2 = \{2.37, 2.20, 8.81, 2.49, 22.7, 3.98, 3.20, 1.68, 3.28\}.$$

Where the spectrum is divided in nine bands: 0~0.4f; 0.4~0.5f; 0.5f; 0.5~1f; 1f; 2f; 3f; (2n-1)f; 2nf (f is the rotating frequency).

7.2 Retrieval Based on Fuzzy Similarity Measure

We use the retrieval method based on spectrum features to the fault feature vector O_1 and O_2.

The results show that the most of possible faults of O_1 are (in descending):

<table>
<tr><td colspan="2" align="center">O_1</td><td colspan="2" align="center">O_2</td></tr>
<tr><td>Fault</td><td>δ</td><td>Fault</td><td>δ</td></tr>
<tr><td>(1) vibration excited deflective backlash</td><td>0.6957</td><td>vibration excited deflective backlash</td><td>0.6138</td></tr>
<tr><td>(2) liquid accumulated inside rotor</td><td>0.5925</td><td>liquid accumulated inside rotor</td><td>0.4917</td></tr>
<tr><td>(3) supporting loosening</td><td>0.5666</td><td>big backlash of radial bearing</td><td>0.430</td></tr>
<tr><td>(4) big backlash of radial bearing</td><td>0.4790</td><td>supporting loosening</td><td>0.4121</td></tr>
</table>

Basing on the rule knowledge in table 3, we know the fault of liquid accumulated inside rotor results in half frequency vibration and the vibration does not increase with charge augment. So probability of the fault (4) may be farther eliminated.

Thereby, we can say that the faults of this steam turbine are mainly resulted from deflective backlash, interaction supporting loosening and big backlash of radial bearing.

7.3 Proof

When the cylinder of steam turbine is uncovered and repaired the big backlashes of high-pressure rotor and middle pressure rotor was found. The averaged backlash of two sides is $0.40\,mm$. The lower bearing bushings of no.2 and no.3 both wore out more than $0.2\,mm$. After adjusting the position of the rotor, replacing new bearing bushings and balancing the shafting, the steam turbine has been normally working all long.

8 DISCUSSION AND CONCLUSIONS

Following work has been done in this paper:

1) The TFMFCA of machine and a case knowledge acquisition method based on TFMFCA are proposed.

2) An ordinary RDB is extended to a MAFRDB. The conceptions of FKC and FKC dictionary are proposed. A representation of the case knowledge based RDB is suggested.

3) Case retrieval models based on FKC are proposed.

Through the example of a 200MW steam turbine, following conclusions could be given:

1) TFMFCA is a fast and effective method.

2) The representation and management of TFMFCA Case knowledge based on RDB is a convenient.

3) The retrieval model and strategy based on FKC are feasible.

ACKNOWLEDGMENT

The financial support from the Ministry of Science and Technology of China under the grant No. PD9521908Z1 is gratefully acknowledged.

REFERENCES

[1] Patterson, D.W.R., Hughes, J.G., Case-based reasoning for fault diagnosis, New Review of Applied Expert Systems v 3 1997 Taylor Graham Pub. London UK p 15-26.

[2] Soumitra D., Berend, W., Arco, D., Case-Based Reasoning Systems: From Automation to Decision-Aiding and Stimulation, IEEE Trans. On Knowledge and Data Engineering, Vol.9, No.6, November/December 1997, p911-921.

[3] Cunningham, P., Smyth, B., Bonzano, A., Incremental retrieval mechanism for case-based electronic fault diagnosis, Knowledge-Based Systems v 11 n 3-4 Nov 12 1998, p 239-248.

[4] Gupta, K. M., Montazemi, A. R., Empirical evaluation of retrieval in case-based reasoning systems using modified cosine matching function, IEEE Transactions on Systems, Man, and Cybernetics Part A: Systems and Humans v 27 n 5 Sep 1997, p 601-612.

[5] Montazemi, A. R., Gupta, K. M., Adaptive agent for case description in diagnostic CBR systems, Computers in Industry, v29 n3 Aug 1 1996, p 209-224.

[6] Costas T., Cheng Q., and Wei H. Y. , Integrating Case-Based Reasoning and Decision Theory, IEEE Expert Systems, 1997.7/8, pp46~55

[7] Hsu, C. C., Ho, C. S., Hybrid case-based medical diagnosis system, Proceedings of the 1998 IEEE 10th International Conference on Tools with Artificial Intelligence, Taipei, China, Nov 10-12 1998, p 359-366.

[8] Lenz, M., Burkhard, H.D., Pirk, P., Auriol, E., Manago, M., CBR for diagnosis and decision support, AI Communications, v9 n3 Sep 1996, p 138-146.

[9] Vingerhoeds, R.A., Janssens, P., Netten, B. D., Fernandez-Montesinos, M. A., Enhancing off-line and on-line condition monitoring and fault diagnosis, Control Engineering Practice, v3 n11 Nov 1995, p 1515-1528.

[10] Zhang Y. Y., A New Strategy Combining Backward Inference with Forward Inference in Monitoring and Diagnosing Techniques for Hydrodynamic Bearing-Rotor Systems, Wear, Vol.173, No.1~2, pp31~37, 1994.4.

[11] D.L Thomas, Vibrational Monitoring Strategy for Large Turbo Generators. The Third int. CONF. On Vibration in Rotating Machinery. Yokshine, Sept.1984.pp91-97.

[12] D.E Bently, A Muszynska, Detection of Rotor Cracks. Proceedings of the 15th Turbo Machinery Symposium. Texas A & M Univ. 1986. 129-139.

[13] Price, C., Taylor, N., Multiple fault diagnosis from FMEA, Proceedings of the 1997 9th Conference on Innovative Applications of Artificial Intelligence Jul 27-31 1997, Providence, RI, USA, AAAI Menlo Park CA USA p 1052-1057.

[14] Boyd, M. A., Khalil, A. A., Herrin, S. A., Real-time automated diagnosis for human-computer based monitoring and control systems, Proceedings of the 1997 Annual Reliability and Maintainability Symposium Jan 13-16 1997, Philadelphia, PA, USA, p 355-360.

[15] Boyd, M. A., Khalil, A. A., Montgomery, T. A., Gebrael, M., Development of automated computer-aided diagnostic systems using FMECA-based knowledge capture methods, Proceedings of the 1998 Reliability and Maintainability Symposium Jan 19-22 1998, Anaheim, CA, USA, p 285-291.

[16] Lehrasab, N., Fararooy, S., Allan, J., Fault detection in intelligent early failure warning sensors system for Train Rotary Door Operator, IEE Colloquium (Digest) Proceedings of the 1996 IEE Colloquium on Intelligent Sensors Sep 19 1996, London, UK, p 1-5.

[17] Cayrac, D., Dubois, D., Prade, H., Handling uncertainty with possibility theory and fuzzy sets in a satellite fault diagnosis application, IEEE Transactions on Fuzzy Systems v4 n3 Aug 1996 p 251-269.

[18] Zhang Jinyu, Zhang Youyun, Xie Youbai, Research on intelligent comprehensive diagnosis processing support center for large rotating machinery, China mechanical engineering, Vol.10, p411-414, Apr. 1999.

[19] Trego, Linda E., Maintenance programs, Aerospace Engineering, 15 1 Jan-Feb 1995 SAE p 15-20.

[20] Hollick, Ludwig J., Nelson, Greg N., Rationalizing scheduled-maintenance requirements using reliability centered maintenance, Proceedings of the Annual Reliability and Maintainability Symposium Jan 16-19 1995 IEEE p 11-17.

About the authors:

J. Y. Zhang received the B. S. Degree in 1986 from Northwest Polytechnical University, Xi'an, China, and the M. S. and Ph. D. Degree in 1989 and 1999, respectively, from Xi'an Jiaotong University. He has done research on maintainability and reliability engineering. He is currently an Associate Professor of mechanical and Electrical Engineering. And he is currently doing research in remote diagnosis, e-maintenance, AI, GDSS and non-stable signal analysis.

Y. Y. Zhang received the B. S. Degree in 1970 from Xi'an Jiaotong University, Xi'an, China, and got M.Sc. Degree in 1981 and Ph.D. in 1989 in the same University. She is a Member of Standing Council, Chinese Society of Mechanical Fault Diagnosis; Committee Member of CAPE Committee of Plant Diagnosis Engineering; Committee Member of Academic Committee of Science and Engineering, Xi'an Jiaotong University; the member of ISO/TW108/SC5.

Y. B. Xie received his B. S. degree in Internal-Combustion Engine from Jiaotong University, Shanghai, in 1955. Subsequently he taught and rescarched at the same University, and moved to Xi'an, Shananxi Province, accompanying with the University in 1957. Prof. Xie has been an academician of Chinese Academy of Engineering since 1994, and is also Proc IMechE (Part J) Journal of Engineering Tribology (UK). His research interests include mechanical design & theory, and the tribological aspects of internal-combustion engine, rotating machine and fault diagnosis.

An Interpolative-Type Reasoning in Sparse Fuzzy Rule Bases

Yan SHI[1] and Masaharu MIZUMOTO[2]
[1]*School of Information Science, Kyushu Tokai University, Kumamoto 862-8652, Japan*
[2]*Faculty of Information Science and Technology*
Osaka Electro-Communication University, Osaka 572-8530, Japan

Abstract. Based on Lagrange's interpolation technique, we present an interpolative-type reasoning approach in sparse fuzzy rule bases. This fuzzy reasoning approach can guarantee the membership function of an inference consequence to be of triangular-type if all of membership functions of fuzzy rules and an observation are given by triangular-type, in a sparse rule-based fuzzy system, in general. Moreover, the efficiency of the presented method shall be shown by a numerical example.

1. Introduction

Fuzzy reasoning plays a very important role in fuzzy applications [8]. To apply some suitable fuzzy reasoning on fuzzy control and expert systems have much succeed in recent decade. One can note that all of the conventional fuzzy reasoning methods are almost based on the compact rule bases [1-3,6,7]. In these fuzzy reasoning approaches the whole input universe of discourse is covered by all the rule bases completely, and when an observation comes, a consequence can be derived by some proper inference rules [1-3,6,7]. However, if a fuzzy rule base is sparse, conventional fuzzy reasoning methods encounter a difficulty, that is that they are not able to obtain a good inference result or, no rule will be fired when an observation comes to the empty space, so that no consequence is derived. To deal with this problem, Koczy and Hirota have proposed a linear interpolative reasoning [4,5], which given a solution for the problem of sparse fuzzy rule bases. But, it has been pointed that the consequence of inference by the linear interpolative reasoning is sometimes an abnormal fuzzy set, and it is very limited to guarantee the membership function of the consequence to be a convex fuzzy set subject to some special conditions suggested by authors [9-11]. In this paper, by means of the Lagrange's interpolation technique, we present a fuzzy reasoning method in sparse fuzzy rule bases, which it can guarantee that the membership function of the consequence to be of triangular-type if all of membership functions of fuzzy rules and an observation are given by triangular-type, in general. Moreover, the efficiency of the presented method shall be illustrated by a numerical example.

2. Fuzzy Reasoning Problem in Sparse Fuzzy Rule Bases

To cope with the fuzzy reasoning problem in sparse rule bases, first, we describe briefly so-called "*Tomato Classification Problem*" proposed by Mizumoto and Zimmerman [1].

According to the generalized modus ponens, we can obtain a proper fuzzy inference consequence for the following inference:

Rule: If a tomato is red then the tomato is ripe
<u>*Observation: This tomato is very red*</u>
Consequence: This tomato is very ripe

However, difficulty arise in the following inference:

Rule 1: If a tomato is red then the tomato is ripe
Rule 2: If a tomato is green then the tomato is unripe
<u>*Observation: This tomato is yellow*</u>
Consequence: ???

What would be the consequence? Intuitively, one would have a consequence, in which the tomato should be half ripe when it is yellow, which can be represented in Fig. 1. But the consequence would be nothing by the conventional fuzzy reasoning methods [1-3,6,7].

Koczy and Hirota have looked into the above problem and, using the idea of gradual rules and analogical reasoning, they proposed a linear interpolative fuzzy reasoning for treating fuzzy rule interpolation when the fuzzy rule base is sparse [4,5]. But, it has been pointed that the consequence of inference by the linear interpolative reasoning is sometimes an abnormal fuzzy set, and it is very limited to guarantee the membership function of the consequence to be a convex fuzzy set subject to some special conditions suggested by authors [9-11].

In order to obtain the solution of this problem, we generally can rewrite the above fuzzy reasoning in the form of *"If...Then..."* fuzzy rule model as follows:

Rule 1: *If x is A_1 then y is B_1*
Rule 2: *If x is A_2 then y is B_2*
<u>Observation: *If x is A^**</u>
Consequence: *y is B^**

where A_1, A_2 and A^* are fuzzy sets on X; B_1, B_2 and B^* are fuzzy sets on Y. And, $A_1 => B_1$, $A_2 => B_2$ are the disjoint fuzzy rules on the universe of discourse $X \times Y$. Let the membership functions of A_1, A_2, B_1, B_2 and an observation A^* be of triangular-type as shown in Fig. 2, where $a_{11} = inf\{A_{1\alpha}\}$, $a_{21} = inf\{A_{2\alpha}\}$, $a_1 = inf\{A^*_\alpha\}$, $a_{13} = sup\{A_{1\alpha}\}$, $a_{23} = sup\{A_{2\alpha}\}$, $a_3 = sup\{A^*_\alpha\}$, $b_{11} = inf\{B_{1\alpha}\}$, $b_{21} = inf\{B_{2\alpha}\}$, $b_{13} = sup\{B_{1\alpha}\}$, $b_{23} = sup\{B_{2\alpha}\}$ at $\alpha = 0$. And $a_{12} = A_{1\alpha}$, $a_{22} = A_{2\alpha}$, $a_2 = A^*_\alpha$, $b_{12} = B_{1\alpha}$, $b_{22} = B_{2\alpha}$ at $\alpha = 1$.

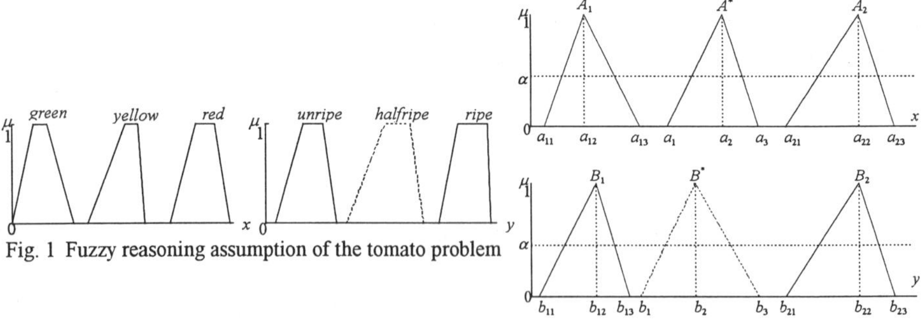

Fig. 1 Fuzzy reasoning assumption of the tomato problem

Fig. 2 Triangular-type membership functions (I)

We consider how to obtain B^*, and make its membership function to be a triangular-type as the same as B_1 and B_2. Using the interval of confidence at level $\alpha \in [0,1]$, fuzzy sets A_1, A_2, B_1, B_2 and A^* can be expressed as the following, respectively:

$$A_{i\alpha} = [inf\{A_{i\alpha}\}, sup\{A_{i\alpha}\}] = [(a_{i2} - a_{i1})\alpha + a_{i1}, -(a_{i3} - a_{i2})\alpha + a_{i3}] \quad i = 1, 2 \qquad (1)$$
$$B_{i\alpha} = [inf\{B_{i\alpha}\}, sup\{B_{i\alpha}\}] = [(b_{i2} - b_{i1})\alpha + b_{i1}, -(b_{i3} - b_{i2})\alpha + b_{i3}] \quad i = 1, 2 \qquad (2)$$
$$A^*_{\alpha} = [inf\{A^*_{\alpha}\}, sup\{A^*_{\alpha}\}] = [(a_2 - a_1)\alpha + a_1, -(a_3 - a_2)\alpha + a_3] \qquad (3)$$

To make B^* to be of triangular-type, here, there should be a form of interval of confidence at level α as

$$B^*_{\alpha} = [inf\{B^*_{\alpha}\}, sup\{B^*_{\alpha}\}] = [(b_2 - b_1)\alpha + b_1, -(b_3 - b_2)\alpha + b_3] \qquad (4)$$

which satisfies conditions if $A^* = A_i$, then $B^* = B_i$ (i=1,2). That is, $b_1 = b_{i1}$, $b_2 = b_{i2}$, $b_3 = b_{i3}$ when $a_1 = a_{i1}$, $a_2 = a_{i2}$, $a_3 = a_{i3}$ (i=1,2). In other words, when α ($\in [0,1]$) is fixed, it should be noted that B^*_{α} is a mapping from A^*_{α} about variables a_1, a_2 and a_3 with the above conditions. If we write such a mapping matrix F, then

$$B^*_{\alpha} = F(A^*_{\alpha}) \qquad (5)$$

Equivalently,

$$[inf\{B^*_{\alpha}\}, sup\{B^*_{\alpha}\}] = [F(inf\{A^*_{\alpha}\}), F(sup\{A^*_{\alpha}\})] \qquad (6)$$

Thus, we have the following two equations:

$$inf\{B^*_{\alpha}\} = F(inf\{A^*_{\alpha}\}) \qquad (7)$$
$$sup\{B^*_{\alpha}\} = F(sup\{A^*_{\alpha}\}) \qquad (8)$$

that is,

$$\alpha b_2 + (1 - \alpha)b_1 = F(\alpha a_2 + (1 - \alpha)a_1) \qquad (9)$$
$$\alpha b_2 + (1 - \alpha)b_3 = F(\alpha a_2 + (1 - \alpha)a_3) \qquad (10)$$

Due to arbitrariness of α ($\in [0,1]$), if we take α as 0 and 1 in (9) and (10) respectively, then we have the following result:

$$b_i = F(a_i) \qquad (i\text{=}1,2,3) \qquad (11)$$

which meets modus ponens [1].

Based on the above analysis, it hits us to search a mapping F which should satisfy (11) subject to the conditions of (1) - (3). If such a mapping F exists, then by (4) one can see that it can guarantee B^* is a triangular-type when A_1, A_2, B_1, B_2 and A^* are defined by triangular-type membership functions, in general. In the next section, by means of Lagrange's interpolation method, we shall give a form of the mapping F above mentioned.

3. Fuzzy Lagrange's Interpolation Reasoning

Generally, in the case of when a fuzzy inference system has n rules in a sparse rule base we have the following fuzzy inference form, and the membership functions of A_i, B_i (i=1,2,...,n) and an observation A^* be all defined by triangular-type as shown in Fig. 3.

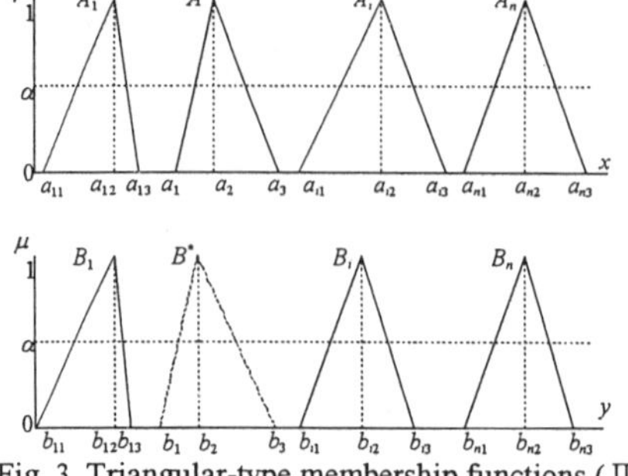

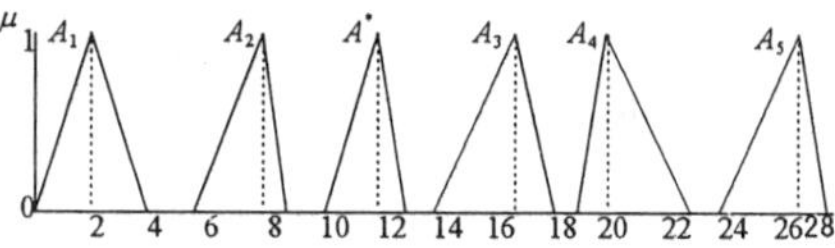

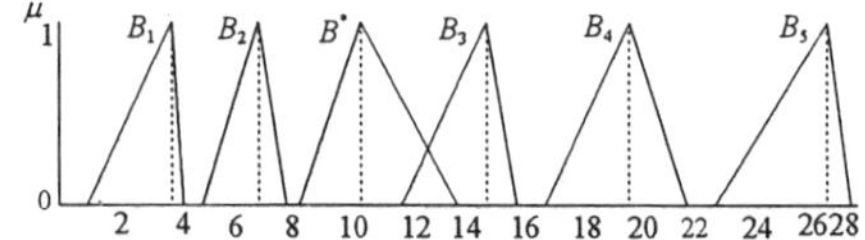

Fig. 3 Triangular-type membership functions (II)   Fig. 4 Fuzzy Lagrange's interpolation with 5 rules

Rule 1: *If x is A_1 then y is B_1*

....

Rule n: *If x is A_n then y is B_n*

Observation: *If x is A^**

Consequence: *y is B^**

To obtain a consequence B^* and make its membership function to be of triangular-type, first, triangular-type membership functions of fuzzy sets A_i, B_i $(i=1,2,...,n)$ and A^* can be expressed in the form of the interval of confidence at α $(\in[0,1])$ level as follows:

$$A_{i\alpha} = [(a_{i2} - a_{i1})\alpha + a_{i1}, -(a_{i3} - a_{i2})\alpha + a_{i3}] \qquad (i=1,2,...,n) \qquad (12)$$

$$B_{i\alpha} = [(b_{i2} - b_{i1})\alpha + b_{i1}, -(b_{i3} - b_{i2})\alpha + b_{i3}] \qquad (i=1,2,...,n) \qquad (13)$$

$$A^*_{\alpha} = [(a_2 - a_1)\alpha + a_1, -(a_3 - a_2)\alpha + a_3] \qquad (14)$$

If third n-order **Vandermonde** determinants made of parameters of $A_{i\alpha} = (a_{i1}, a_{i2}, a_{i3})$ $(i=1,2,...,n)$ satisfy

$$\begin{vmatrix} 1 & a_{1j} & a_{1j}^2 & ... & a_{1j}^{n-1} \\ 1 & a_{2j} & a_{2j}^2 & ... & a_{2j}^{n-1} \\ & & .... & & \\ 1 & a_{nj} & a_{nj}^2 & ... & a_{nj}^{n-1} \end{vmatrix} = \prod_{h>i}(a_{hj} - a_{ij}) \neq 0 \qquad (j=1,2,...,n) \qquad (15)$$

then we can define a n-order polynomial expression mapping $F: A^* \mid \rightarrow B^*$ as

$$b_j = F(a_j) = \sum_{i=1}^{n}\prod_{h\neq i}^{n} \frac{(a_j - a_{hj})}{(a_{ij} - a_{hj})} b_{ij} \qquad (j=1,2,...,n) \qquad (16)$$

Obviously, it can be proved that the definition in (16) satisfies $b_j = b_{ij}$ if $a_j = a_{ij}$ $(j=1,2,3; i=1,2,...,n)$ subject to the conditions in (12) - (14). That is to say, for arbitrary $\alpha \in [0,1]$, if $A^*_{\alpha} = A_{i\alpha}$, then $B^*_{\alpha} = B_{i\alpha}$ $(i=1,2,...,n)$. So that, we have a polynomial expression interpolation formula, which is named a fuzzy Lagrange's interpolation as follows:

$$B^*_{\alpha} = [inf\{B^*_{\alpha}\}, sup\{B^*_{\alpha}\}] = F(A^*_{\alpha}) \qquad (17)$$

$$inf\{B^*_{\alpha}\} = -(b_2 - b_1)\alpha + b_1$$

$$= [\sum_{i=1}^{n}\prod_{h\neq i}^{n} \frac{(a_2 - a_{h2})}{(a_{i2} - a_{h2})} b_{i2} - \sum_{i=1}^{n}\prod_{h\neq i}^{n} \frac{(a_1 - a_{h1})}{(a_{i1} - a_{h1})} b_{i1}]\alpha + \sum_{i=1}^{n}\prod_{h\neq i}^{n} \frac{(a_1 - a_{h1})}{(a_{i1} - a_{h1})} b_{i1} \qquad (18)$$

$$sup\{B^*_{\alpha}\} = -(b_3 - b_2)\alpha + b_3$$

$$= -[\sum_{i=1}^{n}\prod_{h\neq i}^{n} \frac{(a_3 - a_{h3})}{(a_{i3} - a_{h3})} b_{i3} - \sum_{i=1}^{n}\prod_{h\neq i}^{n} \frac{(a_2 - a_{h2})}{(a_{i2} - a_{h2})} b_{i2}]\alpha + \sum_{i=1}^{n}\prod_{h\neq i}^{n} \frac{(a_3 - a_{h3})}{(a_{i3} - a_{h3})} b_{i3} \qquad (19)$$

In (17), one can see that B^*_α is of triangular-type, and its coefficients of one-order term (about α) in the left side is two n-order polynomial expressions about variables a_1 and a_2, respectively, and its constant term is also a n-order polynomial expression about variable a_1. Similarly, we have the same results for the right. If A_i, B_i, and A^* are all given in the form of triangular-type membership functions with the conditions in (12) - (14), then according to (17), we can obtain a function expression of B^*_α, directly.

In the below, we give an illustration to show the above Fuzzy Lagrange's Interpolation method, which there are five rules in the fuzzy inference system. In Fig. 4, the parameters of fuzzy rules are given as $a_{11} = 0$, $a_{12} = 2$, $a_{13} = 4$, $a_{21} = 5.6$, $a_{22} = 8$, $a_{23} = 8.8$, $a_{31} = 14$, $a_{32} = 16.8$, $a_{33} = 18$, $a_{41} = 19$, $a_{42} = 20$, $a_{43} = 23$, $a_{51} = 24$, $a_{52} = 26.8$, $a_{53} = 28$, $a_1 = 10$, $a_2 = 12$, $a_3 = 13$; $b_{11} = 1$, $b_{12} = 4$, $b_{13} = 4.4$, $b_{21} = 5$, $b_{22} = 7$, $b_{23} = 8$, $b_{31} = 12$, $b_{32} = 15$, $b_{33} = 16$, $b_{41} = 17$, $b_{42} = 20$, $b_{43} = 22$, $b_{51} = 23$, $b_{52} = 27$, $b_{53} = 28$.

By (17) - (19), we obtain an inference consequence B^* for any $\alpha \in [0,1]$ as following:

$$B^*_\alpha = (b_1, b_2, b_3) = (8.5, 10.6, 14) = [2.1\,\alpha + 8.5, -3.4\,\alpha + 14]$$

4. Conclusions

As a kind of fuzzy reasoning method in sparse rule bases, we have presented so-called the fuzzy Lagrange's interpolation. Using the method, we can guarantee that in sparse fuzzy rule bases, if the membership functions of all rules and an observation are given in form of triangular-type, then the membership function of the consequence of inference will be also of triangular-type, which show the fuzzy inference to be more intuitive.

References

[1] M. Mizumoto and H.J. Zimmermann, Comparison of fuzzy reasoning methods, *Fuzzy Sets and Systems*, 8(1982), 253-283.

[2] M. Mizumoto, Fuzzy reasoning, *Journal of Japan Society for Fuzzy Theory and Systems* (in Japanese), 4 (1992), 256-264.

[3] M. Mizumoto, Fuzzy reasoning, *Journal of Japan Society for Fuzzy Theory and Systems* (in Japanese), 4 (1992), 35-46.

[4] L.T. Koczy and K. Hirota, Interpolative reasoning with insufficient evidence in sparse fuzzy rules bases, *Information Sciences*, 71 (1993), 169-201.

[5] L.T. Koczy and K. Hirota, Approximate reasoning by linear rule interpolation and general approximation, *International Journal of Approximate Reasoning*, 9 (1993), 197-225.

[6] M.M. Gupta and E. Sanchez (editors), Approximate Reasoning in Decision Analysis, North-Holland Publishing Company, 1982.

[7] E. Sanchez and L.A. Zadeh (editors), Approximate Reasoning in Intelligent Systems, Decision and Control, Pergamon Press, Pergamon Books LtD. 1987.

[8] L.A. Zadeh, Interpolative reasoning in fuzzy logic and neural network theory, IEEE International Conference on Fuzzy Systems (Plenary Talk), San Diego, 1992.

[9] Y. Shi, M. Mizumoto and Z.Q. Wu, Reasoning conditions on Koczy's interpolative reasoning method in sparse fuzzy rule bases, *Fuzzy Sets and Systems*, 75 (1995), 63-71.

[10] Y. Shi and M. Mizumoto, Reasoning conditions on Kóczy's interpolative reasoning method in sparse fuzzy rule bases. Part II, *Fuzzy Sets and Systems*, 87 (1997), 47-56.

[11] Y. Shi and M. Mizumoto, A note on reasoning conditions of Kóczy's interpolative reasoning method, *Fuzzy Sets and Systems*, 96 (1998), 373-379.

Maximum Network Flow Algorithm
With Fuzzy Input Data

Roman Tyshchuk

Information Systems Department, Industrial Association "Kiev-Konti", Donetsk, Ukraine

Abstract. Max network flow algorithm for networks with fuzzy capacities is proposed.
The main problems of classical methods transformation to fuzzy networks are distinguished.
The general scheme of algorithm based on modified procedure of augmenting flow obtaining
and refining procedure is represented. Advantages of the proposed algorithm using are
demonstrated by example.
Keywords: Network flows; Maximum network flow; Fuzzy capacities; Fuzzy flows.

1.   Problem Definition

Let N be a digraph with vertices $V = S \cup I \cup T$, where S represents the source vertex,
T represents the sink vertex, and I represents the intermediate vertices. Let E represents the set
of directed arcs in the digraph N. A capacity function c, $E \rightarrow \Re^+$, assigns a non-negative
number $c(u,v)$ to every arc (u,v) in E. An instance of a max-flow problem consists of a
digraph N and a capacity function c. Given an instance of a max-flow problem, a function, f:
$E \rightarrow \Re^+$, is called a flow function if it is satisfies the following constrains:

$$0 \leq f(u,v) \leq c(u,v), \forall(u,v) \in E \tag{1}$$

$$\sum_{(v,w) \in E} f(v,w) = \sum_{(w,v) \in E} f(w,v), \forall v \in I \tag{2}$$

The max-flow problem is to find an allowable flow, which maximize a linear form (2).

Algorithms for this problem have been studied for over four decades. One of the first
and well-known algorithms is the classical Ford-Fulkerson method with marking procedure
such as a big number of other methods [1, 2]. However the direct applying of this methods on
the networks with fuzzy capacities leads to incorrect intermediate and final results.

First, under the fuzzy input data conditions it is difficult to interpret capacities
constrain because of necessary of comparison of two fuzzy numbers. We have this problem
also in vertices marking procedure.

Second, during the classical algorithm execution we can't in some cases use neither
simple, nor extended arithmetical operations. Therefore in this cases we must define this
operations. Also in some cases we must define a maximum and minimum operations.

Third, in some cases all the scheme of classical algorithm with marking procedure
can't lead us to the final result and allow to find only allowable flow in the network.

The maximum network flow problem in networks with fuzzy capacities and fuzzy
flows was represented in [3-5]. But some of the methods in general can't provide us with the
correct results because of the comparison problem of fuzzy numbers was not described and
because of the decomposition methods of the fuzzy network by the real-valued networks was
used.

2.   The Main Clauses of Maximum Network Flow Algorithm with Fuzzy Input Data

The main idea of the algorithm is following. On the basis of the fuzzy max-flow min-
cut theorem [5] a conclusion about the existence of the augmenting flow have to be drawn.

After that the augmenting flow is determinated based on the direct comparison of the fuzzy numbers. This procedure allows to overcome inconsistencies, which can take a place in the algorithm proposed by Diamond [5]. The last stage of algorithm provides the max network flow obtaining by means of the refining procedure.

2.1. Capacities Constrain Correction

The main problem which occurred during the capacities constrain correction is a necessary of comparison of two fuzzy numbers. However it is not always possible to find the biggest fuzzy interval cause they can be covered with considerable proportions.

Based on basic coefficients of comparison of fuzzy numbers from the possibility theory [6,7] we can determine the following two variants of capacities constrain (Figure 1):

1.   The first variant determines the necessity of the fact, that the lowest values of the capacity $\tilde{c}(e)$ will be greater then the greatest values of the flow $\tilde{f}(e)$ $(\text{Nec}(\underline{c} > \overline{f}) = 1)$.

2. The second variant determines the following 3 conditions:

-    Determining a necessity of the fact, that the lowest values of the capacity $\tilde{c}(e)$ will be not less then the lowest values of the flow $\tilde{f}(e)$ $(\text{Nec}(\underline{c} \geq \underline{f}) = 1)$.

-    Allowing the possibility of the fact, that the greatest values of the flow $\tilde{f}(e)$ will be not less then the lowest values of the capacity $\tilde{c}(e)$ $(0 \leq \text{Pos}(\overline{f} > \underline{c}) \leq 1)$.

-    Exclusion the possibility of the fact, that the greatest values of the flow $\tilde{f}(e)$ will be greater then the greatest values of the capacity $\tilde{c}(e)$ $(\text{Pos}(\overline{f} > \overline{c}) = 0)$.

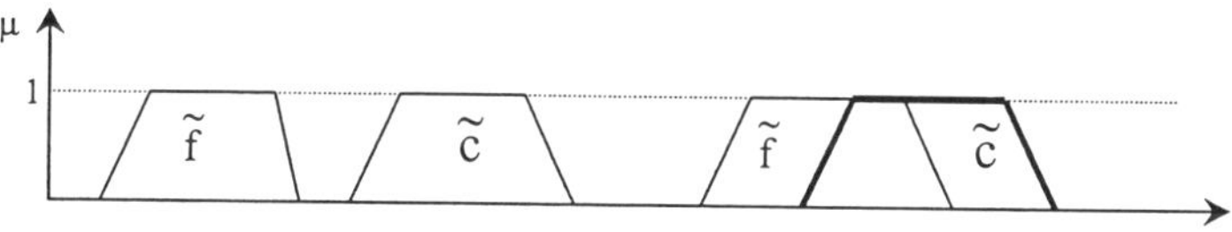

Figure 1. A fuzzy capacity constrains.

Using the first variant we get a determinate max-flow problem, where fuzzy degree of $\tilde{f}$ will reduced to 0 and $\tilde{c}$ will become $\underline{c}$.

2.2. Defining an Addition ε

The flow value ε, which we can add to the current flow in the arc $e(u, v)$ to make this arc F-saturated, in fuzzy case must fulfil the following requirements:

1. $\tilde{\varepsilon}(e) + \tilde{f}(e) = \tilde{c}(e)$. This operation must be the inverse operation to the finding $\tilde{\varepsilon}(e)$ from $\tilde{f}(e)$ and $\tilde{c}(e)$.

2. Fuzzy number $\tilde{\varepsilon}(e)$ must be correct and normal.

3. $\tilde{\varepsilon}(e) \geq 0$ (We must exclude the possibility of the negative flow appearance).

Based on these conditions the value $\tilde{\varepsilon}(e)$ must be defined in the following way:

1. $\underline{\varepsilon}_i = \underline{c}_i - \underline{f}_i$, $\overline{\varepsilon}_i = \overline{c}_i - \overline{f}_i$, where i – number of the α-cut.

2. For i=2 to N (where N – amount of the α-cuts)

    If $\overline{\varepsilon}_i > \overline{\varepsilon}_{i-1}$ then $\overline{\varepsilon}_i = \overline{\varepsilon}_i - (\overline{\varepsilon}_i - \overline{\varepsilon}_{i-1})$

3. If $\underline{\varepsilon}_N > \overline{\varepsilon}_N$ then $\underline{\varepsilon}_N = \underline{\varepsilon}_N - (\underline{\varepsilon}_N - \overline{\varepsilon}_N)$

4. For i = N to 2

    If $\underline{\varepsilon}_{i-1} > \underline{\varepsilon}_i$ then $\underline{\varepsilon}_{i-1} = \underline{\varepsilon}_{i-1} - (\underline{\varepsilon}_{i-1} - \underline{\varepsilon}_i)$

2.3.   A Minimum Operation In Marking Procedure

In fuzzy case we must define a choosing minimum value operation from the previous vertices mark and addition $\tilde{\varepsilon}(e)$. To define a minimum operation in this case we must not use the possibility theory coefficients and describe this operation as following:

Let $\tilde{C}$ = new_min_operation($\tilde{A}, \tilde{B}$). Then $\forall i : \underline{c}_i = \min(\underline{a}_i, \underline{b}_i), \overline{c}_i = \min(\overline{a}_i, \overline{b}_i)$, where i – number of the α-cut.

3.   The General Scheme of Algorithm

The General scheme of algorithm is similar to the general scheme of classical Ford-Fulkerson algorithm [8,9] with the exception of the marking procedure, which we define as following: we can't mark a vertex while all the vertices which can take part in direct or reverse marking of current vertex are not marked. Then we mark the current vertex with the maximum value of all computed marks. To finding a maximum value in this case we can use an integrated superiority coefficients PSE, NSE, PS and NS.

However in general case this method can't lead us to the final result because of following network property in fuzzy conditions: the total sum of the flow, which we can pass through the vertex can be greater than sum of flows, which we passed through vertex using marking procedure. This property we can see on the figure 1.d of Appendix: we can't pass through vertex "d" any more additional flow using marking procedure. However theoretically and practically it is possible to pass through the vertex "d" a total flow with value (4, 10, 2, 2) if the flows $\tilde{f}(Sc)$, $\tilde{f}(cd)$ and $\tilde{f}(dT)$ will be equal to values (3,7,2,1), (3,7,2,1) and (2,3,2,1) (In this example near each arc a value above the line is the flow value, a value below the line is the capacity value, a grey rectangle is representing a vertex mark. All the values are represented as a L-R-type fuzzy numbers). Thus we must define the following refining procedure:

Procedure Refining

Step 1. For each arc e in the network N define a value DIFF(e) = { $\underline{\text{DIFF}}_i, \overline{\text{DIFF}}_i$ }, where $\underline{\text{DIFF}}_i = \underline{c}_i - \underline{f}_i$, $\overline{\text{DIFF}}_i = \overline{c}_i - \overline{f}_i$, i – number of the α-cut.

Step 2. For the network N define a set of paths {P}: $\forall e \in P_i$: any of the values $\underline{\text{DIFF}}_i$ or $\overline{\text{DIFF}}_i$ in the arc e must be not equal to zero.

Step 3. While select the path from {P}

Define for each α-cuts i a non-zero value $\underline{\text{MinDIFF}}_i = \min(\underline{\text{DIFF}}_i(e))$ and a non-zero value $\overline{\text{MinDIFF}}_i = \min(\overline{\text{DIFF}}_i(e))$.

Check if all values $\tilde{f}^{\text{new}}(e)$ on the current path will be a correct fuzzy numbers, where $\underline{f}_i^{\text{new}} = \underline{f}_i + \underline{\text{MinDIFF}}_i$ and $\overline{f}_i^{\text{new}} = \overline{f}_i + \overline{\text{MinDIFF}}_i$ for all e on the current path. If checking is ok, then we correct the flow on the current path: $\underline{f}_i = \underline{f}_i + \underline{\text{MinDIFF}}_i$, $\overline{f}_i = \overline{f}_i + \overline{\text{MinDIFF}}_i$ for each e on the current path.

Wend.

End procedure

Thus let us define a maximum network flow algorithm with fuzzy input data.

Algorithm

Begin

F=0 // defining a starting flow

Label: put all mark = 0

Put ARRAY = NULL

```
            Put LIST = {S}
While LIST is Not NULL
        Let X be a vertex from the LIST
        Delete X from the LIST
        Browse X
        If T is marked then
                Increase the flow along the augmenting path
                Go to Label
        End If
Wend
Refining
End
```

Procedure Browse

Define all the vertices Y_i, from which arcs to X with non-negative flows exists, and to which non-F-saturated arcs from X exists, to ARRAY $(Y_i , X(mark))$.

If for Y_i all the vertices, for which a mark may be define, are entry, then mark Y_i with maximum mark, insert Y_i into the LIST, delete Y_i from the ARRAY.

End procedure

During the solving max flow problem in fuzzy network in some cases failures may take places because a max flow in the arc in a real situation (as a determinate value) may be greater than a capacity of this arc in concrete time (also as a determinate value). This possibility may be computed using an integrated superiority coefficient PSE. In this situation a decision-maker have to find a compromise between max flow value and failure possibility value.

Thus a maximum network flow algorithm with fuzzy input data was proposed based on procedure of the direct comparison of the fuzzy numbers and on the refining procedure to obtain a max network flow. This algorithm uses a fuzzy min-cut max-flow theorem to provide a correct result and allows to overcome inconsistencies in fuzzy numbers representation.

References

[1] A.V. Goldberg, Recent developments in maximum flow algorithms, Technical report #98-045, NEC Research Institute, Inc., 1998.

[2] K.D. Wayne, Generalized maximum flow algorithms, Dissertation, Cornell University, 1999.

[3] J.J.Buckley, Y.Hayashi, Applications of fuzzy chaos to fuzzy simulation, Fuzzy sets and systems, 99 (1998).

[4] Hsu-Shih Shih, E.Stanley Lee, Fuzzy multi-level minimum cost flow problems, Fuzzy sets and systems, 107 (1999).

[5] P.Diamond, A fuzzy max-flow min-cut theorem, Fuzzy sets and systems, 119 (2001).

[6] D. Dubois, H. Prade, Applications a la representation des connaissances en informatique, Masson, 1988.

[7] Fundamentals of fuzzy sets, ed. by D.Dubois and H.Prade, 2000.

[8] M.N.S. Swamy, K. Thulasiraman, Graphs, networks, and algorithms, A Wiley Interscience Publication, 1984.

[9] C.H. Paradimitriou, K. Steiglitz, Combination optimization, Prentice-Hall, Inc., 1982.

[10] L.A. Zadeh, Fuzzy Sets, Inform. and control, 8, 1965.

Appendix A

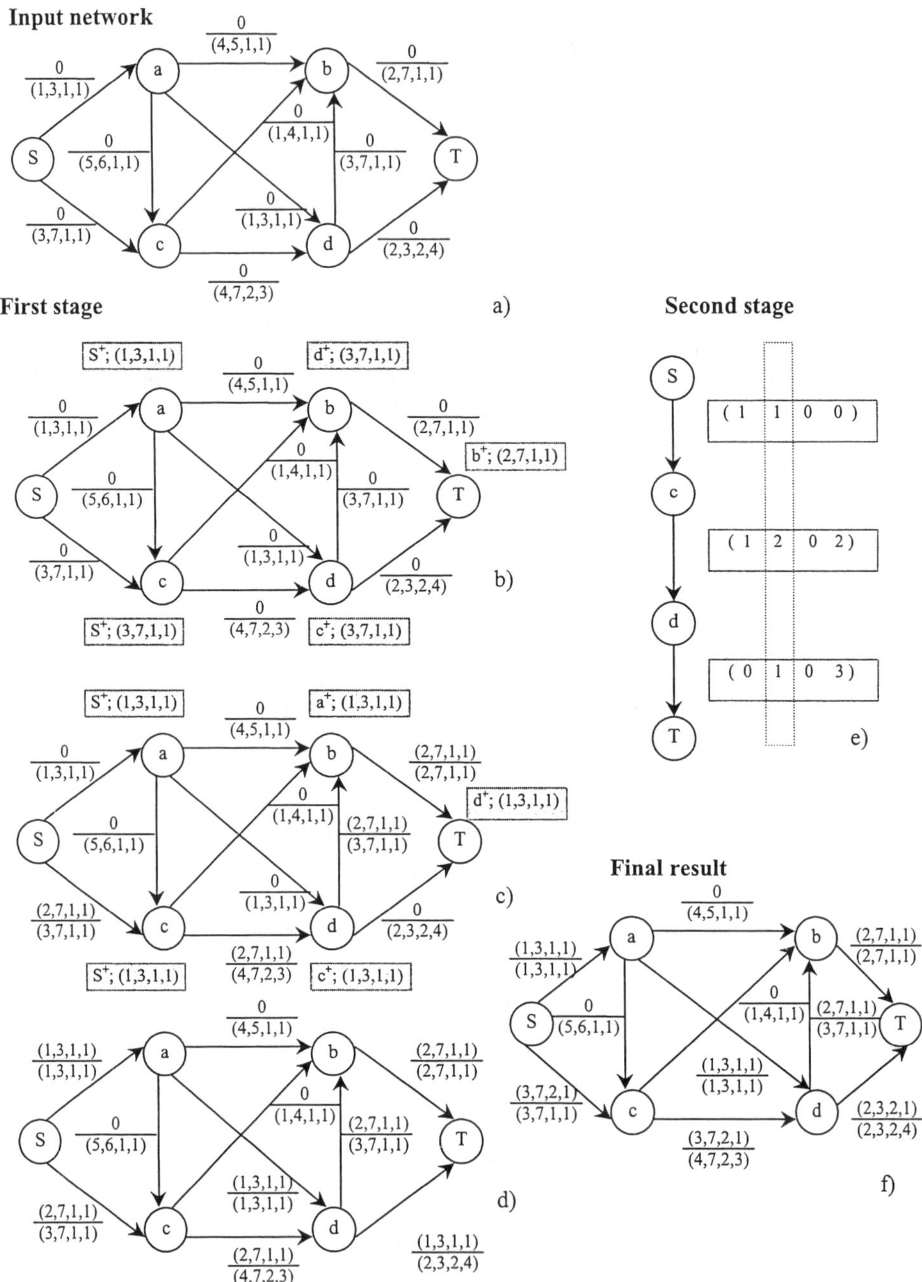

Figure 1. Illustrating example

Fuzzy Concept Formation Based on Context Model

V. N. Huynh and Y. Nakamori
School of Knowledge Science, JAIST
Tatsunokuchi, Ishikawa, 923-1292, JAPAN

Abstract. This paper aims at approaching the problem of mathematical modeling of fuzzy concepts based on the theory of formal concept analysis and the notion of context model. We introduce a notion of fuzzy concepts within a context model and the membership functions associated with these fuzzy concepts. It is shown that fuzzy concepts can be interpreted exactly as the collections of α-cut of their membership functions. An analysis of using the approach in the paper as a simple method of constructing membership functions for fuzzy sets in mining fuzzy association rules from databases is also discussed.

1 Introduction

The mathematical theory of formal concepts and their hierarchical order firstly introduced by R. Wille in 1982 [13] have been successfully developed during the last decades. Recently it becomes clear that most of the mathematical theory of formal concepts can be concentrated in a comprehensive model for conceptual knowledge systems which enables representation, inference, acquisition, and communication of conceptual knowledge to perform [3, 13, 14, 9].

In [7], the authors introduced the context model as an approach to representation, interpretation, and analysis of imperfect data. The short motivation of this approach stems from the observation that the origin of imperfect data is due to situations, where either we are not able to specify an object by an original tuple of elementary characteristics because of incomplete information available, or we only need to use linguistic descriptions about objects, which are more natural and understandable for human beings. In addition, the specific meaning of vague concepts in human thinking and communication is always determined by contexts, by personal views, etc, i.e. their interpretation (and/or meaning) depends on which context they are uttered in.

According to [13], a (crisp) concept is understood as a unit of thoughts consisting of two parts: the *extension* and the *intension*; the extension covers all objects (or entities) belonging to the concept while the intension comprises all attributes (or properties) valid for all those objects.

In this paper, we consider and propose a mathematical modeling of fuzzy concepts based on Kruse's approach within the framework of formal concept analysis (FCA, for short). The main idea is as follows. Firstly, we transform a context model into a many-valued context which can be considered as a further generalization of formal contexts in the sense of Goguen [4]. Then, from the derived many-valued context called L-fuzzy context, we introduce a way of fuzzy concept formation based on the so-called level contexts, and define the notion of membership functions associated with the formed fuzzy concepts. Finally, we present an analysis of using the approach in the paper as a simple method of constructing membership functions for fuzzy sets in mining fuzzy association rules from databases. Due to the limitation of the page number, some illustrative examples will be given in the presentation.

2   The Basics of FCA and Context Model

This section briefly recalls neccessary notions of FCA and context model. For more details as well as illustrated examples, the reader may be referred to, e.g. [3, 13, 14, 7].

2.1   Formal Concept Analysis: A Brief Introduction

FCA begins with the notion of a formal context defined as a triple (O, A, R), where O and A are sets and R is a binary relation between O and A, i.e. $R \subseteq O \times A$. The elements of O and A are called objects and attributes, respectively, while oRa, i.e. $(o, a) \in R$, is read as the object o has the attribute a in the context (O, A, R). Now consider two operations derivated from (O, A, R) as follows:

$$\forall X \subseteq O, \psi(X) = \{a \in A : (o, a) \in R \text{ for all } o \in X\},$$

$$\forall M \subseteq A, \phi(M) = \{o \in O : (o, a) \in R \text{ for all } a \in M\}.$$

These operations form a so-called *Galois connection*[1] between the power sets 2^O and 2^A. In the context (O, A, R), a *(formal) concept* is defined as a pair (X, M) such that $X \subseteq O, M \subseteq A, \phi(M) = X$ and $\psi(X) = M$; X and M are called the *extent* and the *intent* of the concept (X, M), respectively. Denote by $\mathcal{B}(O, A, R)$ the set of all concepts of (O, A, R). The hierarchy of concepts is given by the relation "*subconcept-superconcept*" defined as follows: the concept (X_1, M_1) is a subconcept of the concept (X_2, M_2) if $X_1 \subseteq X_2$ (equivalently, $M_1 \supseteq M_2$), (X_2, M_2) is then a superconcept of (X_1, M_1). Then, the set $\mathcal{B}(O, A, R)$ becomes a complete lattice called *concept lattice* under ordering relation being "subconcept-superconcept" one.

To enlarge the possibilities for expressing conceptual knowledge, the authors in [9] extended the language of concept lattices to the language of algebras of semi-concepts. In the context (O, A, R), a *semi- concept* is defined as a pair (X, M) such that $X \subseteq O, M \subseteq A, \phi(M) = X$ or $\psi(X) = M$. In this paper we use the semi-concepts of the form $(\phi(M), M)$ to form fuzzy concepts generated by a set M of vague characteristics.

2.2   Context Model: Basic Concepts

A context model is defined as a triple $\langle D, C, A_C(D) \rangle$, where D is a nonempty *universe of discourse*, C is a nonempty *finite set of contexts*, and the set $A_C(D) = \{a | a : C \to 2^D\}$ which is called the set of all vague characteristics of D with respect to C.

For $a_1, a_2 \in A_C(D)$, a_1 is said to be *more specific* than a_2 if and only if $(\forall c \in C)(a_1(c) \subseteq a_2(c))$.

If there is a finite measure P_C on the measurable space $(C, 2^C)$, then $a \in A_C(D)$ is called a *valuated vague characteristic* of D w.r.t. P_C. Then we call a quadruple $\langle D, C, A_C(D), P_C \rangle$ a context model valuated by P_C.

Notice that in the case where C is a single-element set, say $C = \{c\}$, a context model becomes a formal context in the sense of Wille as follows. Let $\langle D, C, A_C(D) \rangle$ be a context model such that $|C| = 1$. Then the triple (O, A, R), where $O = D, A = A_C(D)$ and $R \subseteq O \times A$ such that $(o, a) \in R$ iff $o \in a(c)$, is a formal context. Thus, a context model can be considered as a collection of formal contexts. In the following section we will address the problem of how to combine formal contexts from a context model for fuzzy concept formation.

3 Fuzzy Concept and Membership Function

Let $\langle D, C, A_C(D) \rangle$ be a context model. We now define a many-valued context which is called L-fuzzy context in the sense of Goguen [4] as follows.

Let us consider the following mapping

$$R: \quad D \times A_C(D) \longrightarrow 2^C$$
$$(d, a) \longmapsto \{c \in C : d \in a(c)\}.$$

Denote $L = \{\alpha \in 2^C : \exists (d, a) \in D \times A_C(D), R(d, a) = \alpha\} \cup \{C, \emptyset\}$. Then we have a derived many-valued context $\langle D, A_C(D), R, L \rangle$ called L-fuzzy context, where L is a bound poset.

Next we consider a valued context model $\langle D, C, A_C(D), P_C \rangle$, where $D, C, A_C(D)$ are as above and $P_C : 2^C \to [0, 1]$. Then we define the following mapping

$$R: \quad D \times A_C(D) \longrightarrow [0, 1]$$
$$(d, a) \longmapsto P_C\{c \in C : d \in a(c)\},$$

and obtain a fuzzy context in the sense of Zadeh [15], namely $\langle D, A_C(D), R, [0, 1] \rangle$.

Generally, we consider a L-fuzzy context (D, A, R, L), where D is a nonempty universe of discourse, A is the set of all vague characteristics defined over D, L is a poset and R is a mapping from $D \times A$ into L. For each $\alpha \in L$, we define a formal context $FC_\alpha = (D, A, R_\alpha)$ such that $(d, a) \in R_\alpha$ iff $R(o, a) \geq \alpha$, for all $d \in D, a \in A$. For each $M \subseteq A$, denote $\phi_\alpha(M) = \{d \in D : (d, a) \in R_\alpha \text{ for all } a \in M\}$.

Now we are ready to define the notion of a fuzzy concept generated by a set of vague characteristics as follows. For each $M \subseteq A$, by a *fuzzy concept generated from M*, denoted by $\mathcal{F}(M)$, we mean the collection $\{\phi_\alpha(M)\}_{\alpha \in L}$. In particular, for each $\alpha \in L$, the pair $(\phi_\alpha(M), M)$ is a semi-concept in the formal context FC_α.

Let $\mathcal{B}(D, A, R, L) = \{\{\phi_\alpha(M)\}_{\alpha \in L} : M \in 2^A\}$. Then we can define operations $\sqcup, \neg$ in $\mathcal{B}(D, A, R, L)$ as follows[1]: $\forall M, N \in 2^A$,

$$\mathcal{F}(M) \sqcup \mathcal{F}(N) = \mathcal{F}(M \cap N), \text{ and } \neg \mathcal{F}(M) = \mathcal{F}(\neg M),$$

where $\neg M$ mean the complement of M in A. The operation $\sqcap$ is then defined via $\sqcup, \neg$ as usual, i.e. $\mathcal{F}(M) \sqcap \mathcal{F}(N) = \neg(\neg \mathcal{F}(M) \sqcup \neg \mathcal{F}(N))$. In fact, we have the following representation

$$\langle \mathcal{B}(D, A, R, L), \sqcap, \sqcup, \neg \rangle \cong \Pi_{\alpha \in L} \langle \{(\phi_\alpha(M), M) : M \in 2^A\}, \sqcap, \sqcup, \neg \rangle$$

Before defining membership functions associated with fuzzy concepts defined as above, we now recall the notion of the L-ideal in a poset L. Let L be a poset. By any *L-ideal* we mean a non-empty subset I of L such that for any $t, s \in L$, if $t \leq s$ and $s \in I$ then $t \in I$. For any $t \in L$, the set $I(t) = \{s \in L : s \leq t\}$ is a L-ideal and will be called to be a principal ideal generated by t.

By *L-expansion* we mean the family LT consisting of the empty set, and of all L-ideals. It is known [11] that the L-expansion LT is a complete set lattice under partial

---

[1]These operations are defined point-wise based on Luksch & Wille's ones defined on semi-concepts

ordering $\leq$ being the inclusion relation. By a LT-fuzzy set [11] in a universe $U \neq \emptyset$ we mean a function $f : U \to LT$.

Given a L-fuzzy context (D, A, R, L). For each $M \subseteq A$, we define the membership function of the fuzzy concept $\mathcal{F}(M)$ as following

$$\mu_M : \quad D \longrightarrow \mathcal{P}(L)$$
$$d \longmapsto \mu_M(d) = \{\alpha \in L / d \in \phi_\alpha(M)\}.$$

It is easily seen that $\mu_M(d)$, $\forall d \in D$, is a L-ideal. That is μ_M is a LT-fuzzy set in D.

Especially, in the case where $L = [0, 1]$ we can define the membership function of the fuzzy concept $\mathcal{F}(M)$ as follows:

$$\mu_M : \quad D \longrightarrow [0, 1]$$
$$d \longmapsto \mu_M(d) = sup\{\alpha \in [0, 1] / d \in \phi_\alpha(M)\}.$$

Then, each fuzzy concept in $\mathcal{B}(D, A, R, [0, 1])$ is associated with a fuzzy set in D in the sense of Zadeh [15] and, furthermore, each fuzzy concept is exactly the α-cut representation of its membership function. Especially, if M is a single-element set, say $M = \{m\}$, we have simply $\mu_M = R(\cdot, m)$.

In the next section, we will present some discussions about applicability of the proposal in this paper to constructing membership functions for fuzzy sets in mining fuzzy association rules from databases.

4   Discussions

Mining association rules is one of the important problems in data mining. In the past several years, there has been much active work in developing algorithms for mining association rules in databases with binary attributes. However, relational tables in practice have richer attribute types, attributes can be quantitative or categorical, and Boolean attributes can be considered a special case of categorical attributes. Recently, the problem of mining association rules in large relational tables containing both quantitative and categorical attributes [12] has also received considerable attention.

An algorithm has been proposed in [12] for mining quantitative association rules by partitioning the attribute domain, combining adjacent partitions, and then transforming the problem into binary one. Although this algorithm can solve some problems introduced by quantitative attributes, it causes the sharp boundary problem [8, 5]. The algorithm either ignore or over-emphasize the elements near the boundary of the intervals in the mining process.

To overcome the sharp boundary problem, recently, the problem of mining fuzzy association rules has been introduced in [8, 6, 5]. The mined rules are then expressed in the linguistic terms associated with the fuzzy sets, which are natural and understandable for human beings. However, these work assume that the fuzzy sets are given previously. This assumption may not always be realistic. In [2], the authors have proposed a method of finding the fuzzy sets based on clustering techniques. In fact, the authors have used the CLARANS clustering algorithm introduced in [10] to find the k medoids (representative objects) of the database, and then, they defind k triangular fuzzy sets for each quantitative attribute domain on overlapping intervals with normalized points being attribute values of k medoids respectively. This method causes some problems. The first problem is in some applied situations the number of fuzzy sets (considered as vague characteristics defined over quantitative attribute domain) of quantitative attributes may not be the same. The second is the CLARANS clustering algorithm takes the dissimilarity measure considered over all attributes of objects into account, and this may be unsuitable in choosing representative values for non-interactive attributes such as, for example, *age* and *income*.

Now we consider a quantitative attribute domain as the universe of discourse D, and vague characteristics defined on D are linguistic terms used in the mining purpose. We observe that the linguistic terms used in natural language are context-dependent. For example: an income in the interval $[n, m]$ is said to be *high* in the context of a person but not in the context of another person; The students' course score [6] in the interval $[70, 100]$ is *high* in the context of a normal student but in the context of an excellent student, the interval may be $[80, 100]$. Under such an observation, to construct membership functions for linguistic terms we can use a number of partitions corresponding to different contexts associated with weightings. Then the approach in this paper can be used directly as a very simple method of constructing membership functions for linguistic terms in mining fuzzy association rules.

Acknowledgments

The first author is on leave from Quinhon University, Vietnam. He is supported by Inoue Foundation for Science under a postdoctoral fellowship.

References

[1] G. Birkhoff, *Lattice Theory*, Providence, Rhode Island, 1973.

[2] Ada Fu et al., Finding fuzzy sets for the mining of fuzzy association rules for numerical attributes, *Proceedings of the 1st IDEAL'98*, October 1998, pp. 263–268.

[3] B. Ganter & R. Wille, *Formal Concept Analysis: Mathematical Foundations*, Springer-Verlag, Berlin Heidelberg, 1999.

[4] J. A. Goguen, *L*-fuzzy sets, *J. Math. Anal. Appl.* **18** (1967) 145–174.

[5] Attila Gyenesei, *A Fuzzy Approach for Mining Quantitative Association Rules*, Turku Centre for Computer Science, Technical Report No 336, March 2000, 20 pages.

[6] T-P Hong, C-S Kuo & S-C Chi, Mining association rules from quantitative data, *Intelligent Data Analysis* **3** (1999) 363–376.

[7] R. Kruse, J. Gebhardt & F. Klawonn, Numerical and logical approaches to fuzzy set theory by the context model, in: R. Lowen and M. Roubens (Eds.), *Fuzzy Logic: State of the Art*, Kluwer Academic Publishers, Dordrecht, 1993, pp. 365–376.

[8] C. M. Kuok, Ada Fu, M. H. Wong, Mining Fuzzy Association Rules in Databases, *ACM SIGMOD Records* **27** (1998) 41–46.

[9] P. Luksch & R. Wille, A mathematical model for conceptual knowledge systems, in H. H. Bock & P. Ihm (Eds.), *Classification, Data Analysis, and Knowledge Organization*, Springer-Verlag, Heidelberg, 1990, pp. 156–162.

[10] R. T. Ng & J. Han, Efficient and effective clustering methods for spatial data mining, Proceedings of the 20th VLDB Conference, September 1994, pages 144–155.

[11] H. Rasiowa & C. H. Nguyen, *LT*-fuzzy sets, *Fuzzy Sets and Systems* **47** (1992) 323–339.

[12] R. Srikant & R. Agrawal, Mining Quantitative Association Rules in Large Relational Tables, Proceedings of the ACM SIGMOD, Monreal, Canada, June 1996, pp. 1–12.

[13] R. Wille, Restructuring lattice theory: An approach based on hierarchies of concepts, in I. Rival (Ed.), *Ordered Sets*, Reidel, Dordrecht, Boston, 1982, pp. 445–470.

[14] R. Wille, Concept lattices and conceptual knowledge systems, *Computers Math. Appli.* **23** (1992) 493–515.

[15] L. A. Zadeh, Fuzzy sets, *Information and Control* **8** (1965) 338-353.

KES '01
N. Baba et al. (Eds.)
IOS Press, 2001

Soft Computing in E-learning: Towards Intelligent Virtual Learning Milieu

Vesa A. NISKANEN

Univ. of Helsinki, Dept. of Economics & Management, PO Box 27, 00014 Helsinki, Finland

Abstract. This article sketches a system which applies hypermedia, e-learning and soft computing in combination. This provides a basis for virtual learning milieu.

1. E-learning and Virtual Learning Milieu

E-learning means a learning process in which a person may utilize modern technology such as the Internet, mobile phones, computer-aided instruction and multimedia. His/her studies presuppose networking with other students as well as with the instructors. E-learning allows us to construct virtual milieux, such as virtual universities, in which we deal with immaterial things. For example, learning materials are in electronic form and no classrooms are required.

E-learning is often based on hypermedia, and this media may be characterized as follows [1]: (i) Hypermedia comprise units of information such as text, pictures, audios and videos. (ii) These units are connected by links. (iii) By creating, editing and linking the units, users may construct information structures for various purposes. (iv) In the shared systems, several users may simultaneously access the hypermedia database (Fig. 1).

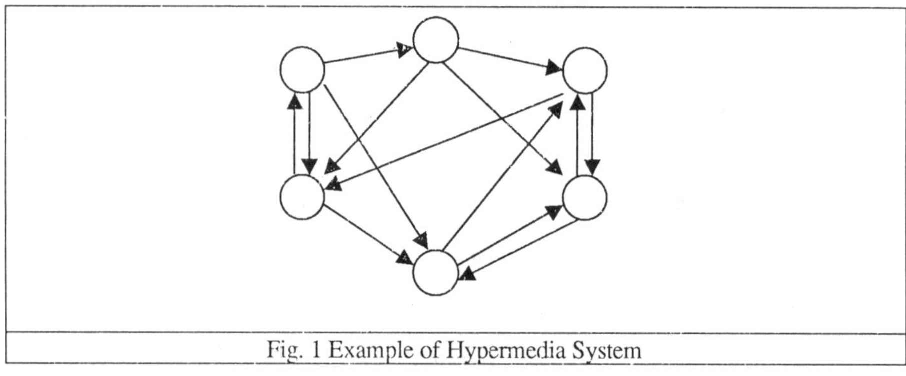

Fig. 1 Example of Hypermedia System

Today, we aim to construct intelligent e-learning milieux, and in this context, soft computing still awaits its golden age (Fig. 2). In an intelligent milieu, the e-learning systems should include appropriate simulations, games, interpretations of text and language, pattern recognition, data mining, data compression, expert systems, intelligent agents, feedback and evaluation services and self-learning services.

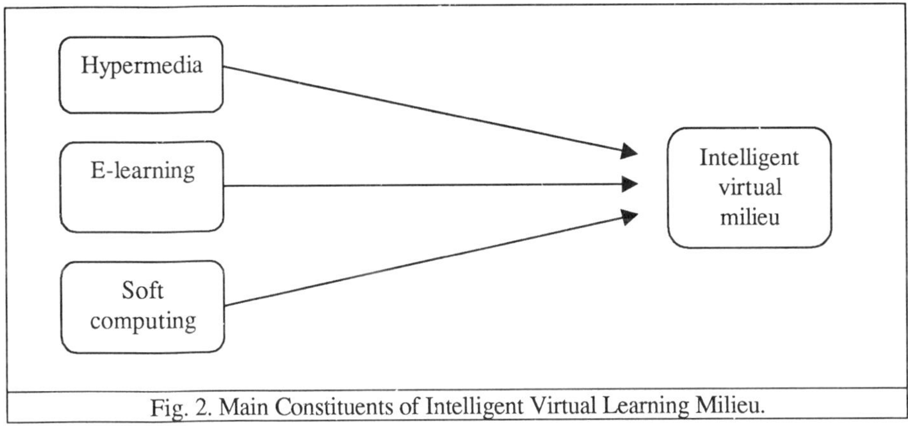

Fig. 2. Main Constituents of Intelligent Virtual Learning Milieu.

Soft computing methods [3,4] are powerful in the e-learning systems, because they can cope with linguistic and imprecise entities and models. In particular, we may apply linguistic rule-based reasoning and fuzzy cognitive maps [2]. Hence, a typical e-learning system may use hypermedia structure, Internet technology and fuzzy cognitive maps in combination. The nodes comprise various learning materials, and the links between the nodes are established with fuzzy rules. We may move in the network according to the instructions, recommendations and feedback of the system. In this manner, we may also construct simulations and other virtual world applications.

2. A Simple Learning System

Figure 3 provides an example of a simplified intelligent virtual system for learning basic arithmetic. This system is a fuzzy cognitive map the nodes of which are information units. The relations between the nodes are fuzzy rule sets. According to the aims, we may attain four basic areas of arithmetic knowledge and the expected knowledge may be good, fairly good etc. The aims also determine the curriculum and the organization of learning in practice. Curriculum and aims, in turn, provide the basis for the appropriate learning materials. The learning process proceeds from addition to division. The user first studies addition. His/her knowledge is evaluated prior to, during and after the learning process according to the given tests and resolved problems. At each stage, the system provides additional knowledge and excercises if the aims of this level have not been achieved. Hence, bad results lead to extra studies, whereas good results lead to the next stage. The teacher supervises and supports the learning process.

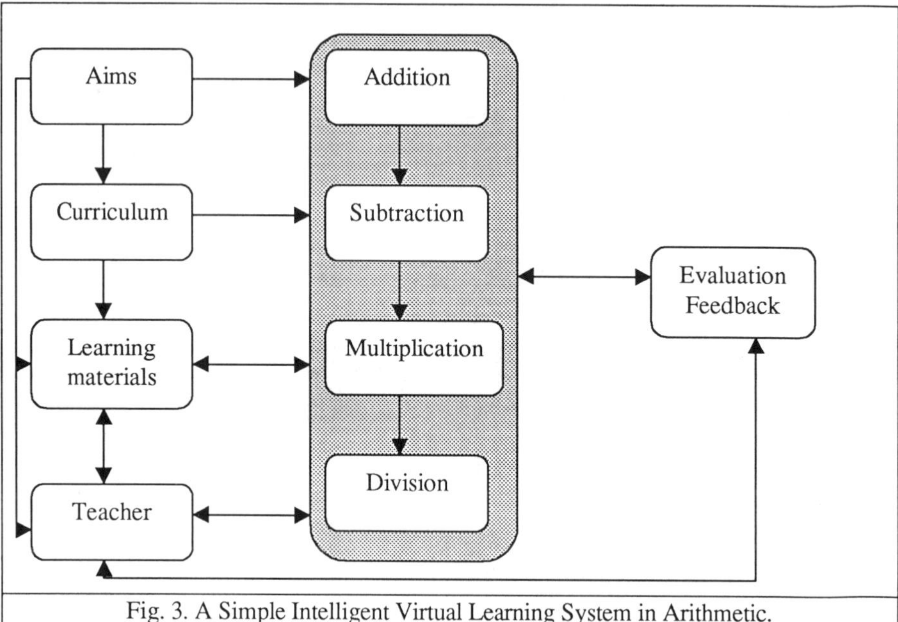

Fig. 3. A Simple Intelligent Virtual Learning System in Arithmetic.

Even this simple system should include several complicated constituents, each of them a fuzzy cognitive map with the hypermedia structure. We thus actually use an augmented map. However, by virtue of fuzzy reasoning and fuzzy cognitive maps, we may construct these systems effortlessly. We also may apply hypermedia, e-learning and soft computing in combination.

References

[1] E. Barrett (ed.), The Society of Text, MIT, 1989.
[2] B. Kosko, Fuzzy Engineering, Prentice Hall, 1997.
[3] L. Zadeh, Fuzzy logic = Computing with words, IEEE Transactions on Fuzzy Systems, vol. 2, (1996) 103-111.
[4] L. Zadeh, Toward a theory of fuzzy information granulation and its centrality in human reasoning and fuzzy logic, Fuzzy Sets and Systems 90/2 (1997) 111-127.

On Field Theory of Λ-Standard Fuzzy Numbers

Jorma K. MATTILA
Lappeenranta University of Technology, Inst. of Appl. Mathematics
P.O. Box 20, FIN-53851 Lappeenranta, Finland

Abstract. The set of fuzzy Λ-standard fuzzy numbers are considered. These fuzzy numbers are convex normal fuzzy sets. Addition, subtraction, multiplication and division are defined so that the set of these fuzzy numbers is closed under these operations. The set of fuzzy natural numbers is a subset of the set of these real fuzzy numbers. These fuzzy numbers are unique, especially in concerning with the shape of their membership functions and the length of the support of the numbers. The axioms of field theory are shown to be satisfied for the set of fuzzy Λ-standard fuzzy numbers.

1. Definitions and Properties of Λ-standard fuzzy numbers

Fuzzy numbers are usually defined to be normal convex fuzzy sets on $\mathbb{R}$, as generally can be seen in the literature concerning fuzzy sets and systems. The membership function of a Λ-shaped fuzzy set A has the general form

$$
(1.1) \qquad A(x) = \Lambda(x;\, a,\, b,\, c) = \begin{cases} 0 & \text{if } x \leq a \\[4pt] \dfrac{x-a}{b-a} & \text{if } a \leq x < b \\[8pt] \dfrac{c-x}{c-b} & \text{if } b \leq x < c \\[6pt] 0 & \text{if } x \geq c \end{cases}
$$

Usually arithmetic operations of fuzzy numbers are defined using extension principle. Same results can be obtained using interval arithmetic of α-cuts (see Kaufmann and Gupta [2]). This kind of arithmetic is sometimes problematic because results of these arithmetic operations do not preserve values of several indices which are used for measuring fuzziness, imprecision, speciality, etc. Sometimes there is no need for changing these values. Also in some special cases multiplication of fuzzy numbers does not preserve even the convexity which should be the presupposition for the result to be a fuzzy number. Such a special case is for instance the situation where the universe if discourse is a discrete set. Also in multiplication, the product of two given Λ-shaped fuzzy numbers is actually no more a Λ-shaped fuzzy number, because the slanting sides of the triangle (which is the non-zero part of the membership function of the product) are curve segments of parabolas. This can be seen very clearly by using interval arithmetic applied to α-cuts.

We will define the usual arithmetic operations for standard fuzzy numbers in such ways that the results are standard fuzzy numbers, too.

We give the definition of standard fuzzy number as follows.

Def. 1.1. A normal fuzzy subset $\tilde{n}$ of $\mathbb{R}$ is a Λ-*standard real fuzzy number* if and only if $\tilde{n}$ satisfies the following properties:

(i)   The membership function of $\tilde{n}$ is of the form (1.1), i.e.

$$\tilde{n}(x) = \Lambda(x;\ a,\ n,\ b),\ a,\ b,\ n \in \mathbb{R}.$$

(ii)   The curve of $\tilde{n}(x) = \Lambda(x;\ a,\ n,\ b)$ is symmetric with respect to the line $x = n$.

(iii)   The first speciality index of $\tilde{n}$ is one, i.e.

$$i_1(\tilde{n}) = \frac{1}{\displaystyle\int_a^b \mu_A(x)dx} = 1.$$

About Yager's speciality index, see Yager, "Decision with Usual Values", In: A. Jones et al (Eds.), *Fuzzy Sets Theory and Applications*, NATO ASI Series, Publisher Reidal, 1986).

Theor. 1.1. $\tilde{n}$ is a standard real fuzzy number if an only if

(1.2) $\tilde{n}(x) = \Lambda(x;\ n-1,\ n,\ n+1) = \begin{cases} x - (n-1) & \text{if } n-1 < x \leq n \\ n+1-x & \text{if } n < x < n+1 \\ 0 & \text{if either } x \leq n-1 \text{ or } x \geq n+1 \end{cases}$

The proof of Theor. 1.1 is quite clear. It has been given in [5].

We adopt the symbol $\tilde{\mathbb{R}}_\Lambda$ standing for the set of Λ-standard fuzzy real numbers.

Corollary. From Def. 1.1 and Theor. 1.1 it immediately follows that

(i) $\ker(\tilde{n}) = n$ for all $\tilde{n} \in \tilde{\mathbb{R}}_\Lambda$;

(ii) $\mathrm{supp}(\tilde{n}) = \,]n - 1,\, n + 1[$;

(iii) $\tilde{n} = \tilde{m}$ if and only if $n = m$;

(iv) $\tilde{n}(x) = 0$ if $x = n \pm k$, $k \geq 1\ (k \in \mathbb{R})$;

(v) $i_0(\tilde{n}) = \frac{1}{2}$ for all $\tilde{n} \in \tilde{\mathbb{R}}_\Lambda$.

The properties (i), (ii), and (iv) are immediately clear, (iii) follows by the fact that $\tilde{n} = \tilde{m}$ if and only if $\tilde{n}(x) = \tilde{m}(x)$ for all $x \in \mathbb{R}$, and (v) follows by Def. 1.1.

It is shown in [5] that a Λ-standard fuzzy number $\tilde{n} \in \tilde{\mathbb{R}}_\Lambda$ is

(i) Λ-standard fuzzy natural number (denoted by $\tilde{n} \in \tilde{\mathbb{N}}_\Lambda$) iff $\ker(\tilde{n}) = n \in \mathbb{N}$;

(ii) Λ-standard fuzzy integer (denoted by $\tilde{n} \in \tilde{\mathbb{Z}}_\Lambda$) iff $n \in \mathbb{Z}$;

(iii) Λ-standard fuzzy rational number (denoted by $\tilde{n} \in \tilde{\mathbb{Q}}_\Lambda$) iff $n \in \mathbb{Q}$;

(iv) Λ-standard fuzzy irrational number (denoted by $\tilde{n} \in \tilde{\mathbb{R}}_\Lambda - \tilde{\mathbb{Q}}_\Lambda$) iff $n \in \mathbb{R} - \mathbb{Q}$.

Arithmetical operations on $\tilde{\mathbb{R}}_\Lambda$ are defined as follows.

Def. 1.2 . For any two Λ-standard numbers $\tilde{m}, \tilde{n} \in \tilde{\mathbb{R}}_\Lambda$ there is a Λ-standard fuzzy number $\tilde{k} \in \tilde{\mathbb{R}}_\Lambda$, such that

$$(1.3) \qquad \tilde{k} = \tilde{m} + \tilde{n} \ \text{ iff } \ k = m + n \qquad \text{(addition),}$$

$$(1.4) \qquad \tilde{k} = \tilde{m} - \tilde{n} \ \text{ iff } \ k = m - n \qquad \text{(subtraction),}$$

$$(1.5) \qquad \tilde{k} = \tilde{m}\,\tilde{n} \ \text{ iff } \ k = m\,n \qquad \text{(multiplication),}$$

$$(1.6) \qquad \tilde{k} = \frac{\tilde{m}}{\tilde{n}} \ \text{ iff } \ k = \frac{m}{n} \qquad \text{(division).}$$

where $k = \ker(\tilde{k})$, $m = \ker(\tilde{m})$, and $n = \ker(\tilde{n})$.

The membership functions for sum, difference, product, and quotient of Λ-standard fuzzy numbers $\tilde{m}$ and $\tilde{n}$ are respectively

$$(1.7) \qquad (\tilde{m} + \tilde{n})(x) = \Lambda(x;\ m+n-1,\ m+n,\ m+n+1)$$

$$(1.8) \qquad (\tilde{m} - \tilde{n}.)(x) = \Lambda(x;\ m-n-1,\ m-n,\ m-n+1),$$

$$(1.9) \qquad (\tilde{m}\,\tilde{n})(x) = \Lambda(x;\ m\,n-1,\ m\,n,\ m\,n+1),$$

$$(1.10) \qquad \frac{\tilde{m}}{\tilde{n}}(x) = \Lambda\left(x;\ \frac{m}{n} - 1,\ \frac{m}{n},\ \frac{m}{n} + 1\right).$$

Addition and multiplication of Λ-standard fuzzy numbers is clearly commutative and associative, and the distributivity law holds similarly as for crisp real numbers.

2. Field Theory of Λ-standard fuzzy numbers

We suppose here that the reader is familiar with first order predicate calculus. We consider the usual axiomatization for field theory which are essentially those given by Tarski in his *Introduction to Logic.*, New York, 1941.

Def. 2.1. Let $\mathbb{P} = \{+, \cdot, -, {}^{-1}, \mathbf{0}, \mathbf{1}\}$ be a predicate calculus where $+$ and $\cdot$ are binary function symbols, $-$ and ${}^{-1}$ are unary function symbols, and $\mathbf{0}$ and $\mathbf{1}$ are individual constants. A theory having the following non-logical axioms is called (*a first order*) *field theory*:

A1. $\forall x \forall y (x + y = y + x)$

A2. $\forall x (x + \mathbf{0} = x)$

A3. $\forall x (x + (-x) = \mathbf{0})$

A4. $\forall x \forall y \forall z ((x + y) + z = x + (y + z))$

A5. $\forall x \forall y \forall z ((x \cdot y) \cdot z = x \cdot (y \cdot z))$

A6. $\forall x (x \cdot \mathbf{1} = x)$

A7. $\forall x (\neg (x = \mathbf{0}) \rightarrow x \cdot x^{-1} = \mathbf{1})$

A8. $\forall x \forall y (x \cdot y = y \cdot x)$

A9. $\forall x \forall y \forall z (x \cdot (y + z) = x \cdot y + x \cdot z)$

A10. $\neg (\mathbf{0} = \mathbf{1})$

A model of a field theory is called a *field*.

As we know, the set of real numbers is a field satisfying these axioms for arithmetical operations of real numbers where $\mathbf{0}$ and $\mathbf{1}$ are the numbers 0 and 1, respectively. The variable symbols refer to crisp real numbers.

Consider Λ-standard fuzzy numbers and the arithmetical operations defined in Section 1. It seems that they fit well to the axioms above, when the variables refer to Λ-standard fuzzy numbers. We choose the individual constants $\mathbf{0}$ and $\mathbf{1}$ to be $\tilde{0}$ and $\tilde{1}$, respectively. We interpret the function symbols $+$, $-$, $\cdot$, and ${}^{-1}$ as arithmetical operations defined above for Λ-standard fuzzy numbers, i.e. we consider the predicate calculus $\mathbb{P}_\Lambda = \{+, \cdot, -, {}^{-1}, \tilde{0}, \tilde{1}\}$ where x^{-1} means the same as $\dfrac{1}{x}$.

Because the commutativity and associativity laws hold for addition and multiplication, and the distributivity law holds on $\tilde{\mathbb{R}}_\Lambda$, the system of Λ-standard fuzzy numbers satisfies the axioms A1, A8, A4, A5, and A9. We show that the axioms A2, A3, A6, A7, and A10 are satisfied by the system $\tilde{\mathbb{R}}_\Lambda$. In the following consideration we suppose that $\tilde{n} \in \tilde{\mathbb{R}}_\Lambda$ is any Λ-standard fuzzy number.

First, we consider the sum of $\tilde{n}$ and $\tilde{0}$. As a result we have
$$(\tilde{n} + \tilde{0})(x) = (\tilde{n} + \tilde{0})(x) = \Lambda(x, n - 1 + 0, n + 0, n + 1 + 0) = \Lambda(x, n - 1, n, n + 1)$$
for any $x \in \mathbb{R}$. Thus A2 is satisfied.

Secondly, we consider the sum of $\tilde{n}$ and $-\tilde{n}$. We have $-\tilde{n}(x) = \Lambda(x, -n - 1, -n, -n + 1)$. Further, $(\tilde{n} + (-\tilde{n}))(x) = \Lambda(x, n - n - 1, n - n, n - n + 1) = \Lambda(x, -1, 0, 1) = \tilde{0}(x)$ for any $x \in \mathbb{R}$. Thus A3 is satisfied. In [5] it is proved that $\tilde{n} - \tilde{n}$ does not depend on $\tilde{n}$.

Thirdly, we consider the product of $\tilde{n}$ and $\tilde{1}$. We have $(\tilde{n} \cdot \tilde{1})(x) = \Lambda(x; n \cdot 1 - 1, n \cdot 1, n \cdot 1 + 1) = \Lambda(x, n - 1, n, n + 1) = \tilde{n}(x)$ for any $x \in \mathbb{R}$. Thus A6 is satisfied.

As the fourth step, suppose $\tilde{n} \neq \tilde{0}$, i.e. $\tilde{n}(x) = \tilde{0}(x)$ does not hold for all $x \in \mathbb{R}$, i.e. $\neg(\tilde{n} = \tilde{0})$. Thus we have $\left(\tilde{n} \cdot \dfrac{1}{\tilde{n}}\right)(x) = \Lambda(x; n \cdot \dfrac{1}{n} - 1, n \cdot \dfrac{1}{n}, n \cdot \dfrac{1}{n} + 1) = \Lambda(x; \dfrac{n}{n} - 1, \dfrac{n}{n}, \dfrac{n}{n} + 1) = \left(\dfrac{\tilde{n}}{\tilde{n}}\right)(x) = \Lambda(x; 0, 1, 2) = \tilde{1}(x)$ for any $x \in \mathbb{R}$. Thus A7 is satisfied.

As the last step, we clearly see that A10 is satisfied, because $\tilde{0}(x) = \tilde{1}(x)$ does not hold for all $x \in \mathbb{R}$.

Our final result is that we have a field theory $\mathbb{P}_\Lambda = \{+, \cdot, -, {}^{-1}, \tilde{0}, \tilde{1}\}$ having the non-logical axioms like those in Def. 2.1, where the variables have their values from $\tilde{\mathbb{R}}_\Lambda$. Thus the system $\tilde{\mathbb{R}}_\Lambda$ is a model of $\mathbb{P}_\Lambda$. i.e. $\tilde{\mathbb{R}}_\Lambda$ is a field.

References

[1] Dirk van Dalen, *Logic and Structures*, Springer-Verlag, Corrected Third Printing, 1989

[2] A. Kaufmann, M. M. Gupta, *Fuzzy Mathematical Models in Engineering and Management Science*, North-Holland, Elsevier Science Publishers, 2th ed., 1991

[3] Jorma. K. Mattila, *A Study on Λ-Shaped Fuzzy Integers*, Research Report no. 76, Lappeenranta University of Technology, Department of Information Technology, Lappeenranta, 2000.

[4] Jorma. K. Mattila, "Diagonal Method for Modifying Fuzzy Numbers", Proceedings of IFSA/NAFIPS 2001 Congress, Vancouver, Canada, July 25-28, 2001

[5] Jorma. K. Mattila, "A Standard System of Λ-Shaped Fuzzy Numbers", submitted for publication

[6] Bruno Poizat, A Course in Model Theory. An Introduction to Contemporary Mathematical Logic, Springer-Verlag, 2000

KES '01
N. Baba et al. (Eds.)
IOS Press, 2001

Natural and Complement Coarsening Operators

Jari Kortelainen
Mikkeli Polytechnic, Mikkeli, Finland

Abstract. The paper presents an approach to coarsening operators. Representation Theorems, presented by Negoita and Ralescu, can be applied to determine a class of coarsenings of an L-set defined by means of a topology on U, as discussed in [5]. Specific coarsening operators, namely, natural and complement coarsening operators are defined, and some properties of those are presented. Especially, a characterization on the existence of a weakening operator as a complement coarsening operator is presented.

1 Introduction

In this paper we continue the work started in [5]. However, we define some preliminary concepts and notations similarly to [5] for convenience.

The fuzzy sets are introduced by Zadeh in [9] and they are generalized to L-sets by Goguen in [2]. As usual, an L-set $\mathcal{A}$ on a non-empty universe U is understood to be a function on U and the class of all L-sets on U is denoted by L^U, where L is a complete lattice satisfying, perhaps, some additional properties. Ordinary subsets of U are denoted by A, B, ..., whenever we need to emphasize that property. The α-level sets of $\mathcal{A} \in L^U$, $\mathcal{A}_\alpha = \{x \in U \mid \mathcal{A}(x) \geq \alpha,\ \alpha \in L\}$, are also ordinary subsets of U. Classes of ordinary sets are denoted by $\mathfrak{A}$, $\mathfrak{B}$, ..., and especially $\mathfrak{P}(U)$ denotes the ordinary power set of U. Any class $\mathfrak{A} \in \mathfrak{P}(\mathfrak{P}(U))$, U is finite, is also referred to as *knowledge about U*. Moreover, $\mathcal{A}$, $\mathcal{B}$, ..., denote ordinary classes of L-sets.

In many cases our knowledge about U is incomplete and this may restrict us to determine, e.g., α-level sets of L-sets. The following Definition is now presented [5]:

Definition 1.1. Let U be a non-empty finite set, $\mathcal{A} \in L^U$ and L be a complete lattice. We say that $\mathcal{A}$ is $\mathfrak{A}$-*definable* if $\forall \alpha \in L$, $\mathcal{A}_\alpha \in \mathfrak{A}$, where $\mathfrak{A} \in \mathfrak{P}(\mathfrak{P}(U))$.

In this paper the universe U is finite, and we are interested in topologies and classes of closed sets on U. If $\mathfrak{T}$ is a topology on U then the associated class of closed sets is denoted by $c(\mathfrak{T})$, thus $c(\mathfrak{T}) = \{S \subset U \mid \bar{S} \in \mathfrak{T}\}$, where the overbar denotes the complementation.

Following [8], L-sets can be represented by means of α-level sets if L satisfies some additional properties. Those properties are the following:

(a) L is a complete lattice,

(b) $\forall a \in L,\ \forall S \subset L,\ a < \bigvee S \ \Rightarrow\ \exists b \in S, a \leq b$, where $\bigvee S = \sup S$.

The paper is organized as follows: In Section 2 we define topologies corresponding to L-sets and define a fineness relation between L-sets as in [3, 4, 5]. In Section 3 we define natural and complement coarsening operators as an application of Representation Theorems presented by Negoita and Ralescu in [8].

2   Topologies Corresponding to L-Sets

In this Section we introduce topologies corresponding to L-sets determined by *compositional modifiers*. A more detailed discussion on these topics can be found, e.g., in [3, 7].

Let us connect L-sets to ordinary binary relations in the following way [3]:

Definition 2.1. Let L satisfy (a), U be a non-empty set and $A \in L^U$. We say that $R^A \subset U \times U$ is *relation corresponding to* A, if

$$\forall x, y \in U, \quad x R^A y \Leftrightarrow A(x) \geq A(y). \tag{1}$$

Clearly, R^A and its converse, denoted in this paper by R^A_{conv}, are both quasi-orderings on U. For any $A \in L^U$ there exists two topologies, denoted by $\mathfrak{T}^A$ and $\mathfrak{T}^A_{\text{conv}}$, which are determined by substantiating compositional modifiers $\left(\mathcal{H}^A \right)^*$ and $\left(\mathcal{H}^A_{\text{conv}} \right)^*$ defined by means of R^A and R^A_{conv}, respectively. For example, $\mathfrak{T}^A = \left\{ \left(\mathcal{H}^A \right)^* (A) \mid A \in \mathfrak{P}(U) \right\}$, where $\left(\mathcal{H}^A \right)^* (A) = \overline{\mathcal{H}^A \left(\overline{A} \right)}$ and $\mathcal{H}^A(A) = \{ y \in U \mid \exists x \in U, A(x) \geq A(y)$ and $x \in A \}$, as discussed in [3]. Interesting enough, $\forall \alpha \in L, A_\alpha \in \mathfrak{T}^A$. We also refer [1, 6] for discussion when R^A is a partial ordering.

Let us define a binary relation on L^U, as follows [3]:

Definition 2.2. Let U be a non-empty set, $A, B \in L^U$ and L satisfies (a). We say that A is *finer than* B, denoted by $A \preceq B$, if $R^A \subset R^B$.

We also say that B is *coarser than* A if $A \preceq B$. Definition 2.2 could be expressed also by means of the corresponding topologies, because $R^A \subset R^B \Leftrightarrow \mathfrak{T}^B \subset \mathfrak{T}^A$ (see [4]).

The following Proposition is proved in [3]:

Proposition 2.3. *Let L satisfy* (a). *Then, each $\mathfrak{T}^A_{\text{conv}}$-closed set is $\mathfrak{T}^A$-open and each $\mathfrak{T}^A$-closed set is $\mathfrak{T}^A_{\text{conv}}$-open.*

The following Corollary is immediately presented:

Corollary 2.4. Each α-level set of $A \in L^U$ is a $\mathfrak{T}^A_{\text{conv}}$-closed set.

Proof. Because $\forall \alpha \in L, A_\alpha \in \mathfrak{T}^A$ then $\overline{(A_\alpha)} \in \mathfrak{T}^A_{\text{conv}}$, by Proposition 2.3. This means that $A_\alpha \in c(\mathfrak{T}^A_{\text{conv}})$, thus, A_α is $\mathfrak{T}^A_{\text{conv}}$-closed. $\quad \square$

The following Proposition is proved in [5],

Proposition 2.5. *Let L satisfy* (a) *and $A, B \in L^U$. Then, $\forall \alpha \in L, B_\alpha \in c(\mathfrak{T}^A_{\text{conv}}) \Leftrightarrow A \preceq B$.*

Now, let us consider that L satisfies (a) and (b), and let $A \in L^U$. We denote a class of those L-sets which are coarser than A by $\mathfrak{M}^A$. Thus, $\mathfrak{M}^A = \{ B \in L^U \mid A \preceq B \}$, and especially $\forall A \in L^U, A \in \mathfrak{M}^A$. This means that an L-set B is $c(\mathfrak{T}^A_{\text{conv}})$-definable if and only if $A \preceq B$, as discussed in [5]. Indeed, we demand that L satisfies (a) and (b), because in this case we can apply Representation Theorem [8] such that $\forall \mathcal{B} \in \mathcal{M}^A$, $\forall x \in U$, $\mathcal{B}(x) = \sup\{\alpha \in L \mid x \in \mathcal{B}_\alpha\} = \sup\{\alpha \in L \mid x \in \mathcal{B}_\alpha, \ \mathcal{B}_\alpha \in c(\mathfrak{T}^A_{\mathrm{conv}})\}$.

Now, let L be such that for each $A \in L^U$ a complement of A, denoted by $\overline{A}$, can be defined by means of a negation function $n: L \longrightarrow L$ such that $\forall \alpha, \beta \in L$,

$$\alpha \geq \beta \ \Leftrightarrow \ n(\beta) \geq n(\alpha), \tag{2}$$

$$n(\mathbf{1}) = \mathbf{0} \text{ and } n(\mathbf{0}) = \mathbf{1}, \tag{3}$$

$$n(n(\alpha)) = \alpha, \tag{4}$$

where $\mathbf{0}$ and $\mathbf{1}$ are the least and the greatest elements of L, and $\forall x \in U$, $\overline{A}(x) = n(A(x))$. To complete this Section we present the following Proposition:

Proposition 2.6. *Let $A \in L^U$. Then,* $\mathfrak{T}^{(\overline{A})}_{\mathrm{conv}} = \mathfrak{T}^A$.

Proof. It is enough to show that $R^{(\overline{A})}_{\mathrm{conv}} = R^A$. Indeed, for any $x, y \in U$, $x R^{(\overline{A})}_{\mathrm{conv}} y \ \Leftrightarrow \ y R^{(\overline{A})} x$. In this case $\overline{A}(y) \geq \overline{A}(x) \ \Leftrightarrow \ A(x) \geq A(y)$. Thus, $R^{(\overline{A})}_{\mathrm{conv}} = R^A$. $\square$

3   Natural and Complement Coarsening Operators

In this Section we study an approach to coarsen L-sets. However, the following Definition is first needed [5]:

Definition 3.1. Let L satisfy (a) and (b), $A \in L^U$ and $\mathfrak{T}$ any topology on U. We say that $\mathcal{M}^A_\mathfrak{T}$ is *class of $\mathfrak{T}$-coarsenings of A,* if

$$\mathcal{M}^A_\mathfrak{T} = \{\mathcal{B} \in L^U \mid \forall \alpha \in L, \mathcal{B}_\alpha \in c(\mathfrak{T}) \cap c(\mathfrak{T}^A_{\mathrm{conv}})\}. \tag{5}$$

Notice that $c(\mathfrak{T}) \cap c(\mathfrak{T}^A_{\mathrm{conv}})$ is a class of all $\mathfrak{T} \cap \mathfrak{T}^A_{\mathrm{conv}}$-closed sets, because $c(\mathfrak{T}) \cap c(\mathfrak{T}^A_{\mathrm{conv}}) = c(\mathfrak{T} \cap \mathfrak{T}^A_{\mathrm{conv}})$. This means that $\forall \mathcal{B} \in \mathcal{M}^A_\mathfrak{T}$, $\forall x \in U$,

$$\mathcal{B}(x) = \sup\{\alpha \in L \mid x \in \mathcal{B}_\alpha, \ \mathcal{B}_\alpha \in c(\mathfrak{T}) \cap c(\mathfrak{T}^A_{\mathrm{conv}})\}. \tag{6}$$

Moreover, $\mathfrak{T}^A_{\mathrm{conv}} \subset \mathfrak{T}$ if and only if $\mathcal{M}^A_\mathfrak{T} = \mathcal{M}^A$. Interesting enough, it is possible to define also another class of L-sets as follows:

Definition 3.2. Let L satisfy (a) and (b), $A \in L^U$ and $\mathfrak{T}$ any topology on U. We say that $\mathcal{Q}^A_\mathfrak{T}$ is *class of complement $\mathfrak{T}$-coarsenings of A,* if

$$\mathcal{Q}^A_\mathfrak{T} = \{\mathcal{B} \in L^U \mid \forall \alpha \in L, \mathcal{B}_\alpha \in c(\mathfrak{T}) \cap c(\mathfrak{T}^A)\}. \tag{7}$$

Proposition 3.3. *Let L satisfy* (a) *and* (b)*, and $A \in L^U$. If R^A is an equivalence relation then $\mathcal{M}^A_\mathfrak{T} = \mathcal{Q}^A_\mathfrak{T}$.*

Proof. Clearly, if R^A is an equivalence relation, then $R^A = R^A_{\mathrm{conv}}$, which means that $\mathfrak{T}^A = \mathfrak{T}^A_{\mathrm{conv}}$. $\square$

Proposition 3.4. *Let L satisfy* (a) *and* (b)*, and $A \in L^U$. Then,* $\mathcal{M}^{(\overline{A})}_\mathfrak{T} = \mathcal{Q}^A_\mathfrak{T}$.

Proof. By Proposition 2.6 and Definition 3.2. $\square$

It is now possible to define operators which coarsen L-sets and their complements, as follows:

Definition 3.5. Let L satisfy (a) and (b), $A \in L^U$, $\mathfrak{T}$ any topology on U, and $f \colon L^U \longrightarrow L^U$. We say that f is $\mathfrak{T}$-*coarsening operator* on L^U if $\forall A \in L^U$, $f(A) \in \mathfrak{M}_{\mathfrak{T}}^A$.

Thus, a $\mathfrak{T}$-coarsening operator f on L^U associates for any L-set A an L-set $f(A)$, which is $c(\mathfrak{T}) \cap c(\mathfrak{T}_{\mathrm{conv}}^A)$-definable.

Definition 3.6. Let L satisfy (a) and (b), $A \in L^U$, $\mathfrak{T}$ any topology on U, and $g \colon L^U \longrightarrow L^U$. We say that g is *complement* $\mathfrak{T}$-*coarsening operator* on L^U if $\forall A \in L^U$, $g(A) \in \mathfrak{Q}_{\mathfrak{T}}^A$.

Proposition 3.7. *Let f be any* $\mathfrak{T}$-*coarsening operator on* L^U. *If* $\mathfrak{T} = \mathfrak{P}(U)$ *then* $\forall A \in L^U$, $f(A)$ *is* $c(\mathfrak{T}_{\mathrm{conv}}^A)$-*definable.*

Proof. Clearly by Definitions 1.1 and 3.5. $\square$

Similarly, we can prove that if g is any complement $\mathfrak{T}$-coarsening operator on L^U and $\mathfrak{T} = \mathfrak{P}(U)$, then $\forall A \in L^U$, $g(A)$ is $c(\mathfrak{T}^A)$-definable. We propose the following specific operators:

Definition 3.8. Let L satisfy (a) and (b), $\mathfrak{T}$ any topology on U, and f a $\mathfrak{T}$-coarsening operator on L^U. We say that f is *natural* $\mathfrak{T}$-*coarsening operator on* L^U if $\forall A \in L^U$, $\forall x \in U$,

$$f(A)(x) = \sup\{\alpha \in L \mid x \in A_\alpha,\ A_\alpha \in c(\mathfrak{T}) \cap c(\mathfrak{T}_{\mathrm{conv}}^A)\}. \tag{8}$$

Moreover, $f(A)$ is called *natural* $\mathfrak{T}$-*coarsening of A*.

Definition 3.9. Let L satisfy (a) and (b), $\mathfrak{T}$ any topology on U, and g a complement $\mathfrak{T}$-coarsening operator on L^U. We say that g is *complement* $\mathfrak{T}$-*coarsening operator on* L^U *associated to f*, if $\forall A \in L^U$, $\forall x \in U$,

$$g(A)(x) = f(\overline{A})(x). \tag{9}$$

Indeed, natural $\mathfrak{T}$-coarsening operator on L^U is well defined because L satisfies (a), (b) and this means that $\sup\{\alpha \in L \mid x \in A_\alpha,\ A_\alpha \in c(\mathfrak{T}) \cap c(\mathfrak{T}_{\mathrm{conv}}^A)\}$ exists uniquely for any $x \in U$. Moreover, $f(A) \in \mathfrak{M}_{\mathfrak{T}}^A$ and $g(A) \in \mathfrak{Q}_{\mathfrak{T}}^A$, and the following Proposition can be presented:

Proposition 3.10. *Let f be the natural* $\mathfrak{T}$-*coarsening operator and g the complement* $\mathfrak{T}$-*coarsening operator on* L^U *associated to f. If* $\mathfrak{T} = \mathfrak{P}(U)$ *then* $\forall A \in L^U$, $f(A) = A$ *and* $g(A) = \overline{A}$.

Proof. If $\mathfrak{T} = \mathfrak{P}(U)$ then $f(A)(x) = \sup\{\alpha \in L \mid x \in A_\alpha,\ A_\alpha \in c(\mathfrak{T}_{\mathrm{conv}}^A)\}$. Because $c(\mathfrak{T}_{\mathrm{conv}}^A)$ is the class of all $\mathfrak{T}_{\mathrm{conv}}^A$-closed sets and $\forall \alpha \in L$, $A_\alpha \in c(\mathfrak{T}_{\mathrm{conv}}^A)$, by Representation Theorem [8] $\forall x \in U$, $f(A)(x) = \sup\{\alpha \in L \mid x \in A_\alpha\} = A(x)$. Thus, $f(A) = A$ and this means also that $g(A) = \overline{A}$. $\square$

Interesting enough, natural $\mathfrak{T}$-coarsening operators on L^U can be understood as *substantiating* operators in the following sense [5]:

Proposition 3.11. *Let f be the natural* $\mathfrak{T}$-*coarsening operator on* L^U. *Then,* $\forall A \in L^U$, $f(A) \subset A$.

Finally, we present the following Proposition:

Proposition 3.12. *Let f be the natural $\mathfrak{T}$-coarsening operator on L^U and g be the complement $\mathfrak{T}$-coarsening operator on L^U associated to f. If $\forall \mathcal{A} \in L^U, \overline{g(\mathcal{A})} \in \mathfrak{Q}_{\mathfrak{T}}^A$, then $g^\star \colon L^U \longrightarrow L^U$, $g^\star(\mathcal{A}) = \overline{g(\mathcal{A})}$, is a complement $\mathfrak{T}$-coarsening operator. Moreover, $\forall \mathcal{A} \in L^U$, $\mathcal{A} \subset g^\star(\mathcal{A})$.*

Proof. By the properties of the negation function and Proposition 3.11, $\forall \mathcal{A} \in L^U$, $\mathcal{A} \subset \overline{f\left(\overline{\mathcal{A}}\right)}$. Thus, $\forall \mathcal{A} \in L^U$, $\mathcal{A} \subset g^\star(\mathcal{A})$. Clearly $g^\star$ is a complement $\mathfrak{T}$-coarsening operator by Definition 3.6. $\square$

Indeed, $g^\star$ can be understood as a *weakening operator* and, in the sense of Proposition 3.12, as a complement $\mathfrak{T}$-coarsening operator also. Clearly, if $\forall \mathcal{A} \in L^U, \overline{g(\mathcal{A})} \in \mathcal{M}_{\mathfrak{T}}^A$, then $g^\star$ is a $\mathfrak{T}$-coarsening operator.

4 Conclusions

In this paper we continued the study started in [5]. We defined a topology $\mathfrak{T}$ dependent operator called $\mathfrak{T}$-coarsening operator on L^U and, especially, a specific operator called natural $\mathfrak{T}$-coarsening operator on L^U. Moreover, a specific complement $\mathfrak{T}$-coarsening operator were also defined. We proved some main results, especially, these specific operators can be understood as substantiating and weakening operators.

Proposition 3.12 gives one characterization on existence of weakening operators as complement coarsening operators. However, there may be special restrictions for the underlying topology, in general.

References

[1] G. Birkhoff, Lattice Theory, volume **XXV**, AMS, Providence, Rhode Island, third edition (1995), Eight printing.

[2] J. A. Goguen, L-fuzzy sets, Journal of Mathematical Analysis and Applications **18** (1967) 145–174.

[3] J. Kortelainen, A Topological Approach to Fuzzy Sets, Ph.D. dissertation, Lappeenranta University of Technology, Lappeenranta, Finland (1999), Acta Universitatis Lappeenrantaensis **90**.

[4] J. Kortelainen, Applying modifiers to knowledge acquisition, To appear in Information Sciences: An International Journal (2001).

[5] J. Kortelainen, On coarsenings of L-sets, Proceedings of IFSA/NAFIPS '01, Vancouver, Canada (2001), To appear.

[6] J. D. Lawson, Order and strongly sober compactifications, In G. M. Reed, A. W. Roscoe and R. F. Wachter, editors, Topology and Category Theory in Computer Science, Oxford Science Publications, Oxford (1991).

[7] J. K. Mattila, On modifier logic, In L. A. Zadeh and J. Kacprzyk, editors, Fuzzy Logic for Management of Uncertainty, John Wiley, New York (1992).

[8] C. V. Negoita and D. A. Ralescu, Representation theorems for fuzzy concepts, In D. Dubois, H. Prade, and R. R. Yager, editors, Readings in Fuzzy Sets for Intelligent Systems, Morgan & Kaufmann Publ, Inc., San Mateo, CA (1993).

[9] L. A. Zadeh, Fuzzy sets, Information and Control **8** (1965) 338–353.

Machine Dreams: Fuzzy Ethical Theory to Cyborg Intelligence

Timo AIRAKSINEN
Department of Moral and Social Philosophy
Box 9, 00014 University of Helsinki, Finland

Abstract. This paper discusses the fact-value distinction which has been one of the main problems of theoretical ethics, originally because of the classic work by David Hume and then G.E. Moore. Also such logical ideas as supervenience and supererogation will be studied. Finally a new, information-age notion, the group mind in the Web, will be discussed as a speculative application of ethics. All these notions present difficult challenges as well to the machine intelligence as to the standard positivistic research methodology. My concluding point is that the technoculture displays, in the beginning of the new millennium, a strong drive towards Cyborg - thought. In other words, the new paradigm of cognitive and emotional life entails the emergence of a Cyborg. This is a surprising and confusing cultural change because it seems to leave ethics out of the picture. I will explain the speculative logic in question and suggest its fuzzy methodological aspects.

1. Introduction to A Problem and Its Solution and Beyond

The human mind is a restless and confused entity. What Lotfi Zadeh writes seems to me to be true: "Among the basic concepts which underlie human cognition there are three that stand out in importance. The three are: granulation, organization and causation." He continues: "the way in which humans employ fuzzy information granulation to make rational decisions in an environment of partial knowledge, partial certainty and partial truth should be viewed as a role model for machine intelligence" [10]. I proceed to study and evaluate his claims in what seems to be the most demanding intellectual context, ethical reasoning. Its challenges have been beyond the capabilities of positivistic methodology.

This is the point: Humans live their lives in the environment which is practically relevant and which they judge and evaluate accordingly. This work is called *ethics*. They analyze the given context, or perform the operations of information granulation. However, after they have reached the ultimate granules, as they see it, they cannot organize what they now know. They are left with some isolated granules, which are, of course, difficult to utilize in practical causal contexts. The analyzed information does not allow for a synthesis. To illustrate: think of a film where you see a stone shattering a window. Now play it back at normal speed. What you see on the screen is a physical impossibility. All the pieces of glass fly together and organize spontaneously to form a perfect panel of glass.

This is just like the situation where we seem to be so often. We know how to create two separate sets of information granules, facts and values, but then we cannot see how they are connected for practical and causal purposes. The *fact-value distinction* has been one of the main problems of theoretical ethics, because of the work by David Hume [3] and then by G.E. Moore [5]. According to the positivistic philosophy facts are simply (a) observable entities and values (b) subjective feelings. In the empirical world only a and b exist but they are neither mutually connected nor organized. Perhaps facts are coded digitally but values, feelings, and emotions are coded analogically, as Robert Nozick has argued [7]. In this way we have lost any possibility of a synthesis and rational organization. Values cannot guide our journey among the myriads of facts. Despite our rationality, or because of it, we are adrift.

2. The Gap between Facts and Values

The relation of facts and values presents an unsoluble problem when reality is understood as containing two distinct categories of entities, exemplified as (i) "Most airplanes are made of metal" and (ii) "Most airplanes are safe to operate". Sentence (i) states a fact. Sentence (ii) entails a human value, safety, which is both needed and desired by most of us.

As Hume said, we cannot logically infer a value-sentence from a set of premises which contain only fact-sentences or statements of facts. For example, we cannot derive (ii) from a set of premises which contain only (i) and logically similar factual sentences. This is a generally accepted conclusion even now. G.E. Moore realized that it is impossible to give a definition of an ordinary language value term like "safety" in terms of fact-terms, like "metallic". Both philosophers agree that some novel elements must be added to facts if we hope to reach the realm of values. Logical positivist, like A.J. Ayer, interpreted this to mean, roughly, that values are emotions [1]. If this is accepted, the human world of knowledge and action is divided into two parts both of which are typically crisp. Moreover they are mutually incompatible. Logically, or syntactically and semantically, no trouble is evident, although pragmatically the case is different. It is indeed an intolerable conclusion that values must be defined and handled in human practical reasoning independently of values, and vice versa. For instance, however detailed the description of human pain and misery is at the factual level, it cannot be concluded that the situation is undesirable or bad. Facts cannot guide human action.

From the pragmatic point of view, we must conclude that there is a fuzzy border area between facts and values. In other words, the standard positivistic fabrication of the duality is artificial and logically viable, but also impossible to accept. We need to fuzzify the fact/value -distinction. Some attempts in this direction has been done. We need to mention here only two. The first mentions socially constituted facts [8] and the second the logical notion of supervenience relation [2]. We have then a sociological and logical solution to the key problem.

The social fuzzification of the fact/value -gap can be found in the following sentence: (iii) "When the ship sunk the Master of the ship left his vessel last, and this was the right thing to do". The reasoning in (iii) to the conclusion that the Master's action was right is valid, even if the antecedents are factual and the conclusion contains a value based norm. The sinking of the ship is a hard fact, but the rightness of the Master's action is a matter of valuation. Why is this so, regardless of what Hume and Moore might have said?

The reason is simple. To say that the Master of the sinking ship was last to abandon it, we refer to the Master and his social role in the game of seafaring. And such a role is defined by certain norms and values without which there are no Masters. One of the defining norms is that the Master will stay on board until everyone has left the ship. This is a socially constructed fact which allows us to perform the reasoning in (iii) and call it valid. Many philosophers would reject such a solution because the concept of a Master is now a value term, which begs the question. Anyway, it is interesting to notice that facts and values have their fuzzy borders in the realm of socially constructed facts. The borders are not crisp.

3. The Supervenience of Values upon Facts

R.M. Hare introduced the notion of supervenience in his studies on ethical language [2]. This notion has then been studied widely and applied to various fields, including cognitive psychology and machine intelligence. Hare argues that values cannot be derived logically from facts, in other words, he agrees with Hume. But, Hare says, facts and values are still connected logically, and the name of their relation is *supervenience*. An example illustrates the situation in question. Go to an art museum and find a painting on the wall. When you look at it, you can freely call it good or bad, because the factual description of the painting can fix no value-based evaluation of it. So, call it good. Then find a painting which resembles the first one as much as possible in all of its essential artistic qualities. You must call it good too, or you are guilty of logical inconsistency.

In the same way, suppose that you observe that your favorite intelligent machine M is in state Z when it seems, according to your free judgement, display intelligent functioning. Every time M or its relative M' is in state Z or in any sufficiently similar state Z', you must call the result intelligent behavior. Any ascription of machine intelligence rests upon M-family's Z-likeness, just like the goodness of a painting in an art gallery rests on the qualities visible in the painting first assessed. Such an analysis of M's Z presupposes that "intelligence" (Z) is an evaluative term and not a matter of mere facts. But such a presupposition may well be warranted. It is also worthwhile to consider the possibility that this is the very reason why it has been so difficult to find any (factual) standards of the applicability of the (normative) concept of intelligence to machines. We now say that *intelligence* is an evaluative property which supervenes upon machines, although intelligence is not logically entailed by any set of the factual descriptions of the machine.

Supervenience seems to be a phenomenon which is easy to handle by means of any standard, positivistic theory of machine intelligence, unlike the idea of socially constructed facts and values. The latter seems ambiguous and unclear indeed, requiring good social knowledge and intuitions to become at all clear. But supervenience looks different. The machine's eye needs only to register a painting and assign a random evaluation to it, or call whatever it observes either good or bad. Then, when the eye meets another painting, it intelligently determines its degree of relative similarity, and assigns the correct evaluative term to it accordingly.

However, the difficulty inherent in the narrative above is clear, when one thinks of it. What do we mean by "relative similarity" between any two observables, such as paintings? If this is not a fuzzy notion, nothing is. The two paintings are never fully similar, or identical, yet they must somehow be similar. The observer needs to apply both his personal and socially defined criteria of similarity, in terms of style, color, shapes, and painting technique. From all of this material he/she must construct the similarity, always keeping in mind that the results should also be socially communicable and even acceptable. It is difficult to see that any intelligent machine could do this without fuzzy operations. Multidimensional comparisons in a vague context are mysterious, but not totally free because they should also be somehow artistically informed. All of this must be based on operations in terms of words.

The main problem might ultimately be that the evaluator needs to convince his/her peers of the sensibility of his/her own views. When a machine performs supervenient operations, they may look easy only if they are arbitrary. In the human world this would be called extreme subjectivity, which is out of place in the machine world. The machine merely calls a selected painting good on a random basis and then selects a similar item on the basis of some irrelevant features. After this it is easy to call the second painting good too. The machine is logically consistent but not really intelligent. Of course this is no news because we all know that the standard machine intelligence is mere logicality.

4. The Idea of Supererogation

When we move deeper into the miraculous world of ethics, we can try our logical skills on the problem of *supererogation*, which was made popular by J.O. Urmson [9]. Suppose you walk by a burning house which is already engulfed in flames. You see a child in the window. You are tempted to rush in and try to save her, but you also realize the extreme danger involved.

You understand that it would be good to try to save the child - but is it also your ethical duty? This kind of a decision situation shows the meaning of supererogation: Some actions are good but they are not required by duty, simply because they are too demanding. In the case of the burning house, the requirement to rush in to the burning house would be cruel, because no one can be blamed of not doing so. In the case of supererogatory actions, the attempt to save the child is laudable, although not doing so cannot be criticized or condemned. It is not one's duty to save the child.

This example shows why ethics is so difficult to comprehend from the point of view of the human mind and machine intelligence alike. The various ambiguities and uncertainties are large and multidimensional. In the case of supererogatory actions the agent must first evaluate the goodness of actions. As we remember, this is a subjective exercise due to Moore's and Hume's logical theories of ethical reasoning. Suppose the action in question is good. Next it must be determined how dangerous the action is. The risks must be evaluated from the agent's own point of view. This is another subjective elements in the process. As usual, a machine intelligence needs to randomize its judgements, but that can certainly be done consistently. A human agent must act differently. He/she must determine whether the benefits are so large that his/her risk is not significant, in which case he/she may call the proposed action his/her ethical duty. If the good is not so weighty, the duty does not materialize, although counterfactually speaking it can still be said that the action would have been good, had he/she done it.

In supererogatory cases the agents are required to compare ethical good/evil and the requirements of duty to each other. To do this successfully, the human mind must be able to determine how much risk is sufficient to override the requirements of duty. In other words, one should quantify both the risk and the duty, and compere the results. Small risk and strong duty lead to action. Serious risk and weak duty prevent action, in supererogatory problem cases. The bad news is that not even the human mind can perform these tasks at the level of exact quantifications and their comparisons. What is needed, and actually applied in real life, is an intuitive approach and method. A mature ethical agent is able to see what to do, without mentioning any conclusive logical reasoning. In the machine world, we can say that one computes with words in a fuzzy context.

5. The Strangeness of Cyborgs and Other Robots

Machine intelligence cannot handle these problems, and so we need to get rid of the idea of a mere robot [6]. Robots are a primitive tribe which has never flourished and which will be extinct soon, except in their last fortresses, factories, where they can be exploited as workers. But they cannot develop outside of that limited realm. They are too clumsy physically and too limited intellectually.

Robot-intelligence is a problem, especially if it is understood in a broad sense including ethics, value-intuitions, and emotional experiences. As I tried to show above, these present some formidable problems. At the same time it is possible to make the idea of intelligence very narrow and include only those aspects in which the binary computational machine excels. This can be done verbally. The only move that is required is to make the machine the paragon of intelligence, instead of the human mind. This narrows down the scope of the concept of intelligence so that it becomes tailor made for machines and their capabilities as intelligent agents. And this seems to be the last opportunity which is available to the traditionally minded AI community in the world of IT. As I have tried to show in this paper, large parts of the human life-world must in this case be forgotten. Certainly fuzzy logic and computing with words will help to solve many of these problems, as Lotfi Zadeh and Vesa A. Niskanen have argued in their various publications. However, more needs to be said.

Robots are dead. They were too isolated and independent, or too far removed from the human world with its deep problems and great intelligibility. The attention turns now to *Cyborgs*, simply because they can retain the mysterious emotions, ethical intuitions, and values which are typical of only human beings. This human component is needed because it contains values which would otherwise be inaccessible to machines. But then Cyborgs will be like humans. They can be given the human cognitive features because they are still partly human. But this is only the first step towards practical and intelligent ethical capability.

However, the human mind may be too limited to be able to solve its own theoretical problems, like those of the ethical theory. Hence no single Cyborg would do any better, because its/his/her mind must depend on the human capabilities at this crucial juncture. Perhaps more fuzzy logic and computing with words help, but it is unreasonable to expect them to do any better than the human mind in the field of practical philosophy. The human mind is a the role model, according to Zadeh. But here we must be talking about its practical capabilities which indeed can overcome the vagaries of ethics. This is the real strength of the human mind and consequently that of the pragmatic interpretation of a Cyborg.

Standard AI in the role of machine intelligence cannot solve the problems of ethics as rational tasks of good reasoning. Too much is left to depend on vague intuitions. This is the critical observation which makes the idea of a *Cyborg* so attractive. It is easy to see why this is true. If intelligent machines can never crack the code of ethical behavior, unlike the human mind, it is perfectly rational to create a Cyborg, instead of a pure robot, to solve this dilemma. The Cyborg has its superior computational machine intelligence and is free of the many limitations of the robot hardware, that is, the body. Human wetware and a nice combination of artificial and human software should do the trick [4]. Yet Zadeh is correct when he says that the human mind is the role model for mere machines. But this must be understood correctly: those problems of ethics which we reviewed above cannot be solved positivistically, neither by computational machine intelligence nor by analytical and scientific philosophy. They can be solved by the practical human mind in action in the course of its normal life, intuitively, so to speak. Such human mind produces results which are valid but also ambiguous, contextual, gradual, and many-values, or fuzzy. This is the aspect of the human reason which is worth simulation. The human mind is not a good problem solver in the theory of ethics, although it is perfectly adapted to ethical practice and its intuitive requirements.

References

[1] Ayer, A.J., *Language, Truth, and Logic*, Gollancz, London, 1936.
[2] Hare, R.M., *The Language of Morals*, Oxford University Press, Oxford, 1952.
[3] Hume, D., *A Treatise of Human Nature*, Clarendon Press, Oxford, [1739-1740] 1978.
[4] Lévy, P., *Collective Intelligence: Mankind's Emerging World in Cyberspace*, tr. R. Bobonno, Perseus Books, New York, 1997.
[5] Moore, G.E., *Principia Ethica*, Cambridge University Press, Cambridge, 1903.
     Moravec, H., *Robot: Mere Machine to Transcendent Mind*, Oxford University Press, New York, 1999.
[6] Nozick, R., "Emotions". In: Robert Nozick: *The Examined Life: Philosophical Meditations*. Simon and Schuster, New York, 1989, pp. 87-98.
[7] Searle, J., "How to Derive 'Ought' from 'Is'?". In: Phillippa Foot (ed.): *Theories of Ethics*, Oxford University Press, Oxford.1967, pp. 101-114.
[8] Urmson, J.O.,"Saints and Heroes". In: Joel Feinberg (ed.): Moral *Concepts*, Oxford University Press, Oxford, 1969, pp. 60-73.
[9] Zadeh, L., "Toward a Theory of Fuzzy Information Granulation and Its Centrality in Human Reasoning and Fuzzy Logic", *Fuzzy Sets and Systems*, Vol. 90, 1997.

KES '01
N. Baba et al. (Eds.)
IOS Press, 2001

Formulation of Regression Model Based on Natural Words [1]

Junzo Watada

Osaka Institute of Technology

School of Industrial Management

Abstract: *Generally when human experts express their ideas and thoughts, human words are basically employed in these expressions. It should be useful to employ fuzzy regression analysis in handling human words and finding the latent structures under these human words, a linguistic regression model is formulated in terms of fuzzy regression analysis and vocabulary matching on the basis of fuzzy numbers.*

Keywords: *Computation of words, Linguistic Regression Model, Fuzzy Regresssion Model*

1   Introduction

As Professor L. A. Zadeh has placed the stress on the importance of the computation of words, fuzzy sets can take a central role in handling words [10, 11]. In this perspective fuzzy logic approach is offen thought as the main and only useful tool to deal with human words. In this paper we intend to present another approach to handle human words instead of fuzzy reasning. That is, fuzzy regression analysis enables us deal with the computation of words.

We intend to abstract the latent structure under the relations between words. Human words can be translated into fuzzy sets such as fuzzy numbers, which is employed in a fuzzy controller. If it is possible such as fuzzy control to formulate a dictionary between fuzzy number and a word, we can build the relations under the data in terms of fuzzy regression analysis.

Consequently, only experts with much professional experiences are capable of making assessment using their intuition and experiences. The measurements and interpretation of these characteristics are taken with uncertainty, because most measured characteristics, analytical result, and field data can be interpreted only intuitively by experts. In such cases, judgments may be expressed by experts with linguistic terms. The difficulty in the direct measurement of certain characteristics makes the estimation of these characteristics imprecise. Such measurements may be dealt with the use of fuzzy set theory [7, 8, 9, 2].

Watada, Fu and Yao [4, 5] proposed a model of damage assessment by using the information given by experts through fuzzy multivariant analysis.

In order to process linguistic variables, we define the vocabulary translation and vocabulary matching which convert linguistic expression into membership functions on the interval [0-1] on the basis of a linguistic dictionary, and vice versa. We employ Fuzzy Regression Analysis [2, 6] in order to deal with the assessment process of experts' from linguistic variables of features and characteristics of an objective into the linguistic expression of the total assessment.

---

[1] This article can be inquired to Professor J. Watada / 5-16-1 Omiya, Asahi, Osaka 535-8585 JAPAN / Phone : 81-6-6952-4327 / FAX : 81-6-6952-6197 / e-mail : junzow@osb.att.ne.jp

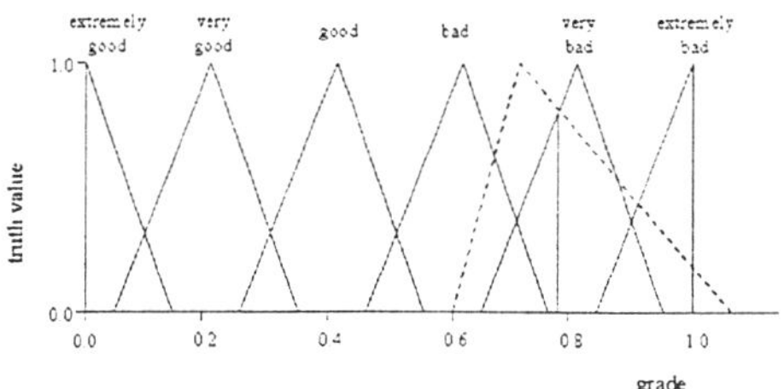

Figure 1. Vocabulary Matching

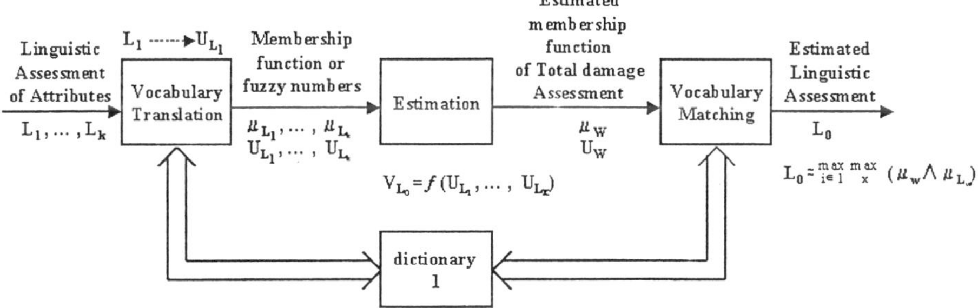

Figure 2. Process of Regression Model Based on Natural Words

2 Linguistic Variable and Vocabulary Matching

In making assessments, experts often (a) evaluate various features and characteristics, and (b) assess the objective in a linguistic form. For instance, though it is possible to measure the production volume, it is difficult to analytically interpret the numerical value in terms of its influence of this amount on the future decision making.

On the other hand, experts may be able to express (a) the effect of a given sales trend as "good" or "not good", and (b) the total assessment or state in a linguistic form such as "very good", "good" or "not good".

Consider the simplest possible value such as production volume. Let $\mathbf{L}$ be a linguistic variable which can have values in the set $\mathbf{X}$. As examples of such a variable, let us consider a production volume as well as the number of visiting customers.

$$\mathbf{L}_{\text{volume}}(\text{production}) \text{ is } /\text{extremely bad}/$$
$$\mathbf{L}_{\text{number}}(\text{visiting customers}) \text{ is } /\text{bad}/$$

where subscripts (number) in $\mathbf{L}_{(}$visiting customers$)$ denote the state of a visiting customers.

These expressions "good", "bad", "extremely bad" can be defined with fuzzy grades on $[0.1]$ such as $\mathbf{U}_{(\text{good})}$, $\mathbf{U}_{(\text{bad})}$, $\mathbf{U}_{(\text{extremely bad})}$. Denote π as a possibility distribution. We can identify the possibility of the state of a sales trend with the degree of its descriptive adjective on $[0.1]$. For example,

$$\pi_{\text{number}}(\text{visiting customer}) \equiv \pi_{(\text{state})}(\text{the number})$$
$$\equiv \mathbf{U}_{(\text{extremely bad})}$$

Table 1. Linguistic data given by experts.

Training Sample	Linguistic variables					Linguistic objective
1	$\mathbf{L}_1(1)$	$\cdots$	$\mathbf{L}_i(1)$	$\cdots$	$\mathbf{L}_K(1)$	$\mathbf{L}_0(1)$
2	$\mathbf{L}_1(2)$	$\cdots$	$\mathbf{L}_i(2)$	$\cdots$	$\mathbf{L}_K(2)$	$\mathbf{L}_0(2)$
3	"good"	$\cdots$	"bad"	$\cdots$	"very bad"	"bad"
$\vdots$	$\vdots$		$\vdots$		$\vdots$	$\vdots$
ω	$\mathbf{L}_1(\omega)$	$\cdots$	$\mathbf{L}_i(\omega)$	$\cdots$	$\mathbf{L}_K(\omega)$	$\mathbf{L}_0(\omega)$
$\vdots$	$\vdots$		$\vdots$		$\vdots$	$\vdots$
n	$\mathbf{L}_1(n)$	$\cdots$	$\mathbf{L}_i(n)$	$\cdots$	$\mathbf{L}_K(n)$	$\mathbf{L}_0(n)$

Define the dictionary of descriptive adjectives in which a descriptive adjective corresponds with its fuzzy grade on $[0,1]$.

The objective of this study is to model the experts' assessment process through which experts might evaluate the possibility of sales trend $\mathbf{L}$ on the basis of states of its features and characteristics $\mathbf{L}_i(i = 1, 2, \ldots, K)$, where these values $\mathbf{L}$ and $\mathbf{L}_i$ are expressed in a linguistic form. In other words, we intend to determine the linguistic assessment process $\mathbf{F}$ of linguistic variable $\mathbf{L}_1, \mathbf{L}_2, \ldots, \mathbf{L}_K$, which produces a value of an objective $\mathbf{Z}$. It can be written in the form

$$\mathbf{Z} = \mathbf{F}(\mathbf{L}_1, \mathbf{L}_2, \ldots, \mathbf{L}_K) \tag{1}$$

We define the descriptive adjectives "extreme", "very", and so on. Let us make a dictionary for corresponding linguistic expressions $\mathbf{L}_i$ and fuzzy grades $\mathbf{U}_{L_i}$. If we obtain the assessment $\mathbf{L}_i(i = 1, 2, \ldots, K)$, $\mathbf{L}_i$ is understood in terms of $\mathbf{U}_{L_i}$ through the dictionary build by experts. Let us divide this linguistic assessment process F into three portion:

(1) translation of attributes form linguistic values L_i into fuzzy grades $\mathbf{U}_{L_i}$,

(2) estimation of total damage by the fuzzy assessment function

$$\mathbf{V} = f(\mathbf{U}_{L_1}, \mathbf{U}_{L_2}, \ldots, \mathbf{U}_{L_K}) \tag{2}$$

which produces a fuzzy grade $\mathbf{V}$ in terms of fuzzy grades of attributes $\mathbf{U}_{L_i}$ where a suffix $\mathbf{L}$ is not attached to $\mathbf{V}$ because it is un known, and

(3) linguistic matching of the fuzzy grade of the objective with the dictionary wherein the linguistic value $\mathbf{Z}$ is decided for the objective.

Let us define the vocabulary matching by using the following minimax calculation

$$\mathbf{Z}_0 \simeq \max_{\mathbf{W}_i \in \mathbf{D}} \{\max_t \mu_{\mathbf{V}}(t) \bigwedge \mu_{W_i}(t)\} \tag{3}$$

where $\mathbf{Z}_0 \simeq \max_{\mathbf{W}_i \in \mathbf{D}} f(\mathbf{W}_i)$ denotes that $\mathbf{Z}_0$ is the word in $\mathbf{D}$ which realizes the maximum value of f. $\mu_{\mathbf{V}}$ denotes a membership function of $\mathbf{V}$, and μ_{W_i} denotes a membership function of a word $\mathbf{W}_i$ as included in the dictionary $\mathbf{D}$. We assign a word $\mathbf{L}_0$ to the fuzzy grade of the total assessment $\mathbf{V}$.

3   Determination of Fuzzy Regression Model

After the model of experts' assessment process is formulated, it is desirable to determine the fuzzy assessment function f of the total assessment as shown in Figure 3. This is a fuzzy function

Table 2. Fuzzy data translated.

Training Sample	Fuzzy grade of variables					Fuzzy grade of the objective
1	$\mathbf{U}_{L_1}(1)$	$\cdots$	$\mathbf{U}_{L_i}(1)$	$\cdots$	$\mathbf{U}_{L_K}(1)$	$\mathbf{V}_{L_0}(1)$
2	$\mathbf{U}_{L_1}(2)$	$\cdots$	$\mathbf{U}_{L_i}(2)$	$\cdots$	$\mathbf{U}_{L_K}(2)$	$\mathbf{V}_{L_0}(2)$
3	$(0.4,0.15,0.15)$	$\cdots$	$(0.6,0.15,0.15)$	$\cdots$	$(0.8,0.15,0.15)$	$(0.6,0.15,0.15)$
$\vdots$	$\vdots$		$\vdots$		$\vdots$	$\vdots$
ω	$\mathbf{U}_{L_1}(\omega)$	$\cdots$	$\mathbf{U}_{L_i}(\omega)$	$\cdots$	$\mathbf{U}_{L_K}(\omega)$	$\mathbf{V}_{L_0}(\omega)$
$\vdots$	$\vdots$		$\vdots$		$\vdots$	$\vdots$
n	$\mathbf{U}_{L_1}(n)$	$\cdots$	$\mathbf{U}_{L_i}(n)$	$\cdots$	$\mathbf{U}_{L_K}(n)$	$\mathbf{V}_{L_0}(n)$

with which K fuzzy grades of attributes, $\mathbf{U}_{L_i}$, can be transformed to one fuzzy grade of its total assessment, $\mathbf{V}$. We must determine this fuzzy regression model on the basis of training data $\omega(\omega = 1, 2, \ldots, n)$ as given by experts in order to mimic experts' assessment process. Let us employ the method proposed as Fuzzy Quantification Theory Type 1 by Watada, Tanaka and Asai [6] for determining this fuzzy regression model f.

Table 3 shows the training data given in linguistic form by experts. This data in Table 3 is translated into data of fuzzy grades in terms of the dictionary (see Table 3). In Table 4, the fuzzy grades of attributes i and of the total damage of sample structures ω are denoted by $\mathbf{U}_{L_i}(\omega)$ and $\mathbf{V}_{L_0}(\omega)$, respectively where $i = 1, 2, \ldots, K$ and $\omega = 1, 2, \ldots, n$.

Assume that all fuzzy grades have triangular shapes which are normal and convex. We employ a fuzzy linear function for a fuzzy regression model f as

$$\begin{aligned} \mathbf{V}_{L_0} &= f(\mathbf{U}_{L_1}, \mathbf{U}_{L_2}, \ldots, \mathbf{U}_{L_K}) \\ &= \sum_{i=1}^{K} \mathbf{A}_i \mathbf{U}_{L_i}(\omega) \quad \omega = 1, 2, \ldots, n \end{aligned} \qquad (4)$$

Using n relation of training samples

$$\mathbf{V}_{L_0}(\omega) = \sum_{i=1}^{K} \mathbf{A}_i \mathbf{U}_{L_i}(\omega) \quad \omega = 1, 2, \ldots, n \qquad (5)$$

We must specify the best fuzzy parameters $\mathbf{A}_i$ in terms of these relations. Two criteria are employed in order to define the goodness of the fuzzy linear function. One criterion is fitness of the fuzzy regression model, h, and the other is fuzziness included in the fuzzy regression model, $\mathbf{S}$.

(i)Fitness.

Assumed that an estimated values $\mathbf{V}_{L_0}(\omega)$ is obtained by the fuzzy linear function f, fitness $b(\omega)$ of $\mathbf{Y}_0(\omega)$ to a sample value $\mathbf{Y}(\omega)$ is $\mathbf{V}_{L_0}(\omega)$ to a sample value $\mathbf{V}_{L_0}(\omega)$ is defined by

$$h(\omega) = \bigvee_{y \in R} \{\mu_{L_0(\omega)(y)} \bigwedge \mu_{L_0(\omega)}(y)\} \qquad (6)$$

(ii)Fuzziness.

The fuzziness $\mathbf{S}^\alpha$ included in the fuzzy function at α-level is defined by

$$\mathbf{S}^\alpha = \sum_{i=1}^{K} (\overline{a_i} - \underline{a_i}) \qquad (7)$$

where $\overline{a_i}$ and $\underline{a_i}$ are number which specify an α-level set $\mathbf{A}_i^\alpha$, i.e.,

$$\mathbf{A}_i^\alpha = [\overline{a_i} - \underline{a_i}] \tag{8}$$

In this paper, we deal with triangular fuzzy number. Note that

$$\mathbf{A}_i^\alpha \equiv [a_i - (1-\alpha)c_i^l, a_i + (1-\alpha)c_i^r]$$

as $\mathbf{A}_i^\alpha$ is defined by

$$\mathbf{A}_i \equiv (a_i, c_i^l, c_i^r)$$

Note that these two indices of fuzziness and fitness are incompatible with each other. The higher fitness we seek, the resulting model will be.

(1) Formulation of the problem.

We formulation a fuzzy regression model by minimizing its fuzziness $\mathbf{S}$ under the constraints that an as estimated fuzzy grade of total assessment of each sample is fit to the fuzzy grade given by experts with the fitness grater than or equal to the given value h, called fitness standard.

[Problem]

If data are given such as that listed in Table 4, the problem is to determine a fuzzy linear function

$$\mathbf{V}_{L_0}(\omega) = \sum_{i=1}^{K} \mathbf{A}_i \cdot \mathbf{U}_{L_i}(\omega) \tag{9}$$

which minimizes the fuzziness

$$\mathbf{S} = \sum_{i=1}^{K} (\overline{a_i} - \underline{a_i}) \tag{10}$$

under the conditions that

$$h(\omega) = \underset{y \in R}{V} \quad \{\mu_{L(\omega)(y)}{\scriptstyle\wedge}\mu_{L_0(\omega)}(y)\} \geq h^0 \tag{11}$$

$$\omega = 1, 2, \ldots, n$$

where $h(\omega)$ indicates the fitness of the estimated value with respect to a sample ω and h^0 denotes the fitness standard, and $\overline{a_i}$ and $\underline{a_i}$ are defined as

$$A_i^{h^0} = [\underline{a_i}, \overline{a_i}] \tag{12}$$

Note that we will employ triangular approximation for calculation of fuzzy number in this paper, although the precise calculation has been obtained in Tanaka, Watada and Asai [3].

The membership function $\mu_{L_0}(y)$ of the fuzzy grade of structural total damage $\mathbf{L}_0$ can be reduced through extension principle to

$$\mu_{\mathbf{L}_0}(y) =$$

$$\underset{\substack{(t_i,u_i)}}{\underset{\substack{|y = \sum_i t_i u_i \\ |0 \leq t_i \leq 1 \\ |0 \leq u_i \leq 1}}{V}} \quad \bigwedge_{i=1}^{K}\{\mu_{\mathbf{A}_i}(t_i)\bigwedge\mu_{\mathbf{L}_i}(u_i)\}$$

$$\tag{13}$$

(2) the fuzzy grade.

In this section, we discuss the heuristic method to determine a fuzzy assessment function by using fuzzy grades of assessment attributes, i.e., U_i are fuzzy numbers in $[0,1]$.

According to the sign of A_i, the production of fuzzy number A_i and U_{L_i} is given in the following three cases:

(i) In this case where $\overline{a}_i \geq \underline{a}_i \geq 0$

$$(A_i U_{L_i})^{h^0} = [\underline{a}_i \underline{u}_i, \overline{a}_i \overline{u}_i] \tag{14}$$

(ii) In this case where $\overline{a}_i \leq \underline{a}_i \leq 0$

$$(A_i U_{L_i})^{h^0} = [\underline{a}_i \overline{u}_i, \overline{a}_i \underline{u}_i] \tag{15}$$

(iii) In this case where $\overline{a}_i \leq \underline{a}_i \leq 0$

$$(U_{L_i})^{h^0} = [\underline{a}_i \overline{u}_i, \overline{a}_i \underline{u}_i] \tag{16}$$

It is difficult to solve analytically this problem. Therefore, we employ the heuristic approach for solving the problem. The procedure is as follows.

An α-level set of the fuzzy degree of a structural attribute $U_{L_i} (i = 1, 2, \ldots, K)$ at h^0 is assumed to be denoted by Eqn. 28.

4   Concluding Remarks

We have discussed the formulation of regression model based on natural words. In assessment, the role of experts is very important because existing structures are extremely complex and the accumulation of professional experience is often required. Experts frequently express their judgment in linguistic form rather than numerical form. Therefore, the linguistic treatment of assessment is central and essential in employing human inspective and subjective judgment into assessment process.

The presented process consists of four portions: (1) vocabulary translation, (2) estimation, (3) vocabulary matching and (4) dictionary. The emphasis of this paper should be placed on the use of natural words. The linguistic judgment of the total assessment is obtained through vocabulary matching on the basis of a dictionary as given by experts. We employed fuzzy quantification theory type 2 for estimating the total assessment in terms of linguistic structural attributes which are obtained from an experts.

References

[1] Dubois, D., and Prade, H.: Fuzzy Sets and systems: Theory and Applications, Academic Press, New York, 1980.

[2] Tanaka, H., Uejima, S., and Asai, K., "Linear Regression Analysis with Fuzzy Model", IEEE, Trans on System., Man. Cybern., Vol. SMC-12, No. 6, pp. 903-907, 1982.

[3] Tanaka, H., Watada, J., and Asai, K.: "Evaluation of Alternatives with Multiple Attribute based on Fuzzy Sets", Systems and Control { The Japan Association of Automatic Control Engineers }, Vol.27, No. 6, pp 403-409, 1983.

[4] Watada, J., Fu, K.S., and Yao, J.T.P.,; "Damage Assessment Using Fuzzy Multivariant Analysis," Technical Report No. CE-STR-84-4, School of Civil Engineering, Purdue University, West Lafayette, In. 1984.

[5] Watada, J., Fu, K.S., and Yao, J.T.P.,; "Fuzzy Classification Approach to Damage Assessment," First International Conference on Fuzzy Information Professing in Hawaii, July 1984.

[6] Watada, J., Tanaka, H., and Asai, K.; "Fuzzy Quantification Theory Type I," The Japanese Journal of Behaviormetrics (The Behaviormetric Society of Japan), Vol. 11, No. 1, pp. 66-73, 1983, in Japanese.

[7] Zadeh, L.A.; "Fuzzy Sets", Int. Control, Vol. 8, p. 338-353, 1965.

[8] Zadeh, L.A.; "The concept of a Linguistic Variable and Its Applications to Approximate Reasoning, Part 1", Int. Sei., Vol. 8, pp. 199-249, 1975.

[9] Zadeh, L.A.; "The concept of a Linguistic Variable and Its Applications to Approximate Reasoning, Part 2", Int. Sei., Vol. 8, p. 301-357, 1975.

[10] Zadeh, L.A.; "What is computing with words," in *Computing with Words in Information/Intelligent Systems, Foundation*, Physica-Verlag, eds. L.A. Zadeh & Janusz Kacpruzyk, pp VIII - IX.

[11] Zadeh, L.A.; "Fuzzy Logic = computing with words," in *Computing with Words in Information/Intelligent Systems, Foundation*, Physica-Verlag, eds. L.A. Zadeh & Janusz Kacpruzyk, pp 3 - 23.

A Stability Test Tool for Parameters Estimation in Portfolio Management

Marina Resta
University of Genova, Italy

Abstract. This work focuses on portfolio management, namely on welfare allocation between a risk free asset, and a risky asset, when the dynamics of this latter is not governed by a standard gaussian process.

In this context the attention is stressed on the problems which arise in the choice of the right model to fit price returns dynamics, with special care to parameters distribution estimation. A solution based on neural networks is here suggested, giving information about the stability of model parameters, as well as on the time occurring between each portfolio re-calibration.

1 Introduction

This paper studies the classical Merton problem of optimal portfolio selection and consumption, when alternatives to standard Brownian Motion are considered to model stock price dynamics.

The problem to be considered may be summarized as follows.

In its simplest form, portfolio management deals on how to allocate investor's wealth between a risk-free asset paying at a constant rate r, and a risky asset, whose dynamics is given by:

$$S_t = S_0 e^{\mu t + \sigma W_t} \tag{1}$$

where μ is the drift coefficient or mean rate, σ is the volatility, and W_t a Brownian motion.

According to Eq. 1, logreturns are normally distributed with mean $\mu \Delta t$, and variance $\sigma^2 \Delta t$, where Δt is a chosen temporal window. Throughout this paper Δt will be measured in days, but generalization are possible both for intraday and weekly, as well as for monthly data.

Although widely used, this model doesn't fit well real log prices return dynamics, as already evidenced by many empirical works, such as in [4].

In recent times Barndorff-Nielsen and others [1], [2], [3] have introduced an alternative model, where log–prices returns dynamics is fitted by means of Levy processes.

Levy Processes are a family of parametric stochastic processes with density $p(x; \alpha, \beta, \gamma, \delta)$, where α has to be intended as a shape controlling parameter, β is a skewness indicator, γ a location parameter, and δ controls the scaling of the distribution.

It has been demonstrated [3] that under proper regularity conditions, it is possible to model price returns dynamics with Levy processes: in such case, Equation 1 becomes:

$$S_t = S_0 e^{\mu t + \sigma L_t} \tag{2}$$

where L is a Levy process, and it is possible to solve the related optimization problem, hence defining the quote of wealth to be invested into the risky asset.

However, a number of observations may be drawn.

- Although Levy processes are a wide class (the Gaussian process itself is a Levy process, for β and γ equal to zero), regularity conditions needed to give to portfolio allocation a closed form does not hold for the whole family of processes [8], so that a considerable number of them which are potential good candidates to fit real data must to be excluded.

  This leads to consider a sub–class of such processes, namely the Generalized Hyperbolic Processes which are tractable under the profile of requirements set in [3], with particular attention to the family of Normal Inverse Gaussian laws or NIG.

- Portfolio management needs a proper fit over price returns dynamics, in order to forecast the quote of wealth to be invested into the risky asset. To this purpose, model parameters must be stable, that is they should assume values into a basin of attraction. However, no empirical evidence exists in such sense, and distribution parameters have to be frequently re-calibrated.

Starting from this point, the purpose of this work is to prove how a neural-network approach may be useful to test the stability of model parameters, as well as to get information about the time occurring between each portfolio re-calibration. To such purpose, the paper is organized as follows.

Section II briefly introduces to the class of Generalized Hyperbolic processes, with particular attention to Normal Inverse Gaussian Processes (NIG since now on). In this context, data from New York Stock Exchange (NYSE) will be used to prove how NIG are well-fitting to real data, as well as to outline the problem of parameters estimation and related portfolio re–calibration.

Section III will introduce to the technique based on Kohonen's Self–Organizing Maps which allows for parameters stability control.

Section IV will end the paper, giving some conclusive remarks and outlines for future works.

2 Generalized Hyperbolic and Normal Inverse Gaussian Processes.

In a probability space $\{\Omega, P, F\}$, with filtration $F = \{F_t\}_{t \in [0, T*]}$ a class of stochastic processes is defined with density given by:

$$p(x; \lambda, \alpha, \beta, \gamma, \delta) = a(\lambda, \alpha, \beta, \delta)(\delta^2 + (x - \gamma^2))^{(\lambda - \frac{1}{2})/2} K_{\lambda - \frac{1}{2}}(\alpha \sqrt{\delta^2 + (x - \gamma^2)}) exp(\beta(x - \gamma)) \tag{3}$$

where $a(\lambda, \alpha, \beta, \delta) = \dfrac{(\alpha^2 - \beta^2)^{\lambda/2}}{\sqrt{2\pi}\alpha^{\lambda - 1/2}\delta^\lambda K_\lambda \delta \sqrt{\alpha^2 - \beta^2}}$ is a normalizing constant, and K_λ indicates the modified Bessel function of the third kind with index λ. An integral representation of K_λ is given by:

Table 1:

K–S	A–D	L1	L2	K–S	A–D	L1	L2
0.0456165	0.131731	1.24862	0.17671	0.112388	0.271647	3.22858	0.466439
0.0639884	0.155276	1.91135	0.26940	0.0700114	0.207496	2.24809	0.317768
0.0690855	0.179434	1.71116	0.25584	0.239862	0.520078	6.41256	1.00794
0.0644726	0.175805	2.25279	0.29083	0.227607	0.538198	6.19596	0.957403
0.0542747	0.174015	2.0469	0.26459	0.307248	0.767039	9.19303	1.3656
0.0750569	0.187395	1.56728	0.25817	0.0676808	0.174233	2.46737	0.323794
0.054155	0.120899	0.891044	0.14646	0.166219	0.335215	4.81112	0.699414
0.110824	0.276452	2.73731	0.39337	0.424273	1.07293	12.28	1.88164
0.1434	0.287271	2.7222	0.46124	0.267207	0.547532	6.19453	1.00736
0.215393	0.450681	3.51081	0.68210	0.552582	1.33667	12.4602	2.1549
0.18231	0.373832	2.75908	0.53529	0.529081	1.14925	12.4466	2.04142
0.165542	0.332999	2.43014	0.44992	0.473839	0.999884	12.1395	1.92395
0.14498	0.298546	2.51144	0.47610	0.318473	0.741078	7.61976	1.28233
0.0599175	0.232676	1.41811	0.20291	0.435769	1.03289	13.5236	1.98772

$$K_\lambda(z) = \int_0^\infty y^{\lambda-1} exp(-\frac{1}{2}z(y + y^{-1}))dy \tag{4}$$

This density depends from the four parameters α, β, γ, and δ, since λ which controls the heaviness of the tails is considered to assume very special values.

In particular, when $\lambda = -\frac{1}{2}$, Eq. 3 becomes:

$$d(x; \alpha, \beta, \gamma, \delta) = \frac{\alpha}{\pi} exp(\delta\sqrt{\alpha^2 - \beta^2} + \beta(x - \gamma))\frac{K_1(\alpha\delta\sqrt{1 + \frac{(x-\gamma)^2}{\delta}})}{\sqrt{1 + \frac{(x-\gamma)^2}{\delta}}} \tag{5}$$

and the corresponding moment generating function is given by:

$$M(u) = exp(\gamma u)(\frac{\alpha^2 - \beta^2}{\alpha^2 - (\beta + u)^2})^{-1}\frac{K_{-\frac{1}{2}}(\delta\sqrt{\alpha^2 - (\beta + u)^2})}{K_{-\frac{1}{2}}(\delta\sqrt{\alpha^2 - \beta^2})} \tag{6}$$

In this way it is possible to completely characterize a family of distributions which is particularly well fitting to real data.

A sketch of this evidence may be given by considering New York Stock Exchange (NYSE) returns since January 2nd 1971 to April 10th 2001. To aims which will be explained later, available data have been divided into blocks of 500 observations shifted of 100 days one to each other. Table 1 shows a comparison between data fitting through NIG (earlier 4 columns) and normal distribution with proper parameters over a number of such blocks. The comparison is run by means of 4 different metrics: Kolmogorov-Smirnov (K-S), Anderson-Darling (A-D), L1, and L2 distances. In all cases NIG provide better data fitting than normal distribution.

However, it can be observed by looking at Figure 1 (where in clockwise sense from top to bottom $\alpha, \beta, \gamma, \delta$ behaviour is shown), that distribution parameters vary sensitively along time, thus requiring a continuous re-calibration of the model.

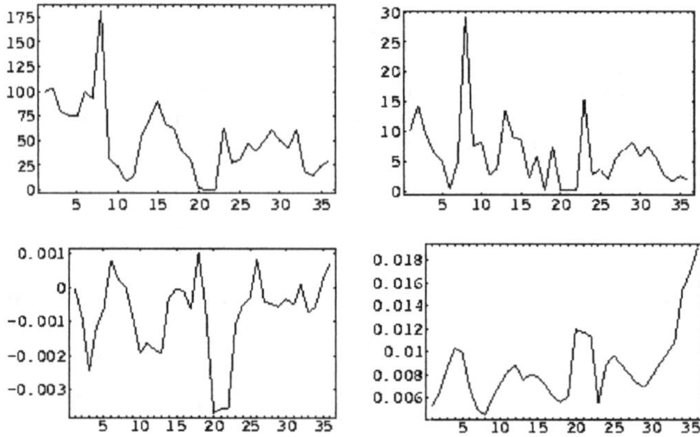

Figure 1: Parameters behaviour for estimated NIG

The technique which will be summarized in the next section tries to overcome such problem.

3   A neural network technique for parameters estimation

The technique we are going to describe is based on Kohonen's Self Organizing Maps. The basic idea is to train a net with vectors consisting of the following components:

- a part of Previous History of the Series (PHS since now on), whose length depends from the embedding dimensions of the data under examination;

- the values for Estimated Distribution Parameters (EDP) fitting well on PHS: since the blocks of data may be overlapping, more that a set of EDP might be necessary

Once the net has been trained, it is possible to verify that clusters of parameters tend to be created (Figures 2–3).

At this point, new vectors made up by the sole part of PHS are presented to the map, that is the impact of new information is evaluated with respect to distribution parameters.

In this way, it is possible to observe where the map locates the new data history, (Figure 4), hence getting out both more probable values for distribution parameters, and their evolution over time.

4   Conclusions

A technique based on artificial neural networks has been suggested in order to estimated well fitting distribution parameters to observed data. This is particularly important in portfolio selection problems, where the model needs to be frequently calibrated, in order to asses to

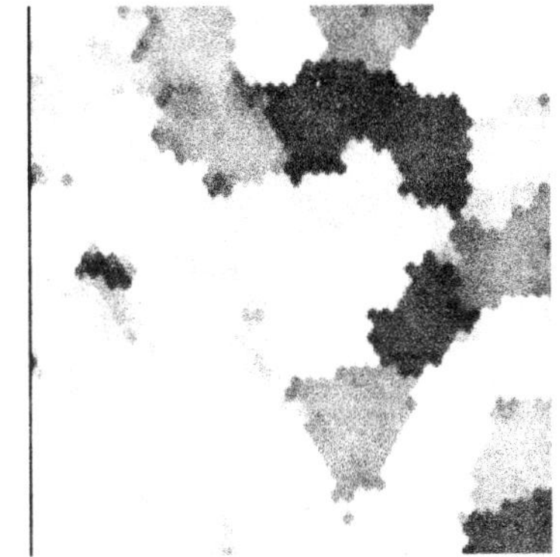

Figure 2: Global Map

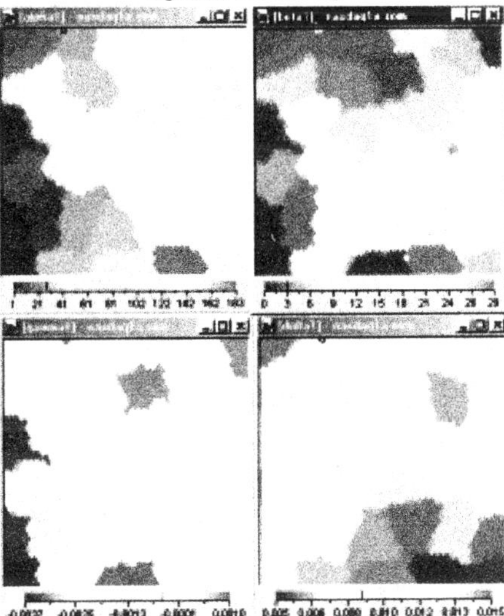

Figure 3: Parameters maps

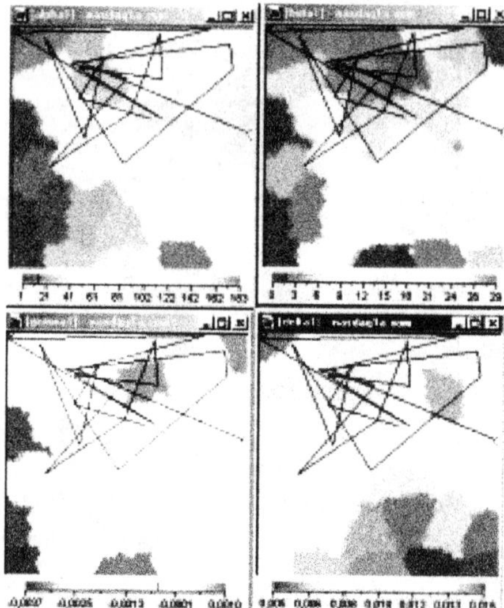

Figure 4: Merging of new vectors

right proportion to risky assets. The technique is aimed to provide information about the stability of parameters, as well as about time occurring between to consequent re-calibrations.

References

[1] O. E. Barndorff-Nielsen, Process of Normal Inverse Gaussian type, Finance and Stochastics (1998) **2**, 41–68.

[2] O. E. Barndorff-Nielsen, Superposition of Ornstein-Uhlenbeck type processes, Theory of Probability and Its Applications (2000).

[3] O. E. Barndorff-Nielsen, N. Shephard, Modelling by Levy Processes for financial econometrics, draft paper, (2000).

[4] F.E. Benth, K. Hvistendhal Karlsen, K. Reikvam, Optimal Portfolio with Transaction Costs and nonlinear integro-differential euqations with gradient constraints: a viscosity solution approach, MaPhySto Research Report nr. 21, University of Aarus, Denmark (1999).

[5] E. Eberlein, U. Keller, Hyperbolic distributions in finance, Bernoulli **1** (1995), 281–289.

[6] I.A. Ibragimov, Y.V. Linnik, Indipendent and Stationary Sequences of Random Variables, Wolters–Noordhoff Publishing Co., Groningen, the Nederlands (1971).

[7] T. Kohonen, Self-Organizing Maps, (1997) Springer Series in Information Science.

[8] M. Resta, Portfolio Selection Models driven by non-Gaussian Price Returns dynamics, draft paper, (2001) University of Genova, Italy.

Rule-based Prediction with Additional Learning in Financial Problems

Hirotaka Nakayama, Takahiro Kotera and Koji Miyazaki

Department of Information Science and Systems Engineering
Konan University
8-9-1 Okamoto, Higashinada-ku, Kobe 658-8501, Japan
E-mail: nakayama@konan-u.ac.jp

Abstract. Recently, machine learning is widely applied to financial problems. One of main features in financial investment problems is that the situation changes very often over time. In applying machine learning techniques under this circumstance, the authors have proved that additional learning plays an effective role, in particular, in financial problems by suggesting several techniques for additional learning using artificial neural networks. However, many practitioners claim that artificial neural networks are not transparent to explain why the result is obtained. In order to overcome this difficulty, in this paper, we consider the effect of additional learning using rule extraction methods. It will be shown that we can analyze the transition of importance of attributes over time by making additional learning in finacial investment problems.

1 Introduction

In many practical problems, e.g., financial investment problems, the situation changes very often over time. In machine learning, therefore, decision rules are needed to adapt for such changeable situations. To this end, additional learning should be made on the basis of new data. One of the authors and his collaborators have reported the effectiveness of additional learning in several machine learning techniques: mathematical programming approach (Nakayama-Kagaku 1998), Potential method (Nakayama-Yoshida 1997) and RBF networks (Nakayama 1999; Nakayama *et al.* 1998). Furthermore, note that the rule for classification becomes more and more complex with only additional learning. Therefore, some appropriate forgetting is also necessary in order to avoid such an overfitting. The authors have also suggested several trials of forgetting in machine learning (Nakayama 1999; Nakayama *et al.* 1998; Nakayama-Yoshii 2000). In particular, they reported that active forgetting which decreases the degree of importance of unnecessary data performs very effectively in stock portfolio problems.

However, the methods stated above are mainly based on artificial neural networks. Unfortunatley, most artificial neural networks are poor at explaining why the result is obtained. On the other hand, rule extraction methods can explain this well. In the following, we show that we can analyze the transition of importance of attributes by making additional learning using such rule extraction methods, in particular, C5.0.

2 Rule Extraction by C5.0 in Stock Investment Problems

There have been several methods for rule extraction so far. Above all, C5.0 (Quinlan 1993), CART (Breiman 1984) and the rough set theory (Pawlak 1991) are applied to a wide range of practical problems. They try to extract as simple rules as possible from given data bases. To this end, C5.0 uses infomation entropy, while CART does Gini index. On the other hand, the rough set theory tries to remove as many unnecessary attributes as possible while maintaining the allowable consistency among data. Although each method was derived from reasonable idea, C5.0 has a good reputation due to its fast computation time. Therefore, we discuss mainly C5.0 in the following.

As an example, consider a stock investment problem as follows: Our problem is to judge whether a stock is to be purchased or not. The data are monthly stock prices of a certain company during 119 terms from January 1985 to November 1994 with seven economic attributes:

1) PER ratio,

2) elasticity ratio,

3) rank correlation (six months),

4) ratio of monthly close price to 3 monthly averaged close price,

5) ratio of monthly volume to 6 months averaged volume,

6) ratio of money supply (M2CD) to GDP,

7) ratio of money supply (M2CD)/ to the one on the same month in the previous year.

Decision value "+1" means to buy the stock, and "-1" means not to buy it. If the highest price in the following 6 terms exceeds a given threshold, then the decision at the term is judged to buy the stock (i.e., "+1").

Note that a bubble economy was in the first half period, and it was broken in the last half as can be seen in the graph of Nikkei-225 in Fig.1. Actually, the number of "+1" is more than that of "-1" in the first half, and the opposit in the last half as can be seen in Fig.2. In cases in which the number of data in two decision categories has some bias, like in this example, we usually impose a cost to each category in applying C5.0. Here, we adopt the cost given by the inverse number of data.

Figure 1: Index of market (Nikkei 225)

test data		51	52	53	54	55	56	57	58	59	60	61	62	63	64	65	66	67	68	69	70	71	72	73
decision		−1	−1	−1	1	−1	1	−1	−1	−1	−1	−1	−1	1	1	−1	−1	−1	−1	1	−1	1	−1	−1
data in	+1	33	33	33	34	34	35	35	35	35	35	35	35	36	37	37	37	37	37	38	38	39	39	39
the past	−1	17	18	19	19	20	20	21	22	23	24	25	26	26	26	27	28	29	30	30	31	31	32	33

test data		74	75	76	77	78	79	80	81	82	83	84	85	86	87	88	89	90	91	92	93	94	95	96
decision		−1	−1	−1	−1	−1	−1	−1	−1	−1	−1	−1	−1	−1	−1	−1	−1	1	1	−1	−1	−1	−1	−1
data in	+1	39	39	39	39	39	39	39	39	39	39	39	39	39	39	39	39	40	41	41	41	41	41	41
the past	−1	34	35	36	37	38	39	40	41	42	43	44	45	46	47	48	49	49	49	50	51	52	53	54

test data		97	98	99	100	101	102	103	104	105	106	107	108	109	110	111	112	113	114	115	116	117	118	119
decision		1	1	1	−1	−1	−1	−1	−1	−1	−1	1	1	−1	−1	1	−1	−1	−1	−1	−1	−1	−1	−1
data in	+1	42	43	44	44	44	44	44	44	44	44	45	46	46	46	47	47	47	47	47	47	47	47	47
the past	−1	54	54	54	55	56	57	58	59	60	61	61	61	62	63	63	64	65	66	67	68	69	70	71

+1: to buy
−1: otherwise

Figure 2: Ingredients of our data (after 51-th term)

In our experiments, data during the first 50 terms are used for the initial learning, while the rest 69 data are used for the prediction test. The result is given in Fig. 3, where black bars show the misclassification. Clearly, additional learning provides a good effect to the correct classification.

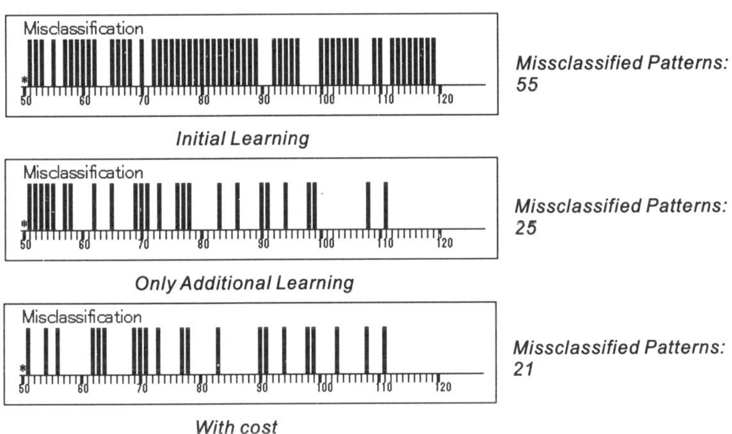

Figure 3: Result by C5.0 (continuous data)

We can analyze the importance of condition attributes in the result by C5.0. Fig. 4 shows the impotant attributes at each term. From this, we can get an information which attributes are important at each term.

At the beginning of test, the number of important attributes is relatively small. This seems because the total number of data is small and rules can be extracted on the basis of relatively small number of condition attributes. However, as data are added, the number of important attributes becomes relatively larger.

test data	51	52	53	54	55	56	57	58	59	60	61	62	63	64	65	66	67	68	69	70	71	72	73
decision	-1	-1	-1	1	-1	1	-1	-1	-1	-1	-1	-1	1	1	-1	-1	-1	-1	1	-1	1	-1	-1
judgment	×	O	O	×	O	×	O	O	O	O	O	×	×	×	O	O	O	O	×	×	×	O	×
conditional attributes	c4	c2	c2 c4 c1 c1 c7 c3 c4 c5	c4 c2 c1 c1 c7 c3 c4 c4	c2	c2	c2	c2	c2	c2 c4 c1 c1 c1 c7 c3 c4 c4	c2 c1 c1 c4 c1 c7 c3 c4 c4	c2 c1 c1 c4 c1 c7 c3 c4 c5	c2 c1 c4 c1 c7 c3 c4 c5	c2 c4 c1 c1	c2 c4	c2 c4	c2 c4	c2 c7 c1 c1 c3	c2 c7 c1 c1	c2 c7	c2 c1 c3 c6 c4 c7 c5 c1	c2	c2 c7 c1 c1

test data	74	75	76	77	78	79	80	81	82	83	84	85	86	87	88	89	90	91	92	93	94	95	96
decision	-1	-1	-1	-1	-1	-1	-1	-1	-1	-1	-1	-1	-1	-1	-1	-1	1	1	-1	-1	-1	-1	-1
judgment	O	O	O	×	×	O	O	O	O	×	O	O	O	O	O	O	×	×	O	O	×	O	O
conditional attributes	c2 c1 c4 c6 c3 c7 c1	c2 c1 c4 c6 c3 c7 c1 c7 c1	c2 c1 c4 c3 c6 c1 c7	c2 c1 c4 c4 c3 c6 c3 c1 c1 c7	c6 c4 c4 c3	c6 c4 c5 c3 c4	c6 c4 c5 c3 c4	c6 c4 c5 c3 c4	c6 c4 c5 c3 c4	c6 c4 c5 c3	c6 c4 c5 c3	c6 c4 c5 c3	c7 c2	c7 c2	c7 c2 c4 c1 c6 c3 c1 c4 c7 c7 c3	c7 c2 c4 c1 c3 c6 c4 c1 c7	c7 c2 c4 c1 c3 c6 c4 c1 c7	c6 c4 c4 c3 c5	c6 c4 c4 c3 c1 c5	c6 c4 c4 c3 c1 c5	c6 c4 c4 c3 c1 c5	c6 c4 c4 c3 c1 c5	c6 c4 c3 c5 c1 c5

test data	97	98	99	100	101	102	103	104	105	106	107	108	109	110	111	112	113	114	115	116	117	118	119
decision	1	1	1	-1	-1	-1	-1	-1	-1	-1	1	1	-1	-1	1	-1	-1	-1	-1	-1	-1	-1	-1
judgment	O	×	×	O	O	O	×	O	O	O	O	×	O	O	×	O	O	O	O	O	O	O	O
conditional attributes	c6 c4 c3 c5 c1	c6 c4 c3 c5 c1	c6 c4 c3 c5 c7	c6 c4 c3 c1 c5	c6 c4 c3 c1 c5	c6 c4 c3 c5 c7	c6 c4 c3 c5 c2 c7 c6	c6 c4 c3 c5 c7	c6 c4 c3 c1 c5	c6 c4 c3 c5 c7	c6 c4 c3 c5 c7	c6 c4 c3 c1 c5	c6 c4 c3 c7 c7	c6 c4 c3 c7 c7	c6 c4 c3 c2 c7	c6 c4 c7 c7 c1 c3	c6 c4 c7 c7 c1 c3	c6 c4 c7 c7 c1 c3	c6 c4 c7 c7 c1 c3	c6 c4 c7 c7 c1 c3	c6 c4 c7 c7 c1 c3	c6 c4 c7 c7 c1 c3	c6 c4 c7 c7 c1 c3

Figure 4: change of important condition attributes

3 Additional Learning by Potential Method

It seems clear that techniques in artificial neural networks can provide a better performance in classification itself than rule extraction methods such as C5.0. Roughly speaking, this is because artificial neural networks approximate nonlinear decision boundaries by smooth curves, but rule extraction methods approximate them by stepwise functions. For reference, we cite the result of additional learning by the potential method which was developed by the authors (Nakayama and Yoshii 2000) in Fig.5.

However, many practitioners claim that artificial neural networks can not explain why/how the result is obtained, even though they usually show a good performance in classification ability. Therefore, it is desirable to use both rule extraction methods and artificial neural networks. One way to combine them is to use rule extraction methods first and then artificial neural networks: we can reduce the number of attributes by using rule extraction methods, and also make a rough prediction in classification. Netly, on the basis of the reduced number of attributes, we apply artificial neural networks. As a result, a high precision of prediction is possible by using artificial neural networks in particular with additional learning and forgetting. Moreover, we can analyze important attributes by rule extraction methods, in other words we can see which attributes are important at the term.

Another way to combine rule extraction methods and artificial neural networks is to use artificial neural networks first and then rule extraction methods. Recall that we should adjust the weight on each data when the number of data in each decision category is biased. It can be expected that adjusting the weight on each data in more detail can increase the prediction precision. Therefore, using the weight to each data generated by artificial neural networks, e.g., the potential method, we can expect to improve the classification ability of rule extraction methods. Experiments along this line is now on-going.

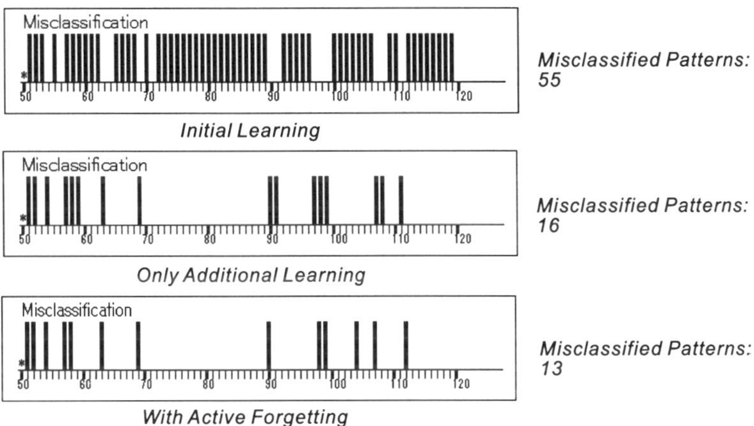

Figure 5: Result by Potential method

4 Concluding Remarks

In this paper, we showed how effectively rule extraction methods can be used jointly with artificial neural networks. In particular, it is important to know which attributes are important at each term in financial problems to analyze the reason (i.e., the economic background). Since artificial neural networks are not good at explaining such reasons, rule extraction methods can be effectively applied to this end. We will report some experimental results by applying rule extraction methods after weighting each data by some artificial neural networks, e.g., the potential method at the conference.

References

[1] Breiman, L., J.H. Friedman, R.A. Olshen and C.J. Stone (1984): *Classification and Regression Trees*, Wadsworth, Inc.,

[2] Quinlan, R.J.,(1993):*C4.5: Programs for Machine Learning*

[3] Nakayama, H. (1999): Growing Learning Mahines and their Applications to Portfolio Problems, *Proc. of the International ICSCC Congress on Computational Intelligence Methods and Aplications (CIMA'99)*

[4] Nakayama, H. and Kagaku, N. (1998): Pattern Classification by Linear Goal Programming and its Extensions, *J. of Global Optimization*, 12, pp.111-126

[5] Nakayama, H. and Yoshida, M. (1997): Additional Learning and Forgetting by Potential Method for Pattern Classification, *Proc. ICNN'97*, Houston/USA, pp. 1839-1844

[6] Nakayama, H., Yanagiuchi, S., Furukawa, K., Araki, Y., Suzuki, S. and Nakata, M. (1998): Additional Learning and Forgetting by RBF Networks and its Application to Design of Support Structures in Tunnel Construction, *Proc. International ICSC/IFAC Symposium on Neural Computation (NC'98)*, pp.544-550

[7] Nakayama, H. and K. Yoshii (2000): Active Forgetting in Machine Learning and its Application to Financial Problems,*Proc. International Joint Conference on Neural Networks*, (in CD ROM)

[8] Pawlak, Z. (1991): *Rough Sets Theoretical Aspects of Reasoning about Data*, Kluwer Academic Publishers (1991)

KES '01
N. Baba et al. (Eds.)
IOS Press, 2001

Knowledge-Based Decision Support Systems for Dealing Nikkei-225 by Soft Computing Techniques

Norio Baba*, Yan Yanjun**, Inoue Naoyuki*,

Xu Lina***, Deng Zhenglong***

** Information Science, Osaka Kyoiku University, Japan*
*** Information Science, Osaka Kyoiku University (On 9 months leave from Control Engineering, Harbin Institute of Technology)*
**** Control Engineering, Harbin Institute of Technology, P.R.China*

Abstract

This article introduces a novel way in dealing Nikkei-225 (a stock price index in Japanese stock market based on the Dow formula). Due to the uncertain nature of the stock market, the traditional approaches such as regression analysis seem to be not so effective any more, so we prefer to use the soft computing techniques including the combination of neural networks (NNs) and genetic algorithms (GAs). After considering several effective impacts on the changeful stock market, we select some as the inputs of NN. And then we tried both the fixed and the shift methods to train the NN in order to learn the patterns of historical data. Simulation shows that: Shift method is more adaptable, whose prediction of Nikkei-225 four weeks in the future presents a high coincidence ratio with the real Nikkei-225 four weeks in the future, and these two values are highly correlated. We construct a Decision Support System (DSS) based on this prediction and further use GAs to optimize the parameters and the rules of dealing strategy. The final DSS can get a considerably big return under the practical condition that the tax and processing fee has been subtracted every time in the dealing.

1 Introduction

Now around the world, people share great interest in the stock market prediction, one aim is to understand more of the stock market from the perspective of economics, the other aim is to get big profit from such a changeful and venturesome market. To achieve such final goals, we should try to get a fairly accurate prediction of certain stock price index and then construct a reliable Decision Support System based on this prediction. Here we select Nikkei-225 as our object of prediction. Different from popular technical analysis, we base our system mainly from the perspective of signal processing techniques. Concerning the natural characters of stock market such as randomness and discontinuity, the traditional approaches seem not so effective any more, while soft computing, distinctive for its power of dealing imprecision, uncertainty and partial truth, seems quite suitable in this problem. So in this article, we try to use soft computing technology such as the combination of NNs and GAs to construct an intelligent DSS for dealing Nikkei-225.

The outline of this article is as following: Based on the current knowledge of stock market, we select several effective impacts on Nikkei-225 as the inputs of NN and then we compare the fixed and shift methods to train it. Simulation shows that the shift method can utilize the latest data more efficiently in the sense that it can master the main patterns of historical data. According to several evaluation standards including coincidence ratio and correlation, the prediction of Nikkei-225 four weeks in the future is very close to the real future Nikkei-225 in four weeks. Thereafter, we construct a DSS based on this prediction, and use GAs to optimize the parameters and dealing rules.

2 Application of NNs to predict Nikkei-225

2.1 Neural Network Model and Input Variables

Figure 1 shows the structure of neural network model. After several trials, we select such a model as nine inputs, one hidden layer with twelve neurons, and two outputs.

The nine inputs include several significant impacts on Nikkei-225, such as the PBR (Price Book-Value Ratio), dealing volume of foreign traders and the ratio of Japanese Yen to US dollar. The inputs can be classified into three groups: The first one is the historical Nikkei-225, the second one is the changing rate of those raw data in different periods, and the third group is the quantified changing trend of some raw data during last four weeks, from which, we can extract some historical information of the market tendency.

Definition 1: changing rate of x during period T

$$changing\ rate\ of\ x = \frac{x(the\ final\ po\mathrm{int}\ of\ T) - x(the\ starting\ point\ of\ T)}{x(the\ starting\ point\ of\ T)}$$

Definition 2: changing trend of x during four weeks

$$t(i) = \sum_{n=0}^{3} 2^{3-n} \cdot hardlimit(x(i-n) - x(i-n-1))$$

In which, the hardlimit function is defined as

$$hardlimit(x) = \begin{cases} 1 & x > 0 \\ 0 & x \le 0 \end{cases}$$

Interested readers are kindly asked to attend our presentation, where we can explain this formula clearly. Input variables are further normalized before they enter the NN, in order to avoid the discrepancy among local training speed, then the training of NN is more efficient.

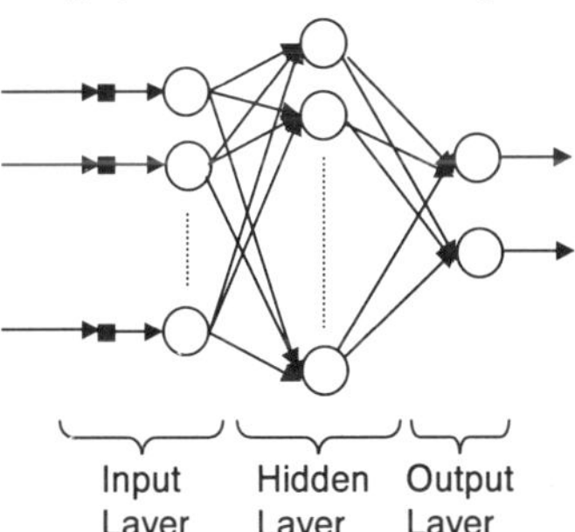

Fig. 1 neural network model with one output

The two outputs are the prediction of the highest and the lowest probable changing rate of Nikkei-225   four weeks in the future. With the prediction of changing rate and current Nikkei-225, we can predict the possible zone of future Nikkei-225 in four weeks. By providing the zone of changing, we can make our prediction more reliable. It's quite interesting to notice that with this two outputs model, the coincidence ratio between prediction and reality can increase a lot from the one output model, which predicts the exact changing rate of future Nikkei-225[1].

2.2 Teacher signals and the Training Algorithm

In a word, we use the modified BP algorithm with momentum. The momentum constant and the learning rate are adapted to the instant error through learning process. And when the gradient of training is too small, the training will automatically stop. In above neural network model, tangential sigmoid function is selected as the activation function of neurons, whose function formula is deduced as following with $\sigma = 2$:

$$f(x) = \frac{1 - e^{-\sigma x}}{1 + e^{-\sigma x}}$$

Because the stock market is changed by so many elements that we can not list them all, so we'd better treat the small force just as noise, and not to stick so exactly to the historical modes. Here we use a validation data set to early stop the training when the generalization ability of the neural networks becomes worse. 20% data are selected randomly from the training period as the validation data set, and the left data is used as training data. We compare the performance of the training data set and this validation data set; once the performance of validation data set is worse than the performance of training data set, early stop the training. This validation process can avoid the over-fitting phenomenon to some extent.

2.3 Fixed and Shift Training Method

Thanks to the provision of data from QUICK Corporation, we can have a general idea of the development of Nikkei-225 from 1985 to 2001. During this time, Nikkei-225 first kept increasing until 1990, then dropped continually until 1992. Only after 1992 did Nikkei-225 appear to change in zigzag. In order to make our system suitable for future utilization, we concentrate our research after 1992, when the changing of Nikkei is somewhat similar to current situation.

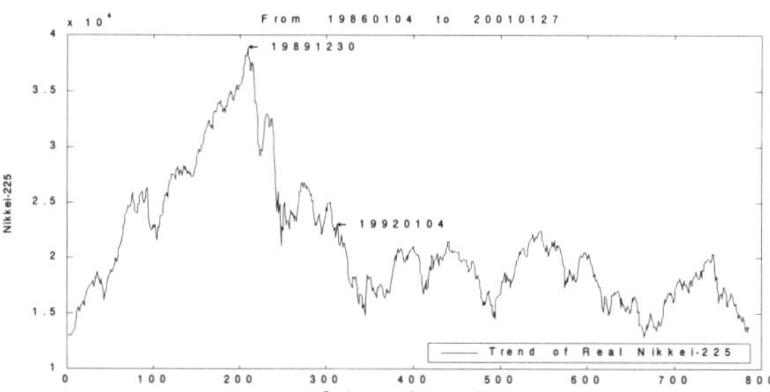

Fig 2 Changing tendency of Nikkei-225

The so-called fixed and shift method is determined by the starting time of training. We can use all the available data from 1992 to train the neural network; this method is called fixed training. While in fact, we can also shift the starting time of training period according to the requirement of prediction on the starting point: For example, we can use the latest two years data for training and predict in the following one year, this method is called shift training. The essence of this shift training is to make the most of the new data.

3. Basic Decision Support System for Dealing Nikkei-225

Based on the prediction, we construct a basic DSS as follows:

1. let α be the positive parameter.

2. Calculate the prediction of the highest value of the Nikkei-225 four weeks in the future and the prediction of the lowest price of Nikkei-225 four weeks in the future. Let PN be the average of the above two predictions, which is the prediction of Nikkei-225 four weeks in the future.

3. If (Current Nikkei-225)$\times(1+\alpha)$ is below PN, then buy. When the Nikkei-225$\times(1-\alpha)$ goes above PN, then sell.

4. If (Current Nikkei-225) $\times(1-\alpha)$ is above PN, then sell. When Nikkei-225$\times(1+\alpha)$ drops below PN, then buy.

5. Otherwise, wait.

We have carried out several computer simulations. Almost all of them have shown that the proposed basic DSS is effective in yielding considerable return with the specific parameter value of α.

4 Utilization of GAs to make the Decision Support System More Effective

Although the basic DSS can profit some, it's not so natural. For one thing, the pattern of Nikkei-225's 'up and down' is not necessarily symmetrical, that means, we can use two different parameters α and β each for selling and buying. Further, the proposed DSS does not specify how to deal in the Nikkei-225 under the general condition that one may use only limited funds. We should consider the possibility that he could have received greater return by dividing dealing according to the magnitude of possible changes from current Nikkei-225. A DSS concerning above ideas is as following:

1. let α and β be the positive parameters.

2. Calculate the prediction of the highest value of the Nikkei-225 four weeks in the future and the prediction of the lowest price of Nikkei-225 four weeks in the future. Let PN be the average of the above two predictions.

3. If (Current Nikkei-225)$\times(1+\alpha)$ is below PN and in some specific range, then buy with appropriate share of money. When the Nikkei-225$\times(1-\beta)$ goes above PN and in some specific range, then sell appropriate share of stocks.

4. If Current Nikkei-225$\times(1-\beta)$ is above PN and in some specific range, then sell appropriate share of stocks. When Nikkei-225$\times(1+\alpha)$ drops below PN and in some specific range, then buy with appropriate share of money.

5. Otherwise, wait.

Concerning that there are so many possible combinations of parameters and dealing rules in some specific range, we try to utilize GAs for the following objectives:

1) To utilize GAs in order to find appropriate parameter values α and β.

2) To utilize GAs in order to find successful way of dealing according to the different ranges of the prediction of future Nikkei-225.

Figure 3 shows the structure of the chromosome used in GA searching. We define six ranges according to the real current Nikkei-225, then deal with different share of stocks in each specific range. In this chromosome, the parameters will occupy 3 bits for each, and the dealing shares will occupy 5 bits for each.

Computer Simulation Results

We have tried to utilize the chromosome (having been found by GAs) for dealing in the Nikkei-225, which succeeded in making a considerably large return in simulation under the practical condition that the tax and processing fee has been subtracted every time in the dealing. Due to limitation of space, we can't go into details, welcome to our presentation or refer to reference [2]-[4].

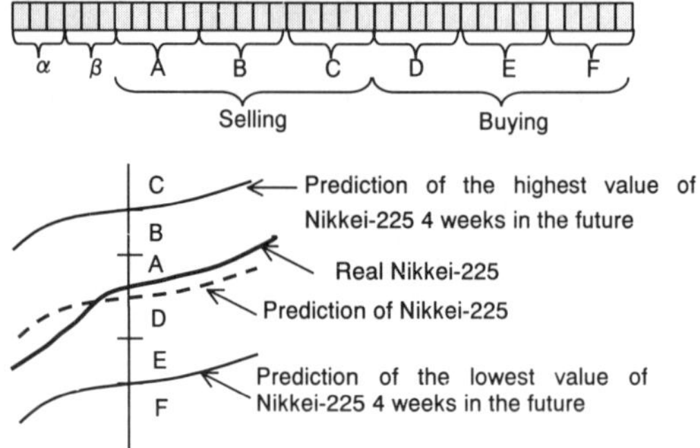

Fig. 3 Structure of Chromosome

5 Concluding Remarks

In our 'out of market' simulation, the final DSS constructed by soft computing technology can get a considerably big return under the practical condition that the tax and processing fee has been subtracted every time in the dealing.

Acknowledgement

The historical data used in this paper is provided by QUICK corporation, Tokyo, Japan.

References

[1] N.Baba, Yan Yanjun, Proceedings of KSCIE'2001, Osaka, Japan, 2001

[2] N.Baba, N.Inoue, and H.Asakawa, Proceedings of The 31st ISCIE International Symposium on Stochastic Systems Theory and its Application, Keio Univ.,1999.

[3] N.Baba, "Intelligent Decision Support Systems for Dealing Stocks", Journal of Systems, Control and Information (in Japanese), 2000: 379~386.

[4] Inoue Naoyuki, Master Graduation Thesis, Osaka Kyoiku University, Feb. 2001

[5] N.Baba et al, Proceedings of IJCNN'2000, Como, Italy, 2000.

[6] N.Baba, H.Suto, European Journal of Operations Reseach, 2000.

[7] N.Baba, Proceedings of the ICSC Congress on CIMA'99, RIT, U.S.A.1999.

[8] N.Baba et al, Proceedings of IJCNN'99, Washington, U.S.A.1999.

[9] N.Baba, MTEC Journal, 1998.

[10] N.Baba, Proceedings of KES'98, 1998.

[11] Xu Lina, Neural Networks Control, Harbin Insititute of Technology, 1998.

Use of conversational agents and dynamic Bayesian nets for message recognition

Julieta NOGUEZ[1], Enrique D. ESPINOSA, Fernando RAMOS
Instituto Tecnológico de Estudios Superiores de Monterrey
Campus Ciudad de México, Campus Cuernavaca
{ jnoguez, eespinos}@campus.ccm.itesm.mx, framos@campus.mor.itesm.mx

Abstract. This work establishes a hypothesis: the use of conversational agents with human shape as tools to improve the interaction in distant collaborative learning will help people who live in rural areas, and lack a computer science culture, to incorporate the use of technology into their business activities. This will them the benefits provided by the e-business global process. Hereby, it is important to explore new motivating and familiar but technologically-devised ways of communication that allow people to come close to the other people, in ways that closely resemble our own behavior: socially, personally and emotionally. As a first stage of the project, we present a simulator of the interpretation of combinations of words, gesture, looks, movement of arm and head using Bayesian Nets. In order to differentiate them from other expressions and to emphasize the importance of key words inside a message.

1. Introduction

Computer science technology is able to perform an important function by helping small enterprises to improve their services and profitability due to the fact that, in a global world, it is of great importance to maintain a level of international competence. The new economy is directed towards digital markets because of many reasons, such as: reduction of margins, efficiencies, and time to market. This reduces timing in the development of products, improves relationship with clients, and increases productivity.

The most successful electronic businesses are connected to organizations with whom them work constantly: clients, providers, designers, associates, financial institutions, etc. [14]. Nevertheless, in Mexico there are a large and specific area of opportunity because small enterprises are far from having an access to the information technologies, just as the Mexican Chamber of Commerce (CONCAMIN) recognizes in a recent declaration [9]. The CONCAMIN agrees with the Mexican Public Education Secretary (SEP) that this challenge is bigger in the rural and margined countryside where communication means are scarce and where there is a widespread lack of an information culture [10].

In the rural areas of countries such as Mexico, an educative, technological, and economic recession can be found. This obstructs the possible interaction between their inhabitants and the modern computer technology [8]. Meanwhile, technology is out from reach of the small producers or the ones who have mini enterprises in far regions.

---

[1] Julieta Noguez is a PhD student at the Monterrey Institute of Technology, and holds a scholarship provided by the Asociación Nacional de Universidades e Instituciones de Educación Superior de México (SUPERA).

Therefore, they will be excluded from global economic processes, without an opportunity of improving their actual situation.

Due to the problems last described, there are important obstacles identified for the use of new technologies in the rural zones, including an additional resistance by ignorance or even by ideological, cultural, or social issues [12]. Information systems, included the ones developed in Web, assume that users have certain experience and previous knowledge in the use of computers. Nevertheless, if learning should be delivered to inhabitants of rural zones it is important to explore new ways of communication that truly permit the interaction between man & machine and that, combined with collaborative learning techniques, permit diffusion and enrichment of knowledge and usage of such systems.

The hypothesis we present states that the use of conversational agents with human shape will improve the interaction in environments of distant collaborative learning. People who live in rural areas, and lack an information technology culture, could achieve better interpersonal communication, incorporating high-tech in their business processes, obtaining access to the benefits of e business, and acquiring abilities of collaborative work and project development to increase the productivity of their enterprises.

To achieve a deeper integration among locals, and between communities, as a result of their increased education, it is important to explore new ways of communication that would motivate their interaction. Because of this we look for new ways to design computer interfaces that consider the conversational behavior as a dialog function and of the demands of social, personal and emotive conventions. The first task we have defined is the study of adaptability of the human being to communicate through computers.

As a first stage of the project, a model simulator that permits the interpretation of gesture combinations: look, arm movement, and head movement was developed achieves a simulation of the exchange of communication between conversational agents. This work uses modules of classification of gestures and messages (words), using Dynamic Bayesian Networks to classify gestures into categories, differentiate expressions, and achieve a recognition of the importance of key words in a message sent through time.

2. Theoretical Framework

To achieve a better communication with a human, his words are considered, as well as his reactions and gestures to different events. These expressions are inherent to her/his own characteristics as a human being. It has been reported that, when there is communication between computing systems and humans, the integral communication is missed, because the emitted and received messages only consider information made up by the words that make up the message [11]. There is no additional information that permits an adequate interpretation of such messages. Actually, new methods to design computing interfaces which consider the conversational behavior as a function of dialog and the demands of social, personal, and emotive conventions are searched for, with the objective to significantly improve the communication between human and computers [2].

Dr. Casell's team has developed an understanding module using the Markov's hidden models [3], which have the disadvantage of representing with a single variable the related information with diverse factors of communication related to each word of the message. The last can generate a wide spectrum of uncertainty, which has to be improved. This work focuses on the communication structure and proposes to consider different aspects of non-verbal signals involved in the effective communication between conversational agents. Thus, using dynamic Bayesian networks permitting the use of time-functional dependencies for gathering significance, or importance, of each word used in a message, with the goal to find its importance for the understanding of such message, is appropriate.

3. Communication by the expression technique

Different studies in communication processes [1] have pointed different signals of great importance in communication. Hereby, those that make a bigger impact in the model to be simulated have been selected:

User interfaces have been constructed that propose a categorization of facial expressions which depend of a communication meaning, and voice systems which recognize voices and generate a response with a display of appropriate facial have been developed [4,5,6]. In other words, models of interaction of the type face to face have been used to generate all the developed behaviors from an information state about the intonation of the communication function of head, look, and gesture in hands.

In the construction of autonomous agents capable of establishing communication using expressions, we have focused in the communication act structure, in which the agent receives a message in a particular situation with certain words, intonation, arm movement, facial expressions which denote emotion and changes in the look from its interlocutor.

4. Probabilistic model for the recognition of expressions

From the analysis of several works on conversational agents [3,4,13], the following model has been inferred. It considers the face-to-face interaction to generate all the developed behaviors from an information state about the importance of the communication functions of intonation, look, and movements of the hands.
Even though, the model of figure 1 represents a high conditional probabilistic incidence over the meaning variable so it was necessary to simplify the model adding the tools of dynamic bayesian nets to determine such variables.

4.1 Dynamic Bayesian Networks

On the other side, the Bayesian nets are a general form of representation of probabilistic, combined, distributions, which use suppositions of conditional independence and chain rules.

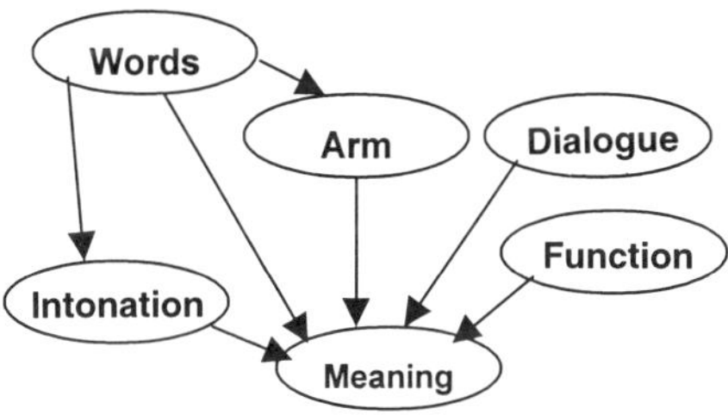

Figure 1. Initial model to determine the degree of significance of a word inside a message

The variables are connected by directed cyclic graphs, whose arcs specify the conditional independence between variables, which its whole probability is given by:

$$P(v_1,...v_n) = \prod_i P(v_i \mid P(v_i))$$

(1)

Where $P(v_i)$ are the parents of v_i in the graph.

The conditional probabilities can be stored in tabular form (matrix) or through a functional representation.

The net also represents the conditional independencies of a variable. This permits the simplification of the representation of knowledge and reasoning. The probabilistic reasoning or probabilistic propagation consists in the propagation of the evident effects through the net to know the a posteriori probabilities of the variables[15].

Figure 2 shows the simplified model of the Bayesian network to determine the significance level of a word in a message.

When the variables represent a temporal sequence, and this has an order in time, the Dynamic Bayesian Network concept is assigned to the networks [7]. These nets maintain values for a collection of variables Xi in each moment of time. Xij represents the value of the i-enth variable in time j. This variables are divided in equivalent collections which share time invariant conditional probabilities.

4.2 Prototype

Although the original design of the system considers the propagation in a complex multi- connected net, the evidence presented of the variable reading allows, in reality, to simplify the model even more. In the first word, the node "meaning" is just affected by the evidence of the word and intonation nodes, and for the following words of the message, the intonation of the last word in conjunction with the word and its own intonation affect the significance of the message.

Diverse runs were made to determine if the significance of every word in the message was obtained accurately. Accordingly to the information from the initial matrixes given by the "expert", very encouraging results were obtained.

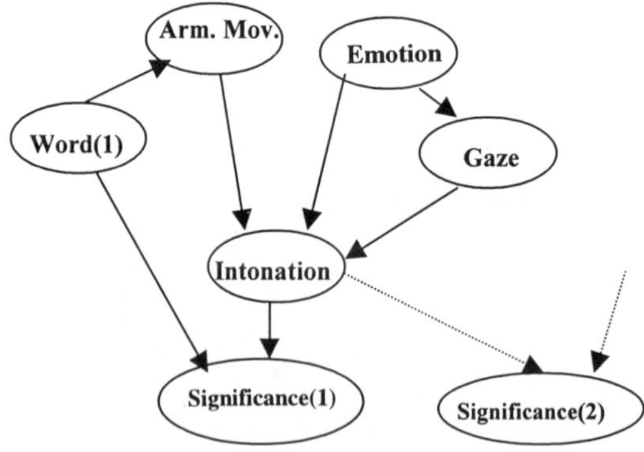

Fig. 2. Dynamic Bayesian Network

5. Conclusions and perspectives

Through this application, it is demonstrated that dynamic Bayesian networks are a good tool to model the behavior of conversational agents which receive messages formed by words, which have an additional information than its content, such as intonation, gaze, emotions, and arm motion.

A future work is the validation of the model of necessary variable dependencies to determine a message meaning. Here, a simplified model is presented based in diverse linguist articles, which model is very complex.

On the other hand, the development for the cases in which there is partial evidence and it is necessary to make the evidence's propagation effect through the model can be continued and its impact verified in the meaning of the message.

References

[1] M. Argyle and M. Cook. Gaze and Mutual Gaze. Cambridge. Cambridge University. Press. 1976.

[2] N. Badler, R. Bindiganalave, J. Allbeck, W. Schuler, L. Zhao and M. Palmer. Parameterized Action Representation for Virtual Human Agents. MIT Press. pp. 256- 284. 2000.

[3] J. Cassell. Embodied Conversation Agents. MIT Press. pp. 5. 2000.

[4] J.Cassell. Embodied Conversational Agents: Representation and Intelligent in User Interface. MIT Media Lab.In press, AI Magazine.

[5] J. Cassell, C.Pelachaud, N. Badler, M. Steedman, B. Achorn, T. Becket, S. Prevost, M. Stone. Animated Conversation: Rule Based Generation of Facial Expression, Gesture & Spoken Intonation for Multiple Conversational Agents. Departament of Computer & Information Science, University of Pennsylvania.

[6] J.Cassell. Fully Embodied Conversational Avatars: Making Comunicaction Behaviors Autonomous. MIT. Media Lab.In press, *AI Magazine*.

[7] T.Dean, K. Kanazawa. A model for reasoning about persistence an causation. *Computation Intelligence*. pp. 142-150, 1989.

[8] Dirección General de Televisión Educativa. U.T.E..-SEP.1999.

[9] Periódico El Financiero (2001) "Compac y Concamin pactan acuerdo de asesoría informática". pp. 67. march 26, 2001.

[10] Instituto Nacional de Estadística e Informática . " Conteo de Población y Vivienda", 1995.

[11] D.Massaro, M. Cohen, J. Beskow and R. Cole . Developing and Evaluating Conversational Agents. MIT Press. pp. 287. 2000.

[12] B. Petrazzini and M. Kibati. The internet in developing countries . Communicacion of the ACM. Vol. **42**, No. **6**, June, 1999.

[13] I. Poggi and C. Pelachaud. Performative Facial Expressions in Animated Faces. MIT Press. pp. 155-188. 2000.

[14] E. Turban, E. McLean and J. Wetherbe. Information Technology for Management. Ed. John Wiley & Sons. 2nd. Ed. 1999.

[15] G. Zweig and S. Russell. Speech Recognition with Dynamic Bayesian Methods. *AAAI*. 1998.

KES '01
N. Baba et al. (Eds.)
IOS Press, 2001

Urban Data Representation and Analysis

David SOL, Humberto GOMEZ & Michel GUENET
CENTIA, Universidad de las Américas-Puebla, Mexico

Abstract: Our paper describes a model and an application to manage urban data. This work presents the catastrophe model. The goal of this model is to represent the dynamic of urban zones. New constructions and buildings can be placed inside an urban area. These constructions can produce a modification and an impact on the land. The urban area needs to keep a structure and a balance between its parameters. The parameters are those related to each polygon inside the urban area: the cost of the land, the size and use, for instance. The model proposes to combine these parameters to generate a unique number. The mechanism used to combine the parameters is the Gödel's codification number. Qualitative data is translated into quantitative data. The application which implements the model can manage important differences between the Gödel´s numbers to show discontinuities in the studied area. The application shows the results of these analysis in 3D and includes modules to store and retrieve data from a database. The spatial data representation is supported by the OpenGIS specification consortium. Our experiment shows how data models help in real applications like urban analysis.

1. Introduction

Land distribution involves all the population. Cities have been planned to give services to its citizens. A heavy concentration of people in a small area can represent a symptom of urban illness. The citizens' services need to be increased and improved to maintain an acceptable quality of life. An urban area can grow, but its land should be used to keep an environment balance. In particular, this phenomenon can be found on urban areas of the third world.

Our work describes a model and an application to help the urbanist to discover patterns that can show unbalanced scenarios in urban areas. The results of these analysis can help to manage and design convenient scenarios and to plan the urban growth. This document describes the catastrophe model implemented by our application, and the operations needed to integrate the data in the model. Finally, we present some tests to validate our work. This work is supported by the Millennium-CONACyT initiative in the project *"Access to High Quality Digital Services and Information for Large Communities of Users"*.

2. Urban Catastrophe Model

Our model is supported by the Thom's catastrophe theory (see [10]). This model was introduced in [5]. The main idea of Thom´s catastrophe theory is concerned to what is known as a catastrophe. A catastrophe is defined as a lack of stability. "Suppose that the stable configuration of a physic system is determined by a function, and suppose that the form of the function suffers a gradual change in time. Stability is lost when the minimum of this function ceases to exist, then we have a catastrophe: the system suffers a rapid sequence of changes, until a new state is achieved" (see [10]). The Thom´s theory is used to understand the general dynamic principles which support the modeling of changes in nature. Thom suggested to use the topology theory of dynamic systems to model the discontinuous changes occurred in natural phenomena, making an emphasis on biology. A mathematical description of Thom's theory can be found in [17]. Our model has two principal objectives:

1. to improve the knowledge about the environmental components to assure a better urban management process and
2. to understand the effects of urban dynamic in a regional context

The first evaluation of our model was made by hand. An expert selected an area and worked with data related to this space. The result of the analysis indicated zones with concentration, dispersion, evasion or grouping. The application described later (see [4]) provides tools to automate this classification.

The second phase of the urban analysis consists of making a 3D representation of the studied area. A centroid and a Gödel's number is computed for each polygon. Each Gödel's number represents a height. The 3D representation can show discontinuities (variations on heights). These discontinuities can be related to the four variables of study.A conflict exists when a dramatic change in height is identified. Remember that the number of Gödel defines the height for each polygon.

3. Gödel's number

A central part of the urban analysis is the Gödel's codification. On this part, the urban analyst has to select a descriptive data subset in order to perform the codification. The Gödel's codification uses a prime number sequence as a mechanism to identify each tuple. The original purpose of Gödel's codification was meant to be used as a mechanism to associate a number to a mathematical expression (see [2], [3], [6], [12] and [13]).

As mentioned above, a centroid on 2D is computed by each polygon. Each of these polygons has several descriptive data associated. A subset is selected to calculate a Gödel's number per polygon. Once we have the Gödel's number, each centroid is related with its corresponded Gödel's number which produces a point in 3D. Every point obtained as described is presented on the screen in order to appreciate the differences on height (defined by each Gödel's number). For instance, if we have the area and functional value of each polygon, shown on table 1, we can compute the Gödel's number for polygon with GID 2 as follows:

Table 1. Descriptive data associated to each polygon.

GID (Geometry ID)	Area	Functional value
2	23	2

$$\text{tuple}(\text{area}_2, \text{functional\_value}_2) = \text{tuple}(23,2) \Rightarrow 2^{23} * 3^2 = 8,388,608 * 9 = 75,497,472$$

In this process, each tuple will be associated to a unique Gödel's number (unless two or more tuples have the same values for each attribute). With this approach we can relate the descriptive and geometric data. Besides, we can decode each Gödel's number to obtain the original values. If we have the number 74,497,472 without knowing the original values that produce it, we can (using successive divisions by prime numbers) obtain the original value.

4. Data Modeling

Spatial data is modeled by using OpenGIS specification. With this specification is possible to relate non spatial data with spatial data. The specification is implemented in a relational model. The model contains a first structure to represent the spatial reference

(SPATIAL_REFERENCE_SYSTEM). This spatial reference is related to other structure (GEOMETRY_COLUMNS) which controls a set of layers which together can represent a real problem. Each layer is represented by one type of feature. The principal features are point, line and polygon. However, the specification models a hierarchy of features. A particular feature is represented in another table. Each type of feature is related to the tables which store all the points needed for the feature. A line needs two points and a polygon needs a sequence of points (at least three points). All the points need to be stored in the database. The non spatial data can be related to the spatial data by means of the GID (Geometry ID) attribute. The data (spatial and non spatial) can be obtained from another database or from a proprietary format.

5. Analysis

We can summarize the steps to make an analysis as follow:

1. Obtain the digital map of the studied area. This map must contains the polygons which represents the studied structures (houses, buildings).
2. For each polygon (house, building) a centroid is calculated.
3. The parameters are selected (e.g. area, cost, date of construction) in order to calculate the number of Gödel for each polygon.
4. The selected parameters are used to estimate the Gödel's number for each polygon. This process is done by implementing the Gödel's codification (see 2,3,6 and 12). This process will allow us to translate qualitative data into quantitative data (area, cost, date of construction).
5. The Gödel's number represents the Z component of each centroid (calculated in point 2), this will produce a 3D representation based both on the geometric data (centroid in 2D) and the parameters associated to each polygon (the descriptive data).
6. The results of these estimations are used to present to the user a 3D visualization. An expert user can interpret the data and the cycle of analysis can conclude.

This process has been designed an implemented in our application. We called this application GISUA (Geographical Information System for Urban Analysis) describe in [4]. It performs the tasks mentioned above. The application consists of six modules which perform operations such as data retrieval, data representation, database access, data presentation, and spatial analysis (figure 1). Our application was developed by using the java programming environment (see [7],[8] and [9]).

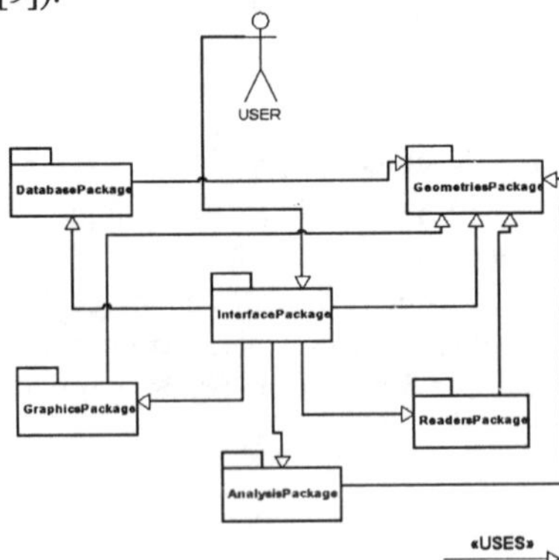

Figure 1. System's architecture

The expert only need to interact with the InterfacePackage module. This module is related with all the modules in the application. For instance if the user needs to access the database, InterfacePackage call the DatabasePackage, which manages several access to the database. Our application is connected with Informix DBMS. Remember that the spatial data is organized by using the OpenGIS specification. Also the spatial data is stored in a binary code format. The InterfacePackage is connected also to the GraphicsPackage which is responsible to manage the design of all the geometric objects on the screen. The GeometriesPackage handles the internal representation of the geometric objects. The ReadersPackage is a very important module because it translates spatial and non spatial data from external formats. Finally, the AnalysisPackage implements the estimation of Gödel's number. We can add new analysis by modifying this module. In [4] is described in detail each component presented in figure 1.

6. Tests

Several tests were performed to the implemented system. A digital map representing the city of Montréal, Canada was used. The original size of this map is 87 megabytes of descriptive information, and 89 megabytes of geometric information. A subset of this map was selected in order to make the tests. The system reads the geographical information using to formats: DBF (descriptive information), and SHP (geometric information). Both formats, DBF and SHP, are used by [1]. In figure 2 we can observe the system's interface used to show the created information during the execution of an urban analysis.

In particular, in the descriptive information section, we present the selected parameters and a column with the Gödel's number computed for this set of parameters. The right-bottom section shows an area with its polygons. For each polygon we can see that a Gödel's number has been computed. An expert can use this interface to interpret relationships between the components of this area and to classify the polygons. It is important the participation of an expert.

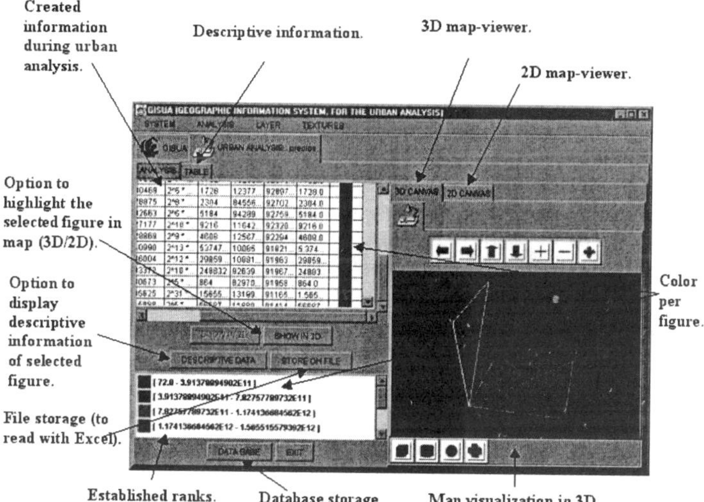

Figure 2. System's interface used to show urban information.

People who are not urbanists selected some parameters and the result shows very divergent values. This type of presentation allows the expert to recognize the region. We can see in the figures that both types of interfaces have buttons to store in the database, to store in a file, to manage the colors and some tools to manage the map (zoom, pan, etc.)

7. Conclusions

As mentioned above, our application is used to assist the urban analyst to perform certain calculations concerning the computation of Gödel´s numbers. The system can also present the digital map in 2D and to compute the centroids in 3D, along with the descriptive information available. The system is not an expert system, because it does not present solutions or interpretations, as this is the responsibility of a human expert on the field of urbanism. Michel Guenet made the interpretations, as he is the expert on urbanism. However, the algorithm implemented to calculate the Gödel´s number is to slow ($O(n^2)$).

During the elaboration of this thesis several achievements were made. The model on urban analysis (see [5]) was implemented in the first step, since the other part corresponds to the Thom´s catastrophe theory. The maps in 2D and 3D are sensitive. The end-user can select any figure on screen in order to obtain descriptive information (figure 2). The database scheme defined by [14] and [15] was implemented, using the binary storage format.

Some extensions in revision are the implementation of Delaunay´s triangulation (see [16]) to join the centroids in 3D, improvement of Gödel´s number calculation algorithm, and the ability to read geographical information from other formats (e.g. [11] format).

Acknowledgements

This project is sponsored by the Mexican National Council of Science and Technology, CONACyT (project 35804-A).

References

[1] D. Sol, A. Razo and E. Loyo. Natural Hazards in the Popocatepetl volcano zone, *Workshop on Advanced Techniques for the Assessment of Natural Hazards in Mountain Areas*, Innsbruck, Austria, 2000, pp. 97-101.

[1] Home page: *http://www.esri.com/software/arcview/*.

[2] F. Ayres. *Algebra Moderna*. McGraw-Hill. México, D. F. 1976.

[3] Dolciani, P. Simon , Berman L. and W. Wooton. *Algebra Moderna y Trigonometría*. Publicaciones Cultural S. A. México, D. F. 1977.

[4] H. Gómez and A. Solís. *SIGAU (Sistema de Información Geográfica para el Análisis Urbano)*. Tesis de licenciatura presentada en Primavera del 2001 en las instalaciones de la Universidad de las Américas – Puebla. 2001.

[5] M. Guenet and J.F. Rotgé. *Expérimentation d'un SIG volumique à partir d'un modèle géométrique générique sur la dynamique urbaine*. Revue Internationale de géomatique. Volume 6 - n° 2-3/1996.

[6] J. E. Hopcroft and J. D. Ullman. *Introducción a la Teoría de Autómatas, Lenguajes y Computación*. Compañía Editorial Continental. México, D. F. 1993.

[7] Home page: *http://java.sun.com/*.

[8] Home page: *http://java.sun.com/products/java-media/3D/*.

[9] Home page: *http://java.sun.com/products/jfc/*.

[10] E. A. Lord and C. B . Wilson. *The Mathematical Description of Shape and Form*. John Wiley & Sons. England. 1984.

[11] Home page: *http://www.mapinfo.com/*.

[12] E. Nagel, J.R. Newman, *El Teorema de Gödel*. Consejo Nacional de Ciencia y Tecnología. México. D. F. 1981.

[13] I. Niven. H. Zuckerman. *Introducción a la teoría de los números*. Editorial Limusa, México DF, 1976.

[14] Home page: *http://www.opengis.org/*.

[15] OpenGIS Simple Features Specification For SQL. *Revision 1.1. OpenGIS Project Document 99-049, Release Date*: May 5, 1999. http://www.opengis.org/.

[16] S. Peterson. *Computing Constrained Delaunay Triangulations in The Plane*. Home page: http://www.geom.umn.edu/~samuelp/del_project.html.

[17] P. Batelli and I. Stewart. *Catastrophe Theory and its applications*. Pitman Publishing Limited. Great Britain, 1978

Improving the Quality of Digital Services and Collections for Large Communities: Research Issues

Gerardo AYALA, J. Alfredo SÁNCHEZ & David SOL
CENTIA, Universidad de las Américas-Puebla, Mexico
{ayalasan,alfredo,sol}@mail.udlap.mx

Abstract. In today's global society, access to knowledge and information determines the competitive advantage of individuals, organizations and countries, In order to address the problems of a digital divide, it is necessary to bring the benefits of information technologies and digital services to large user communities. In this paper we present our proposal and work in progress concerning our research oriented to provide access to high quality digital services and information for large communities of users. The aim of this applied research project is the development of digital collections and services in the context of digital libraries, geographical information systems and distance learning.

1. Introduction

In today's global society, access to knowledge and information determines the competitive advantage of individuals, organizations and countries, In order to address the problems of a *digital divide*, it is necessary to bring the benefits of information technologies and digital services to large user communities. As a research group established in Puebla, Mexico, we believe it is necessary to promote the orientation of our research activities to the development of information technologies that facilitate access to information, knowledge, and decision support for the appropriate use of natural resources and disaster prevention[1].

Our research is oriented to provide access to high quality digital services and information for large and diverse user communities. We develop technology that facilitates access to digital information and services, focusing on areas such as education and decision support concerning disaster management and natural resource usage. These areas have been considered of high priority for applied research in computer science and information technologies, as they are likely to result in direct economical and social benefits [2].

2. High quality digital services and information

In order to enable access to high quality digital services and information for large user communities, it is necessary to conduct research and advance three main areas: personalized user interfaces for digital services available to Internet communities, building high quality digital collections, and distribution of digital collections to wide user communities through high performance networks.

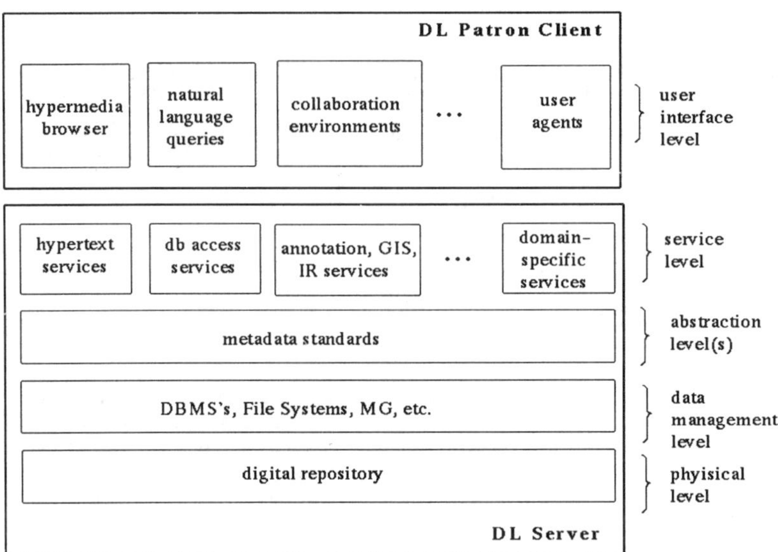

Figure 1. Layered architecture for digital libraries (Sánchez & Leggett [3]).

Our research projects on high quality digital collections and services are framed by the layered architecture for digital libraries proposed by Sánchez & Leggett [3] (see figure 1). Our research in the user interface level includes user agents, collaboration for learning environments, speech recognition interfaces and visual interfaces for geographical information.

At the service level, our research focuses on geographical databases, annotation of documents, image processing, agent essentials and multimedia retrieval. The digital repository (physical level) results both from the digitization (and processing) of existing analog documents and from the integration of digital materials into a comprehensive data model.

3. User interfaces for digital services to wide user communities

3.1 User agents and personal spaces

Research and development on agent technologies imply the formalization of agent modeling and agent cooperation and communication. User agents assist the user by gathering information from distributed repositories [3]. User agents travel among federated collections searching for information in order to generate suggestions according to general user profiles and to respond to specific user queries. Results coming from multiple collections are integrated by the user agents and presented to users in what we refer to as *personal spaces* [4].

The development of user modeling techniques is necessary for the personalization of Internet-based learning environments and digital libraries [5]. We model personal spaces as virtual areas in which individual users organize information units. A user agent in this kind of environment helps individual users handle a vast information space based on personal models held for each of them.

3.2 Speech interfaces in Mexican Spanish

An important component of the interface for access to wide communities in our country implies the research and development of speech interfaces for Mexican Spanish [6]. In this way we seek to extend the means available for accessing digital services through voice interfaces (for example, via telephone). We are developing interfaces that will elicit (from the vocabulary used and the context of the dialogue) what is required to fulfill a task and what the system needs to ask from the user. At the same time the interface gives the user the chance to provide more information than was specifically asked for but the user already knows, speeding up the process of completing the information.

Task specific dialogue and language models for digital services interfaces are required, based on mixed initiative techniques, for access to information through voice interfaces via telephone. Research and development of mechanisms making geographical data accessible to multiple users, together with spatial query languages to access geographical databases, requires the definition of a vocabulary for the Popocatépetl volcano information system, based on a language statistical model.

3.3 Visualization

In order to facilitate visualization and high-performance organization of graphical information for high quality digital services it is necessary to develop visualization algorithms. Among the tools available to the user to facilitate the visualization and organization of a digital collection we are developing an interface designed to aid in the visualization of complex structures (such as multiple taxonomies) which are typically superimposed on information units maintained in digital libraries [14].

Novel structures and algorithms for geographical information systems, supporting decision-making based on geographical information is required [8]. In order to improve visual interfaces our group works on representation models and visualization techniques for realistic computer-generated images, implementing computational models of physical and optical aspects, including the development of algorithms for solid modeling based on computational geometry for virtual reality environments.

3.4 Collaboration and learning environments

Education is a natural application domain for digital libraries, and digital information services for large communities [9]. For any community member, learning is a lifelong, distributed and collaborative activity, based on the active interaction among members of such a community, sharing and constructing knowledge. In this context, it is considered necessary to strengthen the motivation of learning for sustainable communal life and personal development by developing information technologies that provide opportunities of self-directed learning. In a community, learning advances through collaborative social interaction and the social construction of knowledge. In the context of large communities we propose to apply these methodologies from the perspective of lifelong learning [10].

For both collaborative and learning environments, agents are necessary for the configuration of groups in wide communities of users. We are modeling intelligent agents that cooperate in the configuration of groups in internet-based wide communities, promoting collaboration, learning opportunities and the social construction of knowledge in areas considered of the interests of the users. Group configuration and awareness in a large community is important, and this is implemented through the cooperation of agents that work based on their user models.

4. Building high quality digital collections

4.1 Pattern recognition

We propose the modeling of neural networks for pattern recognition in order to construct and classify images and documents in high quality digital collections. Research and development of handwritten character recognition techniques are necessary in order to input and catalogue large amounts of manuscripts for digital repositories. Neural networks applications using handwritten character recognition are applied to input large amounts of information in digital services in a large digital collection of historical documents.

Concretely, our group have results working with the collection of telegrams written by Porfirio Díaz, the president of Mexico who ruled the country before the Mexican Revolution [11]. The system recognizes isolated words taken away from telegrams written by Díaz. Images of the telegrams are obtained from microfilms of the originals.

4.2 Image processing

We are developing image processing algorithms and techniques for the input, enhancement, alignment of images of digital collections, as well as for their manipulation, representation and presentation. Image processing is necessary for the modeling and recovery information repositories for Puebla City, its suburban areas and the Popocatépetl volcano zone, for decision-making support systems. In the context of geographical information systems, it is necessary to design and develop efficient methods for obtaining information from geographical data coping with the problem of 3-dimensionality [12].

5. Distributed and high performance access of digital collections

5.1 High performance information exchange

In the context of our project, we require the development of mechanisms for high performance access of large volumes of information and multimedia data. We are developing communication protocols of Java/Jini technology for real-time digital services [13]. Event-driven task specific servers are necessary to route the events and information to the corresponding server (for example, a video server or an audio server). The use of protocols and standards to exchange geographic data is needed in the context of distributed applications and geographic data. High performance information exchange also requires the formal analysis and design of algorithms and their complexity for information retrieval in large digital repositories and digital services.

5.2 Multimedia data compression

Together with the modeling of high quality and performance multimedia databases for supporting digital services based on audio and video, it is necessary the modeling of software that provide a high performance exchange of multimedia material (audio, video) for distributed learning environments, distance education, digital libraries and speech interfaces. This implies algorithms to compress multimedia data. We are developing a state-of-the-art Unix streaming solution on the Sun platform that will allow universities to use any client for accessing video and audio content seamlessly through their network. This development effort will also pave the way for universities to take advantage of the latest developments in thin-client network devices.

6. Conclusions

Providing access to high quality digital services and information for large communities implies three main research and development components: personalized user interfaces based on agent and speech recognition technologies, construction of high quality digital collections using neural networks and image processing, and high-performance distribution of digital collections based on multimedia data compression and transmission.

Acknowledgements

The authors thank to all the members of the CENTIA who are the researchers participating in the project: Ingrid Kirschning, Pilar Gómez, Antonio Sánchez, Ofelia Cervantes, Juan Manuel Ramírez, Oleg Starostenko, Daniel Vallejo, Mauricio Osorio, Antonio Aguilera and Rogelio Dávila. This project is sponsored by the Mexican National Council of Science and Technology, CONACyT (project 35804-A).

References

[1] D. Sol, A. Razo and E. Loyo. Natural Hazards in the Popocatepetl volcano zone, *Workshop on Advanced Techniques for the Assessment of Natural Hazards in Mountain Areas*, Innsbruck, Austria, 2000, pp. 97-101.

[2] G. Ayala, O. Cervantes and A. Bernat. *Collaboration in Computer Science: Accelerating Progress in Science and Technology*, Final Report of the Third Mexico-U.S.A. Workshop on Computer Science, Universidad de las Américas-Puebla, Cholula, Mexico, 1999.

[3] J. A. Sánchez, and J. A. Leggett. Agent services for users of digital libraries, *Journal of Networks and Computer Applications* 21, 1, 1997, pp.45-58.

[4] L. Fernández, J. A. Sánchez and A. García. MiBiblio: Personal Spaces in a Digital Library Universe, *Proceedings of the Fifth International ACM Conference on Digital Libraries DL'00*, San Antonio, Texas, 2000, pp. 232-233.

[5] J. A. Sánchez, and G. Ayala. User agents in digital libraries and collaborative learning environments, *Memoria del Taller de Inteligencia Artificial TAINA'98*, Mexico City, 1998, pp. 367-369.

[6] I. Kirschning. Research and Development of Speech Technology & Applications for Mexican Spanish at the Tlatoa Group. Proceedings of CHI 2001, Seattle, WA, April, 2001.

[7] C. Proal, J. A. Sánchez, L. Fernández. UVA: 3D representations for visualizing digital collections. *Proceedings of the Third International Conference on Visual Computing* (Visual 2000, Mexico City, Sept. 18-22), 2000, pp. 185-192.

[8] D. Sol and F. Garcia. Geographic Services in a Digital Library, *First NSF-CONACyT workshop on Digital Libraries*, Albuquerque, NM, 1999, pp. 37-41.

[9] G. Ayala, F. Cocón and L. Agosto. BIDACI: Modeling agents for a learning environment based on a digital library, *Proceedings of ED-MEDIA 2000, World Conference on Educational Multimedia, Hypermedia and Telecommunications*, Montreal, CANADA, AACE, 2000, (cd rom version).

[10] G. Ayala. Intelligent Agents Supporting the Social Construction of Knowledge in a Learning Environment, *Human Computer Interaction: Issues and Challenges*, (Qiyang Chen, Editor.) Idea Group Publishing, 2001, pp. 44-63

[11] P. Gómez, S. Linares, S. Spínola and J.M. Ramírez. On the automatic digital storage of historical documents: Recognition of handwritten telegrams of Don Porfirio Diaz, paper submitted to the *Information Systems for the Access to Digital services for a Wide Community special session, (KES'2001)* (to appear in this volume) 2001.

[12] A. Aguilera and D. Ayala. Faster ASV Decomposition for Orthogonal Polyhedra, Using the Extreme Vertices Model (EVM), *WSCG'2000: The 8-th International Conference in Central Europe on Computer Graphics, Visualization and Interactive Digital Media*, Vol. I, Plzen, Czech Republic, 2000, pp. 60-67.

[13] O. Starostenko. Data Acquisition Systems in Distributed Environments: Examples of Java and Jini Technology Implementations, *First International Symposium Educational Informatics and Telecomunications*, IPN, Mexico City, 1999, pp. 51-58 .

KES '01
N. Baba et al. (Eds.)
IOS Press, 2001

Speech Technology to Provide Access to Digital Information in Mexican Spanish

Ingrid KIRSCHNING, Ofelia CERVANTES
Grupo TLATOA (CENTIA), Universidad de las Américas-Puebla
Sta. Catarina Mártir, Cholula, Pue., MEXICO
{ingrid, ocervan}@mail.udlap.mx

Abstract. Thanks to the advances in today's technology in terms of processing speed of computers, storage space and the management of sound and video devices, speech technology is a reality in almost any kind of computerized system. Speech applications are being used in personal computers, cellular phones, etc. Among it's most useful applications we can find telephone-based information services, banking and computer assisted language learning systems. There exist already a large number of commercial products that use speech interfaces, developed mainly for English, German and Japanese. That is why a strong effort is being made to make this technology available also for the Spanish speaking community. Additionally, the fact that a large number of people do not have access to a computer with internet connection or the training to use one, causes a type of "Digital Divide" we need to overcome by developing interfaces that provide digital services to anyone who needs them, regardless of their training in the use of computers and the availability of them in the field.

1. Introduction

The rapid growth of the Internet has produced an extremely large network of information, interconnecting countries and cultures. This global connectivity opens many opportunities for people to have access to data and exchange information in order to solve different problems. However, it is still not there for everybody. Current technology provides ever-improving human-computer interfaces, but there exist large communities of citizens in different countries that cannot use a computer. They do now have access to computers or internet connection, nor do they have the know-how to find information that might be useful. This is caused mainly by economical, educational and technological recessions found in many areas of many countries.

Some of the fundamental problems faced by researchers in this area are the representation, sharing, access, retrieval and translation of information already present on the internet in other languages. Also, the heterogeneous information networks, which differ in design, reliability and performance can pose a difficulty to the communication. However, even when the digital data is available, our concern are the channels by which the access should be provided. Our work is focused on speech technology to provide services (information and education) in Mexican Spanish. As computers are not available anywhere, but a telephone is more commonly found, we base our work on this device to communicate to one central system as the gateway to the world of digital information, trying to breach the digital divide [1].

2. Development of Speech Technology for Mexican Spanish

Alongside the various advances in fundamental research on improving the speech technologies is the consideration of Human-Computer interaction factors, specific to every user's culture and language that need to be included in information systems. Our work has been to perform basic research in the different speech processing techniques, improving the performance of speech recognition (based on both, Artificial Neural Networks or Hidden Markov Models and synthesis in Spanish (using diphone concatenation and Unit Selection). We also work on the analysis of the Spanish language, dialogue structure and perception for the development of conversational speech interfaces and language learning and language acquisition systems[1].

3. Speech Recognition

In the area of speech recognition our research focuses on the improvement of the methods for training and development of speech recognizers, as well as training and testing new versions of our recognizers for Spanish [2, 3]. We have recently trained a general-purpose recognizer based neural networks and another with HMM's using the CSLU Toolkit and with a speech corpus containing more than 500 speakers. The performance has been tested on specific purpose corpora with excellent recognition rates[4]. The following table shows some of the results. The corpus used here contains 50 speakers (at 8000KHz) uttering sequences from 2 to 6 digits and the number 10.

Table 1: Recognition results for the digits corpus.

Recognizer	Accuracy (%)	Comments
Neural Network	98.2 %	Phonemes modeled as context-independent units
Neural Network	98 %	Phonemes modeled as context-dependent units
HMM	71.19 %	Phonemes modeled as context-independent units
HMM	92.14 %	Phonemes modeled as context-dependent units

However, the corpus used to train these recognizers does not reflect all the different dialectal zones of Mexico, thus there is still room for improvement, once more data from different zones of the country becomes available.

4. Speech Synthesis

Until now speech the applications we have developed have used a Mexican Spanish voice developed by one of our members and provided now with the CSLU Toolkit for use with Festival. Although this speech synthesis was later improved by adding a duration estimation module [5], the voice still sounds robotic. Now we have started to develop a new voice with a better quality, using a Unit Selection approach for the synthesis. In this approach a speech corpus, recorded with high quality by a professional speaker is searched

[1] Projects supported by the Mexican National Council of Science & Technology (CONACyT) grants No. CERVO30050-A and No. 35804-A, and CSLR, Colorado University (NSF grant No. IIS-0086107)

to extract large units of speech to concatenate them. These units of speech can be entire phrases, parts of phrases, words, or if none can be found, syllables. The result is a much more natural sounding speech. The program developed can use any pre-recorded corpus, which needs to be completely transcribed and labeled (word- and phoneme-level) [6].

5. Conversational Interfaces

The term "conversational" can have different meanings depending on the context, but in general it refers to an interactive system which works in a restricted domain. Although many speech interfaces are considered conversational, they largely differ in one main aspect, and this is the degree in which the system takes a more active role in the conversation [7].

Conversational systems allow users to interact with an automated system and recover information, perform transactions, or other tasks through a more or less flexible dialogue with the user.

5.1 Architecture of Conversational Interfaces

The elements required in a conversational system (see figure 1) are:
- A speech recognition module to convert speech to text.
- The Natural Language Understanding (NLU) module that obtains the semantic representation of the recognized speech.
- The dialogue manager, which keeps the control of the interaction, and can decide who is to have a more active role in the dialogue, the speaker or the system.
- A language generating module which puts the retrieved information or the questions of the system in understandable and correct sentences / phrases.
- A mechanism to transmit the information to the user, which can include, apart from speech synthesis, other devices to visually display the data as well.

5.2 Conversational Interfaces for Spanish Language

Based on CSLU's robust parser we had initially integrated natural language processing into query systems that allowed students to obtain information about their courses via a speech interface [8]. This is a good approach for developing in a fairly short time some NLU interfaces, using the CSLU Toolkit. There are, however much more powerful architectures, that provide better capabilities for mixed initiative conversational interfaces. In order to provide access to information we are beginning to work with the CU-Communicator System [9].

It is our objective to develop conversational applications that are able to perform a dialogue in Spanish, detecting the degree of expertise of a user with a system and allow a flexible communication where the system can take or relinquish the initiative. This is in order to be able to obtain enough information from the user to perform the required task and at the same time give the user the chance to provide more information than was specifically asked for but the user already knows, speeding up the process of completing the information. Mixed initiative techniques try to give more flexibility to an expert user and guidance to the newcomer in order for both to reach the same goal in a user-friendly environment. It implies the monitoring of the turns taken in the dialogue between the user and the system, both trying to reach the same goal (depending on the task this can be

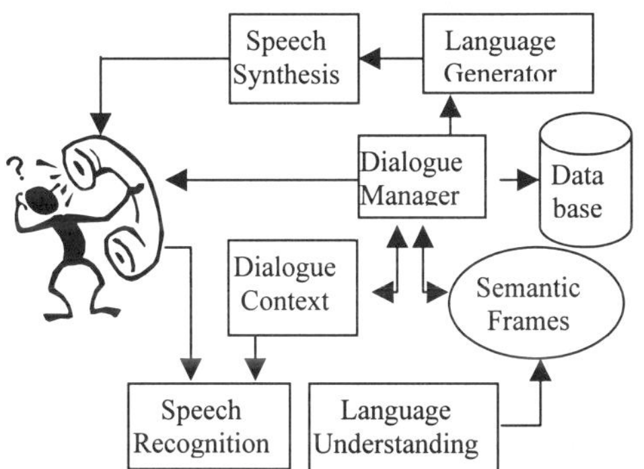

Figure 1. Typical architecture of a conversational interface.

retrieval of information, flight reservation, etc.). The implementation of the necessary modules to maintain a mixed-initiative interaction in Mexican Spanish includes natural language analysis and processing [10] and an user agent that monitors the dialogue, registers he elements relevant in the context (filling the slots of required information) and the data given by the user, until it is enough to form a query to consult a database. The development of mechanisms to make digital data accessible to multiple users, together with query languages to access databases implies the definition of task specific vocabularies, dialogue structures and language models.

6. Speech Application Development

Several small applications have been developed as demonstration systems like a voice mail, a system to access e-mail via the telephone among others [11, 12]. Additionally we collaborated with SpeechWorks Intl. In the development of an auto-attendant for the university, as web as a system that allows the students check on their account status with the university, also via the telephone.

One very interesting application field for speech technology is the education. As part of a larger project to be developed we created a couple of tools for a computer-assisted language-learning environment. These tools are a first prototype for pronunciation verification and a bilingual dictionary, both for Spanish language students, whose native tongue is American English [13,14]. Another very interesting aspect is the use of speech technology based systems to support language acquisition for deaf children. We are developing a system based on the CSLU Toolkit as a first prototype for the Jean Piaget Special Education School in Puebla. The main challenge here is to create a tool that can be used also by the children that do not speak yet (modulating correctly) and design the interface and the content of the lessons in a way that is easy to use for the students and easy to manage by the teachers.

7. Corpus Development

Each and every one of the previous projects could not be done without the existence of a sufficiently complete speech corpus. Our group has recorded a corpus consisting of 550 speakers, mainly from the central area and a little from the north of Mexico, speaking a large variety of words (names, numbers, digits, letters) as well as spontaneous speech. This has been one of the most time-consuming tasks, but now this corpus, completely transcribed and labeled, is available to anyone without cost for educational and research purposes. The recording of speech corpus is a continuing effort, aiming to cover in some near future all the dialectal zones of the country, as well as different ages, including children.

8. Conclusions

Our main interest is first, to produce students with the knowledge and capability to work with speech technology, and the awareness of CHI specific to their environment. Second, we focus on the development of technology that is accessible to Mexican people and also applications that provide a support/aid to the real needs of Mexican society, using speech technology.

References

[1] B. Shneiderman, CUU: Bridging the Digital Divide with Universal Usability. *Interactions*, Special Issue 2001, Vol. **VIII.2**, (2001), pp. 11-15.

[2] M. Espinosa and B. Serridge, Comparación entre redes neuronales y modelos ocultos de Markov para el reconocimiento de voz, utilizando el CSLU Toolkit, in *Proceedings of ENC'99*, Pachuca, Mexico, September 1999.

[3] M.A. Oliver and I. Kirschning, Evaluación de métodos de determinación automáticos de una transcripción fonética, in *Proceedings of ENC'99*, Pachuca, Mexico, September 1999.

[4] E. Clemente, Entrenamiento y Evaluación de reconocedores de Voz de Propósito General basados en Redes Neuronales *feed-forward* y Modelos Ocultos de Markov, Graduate Thesis, Dept. Computer Systems Engineering, UDLAP, June 2001.

[5] H. Meza, Modelos Estadísticos de Duración de los Fonemas en un Corpus de Español Mexicano, in *Proceedings of CONIELECOM 2000* ,Cholula, Mexico, March 2000.

[6] L. Flores, Síntesis de Voz con Unit Selection, Graduate Thesis, Dept. Computer Systems Engineering, UDLAP, June 2001.

[7] J. Glass, Challenges for Spoken Dialogue Systems, *in Proceedings of 1999 IEEE ASRU Workshop*, Keystone, CO, December 1999.

[8] O. Rosas, Sistema de Consultas utilizando Reconocimiento de Voz y Procesamiento de Lenguaje Natural. Master Thesis, Dept. Computer Systems Engineering, UDLAP, June 1999.

[9] B. Pellom, W. Ward and S. Pradhan, The CU Communicator: An Architecture for Dialogue Systems. International Conference on Spoken Language Processing (ICSLP), Beijing China, November 2000.

[10] R.D. Navarrete, R. Davila and A. Sánchez, SAVIA Traductor de un dominio restringido en una biblioteca digital, in *Proceedings of CONIELECOM 2000* , Cholula, Puebla, March 2000.

[11] N. Munive and O. Cervantes, Un Sistema de Correo Electrónico y de Voz usando Reconocimiento de Voz, *Soluciones Avanzadas*, 7, 69, (May 1999), 44-48.

[12] N. Munive, A. Vargas, B. Serridge, O. Cervantes and I. Kirschning, Entrenamiento de un reconocedor Fonético de Dígitos para el Español de México usando el CSLU Toolkit, *Computación y Sistemas*, 3, 2, (1999), 98-104.

[13] I. Kirschning and N. Aguas, Verification of Correct Pronunciation of Mexican Spanish using Speech Technology, in *Proceedings of MICAI 2000: Advances in Artificial Intelligence, Mexican International Conference on Artificial Intelligence*, Springer Verlag, , 493-502, México, April 2000.

[14] I. Kirschning, N. Aguas, and A. Ahuactzin, Aplicación de Tecnología de Voz en la Enseñanza del Español, in *Proceedings of the 1er. Taller Internacional de Tratamiento del Habla, Procesamiento de Voz y el Lenguaje HAVOL 2000*, Mexico, July 2000.

On the automatic digital storage cf historical documents: Recognition of handwritten telegrams of Don Porfirio Diaz

Pilar GÓMEZ-GIL, Sergio LINARES-PÉREZ, Carlos SPÍNOLA-TENORIO, Manuel
RAMÍREZ CORTÉS
Department of Computer Science, CENTIA.
Universidad de las Américas, Puebla 72820. México
pgomez@mail.udlap.mx

Abstract. We present partial results of a project to build a digital collection of a very important set of documents for the History of Mexico: a collection of telegrams written by Porfirio Diaz, a president of the country who ruled at the beginning of 20th century. This article focuses in the recognition of isolated words taken away from telegrams written by Diaz. Images of the telegrams were obtained from microfilms of the originals, hence there is a lot of noise and poor resolution on them. This, added to the fact that old handwritten manuscripts are hard to read, make this task especially difficult for a automatic recognizer. We designed a recognizer based on a back-propagation neural network, that works recognizing each letter in each word. The recognizer receives as input a cleaned word and finds the class of each character in the word. The pre-processing on the words and the segmentation algorithm to get each letter is also described.

1. Introduction

The access to old documents is of high importance to every historian. However, it is mandatory to preserve such documents from being damaged by frequent use, as they must be available for a large number of historians located in very different places. A solution to these problems is to built a digital collection of these documents available through the Web.

At the end of 19th century and beginning of 20th century, General Porfirio Diaz was ruling Mexico. His main and fastest communication medium were telegrams. Most of the telegrams he wrote and received contain important information about the Revolution movement. The originals of these telegrams are located in Mexico City, and the Library of the Universidad de las Americas, Puebla own a microfilm version of them. Our Library wants to create a digital database of the images and transcriptions of these documents, in order to make them available in the Internet to a large community of users. There are around 60,000 letters and telegrams written in Old Spanish in a rather difficult-to-read handwritten style (Figure 1). For these reasons, the translation of the information in the telegrams is a difficult and slow task to be done by a human. Much better could be to have an automatic system to transfer the telegrams from the microfilm to a digital image, and then automatically read them. This system could be able to recognize the writing of one writer, Porfirio Diaz. Figure 2 shows a block diagram of such system.

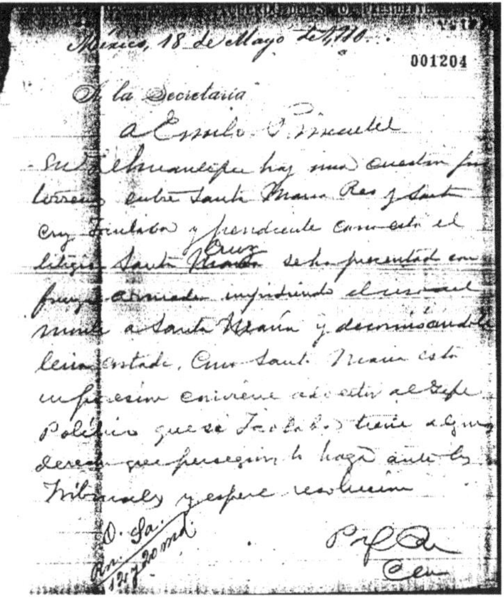

Figure 1. An example of a telegram written by Porfirio Diaz

There are several problems associated to the construction of this digital library, each one belonging to different disciplines, among others: pattern recognition, image processing and databases. Our research focuses on one part of the pattern recognition problem: the recognition of isolated words.

Off-line recognition of handwriting is a challenging problem for a computer. Specially, script characters are less uniform than printed characters; even when they are written by the same person, they present a wide variety of fonts, sizes and leaning, not to mention additional ornaments introduced by the writers. Factors such as ethnic origin, social level, education and age strongly contribute to the lack of uniformity in this kind of writing [1]. Several works using a wide variety of techniques have been proposed to solve this problem. Among them, artificial neural networks have shown good results [1] [2].

In this paper we present some advances in the construction of a word recognizer for this system, supposing that the documents have been digitized, cleaned from ruling lines and some noise, and words have been isolated.

2. Description of the database and preprocessing

Twenty five documents were selected from the collection, taking care that all documents were written by Porfirio Diaz (the collection contains documents written by others). Each document was digitized in a separate file. After that, using Adobe Photoshop, the document was cleaned from noise and horizontal lines. Words were selected and isolated from the documents, in a way that selection could contained patterns belonging to all letters of the alphabet. Each isolated word was stored in a separated file.

During this process it became evident that some letters were written in very different ways. This is a common problem in cursive handwritten [1]. Even when we are dealing with the same writer, the style of each letter may be different in different words. Next, inclination of each word was corrected with respect to its vertical and horizontal axis. The recognizer was

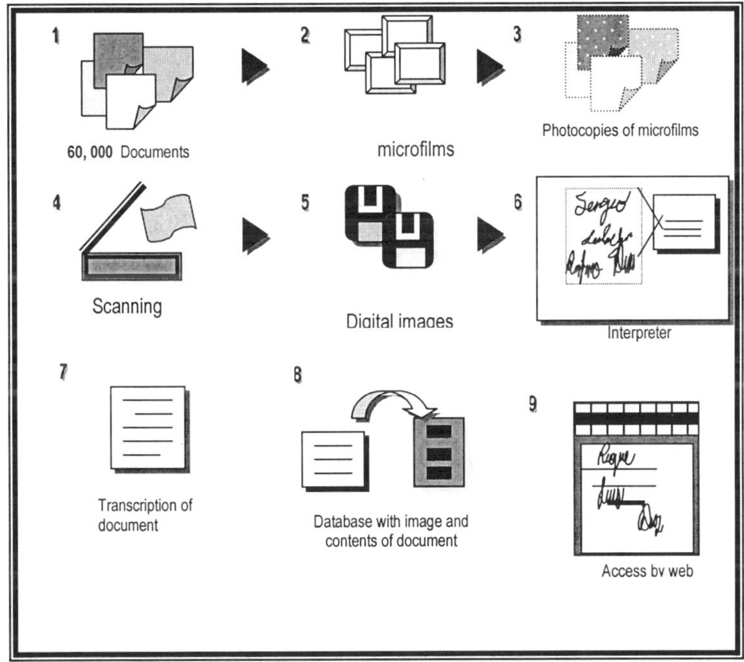

Figure 2. Automatic built of a digital collection of Porfirio Diaz telegrams

tested with both the original and corrected words. The height of all words was normalized to 1.6 cm.

3. The segmentation algorithm

After preprocessing a word, it is segmented into the characters or letters composing it. The segmentation algorithm is based on the one proposed by [2]. The image is stored in a binary matrix, and a vertical histogram of the image is obtained, that is, the number of ones in each column is counted. This histogram is processed to obtain a new histogram with only 2 values: 10 or 0. This threshold histogram is obtained as follows: The average value in the histogram is calculated and is multiplied by 0.8. call this number T. Each value in the original histogram is compared with T. If it is greater than T, 10 is assigned, if it is les or equal, a cero is assigned. Each segment is obtained from this threshold histogram as follows: The histogram is read from right to left. When a zero is found, the next "10" is located, and the segment is composed from the pixels corresponding to the columns where the last "10" was found up to the next "10". If the segment is less than 3 columns in length then it is attached to the previous segment because it is supposed that no letter may be so small. Figure 3 shows an example of the segmentation of a letter. Each segment is normalized in size following the algorithm proposed by [4]. The normalized segments are matrices of 8x27 pixels, a total of 216 binary values. Segments are the feature vector fed to the recognizer.

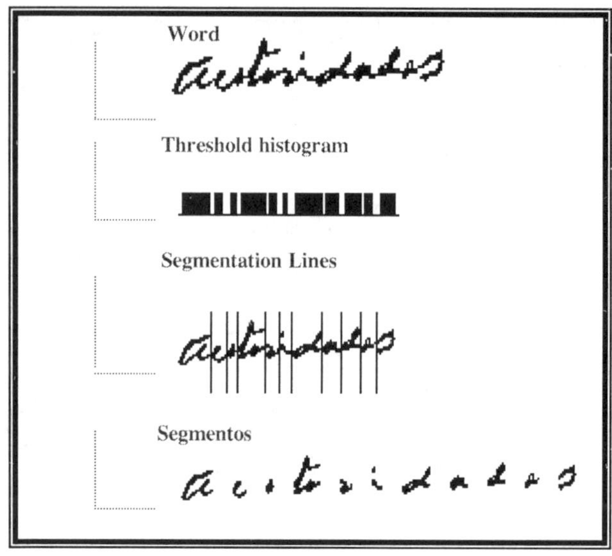

Figure 3. The segmentation process

4. Results

From the selected documents, a set of 90 words was created. Each word was segmented into characters to train the recognizer. Three training files were prepared. The first includes all the letters obtained from the 70 words, even the ones that were erroneously segmented.

Two classes were assigned for letter m, n and u because the segmentation process cuts each of them in 2 segments. The number of patterns for each letter (class) varies from one to 27 depending on the frequency that each letter is presented in the selected words. The second file contains characters obtained from words with and without correction in its inclination, and only 13 letters were considered. The third file contains patterns of 13 different letters obtained from words with correction in the inclination.

The recognizer was built using a feed-forward multi-layer perceptron network. The network was trained using the back propagation algorithm proposed by [5]. Several topologies were tested including fully and partially connected networks. (See Table 1).

Table 1. Topologies of the test networks

Topology	Nodes at Input layer	Nodes at first hidden layer	Nodes at second hidden layer	Output Layer	Total of nodes
Fully connected	216	162	54	26	458
Fully connected	216	162	54	13	445
Partially connected	216	43	11	26	296
Partially connected	216	43	11	13	283

Partial connections were made connecting groups of 10 nodes in the input layer with one node in the hidden layer, overlapping 5 nodes on each group. The three topologies were tested with the three files. For the experiment with 26 classes, the best results were obtained using the partially connected network, getting a recognition of 39% of the characters in a set containing words with 70% of characters that were used for training and 30% new characters. For the experiment with 13 classes, the fully connected network was able to obtain 81% of recognition in a set with 84% of segments used to train and 16% not used to train the network.

5. Conclusions

We have shown preliminary results of a neural net recognizer of a difficult case of handwriting. Even though the percentages of recognition shown here are low, there are more experiments to carry out and tuning of the network topologies is to be tried. The segmentation algorithm worked well for many words. However, it does not work right for an important part of the documents. For this reason, some modifications need to be done to get a right segmentation for such cases. Also a more complicated topology, as the one found in [1] is about to be tested with this database.

Acknowledgments

This project is supported by INIP (Institute of Research and Graduate Studies) at Universidad de las Américas-Puebla and by CONACYT (Consejo Nacional de Ciencia y Tecnología), project No. 35804-A. The Neural Net Shell was partially supported by CONACYT (Consejo Nacional de Ciencia y Tecnología) project No. 132900-A.

References

[1] P. Gomez-Gil, J.M. Ramírez and W. Oldham. On handwritten character recognition through locally connected structural neural networks. *Proceedings of the Second Joint Mexico-US International Workshop on Neural Networks and Neurocontrol Sian Ka'an 97*. Quintana Roo, México, (1997) 251-255

[2] E. M. Kussul and L. M.Kasatkina Neural Network for Continuous handwritten words recognition *Proceedings of the International Joint Conference on Neural Networks*. Washington, DC. July 10 (1999) 22

[3] A. W. Senior, A. J. Robinson. An off-line Cursive Handwriting Recognition System. *IEEE pattern Analysis and machine Intelligence*, **20** No. 3 (1998) 309-321.

[4] A. Güdsen. Quantitative Analysis of preprocessing techniques for Recognition of Hand printed Characters. *Pattern Recognition* **8** (1977) 219-227.

[5] D.E. Rumelhart, G. E. Hinton and R.J. Williams, 1986b. Learning Internal Representation by error propagation In Parallel Distributed Processing: Explorations in the Microstructure of Cognition (D.E. Rumelhart and J.L. McClelland, eds.) Vol. 1, Chapter 8. Cambridge, MA: MIT Press.

KES '01
N. Baba et al. (Eds.)
IOS Press, 2001

Motion Estimation Algorithms of Image Processing Services for Wide Community

Oleg STAROSTENKO, José A. CHÁVEZ A.
CENTIA Research Laboratory, Computer Science Department,
Universidad de las Américas, Cholula, Puebla, 72820, México

Abstract. Actually the development of high quality digital image processing services is important part of support of access to knowledge and information within large communities. In this paper the results of our research regarding algorithmic aspects of low-level motion estimation by flow-based and correspondence-based techniques for limited sequence of frames of dynamic scenes are presented. Two proposed algorithms have been used for constructing motion field, which can be used as supporting the information systems for distributed federation of users. The first algorithm is based on computation of spatial-temporal gradients, another one has been designed for fast processing the principal corners of object in dynamic scene. In order to evaluate algorithm velocity, utility, and efficiency, they have been tested during image retrieval, compression, processing and exchange.

1. Digital Image Processing Services for Large Communities

The important areas of information systems for the access to digital services for wide community are the development of representation models and visualization techniques for realistic computer generated and processed images, for support of virtual reality dynamic environments and for fast and high-performance multimedia data retrieval, compression, interpretation and exchange. That is why, in order to facilitate visualization and high-quality organization of graphical information for digital services within distributed federation of users, it is necessary to develop novel image processing and interpretation techniques, methods, algorithms which can be efficient and fast enough in real-time applications for large communities.

The particular objectives of our research regarding the development of image processing facilities can be presented as it follows:

a. supporting digital image processing services, modeling of software that provide a high performance exchange of multimedia material (particularly images, video) for distributed collaborative and learning environments, distance education, digital libraries;

b. developing the information systems for retrieval, processing and distribution of multimedia data about remote events and collections within impromptu community (for example, digital urban map management, Earth surface corrosion analysis, activity detection monitored volcano);

c. developing the searching facilities for personalized information among federated collections which imply image processing, agent-based personalized information retrieval and exchange (image based navigators, images of personal spaces gathering).

This paper presents some issues devoted to development of image processing algorithms, and taking into account algorithmic aspect of this research, the results may be particularly used as support of image processing procedures during acquisition, retrieval or transmission of information in wide digital environments.

2. Flow and Correspondence Based Image Processing Techniques

The sequence of images contains the necessary information about dynamic scene and usually is defined by optical flow or motion field, which can be estimated by gradient-based or correspondence methods [1]. Gradient-based methods obtain the motion characteristics by analysis of temporal variations of the image brightness using well-known motion constraint equation. The correspondence techniques try to estimate a best match of features or regions within consecutive frames [2]. Normally, the computing of movement characteristics can be obtained more quickly on base of processing the object edges instead of analysis of intensity variations or complete object correlation in consecutive frames [3]. The obtained information about motion can be used as input of different subsequent processes including motion detection, motion compensation, motion-based data compression, 3-D scene reconstruction, autonomous navigation, and analysis of dynamic processes in scientific applications.

For this work two high performance methods have been selected as base for development of novel algorithms [4]. One of them is gradient-based method, which permits fast edge estimation of moving objects in the limited sequence of images. The distance of displacement is not critical for this method but computation is complex enough [2]. Another one is a SUSAN (Smallest Univalue Segment Assimilating Nucleus) method where the principal idea is the manipulations by circular mask with center denominated as a nucleus [6]. The brightness of each pixel within a mask is compared with the brightness of that mask nucleus for USAN area calculation. The USAN area is at a maximum when the nucleus has the same brightness as all pixels, it falls to half of this maximum very near a straight edge, and falls even further when inside a corner. The SUSAN technique uses no image derivatives and it is good in the presence of noise due to method integrating effect.

3. Novel Gradient-Based Algorithm for Edge Detection

The proposed algorithm permits detection of complete edges of all the objects within the set of consecutive frames and then detection of edges in movement. For testing this algorithm, the sequence only of two frames of dynamic scene (Figure 1) has been used.

The proposed algorithm has been implemented on base of Halcon HD Kit that permits the development of interactive real -time software [7]. The algorithm consists of:

a. Normalization of image dimensions;

b. For edge detection the Sobel filter is applied where function $F(x, y)$

$$\nabla F = \begin{pmatrix} G_x \\ G_y \end{pmatrix} = \begin{pmatrix} \partial f / \partial x \\ \partial f / \partial y \end{pmatrix} \tag{1}$$

defines the direction of the maximum variation of the gradient F at the point *(x, y)* [8],

Figure 1. Dynamic scene with object in movement

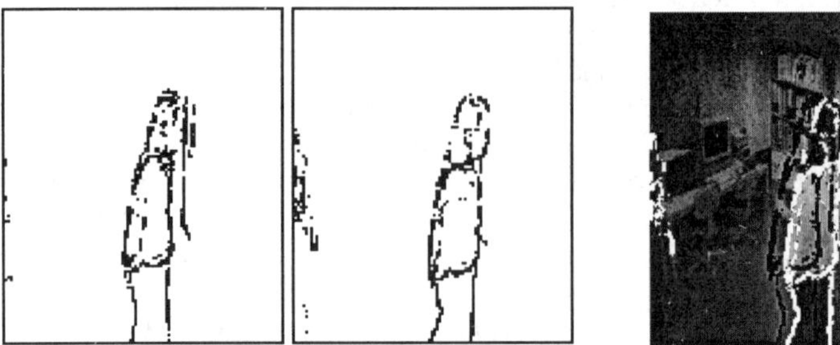

Figure 2. The object edges extraction and composite image with object in movement

In digital form it can be calculated using the masks for direction on axis x and y where each position is represented by variable z_i - gray scale of pixel under the mask:

$$|G_X|=|z_7+2z_8+z_9-(z_1+2z_2+z_3)| \quad \text{and} \quad |G_Y|=|z_8+2z_6+z_9-(z_1+2z_4+z_7)| \tag{2}$$

c. After two frames gradients computing, the edges in movement are estimated as:
1. temporal gradient is obtained by computing differences between first and second frames;
2. the result is multiplied by spatial gradient of the first frame;
3. the gray scale values are normalized in the range from 0 to 255 and obtained images are adjusted to dimensions of initial frames;
4. in case of detection of weak edges two thresholds are used for efficient extraction of the objects from the background. If the selected points lie between the range defined by thresholds for selected object: $T_1 \leq F(x, y) \leq T_2$ they belong to estimated object and obtain value "1", otherwise - $F(x, y) > T_2$ or $F(x, y) < T_1$ they belong to other object or to background (value "0"). After application of thresholds $T_1 = 20$ and $T_2 = 255$ for frames of Figure 1 the edges of moving object are extracted as it shown in Figure 2 and then the composite image with boders in movement is obtained.

4. Novel SUSAN-Based Algorithm for Corner Detection

For estimation of motion the analysis of principal corners of objects in image sometimes is enough. In this case the proposed algorithm based on the property of minimum USAN area near corners of object is quite acceptable. The algorithm consists of following steps:
a. Each point in the input image is used as the nucleus of a circular mask. The best digital approximation for calculation of the mask value is a Gaussian weighting because it is more smoother and stable than square similarity function (versus pixel brightness difference). The usual radius for each mask is 3.4 pixels that gives 37 pixels per mask .
b. Using equation

$$C(r,r_0) = exp-\left(\frac{I(r)-I(r_0)}{t}\right)^6 \tag{3}$$

the brightness function is calculated for each pixel under the mask. In the equation $I(r_0)$, $I(r)$ represent the brightness of nucleus position and the brightness of an other point within the mask respectively; t is the brightness difference threshold which defines minimum contrast of edges and image background. The number of estimated corners directly depends on t value. The 6th power of equation is a theoretical optimum, which permits good balance between stability about the threshold and the sensitivity of method.
c. Then value of USAN area n is calculated as sum of computed C functions.
d. For the detection of corners the USAN area property is used comparing the estimated areas with the threshold $g < n_{max}/2$, where n_{max} is maximum area of the circular mask. The

Figure 3. Composite image with superimposed corners

threshold g defines quality and quantity of the detected corners.

e. Finally, the wrongly detected corners must be rejected as a noise if the mask gravity has a great value (medium of the distance between point and nucleus). The same steps of the algorithm are applied for the second image in order to detect its principal corners. The composite image as result of corner detection procedure is presented in Figure 3.

The corners of the static objects cover each another and can be removed for the following analysis of motion characteristics. Therefore the advantage of this approach is the less amount of processed data for motion field construction. These algorithms directly can be used for high-speed image analysis, automatic compression, complex image skeleton estimation, pattern recognition, manuscript letter interpretation, users image profiles and pattern simplification for agent-based navigation and constructing personalized knowledge.

5. Motion Characteristics Estimation as Digital Processing Services Support

The time interval between two consecutive images of a dynamic scene is a principal characteristic of motion vector computing. Usually, for the estimation of object displacement the correspondence methods are used when the coordinates (x', y') of the center of analyzed pattern are found after pattern translation from coordinates (x, y) during the time interval Δt [2], [9]. This process can be modeled by the analysis of the brightness function of selected pattern. In this approach the same pattern is used within consecutive images as reference segment. It allows overcoming the problem of progressive increment of compared patterns but aggregates accumulative error proportional to time function.

For complete extraction and recognition of reference and analyzed patterns the Segment and Neighbors Matching method proposed by authors in [10] is used due to its high speed and efficient interpretation with presence of noise and distortions usually occurred in dynamic scenes.

a) 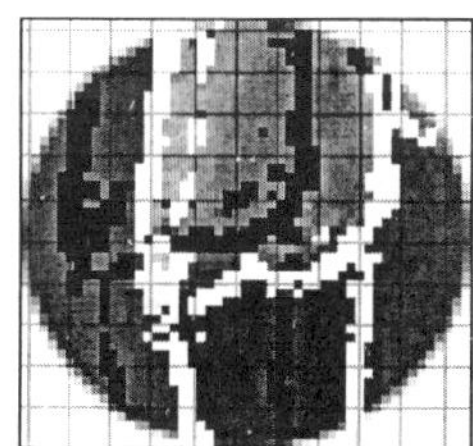   b) 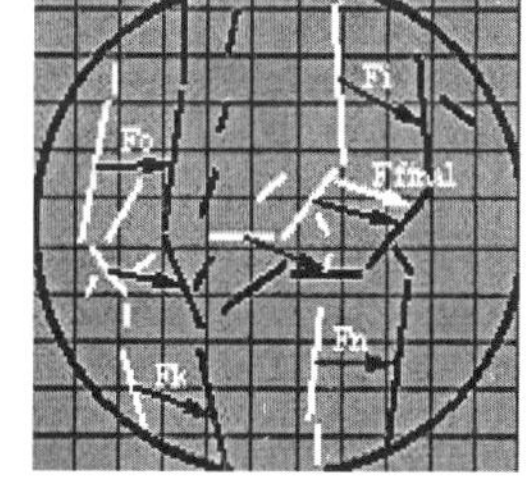

Figure 4. a) Edges in movement, b) fuzzy segments displacement

Figure 5. Corrosion detection of remote monitored volcano's slopes

This method is based on the concept of fuzzy sets and operates with membership grade as a principal characteristic for including analyzed element to the set of fuzzy segments, which compose the pattern. Applying this method to extracted edges of Figure 2, the reduction of similar oriented segments is obtained.

Finally, the fuzzy segments displacement vectors and the total displacement vector F_{final} can be calculated. In this way, the errors aggregated by non-arbitrary segment substitution are reduced considerably. In the Figure 4 the image fragment of Figure 2 and fuzzy segment with corresponding displacement vectors are shown at time t-Δt and t. The testing of the proposed algorithms gives quite simple and acceptable motion estimation in a limited sequence of images.

One example of proposed algorithm application is remote events analysis as it shown in Figure 5, where white and black segments represent the changes of volcano's slopes.

6. Conclusions

The proposed algorithms have been developed for supporting the digital image processing services for large communities, particularly for estimation of motion characteristics in limited sequence of consecutive images to be used by distributed federation of users. The computing of motion characteristics is limited by processing only edges and corners of objects but it is acceptable for the fast construction of the motion field.

Acknowledgments. This research is sponsored by the Mexican National Council of Science and Technology, CONACyT (project 35804-A).

References

[1]   B. Jahne, H. Haußecker, Computer Vision and Applications, Academic Press, USA, 2000.
[2]   F.Laplante, Real-time imaging. Theory, techniques, and applications. IEEE Press, USA, 1996.
[3]   A.Verri, Motion Field and Flow. *IEEE Transactions on Pattern Analysis,*Vol.11, **5** (1989) 258-267
[4]   O.Starostenko, T.Kuzina, Novel algorithms for movement characteristics computing, *Proceeding of International Conference on Computer Systems CIC'2000, IPN, Mexico,* (2000) 215-225.
[5]   R. Jain, R. Kasturi, B. Schunck, Machine vision. McGraw-Hill, Singapore, 1995.
[6]   S.Smith, J.Brady, SUSAN-method, *International Journal of Computer Vision,* Vol.23, **1**, (1997) 45-78.
[7]   Halcon MVTec Software GmbH. MVTec. Germany, January, 2000, www.mvtec.com
[8]   J.Sobel, An Isotropic 3x3 Image Gradient Operator in Machine Vision, Academic Press, USA, 1990.
[9]   I.Lagendijk, Motion estimation. Information and Communication Theory group, Delft University of Technology, www-ict.its.tudelft.nl/, 1999.
[10] O. Starostenko, J.A. Neme, Automatic Complex Glyphs Recognition and Interpretation, *International Journal IEICE Transaction (Institute of Electronics, Information and Communication Engineers),* Japan, Vol. E82-A, **10** (1999) 2154-2160.

Bioinformatics for the Large-Scale Mouse Mutagenesis Project

Yoichi GONDO
Population and Quantitative Genomics Team, Bioinformatics Group
RIKEN Genomic Sciences Center, Tsurumi, Yokohama 230-0045, Japan

Abstract. A large-scale mouse mutagenesis is one of the focuses in the post genome project. A number of mutant mice should provide a useful tool to understand the gene function of the mammalian genome including human. It also becomes an animal model system for the studies of human diseases. To identify mutants, it is necessary to screen a large number of candidate mice with many objective phenotypes, since mutations arise only in a low frequency even if some mutagens are administered. In order to conduct the project in a large scale, an ORACLE-based database system has been introduced in our project at RIKEN Genomic Sciences Center. All the mice in our facilities are bar-coded and all the data, information, messages are transferred back and forth through LAN. All biometrical devices are also directly connected to the database server through LAN with a transporting protocol. Based upon these infrastructures, population and quantitative genomics now becomes feasible to identify the useful mutations from the compiled genotype- and phenotype-database. The expected mutants *in silico* will be examined *in vivo* and *vise versa*. For these process, new data-mining tools are also needed combining with biological and genetical tools. The comparative genomics between human and mouse is also another key to effectively utilize this animal model system to elucidate the mechanism of human diseases and to apply the outcomes to medical and pharmaceutical field.

1. Objectives of the Large-Scale Mouse Mutagenesis

Animal models are an indispensable tool for studying the biological function of genes that are involved in the pathogenesis of human diseases. They are also essential to develop the therapy for and diagnosis of human diseases. In particular, mutant mice have become the ideal model system because of the close similarity in the genomes between the two species and the extensive knowledge of mouse genetics that has been compiled over the last 70 years. In addition, many elaborated manipulation techniques such as transgenesis and gene targeting have recently facilitated a gene-driven analysis of the functions of cloned genes. Our objective is firstly to develop and establish a system to effectively produce a large number of mutant mice. The ultimate goal is to provide a mutant resource encompassing the whole mouse genome.

2. Overview of the Mutagenesis Protocol

One of the most potent chemical mutagen, *N*-ethyl-*N*-nitrosourea (ENU), is used to induce mainly point mutations in the mouse genome. ENU is administered intraperitoneally to male mice (G0). After recovering from an acute sterility, each G0 male will be mated to wild-type females to produce G1 offspring that are the dominant mutant candidates. The mutations are considered to arise only in paternal homolog; thus, any phenotypes identified in G1 mice are usually dominant. We examine more than 100 G1 mice every week with respect to approximately 200 screening parameters. To identify recessive phenotypes, each G1 is mated to wild-type counterpart again to produce G2 progenies. G1 and G2 are mated to produce G3 where ENU-induced mutation has a chance for the first time to become homozygotes. We again examine all the G3 thoroughly for any possible phenotypes.

At the same time, it is crucial to create a preservation system for the mutant mice as a genetic resource for both the basic and applied science communities. All G1 males in our project are preserved for such future use by freezing their sperms in liquid nitrogen tank.

3. Screening the Mutant Phenotypes

Any detectable inherited difference from the original inbred strains is systematically screened. Visible mutations like eye- and coat-color changes, and structural and behavioral anomalies are screened between the 8th and 12th week after birth. Physiological, hematological, biochemical, neurological, and anatomical analyses are also systematically conducted in order to discover any genetic variations that have been caused by mutations. X-ray 2D images are also captured to investigate any possible morphological anomalies. These are categorized as early-onset phenotype screenings. Tumorigenesis susceptibility, senescence anomaly and traits of common diseases like high blood pressure are additional screening categories for the late-onset phenotypes. The late-onset screenings continue up to 18 months after birth. The screening system, therefore, can be referred as "mouse general hospital" to detect any physiological differences from the wild-type counterparts. We also develop novel techniques to detect mouse mutant phenotypes, which should be subsequently applicable for examining the human physiological conditions. New methods and approaches for the genetic diagnosis for human diseases could be facilitated through these renovations in our mouse mutagenesis project.

4. Unique Features

A large-scale mouse mutagenesis project by using ENU has been started at GSF in German and MRC in UK since 1977. In Japan, RIKEN Genomic Sciences Center (GSC) settled Mouse Functional Genomics Research Group on 1999 to conduct a large-scale mouse mutagenesis project. At present, more than ten independent mutagenesis projects have been organized in the world. All the projects aimed to establish animal model resources for the post genome era [1]. The screening schemes are quite similar among these projects for the early-set phenotypes except RIKEN GSC seems to cover the largest numbers of early-onset screening items.

The mouse mutagenesis project at RIKEN GSC has several unique features: i) We conduct late-onset phenotype screening for tumorigenesis, senescence, neurological disorders and common diseases. Current large-scale mouse mutagenesis projects are basically conducted only for early-onset phenotypes up to 12 weeks old. We have extended the observation period up to 18 months. ii) Original Japanese inbred strains like MSM will be used, in addition to conventional laboratory inbred strains that have been developed in Europe and the United States. MSM still represents most of the wild-type phenotypes that have been lost in many laboratory inbred mouse strains. Thus, unique mutant phenotypes related to novel gene functions could be expected from MSM mutagenesis. iii) Synthetic mutant phenotypes or polygenic characteristics with epistatic interaction are analyzed. Many human diseases are known to be influenced not only by environmental factors but also by each individual's genetic background. The understanding of both gene-gene interaction and gene-environment interaction is the key to develop future medical science based upon functional genomics.

5. Infrastructures for the Operation of the Database and Project

Bar-coding. Every week, 200 of G1 mice are produced and all the males, of which sperms are stored in liquid nitrogen tanks, are maintained up to 78 weeks. Therefore, approximately 10,000 G1 mice present in our facility at any given time. G1's parental mice, G2 and G3 mice and other necessary supporting mice (e.g., the mutation mapping scheme requires a number of mice) are also maintained in the specific-pathogen free (SPF) animal facility. Our facility needs 50,000 mouse-cage capacity with a limited number of animal caretakers. A bar-coding system to identify each mouse is the key for the daily husbandry works.

LAN. The mice must be protected from fatal infection by pathogens. Material transfer is, therefore, very restricted in order to maintain the SPF grade. The data transfer also requires no or minimal writing and coping errors. To suffice these conditions, all the data input is conducted only once at the original place directly to computer and transferred into the ORACLE database at the same time. In the mouse facility area, appropriate numbers of client computer machines for the data inputs are set wherever necessary. LAN configuration and maintenance are thus extremely critical, since we minimized the use of pens and paper. Even a minute of LAN interruption could cause serious problems in our tasks.

Biometrical devices. Quite a few biometrical devices are also located at many places for the phenotype screening. Nowadays, such machines are usually controlled by a computer; however, in many cases the controlling computer is set in a stand-alone manner. To transport the data from such stand-alone device effectively, we constructed a data-transfer protocol for each biometrical device when necessary. Occasionally, biometrical device itself needs to be modified, or even newly developed to suffice our purpose. Such renovation of biometrics should contribute a development of new diagnostic systems for human.

Management tools. Daily works and tasks are diversified to many different levels and kinds in the large-scale mouse mutagenesis. Animal caretakers, technical staffs, researchers and other personnel must communicate each other to coordinate the tasks. To organize daily, weekly and long-term plans of this complicated project, some "schedule planner" is necessary. An expert system for the laboratory management and planning is ideal for this purpose. At the same time, the control and maintenance of the facility are indispensable. The supplies of electric power, water, gas, etc. must be concrete and their backup systems are periodically checked. The temperature, humidity, security locks and SPF grade are also monitored all the time and the logs are automatically recorded to the database server through LAN. When an emergency or trouble should occur, it must be able to trigger the alarm and safety systems including the notification to respective staffs and offices.

Genetics. In order to produce 200 G1 mice every week, the mouse husbandry must be well planned way before hand, because one single G1 goes through various examinations for 18 months. When the inheritance test is conducted for the late-onset phenotypes, another 18 months are necessary. It also takes about a year from the ENU injection to G0 males to obtain the G3 offspring. For the mapping and molecular-cloning of identified mutations, additional years will be required. Mutant candidates must be carefully examined to eliminate false-positives and –negatives as many as possible at the very early stage of screenings. Otherwise, the manpower, spaces, and budgets would soon exceed far more than the capacities or the candidates would turn out to be all false after the long screening and examination period. Key genetical setups are: 1) which inbred strain(s) to be used; for the ENU injection, for the G1 production and G2 production, and for the mapping, etc., and 2) effective inheritance test with respect to environmental effects, degree of dominance, penetrance, heterosis, viability, sterility, meiotic drive, segregation distortion, genetic background, genetic drift, and polygenic effects, etc. Obviously the setting of the negative control data is essential for the entire project.

6. Data-mining: Population and Quantitative Genomics

As described above, ENU-induced mutant mice carry single-base substitutions in their genomic DNA. ENU induces many single-base substitutions; therefore, the mutant mice have the causative point mutation for the mutant phenotype as well as many other silent mutations in the genome. In this sense, ENU-induced mutant mice are very similar to the situation of single nucleotide polymorphisms (SNPs) and will be a good animal model for the study of the human SNP project. By using population and quantitative genetics, it is feasible to associate SNPs to genetic traits of diseases and individual's physiological conditions. However, it seems extremely difficult to conclude which SNP(s) is responsible to and causative of the trait in human. The animal model will provide a good experimental tool to investigate the SNP function to the biological function. In the large-scale mouse mutagenesis, biological functions are compiled in a balk of phenotype database. At the same time, SNPs are assorted in the genotype database as described below. We refer this Genome-wide genetical approach with statistics and biometrics to *population and quantitative genomics.*

Phenotype-driven approach. Mutations are firstly detected as a phenotype(s) different from that of the wild-type mice in the large-scale mouse mutagenesis project. Then, the phenotype leads to identify and clone the causative gene(s) and SNP(s). Based on the phenotype, the genetic mapping is conducted first to locate the chromosomal position of the mutation. Then, positional cloning identifies the gene and complete DNA sequencing determines the causative SNP(s). In this approach, the majority of mutations will be dominant phenotypes and the pace of genotype database construction will be slow.

Gene-driven approach. We preserve the sperm samples from all the G1 male independently in the liquid nitrogen tank as described above. Their genomic DNAs also become available within 18 months after their late-onset phenotype screening. From each G1's genomic DNA, target genes are amplified by PCR and directly sequenced to identify all the SNPs in the target regions. After finding SNPs, we investigate *in silico* which SNP could cause biological change in the mouse. Candidate SNPs are examined *in vivo* by retrieving the sperm from the liquid nitrogen and producing the viable mouse. In this gene-driven approach, recessive phenotypes are easily examined because which SNP to be made to homozygotes is determined before hand.

Data mining and concluding remarks. From the phenotype database and genotype database, mutant candidates must be extracted effectively. At the same time, false positives must be carefully eliminated. The expected mutants *in silico* will be examined *in vivo* and *vise versa*. The genotypes are nowadays concretely determined at the DNA sequence level although it is still laborious and expensive for genome-wide determination. On the other hand, the phenotypes are a part of whole biology and physiology and their measurement occasionally very subjective. Therefore, it is necessary to establish the biometrics of the phenotype assessment so that each parameter values are reproducible and reliable. By integrating these fundamental databases with population and quantitative genomics, a knowledge-based database for data mining becomes plausible. For these processes, new data-mining tools are essential combining to biological and genetical tools. The comparative genomics between human and mouse is also another key to effectively utilize this animal model system to elucidate the mechanism of human diseases and to apply the outcomes to medical and pharmaceutical field. It requires a high throughput system that is able to manipulate the huge complicated and diversified database.

References

[1] J.H. Nadeau *et al.*, The International Mouse Mutagenesis Consortium: Annotating Genome Sequences with Biological Functions in Mice, *Science* **291** (2001) 1251-1255.

The output connectivity of randomly constructed genetic networks affects their decision-making capabilities

Chikoo Oosawa[†, §] and Michael A. Savageau[†,¶]

† *Department of Microbiology & Immunology, The University of Michigan Medical School, 5641 Medical Science Building II, Ann Arbor, Michigan 48109-0620, USA*
§ Present address : *Bioinformatics Group, Computational Genomics Team, RIKEN Genomic Sciences Center (GSC), 1-7-22 Suehiro-cho, Tsurumi-ku, Yokohama, Kanagawa, 230-0045, JAPAN*
¶ Corresponding author

Abstract

We have use random Boolean networks as decision-making networks to model regulatory gene networks and to examine their capabilities by comparing their dynamical properties. Four different topologies, which are characterized by different rank distributions of output connections that vary from fairly uniform to highly skewed, are introduced into the standard model. The intermediate topologies include the topology of the standard model, which is included for purposes of comparison, and a topology with a power-law rank distribution, which is based on recent data for the regulatory gene network of the bacterium *Escherichia coli*. Networks with the more uniform rank distributions exhibit longer lengths of attractors and larger numbers of attractors, and networks with the more skewed rank distributions have complementary properties. Either the most uniform or the most skewed rank distributions have disadvantageous properties as decision-making networks. The intermediate rank distribution exhibited by the regulatory gene network of *E. coli* avoids these disadvantages.

1. Introduction

The standard model for random Boolean network (RBN)[1] has a topological structure of a regulatory gene network can be depicted as a directed graph. Each *node* corresponds to a transcriptional unit that includes a regulatory gene. Each *arc* represents an interaction between transcriptional units. The origin of an arc is the output of one transcriptional unit and its destination is the input of a second transcriptional unit. The state of gene expression is either ON or OFF. The number of inputs to each node is exactly K, where K is called the *input connectivity*. The origin of an input is chosen at random from other nodes. Each node is assigned a Boolean function selected at random with equal probability.

The dynamics of the network described above are determined by the following equation

$$X_i(t+1) = B_i(X(t)) \qquad\qquad i = 1,2,\ldots, N$$

where $X_i(t)$ is the binary state, either 0 or 1, of node i at time t, $B_i(\bullet)$ is the Boolean function used to update the state of node i, and $X(t)$ is a binary vector that gives the states of the N nodes in the network. An initial vector $X(0)$ is assigned and successive states of the nodes are updated synchronously.

The dynamical behavior of these networks is represented by the time series of states, which corresponds to the time course of gene expression. The time course of gene expression follows a transient phase from an initial state until a periodic pattern, called an attractor, is eventually established. The length of the transient phase and the length of the attractor are determined by the number of time steps before and after entering the attractor respectively. Different initial states for the same network may lead to the same or different attractors. A large number of different initial states are explored in an effort to determine the number of distinct attractors for a given network. Networks with different numbers of nodes are often used to investigate the dependence of dynamical properties on network size, and many networks are generated and analyzed under fixed K and N to identify the statistical regularities [2].

2. Methods

We define the topology of the network in terms of the rank distribution of input and output connections. Thieffry et al. [3,4] examined the number, type and location of regulatory binding sites in the promoter regions of *E. coli* and concluded that the average input connectivity is low, between one and two. On the bases of these studies we have fixed the input connectivity for all our models at K=2.

We have considered four different network topologies are characterized by their output connectivity. Once the input connectivity K is determined, the total number of connections in a network is also fixed automatically. If a network contains N nodes and each node has K input connections, the total number of connections in the network becomes K*N. These connections are assigned by different methods to the output of each node and the resulting topologies are termed Uniform, Exponential, *E. coli* and Extreme.

2.1. Uniform topology

For this topology all nodes have exactly two output connections. Hence, the rank distribution for the number of output connections is uniform (figure 1). The network in this case is constructed by selecting an unoccupied input at random and connecting it to an unoccupied output selected at random. This procedure is followed until all 2*N connections have been assigned.

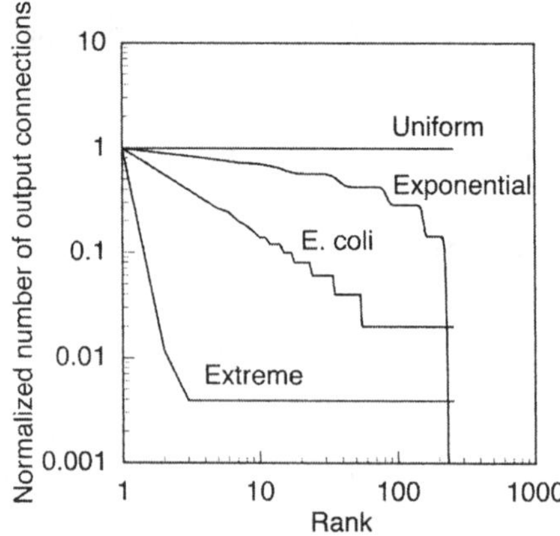

Figure 1. Normalized rank distribution of output connections in random Boolean networks with four different topologies at N=256. The vertical axis corresponds the number of binding sites of a regulatory protein, the horizontal axis means that regulatory proteins with smaller ranks have larger numbers of binding sites.

2.2. Exponential topology

This is the topology of the original RBN, as described in the previous section. The rank distribution of output connections is exponential-like (figure 1). The network in this case is constructed by selecting an unoccupied input at random and connecting it to an output selected at random.

2.3. E. coli topology

We have determined the *E. coli* topology from data in three recent publications [5-7]. We tabulated the number of regulated operons with known binding sites for each known regulator and rank ordered the regulators according to the number of operons with the corresponding target site. The results when plotted give a power-law rank distribution (figure 1). The network in this case is constructed in two stages. First, the *E. coli* rank distribution of output connections is determined with the minimum number of outputs being one. Each node is allotted a number of outputs at random according to this distribution. Second, the connections are made by selecting an unoccupied input at random and connecting it to an unoccupied output selected at random.

2.4. Extreme topology

The opposite of the Uniform rank distribution, for which no node has more connections than another, is the Extreme rank distribution, for which one node has connections to all other nodes in the network and the remaining nodes have only one or a few connections. More specifically, one node has (N-1) output connections, another has three and the rest have only one (figure 1).

2.5. Ensembles of networks

We examined 20 different ensembles of networks corresponding to five different sizes (N=16, 32, 64, 128 and 256) and four different topologies. Each ensemble consisted of 1000 networks randomly constructed as described above. The dynamic behavior of each network was followed from 2000 different initial states that were selected uniformly from among the 2^N possible states.

3. Results & discussion

3.1. RBN as discrete dynamical systems

Our results for the alternative topologies show that smaller numbers of attractors and shorter attractor lengths are characteristic of the more skewed topologies. The numbers of attractors (figure 2) and the attractor lengths, two properties that increase with network size, have very similar distributions that differ markedly from the distribution of transient lengths. They clearly demonstrate that topology is another critical factor influencing the dynamical behavior of an RBN.

3.2. Regulatory genetic networks as decision-making networks

A striking property of regulatory gene networks in cells is the ability to switch expression of specific genes ON or OFF according to circumstances. Since the resulting patterns of gene expression can be viewed as the outcome of a set of decisions executed by the network, the regulatory gene network of an organism must be capable of making decisions that produce appropriate patterns of gene expression in response to different inputs from the external environment.

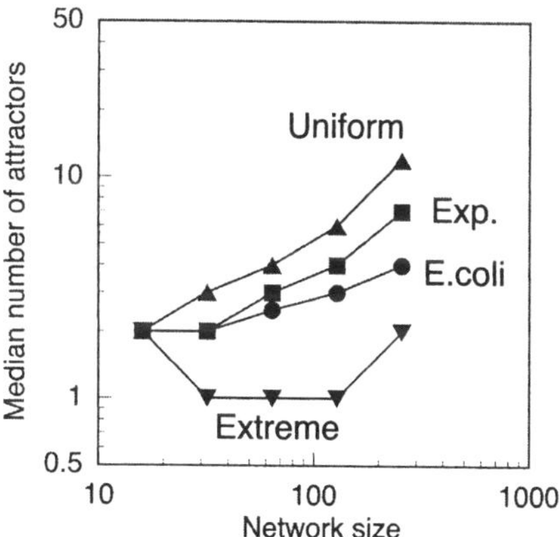

Figure 2. Median number of attractors in random Boolean networks with four different topologies and various network sizes.

The regulatory gene network of an organism can be viewed as making decisions that correspond to different attractors. For example, the regulatory gene network of bacteriophage lambda makes decisions corresponding to two attractors for lytic and lysogenic growth. The decision-making process is based on interactions among regulatory elements of the phage and the external environment provided by its host.

If an organism is to survive in many different environmental conditions, then its regulatory gene network must be capable of making a large number of decisions that correspond to a large number of appropriate attractors. Moreover, these decisions must be made quickly in order to successfully establish the organism's response to stresses such as heat, radiation and starvation. Thus, we might expect organisms to evolve toward large numbers of attractors with short attractor lengths. This appears to be the case for the *E. coli* topology, which has more patterns of decisions (more different attractors) than the Extreme topology and shorter times to establish the patterns (shorter attractor lengths) than the Uniform topology.

3.3. Extension of the results

The *E. coli* topology is the first biologically-relevant distribution of connections to be to incorporate into a random Boolean network. This distribution is based only on experimental data [5,7] and theoretical predictions of regulator-regulon relationships for *E. coli* [6]. However, the complete genome sequence has been determined more than 40 other microbial species (http://www.tigr.org/tdb/mdb/mdbcomplete.html), so we can expect that data on the full connectivity of regulatory gene networks for several organisms will soon become available. These data will permit a more detailed exploration of the generality of the results presented here.

References

[1]   S.A. Kauffman, Metabolic stability and epistasis in randomly constructed genetic nets, *J. Theoret. Biol.* **22** (1969) 437-467.

[2]   S.A. Kauffman, Origin of Order, Oxford University Press, Oxford, 1993.

[3]   D. Thieffry, A.M. Huerta, E. Pérez-Rueda, J. Collado-Vides, From specific gene regulation to genomic networks: A global analysis of transcriptional regulation in *Escherichia coli*, *BioEssays* **20** (1998) 433-440.

[4]   D. Thieffry, H. Salgado, A.M. Huerta, J. Collado-Vides, Prediction of transcriptional regulatory sites in the complete genome sequence of *Escherichia coli* K-12, *BioInformatics* **14** (1998) 391-400.

[5]   J. Collado-Vides, B. Magasanik, J.D. Gralla, Control site location and transcriptional regulation in *Escherichia coli*, *Microbiological Review* **55** (1991) 371-394.

[6]   J. Otsuka, H. Watanabe, K.T. Mori, Evolution of transcription regulation system through promiscuous coupling of regulatory proteins with operons; Structure from protein sequence similarities, *J. Theoret. Biol.* **178** (1996) 183-204.

[7]   A.M. Huerta, H. Salgado, D. Thieffry, J. Collado-Vides, RegulonDB: A database on transcriptional regulon in *Escherichia coli*, *Nucle. Acids Res.* **26** (1998) 55-59.

KES '01
N. Baba et al. (Eds.)
IOS Press, 2001

A Framework for Quick-and-Pinpoint Data Mining and its Application to Heterogeneous Genome Databases

Kenji Satou[1], Yoshiki Fuseda[2], Akihiko Konagaya[1], and Toshihisa Takagi[3]

1 Japan Advanced Institute of Science and Technology, 1-1 Asahidai, Tatsunokuchi, Ishikawa 923-1292, Japan

2 Alphanet.Inc., 1-11-8, Kitano BLDG., Ooyodo-minami, Kita, Osaka, Japan

3 Human Genome Center, Institute of Medical Science, The University of Tokyo, 4-6-1 Shiroganedai, Minato-ku, Tokyo 108-8639, Japan

Abstract. This paper proposes a new framework for knowledge discovery from large and heterogeneous genome databases. By regarding a set of entries as user's interest, the system drastically cuts off irrelevant items and transactions to allow "quick-and-pinpoint" data mining. Two examples of practical applications showed that the framework is effective for knowledge discovery from genome databases via the WWW.

1. Introduction

Today, major sites of genome database services, including GenomeNet [1] in Japan, are gathering and updating over 25 millions of heterogeneous database entries. Furthermore, most of them equip retrieval functions and analysis tools for processing such data. This situation is similar to some sort of data warehouses since heterogeneous and large amount of data are gathered for trans-database searches and analyses. On the other hand, there are several application studies, which use some kind of data mining algorithms for finding useful knowledge from genome databases. Data mining is one of the most actively studied area of computer science. Also in the field of bioinformatics, data mining against DNA, protein, and disease data are highly noticed as a promising approach to scientific discovery. However, since most of data mining algorithms are combinatorial (it means high computational complexity), in typical cases, they use only small amount of static subset of genome database. There is one more reason of it, that is, data mining technology still remains the area of computer science (it means biologists could not utilize data mining tools without aid of a computer scientist or a skilful programmer).

For aiding knowledge discovery from heterogeneous genome databases, first we propose a framework, which we call ``quick-and-pinpoint data mining''. Though its core algorithm is common and traditional (i.e. association rule finding), by assuming an extra input, which represents a user's interest, we can rapidly extract a desired subset of association rules from huge amount of data. Then, two examples are shown for demonstrating the usefulness of the framework.

2. Limiting the discovered knowledge to the one desired

Association rule finding was first proposed by Agrawal *et al.* in 1993 [2,3]. Roughly speaking, it accepts a table of bit vectors and searches sets of items (columns) frequently hold together in many transactions (rows). Since its flexibility and adaptability, the technique has been applied to various real problems with excellent results. In bioinformatics, for instance, we tried find correlation among the sequential, structural, and functional features of proteins using the technique [4,5]. However, since association rule finding is combinatorial approach, it has essential limit in scalability. Therefore, in general, a small amount of static data set, which was specially selected and tailored for one-off data mining job, is used as a throwaway input.

To solve the problem above, we focused on user's interest. Suppose that there is a frequently updated table with millions of items and transactions, it is all too audacious to find all the association rules from the table. However, if there is a small subset of transactions in which a user is interested, we can limit the discovered rules to the ones related to the transactions by selecting the items and transactions in the huge table. More precise description of our framework of quick and pinpoint data mining is as follows.

Input) Usual bit vector table for association rule finding. In addition, a subset of transactions in the table is given. Here we call them ``target transactions", which represent user's interest.

Step1) Select all the items hold in target transactions. Here we call them ``candidate items".

Step2) Select all the transactions in which at least one candidate item holds. Here we call them ``potential transactions". Of course, target transactions are included in them.

Step3) Quarry a smaller bit vector table from input by using candidate items and potential transactions. Additionally insert a special item, which holds only for target transactions. Here we call it ``target item".

Step4) Using the table generated in Step3, perform association rule finding.

Step5) From the discovered association rules, discard all the rules which do not include the target item.

Output) Return the survived association rules to a user.

Since in association rule finding, highly sparse bit vector table is processed in general, our framework can drastically reduce the cost of computation.

3. Application Examples

Today, in the field of genome and protein analysis, it is inevitable for researchers to access to and search public databases on a daily basis. Then, in many situations, they obtain a set of entries in a database. For instance, keyword, homology, and motif search return a set of entries as a result of search. By regarding the entries as a user's interest, we can position the framework of quick-and-pinpoint data mining to postprocessing of search results. In other words, we can summarize a set of entries by discovering common and specific features to them.

On the other hand, what type of features should be used for knowledge discovery from large and heterogeneous genome databases? In this study, we used LinkDB[6] which stores most of the direct and indirect cross-reference information among 25 millions of database entries in GenomeNet. Since the cross-reference information in LinkDB is represented as binary relation between two entries, it can be regarded as a bit vector table with 25 millions of items and transactions. Thus, we developed a system for discovering the cross-references, which are common and specific to target entries, from LinkDB. Then, we connected the system to two existing search and analysis services.

3.1 Postprocessing the Result of Keyword Search

STAG[7] is a simple search service constructed on GenomeNet. It provides trans-database and full-text search against all the genome databases in GenomeNet. After the keyword search by STAG, a user can jump to an input form for data mining by following the hyperlink. If there are cross-references common and specific to a set of entries (i.e. a result of keyword search), the system returns them as a result of quick-and-pinpoint data mining. Fig.1 shows that, starting from keyword search by BSE (bovine spongiform encephalopathy), finally a user could reach to the knowledge that 41 entries out of 45 target entries in SWISS-PROT had cross-references to two entries in PROSITE (PS00291 and PS00706), which were motifs of prion proteins.

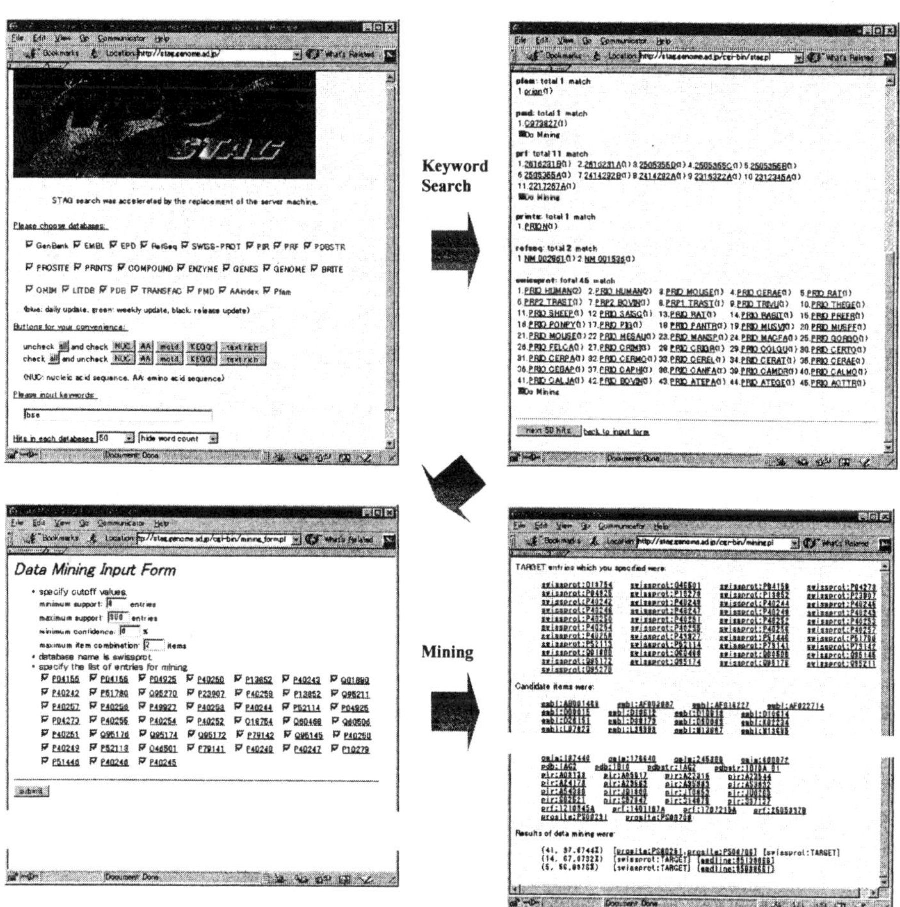

Fig.1 Mining from a result of keyword search

3.2 Postprocessing the Result of Structural Similarity Search

WebPACADE is an integrated system for structural similarity search[8] and visualization of proteins. In Fig.2, first a user searched for substructures similar to a specific substructure in a protein 5TNC (Troponin-C), where 5TNC is an entry in PDB. Then, by the same operation as the one in the previous subsection, the user found that 23 entries out of 24 target entries in PDB (which have similar substructures) were related to an entry in PROSITE (PS00018), which was a motif of calcium binding proteins (EF-HAND motif).

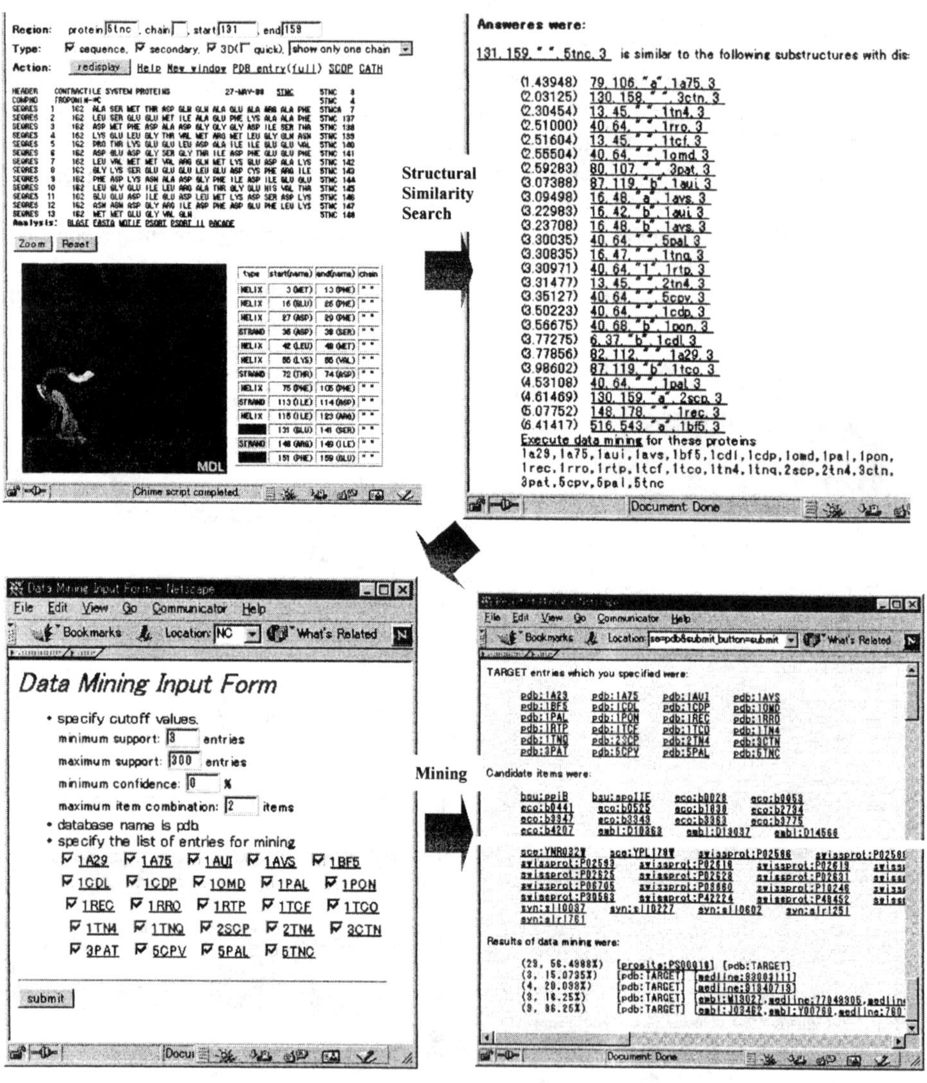

Fig.2 Mining from a result of structural similarity search

4. Concluding Remarks

In this paper, we proposed a new framework for quick discovery from large amount of databases. With two application examples, effectiveness and practicality of the framework were demonstrated. Since the framework itself is not dependent on genome databases the following future work may arise:

- Connecting the quick-and-pinpoint data mining system to the other search and analysis services (e.g. homology and motif searches).

- Applying the framework to usual WWW search engines (cross-references in genome databases and hyperlinks in WWW pages are essentially the same).

Acknowledgements

This work was supported by Grant-in-Aid for Scientific Research on Priority Areas (C) "Genome Information Science" from the Ministry of Education, Culture, Sports, Science, and Technology of Japan.

Rererences

[1] http://www.genome.ad.jp/

[2] Agrawal, R., Imielinski, T., and Swami, A.: Mining Association Rules between Sets of Items in Large Databases, Proc. of ACM SIGMOD, pp.207-216, 1993.

[3] Agrawal, R. and Srikant, R.: Fast Algorithms for Mining Association Rules, Proc. of VLDB, pp.487-499, 1994.

[4] Satou,K., Shibayama,G., Ono,T., Yamamura,Y., Furuichi,E., Kuhara,S., and Takagi,T.: Finding Association Rules on Heterogeneous Genome Data, Proc. of the Pacific Symposium on Biocomputing '97 (PSB'97), pp.397-408, 1997.

[5] Satou,K., Ono,T., Yamamura,Y., Furuichi,E., Kuhara,S., and Takagi,T.: Extraction of Substructures of Proteins Essential to their Biological Functions by a Data Mining Technique, Proc. of the Fifth International Conference on Intelligent Systems for Molecular Biology (ISMB'97), pp.254-257, 1997.

[6] Fujibuchi,W., Goto,S., Migimatsu,H., Uchiyama,I., Ogiwara,A., Akiyama,Y., and Kanehisa,M.: DBGET/LinkDB: an Integrated Database Retrieval System, Pacific Symposium on Biocomputing'98 (PSB'98), pp.683-694, 1998.

[7] http://stag.genome.ad.jp/

[8] Tsukamoto,Y., Takiguchi,K., Satou,K., Furuichi,E., Takagi,T., and Kuhara,S.: Application of a deductive database system to search for topological and similar three-dimensional structures in protein, Computer Applications in the Biosciences (CABIOS), Vol.13, No.2, pp.183-190, Apr. 1997.

KES '01
N. Baba et al. (Eds.)
IOS Press, 2001

Determining Protein Structures
by A Real-Coded Genetic Algorithm

Isao Ono*, Hiroshi Fujiki*, Norihiko Ono* and Shin-ichi Tate**
The University of Tokushima, 2-1 Minamijosanjima, Tokushima, 770-8506, Japan. *
Japan Advance Institute of Science and Technology, 1-1 Asahidai, Tatsunokuchi,Ishikawa,
932-1292, Japan. **

Abstract. Protein structure determination is one of the most important problems in molecular biology. Nuclear Magnetic Resonance (NMR) spectroscopy and X-Ray crystallography currently are the only techniques capable of determining the three-dimensional structures of proteins. In determining protein structures by using NMR spectroscopy, Nuclear Overhauser Effect (NOE) signal assignment is the most laborious and time-consuming process. In this paper, we propose a new automatic assignment method of NOE signals based on a real-coded genetic algorithm and examine its effectiveness by applying it to determining the structure of α-helix, which is a well-known common sub-structure of proteins.

1. Introduction

Protein structure determination is one of the most important problems in molecular biology. Nuclear Magnetic Resonance (NMR) spectroscopy and X-Ray crystallography currently are the only techniques capable of determining the three dimensional structures of proteins at atomic resolution.

The basis of current techniques of determining three-dimensional structures of proteins by NMR spectroscopy was established by Wüthrich in 1980s [7]. The process of determining protein structures by NMR spectroscopy mainly consists of 1) sequential assignments and 2) NOE (Nuclear Overhauser Effect) assignments and structure determination. In the phase of sequential assignments, all chemical shifts of H, N and C atoms in a protein are identified by analysing NMR spectra such as the HNCA spectrum and the HCACO spectrum [2]. A chemical shift is unique to the corresponding atom. In the phase of NOE assignments and structure determination, the three dimensional structure of the protein is determined by assigning NOE signals to corresponding H-H pairs. NOE signals are useful information to determine a three dimensional structure because, if an NOE signal is observed from a H -H pair, the distance of the H-H pair is in the range between approximately 5 Å and 10 Å.

The current technique of determining three-dimensional structures of proteins by NMR spectroscopy has a serious problem that the process of NOE assignments and structure determination is very laborious and time-consuming. A human expert has to gradually assign NOE signals and construct a three-dimensional structure by trial and error, using the domain knowledge on protein structures. This is because many NOE signals overlap each other and cannot be assigned to H atoms uniquely. In many cases, it takes several months to determine a single protein structure.

There have been proposed some methods for automatically assigning NOE signals and determining protein structures [4, 5]. However, these methods have not succeeded in automating the process of NOE assignments and structure determination completely.

In this paper, we propose a new method that determines a protein structure by searching a protein structure that can predict as many observed NOE signals as possible by a real-coded GA and examine its effectiveness through some experiments.

2. NOE Assignments in Protein Structure Determination by NMR Spectroscopy

Protein structure determination by NMR spectroscopy is based on minimizing the NOE violation and the van der Waals violation. The NOE violation is a degree of violating distance constraints between two H atoms obtained by assigning NOE signals to H-H pairs. The van der Waals violation is a degree of overlaps of atoms in the protein.

If an NOE signal is observed from a pair of H atoms, the distance of the two H atoms is in the range between approximately 5 Å and 10 Å. An NOE signal is given by a combination of three chemical shifts of an N atom, an NH atom that is the H atom directly connecting to the N atom and an arbitrary H atom. As shown in Fig. 1, an NOE signal is assigned to the combination of N, NH and H atoms whose chemical shifts, which are identified in the process of sequential assignments in advance, coincide with the NOE signal.

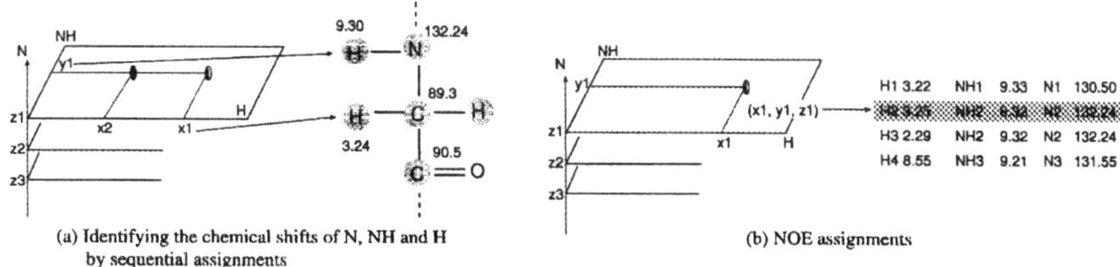

Fig. 1: Sequential assignments and NOE assignments

However, practically, many NOE signals cannot be assigned to unique combinations of N, NH and H atoms because many chemical shifts overlap, which is caused by measurement errors and various noises in measuring NMR signals. In many cases, the number of alternatives of the combinations of H, NH and H atoms to which an NOE signal is assigned is about ten. As the size of protein increases and, as the result, the number of NOE signals to be assigned increases, the number of the combinations of NOE assignments increases exponentially. This makes NOE assignments and protein structure determination very difficult.

Most of experts of determining protein structures by NMR spectroscopy take the following procedures by using some software tools such as DYANA [3] and X-PLOR [1]:

1. In order to obtain an NOE violation, assign some NOE signals which are easy to distinguish because they are in regions where the distribution of chemical shifts is sparse. Find some initial structures by minimizing the NOE violation and the van der Waals violation by Simulated Annealing (SA).

2. Assign an NOE signal to the combination of N, NH and H whose chemical shifts are close to those of the NOE signal and in which the distance between the NH and H atoms is close in the structure found by SA, referring various typical protein structures that have been already determined in databases. Repeat this processes multiple times to obtain a new NOE violation.

3. Find some structures by minimizing the NOE violation and the van der Waals violation by SA. Here, the processes of optimizing structures are performed independently 20 times.

4. Repeat step 2 and step 3 until almost all NOE signals are assigned, van der Waals violation is almost zero and the 20 structures found by SA are almost same. If more than 5 % of NOE signals that cannot be assigned remain or a strange structure from the

viewpoint of molecular biology is obtained, NOE assignments are wrong and do over again from the beginning.

As described above, in the current method of determining protein structures, a human has to construct a protein structure gradually, assigning NOE signals by trail an error. The human has to be an expert because expertise and experience in protein structure determination are required to distinguish strange structures from the view point of molecular biology and to assign NOE signals appropriately. It takes more than several months to determine a single protein structure.

3. A Genetic Algorithm for Determining Protein Structures

In this paper, in order to solve the problems of the current method of determining protein structure, we propose a new method based on a real-coded genetic algorithm (GA). This method searches a three-dimensional structure that can predict as many NOE signals as possible. The basic idea is as follows:

- A three-dimensional structure is generated by a real-coded GA.
- The NOE signals of the structure generated by the GA are predicted.
- The observed NOE signals are assigned to the predicted NOE signals.
- The structure is evaluated from the viewpoint of the NOE violation and the van der Waals violation.

3.1 Designing A Real-Coded GA

[Representation] We employ a real-number vector consisting of dihedral angles as a representation.

[Crossover Operator] We employ the Uniform Crossover (UX) as a crossover operator as shown in Fig. 2 because hundreds of variables have to be optimised and interactions of variables would be weak in this problem.

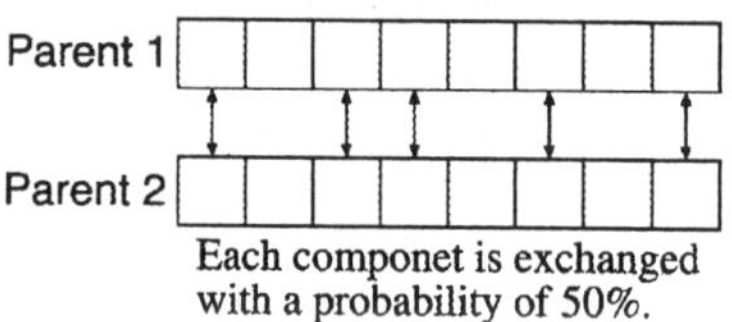

Fig. 2: Uniform Crossover

[Mutation Operator] We employ a mutation operator that adds a uniformly random number with the range $[-\pi/180, +\pi/180]$ to the variables which are chosen with a mutation probability of 0.01.

[Generation-alternation Model] We employ the Minimal Generation Gap (MGG) [6] as shown in Fig. 3.

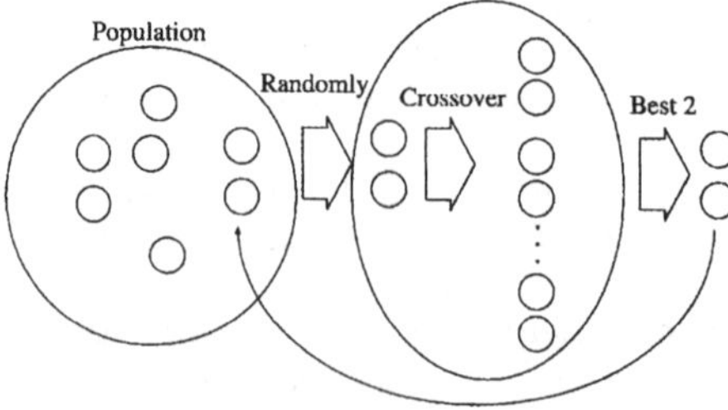

Fig. 3: Minimal Generation Gap (MGG) [6]

3.2 Designing An Evaluation Function

The procedure of evaluating a structure generated by the GA is as follows:

1. Make a table of the predicted NOE signals as follows:
   (a) Calculate the distance of all pairs of NH and H in the structure.
   (b) Add the pairs whose distance are less than 8.0 Å to the table.
2. Apply the following procedures to all observed NOE signals:
   (a) Choose the predicted NOE signals which meets the following conditions:
   $$|p_N - o_N| < T_N, |p_{NH} - o_{NH}| < T_{NH}, |p_H - o_H| < T_H$$
   where p_N, p_{NH} and p_H are the chemical shifts of N, NH and H of the predicted NOE signal respectively, o_N, o_{NH} and o_{NH} are the chemical shifts of N, NH and H of the observed NOE signal respectively, and T_N, T_{NH} and T_H are the thresholds of N, NH and H given by a user in advance respectively.
   (b) Sort the predicted NOE signals chosen in (a) by d_{shift} so that the predicted NOE signal with the smallest d_{shift} is placed at the top and choose three predicted NOE signals from the top, where d_{shift} is given by
   $$d_{shift} = (4|p_H - o_H|/T_H + |p_{NH} - o_{NH}|/T_{NH} + |p_N - o_N|/T_N)/6.$$
   (c) Choose the predicted NOE signal with the smallest v_{NOE} from the three NOE signals chosen in (b) and assign the observed NOE signal to the combination of N, HH and H corresponding to the chosen predicted NOE signal. Delete the chosen predicted NOE signal from the table. v_{NOE} is given by
   $$v_{NOE} = \begin{cases} 1/(d - d_{min} + 1) & if\, d < d_{min} \\ 0 & if\, d_{min} \le d \le d_{max} \\ 1/(d - d_{max} - 1) & if\, d_{max} < d \end{cases}$$
   where d is the distance of the NH-H pair of the predicted NOE signal and d_{min} and d_{max} are the minimum and maximum of the distance calculated from the observed NOE signal, respectively.
3. Calculate the following evaluation value:
   $$E_{structure} = V_{NOE} + V_{vdw},$$
   $$V_{NOE} = \sum v_{NOE}, \quad V_{vdw} = \sum_i \sum_j (d(a_i, a_j) - 0.8r_i - 0.8r_j),$$

where $d(a_i, a_j)$ is the distance of the centers of atoms, a_i and a_j, and r_i is the radius of atom a_i.

4. Experiments

We made an experiment in order to examine whether the proposed method can determine an α-helix structure which is a part of a protein called "hmg2b". In this experiment, the number of observed NOE signals is 173 and the number of dihedral angles to be determined is 76. We set the population size to 500. Five independent trials were done.

Figure 4 shows the transitions of the best evaluation value in the population in five trials. Figure 5 shows the structures determined by the proposed method in five trials. Figure 6 shows a structure determined by a human expert using the conventional method described in section 2.

Figure 5 shows that the proposed method succeeded in determining the α-helix structures in all trials. In Fig. 5, the structures of the parts near both ends are different from each other because the number of NOE signals included in the parts near both ends is less than other parts. The quality of the structures determined by the proposed method in Fig. 5 is slightly worse than that of the structure determined by a human expert in Fig. 6. We believe that this is because the proposed method does not take account of hydrogen-bond energies and information on the side-chain torsion angles that a human expert used to determine the structure in Fig. 6.

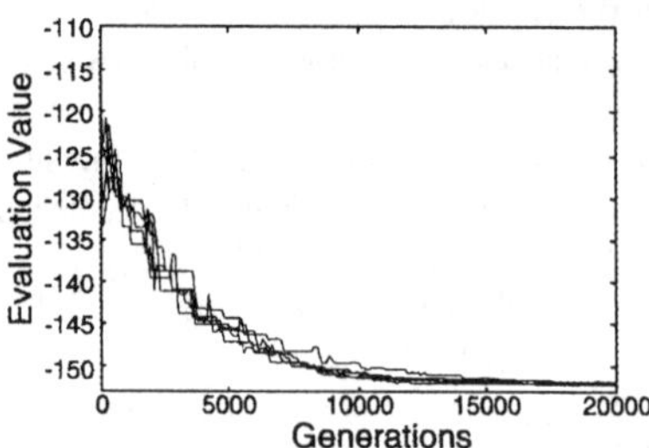

Fig. 4: Online curves

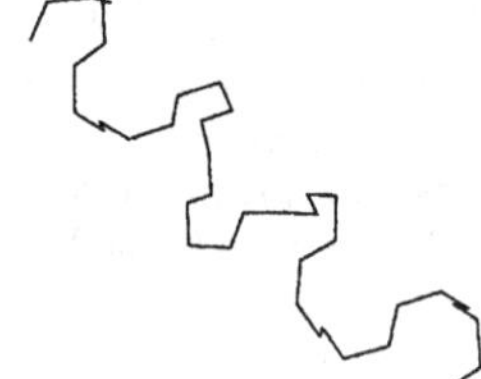

Fig.5: α-helix structure determined
by the proposed method

Fig. 6: α-helix structures determined
by an expert

5. Conclusion

In this paper, we proposed a new method based on a real-coded GA that searches a three dimensional structure which can predict as many NOE signals as possible. We applied the proposed method to the determination of the α-helix structure and succeeded in obtaining α-helix structures in all trials. It took about five hours for the proposed method to determine an α-helix structure on a single computer with 1GHz AMD Athlon processor. This suggests that the proposed method can greatly reduce time required to determine three-dimensional structures of proteins by NMR spectroscopy.

In future work, we have a plan to improve the evaluation function. We also would like to apply the proposed method to larger structures in order to show the effectiveness of the proposed method.

References

[1] Brünger, A. T.: X-PLOR version 3.1. A system for X-ray crystallography and NMR, Yale University Press, New Haven (1992).

[2] Evans, J. N. S.: Biomolecular NMR Spectroscopy, Oxford University Press, New York (1995).

[3] Güntert, P., Braun, W. & Wüthrich, K.: Efficient computation of three-dimensional protein structures in solution from nuclear magnetic resonance data using the program DIANA and the supporting programs CALIBA, HABAS and GLOMSA. J. Mol. Biol. 217, pp.517-530 (1991).

[4] Mumenthaler, C., Güntert, P., Braun, W. and Wüthrich, K.: Automated combined assignment of NOESY spectra and three-dimensional protein structure determination, J. Biomol. NMR, 10, pp.351-362 (1997).

[5] Nilges, M. and O'Donoghue, S. I.: Ambiguous NOEs and automated NOE assignment, Progress in Nuclear Magnetic Resonance Spectroscopy, 32, pp.107-139 (1998).

[6] Sato, H., Yamamura, M. and Kobayashi, S.: Minimal Generation Gap Model for GAs Considering Both Exploration and Exploitation, Proc. IIZUKA'96, pp.494-497 (1996).

[7] Wüthrich, K.: NMR of Proteins and Nucleic Acids, John Wiley & Sons, New York (1986).

KES '01
N. Baba et al. (Eds.)
IOS Press, 2001

An Approach for High Speed Homology Search with FPGAs

Yoshiki YAMAGUCHI [1], Tsutomu MARUYAMA [1] and Akihiko KONAGAYA [2,3]

[1] *Institute of Engineering Mechanics and Systems, University of Tsukuba,*
1-1-1 Ten-ou-dai Tsukuba Ibaraki 305-8573, Japan
[2] *Japan Advanced Institute of Science and Technology,*
1-1 Asahidai Tatsunokuchi Ishikawa 923-1292, Japan
[3] *Japan Riken Genomic Sciences Center,*
1-7-22 Suehiro Tsurumi Yokohama Kanagawa 230-0045, Japan
Email: [1] *{yoshiki, maruyama}@darwin.esys.tsukuba.ac.jp,* [2,3] *kona@jaist.ac.jp*

Abstract. In this paper, we show that we can achieve high performance in homology search by only adding one off-the-shelf PCI board with one Field Programmable Gate Array (FPGA) to a Pentium based computer system in use. By using off-the-shelf boards, we can easily obtain latest FPGAs at low costs, which is very important because the performance is almost proportional to the size of the FPGA, and FPGAs are becoming larger and larger following Moore's law. The performance can be furthermore accelerated by using more number of FPGA boards. In our approach, the search is divided into two phases, and different circuits are configured on the FPGA in each phase, in order to make up limited hardware resources of the off-the-shelf boards.

The performance is almost comparable with small to middle class dedicated hardware systems when we use a board with one of the latest FPGAs. The computation time is almost proportional to the size of the query sequence. The time for comparing a query sequence of 2,048 elements with a database sequence of 64 millions elements by the Smith-Waterman algorithm is about 34 sec, which is about 330 times faster than a desktop computer with a 1GHz Pentium-III.

1 Introduction

Many pattern matching problems in bioinformatics are very time consuming, and many algorithms and dedicated hardware systems have been developed [6][7][8]. The obtained results by the algorithms and the systems are under tradeoffs of quality, time and costs. With desktop computer systems, it is unrealistic to check all pattern matching possibilities within a reasonable time. Therefore, simplified (but still very effective) algorithms have been designed and used on the systems. With dedicated hardware systems, the computation time can be drastically improved, and all the possibilities can be checked, because most of the pattern matching problems have many parallelism in them. However, the costs of the systems are expensive.

Field Programmable Gate Array (FPGA) is a reconfigurable device designed for rapid prototyping, and any kinds of circuits can be realized on the FPGA in a moment by downloading configuration data from host computers or dedicated memories. In the past years, the improvement of the size of FPGAs have been drastic and FPGAs are increasingly used as accelerators in many application areas [3][4][5]. (FPGAs are also used in some dedicated hardware systems [6][8] for bioinformatics). Many optional boards with FPGAs for Pentium based systems are now being shipped from many companies[2], and we can obtain many kinds of these boards at low costs. These boards are designed mainly for rapid prototyping or digital signal processing. However, in general, they do not have enough hardware resources for the pattern matching problems in bioinformatics.

In this paper, we show that we can achieve high performance in homology search by only adding one off-the-shelf FPGA board to a Pentium based computer system in use. The performance can be furthermore accelerated by using higher numbers of FPGA boards. In our approach, the search is divided into two phases, and different configuration data (namely different circuits) are downloaded from the host computer in each phase in order to make up for the limited hardware resources. By using off-the-shelf boards, we can easily obtain the latest FPGAs at low costs, which is very important because the total performance is almost proportional to the size of the FPGAs, and FPGAs are becoming larger and larger following Moore's low (the number of transistors in a fixed size (namely the size of FPGAs) become twice in every 18 months). The configuration data can be easily modified for new FPGA boards, because they are generated from the programs written in hardware description languages without assuming any special hardware resources on the FPGA boards.

This paper is organized as follows. Section 2 describes the overview of our approach. Then, experimental results of the approach are shown in section 3. In section 4, current status and future works are given.

2 Overview of the Approach

2.1 Target Problems

Our current target problems are homology search problems shown in Table 1. In our approach, the Smith-Waterman algorithm[1] is used in all comparison of sequences shown in the Table 1.

Table 1: Target Problems in Homology Search

Query sequence	Database sequence
amino acid	amino acid
amino acid	translated nucleotide
nucleotide	nucleotide

2.2 Hardware and Software for our Approach

We need the followings as components of the hardware platform (Figure 1).

1. one off-the-shelf FPGA board (with PCI bus interface), and
2. one host computer (a Pentium based computer, because driver programs for most FPGA boards run only under Windows or Linux).

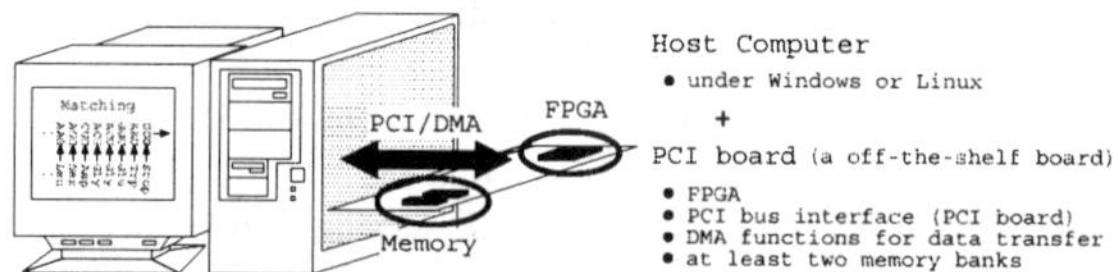

Figure 1: Required Components of Hardware Platform

The softwares which are necessary for our approach are

1. drivers programs to control FPGA boards from the host computer, which are developed by the board maker and attached to the FPGA boards, and

2. CAD tools for the FPGA, only if we need to modify the configuration data for new FPGA boards.

Among the components shown above, what we have developed are

1. the programs for the circuits which are implemented on the FPGA, and

2. interface programs which run on the host computer.

In our programs for the circuits, only the two memory banks to transfer data between the FPGA and the host computer are assumed. Most FPGA boards have at least two memory banks in order to receive data from the host computer while the FPGA is running using another memory bank. Therefore, configuration data for new FPGA boards can be easily generated by only changing some parameters in the programs (FPGA size, memory size in FPGA, I/O pin assignment and so on).

The interface programs need to control FPGA boards using the driver programs. The structure of driver programs depends on the boards, and we need to modify a part of the programs for each board.

2.3 Outline of the Homology Search

In our approach, the search consists of two phases in order to make up for the limited hardware resources, and different configuration data is downloaded from the host computer in each phase.

In the first phase, the query sequence is compared with database sequences by the Smith-Waterman algorithm (Figure 2-(1)(2)), but only the positions of fragments which are similar to the query sequence (thresholds can be given by the users) are output, because of the following reasons.

1. In general, the memory bandwidth of the FPGA is not enough to output all the results by the Smith-Waterman algorithm at the speed the results are generated by the FPGA (at least $2 \times p$ bits memory width is necessary when p cells on the comparison array can be processed at once).

2. The performance is proportional to the number of the cells (p) which can be processed at once. Therefore, it is very important to process more cells at once.

3. The improvement of the memory bandwidth is very slow compared with the improvement of the size of FPGAs.

In this phase, the database sequences are divided into subsequences with fixed size (the size is decided based on the size of the internal memory of the FPGA), and compared with the query sequence. Suppose that the FPGA can compare p elements at a time (Figure 2-(3)), and the length of the query sequence (m) is longer than p ($m > p$). Then, the first p elements of the query sequence are compared with database sequences first, and the intermediate results are stored in the internal memory of the FPGA. Then, next p elements are compared with the database sequences using the intermediate results. However, the length of each database sequence is very long, and it is impossible to store all the intermediate results in the internal memory of the FPGA. Therefore, each database sequence is divided, and compared with the query sequence. In this division, the first and the last parts of each subsequence (their length is a fraction of the length of the query sequence) are overlapped. These overlapped areas are processed twice, which becomes the major overhead in our approach. The overhead is almost proportional to the length of the overlapped area, and becomes relatively smaller by using larger FPGAs because the internal memory size becomes larger according to the size of the FPGAs.

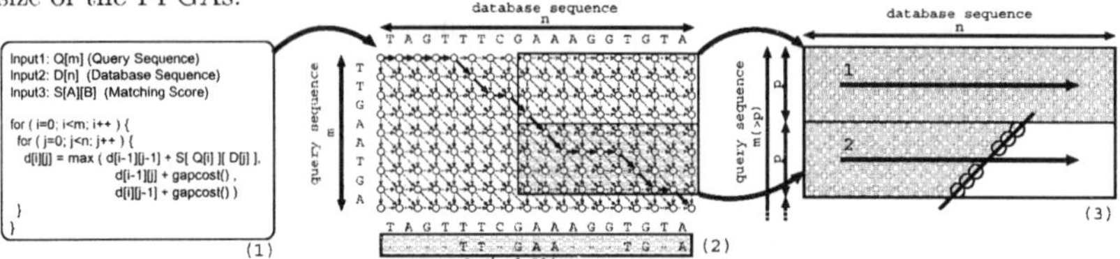

Figure 2: Smith-Waterman Algorithm

In the second phase, the detailed results by the Smith-Waterman algorithm for the fragments are output. The performance of this phase is fixed by the memory bandwidth of the FPGA board. However, the performance of this phase is not so important because the computation time of the second phase is much smaller than that of the first phase.

3 Experiments

Figure 3 shows a block diagram of the FPGA board (RC1000-PP PCI board by Celoxica with FPGA XCV2000E by Xilinx) which we used to evaluate the performance on desktop environment. The board has four memory banks, and two of them are used for data transfer between the FPGA board and the host computer. The FPGA (Xilinx XCV2000E) on the board is one of the largest FPGAs that we can obtain now. We could implement 144 processing elements for the first phase of the homology search, and they run at 40 MHz. The size of the internal memory of the FPGA is 640 Kbits, and the length of the subsequence becomes 32768 elements. Therefore, the overhead caused by the overlapped area becomes about 5 - 10% when the length of the query sequence is 2048 and the size of the fragment is several times of the query sequence.

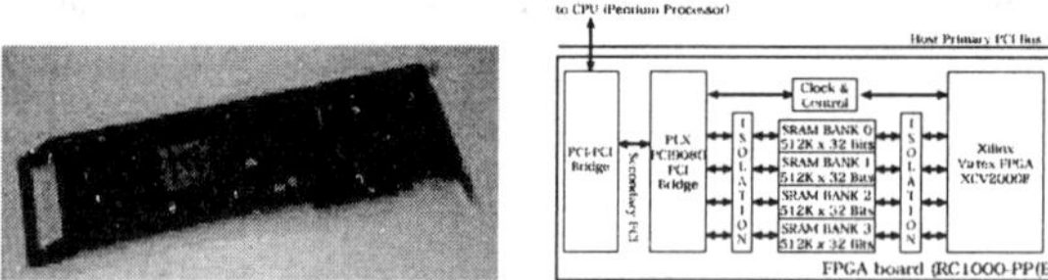

Figure 3: a PCI Board with FPGA and the Block Diagram

Figure 4 shows the relation between the time of the first phase and the length of the query sequence, when the length of the database sequence is 64 millions. The slope of the search time becomes slightly larger as the length of the query sequence becomes larger, because the percentage of the overhead by the overlapped area will gradually increase. The speedup compared with a Pentium-III 1 GHz and 1GB physical memory under Linux kernel version 2.2.5 and gcc-2.91.66 is 327 times when the length of the query sequence is 2048.

4 Current Status and Future Works

We have developed the circuits for homology search, and showed that we can achieve high performance using off-the-shelf FPGA boards. The performance is almost compa-

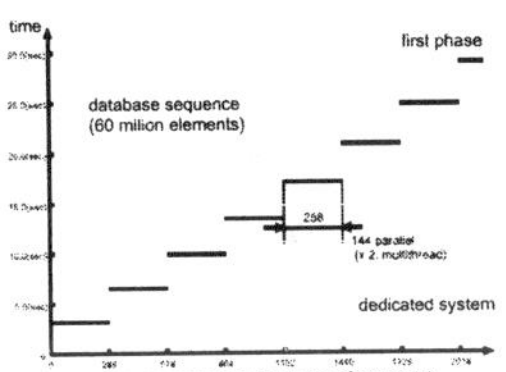

Figure 4: Computation time of the First Phase

rable with small to middle class dedicated hardware systems when we use one board with one of the latest FPGAs (Xilinx XCV2000E). The time for comparing a query sequence of 2048 elements with a database sequence of 64 millions elements by the Smith-Waterman algorithm is about 34 sec, which is about 330 times faster than a desktop computer with a 1GHz Pentium-III. We can also accelerate the performance of a note computer using one PC card with one FPGA (Xilinx XCV300). The time for comparing a query sequence (1024) with database sequence (64 millions) is about 185 sec, which is about 30 times faster than the desktop computer.

We are now evaluating the performance for the translated nucleotides. When we need to translate the sequences during the comparison, the size of each unit on the FPGA becomes about 10% larger and the parallelism in the first phase will go down to 120 from 144 (about 20% performance down). We are now improving the circuits of the unit to achieve higher performance.

Some parts of the programs for the homology search are still under development, and we also need to improve other parts. We are also developing software for parallel processing of the homology search with higher number of pairs of FPGAs and host computers connected by Ethernet.

We are also planning to accelerate other pattern matching problems in bioinformatics with FPGAs.

Acknowledgement

This work was partly supported by Japan Society for the Promotion of Science (JSPS) Research Fellowships for Young Scientists (#5304) and a Grant-in-Aid for Scientific Research on Priority Areas 'Genome Science' from the Ministry of Education, Science, Sports and Culture of Japan.

References

[1] Smith T. F. and Waterman M. S.: *Identification of common molecular subsequences*, Journal of Molecular Biology 147, pp.195-197, (1981).

[2] http://www.optimagic.com/boards.html

[3] http://www.fccm.org
(IEEE Symposium on Field-Programmable Custom Computing Machines)

[4] http://www.ecs.umass.edu/ece/fpga2002/index.html
(ACM International Symposium on Field-Programmable Gate Arrays)

[5] http://xputers.informatik.uni-kl.de/fpl/index_fpl.html
(the International Conference on Field-Programmable Logic and Applications)

[6] http://www.compugen.com

[7] http://www.paracel.com/index.html

[8] http://www.timelogic.com

KES '01
N. Baba et al. (Eds.)
IOS Press, 2001

Fuzzy Multiobjective Programming Problems with Possibility Measures

Hiroaki KUWANO

Faculty of Business Administration and Information Science,
Kanazawa Gakuin University, Kanazawa 920-1392, Japan

Abstract. In this article, we discuss about one of four comparison indices based Possibility theory for a pair of multi-dimensional fuzzy sets. And we refer the difficulty of fuzzy multiobjective programming problems containing interactive fuzzy numbers.

1. Introduction

In 1985, Ramík and Řimánek proposed a binary relation between fuzzy sets, which is called *fuzzy max order*, as a natural fuzzified extension of an ordinary inequality relation "*max*" defined on a set of all real numbers $\mathbb{R}$([3]). Using this relation, they reduced a linear programming problem with constraints those coefficients are *L-R* type fuzzy numbers to a substitute linear mathematical programming problem. On the other hand, in [1], Dubois and Prade introduced four comparison indices of fuzzy numbers in the setting of Possibility Theory established by Zadeh([5]) to fuzzy mathematical programming problems. These indices are defined as the grade of possibility or necessity of the assertion about the ranking of two fuzzy numbers. Essentially, fuzzy max order and four comparison indices are applicable to a pair of fuzzy numbers. Therefore, these are useful in single-objective fuzzy mathematical programming problems.

Recently, Kurano et al.[2] gave the definition of fuzzy max order for multi-dimensional convex fuzzy sets. Hence, we can treat fuzzy multiobjective mathematical programming problems by applying their relation. In this article, we focus to discussion about one of four comparison indices based Possibility theory for a pair of multi-dimensional fuzzy sets. In last section, we refer the difficulty of fuzzy multiobjective programming problems containing interactive fuzzy numbers.

2. Preliminaries

The aim of this section is to remember basic definitions on comparison indices of fuzzy numbers, briefly.

Definition 1. Let $\mathbb{R}^n$ be n-dimensional Euclidean space. Then, a normal fuzzy set $\widetilde{a}$ on $\mathbb{R}^n$ is an n-dimensional fuzzy vector (or a fuzzy vector, simply), if its membership function $\mu_{\widetilde{a}}$ is upper-semicontinuous, quasi-concave, and $\widetilde{a}$ is unique modal and has a compact support. In particular, when $n = 1$, a fuzzy vector $\widetilde{a}$ is denoted by $\widetilde{a}$, and called a fuzzy number.

The α-level set of a fuzzy vector $\widetilde{a}$ is defined by $[\widetilde{a}]^\alpha = \{x \in \mathbb{R}^n \,|\, \mu_{\widetilde{a}}(x) \geq \alpha\}$, if $\alpha \in (0, 1]$, and $[\widetilde{a}]^0 = \mathrm{cl}\,\{x \in \mathbb{R}^n \,|\, \mu_{\widetilde{a}}(x) > 0\}$, where cl denotes the closure of the set.

Definition 2 ([1]). For a fuzzy number $\widetilde{a}$, two fuzzy intervals $[\widetilde{a}, \infty)$ and $(-\infty, \widetilde{a}]$ are characterized by $\mu_{[\widetilde{a},\infty)}(y) = \sup_{x:x \leq y} \mu_{\widetilde{a}}(x)$, $\mu_{(-\infty,\widetilde{a}]}(y) = \inf_{x:x \leq y} \{1 - \mu_{\widetilde{a}}(x)\}$ respectively.

Definition 3 (cf. [5]). Given a fuzzy number $\widetilde{b}$, the possibility measure $\Pi_{\widetilde{b}}$ is defined by $\Pi_{\widetilde{b}}(\widetilde{a}) = \sup_{x \in \mathbb{R}} \min \left\{\mu_{\widetilde{a}}(x), \mu_{\widetilde{b}}(x)\right\}$ for each fuzzy set $\widetilde{a}$ on $\mathbb{R}$. For a fuzzy vector $\widetilde{b}$, the possibility measure $\Pi_{\widetilde{b}}$ is defined in the same way.

Definition 4 ([1]). Let $\widetilde{a}$, $\widetilde{b}$ be fuzzy numbers. Then the index of the grade of possibility of the proposition "$\widetilde{a} \leq \widetilde{b}$" is defined by $\mathrm{Pos}\!\left(\widetilde{a} \leq \widetilde{b}\right) = \Pi_{\widetilde{b}}([\widetilde{a}, \infty))$. That is,

$$\mathrm{Pos}\!\left(\widetilde{a} \leq \widetilde{b}\right) = \sup_{x,y:x \leq y} \min \left\{\mu_{\widetilde{a}}(x), \mu_{\widetilde{b}}(y)\right\}.$$

If two fuzzy numbers $\widetilde{a}$ and $\widetilde{b}$ are real, that is, $\mu_{\widetilde{a}} = I_{\{a\}}$ and $\mu_{\widetilde{b}} = I_{\{b\}}$, then $\mathrm{Pos}\!\left(\widetilde{a} \leq \widetilde{b}\right) = 1$ ($a \leq b$) and $= 0$ ($a > b$), where $I_{\{\cdot\}}$ is an indicator function of the real number.

Dubois and Prade[1] defined three other indices like as the above expression and investigated their properties in more detail.

3. The comparison index

In this section, we discuss an extension of the comparison index of fuzzy numbers to the case of fuzzy vectors.

Let K be a pointed and closed convex ordered cone on $\mathbb{R}$, and the binary relation $x \leqslant_K y$ is defined by $y - x \in K$ for a pair of real numbers $x, y \in \mathbb{R}^n$.

For a fuzzy vector $\widetilde{a}$, a fuzzy set $[\widetilde{a}, \infty)_K$ is defined as $\widetilde{a} + K$. In other words, $[[\widetilde{a}, \infty)_K]^\alpha = [\widetilde{a}]^\alpha + K$, for any α belonging an unit interval $[0, 1]$. Clearly, a fuzzy set $[\widetilde{a}, \infty)_K$ is normal convex, furthermore any α-level set of it is closed. Therefore the membership function is upper-semicontinuous. Then the following property is derived.

Proposition 1. *Let $\widetilde{a}$ be a fuzzy vector on $\mathbb{R}^n$. The membership function of the fuzzy set $[\widetilde{a}, \infty)_K$ is expressed by using a possibility measure $\Pi_{\widetilde{a}}$, as follows;*

$$\mu_{[\widetilde{a},\infty)_K}(x) = \Pi_{\widetilde{a}}(\{x\} - K), \quad \text{for each } x \in \mathbb{R}^n.$$

Proof. Let $x \in \mathbb{R}^n$ be a arbitrarily fixed point. From the above statement, we know that $\mu_{[\widetilde{a},\infty)_K}(x) = \mu_{\widetilde{a}+K}(x)$. Therefore, by Zadeh's extension principle(e.g. [4])

$$\mu_{[\widetilde{a},\infty)_K}(x) = \sup_{y,z:x=y+z} \min \{\mu_{\widetilde{a}}(y), I_K(z)\} = \sup_{y \in \mathbb{R}^n} \min \{\mu_{\widetilde{a}}(y), I_K(x - y)\} = \sup_{y \in \mathbb{R}^n} \min \{I_{\{x\}-K}(y), \mu_{\widetilde{a}}(y)\}.$$

From Definition 3, we get $\mu_{[\widetilde{a},\infty)_K}(x) = \Pi_{\widetilde{a}}(\{x\} - K)$ for each $x \in \mathbb{R}^n$. $\qquad\square$

Remark 1. If $n = 1$ and $K = [0, \infty)$, the previous result coincides with $\mu_{[\widetilde{a},\infty)}(x) = \Pi_{\widetilde{a}}((-\infty, x])$ on Dubois and Prade[1].

Definition 5. Let $\widetilde{a}$, $\widetilde{b}$ be fuzzy vectors. Then the index of the grade of possibility of the proposition "$\widetilde{a} \leq_K \widetilde{b}$" is defined by $\mathrm{Pos}\!\left(\widetilde{a} \leq_K \widetilde{b}\right) = \Pi_{\widetilde{b}}([\widetilde{a}, \infty)_K)$.

Proposition 2. *Let $\widetilde{a}$, $\widetilde{b}$ be fuzzy vectors. Then, the following expression is valid.*

$$\mathrm{Pos}\!\left(\widetilde{a}\le_K \widetilde{b}\right) = \sup_{x,y:x\le_K y}\ \min\left\{\mu_{\widetilde{a}}(x),\ \mu_{\widetilde{b}}(y)\right\}$$

Proof. From Definition 3 and 5,

$$\mathrm{Pos}\!\left(\widetilde{a}\le_K \widetilde{b}\right) = \Pi_{\widetilde{b}}\left(([\widetilde{a},\infty)_K\right) = \sup_{y\in\mathbb{R}^n}\min\left\{\mu_{[\widetilde{a},\infty)_K}(y),\mu_{\widetilde{b}}(y)\right\}.$$

And

$$\mu_{[\widetilde{a},\infty)_K}(y) = \sup_{x,z:y=x+z}\ \min\{\mu_{\widetilde{a}}(x),I_K(z)\} = \sup_{x\in\mathbb{R}^n}\min\{\mu_{\widetilde{a}}(x),I_K(y-x)\} = \sup_{x\in\mathbb{R}^n:y-x\in K}\mu_{\widetilde{a}}(x).$$

Therefore, because $\mu_{\widetilde{a}}$ is upper-semicontinuous and $\widetilde{a}$ has a compact support,

$$\mathrm{Pos}\!\left(\widetilde{a}\le_K \widetilde{b}\right) = \sup_{y\in\mathbb{R}^n}\min\left\{\sup_{x\in\mathbb{R}^n:y-x\in K}\mu_{\widetilde{a}}(x),\mu_{\widetilde{b}}(y)\right\} = \sup_{x,y:x\le_K y}\min\left\{\mu_{\widetilde{a}}(x),\mu_{\widetilde{b}}(y)\right\}.$$

$\square$

Theorem 3. *Let $\widetilde{a}$, $\widetilde{b}$ be fuzzy vectors. Then, for each $\alpha \in (0,1]$, the index of the grade of possibility of the proposition "$\widetilde{a} \le_K \widetilde{b}$", $\mathrm{Pos}\!\left(\widetilde{a}\le_K \widetilde{b}\right)$, is greater than or equal to α, if and only if $[\widetilde{a}]^\alpha + K$ and $[\widetilde{b}]^\alpha$ have nonempty intersection.*

Proof. Let $\alpha \in (0,1]$ be an arbitrary fixed grade. If $\mathrm{Pos}\!\left(\widetilde{a}\le_K \widetilde{b}\right) \ge \alpha$, then there exist $x_0, y_0 \in \mathbb{R}^n$ such that $x \le_K y$, $\mu_{\widetilde{a}}(x) \ge \alpha$ and $\mu_{\widetilde{b}}(y) \ge \alpha$, from Proposition 2. Setting $z = y - x$, we get $y = x + z \in [\widetilde{a}]^\alpha + K$. Hence, $([\widetilde{a}]^\alpha + K) \cap [\widetilde{b}]^\alpha \ne \emptyset$. The inverse assertion is derived similarly. $\square$

Proposition 2 and the previous theorem show that the comparison index $\mathrm{Pos}\!\left(\widetilde{a}\le_K \widetilde{b}\right)$ is a natural extension of Dubois and Prade's $\mathrm{Pos}\!\left(\widetilde{a} \le \widetilde{b}\right)$.

4. An application to a fuzzy multiobjective programming problem and Conclusion

Consider the mathematical programming problem

$$\begin{cases} \text{maximize} & \widetilde{C}x, \\ \text{subject to} & \widetilde{A}x \le \widetilde{b} \\ & x \ge 0, \end{cases} \qquad\qquad \text{(FMOLP)}$$

where $\widetilde{C}x = (\langle\widetilde{c}_1,x\rangle_F,\langle\widetilde{c}_2,x\rangle_F,\ldots,\langle\widetilde{c}_\ell,x\rangle_F)^T$, $\langle\widetilde{c}_k,x\rangle_F = \sum_{j=1}^n \widetilde{c}_{jk}x_j$, $k = 1,2,\ldots,\ell$, and $\widetilde{A}x = (\langle\widetilde{a}_1,x\rangle_F,\langle\widetilde{a}_2,x\rangle_F,\ldots,\langle\widetilde{a}_m,x\rangle_F)^T$. And $\widetilde{c}_k, \widetilde{a}_i$ ($i = 1,2,\ldots,m; k = 1,2,\ldots,\ell$) are n-dimensional fuzzy vectors, $\widetilde{b}$ is m-dimensional fuzzy vector.

Definition 6. A feasible set of FMOLP for a given possibility $\alpha \in [0,1]$ is defined as $X(\alpha) = \left\{x \ge 0 \,\middle|\, \mathrm{Pos}\!\left(\widetilde{A}x\le_K \widetilde{b}\right) \ge \alpha\right\}$

From Proposition 2,

$$\text{Pos}\left(\widetilde{A}x \leq_K \widetilde{b}\right) = \sup_{y,z:y \leqslant_K z} \min \left\{\mu_{\widetilde{Ax}}(y), \mu_{\widetilde{b}}(z)\right\}.$$

Therefore, we can get reformulated constraints on FMOLP by possibilistic programming approaches. However, if there exists at least one pair of the parameters of FMOLP are interactive, this problem can not be solvable substantially. In fact, if $x \in X(\alpha)$ if and only if there exists an optimal solution of parametric programming problem $\sup \left\{\gamma \,\middle|\, \mu_{\widetilde{Ax}}(y) \geq \gamma, z \in [\widetilde{b}]^\gamma, y \leqslant_K z\right\}$. Besides, we have to calculate $\mu_{\widetilde{Ax}}(y)$.

In this article, we discuss about one of four comparison indices based Possibility theory for a pair of n-dimensional fuzzy vector. Our definition is a natural extension of fuzzy numbers case. However, an application to fuzzy multiobjective programming problems is terribly difficult. The further researches are anxious.

References

[1] D. Dubois and H. Prade, Ranking fuzzy numbers in the setting of possibility theory, *Information Sciences* **30** (1983) 183-224.

[2] M. Kurano, M. Yasuda, J. Nakagami and Y. Yoshida, Ordering of convex fuzzy sets – A brief survey and new results, *J. Operations Research Society of Japan* **43** (2000) 138-148.

[3] J. Ramík and J. Řimánek, Inequality relation between fuzzy numbers and its use in fuzzy optimization, *Fuzzy Sets and Systems* **16** (1985) 123-138.

[4] L.A. Zadeh, The concept of a linguistic variable and its applications to approximate reasoning Part 1, *Information Science* **8** (1975) 199-239.

[5] L.A. Zadeh, Fuzzy sets as a basis for a theory of possibility, *Fuzzy Sets and Systems* **1** (1978) 3-28.

KES '01
N. Baba et al. (Eds.)
IOS Press, 2001

An Optimal Path Problem with Threshold Probability

Toshiharu FUJITA

Department of Electric, Electronic and Computer Engineering, Faculty of Engineering
Kyushu Institute of Technology, Kitakyushu 804-8550, Japan
tel & fax: +81(93)884-3256, email: fujita@comp.kyutech.ac.jp

Abstract. The purpose of this paper is to find the optimal paths for reaching goal in time. In our problem, a network in which a time function taking value with randomness is given on the arc set is considered. Then we discuss how to find the optimal paths from one node to another node which maximize the probability that the sum of time on a path is less than or equal to a given limit. A difficulty of this type of problem is caused by the fact that the optimal paths are not always Markov. In other words, a part of the optimal path is not optimal unlike fundamental shortest paths problems. That is why we need invariant imbedding technique. First, we imbed the problem into a parameterized problem and define a family of subproblems. Then we derive a recursive formula and give an algorithm.

1. Introduction

In this paper we are concerned with a certain network optimization problem. Our aim is to find the optimal paths (or policy) which maximize the probability of reaching goal in time. For example, in the case where we want to avoid missing the train or being late for a meeting, we prefer a path which gives stronger possibility that we are in time. In the problem, a network in which a time function taking value with randomness is given on the arc set is considered. Then we discuss how to find the optimal paths from one node to another node which maximize the probability that the sum of time on a path is less than or equal to a given limit. The objective function of the problem is expressed by a threshold probability. A difficulty of this type of problem is caused by the fact that optimal paths is not always Markov. In other words, a part of the optimal path is not optimal unlike fundamental shortest paths problems. Similar difficulties are often occurred in dynamic programming theory ([2] – [7]), when problems have non-additive criteria. Our solution for this difficulty is invariant imbedding technique. First the problem is formulated in section 2. In section 3, we introduce an additional parameter and imbed the problem into a parameterized problem. Then we define a family of subproblems and derive a recursive formula. By using this result, we give an algorithm. The basic idea is how to use label correcting algorithms ([1]). Finally, section 4 give a numerical example.

2. Problem Formulation

The purpose here is to formulate an optimal path problem with threshold probability. We consider it on the undirected graph $G = (N, A)$, where we denote by N a finite node set and by $A \subset N \times N$ a finite arc set. We call an ordered set of nodes

$$P = \{x_1, x_2, \ldots, x_l\}, \quad (x_i, x_{i+1}) \in A$$

(include the case $P = \{x_1\}$ $(x_1 \in N)$) path. Especially, in case we emphasize the origin and the destination of the path, we call P x_1-x_n-path.

For the graph G, we define a time function $T(\geq 0)$ (which is so-called a cost function). So $T(x, y)$ expresses time for each arc $(x, y) \in A$. In addition, we suppose that $T(x, y)$ is discrete random variable which takes values $t_i(x, y)(i = 1, 2, \ldots m)$ with corresponding probabilities $p_i(x, y)$. Then, for an x_1-x_n-path : $\{x_1, x_2, \ldots, x_{n-1}, x_n\}$, total time is expressed by

$$T(x_1, x_2) + T(x_2, x_3) + \cdots + T(x_{n-1}, x_n).$$

Now we discuss the problem of finding the optimal paths between two specified nodes $S, G \in N$, which is to maximize the probability that total time is less than or equal to a given limit M. We assume that all nodes in N are reachable from goal node G. Thus, for the given start node S and goal node G, the optimal path problem is formulated as follows:

$$(P) \quad \text{Maximize } P[T(S, x_1) + T(x_1, x_2) + \cdots + T(x_{n-1}, x_n) + T(x_n, G) \leq M]$$

$$\text{subject to} \quad \{S, x_1, x_2, \ldots, x_n, G\} : S\text{-}G\text{-path}.$$

It must be noted that the path length n is not fixed in above problem.

We introduce the characteristic function of the interval $(-\infty, M]$:

$$\chi_{(-\infty, M]}(x) = \begin{cases} 1 & (x \leq M) \\ 0 & (x > M) \end{cases}$$

Then the objective function of the problem (P) is transformed into an expected value :

$$P[T(S, x_1) + T(x_1, x_2) + \cdots + T(x_{n-1}, x_n) + T(x_n, G) \leq M]$$

$$= \sum_{i_1, i_2, \ldots, i_{n+1}} \chi_{(-\infty, M]}(t_{i_1}(S, x_1) + t_{i_2}(x_1, x_2) + \cdots + t_{i_n}(x_{n-1}, x_n) + t_{i_{n+1}}(x_n, G))$$
$$\times \{p_{i_1}(S, x_1) \times p_{i_2}(x_1, x_2) \times \cdots \times p_{i_n}(x_{n-1}, x_n) \times p_{i_{n+1}}(x_n, G)\}$$

$$= E[\chi_{(-\infty, M]}(T(S, x_1) + T(x_1, x_2) + \cdots + T(x_{n-1}, x_n) + T(x_n, G))].$$

Hence the problem (P) is equivalent to the following problem :

$$(P_0) \quad \text{Maximize } E[\chi_{(-\infty, M]}(T(S, x_1) + T(x_1, x_2) + \cdots + T(x_{n-1}, x_n) + T(x_n, G))]$$

$$\text{subject to} \quad \{S, x_1, x_2, \ldots, x_n, G\} : S\text{-}G\text{-path}.$$

3. Recursive Equation

For the shortest paths problem, we usually define subproblems which are to find the shortest paths from one node to all other node. But this idea is not applied to our problem, because the problem(P_0) (or (P)) has the property that a part of the optimal path is not always optimal. Thus we must use invariant imbedding technique. In this case we need to introduce a new parameter $\lambda \in \mathbf{R}$ (the letter $\mathbf{R}$ denotes the real number system). and to imbed the problem(P_0) into the parameterized problem defined as follows :

(P_λ) Maximize $E[\chi_{(-\infty,M]}(\lambda + T(S,x_1) + T(x_1,x_2) + \cdots + T(x_{n-1},x_n) + T(x_n,G))]$

$$\text{subject to} \quad \{\, S,\ x_1,\ x_2,\ \ldots,\ x_n,\ G \,\} : S\text{-}G\text{-path.}$$

If λ is equal to 0 in the problem above, it is clear that this imbedded problem(P_λ) is equivalent to the problem(P_0). For the imbedded problem, we consider the following subproblems :

$$w(G;\lambda) \;=\; E[\chi_{(-\infty,M]}(\lambda)]$$

$$w(x;\lambda) \;=\; \underset{\substack{x,\ x_1,\ \ldots,\ x_l,G:\\ x\text{-}G\text{-path}}}{\mathrm{Max}}\ E[\chi_{(-\infty,M]}(\lambda + T(x,x_1) + \cdots + T(x_n,G))] \quad (x \in N\backslash\{G\})$$

Then we have the following theorem.

Theorem 3.1 (recursive formula)

$$w(G;\lambda) \;=\; \begin{cases} 1 & \lambda \le M \\ 0 & \lambda < M \end{cases}$$

$$w(x;\lambda) \;=\; \underset{(x,y)\in A}{\mathrm{Max}} \sum_i w(y;\lambda + t_i(x,y)) \times p_i(x,y) \qquad (x \in N).$$

Further, for each $x \in N$, we denote by $\pi^*(x;\lambda)$ the node which is the maximizer of $w(x;\lambda)$. Thus we can get the optimal value $w(S;0)$ and the corresponding optimal paths : $P^* = \{\, S,\ x_1^*,\ x_2^*,\ \ldots,\ x_n^*,\ G \,\}$ generated as follows :

$$\begin{aligned}
\lambda_0 &= 0 \\
x_1^* &= \pi^*(S;\lambda_0), \quad \lambda_1 = \lambda_0 + t_i(S,x_1^*) \\
x_2^* &= \pi^*(x_1^*;\lambda_1), \quad \lambda_2 = \lambda_1 + t_i(x_1^*,x_2^*) \\
&\;\;\vdots \\
x_{n-1}^* &= \pi^*(x_{n-2}^*;\lambda_{n-2}), \quad \lambda_{n-1} = \lambda_{n-2} + t_i(x_{n-2}^*,x_{n-1}^*) \\
x_n^* &= \pi^*(x_{n-1}^*;\lambda_{n-1}).
\end{aligned}$$

Algorithm

By Theorem 3.1 we now get the following algorithm for finding the optimal solutions of the optimal path problem(P_λ).

Step 1: For each $x \in N\backslash\{G\}$, set

$$w(x;\lambda) = 0 \quad \text{and} \quad w(T;\lambda) = \begin{cases} 1 & \lambda \le M \\ 0 & \lambda > M. \end{cases}$$

Step 2: For each $x \in N\backslash\{G\}$, compute

$$w'(x;\lambda) = \underset{y;(x,y)\in N}{\mathrm{Max}}\ E[w(y;\lambda + T(x,y)]$$

and denote by $Y_x^*(\lambda)$ the maximizer of the right-hand side.

Step 3:   If $w'(x;\lambda) \le w(x;\lambda)$ for all $(x,\lambda) \in N \times \mathbf{R}$, then stop.

Step 4:   For $(x,\lambda) \in N \times \mathbf{R}$ such that $w'(x;\lambda) > w(x;\lambda)$, set

$$w(x;\lambda) = w'(x;\lambda), \quad \pi^*(x;\lambda) = Y_x^*(\lambda)$$

and go to Step 2.

$(x,\ y)$	$(t_1(x,\ y),\ p_1(x,\ y))$	$(t_2(x,\ y),\ p_2(x,\ y))$
$(S,\ A)$	$(7,\ 0.40)$	$(8,\ 0.60)$
$(A,\ B)$	$(5,\ 0.50)$	$(8,\ 0.50)$
$(A,\ C)$	$(4,\ 0.70)$	$(8,\ 0.30)$
$(B,\ G)$	$(4,\ 0.50)$	$(6,\ 0.50)$
$(C,\ G)$	$(5,\ 0.70)$	$(6,\ 0.30)$

Table 1: Numerical Data

4. Numerical Example

Let us consider the graph $G = (N, A)$ with

$$N = \{S, A, B, C, G\}, \quad A = \{(S, A), (A, B), (A, C), (B, G), (C, G)\}.$$

and pairs of time and probability on the each arc given in Table 1. We apply the algorithm given in the preceding section to the optimal path problem with a limit $M = 20$.

To begin with, we set

$$w(x; \lambda) = 0.0 \ \ (x = S, A, B, C) \quad \text{and} \quad w(T; \lambda) = \begin{cases} 1.0 & \lambda \leq M \\ 0.0 & \lambda > M. \end{cases}$$

Next, for each $x \in N$, we have to compute $w'(x; \lambda)$ and, if necessary, update $w(x; \lambda)$. For example, in case $x = B$,

$$\begin{aligned} w'(B; \lambda) &= \underset{y;(B,y)\in N}{\text{Max}} E[w(y; \lambda + T(B, y)] \\ &= \text{Max}(E[w(A; \lambda + T(B, A)], E[w(G; \lambda + T(B, G)]) \end{aligned}$$

$$E[w(G; \lambda + T(B, G)] = \begin{cases} 1.0 & (\lambda + 4 \leq 20) \\ 0.0 & (\lambda + 4 > 20) \end{cases} \times 0.5 + \begin{cases} 1.0 & (\lambda + 6 \leq 20) \\ 0.0 & (\lambda + 6 > 20) \end{cases} \times 0.5$$

$$= \begin{cases} 0.5 & (\lambda \leq 16) \\ 0.0 & (\lambda > 16) \end{cases} + \begin{cases} 0.5 & (\lambda \leq 14) \\ 0.0 & (\lambda > 14) \end{cases} = \begin{cases} 1.0 & \lambda \leq 14 \\ 0.5 & 14 < \lambda \leq 16 \\ 0.0 & 16 < \lambda \end{cases},$$

$$E[w(A; \lambda + T(B, A)] = 0$$

Therefore we obtain

$$w'(B; \lambda) = \begin{cases} 1.0 & \lambda \leq 14 \\ 0.5 & 14 < \lambda \leq 16 \\ 0.0 & 16 < \lambda \end{cases}, \qquad Y_B^*(\lambda) = \begin{cases} G & \lambda \leq 14 \\ G & 14 < \lambda \leq 16 \\ A, G & 16 < \lambda \end{cases}.$$

Since $w'(B; \lambda) > w(B; \lambda)$, we set

$$w(B; \lambda) = w(B; \lambda), \ \ \pi^*(B; \lambda) = Y_B^*(\lambda)$$

We similarly repeat step 2, step 3 and step 4 until the condition of step 3 is satisfied. Then we get

$$w(B; \lambda) = \begin{cases} 1.0 & \lambda \leq 14 \\ 0.5 & 14 < \lambda \leq 16 \\ 0.0 & 16 < \lambda \end{cases}, \qquad \pi^*(B; \lambda) = \begin{cases} G & \lambda \leq 14 \\ G & 14 < \lambda \leq 16 \\ A, G & 16 < \lambda \end{cases}$$

$$w(C;\lambda) = \begin{cases} 1.0 & \lambda \le 14 \\ 0.7 & 14 < \lambda \le 15 \\ 0.0 & 15 < \lambda \end{cases}, \qquad \pi^*(C;\lambda) = \begin{cases} G & \lambda \le 14 \\ G & 14 < \lambda \le 15 \\ A,G & 15 < \lambda \end{cases}$$

$$w(A;\lambda) = \begin{cases} 1.00 & \lambda \le 6 \\ 0.91 & 6 < \lambda \le 7 \\ 0.75 & 7 < \lambda \le 8 \\ 0.70 & 8 < \lambda \le 10 \\ 0.49 & 10 < \lambda \le 11 \\ 0.00 & 11 < \lambda \end{cases}, \qquad \pi^*(A;\lambda) = \begin{cases} B,C & \lambda \le 6 \\ C & 6 < \lambda \le 7 \\ B & 7 < \lambda \le 8 \\ C & 8 < \lambda \le 10 \\ C & 10 < \lambda \le 11 \\ B,C & 11 < \lambda \end{cases}$$

$$w(S;\lambda) = \begin{cases} 1.000 & \lambda \le -2 \\ 0.946 & -2 < \lambda \le -1 \\ 0.814 & -1 < \lambda \le 0 \\ 0.720 & 0 < \lambda \le 1 \\ 0.700 & 1 < \lambda \le 2 \\ 0.574 & 2 < \lambda \le 3 \\ 0.196 & 3 < \lambda \le 4 \\ 0.000 & 4 < \lambda \end{cases}, \qquad \pi^*(S;\lambda) = A.$$

Hence, we get the maximum probability $w(S;0) = 0.814$ and the optimal paths are given as follows:

$$\lambda_0 = 0$$
$$x_1^* = \pi^*(S;\lambda_0) = \pi^*(S;0) = A, \quad \lambda_1 = 0 + t_i(S,A) \quad (i = 1,2)$$
$$x_2^* = \pi^*(x_1^*;\lambda_1) = \begin{cases} C & \lambda_1 = 7 \\ B & \lambda_1 = 8 \end{cases}, \quad \lambda_2 = \lambda_1 + T(A,x_2^*)$$
$$x_3^* = \pi^*(x_2^*;\lambda_2) = G$$

This result about the optimal path means that if we take 7 from S to A, then the optimal path is { S, A, C, T}, and that if we take 8, then { S, A, B, T}. In general, the solution yielded by our method proposes that we should dynamically select the optimal path by using information gotten until then.

References

[1] R.K. Ahuja, T.L. Magnanti and J.B. Orlin: Handbooks in Operations Research and Management Science, **1**, Optimization (Eds. G.L. Namhauser et al.), North-Holland, **17**, 1989, pp. 211-369.

[2] R.E. Bellman and L.A. Zadeh: Decision-making in a fuzzy environment, Management Sci., **17**, 1970, pp. B141-B164.

[3] T. Fujita: Stochastic Optimization of Multiplicative Functions with Negative Value, J. Operations Res. Soc. Japan, **41**, 1998, pp.351-373

[4] T. Fujita: An Optimal Path Problem with Fuzzy Expectation, Proceedings of 8th Bellman Continuum, 2000, pp. 191-195.

[5] S. Iwamoto: Maximizing Threshold Probability Through Invariant Imbedding, Proceedings of 8th Bellman Continuum, 2000, pp. 17-18.

[6] S. Iwamoto and T. Fujita: Stochastic decision-making in a fuzzy environment, J. Operations Res. Soc. Japan, **38**, No.4, 1995, pp.467-482

[7] S. Iwamoto, K. Tsurusaki and T. Fujita: On Markov Policies for Minimax Decision Processes, J. Math. Anal. Appl., **253**, 2001, pp. 58-78

Fuzzy Decision-Making
Under
Threshold Membership Criterion

Seiichi IWAMOTO

Department of Economic Engineering, Graduate School of Economics
Kyushu University 27, Hakozaki 6–19–1, Fukuoka 812–8581, Japan
tel&fax: +81(92)642–2488, e-mail: `iwamoto@en.kyushu-u.ac.jp`

Abstract. We propose a "threshold probability" decision-making in fuzzy environment as a multi-stage stochastic control procees. The objective value is the threshold probability that *total* membership function is greater than or equal to a given lower grade. Under the controlled Markov chain we optimize the threshold probability not in *large* general policy class but in *small* Markov policy class. This choice turns out to lead a valid recursive formula. We show that there exists an optimal policy in Markov class.

1 Introduction

Since Bellman and Zadeh has proposed three – *deterministic, stochastic* and *fuzzy* – systems on multistage decision processes in fuzzy environment [4, §4,5], an intensive study on fuzzy decision making under uncertainty has been developed both in theory and in its wide applications ([1,5,16] and others). In Markov decision processes [6,18], it has been tacitly known that there exists an optimal policy which is *Markov* for the *additive* criterion, where Markov policy takes decision on the basis of only today's state. Recently, from a stochastic control theory, Iwamoto has developed Bellman and Zadeh's fuzzy decision-making on the stochastic system for a *non-additive (minimum)* criterion. He has shown that an optimal policy does not exist in Markov class for minimum criterion but does exist in *general* class, where general policy depends on state-sequence up until today [7–15, 19]. His tool is identical twins – both dynamic programming [2] and invariant imbedding [3,17] – for non-additive criterion in stochastic system [7,9,10].

In this paper, we consider a "threshold probability" decision-making in fuzzy environment. On the multi-stage stochastic control process, we evaluate the threshold probability that the minimum criterion exceeds a lower membership-degree. The minimum criterion denotes a *total* membership function of the multistage fuzzy decison process with *stage-wise* membership functions and a *goal* membership function. It is the membership function of intersection of the underlying fuzzy sets [4, p.144, §4,5]. Under the controlled Markov chain we optimize the threshold probability not in general class but in Markov class. We show that this choice will be successful; there exists an optimal policy in Markov class. We also derive the recursive relation for the threshold probability. We use the notations and terminology in [4,9,10].

2   Decision Process with Threshold Probability

Let us consider an N-stage ($N \geq 2$) stochastic decision process $\{(X_n, U_n)\}_0^N$ on a finite state space X and decision space U, which is governed by a Markov transition law $p = \{p(\cdot|\cdot, \cdot)\}$:

$$p(y|x, u) \geq 0, \quad \sum_{y \in X} p(y|x, u) = 1.$$

Thus $p(y|x, u)$ is a conditional probability that the next state X_{n+1} will be y when the current state X_n is x and current decision U_n is u :

$$P(X_{n+1} = y \,|\, X_n = x, U_n = u) = p(y|x, u).$$

This transition is expressed as $X_{n+1} \sim p(\,\cdot\,|x, u)$.

We begin to introduce a large class of policies, which depend not only on today's state but also on state-to-date. Let $X^n := X \times X \times \cdots \times X$ be direct product of n state spaces X. A mapping $\sigma_n : X^{n+1} \to U$ is called *n-th general decision function*, whose sequence $\sigma = \{\sigma_0, \sigma_1, \dots, \sigma_{N-1}\}$ constitutes a *general policy*. The set of all general policies Π_g is called *general class*. When each general decision function σ_n depends only on the last (= current) state, the general policy reduces to a *Markov policy* $\pi = \{\pi_0, \pi_1, \dots, \pi_{N-1}\}$. Let *Markov class* Π denote the set of all Markov policies. Thus we have an inclusion relation : $\Pi \subset \Pi_g$.

Further, given an *n-th membership function* $\mu_n : X \times U \to [0, 1]$ ($0 \leq n \leq N\text{--}1$) and a *goal membership function* $\mu_G : X \to [0, 1]$, the random variables $\mu_n = \mu_n(X_n, U_n)$, $\mu_G = \mu_G(X_N)$ denote the resulting grade of membership [4].

Now we consider the problem of maximizing a threshold probability that *total membership* is greater than or equal to a given *lower grade* $\alpha \in [0, 1]$:

$$\text{Maximize } P_{x_0}^\pi(\, \mu_0 \wedge \mu_1 \wedge \cdots \wedge \mu_{N-1} \wedge \mu_G \geq \alpha \,)$$

$$\mathrm{P}_0(x_0) \qquad \text{subject to (i)}_n \ X_{n+1} \sim p(\,\cdot\,|x_n, u_n) \qquad 0 \leq n \leq N\text{--}1$$

$$\text{(ii)}_n \ u_n \in U$$

where $P_{x_0}^\pi$ is the (discrete) probability measure on history space

$$H_N := X \times U \times X \times U \times \cdots \times U \times X \quad (2N + 1)\text{-factors}$$

induced through an initial state x_0, the Markov transition law p and a Markov policy $\pi(\in \Pi)$.

We dare to maximize the threshold probability over Markov class Π. We do not optimize it over general class Π_g. This choice will be turned to generate a valid recursive equation. Any Markov policy $\pi(\in \Pi)$ determines the threshold probability in $\mathrm{P}_0(x_0)$, which is a "partial" multiple sum :

$$P_{x_0}^\pi(\, \mu_0 \wedge \mu_1 \wedge \cdots \wedge \mu_{N-1} \wedge \mu_G \geq \alpha \,) \;=\; \sum \sum \cdots \sum_{(x_1, x_2, \dots, x_N) \in (*)} p_0 p_1 \cdots p_{N-1} \tag{1}$$

$$(\, p_n = p(x_{n+1}|x_n, u_n) \,)$$

where the domain $(*)$ is the set of all $(x_1, x_2, \dots, x_N) \in X^N$ satisfying

$$\mu_0(x_0, u_0) \wedge \mu_1(x_1, u_1) \wedge \cdots \wedge \mu_{N-1}(x_{N-1}, u_{N-1}) \wedge \mu_G(x_N) \;\geq\; \alpha. \tag{2}$$

Here the sequence of decisions $\{u_0, u_1, \ldots, u_{N-1}\}$ in (1),(2) is uniquely determined through Markov policy $\pi = \{\pi_0, \ldots, \pi_{N-1}\}$:

$$u_0 = \pi_0(x_0), \ u_1 = \pi_1(x_1), \ \ldots, \ u_{N-1} = \pi_{N-1}(x_{N-1}). \tag{3}$$

As for controlling threshold probability on the Markov chain $\{(X_n, U_n)\}$ with reward functions $\{\{r_n\}_0^{N-1}, \ r_N\}$ and a lower level value c, Markov class Π is not enough for *additive* criteria :

$$Q_0(x_0) \qquad \begin{array}{l} \text{Maximize} \ P_{x_0}^{\sigma}(\, r_0 + r_1 + \cdots + r_{N-1} + r_N \ \geq \ c\,) \\[2mm] \text{subject to } (\mathrm{i})_\mathrm{n}, \ (\mathrm{ii})_\mathrm{n} \qquad 0 \leq n \leq N\text{--}1 \end{array}$$

but general class Π_g is enough [9,19]. However, in this paper, we dare to maximize the threshold probability for *minimum* criteria over *Markov class*.

Thus our problem $P_0(x_0)$ is to find the *maximum value function* $v_0 = v_0(x_0)$ and an *optimal policy* $\pi^*(\in \Pi)$ which attains the maximum :

$$\begin{aligned} v_0(x_0) &= P_{x_0}^{\pi^*}(\, \mu_0 \wedge \cdots \wedge \mu_{N-1} \wedge \mu_G \geq \alpha \,) \\ &= \operatorname*{Max}_{\pi \in \Pi} P_{x_0}^{\pi}(\, \mu_0 \wedge \cdots \wedge \mu_{N-1} \wedge \mu_G \geq \alpha \,). \end{aligned} \qquad x_0 \in X \tag{4}$$

3 Recursive formula

We consider the subproblem starting at state $x_n(\in X)$ on n-th stage and terminating on the final N-th stage $(0 \leq n \leq N - 1)$:

$$P_n(x_n) \qquad \text{Max } P_{x_n}^{\pi}(\, \mu_n \wedge \cdots \wedge \mu_{N-1} \wedge \mu_G \geq \alpha \,) \quad \text{s.t. } (\mathrm{i})_\mathrm{m}, \ (\mathrm{ii})_\mathrm{m} \quad n \leq m \leq N\text{--}1$$

where $\pi = \{\pi_n, \ldots, \pi_{N-1}\}$ is taken over *Markov class from n-th stage on* $\Pi(n)$.

Let $v_n(x_n)$ be the maximum value, where

$$v_N(x_N) \overset{\triangle}{=} P(\mu_G \geq \alpha \,|\, X_N = x_N) = \begin{cases} 1 & x_N \geq \alpha \\ 0 & \text{otherwise} \end{cases} \qquad x_N \in X. \tag{5}$$

Lemma 3.1 *We have for any* $0 \leq n \leq N\text{--}1$, $x_n \in X$ *and* $\pi = \{\pi_n, \ldots, \pi_{N-1}\} \in \Pi(n)$

$$P_{x_n}^{\pi}(\, \mu_n \wedge \cdots \wedge \mu_G \geq c\,) \ = \ \begin{cases} \displaystyle\sum_{x_{n+1} \in X} P_{x_{n+1}}^{\pi'}(\, \mu_{n+1} \wedge \cdots \wedge \mu_G \geq c)p(x_{n+1}|x_n, u_n) \\ \qquad\qquad\qquad\qquad\qquad \textit{if } \ \mu_n(x_n, u_n) \geq \alpha \\[3mm] \qquad 0 \qquad\qquad\qquad\qquad\qquad\quad \textit{otherwise} \end{cases}$$

where $u_n = \pi_n(x_n)$, $\pi' = \{\pi_{n+1}, \ldots, \pi_{N-1}\}$ *and* $P_{x_N}^{\pi'} := P$ *in (5) for* $\pi = \{\pi_{N-1}\}$. *Equivalently, in terms of multiple sum, we get*

$$\sum_{(x_{n+1},x_{n+2},\ldots,x_N) \in (*)} \sum \cdots \sum p_{n+1}p_{n+2} \cdots p_N \ = \ \begin{cases} \displaystyle\sum_{x_{n+1} \in X}\left[\sum_{(x_{n+2},\ldots,x_N) \in (\star)}\sum \cdots \sum p_{n+2} \cdots p_N\right]p(x_{n+1}|x_n, u_n) \\ \qquad\qquad\qquad\qquad\qquad\qquad \textit{if } \ \mu_n(x_n, u_n) \geq \alpha \\[3mm] \qquad\qquad 0 \qquad\qquad\qquad\qquad\qquad \textit{otherwise} \end{cases}$$

where $p_m = p(x_m|x_{m-1}, u_{m-1})$, $u_m = \pi_m(x_m)$, $(*)$ *denotes the partial multiple sum over* $(x_{n+1}, \ldots, x_N) \in X \times \cdots \times X$ *satisfying* $\mu_n(x_n, u_n) \wedge \cdots \wedge \mu_G(x_N) \geq \alpha$, *and* $(\star)$ *denotes* $(x_{n+2}, \ldots, x_N)$ *satisfying* $\mu_{n+1}(x_{n+1}, u_{n+1}) \wedge \cdots \wedge \mu_G(x_N) \geq \alpha$.

Thus we have the backward recursive relation :

Theorem 3.1 *(Recursive Equation)*

$$v_N(x) = \begin{cases} 1 & if \ \mu_G(x) \geq \alpha \\ 0 & otherwise \end{cases} \qquad x \in X$$

$$v_n(x) = \begin{cases} \underset{u;\mu_n(x,u)\geq c}{\text{Max}} \sum_{y\in X} v_{n+1}(y)p(y|x,u) & if \ \exists \ u \ ; \ \mu_n(x,u) \geq \alpha \\ 0 & otherwise \end{cases} \qquad (6)$$

$$x \in X, \ \ 0 \leq n \leq N{-}1.$$

Now let us take any pair (n, x). If it satisfies $\mu_n(x, u) \geq \alpha$, then let $\pi_n^*(x)$ denote a $u^* \in U$ which attains the maximum in (6). Otherwise, let $\pi_n^*(x)$ denote any $u \in U$. Then we have an optimal n-th decision function $\pi_n^* : X \to U$. Thus we construct an optimal policy $\pi^* = \{\pi_0^*, \ldots, \pi_{N-1}^*\}$ in Markov class II.

4   An Illustrative Model

Let us consider maximizing the threshold probability with lower membership-degree $\alpha = 0.7$ on Bellman and Zadeh's model [4, pp.B154] :

$$\text{Max} \ \ P_{x_0}^\pi(\ \mu_0(U_0) \wedge \mu_1(U_1) \wedge \mu_G(X_2) \geq 0.7\ ) \ \ \ \text{s.t.} \ \ \ \text{(i)}_n, \ \text{(ii)}_n \ \ \ n = 0, 1$$

where the numerical data is theirs :

$$\mu_0(a_1) = 0.7 \quad \mu_0(a_2) = 1.0; \qquad \mu_1(a_1) = 1.0 \quad \mu_1(a_2) = 0.6$$
$$\mu_G(s_1) = 0.3 \quad \mu_G(s_2) = 1.0 \quad \mu_G(s_3) = 0.8$$

	$p(x_{n+1}|x_n, a_1)$			$p(x_{n+1}|x_n, a_2)$		
$x_n \backslash x_{n+1}$	s_1	s_2	s_3	s_1	s_2	s_3
s_1	0.8	0.1	0.1	0.1	0.9	0.0
s_2	0.0	0.1	0.9	0.8	0.1	0.1
s_3	0.8	0.1	0.1	0.1	0.0	0.9

The backward recursion (6) yields optimal solution in Markov class II — a pair of a sequence of optimal value fuctions $v_0 = v_0(x_0)$, $v_1 = v_1(x_1)$, $v_2 = v_2(x_2)$ and an optimal policy $\pi^* = \{\pi_0^*(x_0), \ \pi_1^*(x_1)\}$. The optimal solution is tabulated as

x_n	$v_2(x_2)$	$v_1(x_1)$	$\pi_1^*(x_1)$	$v_0(x_0)$	$\pi_0^*(x_0)$
s_1	0	0.2	a_1	0.92	a_2
s_2	1	1.0	a_1	0.28	a_1, a_2
s_3	1	0.2	a_1	0.28	a_1

Table 1: Optimal Solution

References

[1] J.F. Baldwin and B.W. Pilsworth, "Dynamic programming for fuzzy systems with fuzzy environment," *J. Math. Anal. Appl.*, vol.85,no.1, pp.1-23, 1982.

[2] R.E. Bellman, *Dynamic Programming*, NJ: Princeton Univ. Press, 1957.

[3] R.E. Bellman and E.D. Denman, "Invariant Imbedding," *Lect. Notes in Operation Research and Mathematical Systems*, vol.52, Berlin: Springer-Verlag, 1971.

[4] R.E. Bellman and L.A. Zadeh, "Decision-making in a fuzzy environment," *Management Science*, vol.17, no.4, pp.B141-B164, 1970.

[5] A.O. Esogbue and R.E. Bellman, "Fuzzy dynamic programming and its extensions," *TIMS/Studies in the Management Sciences*, vol.20, pp.147-167, 1984.

[6] R.A. Howard, *Dynamic Programming and Markov Processes*, Mass.: MIT Press, 1960.

[7] S. Iwamoto, "Associative dynamic programs," *J. Math. Anal. Appl.*, vol.201, no.2, pp.195-211, 1996.

[8] S. Iwamoto, "Decison-making in fuzzy environment: a survey from stochastic decision process," *Proc. 2nd Intl Conf. Knowledge-based Intelligent Electronics Systems (KES '98)*, Adelaide, AUS, Apr. 1998, pp.542-546.

[9] S. Iwamoto, "Maximizing threshold probability through invariant imbedding," *Proc. Intl Workshop on Intelligent Systems Resolutions*; The 8-th Bellman Continuum, Hsinchu, ROC, Dec. 2000, pp.17-22.

[10] S. Iwamoto, "Fuzzy decision-making through three dynamic programming approaches," *Proc. Intl Workshop on Intelligent Systems Resolutions*; The 8-th Bellman Continuum, Hsinchu, ROC, Dec. 2000, pp.23-28.

[11] S. Iwamoto and T. Fujita, "Stochastic decision-making in a fuzzy environment," *J. Oper. Res. Soc. Japan*, vol.38, no.4, pp.467-482, 1995.

[12] S. Iwamoto and M. Sniedovich, "Sequential decision making in fuzzy environment," *J. Math. Anal. Appl.*, vol.222, no.2, pp.208-224, 1998.

[13] S. Iwamoto, K. Tsurusaki and T. Fujita, "Conditional decision-making in a fuzzy environment," *J. Oper. Res. Soc. Japan*, vol.42, no.2, pp.198-218, 1999.

[14] S. Iwamoto, K. Tsurusaki and T. Fujita, "On Markov policies for minimax decision processes," *J. Math. Anal. Appl.*, vol.253, no.1, 58-78, 2001.

[15] S. Iwamoto, T. Ueno and T. Fujita, "Controlled Markov chains with utility functions," *Proc. of Intl Workshop on Markov Processes and Controlled Markov Chains*, Changsha, China, Aug. 1999.

[16] J. Kacprzyk and A.O. Esogbue, "Fuzzy dynamic programming: Main developments and applications," *Fuzzy Sets and Systems*, vol.81, no.1, pp.31-45, 1996.

[17] E.S. Lee, *Quasilinearization and Invariant Imbedding*, New York: Academic Press, 1968.

[18] M.L. Puterman, *Markov Decision Processes : discrete stochastic dynamic programming*, New York: Wiley & Sons, 1994.

[19] T. Ueno and S. Iwamoto, "A dual approach for optimizing threshold probabilities," under consideration.

KES '01
N. Baba et al. (Eds.)
IOS Press, 2001

Stopping Game Problem for Fuzzy Stochastic Systems

Yuji YOSHIDA[§], Masami YASUDA[†], Jun-ichi NAKAGAMI[†] and Masami KURANO[‡]

[§]*Faculty of Economics and Business Administration, the University of Kitakyushu,*
4-2-1 Kitagata, Kokuraminami, Kitakyushu 802-8577, Japan
[†]*Faculty of Science,* [‡]*Faculty of Education, Chiba University,*
1-33 Yayoi-cho,Inage-ku,Chiba 263-8522, Japan

Abstract. Observing a sequence of fuzzy-valued random variables(fuzzy RV), a zero-sum stopping game is discussed. Firstly the fuzzy RV's by the classical stopping times are defined, and then the fuzzy RV's by the fuzzy stopping times are constructed using the α-cut technique. The observed fuzzy RV's are evaluated by probabilistic expectation and scalarization functions, that is, linear ranking functions. A corporation with a cone-saddle point in the multi-objective programming theory makes it possible to consider the model by an order preference. Under a regularity condition, the saddle points for the fuzzy stopping games are given.

1. Introduction and notations

A zero-sum stopping game for sequence of random variables is known as Dynkin's stopping problem(Dynkin [1]) and it is the two-person zero-sum game where each player stops the game so as to maximize/minimize his own expected payoff in the gambling. In this paper, we extend the classical game in the probability theory ([1]) and we consider a stopping game for sequence of fuzzy-valued random variables(fuzzy RV's), which are random variables taking values in fuzzy numbers. Fuzzy RV were first studied by Puri and Ralescu [6] and have been discussed by many authors. Our objective is to discuss the game when the fuzzy RV's are evaluated by probabilistic expectation and scalarization functions, that is, linear ranking functions.

This paper treats a problem of fuzzy RV's and fuzzy stopping times. *Fuzzy stopping times* are introduced in dynamic fuzzy systems by Kurano et al. [4]. In the stopping model, by an order preference defined in the class of fuzzy numbers, we consider an equilibrium point, i.e. a saddle point in this case, corporate with the concept of extreme point in mathematical programming. Then the saddle point for fuzzy stopping games is given under a regularity condition.

In the rest of this section, we give the notations of fuzzy random variables. Let $(\Omega, \mathcal{M}, P)$ be a non-atomic probability space, where $\mathcal{M}$ is a σ-field and P is a probability measure. Let $\mathbb{R} := (-\infty, \infty)$ and $\mathbb{N} := \{0, 1, 2, 3, \cdots\}$. A fuzzy number is denoted by its membership function $\tilde{a} : \mathbb{R} \mapsto [0, 1]$ which is normal, upper-semicontinuous, fuzzy convex and has a compact support. $\mathcal{R}$ denotes the set of all fuzzy numbers. The α-cut of a fuzzy number $\tilde{a}(\in \mathcal{R})$ is given by $\tilde{a}_\alpha := \{x \in \mathbb{R} \mid \tilde{a}(x) \geq \alpha\}$ $(\alpha \in (0, 1])$ and $\tilde{a}_0 := \text{cl}\{x \in \mathbb{R} \mid \tilde{a}(x) > 0\}$, where cl denotes the closure of an interval. In this paper, we write the closed intervals by $\tilde{a}_\alpha := [\tilde{a}_\alpha^-, \tilde{a}_\alpha^+]$ for $\alpha \in [0, 1]$. A map $\tilde{X} : \Omega \mapsto \mathcal{R}$ is called a *fuzzy random variable* if the maps $\omega \mapsto \tilde{X}_\alpha^-(\omega)$ and $\omega \mapsto \tilde{X}_\alpha^+(\omega)$ are measurable for all $\alpha \in [0, 1]$, where $\tilde{X}_\alpha(\omega) = [\tilde{X}_\alpha^-(\omega), \tilde{X}_\alpha^+(\omega)] := \{x \in \mathbb{R} \mid \tilde{X}(\omega)(x) \geq \alpha\}$ is the α-cut of the fuzzy number $\tilde{X}(\omega)$ for $\omega \in \Omega$.

2. A stopping game of fuzzy random variables

In this section, we formulate a stopping game for sequences of fuzzy random variables. Firstly a 'fuzzy' stopping game of the classical stopping time are discussed. By constructing a random variable from the integrand of a linear ranking function for α-cut of a fuzzy RV's, and an existence of saddle point for the problem is shown. Then a 'fuzzy' stopping game of 'fuzzy' stopping time will be defined in the next section.

Let $\{\tilde{W}_n\}_{n=0}^{\infty}$, $\{\tilde{X}_n\}_{n=0}^{\infty}$ and $\{\tilde{Y}_n\}_{n=0}^{\infty}$ be three sequences of integrable bounded fuzzy random variables such that $E(\sup_n \tilde{X}_{n,0}^-) > -\infty$ and $E(\sup_n \tilde{Y}_{n,0}^+) < \infty$, where $\tilde{Y}_{n,0}^+(\omega)$ $(\tilde{X}_{n,0}^-(\omega))$ is the right-end (left-end) of the 0-cut of the fuzzy number $\tilde{Y}_n(\omega)$ $(\tilde{X}_n(\omega)$ resp.). We assume $\tilde{X}_n \preceq \tilde{W}_n \preceq \tilde{Y}_n$ almost surely for all $n = 0, 1, 2, \cdots$ and $P(\Lambda \cup \Gamma) = 1$, where $\Lambda := \{\omega \mid \liminf_{n\to\infty} \tilde{Y}_{n,\alpha}^{\pm}(\omega) \leq \limsup_{n\to\infty} \tilde{X}_{n,\alpha}^{\pm}(\omega)$ for all $\alpha \in [0,1]\}$ and $\Gamma := \bigcup_{m=0}^{\infty}\{\omega \mid \tilde{Y}_n(\omega) = \tilde{X}_n(\omega)$ for all $n \geq m\}$, and $\preceq$ means the *fuzzy max order* ([3,4,6]), i.e. for fuzzy numbers $\tilde{a}, \tilde{b} \in \mathcal{R}$, $\tilde{a} \preceq \tilde{b}$ means that $\tilde{a}_\alpha^- \leq \tilde{b}_\alpha^-$ and $\tilde{a}_\alpha^+ \leq \tilde{b}_\alpha^+$ for all $\alpha \in [0,1]$. Then there exist $\lim_{n\to\infty} \tilde{W}_{n,\alpha}^{\pm}$ almost surely for all $\alpha \in [0,1]$. Therefore we can define a fuzzy random variable $\tilde{W}$ such that

$$\tilde{W}_{\infty,\alpha}^{\pm}(\omega) := \begin{cases} \lim_{\alpha'\uparrow\alpha} \lim_{n\to\infty} \tilde{W}_{n,\alpha'}^{\pm}(\omega) & \text{if } \alpha > 0 \\ \lim_{\alpha'\downarrow 0} \lim_{n\to\infty} \tilde{W}_{n,\alpha'}^{\pm}(\omega) & \text{if } \alpha = 0 \end{cases} \quad \text{for } \omega \in \Omega. \tag{1}$$

$\mathcal{M}_n$ denotes the smallest σ-field on Ω generated by $\{\tilde{W}_{k,\alpha}^{\pm}, \tilde{X}_{k,\alpha}^{\pm}, \tilde{Y}_{k,\alpha}^{\pm} \mid k = 0, 1, 2, \cdots, n; \alpha \in [0,1]\}$ for $n = 0, 1, 2, \cdots$. A map $\tau : \Omega \mapsto \mathbb{N} \cup \{\infty\}$ is called a *stopping time* if $\{\omega \mid \tau(\omega) = n\} \in \mathcal{M}_n$ for all $n = 0, 1, 2, \cdots$. Let $\omega \in \Omega$. For player 1's stopping time τ and player 2's stopping time σ, we define a fuzzy random variable by

$$\tilde{Z}[\tau,\sigma](\omega) := \begin{cases} \tilde{X}_\tau(\omega) & \text{if } \tau(\omega) < \sigma(\omega) \\ \tilde{W}_\tau(\omega) & \text{if } \tau(\omega) = \sigma(\omega) \\ \tilde{Y}_\sigma(\omega) & \text{if } \tau(\omega) > \sigma(\omega). \end{cases} \tag{2}$$

Put a map $g : \mathcal{I} \mapsto \mathbb{R}$ by

$$g([a,b]) = \lambda a + (1-\lambda)b \tag{3}$$

for closed intervals $[a,b]$, where $\mathcal{I}$ denotes the set of all bounded closed sub-intervals of $\mathbb{R}$ and λ is a constant$(0 \leq \lambda \leq 1)$ and means a balance factor of their estimation regarding the game. This function is called a *linear ranking function* which is used for the evaluation of fuzzy numbers(Fortemps and Roubens [2]).

Let (τ, σ) be a pair of finite stopping times. Now we consider the evaluation of the fuzzy random variable $\tilde{Z}[\tau,\sigma]$. The expectation is given by the closed interval $E(\tilde{Z}[\tau,\sigma])_\alpha$ and its evaluation is $g(E(\tilde{Z}[\tau,\sigma])_\alpha)$. For a pair of finite stopping times (τ,σ), we define a random variable

$$\tilde{Z}_g[\tau,\sigma](\omega) := \int_0^1 g(\tilde{Z}[\tau,\sigma]_\alpha(\omega)) \, d\alpha, \quad \omega \in \Omega. \tag{4}$$

Lemma 1. *For a pair of finite stopping times (τ,σ), it holds that*

$$\int_0^1 g(E(\tilde{Z}[\tau,\sigma])_\alpha) \, d\alpha = \int_0^1 E(g(\tilde{Z}[\tau,\sigma]_\alpha)) \, d\alpha = E(\tilde{Z}_g[\tau,\sigma]). \tag{5}$$

The expectation

$$J[\tau, \sigma] := E(\tilde{Z}_g[\tau, \sigma]) \tag{6}$$

means player 1's expected reward for a pair of the player's stopping times (τ, σ). Now we discuss a stopping game such that player 1 maximizes $J[\tau, \sigma]$ with respect to stopping time τ and player 2 minimizes it with respect to stopping time σ respectively.

Problem 1. Find a pair of the player's stopping times (τ^*, σ^*) such that

$$J[\tau, \sigma^*] \leq J[\tau^*, \sigma^*] \leq J[\tau^*, \sigma]. \tag{7}$$

for all stopping times τ and σ. Then, (τ^*, σ^*) is called a 'g-saddle point' for the zero-sum stopping game with respect to a linear ranking function g.

3. A fuzzy stopping game for fuzzy stopping times

The model of a fuzzy stopping game for fuzzy stopping times, that is, both of fuzzy situations, will be constructed in this section which is based on the previous results for the classical stopping time. The keypoint is that the model of the game is defined by using α-cut of a fuzzy stopping time.

Definition 1. A fuzzy stopping time is a map $\tilde{\tau} : \mathbb{N} \times \Omega \mapsto [0, 1]$ satisfying the following (i) and (ii):

(i) For each $n = 0, 1, 2, \cdots$, the map $\omega \mapsto \tilde{\tau}(n, \omega)$ is $\mathcal{M}_n$-measurable.

(ii) For each $\omega \in \Omega$, the map $n \mapsto \tilde{\tau}(n, \omega)$ is non-increasing.

Regarding the membership grade of fuzzy stopping times, $\tilde{\tau}(n, \omega) = 0$ means 'to stop at time n' and $\tilde{\tau}(n, \omega) = 1$ means 'to continue at time n' respectively. And the intermediate value '$0 < \tilde{\tau}(n, \omega) < 1$' is a notion of 'fuzzy stopping'. The following lemmas imply the properties of fuzzy stopping times.

Lemma 2.

(i) *Let $\tilde{\tau}$ be a fuzzy stopping time. Define a map $\tilde{\tau}_\alpha : \Omega \mapsto \mathbb{N}$ by*

$$\tilde{\tau}_\alpha(\omega) := \inf\{n \mid \tilde{\tau}(n, \omega) < \alpha\} \quad (\omega \in \Omega) \quad \text{for } \alpha \in (0, 1], \tag{8}$$

where the infimum of the empty set is understood to be $+\infty$. Then, we have:

(a) $\{\omega \mid \tilde{\tau}_\alpha(\omega) \leq n\} \in \mathcal{M}_n \quad (n = 0, 1, 2, \cdots)$;

(b) $\tilde{\tau}_\alpha(\omega) \leq \tilde{\tau}_{\alpha'}(\omega) \quad (\omega \in \Omega) \quad \text{if } \alpha \geq \alpha'$;

(c) $\lim_{\alpha' \uparrow \alpha} \tilde{\tau}_{\alpha'}(\omega) = \tilde{\tau}_\alpha(\omega) \quad (\omega \in \Omega) \quad \text{if } \alpha > 0$.

(ii) *Let $\{\tilde{\tau}_\alpha\}_{\alpha \in [0,1]}$ be maps $\tilde{\tau}_\alpha : \Omega \mapsto \mathbb{N}$ satisfying the above (a) and (b). Define a map $\tilde{\tau} : \mathbb{N} \times \Omega \mapsto [0, 1]$ by*

$$\tilde{\tau}(n, \omega) := \sup_{\alpha \in [0,1]} \min\{\alpha, 1_{\{\omega \mid \tilde{\tau}_\alpha(\omega) > n\}}(\omega)\}, \quad n = 0, 1, 2, \cdots, \ \omega \in \Omega. \tag{9}$$

Then $\tilde{\tau}$ is a fuzzy stopping time.

Now we consider the definition of the fuzzy RV $\tilde{Z}[\tilde{\tau}, \tilde{\sigma}]$ associated with a pair of fuzzy stopping times $(\tilde{\tau}, \tilde{\sigma})$. Here assume that each fuzzy stopping time is finite, i.e. such as $\tilde{\tau}_0 := \lim_{\alpha \downarrow 0} \tilde{\tau}_\alpha < \infty$ almost surely. However the extension from a classical stopping time to a fuzzy stopping time in the composition for fuzzy RV does not so straightforward. We consider the α-cut of the fuzzy RV so that we construct the fuzzy RV $\tilde{Z}(\tilde{\tau}, \tilde{\sigma})$ by the usual technique, and the evaluation proceeds after that.

Let $\omega \in \Omega$ of underlaying probability space. From Lemma 2(i), each α-cut of the the fuzzy stopping times $\tilde{\tau}$ and $\tilde{\sigma}$ is a classical stopping time, so it is possible to define $\tilde{Z}[\tilde{\tau}_\alpha, \tilde{\sigma}_\alpha](\omega)$, and therefore, the expectation is given by the closed interval

$$K_\alpha[\tilde{\tau}, \tilde{\sigma}] := E(\tilde{Z}[\tilde{\tau}_\alpha, \tilde{\sigma}_\alpha](\cdot)). \tag{10}$$

For each α, define $\max K_\alpha, \min K_\alpha$ by fuzzy max order $\prec$ respectively as follows:

$$\tilde{t} \in \max K_\alpha[\cdot, \tilde{\sigma}] \quad \text{iff} \quad \{\tilde{\tau} \mid K_\alpha[\tilde{t}, \tilde{\sigma}] \prec K_\alpha[\tilde{\tau}, \tilde{\sigma}], \; \tilde{t} \neq \tilde{\tau}\} = \emptyset \quad \text{with fixed } \tilde{\sigma},$$

$$\tilde{s} \in \min K_\alpha[\tilde{\tau}, \cdot] \quad \text{iff} \quad \{\tilde{\sigma} \mid K_\alpha[\tilde{\tau}, \tilde{\sigma}] \prec K_\alpha[\tilde{\tau}, \tilde{s}], \; \tilde{s} \neq \tilde{\sigma}\} = \emptyset \quad \text{with fixed } \tilde{\tau},$$

where $K_\alpha[\cdot, \tilde{\sigma}] := \cup_{\tilde{\tau}} K_\alpha[\tilde{\tau}, \tilde{\sigma}]$ and $K_\alpha[\tilde{\tau}, \cdot] := \cup_{\tilde{\sigma}} K_\alpha[\tilde{\tau}, \tilde{\sigma}]$. To analyze the following Problem 2, we consider the next subproblem:

Problem 2α. Let $\alpha \in [0, 1]$. Find a pair of the player's stopping times $(\tilde{\tau}^*, \tilde{\sigma}^*)$ such that

$$K_\alpha(\tilde{\tau}^*, \tilde{\sigma}^*) \in \min K_\alpha(\tilde{\tau}^*, \cdot) \cap \max K_\alpha(\cdot, \tilde{\sigma}^*). \tag{11}$$

The pair of fuzzy stopping times $(\tilde{\tau}^*, \tilde{\sigma}^*)$ is called an 'α-saddle point'.

Let $\alpha \in [0, 1]$. For $n = 0, 1, 2, \cdots$, we define

$$\overline{V}_{n,\alpha} := \text{ess inf}_{\tilde{\sigma} \geq n} \, \text{ess sup}_{\tilde{\tau} \geq n} \, E(g(\tilde{Z}[\tilde{\tau}_\alpha, \tilde{\sigma}_\alpha]_\alpha) | \mathcal{M}_n),$$

$$\underline{V}_{n,\alpha} := \text{ess sup}_{\tilde{\tau} \geq n} \, \text{ess inf}_{\tilde{\sigma} \geq n} \, E(g(\tilde{Z}[\tilde{\tau}_\alpha, \tilde{\sigma}_\alpha]_\alpha) | \mathcal{M}_n), \tag{12}$$

$$X_n(\omega) := g(\tilde{X}_{n,\alpha}(\omega)) \quad \text{and} \quad Y_n(\omega) := g(\tilde{Y}_{n,\alpha}(\omega)) \tag{13}$$

for $n = 0, 1, 2, \cdots, \omega \in \Omega$. Then we have the following lemma, by applying Ohtsubo [5].

Lemma 3.

(i) $\overline{V}_{n,\alpha} = \underline{V}_{n,\alpha}$ *for all* $n = 0, 1, 2, \cdots$. *So, we put* $V_{n,\alpha} := \overline{V}_{n,\alpha} = \underline{V}_{n,\alpha}$ *for* $n = 0, 1, 2, \cdots$. *The sequence of random variables* $\{V_{n,\alpha}\}_{n=0}^{\infty}$ *satisfies*

$$V_{n,\alpha} = \min\{\max\{g(\tilde{X}_{n,\alpha}), E(V_{n+1,\alpha} | \mathcal{M}_n)\}, g(\tilde{Y}_{n,\alpha})\}, \quad n = 0, 1, 2, \cdots. \tag{14}$$

(ii) *Define*

$$\tau_\alpha^*(\omega) := \inf\{k \geq n \mid V_{k,\alpha}(\omega) \leq g(\tilde{X}_{k,\alpha}(\omega))\}, \quad \omega \in \Omega, \tag{15}$$

$$\sigma_\alpha^*(\omega) := \inf\{k \geq n \mid V_{k,\alpha}(\omega) \geq g(\tilde{Y}_{k,\alpha}(\omega))\}, \quad \omega \in \Omega, \tag{16}$$

where the infimum of the empty set is understood to be $+\infty$. *If* $\min\{\tau_\alpha^*, \sigma_\alpha^*\} < \infty$, *then* $(\tau_\alpha^*, \sigma_\alpha^*)$ *is an* α-saddle point of Problem 2α *and* $E(V_{0,\alpha}) = E(g(\tilde{Z}[\tau_\alpha^*, \sigma_\alpha^*]_\alpha))$.

Assumption A. The maps $\alpha \mapsto \tau_\alpha^*(\omega)$ and $\alpha \mapsto \sigma_\alpha^*(\omega)$ are non-increasing for $\omega \in \Omega$.

In order to construct a saddle point from α-saddle points $\{(\tau_\alpha^*, \sigma_\alpha^*)\}_{\alpha \in [0,1]}$, we need Assumption A(a regularity condition). Then, we can define maps $\tilde{\tau}^*, \tilde{\sigma}^* : \mathbb{N} \times \Omega \mapsto [0,1]$ by

$$\tilde{\tau}^*(n,\omega) := \sup_{\alpha \in [0,1]} \min\{\alpha, 1_{\{\omega | \tau_\alpha^*(\omega) > n\}}(\omega)\}, \tag{17}$$

$$\tilde{\sigma}^*(n,\omega) := \sup_{\alpha \in [0,1]} \min\{\alpha, 1_{\{\omega | \sigma_\alpha^*(\omega) > n\}}(\omega)\} \tag{18}$$

for $n = 0, 1, 2, \cdots$, $\omega \in \Omega$. Remark that, for the α-cuts of $\tilde{\tau}^*(n,\omega)$ and $\tilde{\sigma}^*(n,\omega)$ defined by $\tilde{\tau}_\alpha^*(\omega)$ and $\tilde{\sigma}_\alpha^*(\omega)$ similar to (8), $\tilde{\tau}_\alpha^*(\omega) \neq \tau_\alpha^*(\omega)$ or $\tilde{\sigma}_\alpha^*(\omega) \neq \sigma_\alpha^*(\omega)$ may occur at most countable many $\alpha \in (0,1]$. Similarly to Section 2, the evaluation of the fuzzy RV is induced by the integral of a linear ranking function and its probabilistic expectation. For a pair of finite fuzzy stopping times $(\tilde{\tau}, \tilde{\sigma})$, we define a random variable

$$\tilde{Z}_g[\tilde{\tau}, \tilde{\sigma}](\omega) := \int_0^1 g(\tilde{Z}[\tilde{\tau}_\alpha, \tilde{\sigma}_\alpha]_\alpha(\omega))\, d\alpha, \quad \omega \in \Omega. \tag{19}$$

Similarly to (6), the player 1's expected reward in the fuzzy stopping game is set by

$$J[\tilde{\tau}, \tilde{\sigma}] := E(\tilde{Z}_g[\tilde{\tau}, \tilde{\sigma}]) \tag{20}$$

for a pair of the player's stopping times $(\tilde{\tau}, \tilde{\sigma})$.

Lemma 4. *For a pair of finite fuzzy stopping times $(\tilde{\tau}, \tilde{\sigma})$, it holds that*

$$J[\tilde{\tau}, \tilde{\sigma}] = \int_0^1 g(K_\alpha[\tilde{\tau}, \tilde{\sigma}])\, d\alpha = \int_0^1 E(g(\tilde{Z}[\tilde{\tau}_\alpha, \tilde{\sigma}_\alpha]_\alpha))\, d\alpha. \tag{21}$$

Now we consider the fuzzy stopping game for the sequence of fuzzy random variables.

Problem 2. Find a pair of the player's fuzzy stopping times $(\tilde{\tau}^*, \tilde{\sigma}^*)$ such that

$$J[\tilde{\tau}, \tilde{\sigma}^*] \leq J[\tilde{\tau}^*, \tilde{\sigma}^*] \leq J[\tilde{\tau}^*, \tilde{\sigma}] \tag{22}$$

for all fuzzy stopping times $\tilde{\tau}$ and $\tilde{\sigma}$. Then $(\tilde{\tau}^*, \tilde{\sigma}^*)$ is called a 'fuzzy saddle point' for the stopping game.

Theorem 1. *Under Assumption A, $(\tilde{\tau}^*, \tilde{\sigma}^*)$ is a fuzzy saddle point for Problem 2.*

References

[1] E.B. Dynkin, Game variant of a problem on optimal stopping, *Soviet Math. Dokl.* **10** (1965) 270-274.

[2] P. Fortemps and M. Roubens, Ranking and defuzzification methods based on area compensation, *Fuzzy Sets and Systems* **82** (1996) 319-330.

[3] M. Kurano, M. Yasuda, J.N akagami and Y. Yoshida, Ordering of fuzzy sets — A brief survey and new results, *J. Oper. Res. Soc. of Japan* **43** (2000) 137-148.

[4] M. Kurano, M. Yasuda, J. Nakagami and Y. Yoshida, An approach to stopping problems of a dynamic fuzzy system, preprint.

[5] Y, Ohtsubo, Optimal stopping sequential games with or without a constraint of always terminating, *Math. Oper. Res.* **11** (1986) 591-607.

[6] M.L. Puri and D.A. Ralescu, Fuzzy random variables, *J. Math. Anal. Appl.* **114** (1986) 409-422.

Option Pricing with Uncertainty
in Financial Engineering

Yuji YOSHIDA

Faculty of Economics and Business Administration, the University of Kitakyushu
4-2-1 Kitagata, Kokuraminami, Kitakyushu 802-8577, Japan

Abstract. A discrete-time mathematical model for American put option with uncertainty is presented and the randomness and fuzziness are evaluated by both probabilistic expectation and fuzzy expectation defined by a possibility measure from the viewpoint of fuzzy expectation. It is shown that the optimal fuzzy price is a solution of the optimality equation under a reasonable assumption. The writer's (seller's) permissible range of optimal expected price in the American put option is presented.

1. Introduction and notations

This paper deals with a discrete-time mathematical model for American put option with uncertainty. It is not easy in general to analyze American put option in continuous-time, and so the study of the discrete-time case is one of the important approaches to investigate the continuous-time model through approximation. To analyze American put option, we need to discuss an optimal stopping problem in stochastic processes. The optimal stopping problem for a sequence of random variables has a long history and has been studied by many authors in probability theory. For example, financial engineering(Pliska [5], Ross [6]) and so on. The aim of this paper is to discuss the optimal stopping problem with uncertainty(randomness and fuzziness) from the viewpoint of fuzzy expectation, taking account of human subjective judgment. In order to describe an optimal stopping model with fuzziness, we need to extend real-valued random variables in the classical probability theory to *fuzzy random variables* which are random variables with fuzzy number values. Fuzzy random variables were first studied by Puri and Ralescu [4] and have been studied by many authors. This paper derives an optimality equation for the *fuzzy stochastic process* and gives a method to solve the stopping problem.

2. American put option in uncertain environment

First we give some mathematical notations regarding fuzzy numbers. $\mathbb{R}$ denotes the set of all real numbers. A fuzzy number is denoted by its membership function $\tilde{a} : \mathbb{R} \mapsto [0, 1]$ which is normal, upper-semicontinuous, fuzzy convex and has a compact support ([8]). $\mathcal{R}$ denotes the set of all fuzzy numbers. In this paper, we identify fuzzy numbers with its corresponding membership functions. The α-cut of a fuzzy number $\tilde{a}(\in \mathcal{R})$ is given by $\tilde{a}_\alpha := \{x \in \mathbb{R} \mid \tilde{a}(x) \geq \alpha\}$ $(\alpha \in (0, 1])$ and $\tilde{a}_0 := \mathrm{cl}\{x \in \mathbb{R} \mid \tilde{a}(x) > 0\}$, where cl denotes the closure of an interval. We write the closed intervals by $\tilde{a}_\alpha := [\tilde{a}_\alpha^-, \tilde{a}_\alpha^+]$ for $\alpha \in [0, 1]$.

Next we describe a discrete-time fuzzy stochastic system and an American option model with fuzziness. Let $(\Omega, \mathcal{M}, P)$ be a probability space, where $\mathcal{M}$ is a σ-field and P is a non-atomic probability measure. Let T $(T > 0)$ be an *expiration date* and let

$\mathbb{T} := \{0, 1, 2, \cdots, T\}$ be the time space. Let r $(r > 0)$ be an *interest rate* of a bond price, which is riskless asset, and let a *discount rate* $\beta = 1/(1+r)$. Define a stock price process $\{S_t\}_{t=0}^{T}$ as follows: S_0 is a positive constant and

$$S_t := S_0 \prod_{s=1}^{t} (1 + Y_s) \quad \text{for } t = 1, 2, \cdots, T, \tag{1}$$

where $\{Y_t\}_{t=1}^{T}$ is a uniform integrable sequence of independent, identically distributed real random variables on $[r-1, r+1]$ such that $E(Y_t) = r$ for all $t = 1, 2, \cdots, T$. Let $\mathcal{M}_0 := \{\emptyset, \Omega\}$ and $\mathcal{M}_t (t = 1, 2, \cdots, T)$ denote the σ-fields generated by $\{Y_1, Y_2 \cdots Y_t\}$.

Let $\{a_t\}_{t=0}^{T}$ be an $\mathcal{M}_t$-adapted stochastic process such that $0 < a_t(\omega) \le S_t(\omega)$ for almost all $\omega \in \Omega$. We consider a finance model where the stock price process $\{S_t\}_{t=0}^{T}$ takes fuzzy values. We give fuzzy values by triangular fuzzy numbers for simplicity. Then, stock price process with fuzzy values are represented by a *fuzzy stochastic process* $\{\tilde{S}_t\}_{t=0}^{T}$ of fuzzy random variables:

$$\tilde{S}_t(\omega)(x) := L((x - S_t(\omega))/a_t(\omega)) \tag{2}$$

for $t \in \mathbb{T}$, $\omega \in \Omega$ and $x \in \mathbb{R}$, where $L(x) := \max\{1 - |x|, 0\}$ $(x \in \mathbb{R})$ is the triangle-shape function and $\{a_t\}_{t=0}^{T}$ is a sequence of random variables with positive values. Hence, $a_t(\omega)$ is a spread of triangular fuzzy numbers $\tilde{S}_t(\omega)$ and corresponds to the amount of fuzziness in the process. The fuzziness in the process increases as $a_t(\omega)$ becomes bigger, and $a_t(\omega)$ should be an increasing function of the stock price $S_t(\omega)$ (see Assumption S in this section).

Let K $(K > 0)$ be a *strike price*. By using stopping times, we maximize the expected values of the price process $\{\tilde{P}_t\}_{t=0}^{T}$ in American put option:

$$\tilde{P}_t(\omega) := \beta^t (1_{\{K\}} - \tilde{S}_t(\omega)) \vee 1_{\{0\}} \quad \text{for } t \in \mathbb{T}, \tag{3}$$

where $1_{\{K\}}$ and $1_{\{0\}}$ denote the crisp number K and zero respectively, and $\vee$ is the maximum with respect to the *fuzzy max order*. For fuzzy numbers $\tilde{a}, \tilde{b} \in \mathcal{R}$, the α-cuts of the maximum $\tilde{a} \vee \tilde{b}$ are defined by $(\tilde{a} \vee \tilde{b})_\alpha = [\max\{\tilde{a}_\alpha^-, \tilde{b}_\alpha^-\}, \max\{\tilde{a}_\alpha^+, \tilde{b}_\alpha^+\}]$, $\alpha \in [0, 1]$.

Now, we introduce a valuation method of fuzzy prices, taking into account of decision maker's subjective judgment. Let $\mathbb{R}_+ := [0, \infty)$ and give a fuzzy goal by a fuzzy set $\varphi : \mathbb{R} \mapsto [0, 1]$ which is a continuous and increasing function with $\varphi(0) = 0$ and $\lim_{x \to \infty} \varphi(x) = 1$. Then we note that the α-cut is $\varphi_\alpha = [\varphi_\alpha^-, \infty)$ for $\alpha \in (0, 1)$. A map $\tau : \Omega \mapsto \mathbb{T}$ is called a stopping time ([3]) if $\{\omega \in \Omega \mid \tau(\omega) = t\} \in \mathcal{M}_t$ for all $t \in \mathbb{T}$. For a stopping time τ, we define a *fuzzy expectation* of the fuzzy numbers $E(\tilde{P}_\tau)$ by

$$\tilde{E}(E(\tilde{P}_\tau)) := \fint_{\mathbb{R}_+} E(\tilde{P}_\tau)(x) \, d\tilde{m}(x) = \sup_{x \in \mathbb{R}_+} \min\{E(\tilde{P}_\tau)(x), \varphi(x)\}, \tag{4}$$

where $\tilde{m}$ is the possibility measure generated by the density φ and $\fint d\tilde{m}$ denotes Sugeno integral ([7]). The fuzzy number $E(\tilde{P}_\tau)$ means a *fuzzy price*, and the fuzzy expectation (4) implies the degree of writer's (seller's) satisfaction regarding fuzzy prices $E(\tilde{P}_\tau)$. Then the fuzzy goal $\varphi(x)$ means a kind of utility function for expected prices x in (4), and it represents a writer's (seller's) subjective judgment from the idea of Bellman and Zadeh [1] (see [9,10]). We define an optimal fuzzy price $\tilde{V}$ as follows: Consider

$$\tilde{V} := \bigvee_{\tau:\ \text{stopping times},\, \tau \leq T} E(\tilde{P}_\tau), \tag{5}$$

where $\vee$ means the supremum with respect to the fuzzy max order (see (4)). Here we can construct a fuzzy number $\tilde{V}$ from (5), by considering left continuous versions of maps $\alpha(\in (0,1]) \mapsto \sup_{\tau:\tau \leq T} E(\tilde{P}_\tau)_\alpha^\pm$ (see the proof of Yoshida [8,Theorem 2.1]). Now this paper discusses the following optimal stopping problem regarding American put option with fuzziness, which is described as a fuzzy-number-valued extension of the stopping models in [3,Chapter 4].

Problem P. Find a stopping time $\tau^*(\tau^* \leq T)$ such that

$$\tilde{E}(E(\tilde{P}_{\tau^*})) = \tilde{E}(\tilde{V}), \tag{6}$$

where $\tilde{V}$ is given by (5).

Then, τ^* is called an *optimal exercise time* and a real number $x^*(\in \mathbb{R}_+)$ is called an *optimal expected price* under the fuzzy expectation generated by possibility measures if it attains the supremum of the fuzzy expectation (4)

Next, to analyze the optimal fuzzy price $\tilde{V}$, we introduce some notations. Let $t \in \mathbb{T}$. Define

$$Z_t(\omega) := \bigvee_{\tau:\ \text{stopping times},\, t \leq \tau \leq T} E(\tilde{P}_\tau | \mathcal{M}_t)(\omega) \tag{7}$$

for $\omega \in \Omega$, where $\vee$ means the essential supremum with respect to (4). Here we can construct a fuzzy random variable $\tilde{Z}_t$ from (7) similarly to the method to obtain the fuzzy number $\tilde{V}$ from (5). The fuzzy random variables $\tilde{Z}_t$ correspond to Snell's envelope in probability theory([3]). Then, by modifying the proof of Yoshida [8,Theorem 2.2], we obtain the following optimality equation for the fuzzy stochastic process $\{\tilde{Z}_t\}_{t=0}^T$ regarding the optimal fuzzy price $\tilde{V}$.

Theorem 1 (Optimality equation). *The following (i) and (ii) hold.*

(i)　$\tilde{V} = E(\tilde{Z}_0)$.

(ii)　$\tilde{Z}_t(\omega) = \tilde{P}_t(\omega) \vee E(\tilde{Z}_{t+1} | \mathcal{M}_t)(\omega)$　*almost all* ω　*for* $t = 0, 1, \cdots, T - 1$.

3. The optimal expected price and the optimal exercise time

In this section, we discuss the fuzzy expectation of the optimal fuzzy price $\tilde{V}$. Now we introduce a reasonable assumption.

Assumption S. The stochastic process $\{a_t\}_{t=0}^T$ is represented by

$$a_t(\omega) := cS_t(\omega), \quad t = 0, 1, \cdots, T - 1, \ \omega \in \Omega,$$

where c is a constant satisfying $0 < c < 1$

Assumption S is reasonable since $a_t(\omega)$ means a size of fuzziness and it should depend on the volatility and the stock price $S_t(\omega)$ because one of the most difficulties is estimation of the volatility in actual cases ([6,Sect.7.5.1]). In this model, we represent by c the fuzziness of the volatility, and we call c a fuzzy factor of the process. From now on, we suppose that Assumption S holds. By putting $b^{\pm}(\alpha) := 1 \pm (1-\alpha)c$ $(\alpha \in [0,1])$, from (2) we obtain

$$\tilde{S}^{\pm}_{t,\alpha}(\omega) = b^{\pm}(\alpha)S_0 \prod_{s=1}^{t}(1 + Y_s(\omega)), \quad \omega \in \Omega. \tag{8}$$

for $t \in \mathbb{T}$ and $\alpha \in [0,1]$. Then we have the fuzzy price process:

$$\tilde{V}^{\pm}_{\alpha} = \sup_{\tau \leq T} E(\beta^{\tau} \max\{K - b^{\mp}(\alpha)S_{\tau}, 0\} \mid S_0 = y) \tag{9}$$

for an initial stock price y $(y > 0)$. Define a grade α^* by

$$\alpha^* := \sup\{\alpha \in [0,1] \mid \varphi^-_{\alpha} \leq \tilde{V}^+_{\alpha}\}, \tag{10}$$

where $\varphi_{\alpha} = [\varphi^-_{\alpha}, \infty)$ for $\alpha \in (0,1)$, and the supremum of the empty set is understood to be 0. From (9) and the continuity of φ and $\tilde{V}$, we can easily check $\varphi^-_{\alpha^*} = \tilde{V}^+_{\alpha^*}$. The following results are obtained by a modification of the results in Yoshida [8] under the fuzzy expectation generated by possibility measures (4).

Theorem 2. *It holds that*

$$\alpha^* = \tilde{E}(\tilde{V}) = \sup_{\tau:\ stopping\ times,\, \tau \leq T} \tilde{E}(E(\tilde{P}_{\tau})). \tag{11}$$

The value α^* means the optimal grade of the fuzzy expectation of the fuzzy prices. Next, we give an optimal exercise time for Problem P. Define a stopping time

$$\sigma^*(\omega) := \min\{t \in \mathbb{T} \mid Z^+_{t,\alpha^*}(\omega) = \tilde{P}^+_{t,\alpha^*}(\omega)\}, \quad \omega \in \Omega, \tag{12}$$

where the infimum of the empty set is understood to be T.

Theorem 3. *The stopping time σ^* is an optimal exercise time for Problem P, and it is the shortest in the class of optimal exercise times. Further, under the fuzzy expectation generated by possibility measures (4), the optimal expected price is $x^* = \varphi^-_{\alpha^*}$.*

To analyze the optimal fuzzy price $\tilde{V}^{\pm}_{\alpha}$, we put

$$\tilde{V}^{\pm}_{t,\alpha}(y) = \sup_{\tau:t \leq \tau \leq T} E(\beta^{\tau-t} \max\{K - b^{\mp}(\alpha)S_{\tau}, 0\} \mid S_t = y) \tag{13}$$

for $t \in \mathbb{T}$, a stock price y $(y > 0)$ and a strike price K $(K > 0)$. Then we note that $\tilde{V}^{\pm}_{0,\alpha}(y) = \tilde{V}^{\pm}_{\alpha}$. Since the price process is Markov (see (1)), by (9) we can reduce from Theorem 1(ii) to the following recursive *optimality equation*:

$$\tilde{V}^{\pm}_{t,\alpha}(y) = \max\{\beta E(\tilde{V}^{\pm}_{t+1,\alpha}(y(1 + Y_1))), f^{\pm}_{\alpha}(y)\}, \quad t = 0, 1, \cdots, T-1,\ y > 0, \tag{14}$$

$$\tilde{V}_{T,\alpha}^{\pm}(y) = f_{\alpha}^{\pm}(y), \quad y > 0, \tag{15}$$

where $f_{\alpha}^{\pm}(y) := \max\{K - b^{\mp}(\alpha)y, 0\}$ $(y > 0)$. From (15), the grade α^* of the fuzzy expectation of the optimal fuzzy price is given by $\varphi_{\alpha^*}^- = \tilde{V}_{0,\alpha^*}^+(y) = \tilde{V}_{\alpha^*}^+(y)$. By Theorem 3, the corresponding optimal expected price is

$$x^* = \varphi_{\alpha^*}^-, \tag{16}$$

and the corresponding optimal exercise time (12) is reduced to

$$\sigma^*(\omega) = \inf\{t \in \mathbb{T} \mid \tilde{V}_{t,\alpha^*}^+(S_t(\omega)) = f_{\alpha^*}^+(S_t(\omega))\}, \quad \omega \in \Omega. \tag{17}$$

Since the fuzzy expectation (4) is defined by possibility measures, (16) gives an upper bound on optimal expected prices of American put option. Therefore, similarly to (10) we can define another grade, which gives a lower bound on optimal expected prices of American put option as follows:

$$x_* = \varphi_{\alpha_*}^-, \tag{18}$$

where α_* is defined by $\alpha_* := \sup\{\alpha \in [0,1] \mid \varphi_{\alpha}^- \leq \tilde{V}_{\alpha}^-(y)\}$. and it satisfies $\varphi_{\alpha_*}^- = \tilde{V}_{\alpha_*}^-(y)$. Then, its corresponding optimal exercise time is given by

$$\sigma_*(\omega) = \inf\{t \in \mathbb{T} \mid \tilde{V}_{t,\alpha_*}^-(S_t(\omega)) = f_{\alpha_*}^-(S_t(\omega))\}, \quad \omega \in \Omega. \tag{19}$$

Hence, from (10), (16), (18) and (24), we can easily check the interval $[x_*, x^*]$ is written as

$$[x_*, x^*] = \{x \in \mathbb{R} \mid \tilde{V}(x) \geq \varphi(x)\}, \tag{20}$$

which is the range of prices x such that the reliability degree of the optimal expected price, $\tilde{V}(x)$, is greater than the degree of writer's satisfaction, $\varphi(x)$. Therefore, $[x_*, x^*]$ means *writer's permissible range of expected prices* under his fuzzy goal φ.

References

[1] R.E. Bellman and L.A. Zadeh, Decision-making in a fuzzy environment, *Management Sci. Ser. B.* **17** (1970) 141-164.

[2] M. Kurano, M. Yasuda, J. Nakagami and Y. Yoshida, Markov-type fuzzy decision processes with a discounted reward on a closed interval, *Europ. J. Oper. Res.* **92** (1996) 649-662.

[3] J. Neveu, Discrete-Parameter Martingales, North-Holland, New York, 1975.

[4] M.L. Puri and D.A. Ralescu, Fuzzy random variables, *J. Math. Anal. Appl.* **114** (1986) 409-422.

[5] S.R. Pliska Introduction to Mathematical Finance: Discrete-Time Models, Blackwell Publ., New York, 1997.

[6] S.M. Ross, An Introduction to Mathematical Finance, Cambridge Univ. Press, Cambridge, 1999.

[7] M. Sugeno, Theory of fuzzy integrals and its applications, *Doctoral Thesis, Tokyo Institute of Technology* (1974).

[8] Y. Yoshida, An optimal stopping problem in dynamic fuzzy systems with fuzzy rewards, *Computers Math. Appl.* **32** (1996) 17-28.

[9] Y. Yoshida, Fuzzy decision processes with expected fuzzy rewards, In: B.M. Ayyub and M.M. Gupta (eds.), Uncertainty Analysis in Engineering and the Sciences, Kluwer Academic Publishers, Boston, 1997, pp. 313-323.

[10] Y. Yoshida, A time-average fuzzy reward criterion in fuzzy decision processes, *Information Sciences* **110** (1998) 103-112.

Author Index